图解机械工程英语

朱派龙　主编　　刘晓初　雷宇航　副主编

Illustrated English of Mechanical Engineering

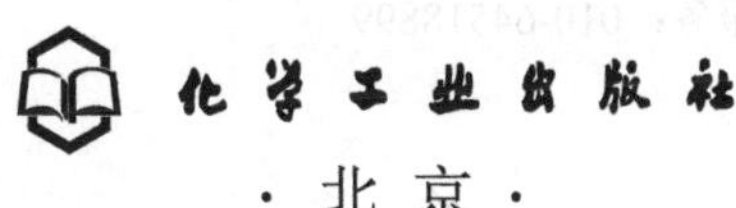

化学工业出版社
·北京·

内容简介

以“看图识词”的方式编写专业英语，将专业内容表达地更为直观、具体、生动。

本书内容全面，取材广泛，囊括了机械工程的方方面面，包括机械设计、机械制造、机械零部件、专业机械几大类。对于已经学习过机械专业知识的读者，可以温故而知“英”，扩大知识面，开拓专业技术视野；对于非机类专业人士，可以简单学习专业术语及名称，并认识各类机械零件、结构、设备和加工方法，便于自学和应用于实际工作。

本书作为一本涵盖面广、编写形式新颖的工具书，不仅可供大中专院校机械专业师生查阅参考，也可供机类及非机类专业从事技术翻译、技术交流、进出口贸易等工作的各界涉外人士、工矿企业的专业技术人员学习参考。

图书在版编目（CIP）数据

图解机械工程英语 / 朱派龙主编. —北京：化学工业出版社，2017.3（2025.1 重印）

ISBN 978-7-122-28943-8

Ⅰ. ①图… Ⅱ. ①朱… Ⅲ. ①机械工程-英语-图解 Ⅳ. ①TH-64

中国版本图书馆 CIP 数据核字（2017）第 016206 号

责任编辑：贾 娜　　文字编辑：项 溦
责任校对：宋 夏　　装帧设计：王晓宇

出版发行：化学工业出版社（北京市东城区青年湖南街 13 号　邮政编码 100011）
印　　装：北京盛通数码印刷有限公司
787mm×1092mm　1/16　印张 $31^3/_4$　字数 856 千字　2025 年 1 月北京第 1 版第 3 次印刷

购书咨询：010-64518888　　售后服务：010-64518899
网　　址：http://www.cip.com.cn
凡购买本书，如有缺损质量问题，本社销售中心负责调换。

定　　价：138.00 元

前言

FOREWORD

机械工业在国民经济中占有非常重要的地位。随着机械行业的发展，掌握一定的机械专业英语成为机械领域从业人员的必备技能。我国高校陆续开设了机械专业英语课程并编制了相应的教材，这些教材基本上都是以科技文章为主。近年来，也有英汉或汉英机械专业词典出版。

机械工业对应的机械工程是一个看得见、摸得着、具体、形象的行业，从事机械工程相关工作的专家、教师、工程师或技术员在讨论技术问题或进行技术交流时，经常借助于画图或照片以直接说明或辅助理解，所以说，工程图是工程师交流的语言。根据机械工业的专业特点，通过图形与英汉词汇相结合，可使读者更加直观、生动地学习专业英语，并进一步巩固、加深对机械本身专业术语的理解。因此，我们编写了《图解机械工程英语》一书。

本书内容全面，囊括了机械工程的方方面面，包括机械设计、机械制造、机械零部件、专业机械等，诸多内容和图片是国内图书尚未呈现过的；取材广泛，确保专业词汇的英文翻译正宗、原味；编写方式新颖，以“专业图＋中文＋英文”的方式来表达较为枯燥的专业英语，使专业内容变得直观、生动；内容编排有序，体现由浅入深、由简到繁、由普遍性到特殊性的逻辑递进关系。

本书由朱派龙主编，刘晓初、雷宇航副主编。唐电波、李平凡、颜学定、张铠锋、范瑞、沈健、张东峰、吴峥强、王康、潘朝参与了编写。其中，范瑞编写 Unit 1；张东峰编写 Unit 2、Unit 10、Unit 11、Unit 13；王康编写 Unit 3、Unit 6；颜学定编写 Unit 4、Unit9；吴峥强编写 Unit 5；朱派龙编写 Unit 7、Unit 8 及附录并负责统稿；刘晓初编写 Unit 12、Unit 14；潘朝编写 Unit 15 ～ Unit 17；雷宇航编写 Unit 18、Unit 21、Unit 22；唐电波编写 Unit 19；张铠锋编写 Unit 20 中的 20.1 ～ 20.3；李平凡编写 Unit 20 中的 20.4 ～ 20.6；沈健编写 Unit 23。

由于编者水平所限，书中不足之处在所难免，敬请各位读者不吝赐教，以求日臻完善，使本书发挥其应有的作用。

主　编

目录
CONTENTS

Part 1 Mechanical Design 机械设计

Unit 1 Engineering Materials 工程材料 / 1

1.1 Material properties and tests 材料性能与测试 / 1

1.2 Metals 金属材料 / 7

1.3 Non-metals 非金属材料 / 13

1.4 Composite materials and nano materials 复合材料和纳米材料 / 15

Unit 2 Drawing and Tolerance 制图与公差 / 21

2.1 Engineering drawing 工程制图 / 21

2.1.1 Types of line and notation 线型和标注 / 21

2.1.2 CAD presentation 计算机辅助设计绘图 / 25

2.1.3 On engineering drawing presentation 机械制图 / 25

2.2 Tolerance and fits 公差与配合 / 27

Unit 3 Technological Appropriateness for Machine Parts and Assemblies 机械零部件结构工艺性 / 34

3.1 Technological appropriateness for castings, forgings, weldings and heat treatment 铸造、锻压、焊接和热处理结构工艺性 / 34

3.2 Technological appropriateness for stampings, metal cutting 冲压、切削加工结构工艺性 / 38

3.3 Technological appropriateness for plastic parts and powder metallurgy parts 塑件、粉末冶金件的结构工艺性 / 48

3.4 Technological appropriateness for machine assembly 机器装配的结构工艺性 / 48

Unit 4 Design of Mechanisms 机构设计 / 50

4.1 Design of linkage mechanisms 连杆机构设计 / 50

4.2 Design of spatial linkage mechanisms 空间机构设计 / 51

4.3 Design of cam mechanisms 凸轮机构设计 / 52

4.4 Design of common variated mechanisms 常用变异机构设计 / 55

Unit 5 Transmission Design 传动设计 / 65

5.1 Design of gear transmission 齿轮传动设计 / 65

5.2 Design of belt and chain drive 带和链传动设计 / 73
5.2.1 Belt drive 带传动 / 73
5.2.2 Chain drive 链传动 / 77
5.3 Design of hydraulic power transmission 液压传动设计 / 79

Unit 6 Mo(u)ld and Die Structure Design 模具结构设计 / 96

6.1 Mo(u)ld and die for metal forming 冲压模具 / 96
6.2 Die and mo(u)ld for plastics 塑料模具 / 109
6.3 Mo(u)ld for die cast 压铸模具 / 115
6.4 Other mo(u)lds 其他模具 / 115
6.4.1 Powder metallurgy mo(u)ld 粉末冶金模具 / 115
6.4.2 Sand mo(u)ld 翻砂模具 / 118
6.4.3 Forging mo(u)ld 锻压模具 / 121

Part 2 Mechanical Manufacture 机械制造

Unit 7 Methods of Mechanical Manufacturing 机械制造技术 / 122

7.1 Manual operations 手工操作 / 123
7.2 Heat process for metals 金属的热加工工艺 / 130
7.2.1 Casting 铸造 / 130
7.2.2 Forging 锻压 / 138
7.2.3 Welding 焊接 / 144
7.2.4 Heat treatment 热处理 / 153
7.3 Cutting principle 切削原理 / 154
7.4 Traditional mechanical machining methods 传统机械加工技术 / 158
7.4.1 Turning on lathe 车削加工 / 158
7.4.2 Milling 铣削加工 / 160
7.4.3 Boring process 镗削加工 / 163
7.4.4 Drilling and ream process 钻削、铰削加工 / 163
7.4.5 Reciprocating machining process 往复式加工 / 164
7.4.6 Abrasives process 磨削加工 / 164
7.4.7 Coated abrasive belt grinding 砂带磨削加工 / 167
7.4.8 Super finishing 超精加工 / 167
7.4.9 Gear cutting 齿轮加工 / 168
7.5 Non-traditional processes 非传统加工（特种加工）工艺方法 / 170
7.5.1 EDM 电火花加工 / 171
7.5.2 ECM 电化学加工 / 173
7.5.3 USM 超声加工 / 175
7.5.4 CHM 化学加工 / 176
7.5.5 EBM 电子束加工 / 177
7.5.6 IBM 离子加工 / 177
7.5.7 LBM 激光束加工 / 179

7.5.8 RP 快速成形 / 180
7.5.9 AJM/WJM/AWJM 磨料射流 / 水射流 / 磨料水射流加工 / 181
7.5.10 MAM 磁性磨料加工 / 183
7.5.11 MEMS 微电子制造 / 183

Unit 8 Manufacturing Processes and Toolings 机械制造装备 / 186

8.1 Cutting tools 切削刀具 / 186
8.1.1 Turning tools 车刀 / 186
8.1.2 Milling tools 铣刀 / 187
8.1.3 Boring tools 镗刀 / 188
8.1.4 Hole-making tools 孔加工刀具 / 189
8.1.5 Reciprocating process tools 往复运动加工刀具 / 195
8.1.6 Abrasives 磨具 / 197
8.1.7 Gear cutting tools 齿轮刀具 / 201
8.2 Jigs and fixtures, toolings 工装夹具 / 202
8.2.1 General purpose jigs and fixtures 通用夹具 / 202
8.2.2 Special purpose jigs and fixtures 专用夹具 / 204
8.3 Various types of machine tools 各种机械加工机床 / 212
8.3.1 Lathes 车床 / 214
8.3.2 Milling machines 铣床 / 216
8.3.3 Boring machines 镗床 / 218
8.3.4 Drilling machines 钻床 / 218
8.3.5 Reciprocating machines 往复式加工机床 / 219
8.3.6 Grinding machines 磨床 / 221
8.3.7 Gear cutting machines 齿轮加工机床 / 224
8.3.8 Automatic screw machines 自动螺纹加工机床 / 225
8.4 Non-traditional machine tools 非传统（特种）加工机床 / 226
8.5 CAD/CAM/CAPP/FMS/CIMS 与计算机相关的先进系统 / 227

Unit 9 NC Machining and NC Machine Tools 数控加工与数控机床 / 232

9.1 Basic knowledge on numerical control 数控基本知识 / 232
9.2 NC machining tools 数控加工机床 / 235
9.3 NC functional components and appendixes 数控机床功能部件及附件 / 240
9.4 NC programming 数控加工编程 / 242

Part 3 Machine Elements 机械零部件

Unit 10 Joints and Fasteners 连接件与紧固件 / 246

10.1 Screws and threads 螺纹 / 246
10.2 Washers and retaining rings 垫片和保持环 / 250
10.3 Keys 键 / 252

10.4 Pins 销 / 253
10.5 Splines 花键 / 254
10.6 Rivets 铆钉 / 254
10.7 Bonding and other fasteners 粘接和其他紧固件 / 256

Unit 11 Shafts 轴 / 258

11.1 Shafts 直轴 / 258
11.2 Other shafts 其他类型的轴 / 260

Unit 12 Bearings 轴承 / 261

12.1 Plain bearings 滑动轴承 / 261
12.2 Rolling bearings 滚动轴承 / 265

Unit 13 Springs and Flywheels 弹簧和飞轮 / 273

13.1 Helical springs 螺旋弹簧 / 273
13.2 Other springs 其他弹簧 / 275
13.3 Flywheels 飞轮 / 276

Unit 14 Lubrication Systems and Sealings 润滑系统与密封件 / 277

14.1 Lubrication systems 润滑系统 / 277
14.2 Sealings 密封件 / 281

Unit 15 Beds, Frames and Guideways 床身、支架及导轨 / 287

15.1 Beds and frames 床身和支架 / 287
15.2 Guideways 导轨 / 288

Unit 16 Couplings, Clutches and Brakes 联轴器、离合器和制动器 / 291

16.1 Couplings 联轴器 / 292
16.2 Clutches 离合器 / 296
16.3 Brakes 制动器 / 299

Unit 17 Reductors and Speed Changers 减速器与变速器 / 303

17.1 Reductors 减速器 / 303
17.2 Speed changers 变速器 / 304

Part 4 Mechanical Equipment for Special Industry 专业机械

Unit 18 Power Generation Equipments 动力设备 / 307

18.1 Power generation equipments 多种发电设备 / 307
18.2 Boilers 锅炉 / 309
18.3 Steam turbines 蒸汽轮机 / 316
18.4 Gas turbines 燃气轮机 / 318

18.5 Hydraulic turbines 水轮机 / 321
18.6 Compressors, fans and blowers 压缩机、通风机和鼓风机 / 322

Unit 19 Materials Handling Equipments 运输机械 / 327

19.1 Elevators 电梯 / 327
19.1.1 The types of elevator 电梯分类 / 327
19.1.2 The lift machine 电梯主机 / 329
19.1.3 The car frame and counterweight 轿架和对重 / 330
19.1.4 The safety device 安全装置 / 331
19.1.5 The hoistway 井道件 / 333
19.2 Hoisting machinery 起重机械 / 333
19.3 Materials handling systems 物料搬运系统 / 339
19.4 Industrial handling trucks 工业搬运车辆 / 344

Unit 20 Automobile Constructs 汽车结构 / 346

20.1 Engine 发动机 / 346
20.2 Body 车身 / 359
20.3 Chassis 底盘 / 360
20.4 Functional components 功能部件 / 367
20.5 Special vehicles 特种车辆 / 375
20.6 Electric vehicles 电动车 / 377

Unit 21 Construction Machinery 建筑机械 / 378

21.1 Excavators 挖掘机 / 378
21.2 Earthmoving machinery 铲土运输机 / 379
21.3 Compacting machinery 压实机械 / 380
21.4 Pavement machinery 铺路机械 / 381
21.5 Pile-driving machinery 桩工机械 / 383
21.6 Concrete reinforcing machinery 钢筋混凝土机械 / 384

Unit 22 Petroleum Drilling Machinery and Refining Equipments 石油钻采与炼制机械 / 391

22.1 Petroleum drilling rigs 石油钻机 / 391
22.2 Downhole drilling tools 井下钻具 / 392
22.3 Petroleum extraction equipments 采油设备 / 394
22.4 Offshore petroleum drilling and extraction equipments 海洋石油钻采设备 / 398
22.5 Petroleum refining equipments 石油炼制设备 / 401

Unit 23 Food and Packaging Machinery 食品与包装机械 / 405

23.1 Grading and sorting machinery 分级分选机械 / 405
23.2 Cleaning machines and devices 清洗机械与设备 / 409
23.3 Drying equipments 干燥设备 / 411
23.4 Crushing and cutting machinery 粉碎切割机械 / 413

23.5 Hot and cold exchange processing machinery 冷热交换处理机械 / 415
23.6 Sterilization machinery 杀菌机械 / 417
23.7 Refrigeration machinery and equipments 冷冻机械与设备 / 418
23.8 Forming machines 成形机械 / 421
23.9 Material conveying machinery 物料输送机械 / 423

Vocabulary with Figure Index 词汇及图形索引（英中对照）/426

Vocabulary with Figure Index 词汇及图形索引（中英对照）/455

附录 / 484

附录 1 One-hand alphabet 单手势字母表 / 484
附录 2 Deaf-and-dump alphabet 手语字母表 / 485
附录 3 Mathematical signs and symbols 数学标识和符号 / 486

参考文献 / 487

Part 1 Mechanical Design

机械设计

Unit 1 Engineering Materials 工程材料

1.1 Material properties and tests 材料性能与测试

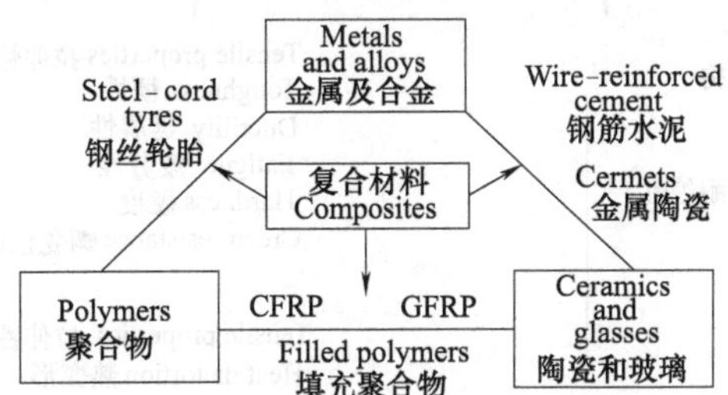

Fig1.1 The classes of engineering materials from which articles are made 按照制造物品的工程材料分类

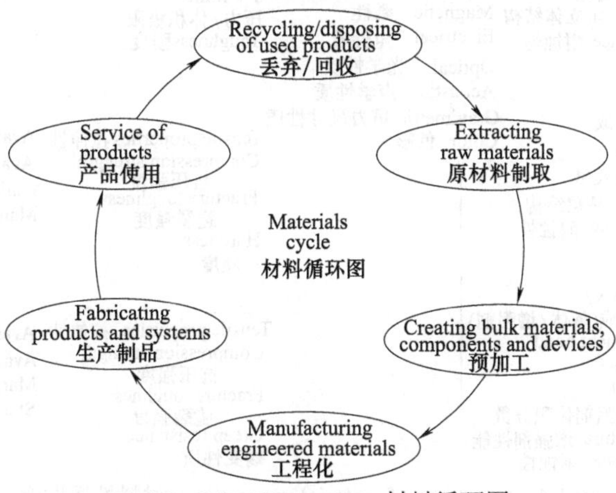

Fig1.2 Materials cycle 材料循环图

Properties
特性
Intrinsic
固有特性
Attributive
属性
Bulk mechanical properties
宏观力学性能
Bulk non-mechanical properties
宏观非力学性能
Surface properties
表面特性
Design
设计
Price and availability
价格和来源
Production properties-ease of manufacture、fabrication、joining、finishing
生产特性：易于制造、加工连接、精加工
Aesthetic properties-appearance、texture、feel
美学特性：外观、组织和手感

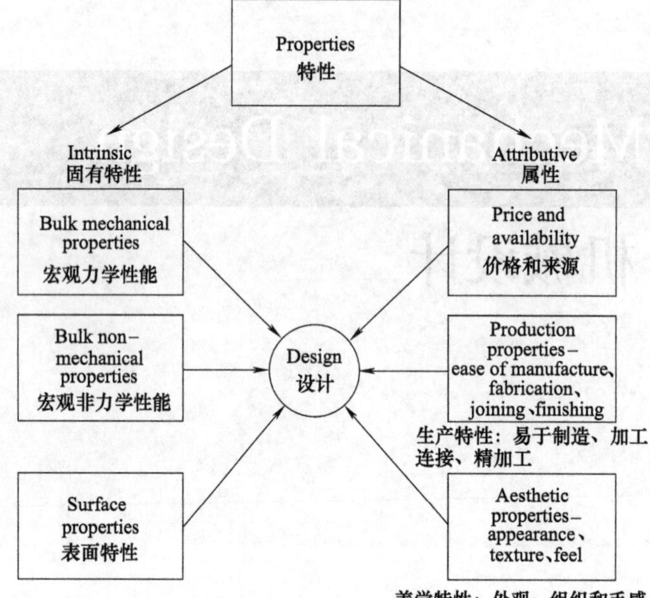

Fig1.3 How the properties of engineering materials affect the way in which products are designed
材料特性如何影响产品设计

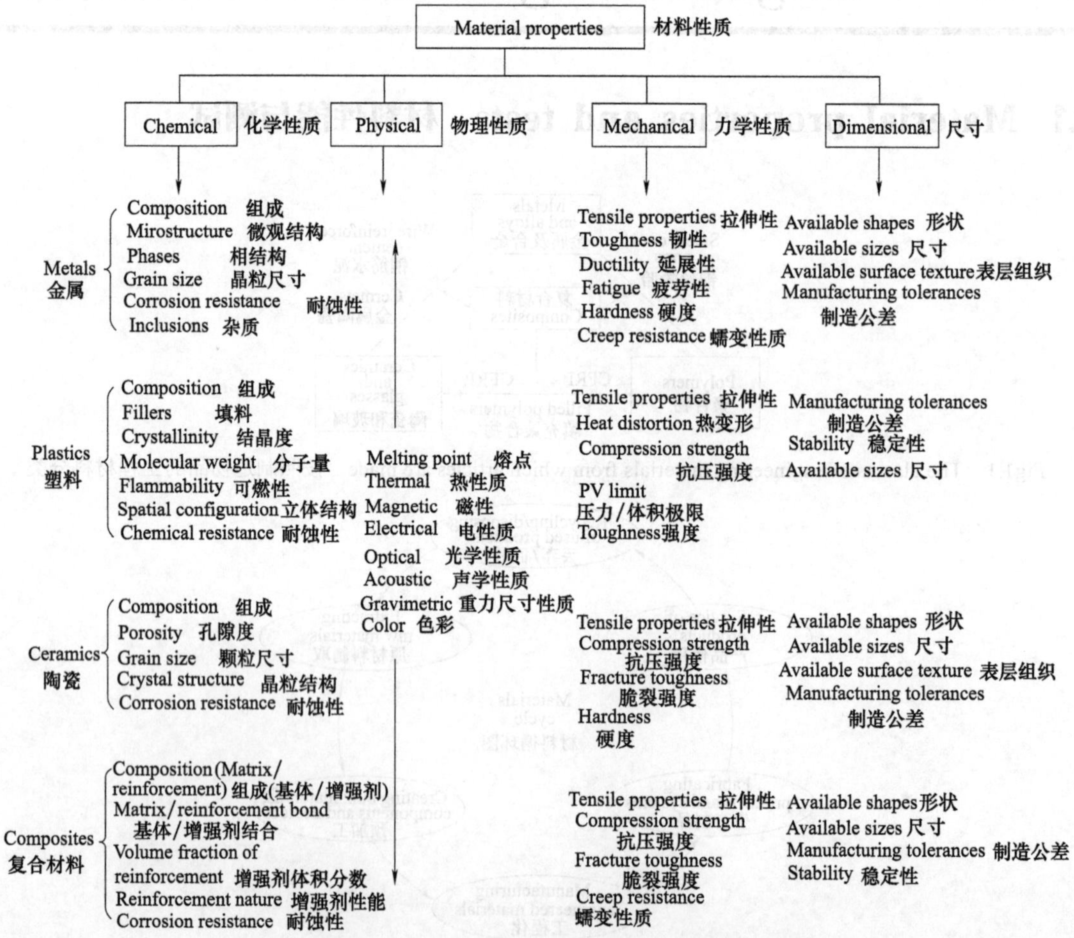

Material properties 材料性质
Chemical 化学性质
Physical 物理性质
Mechanical 力学性质
Dimensional 尺寸
Metals 金属
Composition 组成
Mirostructure 微观结构
Phases 相结构
Grain size 晶粒尺寸
Corrosion resistance 耐蚀性
Inclusions 杂质
Tensile properties 拉伸性
Toughness 韧性
Ductility 延展性
Fatigue 疲劳性
Hardness 硬度
Creep resistance 蠕变性质
Available shapes 形状
Available sizes 尺寸
Available surface texture 表层组织
Manufacturing tolerances 制造公差
Plastics 塑料
Composition 组成
Fillers 填料
Crystallinity 结晶度
Molecular weight 分子量
Flammability 可燃性
Spatial configuration 立体结构
Chemical resistance 耐蚀性
Melting point 熔点
Thermal 热性质
Magnetic 磁性
Electrical 电性质
Optical 光学性质
Acoustic 声学性质
Gravimetric 重力尺寸性质
Color 色彩
Tensile properties 拉伸性
Heat distortion 热变形
Compression strength 抗压强度
PV limit 压力/体积极限
Toughness 强度
Manufacturing tolerances 制造公差
Stability 稳定性
Available sizes 尺寸
Ceramics 陶瓷
Composition 组成
Porosity 孔隙度
Grain size 颗粒尺寸
Crystal structure 晶粒结构
Corrosion resistance 耐蚀性
Tensile properties 拉伸性
Compression strength 抗压强度
Fracture toughness 脆裂强度
Hardness 硬度
Available shapes 形状
Available sizes 尺寸
Available surface texture 表层组织
Manufacturing tolerances 制造公差
Composites 复合材料
Composition (Matrix/reinforcement) 组成(基体/增强剂)
Matrix/reinforcement bond 基体/增强剂结合
Volume fraction of reinforcement 增强剂体积分数
Reinforcement nature 增强剂性能
Corrosion resistance 耐蚀性
Tensile properties 拉伸性
Compression strength 抗压强度
Fracture toughness 脆裂强度
Creep resistance 蠕变性质
Available shapes 形状
Available sizes 尺寸
Manufacturing tolerances 制造公差
Stability 稳定性

Fig1.4 Spectrum of material properties 材料性质图谱

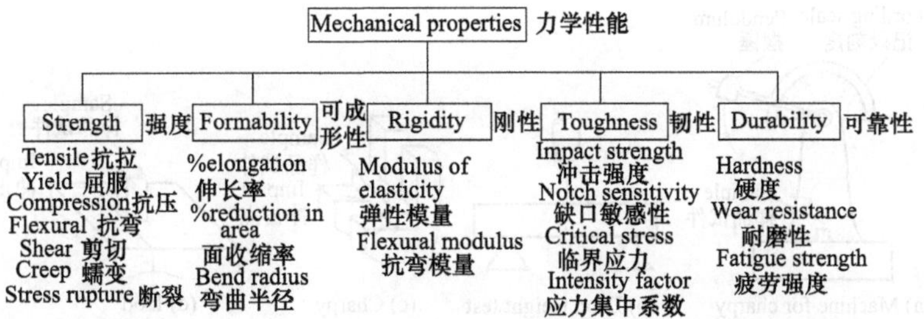

Fig 1.5 Spectrum of mechanical properties 力学性能图谱

HB 布氏硬度
HV 韦氏硬度
洛氏硬度HRA
洛氏硬度HRB
洛氏硬度HRC
回跳(塞勒)硬度计Seleroscope
努氏硬度HK
莫氏刻度 Mohs scale
极限抗拉强度UTS (steels) 钢 psi×10³

10 Diamond 金刚石
9 Corundum 刚玉(氧化铝)
8 Topaz 黄玉
7 Quartz 石英
6 Orthoclase 正长石
5 Apatite 磷灰石
4 Fluorite 萤石
3 Calcite 方解石
2 Gypsum 石膏
1 Talc 滑石

Diamond 金刚石
Cubic boron nitride 立方氮化硼
Boron carbide 碳化硼
Titanium carbide 碳化钛
Tungsten carbide 碳化钨
Cementite 碳化铁
Hard chromium plate 硬铬板
Martensite 马氏体钢
Fully hardened tool steel 淬透工具钢
Glass 玻璃
冷拉0.5C碳钢
Cold drawn 0.5C steel
Cold drawn 0.25C steel 冷拉0.25C碳钢
Titanium 钛
Gray iron 灰铸铁
Malleable iron 可锻铸铁
Annealed mild steel 退火中碳钢
70-30 brass 黄铜
Nickel 镍
Pure iron 纯铁
Bakelite 胶木
Zinc、gold、magnesium 锌、金、镁
Silver 银
Polystyrene 聚苯乙烯
Pure aluminum 纯铝
Vinyl resin 乙烯基树脂
Graphite 石墨
Tin 锡
Lead 铅

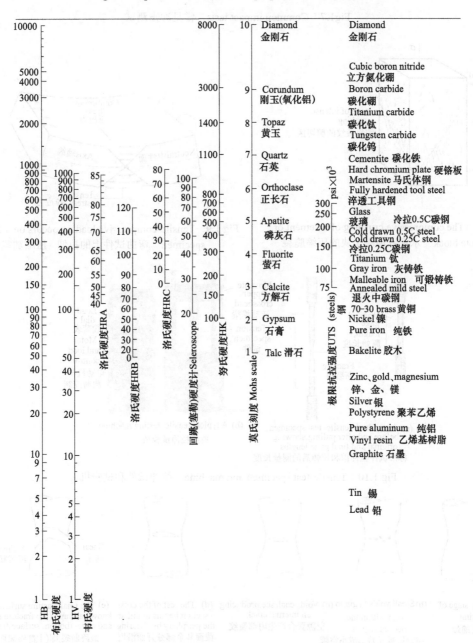

Fig1.6 Hardness conversion chart 硬度换算图

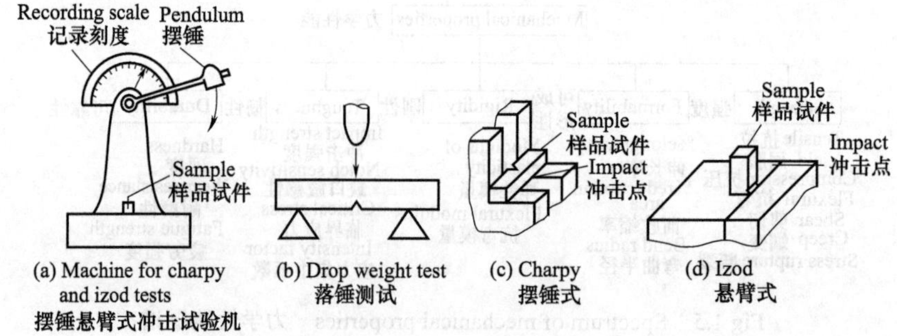

Fig1.7 Common impact tests 常见冲击测试

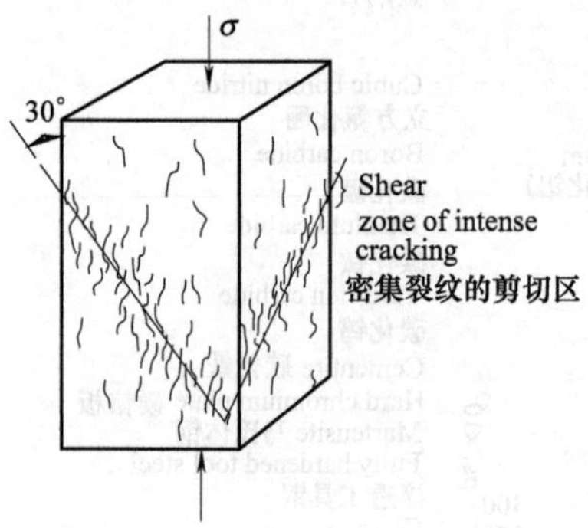

Fig 1.8 The compressive crushing of a cement or concrete block 水泥或混凝土块的压缩脆裂

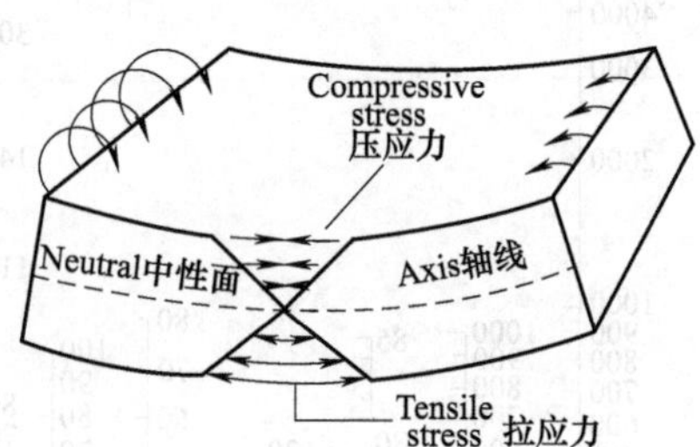

Fig 1.9 Tensile-compressive loading occurs on a flexural specimen 弯曲试件上的拉伸 - 压缩加载

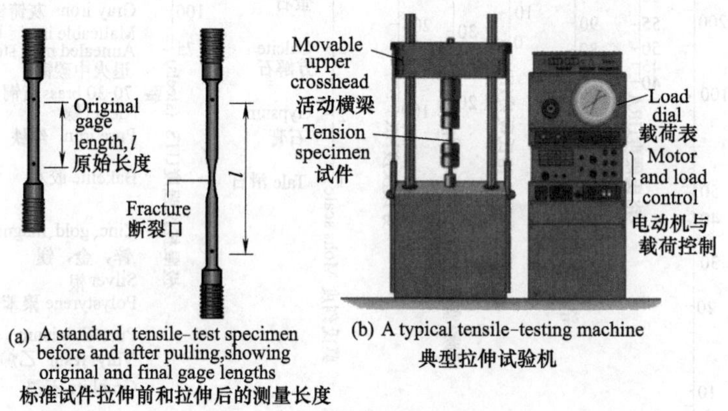

Fig 1.10 Tensile-test specimen and machine 拉伸试件和试验机

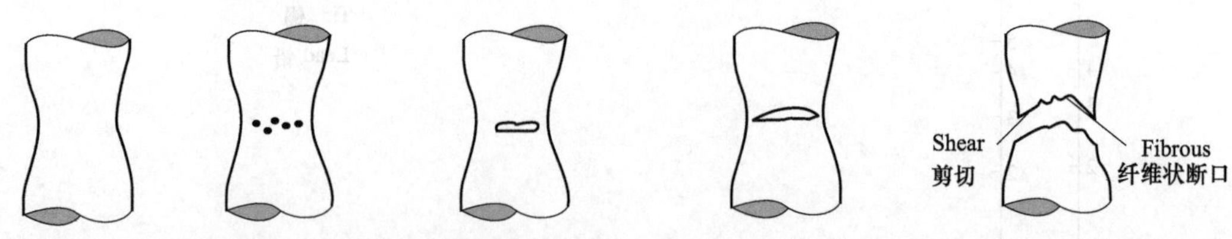

(a) Early stage of necking
初期缩颈

(b) Small voids begin to form within the necked region
缩颈区域开始形成小空隙

(c) Voids coalesce,producing an internal crack
空隙聚合产生内部裂纹

(d) The rest of the cross section begins to fail at the periphery,by shearing
截面其余部分开始沿周边剪切断裂

(e) The final fracture surfaces, known as cup -(top fracture surface) and cone-(bottom surface)fracture
最终断裂,即(上部断裂面)杯形和(底面)锥面断裂

Fig 1.11 Sequence of events in necking and fracture of a tensile-test specimen 拉伸试件缩颈、断裂顺序

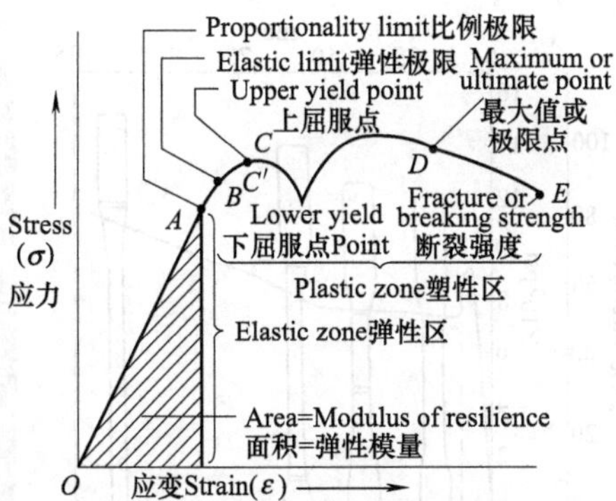

Fig 1.12 σ-ε Diagram for ductile material 塑性材料的应力 / 应变图

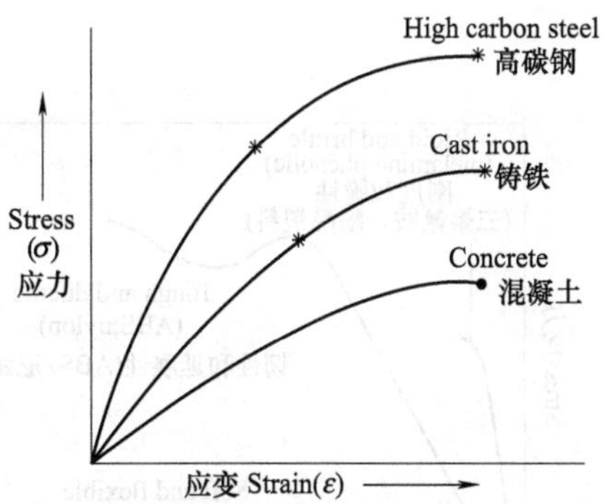

Fig 1.13 σ-ε Diagram for brittle material 脆性材料的 σ-ε 应力应变图

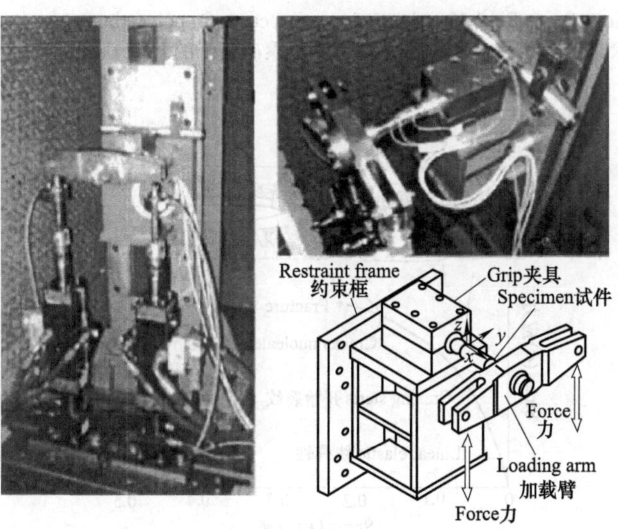

Fig 1.14 Biaxial test machine 双轴向试验机

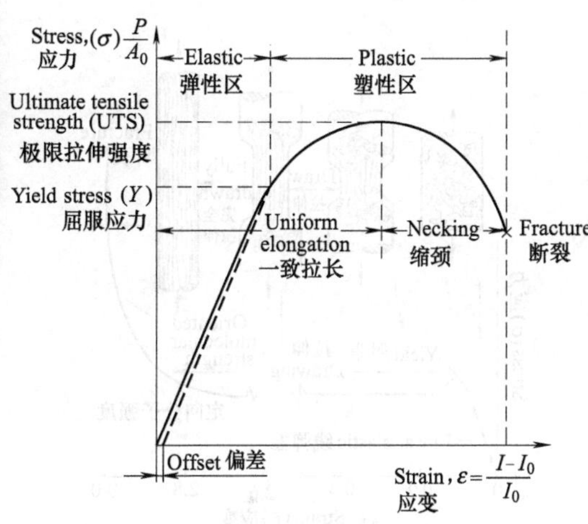

Fig 1.15 A typical stressstrain curve obtained from a tension test, showing various featuress 拉伸试验获得的应力应变曲线

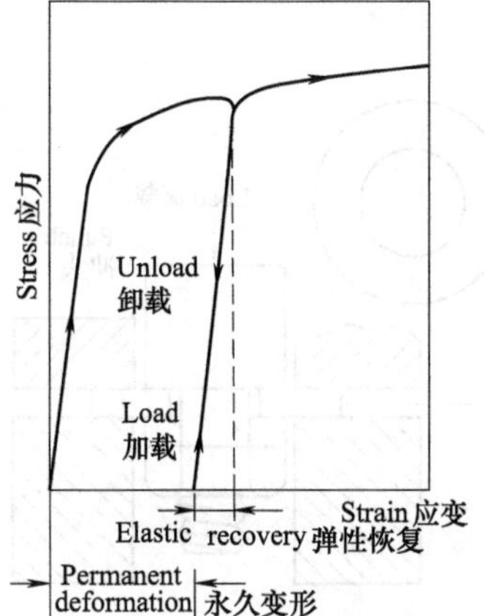

Fig 1.16 Loading and the unloading of a tensile test specimen 拉伸试件的加载和卸载

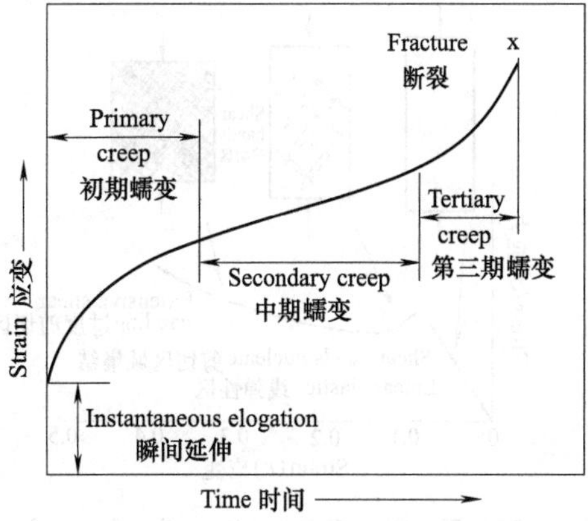

Fig 1.17 Creep test curve 蠕变测试曲线

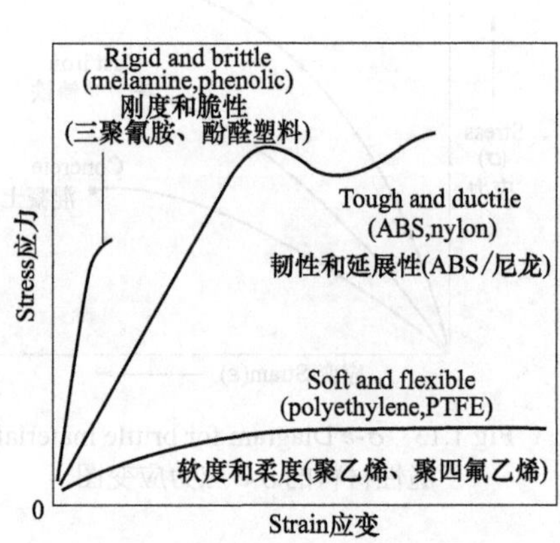

Fig 1.18 General terminology describing the behavior of three types of plastic 描述三种塑料性能的通用术语

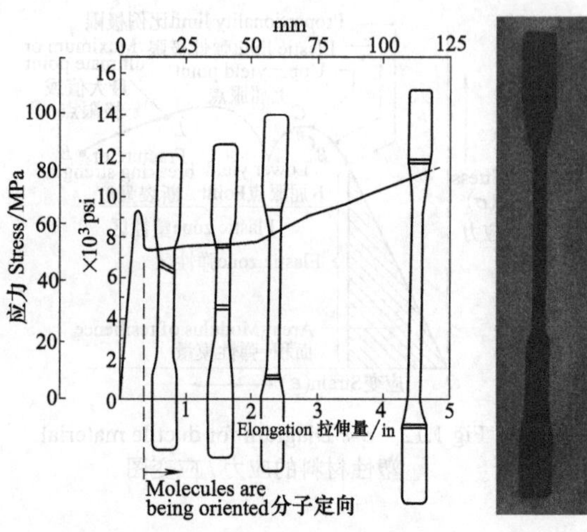

Fig 1.19 Load elongation curve for polycarbonate, a thermoplastic 热塑性材料聚碳酸酯的载荷 / 拉伸曲线

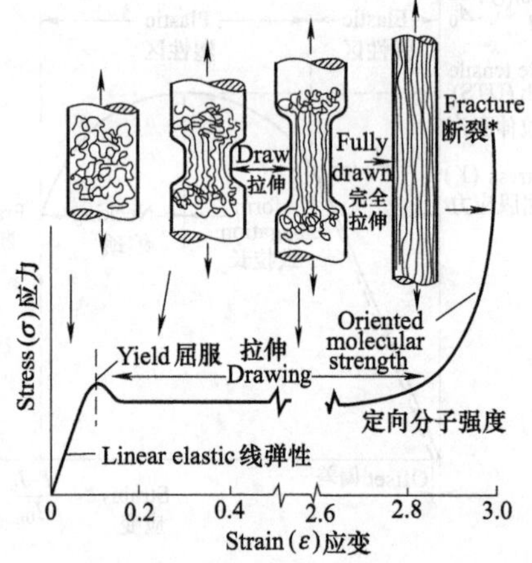

Fig 1.20 Cold-drawing of a linear polymer 线性聚合物的冷拉伸

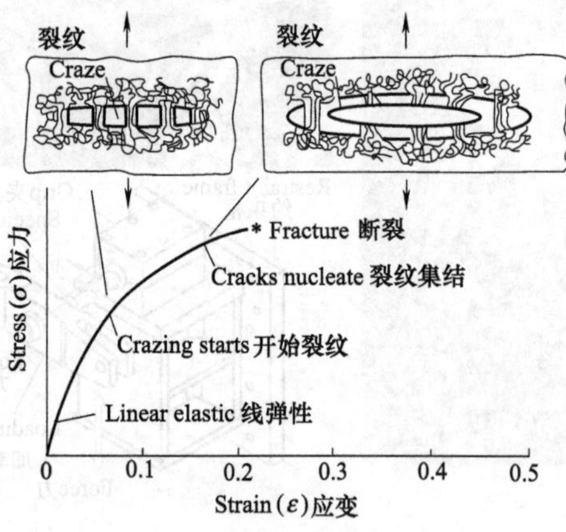

Fig 1.21 Crazing in a linear polymer molecules are drawn out but on a much smaller scale, giving strong strands which bridge the microcracks 线性聚合物裂纹生成图，尽管程度较低，但是裂纹带来的坚韧线头会弥合微裂纹

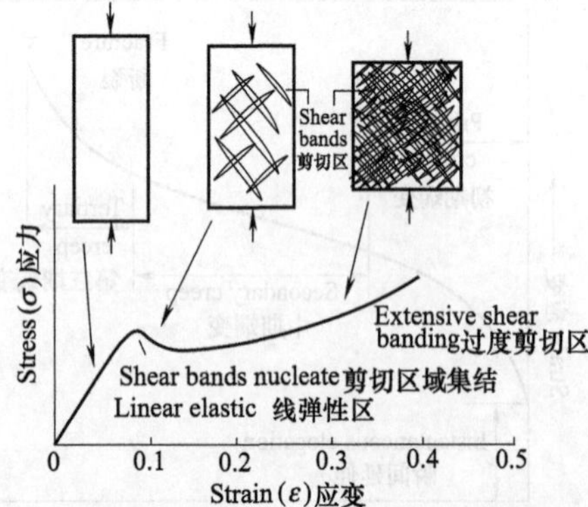

Fig 1.22 Shear banding an alternative form of polymer plasticity which appears in compression 剪切区域：聚合物压缩时呈现的不同形式的塑性

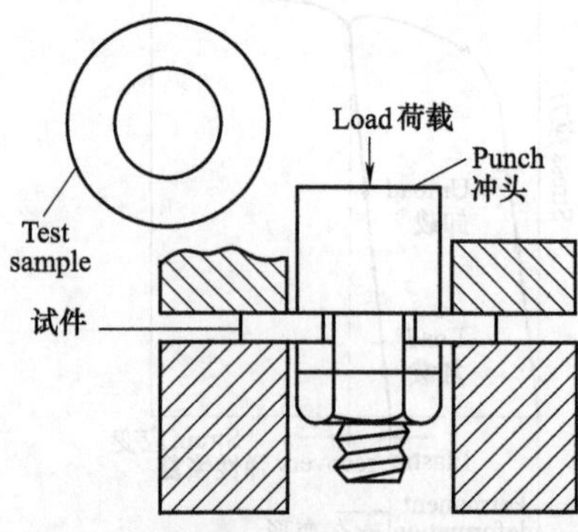

Fig 1.23 Direct shear test used in the plastics industry 塑料工业采用的直接测试剪切应力方法

1.2 Metals 金属材料

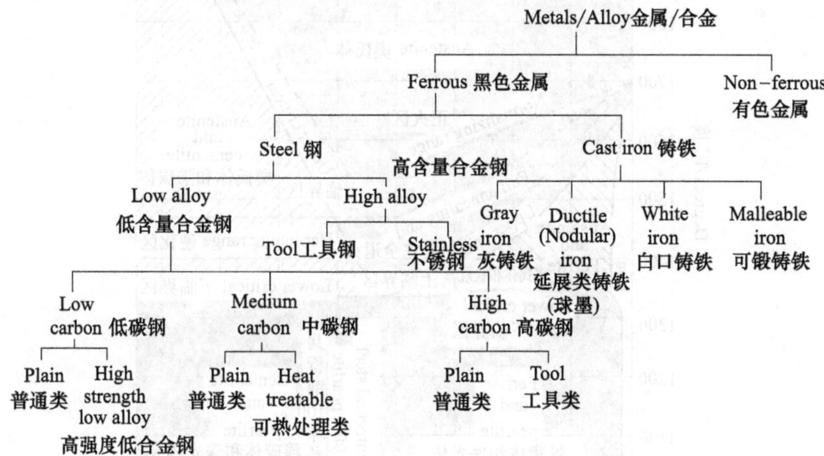

Fig 1.24 Ferrous material 铁族金属（黑色金属）

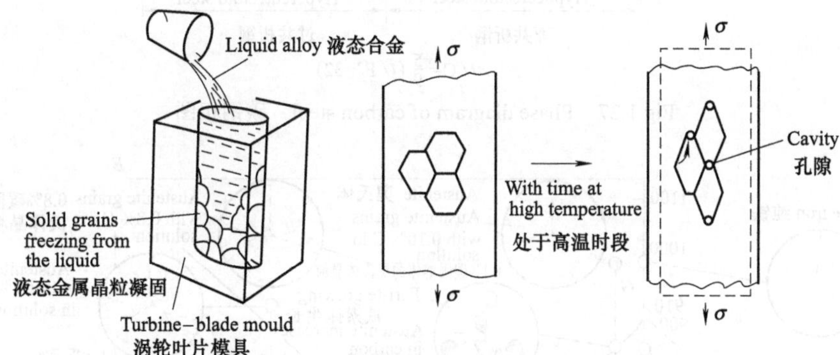

Fig 1.25 Investment casting of turbine blades 消失模铸造涡轮叶片

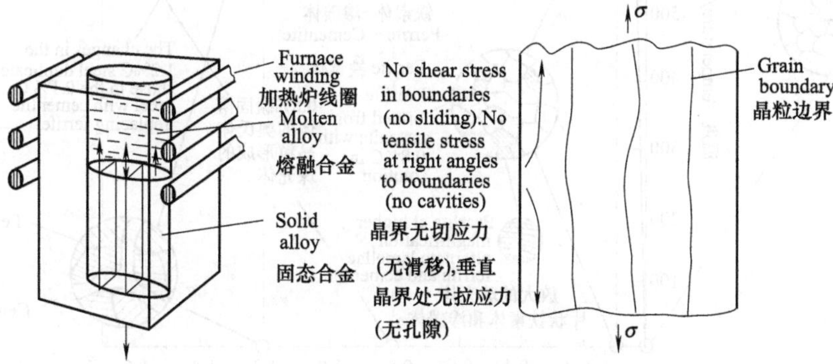

Fig 1.26 Directional solidification(DS)of turbine blade 涡轮叶片的定向固化

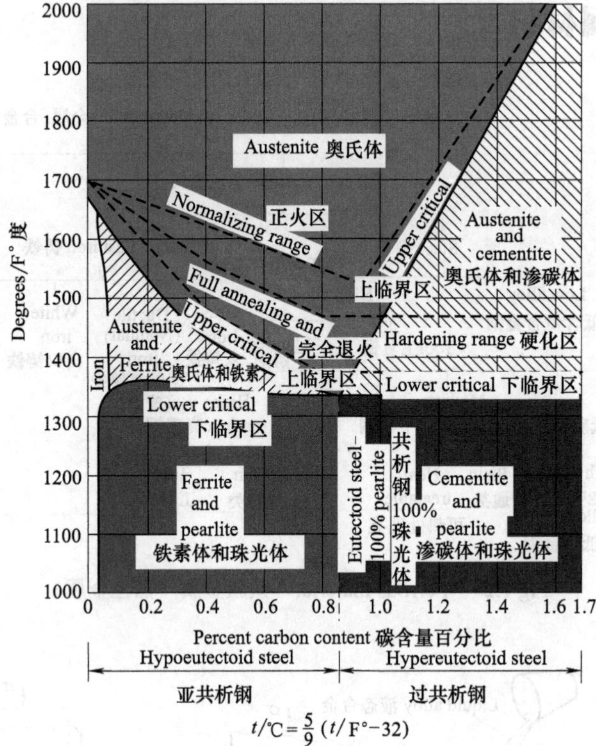

$$t/℃=\frac{5}{9}(t/F°-32)$$

Fig 1.27 Phase diagram of carbon steel 碳钢相图

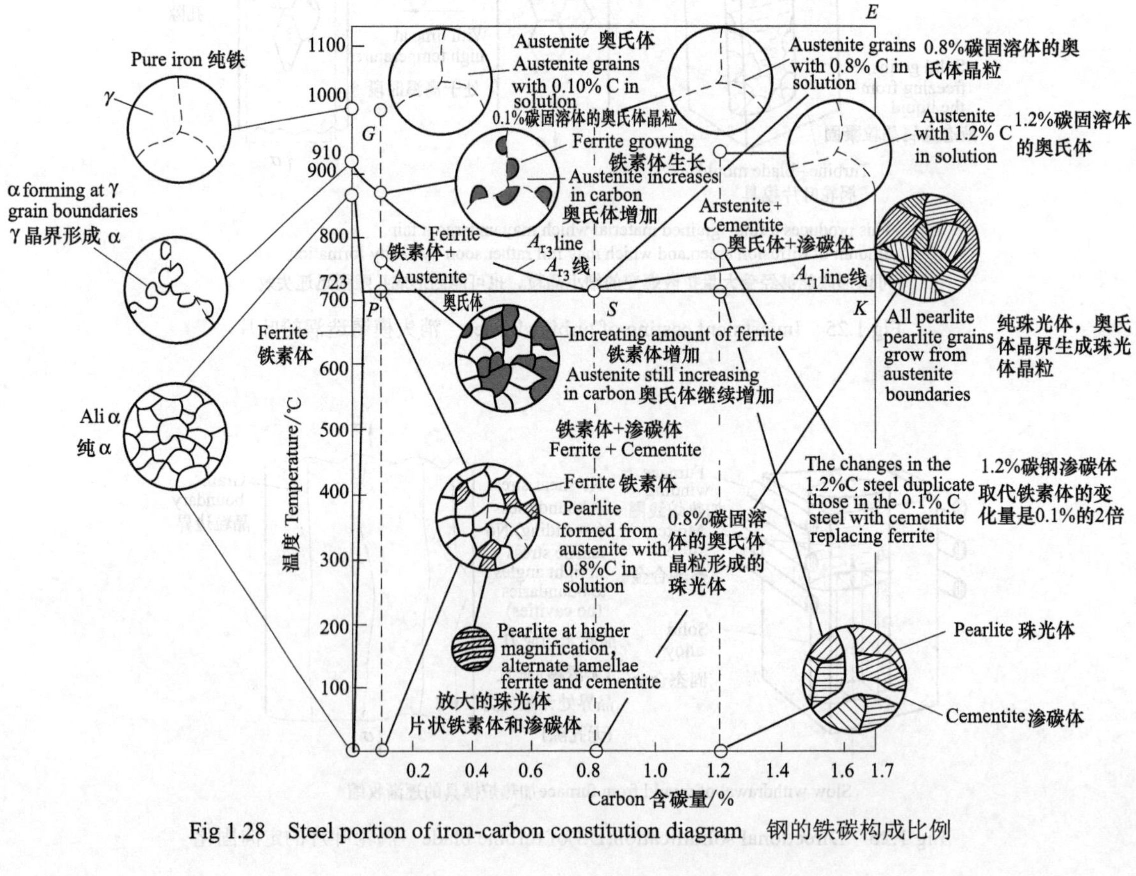

Fig 1.28 Steel portion of iron-carbon constitution diagram 钢的铁碳构成比例

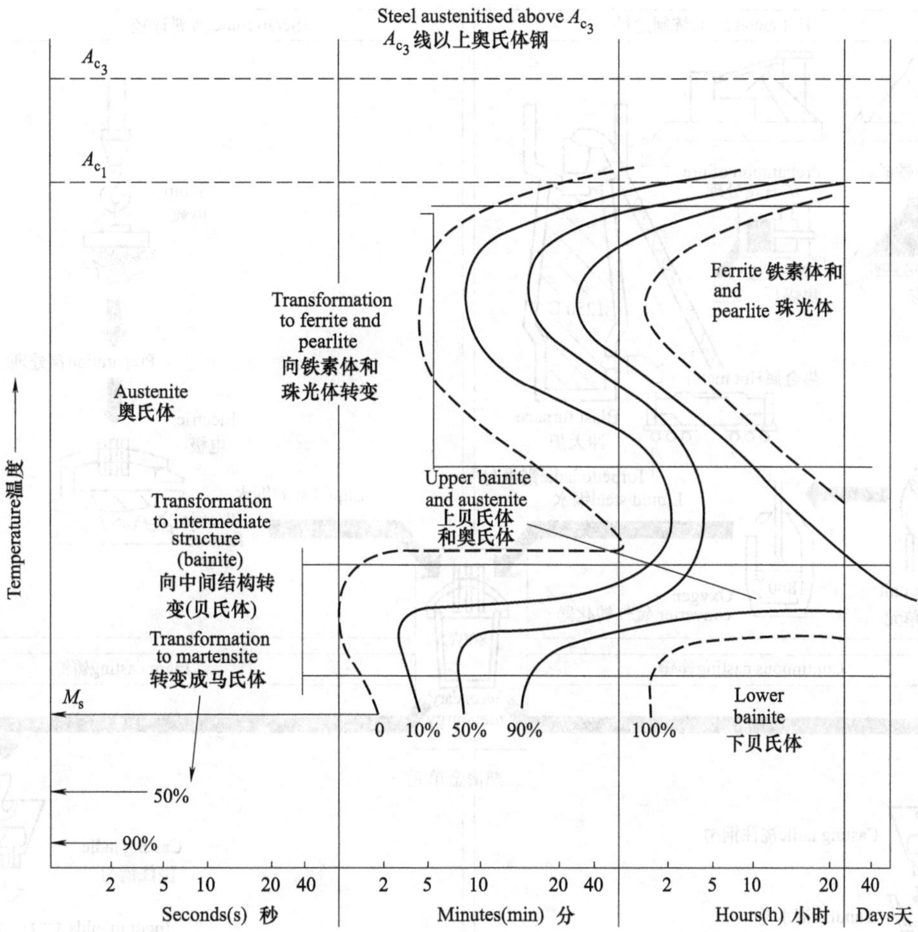

Fig 1.29 Schematic isothermal transformation diagramm 等温转变图

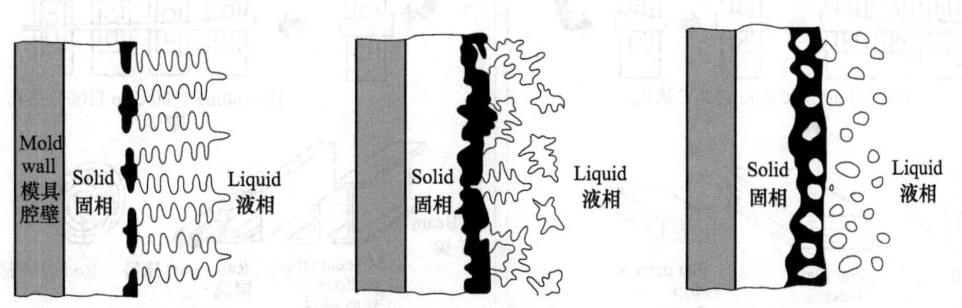

Fig 1.30 Three basic types of cast structures 三种基本晶粒铸造件结构

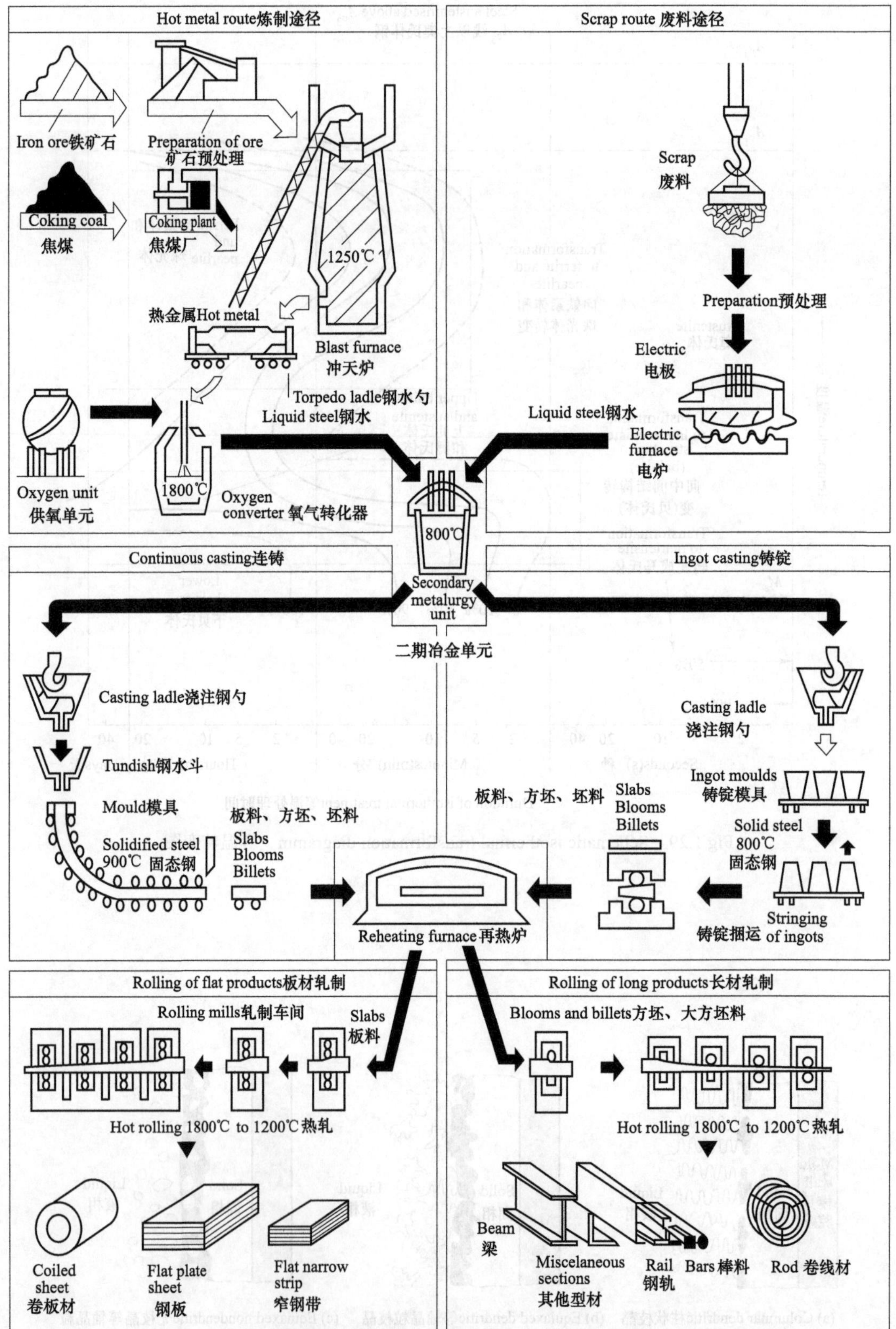

Fig 1.31 Steel production process 钢材生产工艺

Fig 1.32 Basic-oxygen process 基本氧气炼钢法

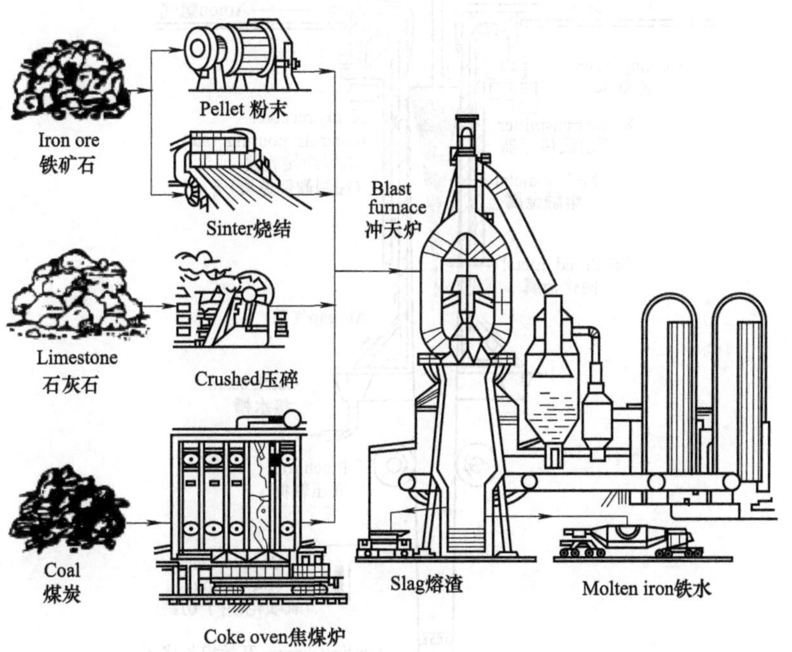

Fig 1.33 Blast furnace 冲天炉

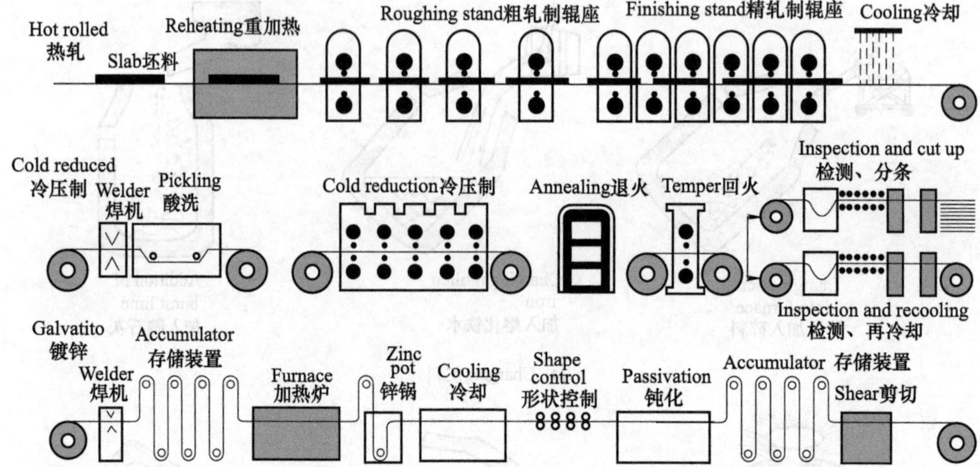

Fig 1.34 Mill trains for producing hot and cold-rolled strip 冷轧制和热轧制钢带的加工流程

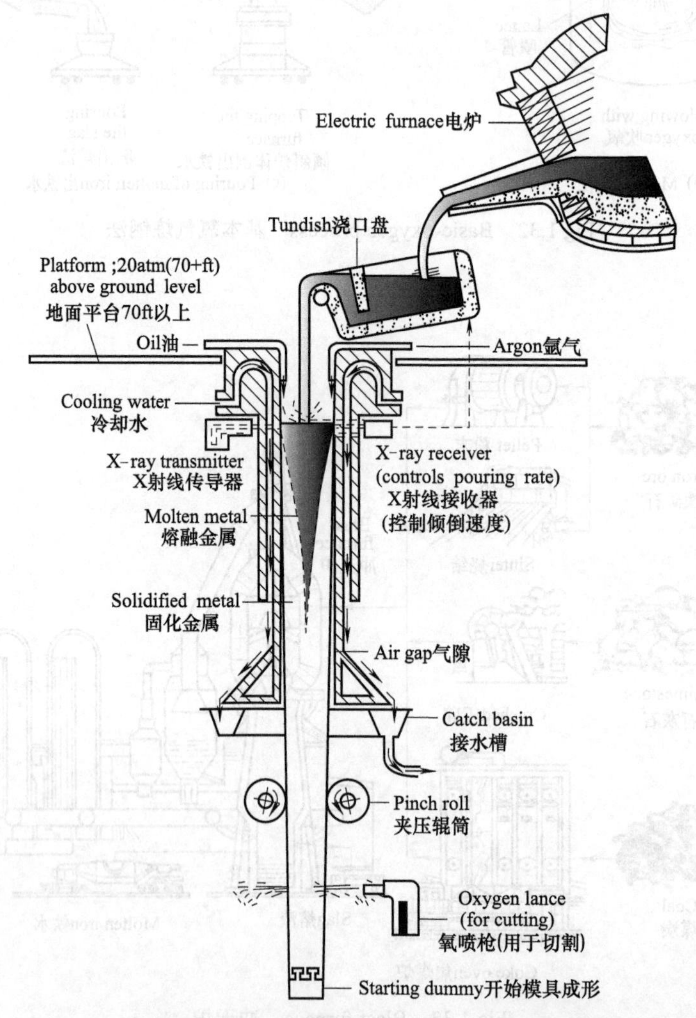

Fig 1.35 The continuous-casting process for steel 钢的连铸工艺

1.3 Non-metals 非金属材料

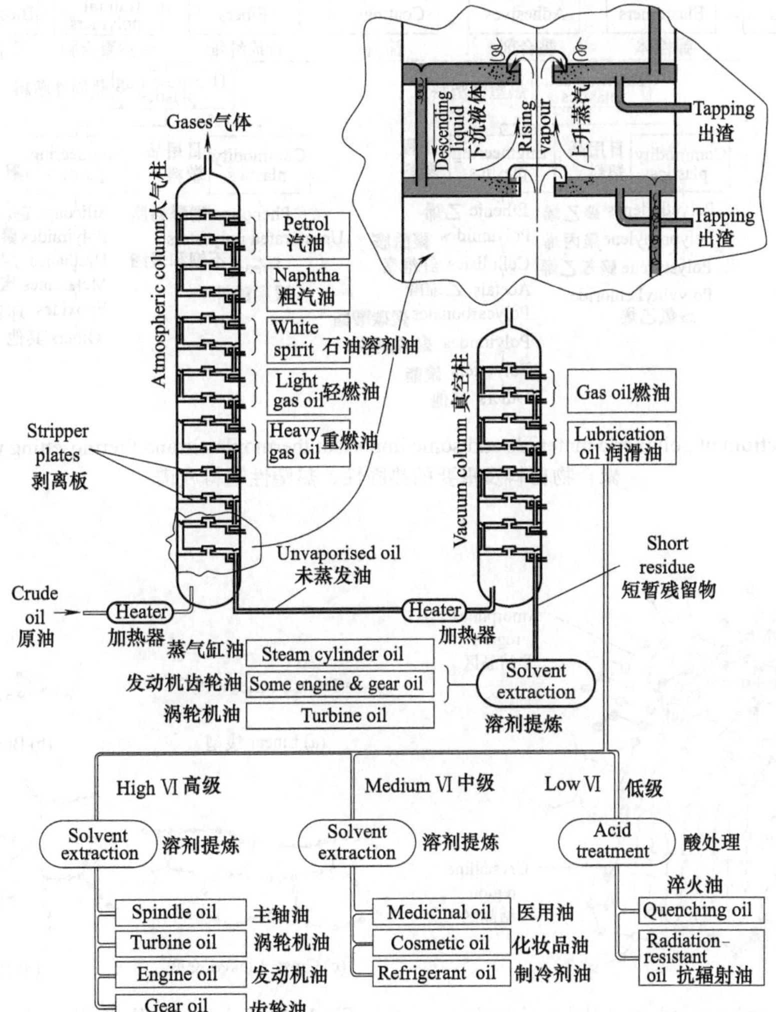

Fig 1.36 Schematic diagram of crude oil distillation process 原油提炼过程图解

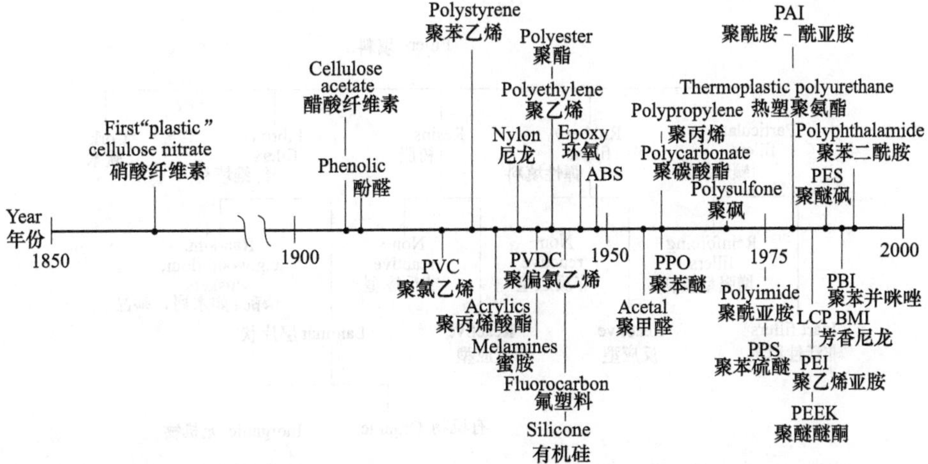

Fig 1.37 Chronological development of important engineering polymers 重要工程聚合物产生的年份

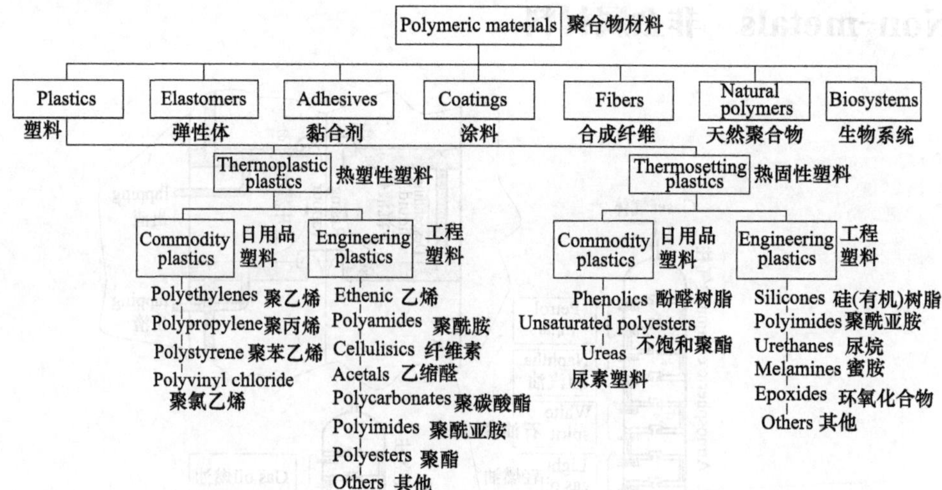

Fig 1.38 Spectrum of polymeric materials and some important thermoplastic and thermosetting plastic families 聚合物材料及重要的热固性、热塑性塑料族谱

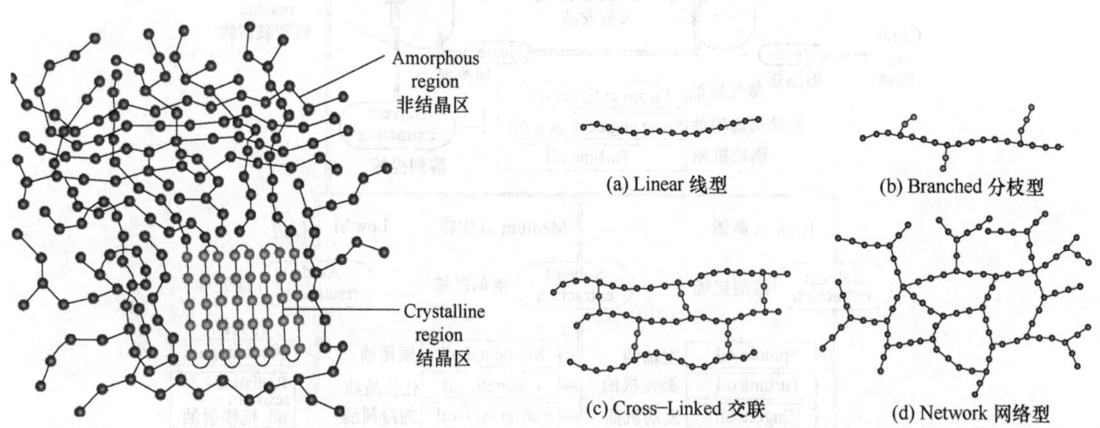

Fig 1.39 Amorphous and crystallne regions in a polymer 聚合物的非结晶区和结晶区

Fig 1.40 Schematic illustration of polymer chains 聚合物链型类别

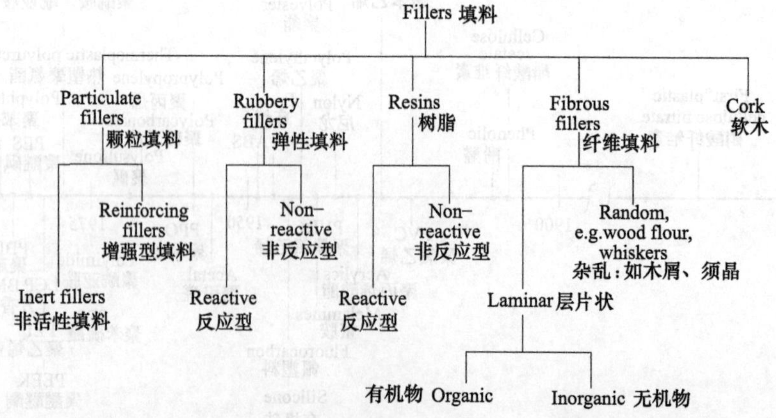

Fig 1.41 Classification of fillers in polymer compounds 混合聚合物填料的类别

Growth ring 年轮
Ray 放射状
Axial轴向
Radial 径向
Tangential 切向
Axial 轴向
Growth ring 年轮
Early wood早期木质
Late wood晚期木质
Cambium新生组织
Bark 树皮
Ray 放射状
Tangential 切向
Radial 径向
Cell 木细胞

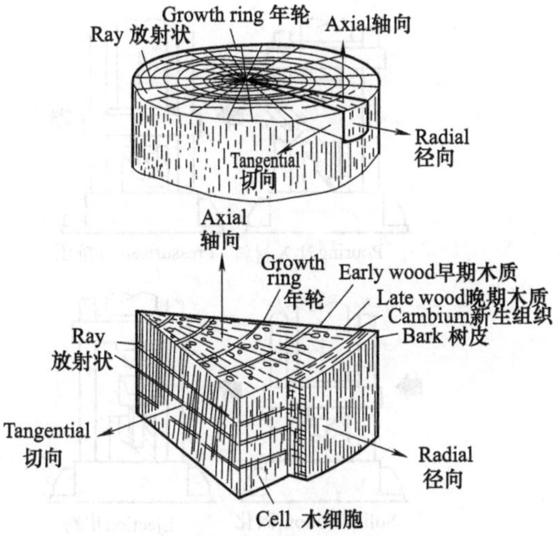

Fig 1.42 The macrostructure of wood 木质的宏观结构

Radial 径向
Grain direction 颗粒方向
Longitudinal 纵向
Tangential 切向

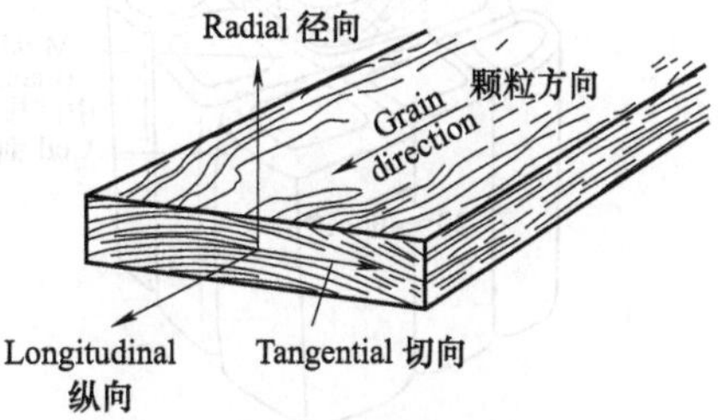

Fig 1.43 Wood is an orthotropic material 木材的正交各向异性

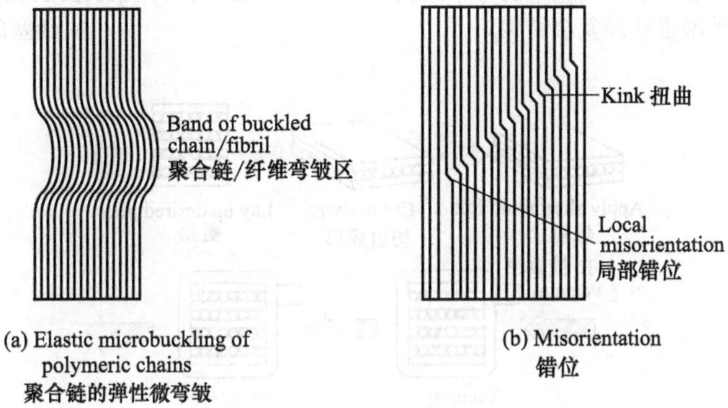

Band of buckled chain/fibril
聚合链/纤维弯皱区
Kink 扭曲
Local misorientation
局部错位
(a) Elastic microbuckling of polymeric chains
聚合链的弹性微弯皱
(b) Misorientation
错位

Fig 1.44 Two compressive failure models 两种压缩失效模型

1.4 Composite materials and nanomaterials 复合材料和纳米材料

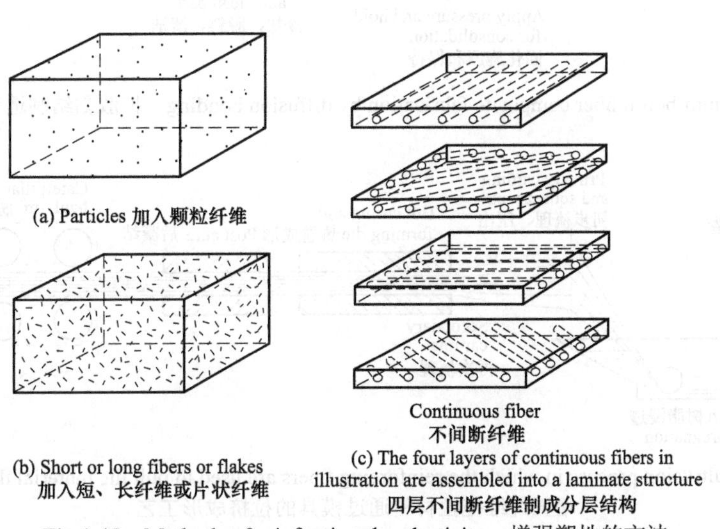

(a) Particles 加入颗粒纤维
(b) Short or long fibers or flakes
加入短、长纤维或片状纤维
Continuous fiber
不间断纤维
(c) The four layers of continuous fibers in illustration are assembled into a laminate structure
四层不间断纤维制成分层结构

Fig 1.45 Methods of reinforcing the plasticity 增强塑性的方法

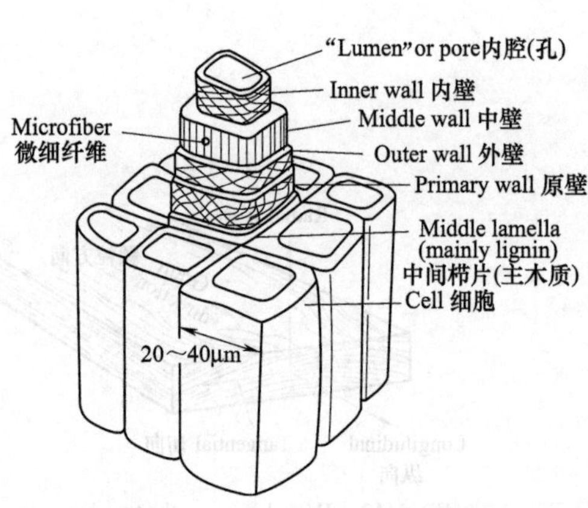

Fig 1.46 The molecular structure of a cell wall (It is a fiber-reinforced composite) 细胞壁的分子结构（它是一种增强纤维复合结构）

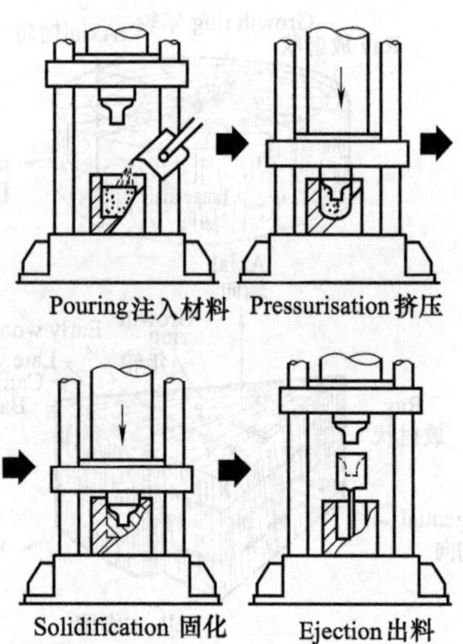

Fig 1.47 Production of a piston with fibrous inserts by squeeze casting 挤压铸造纤维填料活塞的生产

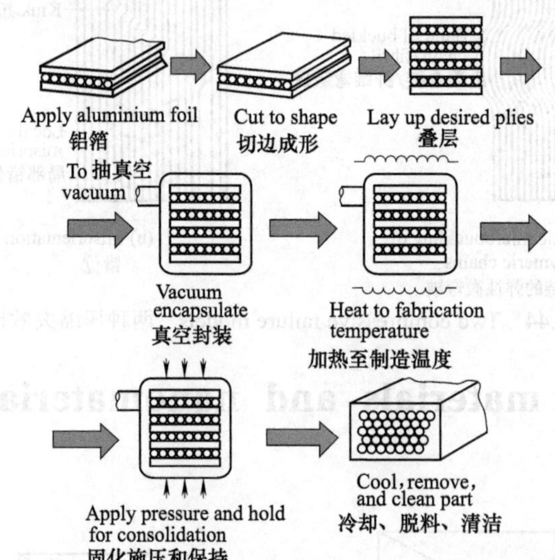

Fig 1.48 Aluminium-boron fiber composite fabrication by diffusion bonding 扩散黏结制造铝硼纤维复合材料

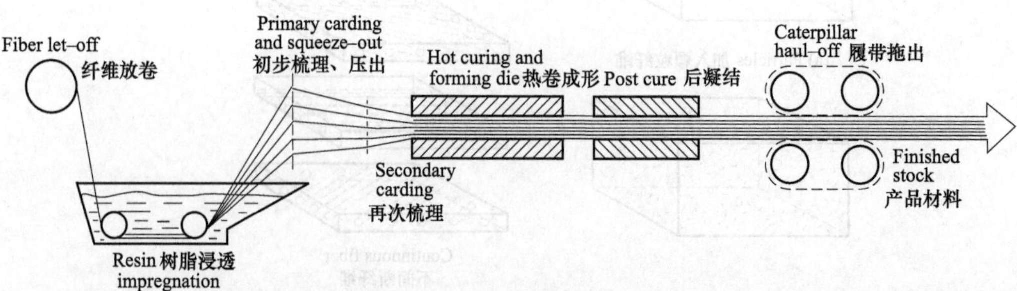

Fig 1.49 Pultrusion process in which the reinforcing fibers are used to pull the material through the die 增强纤维拖拉材料通过模具的拉挤成形工艺

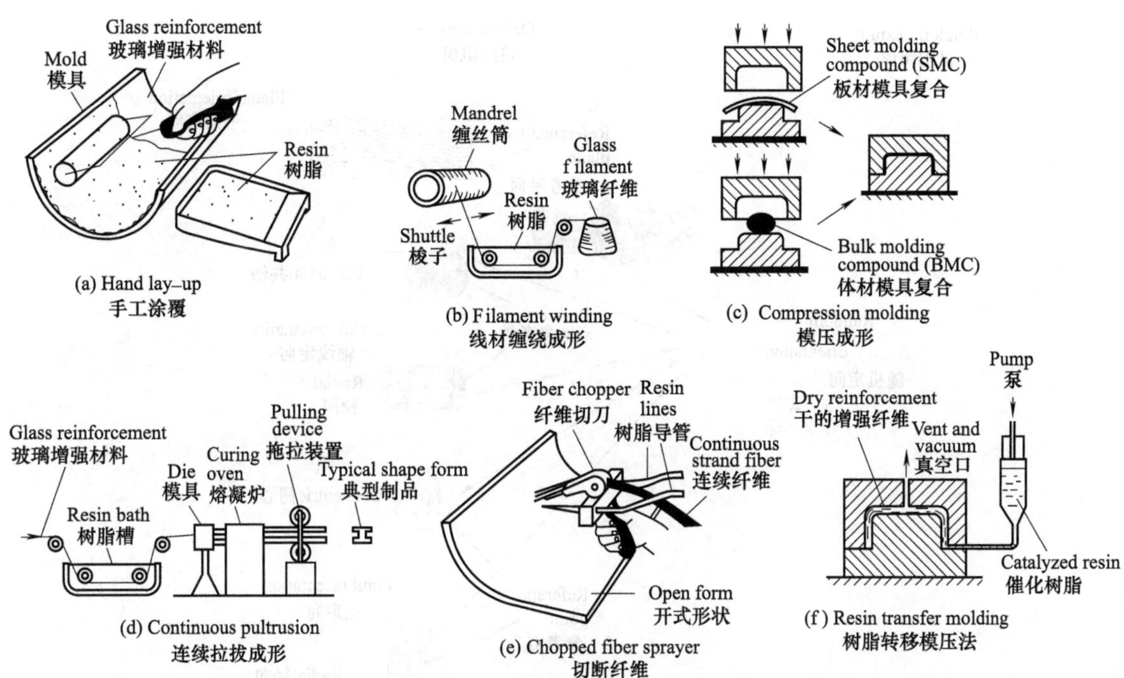

Fig 1.50 Techniques for the fabrication of fiber-reinforced composites 纤维增强复合材料的制作

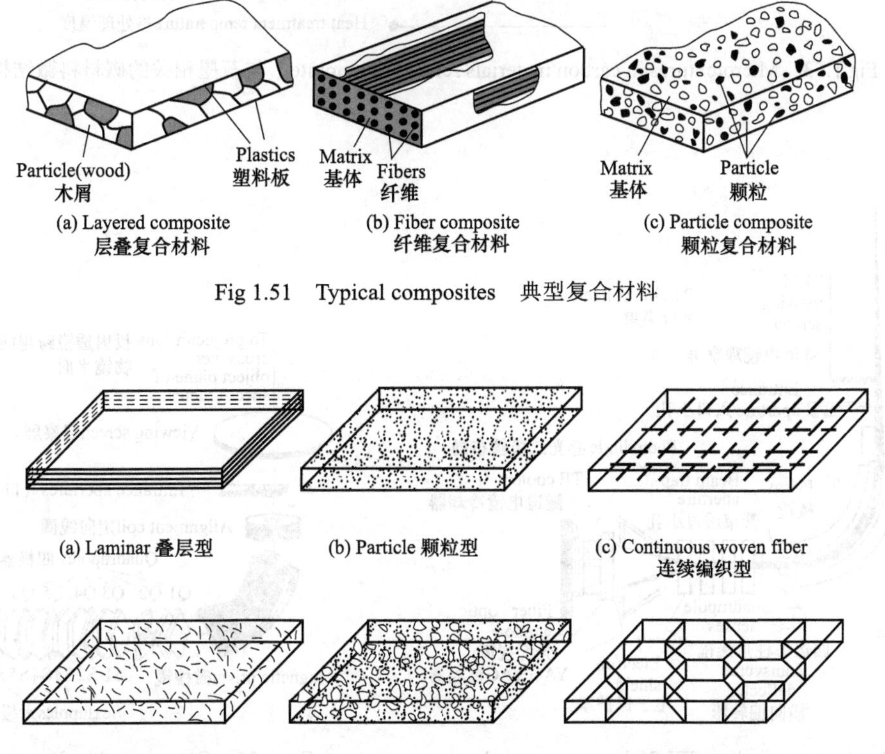

Fig 1.51 Typical composites 典型复合材料

Fig 1.52 Reinforcements used in polymer composites 聚合物复合材料的增强结构

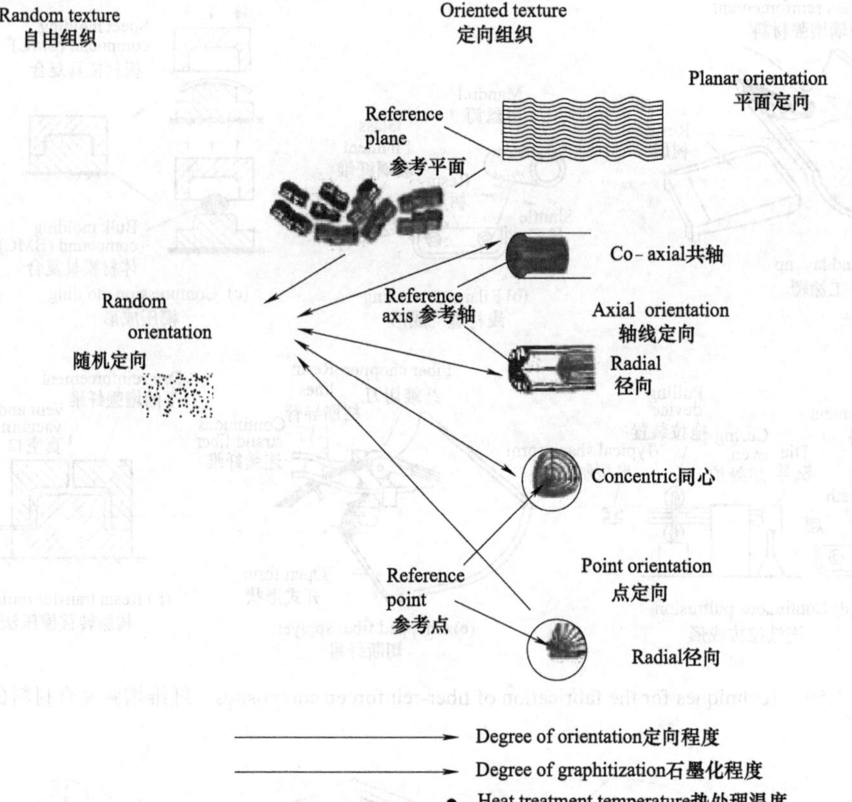

Fig 1.53 Microtextures in carbon materials related to graphite 与石墨相关的碳材料微结构

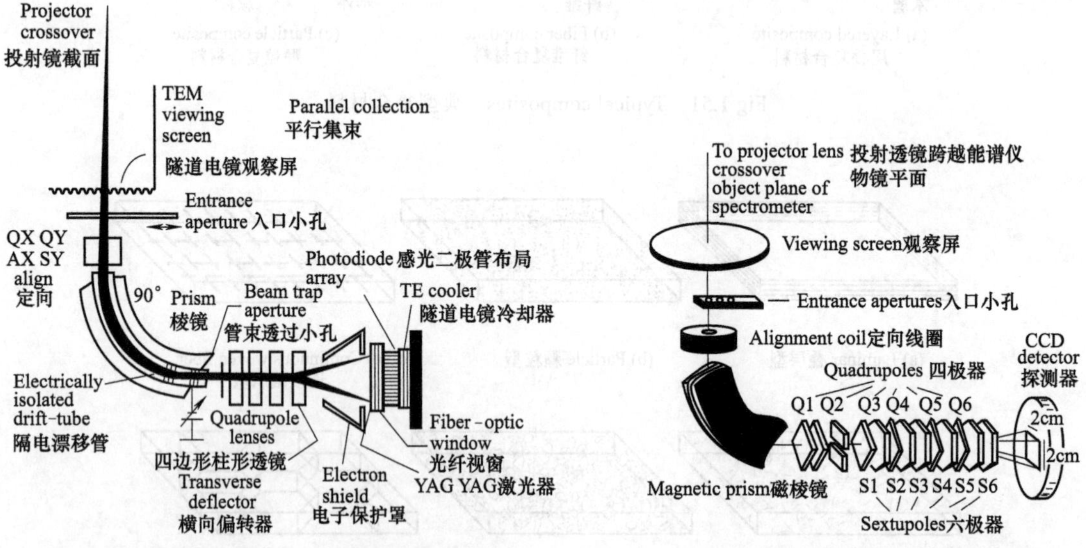

Fig 1.54 A representative EELS(electron energy-loss spectroscopy) acquisition mode, parallel collection 失能电子能谱典型的拾取模式

Fig 1.55 Gatan imaging filer using CCD detector CCD 记录的嘎坦影像过滤

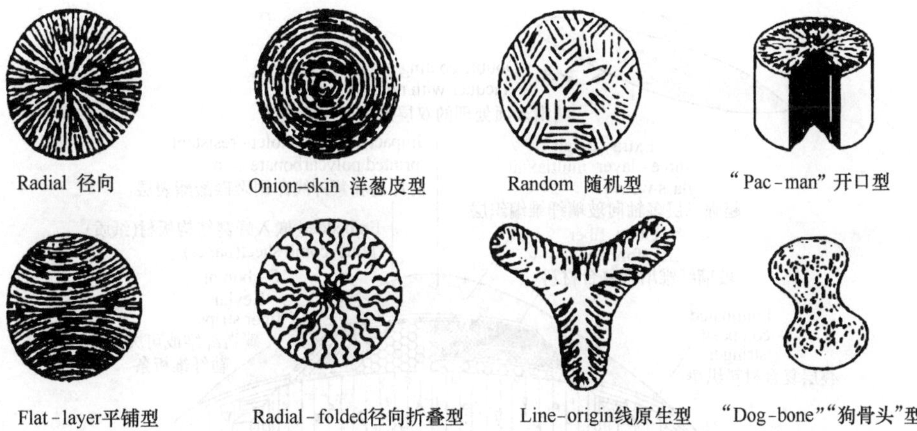

Fig 1.56 Summary of possible carbon fiber morphologies 可能的碳纤维形状汇总

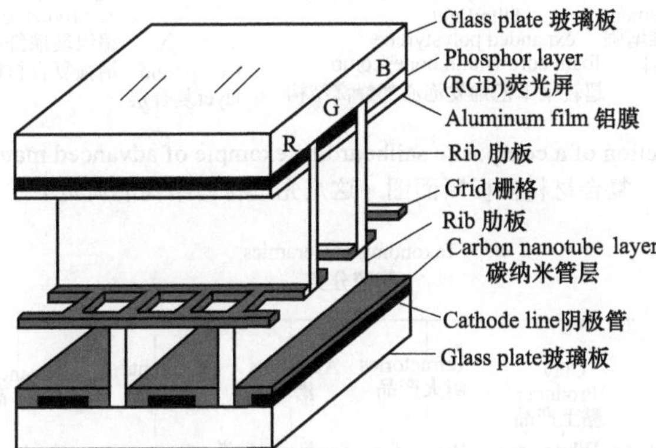

Fig 1.57 Structure of a triode-type vacuum fluorescent display (VFD)-like flat panel display with carbon nanotube cathodes 碳纳米管制成的三极管真空平板荧光屏结构

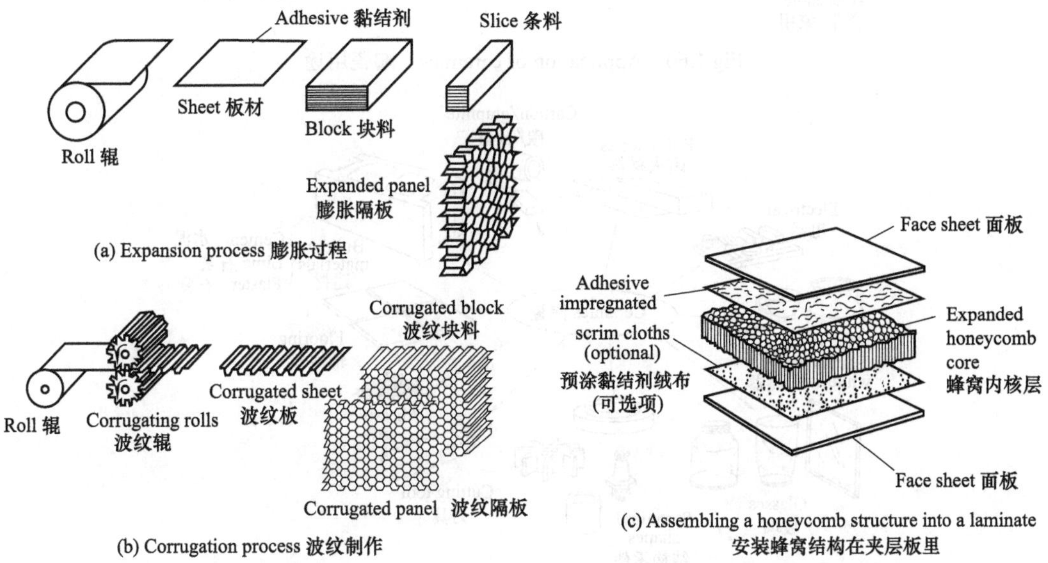

Fig 1.58 Methods of manufacturing honeycomb structures 蜂窝夹层结构制造方法

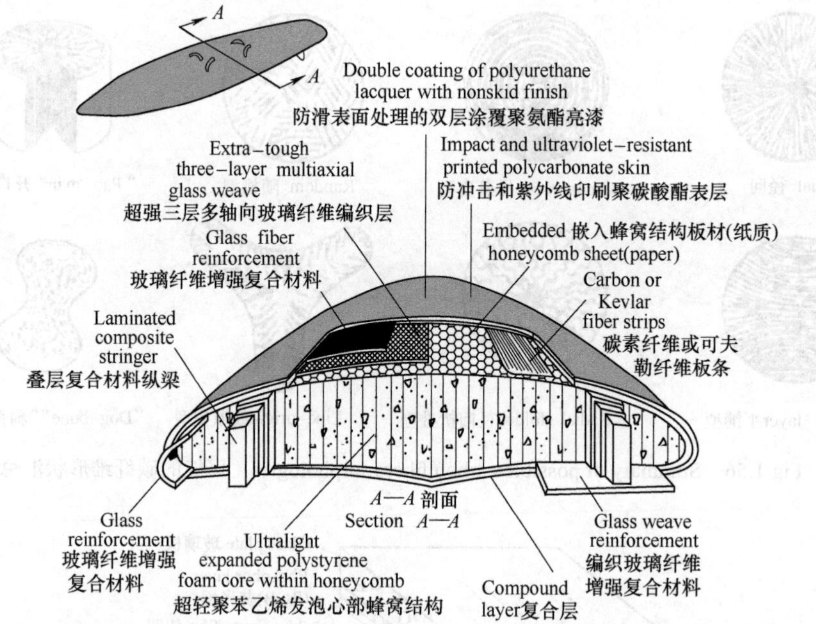

Fig 1.59 Cross-section of a composite sailboard,an example of advanced materials construction 复合材料帆板断面图（这是先进材料结构的典范）

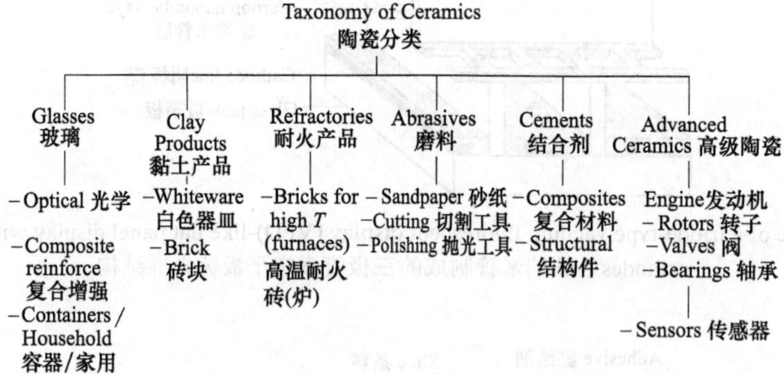

Fig 1.60 Application of ceramics 陶瓷用途

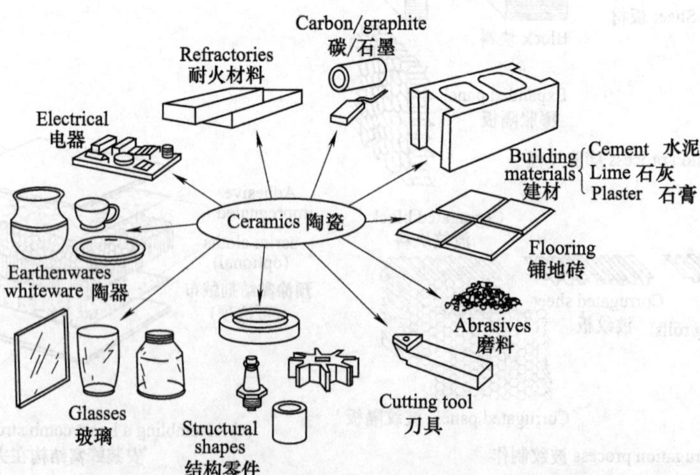

Fig 1.61 Spectrum of ceramic uses 陶瓷材料应用谱图

Unit 2 Drawing and Tolerance 制图与公差

2.1 Engineering drawing 工程制图

2.1.1 Types of line and notation 线型和标注

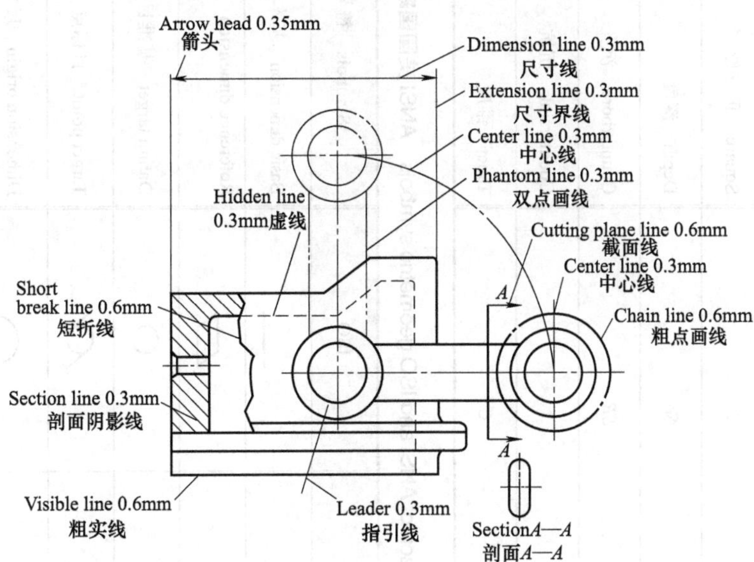

Fig 2.1 Application of different line types 各种线型的应用

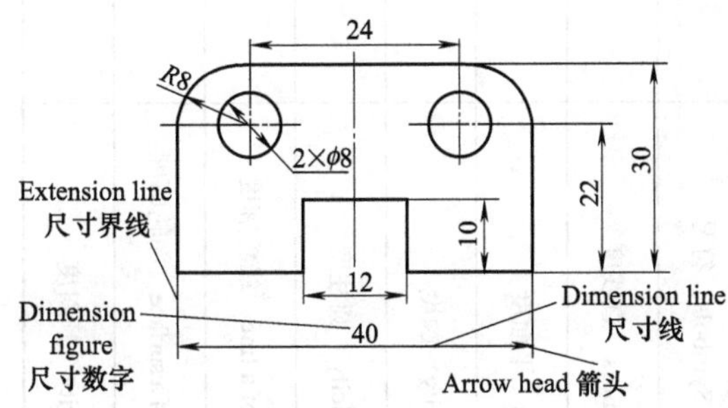

Fig 2.2 Components of dimension 尺寸的组成

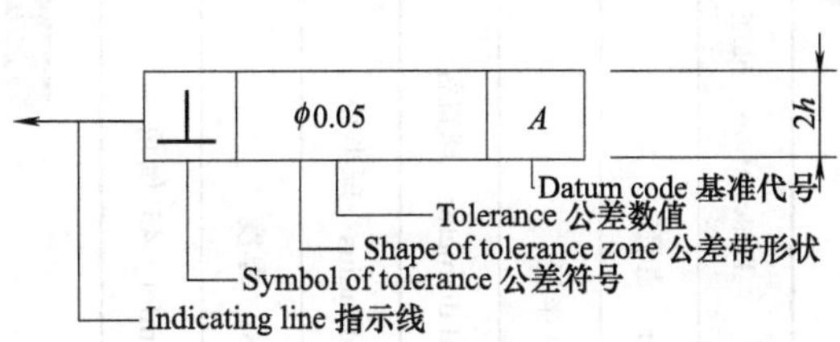

Fig 2.3 Code for shape and location tolerance 形位公差代号

Table 2.1 Commonly used symbols and abbreviations in dimensioning 常用尺寸标注符号和缩写

Description 描述	Symbols and abbreviation 符号缩写	Description 描述	Symbols and abbreviation 符号缩写
Diameter 直径	ϕ	Equally spaced 等距、均布	*EQS*
Radius 半径	*R*	Square 正方形	□
Spherical diameter 球直径	$S\phi$	Depth 深度	↧
Spherical radius 球半径	*SR*	Counterbore 沉孔	⌴
Thickness 厚度	*t*	Countersink 沉坑	⌵
45° Chamfer 45°倒角	*C*	Taper 锥度	⌲

Table 2.2 Comparison of ANSI and ISO geometric symbols ANSI 美国国家标准和 ISO 国际标准几何符号的比较

Symbols 符号	ANSI Y14.5M	ISO	Symbols 符号	ANSI Y14.5M	ISO
Straightness 直线度	⏤	⏤	Basic dimension 基本尺寸	$\boxed{50}$	$\boxed{50}$
Flatness 平面度	⏥	⏥	Reference dimension 参考尺寸	(50)	(50)
Circularity 圆度	○	○	Datum target 基准目标	(φ6/A1)	(φ6/A1)
Cylindricity 圆柱度	⌭	⌭	Target point 目标点	×	×
Profile of a line 线轮廓度	⌒	⌒	Dimension origin 尺寸起点	⌀→	⌀→
Profile of a surface 面轮廓度	⌓	⌓	Spherical radius 球半径	*SR*	*SR*
Angularity 倾斜度	∠	∠	Controlled radius 受控半径	*CR*	None无

续表

Symbols 符号	ANSI Y14.5M	ISO	Symbols 符号	ANSI Y14.5M	ISO
Perpendicularity 垂直度	⊥	⊥	Feature control frame 特征控制框	⌖ | ϕ0.5 Ⓜ | A | B | C	⌖ | ϕ0.5 Ⓜ | A | B | C
Parallelism 平行度	//	//	Datum feature① 基准特征	[A]	[A] 或 [A]
Position 位置度	⌖	⌖	All around-profile 周围轮廓	⊙	⊙ (Proposed) 推荐
Concentricity/coaxiality 同心度 / 同轴度	◎	◎	Conical taper 锥度	▷	▷
Symmetry 对称度	⌯	⌯	Slope 斜度	◺	◺
Radius 半径	*R*	*R*	Counterbore/spotface 沉孔 / 点锪	⌴	⌴ (Proposed) 推荐
Between① 两者之间	↔	None 无	Countersink 锪锥形沉孔	⌵	⌵ (Proposed) 推荐
Circular runout① 圆跳动	↗	↗	Depth/deep 深度	↧	↧ (Proposed) 推荐
Total runout① 全跳动	⌰	⌰	Square(Shape) 正方形	□	□
At maximum material condition 最大实体条件	Ⓜ	Ⓜ	Dimension not to scale 尺寸不成比例	<u>15</u>	<u>15</u>
At least material condition 最小实体条件	Ⓛ	Ⓛ	Number of times/Places 次数 / 地点数	8X	8X
Regardless of feature size 不计特征尺寸	None 无	None 无	Arc length 弧长	⌒105	⌒105
Projected tolerance zone 延伸公差带	Ⓟ	Ⓟ	Spherical diameter 球直径	$S\phi$	$S\phi$
Diameter 直径	∅	∅	Statical tolerance 静公差	⟨ST⟩	None无

① Arrowheads may be filled in. 箭头可以填黑。

Table 2.3 Application of geometric control symbols 几何控制符号的应用

Type 类别	Geometric characteristics 几何特征	Pertains to 相关要素	Basic dimensions 基本尺寸	Feature modifier 特征修正	Datum modifier 基准修正
Form 形状	Straightness 直线度 Circularity 圆度 Flatness 平面度 Cylindricity 圆柱度	Only individual feature 只需自身特征		Modifier not applicable 不需修正	No datum 无基准
Profile 轮廓	Profile(Line) 轮廓（线） Profile(Surface) 轮廓（面）	Individual or related 独立或相关	Yes if related 相关则有		RFS implied unless MMC or LMC is stated 不计特征尺寸，除非注明了MMC或LMC
Orientation 定位	Angularity 倾斜度 Perpendicularity 垂直度 Parallelism 平行度	Always related feature(s) 总需相关特征	Yes 有	不计特征尺寸 RFS implied unless MMC or LMC is stated 注明MMC或LMC	
Location 位置	Position 位置度 Concentricity 同轴度 Symmetry 对称度		Yes 有		
Runout 跳动	Circular runout 圆跳动 Total runout 全跳动			Only RFS 不计特征尺寸	Only RFS 不计特征尺寸

注：RFS——regardless of feature size 不计特征尺寸。

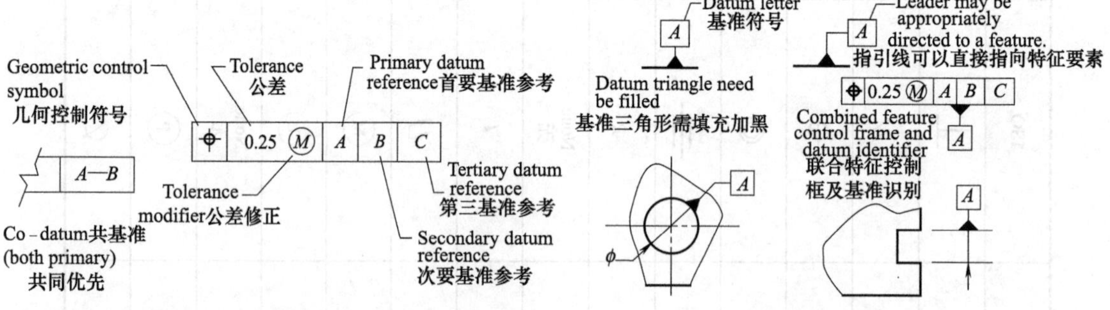

Fig 2.4 Feature control frame and datum order of precedence 特征控制框和基准优先顺序

Fig 2.5 Datum feature symbol 基准特征符号

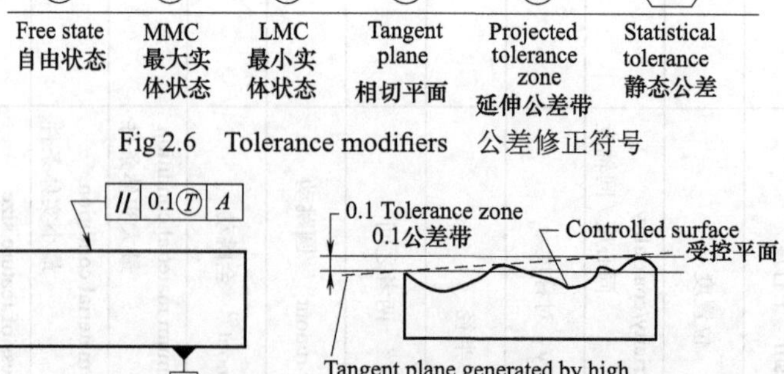

Fig 2.6 Tolerance modifiers 公差修正符号

Fig 2.7 Tangent plane modifier 相切平面修正符号

2.1.2 CAD presentation 计算机辅助设计绘图

2D Lateral model	2D Profile body	2D Rotating body	3D Wire-frame model	3D Surface model	3D Volume model

(a) 二维边界模型 (b) 二维线框模型 (c)二维回转体线框模型 (d) 三维线框模型 (e) 三维表面模型 (f) 三维立体模型

Fig 2.8 Various types of modeling for CAD 几种 CAD 建模

2.1.3 On engineering drawing presentation 机械制图

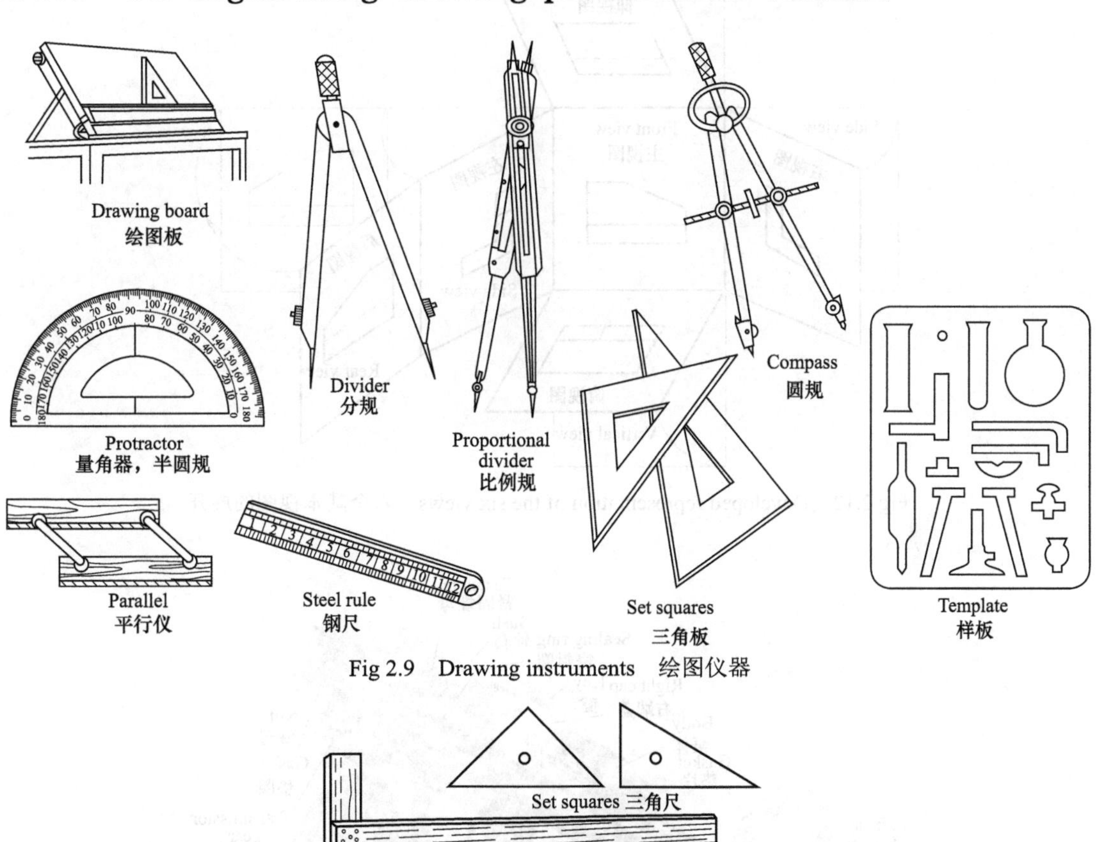

Fig 2.9 Drawing instruments 绘图仪器

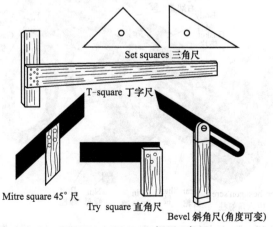

Set squares 三角尺

T-square 丁字尺

Mitre square 45° 尺

Try square 直角尺

Bevel 斜角尺(角度可变)

Fig 2.10 Square 规尺 / 角尺

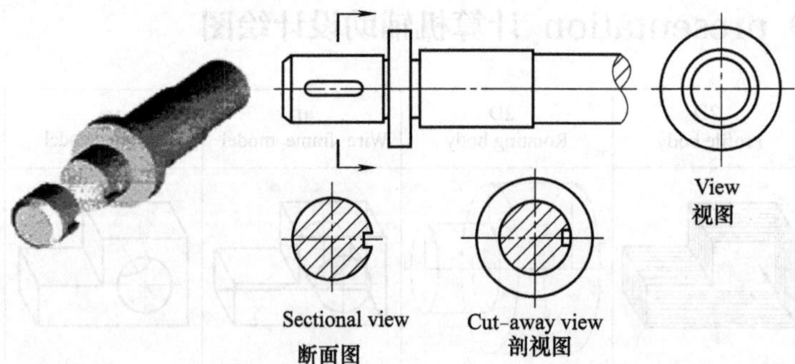

Fig 2.11 Comparison between sectional view and cut-away view 断面图与剖视图的比较

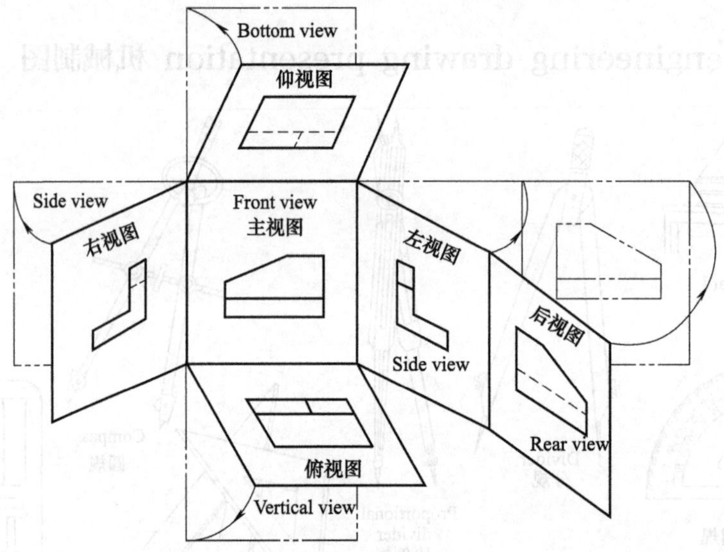

Fig 2.12 Developed representation of the six views 六个基本视图的展开

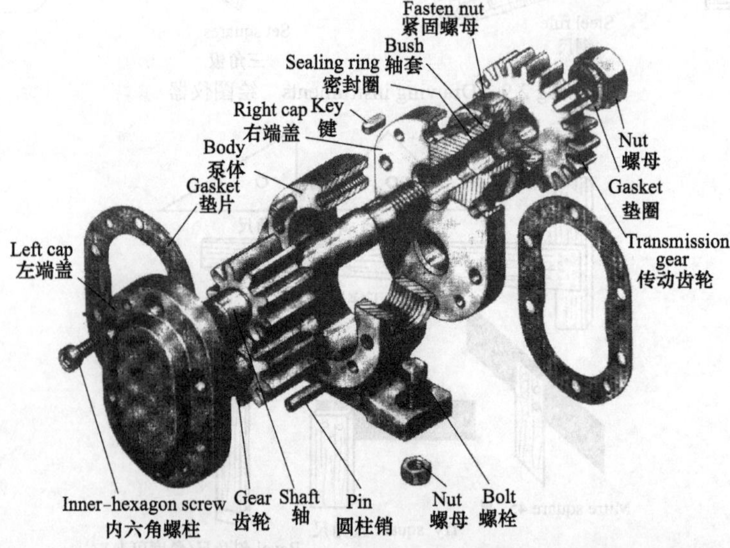

Fig 2.13 Exploded drawing of gear pump 齿轮泵（分解）爆炸图

2.2 Tolerance and fits 公差与配合

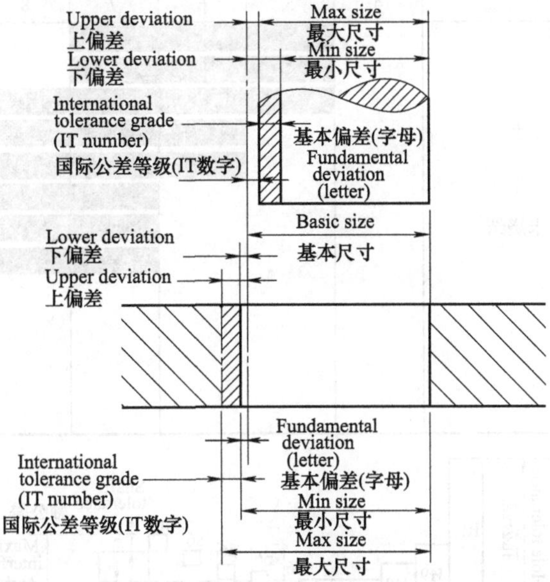

Fing 2.14 Illustration of definitions of limits and fits 极限与配合定义图解

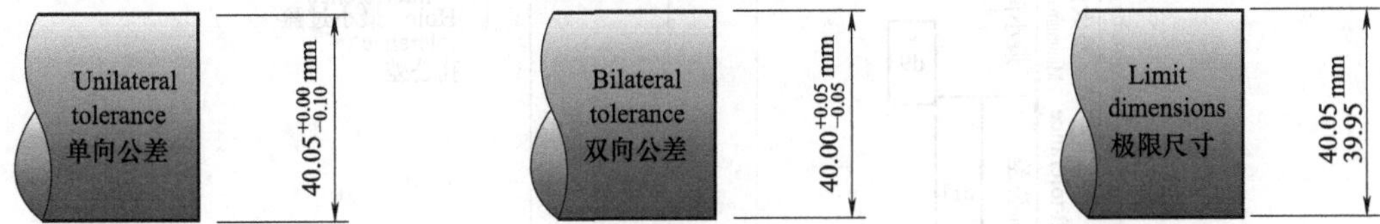

Fig 2.15 Various methods of assigning tolerance on a shaft 轴上公差分配的几种标注方法

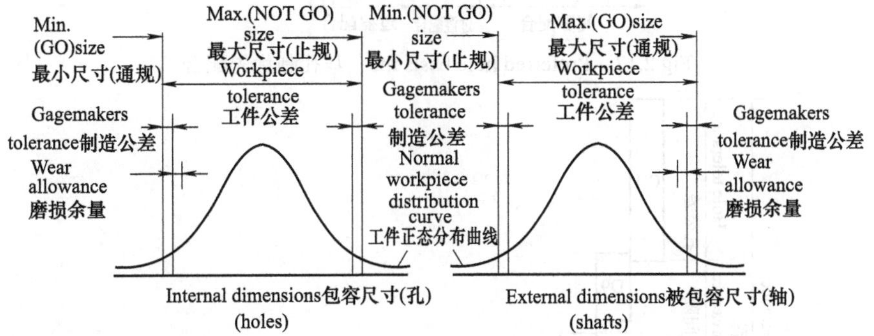

Fig 2.16 Gagemakers tolerance, wear allowance and workpiece tolerance 极限量规制造公差、磨损余量与工件公差

Table 2.4 Machining processes and IT tolerance grades 加工工艺与公差等级

Machining processes 加工工艺	IT grades 公差等级							
	4	5	6	7	8	9	10	11
Lapping & Honing 研磨 / 珩磨	■	■						
Cylindrical grinding 外圆磨削		■	■	■				
Surface grinding 平面磨削		■	■	■	■			

续表

Machining processes 加工工艺	IT grades 公差等级							
	4	5	6	7	8	9	10	11
Diamond turning 金刚石车削		■	■	■				
Diamond boring 金刚石镗削		■	■	■				
Broaching 拉削		■	■	■	■			
Powder metal sizes 金属粉末尺寸			■	■	■			
Reaming 铰孔			■	■	■	■	■	
Turning 车削				■	■	■	■	■
Powder metal sintered 金属粉末烧结				■	■	■	■	
Boring 镗孔				■	■	■	■	■
Milling 铣削							■	■
Planing & Shaping 刨削							■	■
Drilling 钻削							■	■
Punching 冲压							■	■
Die casting 压铸								■

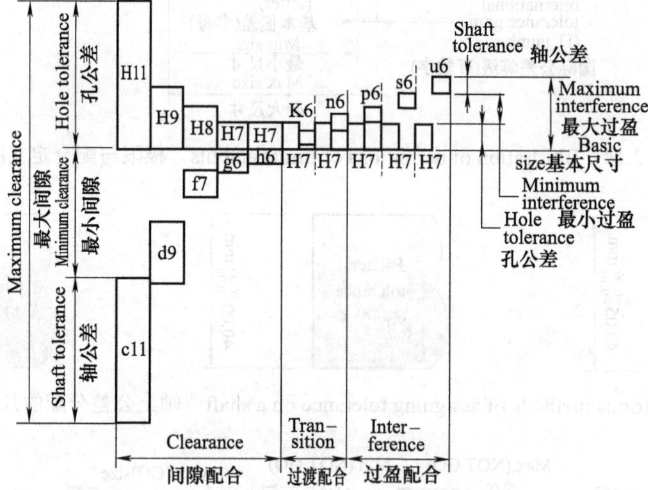

Fig 2.17 Preferred hole basis fits 基孔制优先配合

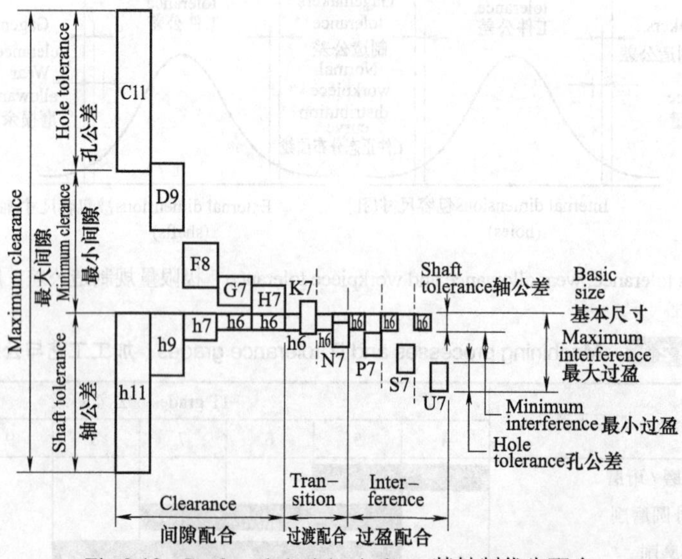

Fig 2.18 Preferred shaft basis fits 基轴制优先配合

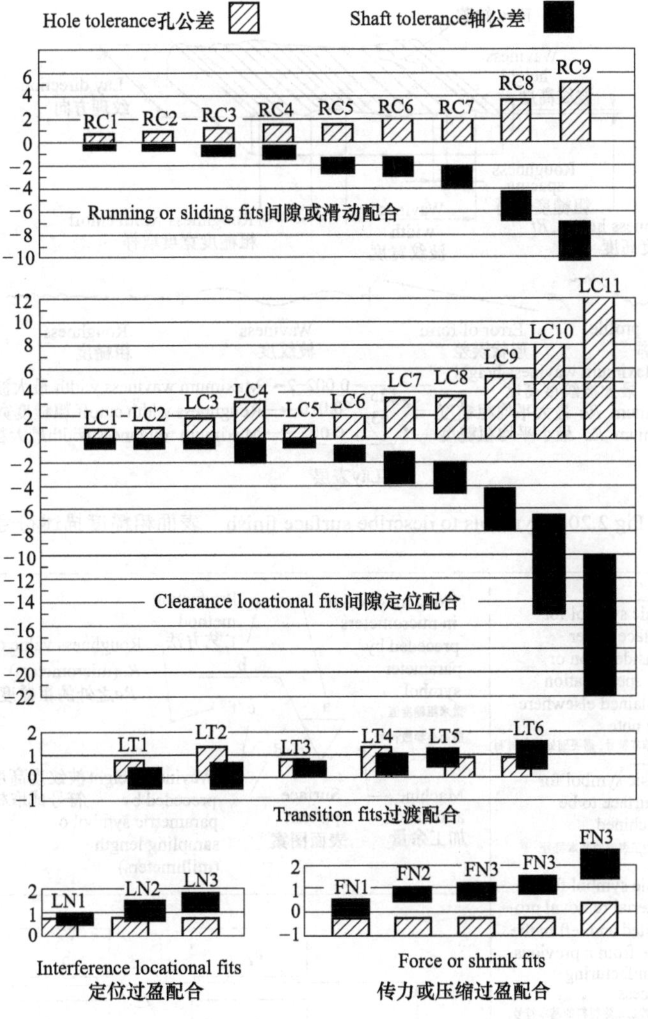

Fig 2.19 Graphical representation of standard limits and fits 标准极限与配合图解

Table 2.5 ISO surface parameter symbols 表面参数符号

Rp=max height profile 最大高度轮廓	*Rδc*=profile section height difference 轮廓剖面高度差
Rv=max profile valley depth 最大轮廓谷底深度	*lp*=sampling length-primary profile 取样长度初始轮廓
Rz=max height of the profile 最大轮廓高度	*lw*=sampling length-waviness profile 取样长度波纹轮廓
Rc=mean height of the profile 轮廓平均高度	*lr*=sampling lenght-roughness profile 取样长度粗糙度轮廓
Rt=total height of the profile 轮廓总高度	*ln*=evaluation length 评价长度
Ra=arithmetic mean deviation of the profile 轮廓算术平均偏差	*Z(x)*=ordinate value 坐标值
Rq=root mean square deviation of the profile 轮廓均方差	d*Z*/d*x*=local slope 局部倾斜
Rsk=skewness of the profile 轮廓畸变	*Zp*=profile peak height 轮廓峰值高度
Rku=kurtosis of the profile 轮廓凸出度	*Zv*=profile valley depth 轮廓谷底高度
RSm=mean width of the profile 轮廓平均宽度	*Zt*=profile element height 轮廓元件高度
RΔq=root mean square slope of the profile 轮廓倾斜均方根	*Xs*=profile element width 轮廓元件宽度
Rmr=material ration of the profile 轮廓材料的分配	*Ml*=material length of profile 轮廓材料长度

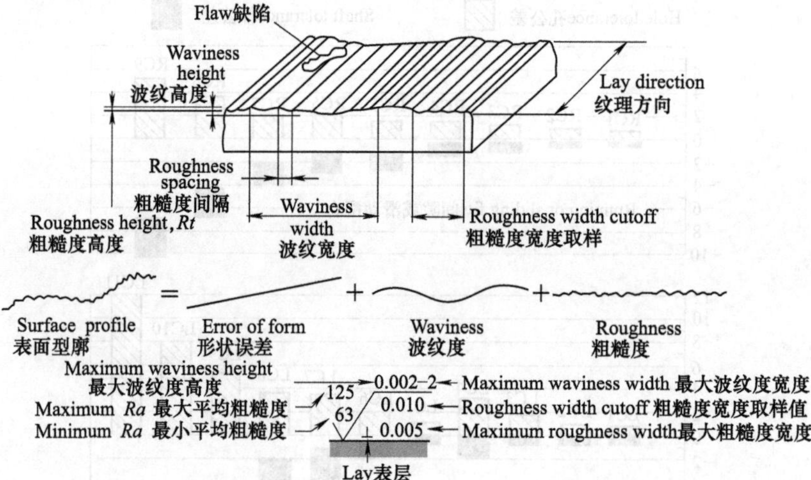

Fig 2.20 Symbols to describe surface finish 表面粗糙度描述符号

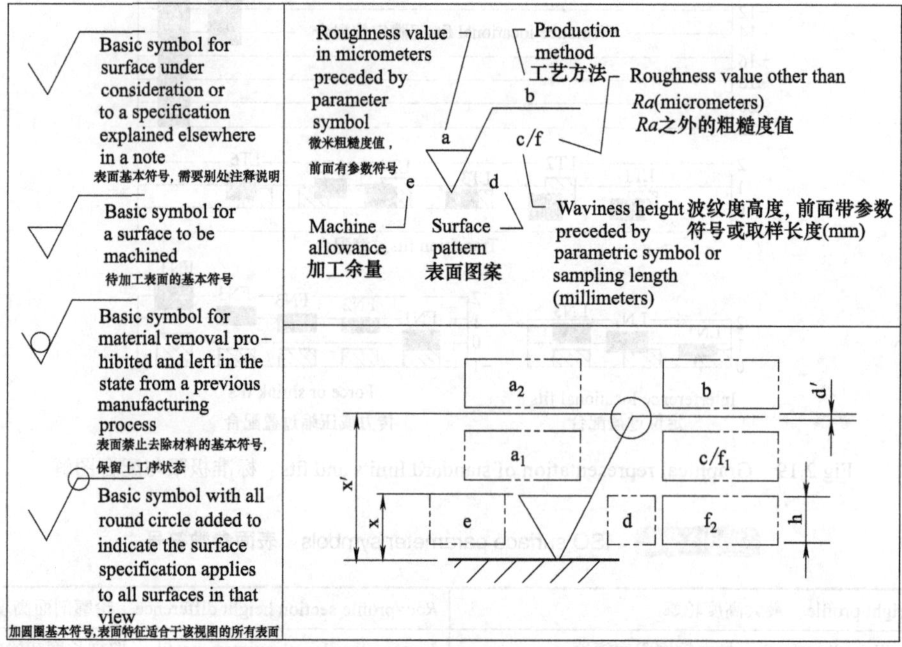

Fig 2.21 ISO surface finish symbol ISO 表面粗糙度符号

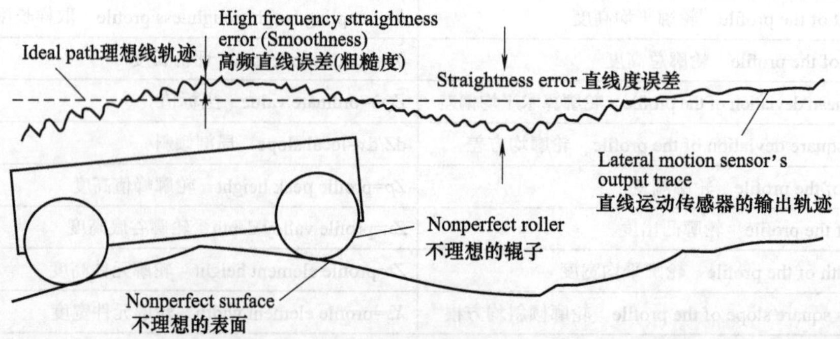

Fig 2.22 Straightness errors caused by surface form and finish errors 表面形状和粗糙度误差引起的直线度误差

Table 2.6 Surface roughness produced by common production methods 常见加工方法获得的粗糙度

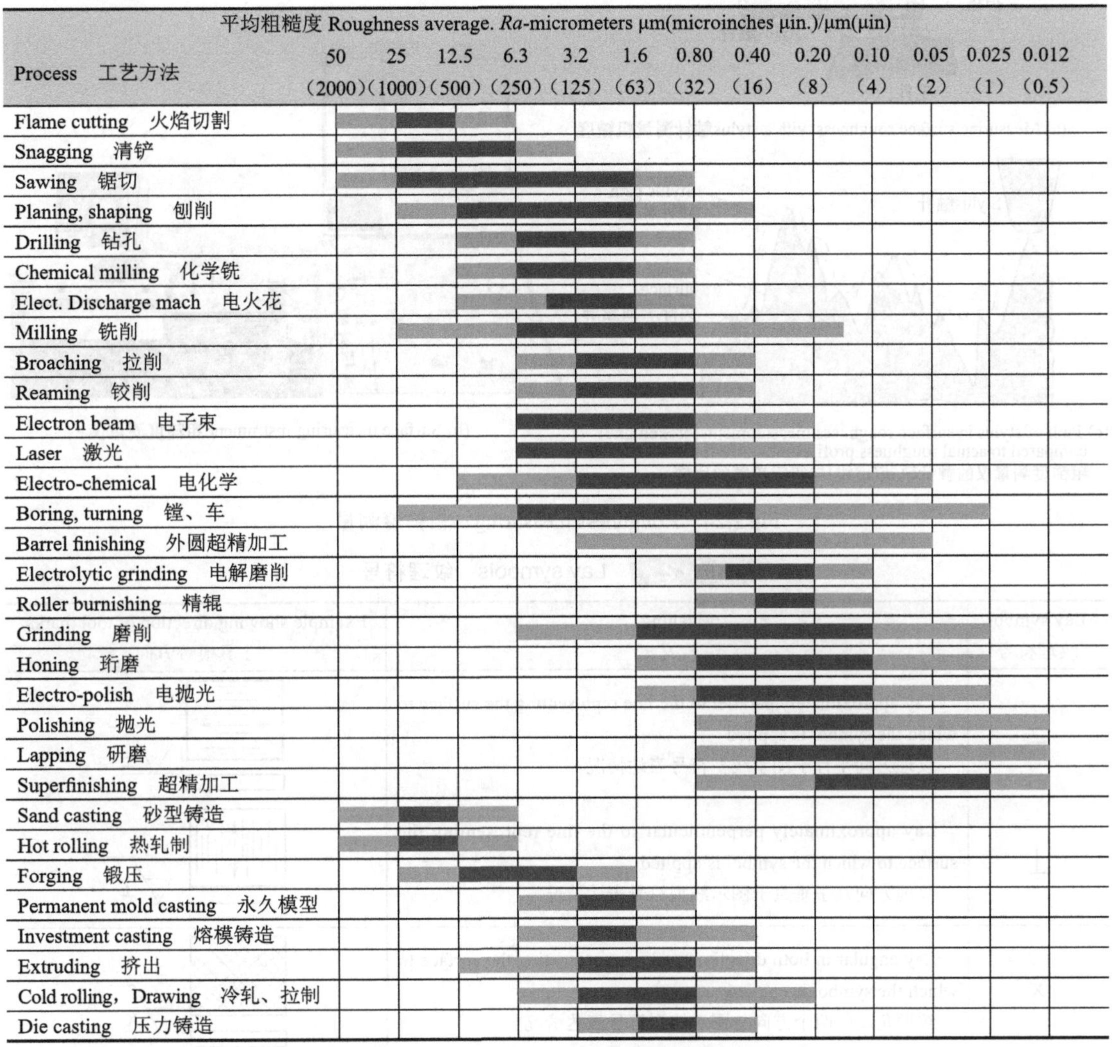

平均粗糙度 Roughness average. Ra-micrometers μm(microinches μin.)/μm(μin)
Process 工艺方法
50 (2000)
25 (1000)
12.5 (500)
6.3 (250)
3.2 (125)
1.6 (63)
0.80 (32)
0.40 (16)
0.20 (8)
0.10 (4)
0.05 (2)
0.025 (1)
0.012 (0.5)
Flame cutting 火焰切割
Snagging 清铲
Sawing 锯切
Planing, shaping 刨削
Drilling 钻孔
Chemical milling 化学铣
Elect. Discharge mach 电火花
Milling 铣削
Broaching 拉削
Reaming 铰削
Electron beam 电子束
Laser 激光
Electro-chemical 电化学
Boring, turning 镗、车
Barrel finishing 外圆超精加工
Electrolytic grinding 电解磨削
Roller burnishing 精辊
Grinding 磨削
Honing 珩磨
Electro-polish 电抛光
Polishing 抛光
Lapping 研磨
Superfinishing 超精加工
Sand casting 砂型铸造
Hot rolling 热轧制
Forging 锻压
Permanent mold casting 永久模型
Investment casting 熔模铸造
Extruding 挤出
Cold rolling，Drawing 冷轧、拉制
Die casting 压力铸造

注：1. The ranges shown above are typical of the processes listed 所列范围为典型工艺。

2. Higher or lower values may be obtained under special conditions 特殊条件下获得值会变化。

Normal force法向力(正压力)
Angular velocity 角速度
Reaction force 反作用力
Net contact force 纯接触力
Resistance 阻力 force
Coefficient of rolling friction 滚动摩擦系数 = Resistance force 阻力 / Normal force 法向力
Normal force 法向力
Braking torque 制动转矩
Hysteresis of rolling material 滚动材料滞留
Angular velocity 角速度
Severe deformation 严重变形
Sticking or adhesion of rolling material 滚动材料的附着/黏滞
Very large contact area 很大接触面积
High coefficient of rolling friction 滚动摩擦系数很大
Normal force 法向力
Braking torque 制动转矩
Net contact force 纯接触力
Angular velocity角速度
Possible skidding 可能滑移
Reaction force 反作用力
Traction force牵引力
Coefficient of adhesion 黏滞摩擦系数 = Traction force 牵引力 / Normal force 法向力

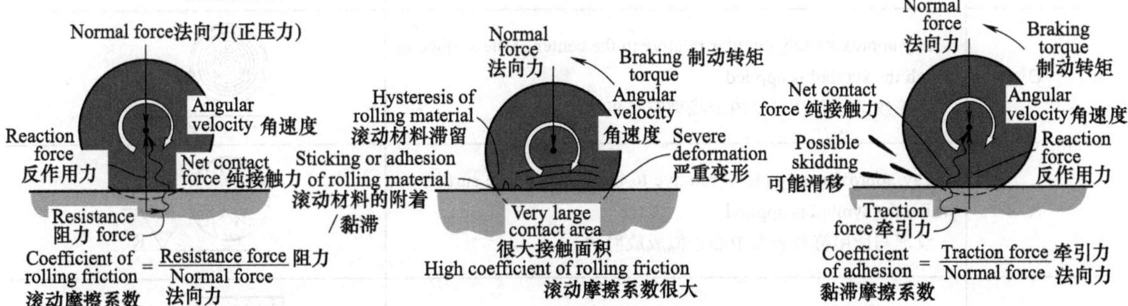

Fig 2.23 The difference between coefficient of rolling friction and coefficient of adhesion and the role of hardness and adhesion in controlling the coefficient of rolling friction 滚动摩擦系数和黏滞摩擦系数的差异及控制滚动摩擦系数时硬度和黏滞的作用

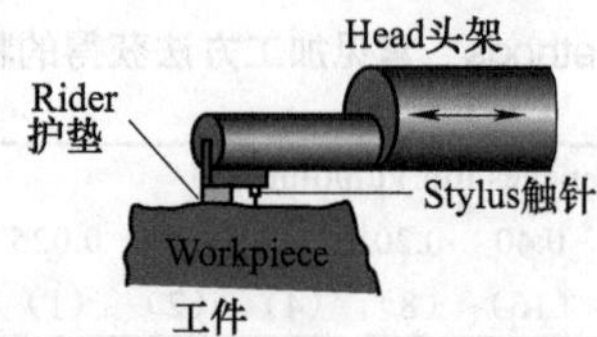

(a) Measuring surface roughness with a stylus触针测量粗糙度

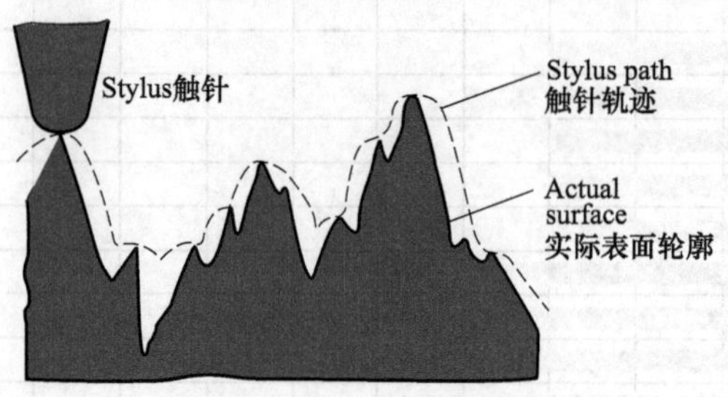

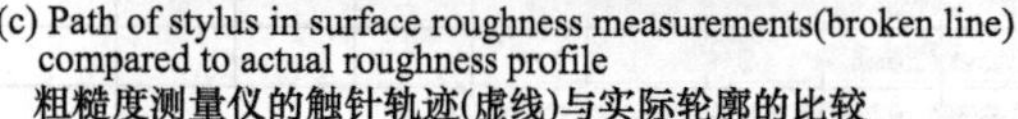
(c) Path of stylus in surface roughness measurements(broken line) compared to actual roughness profile
粗糙度测量仪的触针轨迹(虚线)与实际轮廓的比较

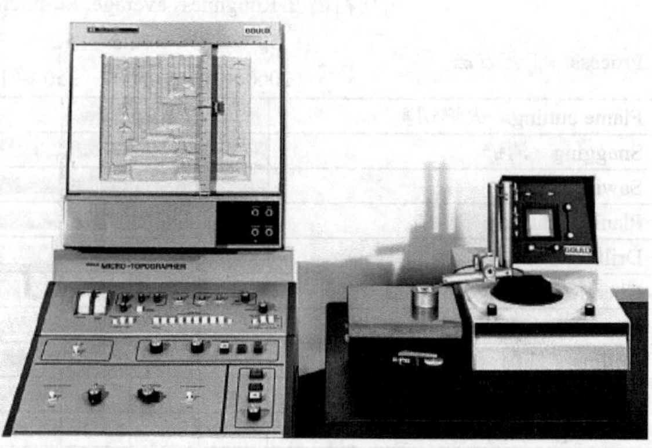
(b) Surface measuring instrument 粗糙度测量仪

Fig 2.24 Roughness measuring 粗糙度测量

Table 2.7 Lay symbols 纹理符号

Lay symbol 纹理符号	Meaning 含义	Example showing direction of tool marks 工具痕迹方向
=	Lay approximately parallel to the line representing the surface to which the symbol is applied 纹理方向平行于图示线型符号表达情况	=
⊥	Lay approximately perpendicular to the line representing the surface to which the symbol is applied 纹理方向几乎垂直于图示线型符号表达情况	⊥
X	Lay angular in both directions to line representing the surface to which the symbol is applied 纹理角度的两个方向与图示线型符号表达情况	X
M	Lay multidirectional 多个方向	M
C	Lay approximately circular relative to the center of the surface to which the symbol is applied 圆形纹理方向几乎与图示线型符号表达情况一致	C
R	Lay approximately radial relative to the center of the surface to which the symbol is applied 纹理与所用符号表面中心近似成放射状	R
P	Lay particulate, non-directional, or protuberant 离子散粒，无方向，或凸起	P

Table 2.8 Description of preferred fits 优先级配合的描述

ISO symbol 国际标准代号			Description 描述	
	Hole basis 基孔制	Shaft basis 基轴制		
Clearance fits 间隙配合	H11/c11	C11/h11	*Loose-running* fit for wide commercial tolerance or allowance on external members 松配合：适于极大的经济公差或外部构件允差较大	↑ More clearance 间隙增大
	H9/d9	D9/h9	*Free-running* fit not for where accuracy is essential, but good for large temperature variations, high running speed and journal pressures 自由运转配合：不适合于有精度要求的地方，适于温度变化大、高速和滑动轴承的场合	
	H8/f7	F8/h7	*Close-running* fit for running on accurate machines, and for accurate location at moderate speeds and journal pressures 精密运转配合：适于精磨机械、中等速度精确定位及滑动轴承	
	H7/g6	G7/h6	*Slide fit* not intended to run freely, but to move and turn freely and locate accurately 滑移配合：不适合于高速自由运行，而适于自由移动或转动及精确定位	
	H7/h6	H7/h6	*Locational Clearance* fit provides snug fit for locating stationary parts; but can be freely assembled or dissembled 定位间隙：对于静止定位元件产生阻滞，能够自由装拆	
Transition fits 过渡配合	H7/k6	K7/h6	*Locational transition* fit for accurate location, a compromise between clearance and interference 定位过渡：对于精确定位，间隙和过盈的适当配合	
	H7/n6	N7/h6	*Locational transition* fit for more accurate location where greater interference is permissible 定位过渡：对于较精确定位可以有较大的过盈量	More interference 过盈增大 ↓
Interference fits 过盈配合	H7/p6	P7/h6	*Locational interference* fit for parts requiring rigidity and alignment with prime accuracy of location but without special bore pressure requirement 定位过盈：对于静止定位适于有刚性和对位的精确定位，但不能有孔压力的特殊要求	
	H7/s6	S7/h6	*Medium drive* fit for ordinary steel parts or shrink fits on light sections, the tightest fit usable with cast iron 中等传动：适于普通钢件或轻型缩颈配合，铸铁件的最紧配合	
	H7/u6	U7/h6	*Force* fit suitable for parts which can be highly stressed or for shrink fit where the heavy pressing forces required are impractical 重载：适于可高度压缩零件或缩颈而压力无法实施的配合	

Unit 3 Technological Appropriateness for Machine Parts and Assemblies 机械零部件结构工艺性

3.1 Technological appropriateness for castings, forgings, weldings and heat treatment 铸造、锻压、焊接和热处理结构工艺性

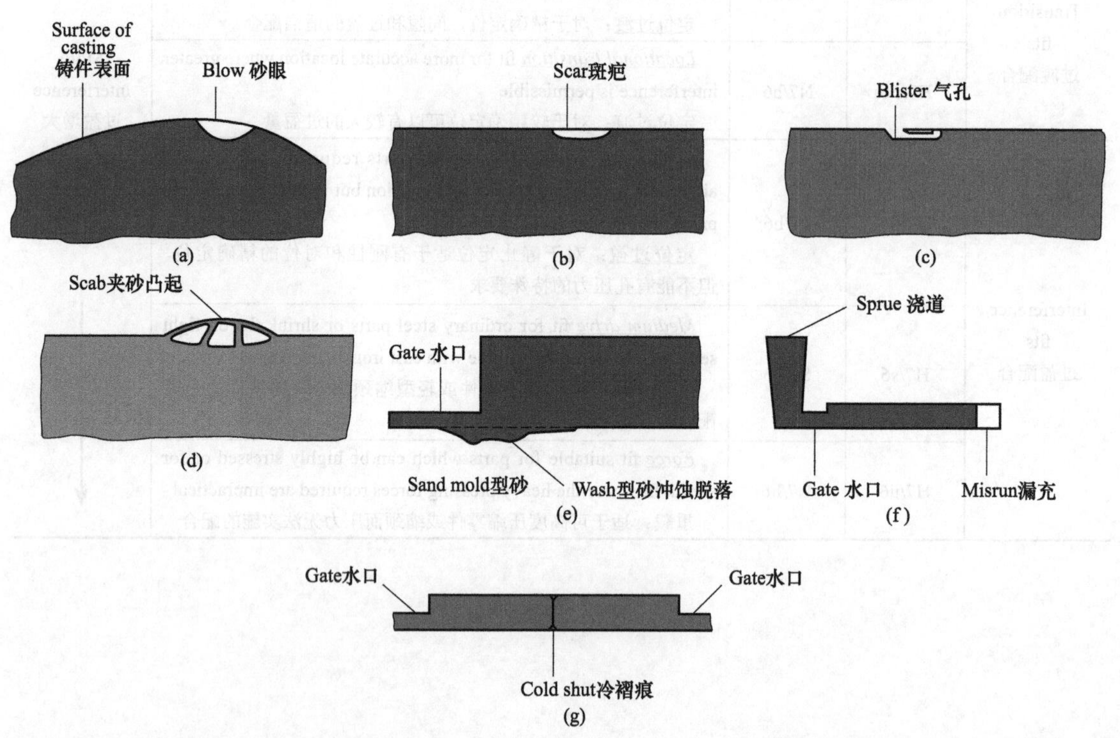

Fig 3.1 Examples of common defects in castings 铸件常见缺陷

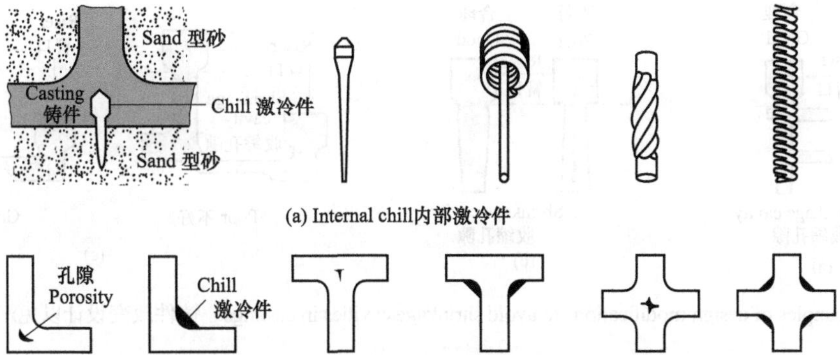

(a) Internal chill内部激冷件

(b) External chills(dark areas at corners)外部激冷件(拐角黑色处) ,used in castings to eliminate porosity caused by shrinkage 用于减少铸件因收缩产生的孔隙

(c) Chills are placed in regions where there is a larger volume of metals 激冷件布置在体积较厚的地方

Fig 3.2 Various types of internal and external chills 各种内、外激冷件

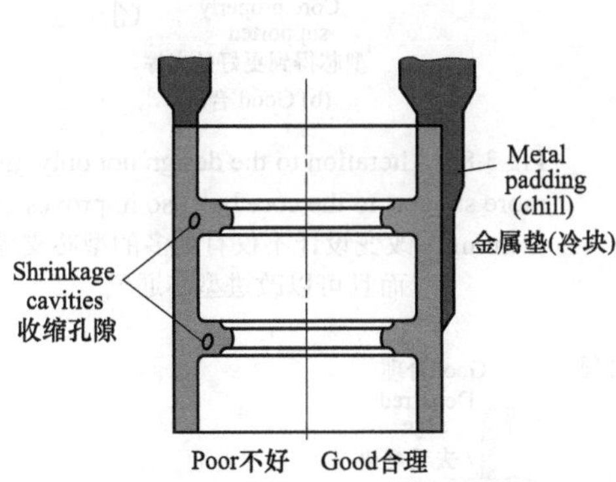

Fig 3.3 The use of metal padding (chills) to increase the rate of cooling in thick regions in a casting to avoid shrinkage cavities
铸件厚壁处采用冷块增加冷却速度可以避免收缩孔隙

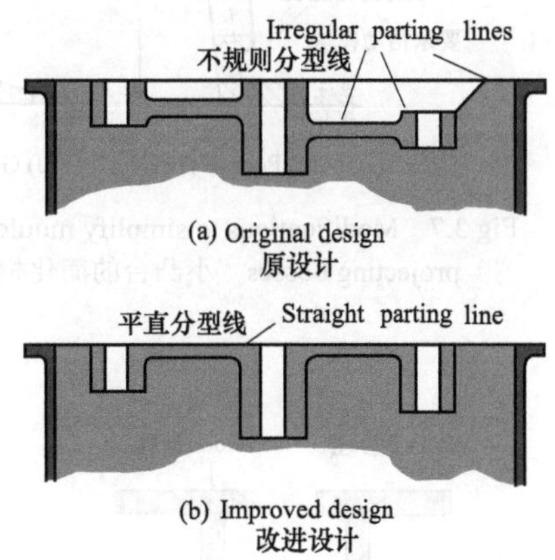

Fig 3.4 Redesign of a casting by making the parting line straight to avoid defects
重新设计铸件的分型线为直线可避免缺陷

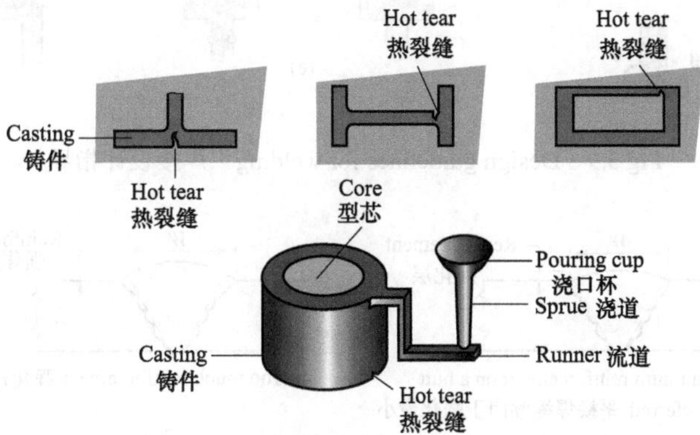

Fig 3.5 Examples of hot tears in castings 铸件热裂缝示例

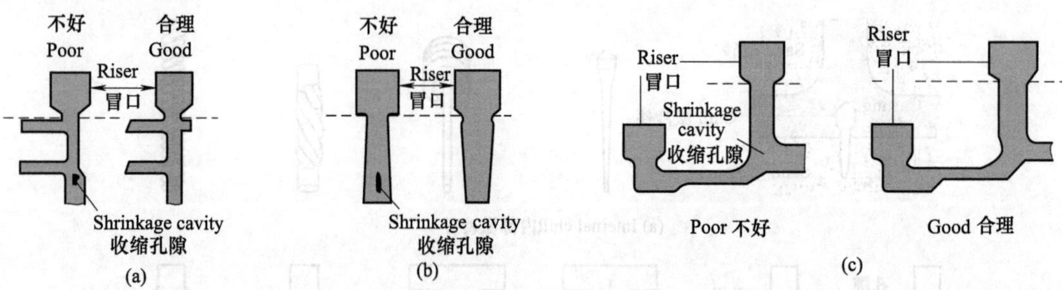

Fig 3.6 Examples of design modifications to avoid shrinkage cavities in castings 铸件改变设计以免产生收缩孔隙

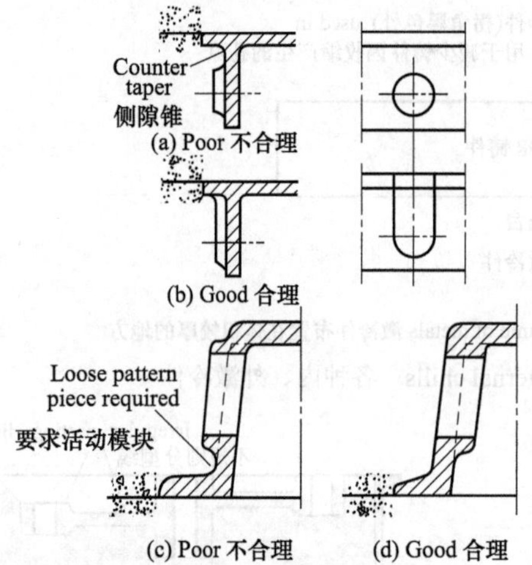

Fig 3.7 Modifications to simplify moulding of small projecting bosses 小凸台的简化模具改进

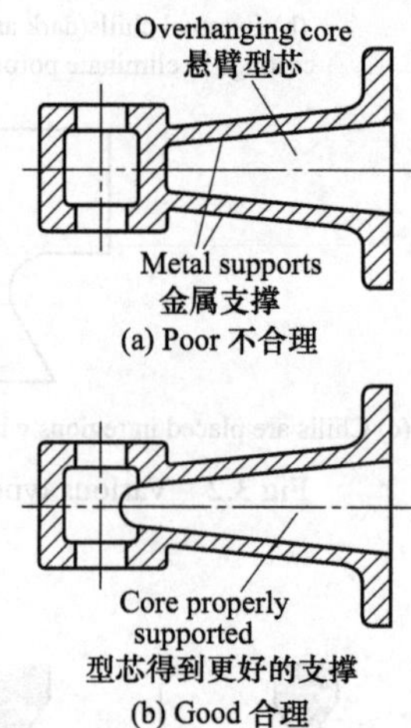

Fig 3.8 Alteration to the design not only gives more support to the core but also improves core venting 改变设计不仅有更多的型芯支撑，而且可以改进型芯通气

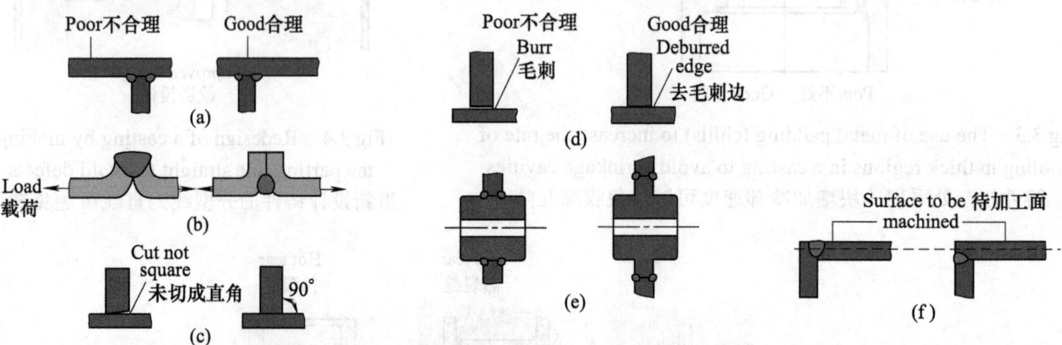

Fig 3.9 Design guidelines for welding 焊接设计指导

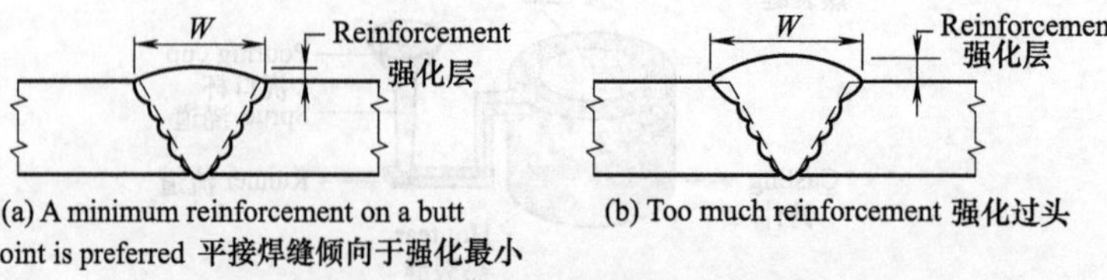

Fig 3.10 Reinforcement 强化层

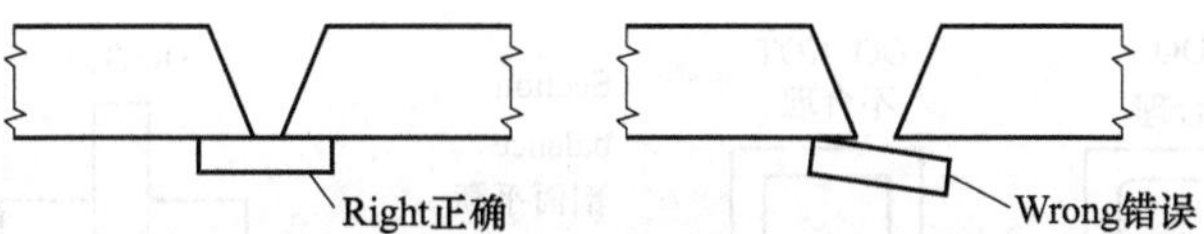

Fig 3.11 The backup strip should be in intimate contact with both edges of the plate
支撑条应当与焊接板两边紧密接触

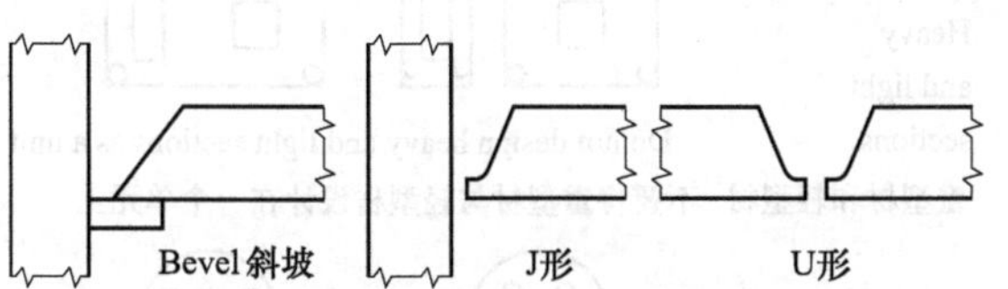

Fig 3.12 A bevel preparation with a backup strip may be more economical than a J or U groove
带背衬条的斜坡口可以比J形和U形开槽更为经济

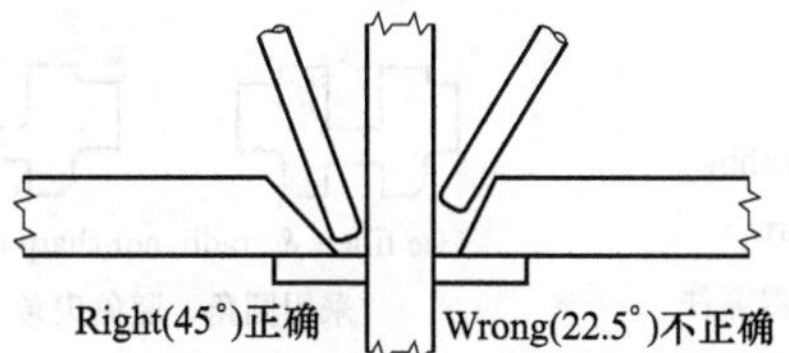

Fig 3.13 Degree of bevel may be dictated by the need for maintaining proper electrode angle
锥角的开设要考虑能够保持合理的电极角度

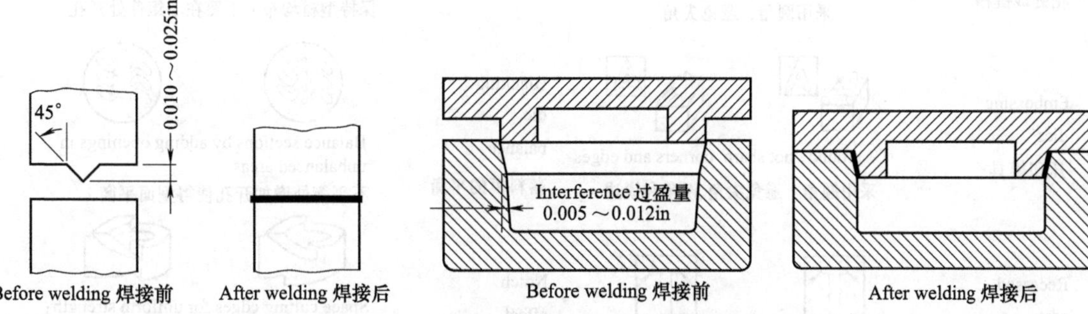

Fig 3.14 Energy director types of ultrasonic weld joint designs for assembly of plastics moldings: and (right) typical shear interference 塑件制品超声焊接设计和（右图）典型的坡度过盈

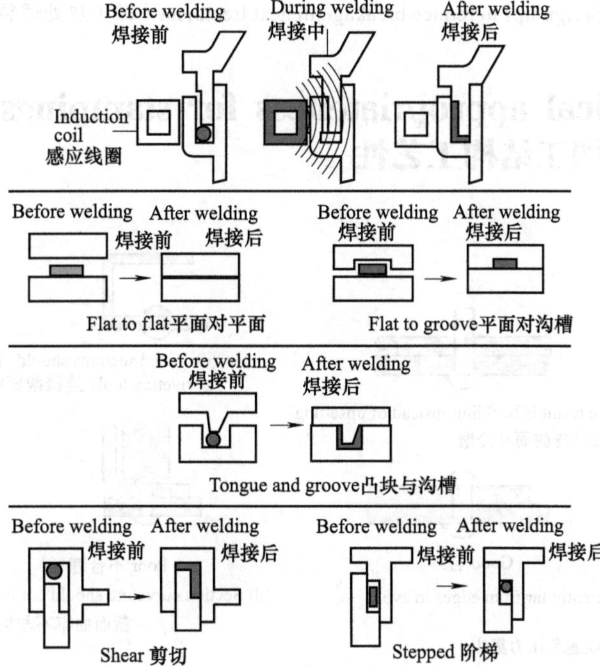

Fig 3.15 Typical joint designs used in induction welding of plastics materials 感应焊接塑料材料的典型接头设计

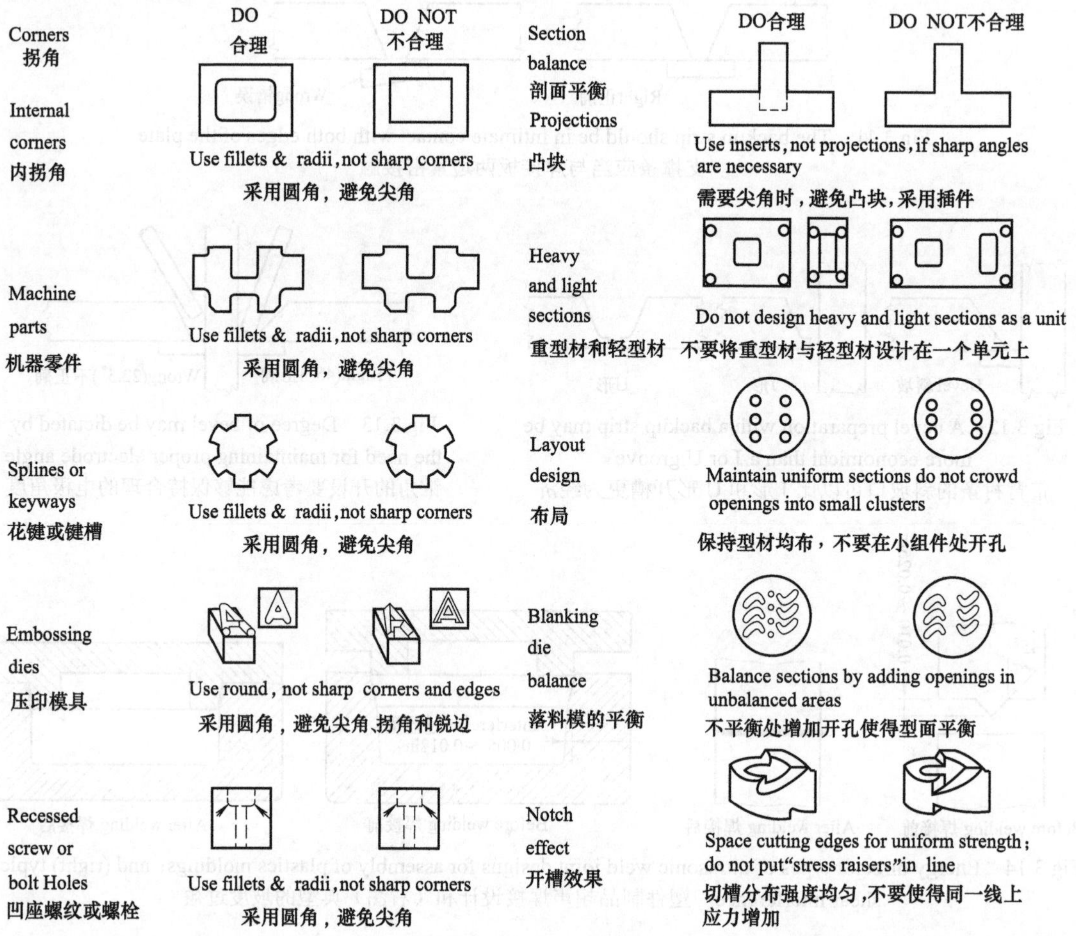

	DO 合理	DO NOT 不合理		DO合理	DO NOT不合理
Corners 拐角 Internal corners 内拐角	Use fillets & radii, not sharp corners 采用圆角，避免尖角		Section balance 剖面平衡 Projections 凸块	Use inserts, not projections, if sharp angles are necessary 需要尖角时，避免凸块，采用插件	
Machine parts 机器零件	Use fillets & radii, not sharp corners 采用圆角，避免尖角		Heavy and light sections 重型材和轻型材	Do not design heavy and light sections as a unit 不要将重型材与轻型材设计在一个单元上	
Splines or keyways 花键或键槽	Use fillets & radii, not sharp corners 采用圆角，避免尖角		Layout design 布局	Maintain uniform sections do not crowd openings into small clusters 保持型材均布，不要在小组件处开孔	
Embossing dies 压印模具	Use round, not sharp corners and edges 采用圆角，避免尖角、拐角和锐边		Blanking die balance 落料模的平衡	Balance sections by adding openings in unbalanced areas 不平衡处增加开孔使得型面平衡	
Recessed screw or bolt Holes 凹座螺纹或螺栓	Use fillets & radii, not sharp corners 采用圆角，避免尖角		Notch effect 开槽效果	Space cutting edges for uniform strength; do not put "stress raisers" in line 切槽分布强度均匀，不要使得同一线上应力增加	

Fig 3.16 Tool and die design tips to reduce breakage in heat treatment 减少热处理裂纹的工模具设计要点

3.2 Technological appropriateness for stampings, metal cutting 冲压、切削加工结构工艺性

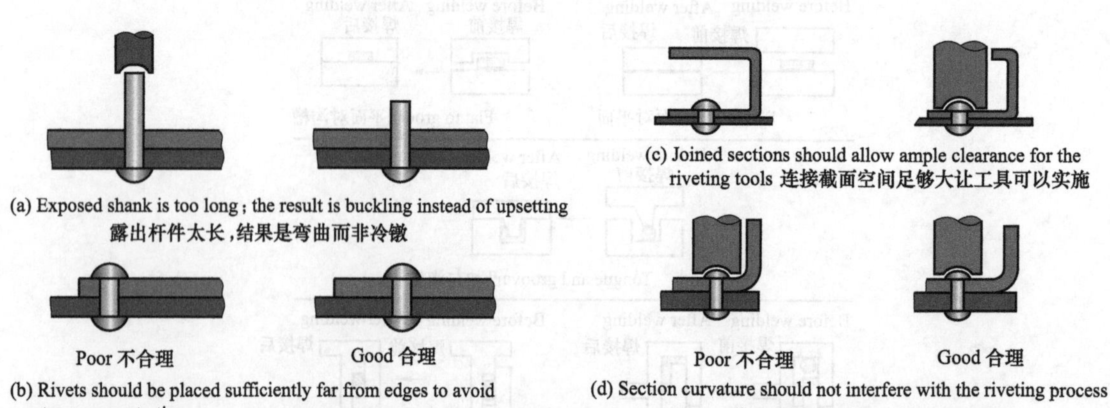

(a) Exposed shank is too long; the result is buckling instead of upsetting 露出杆件太长，结果是弯曲而非冷镦

(b) Rivets should be placed sufficiently far from edges to avoid stress concentrations 铆钉距离边口应该足够远，以避免压力集中

(c) Joined sections should allow ample clearance for the riveting tools 连接截面空间足够大让工具可以实施

(d) Section curvature should not interfere with the riveting process 截面曲率不与铆接工艺实施干涉

Fig 3.17 Design guidelines for riveting 铆接设计指导

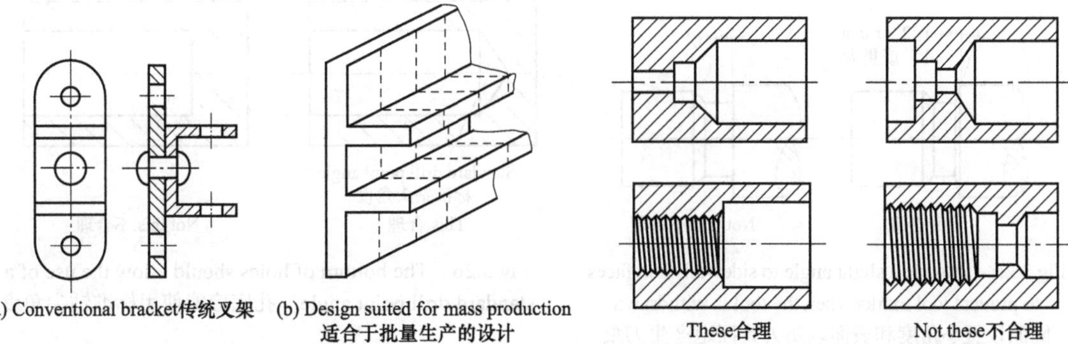

Fig 3.18 Bracket 叉架

Fig 3.19 Operations should be finished without the need for rechucking 加工完成不需再次装夹

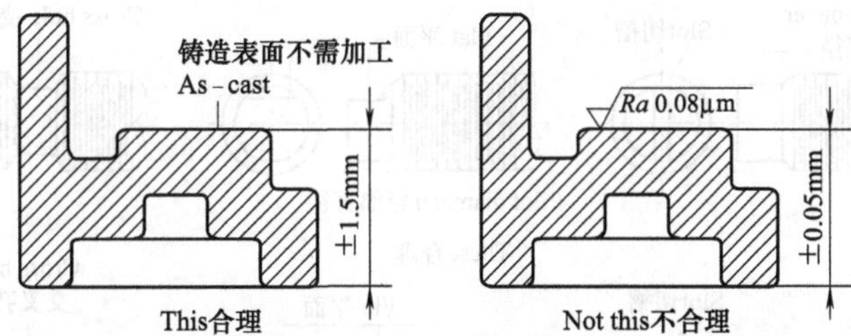

Fig 3.20 Avoid tolerances that necessitate machining if as-cast, as-forged, or as-formed dimensions and surface finishes are satisfactory for the parts function

如果铸造、锻压、成形的尺寸精度和表面粗糙度满足使用需求，避免需要加工的公差标注

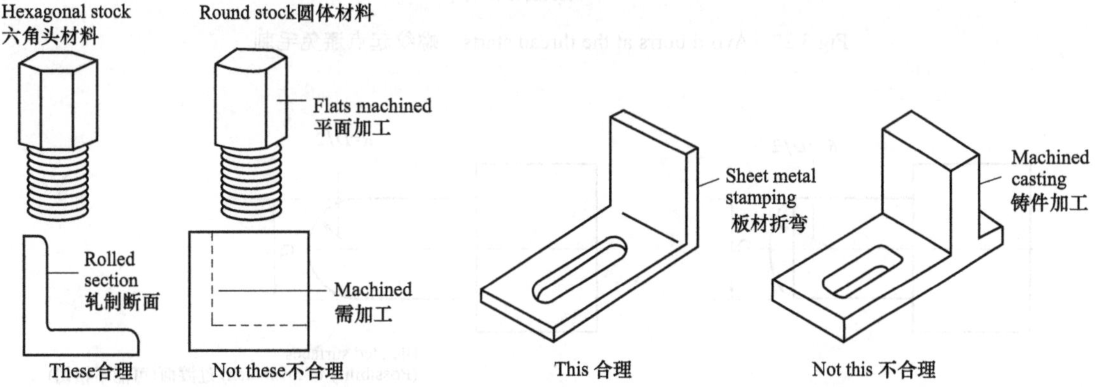

Fig 3.21 Use stock dimensions and minimize the machining allowance 利用材料尺寸减小加工允差

Fig 3.22 Metal-formed parts are better than machined castings 金属成形件好过铸件加工

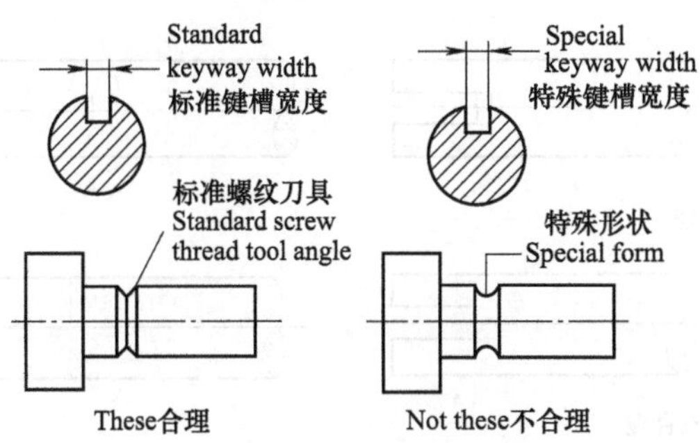

Fig 3.23 Design parts to be machined by standard tools 设计零件采用标准刀具

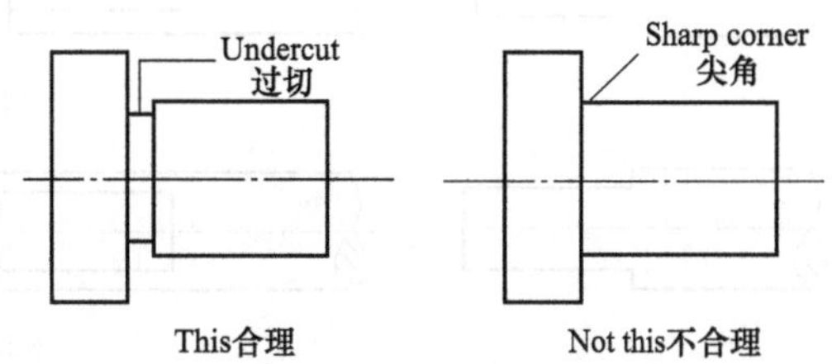

Fig 3.24 Undercuts avoid the problems of sharp corners 过切避免尖角问题

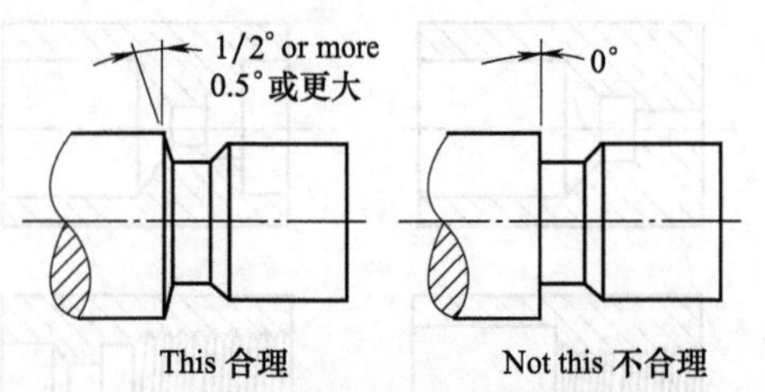

Fig 3.25 Provide a slight angle to sidewalls and faces to prevent tool marks when the tool is withdrawn 侧面设置小角度和表面以防刀具回退产生刀痕

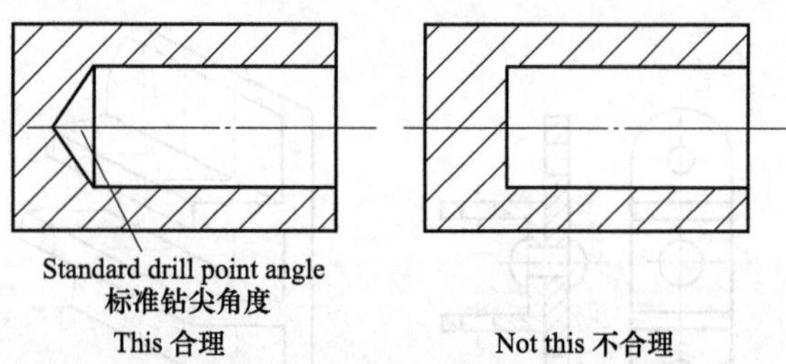

Fig 3.26 The bottom of holes should allow the use of a standard drill point angle 孔底应能使用标准钻尖角度

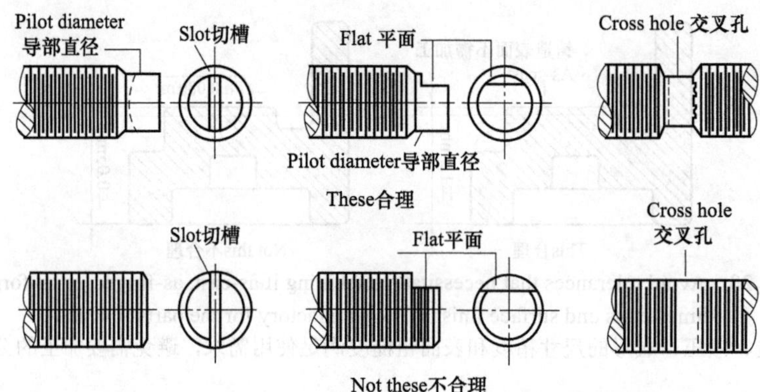

Fig 3.27 Avoid burrs at the thread starts 螺纹起点避免毛刺

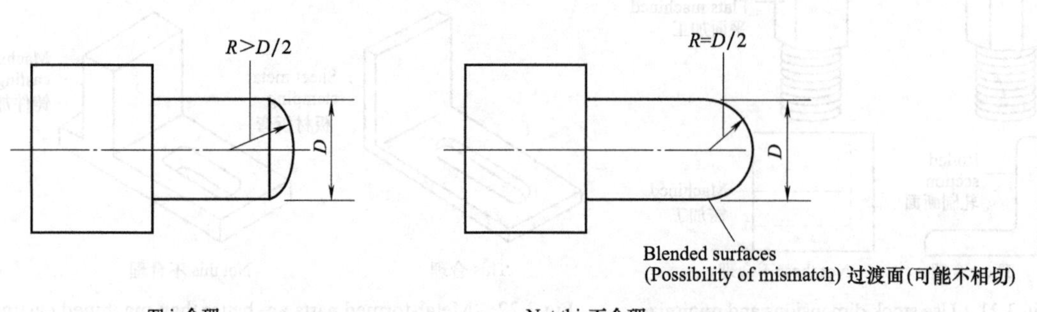

Fig 3.28 Avoid blended surfaces formed by a separate cutter 避免多种刀具加工过渡面

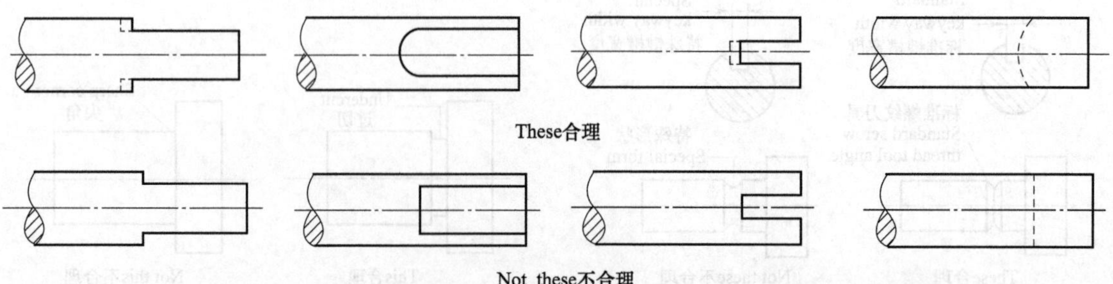

Fig 3.29 Permit curved bottoms of slots and flats if possible 可能的话，槽底为曲面

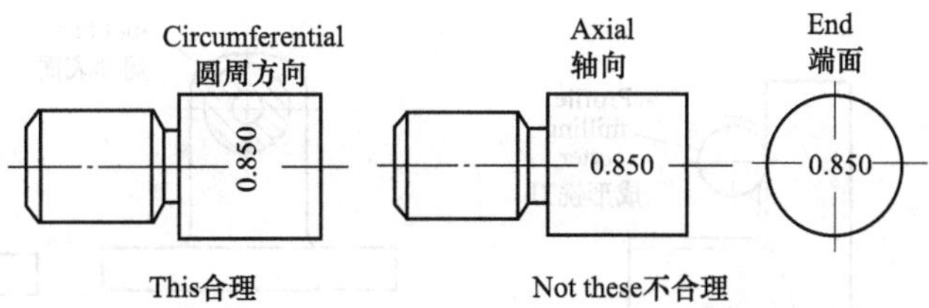

Fig 3.30 Marking should allow the use of roller tools 打标记要能使用滚压刀

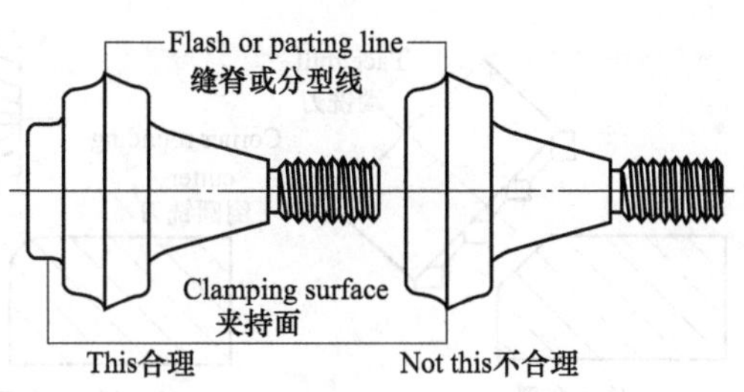

Fig 3.31 Avoid designs that require clamping on parting lines or flash areas 避免夹紧分型线或缝脊

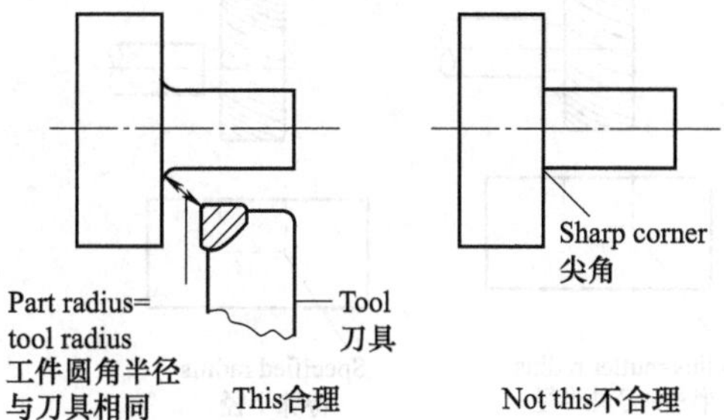

Fig 3.32 Avoid sharp corners 避免尖角

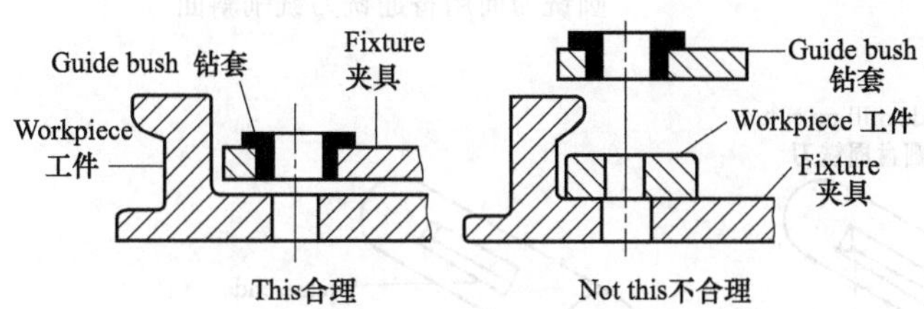

Fig 3.33 Allow room for drill bushes close to the drilled surface 钻套应该靠近钻削表面

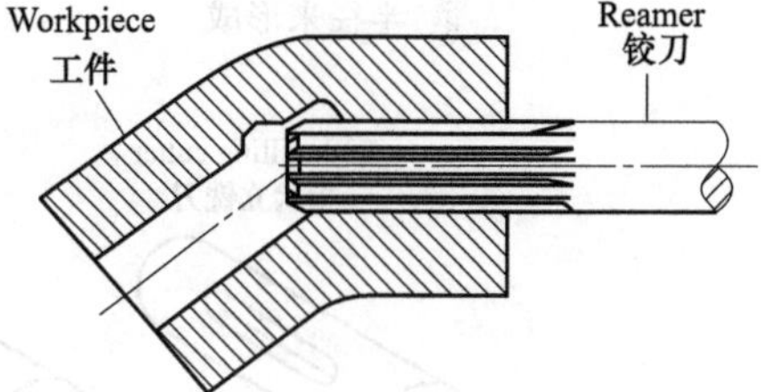

Fig 3.34 Avoid intersecting drilled and reamed holes 避免铰孔和钻孔交叉

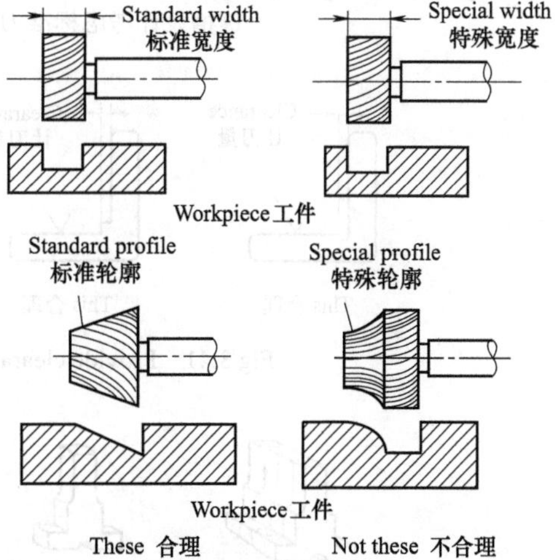

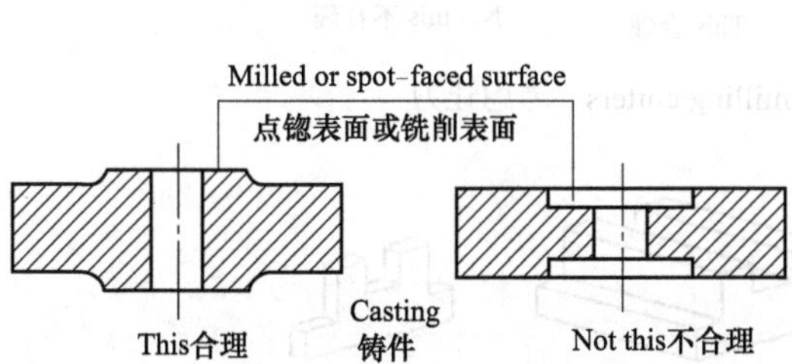

Fig 3.35 A low boss simplifies the machining of a flat surface 小凸台使得平面加工简单

Fig 3.36 Allow the use of standard cutter shapes and sizes rather than special ones 尽量使用标准形状和尺寸的刀具

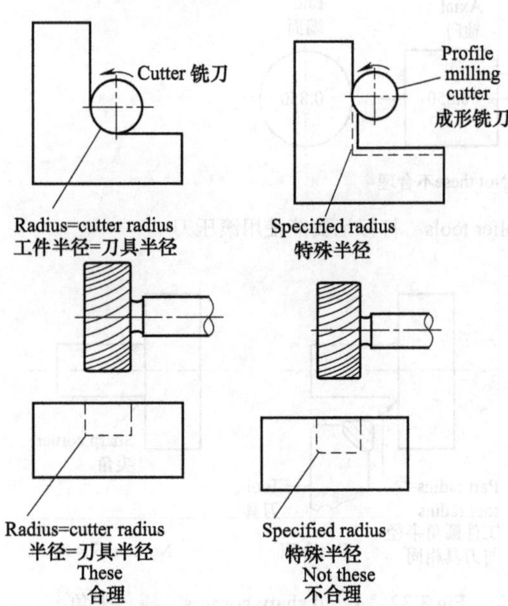

Fig 3.37 Allow the use of radii generated by the milling cutter 拐角半径尽量采用铣刀半径来形成

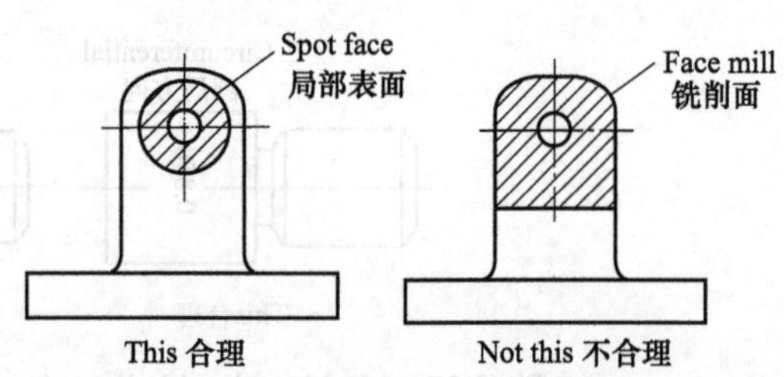

Fig 3.38 Spot facing of small surfaces is preferred over face milling 小面积局部表面优于整个表面

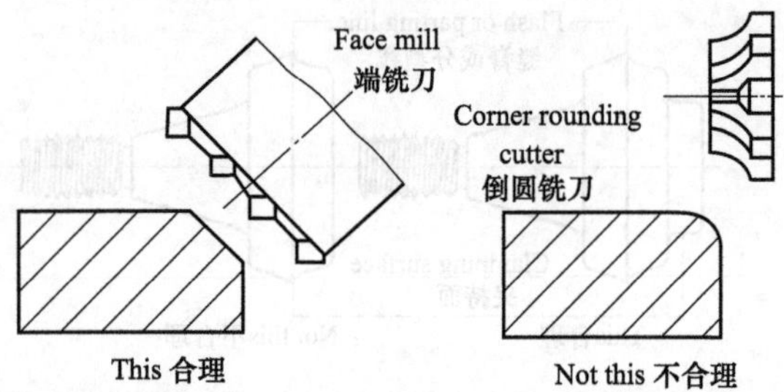

Fig 3.39 Allow beveled rather than rounded corners for economical milling 为了降低刀具成本，不要采用倒圆铣刀而用普通铣刀铣削斜面

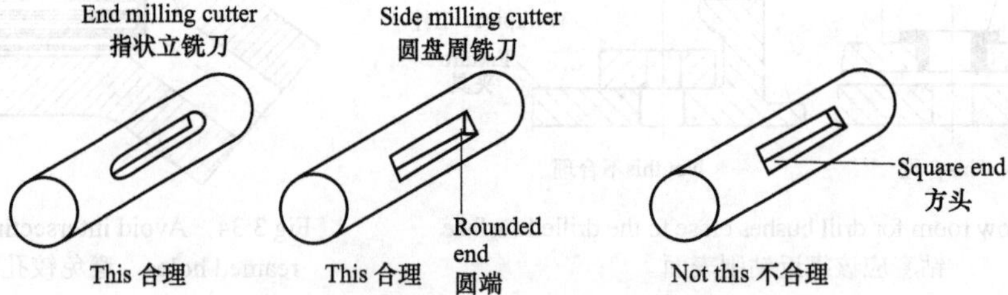

Fig 3.40 Design keyways so that a standard milling cutter finishes its sides and ends in one operation 键槽设计考虑标准刀具一次加工完成其侧面和端部

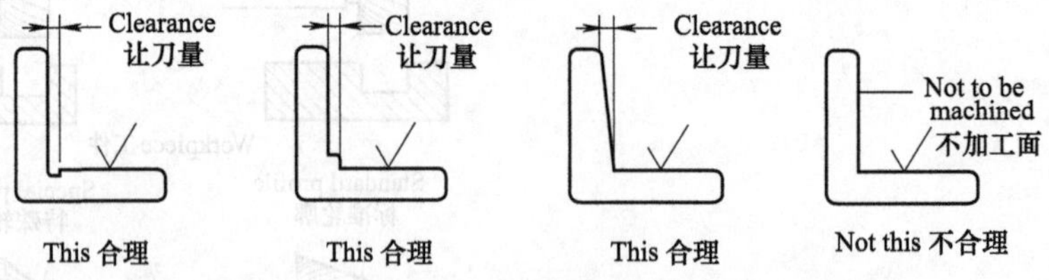

Fig 3.41 Provide clearances for milling cutters 考虑让刀

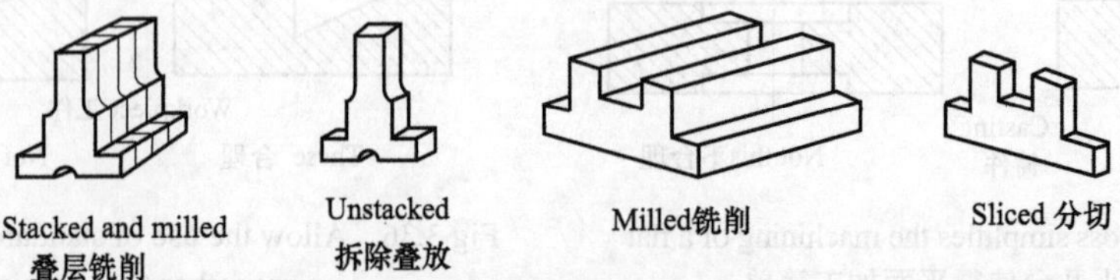

Fig 3.42 Designs that permit stacking or slicing are more economical 考虑叠放加工或分片更为经济

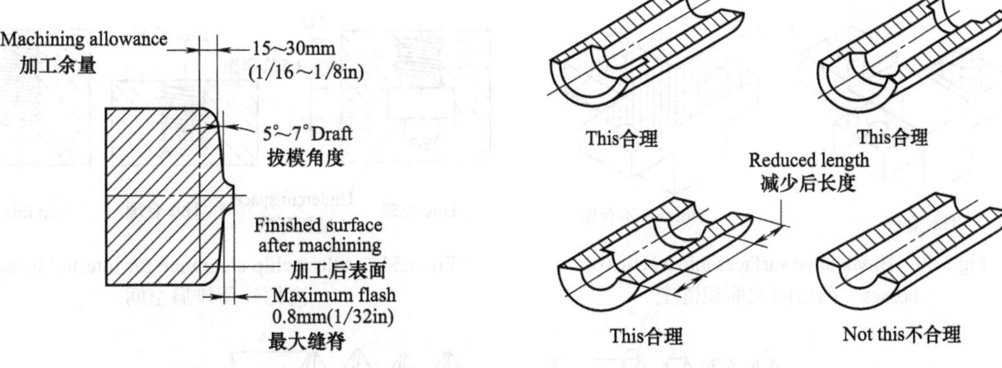

Fig 3.43 Stock allowance for broaching a forged part
锻件拉削的材料余量

Fig 3.44 Long holes should be recessed
深孔应当有阶梯

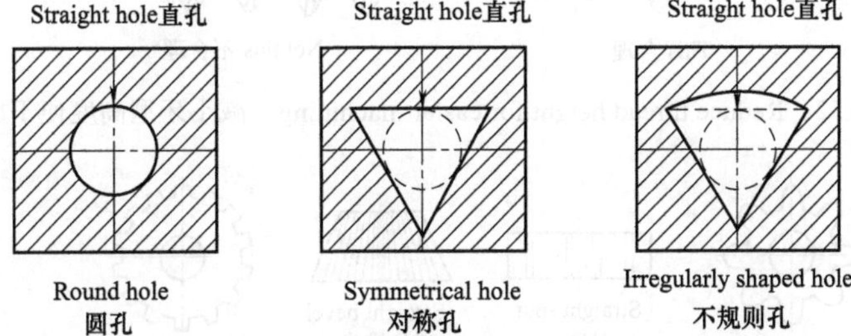

Fig 3.45 Irregularly shaped broached holes are started from round holes 不规则形状的拉削都是始于圆孔

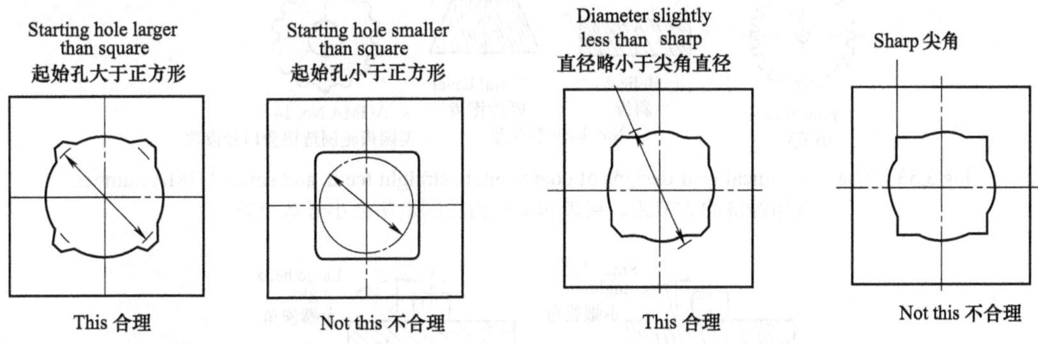

Fig 3.46 Use a slightly oversized starting hole
起始孔略大于正方形孔

Fig 3.47 Avoid sharp corners on major diameters
大径避免尖角

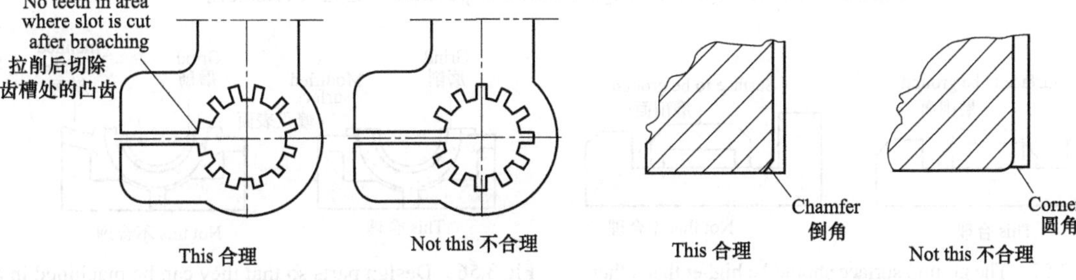

Fig 3.48 Allow room for the burr produced by the saw cut
允许锯切毛刺有去除空间

Fig 3.49 Chamfer outer corners
外部拐角最好倒角

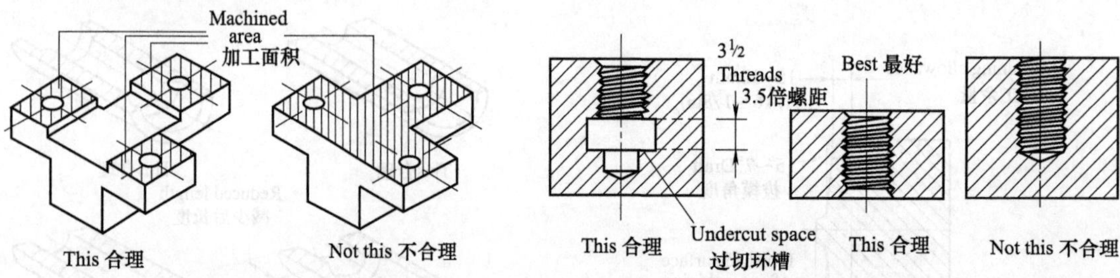

Fig 3.50 Break large surfaces into a series of bosses 凸台间大面积落空

Fig 3.51 Allow chip clearance for internal threads 内螺纹留排屑空间

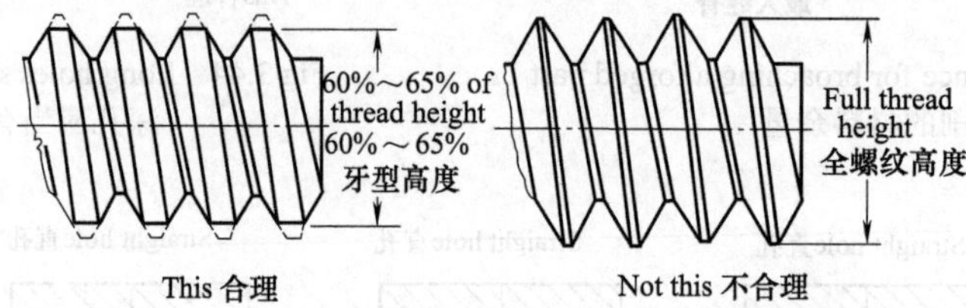

Fig 3.52 Reduce thread height for easier machining 减小牙型高度便于加工

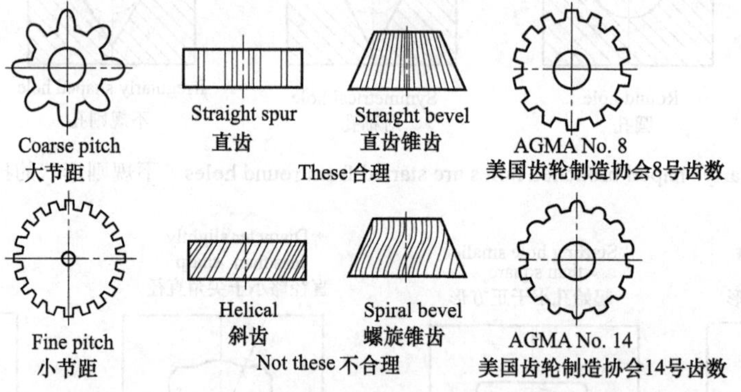

Fig 3.53 Use economical gear designs of coarse pitch, straight teeth, and small AGMA number 采用经济的大节距、直齿和美国齿轮制造协会小号数齿数

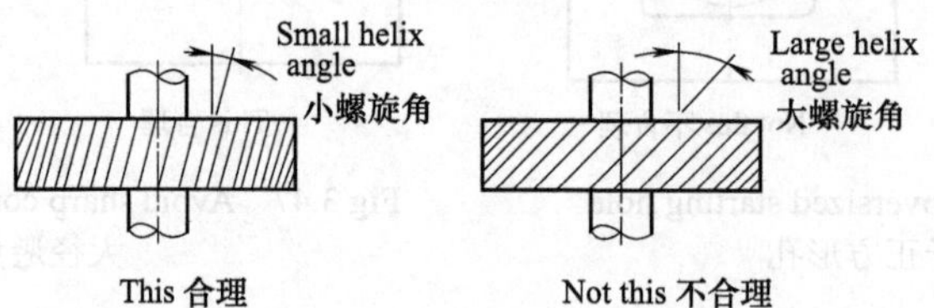

Fig 3.54 Avoid large helix angles whenever possible 避免大的螺旋角

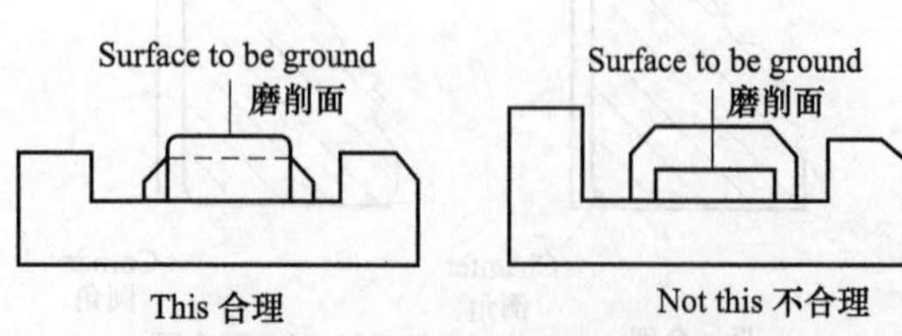

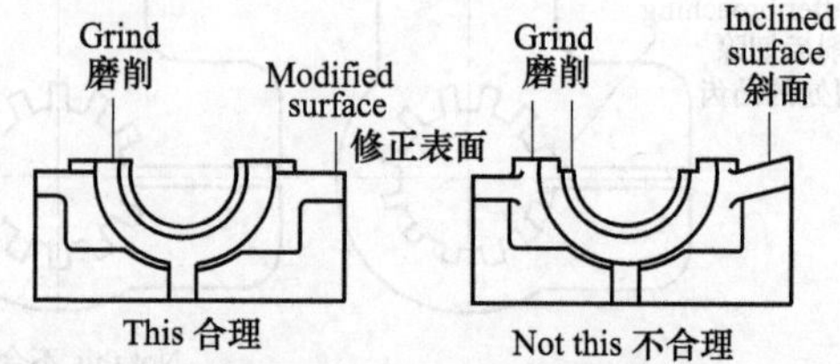

Fig 3.55 The ground surface should be higher than other surfaces to avoid wheel obstruction 磨削面应当高于其他面以免砂轮碰撞

Fig 3.56 Design parts so that they can be machined in a single setup without wheel obstruction 零件设计考虑一次装夹可以加工而没有砂轮碰撞

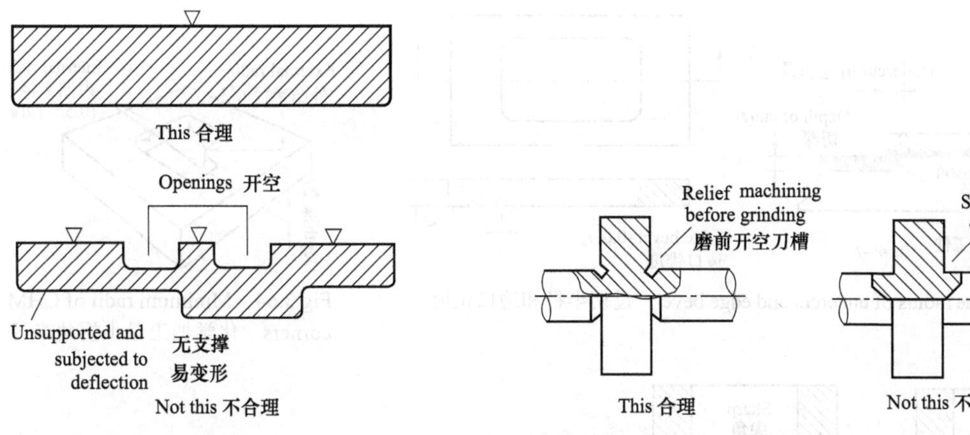

Fig 3.57 Avoid openings and unsupported surfaces 避免落空无支撑

Fig 3.58 Machine or cast a relief at the junction of two surfaces before grinding 磨前两个表面交汇处加工空刀槽

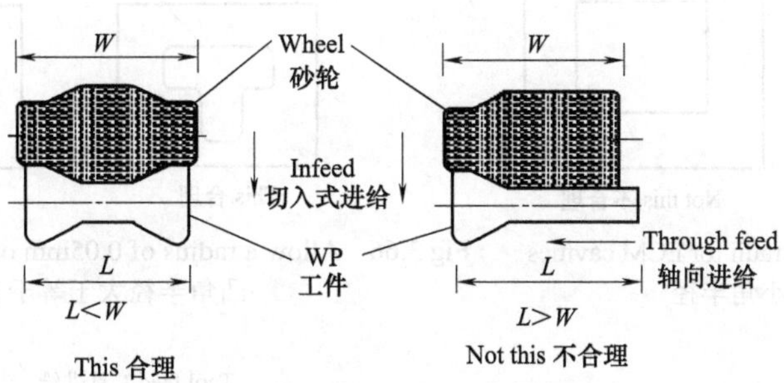

Fig 3.59 Avoid long parts of irregular surfaces so that infeed can be applied only to the wheel 工件不规则表面不宜过长，这样只需砂轮进给

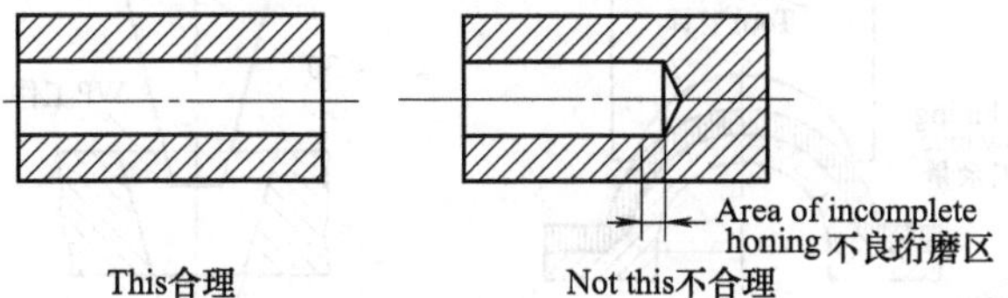

Fig 3.60 Design recommendations for honing of internal holes 内孔珩磨设计推荐

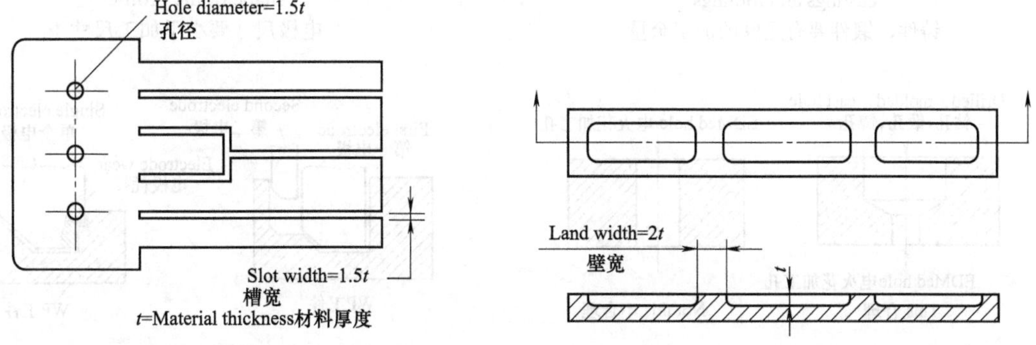

Fig 3.61 Maximum hole diameter and slot width by CHM 化学加工最大孔径和最大槽宽

Fig 3.62 Minimum land widths 最小壁宽

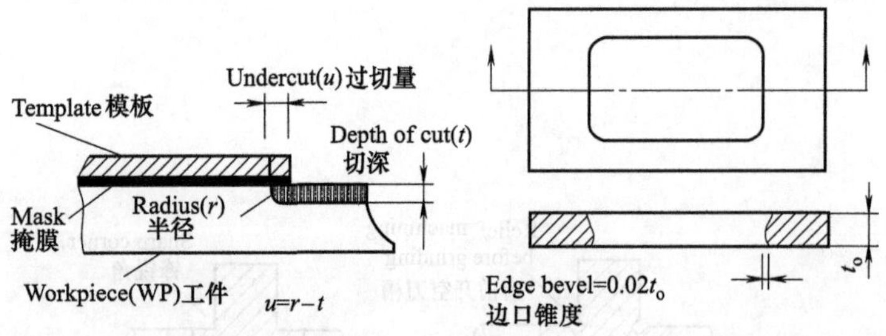

Fig 3.63 The radius of undercut and edge bevel 过切半径和边口锥度

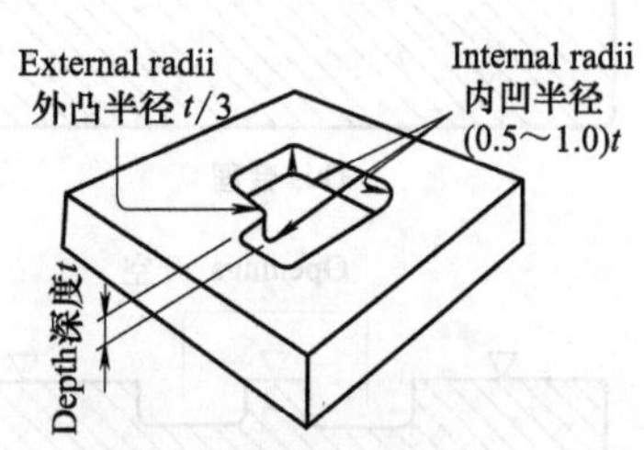

Fig 3.64 Minimum radii of CHM corners 化学加工最小拐角半径

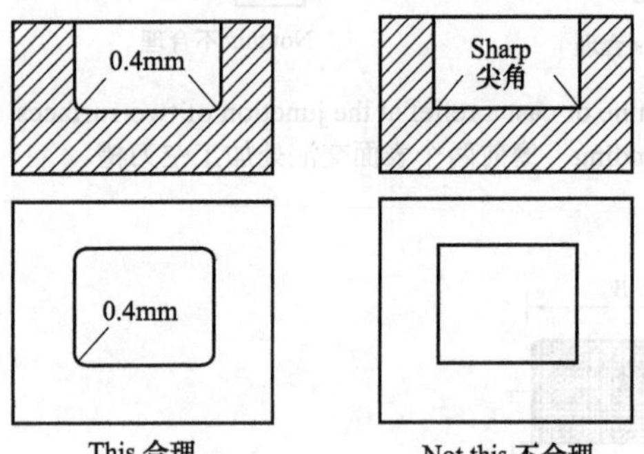

Fig 3.65 Minimum corner radii for ECM cavities 电化学加工最小角半径

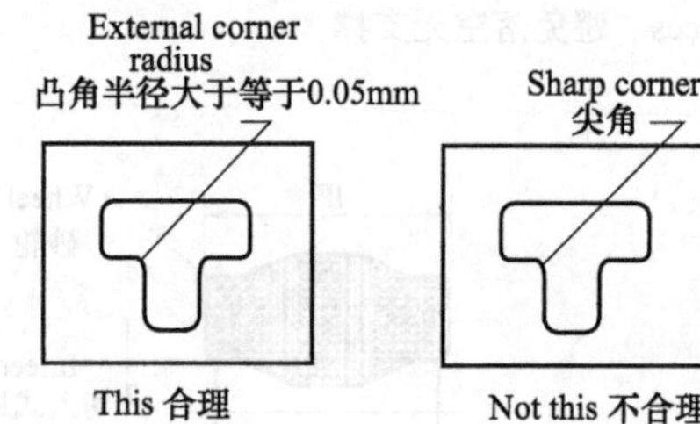

Fig 3.66 Allow a radius of 0.05mm or more for external corners 凸角半径大于等于 0.05mm

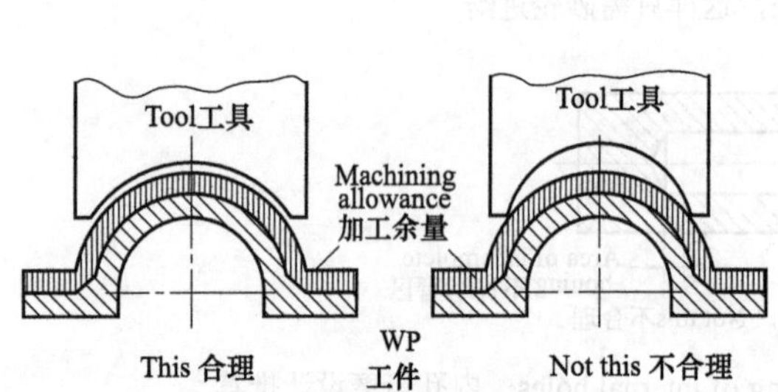

Fig 3.67 Allow sufficient machining allowance on castings and forgings 铸件、锻件要有足够的加工余量

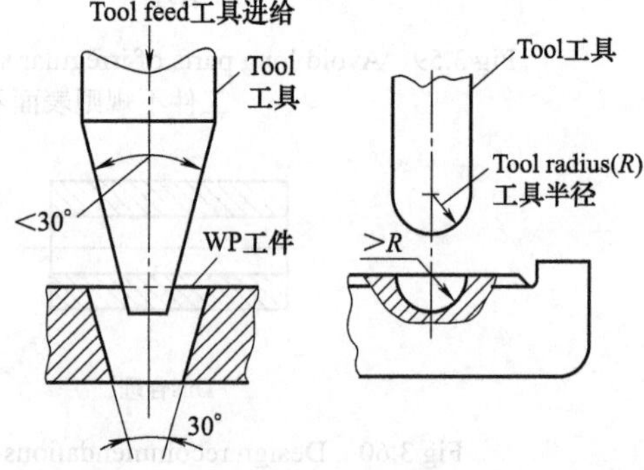

Fig 3.68 The electrode tool is less than the true shape of the profile 电极尺寸要小于加工尺寸

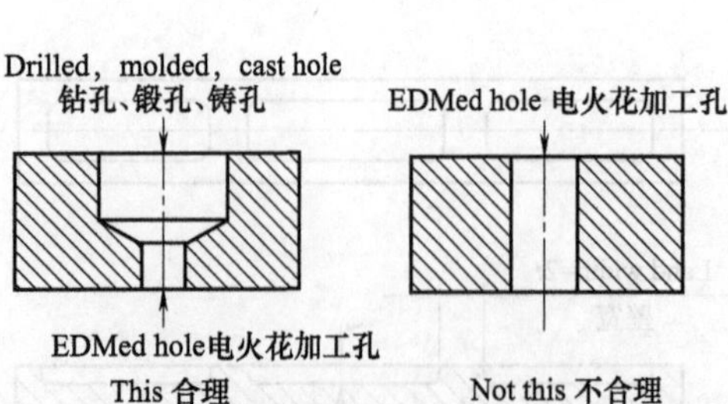

Fig 3.69 Perform maximum conventional machining, molding,or casting before the slow EDM 低效率的电火花加工之前尽量采用传统加工、模具成形或铸造工艺

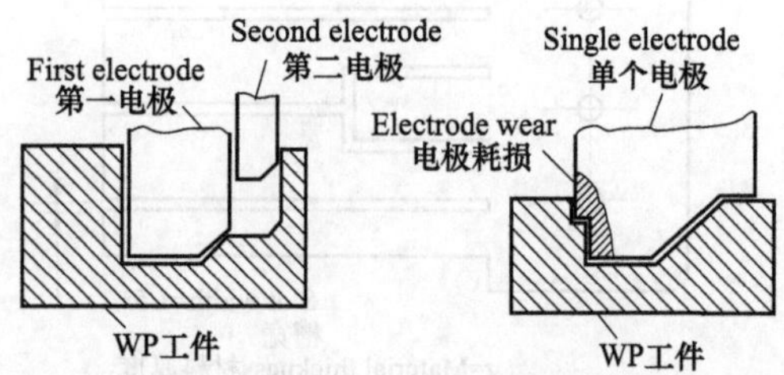

Fig 3.70 Complex shapes require special or multiple electrodes 复杂形状需要特殊电极或多个电极

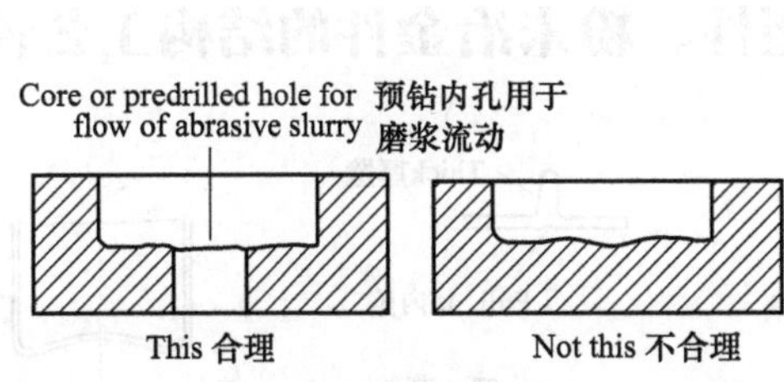

Fig 3.71 Provide through passage for the abrasive slurry 考虑磨浆通道

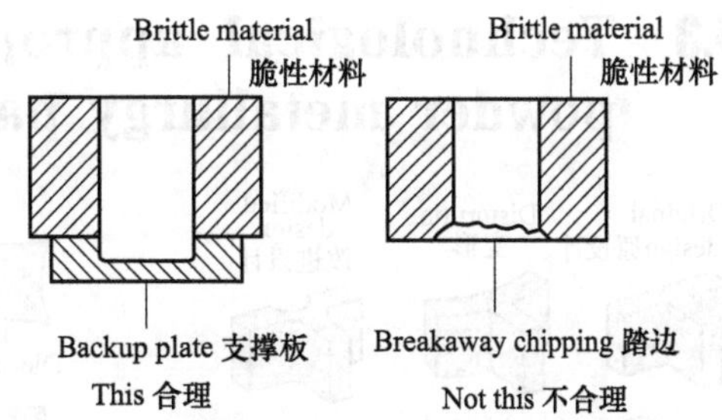

Fig 3.72 Avoid breakaway chipping at the exit surface of USM cavities 超声加工避免踏边

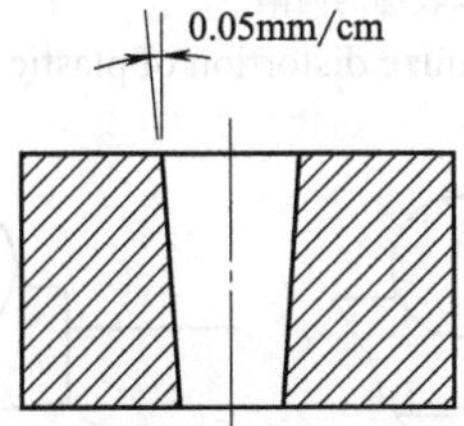

Fig 3.73 Allow taper for sidewalls of USM cavities 超声加工孔有斜度

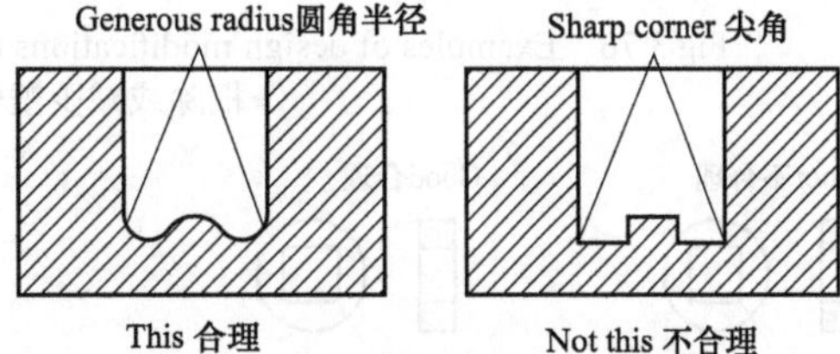

Fig 3.74 Allow generous radii at machined corners 加工拐角有圆角

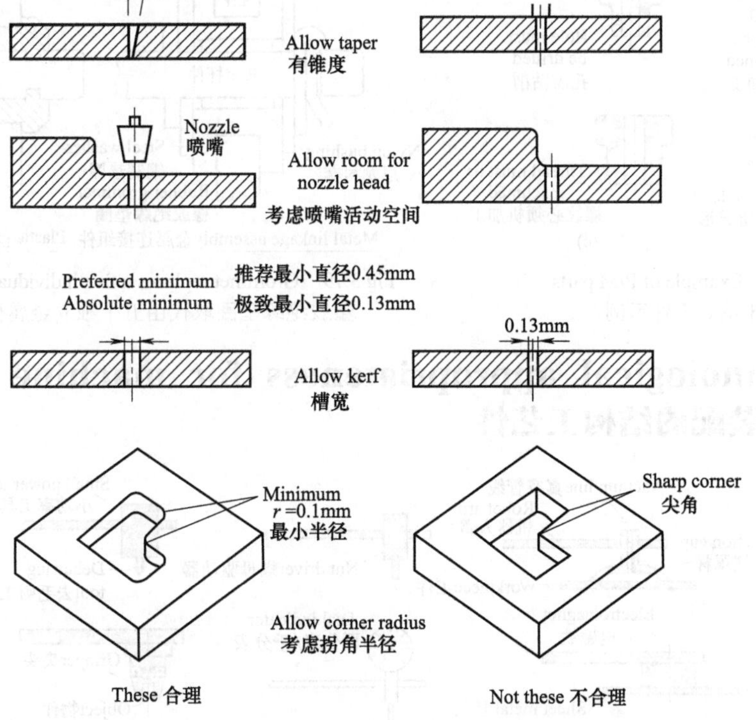

Fig 3.75 Design recommendations for AJM 磨料喷射加工设计推荐

3.3 Technological appropriateness for plastic parts and powder metallurgy parts 塑件、粉末冶金件的结构工艺性

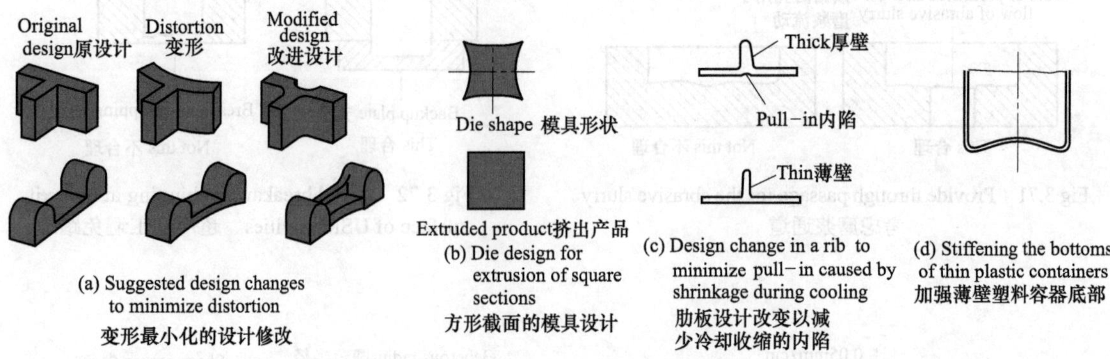

Fig 3.76 Examples of design modifications to eliminate or minimize distortion of plastic parts
根除或减少塑性变形的设计改变

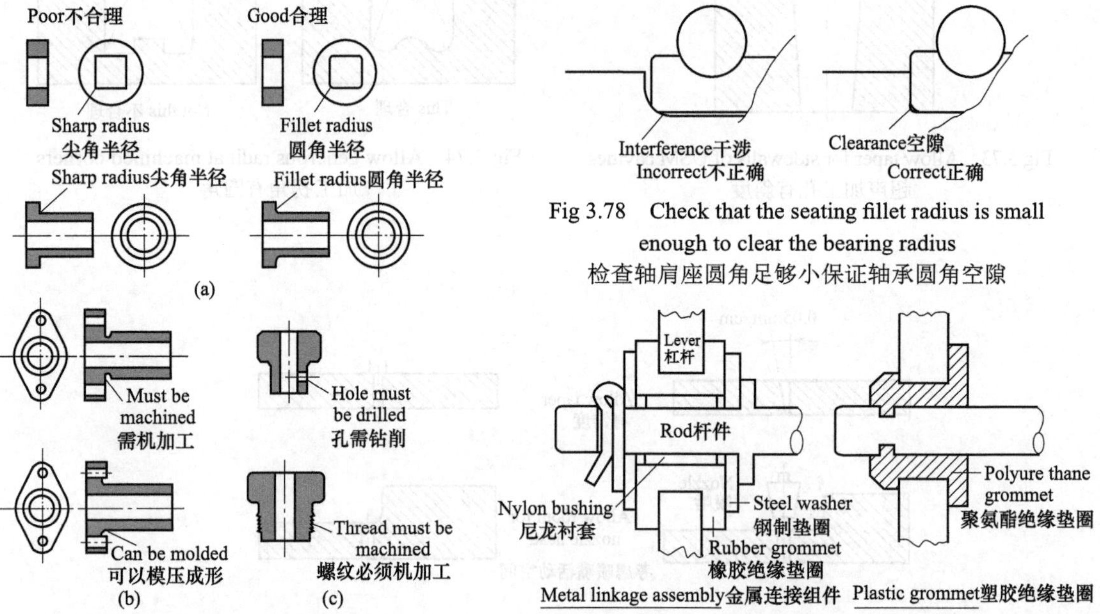

Fig 3.77 Example of P/M parts
粉末冶金零件示例

Fig 3.78 Check that the seating fillet radius is small enough to clear the bearing radius
检查轴肩座圆角足够小保证轴承圆角空隙

Fig 3.79 Grommet replaces a five individual metal assembly
塑胶绝缘垫圈取代由五个独立金属构成的组件

3.4 Technological appropriateness for machine assembly 机器装配的结构工艺性

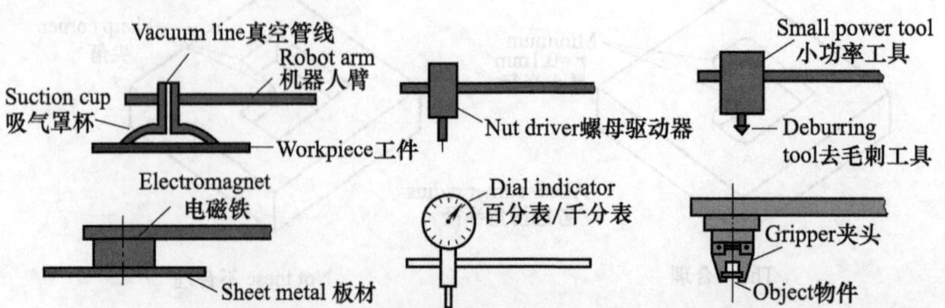

Fig 3.80 Various devices and tools attached to end effectors to perform a variety of operations
末端操控装置附加各种工具完成多种操作任务

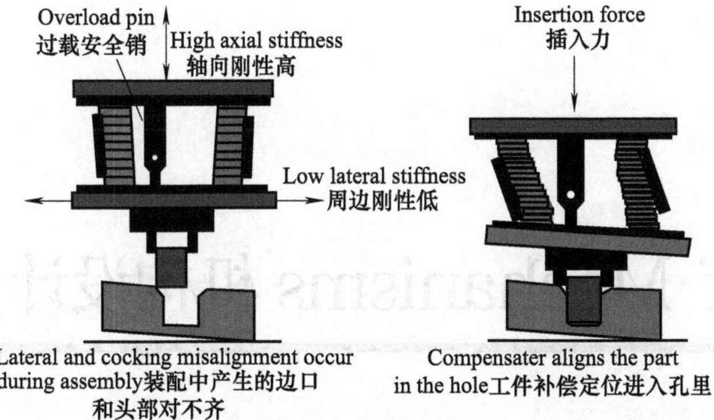

Fig 3.81 A system of compensating for misalignment during automated assembly 自动装配中定位偏差的自动补偿

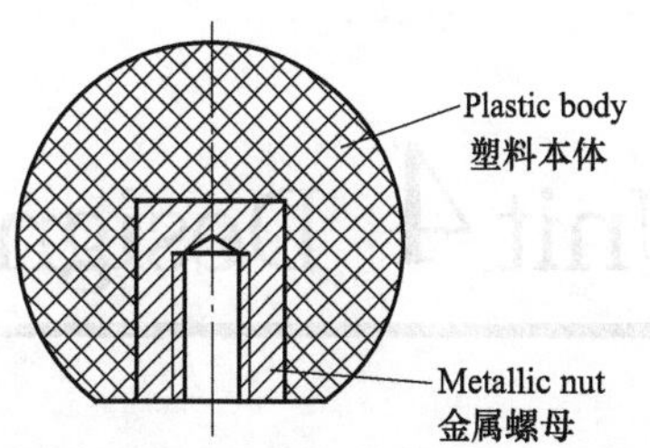

Fig 3.82 Handle produced simultaneously with its assembly on a metal nut for fastening 生产时同时装配金属固定螺母手把

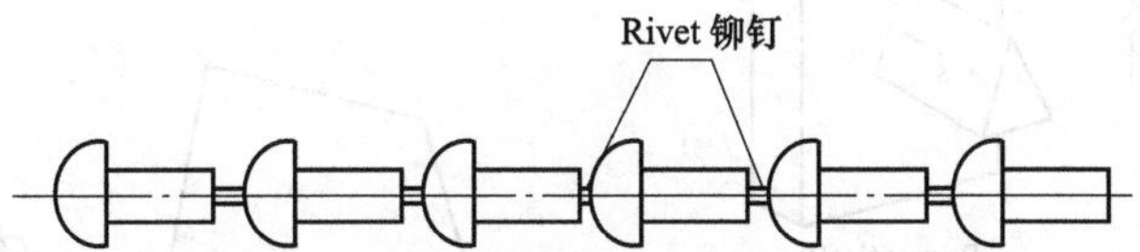

Fig 3.83 Example of continuous chain of rivets for more effective assembly 连续铆钉链可以使得装配更加高效

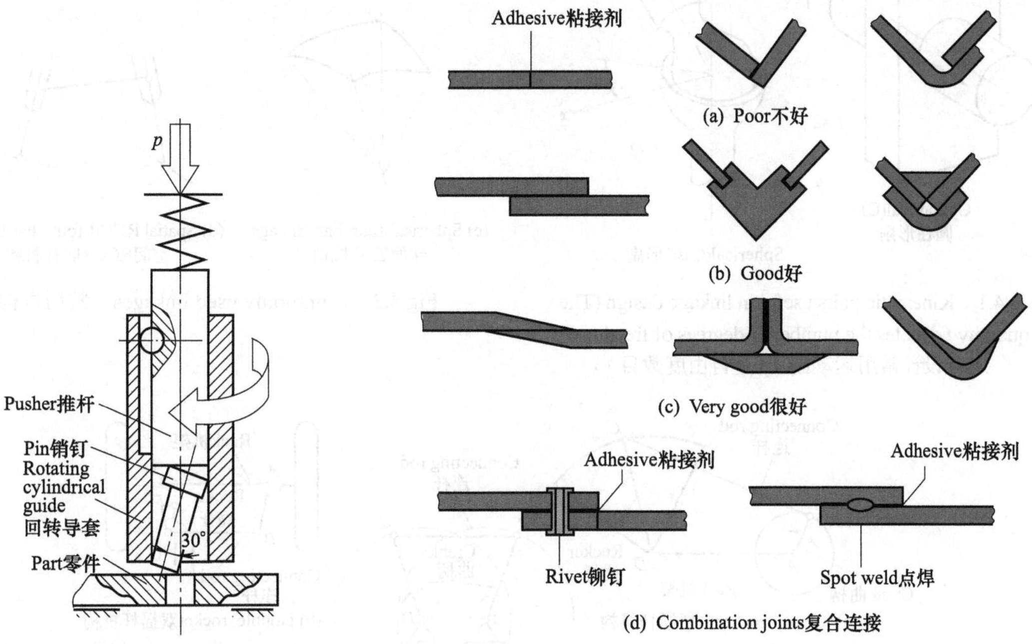

Fig 3.84 Rotating assembly device 回转装配装置

Fig 3.85 Various joint designs in adhesive bonding 粘接的几种接头设计

Unit 4 Design of Mechanisms 机构设计

4.1 Design of linkage mechanisms 连杆机构设计

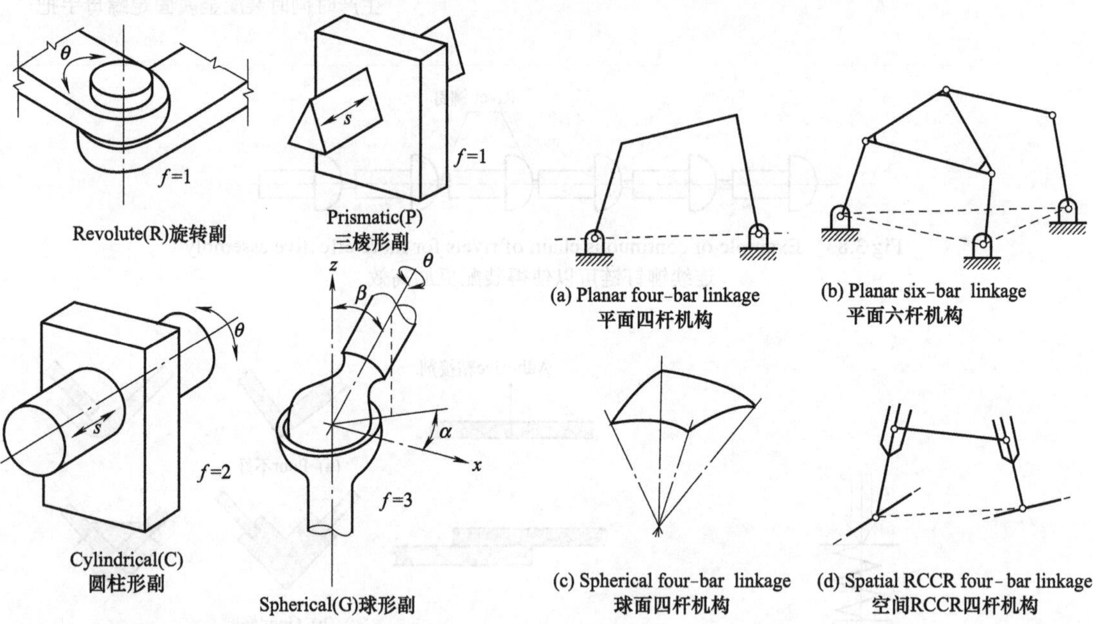

Fig 4.1 Kinematic pairs useful in linkage design (The quantity f denotes the number of degrees of freedom) 连杆设计常用运动副（f是自由度数目）

Fig 4.2 Commonly used linkages 常用机构

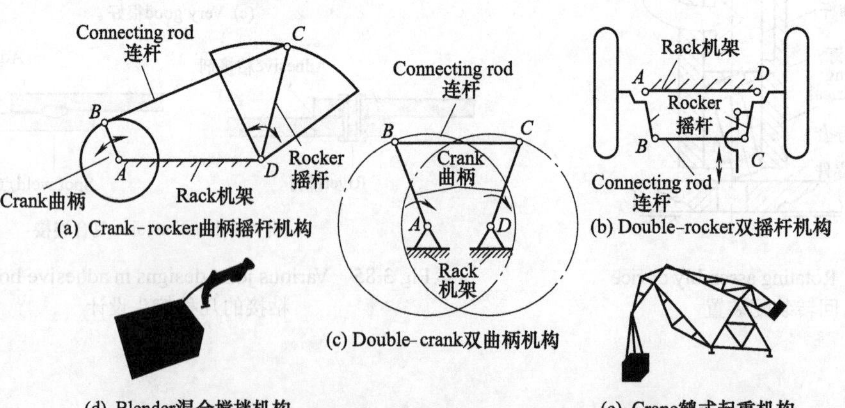

Fig 4.3 Types of Four-bar mechanism 四杆机构的类型

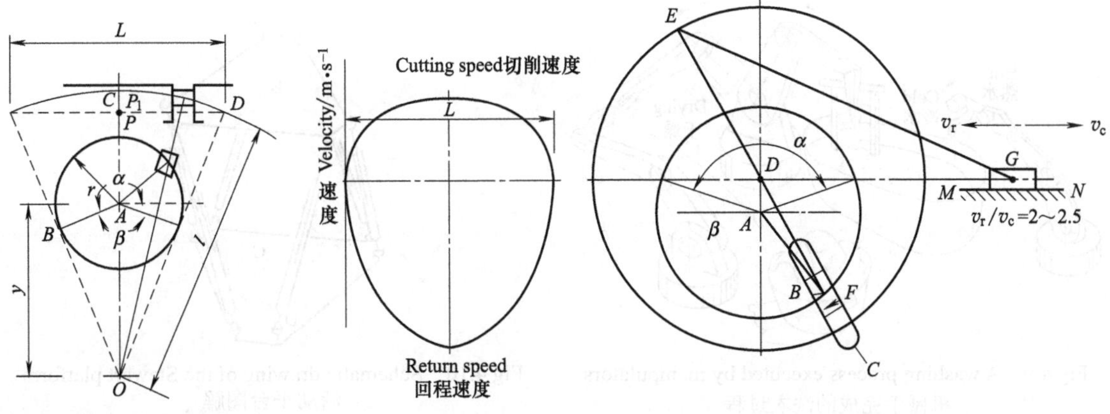

Fig 4.4 The quick-return mechanism 快回机构

Fig 4.5 Whitworth quick-return mechansim 惠氏快回机构

4.2 Design of spatial linkage mechanisms 空间机构设计

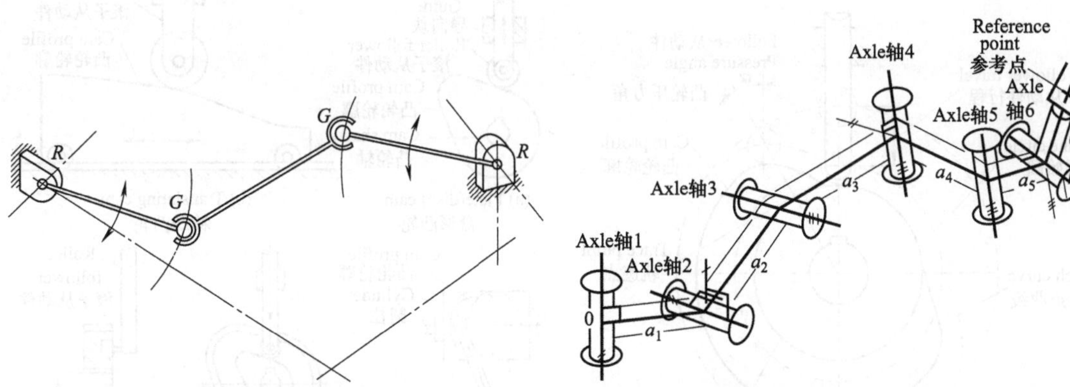

Fig 4.6 RGGR spatial linkage（R designates a revolute joint；G designates a spherical joint）
RGGR 空间连杆（R 代表回转铰链；G 代表球形铰链）

Fig 4.7 Manipulator chain with no offsets and with axles parallel or orthogonal
无偏置链条机械手，其轴心线或是平行或是垂直

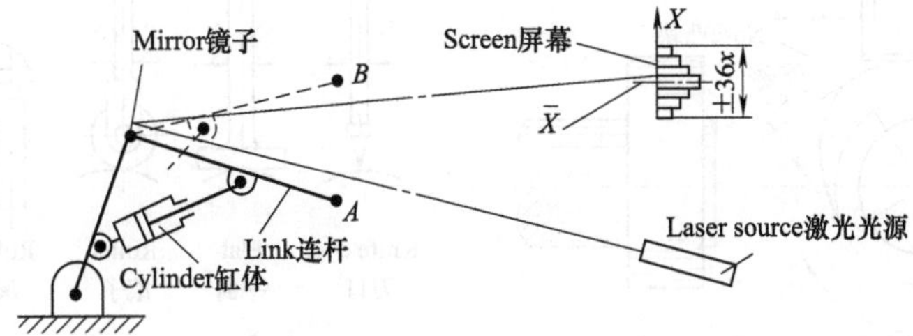

Fig 4.8 Reproducibility of manipulator link movement 机械手连杆运动的复制能力

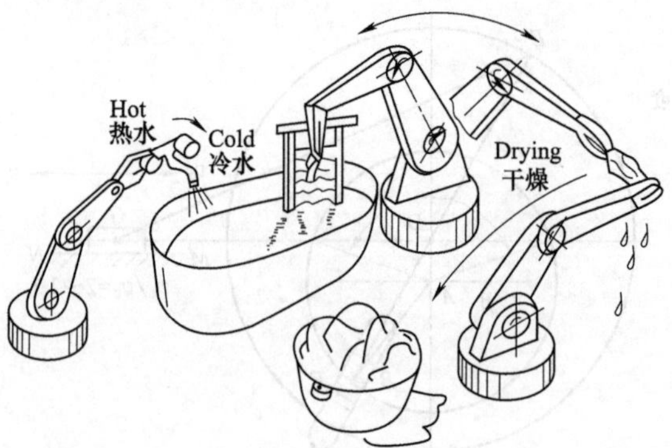

Fig 4.9 A washing process executed by manipulators 机械手完成的洗衣过程

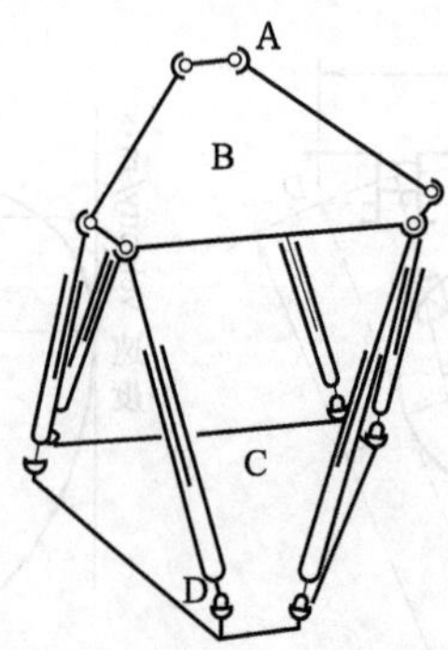

Fig 4.10 Schematic drawing of the Stewart platform 活动平台图解
(The movable platform is supported on six legs 活动平台由六杆支撑)
A—Ball-and-socket joint 球型铰连；B—Movable platform 活动平台；C—Fixed base 固定座；D—Hooke joint 万向节

4.3 Design of cam mechanisms 凸轮机构设计

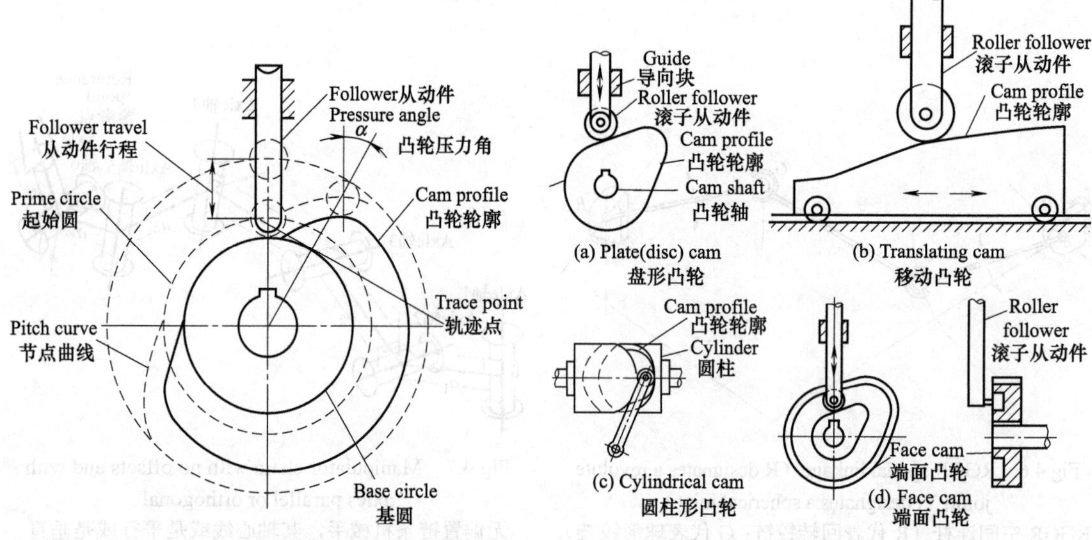

Fig 4.11 Basic components of cam mechanism 凸轮机构的基本构件

Fig 4.12 Types of cams 凸轮类型

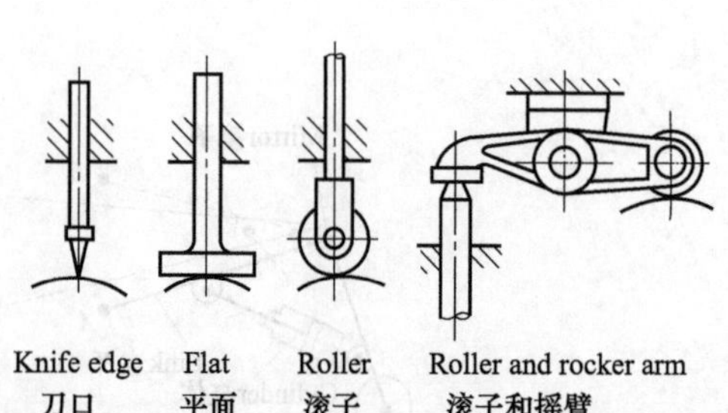
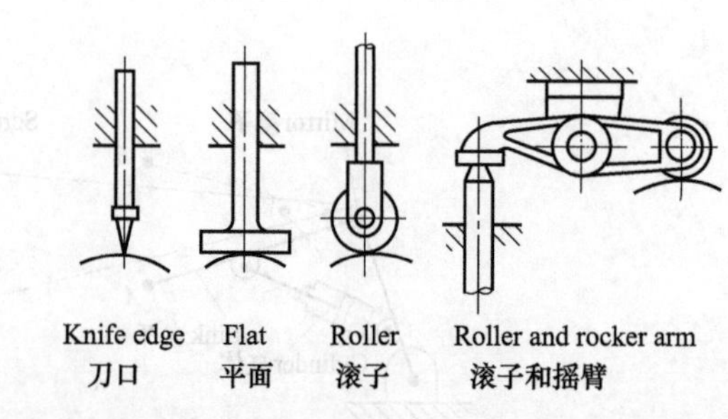

Fig 4.13 Four types of radial cams 四种径向凸轮

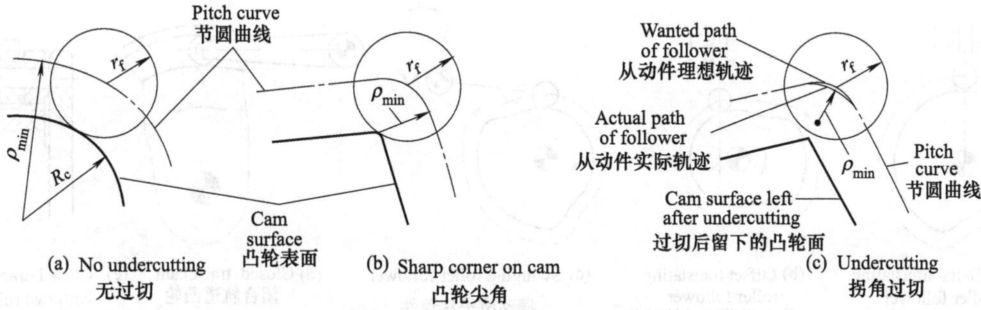

Fig 4.14 The minimum radius of curvature of a cam should be kept as large as possible
凸轮的最小曲率半径保持尽可能大

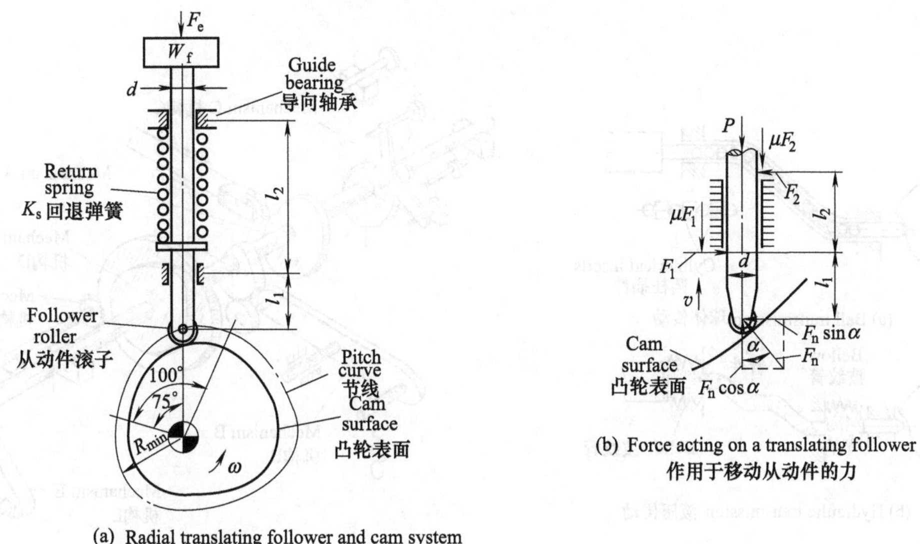

Fig 4.15 Cam forces, contact stresses 凸轮的力和应力

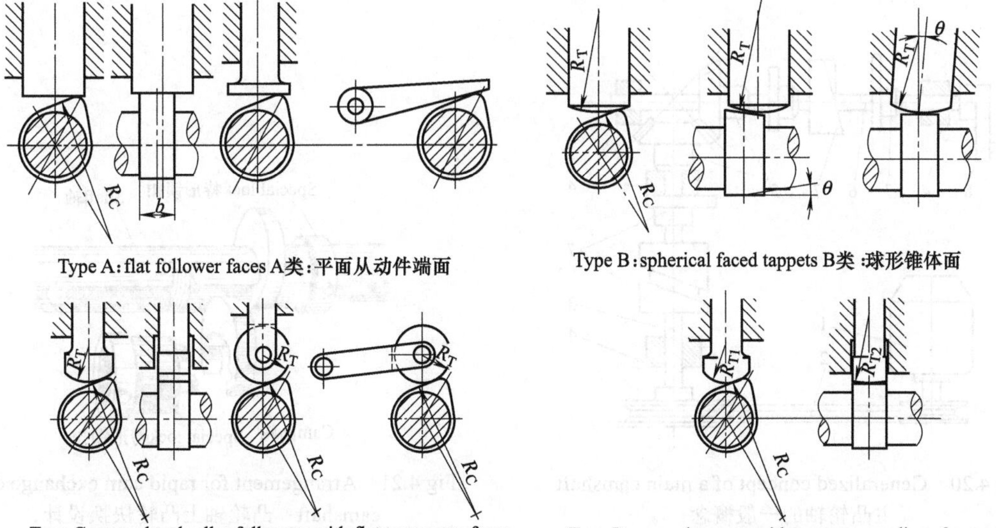

Fig 4.16 Classification of cams and tappets for determination of contact stresses 决定接触应力的凸轮挺杆的分类

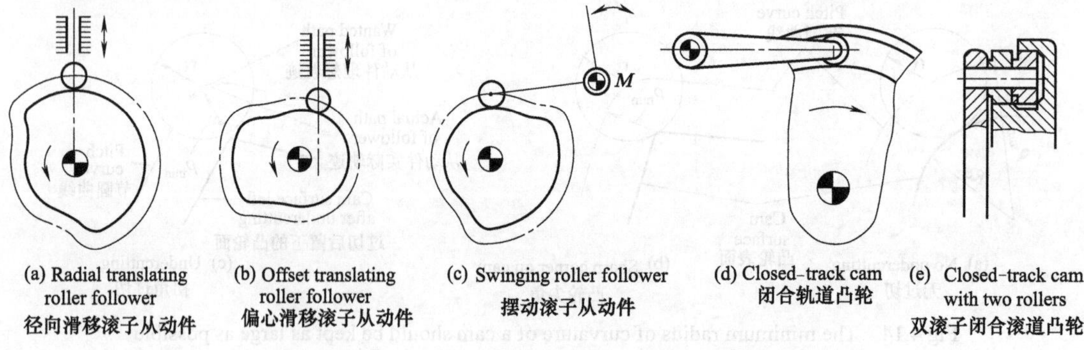

(a) Radial translating roller follower 径向滑移滚子从动件
(b) Offset translating roller follower 偏心滑移滚子从动件
(c) Swinging roller follower 摆动滚子从动件
(d) Closed-track cam 闭合轨道凸轮
(e) Closed-track cam with two rollers 双滚子闭合滚道凸轮

Fig 4.17 Cam Follower Systems 凸轮从动件系统

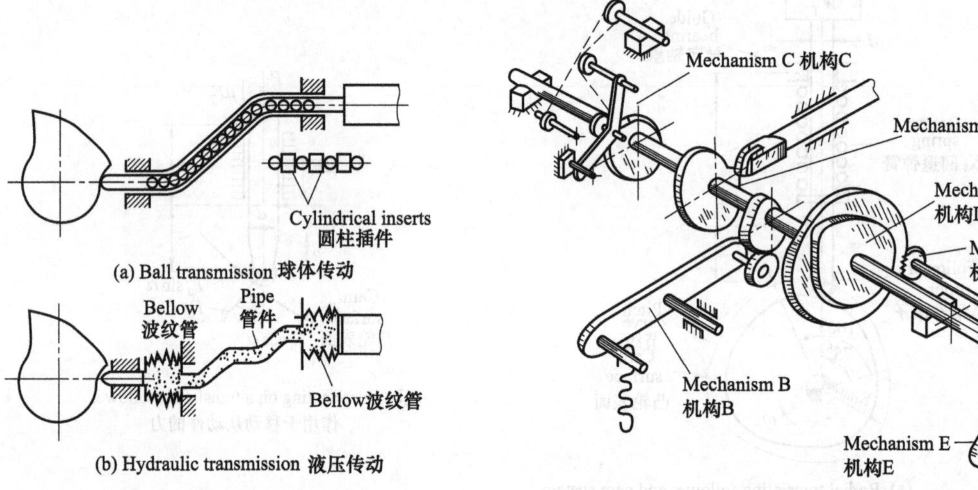

(a) Ball transmission 球体传动

(b) Hydraulic transmission 液压传动

Fig 4.18 Models of a“flexible”transmission of the cam follower's motion
凸轮从动件的柔性传动模型

Fig 4.19 Camshaft as a mechanical program carrier main camshaft 凸轮轴用作程序载体：主凸轮轴

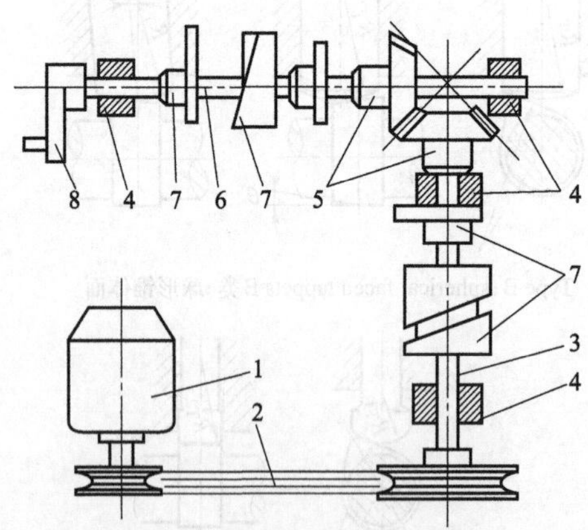

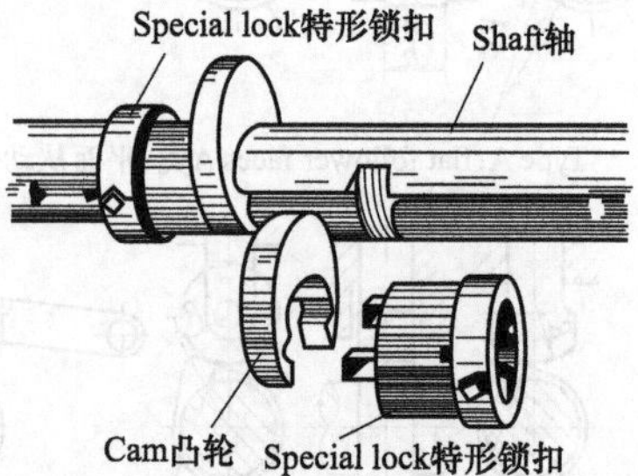

Fig 4.20 Generalized concept of a main camshaft
主凸轮轴的一般概念

1—Motor 电动机；2—Belt drive 带传动；3—Camshaft 凸轮轴；
4—Bearings 轴承；5—Bevel gears 锥齿轮；6—Shaft 轴；
7—Cams 凸轮；8—Crank 曲柄

Fig 4.21 Arrangement for rapid cam exchange on a camshaft 凸轮轴上凸轮快换设计

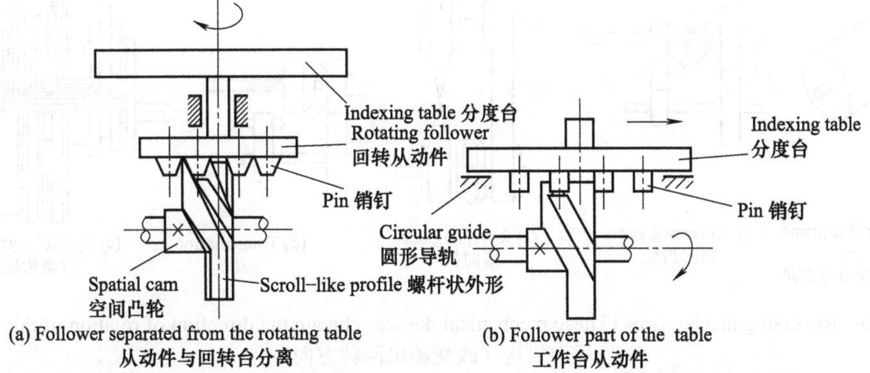

(a) Follower separated from the rotating table 从动件与回转台分离 (b) Follower part of the table 工作台从动件

Fig 4.22 Spatial cam drives for a circular transporting device 圆形传输装置的空间凸轮机构

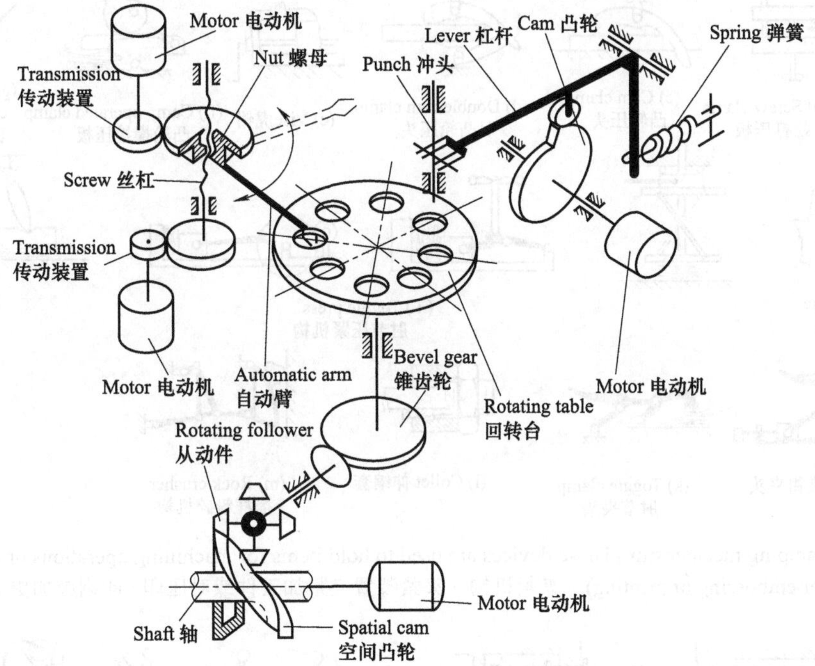

Fig 4.23 Layout of an automatic machine with autonomous, independent drive of mechanisms 带独立驱动机构的自动机设计

4.4 Design of common variated mechanisms 常用变异机构设计

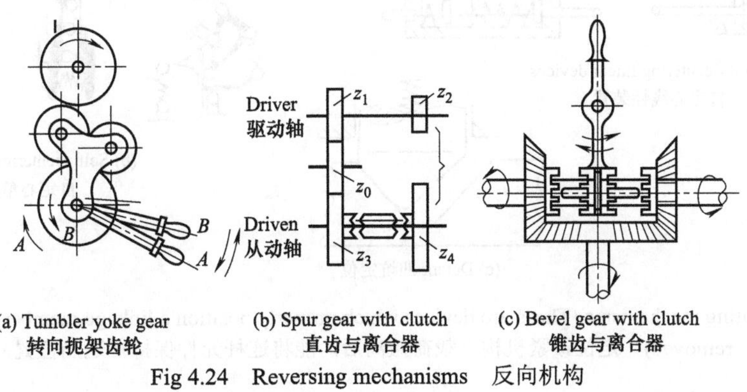

(a) Tumbler yoke gear 转向扼架齿轮 (b) Spur gear with clutch 直齿与离合器 (c) Bevel gear with clutch 锥齿与离合器

Fig 4.24 Reversing mechanisms 反向机构

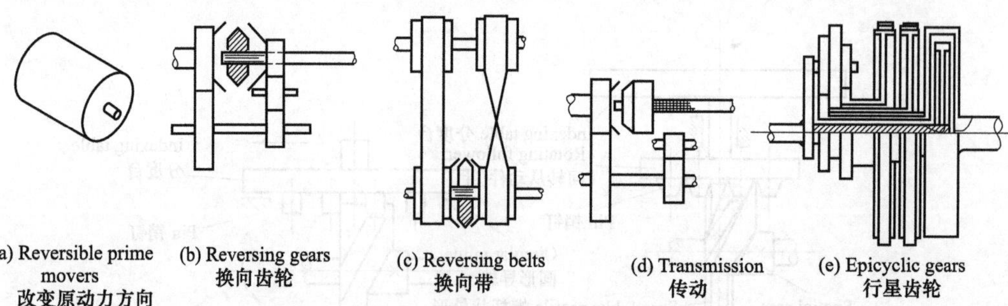

Fig 4.25 Reversing mechanisms (These mechanical devices change the direction of rotation of the output) 反向机构（改变输出回转方向）

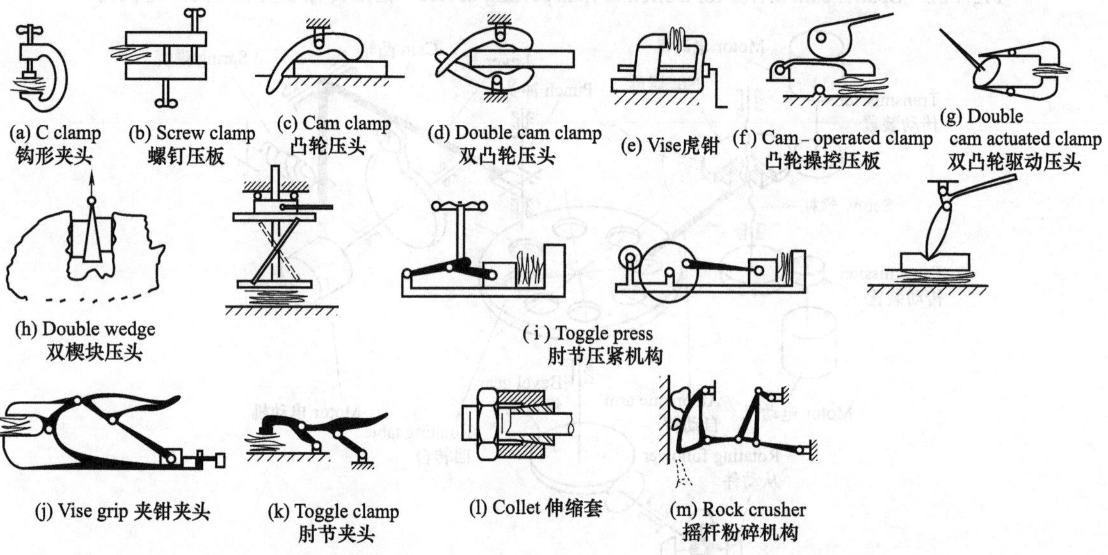

Fig 4.26 Clamping mechanisms (These devices are used to hold items for machining operations or to exert great forces for embossing or printing) 夹紧机构（夹紧装置夹紧加工件或对压印 / 印刷施加更大的力）

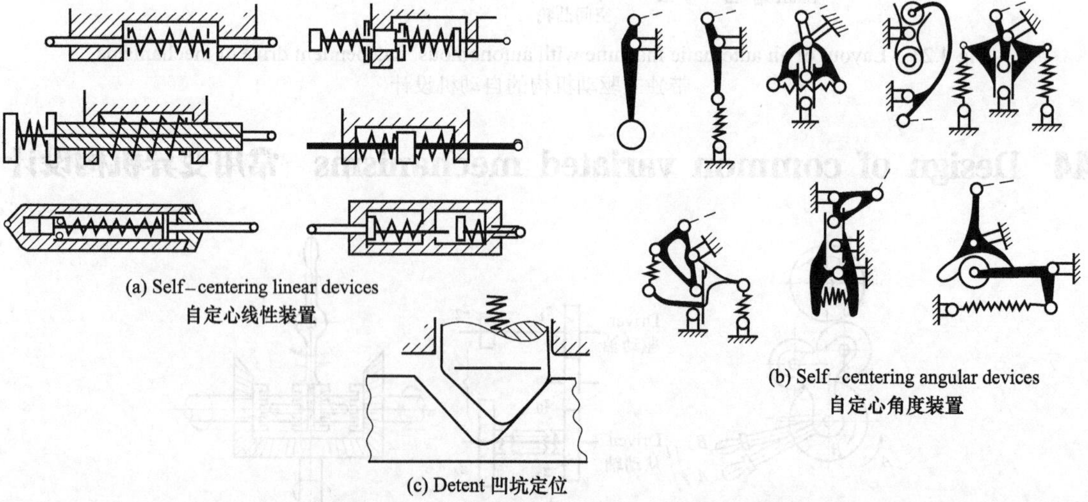

Fig 4.27 Locating mechanisms (These are devices which properly position a linkage member when the load is removed) 定位锁紧机构（载荷撤离后，能将连杆元件保持在适当位置）

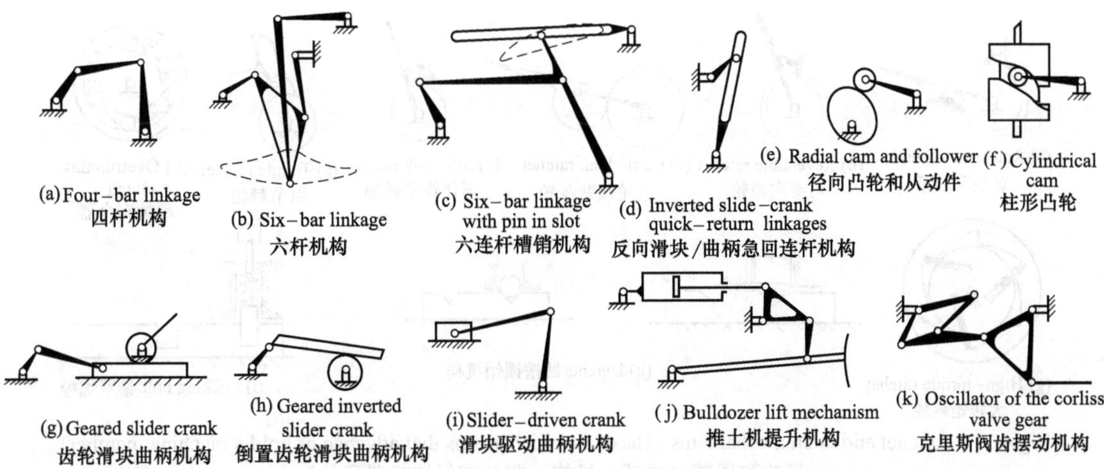

Fig 4.28 Oscillating mechanisms Ⅰ (These mechanisms cause an output to repeatedly swing through a preset angle) 摆动机构Ⅰ（输出预定角度的重复摆动运动）

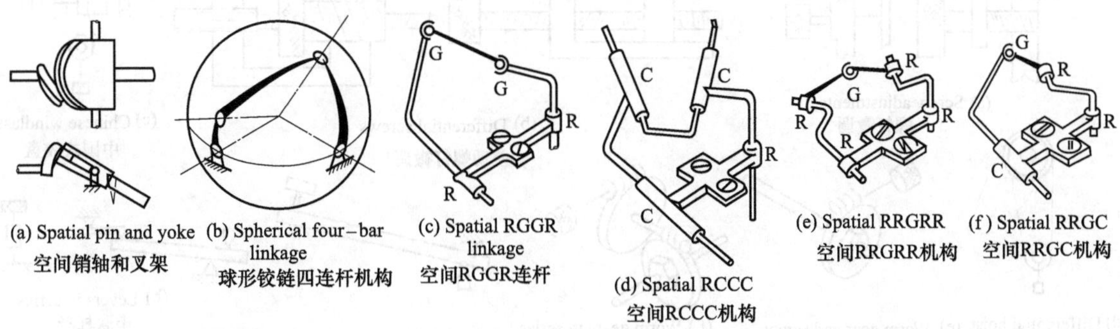

Fig 4.29 Oscillating mechanisms Ⅱ (These all use spatial linkages) 摆动机构Ⅱ（利用空间连杆机构）

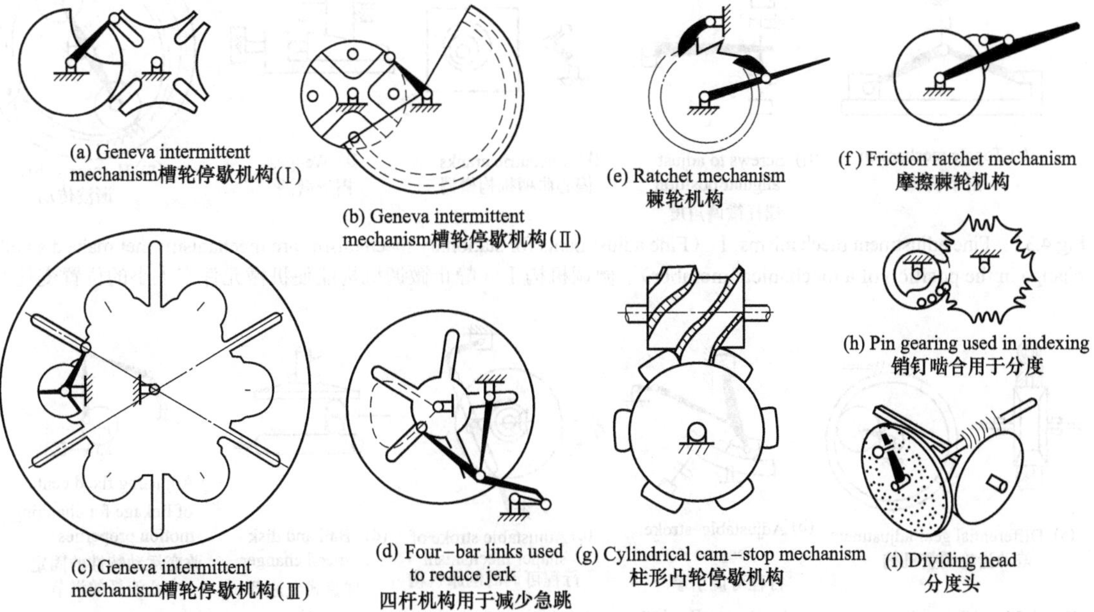

Fig 4.30 Indexing mechanisms (These mechanical devices advance a body to a specific position, hold it there for a period, and then advance it again) 分度机构（驱动从动件到达特定位置并保持一定时间，然后再驱动）

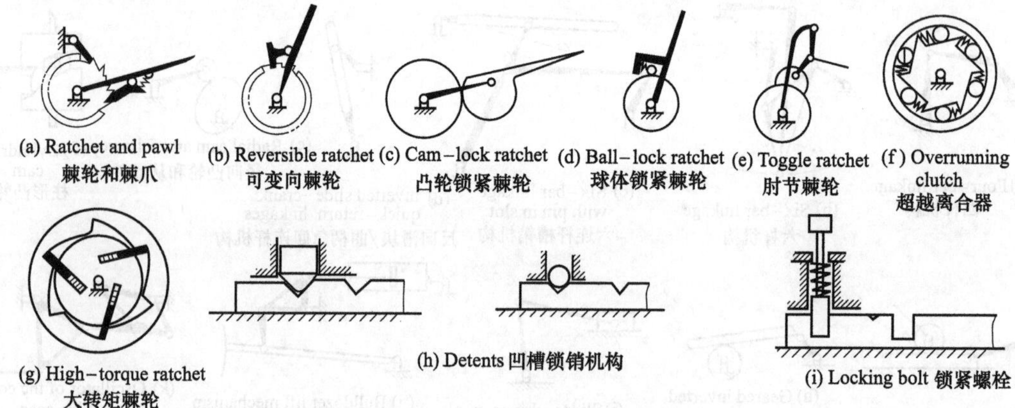

Fig 4.31 Ratchet and latch mechanisms (These are mechanisms that advance or hold a machine member)
棘轮和插销（碰锁）机构（驱动或保持机器元件）

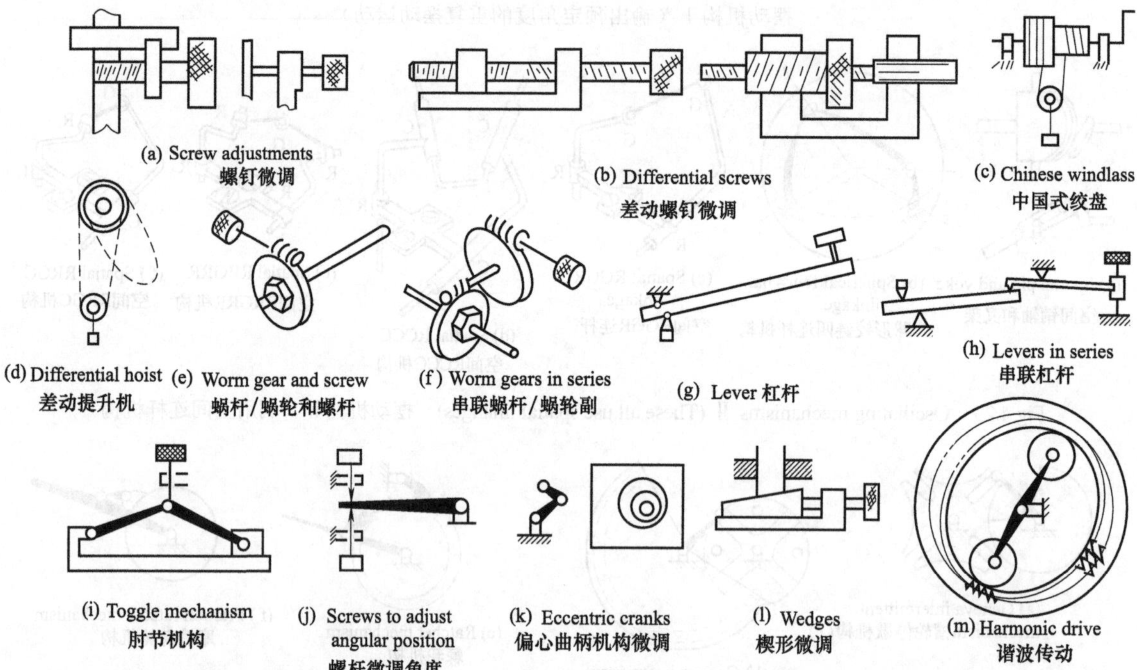

Fig 4.32 Fine adjustment mechanisms Ⅰ（Fine adjustments for stationary mechanisms are mechanisms that make a small change in the position of a mechanical member） 微调机构Ⅰ（静止微调机构能使机械元件产生小的位置变化）

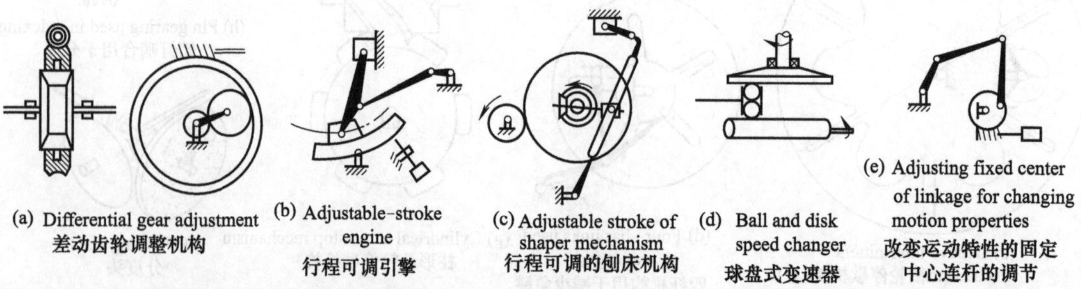

Fig 4.33 Fine adjustment mechanisms Ⅱ（Fine adjustments for moving mechanisms are adjusting devices which control the motion of linkages such as stroke, etc., while the mechanism is in motion）
微调机构Ⅱ（机构在运动过程微调，这些装置控制连杆运动，如行程）

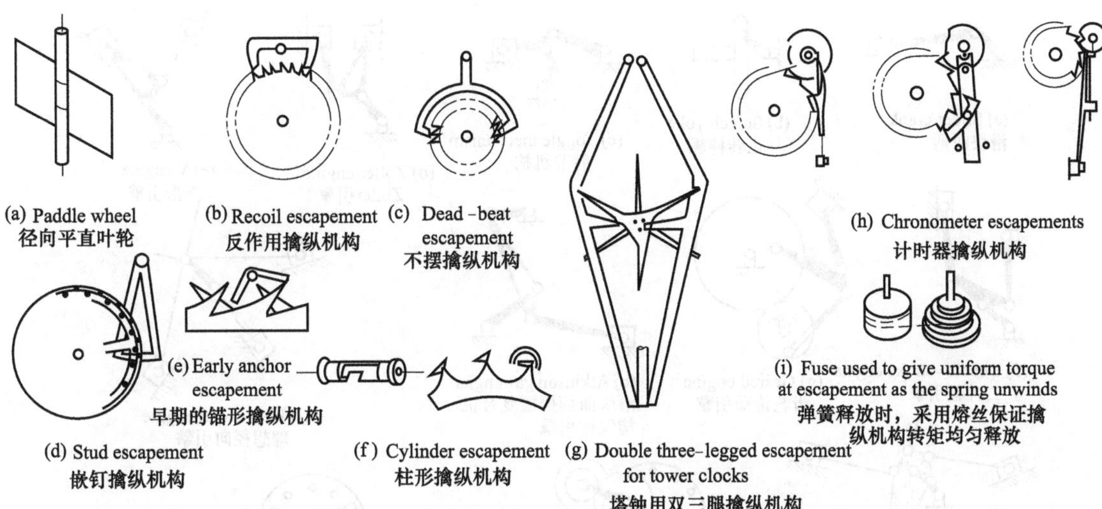

Fig 4.34 Escapements mechanisms (These devices slowly release the potential energy stored in a spring to control devices such as clocks) 擒纵机构（能慢速释放弹簧储存的能量以控制装置，如时钟）

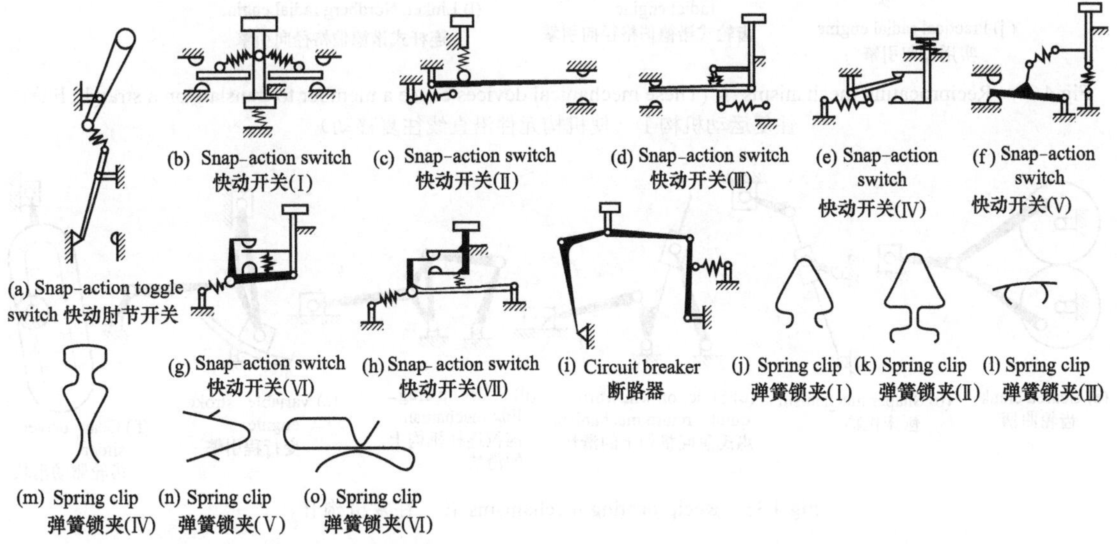

Fig 4.35 Snap-action mechanisms (These mechanisms are bistable elements in machines. They are used in switches to quickly make and break electric circuits and for fastening items)
快动机构（这些机构是机器中的双稳元件，用于电路的快速接通和断开或锁紧功能）

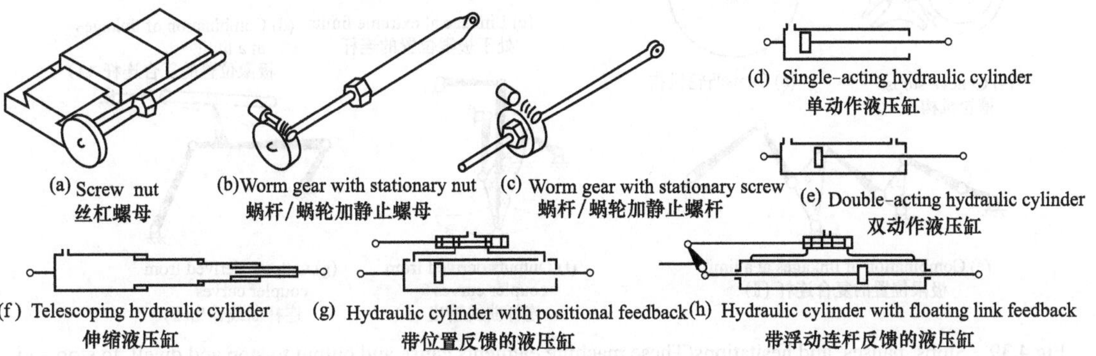

Fig 4.36 Linear actuators (These are devices that cause a straight-line displacement between two machine elements)
线性执行元件（执行装置能够产生两个元件之间的直线位移）

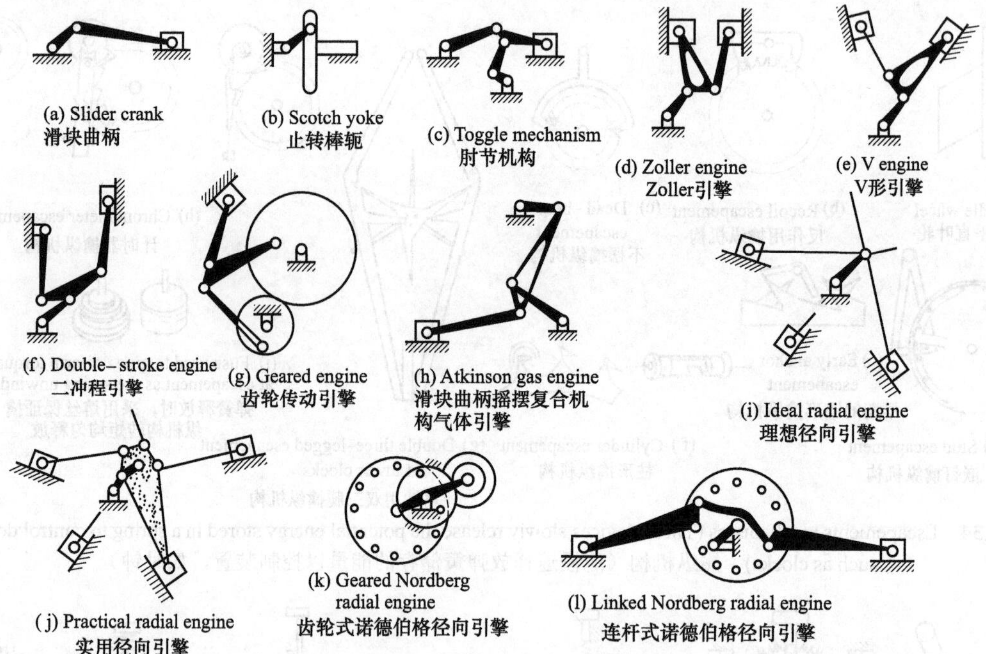

Fig 4.37 Reciprocating mechanisms Ⅰ (These mechanical devices cause a member to translate on a straight line)
往复运动机构Ⅰ（使机构元件沿直线往复移动）

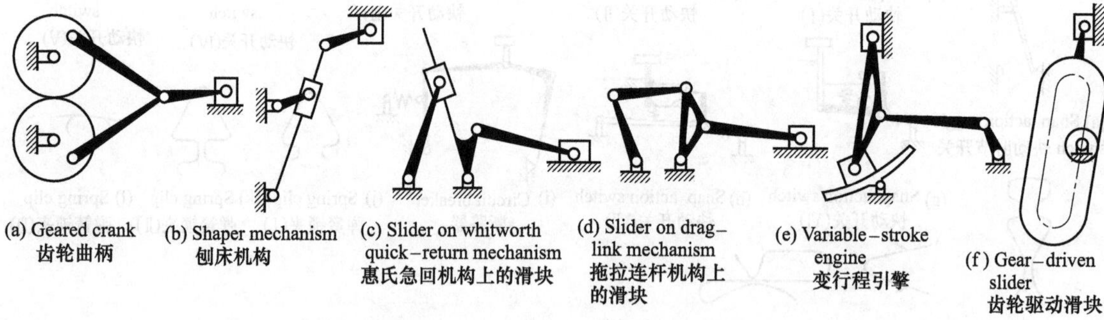

Fig 4.38 Reciprocating mechanisms Ⅱ 往复机构Ⅱ

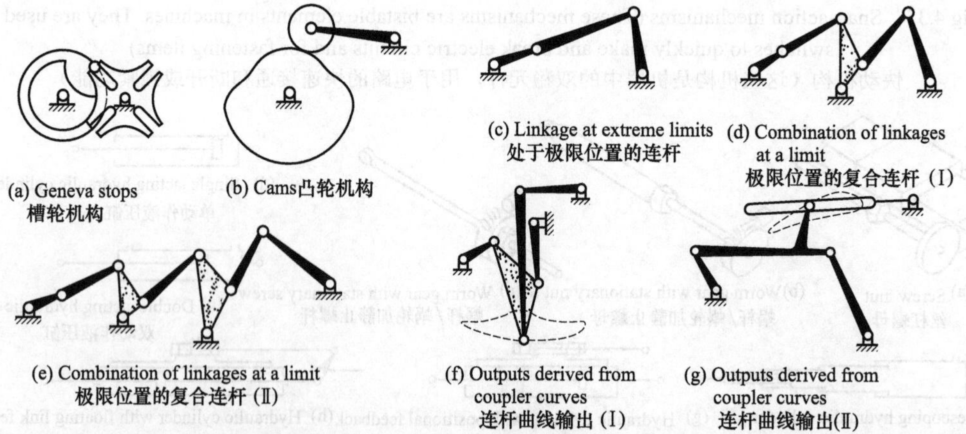

Fig 4.39 Stops, pauses, and hesitations(These machine elements cause and output to stop and dwell, to stop and return to stop and advance etc. The derivatives of the motion at the stop determine which category the motion fits)
停止、暂停和延迟装置（使运动输出停止、暂停、停止、返回、停止、前进等。输出运动决定其适合的类别）

(a) Four-bar film advance 四杆胶片推进机构

(b) Circular motion transport 圆周运动传输

(c) Coupler curve transport 连接件曲线传输（一）

(d) Coupler curve transport 连接件曲线传输（二）

(e) Geared linkage transport 齿轮连杆传输

(f) Fishing-reel feed 鱼竿进给

Fig 4.40 Transportation devices(These mechanisms move one or more objects a discrete distance in stepped motion) 传输装置（将一个或多个物品以步进运动方式移动一定距离）

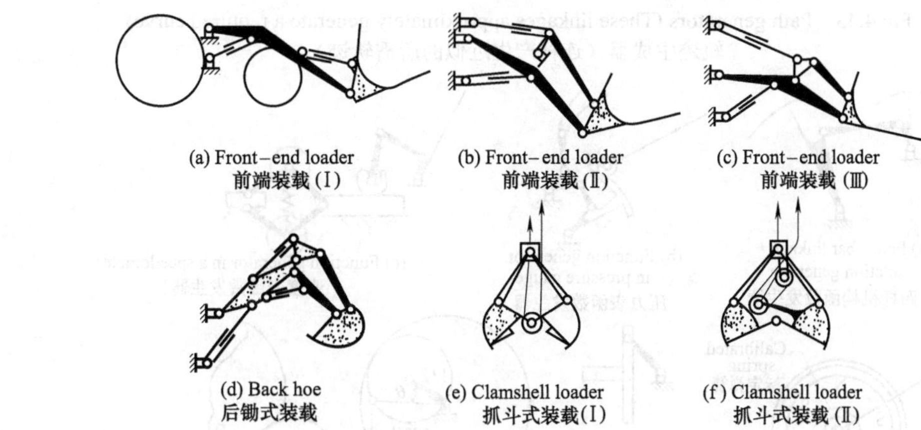

(a) Front-end loader 前端装载（Ⅰ）

(b) Front-end loader 前端装载（Ⅱ）

(c) Front-end loader 前端装载（Ⅲ）

(d) Back hoe 后锄式装载

(e) Clamshell loader 抓斗式装载（Ⅰ）

(f) Clamshell loader 抓斗式装载（Ⅱ）

Fig 4.41 Loading and unloading mechanisms Ⅰ (These mechanisms pick up material and transport it to another location) 装卸载机构Ⅰ（拾取材料并传输到另一个位置）

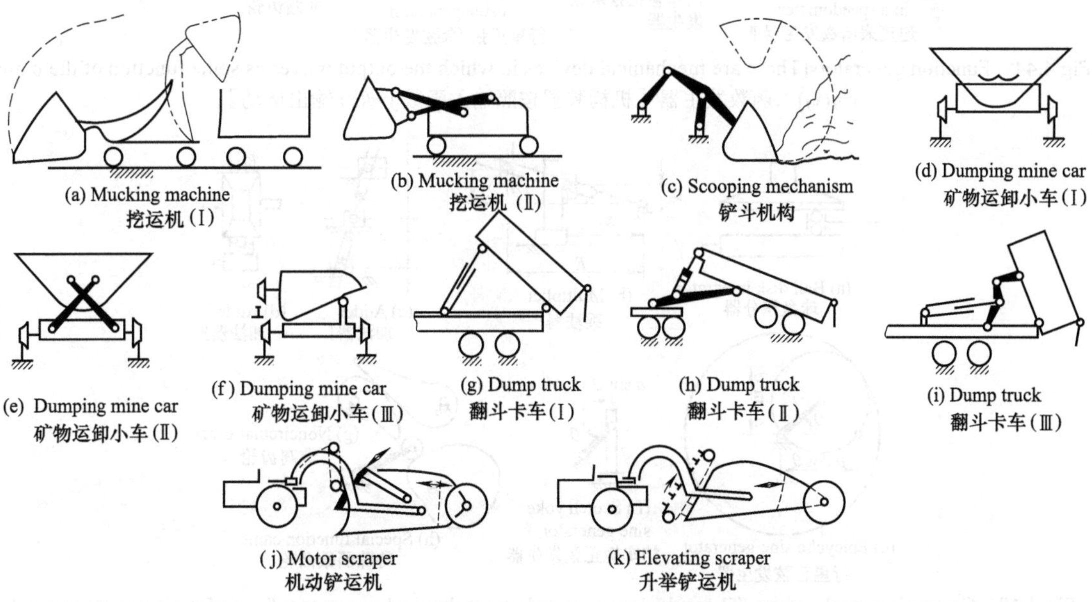

(a) Mucking machine 挖运机（Ⅰ）

(b) Mucking machine 挖运机（Ⅱ）

(c) Scooping mechanism 铲斗机构

(d) Dumping mine car 矿物运卸小车（Ⅰ）

(e) Dumping mine car 矿物运卸小车（Ⅱ）

(f) Dumping mine car 矿物运卸小车（Ⅲ）

(g) Dump truck 翻斗卡车（Ⅰ）

(h) Dump truck 翻斗卡车（Ⅱ）

(i) Dump truck 翻斗卡车（Ⅲ）

(j) Motor scraper 机动铲运机

(k) Elevating scraper 升举铲运机

Fig 4.42 Loading and unloading mechanisms Ⅱ 装卸载机构Ⅱ

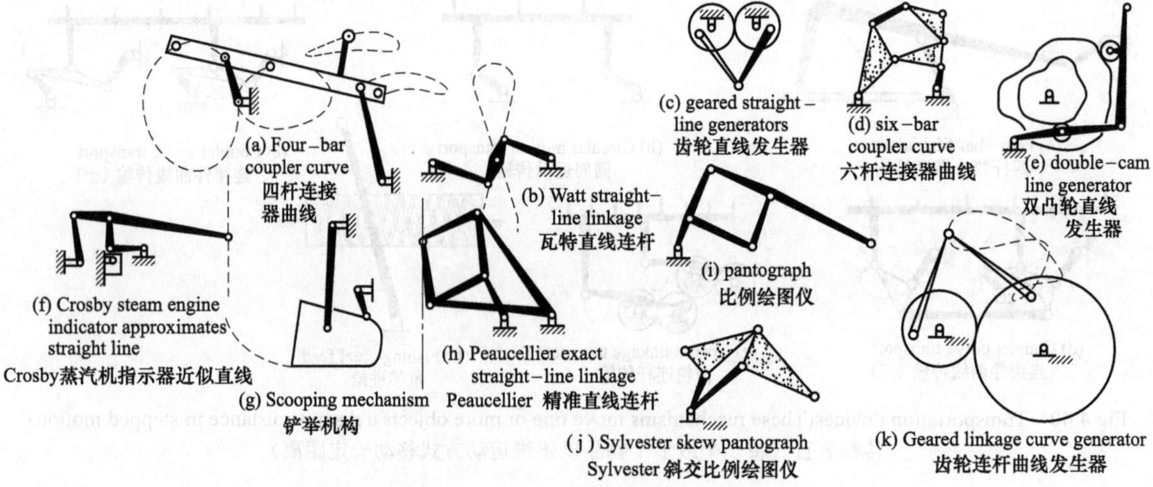

Fig 4.43 Path generators (These linkages approximately generate a required curve) 轨迹生成器（连杆产生近似的所需轨迹）

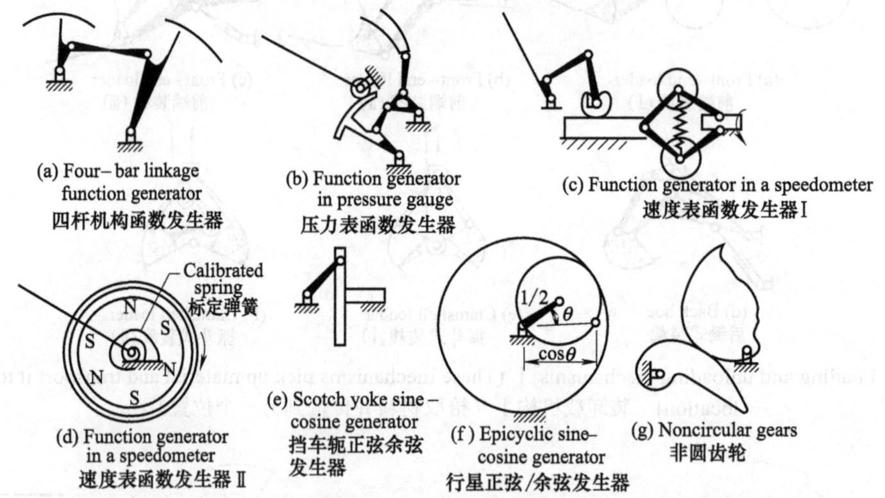

Fig 4.44 Function generators[These are mechanical devices in which the output moves as some function of the input $y=f(x)$] 函数发生器［机构装置按照输入函数 $y=f(x)$ 输出运动］

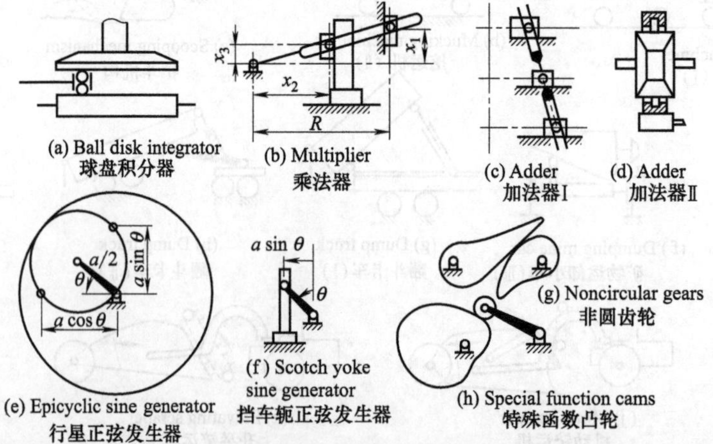

Fig 4.45 Computing mechanisms (These devices are used on mechanical computers for performing mathematical operations) 计算机构（用于机械式计算机完成数学计算）

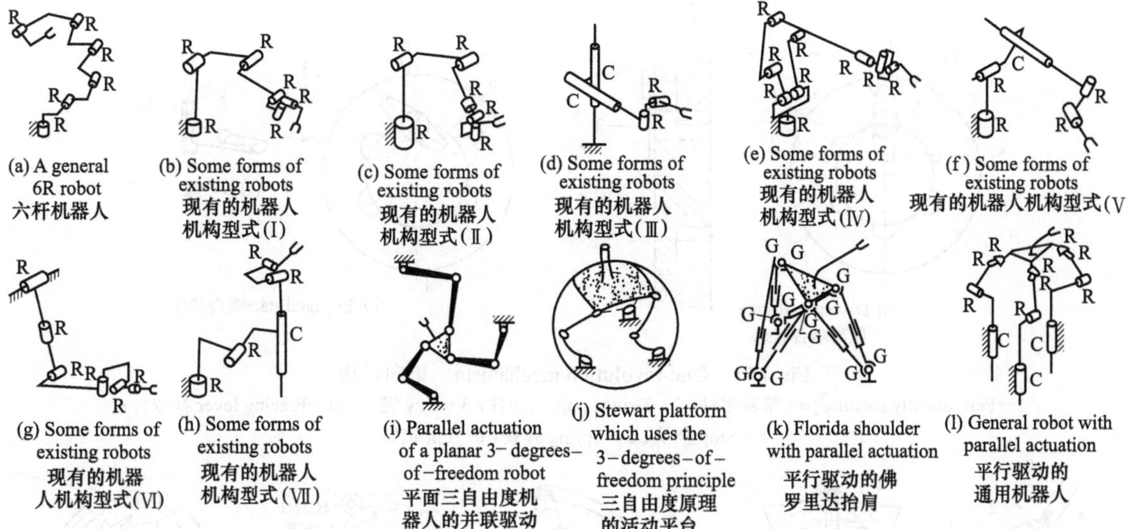

Fig 4.46 Robots mechanisms (These are multidegree-of-freedom devices used for positioning or assembly of items. They usually have some degree of machine intelligence and work under computer control) 机器人机构（多自由度装置用于零部件的定位和装配，通常有一定程度的机器智能，并在计算机控制下作业）

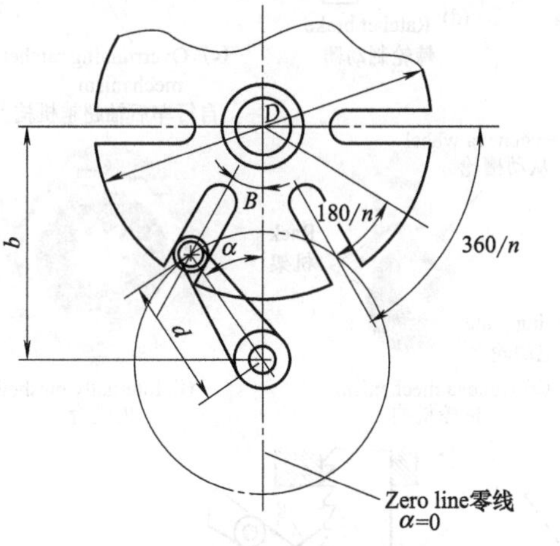

Fig 4.47 External geneva mechanism 外槽轮机构

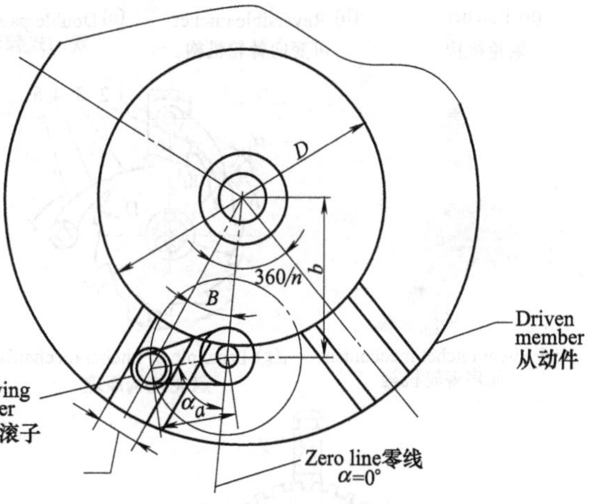

Fig 4.48 Internal geneva mechanism(six-slot internal geneva wheel) 内槽轮机构（六槽）

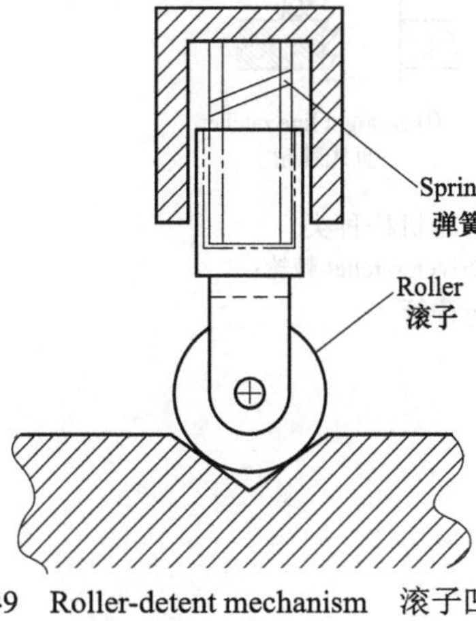

Fig 4.49 Roller-detent mechanism 滚子凹槽机构

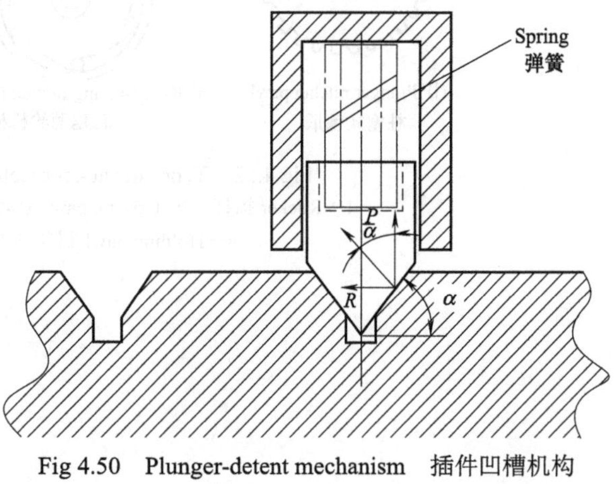

Fig 4.50 Plunger-detent mechanism 插件凹槽机构

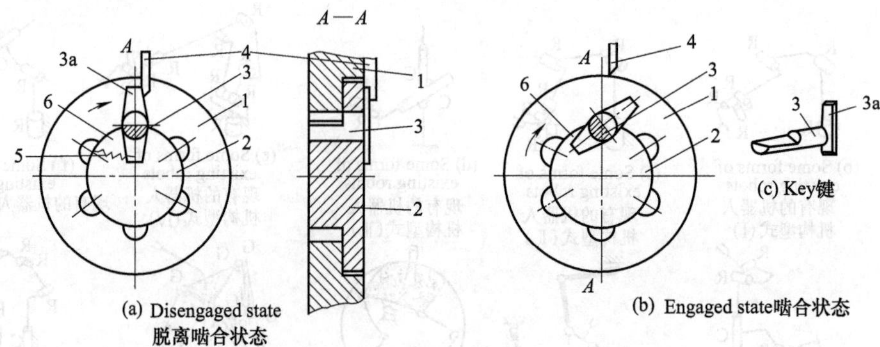

(a) Disengaged state 脱离啮合状态　(b) Engaged state啮合状态　(c) Key键

Fig 4.51　One-revolution mechanism　单向机构

1—Permanently rotating part 常转零件；2—Driven part 从动件；3—Key 键；　3a—Freeing lever 释放杆；
4—Stop 挡块；5—Spring 弹簧；6—Pin 销钉

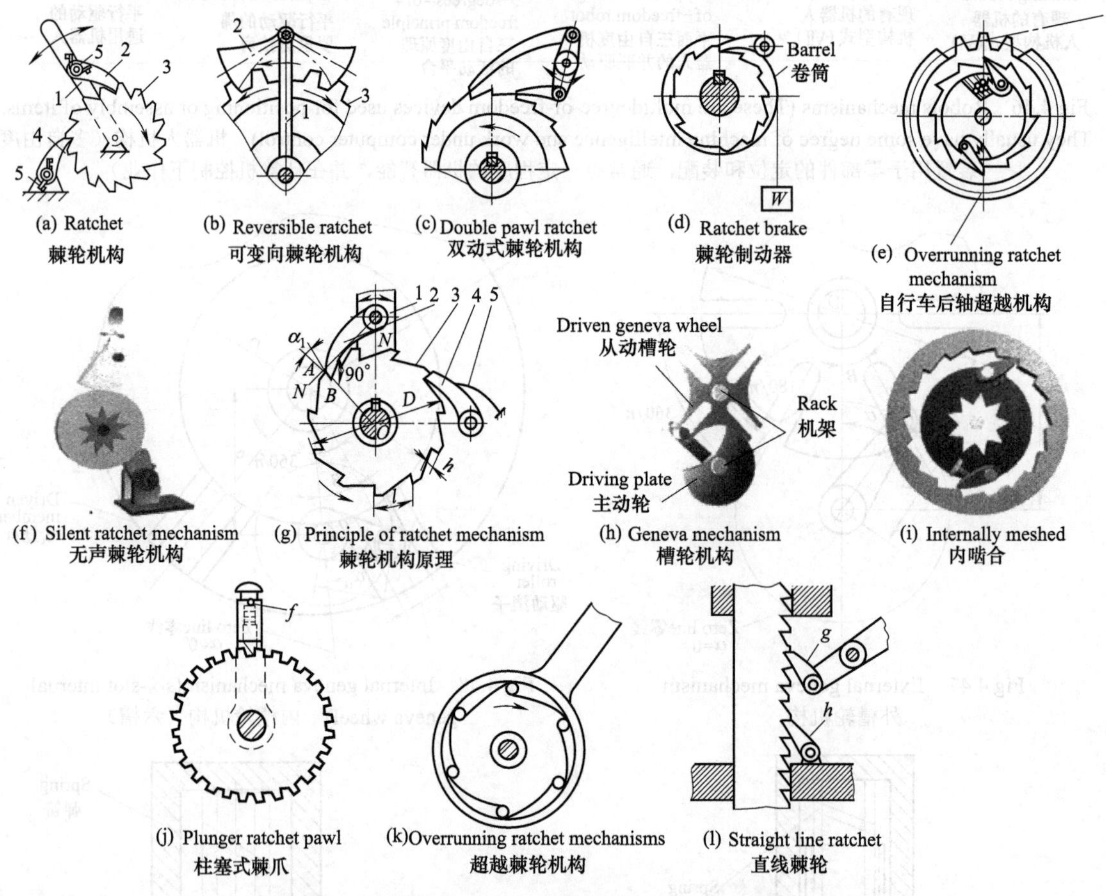

(a) Ratchet 棘轮机构　(b) Reversible ratchet 可变向棘轮机构　(c) Double pawl ratchet 双动式棘轮机构　(d) Ratchet brake 棘轮制动器　(e) Overrunning ratchet mechanism 自行车后轴超越机构

(f) Silent ratchet mechanism 无声棘轮机构　(g) Principle of ratchet mechanism 棘轮机构原理　(h) Geneva mechanism 槽轮机构　(i) Internally meshed 内啮合

(j) Plunger ratchet pawl 柱塞式棘爪　(k) Overrunning ratchet mechanisms 超越棘轮机构　(l) Straight line ratchet 直线棘轮

Fig 4.52　Types of racket mechanisms　棘轮机构种类

1—Rocker 摇杆；2—Driving pawl 驱动棘爪；3—Driven ratchet 棘轮；
4—Holding pawl 制动棘爪；5—Spring 弹簧

Unit 5 Transmission Design 传动设计

5.1 Design of gear transmission 齿轮传动设计

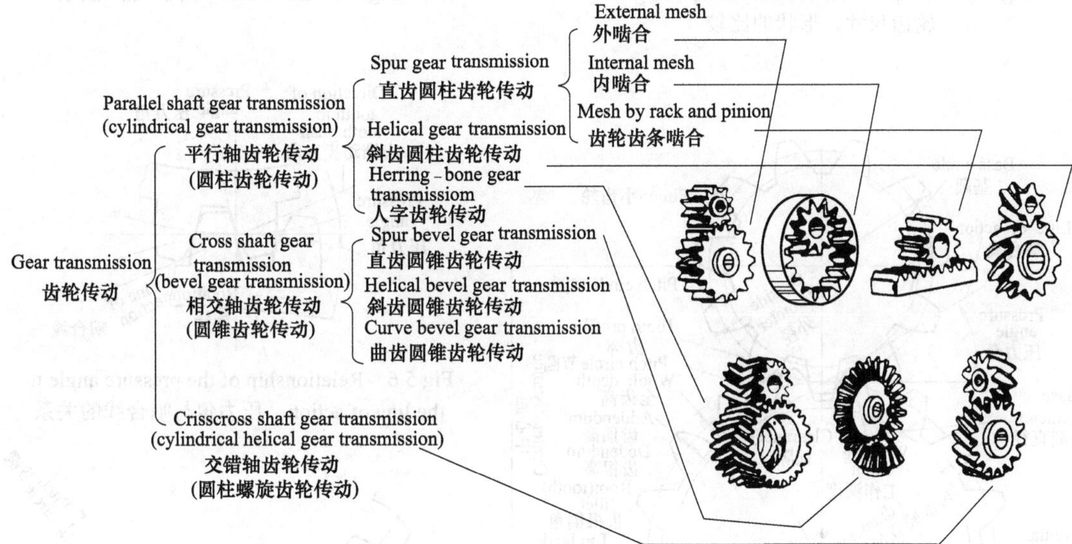

Fig 5.1 Types of gear drive 齿轮传动的类型

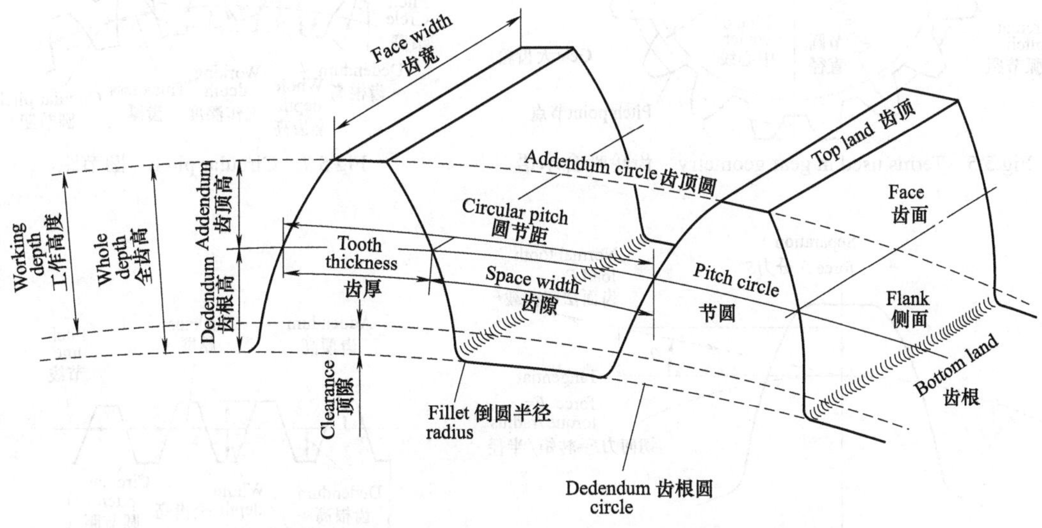

Fig 5.2 Nomenclature for an involute spur gear 渐开线直齿齿轮术语

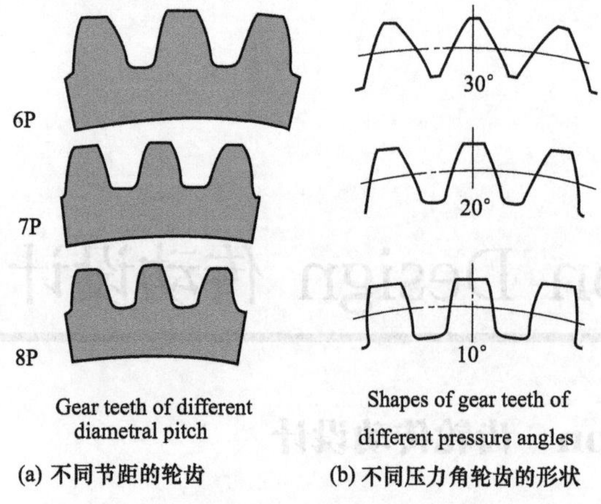

(a) 不同节距的轮齿　(b) 不同压力角轮齿的形状

Fig 5.3　Comparative sizes and shape of gear 轮齿尺寸、形状的比较

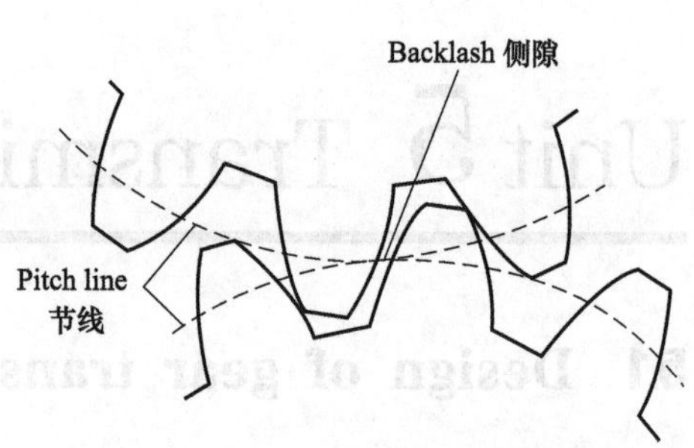

Fig 5.4　Backlash　侧隙或后退间隙

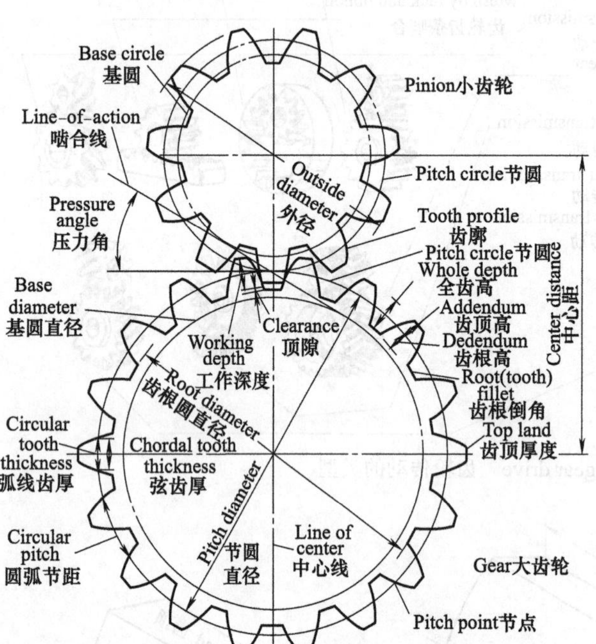

Fig 5.5　Terms used in gear geometry　齿轮啮合术语

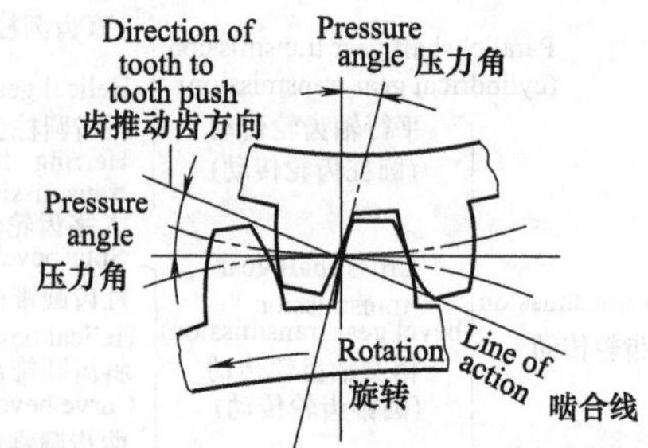

Fig 5.6　Relationship of the pressure angle to the line of action　压力角与啮合线的关系

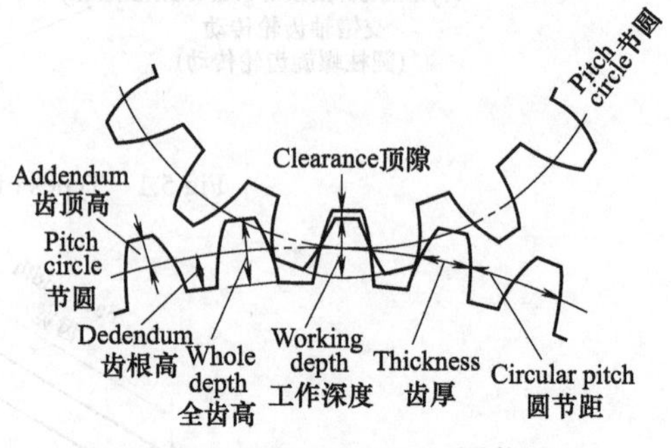

Fig 5.7　Circular pitch　圆节距

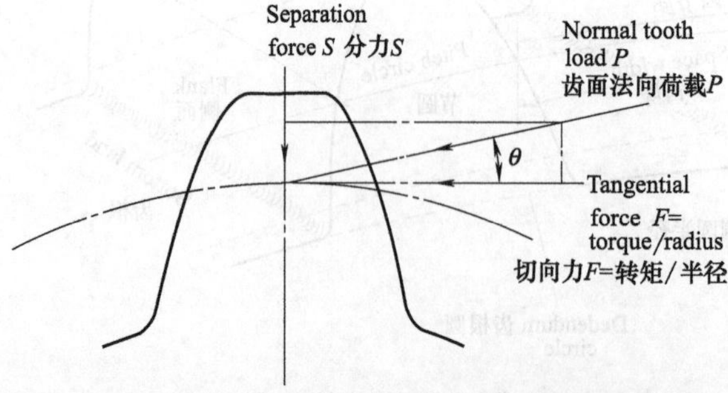

Fig 5.8　Resolution of tooth loads　轮齿荷载的分解

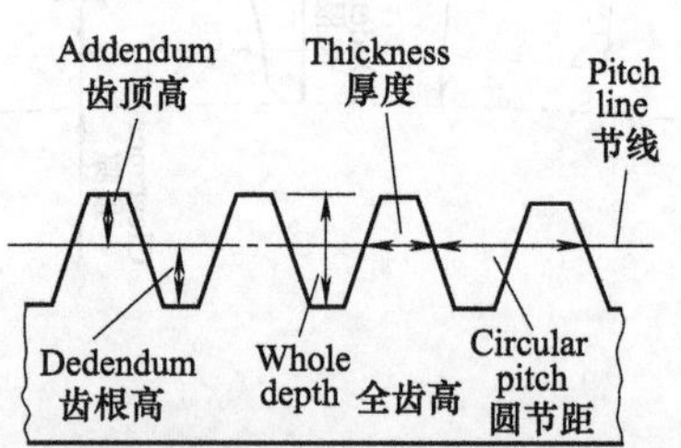

Fig 5.9　Names of rack parts　齿条零件的名称

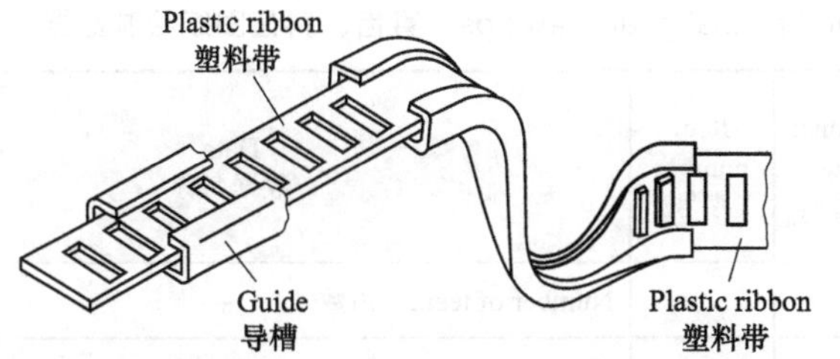

Fig 5.10 Flexible toothed rack for motion transmission 传递运动的柔性齿条

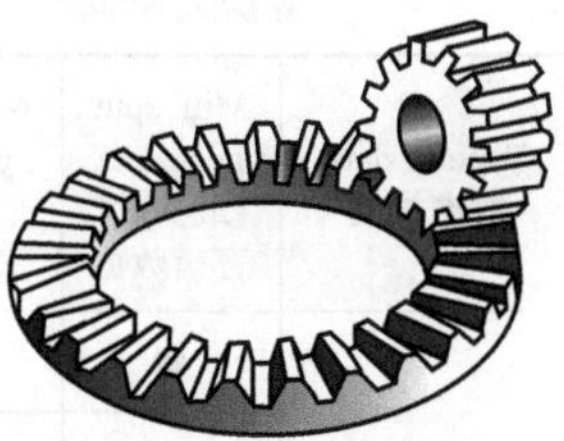

Fig 5.11 Face gear 端齿盘

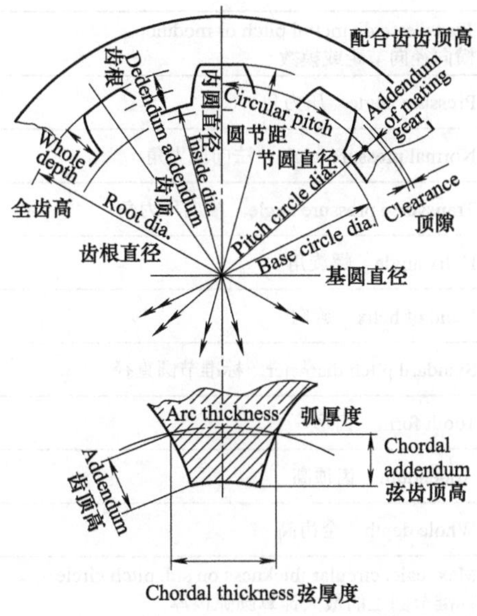

Fig 5.12 Parts of an internal spur gear 内齿直齿轮局部

Fig 5.13 Helical gears 螺旋齿轮

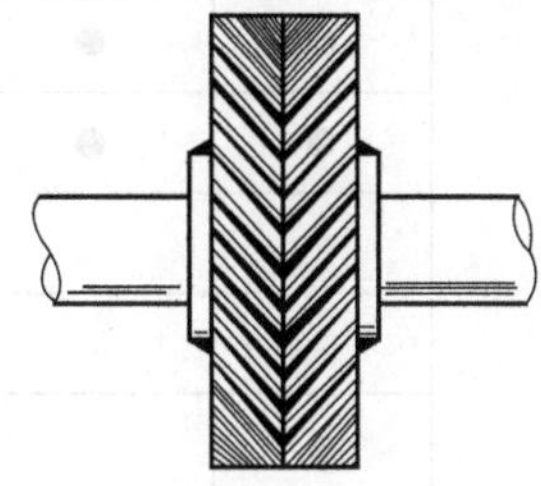

Fig 5.14 Herring-bone gear 人字齿轮

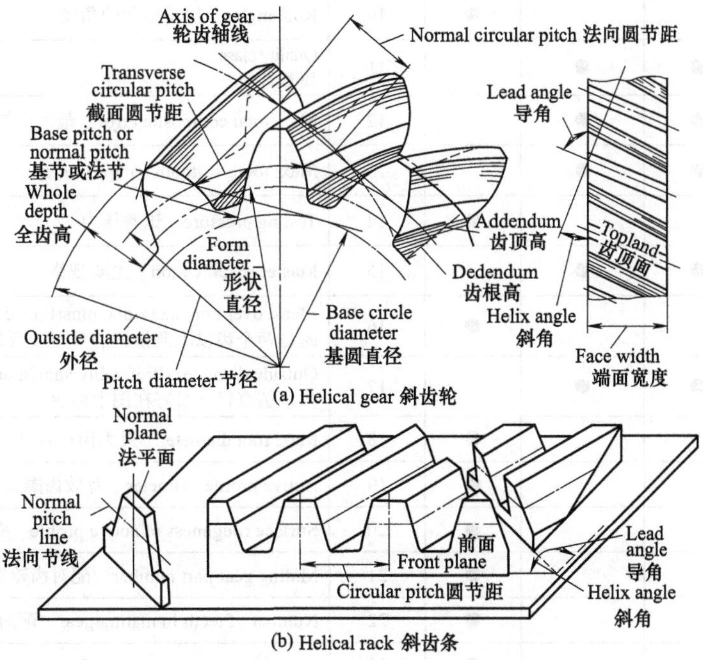

Fig 5.15 Helical gear and rack terminology 斜齿轮和斜齿条术语

Table 5.1 Data for spur and helical gear drawings 直齿、斜齿图纸绘制数据

Type of data 数据类别	Min. spur gear data 最少的直齿数据	Min. helica gear-data 最少的斜齿数据	Optional data 可选项	Item number 序号	Data 数据名称
Basic specifications 基本规格	●	●		1	Number of teeth 齿数
	●			2	Diametral pitch or module 径向节距或模数
		●		2a	Normal diametral pitch or module 法向节距或模数
			●	2b	Transverse diametral pitch or module 横向径向节距或模数
	●			3	Pressure angle 压力角
		●		3a	Normal pressure angle 法向压力角
			●	3b	Transverse pressure angle 横向压力角
		●		4	Helix angle 螺旋角
		●		4a	Hand of helix 旋向
	●	●		5	Standard pitch diameter 标准节圆直径
	●	●		6	Tooth form 齿形
			●	7	Addendum 齿顶高
			●	8	Whole depth 全齿高
	●			9	Max. calc. circular thickness on std. pitch circle 标准节圆上的最大计算圆弧齿厚
		●		9a	Max. calc. normal thickness on std. pitch circle 标准节圆上的最大计算法向齿厚
Manufacturing and inspection 制造与检验			●	10	Roll angles 检验滚子的角度
	●	●		11	Quality class 质量等级
	●	●		12	Max. total composite error 最大的综合误差
	●	●		13	Max. tooth-to-tooth composite error 最大的齿间综合误差
			●	14	Testing pressure 试验压力
	●	●		15	Master specification 主要规格
			●	16	Meas. over two. xxxx dia. pins(For setup only) 跨越两个齿以上测量销的节距（测量装置用）
	●	●		17	Outside diameter (Preferably shown on drawing of gear) 齿顶圆直径（最好在图上标明）
			●	18	Max. root diameter 最大齿根直径
			●	19	Active profile diameter 有效齿廓直径
			●	20	Surface roughness of active profile 有效齿廓表面粗糙度
Engineering reference 工程资料			●	21	Mating gear part number 配对齿轮零件号
			●	22	Number of teeth in mating gear 配对齿轮齿数
			●	23	Minimum operating center distance 运转的最小中心距

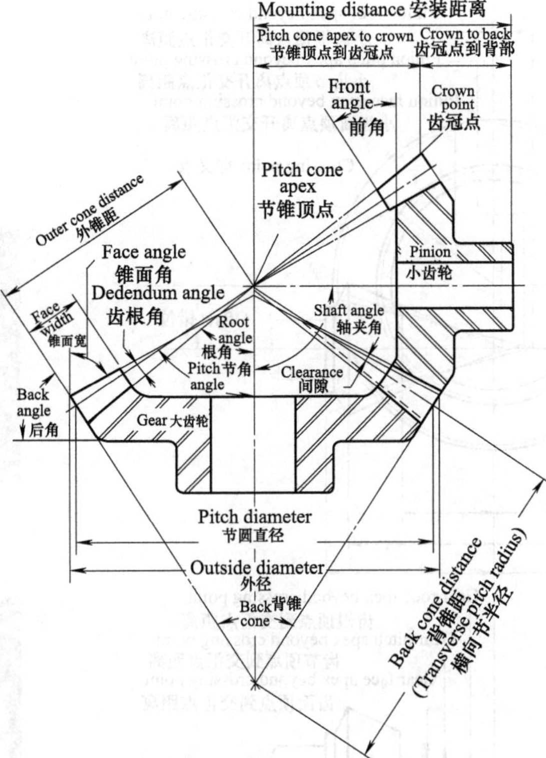

Fig 5.16 Bevelgear nomenclature (axial plane) 锥齿轮术语（轴剖面）

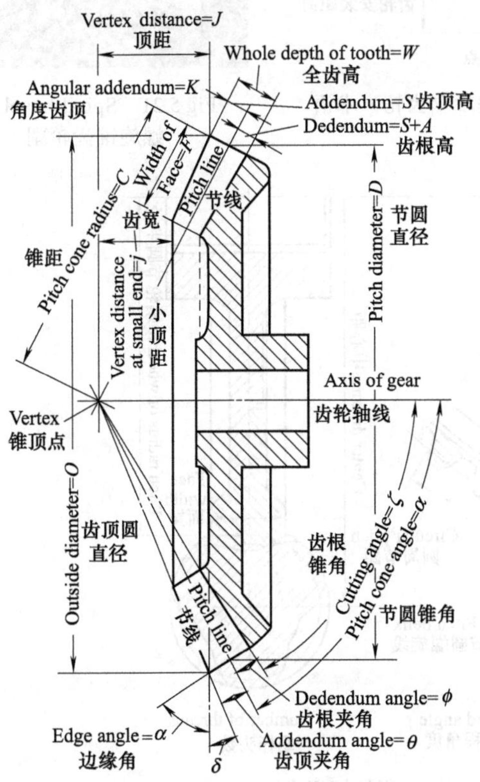

Fig 5.17 Bevel gear nomenclature 锥齿轮术语

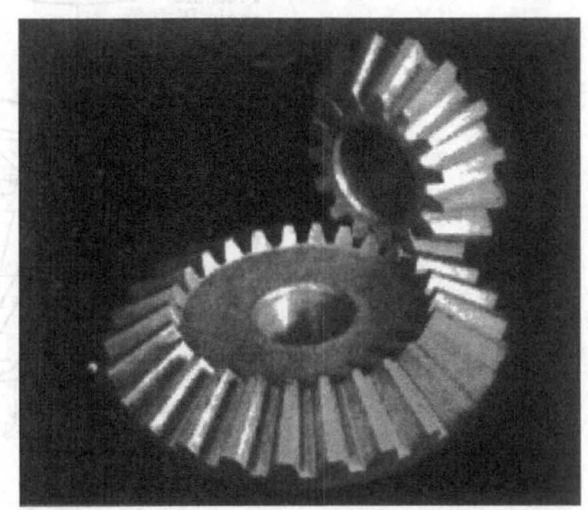

Fig 5.18 Straight bevel gears 直齿锥齿轮

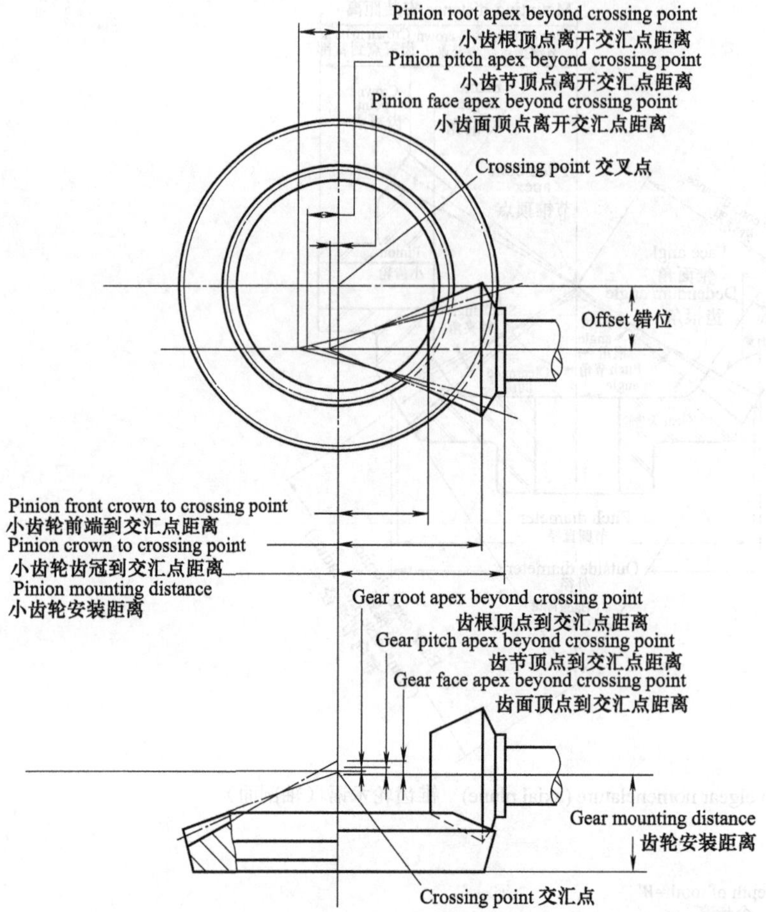

Fig 5.19 Hypoid gear nomenclature 准双曲面齿轮（偏轴伞齿轮）术语

Fig 5.20 Hypoid bevel gears
准双曲面锥齿轮

Fig 5.21 Spiral-bevel set
螺旋锥齿轮副

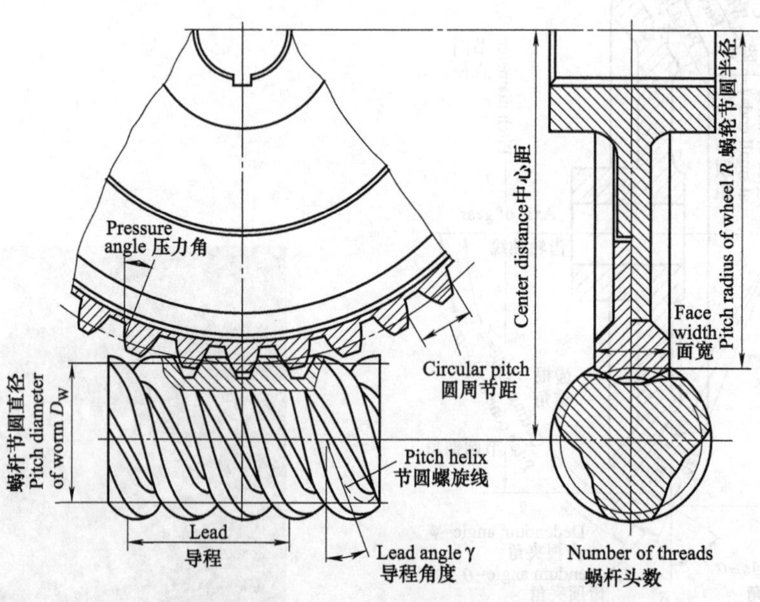

Fig 5.22 Worm gear terminology 蜗轮蜗杆术语

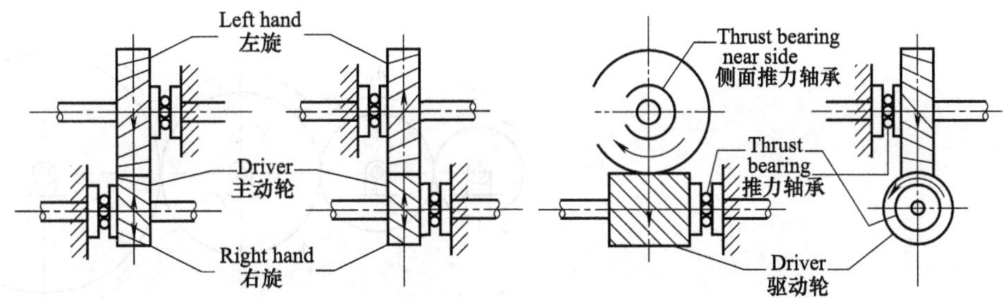

Fig 5.23 Directions of rotation and resulting thrust for parallel shaft and 90 degree shaft angle helical gears
平行轴和 90°轴夹角螺旋齿轮传动的轴向力和回转方向

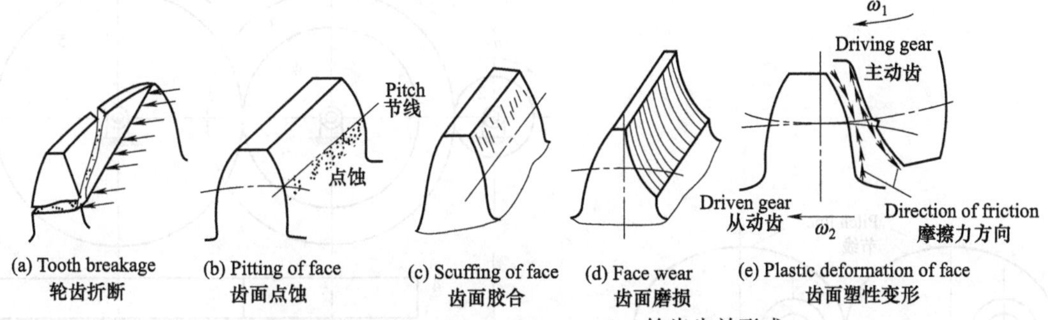

Fig 5.24 Failure forms of gear tooth 轮齿失效形式

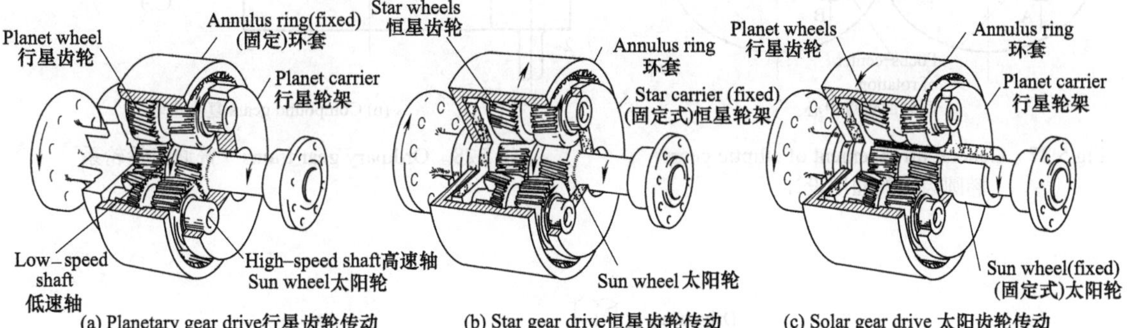

Fig 5.25 Three double helical drives 三种双斜齿轮传动

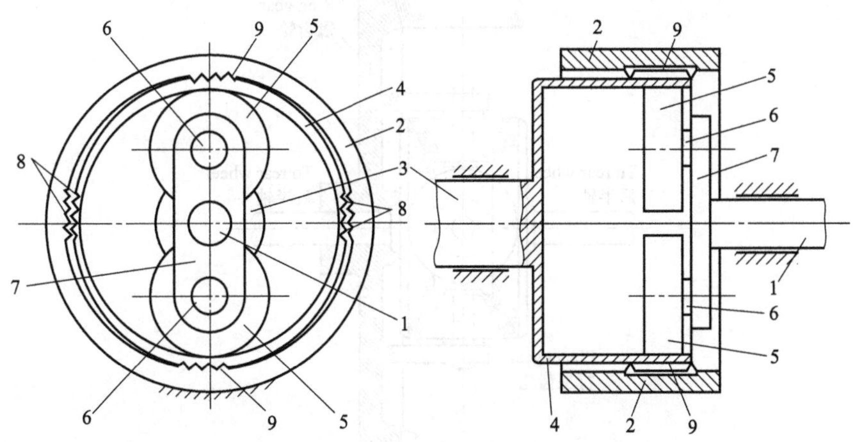

Fig 5.26 Layout of a harmonic drive 谐波传动设计

1—Driving shaft 驱动轴；2—Rigid ring 刚性环；3—Driven shaft 从动轴；4—Elastic ring 弹性环；5—Roller 滚子；6—Axes 轴；7—Transverse 横跨板；8—Disengaged 脱离啮合；9—Engaged location 啮合位置

Pitch lines
节线
A
B
Focus-center
of rotation
回转的焦点中心

Fig 5.27 General arrangement of elliptic gears 椭圆齿轮的通用设计

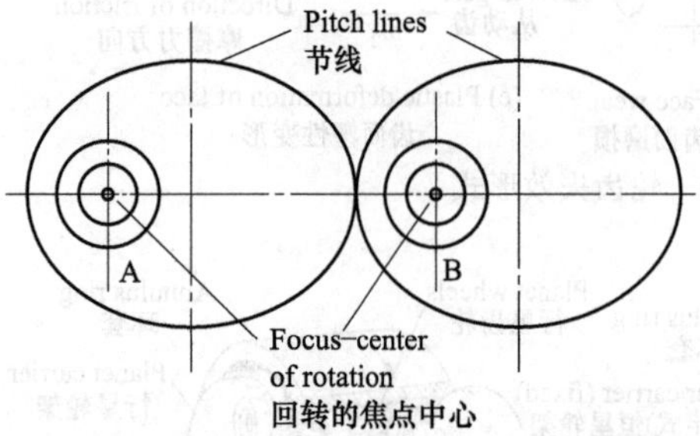

n_2
n_3
n_4
n_5
2
3
4
5
(a) Simple gears 简单齿轮
n_3,n_4
n_5
n_2
2
3
4
5
4
5
2
3
(b) Compound gears 复合齿轮

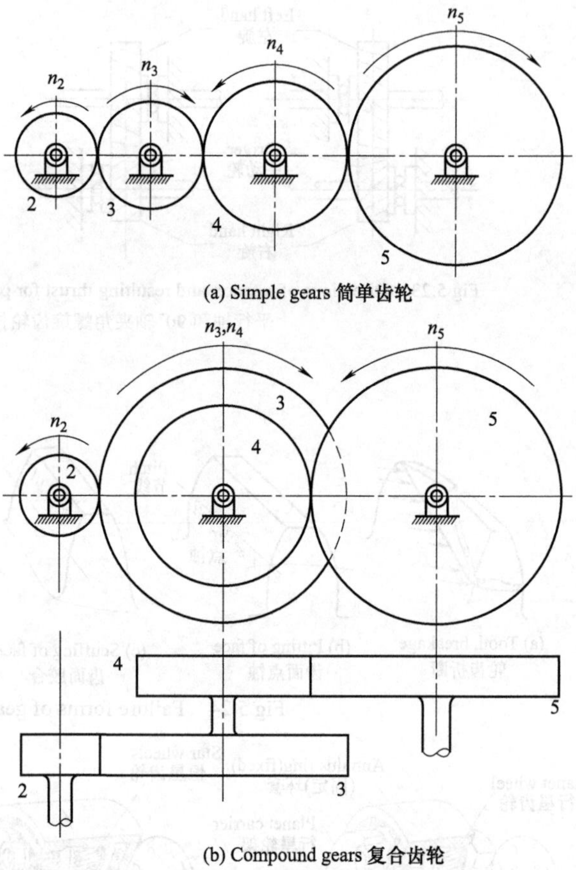

Fig 5.28 Ordinary gear trains 普通齿轮轮系

Drive shaft
驱动轴
Ring gear
冠齿轮
To rear wheel
后半轴
To rear wheel
后半轴
Planet gear
行星齿轮

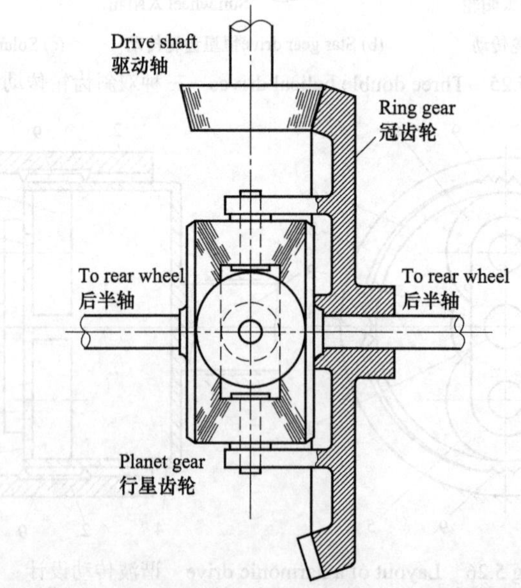

Fig 5.29 Schematic drawing of a bevel-gear automotive differential 汽车差动锥齿轮图解

5.2 Design of belt and chain drive 带和链传动设计

5.2.1 Belt drive 带传动

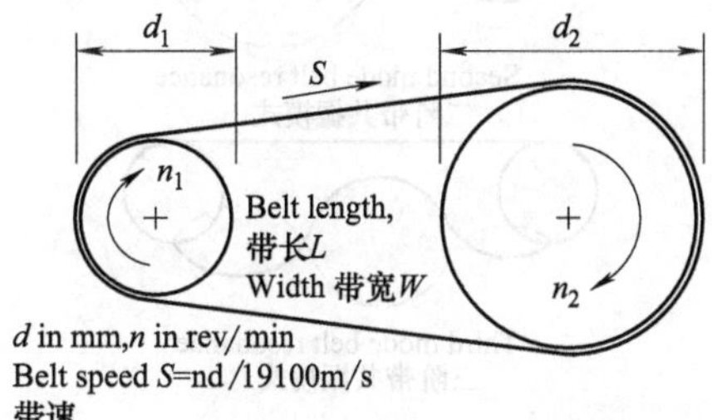

(a) Basic 2-shaft drive 基本两轴传动

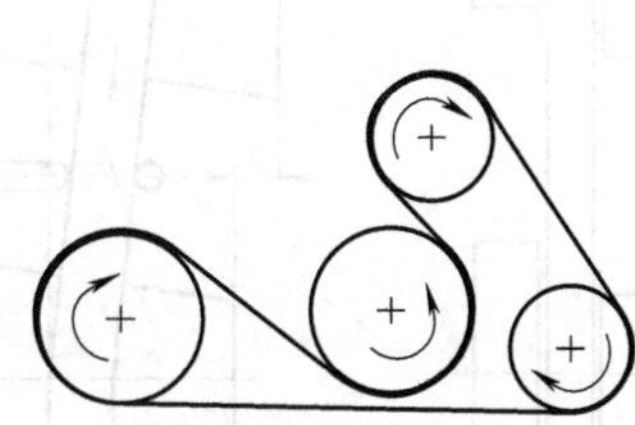

(c) Serpentine layout to drive more than one shaft or reverse direction of roation
蜿蜒形设计以达到多轴驱动,或改变转动方向

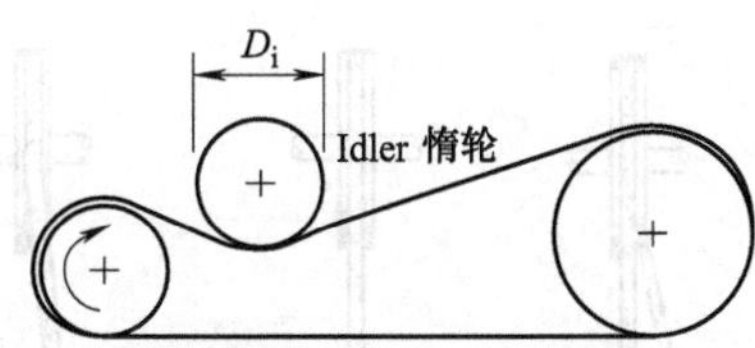

(b) Idler added to increase arc of contact or control belt vibration
附加惰轮增加接触弧长度或控制带的振动

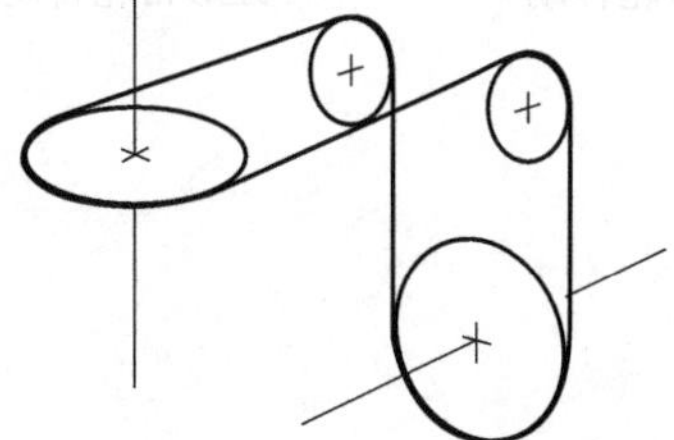

(d) Out-of-plane layout异面布局

Fig 5.30 Some belt drive layouts 几种皮带传动设计

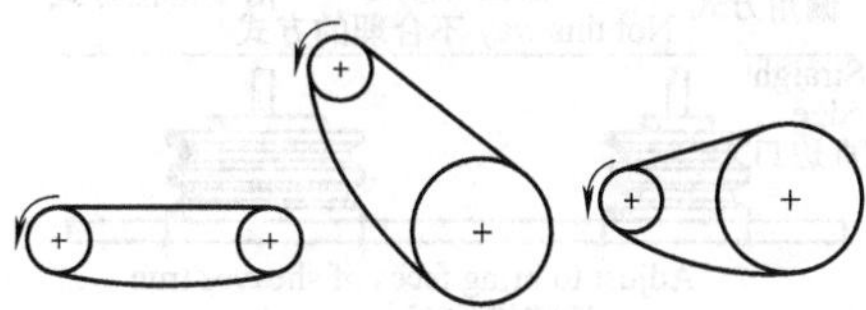

Recommended 推荐方案

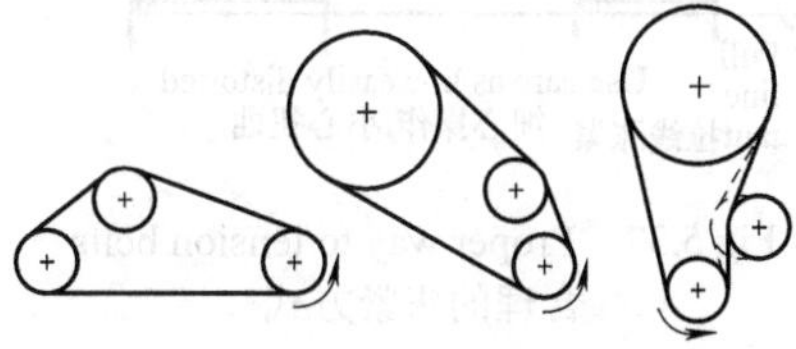

Acceptable 可接受方案

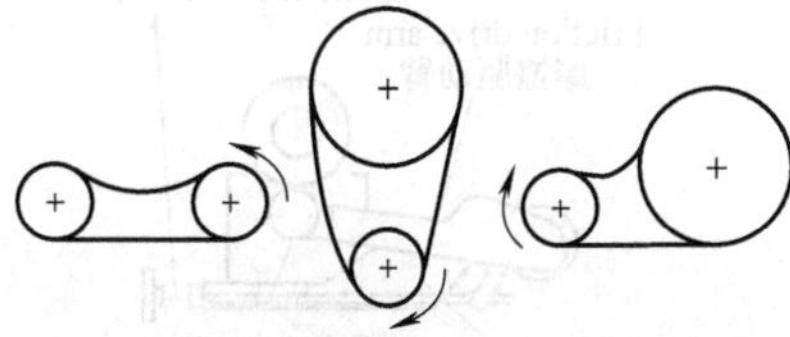

Not recommended 不推荐方案

Fig 5.31 Drive arrangements 传动方案设计

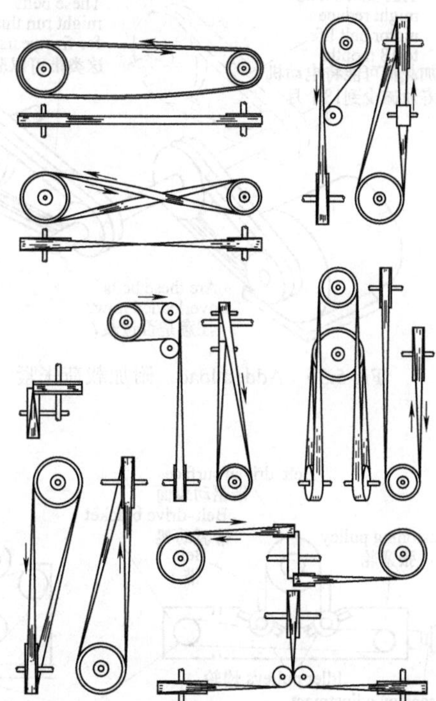

Fig 5.32 Various ways in which a belt drive may be used 可以实施的带传动方式

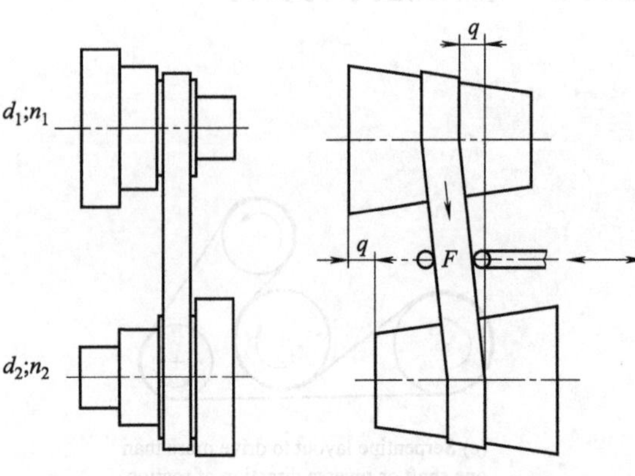

Fig 5.33 Step pulleys drive 阶梯带轮传动

Fig 5.34 Cone-pulley drive 锥形带轮传动

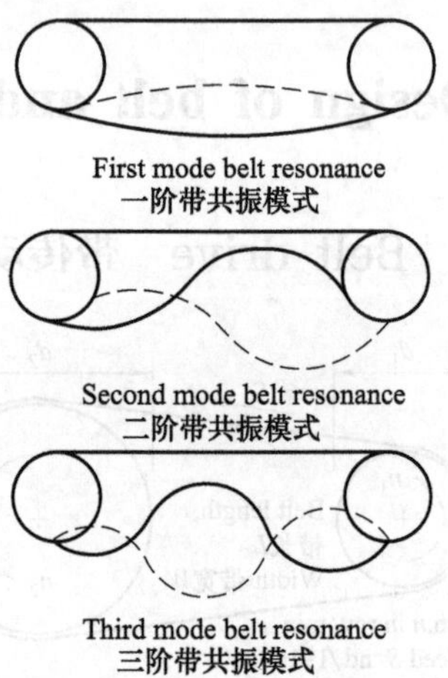

Fig 5.35 Examples of mode resonance in a belt span 带跨距的共振模式示例

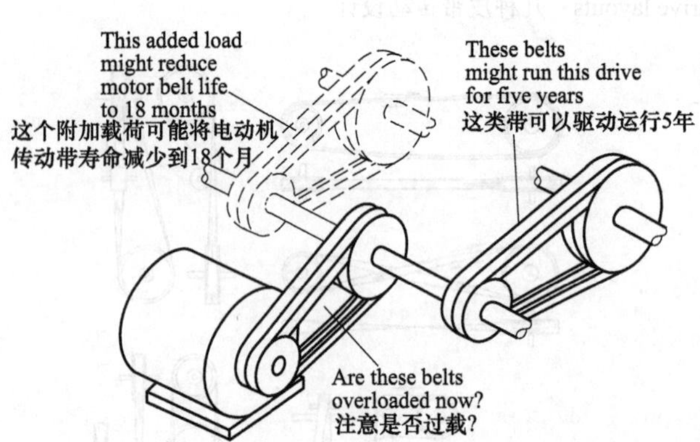

Fig 5.36 Added load 附加载荷张紧

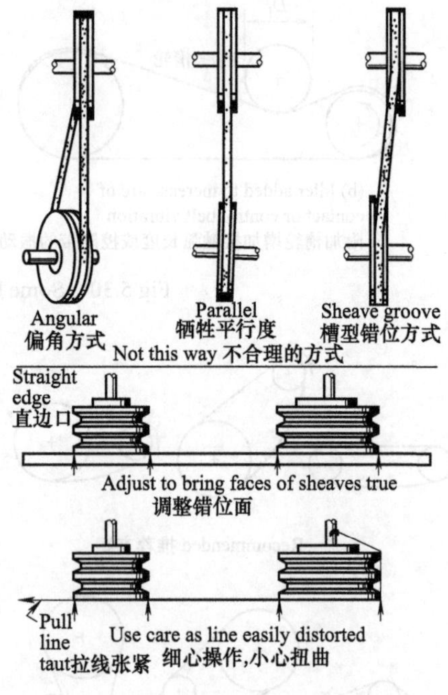

Fig 5.37 Proper way to tension belts 合理的张紧方式

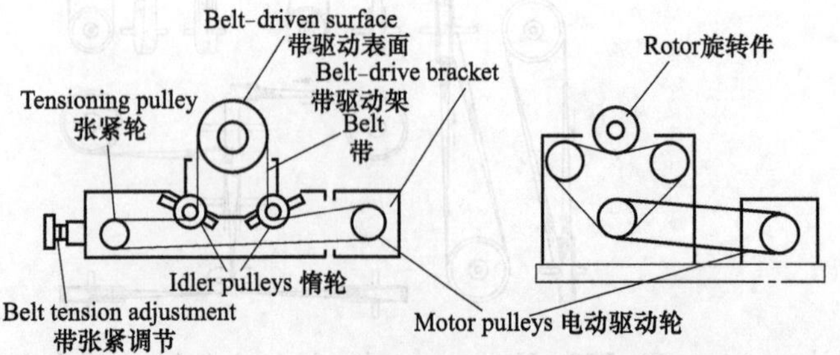

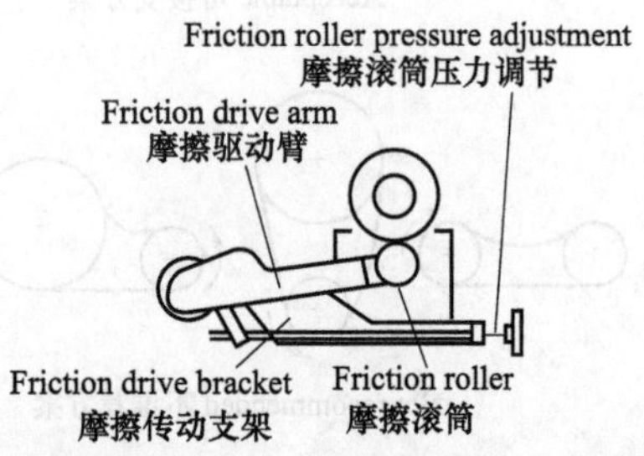

Fig 5.38 Belt and other drives 带和其他传动

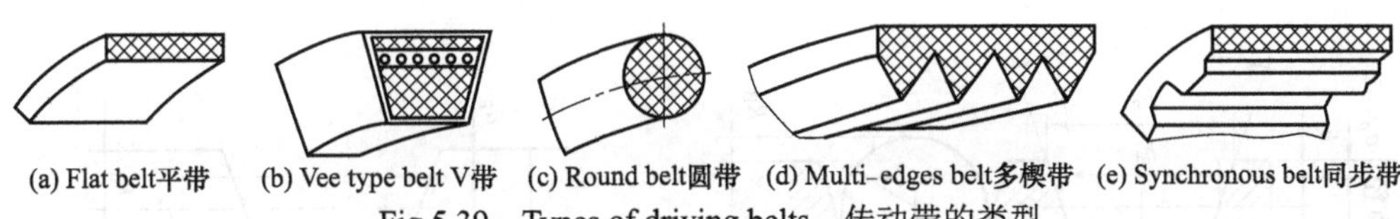

Fig 5.39 Types of driving belts 传动带的类型

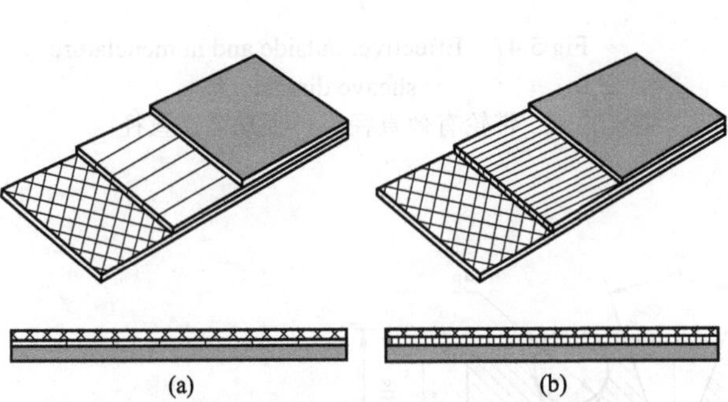

Fig 5.40 Multiple-ply belt 多层带

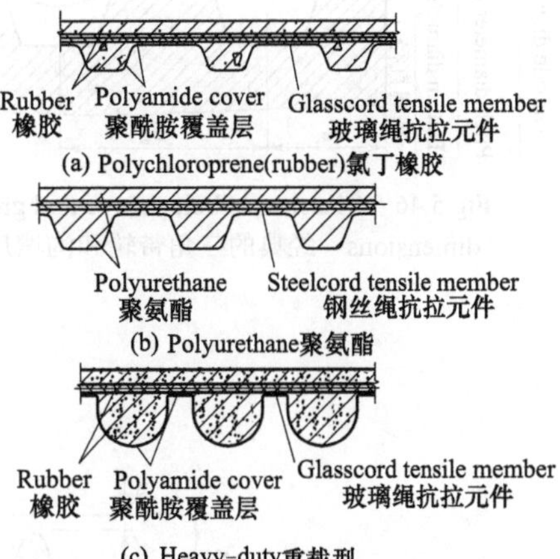

Fig 5.41 Construction and differences of timing belts 不同类别的同步带结构

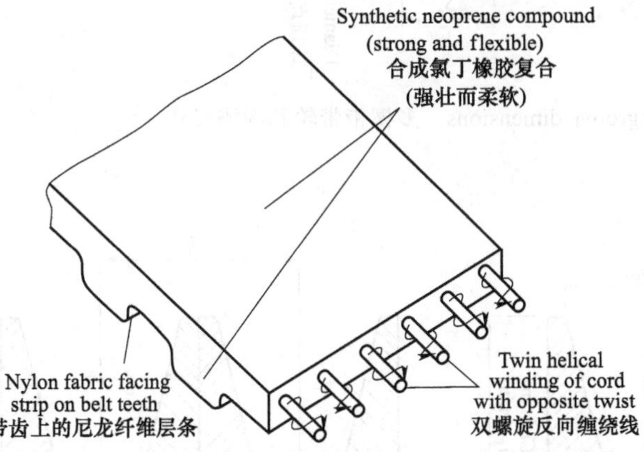

Fig 5.42 Illustrates the typical components of a timing belt 正时带结构

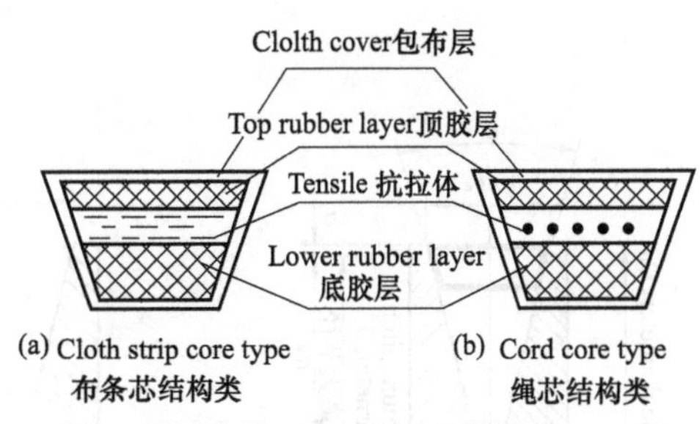

Fig 5.43 Construct of V-belt V 带结构

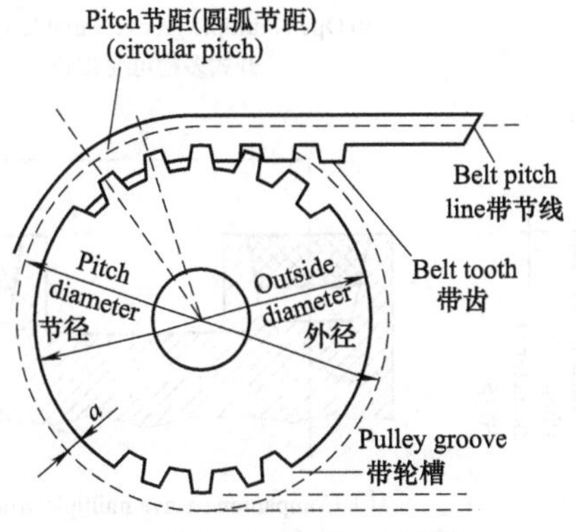

Fig 5.44 Synchronous belt pulley dimensions 同步带轮尺寸

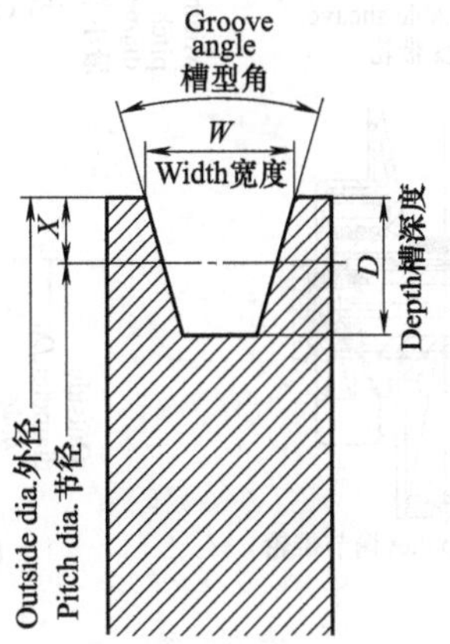

Fig 5.45 Typical pulley dimensions 典型带轮尺寸

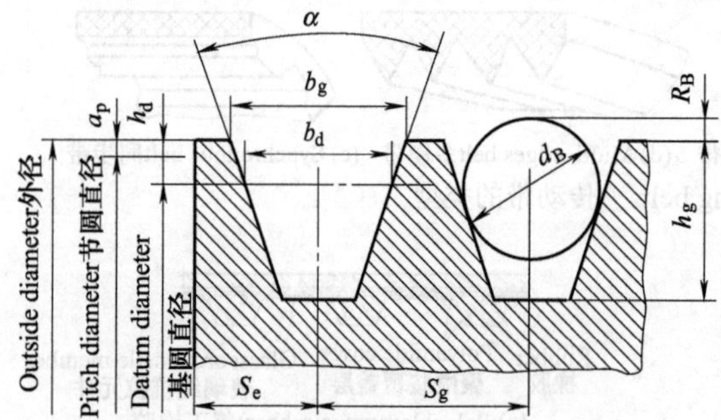

Fig 5.46 Classical V-belt sheave and groove dimensions 经典的三角带轮和沟槽尺寸

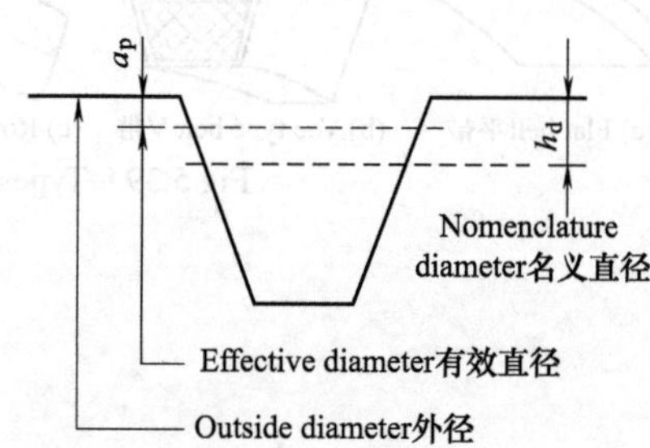

Fig 5.47 Effective, outside and nomenclature sheave diameters 带轮有效直径、外径和名义直径

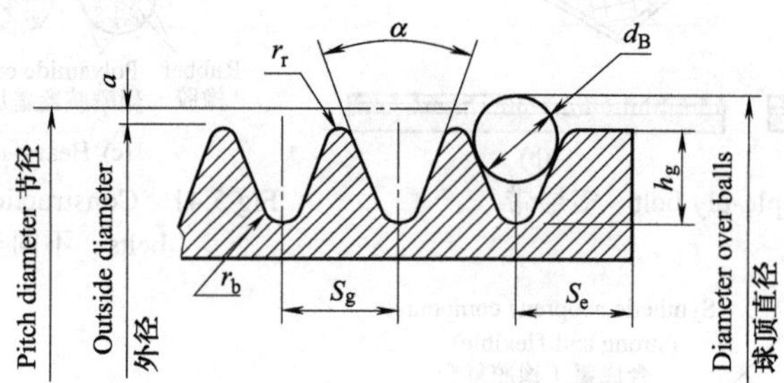

Fig 5.48 V-ribbed belt sheave and groove dimensions 多楔带带轮和沟槽尺寸

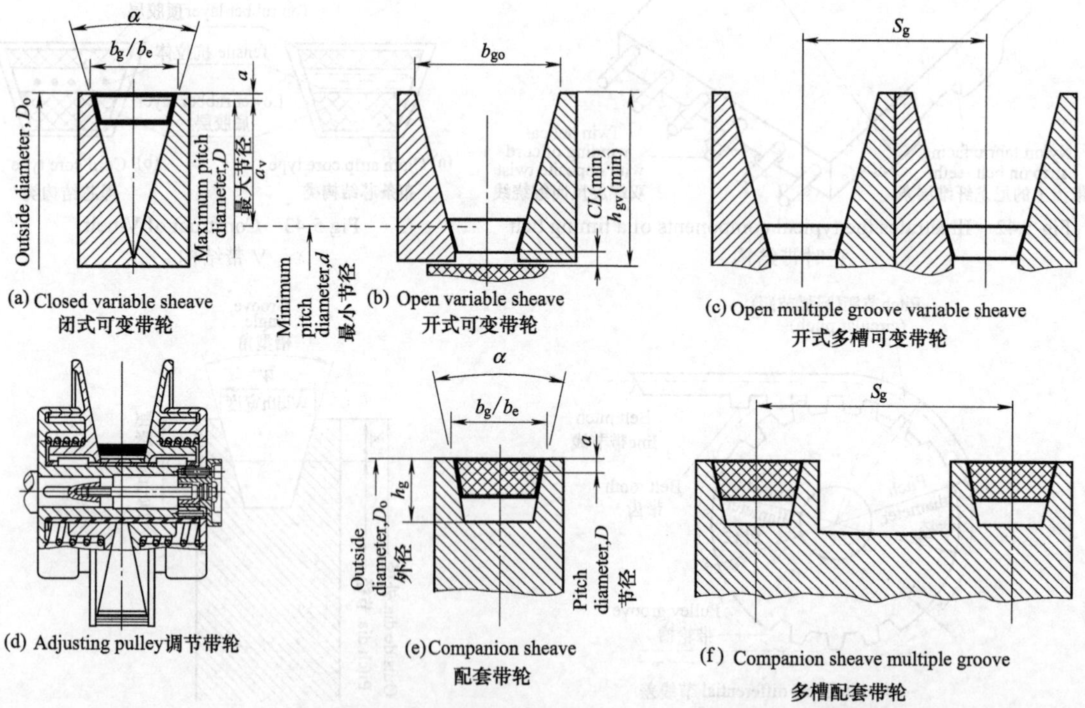

Fig 5.49 Companion sheaves for variable speed sheave 变速带轮的配套带轮

5.2.2 Chain drive 链传动

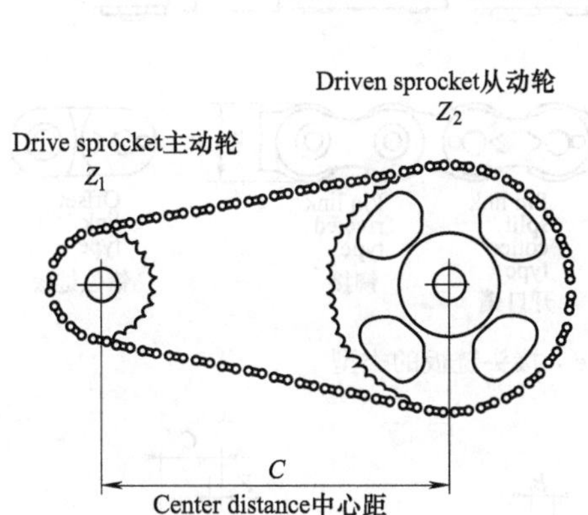

Fig 5.50 Roller chain drives 滚子套筒链传动

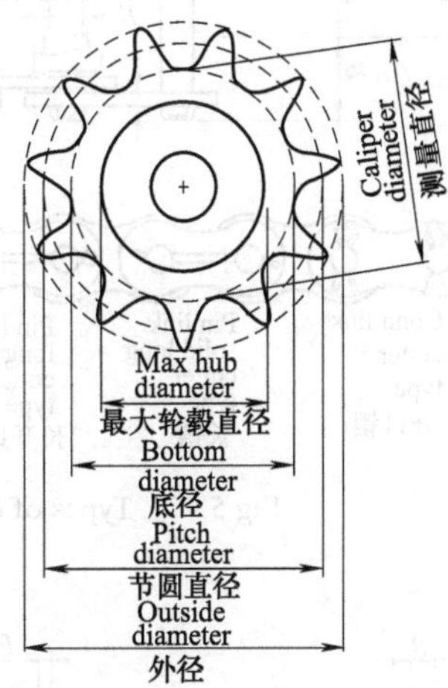

Fig 5.51 Roller chain sprocket diameters 滚子链链轮直径

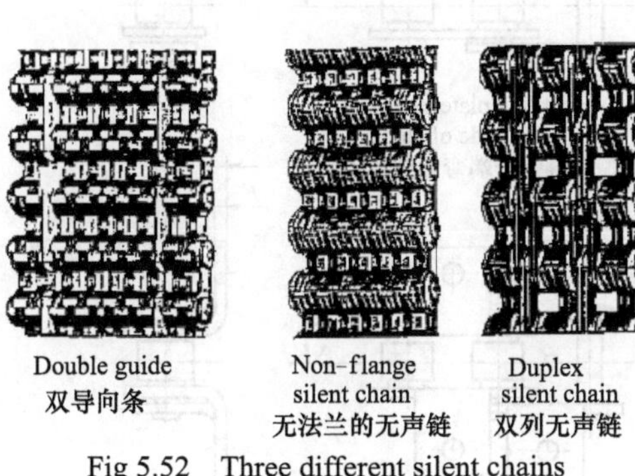

Fig 5.52 Three different silent chains 三种类型的无声链

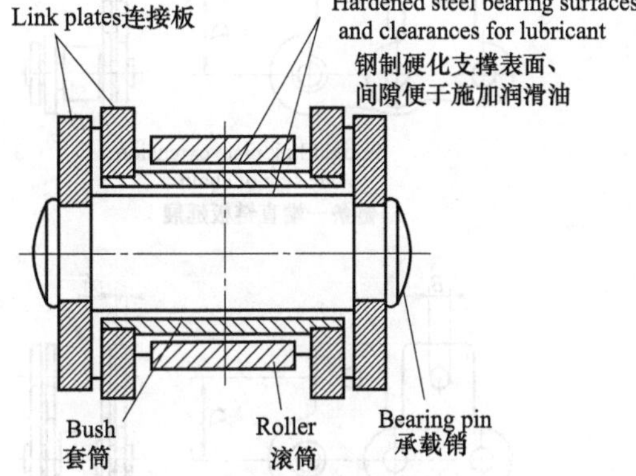

Fig 5.53 Typical section through roller chain 典型的滚筒链剖面

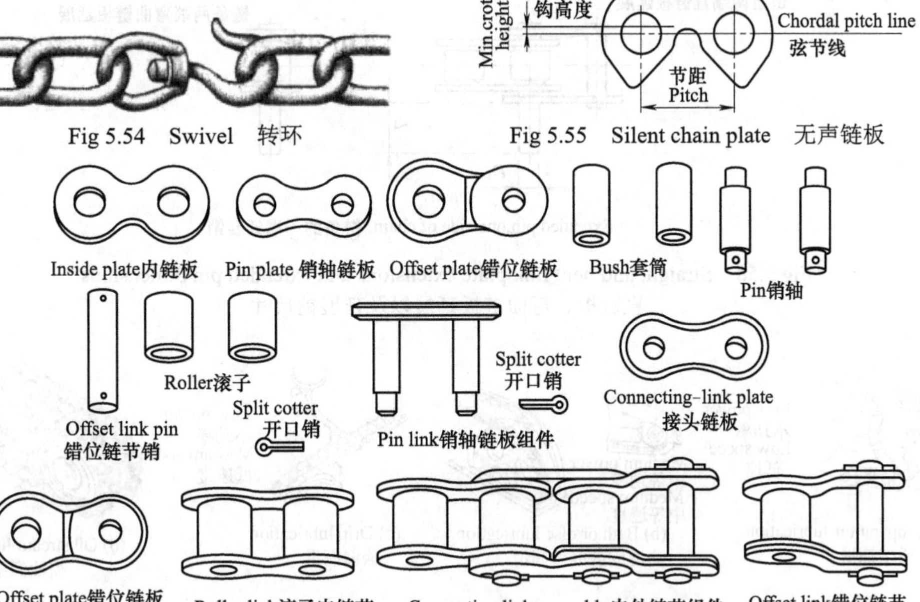

Fig 5.54 Swivel 转环

Fig 5.55 Silent chain plate 无声链板

Fig 5.56 Nomenclature for roller chain parts 滚子链部件术语

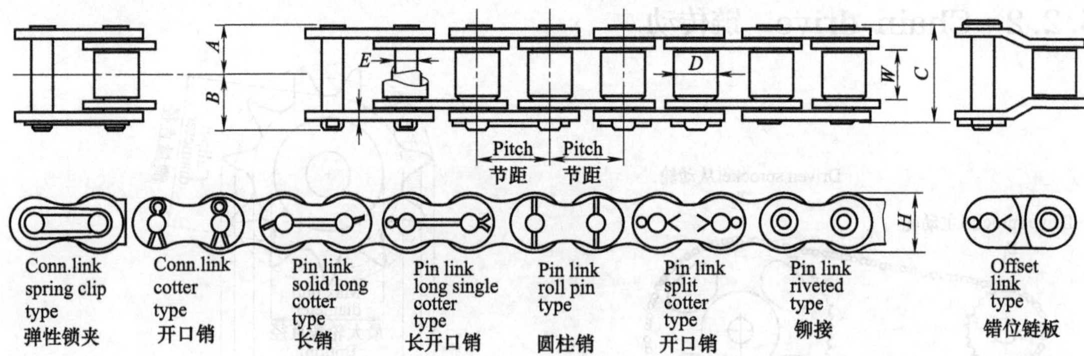

Fig 5.57 Types of connecting-link plate 接头链板的类型

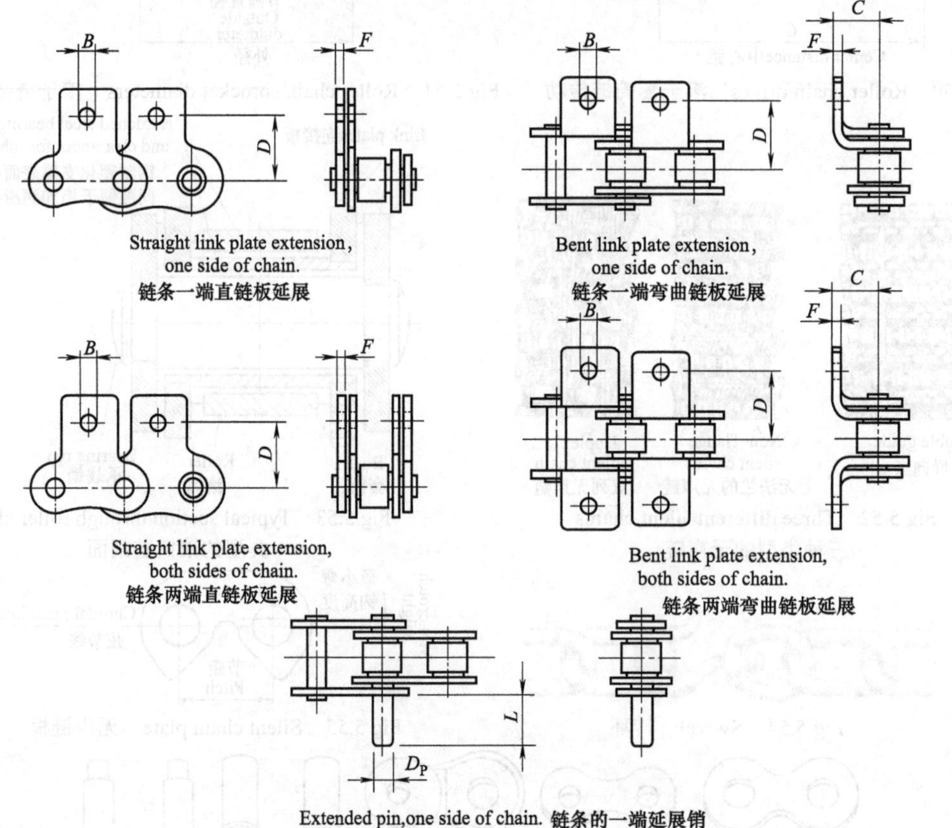

Fig 5.58 Straight and bent link plate extensions and extended pin dimensions
直链板、弯曲链板延展以及延展销尺寸

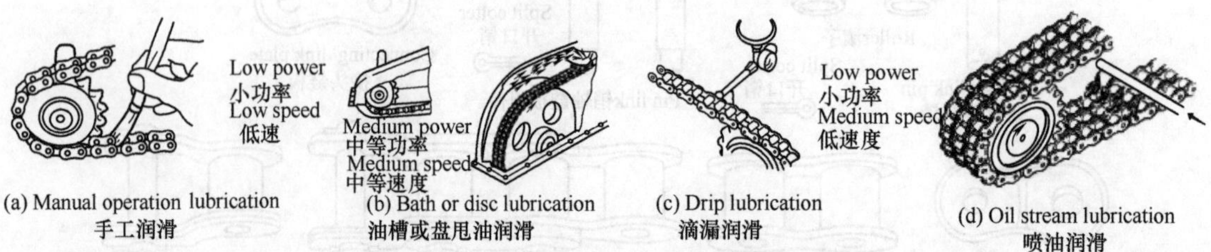

Fig 5.59 Lubrications for chains 链条的润滑

5.3 Design of hydraulic power transmission 液压传动设计

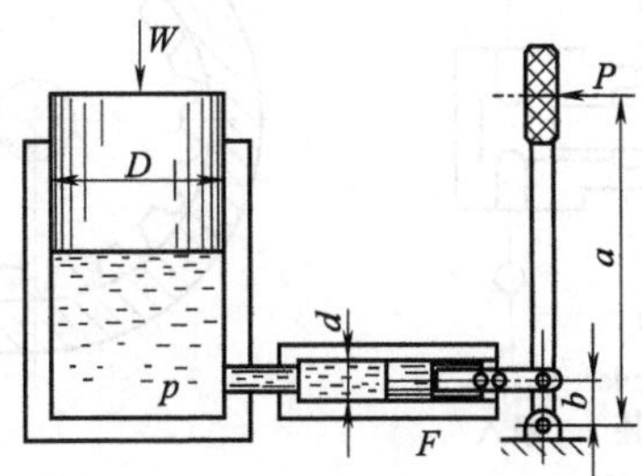

Fig 5.60 Hydraulic jack 液压千斤顶

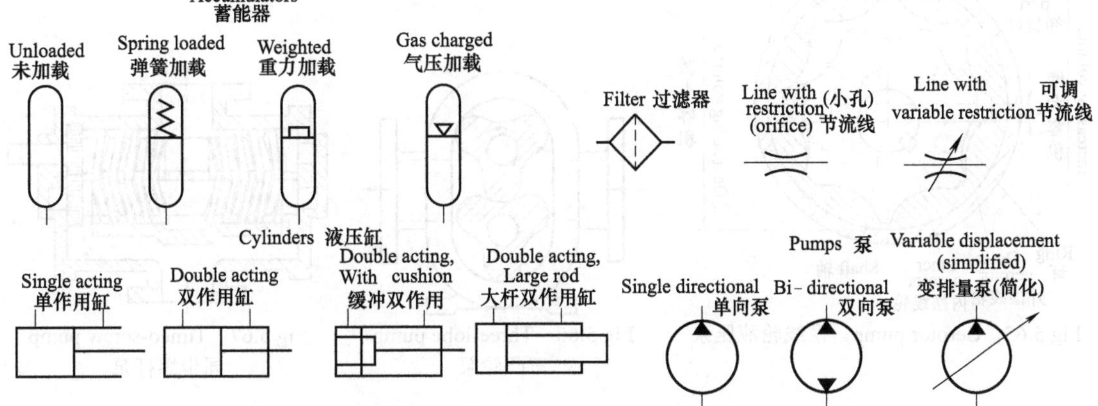

Fig 5.61 Activation symbols 执行件符号

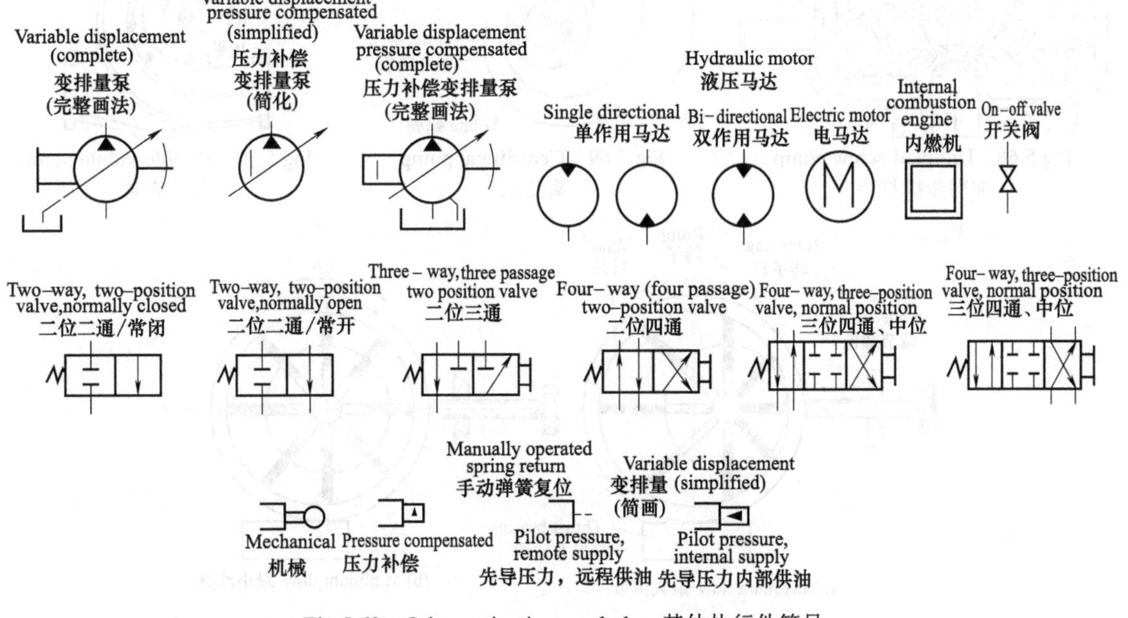

Fig 5.62 Other activation symbols 其他执行件符号

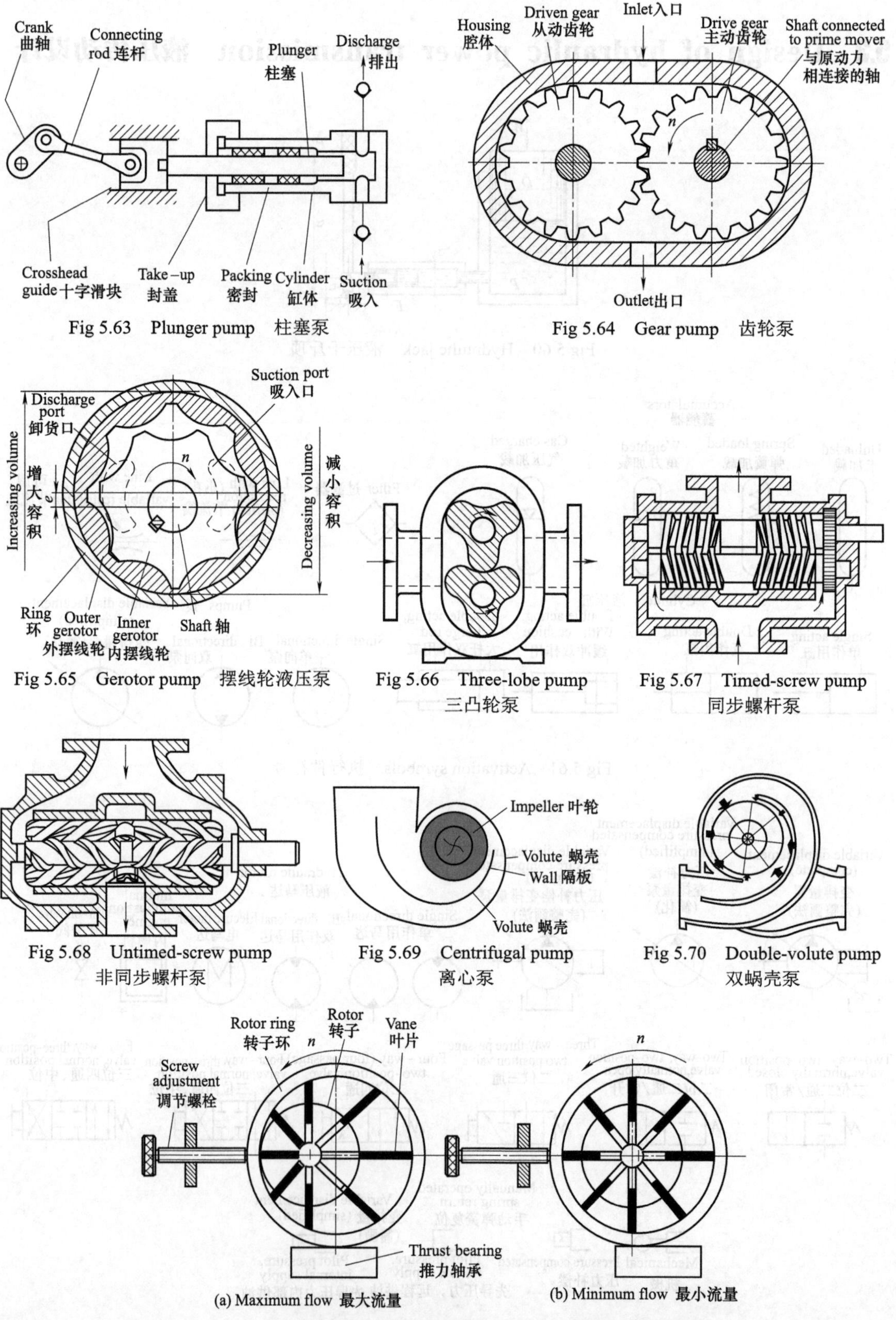

Fig 5.63 Plunger pump 柱塞泵

Fig 5.64 Gear pump 齿轮泵

Fig 5.65 Gerotor pump 摆线轮液压泵

Fig 5.66 Three-lobe pump 三凸轮泵

Fig 5.67 Timed-screw pump 同步螺杆泵

Fig 5.68 Untimed-screw pump 非同步螺杆泵

Fig 5.69 Centrifugal pump 离心泵

Fig 5.70 Double-volute pump 双蜗壳泵

Fig 5.71 Cross-sectional schematic variable vane pump 变量叶片泵横截面示意图

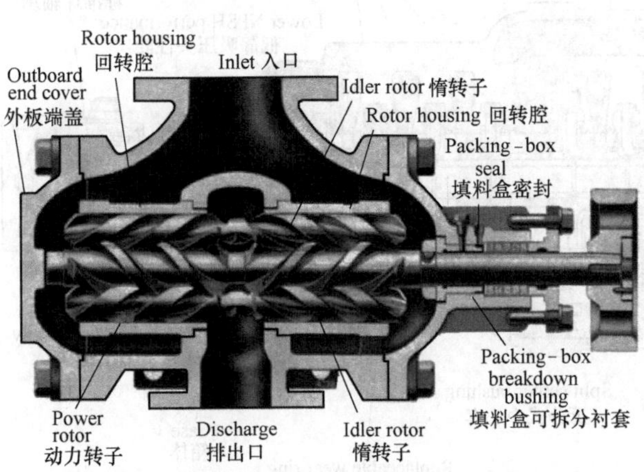

Fig 5.72 Cutaway view of double-end IMO pump
双端三螺杆泵剖视图

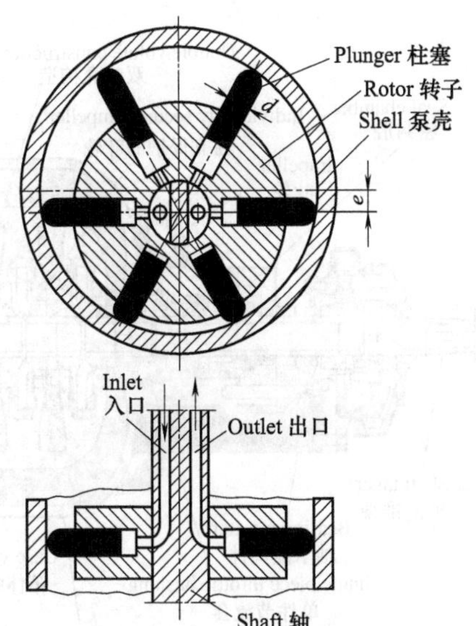

Fig 5.73 Sectional views of radial piston pump
径向柱塞泵截面图

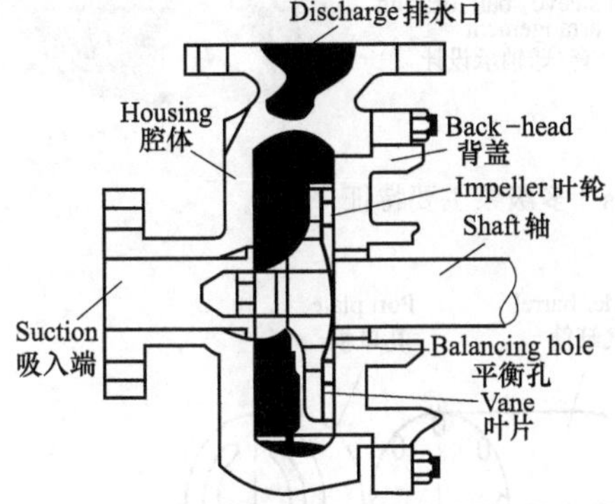

Fig 5.74 Cross-section of a typical end-suction centrifugal 典型端部吸入离心泵剖开图

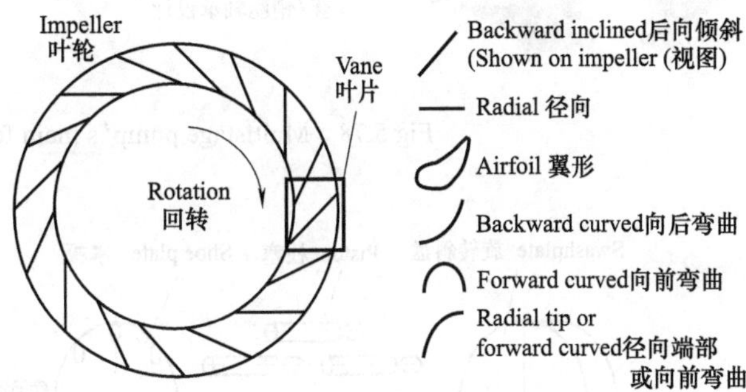

Fig 5.75 A variety of vane types that might be used on a centrifugal fan 离心泵上可能应用的叶片类别

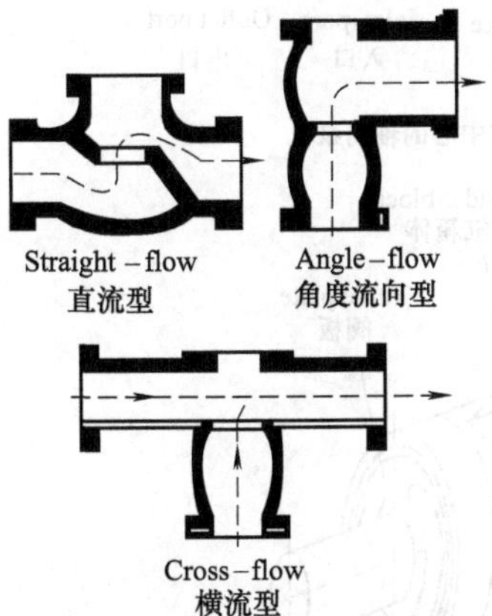

Fig 5.76 Three globe valve configurations: straight-flow, angle-flow, and cross-flow 三种球阀构造：直流型、角度流向型和横流型

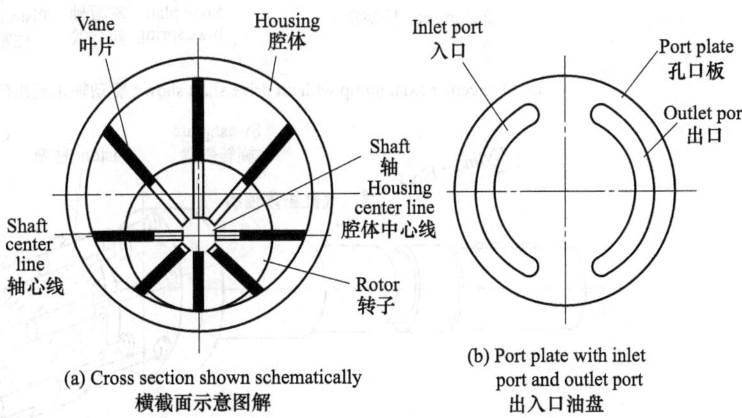

Fig 5.77 Vane pump 叶片泵

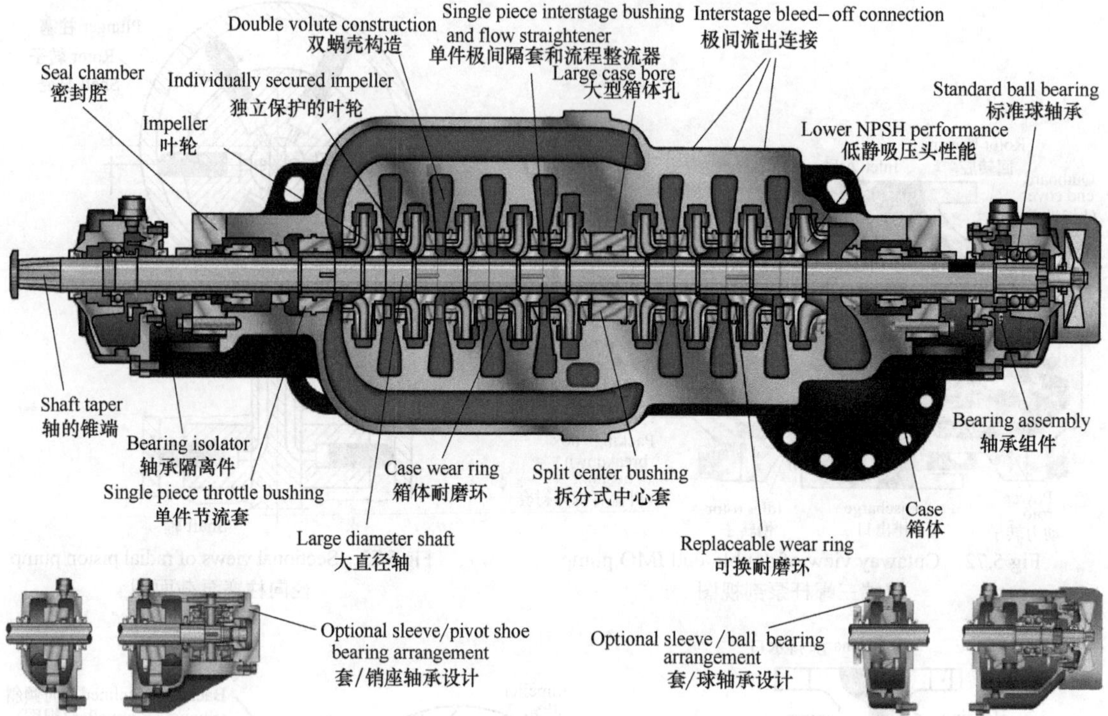

Fig 5.78 Multistage pump's main features 多级泵主要特征

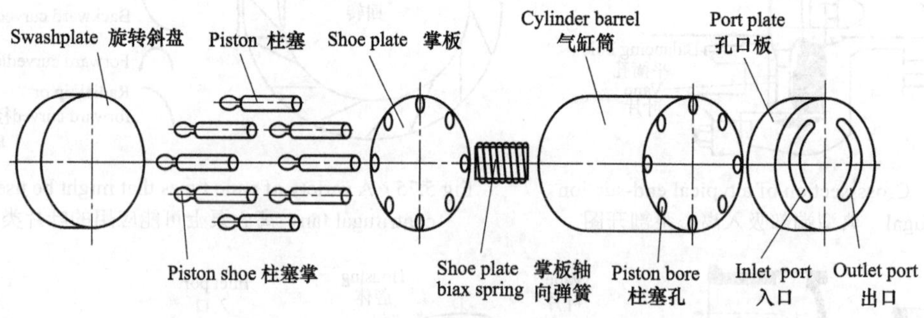

(a) Overcenter axial pump without drive shaft shown 驱动轴未画出的过中心的轴向泵

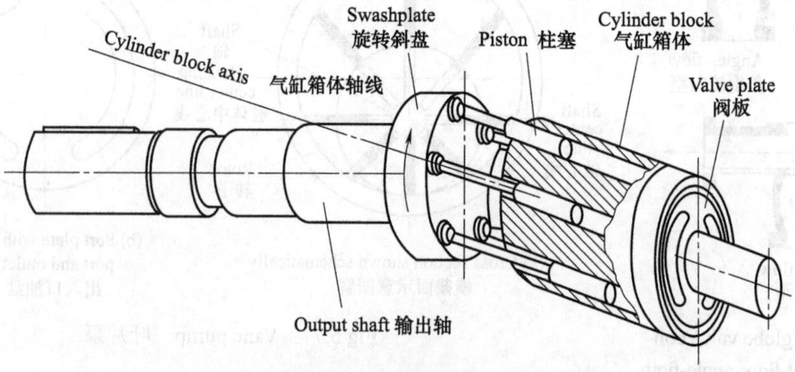

(b) Basic parts for axial piston pump 轴向柱塞泵的基本零部件

Fig 5.79 Axial piston pump 轴向柱塞泵

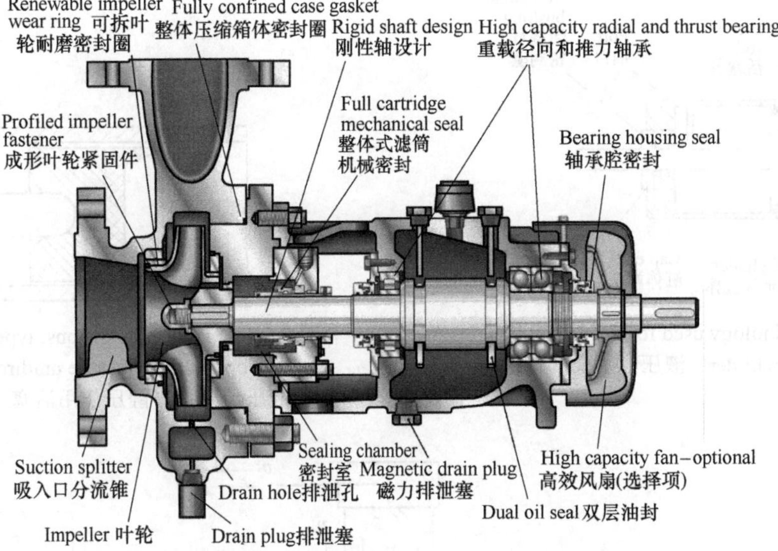

Fig 5.80 Pump's main feature 泵的主要特征

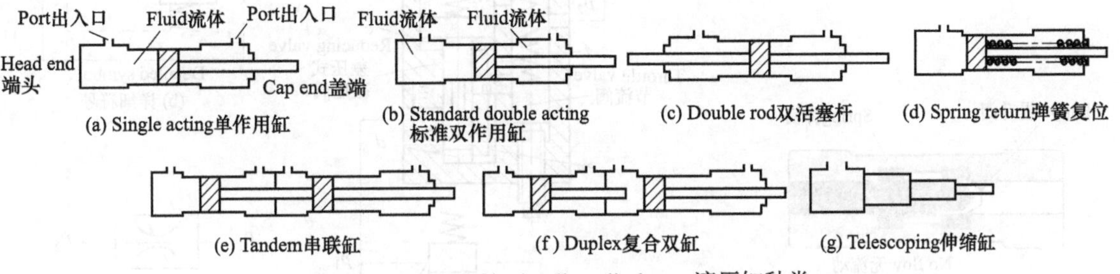

Fig 5.81 Types of hydraulic cylinders 液压缸种类

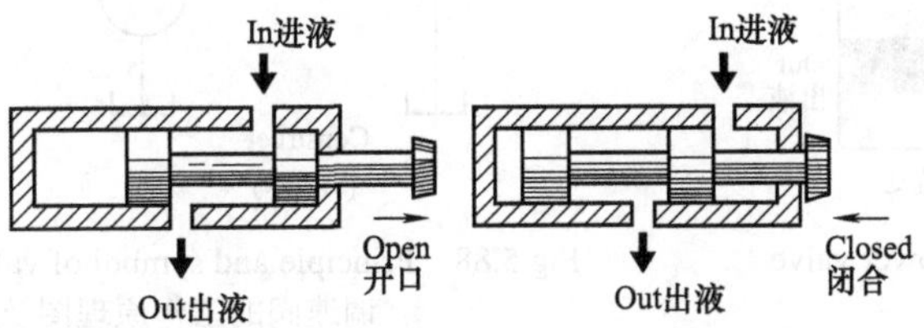

Fig 5.82 Two-way, fluid-power valve 二通流体动力阀

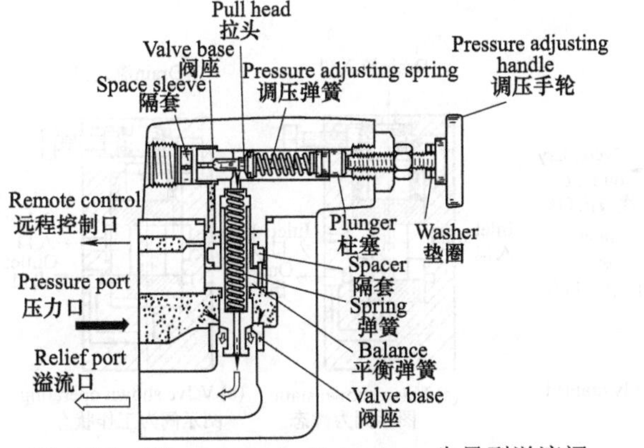

Fig 5.83 Pilot operated relief valve 先导型溢流阀

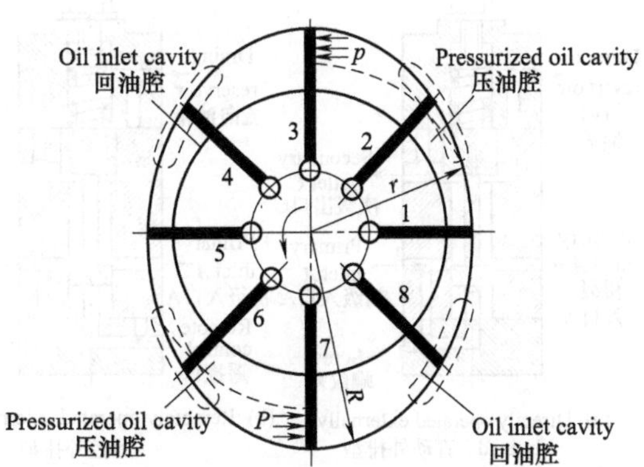

Fig 5.84 Principle of hydraulic vane motor 叶片式液压马达工作原理图

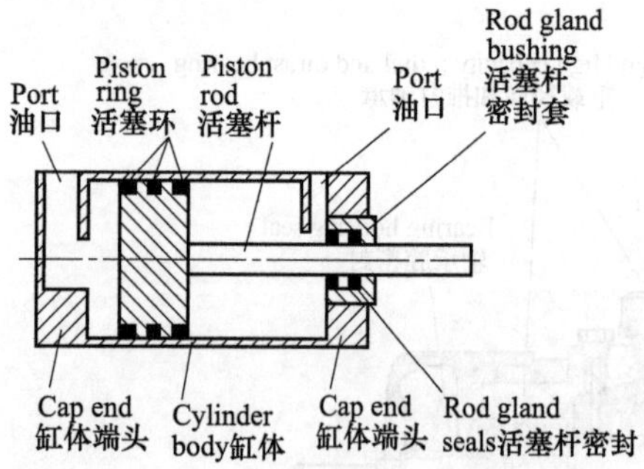

Fig 5.85 Terminology used for major components of hydraulic cylinder 液压缸主要部件术语

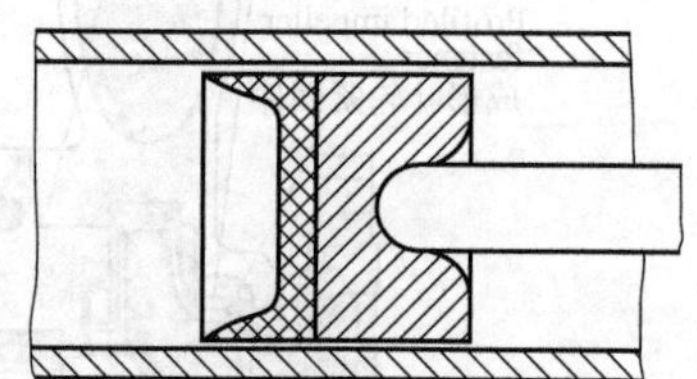

Fig 5.86 Hydraulic pistons: type A-Semi-static operation (pressure unidirectional)
液压活塞：半静压作用活塞（单向压力）

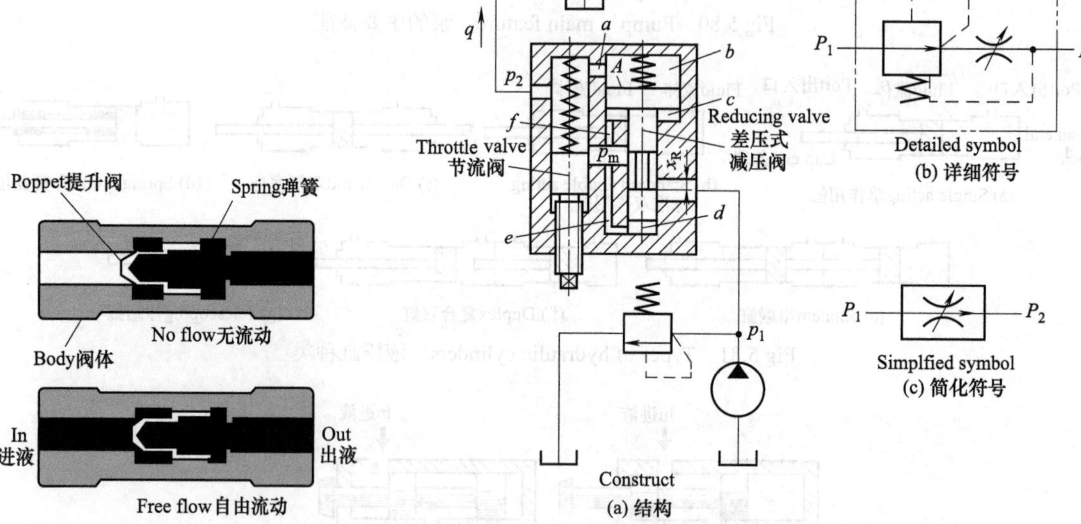

Fig 5.87 One-way, fluid-power valve
单向流体动力阀

Fig 5.88 Principle and symbol of valve for speed regulation
调速阀的工作原理图及图形符号

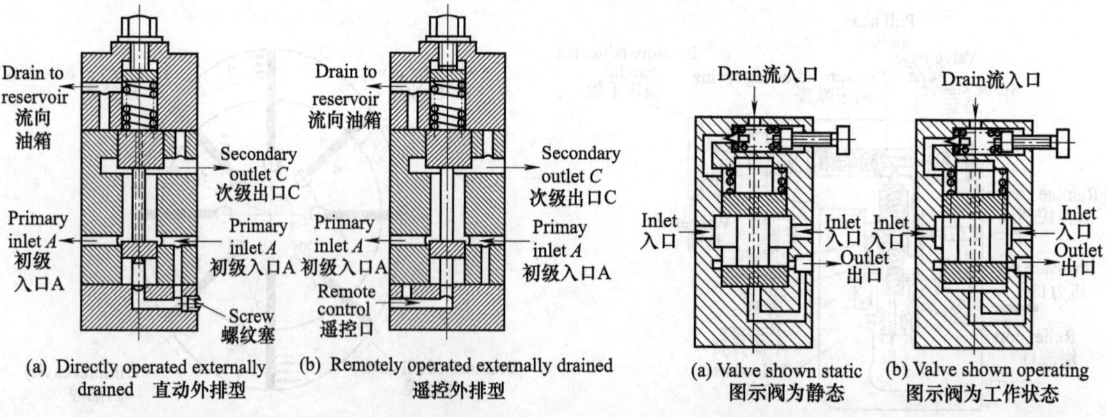

Fig 5.89 Sequence valve 顺序阀

Fig 5.90 Pressure-reducing valve 减压阀

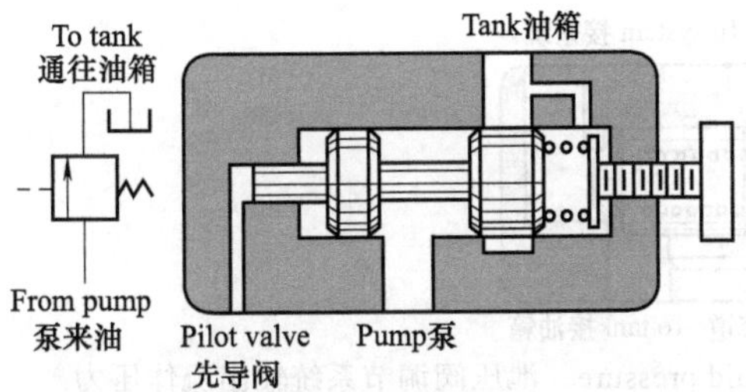

Fig 5.91 Pressure unloading valve unloads pump output to the tank at low pressure when high-pressure flow is not required 不需要高压时，卸荷阀对泵输出卸荷并低压流回油箱

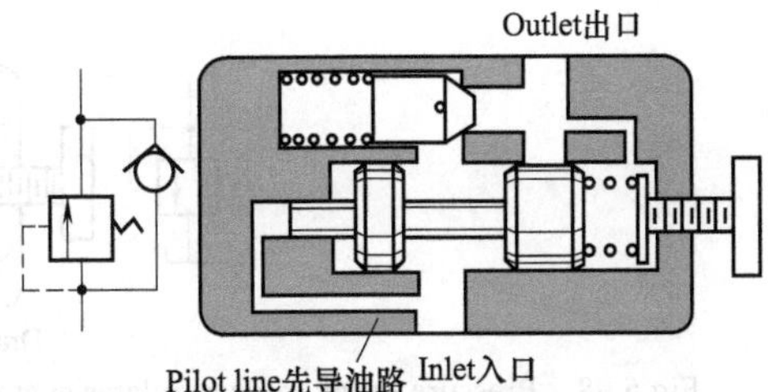

Fig 5.92 Counterbalance valve holds fluid pressure in part of a circuit to counterbalance weight on the external force 平衡阀保持局部油路压力以平衡外力上的重力

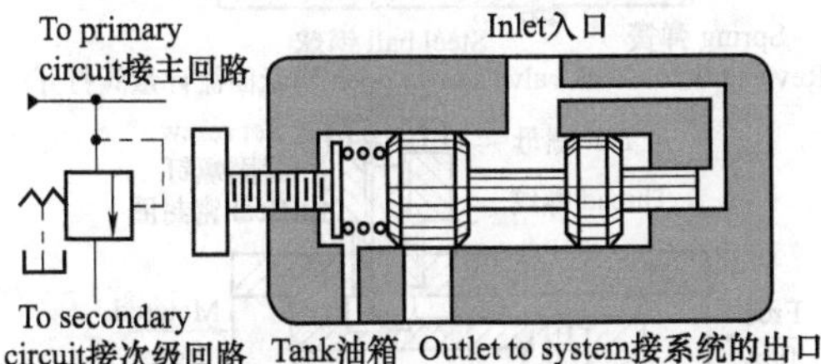

Fig 5.93 Sequence valve prevents fluid from entering one branch of a circuit before a preset pressure is reached in the main circuit 主回路预设压力达到之前，顺序阀阻止流体流入支路

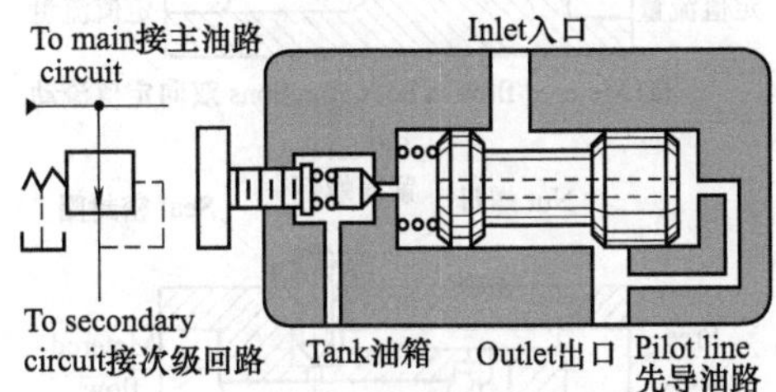

Fig 5.94 Pressure-reducing valve allows one branch of a circuit to operate at a lower pressure than the main sysytem 减压阀使得支路可以在比主油路压力低的情况下工作

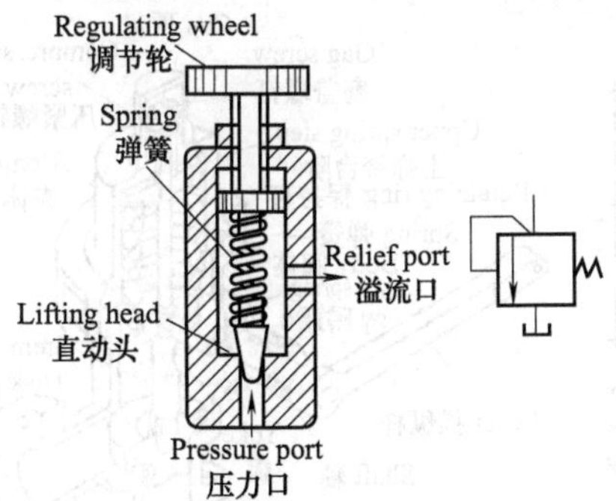

Fig 5.95 Spring loaded type relief valve 弹簧直动型溢流阀

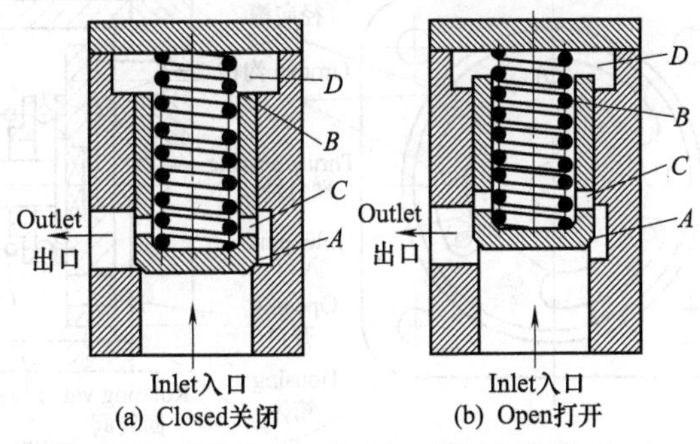

Fig 5.96 Simple check valve, Right-angle check valve 简易计量阀，直角计量阀

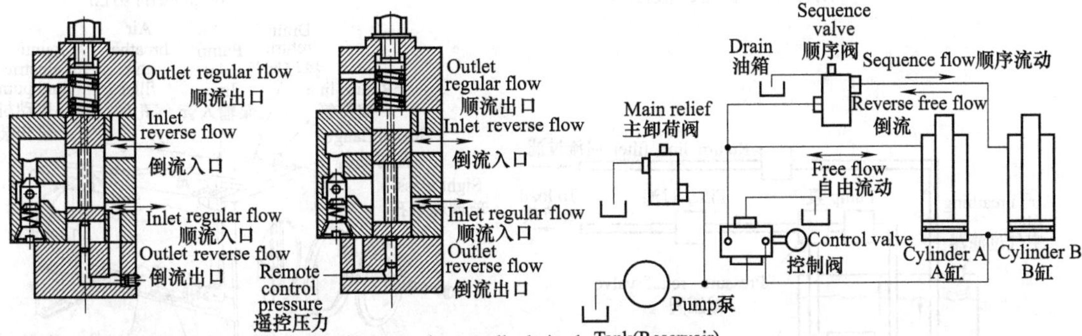

(a) Directly operated ,externally drained, Valve shown open in sequence position 直动外排型，图示顺序位置打开 (b) Remotely operated,externally drained, Valve shown close to regular flow 遥控外排型，图示正常流动关闭 (c) Application drawing showing typical circuit with sequence valve 顺序阀典型回路应用图

Fig 5.97 Sequence valve for reverse free flow 顺序阀用于自由回流

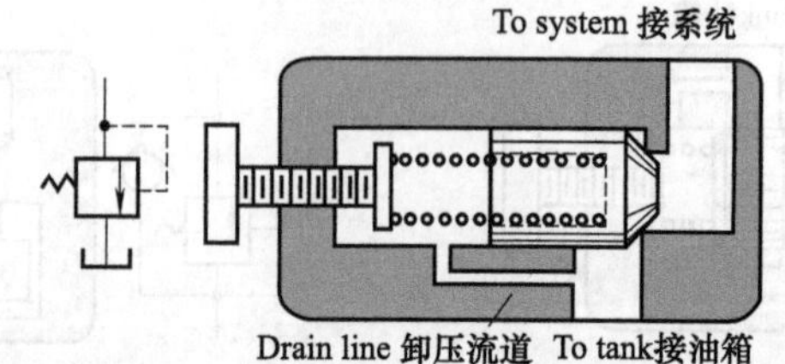

Fig 5.98 Pressure relief valve regulates system output fluid pressure 泄压阀调节系统输出流体压力

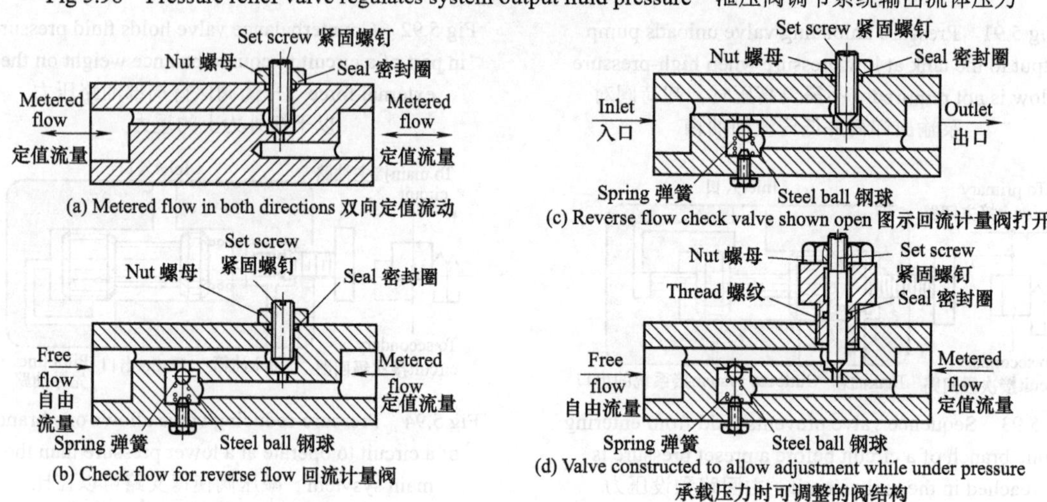

Fig 5.99 Single-needle valve flow control 单一针阀的流量控制

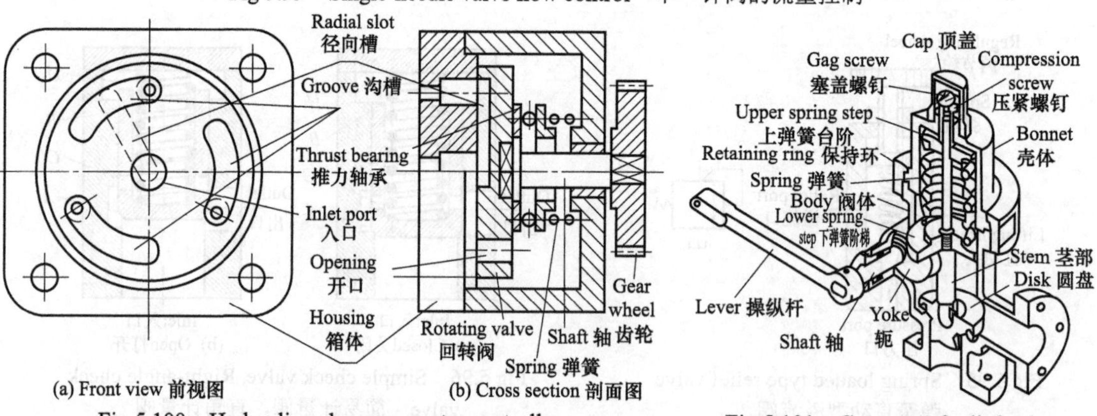

Fig 5.100 Hydraulic valve as a master controller
用作主控制器的液压阀

Fig 5.101 Cutaway of relief valve
泄压阀剖切图

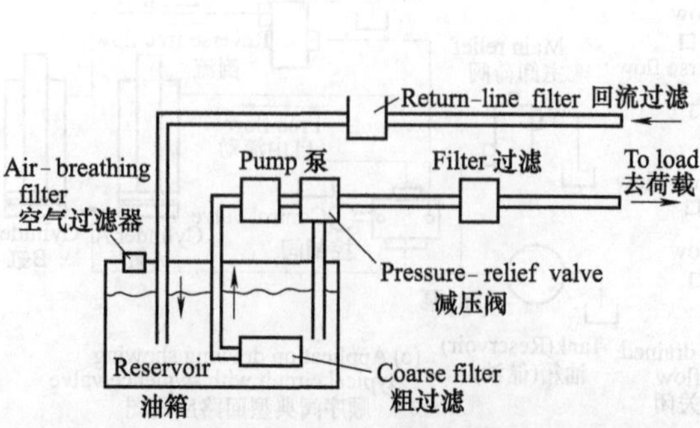

Fig 5.102 Some of the auxiliary components used in a practical hydraulic system 实际液压系统使用的附件

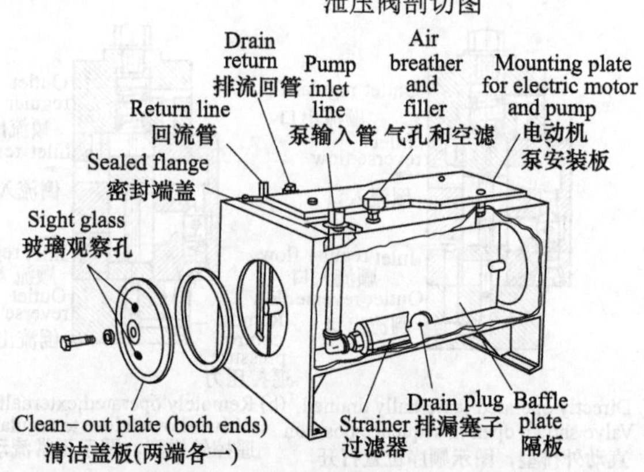

Fig 5.103 Non-pressurized reservoir
无压力油箱

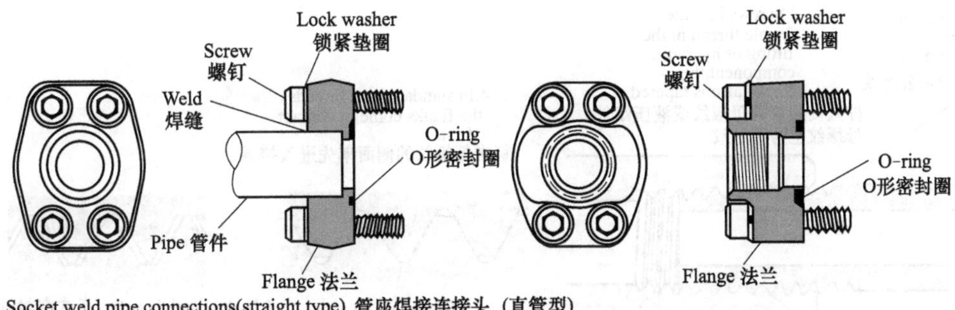

Lock washer 锁紧垫圈
Screw 螺钉
Weld 焊缝
O-ring O形密封圈
Pipe 管件
Flange 法兰
Socket weld pipe connections(straight type) 管座焊接连接头（直管型）
Lock washer 锁紧垫圈
Screw 螺钉
O-ring O形密封圈
Flange 法兰

Fig 5.104 Four bolt flange connectors 四螺栓孔法兰连接器

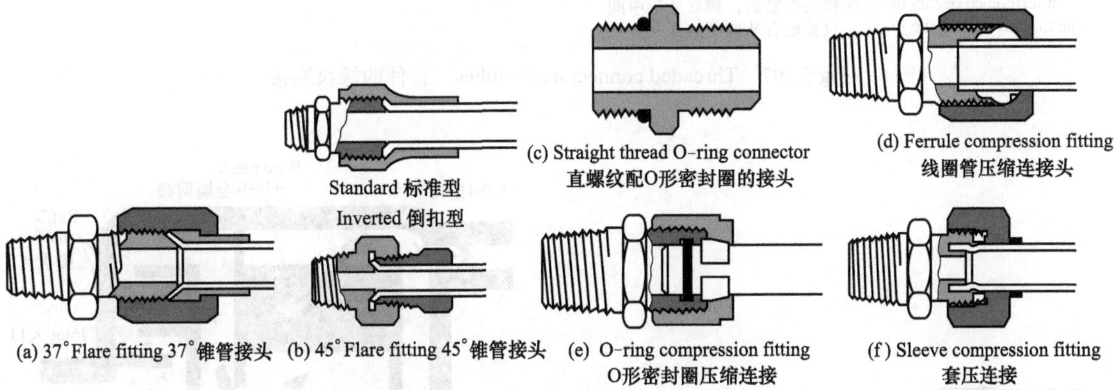

Standard 标准型
Inverted 倒扣型
(c) Straight thread O-ring connector 直螺纹配O形密封圈的接头
(d) Ferrule compression fitting 线圈管压缩连接头
(a) 37° Flare fitting 37°锥管接头
(b) 45° Flare fitting 45°锥管接头
(e) O-ring compression fitting O形密封圈压缩连接
(f) Sleeve compression fitting 套压连接

Fig 5.105 Flared tube fitting 锥管头连接

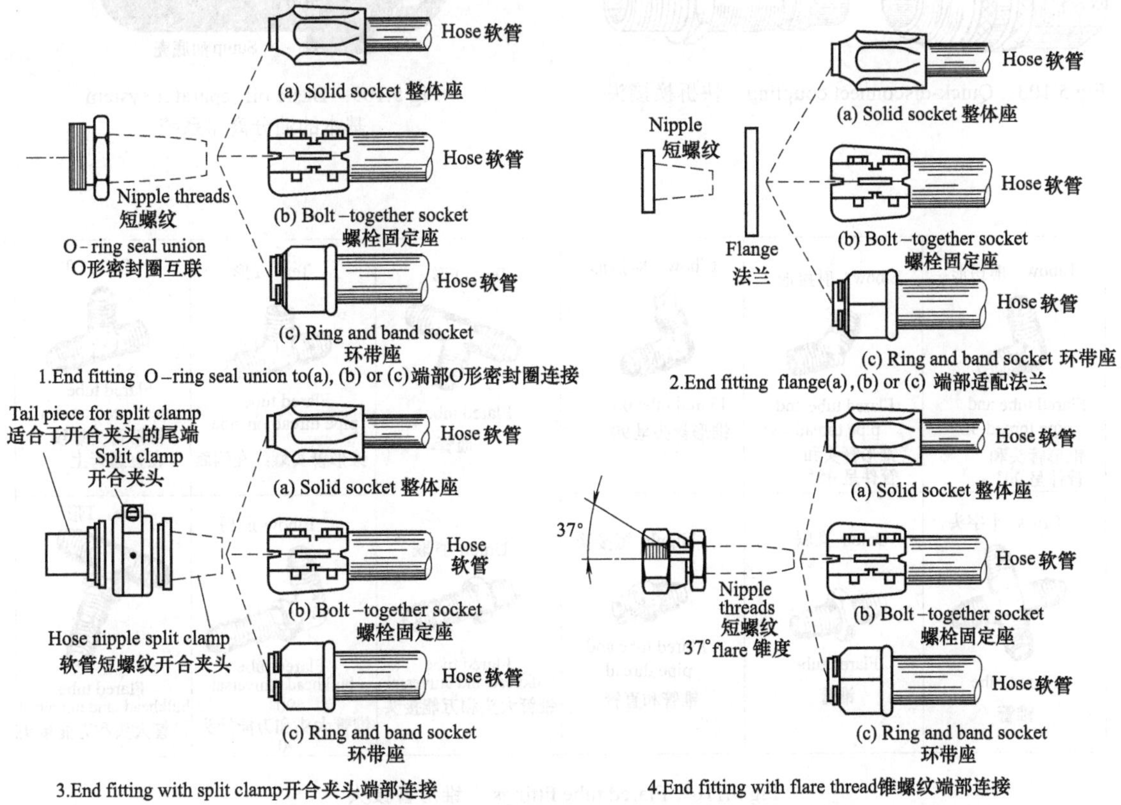

Hose 软管
(a) Solid socket 整体座
Hose 软管
Nipple threads 短螺纹
(b) Bolt-together socket 螺栓固定座
O-ring seal union O形密封圈互联
Hose 软管
(c) Ring and band socket 环带座
1.End fitting O-ring seal union to(a), (b) or (c)端部O形密封圈连接
Hose 软管
(a) Solid socket 整体座
Nipple 短螺纹
Hose 软管
Flange 法兰
(b) Bolt-together socket 螺栓固定座
Hose 软管
(c) Ring and band socket 环带座
2.End fitting flange(a), (b) or (c) 端部适配法兰
Tail piece for split clamp 适合于开合夹头的尾端
Split clamp 开合夹头
Hose 软管
(a) Solid socket 整体座
Hose 软管
(b) Bolt-together socket 螺栓固定座
Hose nipple split clamp 软管短螺纹开合夹头
Hose 软管
(c) Ring and band socket 环带座
3.End fitting with split clamp开合夹头端部连接
Hose 软管
(a) Solid socket 整体座
37°
Hose 软管
Nipple threads 短螺纹
37° flare 锥度
(b) Bolt-together socket 螺栓固定座
Hose 软管
(c) Ring and band socket 环带座
4.End fitting with flare thread锥螺纹端部连接

Fig 5.106 End fittings and hose fittings 端部连接和软管连接

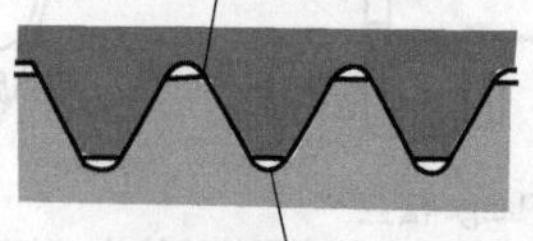

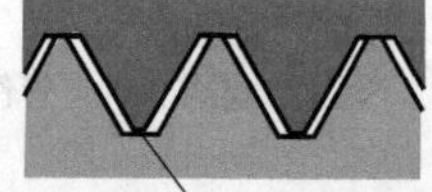

Fig 5.107 Threaded connectors for tubes 管件的螺纹连接

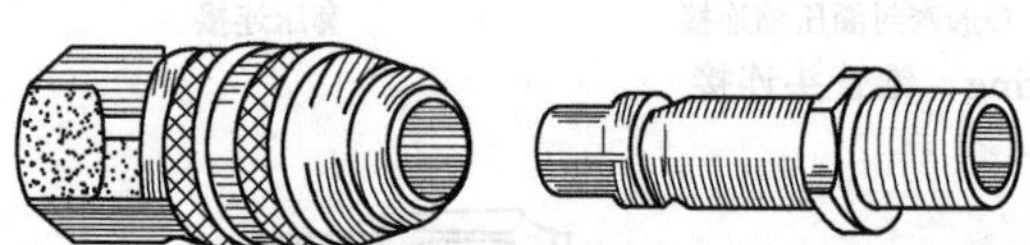

Fig 5.108 Quick-disconnect coupling 快拆换接头

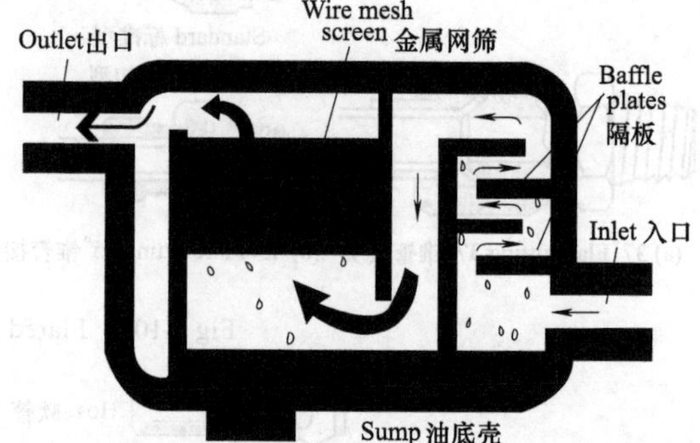

Fig 5.109 Basic oil separator system 基本的油分离器系统

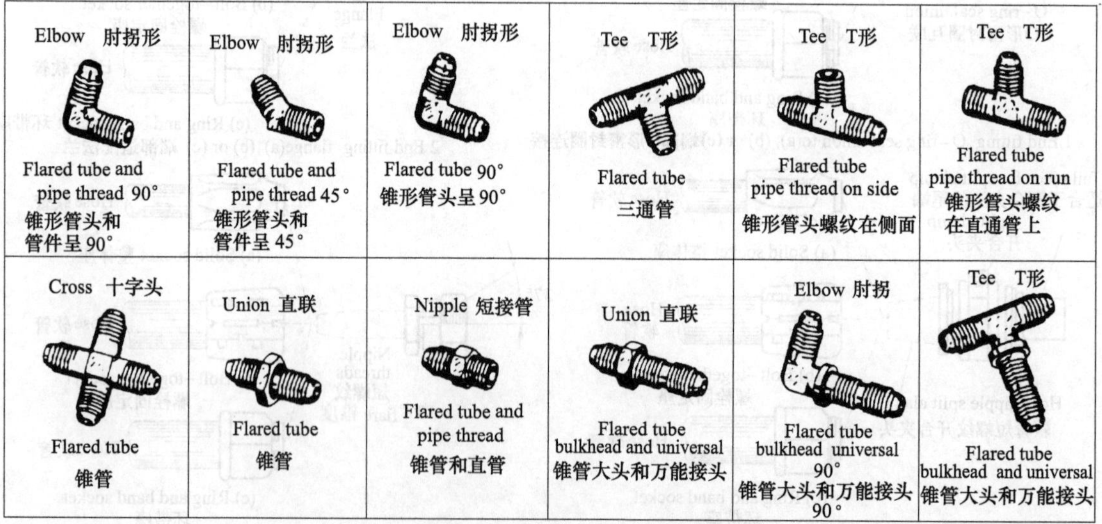

Elbow 肘拐形	Elbow 肘拐形	Elbow 肘拐形	Tee T形	Tee T形	Tee T形
Flared tube and pipe thread 90° 锥形管头和管件呈90°	Flared tube and pipe thread 45° 锥形管头和管件呈45°	Flared tube 90° 锥形管头呈90°	Flared tube 三通管	Flared tube pipe thread on side 锥形管头螺纹在侧面	Flared tube pipe thread on run 锥形管头螺纹在直通管上
Cross 十字头 Flared tube 锥管	Union 直联 Flared tube 锥管	Nipple 短接管 Flared tube and pipe thread 锥管和直管	Union 直联 Flared tube bulkhead and universal 锥管大头和万能接头	Elbow 肘拐 Flared tube bulkhead universal 90° 锥管大头和万能接头 90°	Tee T形 Flared tube bulkhead and universal 锥管大头和万能接头

Fig 5.110 Flared tube fittings 锥形管接头

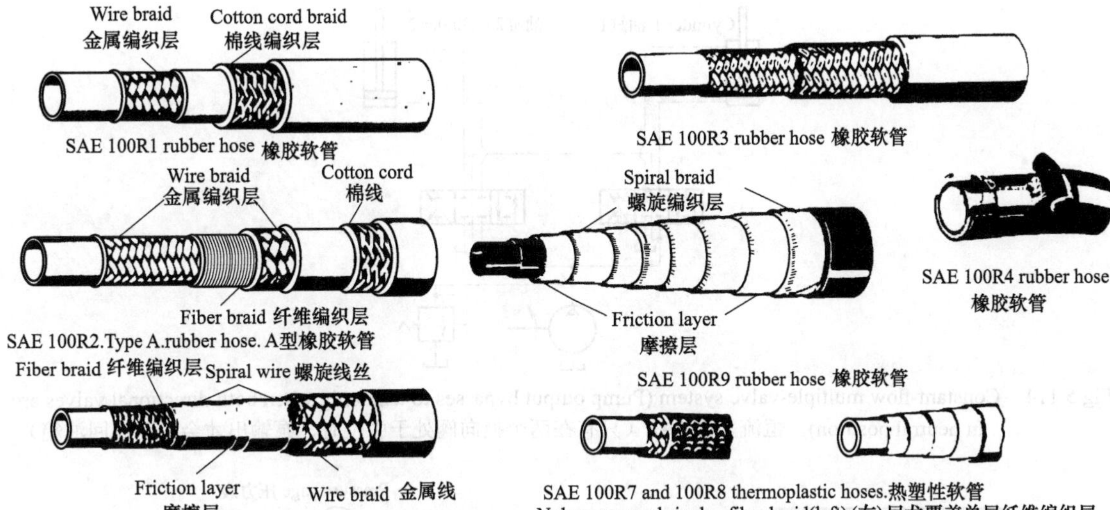

Fig 5.111 Types of flexible hose 软管的种类

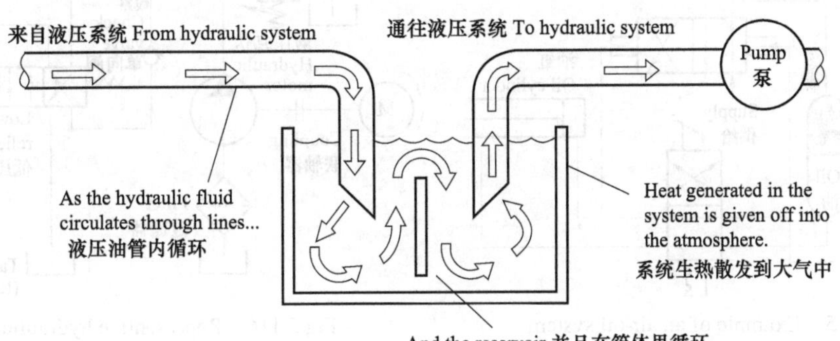

Fig 5.112 Circulating oil cools system 循环油冷却系统

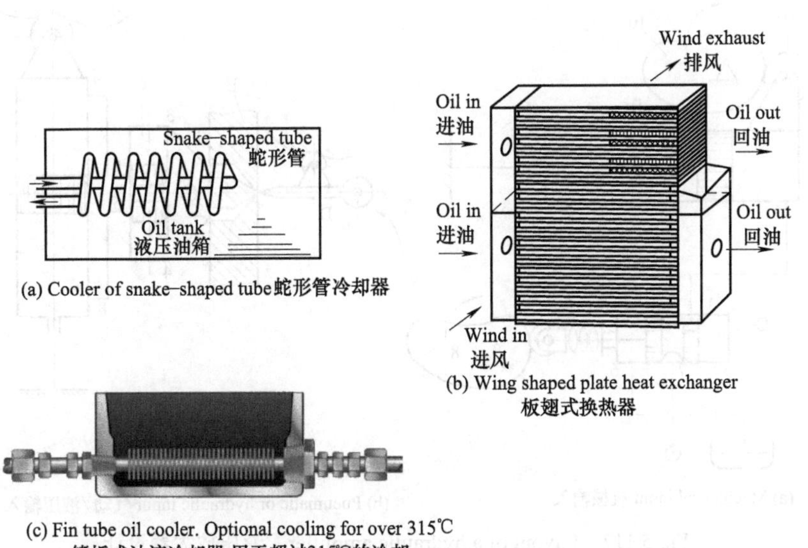

Fig 5.113 Coolers 冷却器

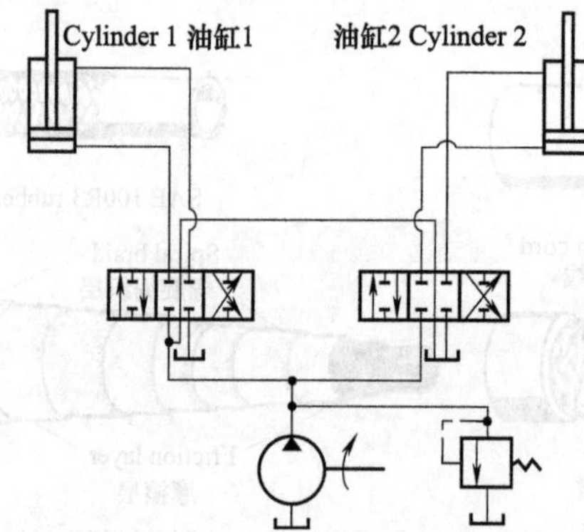

Fig 5.114 Constant-flow multiple-valve system (Pump output bypasses to tank only when both directional valves are in neutral position) 恒流多阀系统（只有在两个换向阀处于中位时，泵输出才会旁路流回油箱）

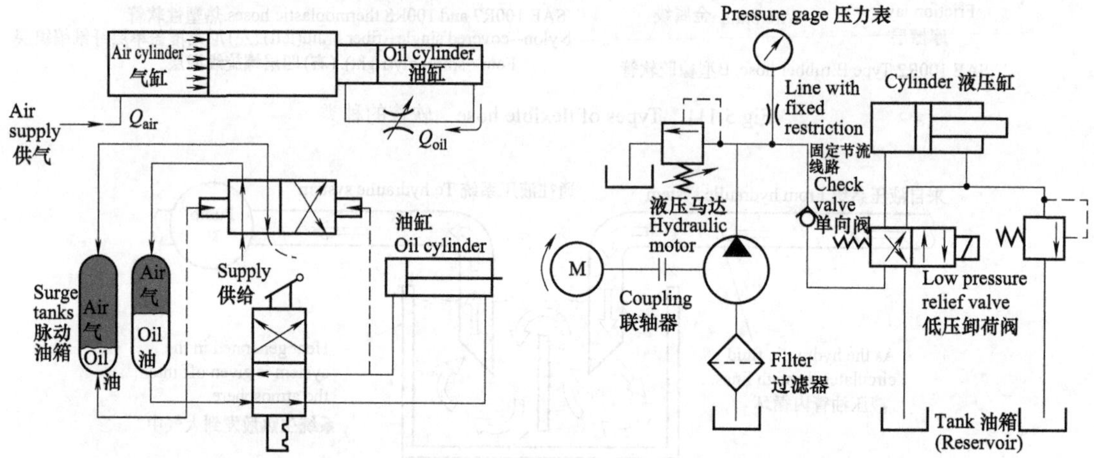

Fig 5.115 Example of an air-oil system 气液系统示例

Fig 5.116 Representive hydraulic circuit 典型液压回路

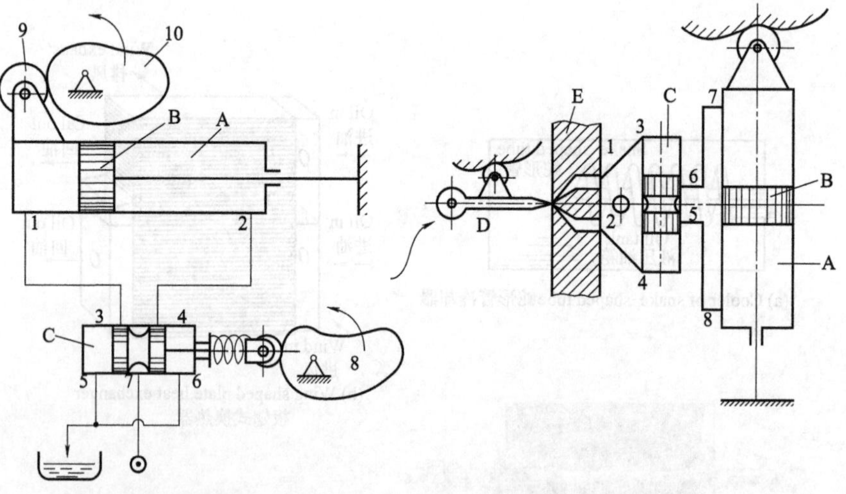

(a) Mechanical input 机械输入

(b) Pneumatic or hydraulic input 气动/液压输入

Fig 5.117 Layout of a hydraulic amplifier 液压放大器设计

1, 2—Cylinder port 缸口；3—Outlet 出口；4—Inlet 入口；5 ～ 7—Port 阀口；8, 10—Cam 凸轮；9—Rotor 转子；A—Cylinder 缸体；B—Piston 活塞；C—Slide valve 滑阀；D—Nozzle 喷嘴；E—Opposite partition 反向分离块

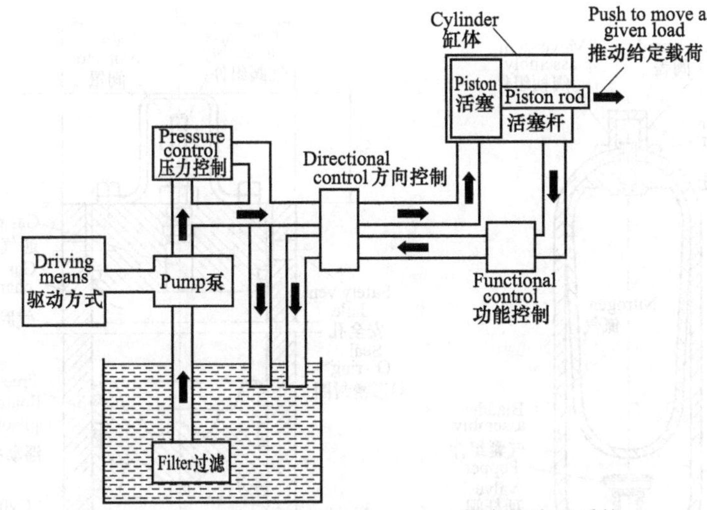

Fig 5.118 Basic hydraulic system 基本的液压系统

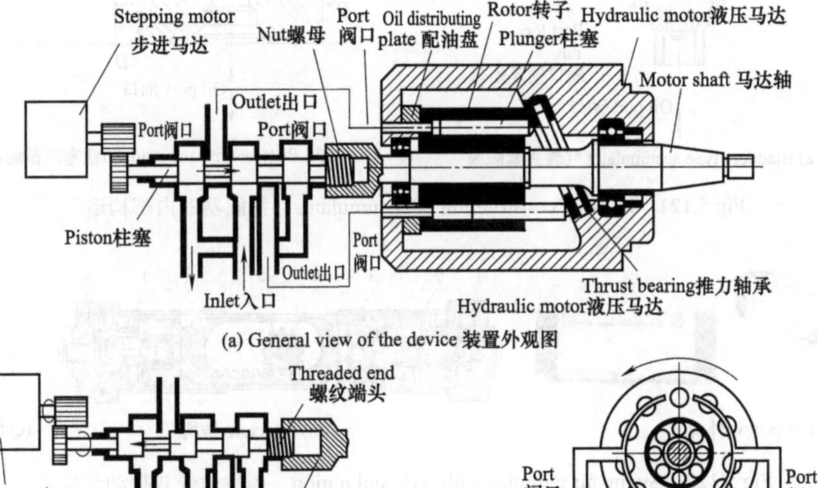

(a) General view of the device 装置外观图

(b) Leftward movement of the valve's piston 阀片活塞向左运动 (c) Cross section of the oil distributor 配硬盘剖面

Fig 5.119 Layout of hydraulic pulse motor-stepmotor combination with a hydraulic servomotor
液压脉冲伺服马达与液压伺服马达的复合设计

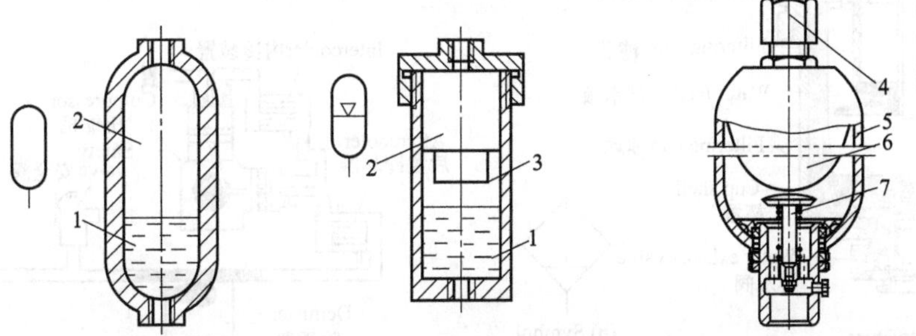

(a) Gas bottle type 气瓶式蓄能器 (b) Piston type 活塞式蓄能器 (c) Gas bag type 气囊式蓄能器

Fig 5.120 Accumulator with gas loading 气体加载式蓄能器

1—Oil 液压油； 2—Gas 气体； 3—Piston 活塞； 4—Valve for air in 充气阀； 5—Shell 壳体；
6—Bag 皮囊； 7—Valve for oil in 进油阀

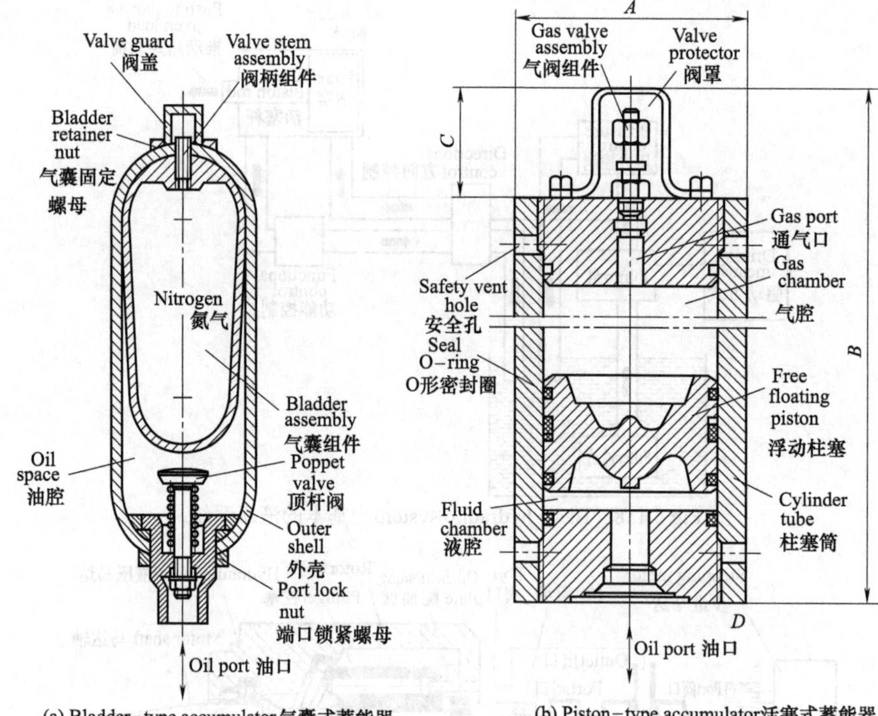

(a) Bladder-type accumulator气囊式蓄能器　(b) Piston-type accumulator活塞式蓄能器

Fig 5.121　Internal construction of accumulator　蓄能器的内部构造

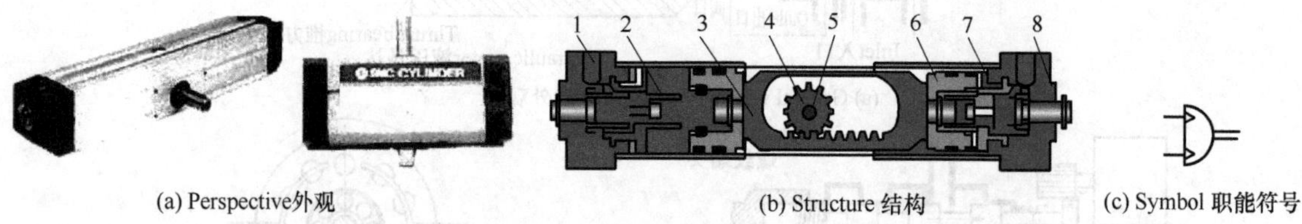

(a) Perspective外观　(b) Structure 结构　(c) Symbol 职能符号

Fig 5.122　Swing air cylinder with rack and pinion　齿轮齿条式摆动气缸

1—Buffer throttle 缓冲节流阀；2—Buffer plunger 缓冲柱塞；3—Rack 齿条组件；4—Pinion 齿轮；5—Output shaft 输出轴；6—Piston 活塞；7—Cylinder 缸体；8—Side plate 端盖

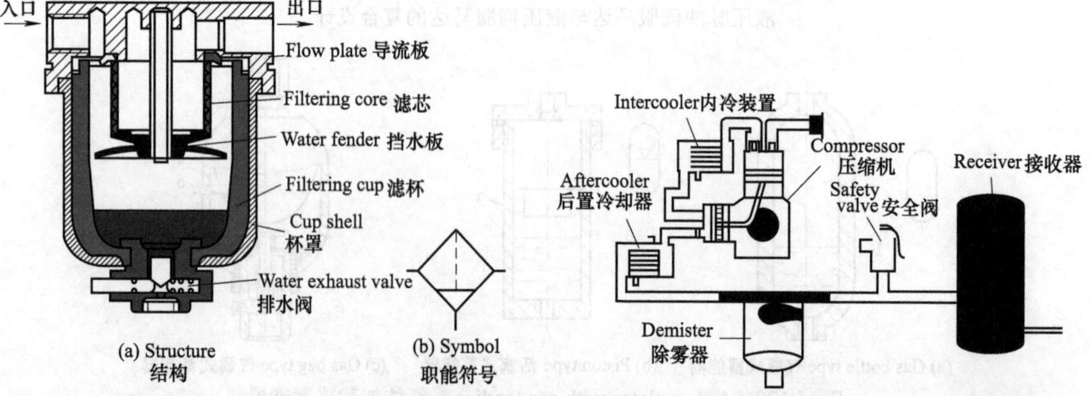

(a) Structure 结构　(b) Symbol 职能符号

Fig 5.123　Standard air filter　标准气体过滤器

Fig 5.124　Illustrates how a safety valve functions　安全阀工作图解

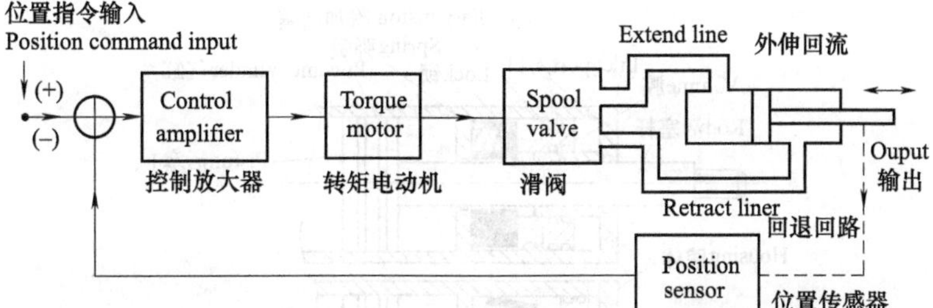

Fig 5.125 A pneumatic closed-loop linear control system 气动闭环线性控制系统

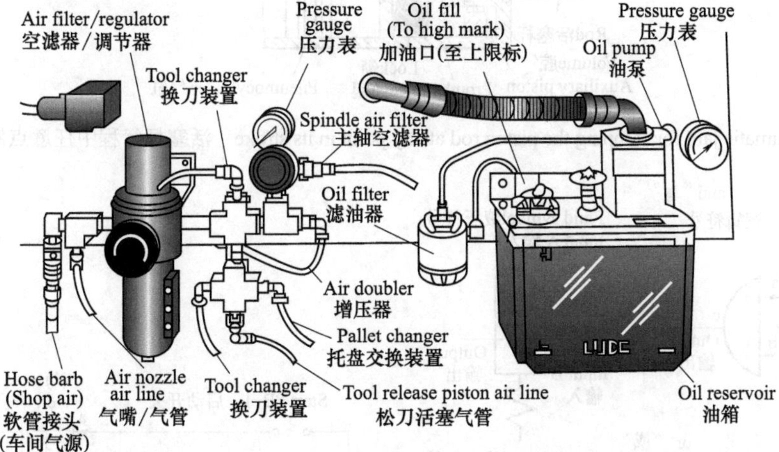

Fig 5.126 Lubricant way for pneumatic system 气动系统润滑回路

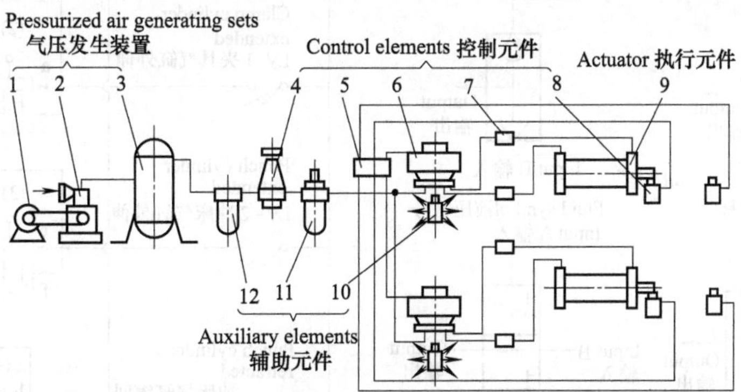

Fig 5.127 Construct of pneumatic system 气动系统的组成

1—Motor 电动机；2—Air compressor 空压机；3—Gas tank 储气罐；4—Pressure control valve 压力控制阀；5—Logic element 逻辑元件；6—Direction control valve 方向控制阀；7—Flowrate control valve 流量控制阀；8—Mechanical control valve 机控阀；9—Air cylinder 气缸；10—Noise absorber 消声器；11—Oil spray 油雾器；12—Air filter 空气滤清器

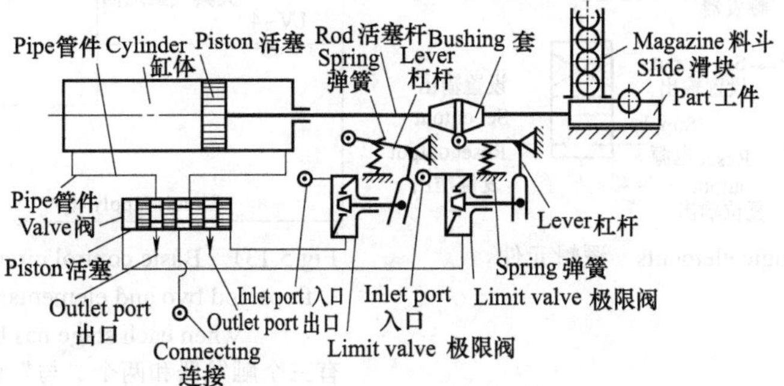

Fig 5.128 Pneumatic circuit controlled by a limit valve 极限阀控制的气动回路

Auxiliary piston 附加活塞
Spring 弹簧
Inlet port 入口
Lock 锁
Pneumocylinder 气缸
Volume 腔
Rod 活塞杆
Piston 活塞Ⅰ
Housing 腔体
Piston 活塞Ⅱ
Rod 活塞杆
Volume 腔
Lock 锁
Auxiliary piston 附加活塞
Protuberance 凸起
Pneumocylinder 气缸

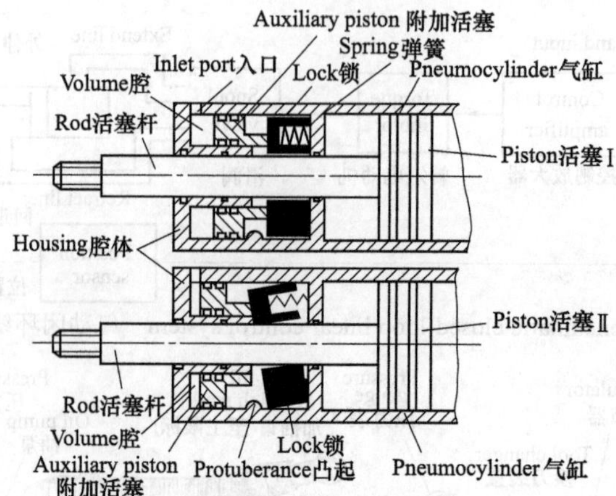

Fig 5.129 Pneumatic jam for stopping the piston rod at any point in its stroke 活塞杆行程中任意点停止用气动夹头

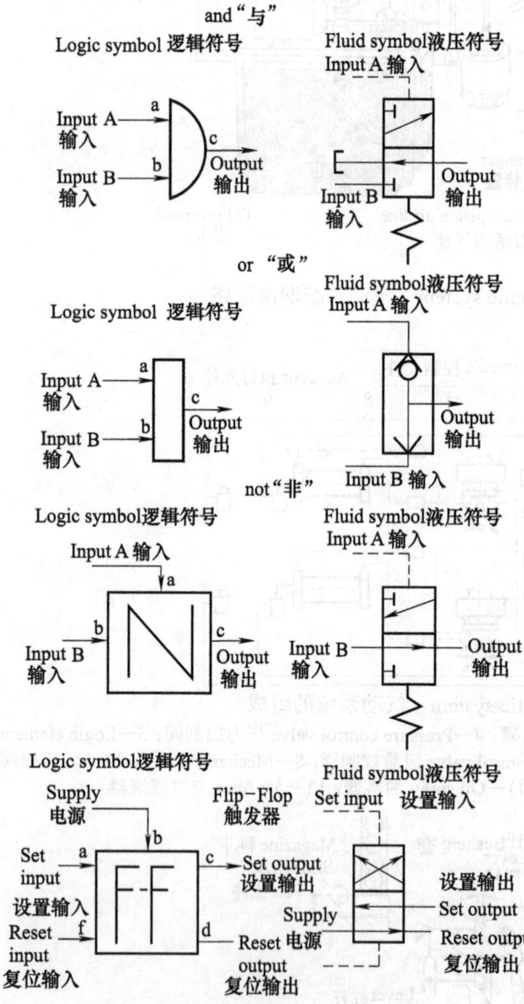

and "与"
Logic symbol 逻辑符号
Fluid symbol 液压符号
Input A 输入
Input A 输入
Input B 输入
Output 输出
Input B 输入
Output 输出
or "或"
Logic symbol 逻辑符号
Fluid symbol 液压符号
Input A 输入
Input A 输入
Input B 输入
Output 输出
Output 输出
Input B 输入
not "非"
Logic symbol 逻辑符号
Fluid symbol 液压符号
Input A 输入
Input A 输入
Input B 输入
Output 输出
Input B 输入
Output 输出
Logic symbol 逻辑符号
Fluid symbol 液压符号
Supply 电源
Flip-Flop 触发器
Set input 设置输入
Set input 设置输入
Set output 设置输出
Reset input 复位输入
Reset output 复位输出
Supply
Reset 电源
设置输出
Set output
Reset output
复位输出

Fig 5.130 Logic elements 逻辑元件

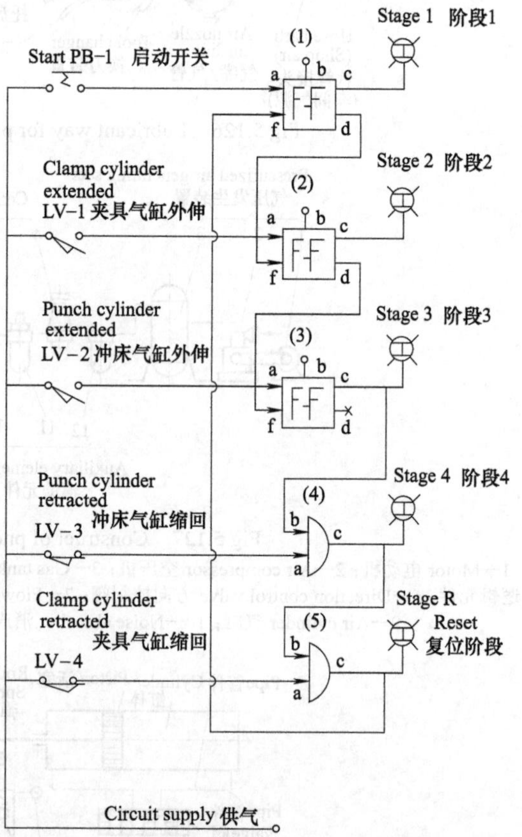

Start PB-1 启动开关
Stage 1 阶段1
Clamp cylinder extended LV-1 夹具气缸外伸
Stage 2 阶段2
Punch cylinder extended LV-2 冲床气缸外伸
Stage 3 阶段3
Punch cylinder retracted LV-3 冲床气缸缩回
Stage 4 阶段4
Clamp cylinder retracted LV-4 夹具气缸缩回
Stage R Reset 复位阶段
Circuit supply 供气

Fig 5.131 Basic control circuit consists of three flip-flops and two and elements. Indicator lights show when each stage has been completed 有三个触发器和两个"与"逻辑元件构成的基本控制回路，每个阶段结束时指示灯点亮

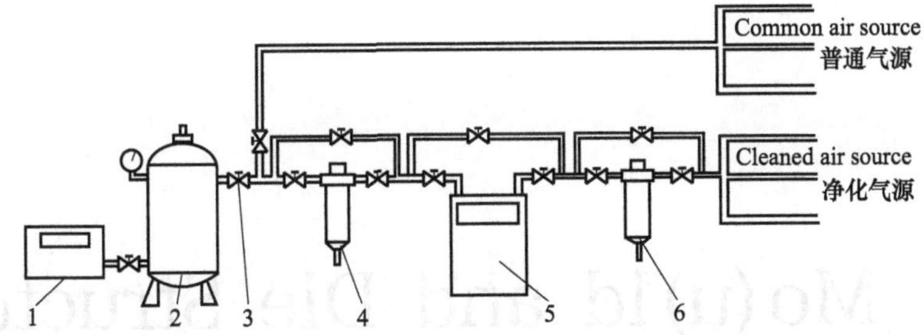

Fig 5.132 Air source and its air cleaning system 气源及空气净化处理系统

1—Air compressor 空压机；2—Gas tank 气体罐；3—Valve(switch) 阀门；4—Main tube filter 主管过滤器；
5—Air drier 干燥机；6—Filter of main tube 主管过滤器

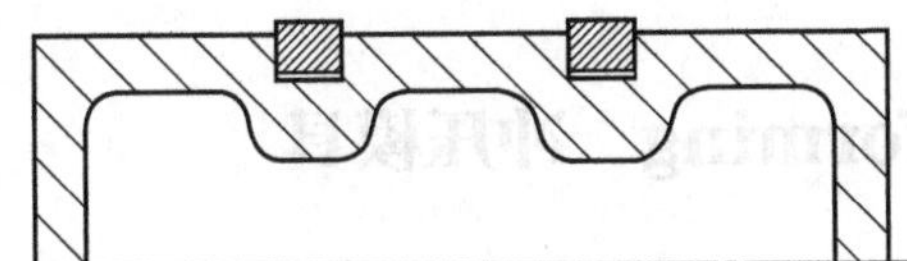

Double-acting compressor piston(lubricated)with plain piston rings
(带润滑的)普通活塞环的双作用压缩机活塞

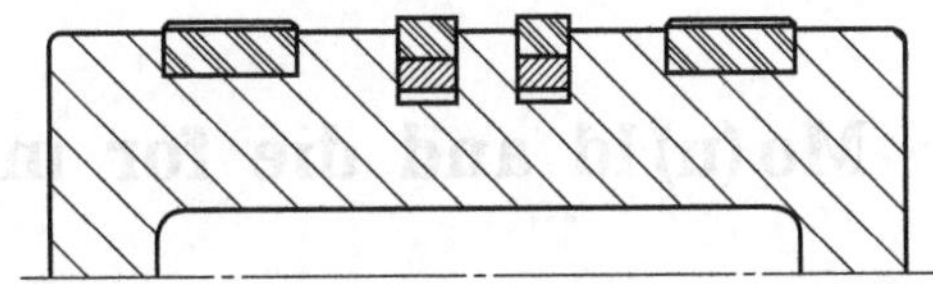

Double-acting compressor piston(non-lubricated)with PTFE sealing rings and bearer bands
有承载环的聚四氟乙烯密封环(非润滑)双作用压缩机活塞

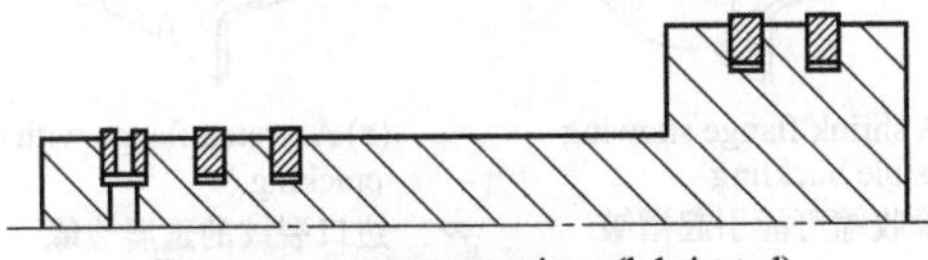

Two-stage compressor piston(lubricated)
(带润滑的)两级压缩机活塞

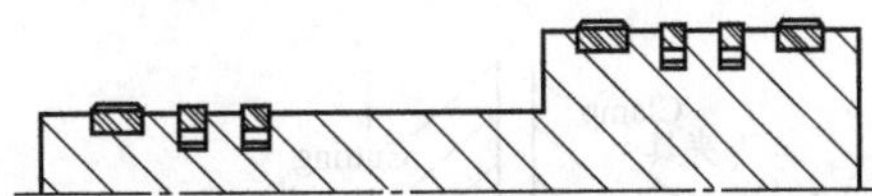

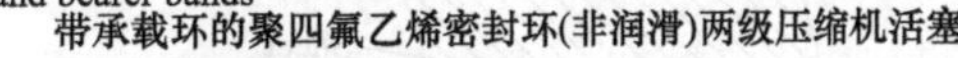

Two-stage compressor piston(non-lubricated)with PTFE sealing rings and bearer bands
带承载环的聚四氟乙烯密封环(非润滑)两级压缩机活塞

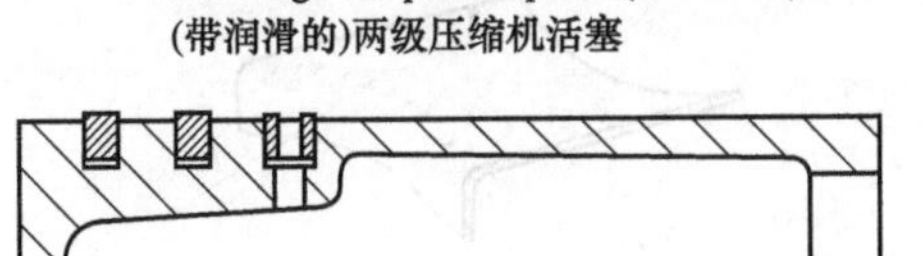

Simple single-stage piston (lubricated)润滑型简易单级活塞

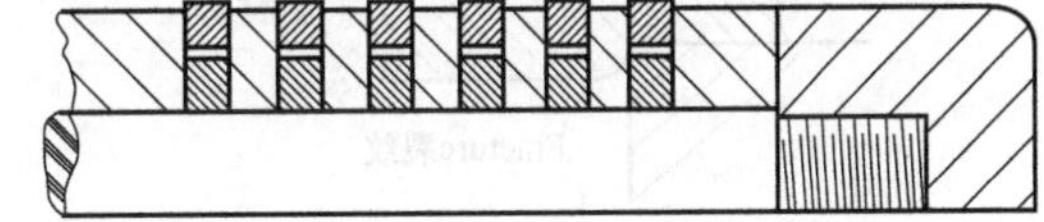

Built-up piston for high-pressure compressor高压压缩机用组合密封

Fig 5.133 Pistons for air and gas compressors 空气或气体压缩机用活塞

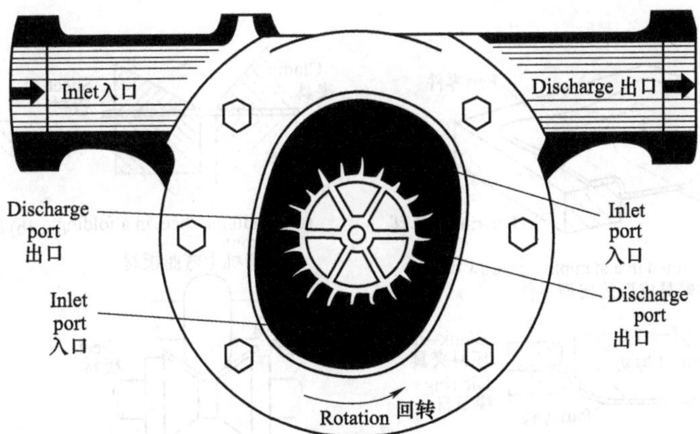

Fig 5.134 Liquid seal ring rotary air compressor 液体密封回转型空气压缩机

Unit 6 Mo(u)ld and Die Structure Design 模具结构设计

6.1 Mo(u)ld and die for metal forming 冲压模具

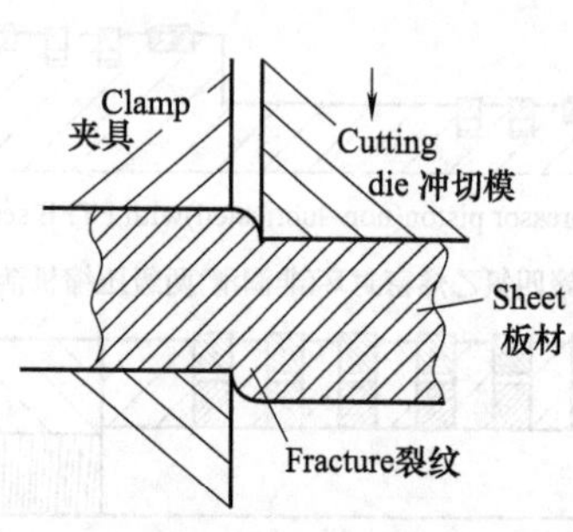

Fig 6.1 Magnified section of blanking a sheet showing plastic deformation and cracking
板材冲裁放大剖面可见的塑性变形和裂纹

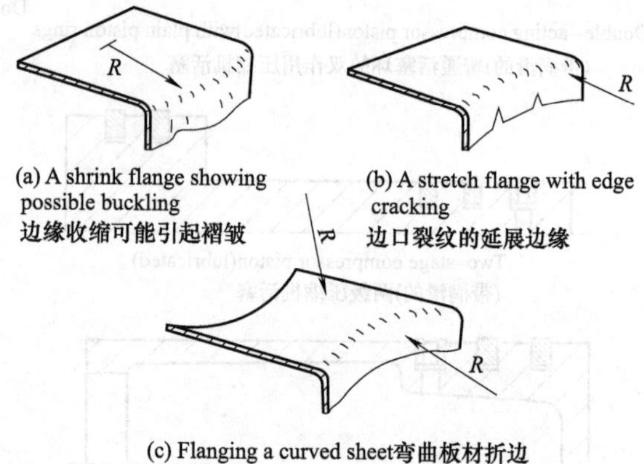

Fig 6.2 Plastic deformation occurs in the adjoining sheet
相接板件发生的塑性变形

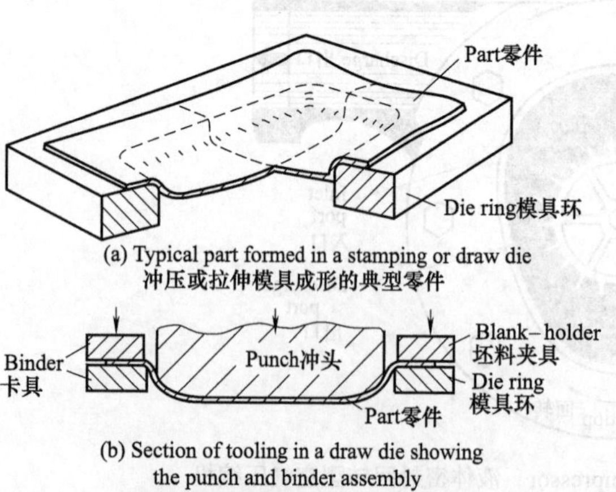

Fig 6.3 Stretching over a punch of more complicated shape in a draw die 拉伸模里相对复杂形状冲头拉伸成形

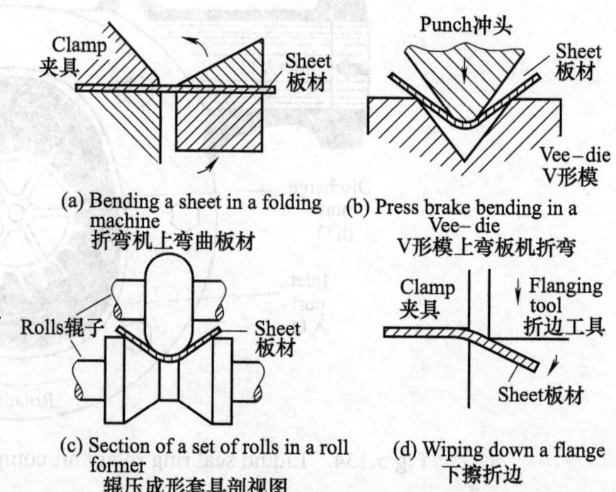

Fig 6.4 Various ways of bending along a straight line
沿直线的几种折弯方式

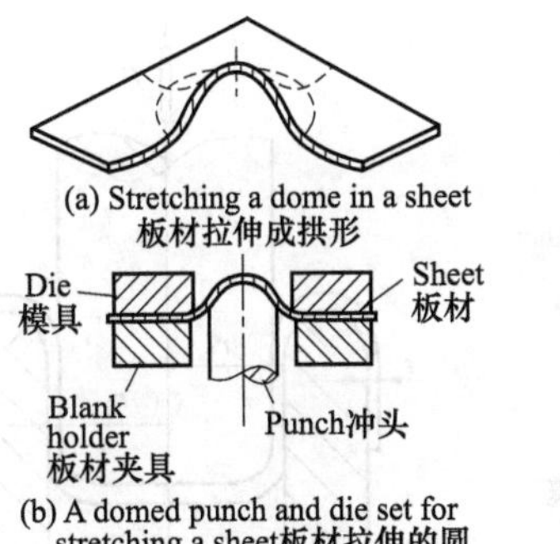

Fig 6.5 Stretching process
拉伸工艺

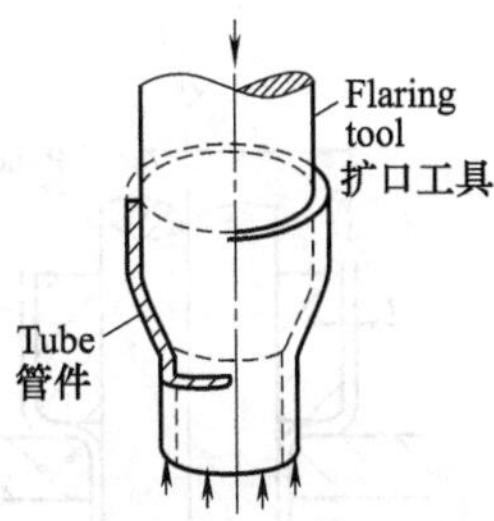

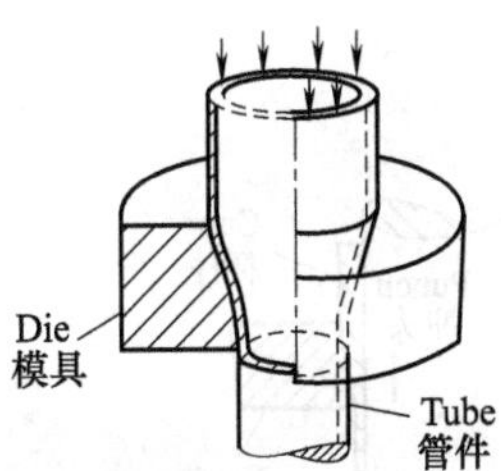

Fig 6.6 Tube forming
管径成形

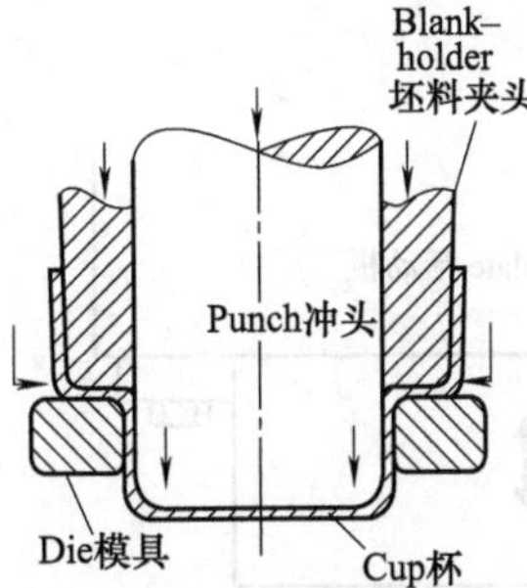

Fig 6.7 Section of tooling for forward redrawing of a cylindrical cup 向前并反拉圆柱形杯的工装剖面图

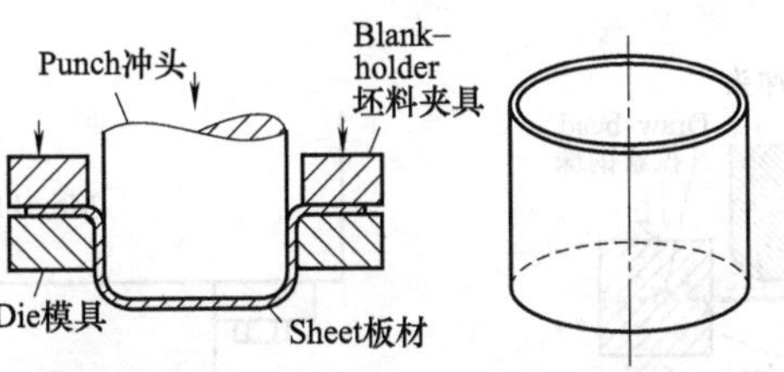

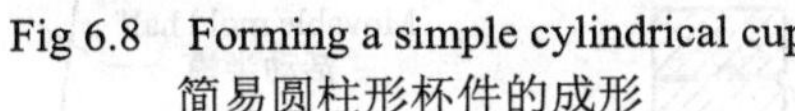

Fig 6.8 Forming a simple cylindrical cup
简易圆柱形杯件的成形

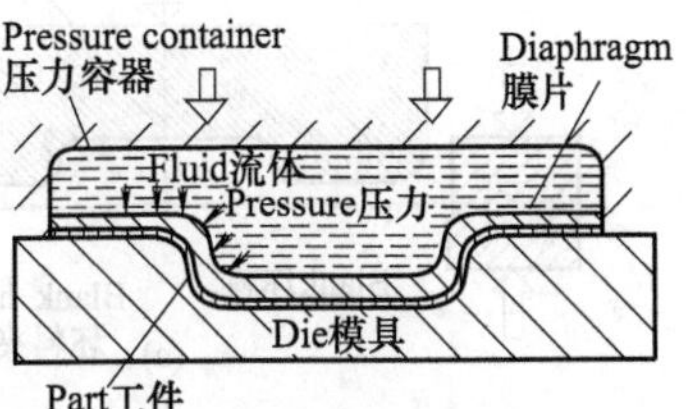

Fig 6.9 Using fluid pressure (hydroforming) to form a shallow part 采用流体压力（液压成形）对深度较浅的工件成形

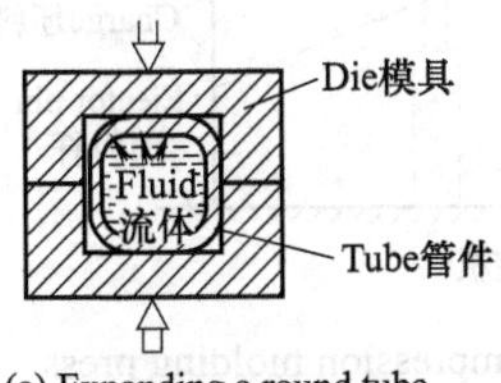

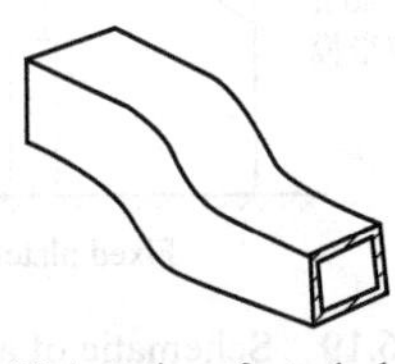

Fig 6.10 A square section are made in a closed die
闭合模具制造正方形截面管件

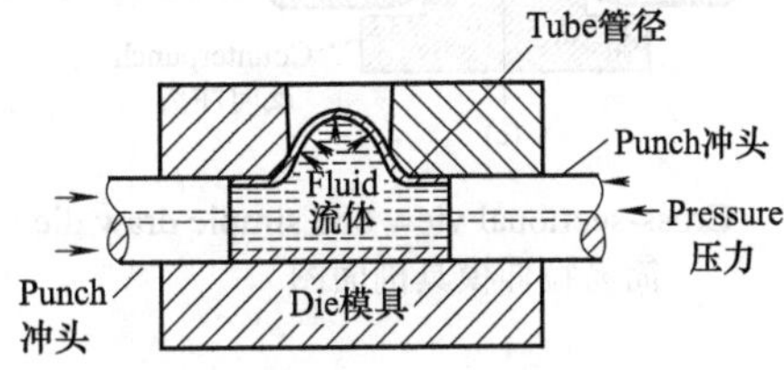

Fig 6.11 Using combined axial force and fluid pressure to form a plumbing fitting (tee joint)
轴向力与液压力联合成形制备隆起配件（T 形接头）

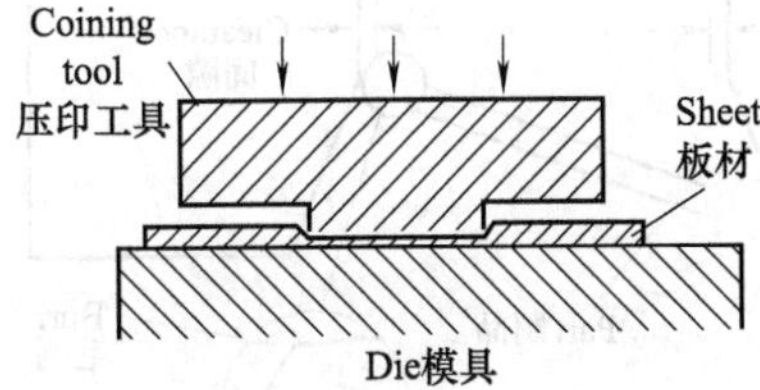

Fig 6.12 Thinning a sheet locally using a coining tool 采用压印工具局部压印板材

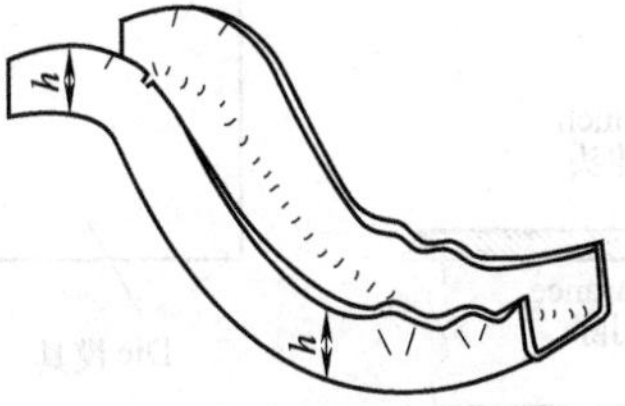

Fig 6.13 Inside and outside bends in a channel section
渠道型截面的内外弯曲

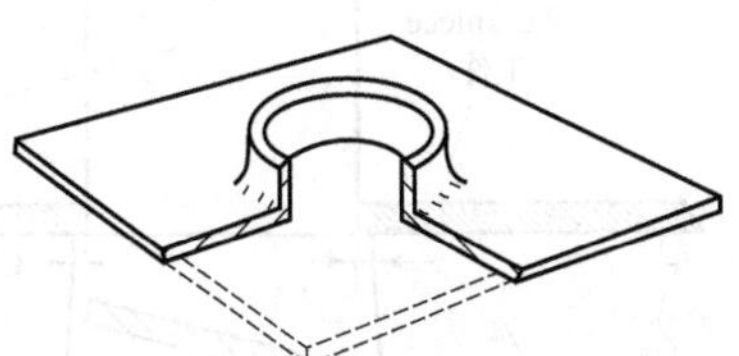

Fig 6.14 Extrusion of a punched hole
冲孔的挤出翻边

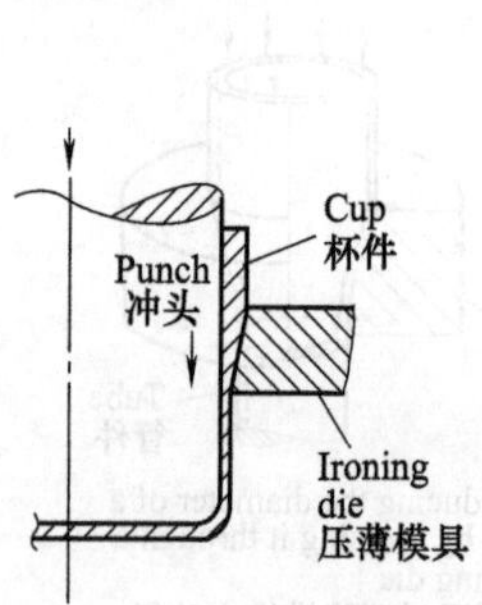

Fig 6.15 Thinning the wall of a cylindrical cup by passing it through an ironing die 通过压薄模具圆柱形杯壁变薄

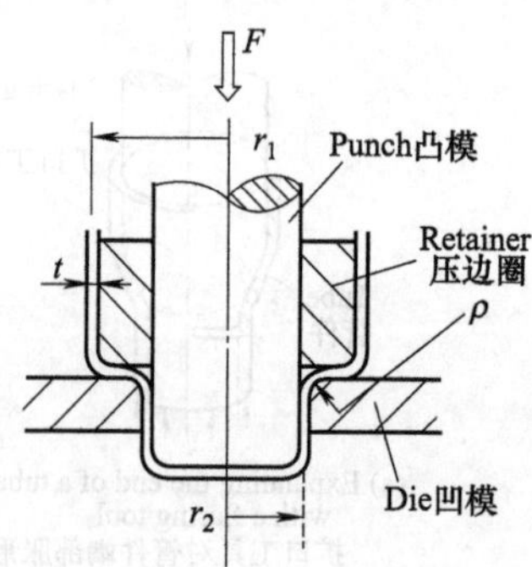

Fig 6.16 Forward redrawing of a deep-drawn cup 已拉深杯体的正向二次拉深

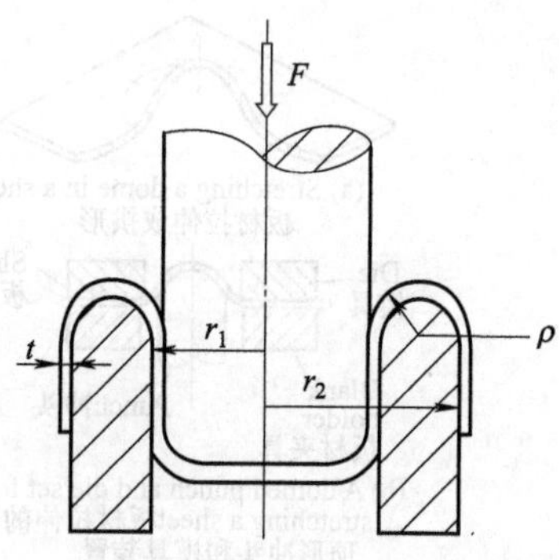

Fig 6.17 Reverse redrawing of a cylindrical cup 圆柱形杯筒的二次反向拉深

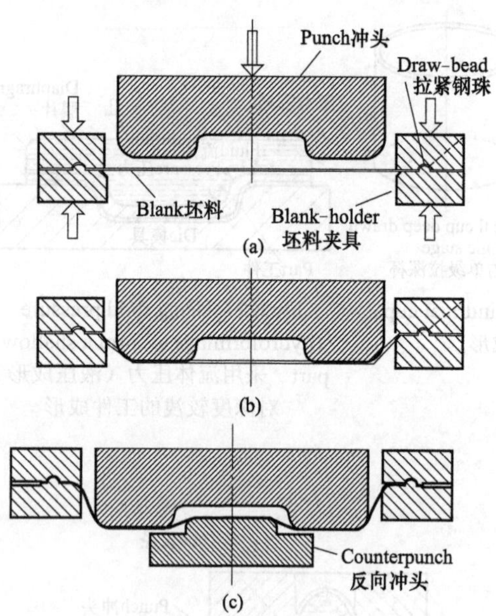

Fig 6.18 Cross-sectional view of a simple draw die 简易拉伸模具剖面图

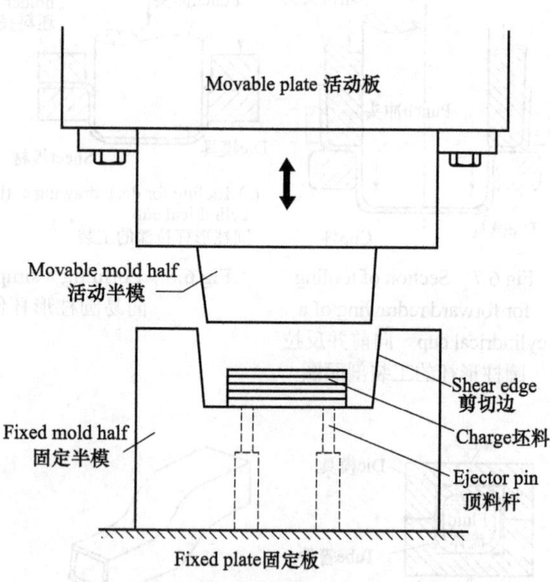

Fig 6.19 Schematic of a compression molding press 压缩模具设备图解

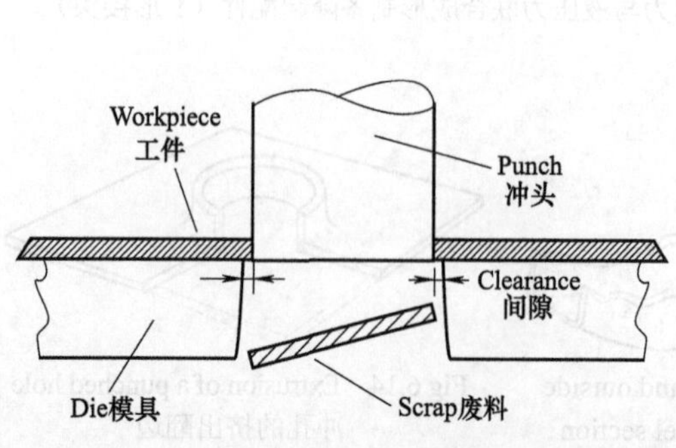

Fig 6.20 Piercing 冲孔

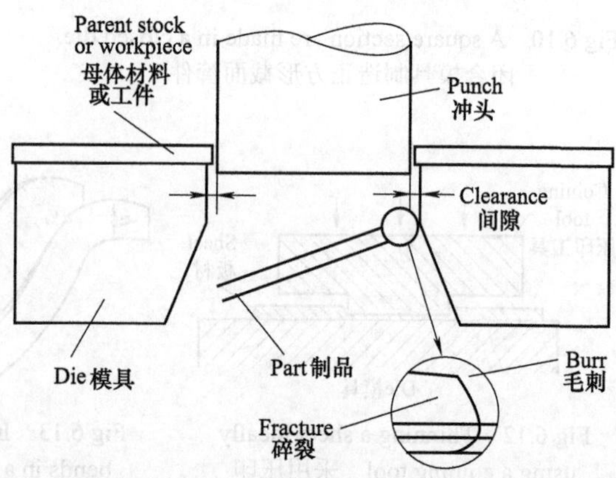

Fig 6.21 Blanking 落料

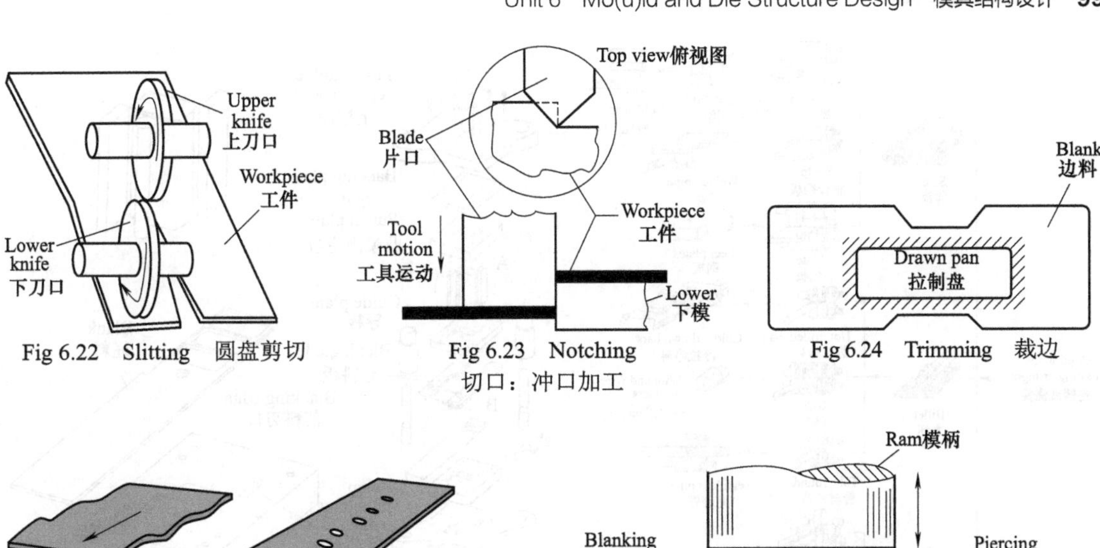

Fig 6.22 Slitting 圆盘剪切

Fig 6.23 Notching 切口：冲口加工

Fig 6.24 Trimming 裁边

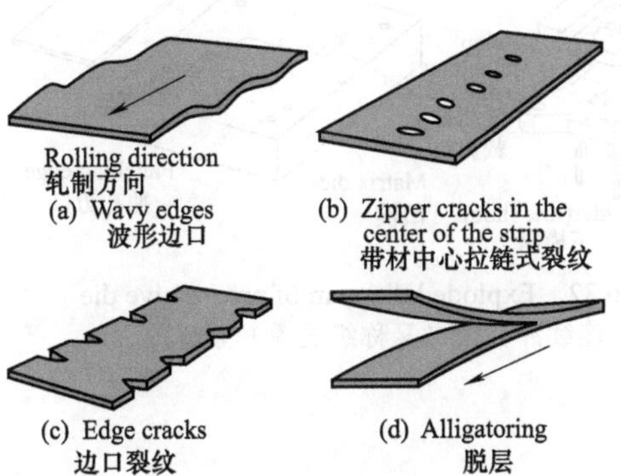

Fig 6.25 Typical defects in flat rolling 平面轧制典型缺陷

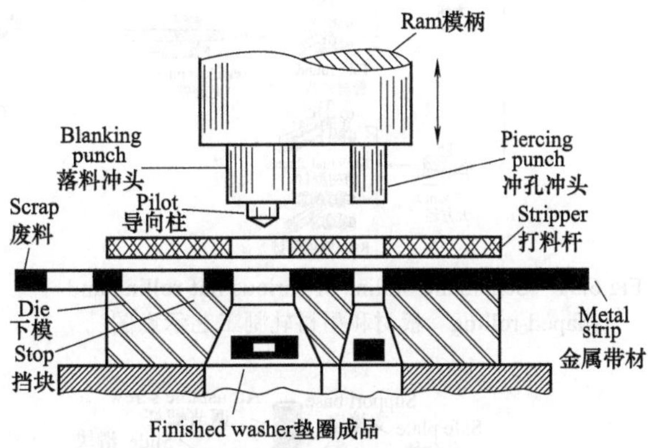

Fig 6.26 Washer production in progressive die 级进模冲制垫圈

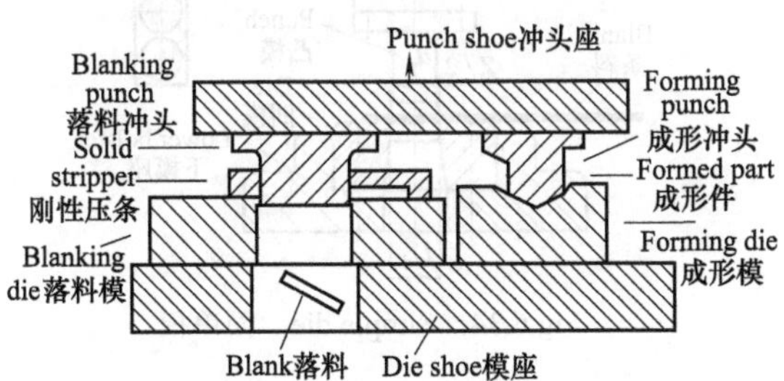

Fig 6.27 Combination die 复合模具

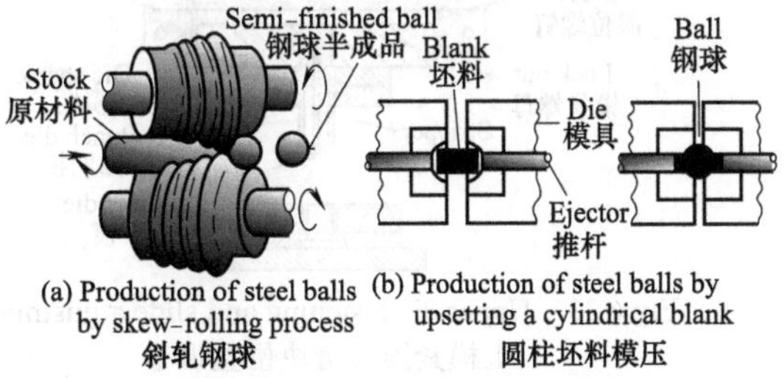

Fig 6.28 Steel ball machined by dies 模具加工钢球

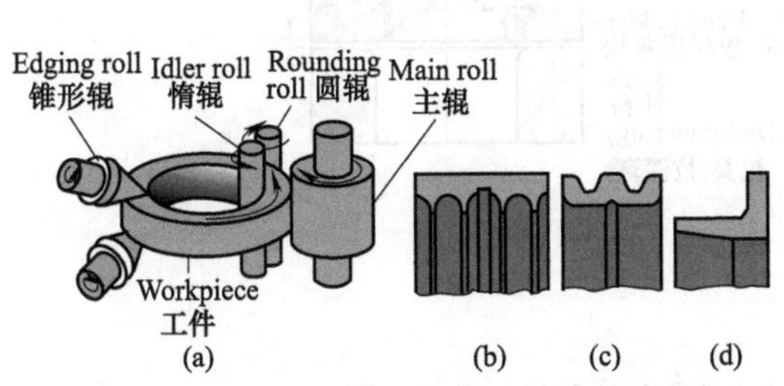

Fig 6.29 Ring rolling operations 环件的滚压加工

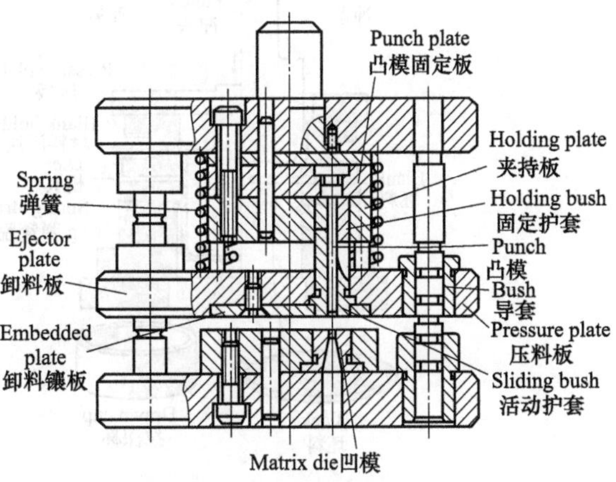

Fig 6.30 Piercing dies for small holes 小孔冲裁模

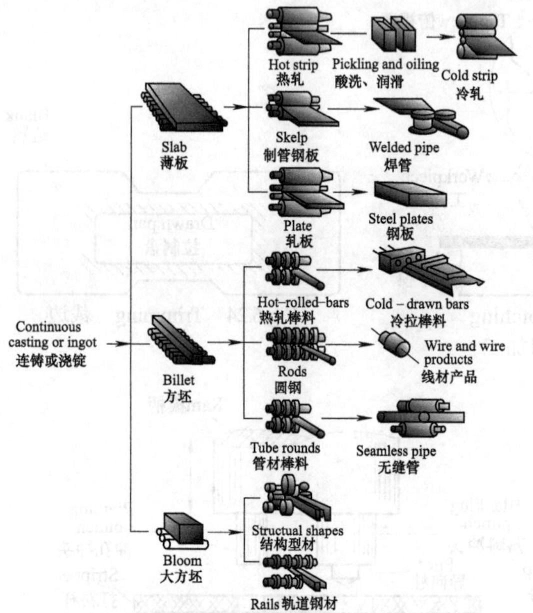

Fig 6.31 Schemalic outline of various flat-rolling and shaped-rolling 板材和型材轧制工艺示意图

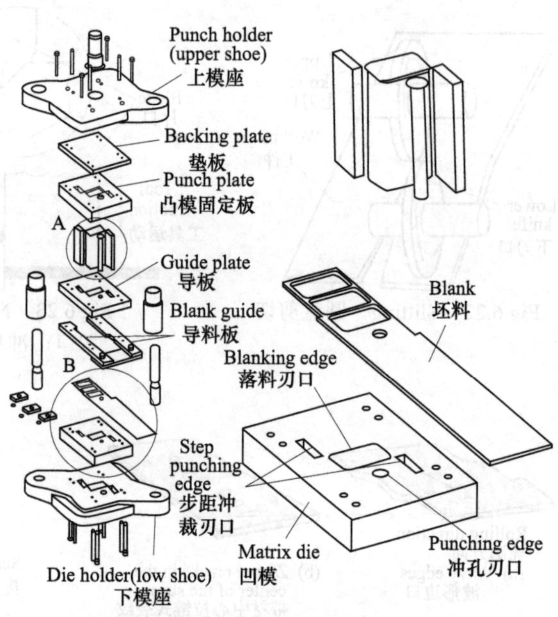

Fig 6.32 Exploded diagram of progressive die 连续冲裁模（又称级进模）分解图

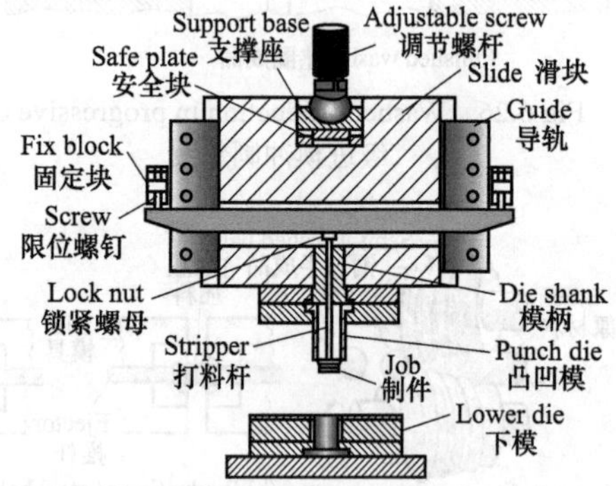

Fig 6.33 Upper die fastening and slide adjustment 上模紧固及滑块位置调节

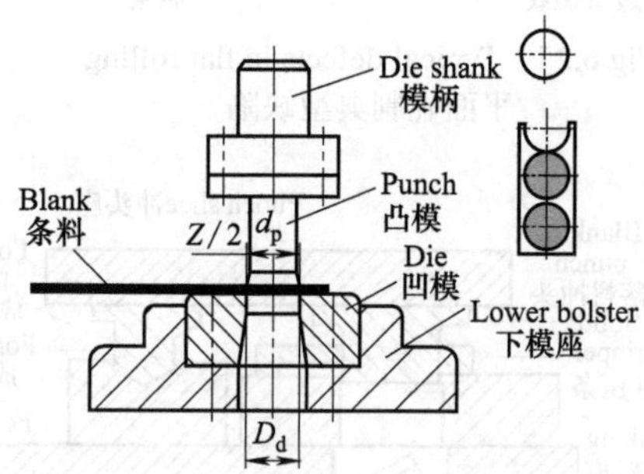

Fig 6.34 Simple die 简单模

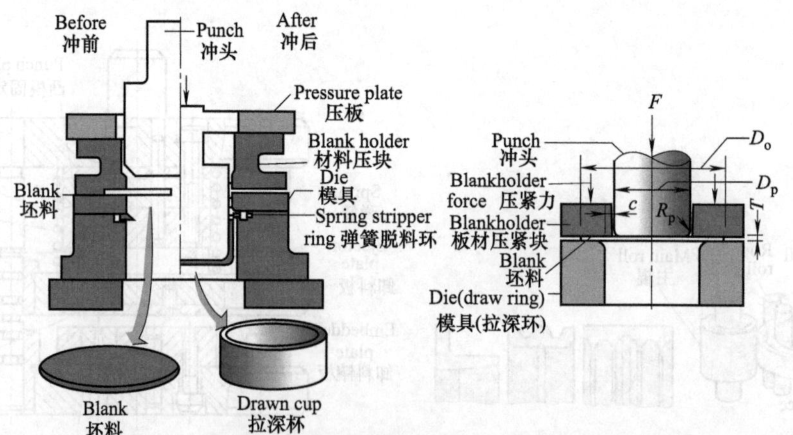

Fig 6.35 Illustration of the deep-drawing process 拉深工艺示意图

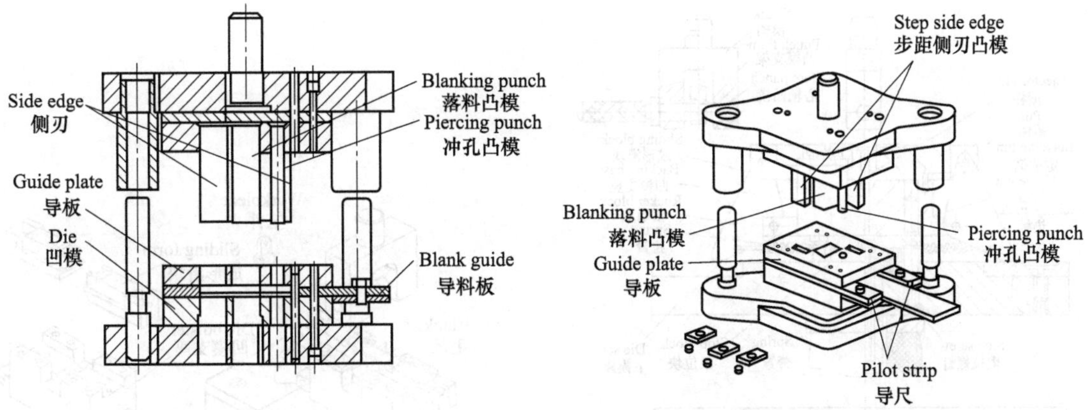

Fig 6.36 Progressive die and its model 连续冲裁模（又称级进模）及模型

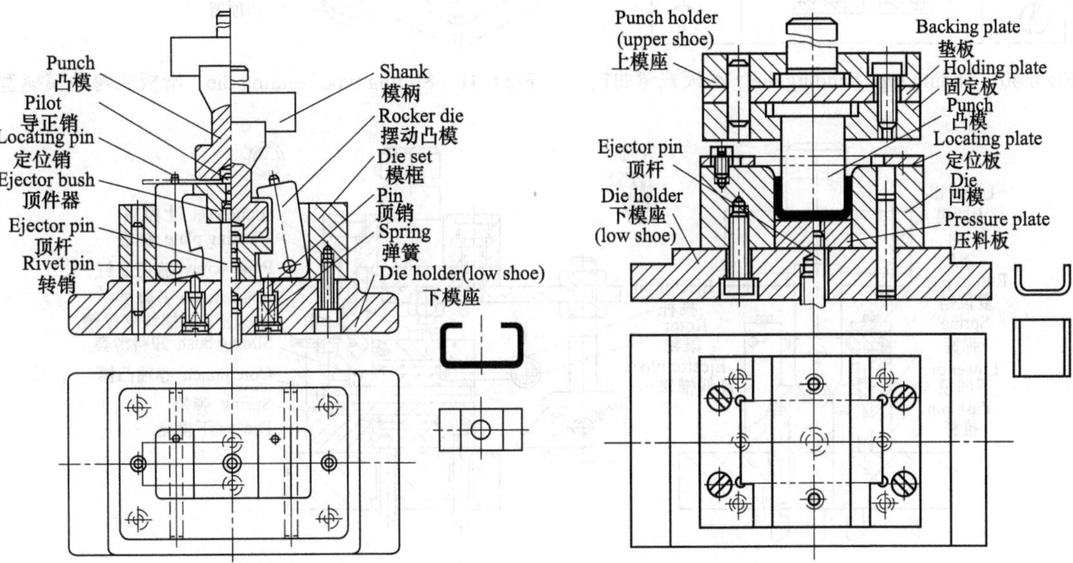

Fig 6.37 Complex bending dies 复杂弯曲模 Fig 6.38 Bending die for U-type job U形件弯曲模

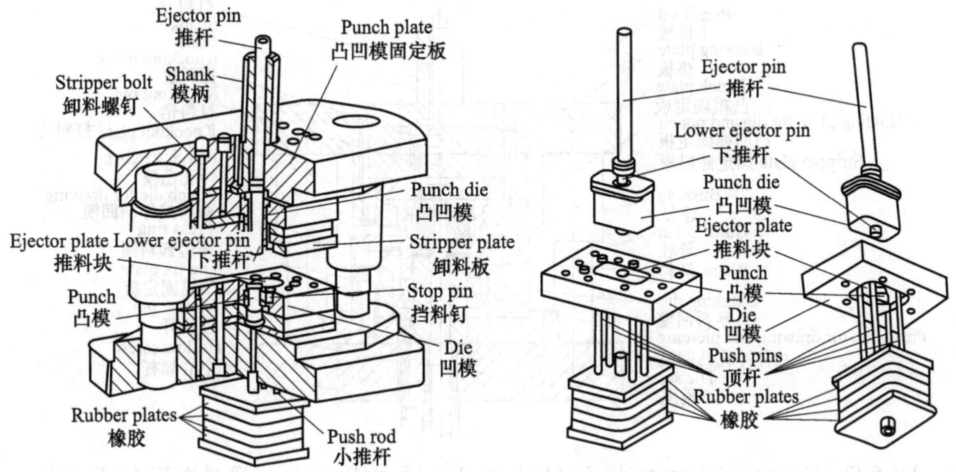

Fig 6.39 Compound die 复合模

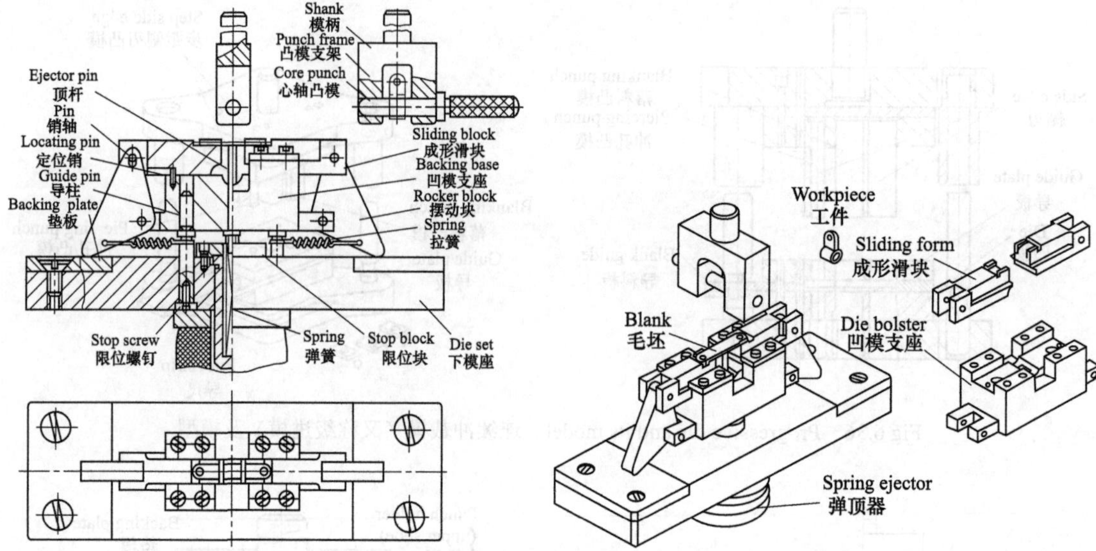

Fig 6.40 Sliding type bending die 滑板式弯曲模

Fig 6.41 Sliding type bending die 滑板式弯曲模模型

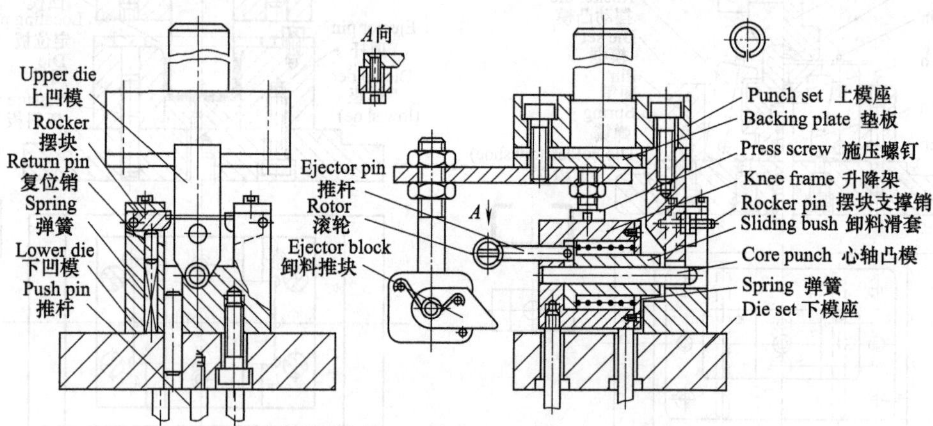

Fig 6.42 Bending die for circular parts 圆形件自动卸料弯曲模

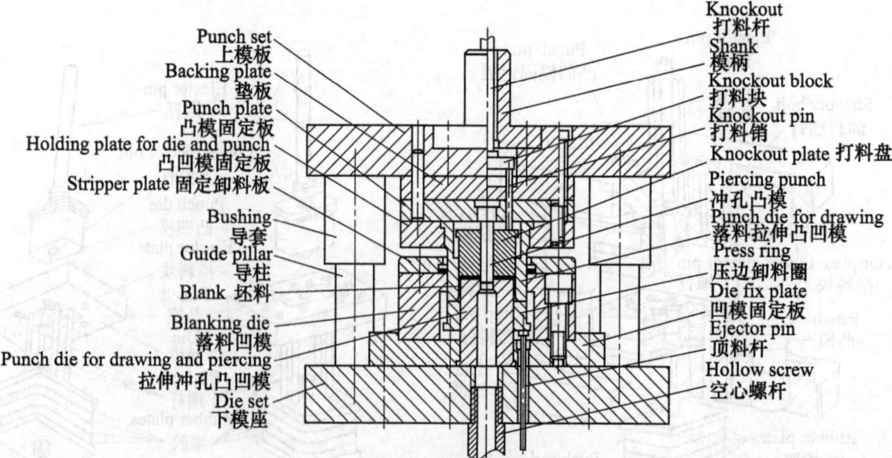

Fig 6.43 Compound drawing die for blanking, drawing and piercing 落料拉深冲孔复合模

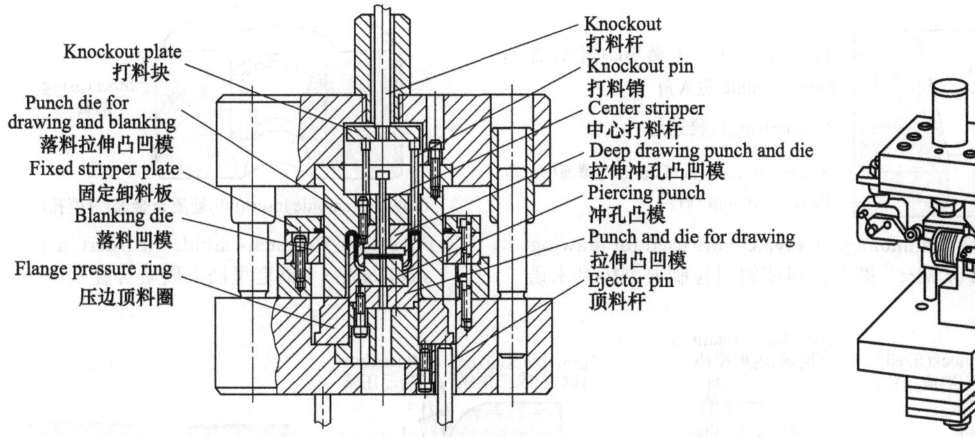

Fig 6.44 Compoungd drawing die for blanking, drawing, reverse drawing, piercing and flanging 落料、正反拉深、冲孔翻边复合模

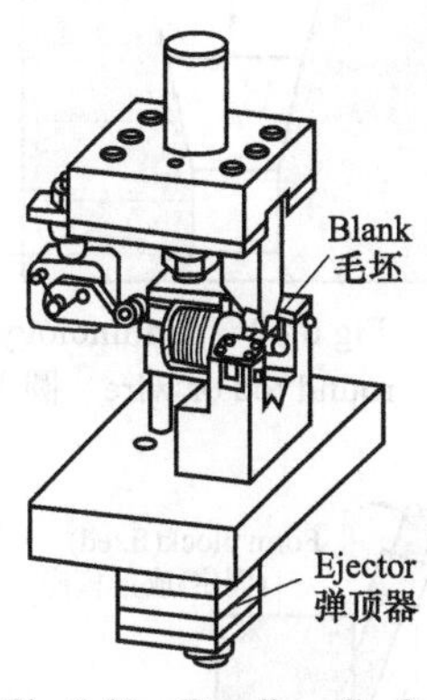

Fig 6.45 Bending die for circular parts 圆形件弯曲模模型

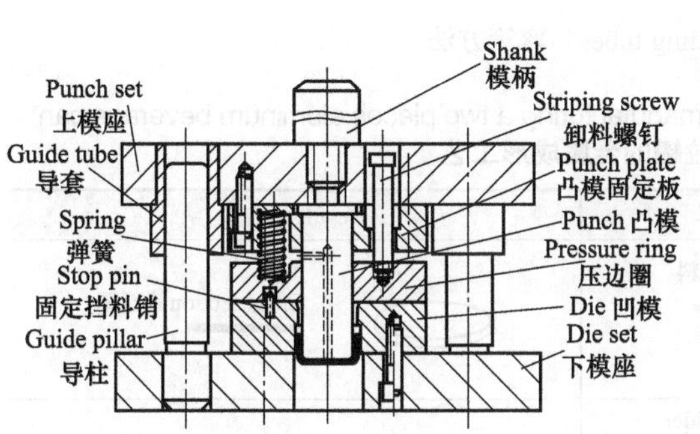

Fig 6.46 Drawing die with blank holding device 有弹性压边装置的拉深模

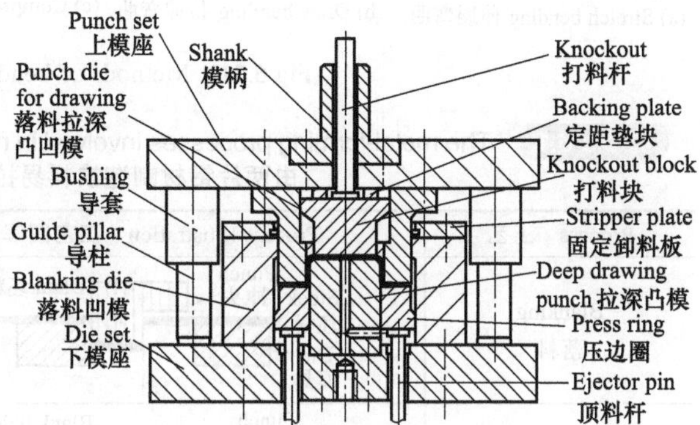

Fig 6.47 Compound drawing die for part with flange 带凸缘制件落料拉深复合模

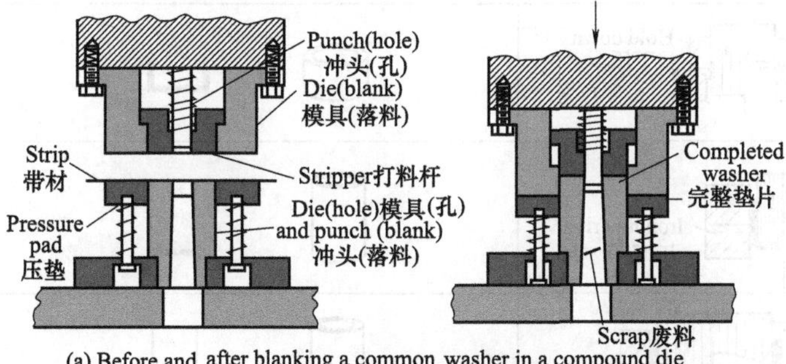

(a) Before and after blanking a common washer in a compound die 复合模具对普通垫片落料前、后图

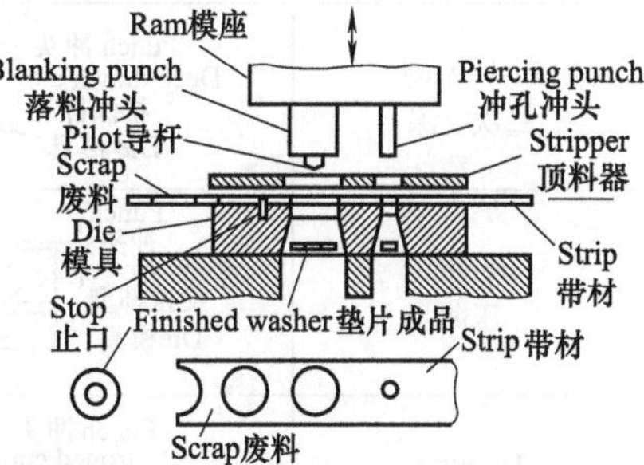

(b) Schematic illustration of making a washer in a progressive die级进模加工垫片图解

(c) Forming of the top piece of an aerosol spray can in a progressive die 级进模上喷雾罐顶片的成形

Fig 6.48 Compound die and progressive die 复合模与级进模

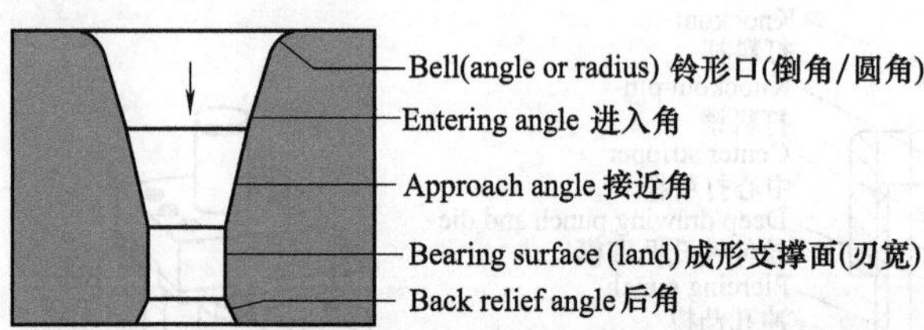

Fig 6.49 Terminology of a typical die used for drawing round rod or wire 圆形棒料或线材拉制典型模具术语

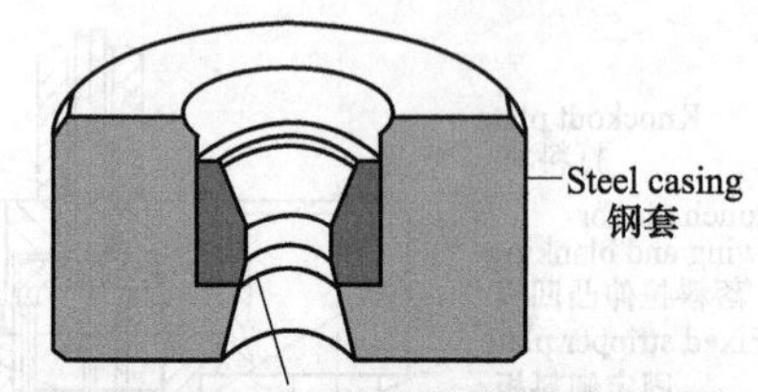

Fig 6.50 Tungsten-carbide die insert in a steel casing 钢套里嵌入硬质合金模具

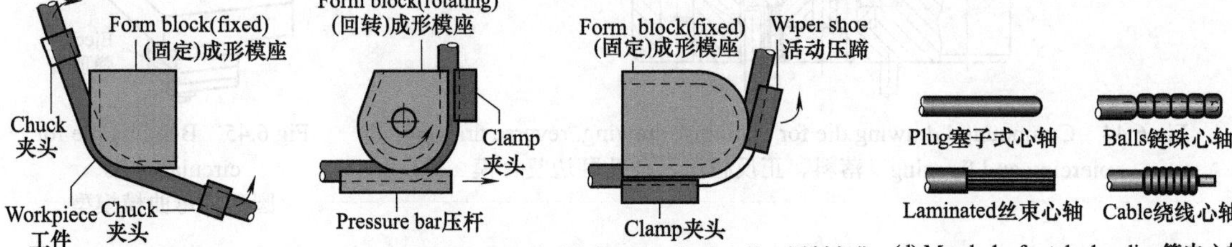

(a) Stretch bending 伸展弯曲 (b) Draw bending 拉伸弯曲 (c) Compression bending 压制弯曲 (d) Mandrels for tube bending管内心轴

Fig 6.51 Methods of bending tubes 弯管方法

Table 6.1 The metal-forming processes involved in manufacturing a two pieces aluminum beverage can 由两片铝材制造饮料易拉罐的金属成形工艺

Process 工艺	Process Illustration 工艺图解	Result 结果
Blanking 落料	Punch 冲头; Stock 坯材; Die 模具; Blank坯料	Cross section截面图
Deep drawing 拉深	Punch 冲头; Blank 坯料; Die 模具; Blank holder 材料压头	
Redrawing 二次拉深	Punch 冲头; Deep drawn cup 拉深杯; Die 模具; Hold down 下压	
Ironing 压薄	Punch 冲头; Redrawn cup 反拉杯型; Die模具; Ironing-ring 压薄环	
Doming 圆顶	Punch 冲头; Ironed cup 压薄杯型; Die模具	
Necking 缩口	Domed cup 圆顶杯; Supports 支件; Spinning tool 旋压工具	
Seaming 接缝	Chuck 夹盘; Lid 盖子; Roller 滚子; Can body 罐体; Before 接缝前; After 接缝后	

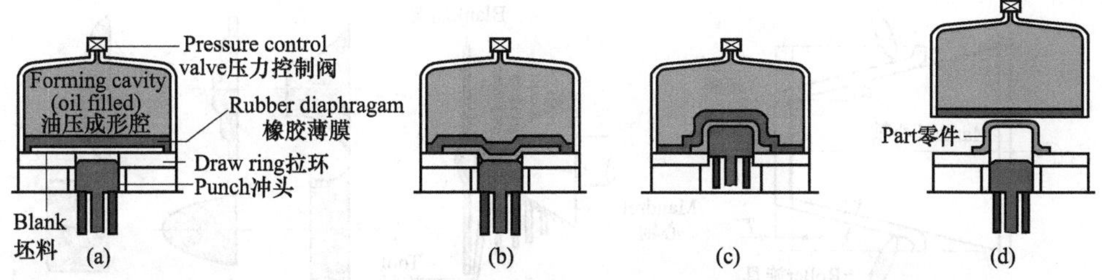

Fig 6.52 The hydroform (or fluid forming) process 液压（流体）成形工艺

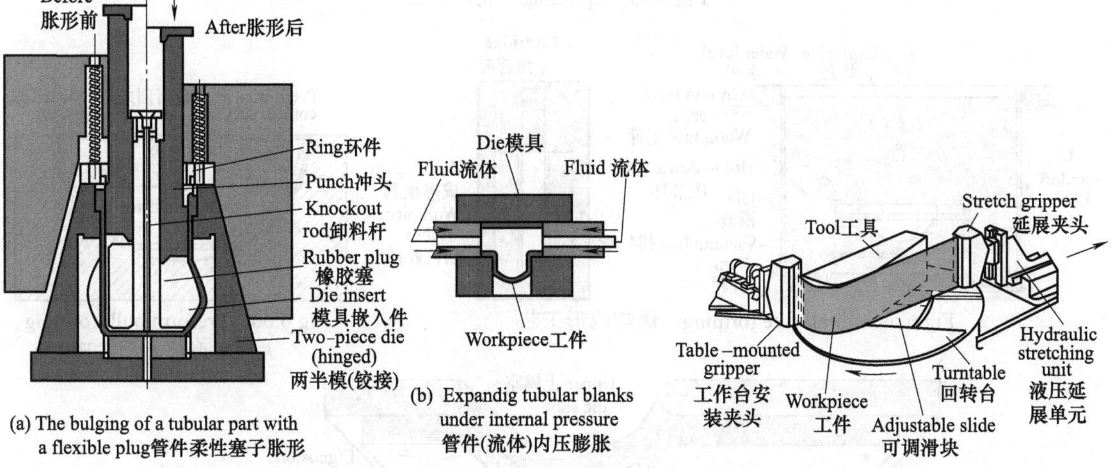

(a) The bulging of a tubular part with a flexible plug管件柔性塞子胀形

(b) Expandig tubular blanks under internal pressure 管件(流体)内压膨胀

Fig 6.53 Bulging 胀芯成形

Fig 6.54 Stretch bending 延展折弯

(a) Comparison of sheared edges produced by conventional (left) and by fine-blanking (right) techniques 传统方法落料(左图)和精密落料(右图)的剪切边口比较

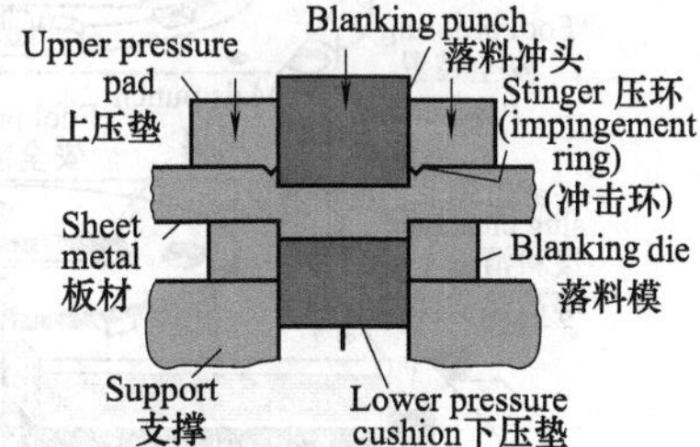

(b) Schematic illustration of one setup for fine blanking 精密落料装置图

Fig 6.55 Fine blanking 精密落料

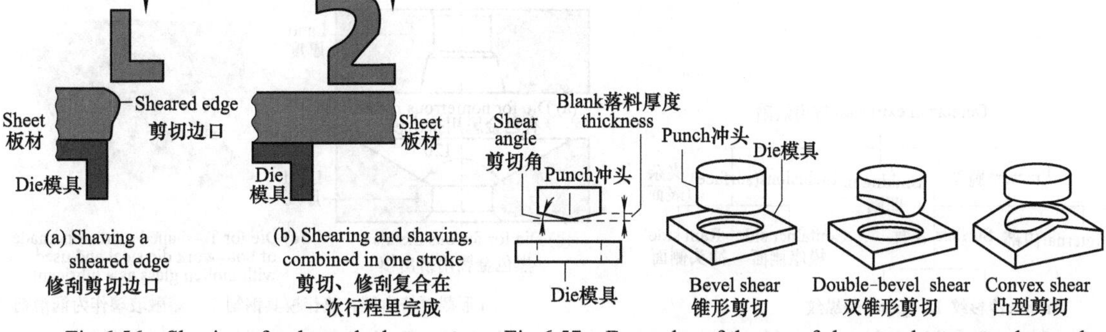

(a) Shaving a sheared edge 修刮剪切边口

(b) Shearing and shaving, combined in one stroke 剪切、修刮复合在一次行程里完成

Fig 6.56 Shaving of a sheared edge 剪切边口的修刮

Fig 6.57 Examples of the use of shear angles on punches and dies 冲头和模具剪切角度的应用例子

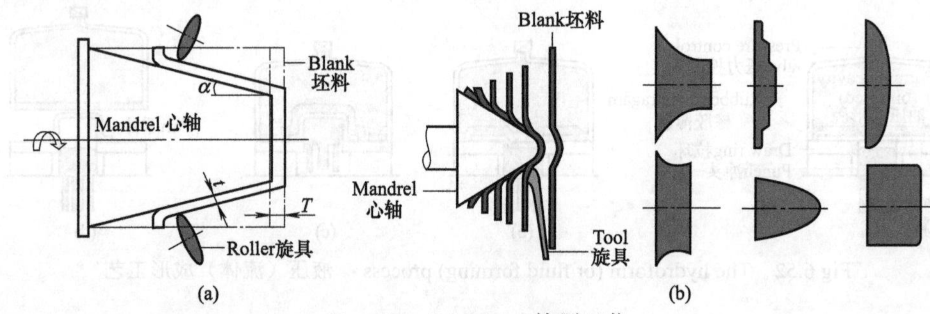

Fig 6.58 Spinning 旋压工艺

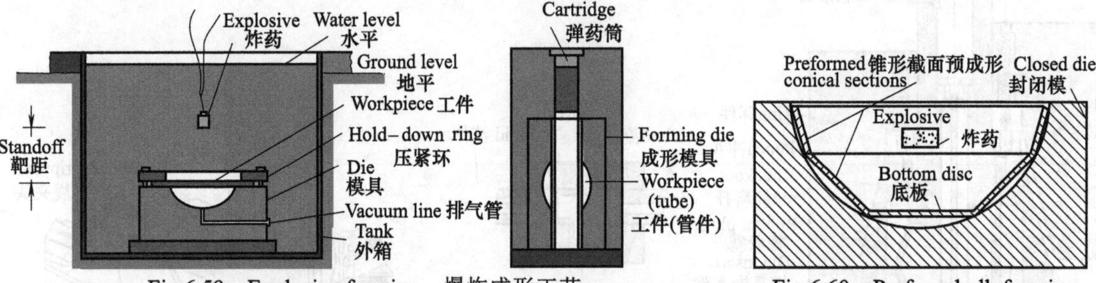

Fig 6.59 Explosive forming 爆炸成形工艺

Fig 6.60 Preform bulk forming 大件成形前的预成形

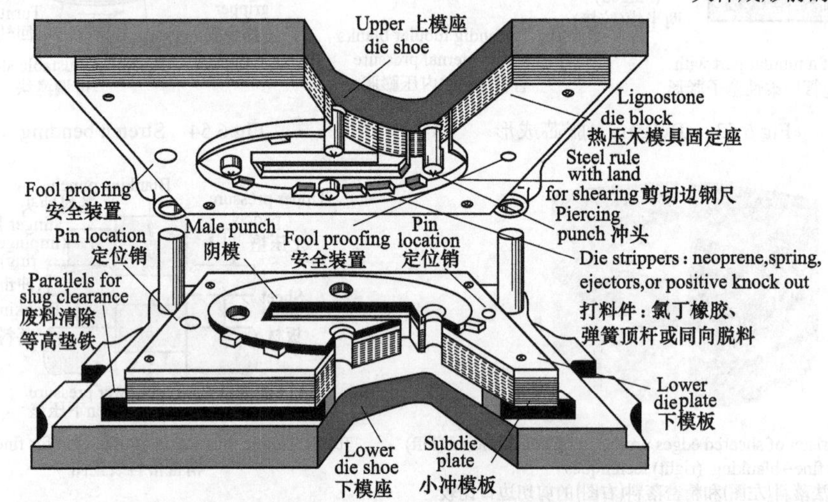

Fig 6.61 Steel rule die for cutting a circular shape, sectioned to show the construction
钢尺模用于切制圆形，剖面展示其结果

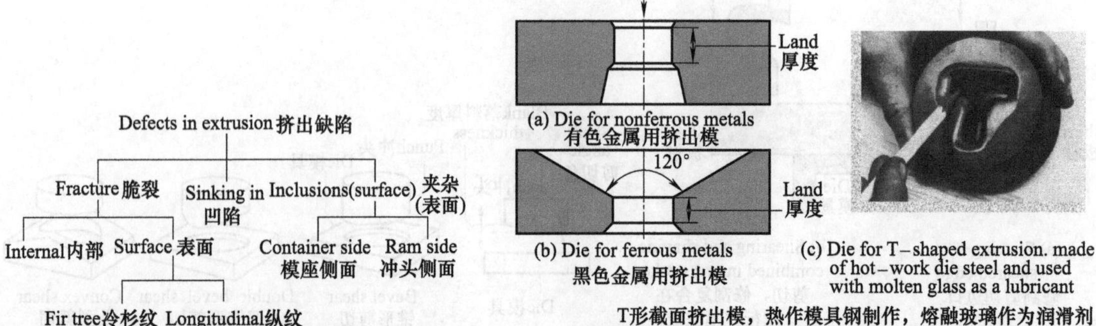

Fig 6.62 Possible defects caused by extrusion
挤出可能产生的缺陷

Fig 6.63 Typical extrusion-die configurations
典型挤出模结构

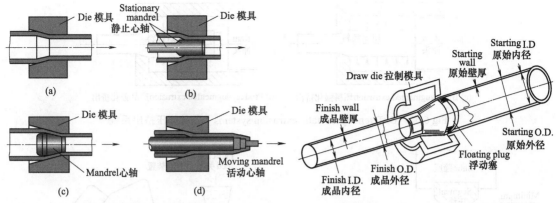

Fig 6.64 Examples of tubedrawing operations,with and without an internal mandrel
内部有心轴和无心轴的管材拉制示例

Fig 6.65 Tube Drawing 管件拉制

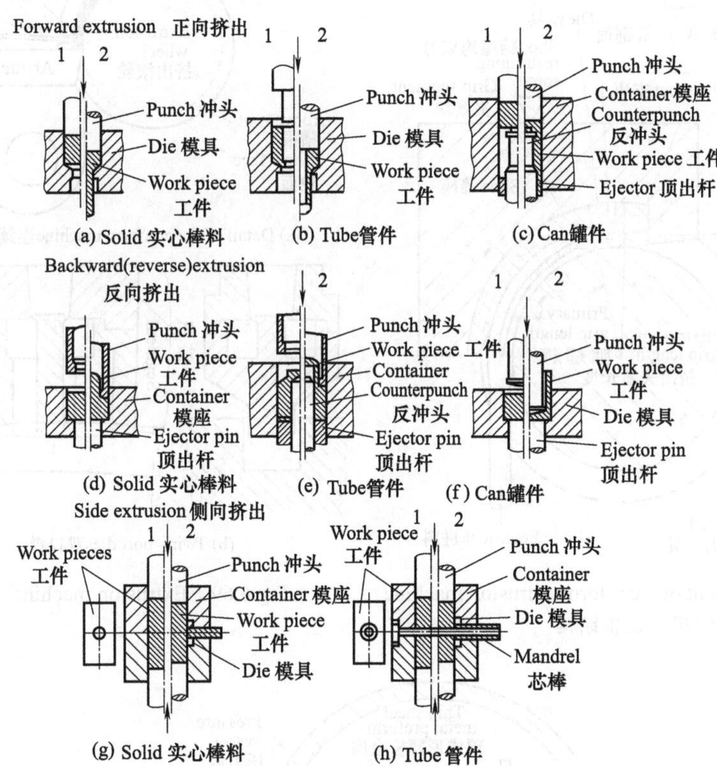

Fig 6.66 Extrusion 挤出

1—Pre-extrusion condition 预挤出条件；2—Post-extrusion condition 终挤出条件

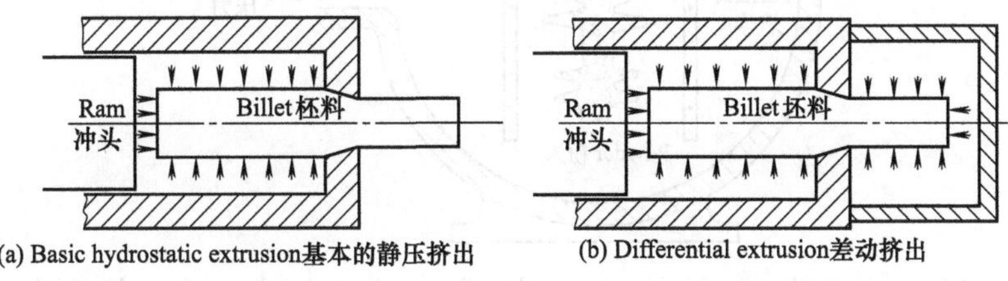

Fig 6.67

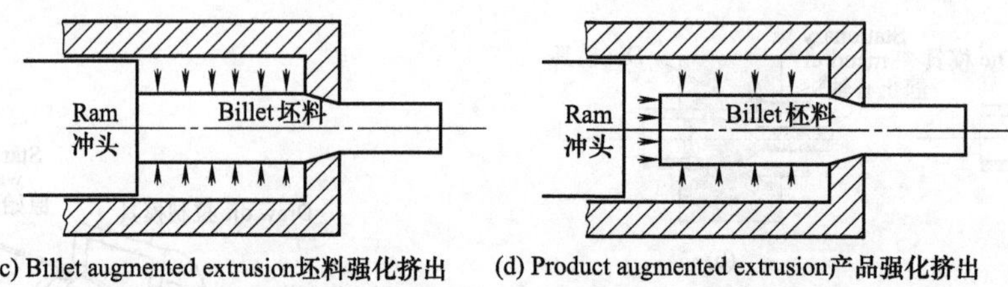

Fig 6.67 Different hydrostatic extrusion systems 各种静压挤出系统

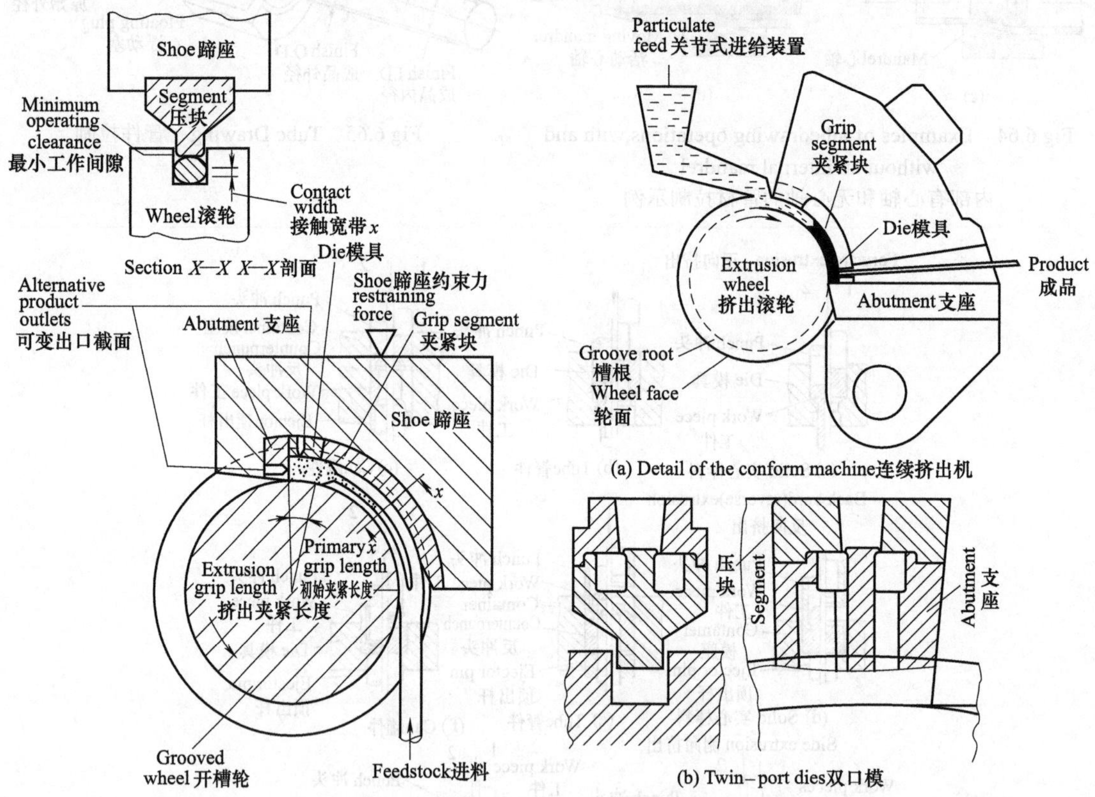

Fig 6.68 Arrangement of a conform extrusion machine 连续挤出设备结构

Fig 6.69 Extrusion machine 挤出设备

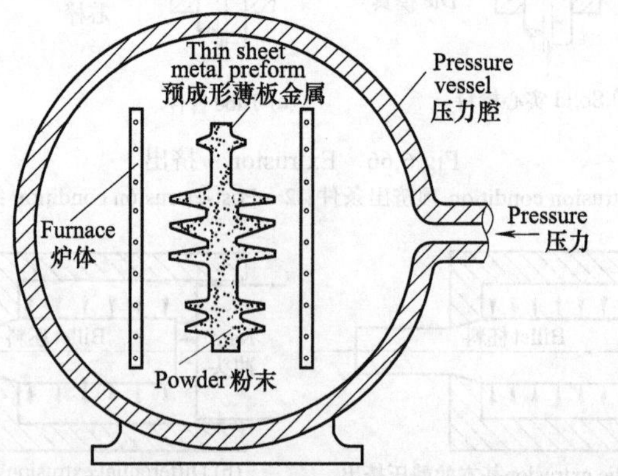

Fig 6.70 Hot-isostatic pressing (the powder, in a thin steel preform, is heated and compressed by high-pressure argon) 热等静压（预成形薄板金属被加热并承受高压氩气）

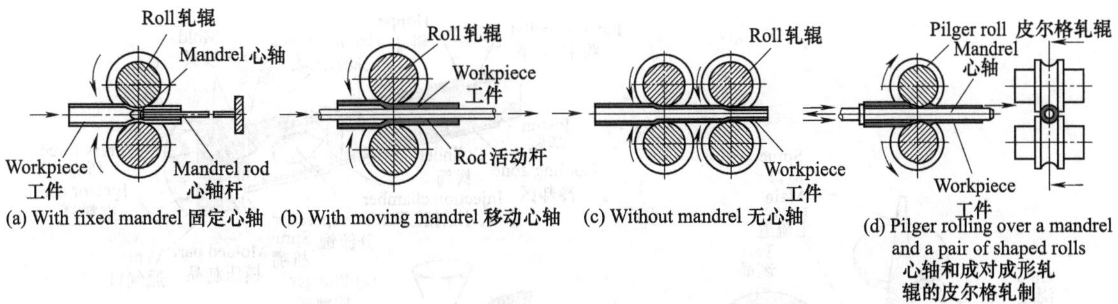

Fig 6.71 Various tube-rolling processes 几种管材轧制工艺

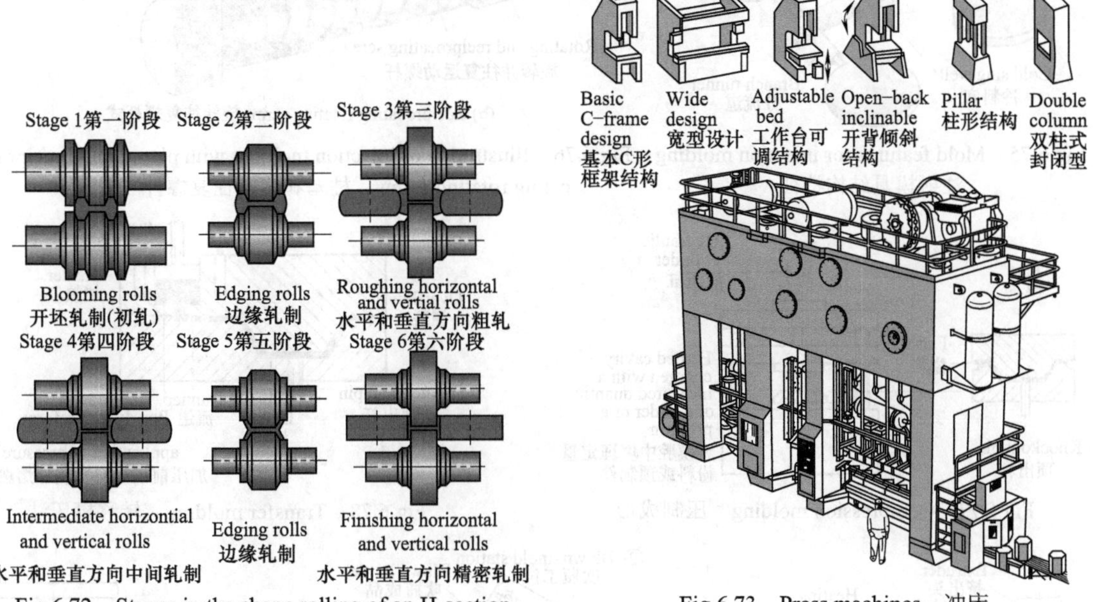

Fig 6.72 Stages in the shape rolling of an H-section part H 形截面型材形状轧制的各个阶段

Fig 6.73 Press machines 冲床

6.2 Die and mo(u)ld for plastics 塑料模具

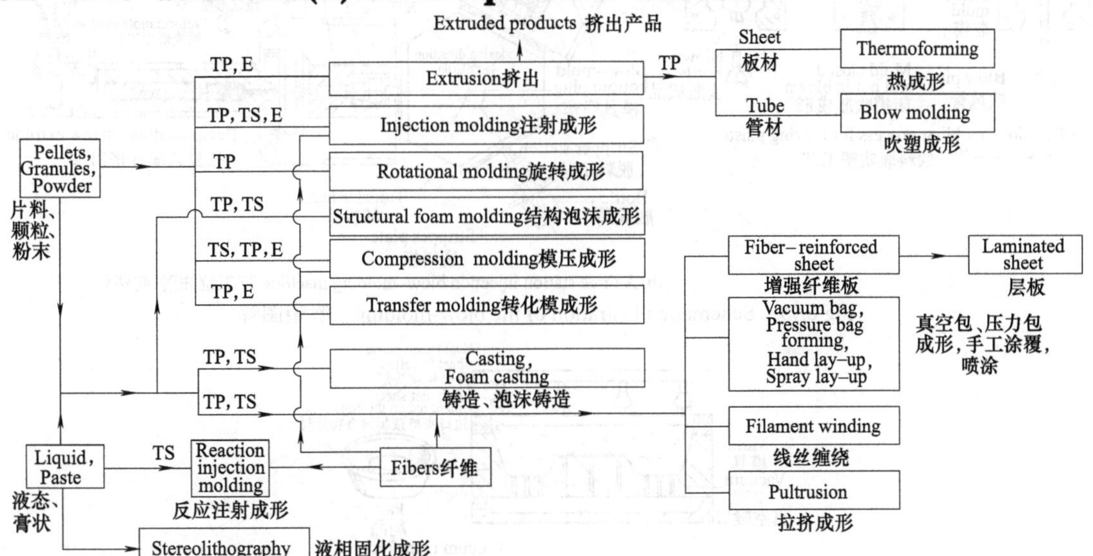

Fig 6.74 Outline of forming and shaping processes for plastics, elastomers, and composite materials 塑性、弹塑性、复合材料的成形工艺概览

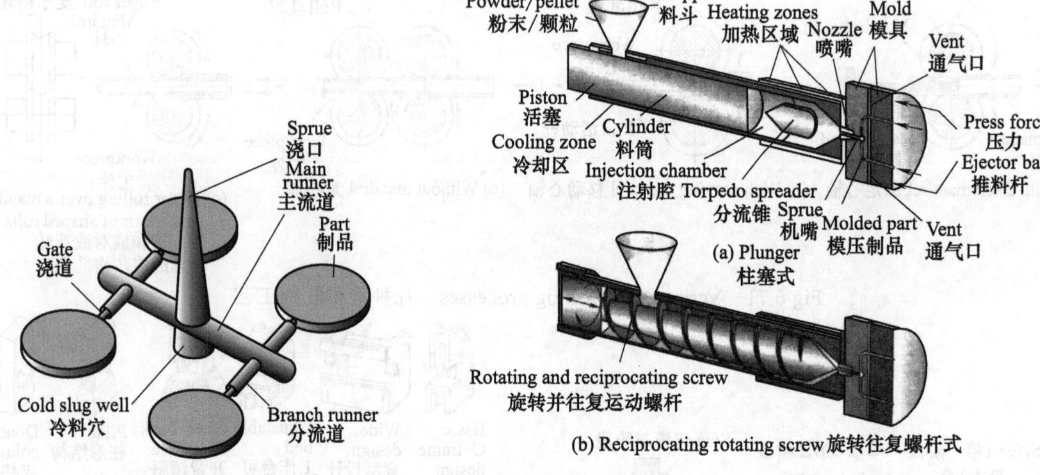

Fig 6.75 Mold features for injection molding 注射模具结构特征

Fig 6.76 Illustration of injection molding with plunger and reciprocating rotating screw 柱塞和旋转往复螺杆注射模具

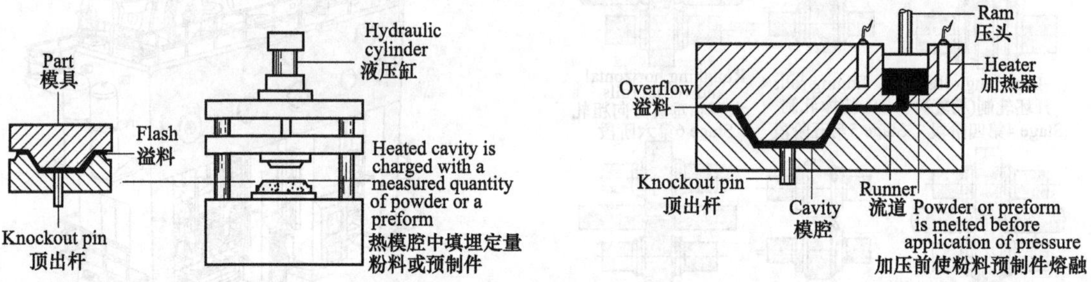

Fig 6.77 Compression molding 压制成形

Fig 6.78 Transfer molding 转移模压法

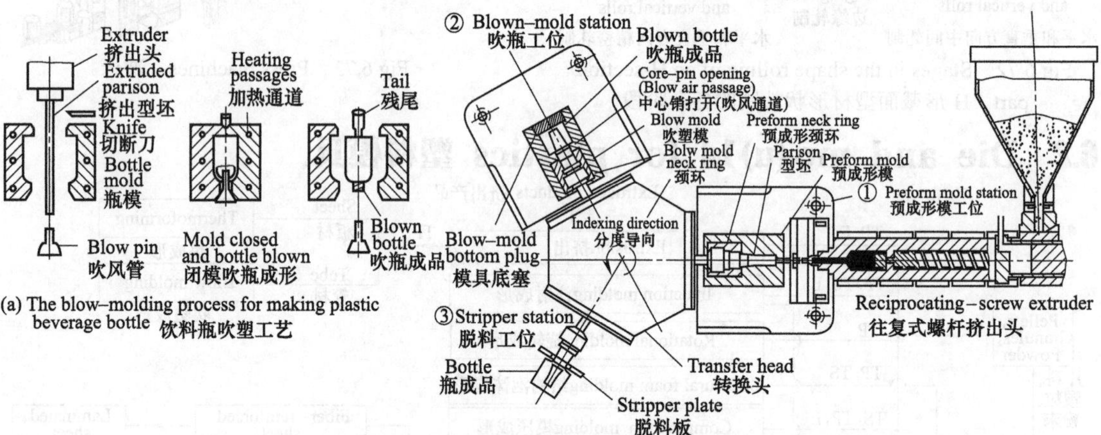

Fig 6.79 Schematic illustration of the blow-molding 吹塑图解

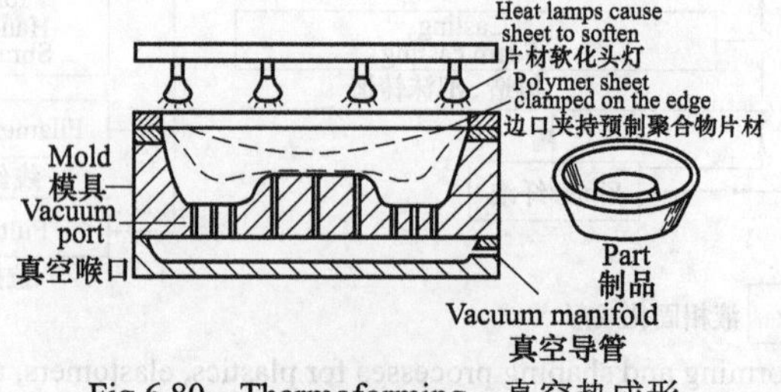

Fig 6.80 Thermoforming 真空热成形

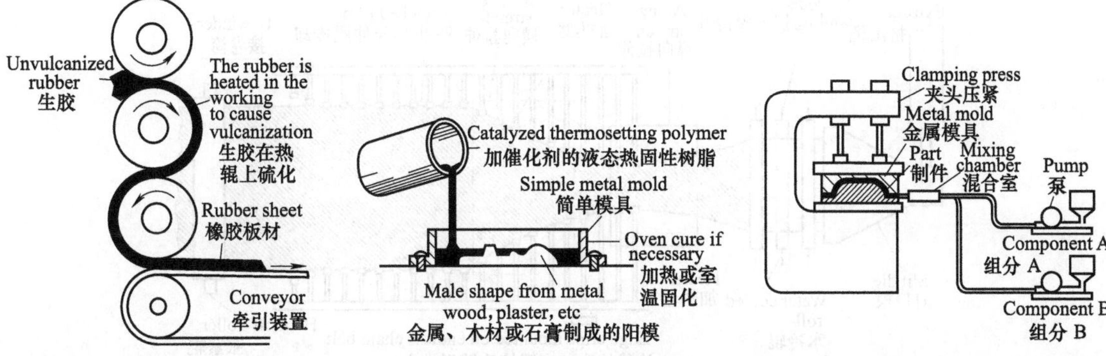

Fig 6.81 Calendering 压延

Fig 6.82 Casting 浇铸成形

Fig 6.83 Reaction injection molding 反应注射模具

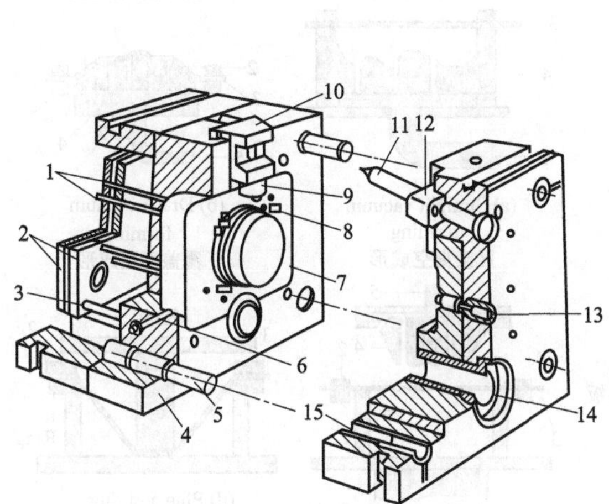

Fig 6.84 Typical die assembly 典型模具装配

1—Ejector pin 推杆；2—Ejector plate 推板；3—Ejector return pin 复位杆；4—Base support 支承底板；5—Guide pillar 导柱；6—Die insert waterway 镶块冷却水道；7—Die insert 镶块；8—Fixed core 定模型芯；9—Moving core 活动型芯；10—Moving core holder 活动型芯固定板；11—Angle pin 斜导柱；12—Core locking wedge 型芯楔压块；13—Cascade waterway 串联水道；14—Plunger bush 衬套；15—Guide bush 导套

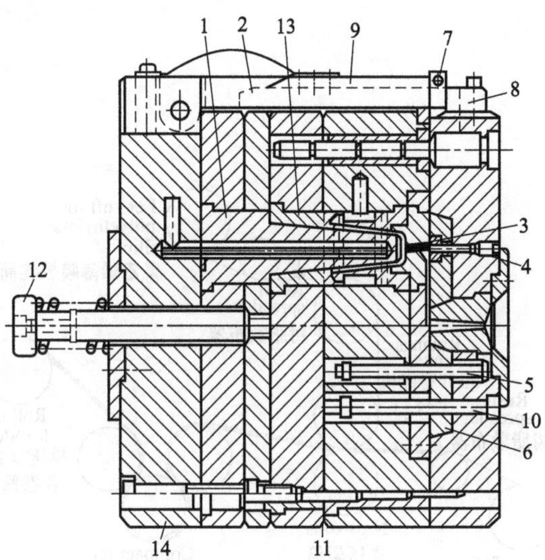

Fig 6.85 Typical degating tool 典型注塑模具

1—Sprue separating area 浇道分型面；2—Latch 插销；3—Sprue 直浇道；4—Retainer pin 固定销；5，10—Bolt head 插销头；6—Sprue-stripping plate 注料口分模板；7—Pin 销；8—Guide strip 导轨；9—Waterway 冷却水道；11—Main parting line 主要分型线；12—Ejector rod 预料杆；13—Stripper sleeve 分型套；14—Plate assembly 模板装置

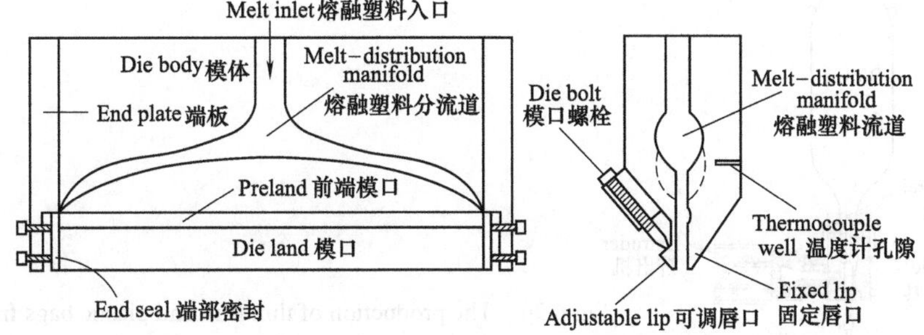

Fig 6.86 Die geometry for extruding sheet 板材挤出模具几何形状

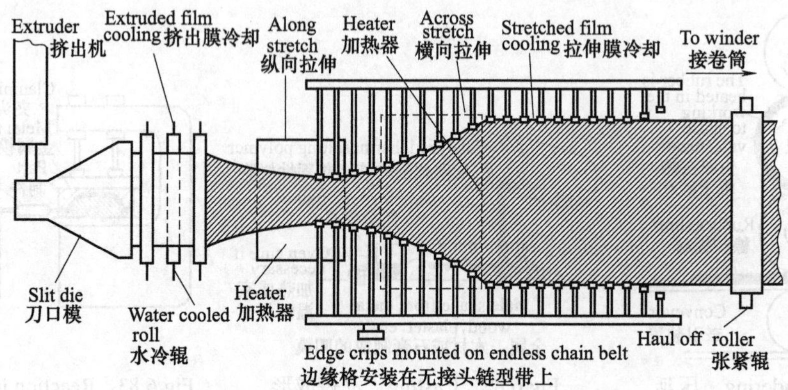

Fig 6.87 Process for manufacture of biaxially stretched polystyrene film 双向拉伸聚苯乙烯薄膜的制造工艺

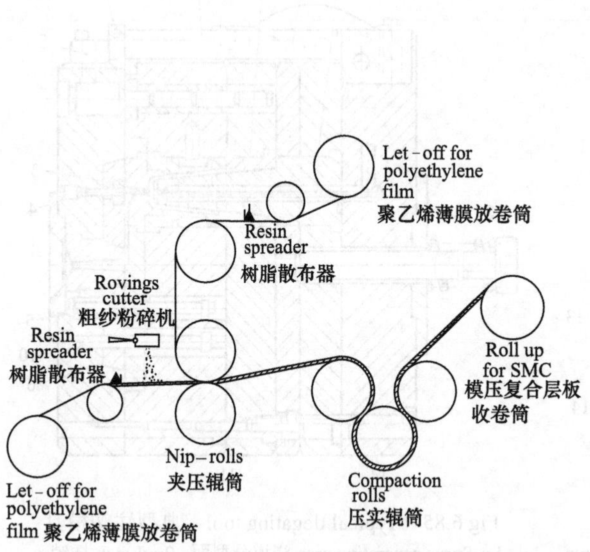

Fig 6.88 Outline of machine for preparing sheet moulding compounds

制备模压成形复合板材的机器布局

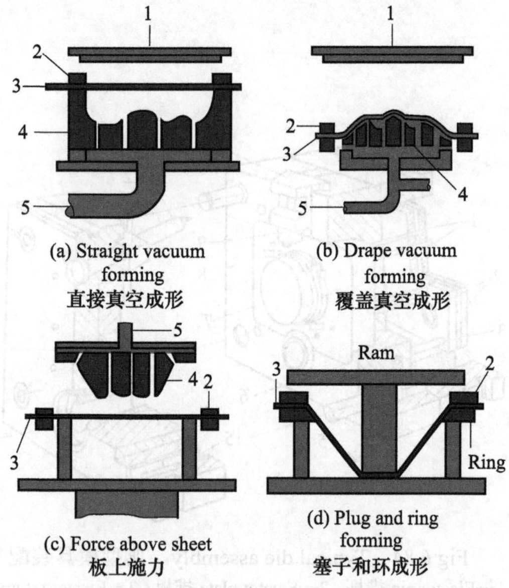

Fig 6.89 Various thermoforming processes for thermoplastic sheet

热塑性板材的几种热塑性成形方法

1—Heater 加热器；2—Clamp 夹头；3—Plastic sheet 塑料板；4—Mold 模具；5—Vacuum line 真空管

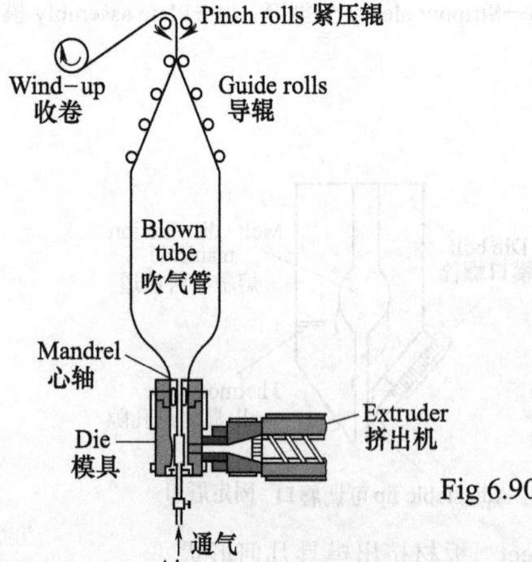

Fig 6.90 The production of thin film and plastic bags from tube first produced by an extruder and then blown by air

先挤出塑管，进而吹气生产薄膜或塑料袋

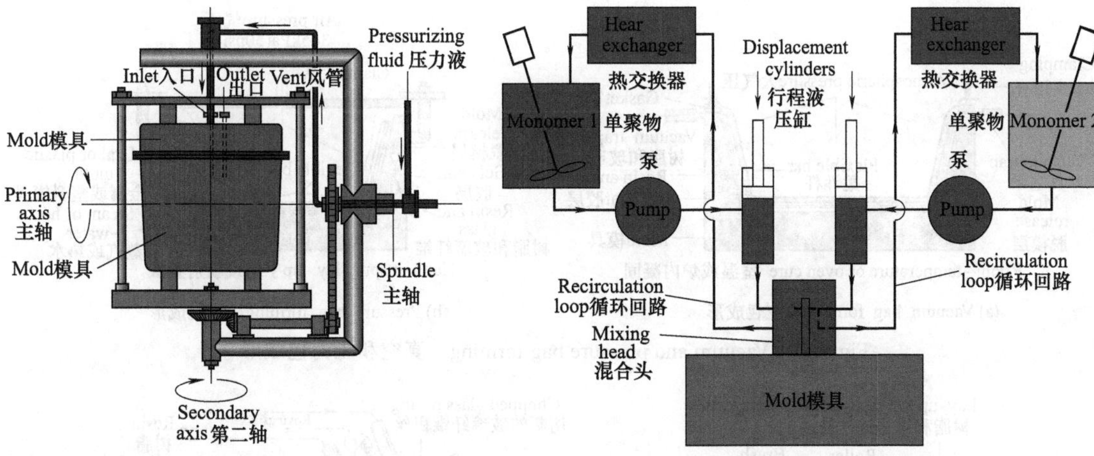

Fig 6.91 The rotational molding (rotomolding or rotocasting) process 回转模具形成工艺

Fig 6.92 Reaction-injection molding process 反应注射模具成形工艺

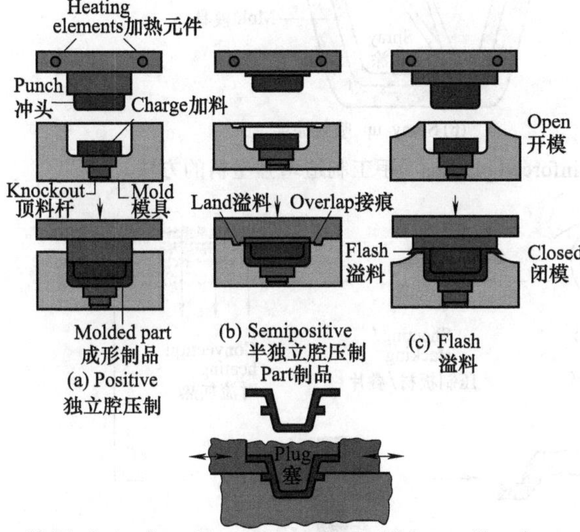

Fig 6.93 Types of compression molding, a process similar to forging 模具压制成形的类别，方法类似于锻压

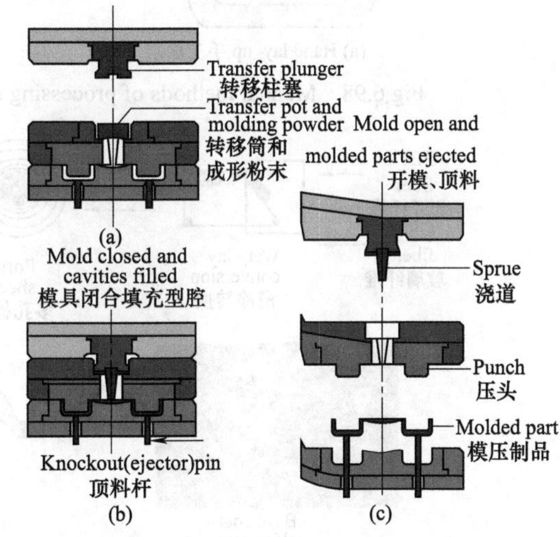

Fig 6.94 Sequence of operations in transfer molding for thermosetting plastics. This process is particularly suitable for intricate parts with varying wall thickness 热固性塑料的转移模具成形，特别适合于壁厚不均的复杂零件

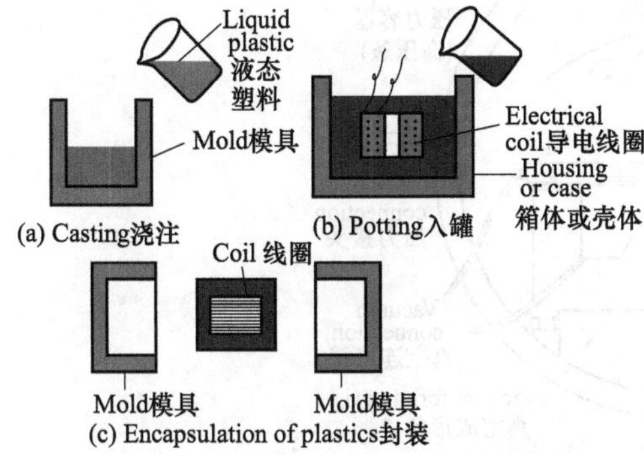

Fig 6.95 Encapsulation process 封装过程

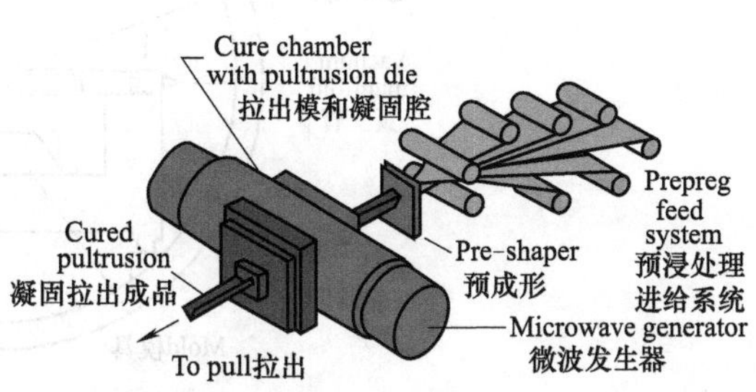

Fig 6.96 Pultrusion process 拉出工艺

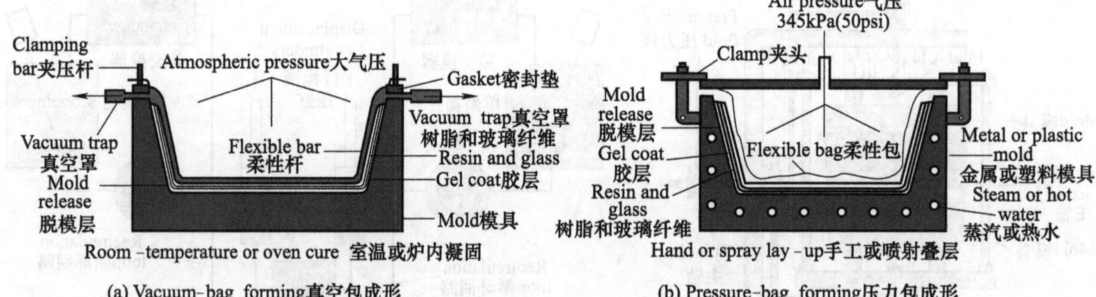

Fig 6.97 Vacuum and pressure bag forming 真空和压力包成形

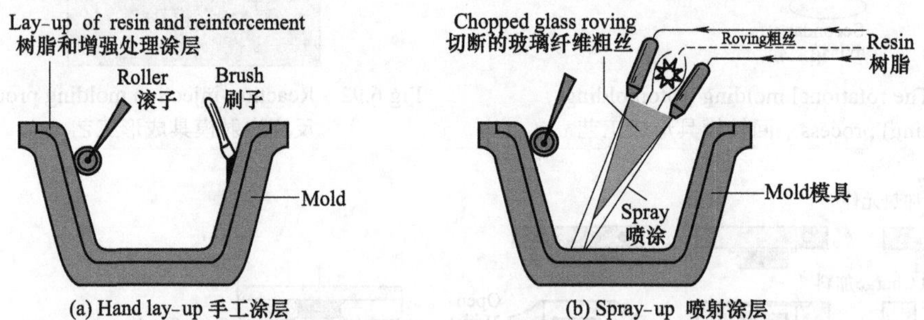

Fig 6.98 Manual methods of processing reinforced plastics 手工制造增强塑料的方法

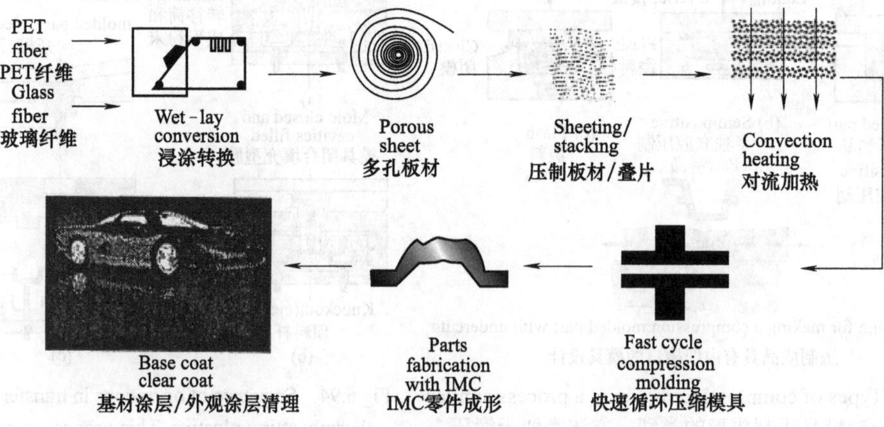

Fig 6.99 Compression molding 压缩模具成形

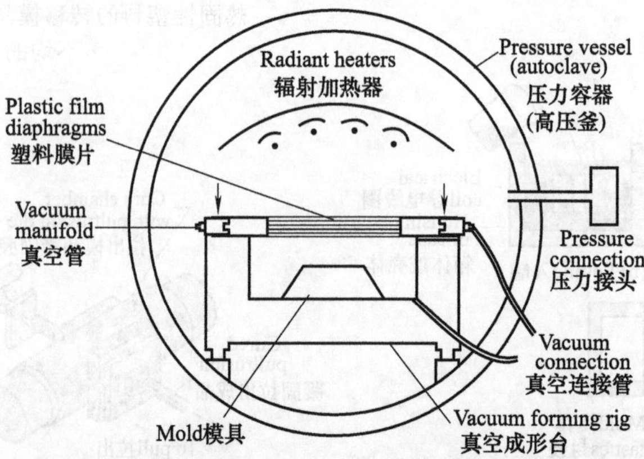

Fig 6.100 Vacuum forming in autoclave 高压釜真空成形

6.3 Mo(u)ld for die cast 压铸模具

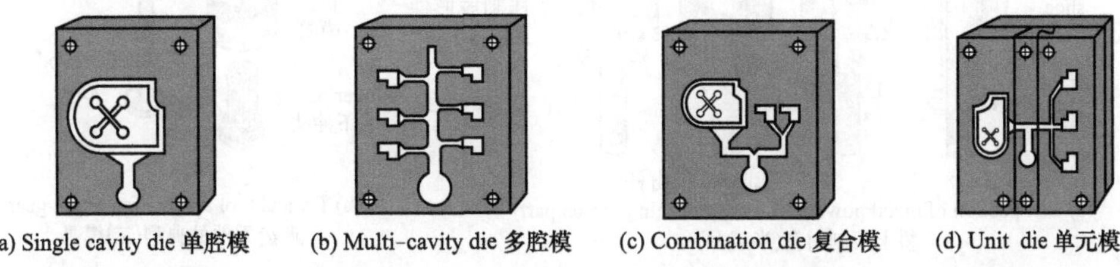

Fig 6.101 Various types of cavities in a die-casting mould 压铸模具的不同型腔结构

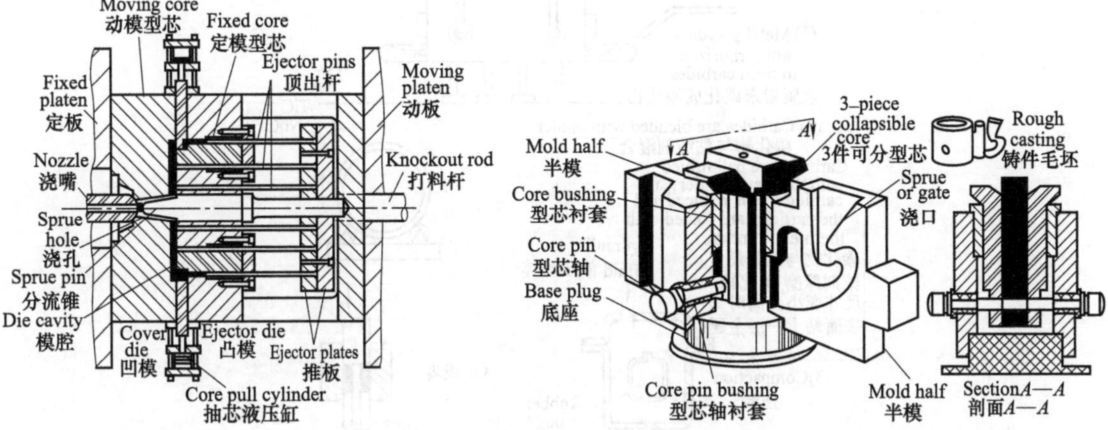

Fig 6.102 Sectional drawing of a typical die cast 典型压铸模具剖面图

Fig 6.103 Typical permanent mold arrangement for nonferrous casting 典型黑色金属永久型铸造

6.4 Other mo(u)lds 其他模具

6.4.1 Powder metallurgy mo(u)ld 粉末冶金模具

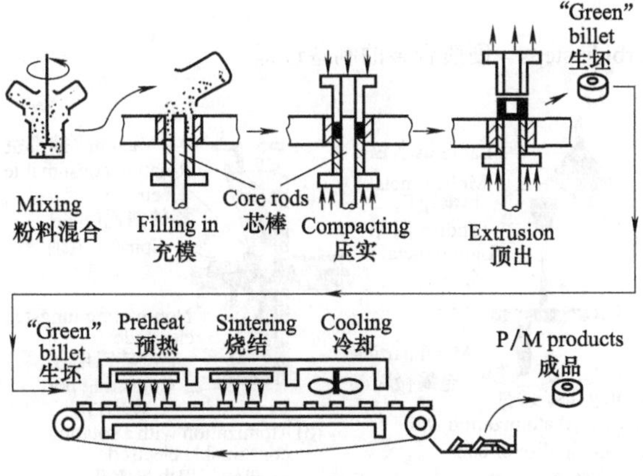

Fig 6.104 Flow diagram of P/M process 粉末冶金的基本工序

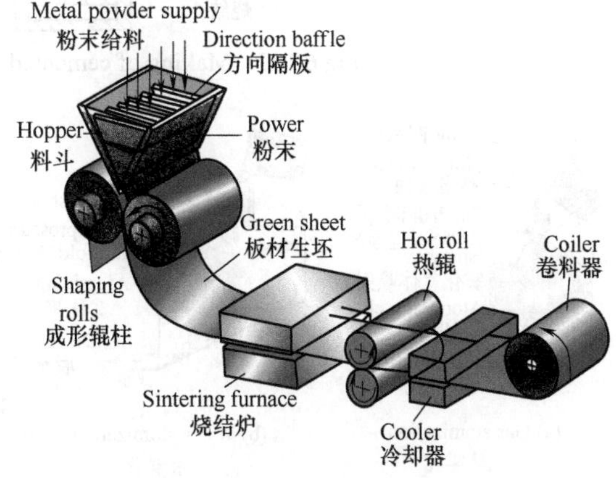

Fig 6.105 Illustration of powder rolling 粉末滚压成形

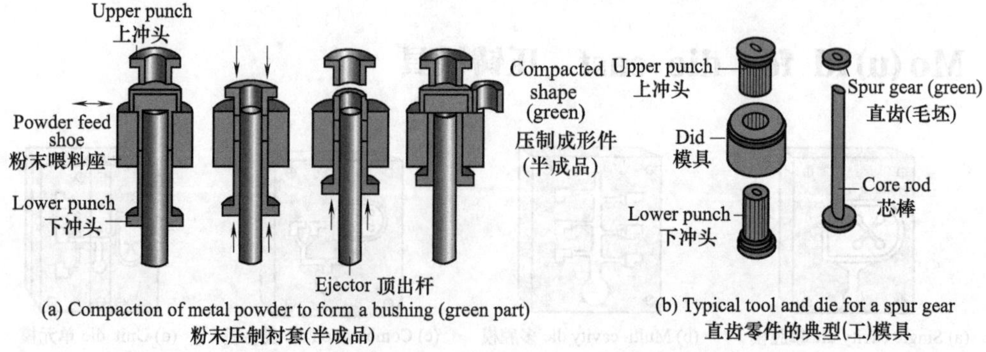

(a) Compaction of metal powder to form a bushing (green part)
粉末压制衬套(半成品)

(b) Typical tool and die for a spur gear
直齿零件的典型(工)模具

Fig 6.106 Powder metallurgy process 粉末冶金成形工艺

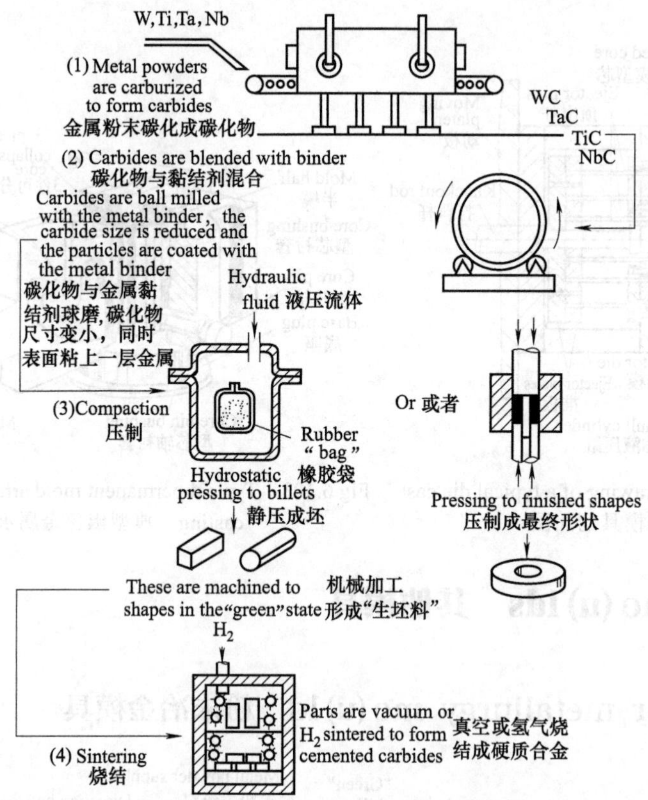

Fig 6.107 Making of cemented carbide steps 硬质合金的制备过程

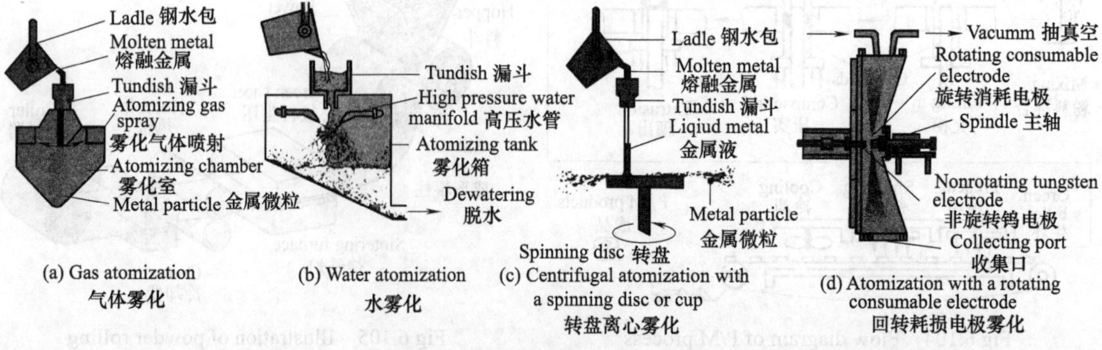

(a) Gas atomization
气体雾化

(b) Water atomization
水雾化

(c) Centrifugal atomization with a spinning disc or cup
转盘离心雾化

(d) Atomization with a rotating consumable electrode
回转耗损电极雾化

Fig 6.108 Methods of meatal powder production by atomization 雾化制粉技术方法

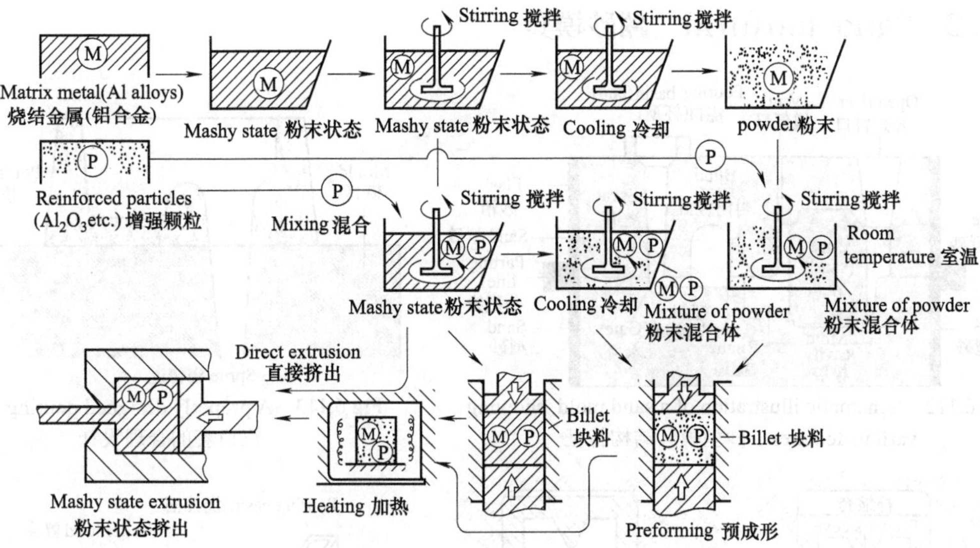

Fig 6.109 Production of particle-reinforced metals in the mashy state 粉末颗粒增强金属的制备

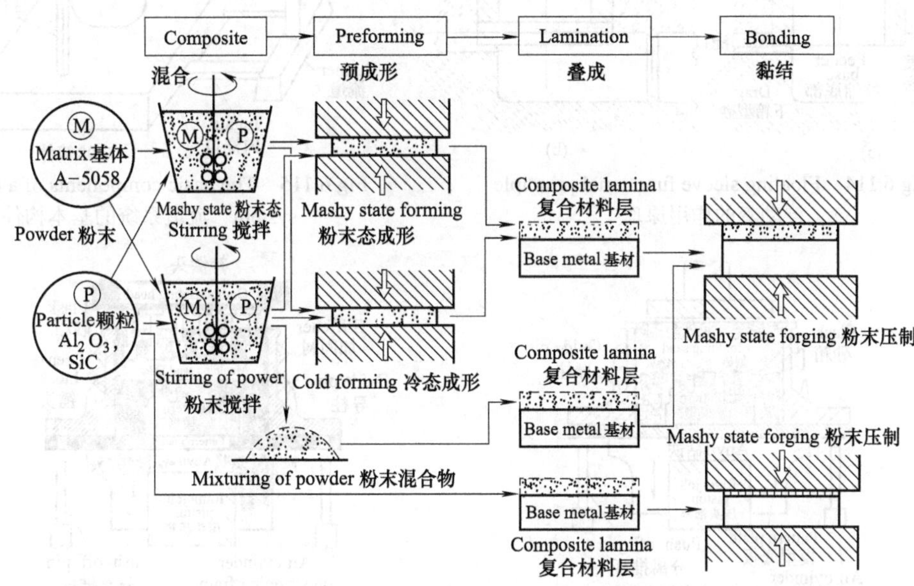

Fig 6.110 The manufacturing techniques for particle-reinforced cladding metals 粉末增强包覆金属制备技术

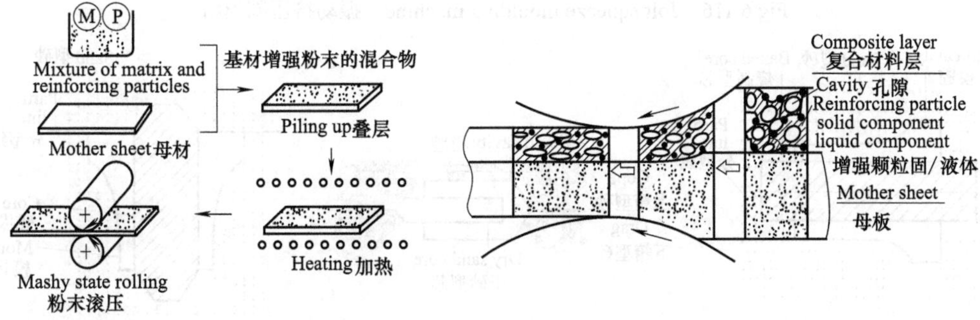

Fig 6.111 The mashy state manufacture of composite sheet 复合板材的粉末压制

6.4.2 Sand mo(u)ld 翻砂模具

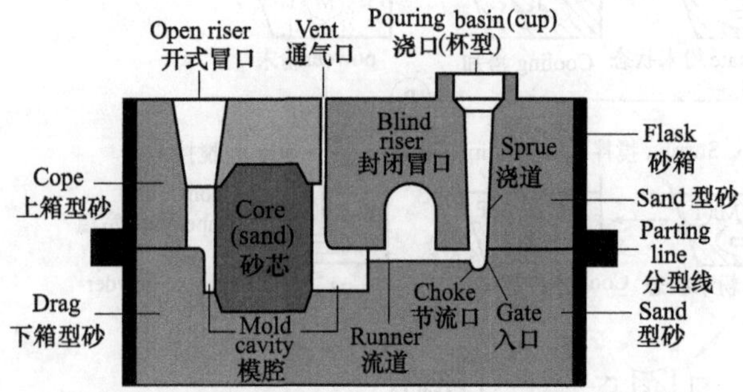

Fig 6.112 Schematic illustration of a sand mold, showing various features 砂型模具结构特征

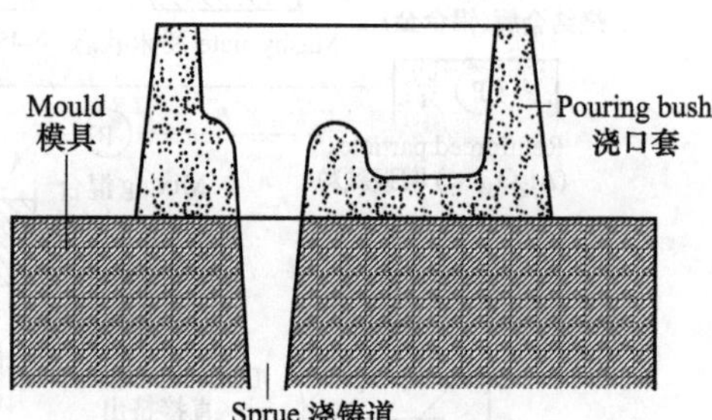

Fig 6.113 A properly designed pouring bush 浇口套的合理设计

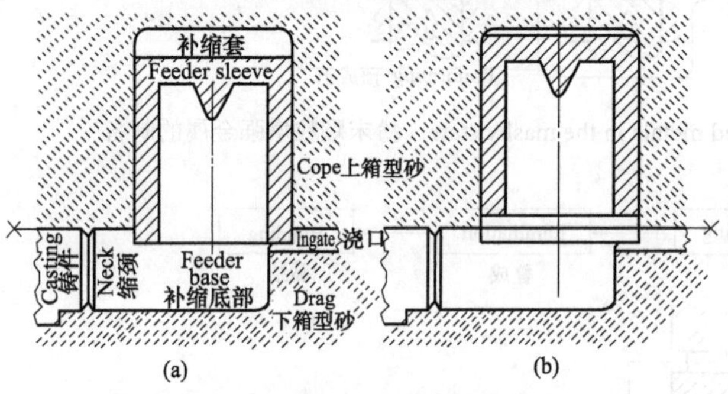

Fig 6.114 Floating sleeve functional principle 浮动套的作用原理

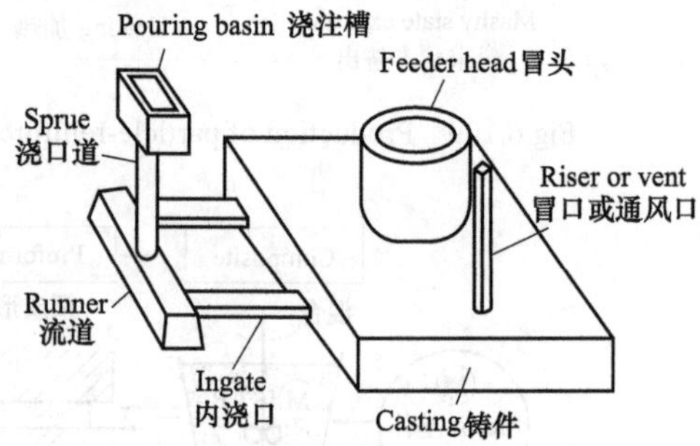

Fig 6.115 The basic components of a running system 流道系统的基本构件

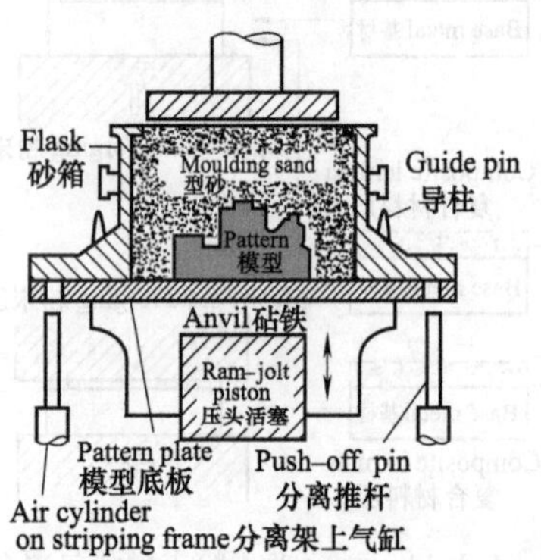

(a) Jolt squeeze moulding machine with solid squeeze heads 采用实体挤压头的振动挤压模机

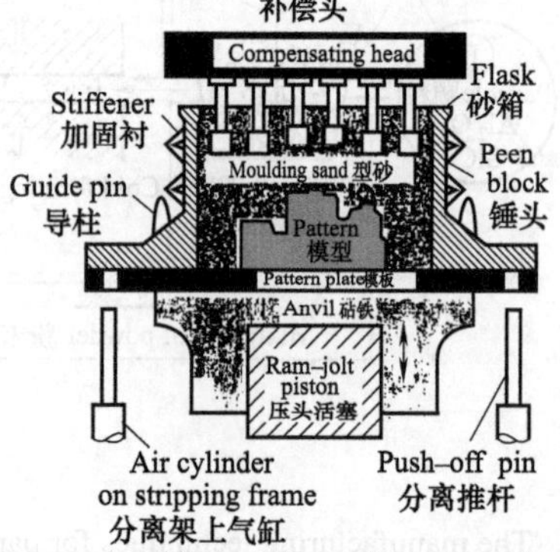

(b) Jolt squeeze moulding machine with compensating heads 采用补偿挤压头的振动挤压制模机

Fig 6.116 Jolt squeeze moulding machine 振动挤压制模机

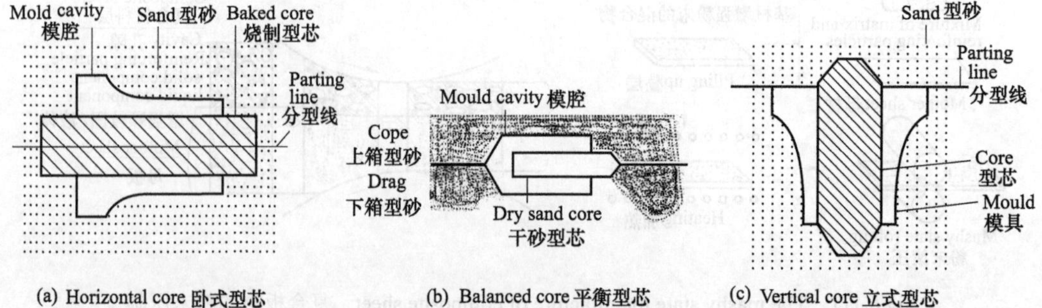

(a) Horizontal core 卧式型芯 (b) Balanced core 平衡型芯 (c) Vertical core 立式型芯

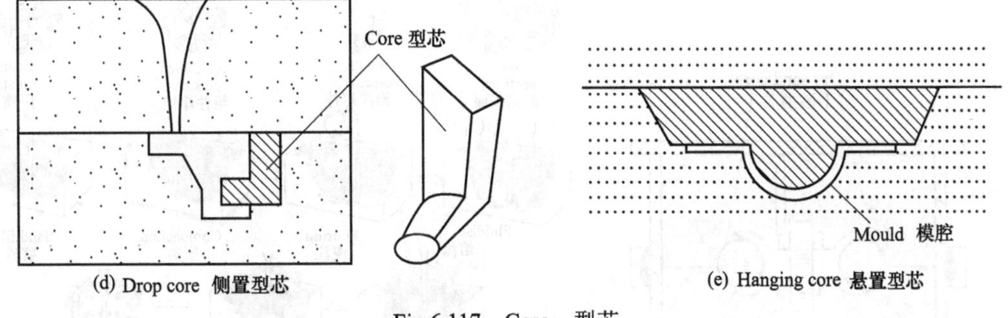

Fig 6.117 Core 型芯

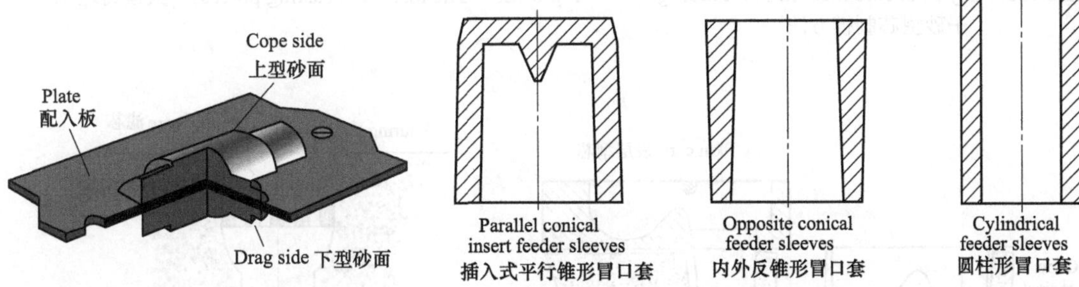

Fig 6.118 A typical metal match-plate pattern used in sand casting
砂型铸造采用的典型金属配入板图样

Fig 6.119 Feeder sleeve types 冒口套的种类

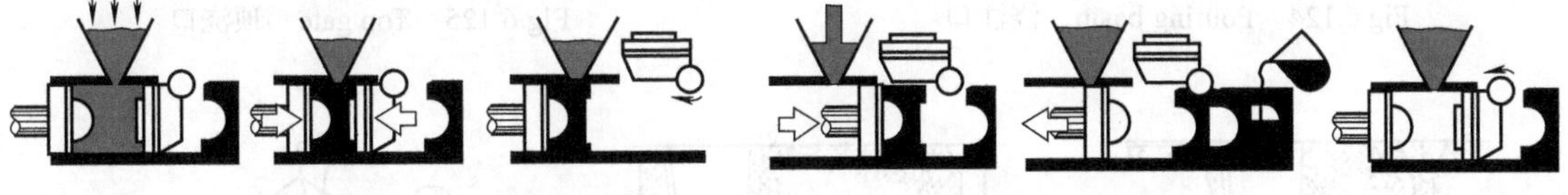

Fig 6.120 Vertically parted flaskless moulding machine 立式进料无砂箱制模机

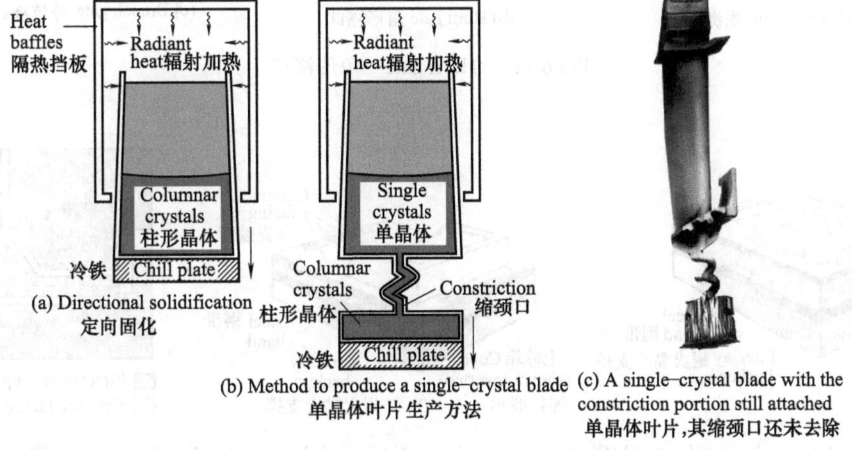

Fig 6.121 Methods of casting turbine blades 涡轮叶片铸造方法

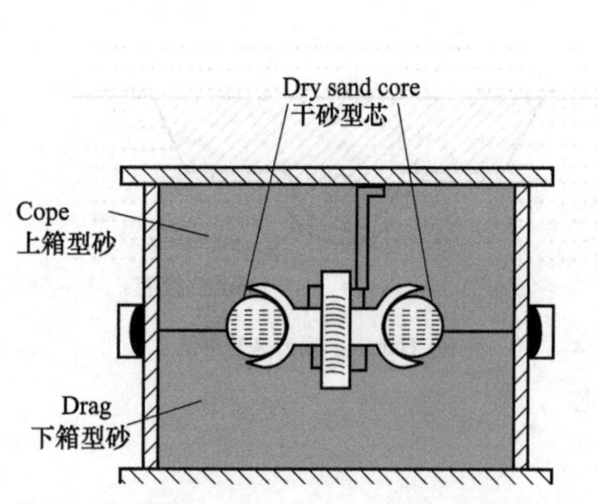

Fig 6.122 Dry sand core method of moulding 干砂型芯制模方法

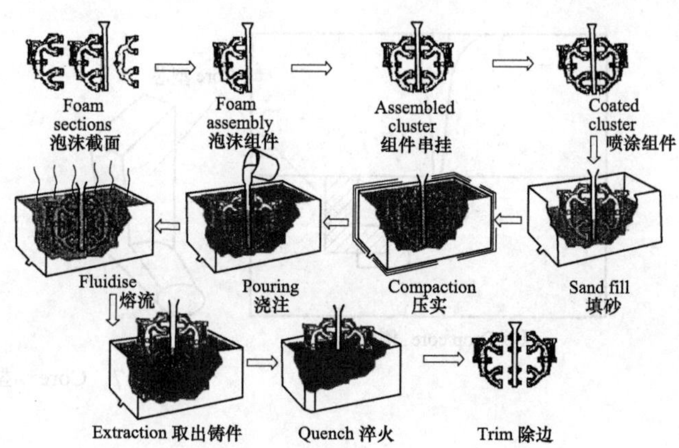

Fig 6.123 The lost foam casting process 失模铸造工艺

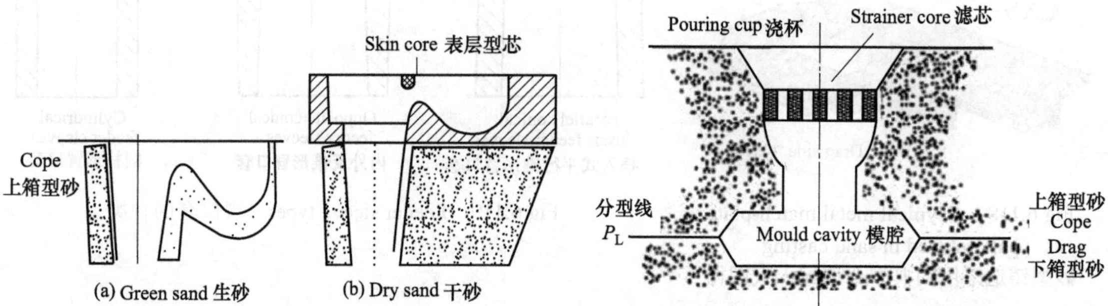

Fig 6.124 Pouring basin 浇注口

Fig 6.125 Top gate 顶浇口

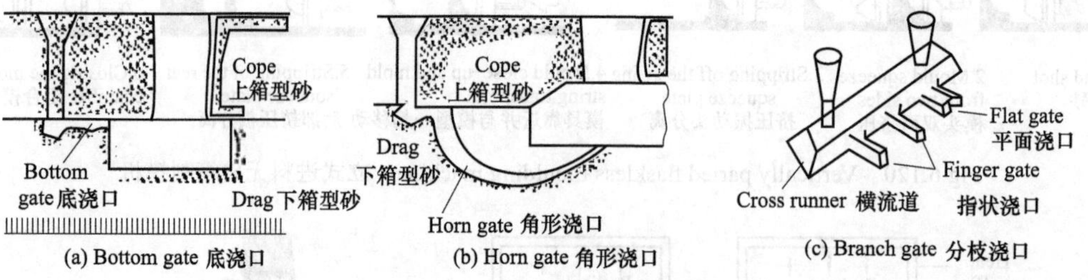

Fig 6.126 Horn gate 角形浇口

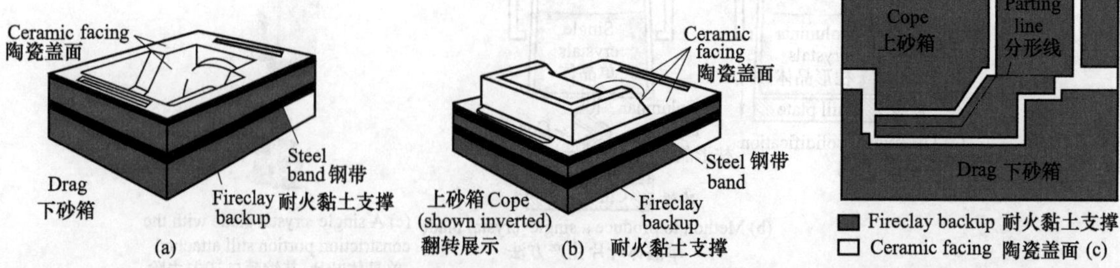

Fig 6.127 A typical ceramic mold (Shaw process) for casting steel dies used in hot forging 热锻铸钢陶瓷模

6.4.3 Forging mo(u)ld 锻压模具

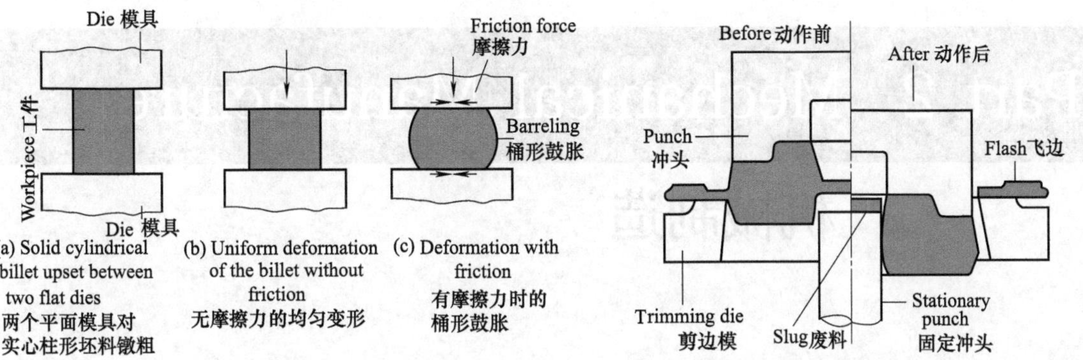

Fig 6.128 Upsetting 镦粗

Fig 6.129 Trimming flash from a forged part (The thin material at the center is removed by punching)
锻件飞边去除（中心薄片材料靠冲头去除）

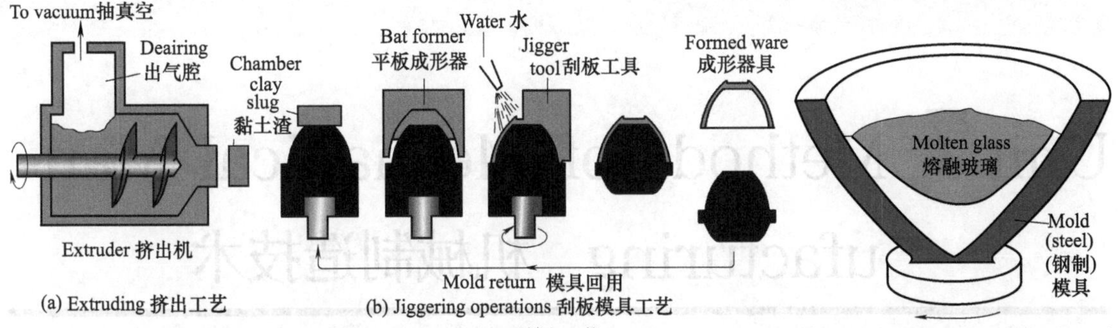

Fig 6.130 Jiggering operation 刮板工艺

Fig 6.131 Centrifugal casting of glass 玻璃的离心铸造

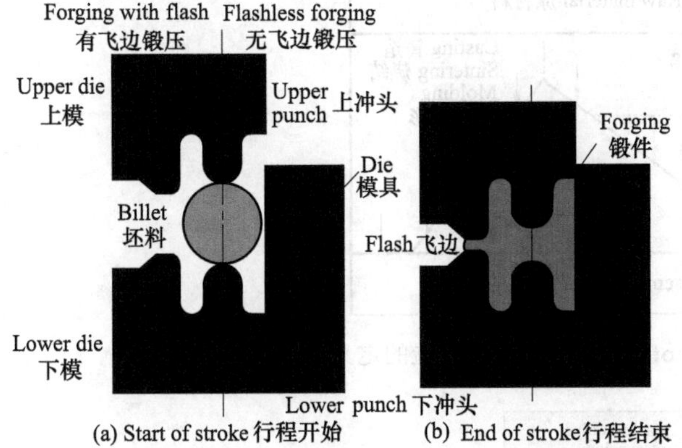

Fig 6.132 Comparison of closed-die forging to precision or flashless forging of a cylindrical billet
圆柱坯料闭合模具精锻与无飞边锻压的比较

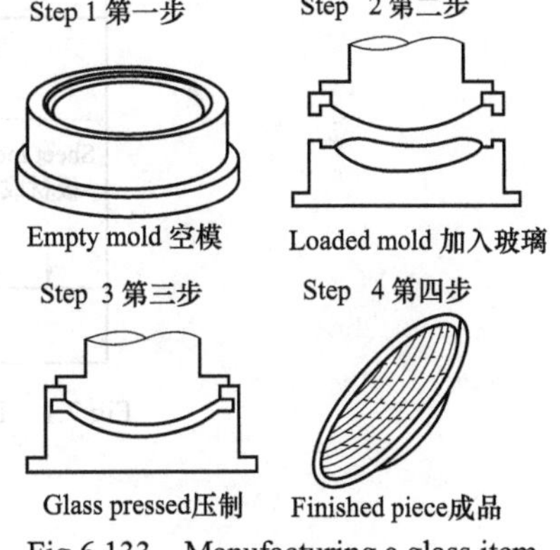

Fig 6.133 Manufacturing a glass item by pressing glass in a mold
模具压制玻璃器件

Step 1 第一步

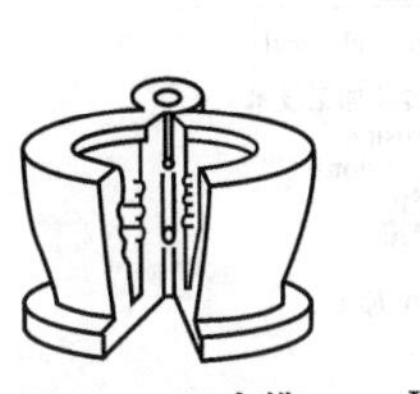

Empty mold 空模

Step 2 第二步

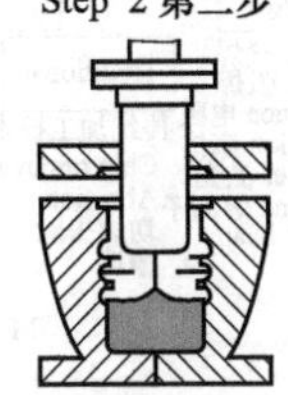

Loaded mold 模具里加入玻璃

Step 3 第三步

Glass pressed 压制

Step 4 第四步

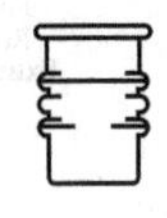

Finished product 成品

Fig 6.134 Pressing glass in a split mold 开合模具压制玻璃

Part 2 Mechanical Manufacture

机械制造

Unit 7 Methods of Mechanical Manufacturing 机械制造技术

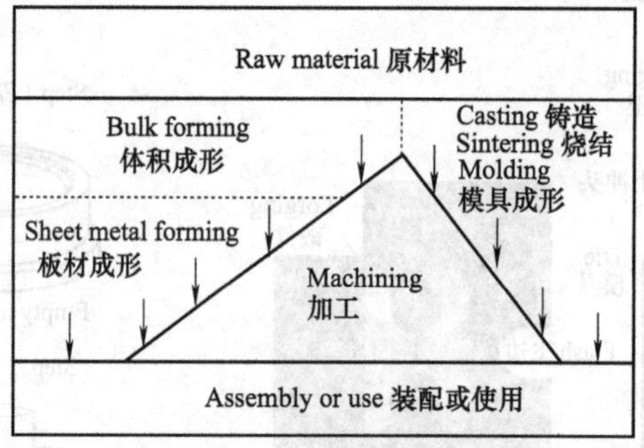

Fig 7.1 Definition of manufacturing 制造的定义

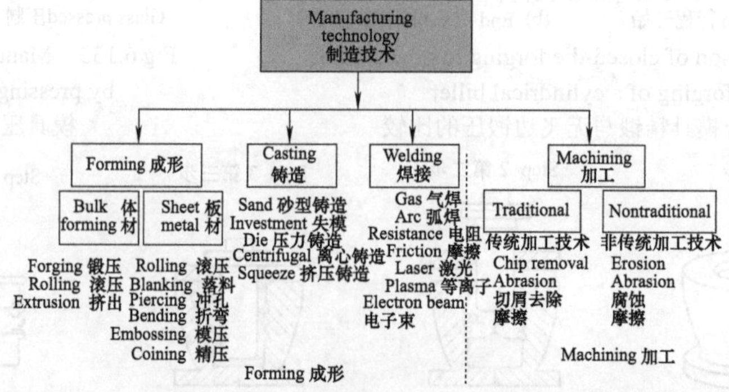

Fig 7.2 Classification of manufacturing processes 制造技术分类

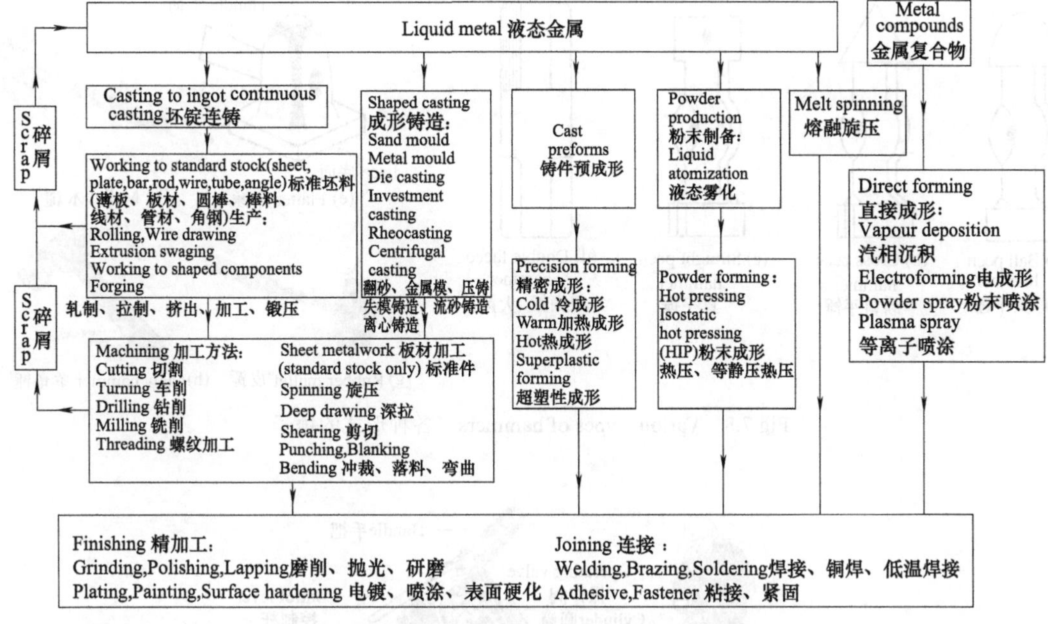

Fig 7.3 Processing routes for metals 金属加工方法

7.1 Manual operations 手工操作

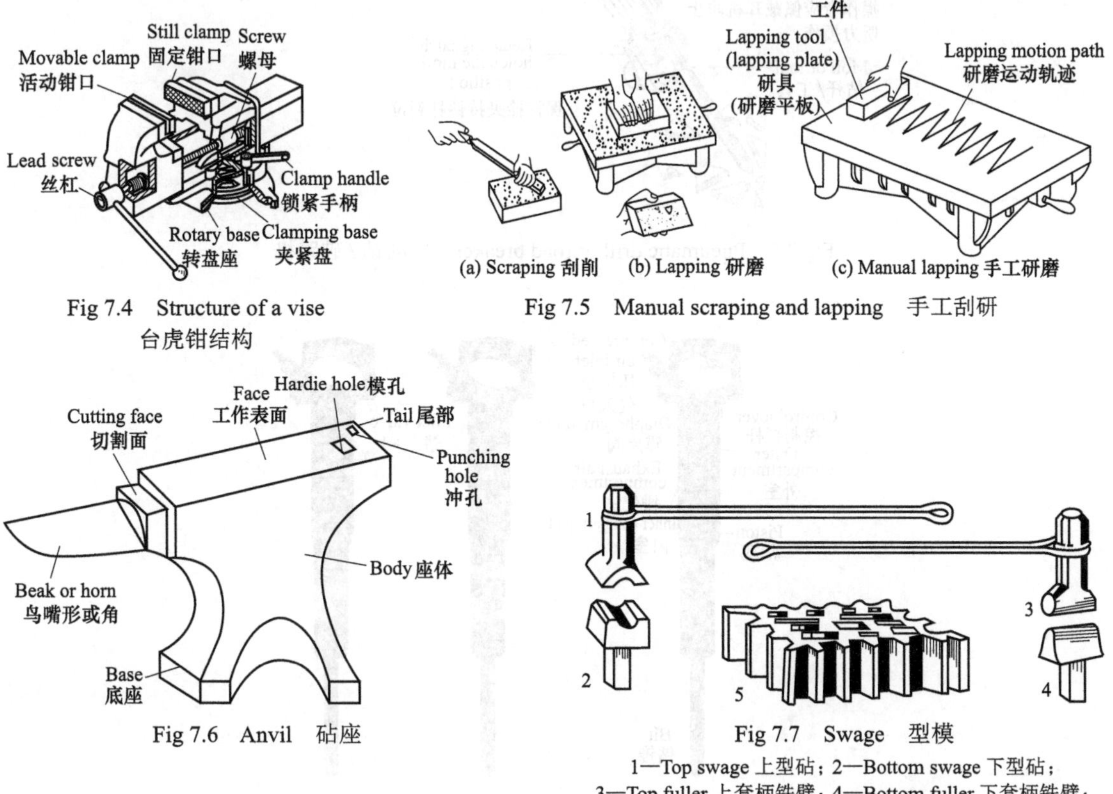

Fig 7.4 Structure of a vise 台虎钳结构

Fig 7.5 Manual scraping and lapping 手工刮研

Fig 7.6 Anvil 砧座

Fig 7.7 Swage 型模

1—Top swage 上型砧；2—Bottom swage 下型砧；
3—Top fuller 上套柄铁劈；4—Bottom fuller 下套柄铁劈；
5—Swage block 型模块

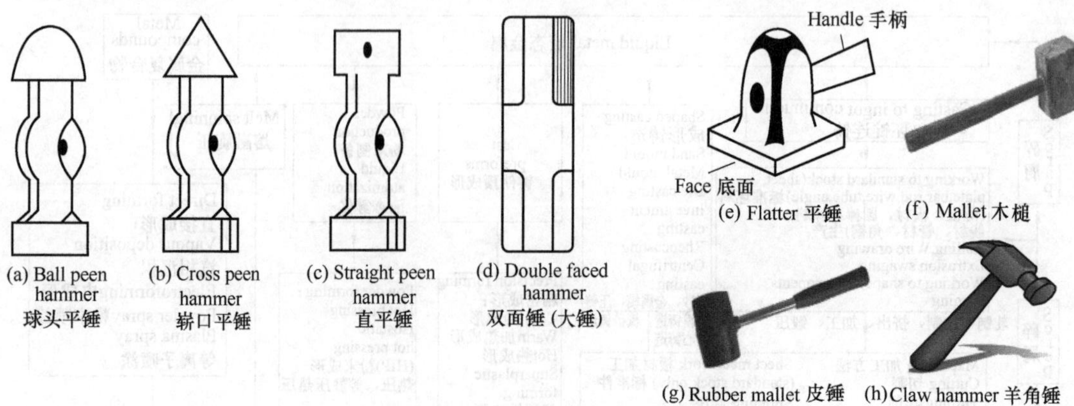

Fig 7.8 Various types of hammers 各种形式的锤子

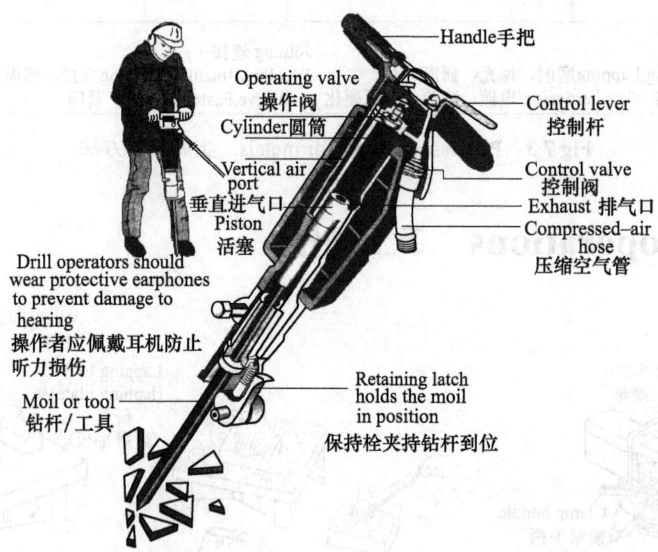

Fig 7.9 Pneumatic drill or road breaker 气动钻 / 破路机

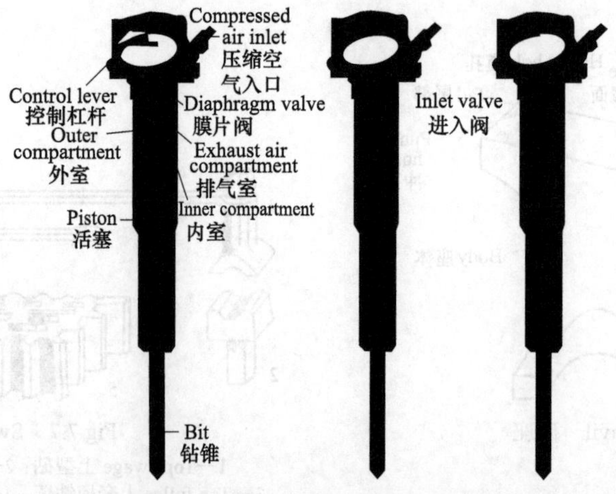

Fig 7.10 Pneumatic hammer 气动锤

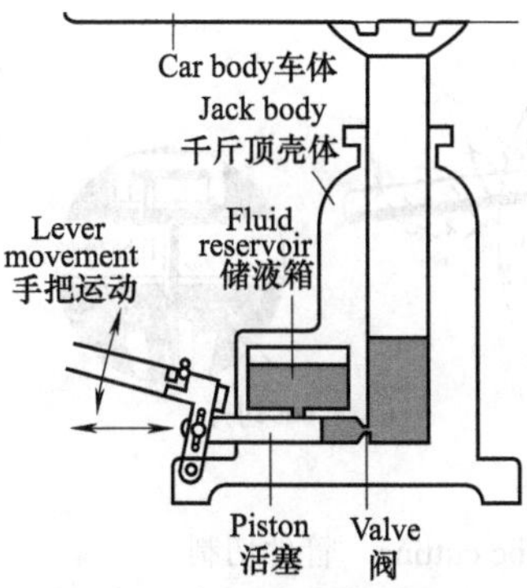

Car body车体
Jack body
千斤顶壳体
Fluid reservoir
储液箱
Lever movement
手把运动
Piston
活塞
Valve
阀

Fig 7.11 Hydraulic jack
液压千斤顶

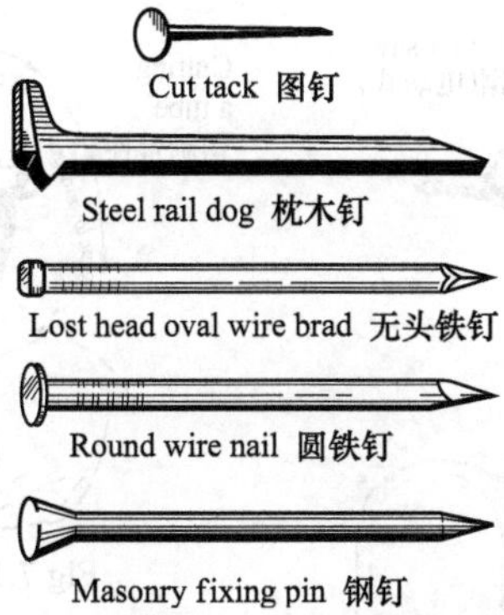

Cut tack 图钉
Steel rail dog 枕木钉
Lost head oval wire brad 无头铁钉
Round wire nail 圆铁钉
Masonry fixing pin 钢钉

Fig 7.12 Nail 钉子

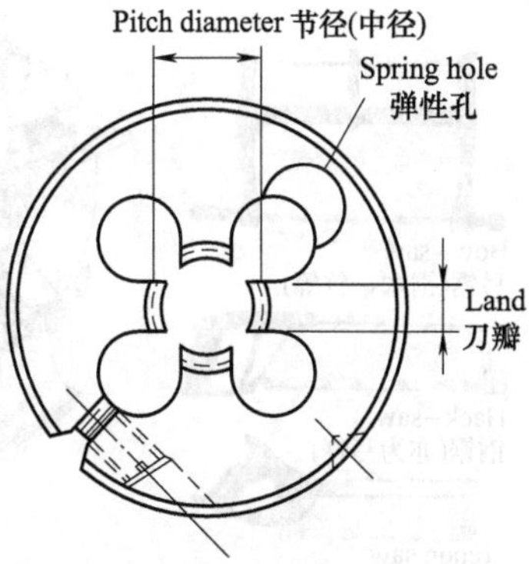

Pitch diameter 节径(中径)
Spring hole
弹性孔
Land
刀瓣

Fig 7.13 Solid screw-adjustable die
整体可调板牙

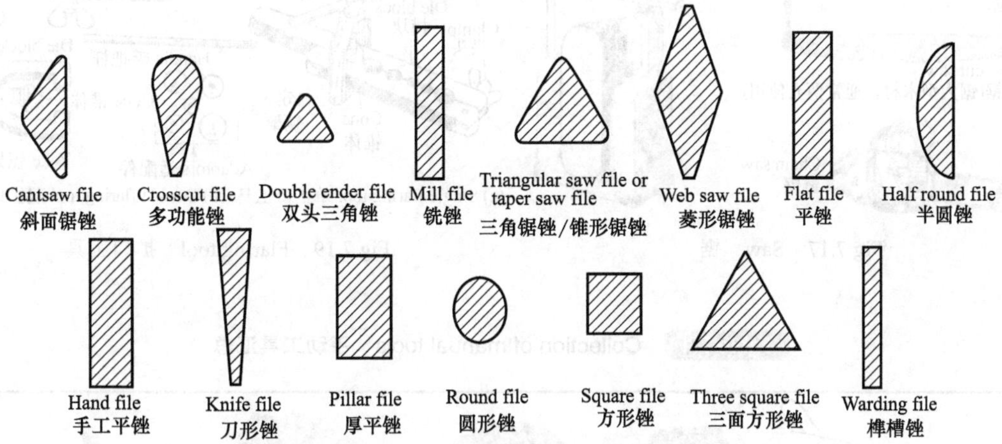

Cantsaw file
斜面锯锉
Crosscut file
多功能锉
Double ender file
双头三角锉
Mill file
铣锉
Triangular saw file or taper saw file
三角锯锉/锥形锯锉
Web saw file
菱形锯锉
Flat file
平锉
Half round file
半圆锉
Hand file
手工平锉
Knife file
刀形锉
Pillar file
厚平锉
Round file
圆形锉
Square file
方形锉
Three square file
三面方形锉
Warding file
榫槽锉

Fig 7.14 Styles of mills or saw files 铣锉或锯锉类别

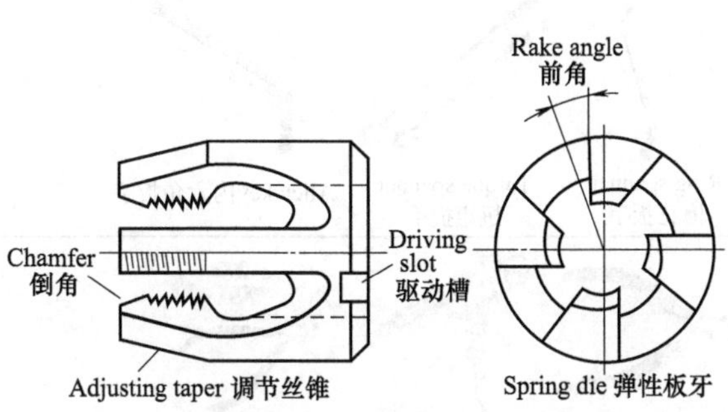

Rake angle
前角
Chamfer
倒角
Driving slot
驱动槽
Adjusting taper 调节丝锥
Spring die 弹性板牙

Fig 7.15 Spring type collet-adjustable die and holder
弹性类收口式可调丝锥和板牙

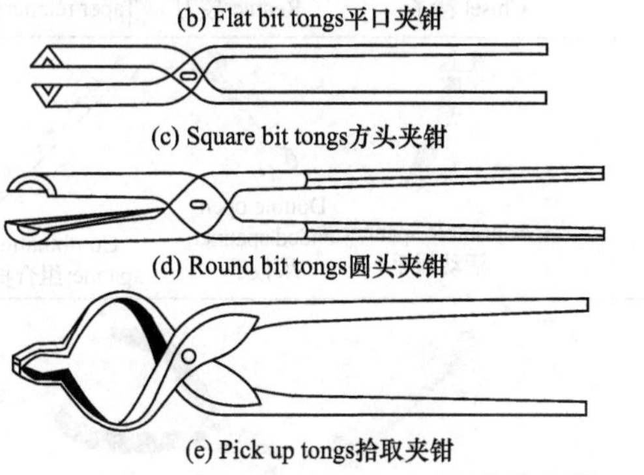

(a) Goose-neck tongs鹅颈夹钳
(b) Flat bit tongs平口夹钳
(c) Square bit tongs方头夹钳
(d) Round bit tongs圆头夹钳
(e) Pick up tongs拾取夹钳

Fig 7.16 Various types of tongs 各种夹钳

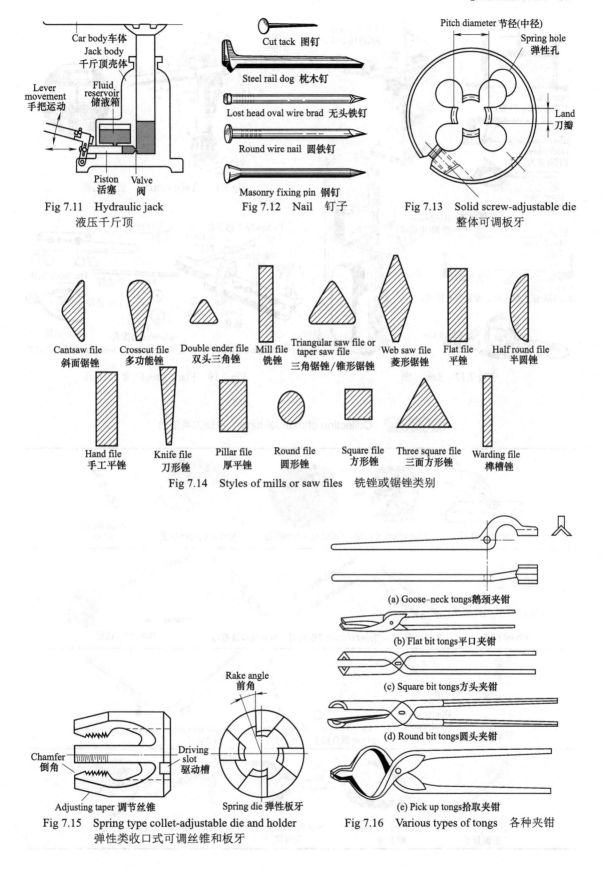

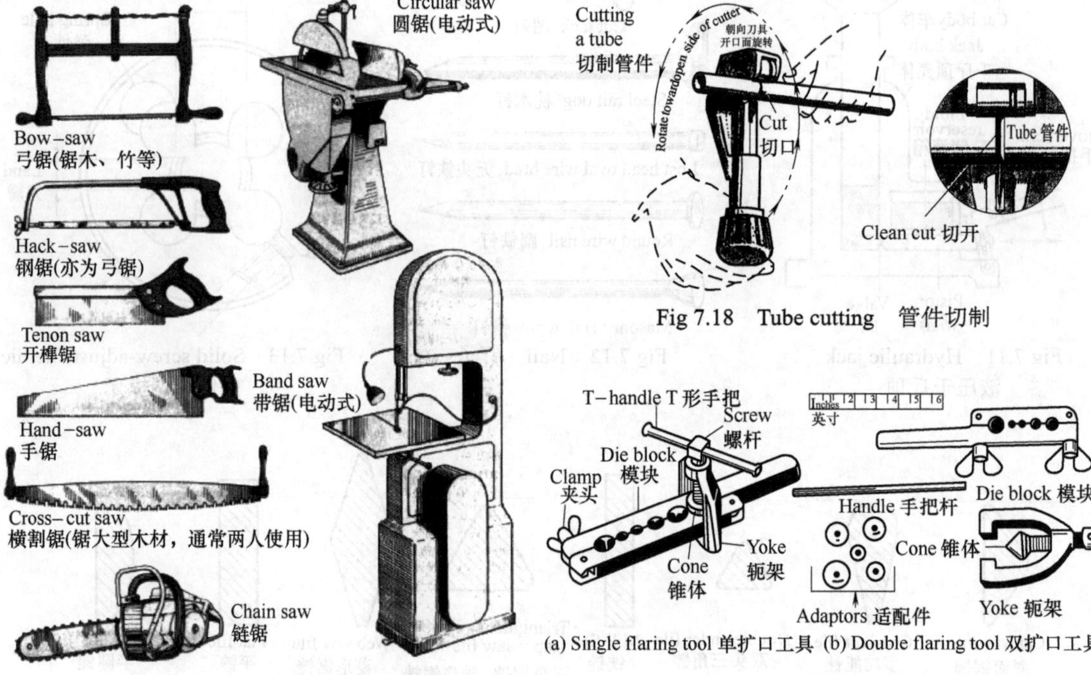

Fig 7.17 Saw 锯

Fig 7.18 Tube cutting 管件切制

Fig 7.19 Flaring tool 扩口工具

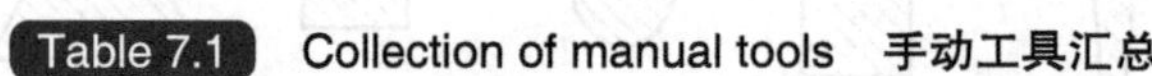
Table 7.1 Collection of manual tools 手动工具汇总

Oil stone油石	Flap wheel飞翼轮	Grinding wheel砂轮	Abrasive paper砂纸	Point punch 样冲	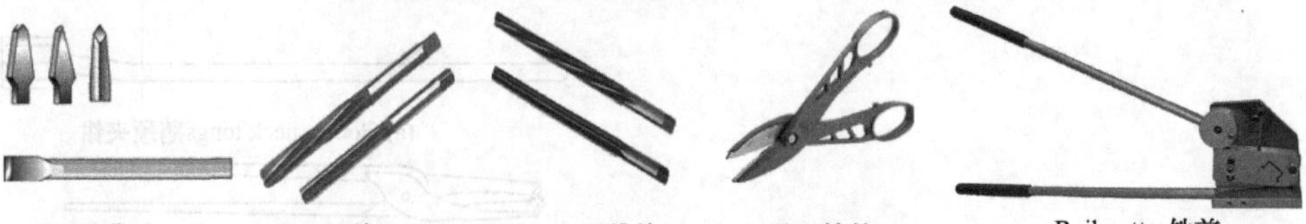
Chisel 凿子	Reamer铰刀	Taper reamer圆锥铰刀	Snip马口铁剪刀	Rail cutter铁剪	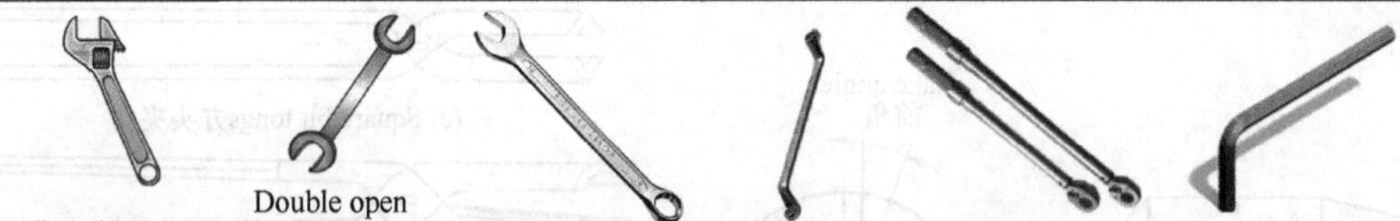
Adjustable spanner 活动扳手	Double open ended spanner 呆扳手	Combination spanner组合扳手	Ring spanner 梅花扳手	Torque spanner 扭矩扳手	Allen key内六角扳手
Box spanner 套筒扳手	Combination plier 组合钳	Needle nose plier 尖嘴钳	Locking plier 锁紧钳	Cutting plier 切断钳	

续表

Die板牙	 Die-stock 板牙扳手	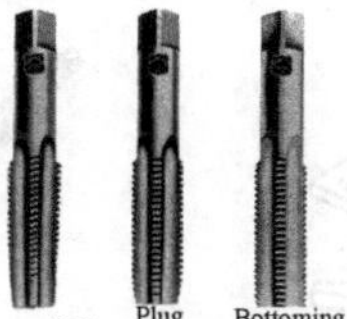 Taper锥形 Plug柱塞形 Bottoming平底形 Straight flute hand taps 直槽手动丝锥	 Tap wrench 丝锥扳手	 Philips screwdriver 十字螺丝刀 (螺钉旋具)
 Slotted screwdriver 一字螺丝刀	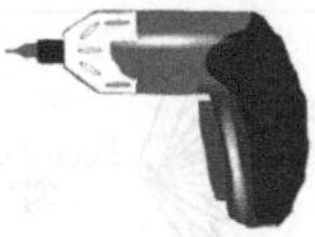 Electric screwdriver 电动螺丝刀	 Socket screwdriver 套筒螺丝刀	 Electric drill (手)电钻	 Pneumatic drill 风钻
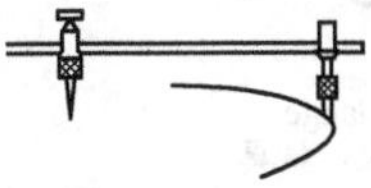 Beam compass 长臂划规	 Breast drill 胸压手摇钻	 Brace 弓形钻	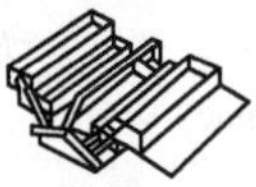 Toolbox 工具箱	
 Tommy bar 套筒扳手旋转手把	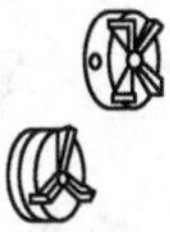 Chuck 卡盘	 Grinding machine 砂轮机、磨床	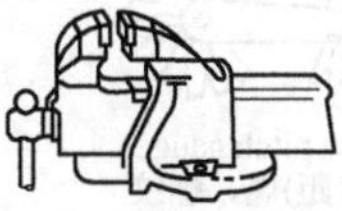 Vise 钳工台虎钳	 Try square 直角尺、矩尺
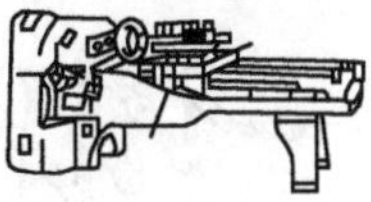 Capstan/turret lathe 六角、转塔车床	 Workbench 工作台	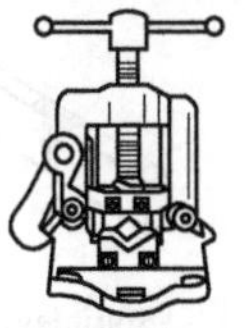 Pipe vise 管子台虎钳	 Gouge 弧口凿、半圆凿	
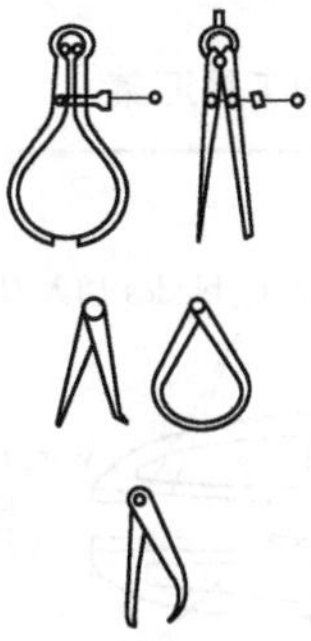 Callipers 卡钳	 Coil spring 螺旋弹簧	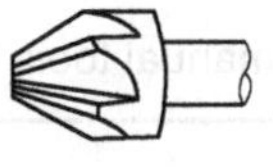 Countersink bit 锪钻、沉头钻	 Instrument screw driver 手捻、仪表起子 (旋具)	 Keyhole saw 键孔锯
	 Jack 千斤顶	 Hydraulic jack 液压千斤顶	 Knurling 滚花	 Hook spanner 钩形扳手
 G clamp 弓形钩、夹钳	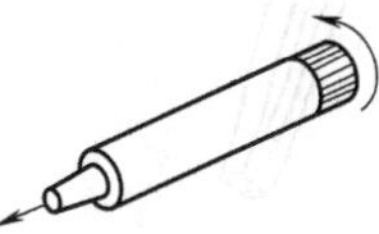 Kaulking gun 填缝枪	 Right-angled screwdriver 直角改锥(旋具)	 Nail puller 起钉器	 Sander 打磨机、磨光机

续表

Awl 木工锥子	 Firmer chisel 榫孔凿	 Microscope 显微镜	 Half-round file 半圆锉	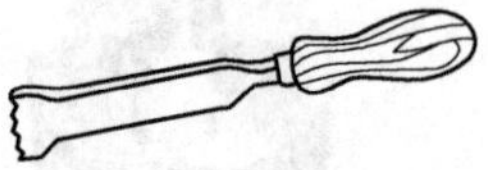 Chaser 螺纹梳刀	
 Drilling machine 钻床	 Magnifying glass 放大镜	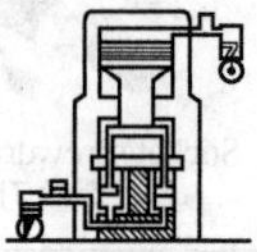 Hydraulic press 水压机 Hydraulic ram 液压夯锤	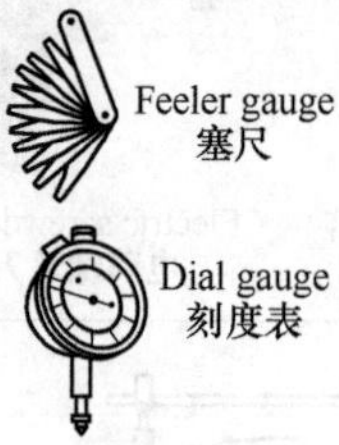 Feeler gauge 塞尺 Dial gauge 刻度表 Gauge 量仪、量具	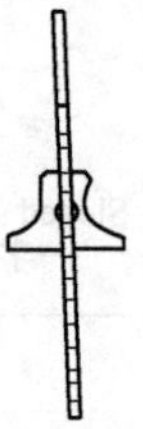 Depth gauge 深度尺	
 Screw extractor 起螺丝器	 Screw pitch gauge 螺纹(距)规、螺纹样板规	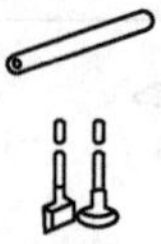 Punch 冲头、戳子	 Scribing block 划线盘	 Pipe wrench 管扳手、管子钳	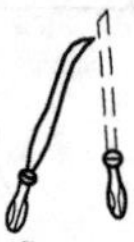 Scraper 刮刀、铲运机
 Ratchet screwdriver 棘轮式改锥	 Polishing machine 抛光机	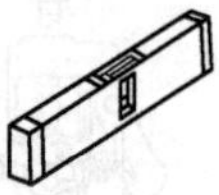 Spirit level 水平仪(尺)	 Stopcock 旋塞阀	 Oil can 油壶	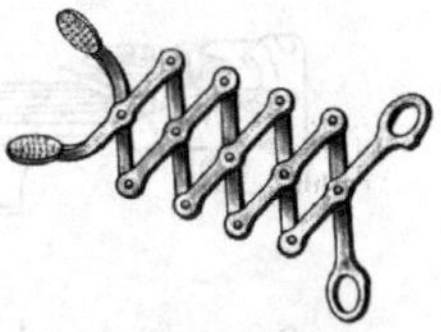 Lazy-tongs伸缩夹钳

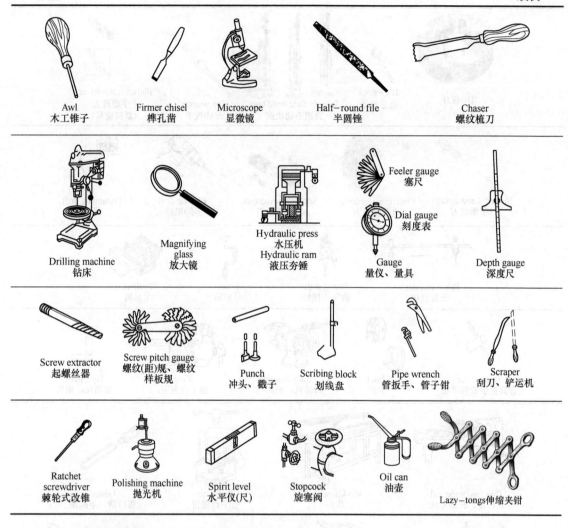

Table 7.2 Collection of manual tools for electrical uses 电工手动工具汇集

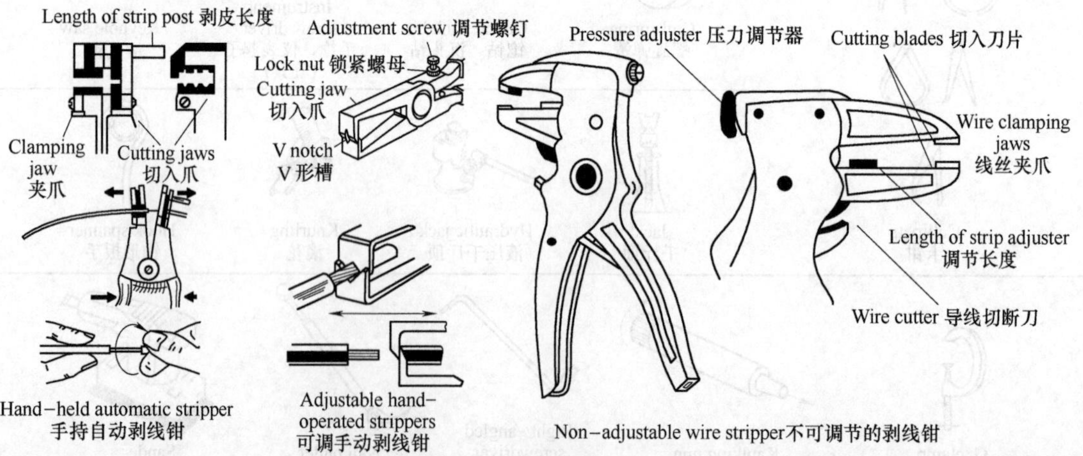

Hand-held automatic stripper 手持自动剥线钳

Adjustable hand-operated strippers 可调手动剥线钳

Non-adjustable wire stripper不可调节的剥线钳

续表

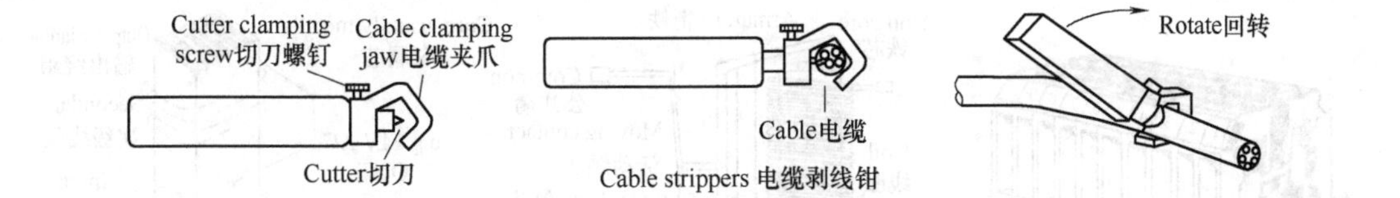

Cutter clamping screw切刀螺钉
Cable clamping jaw电缆夹爪
Cutter切刀
Cable电缆
Rotate回转

Cable strippers 电缆剥线钳

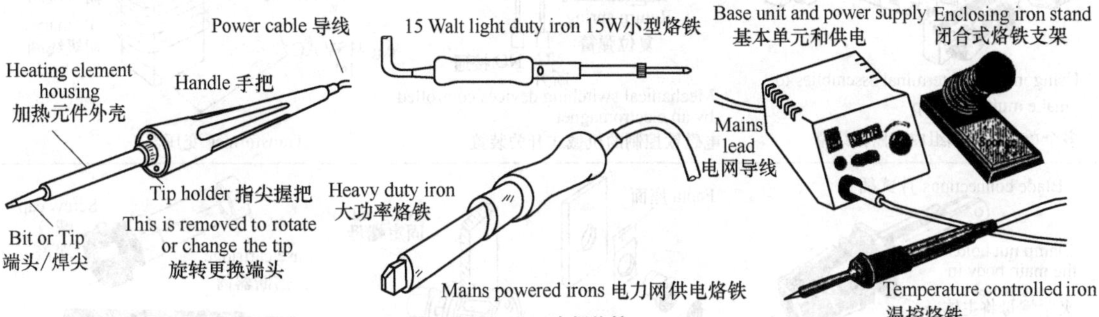

Power cable 导线
Heating element housing 加热元件外壳
Handle 手把
Tip holder 指尖握把
This is removed to rotate or change the tip 旋转更换端头
Bit or Tip 端头/焊尖
15 Walt light duty iron 15W小型烙铁
Heavy duty iron 大功率烙铁
Base unit and power supply 基本单元和供电
Enclosing iron stand 闭合式烙铁支架
Mains lead 电网导线
Mains powered irons 电力网供电烙铁
Temperature controlled iron 温控烙铁

The soldering iron电焊烙铁

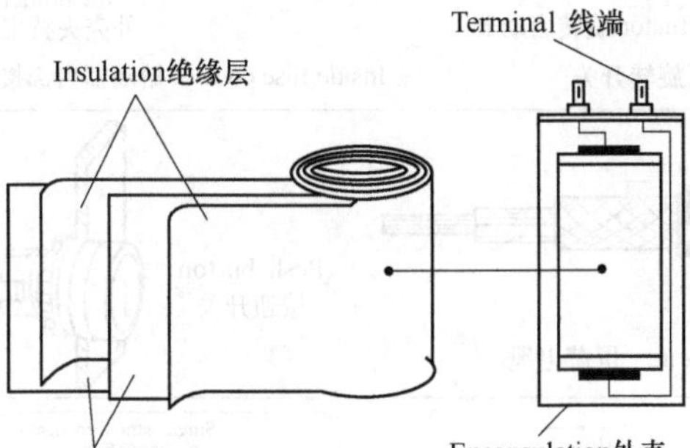

Insulation绝缘层
Terminal 线端
Conductive foils导电薄膜
Encapsulation外壳

Construction of capacitors电容器结构

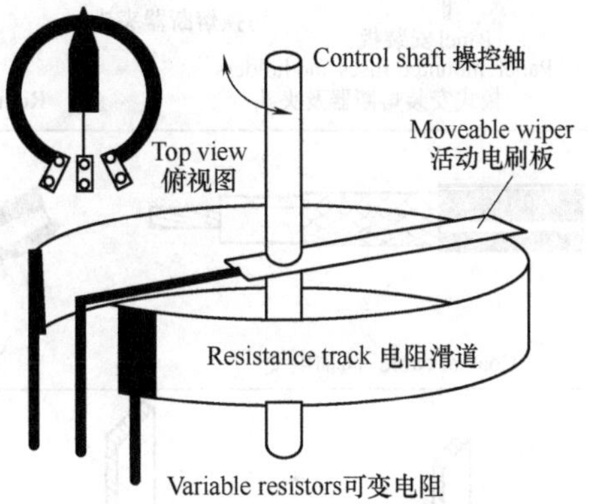

Top view 俯视图
Control shaft 操控轴
Moveable wiper 活动电刷板
Resistance track 电阻滑道

Variable resistors可变电阻

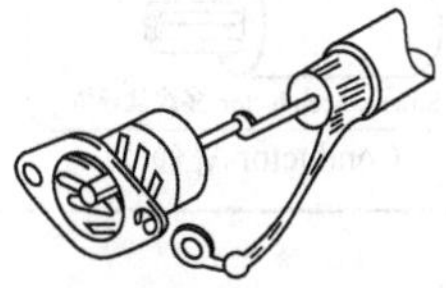

Fitting a coaxial socket 安装同轴电缆座

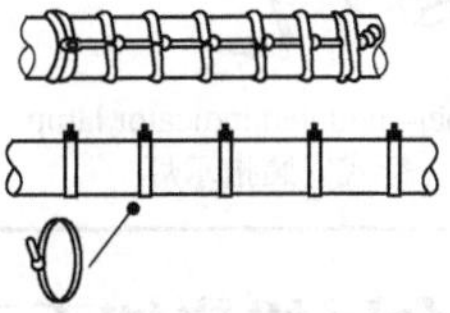

Cableform binding 电缆锁夹

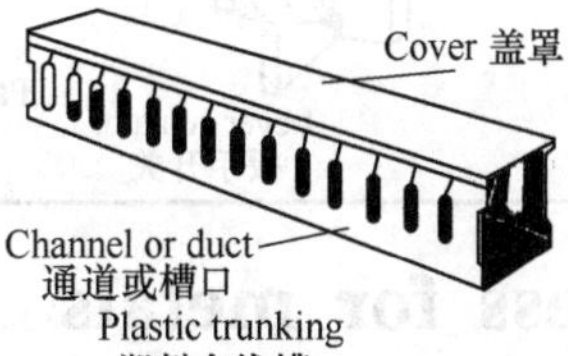

Cover 盖罩
Channel or duct 通道或槽口

Plastic trunking 塑料布线槽

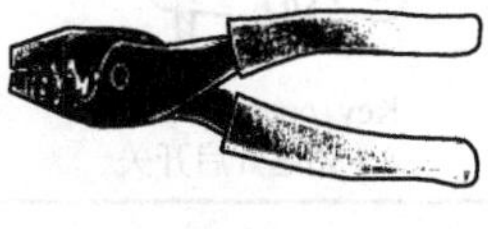

Hand-operated crimping tools 手动卷边工具

Terminal clamp screw 线座夹持螺钉
Wire clamp screw 夹线螺钉
Metal lever clamps to the rail 固定到支架的金属槽杆
Wire entry 导线入口

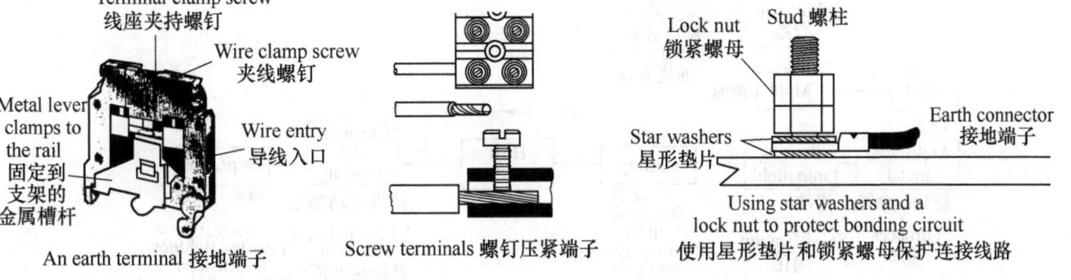

An earth terminal 接地端子

Screw terminals 螺钉压紧端子

Lock nut 锁紧螺母
Stud 螺柱
Star washers 星形垫片
Earth connector 接地端子

Using star washers and a lock nut to protect bonding circuit 使用星形垫片和锁紧螺母保护连接线路

续表

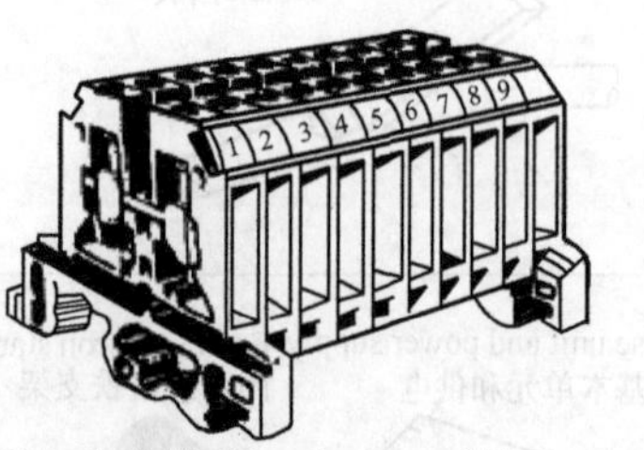

Using individual terminal assemblies to make multiway strip 多个单独接线端组成多路接线槽	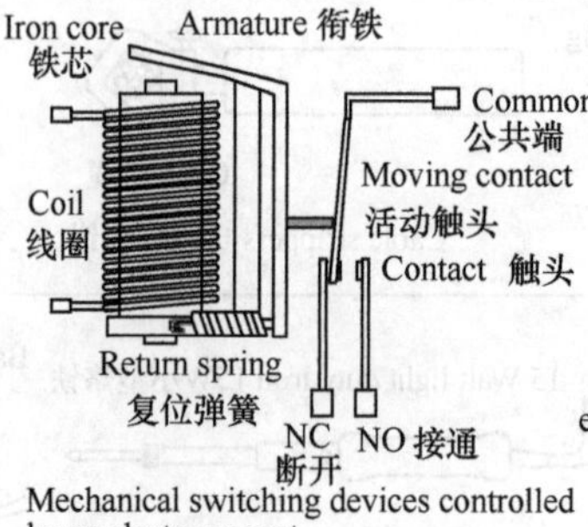 Iron core 铁芯; Armature 衔铁; Common 公共端; Moving contact 活动触头; Coil 线圈; Contact 触头; Return spring 复位弹簧; NC 断开; NO 接通 Mechanical switching devices controlled by an electromagnet 电磁铁控制的机械式开关装置	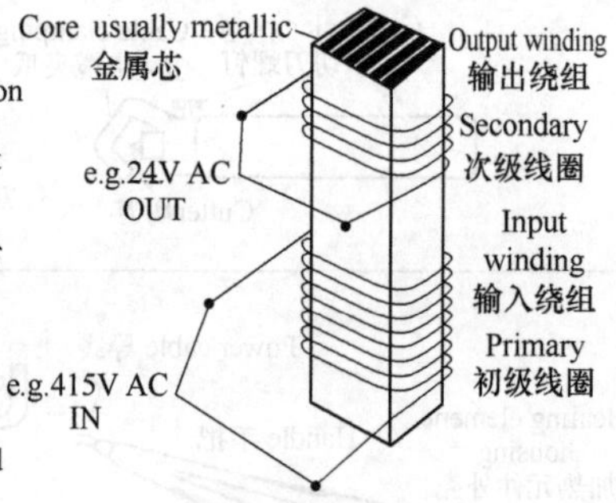 Core usually metallic 金属芯; Output winding 输出绕组; Secondary 次级线圈; e.g.24V AC OUT; Input winding 输入绕组; Primary 初级线圈; e.g.415V AC IN Transformer 变压器
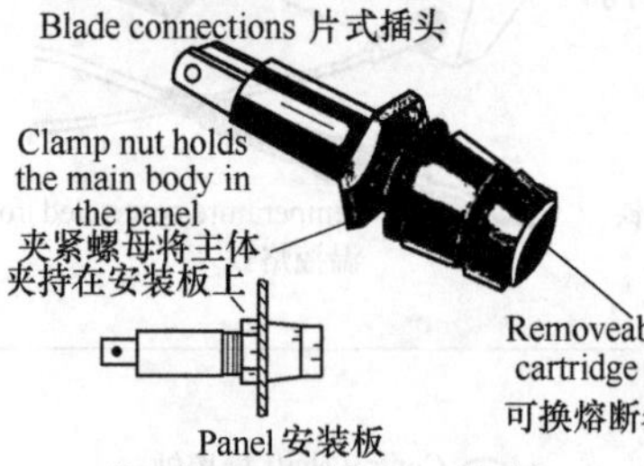 Blade connections 片式插头; Clamp nut holds the main body in the panel 夹紧螺母将主体夹持在安装板上; Removeable fuse cartridge holder 可换熔断器夹头; Panel 安装板 Panel-mounted fuses and holders 板式安装熔断器及夹头	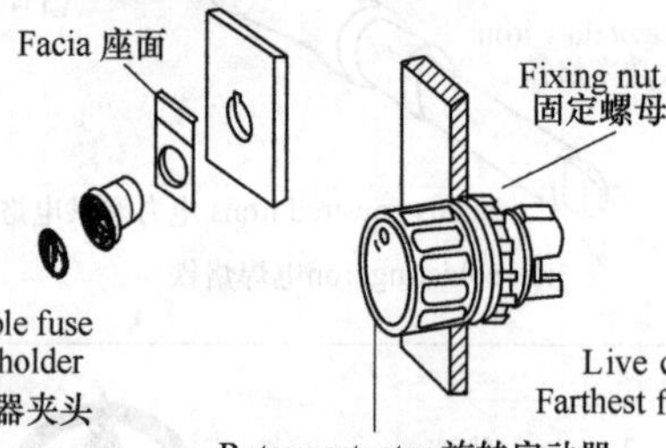 Facia 座面; Fixing nut 固定螺母; Rotary actuator 旋转启动器 Rotary switch 旋转开关	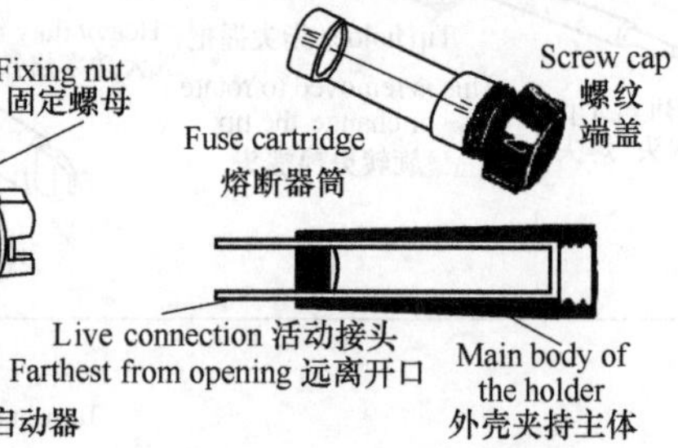 Screw cap 螺纹端盖; Fuse cartridge 熔断器筒; Live connection 活动接头; Farthest from opening 远离开口; Main body of the holder 外壳夹持主体 Inside fuse contact 熔断器内部接触
Coaxial cable 同轴电缆	Screened wire 屏蔽电缆	Push-button 按钮开关

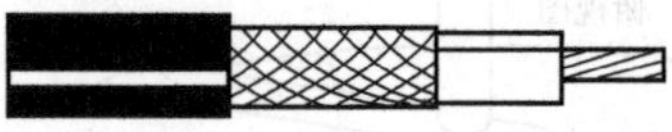

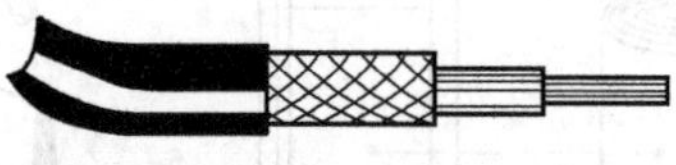

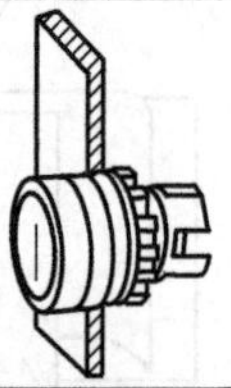

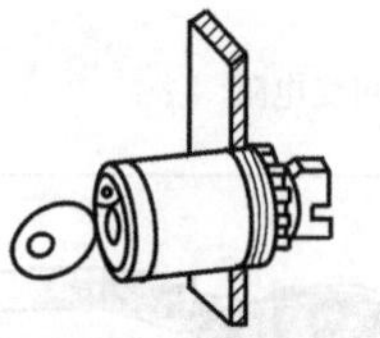

Key-operated actuator 钥匙开启开关	Lever actuator 扳手开关	Panel-mounted indicator lamp 板式安装指示灯	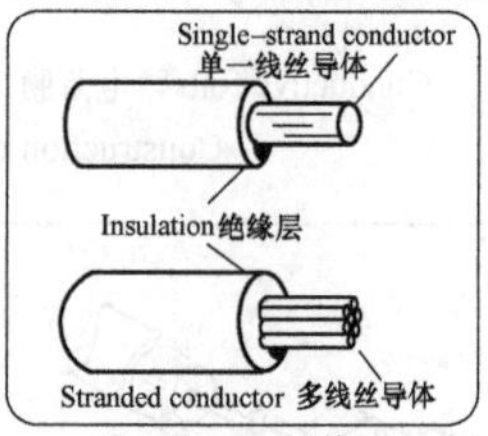 Single-strand conductor 单一线丝导体; Insulation 绝缘层; Stranded conductor 多线丝导体 Conductors 导线

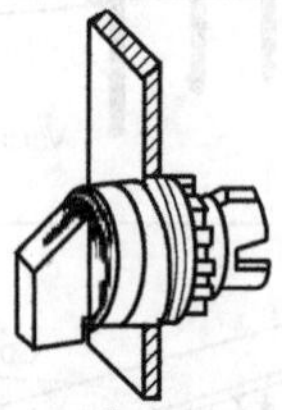

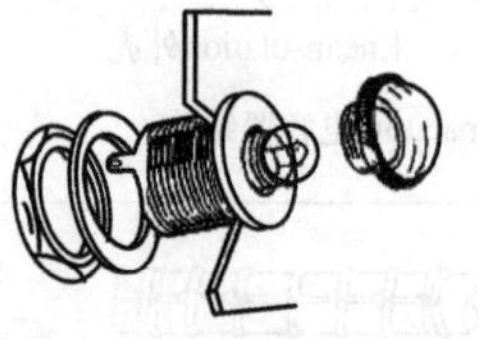

7.2 Heat process for metals 金属的热加工工艺

7.2.1 Casting 铸造

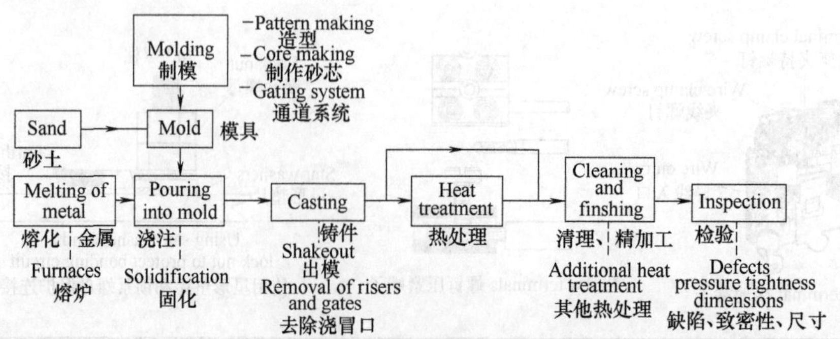

Fig 7.20 Outline of production steps in a typical sand-casting operation 砂芯铸造典型生产流程

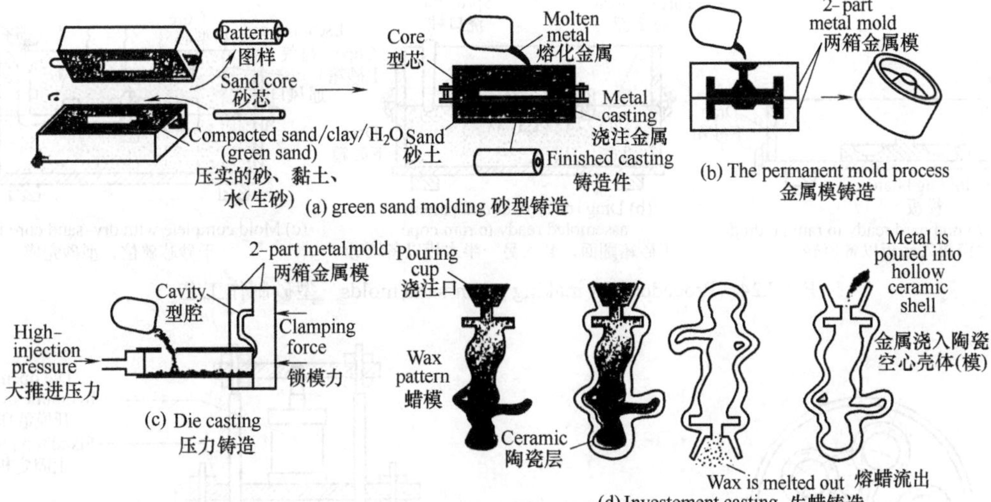

Fig 7.21 Typical casting process 四类典型铸造工艺

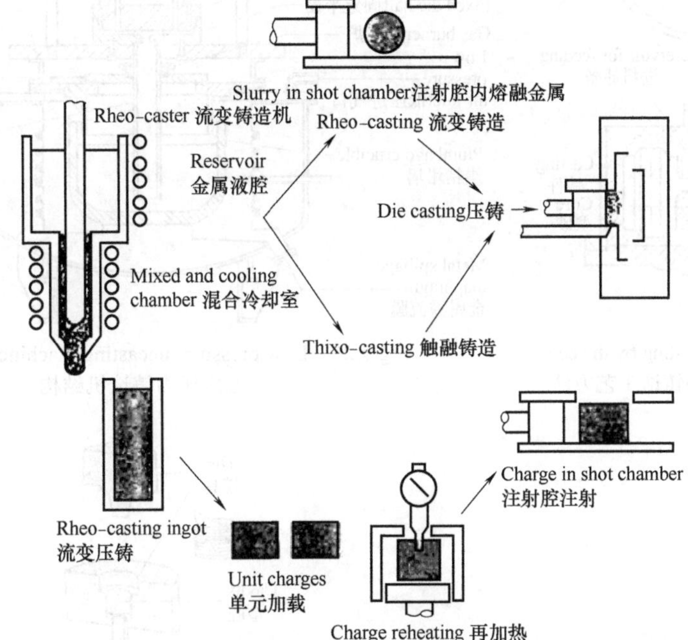

Fig 7.22 Rheo-casting and thixo-casting processes 流变压铸和触融压铸

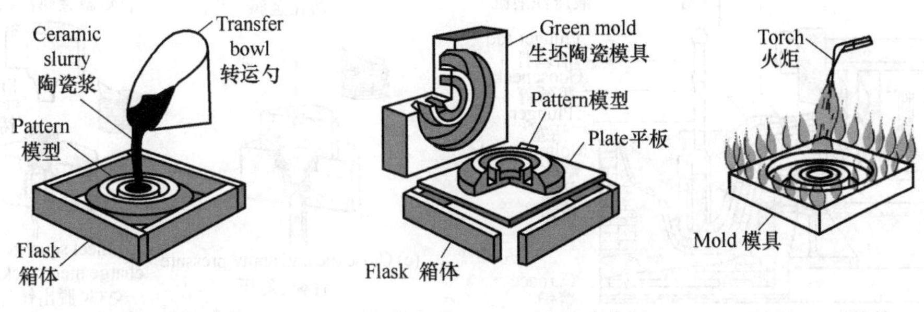

Fig 7.23 Sequence of operations in making a ceramic mold 陶瓷模具制作工序

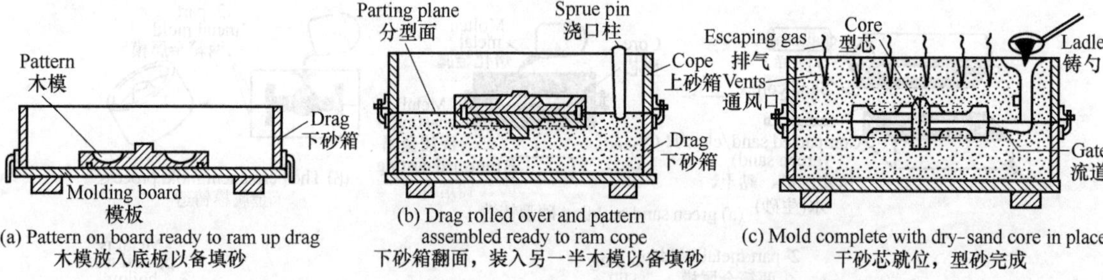

(a) Pattern on board ready to ram up drag 木模放入底板以备填砂
(b) Drag rolled over and pattern assembled ready to ram cope 下砂箱翻面，装入另一半木模以备填砂
(c) Mold complete with dry-sand core in place 干砂芯就位，型砂完成

Fig 7.24 Procedure for making green-sand molds 型砂制作工艺

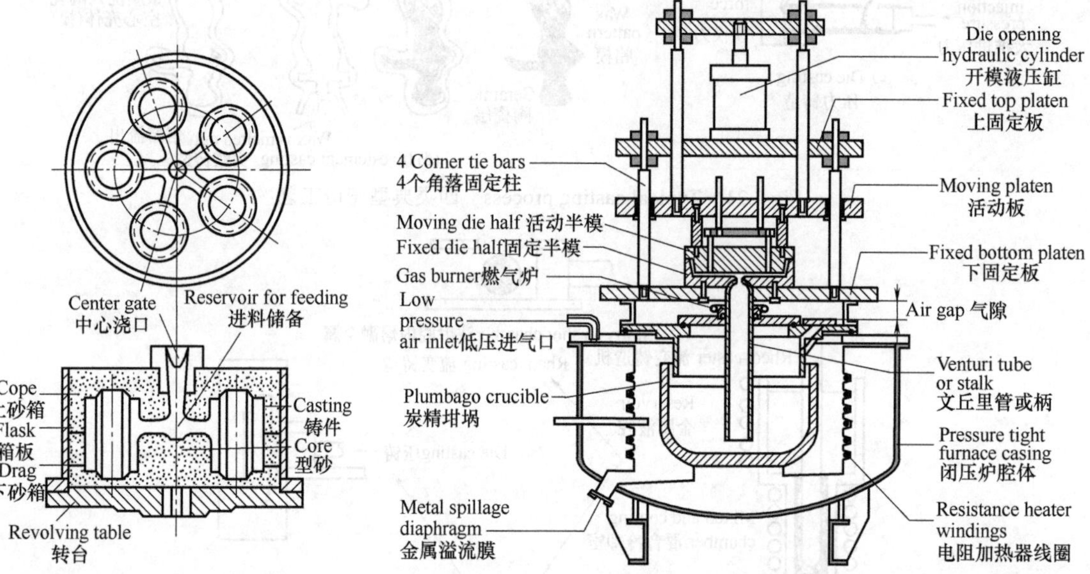

Fig 7.25 Method of casting by the centrifugal process 离心铸造工艺方法

Fig 7.26 Low pressure diecasting machine construction 低压压力铸造机结构

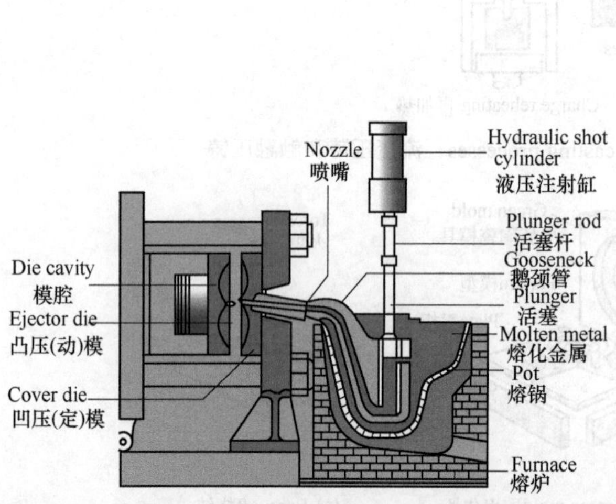

Fig 7.27 Illustration of the hot-chamber 热室压铸工艺

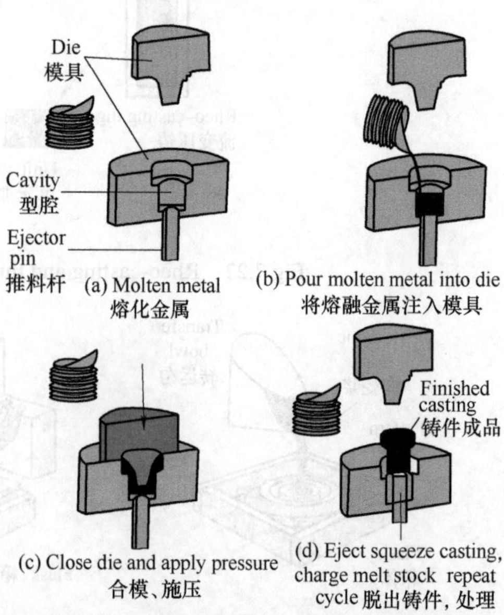

(a) Molten metal 熔化金属
(b) Pour molten metal into die 将熔融金属注入模具
(c) Close die and apply pressure 合模、施压
(d) Eject squeeze casting, charge melt stock repeat cycle 脱出铸件，处理

Fig 7.28 Operations of the squeeze casting process 挤压铸造工艺操作过程

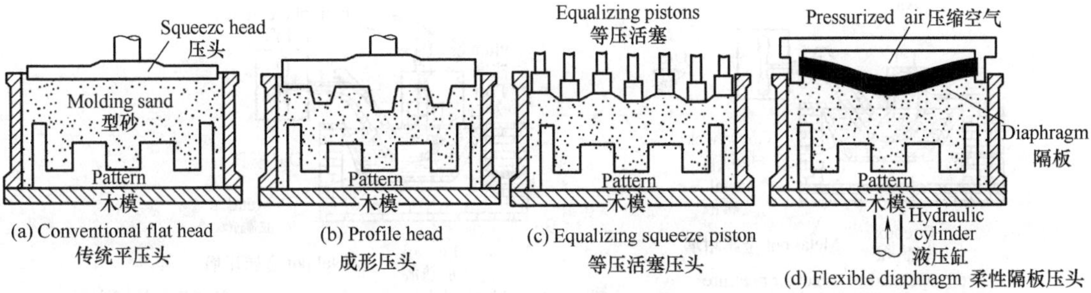

Fig 7.29 Various designs of squeeze heads for mold making 砂型制模的几种压头设计

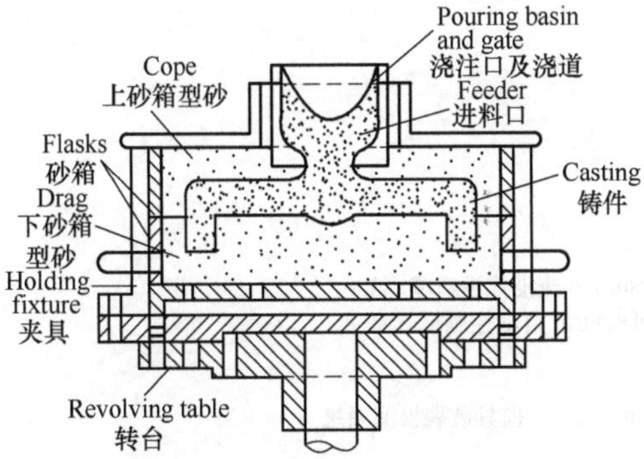

Fig 7.30 Semi-centrifugal casting 半离心铸造工艺

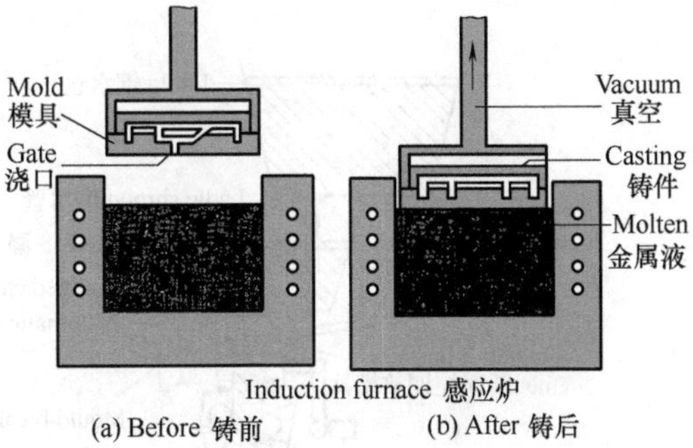

Fig 7.31 Vacuum casting process 真空铸造

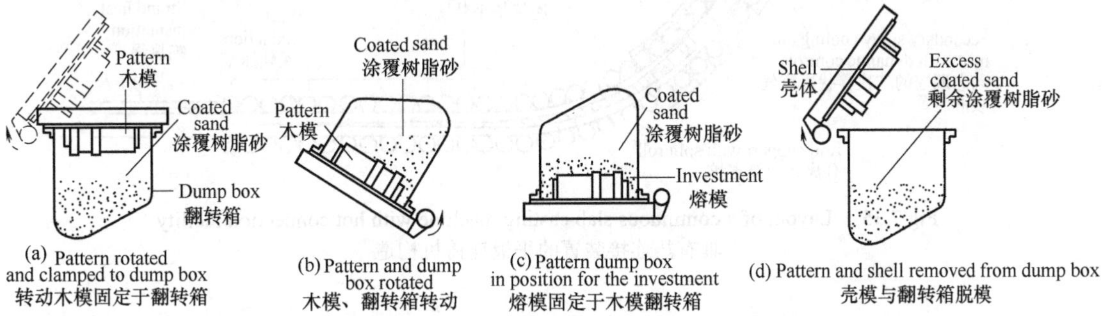

Fig 7.32 Common method of making shell molds 壳模制备的常用方法

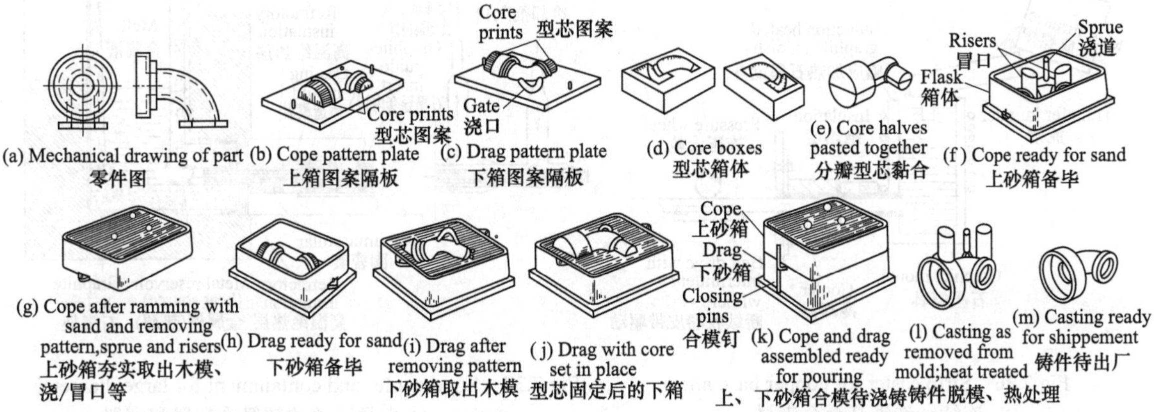

Fig 7.33 Schematic illustration of the sequence of operations for sand casting 砂型铸造操作顺序

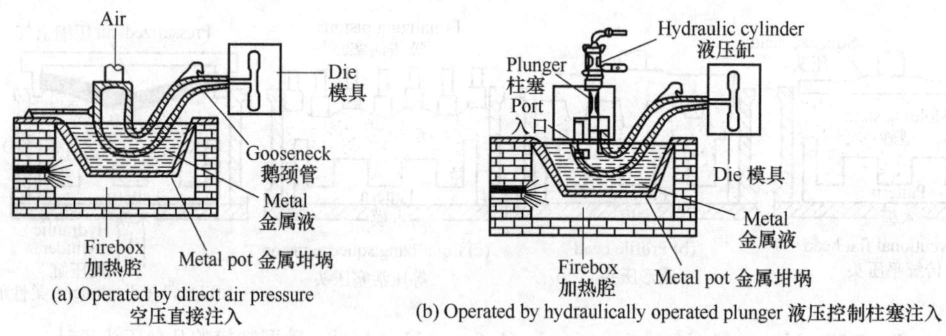

Fig 7.34 Hot chamber die casting machine 热室压铸机

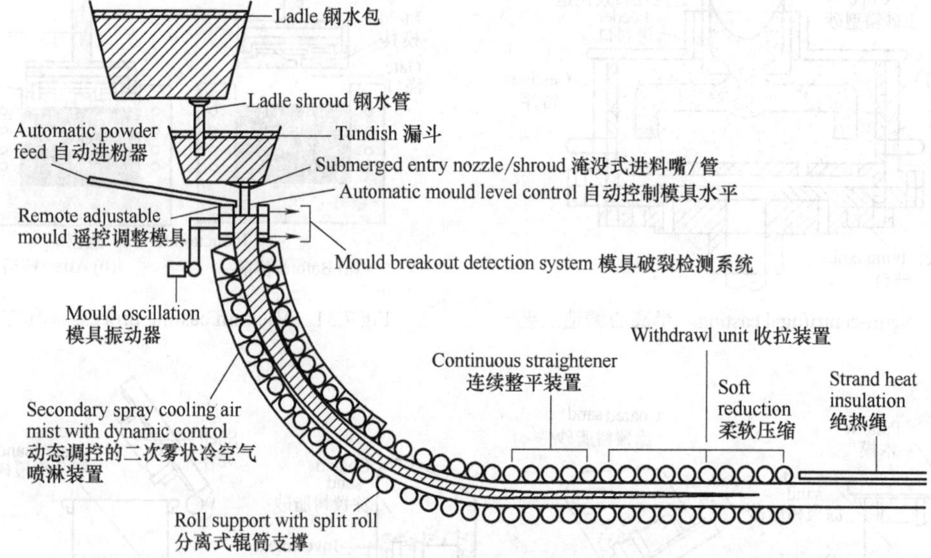

Fig 7.35 Layout of a continuous slab casting machine with hot connection facility
具有热连接装置的平板连铸机构造

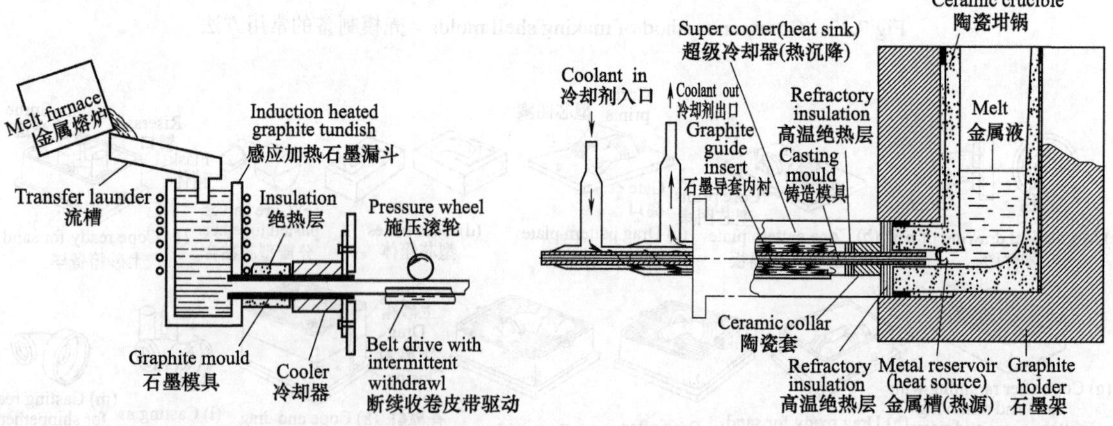

Fig 7.36 Strip caster for copper base alloy
连续铸造铜基合金带材

Fig 7.37 Supercooler and containment for large diameter casting 大直径连铸的超级冷却器和容器

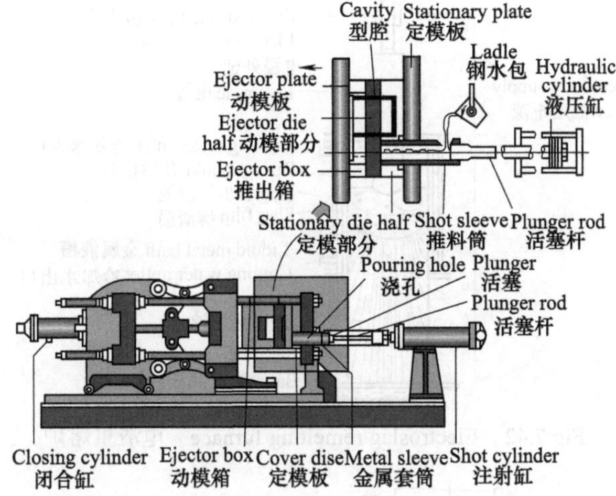

Fig 7.38 Illustration of the cold-chamber die-casting process 冷室压铸机及其工艺

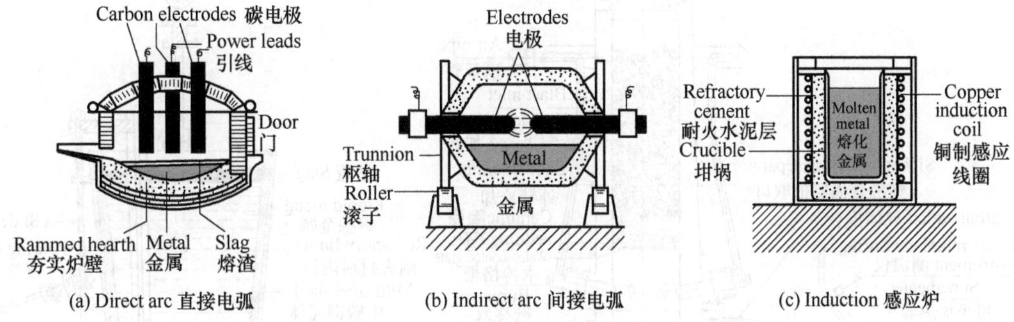

Fig 7.39 Types of electric furnaces 电炉类型

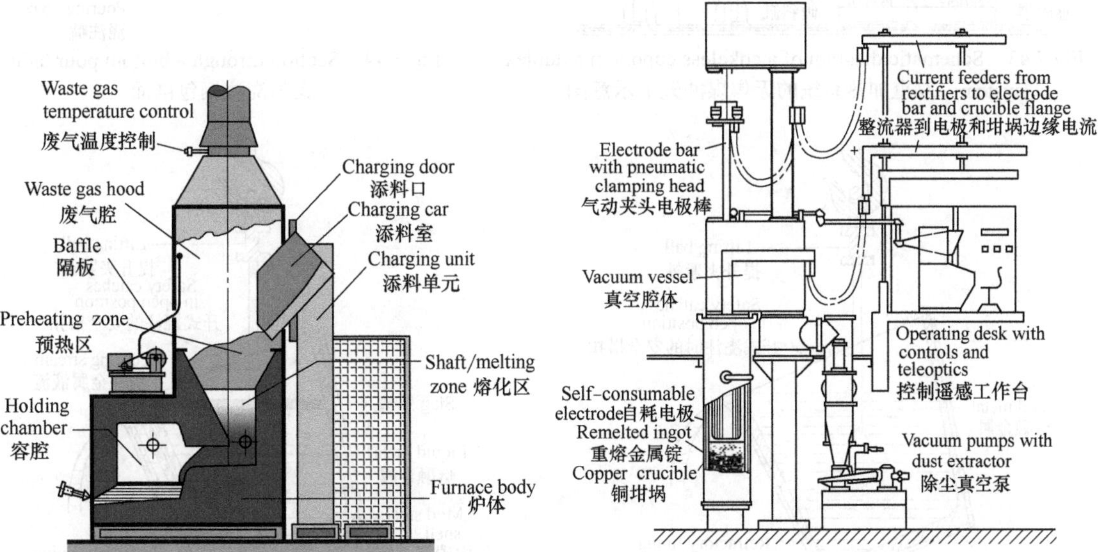

Fig 7.40 Gas-fired shaft furnace 燃气塔炉

Fig 7.41 Vacuum arc remelting furnace 真空自耗电弧炉

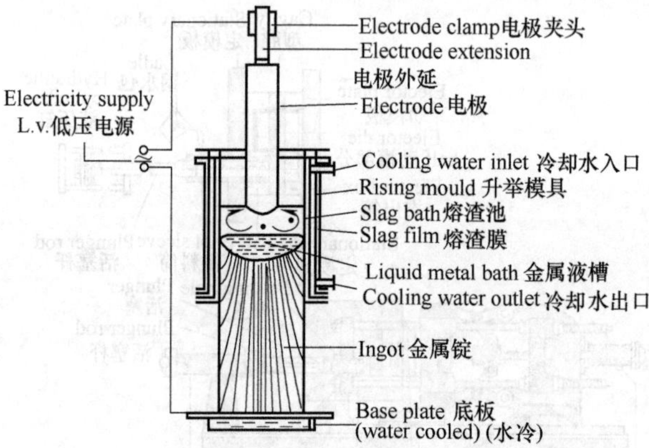

Fig 7.42 Electroslag remelting furnace 电渣重熔炉

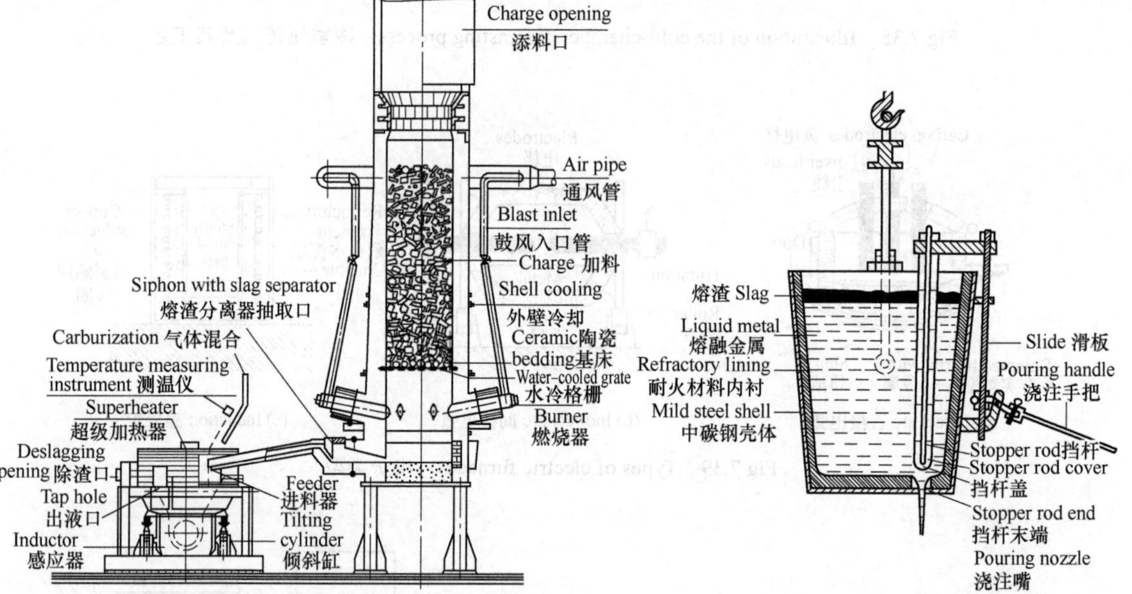

Fig 7.43 Schematic diagram of a cokeless cupola in a duplex system 双重加热系统的无焦煤冲天炉示意图

Fig 7.44 Section through a bottom pour ladle 底部浇注钢包剖面

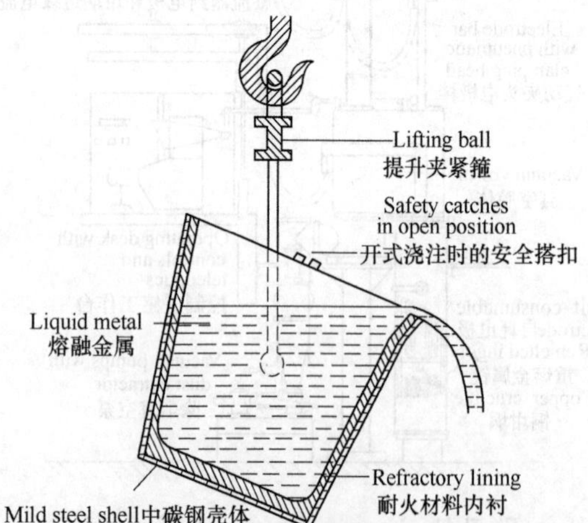

Fig 7.45 Section through a lip pour ladle 突出边缘的钢包

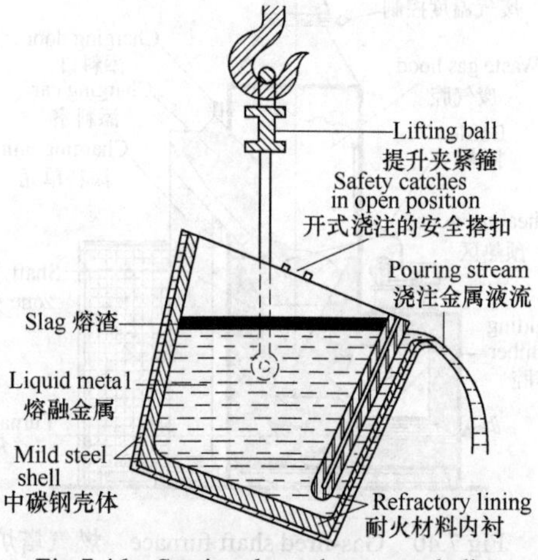

Fig 7.46 Section through a teapot ladle 茶壶形钢包剖面

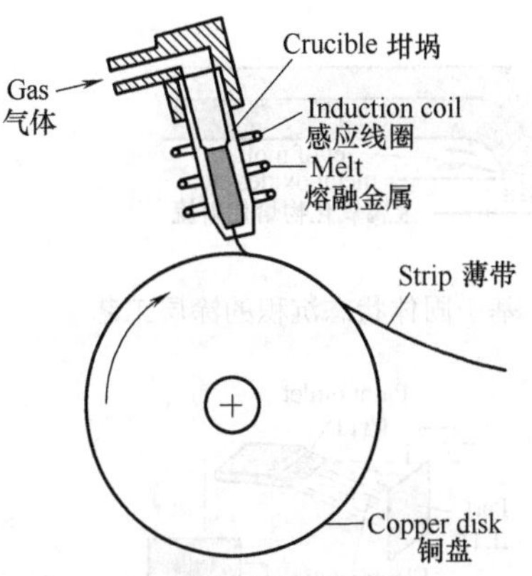

Fig 7.47 Melt-spinning to produce thin strips of amorphous metal
非晶体金属薄带的熔旋制造方法

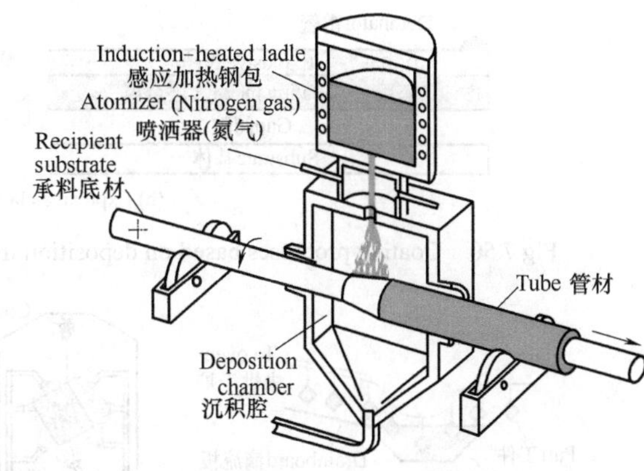

Fig 7.48 Spray casting (in which molten metal is sprayed over a rotating mandrel to produce seamless tubing and pipe)
喷射铸造（熔融金属喷射到回转心轴上铸造无缝钢管）

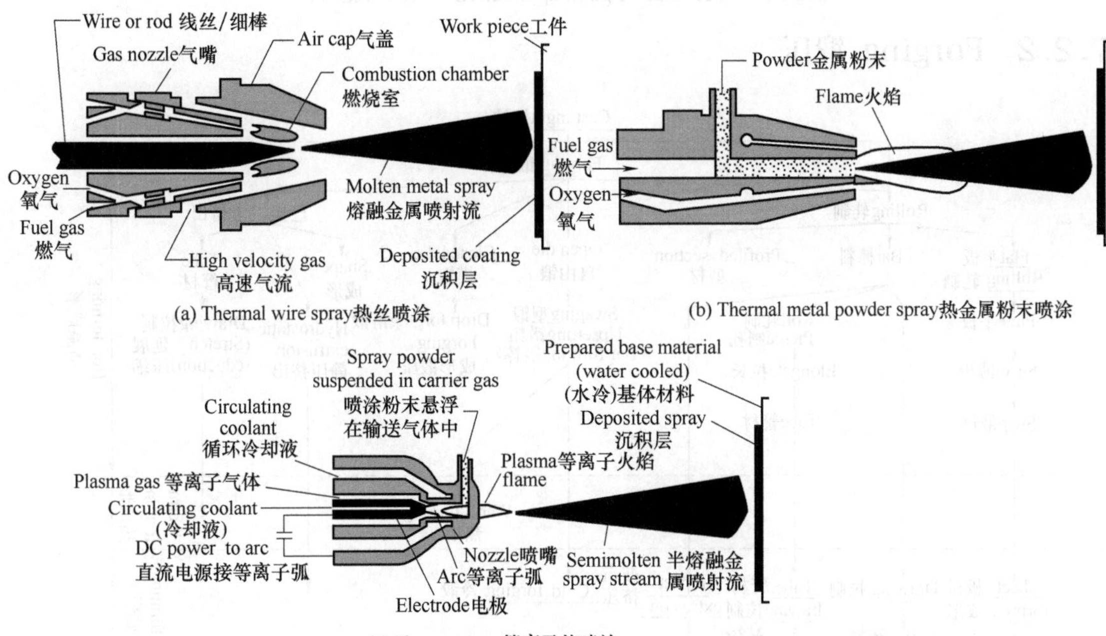

Fig 7.49 Thermal spray operations 热喷涂工艺

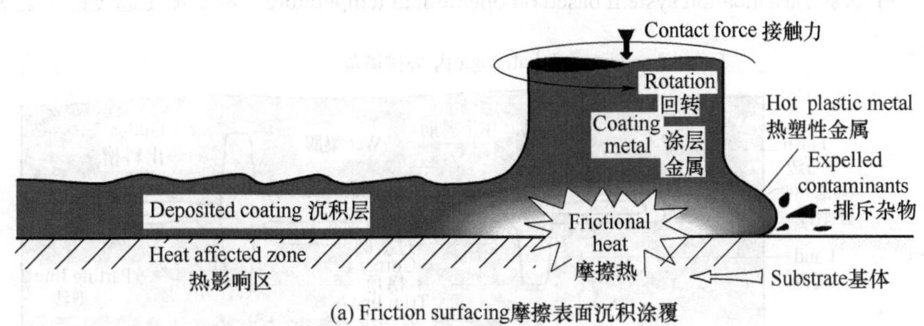

(a) Friction surfacing摩擦表面沉积涂覆

Fig 7.50

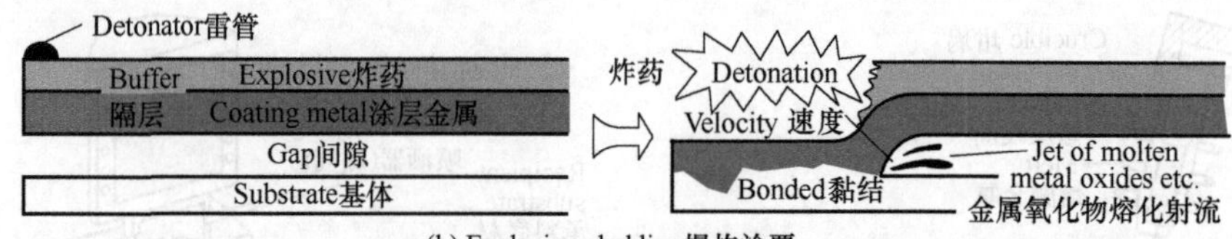

(b) Explosive cladding爆炸涂覆

Fig 7.50 Coating processes based on deposition in the solid state 基于固体状态沉积的涂层工艺

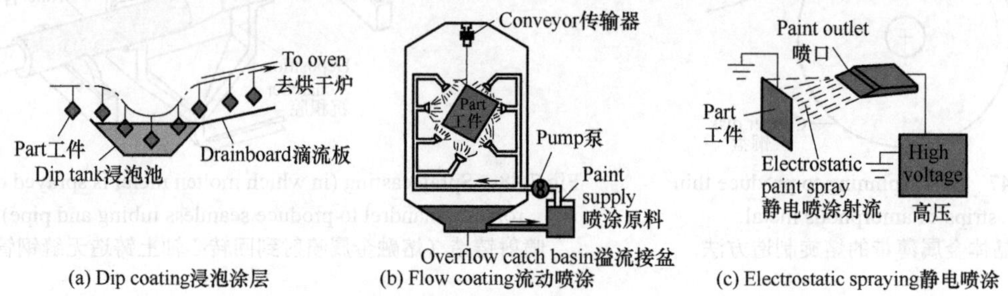

(a) Dip coating浸泡涂层 (b) Flow coating流动喷涂 (c) Electrostatic spraying静电喷涂

Fig 7.51 Methods of paint application 喷涂实施方法

7.2.2 Forging 锻压

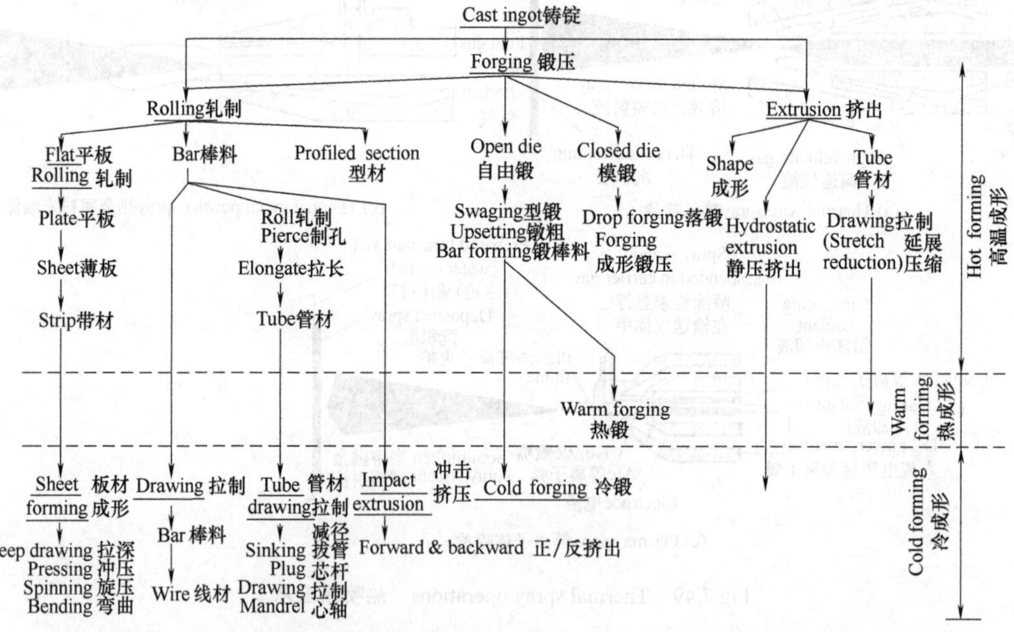

Fig 7.52 Process classification system based on operational temperature 基于加工温度的工艺分类系统

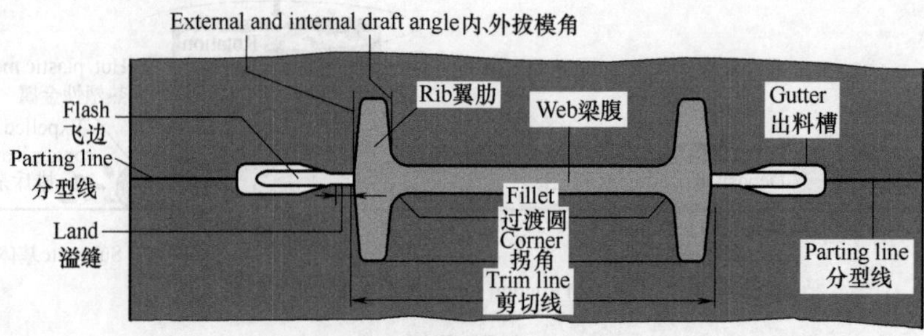

Fig 7.53 Die forging 模锻

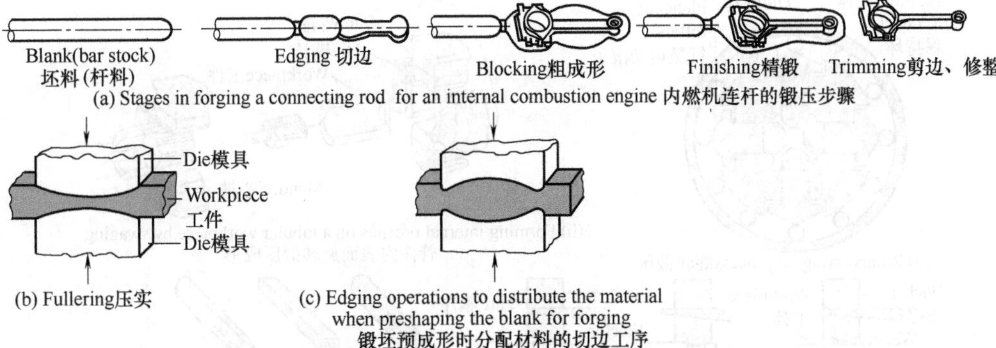

Fig 7.54 Forging a connecting rod 连杆的锻压

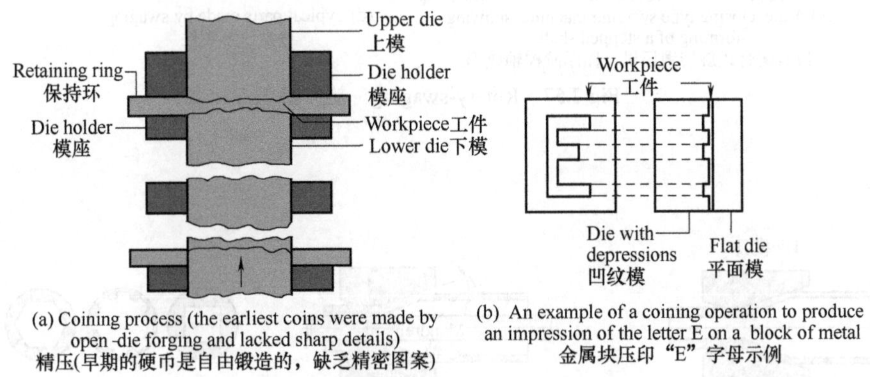

Fig 7.55 Coining 精压、压印

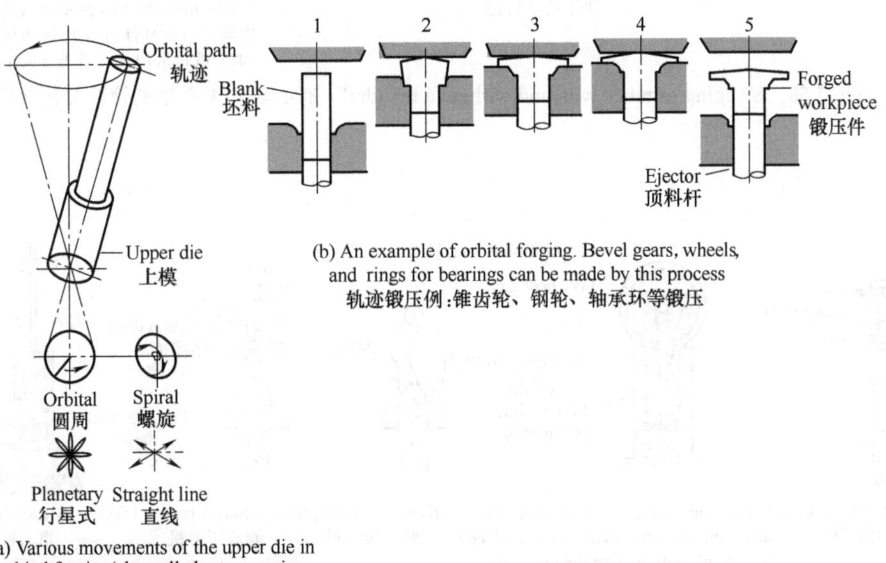

Fig 7.56 Rocking-die forging 摇头锻压

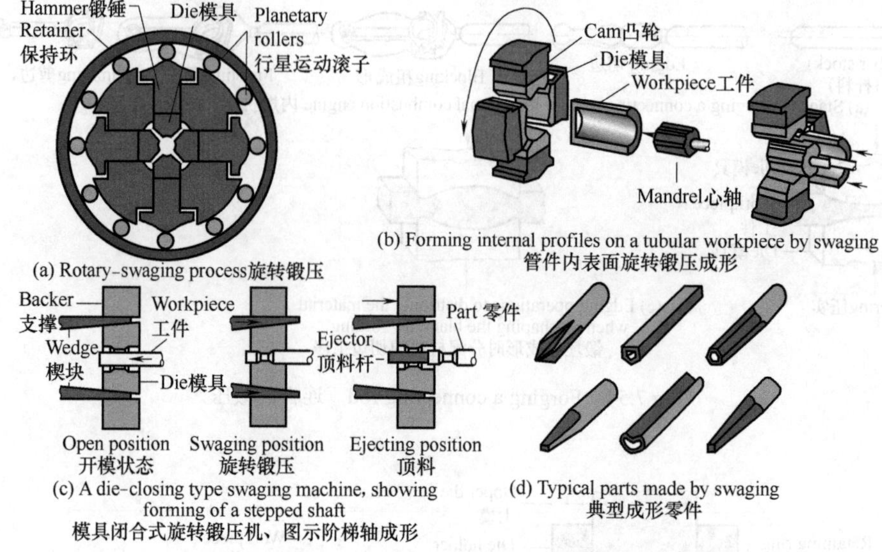

(a) Rotary-swaging process旋转锻压

(b) Forming internal profiles on a tubular workpiece by swaging 管件内表面旋转锻压成形

(c) A die-closing type swaging machine, showing forming of a stepped shaft 模具闭合式旋转锻压机、图示阶梯轴成形

(d) Typical parts made by swaging 典型成形零件

Fig 7.57 Rotary-swaging 旋转锻压

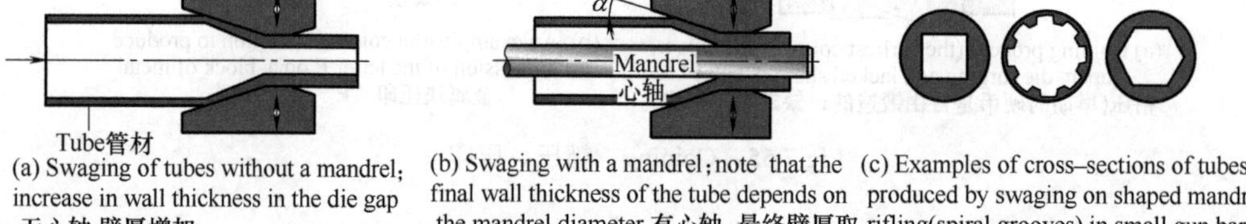

(a) Swaging of tubes without a mandrel; increase in wall thickness in the die gap 无心轴、壁厚增加

(b) Swaging with a mandrel; note that the final wall thickness of the tube depends on the mandrel diameter 有心轴；最终壁厚取决于心轴直径

(c) Examples of cross-sections of tubes produced by swaging on shaped mandrels rifling(spiral grooves) in small gun barrels can be made by this process 旋转锻压管内截面形状样件,小型枪管来复线(螺旋沟槽)就是采用旋转锻压来成形

Fig 7.58 Swaging of tubes with and without a mandrel 有心轴和无心轴的管件旋转锻压

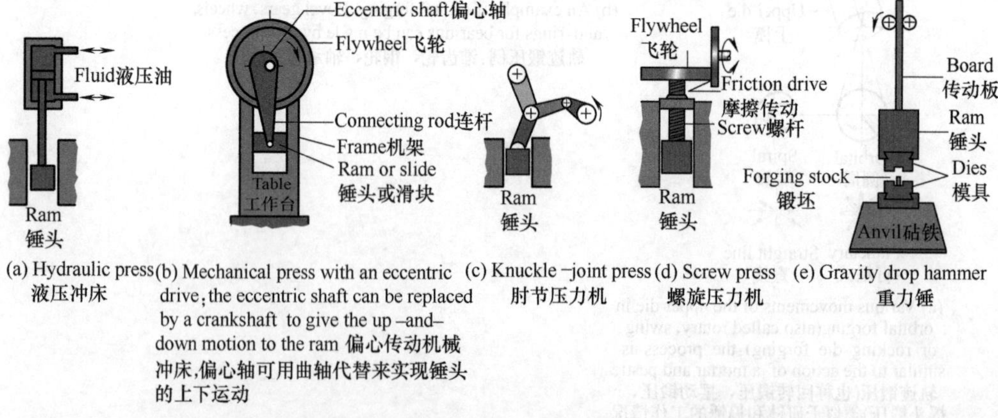

(a) Hydraulic press 液压冲床

(b) Mechanical press with an eccentric drive; the eccentric shaft can be replaced by a crankshaft to give the up-and-down motion to the ram 偏心传动机械冲床,偏心轴可用曲轴代替来实现锤头的上下运动

(c) Knuckle-joint press 肘节压力机

(d) Screw press 螺旋压力机

(e) Gravity drop hammer 重力锤

Fig 7.59 Principles of various forging machines 几种锻压机床原理

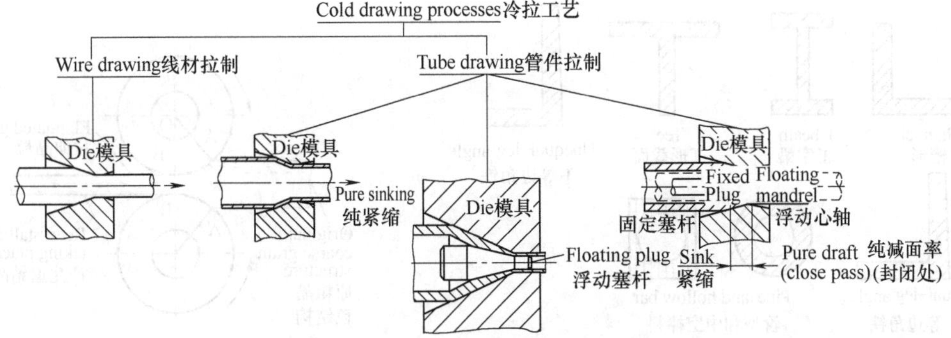

Fig 7.60 Cold drawing processes 冷拉工艺

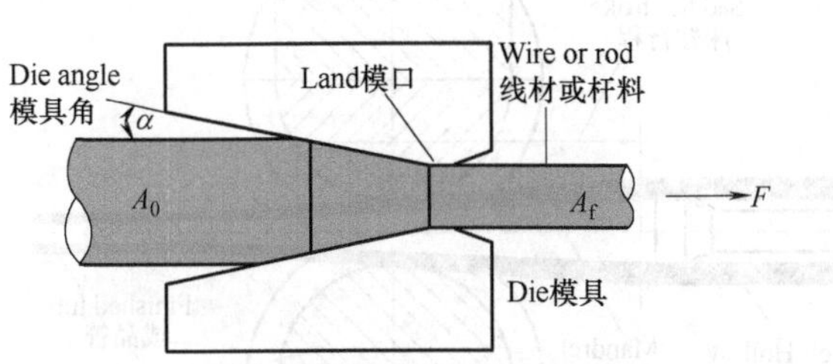

Fig 7.61 Process variables in wire drawing 线材拉制工艺参数

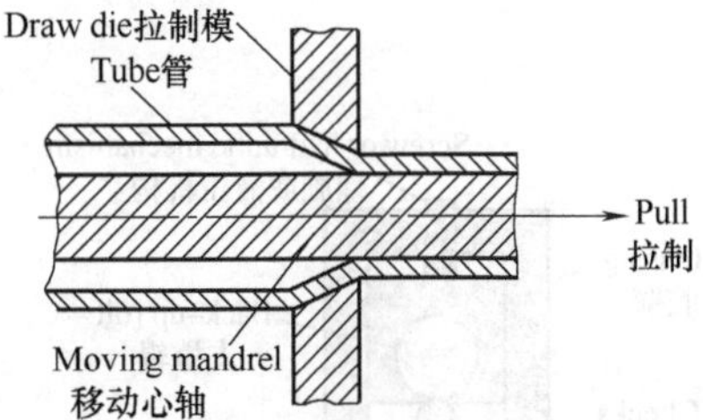

Fig 7.62 Tube drawing with a moving mandrel 活动心轴管材拉制

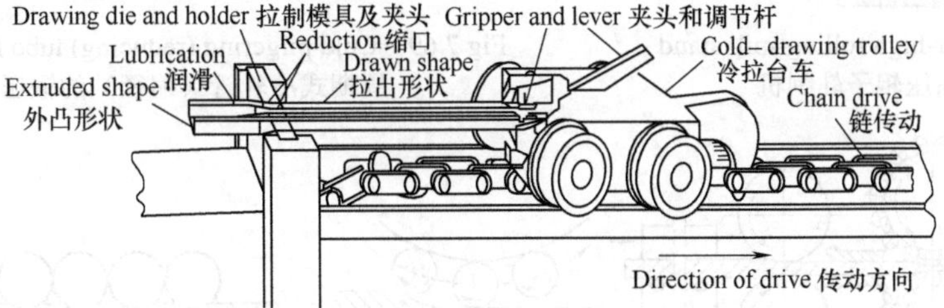

Fig 7.63 Cold drawing of an extruded channel on a draw bench, to reduce its cross-section 拉制台上冷拉外凸管件以减少断面尺寸

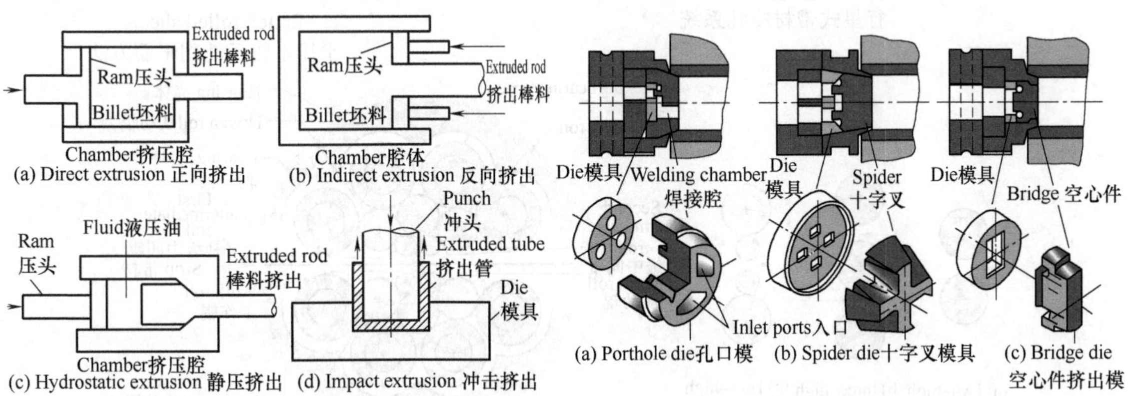

Fig 7.64 Extrusion process 挤出工艺

Fig 7.65 Components for extruding hollow shapes 空心挤出件

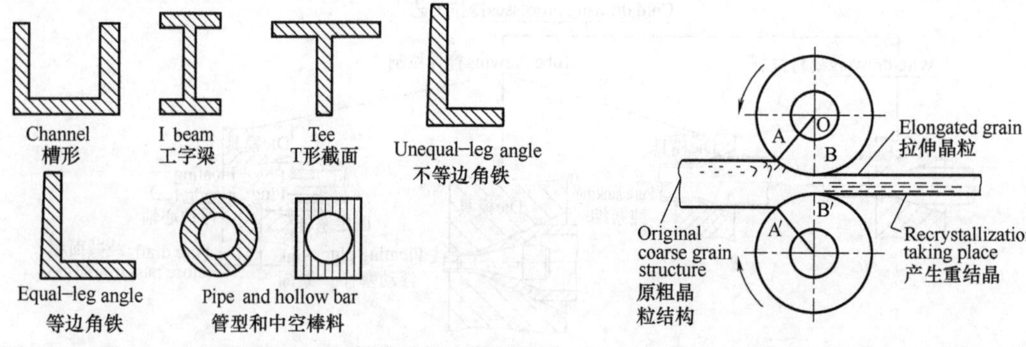

Fig 7.66 Standard extruded shapes 标准的挤出截面形状

Fig 7.67 Hot-rolling process 热轧工艺

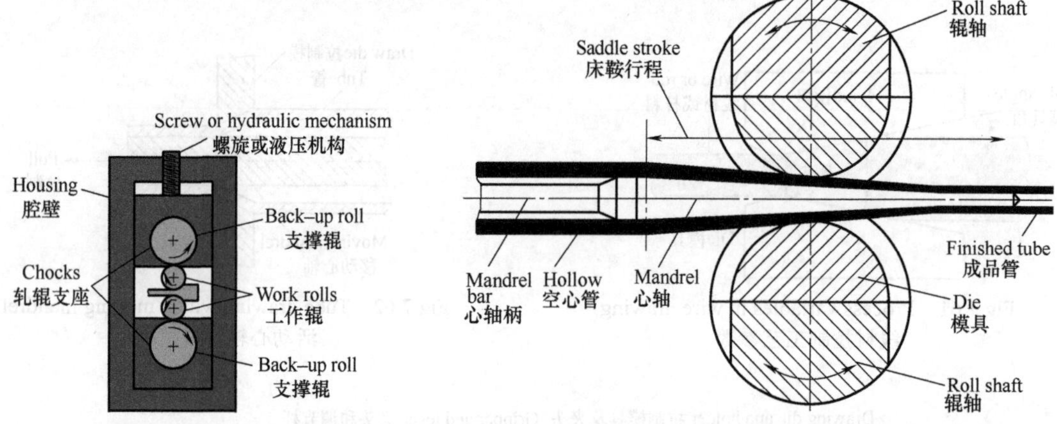

Fig 7.68 Four-high rolling-mill stand
四个高压辊子轧制机

Fig 7.69 Cold pilgering (reducing) tube forging process
周期式冷轧管（缩径）的锻压过程

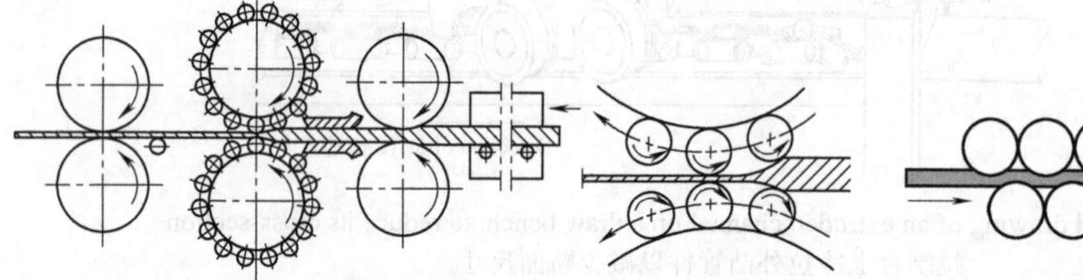

Fig 7.70 System incorporating a planetary mill for cold strip rolling
行星式带材冷轧系统

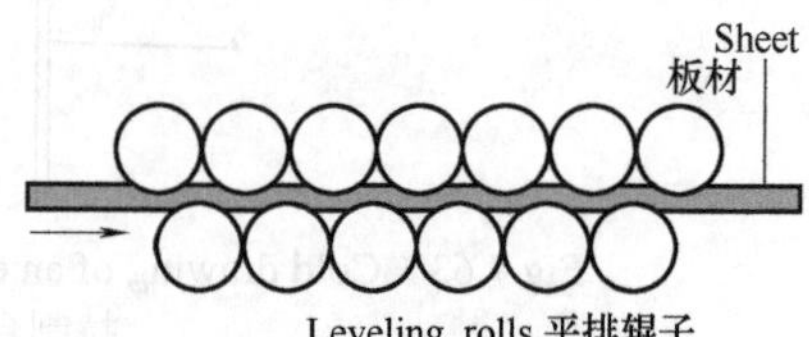

Fig 7.71 A method of roller leveling to flatten rolled sheets
平排多辊子轧制平面方法

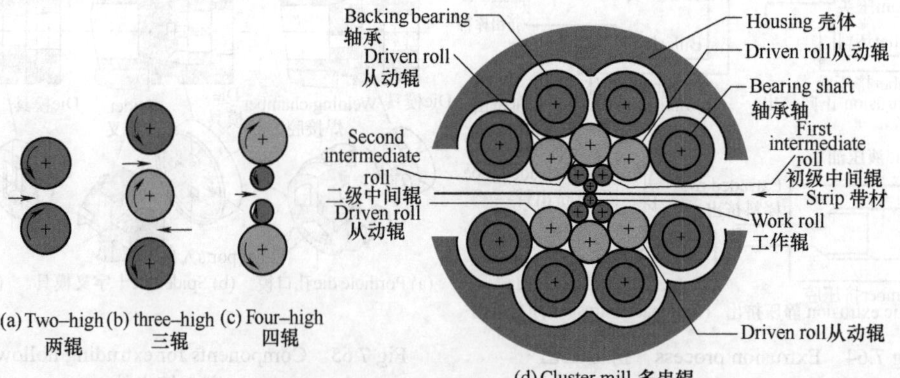

Fig 7.72 Various roll arrangements 几种轧辊布局

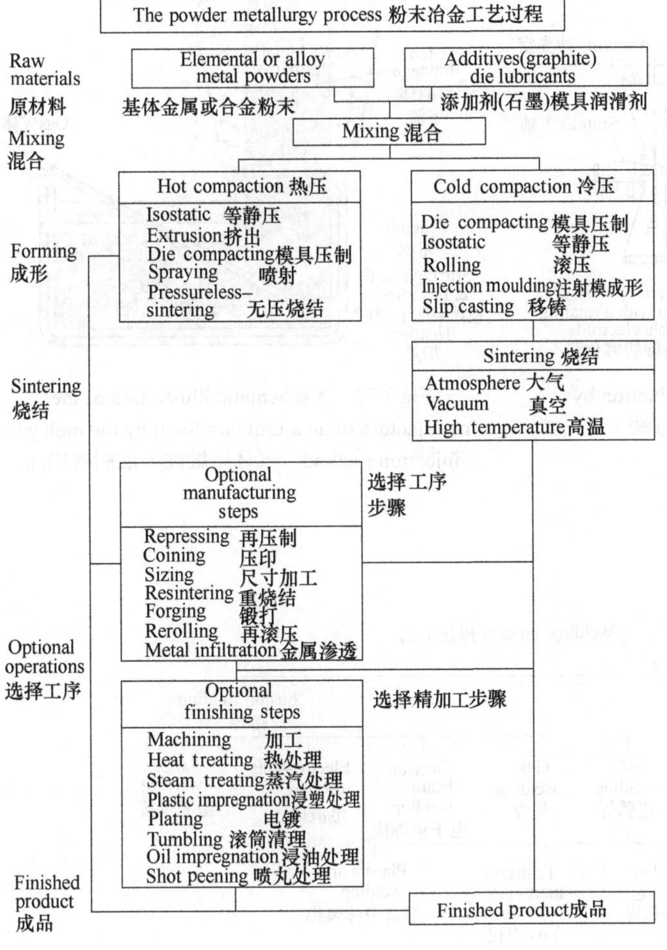

Fig 7.73 Powder metallurgy process 粉末冶金工艺

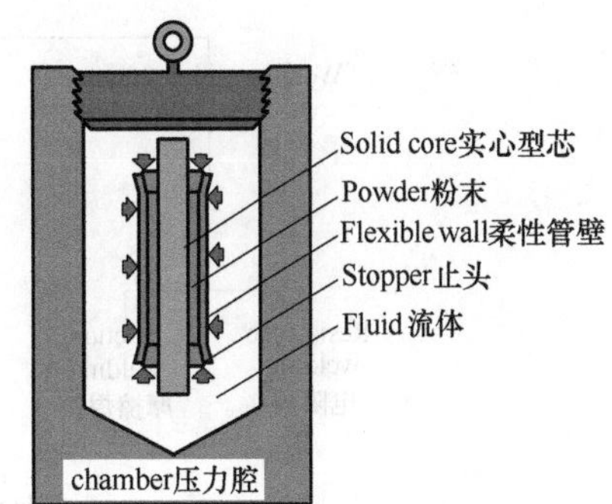

Fig 7.74 Cold isostatic pressing, as applied to forming a tube
冷等静压，图示用于管件成形

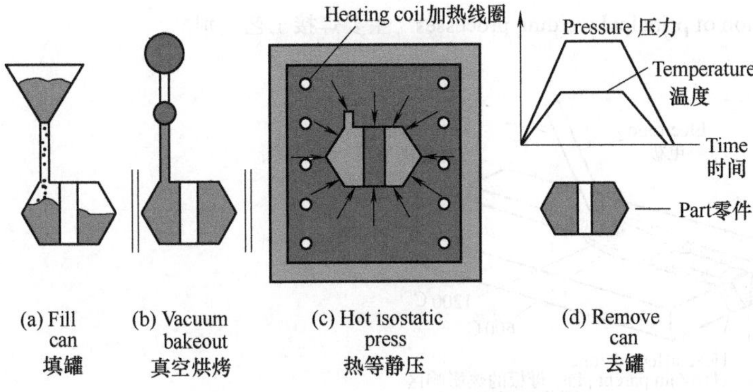

Fig 7.75 Hot isostatic pressing 热等静压压制成形

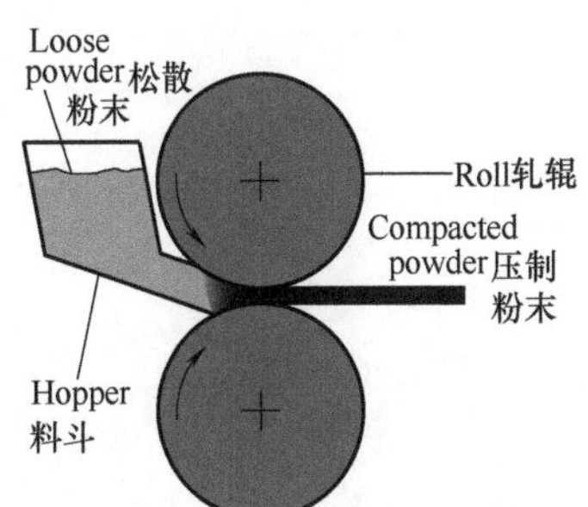

Fig 7.76 An example of powder rolling 粉末轧制示例

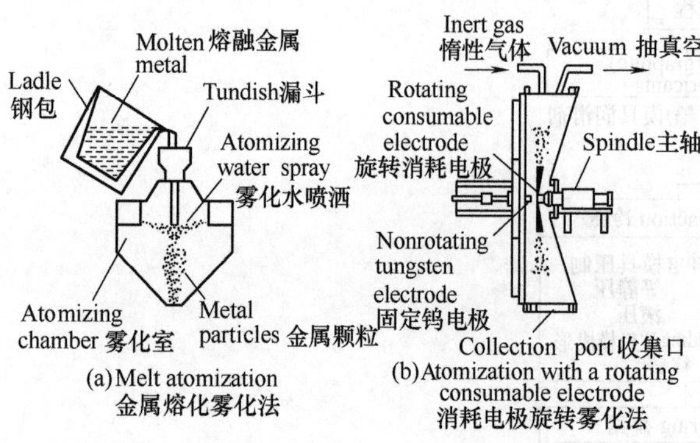

Fig 7.77 Methods of metal-powder production by atomization 金属粉末雾化制备方法

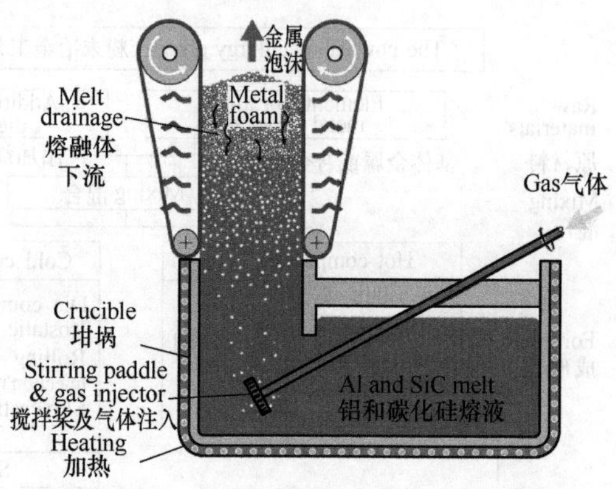

Fig 7.78 A schematic illustration of the manufacture of an aluminum foam by the melt gas injection method 气体熔融注入法制备铝沫

7.2.3 Welding 焊接

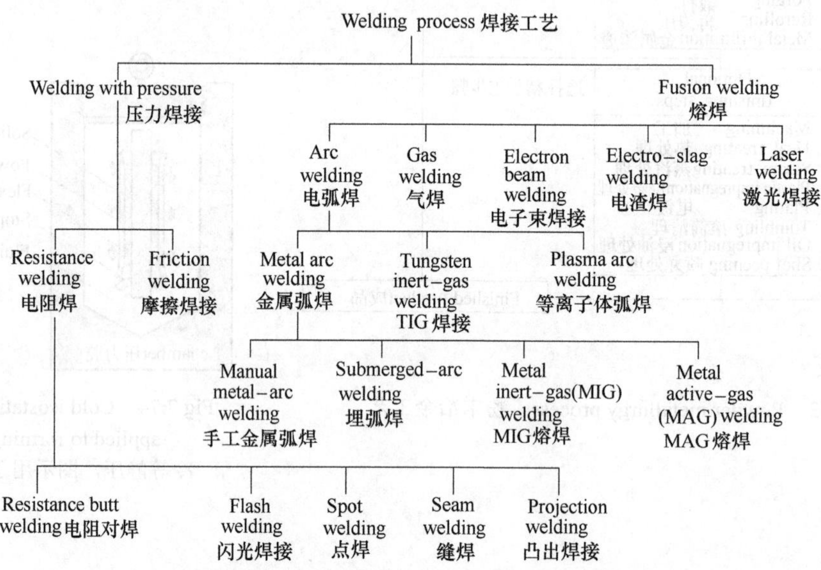

Fig 7.79 Classification of principal welding processes 主要焊接工艺类别

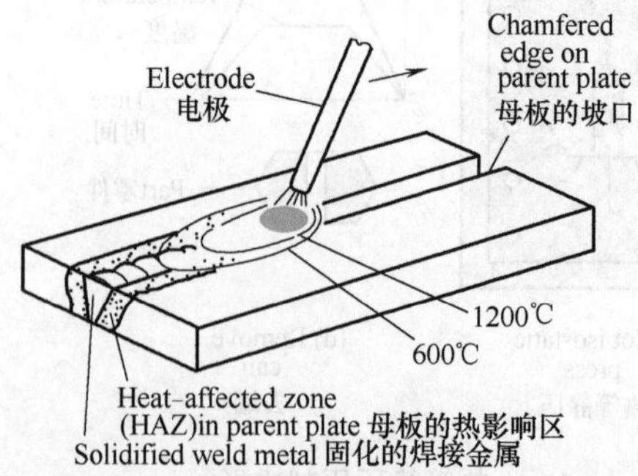

Fig 7.80 Schematic of a typical welding operation 典型焊接操作图解

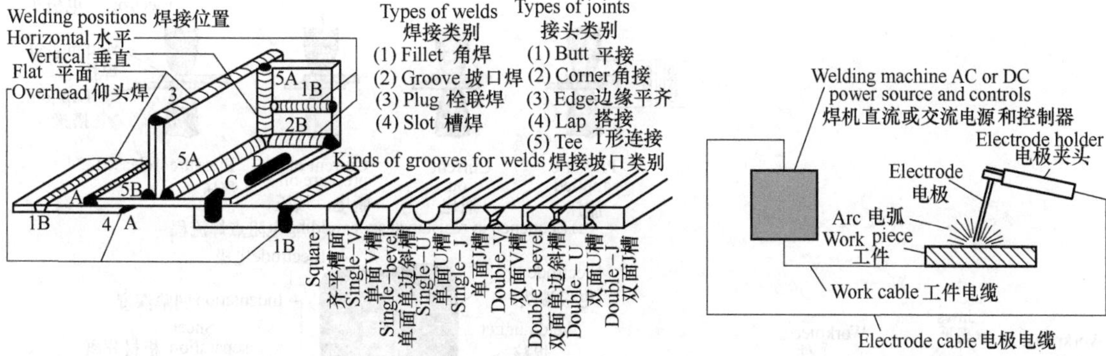

Fig 7.81 Types of welded joints 焊接接口种类

Fig 7.82 The basic arc-welding circuit 基本弧焊回路

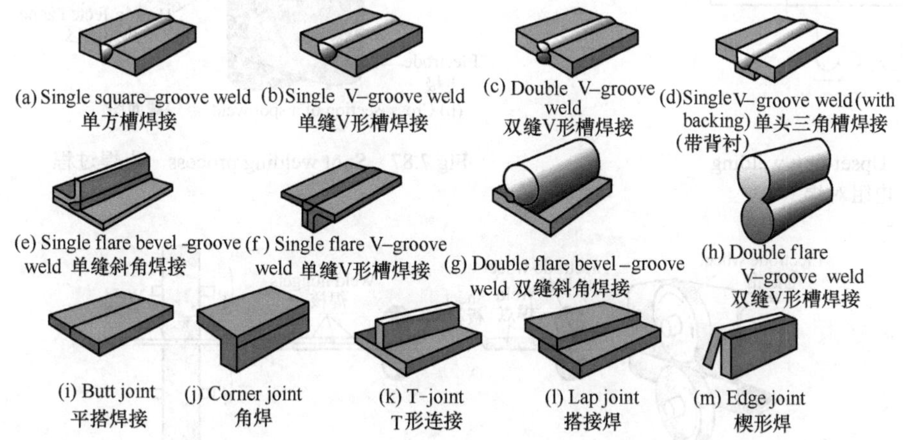

Fig 7.83 Examples of welded joints and their terminology 焊接头形式及其术语

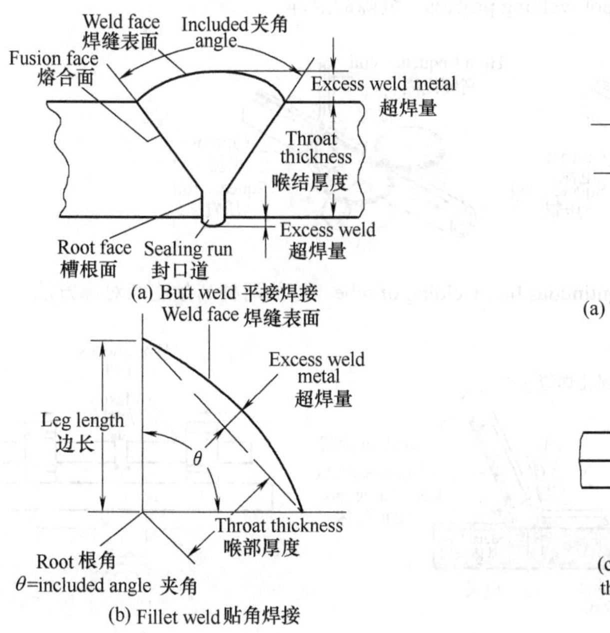

Fig 7.84 Welding terminology 焊接术语

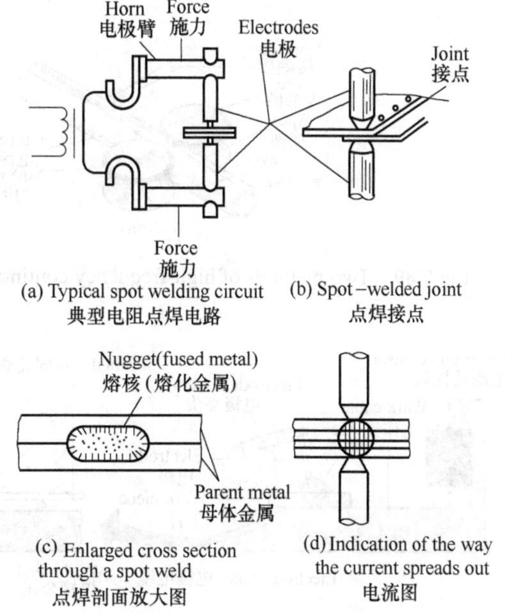

Fig 7.85 Process of spot resistance welding 电阻点焊工艺

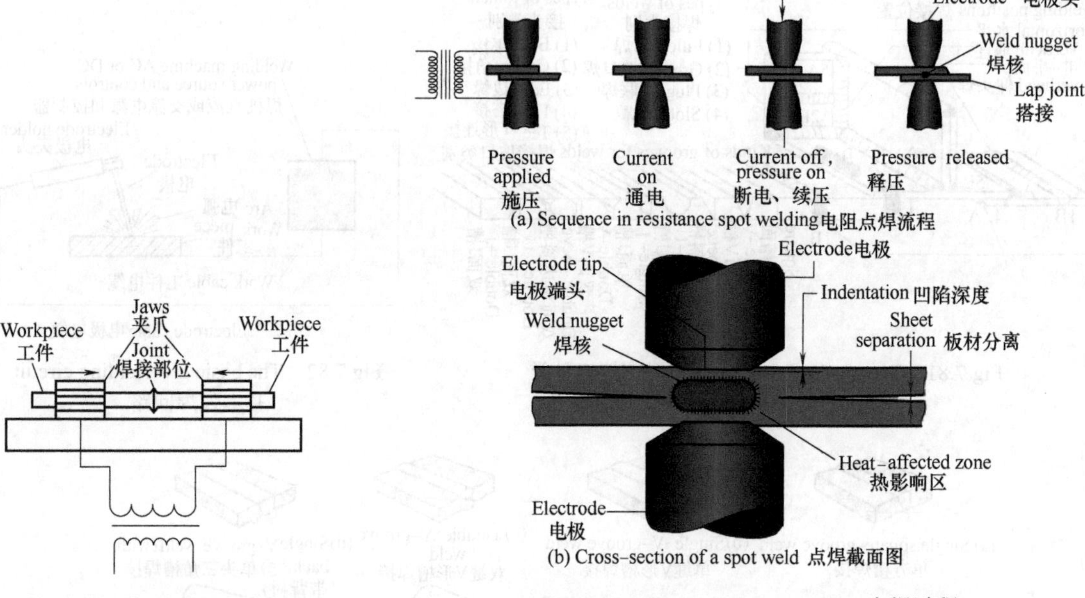

Fig 7.86 Upset-butt welding 电阻对焊

Fig 7.87 Spot welding process 点焊过程

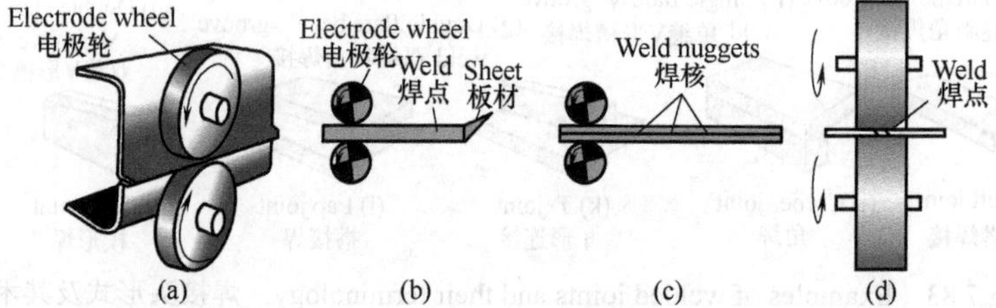

Fig 7.88 Seam spot-welding process 缝隙的点焊

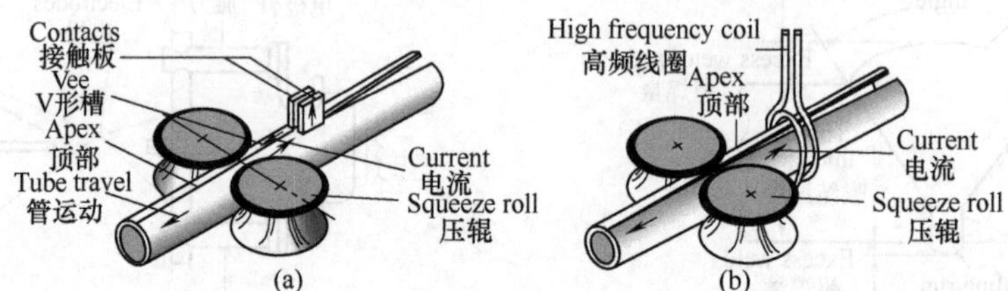

Fig 7.89 Two methods of high-frequency continuous butt welding of tube 管的两种高频连续对焊方法

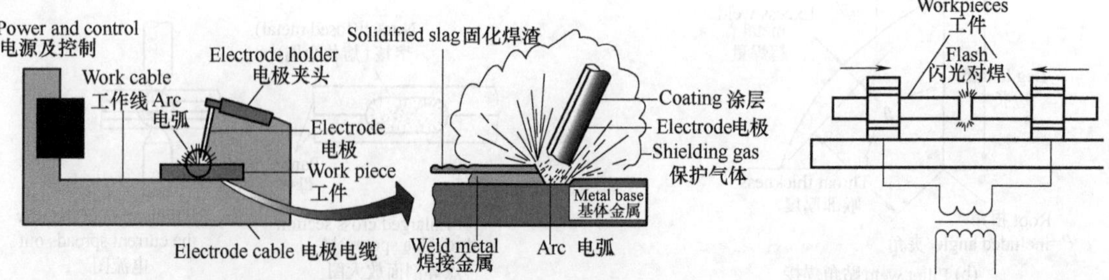

Fig 7.90 Illustration of shielding metal-arc welding process 金属引弧保护焊

Fig 7.91 Flash-butt welding 闪光对焊

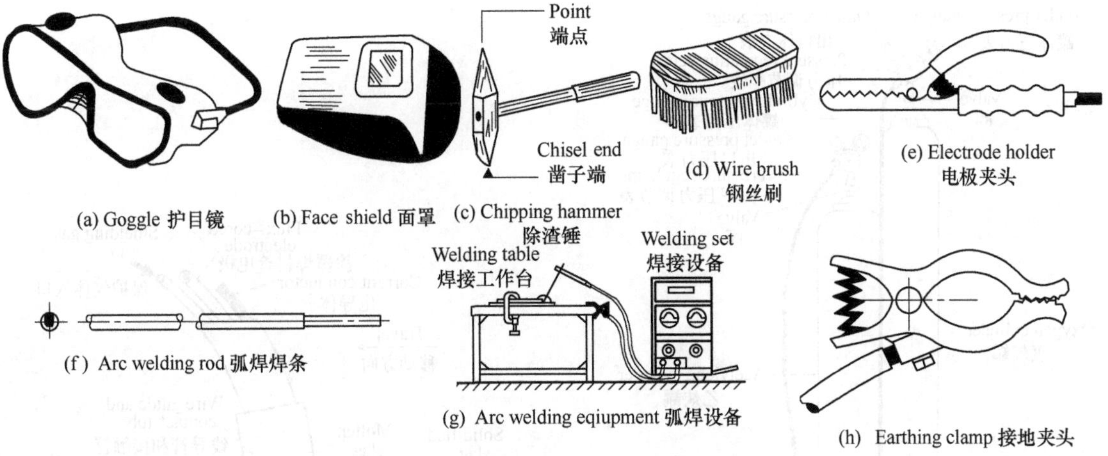

Fig 7.92 Other accessories 焊接用其他附件

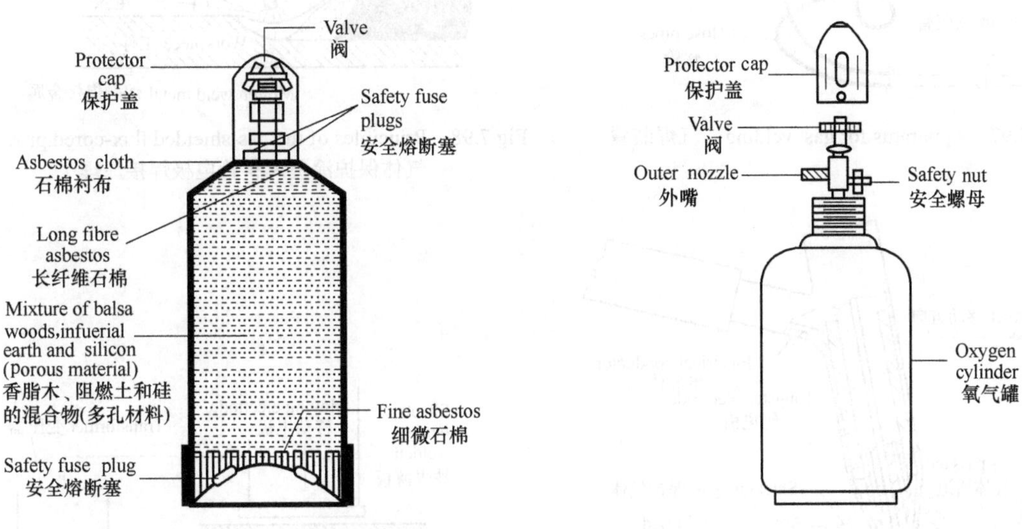

Fig 7.93 Acetylene cylinder 乙炔罐

Fig 7.94 Oxygen cylinder 氧气罐

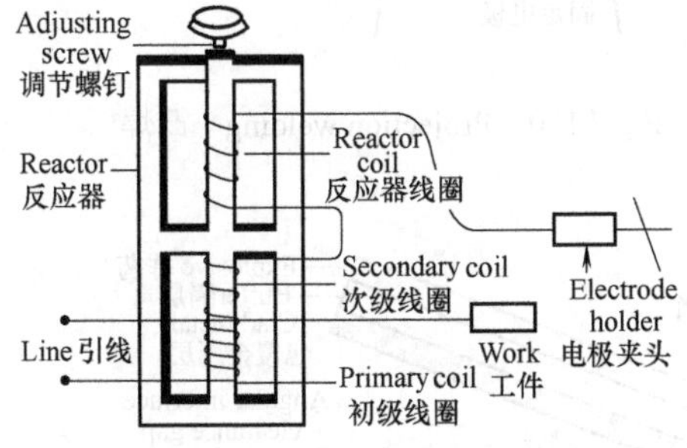

Fig 7.95 Welding transformer set
焊接变压器

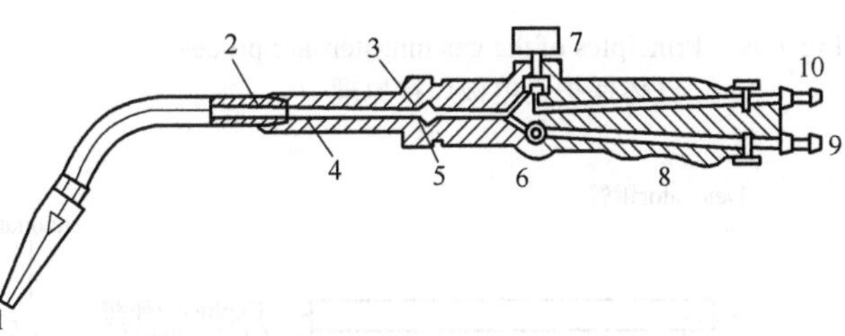

Fig 7.96 Welding torch 焊炬

1—Torch mouth 火焰口；2—Mixing tube 混合管；3—Injector 注入口；
4—Mixing nozzle 混合喷嘴；5—Pressure nozzle 压力喷嘴；
6—Acetylene valve 乙炔阀；7—Oxygen cylinder 氧气阀；
8—Grip 握柄；9—Acetylene entrance 乙炔入口；
10—Oxygen entrance 氧气入口

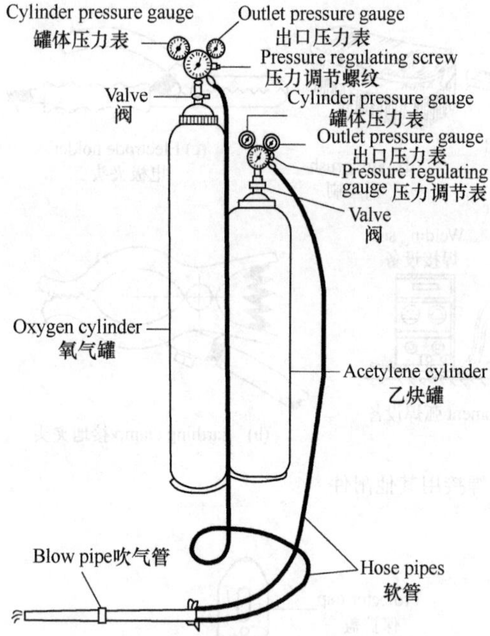

Fig 7.97 Apparatus for gas welding 气焊装置

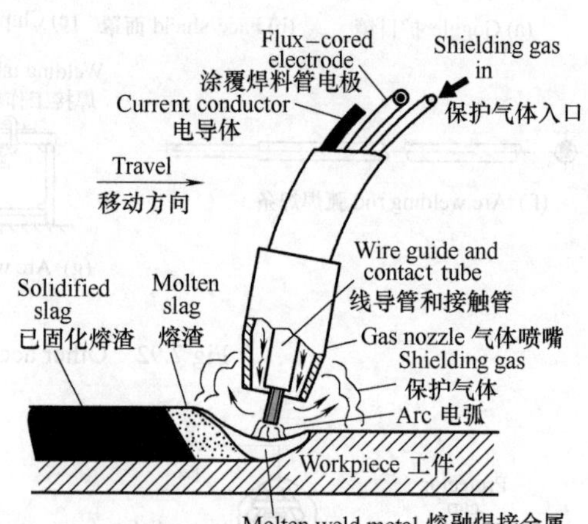

Fig 7.98 Principles of the gas-shielded flux-cored process
气体保护涂覆焊料管电极焊接原理

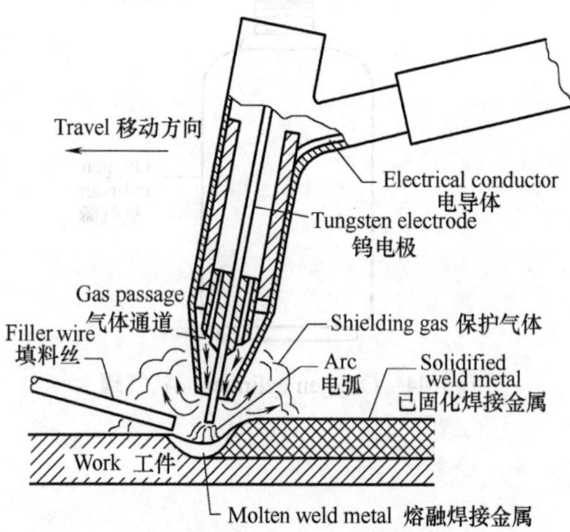

Fig 7.99 Principles of the gas tungsten-arc process
气体保护钨电极焊接原理

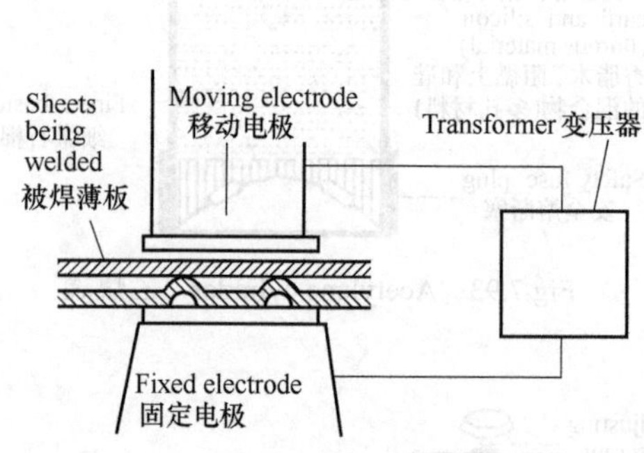

Fig 7.100 Projection welding 凸焊

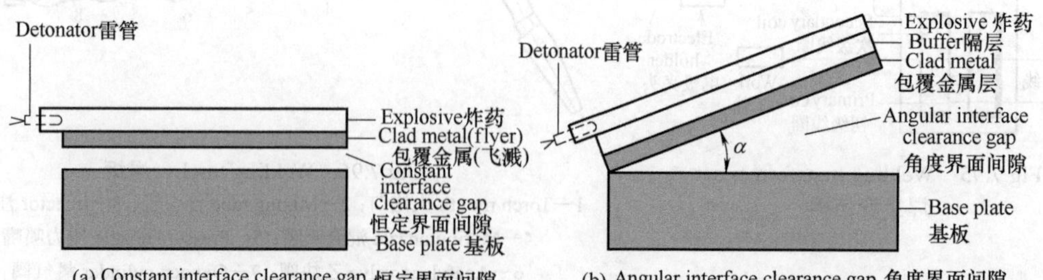

(a) Constant interface clearance gap 恒定界面间隙
(b) Angular interface clearance gap 角度界面间隙

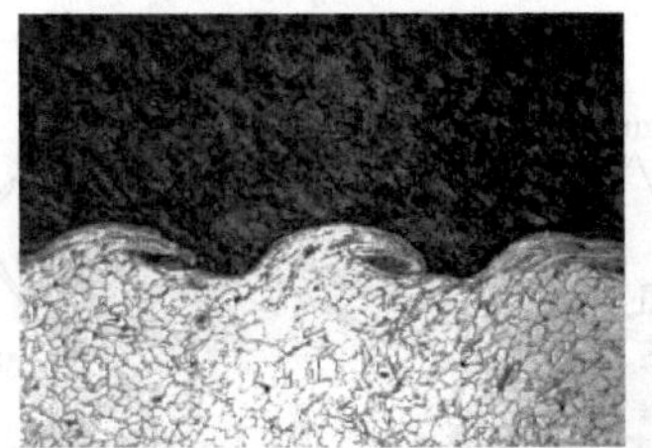

(c) Titanium(top piece)on low- carbon steel(bottom)
低碳钢(下板)上焊接钛板(上板)

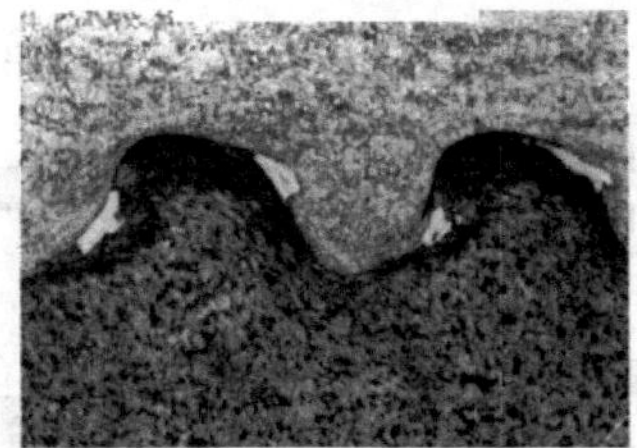

(d) Incoloy 800(an ironnickel-based alloy)on low carbon steel
低碳钢(下板)上焊接铁镍基合金板

Fig 7.101 Explosion welding process 爆炸焊接

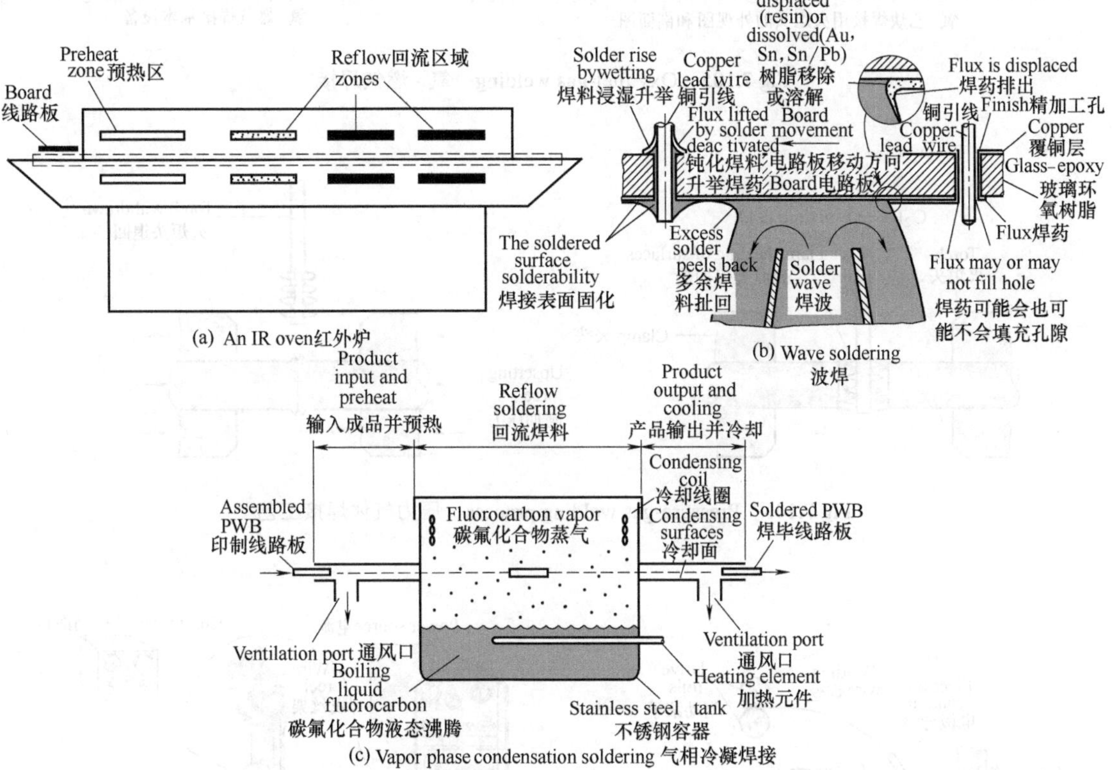

(a) An IR oven 红外炉

(b) Wave soldering 波焊

(c) Vapor phase condensation soldering 气相冷凝焊接

Fig 7.102 Three commonly used soldering (or reflow soldering) facilities
三种常用（低温）焊接（回流焊接）设备

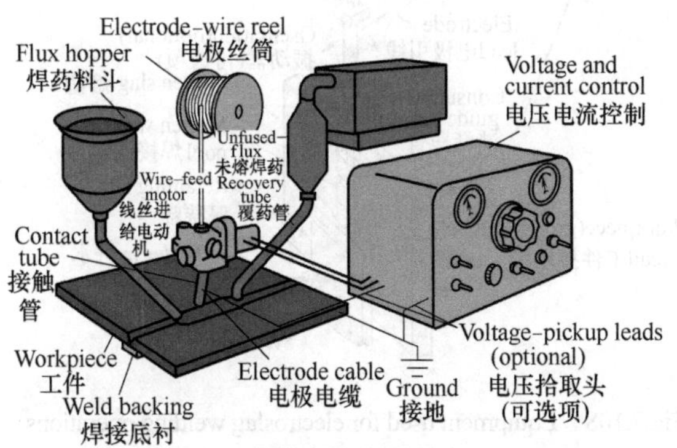

Fig 7.103 Submerged-arc welding process and equipment 埋弧焊接工艺和设备

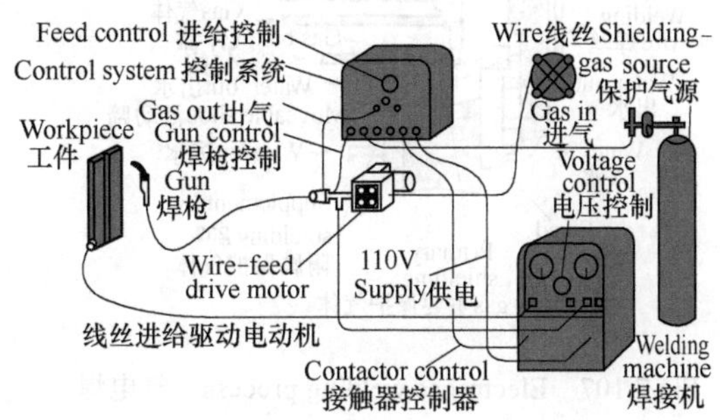

Fig 7.104 Basic equipment used in gas metal-arc welding operations 惰性气体金属保护焊的基本设备

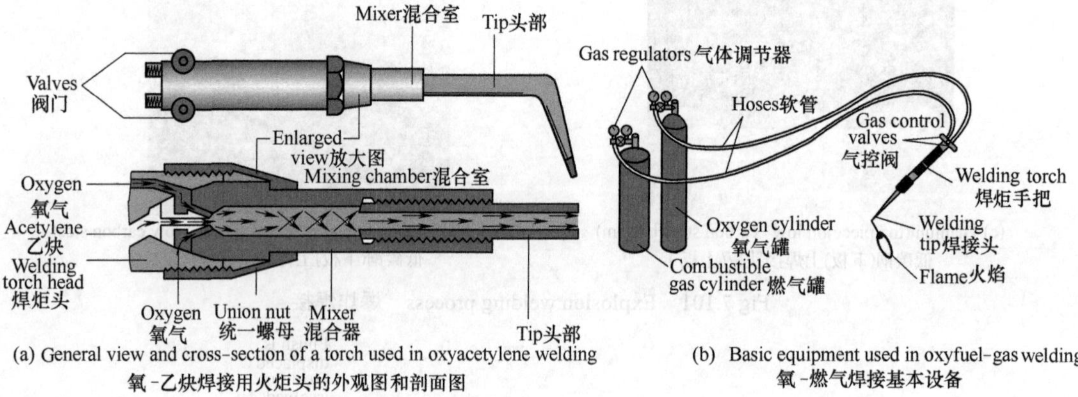

(a) General view and cross-section of a torch used in oxyacetylene welding
氧-乙炔焊接用火炬头的外观图和剖面图

(b) Basic equipment used in oxyfuel-gas welding
氧-燃气焊接基本设备

Fig 7.105 Oxy fuel-gas welding 氧-燃气焊接

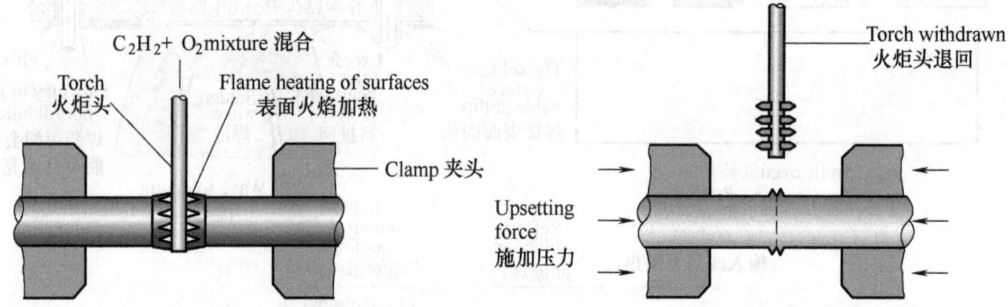

Fig 7.106 Pressure-gas welding process 压力气体焊接工艺

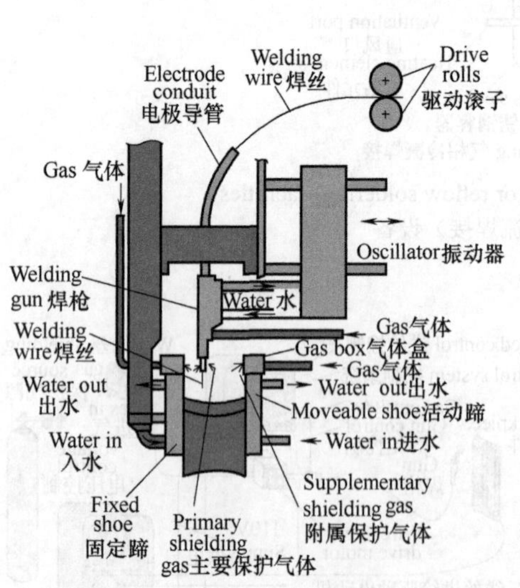

Fig 7.107 Electrogas welding process 气电焊

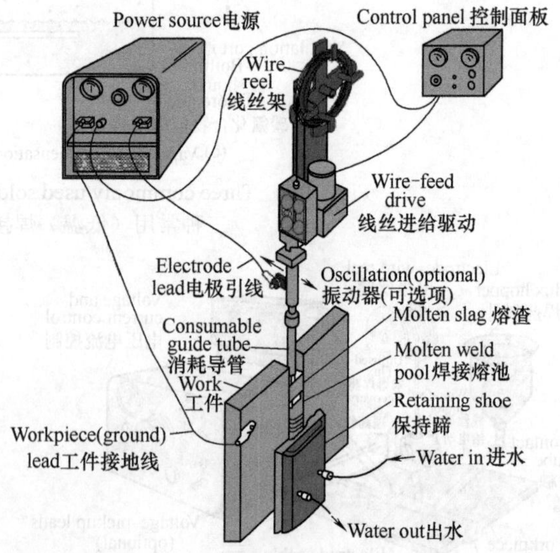

Fig 7.108 Equipment used for electroslag welding operations
电熔渣焊基本设备

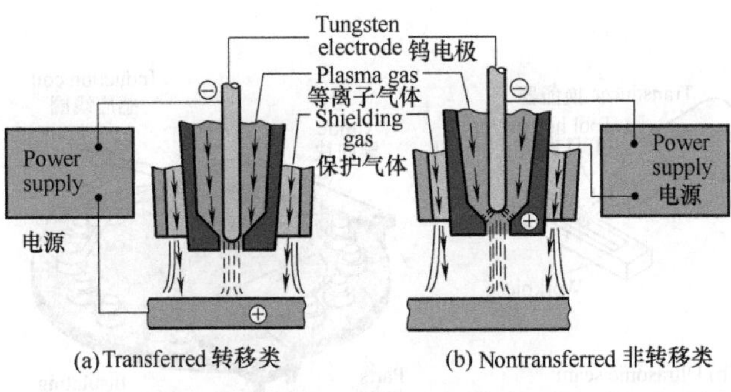

Fig 7.109 Two types of plasma-arc welding processes
两类等离子弧焊接

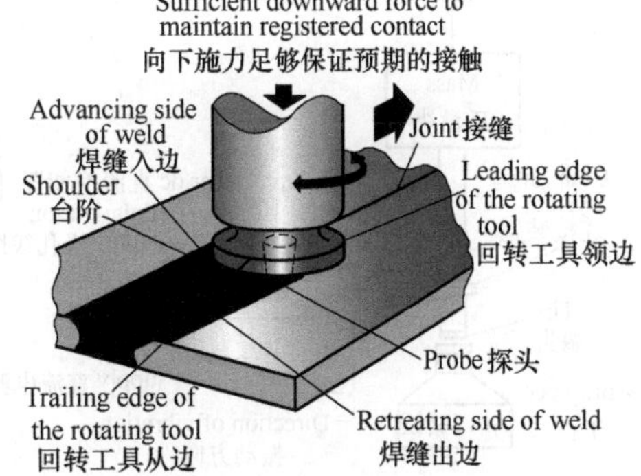

Fig 7.110 Friction stir welding
搅拌摩擦焊接

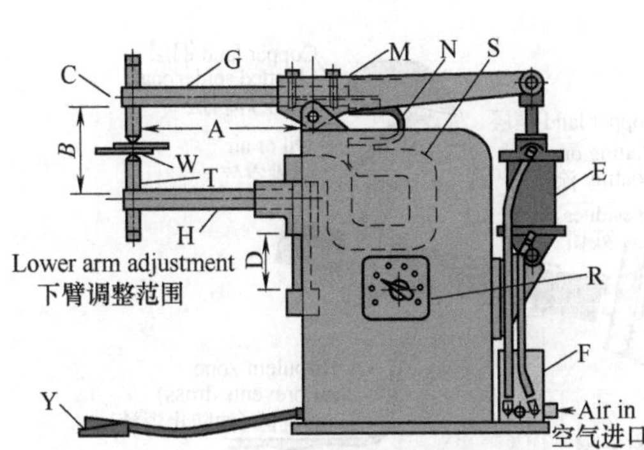

Fig 7.111 An air-operated rocker-arm spot welding machine 气动摇臂式点焊机

A—Throat depth 喉部深度；B—Horn spacing 电极头开度；C—Centerline of rocker arm 摇杆中心线；E—Air cylinder 气缸；F—Air valve 气阀；G—Upper horn 上臂；H—Lower horn 下臂；M—Rocker arm 摇臂；N—Secondary flexible conductor 第二柔性导体；R—Current regulator (tap switch) 电流调节（旋钮）；S—Transformer secondary 变压器次级线圈；T—Electrode holder 电极夹头；W—Electrode 电极；Y—Foot control 脚踏

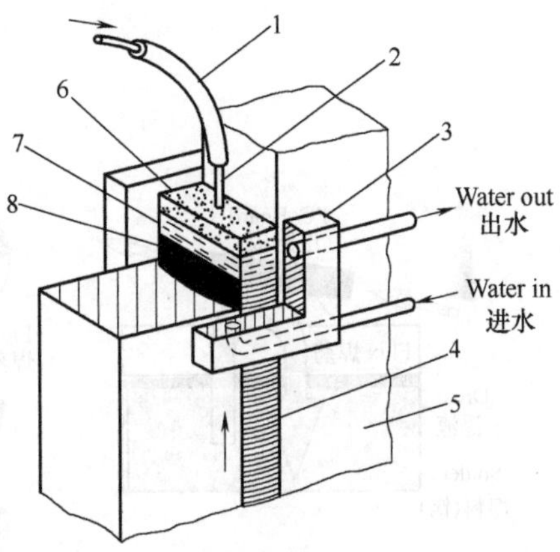

Fig 7.112 Schematic sketch of electroslag welding
电熔渣焊接图解

1—Electrode guide tube 电极导管；2—Electrode 电极；3—Water-cooled copper shoes 水冷铜座；4—Finished weld 已经焊接部分；5—Base metal 基体金属；6—Molten slag 熔渣；7—Molten weld metal 熔融焊接金属；8—Solidified weld metal 已经固化焊接金属

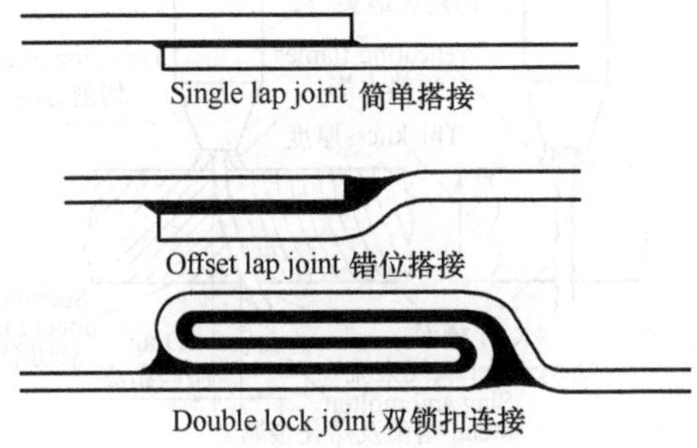

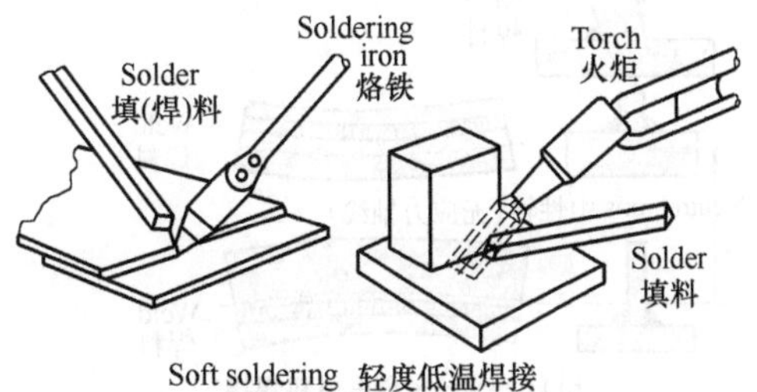

Fig 7.113 Soldered joints 低温焊接

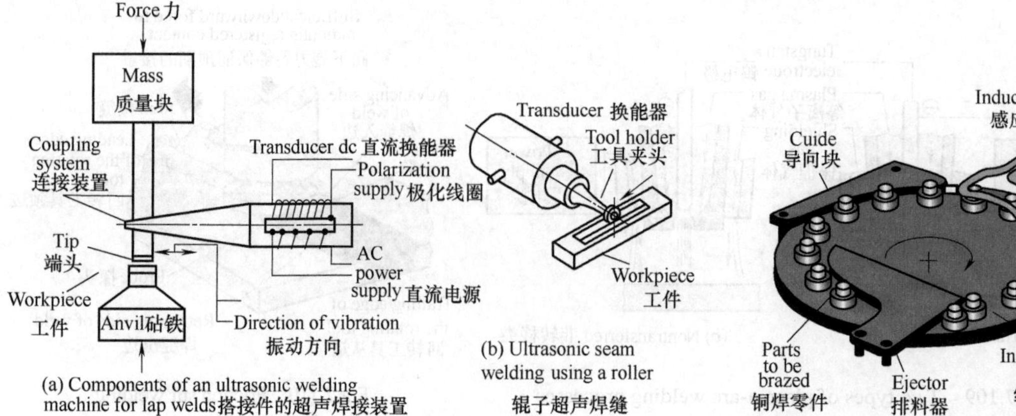

(a) Components of an ultrasonic welding machine for lap welds 搭接件的超声焊接装置

(b) Ultrasonic seam welding using a roller 辊子超声焊缝

Fig 7.114　Ultrasonic welding　超声焊接

Fig 7.115　A continuous induction-brazing setup, for increased productivity
连续感应加热铜焊装置用于提高效率

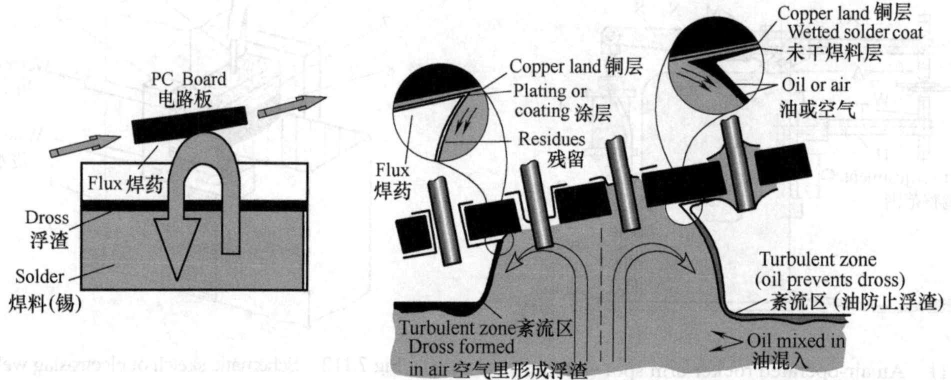

Fig 7.116　Wave soldering process　波形焊工艺

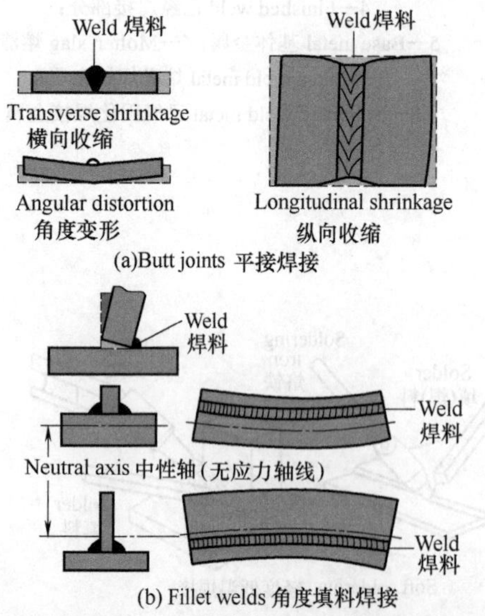

(a) Butt joints 平接焊接

(b) Fillet welds 角度填料焊接

Fig 7.117　Distortion of parts after welding　焊后变形

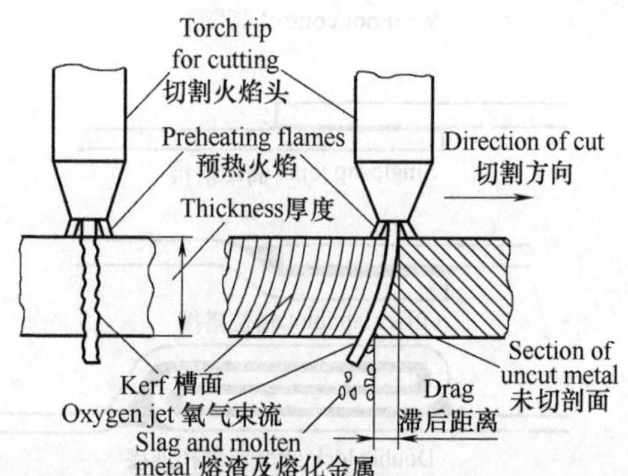

Fig 7.118　Principles of flame cutting　火焰切割原理

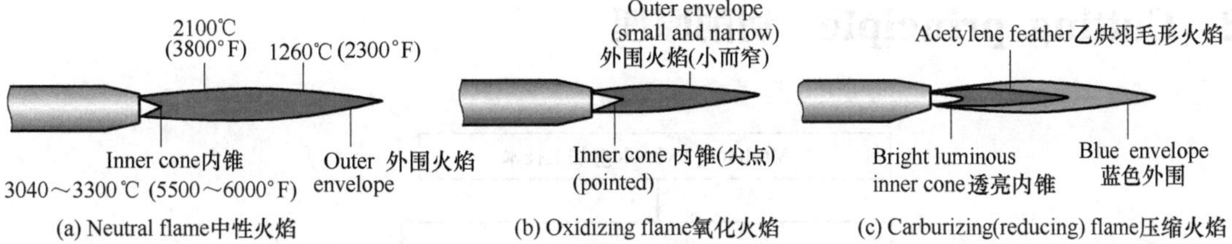

Fig 7.119 Three basic types of oxyacetylene flames used in oxyfuel-gas welding and cutting operations
用于氧 - 乙炔焊接和切割的三种基本火焰类型

7.2.4 Heat treatment 热处理

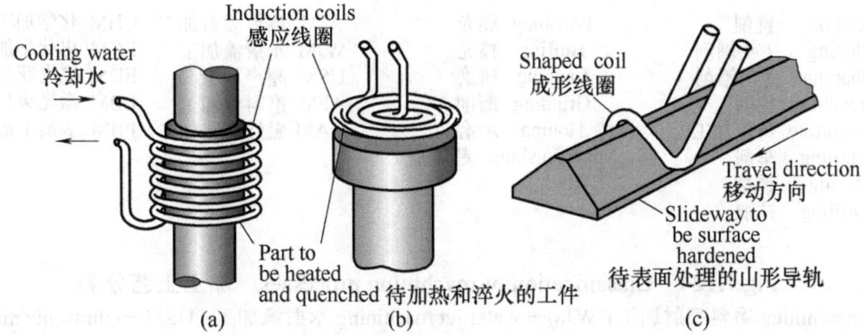

Fig 7.120 Types of coils used in induction heating of various surfaces of parts 几种工件表面感应加热线圈型式

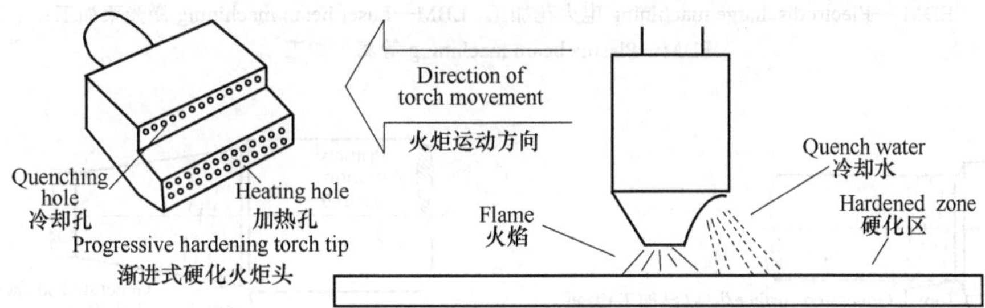

Fig 7.121 Progressive hardening 渐进式硬化

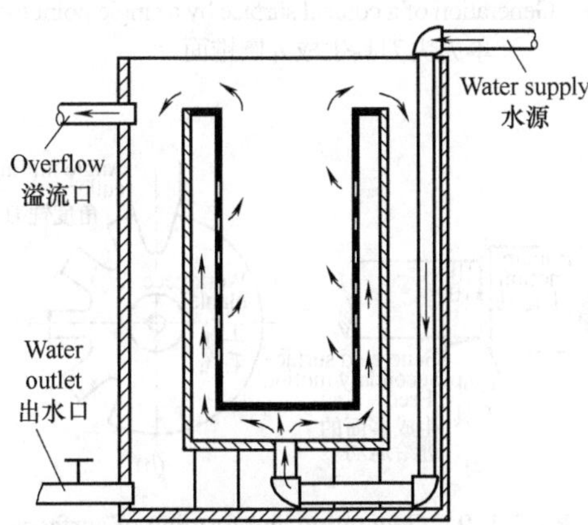

Fig 7.122 A water or brine tank for quenching baths
淬火用的水箱或盐水箱

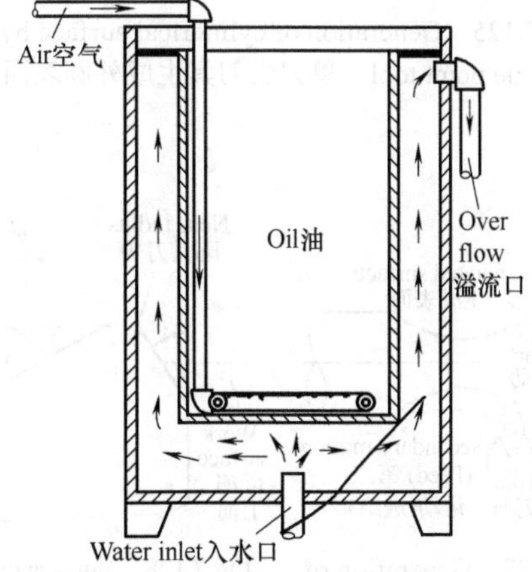

Fig 7.123 An oil-quenching tank in which water is circulated in an outer surrounding tank to keep the oil bath cool
淬火用油箱，水绕油箱外围循环冷却油液

7.3 Cutting principle 切削原理

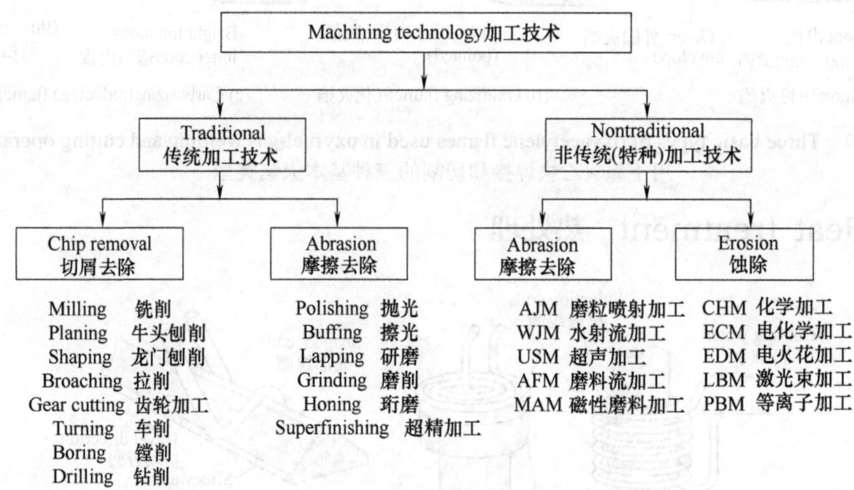

Fig 7.124 Classification of machining processes 加工工艺分类

AJM—Abrasive jet machining 磨料喷射加工；WJM—Water jet machining 水射流加工；USM—Ultrasonic machining 超声加工；
AFM—Abrasive flow machining 磨料流加工；MAM—Magnetic abrasive machining 磁性磨料加工；
CHM—Chemical machining 化学加工；ECM—Electrochemical machining 电化学加工；
EDM —Electrodischarge machining 电火花加工；LBM—Laser beam machining 激光束加工；
PBM—Plasma beam machining 等离子加工

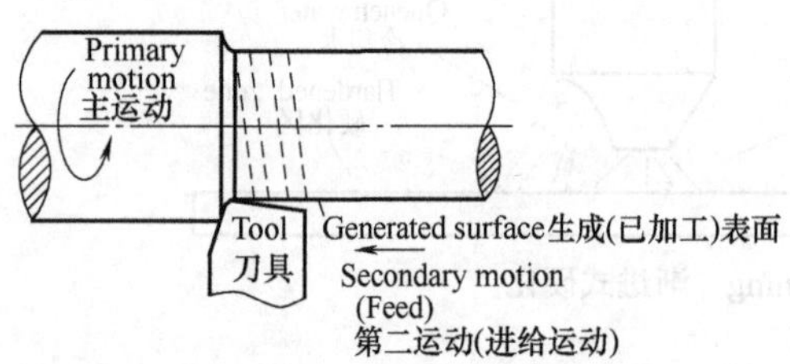

Fig 7.125 Generation of cylindrical surface by a single point tool 单刀尖刀具生成外圆表面

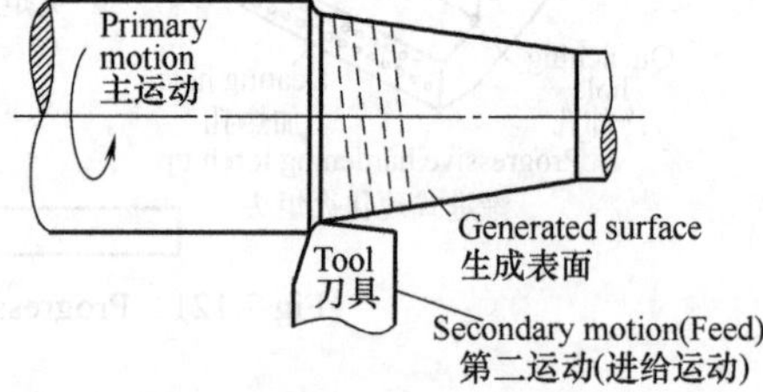

Fig 7.126 Generation of a conical surface by a single point tool 单刀尖刀具生成外圆锥面

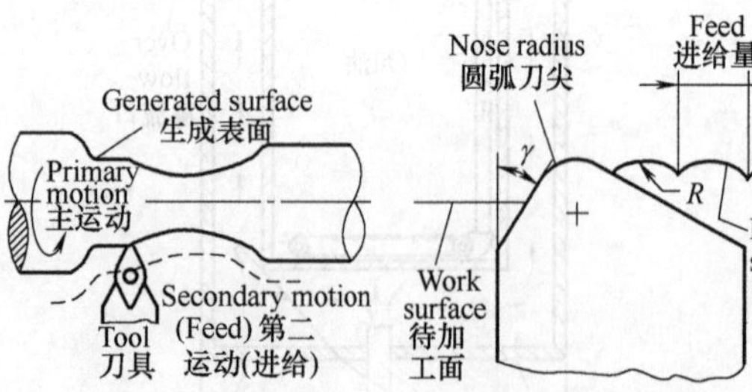

Fig 7.127 Generation of a contoured surface with a single point tool 单刀尖刀具生成外轮廓面

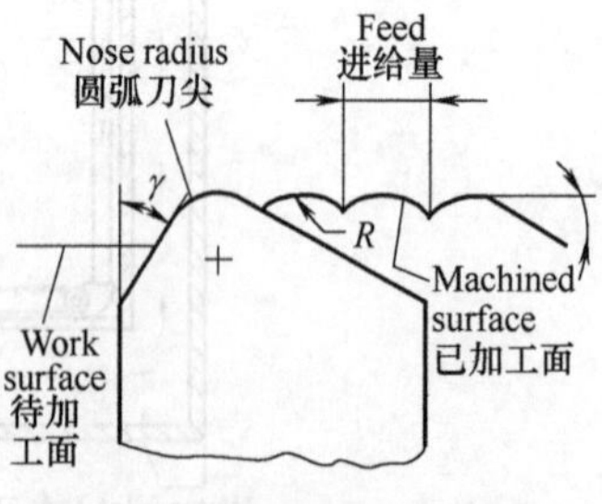

Fig 7.128 Surface profile as produced by turning with a cutting tool having a nose radius 圆弧车刀车削表面形貌

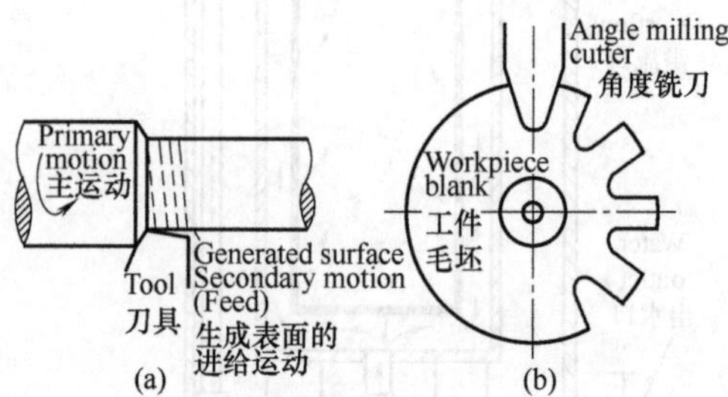

Fig 7.129 Generating and forming of surfaces 展成法和成形法生成加工表面

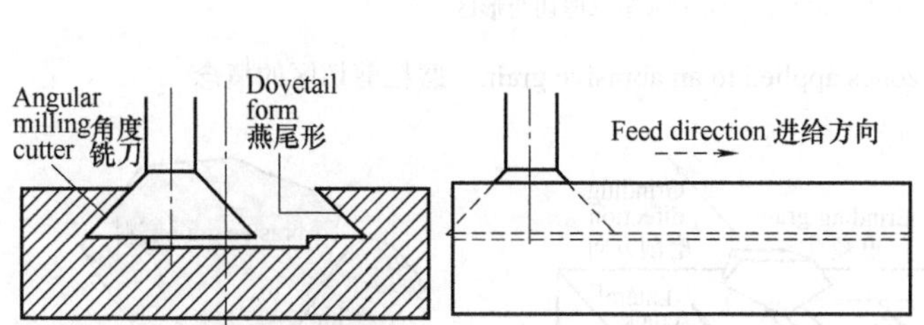

Fig 7.130 Generation of surfaces 加工表面的形成

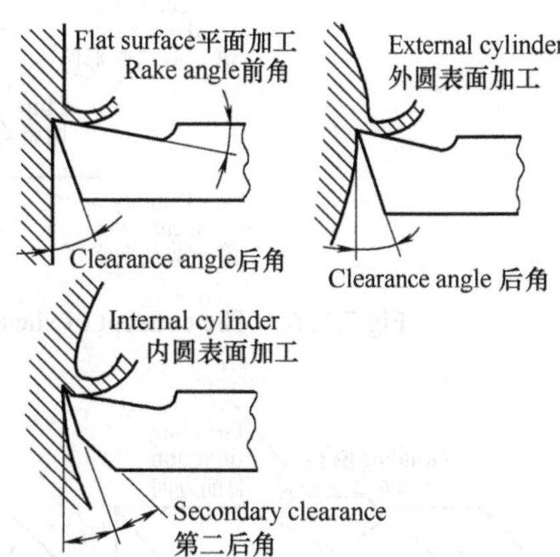

Fig 7.131 Rake angle and clearance angle for cutting tool 刀具的前角和后角

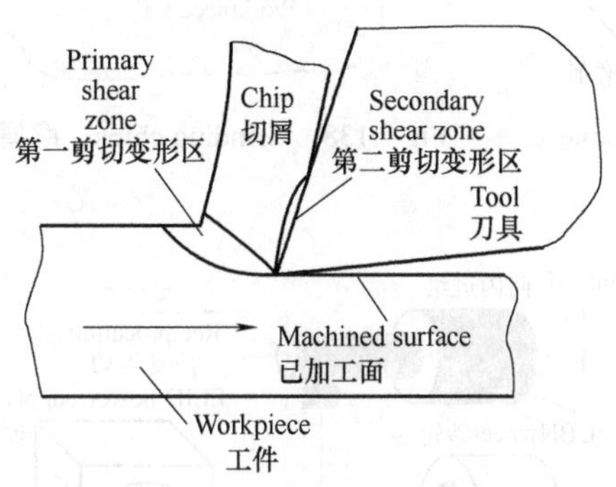

Fig 7.132 The possible deformation in metal cutting 金属切削过程可能的变形

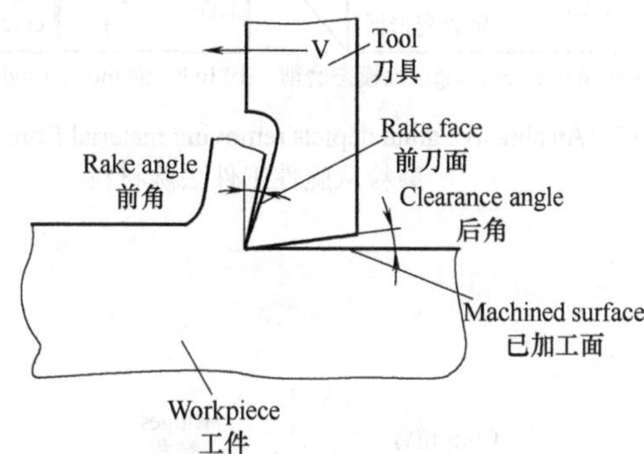

Fig 7.133 The general characteristics of a metal cutting tool 金属切削刀具的通用特征

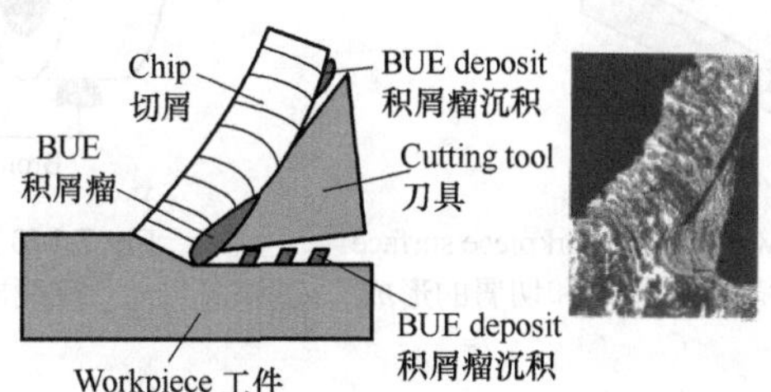

Fig 7.134 Formation of Built-up Edge (BUE) 积屑瘤的形成

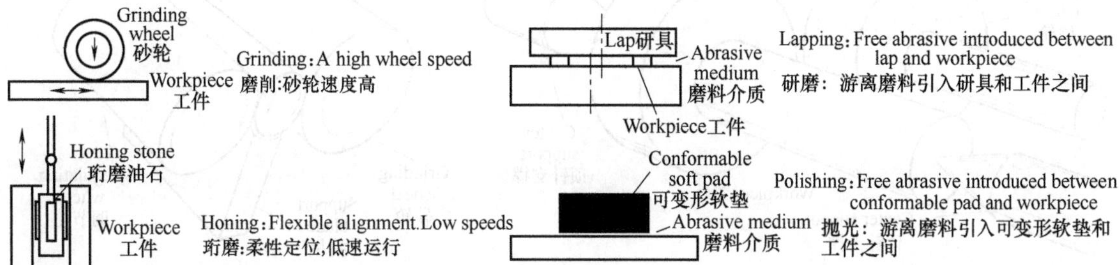

Fig 7.135 Basic principles of grinding, honing, lapping, and polishing 磨削、珩磨、研磨和抛光的基本原理

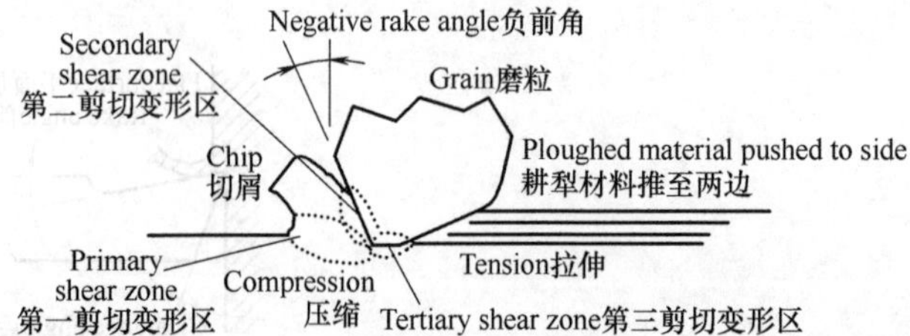

Fig 7.136 The concept of shear zones applied to an abrasive grain 磨粒剪切区的概念

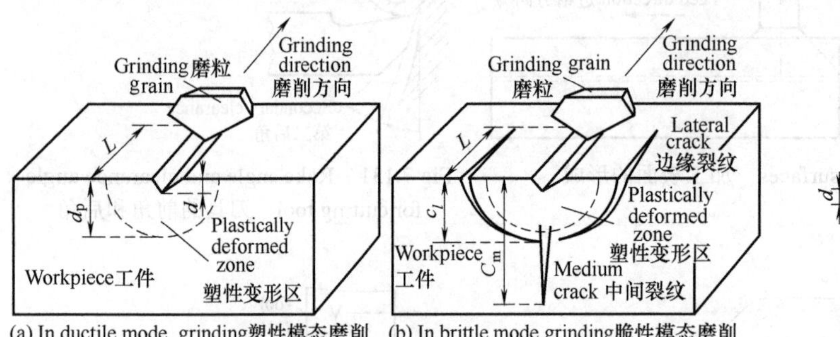

Fig 7.137 An abrasive grain depicts removing material from a brittle workpiece 磨粒从脆性工件去除材料

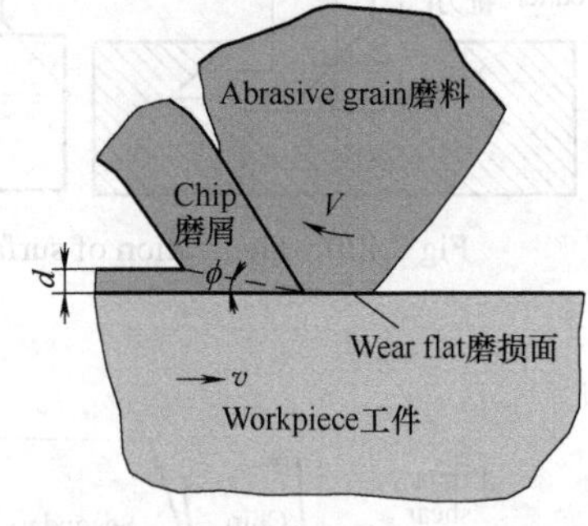

Fig 7.138 Grinding chips 磨屑

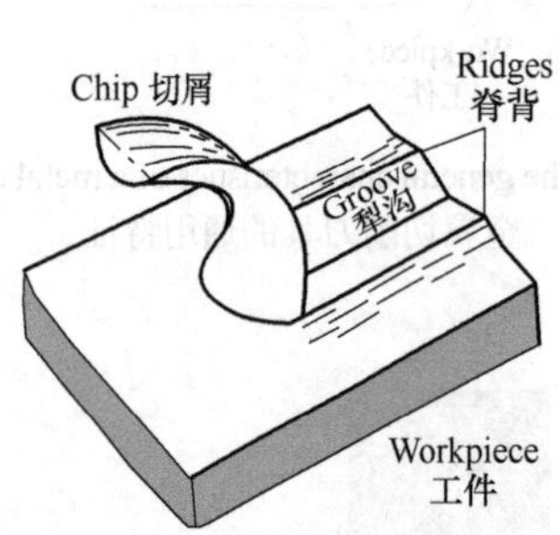

Fig 7.139 Chip formation and plowing of the workpiece surface by an abrasive grain 磨粒对工件表面的根犁和切屑的形成

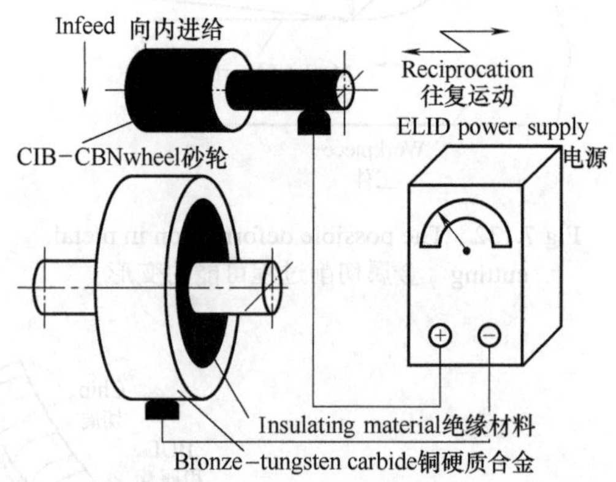

Fig 7.140 ELID truing mechanism 在线电解修形磨削的机理

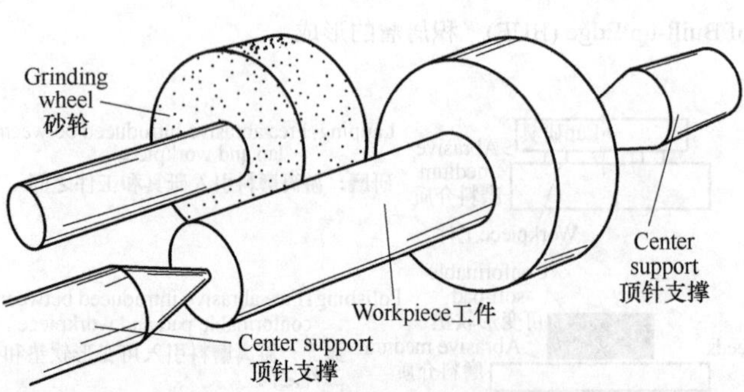

Fig 7.141 Centered grinding 中心磨削

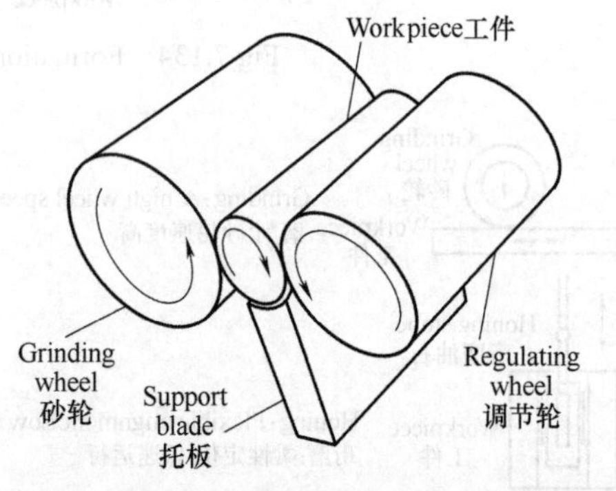

Fig 7.142 Centerless grinding 无心磨削

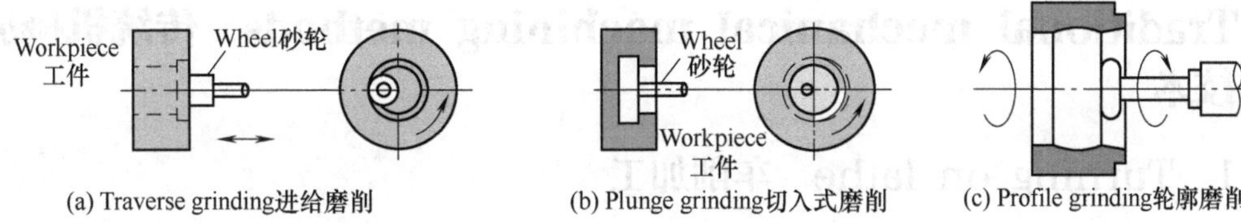

Fig 7.143 Internal grinding operations 内圆磨削

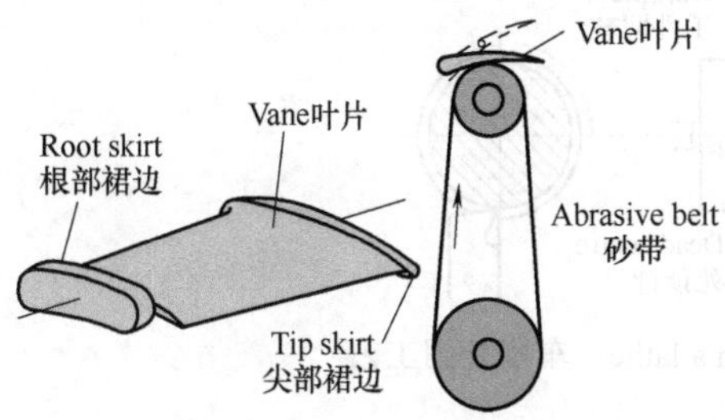

Fig 7.144 Belt Grinding of turbine nozzle vanes
砂带磨削涡轮喷嘴叶片

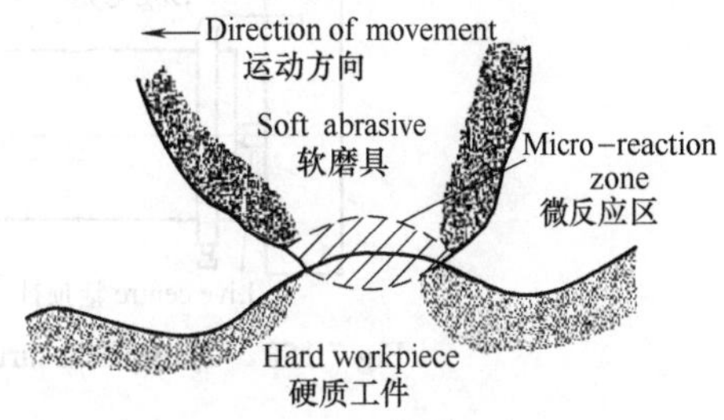

Fig 7.145 The chemo-mechanical action between abrasive, work material, and environment
磨粒、工件及环境间的化学 / 机械作用

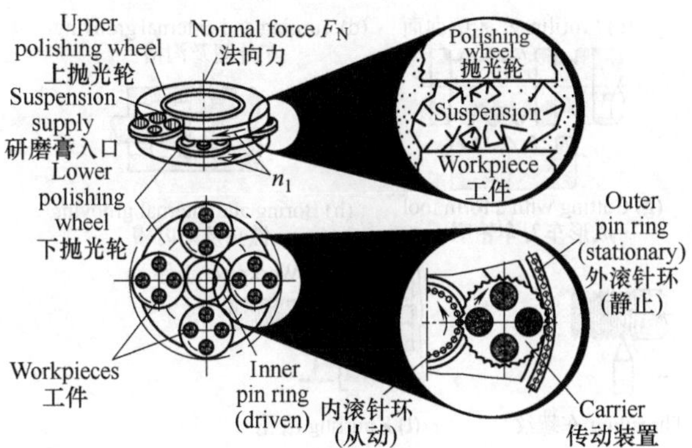

Fig 7.146 Principle of two-wheel polishing with guided workpieces
带工件导向的双轮抛光原理

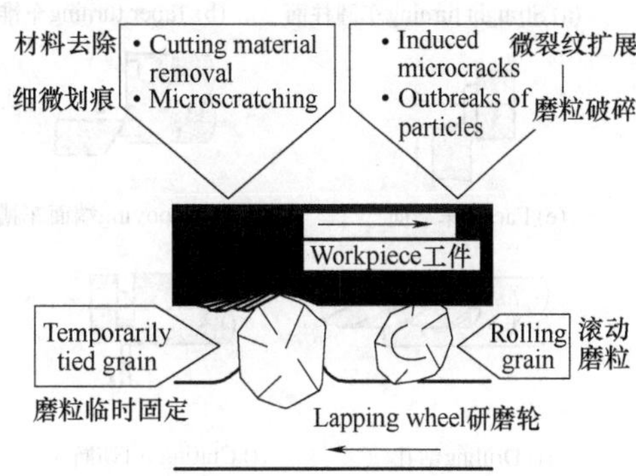

Fig 7.147 Polishing of a plane surface
平面抛光

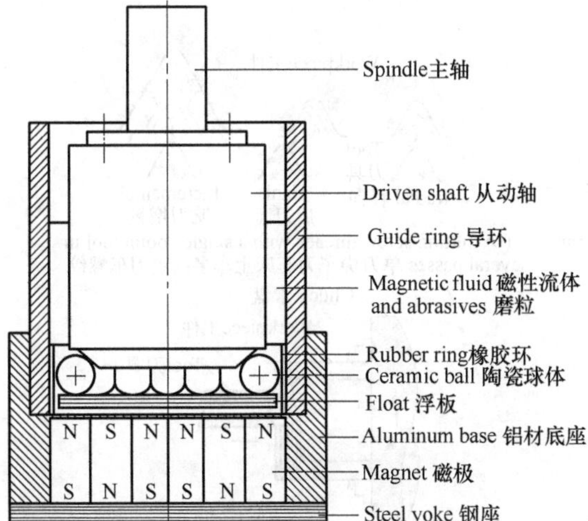

Fig 7.148 The magnetic floating polishing apparatus used for polishing Si_3N_4 balls 用于抛光 Si_3N_4 球体的磁力浮动抛光装置

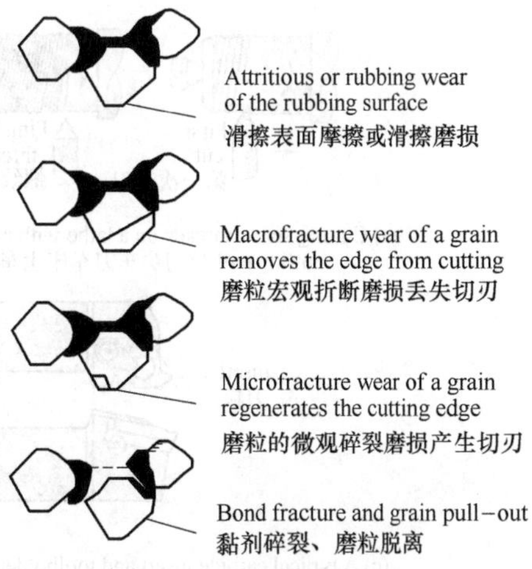

Fig 7.149 Types of grain wear
磨粒磨损的类型

7.4 Traditional mechanical machining methods 传统机械加工技术

7.4.1 Turning on lathe 车削加工

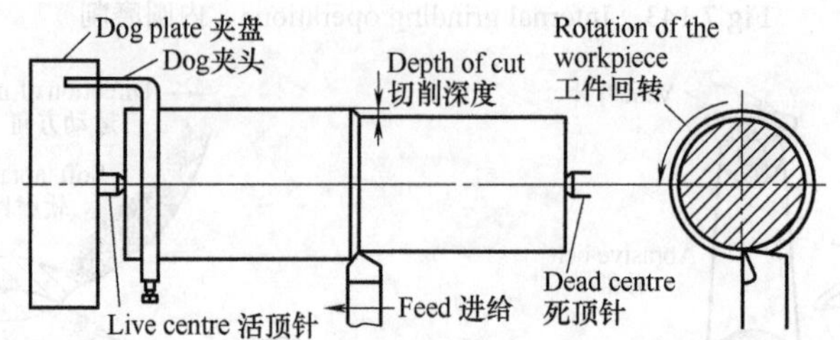

Fig 7.150 Cylindrical turning operation in a lathe 车削外圆工艺

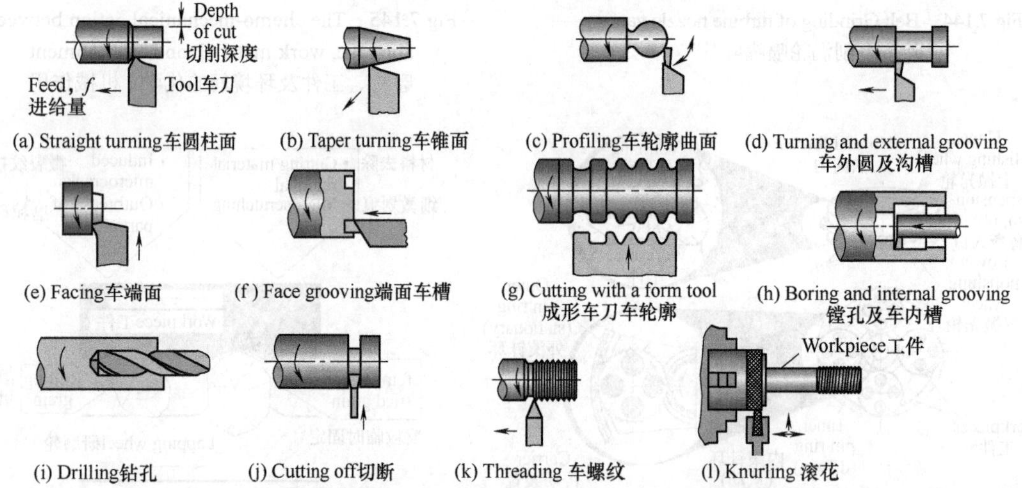

(a) Straight turning车圆柱面 (b) Taper turning车锥面 (c) Profiling车轮廓曲面 (d) Turning and external grooving 车外圆及沟槽

(e) Facing车端面 (f) Face grooving端面车槽 (g) Cutting with a form tool 成形车刀车轮廓 (h) Boring and internal grooving 镗孔及车内槽

(i) Drilling钻孔 (j) Cutting off切断 (k) Threading 车螺纹 (l) Knurling 滚花

Fig 7.151 Various cutting operations that can be performed on a lathe (Note that all parts have circular symmetry) 车床上可实施的各种操作（所有工件为圆对称回转件）

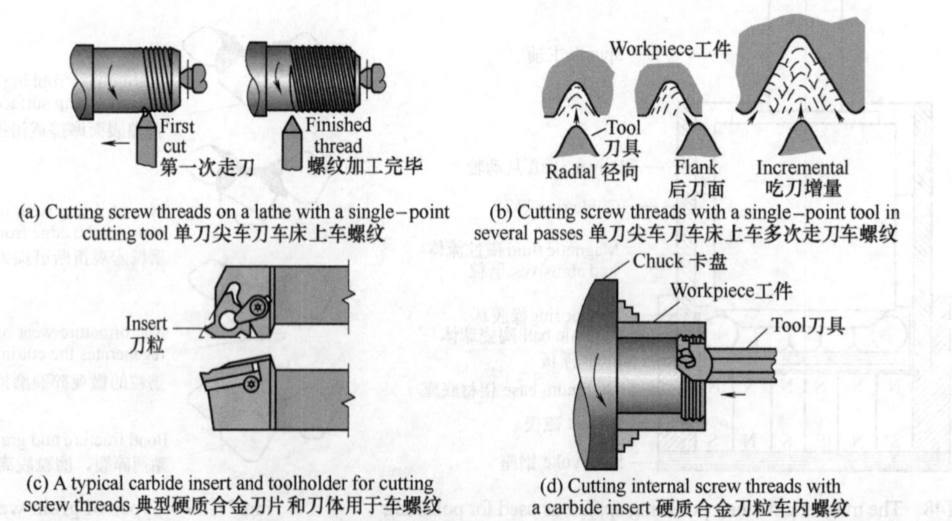

(a) Cutting screw threads on a lathe with a single-point cutting tool 单刀尖车刀车床上车螺纹

(b) Cutting screw threads with a single-point tool in several passes 单刀尖车刀车床上车多次走刀车螺纹

(c) A typical carbide insert and toolholder for cutting screw threads 典型硬质合金刀片和刀体用于车螺纹

(d) Cutting internal screw threads with a carbide insert 硬质合金刀粒车内螺纹

Fig 7.152 Screw threads cutting 螺纹车削

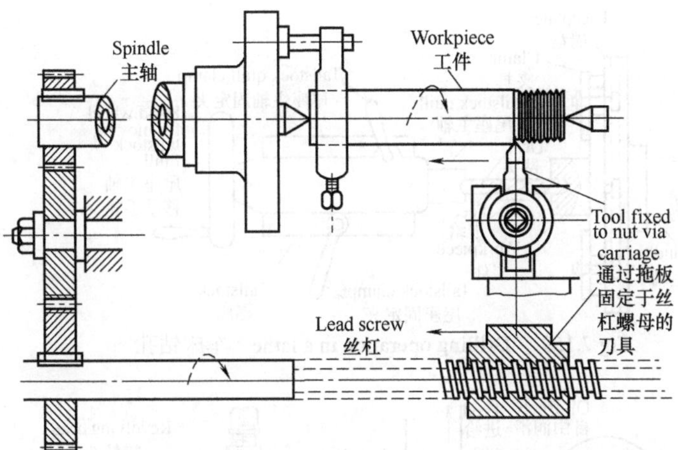

Fig 7.153 Thread cutting using compound slide 使用复合溜板加工螺纹

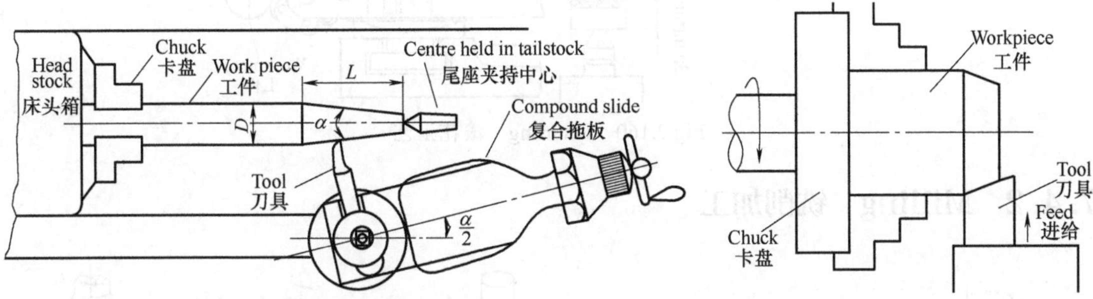

Fig 7.154 Compound slide method for taper turning 复合拖板车削锥度

Fig 7.155 Taper turning using form tools 成形刀具车削锥度

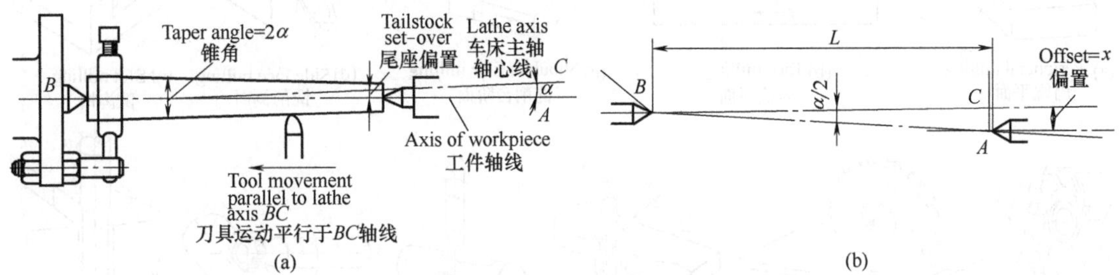

Fig 7.156 Tail stock offset 尾座偏移车削锥度

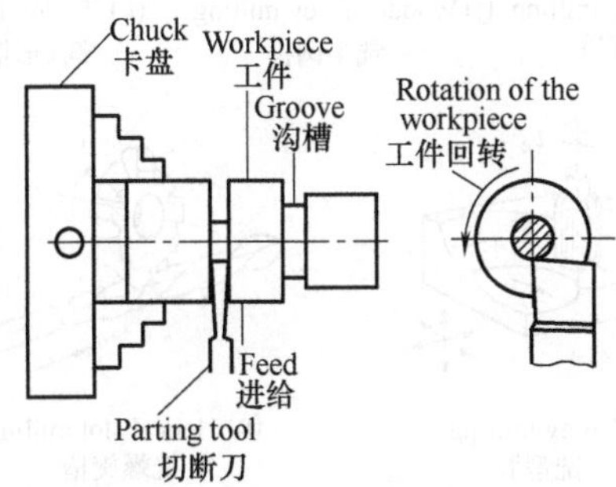

Fig 7.157 Parting tool in operation 切断工作

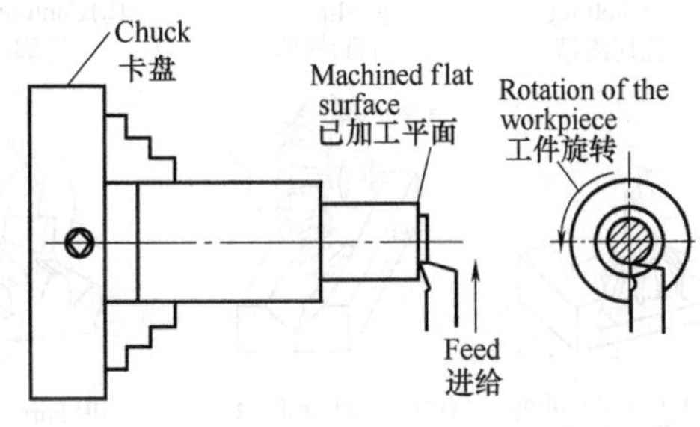

Fig 7.158 Facing 车削端面

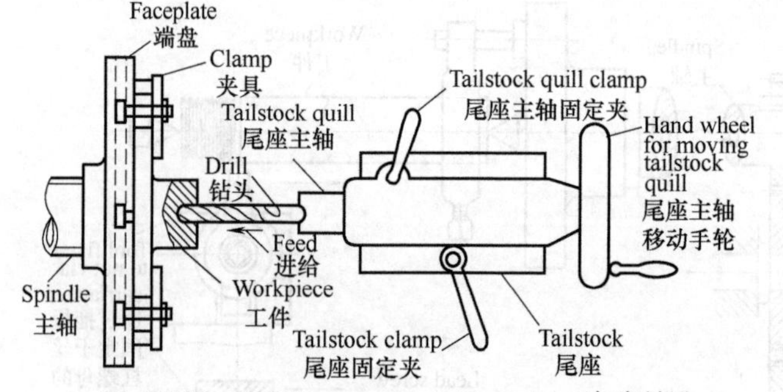

Fig 7.159 Drilling operation in a lathe 车床钻孔

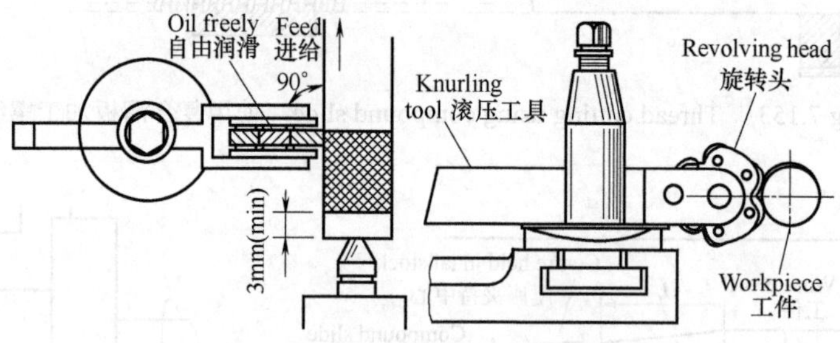

Fig 7.160 Knurling 滚花工艺

7.4.2 Milling 铣削加工

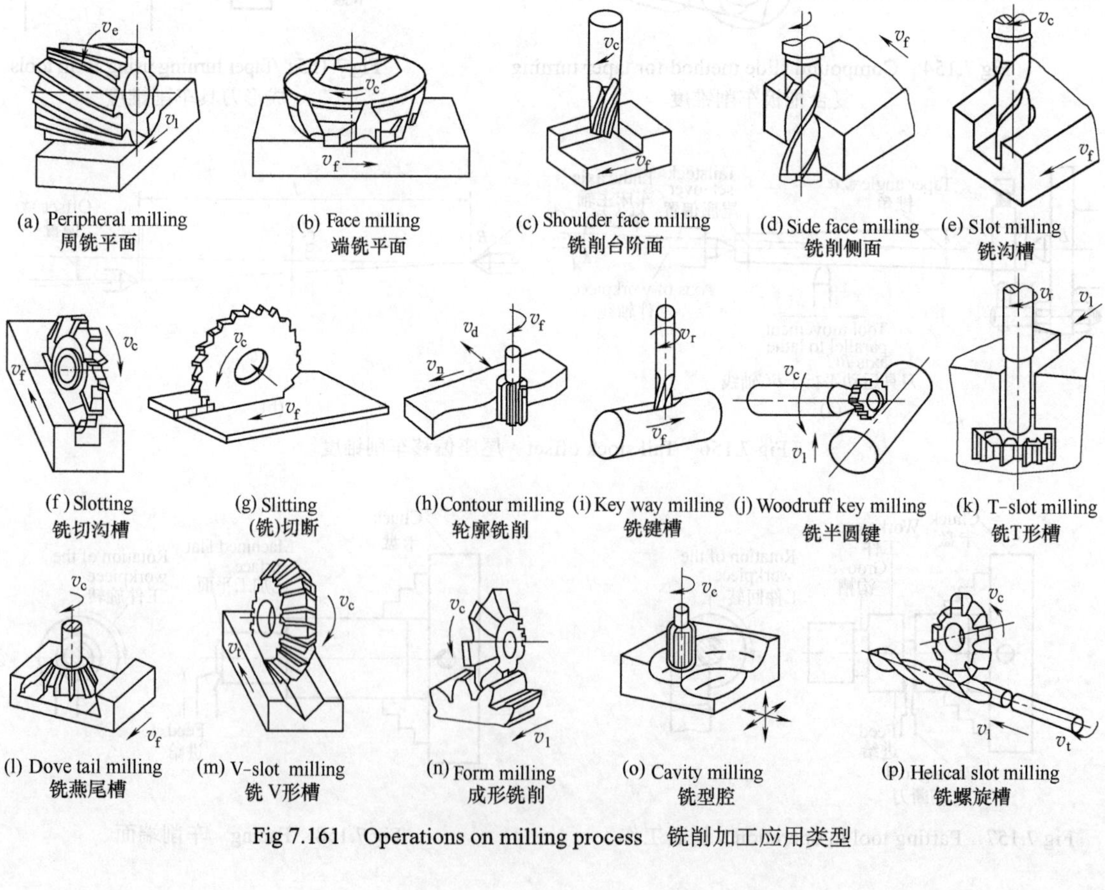

(a) Peripheral milling 周铣平面

(b) Face milling 端铣平面

(c) Shoulder face milling 铣削台阶面

(d) Side face milling 铣削侧面

(e) Slot milling 铣沟槽

(f) Slotting 铣切沟槽

(g) Slitting (铣)切断

(h) Contour milling 轮廓铣削

(i) Key way milling 铣键槽

(j) Woodruff key milling 铣半圆键

(k) T-slot milling 铣T形槽

(l) Dove tail milling 铣燕尾槽

(m) V-slot milling 铣 V形槽

(n) Form milling 成形铣削

(o) Cavity milling 铣型腔

(p) Helical slot milling 铣螺旋槽

Fig 7.161 Operations on milling process 铣削加工应用类型

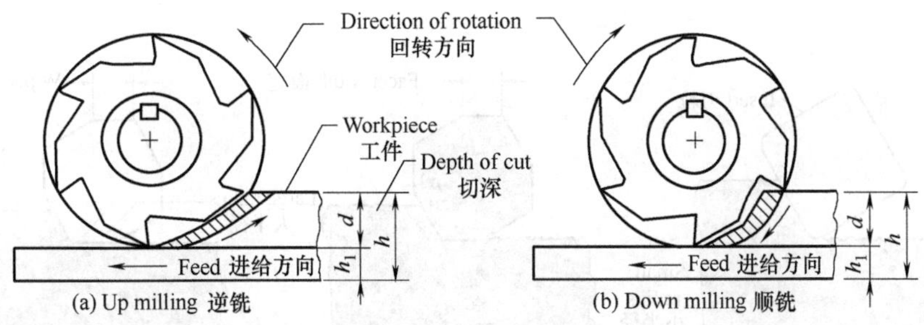

Fig 7.162 Up milling and down milling 顺铣和逆铣

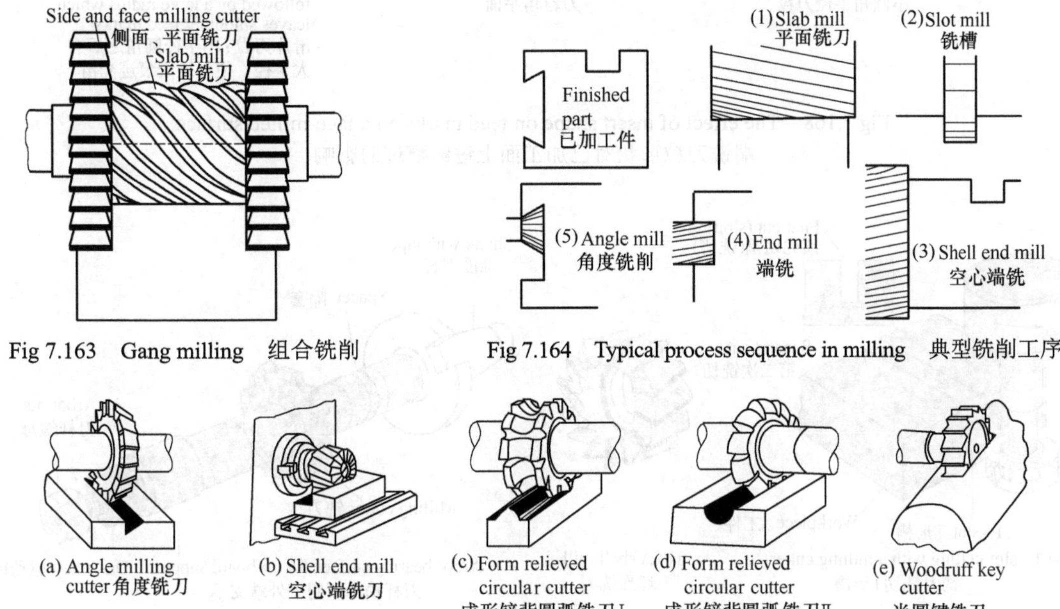

Fig 7.163 Gang milling 组合铣削

Fig 7.164 Typical process sequence in milling 典型铣削工序

Fig 7.165 Special forms of arbor mounted milling cutter 安装刀杆的特型铣刀

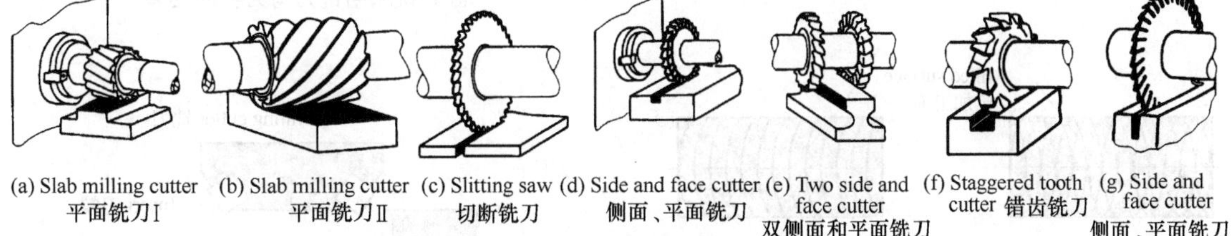

Fig 7.166 Arbor mounted milling cutters for general purpose 刀杆安装通用铣刀

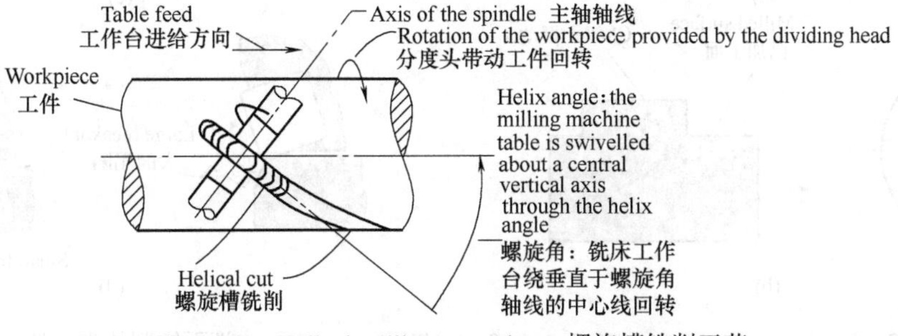

Fig 7.167 Helical milling operation 螺旋槽铣削工艺

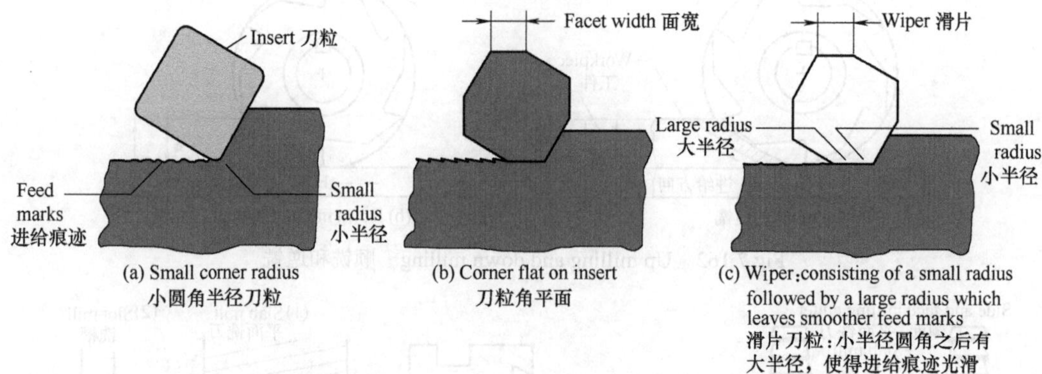

(a) Small corner radius
小圆角半径刀粒

(b) Corner flat on insert
刀粒角平面

(c) Wiper:consisting of a small radius followed by a large radius which leaves smoother feed marks
滑片刀粒:小半径圆角之后有大半径，使得进给痕迹光滑

Fig 7.168 The effect of insert shape on feed marks on a face-milled surface
端铣刀粒形状对已加工面上进给痕迹的影响

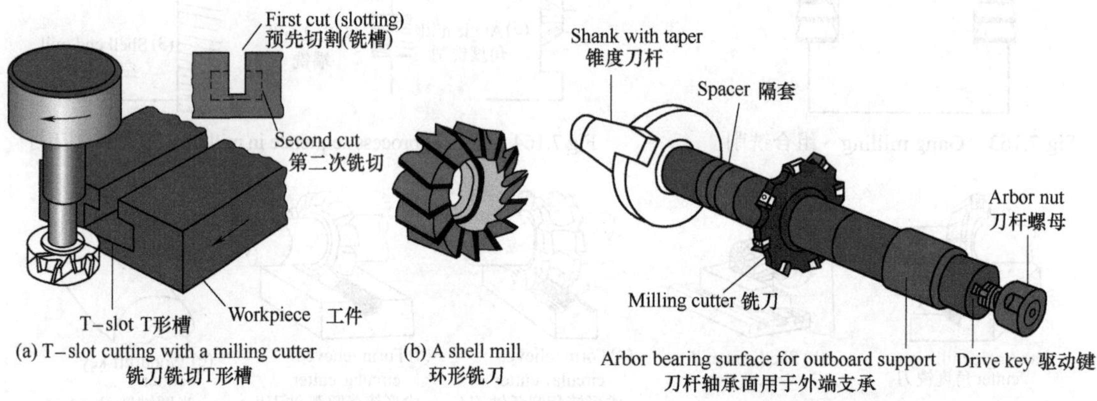

(a) T-slot cutting with a milling cutter
铣刀铣切T形槽

(b) A shell mill
环形铣刀

Fig 7.169 T-slot milling 铣 T 形槽

Fig 7.170 Mounting a milling cutter on an arbor for use on a horizontal milling machine
卧式铣床用铣刀与刀杆的安装

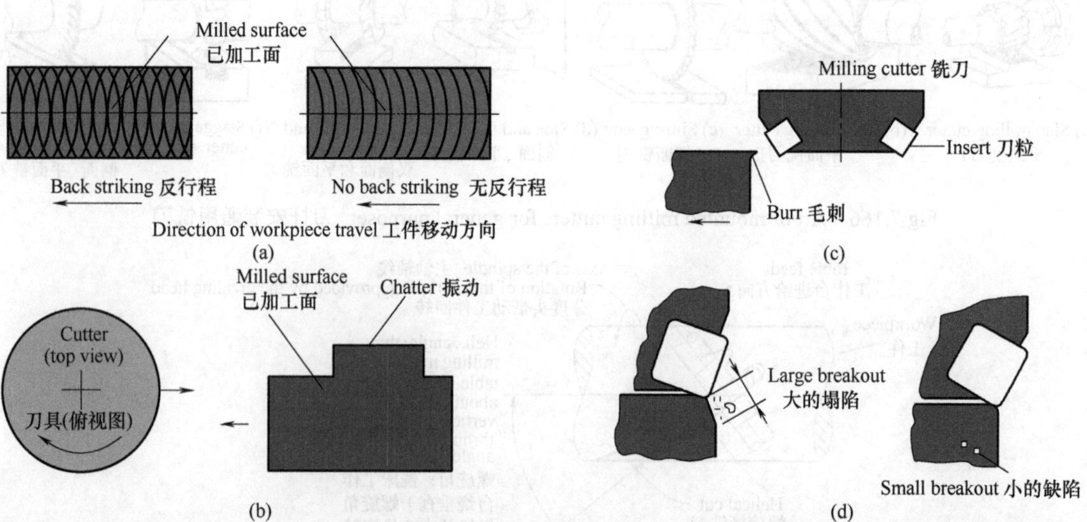

Fig 7.171 Surface features and corner defects in face milling operations 端面铣削的表面特征和拐角缺陷

7.4.3 Boring process 镗削加工

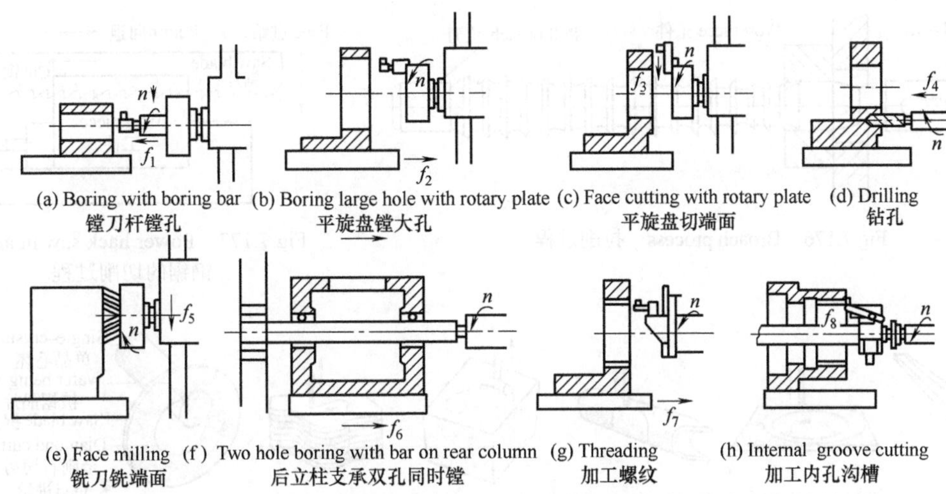

Fig 7.172 Main operations for horizontal boring machine 卧式镗床主要加工方法

7.4.4 Drilling and ream process 钻削、铰削加工

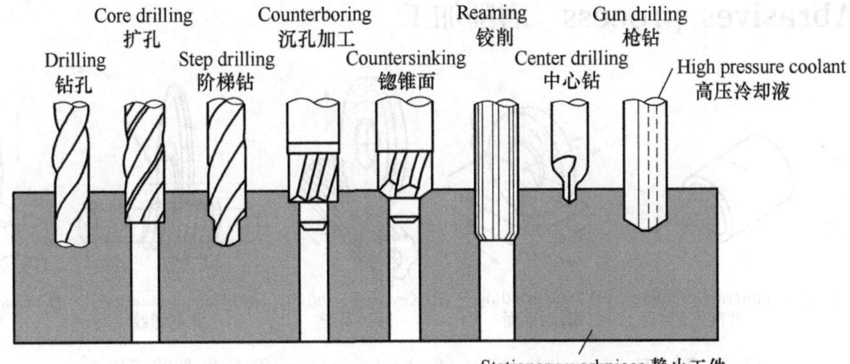

Fig 7.173 Drilling and drilling allied operations 钻削及关联操作

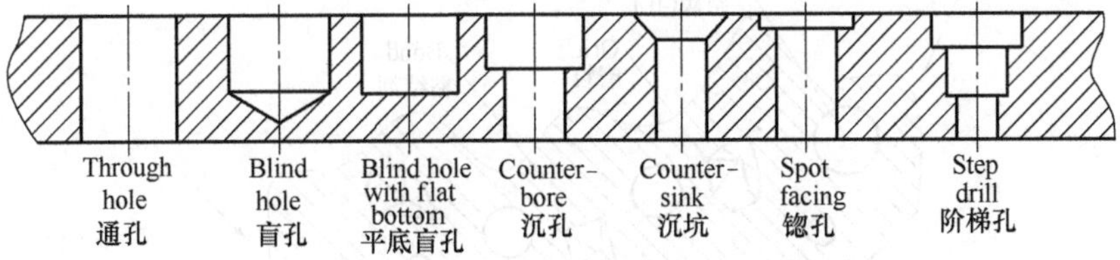

Fig 7.174 Various types of holes 孔的类型

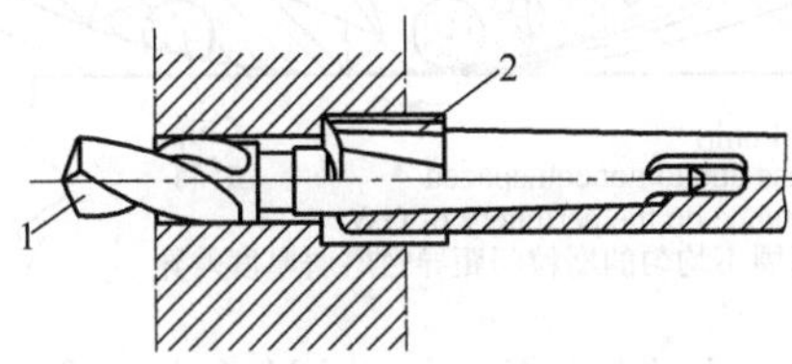

Fig 7.175 Complex tool for drilling and enlarging 钻扩复合加工

1—Twist drill 麻花钻头；2—Enlarging tool 扩孔刀

7.4.5 Reciprocating machining process 往复式加工

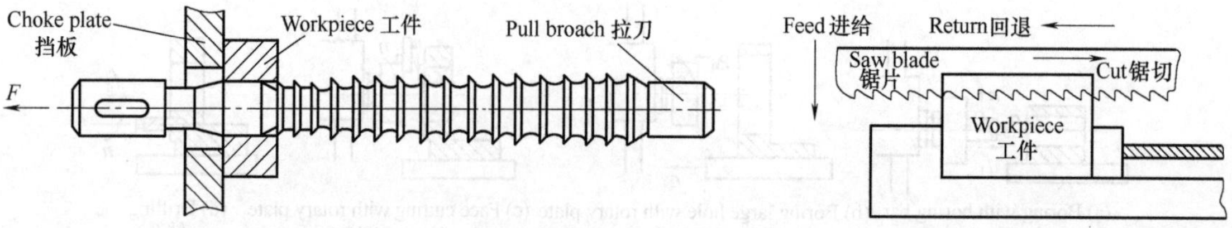

Fig 7.176 Broach process 拉削过程

Fig 7.177 Power hack saw in action 钢锯的切削过程

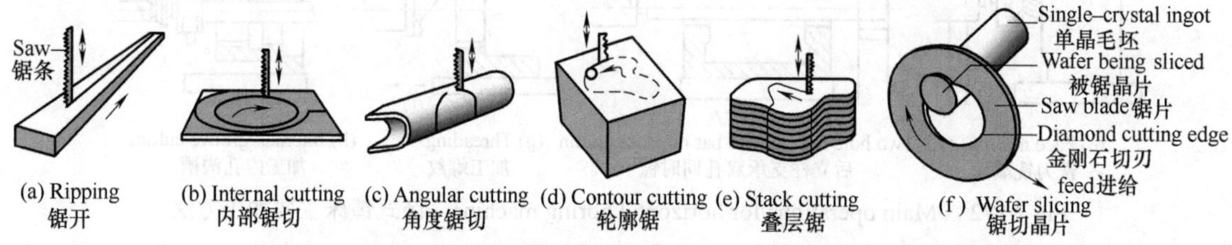

Fig 7.178 Various examples of sawing operations 锯切操作实例

7.4.6 Abrasives process 磨削加工

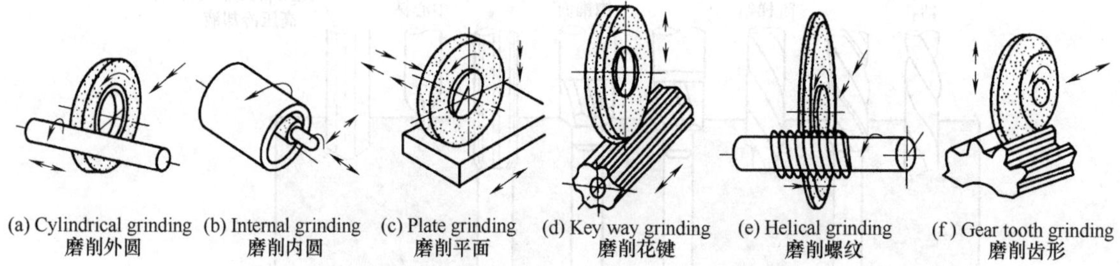

Fig 7.179 Various applications of wheel grinding 砂轮磨削的应用

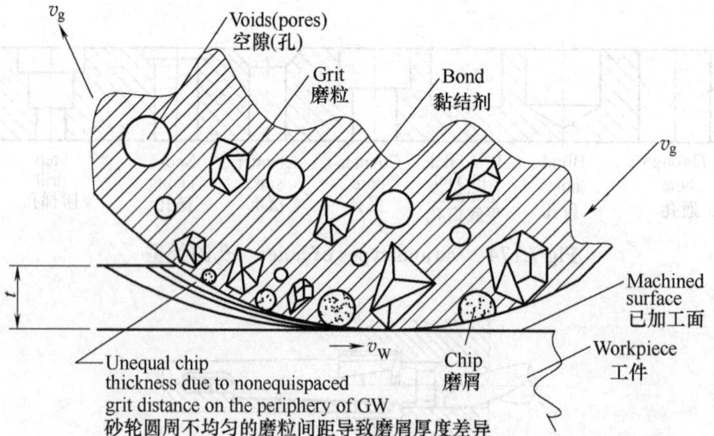

Fig 7.180 Cutting principles and main variables of a surface grinding process
平面磨削工艺加工原理和主要参数

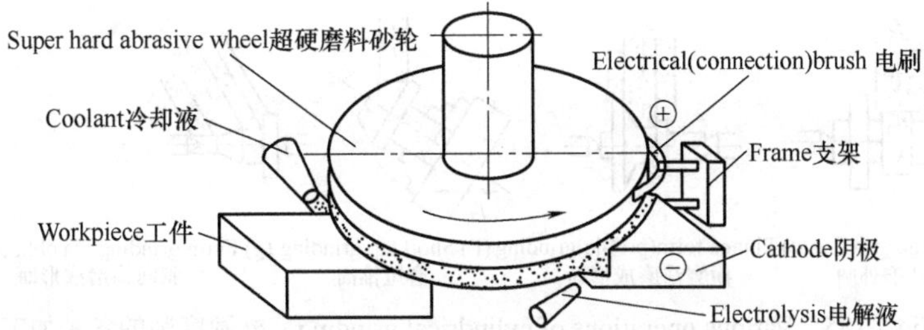

Fig 7.181 Electrolytic in-process dressing (ELID) 电解在线砂轮修正

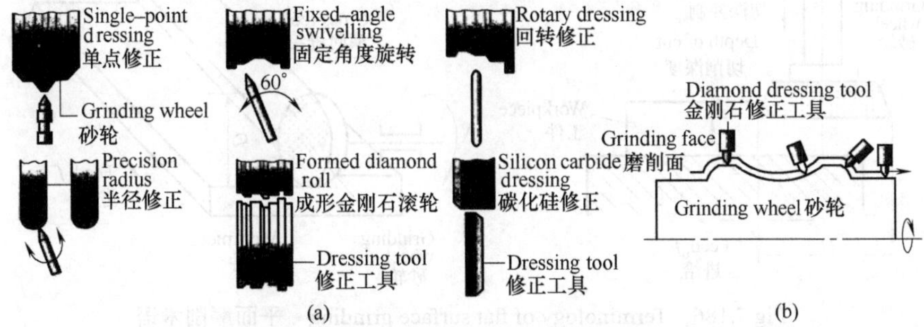

Fig 7.182 Forms of grinding wheel dressing 砂轮修正方式

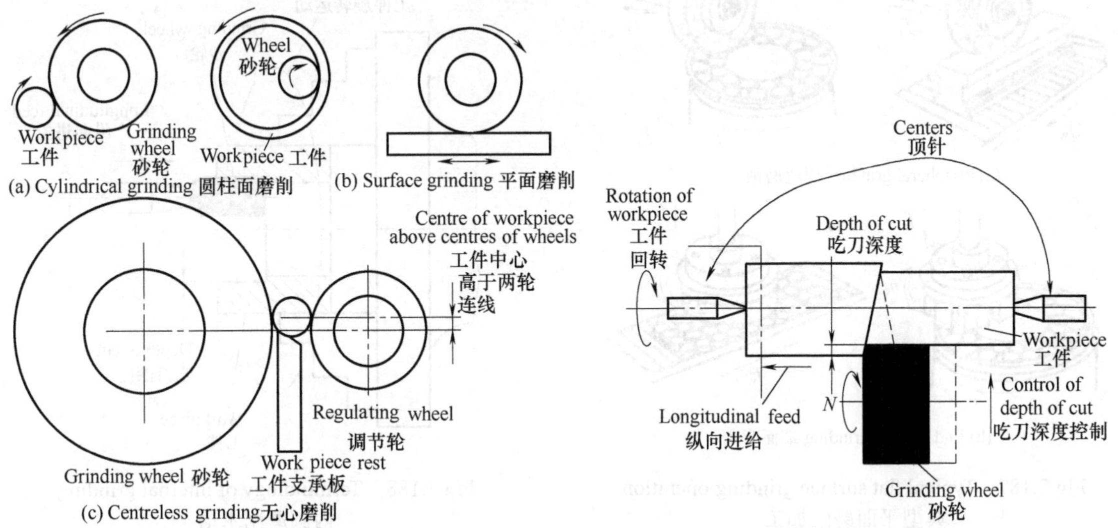

Fig 7.183 Typical grinding operations 典型磨削加工工艺

Fig 7.184 Terminology of cylindrical grinding 外圆磨削术语

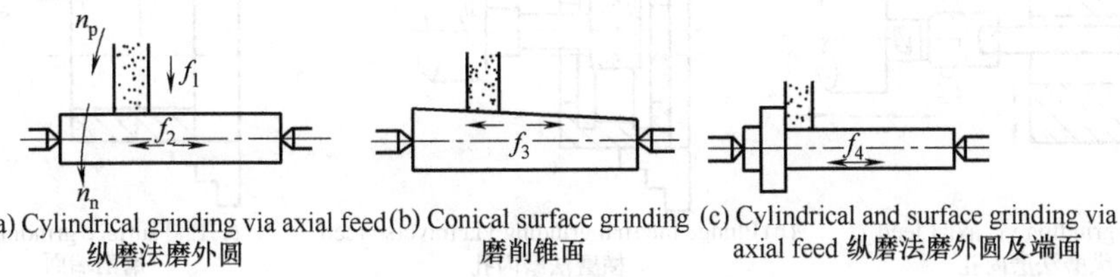

Fig 7.185

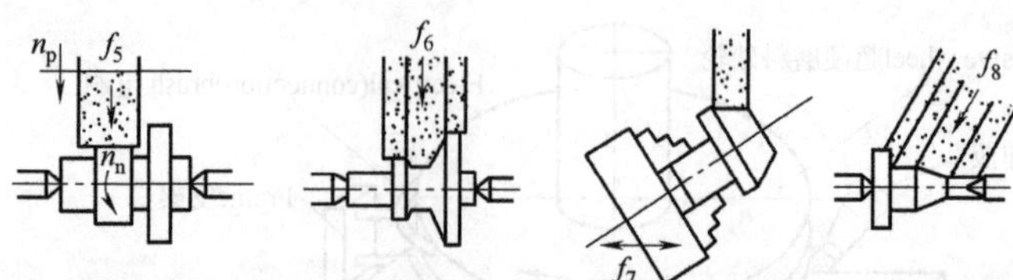

(d) Plunge grinding 横磨法磨外圆 (e) Plunge form(profile)grinding 横磨法磨成形面 (f) Short tap grinding 磨短锥面 (g) Form grinding via oblique feed 斜向横磨成形面

Fig 7.185 Various operations of cylindrical grinding 外圆磨削的各种加工

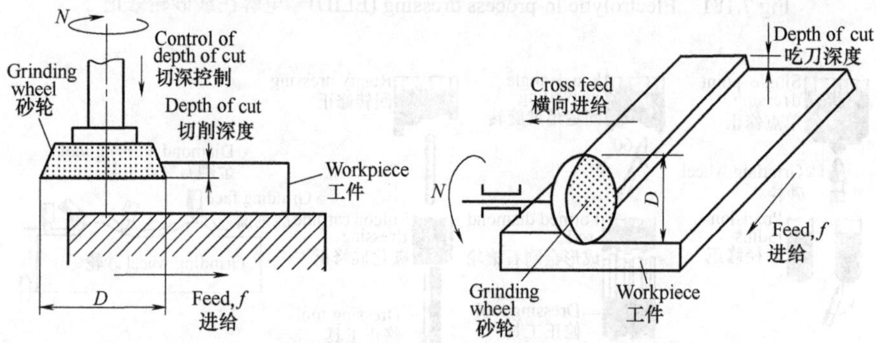

Fig 7.186 Terminology of flat surface grinding 平面磨削术语

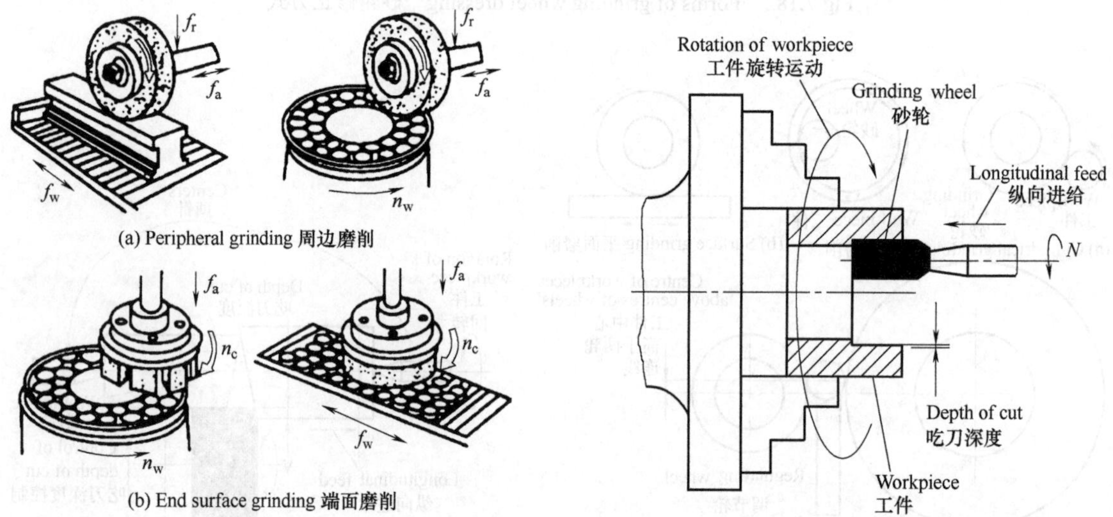

Fig 7.187 Typical flat surface grinding operation 典型平面磨削加工

Fig 7.188 Terminology of internal grinding 内圆磨削术语

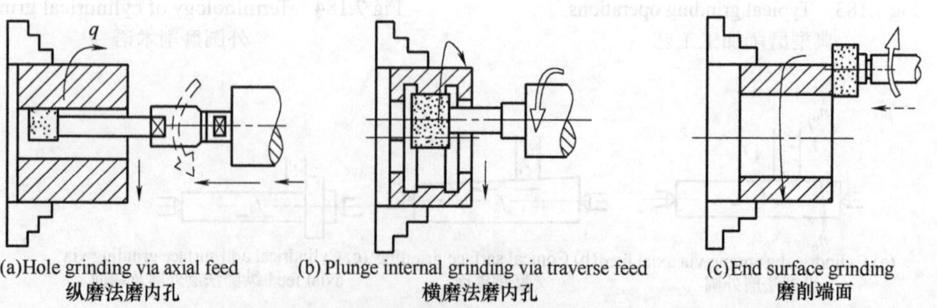

(a)Hole grinding via axial feed 纵磨法磨内孔 (b) Plunge internal grinding via traverse feed 横磨法磨内孔 (c)End surface grinding 磨削端面

Fig 7.189 Operations for inner surface grinding machine 内圆磨削方法

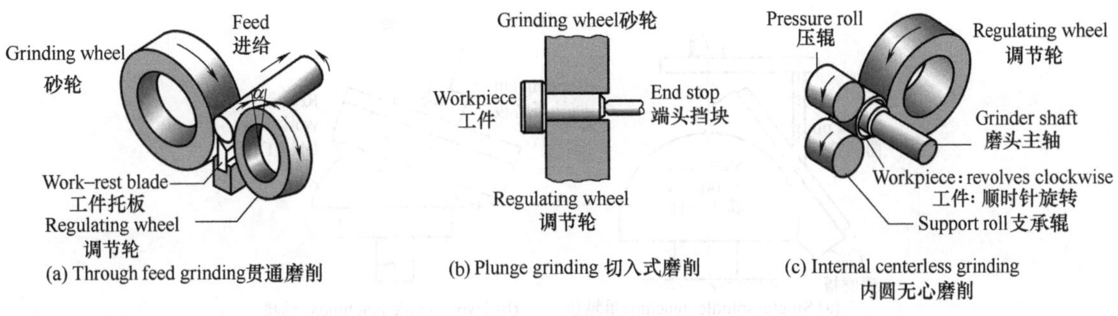

Fig 7.190 Forms of centreless grinding 无心磨削的形式种类

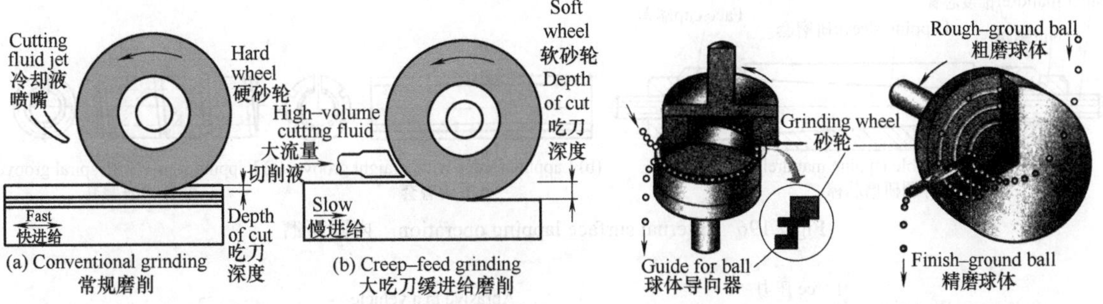

Fig 7.191 Creep feed grinding operation 缓进给磨削

Fig 7.192 Steel-ball grinding 钢球磨削

7.4.7 Coated abrasive belt grinding 砂带磨削加工

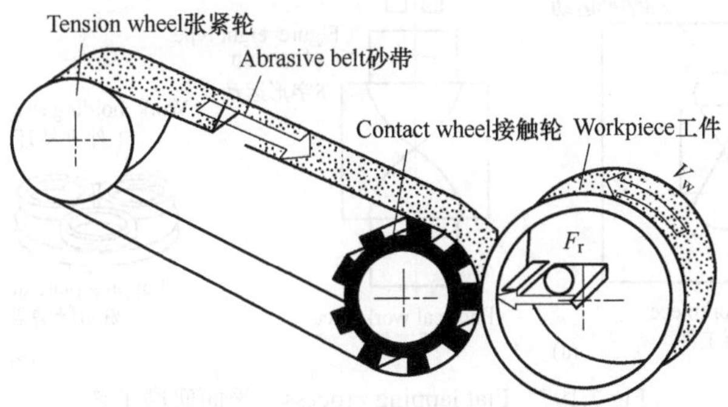

Fig 7.193 Coated abrasive belt grinding set structure 砂带磨削装置构成

7.4.8 Super finishing 超精加工

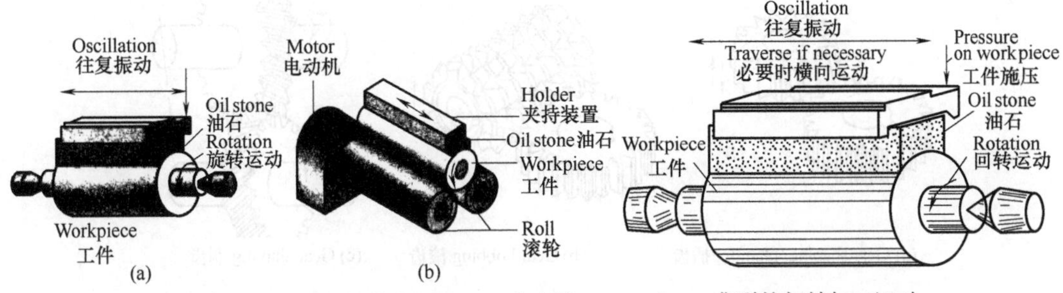

Fig 7.194 Typical motions in a super finishing operation 典型的超精加工运动

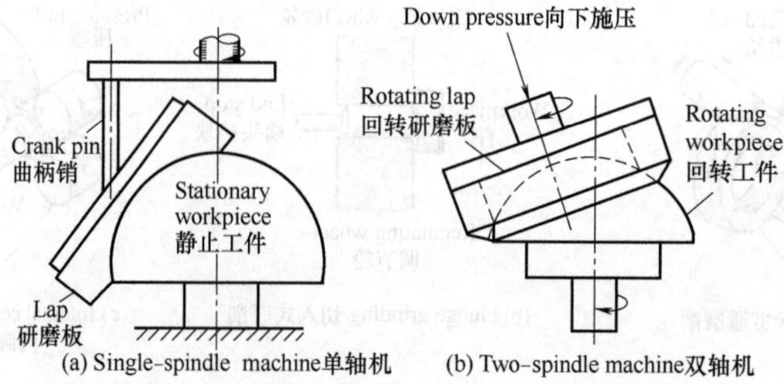

(a) Single-spindle machine单轴机 (b) Two-spindle machine双轴机

Fig 7.195 Lapping of spherical surface 球面研磨机

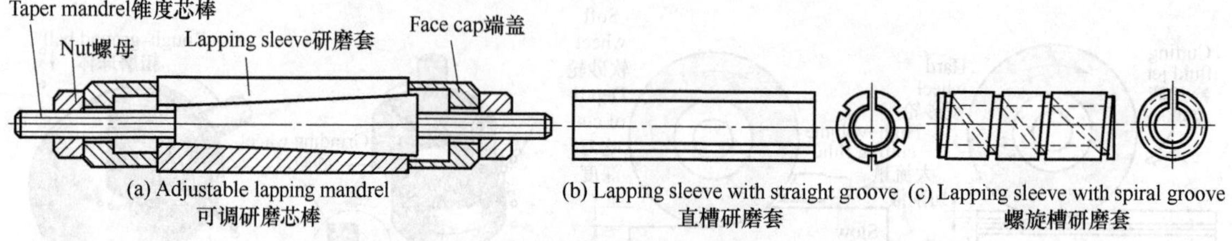

(a) Adjustable lapping mandrel 可调研磨芯棒 (b) Lapping sleeve with straight groove 直槽研磨套 (c) Lapping sleeve with spiral groove 螺旋槽研磨套

Fig 7.196 Internal surface lapping operation 内圆研磨

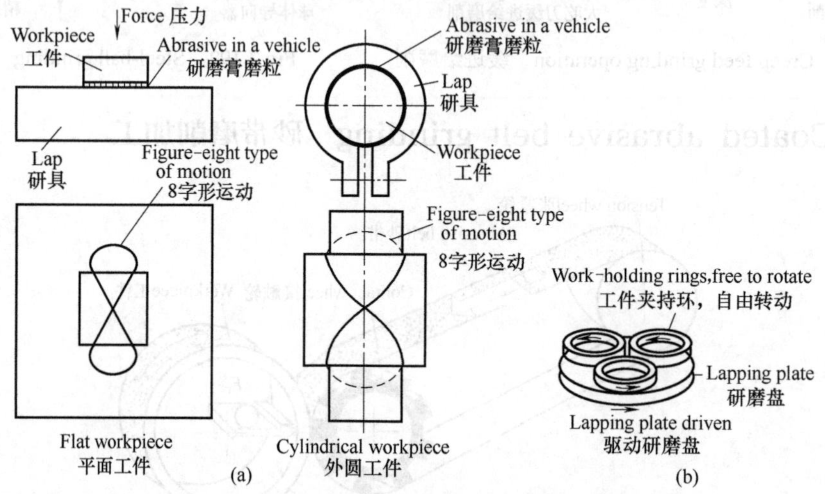

Fig 7.197 Flat lapping process 平面研磨工艺

7.4.9 Gear cutting 齿轮加工

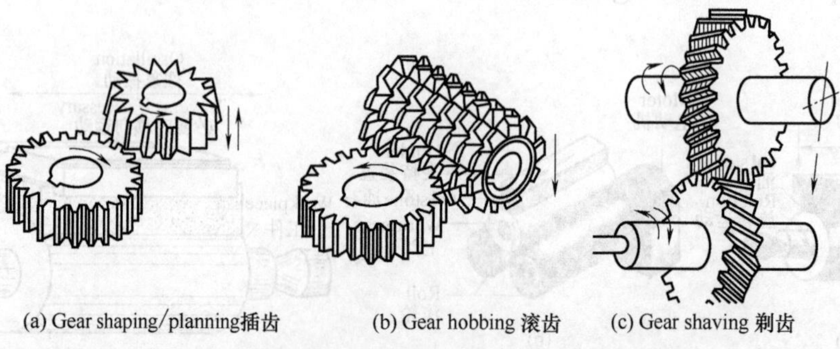

(a) Gear shaping/planning插齿 (b) Gear hobbing 滚齿 (c) Gear shaving 剃齿

Fig 7.198 Gear generating machining 齿轮展成加工

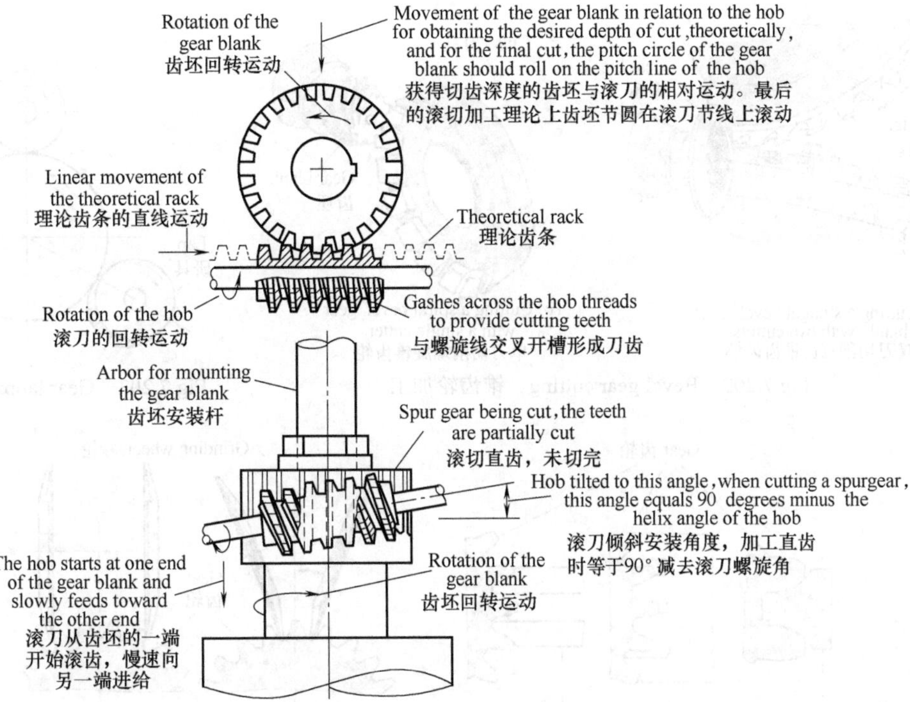

Fig 7.199 Schematic representation of a gear hobbing operation 滚齿加工示意图解

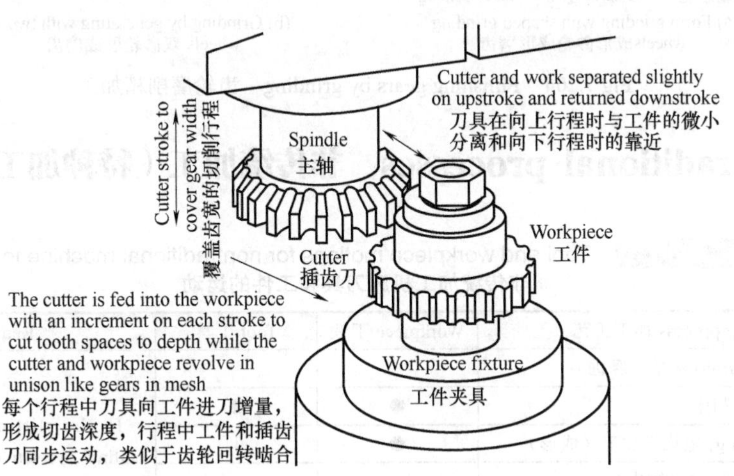

Fig 7.200 Operation zone of a gear shaper 插齿刀工作区域

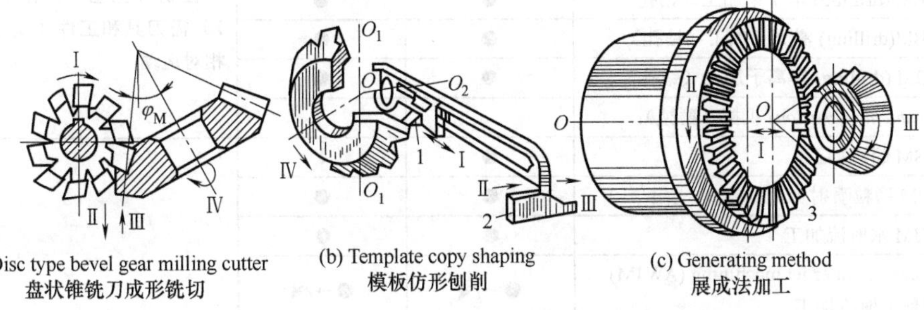

Fig 7.201 Bevel gear cutting 锥齿轮加工图解

Fig 7.202 Bevel gear cutting 锥齿轮加工

Fig 7.203 Gear lapping 研磨齿轮

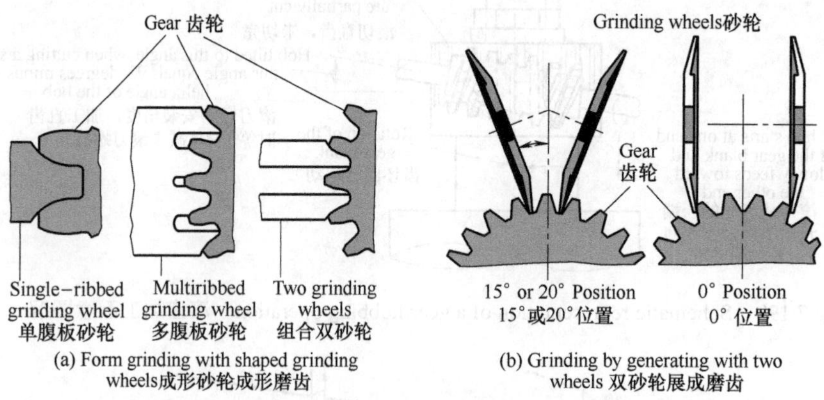

Fig 7.204 Finishing gears by grinding 齿轮磨削精加工

7.5 Non-traditional processes 非传统加工（特种加工）工艺方法

Table 7.3 Tool and workpiece motions for nontraditional machine tools
非传统加工机床刀具和工件的运动

Machining process 加工工艺	Workpiece 工件	Tool 工具	Remark 备注
Chemical (*erosion*) 化学（腐蚀）:			
CHM 化学铣削	●	●	In the slitting processes (plate cutting), a relative motion between tool and WP (traverse speed v_t) is imparted in horizontal directions (X,Y) 在切片工艺中，在水平方向（X, Y）需刀具和工件（横向进给 v_t）的相对运动
ECM (sinking) 电化学加工（成形）	●	↓	
Themal (*erosion*) 热（蚀出）:			
EDM (sinking) 电火花加工（成形）	●	↓ ↷	
EBM (drilling) 电子束加工（钻孔）	●	●	
LBM(drilling) 激光束加工（钻孔）	●	●	
PBM (drilling) 等离子加工（钻孔）	●	●	
Mechanical (*abrasion*) 机械（磨蚀）:			
USM 超声加工	●	↓	
AJM 磨粒喷射加工	●	●	
WJM 水射流加工	●	●	
Abrasive water jet machining (AWJM) 磨料水射流加工	●→↷	●→↷	

Note（注）：↷，Rotation 回转；●，Stationary 静止；→，Linear motion 线性运动。

7.5.1 EDM 电火花加工

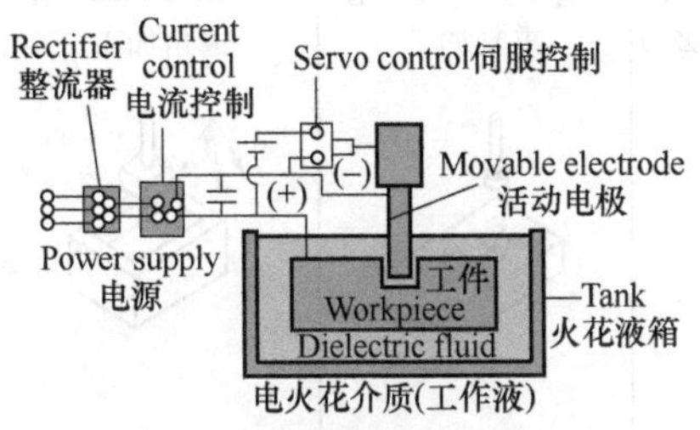

Fig 7.205 Illustration of EDM 电火花加工图解

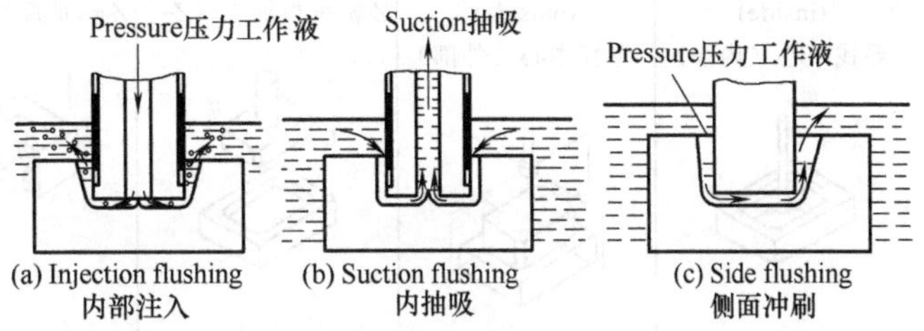

Fig 7.206 Dielectric flushing modes 电火花工作液冲刷方式

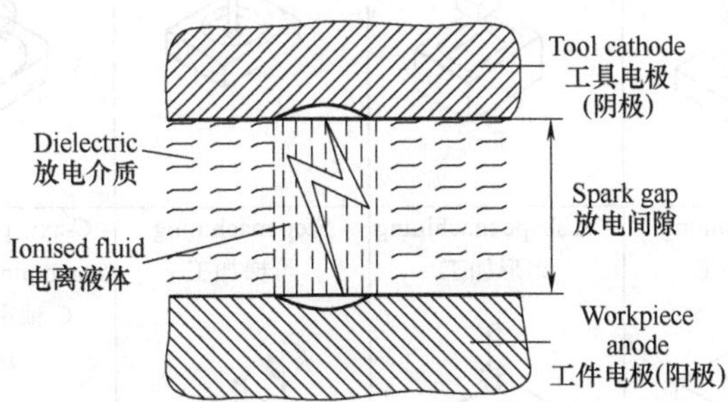

Fig 7.207 Schematic diagram of arc formation in EDM process 电火花弧的形成

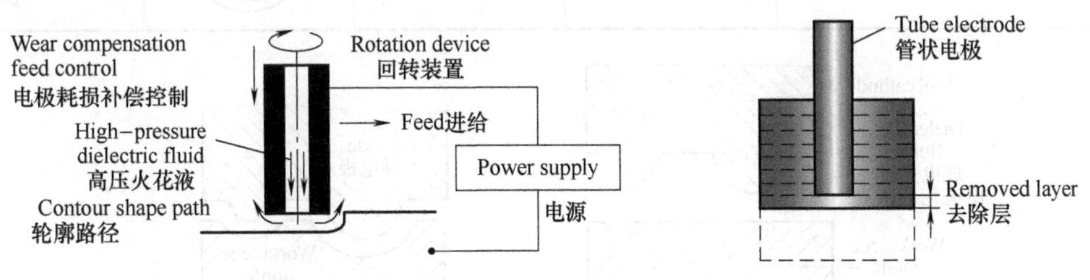

Fig 7.208 Theory of EDM milling 电火花铣削原理

Fig 7.209 EDSCAN machining 电火花电极平动加工

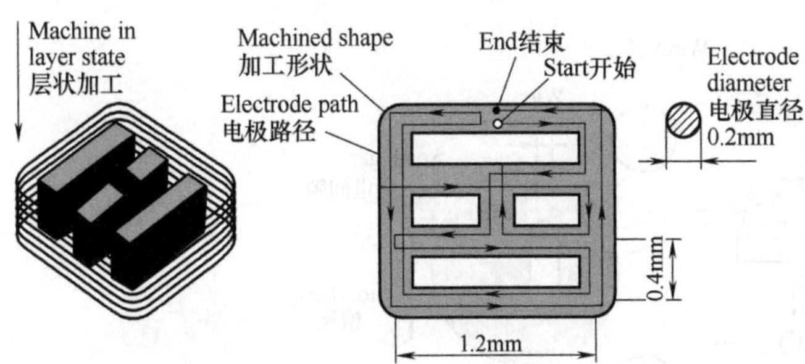

Fig 7.210 Micro EDM 细微电火花加工

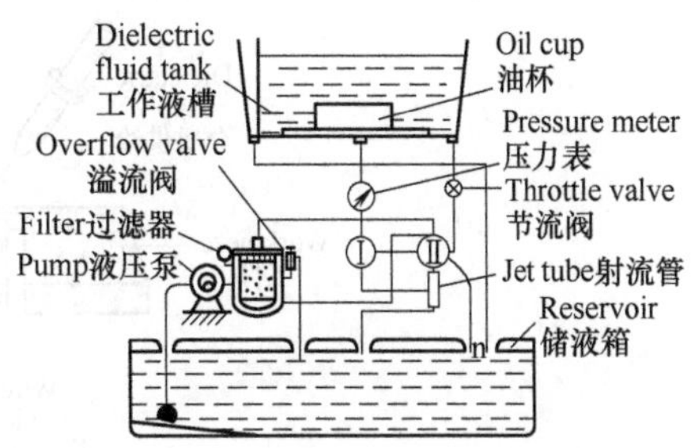

Fig 7.211 Dielectric recycling system 工作液循环系统

Table 7.4 Applications in EDM 电火花加工的应用

Center searching (inside) 寻找中心（内部）	Center searching (outside) 寻找中心（外部）	*Z*-axis machining *Z* 轴垂直加工（*Z*–）	*Z*-axis machining *Z* 轴垂直加工（*Z*+）	Repeat machining 重复加工	Corner machining 清角加工
Sliding machining 平动加工	Rotation machining 旋转加工Ⅰ	Rotation machining 旋转加工Ⅱ	Rotation machining 旋转加工Ⅲ	Rotation machining 旋转加工Ⅳ	Rotation machining 旋转加工Ⅴ
Sliding machining 平动加工	Side machining 侧面加工	Fan shape machining 扇形加工	Step machining 阶梯加工	*C*-axis positioning machining(optional) *C* 轴定位加工（选项）	*C*-axis helical gear machining(optional) *C* 轴螺旋齿轮加工（选项）

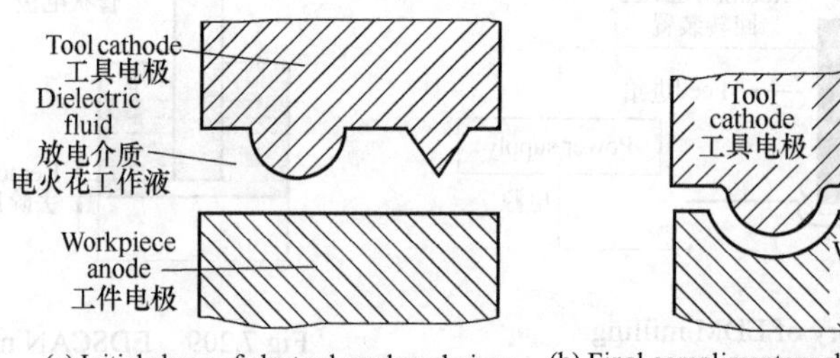

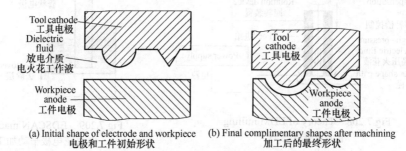

Fig 7.212 Typical surface generation in EDM process 电火花加工典型表面

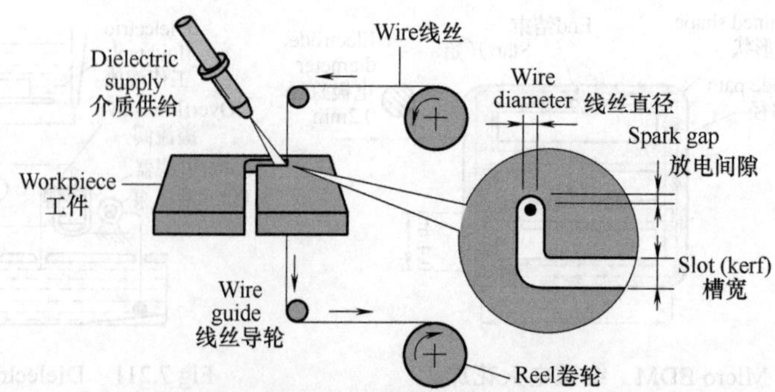

Fig 7.213 Wire EDM 电火花线切割

7.5.2 ECM 电化学加工

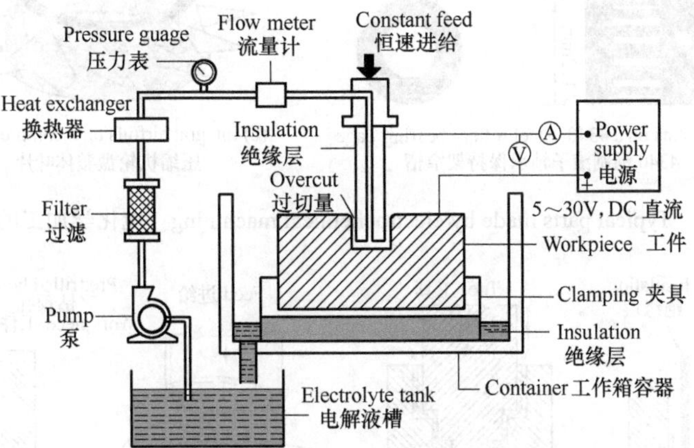

Fig 7.214 ECM setup 电化学加工装置

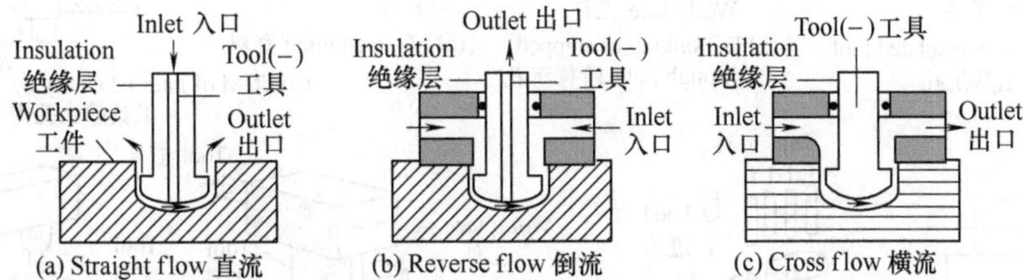

Fig 7.215 Methods of elcctrolyte feeding in ECM 电解液供给方法

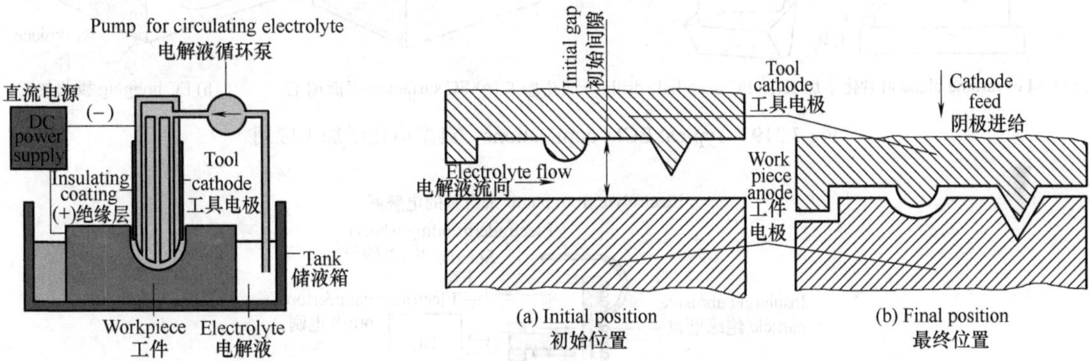

Fig 7.216 Electrochemical machining 电化学加工

Fig 7.217 Complimentary shape produced by ECM 电解加工获得的形状

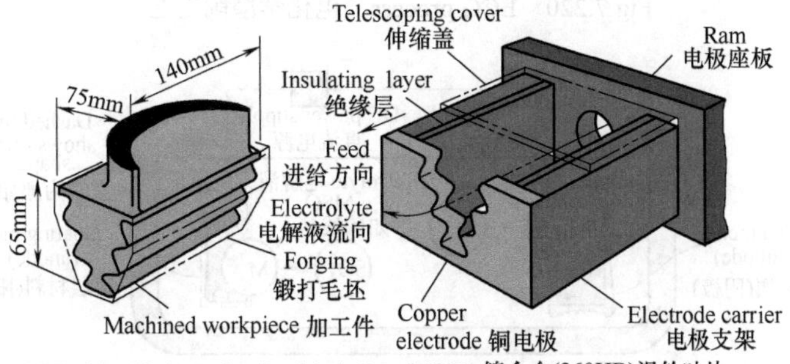

(a) Turbine blade made of a nickel alloy: 360HB 镍合金(360HB)涡轮叶片

Fig 7.218

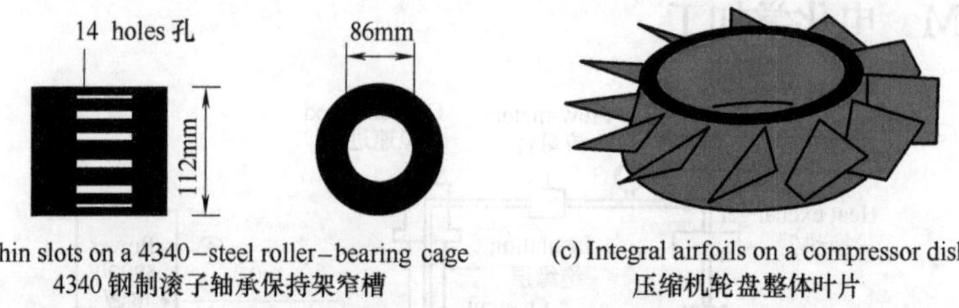

(b) Thin slots on a 4340-steel roller-bearing cage 4340钢制滚子轴承保持架窄槽

(c) Integral airfoils on a compressor disk 压缩机轮盘整体叶片

Fig 7.218 Typical parts made by electrochemical machining 电化学加工的典型零件

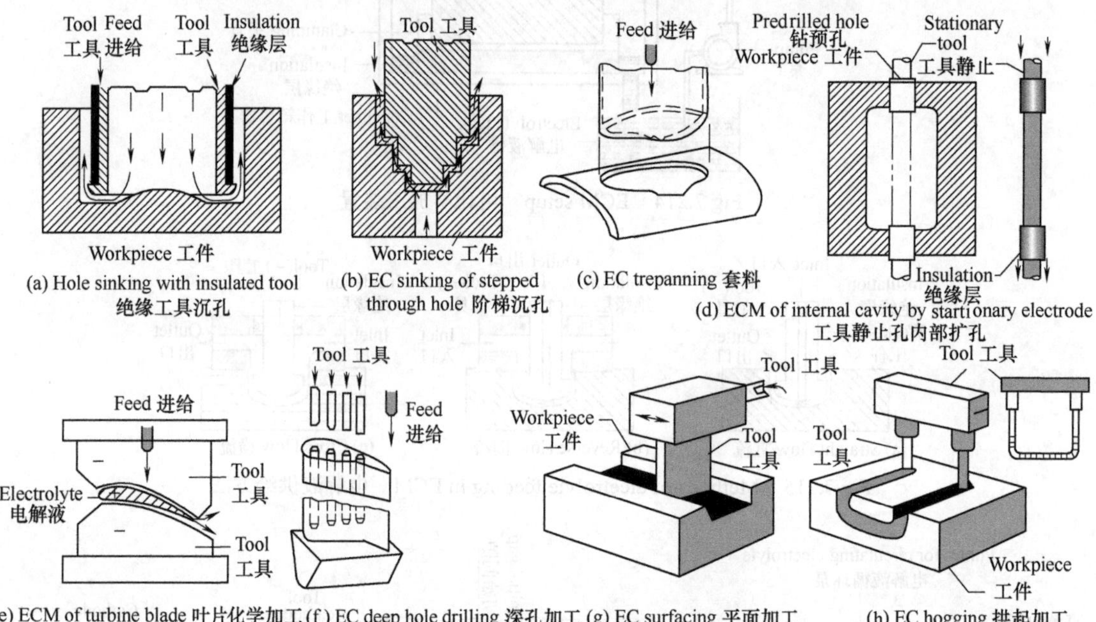

(a) Hole sinking with insulated tool 绝缘工具沉孔

(b) EC sinking of stepped through hole 阶梯沉孔

(c) EC trepanning 套料

(d) ECM of internal cavity by starti onary electrode 工具静止孔内部扩孔

(e) ECM of turbine blade 叶片化学加工 (f) EC deep hole drilling 深孔加工 (g) EC surfacing 平面加工 (h) EC hogging 拱起加工

Fig 7.219 Typical ECM applications 典型电化学加工应用

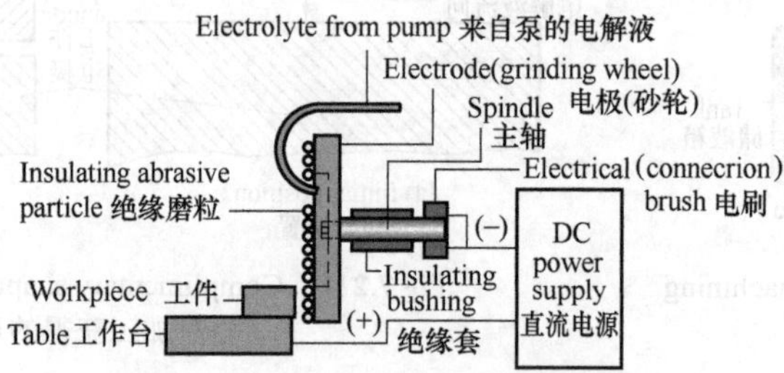

Fig 7.220 ECG process 电化学磨削工艺

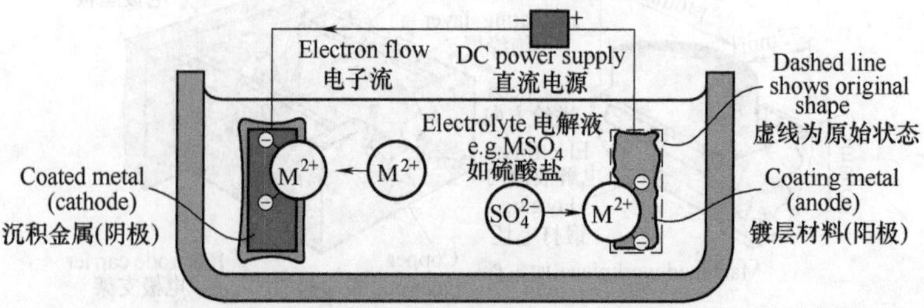

Fig 7.221 Electroplating process 电镀工艺

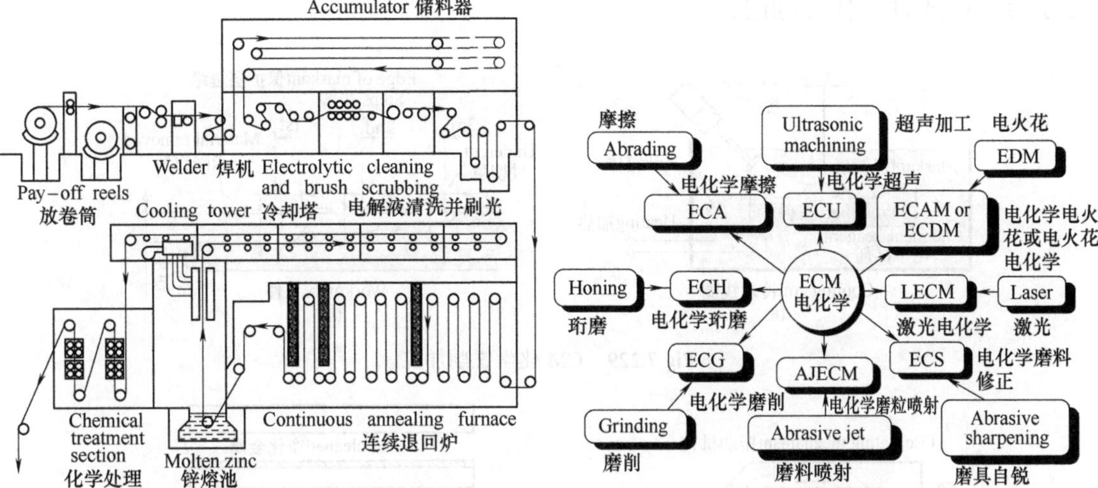

Fig 7.222 Flowline for continuous hot-dip galvanizing of sheet steel 钢板连续热浸镀锌工艺流程

Fig 7.223 Hybrid NTMP integrated with ECM 与电化学方法结合的复合非传统加工

7.5.3 USM 超声加工

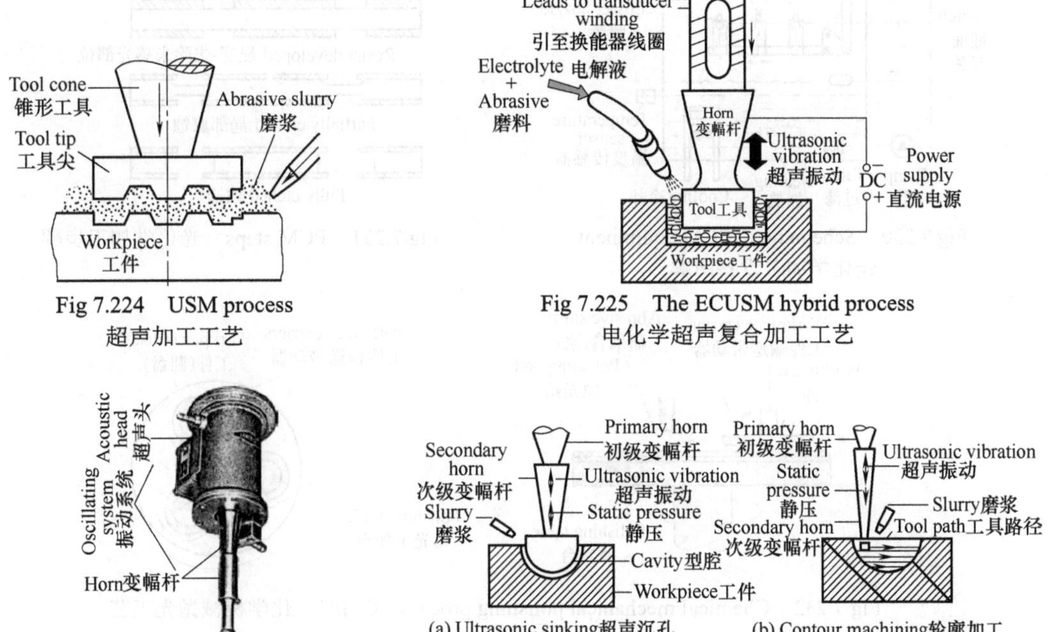

Fig 7.224 USM process 超声加工工艺

Fig 7.225 The ECUSM hybrid process 电化学超声复合加工工艺

Fig 7.226 Oscillating system of USM equipment 超声加工设备的振动系统

Fig 7.227 Snking (conventional) and contouring USM 超声（常规）成形加工和轮廓加工

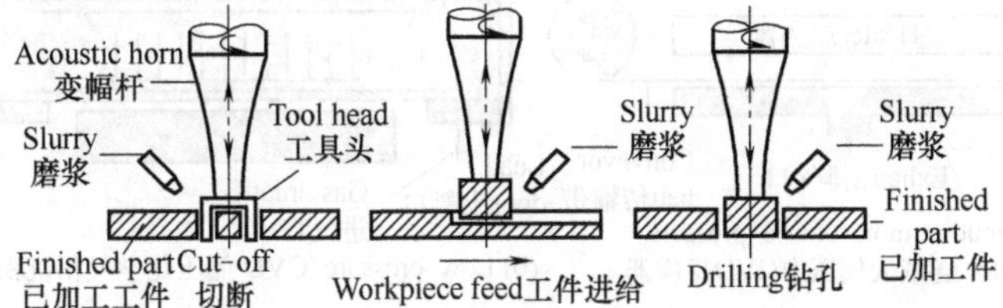

Fig 7.228 Rotary ultrasonic machining (RUM) 回转超声加工

7.5.4 CHM 化学加工

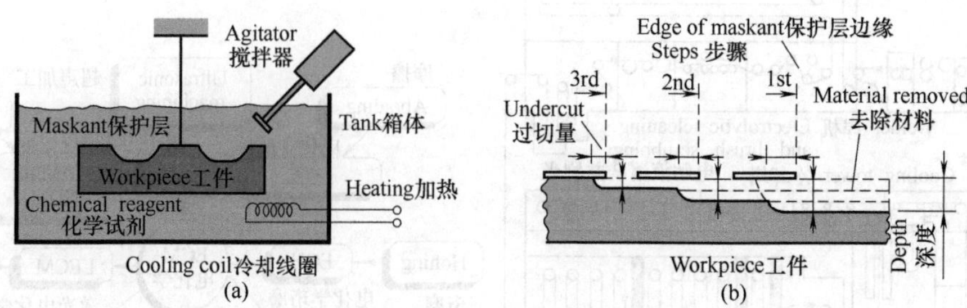

Fig 7.229 CM 化学铣削加工

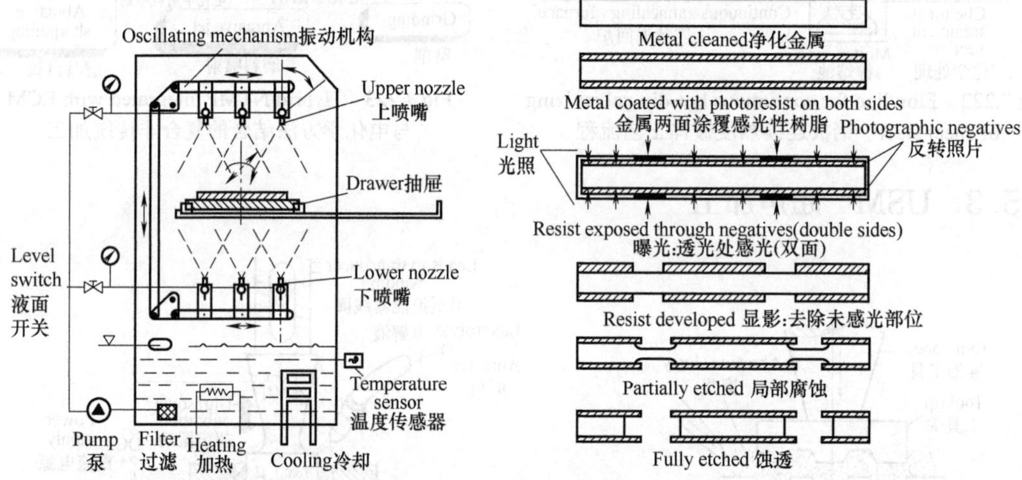

Fig 7.230 Schematic of PCM equipment 光化学加工设备图解

Fig 7.231 PCM steps 光化学加工步骤

Fig 7.232 Chemical mechanical polishing process（CMP） 化学机械抛光工艺

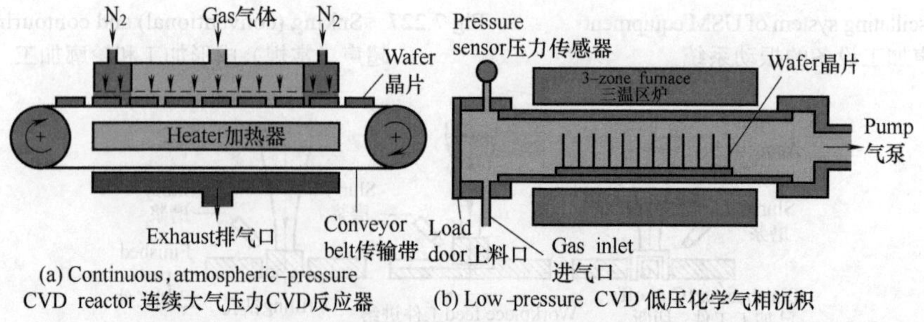

Fig 7.233 Chemical vapor deposition（CVD） 化学气相沉积

7.5.5 EBM 电子束加工

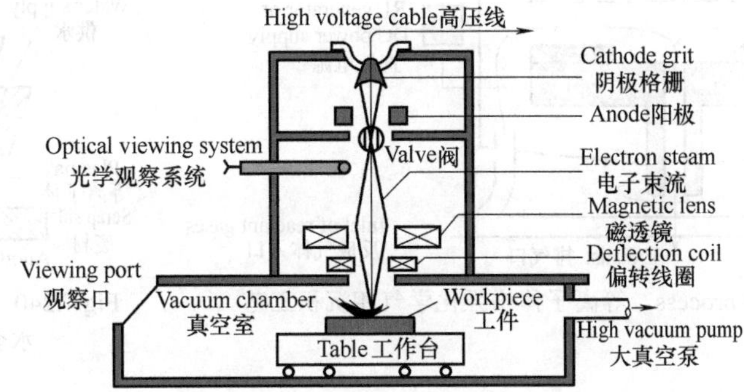

Fig 7.234 Schematic illustration of the EBM 电子束加工示意图

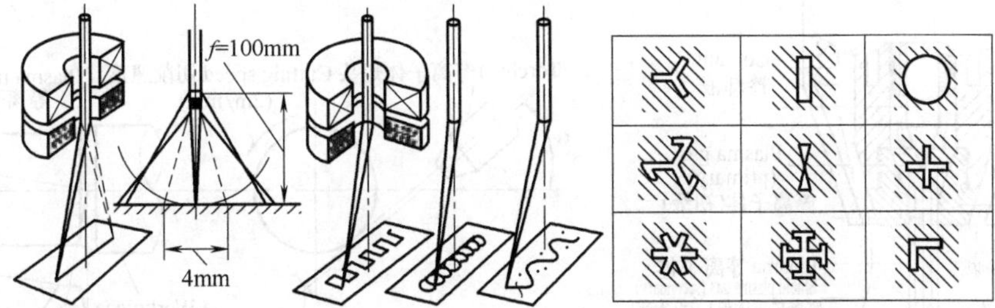

Fig 7.235 Standard configurations produced by EB without WP manipulation 无操控工件的电子束加工标准零件

7.5.6 IBM 离子加工

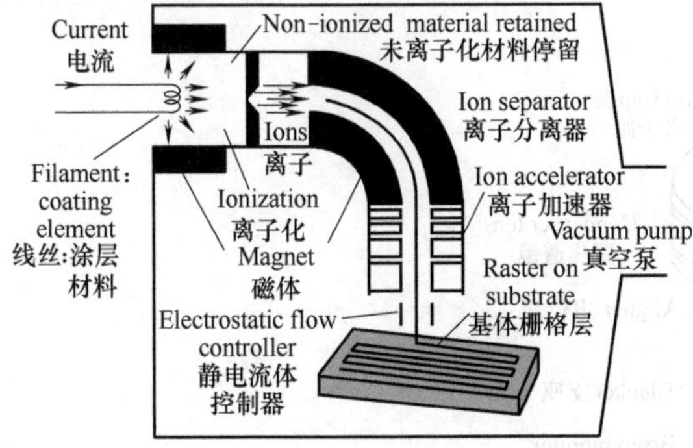

Fig 7.236 Ion implantation process 离子注入工艺

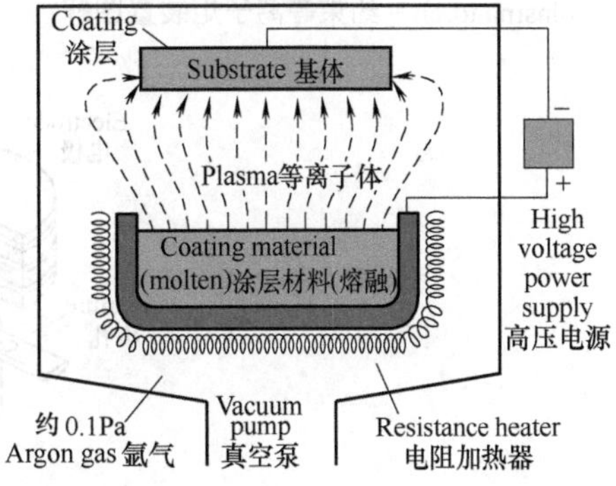

Fig 7.237 Ion-plating process 离子镀过程

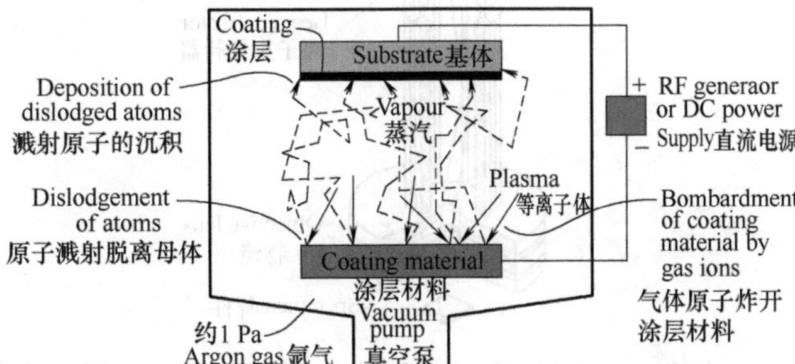

Fig 7.238 Sputtering process 溅射涂覆工艺

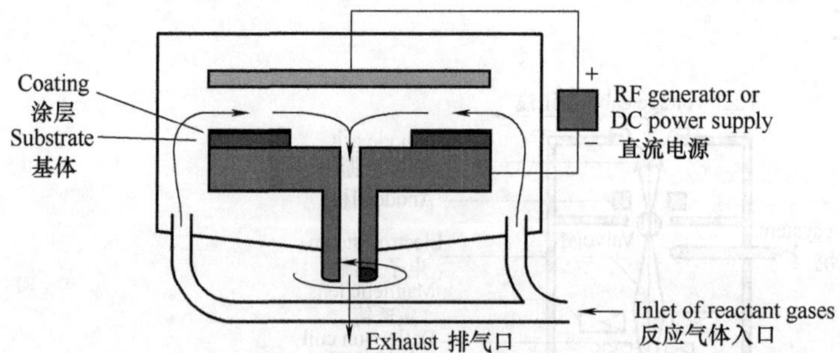

Fig 7.239 PECVD process 等离子体增强化学气相沉积工艺

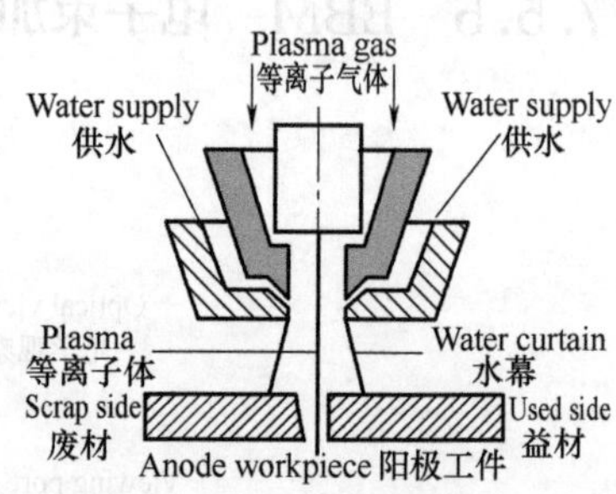

Fig 7.240 Water-Shielded plasma 水套保护等离子体

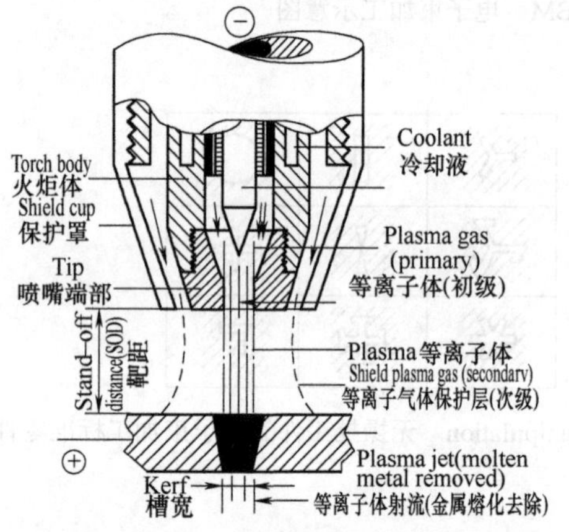

Fig 7.241 Constructional assembly of transferred plasma torch 约束等离子炬装置图解

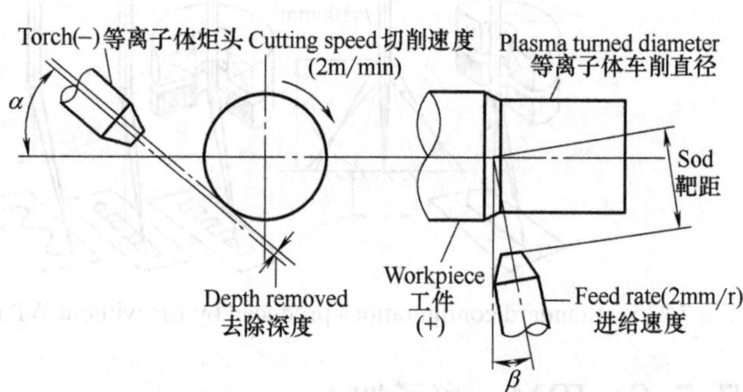

Fig 7.242 Plasma arc turning(PAT) 等离子弧车削

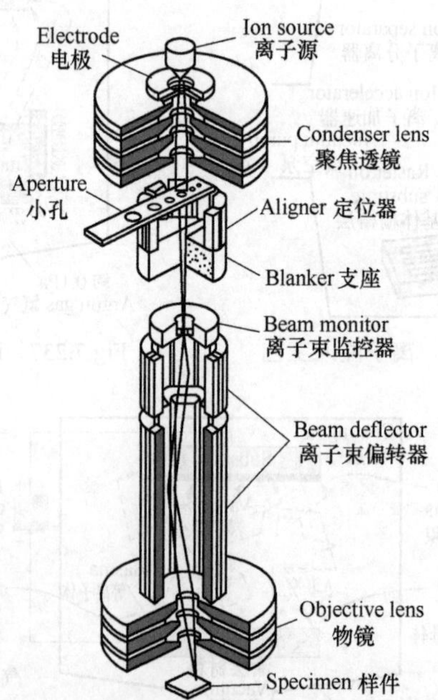

Fig 7.243 Schematic diagram of the focused ion beam (FIB) system 聚焦离子束系统图解

7.5.7 LBM 激光束加工

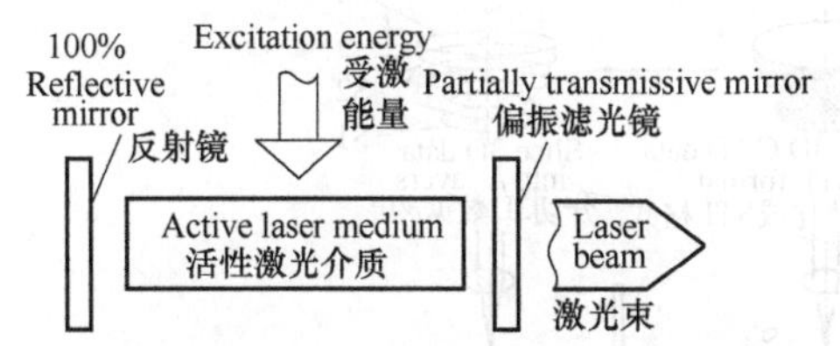

Fig 7.244 Basic components of a laser 激光的基本组成

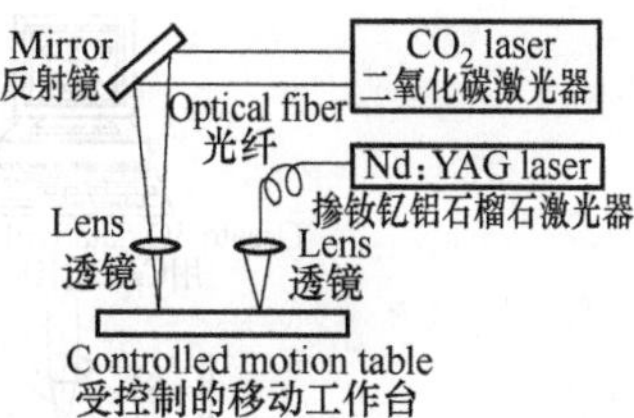

Fig 7.245 Typical laser system 典型激光系统

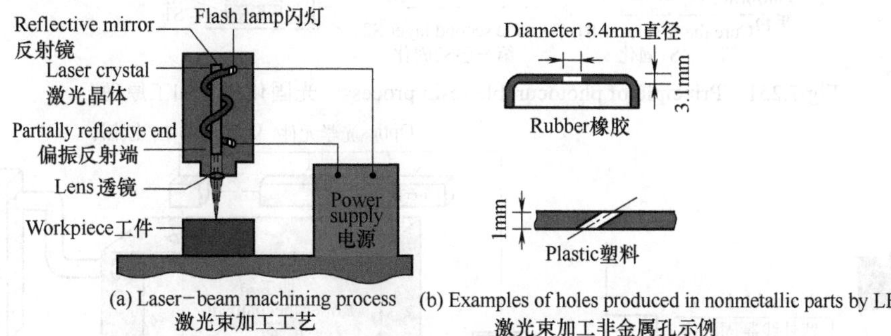

(a) Laser-beam machining process 激光束加工工艺 (b) Examples of holes produced in nonmetallic parts by LBM 激光束加工非金属孔示例

Fig 7.246 Illustration of LBM 激光束加工图

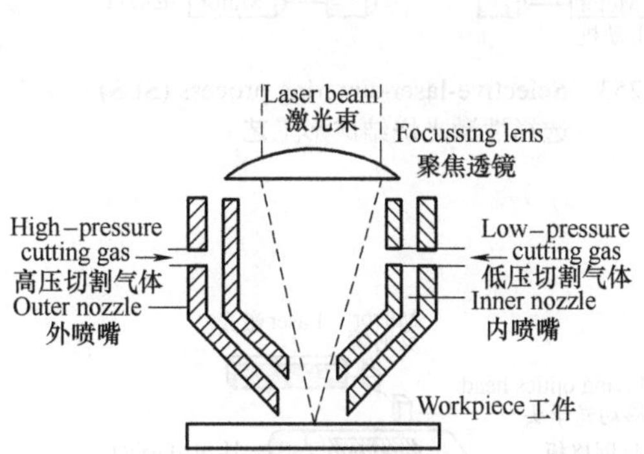

Fig 7.247 Laser gas cutting nozzle for steel 激光气体切割钢材喷嘴

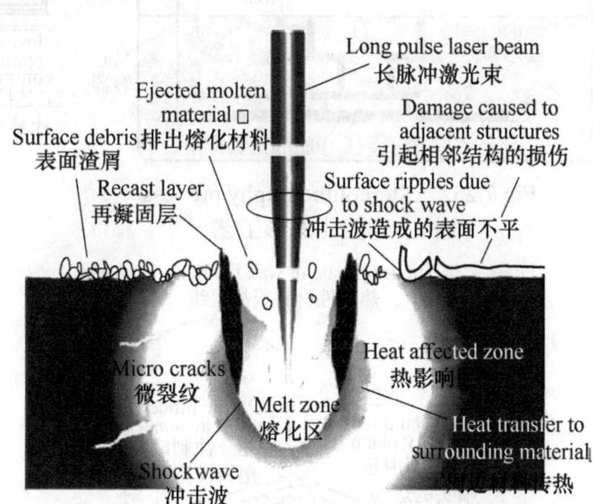

Fig 7.248 Laser ablation 激光烧蚀加工

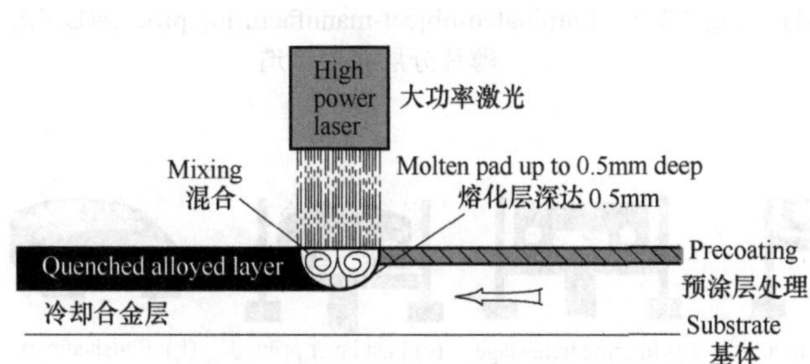

Fig 7.249 Laser surface alloying process 激光表面合金化处理工艺

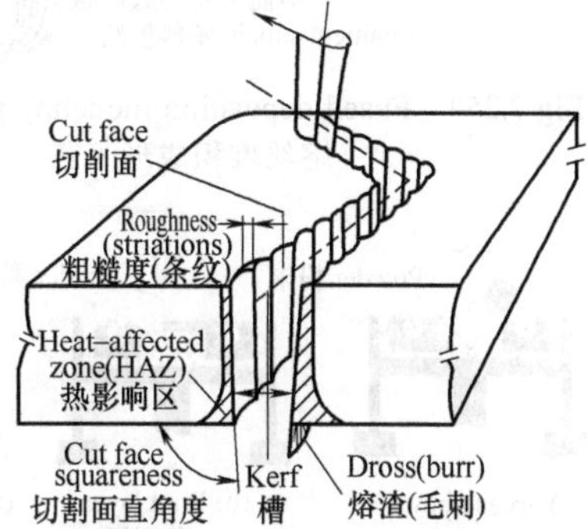

Fig 7.250 Factors in laser cutting 激光切割要素

7.5.8 RP 快速成形

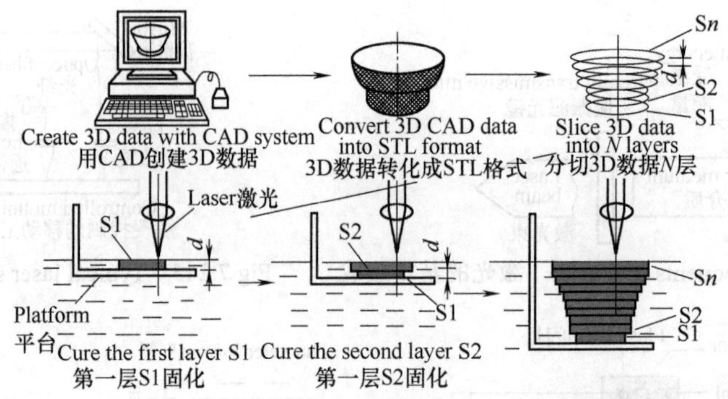

Fig 7.251 Principle of photocurable resin process 光固化树脂加工原理

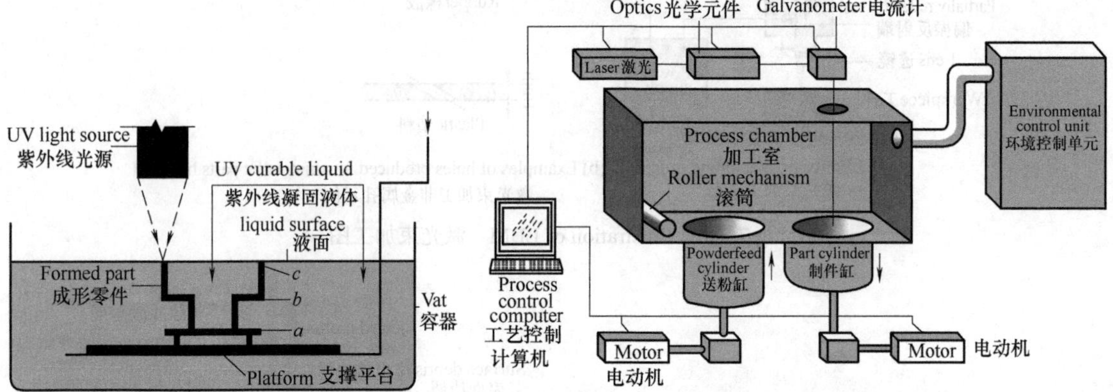

Fig 7.252 Stereolithography process
液相立体固化成形工艺

Fig 7.253 Selective-laser-sintering process (SLS)
选择性激光烧结成形工艺

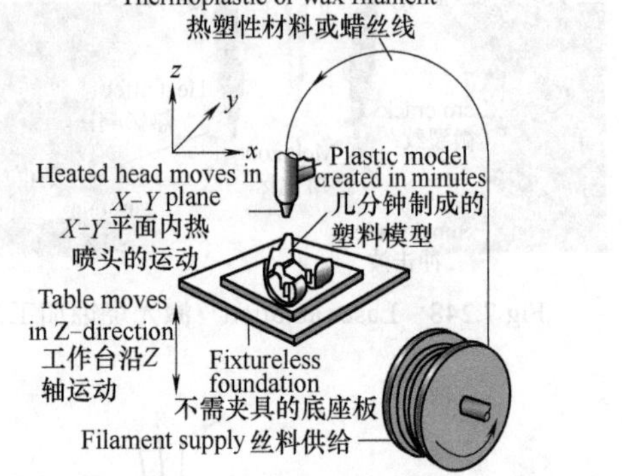

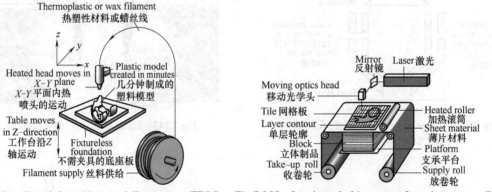

Fig 7.254 Fused-deposition-modeling process(FDM)
熔丝堆积成形工艺

Mirror
反射镜
Laser激光
Moving optics head
移动光学头
Tile 网格板
Layer contour
单层轮廓
Block
立体制品
Take-up roll
收卷轮
Heated roller
加热滚筒
Sheet material
薄片材料
Platform
支承平台
Supply roll
放卷轮

Fig 7.255 Laminated-object-manufacturing process(LOM)
薄片分层叠层制造

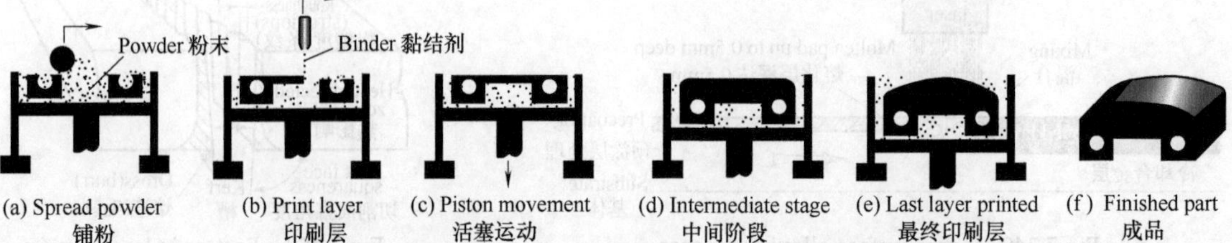

(a) Spread powder 铺粉
(b) Print layer 印刷层
(c) Piston movement 活塞运动
(d) Intermediate stage 中间阶段
(e) Last layer printed 最终印刷层
(f) Finished part 成品

Fig 7.256 Three dimensional printing process 立体印刷工艺

7.5.9 AJM/WJM/AWJM 磨料射流 / 水射流 / 磨料水射流加工

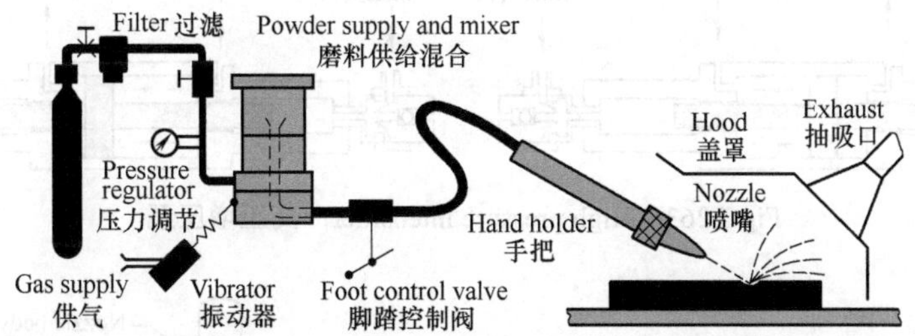

Fig 7.257 Abrasive jet (sand blasting) machining (AJM) 磨料射流（喷砂）加工

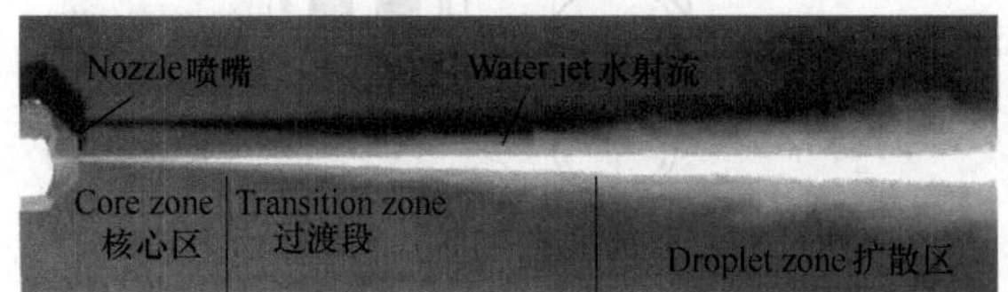

(a) Structure of a high-speed water jet 高速水射流的结构

(b) Structure of an ultrasonically modulated water jet 超声调制水射流结构

Fig 7.258 Structure of water jet 水射流结构

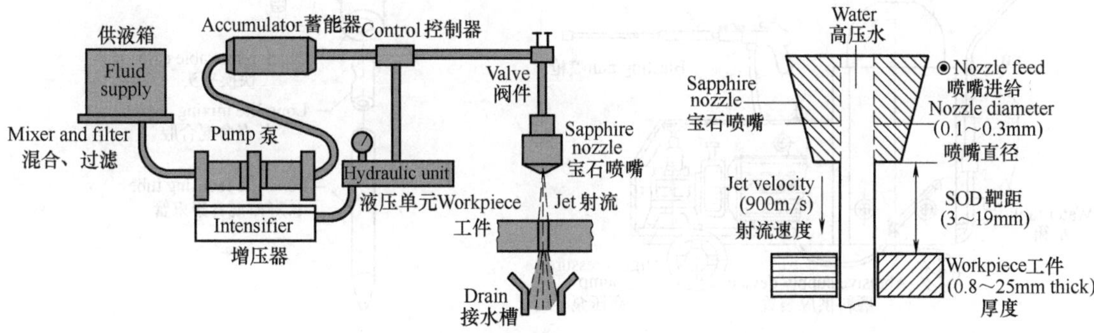

Fig 7.259 Water jet machining (WJM) 水射流加工

Fig 7.260 WJM terms 水射流加工术语

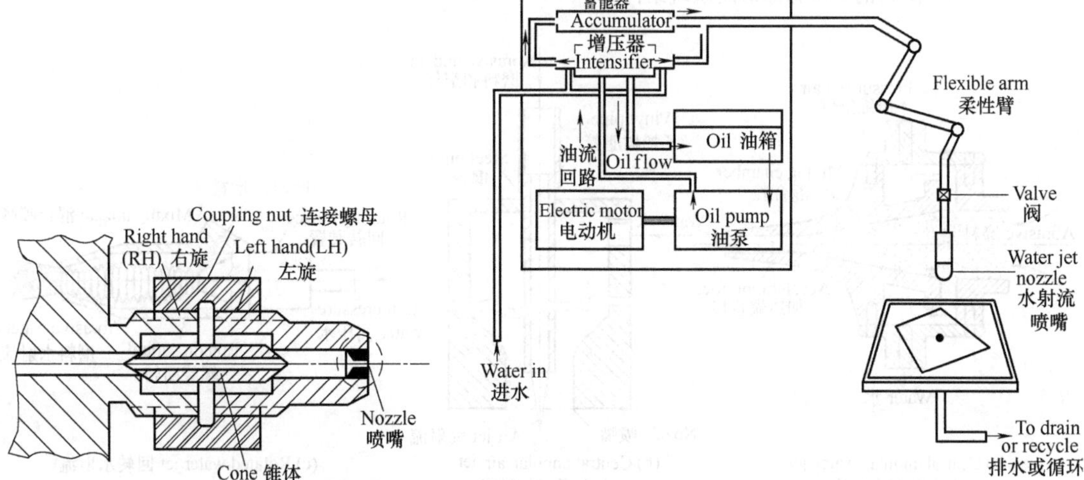

Fig 7.261 Nozzle assembly of WJM equipment 水射流加工设备喷嘴组件

Fig 7.262 Schematic illustration of WJM equipment 水射流设备图解

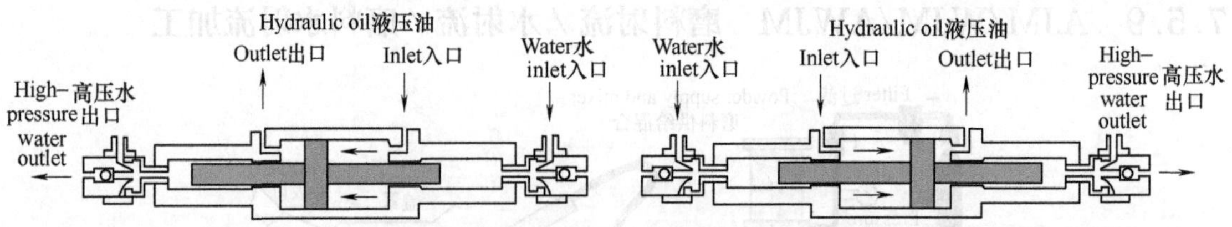

Fig 7.263 High pressure intensifier 高压增压泵

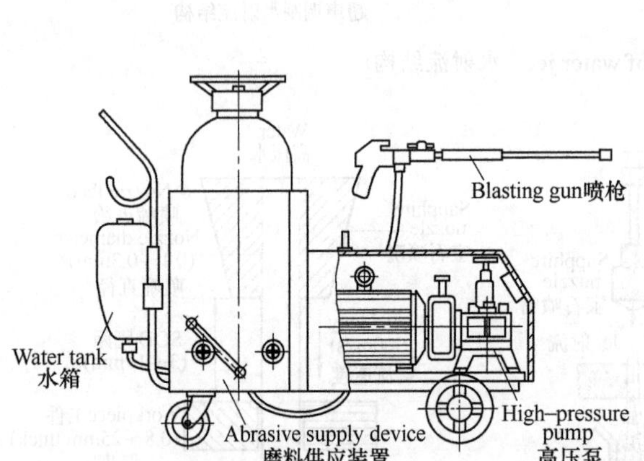

Fig 7.264 Structure of a hydro-abrasive suspension jet system for rust removal 用于除锈的磨浆悬浮射流系统结构

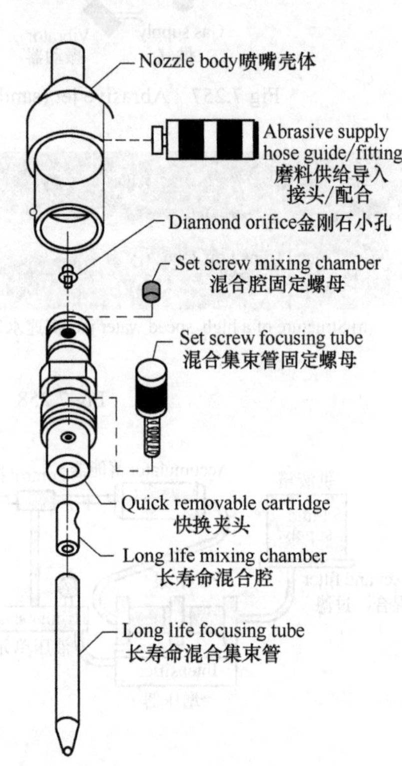

Fig 7.265 Assembly chart of jet former AWJM 磨料水射流加工喷嘴前端组件图

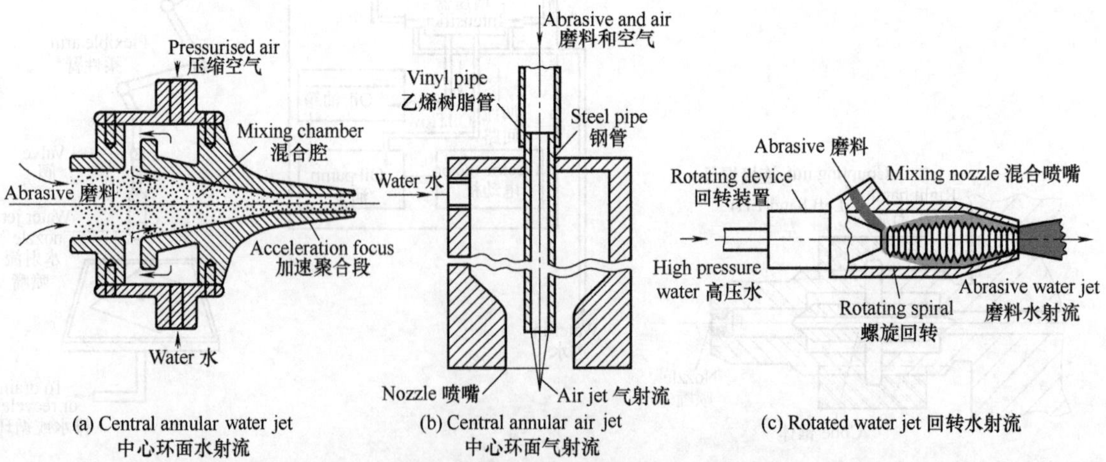

Fig 7.266 Alternative abrasive mixing principles 不同的磨料混合原理

7.5.10 MAM 磁性磨料加工

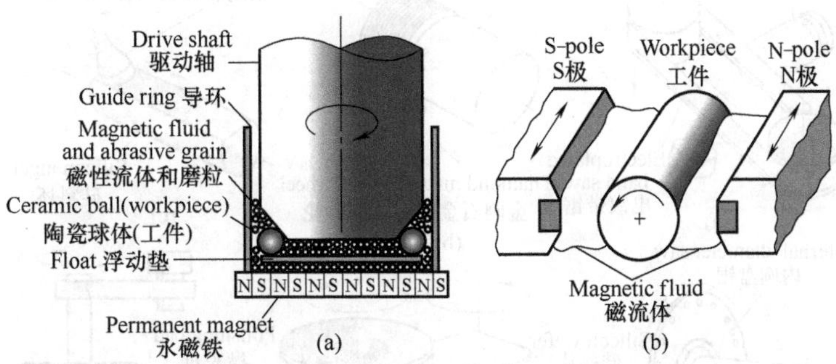

Fig 7.267 Magnetic fluid and abrasive polishing 磁性流体磨拉抛光

7.5.11 MEMS 微电子制造

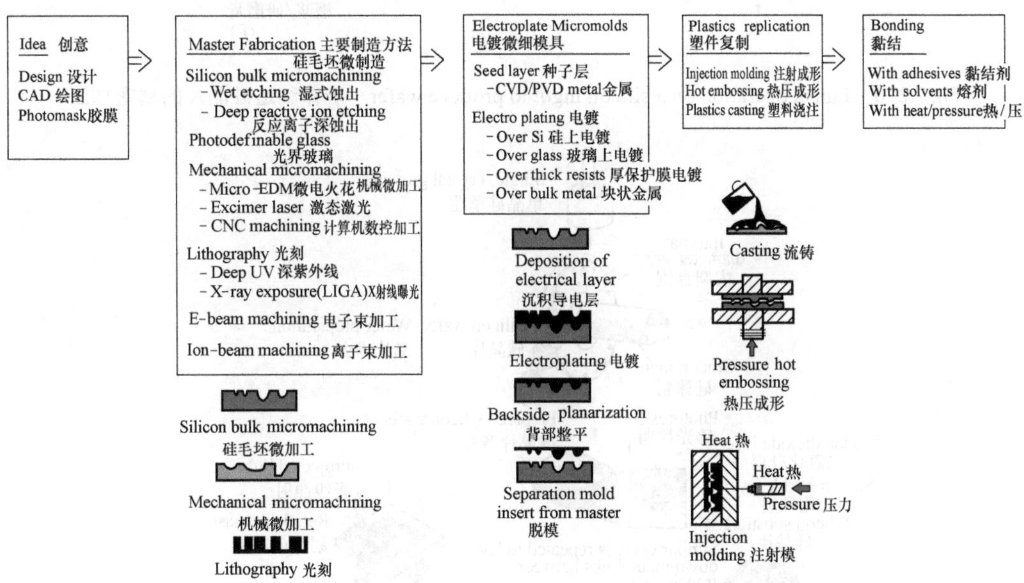

Fig 7.268 Process flow for plastic microfabrication 塑料微制造工艺流程

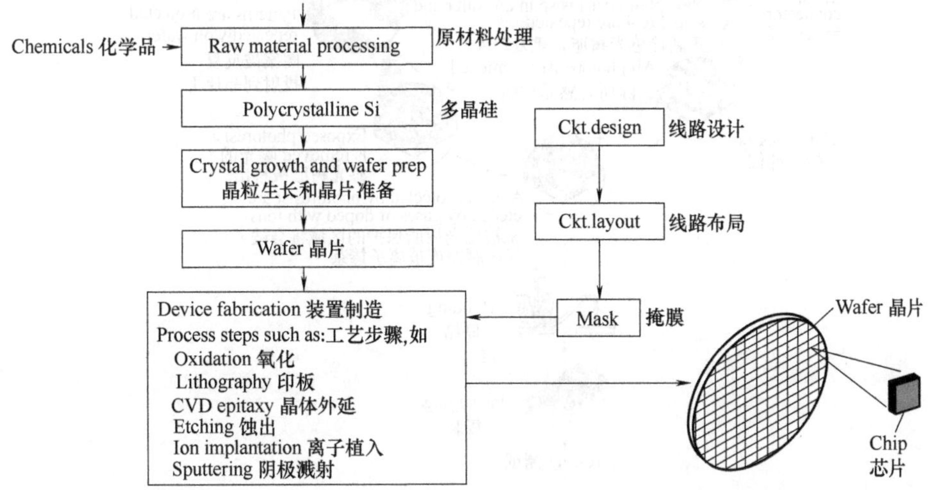

Fig 7.269 Fabrication sequence from raw material to chip 原材料制造芯片的工序

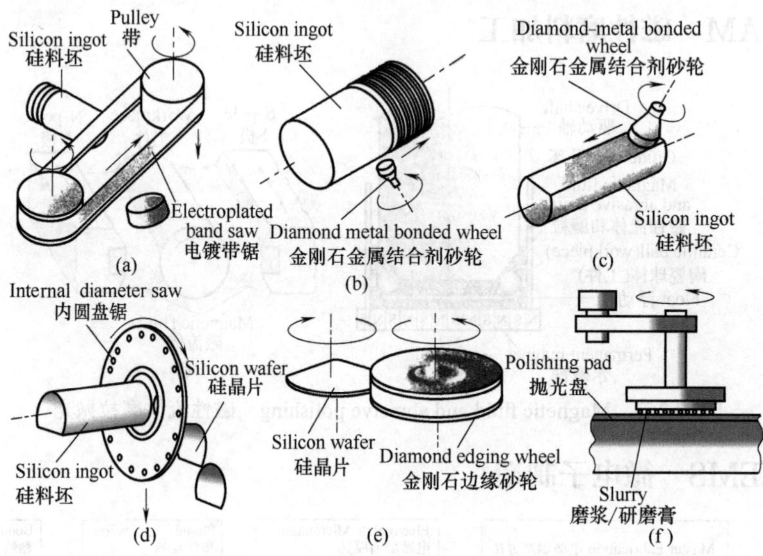

Fig 7.270 Finish operation on a silicon ingot to produce wafer 硅棒制造硅晶片的精密加工

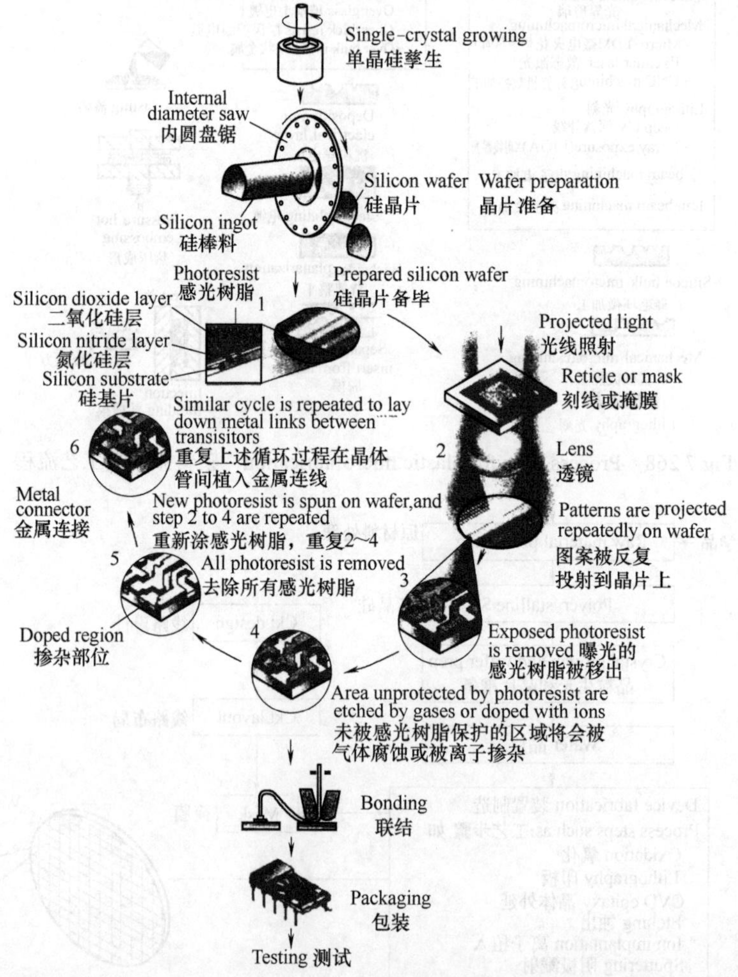

Fig 7.271 The general fabrication sequence for integrated circuit 集成电路的通用制作方法

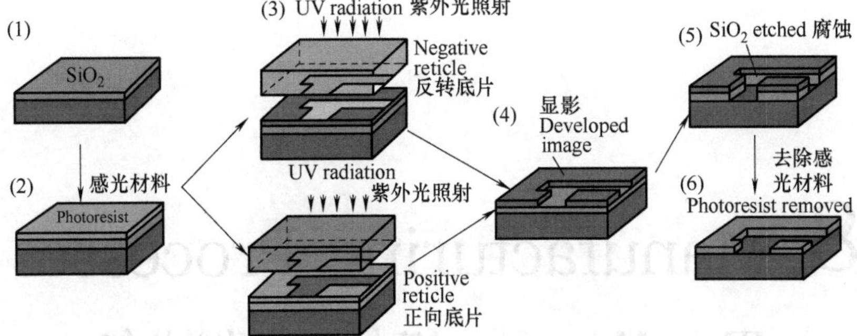

Fig 7.272 Pattern transfer by lithography 制版图形转移

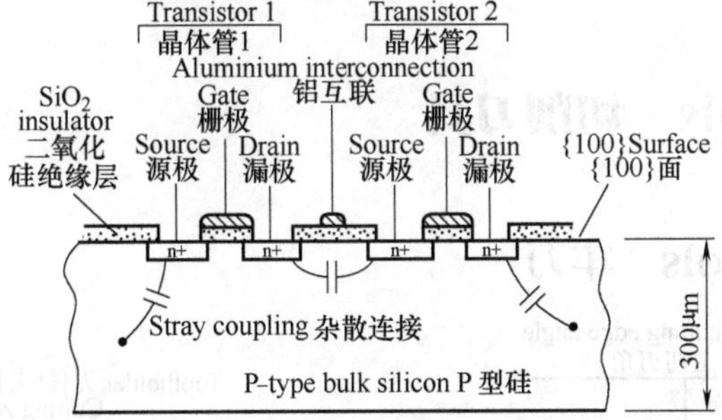

Fig 7.273 A typical integrated circuit. The silicon wafer is cut from a large single crystal using a chemical saw, mechanical sawing would introduce too many dislocations

典型的集成电路：采用化学锯切将大的单晶体切成硅晶片，因为机械切割会引起晶粒错位

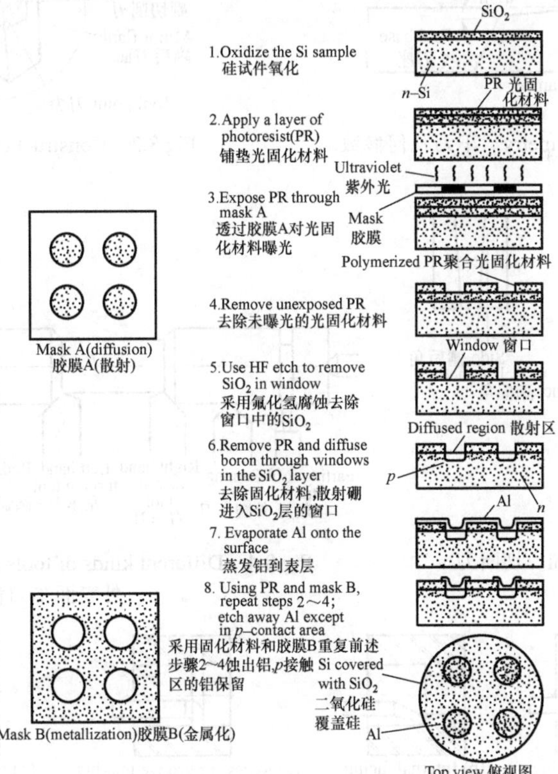

Fig 7.274 Outline of steps in the fabrication of diodes 二极管制造步骤概览

Unit 8 Manufacturing Processes and Toolings 机械制造装备

8.1 Cutting tools 切削刀具

8.1.1 Turning tools 车刀

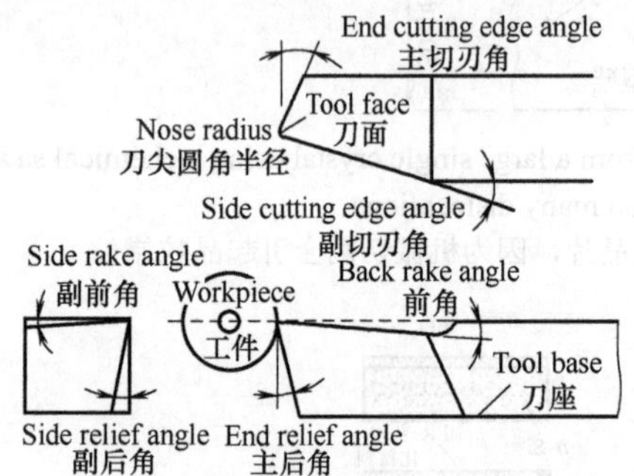

Fig 8.1 Turning tool geometry 车刀几何参数

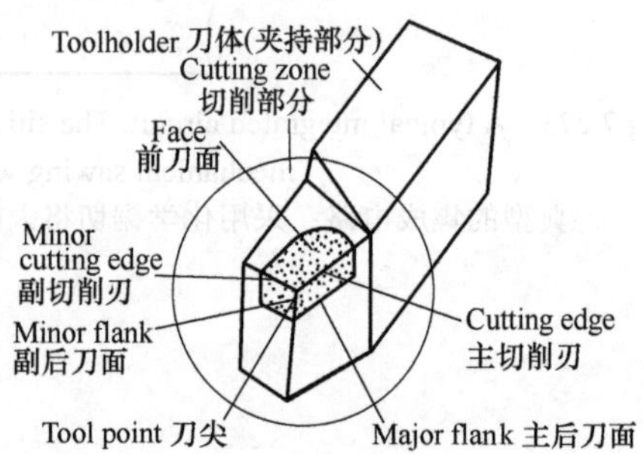

Fig 8.2 Construct of a cutting tool 刀具结构

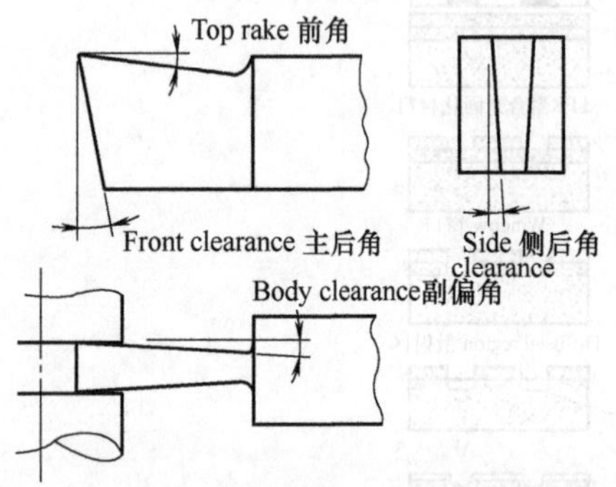

Fig 8.3 Parting-off tool 切断刀

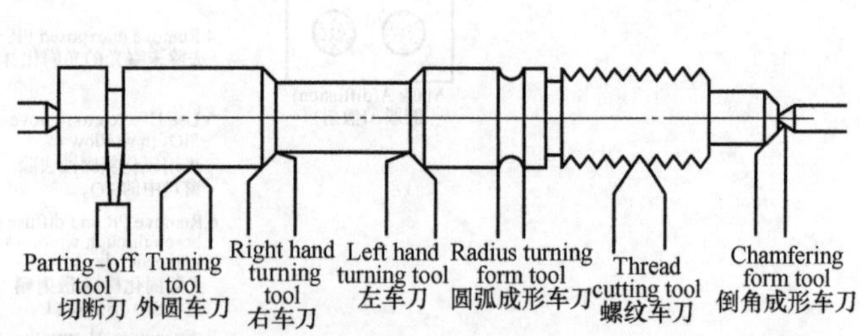

Fig 8.4 Different kinds of tools used for external surface 外表面车刀种类

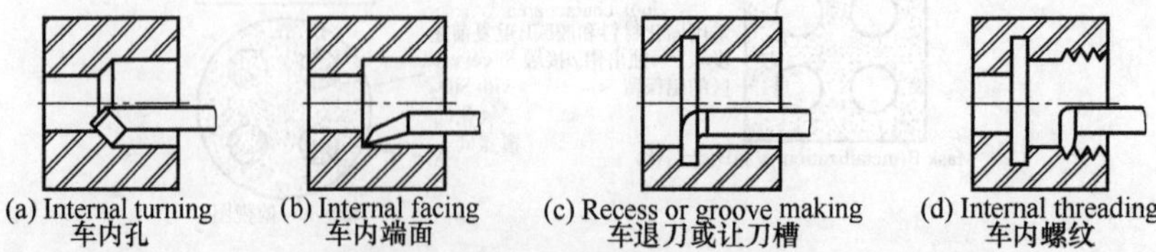

Fig 8.5 Different kinds of tools used for internal surface 内表面车刀种类

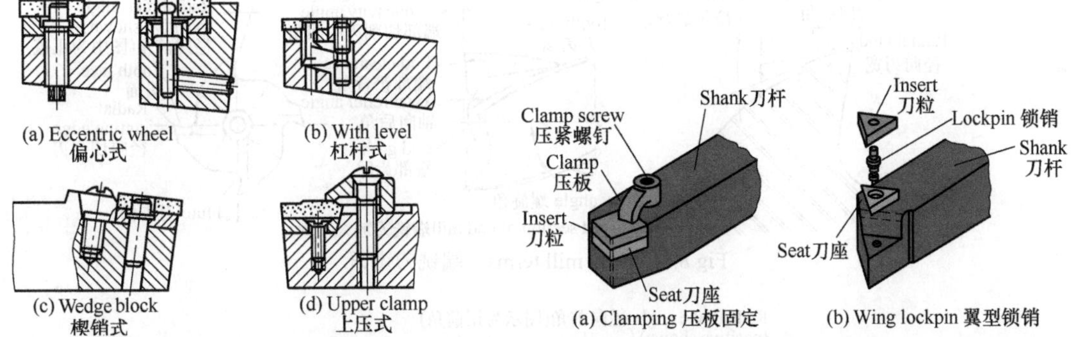

Fig 8.6 Various types of fixing inserts into tool body 可转位车刀的结构

Fig 8.7 Methods of attaching inserts to toolholders 刀粒固定在刀座上的方法

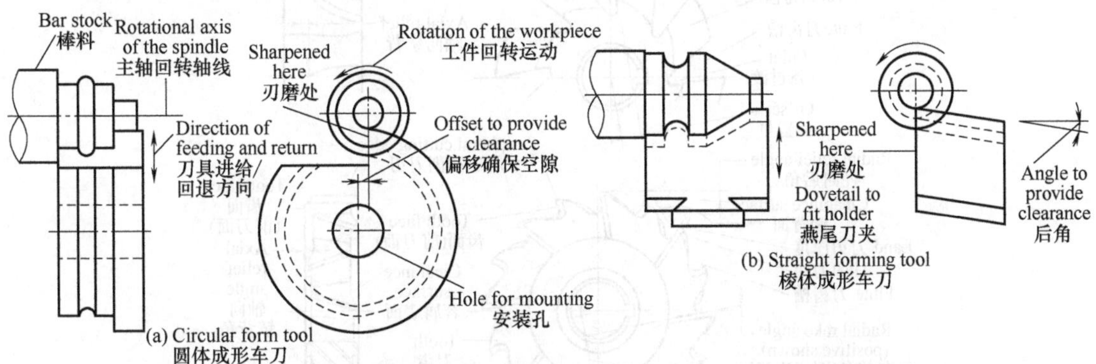

Fig 8.8 Form tool types used in centre lathe 中心车床用成形车刀种类

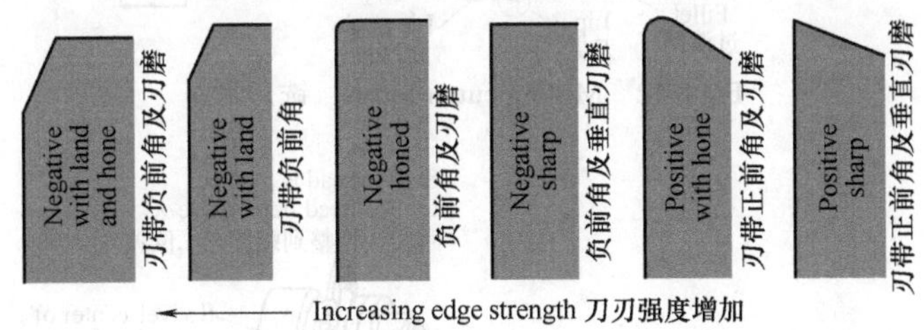

Fig 8.9 Edge preparation of inserts to improve edge strength 提高刀刃强度的刀粒刃口改变

8.1.2 Milling tools 铣刀

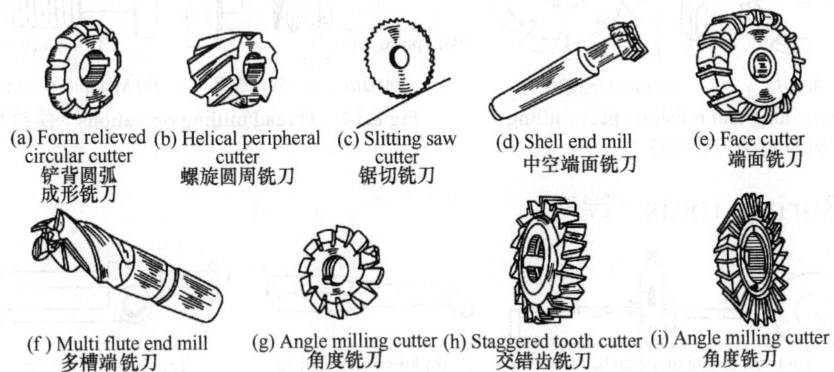

Fig 8.10 Various types of milling cutters 各种形式的铣刀

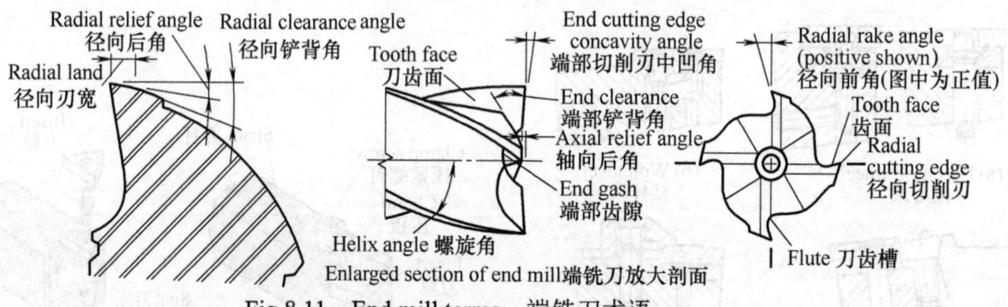

Fig 8.11　End mill terms　端铣刀术语

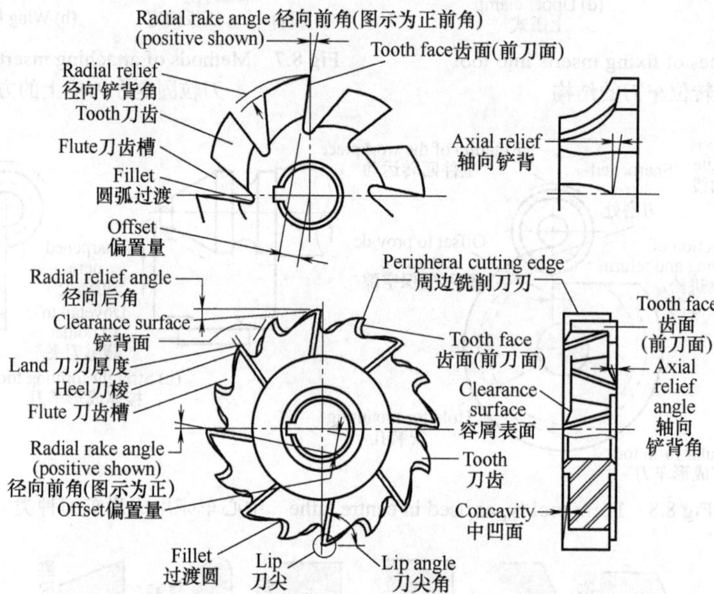

Fig 8.12　Milling cutter terms　铣刀术语

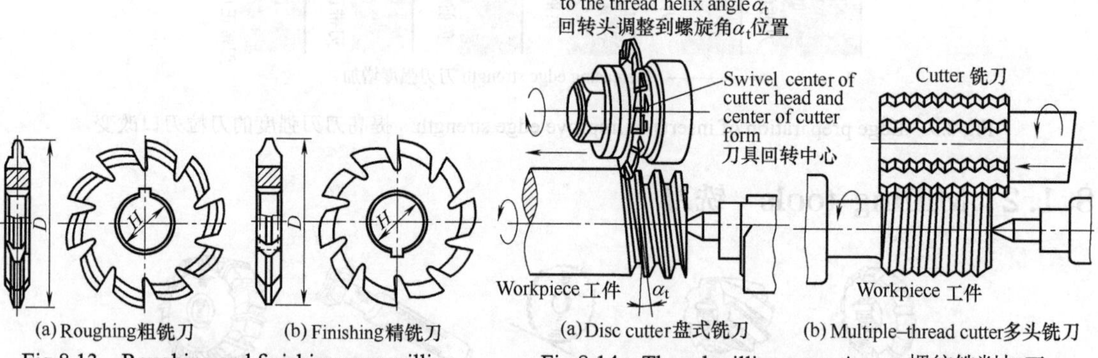

Fig 8.13　Roughing and finishing gear milling 齿轮粗铣刀和精铣刀

Fig 8.14　Thread milling operations　螺纹铣削加工

8.1.3　Boring tools　镗刀

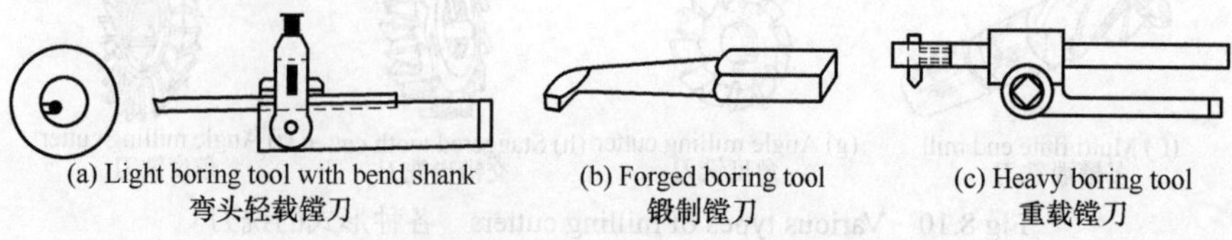

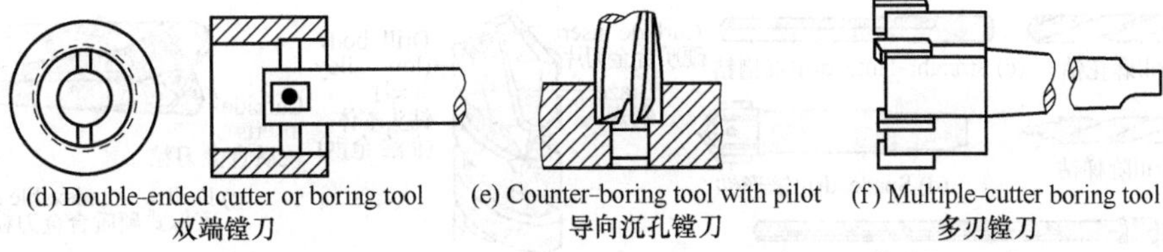
(d) Double-ended cutter or boring tool 双端镗刀 (e) Counter-boring tool with pilot 导向沉孔镗刀 (f) Multiple-cutter boring tool 多刃镗刀

Fig 8.15 Types of boring tools 镗刀种类

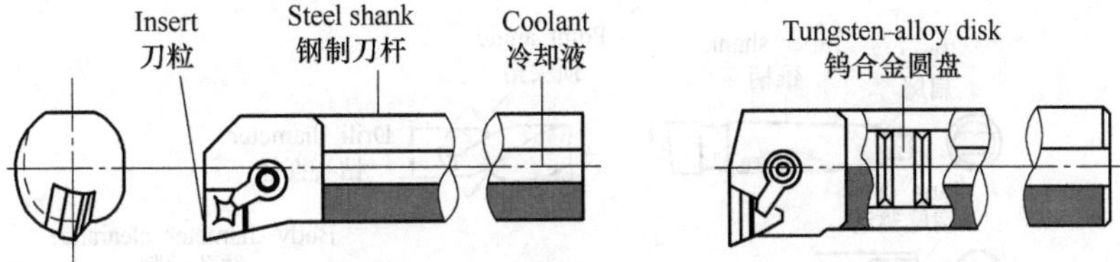

Fig 8.16 Boring bar structure 镗刀杆结构（一）

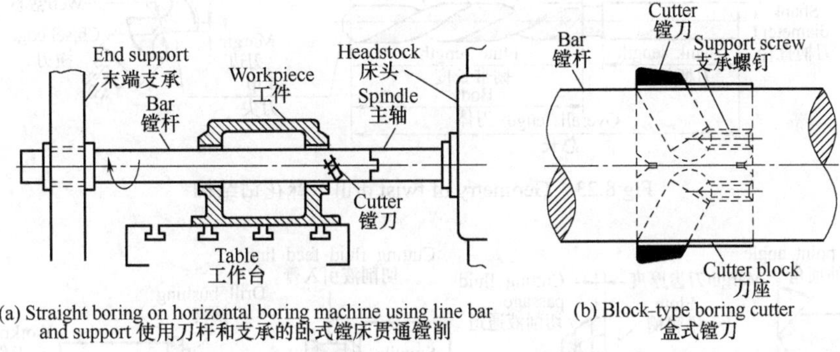

(a) Straight boring on horizontal boring machine using line bar and support 使用刀杆和支承的卧式镗床贯通镗削 (b) Block-type boring cutter 盒式镗刀

Fig 8.17 Boring bar structure 镗刀杆结构（二）

8.1.4 Hole-making tools 孔加工刀具

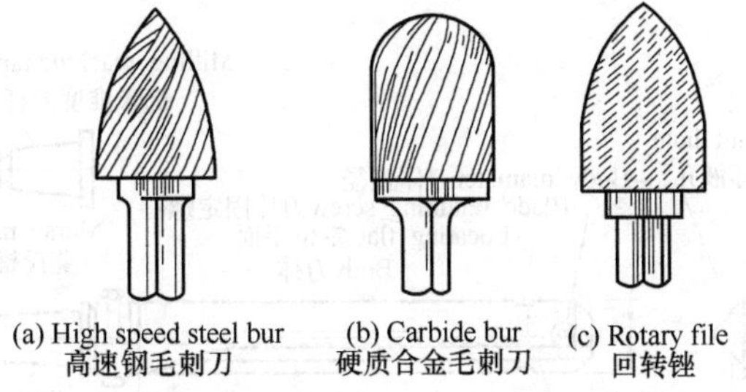
(a) High speed steel bur 高速钢毛刺刀 (b) Carbide bur 硬质合金毛刺刀 (c) Rotary file 回转锉

Fig 8.18 Various types of burs 去毛刺刀类别

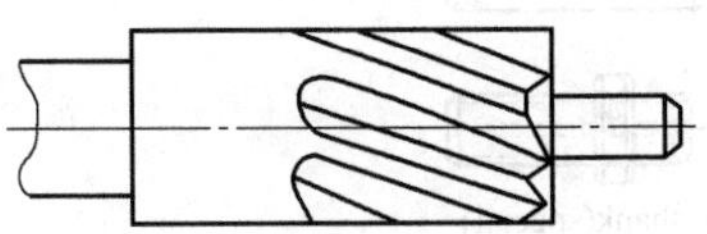
Fig 8.19 Counter boring drill 沉孔锪钻

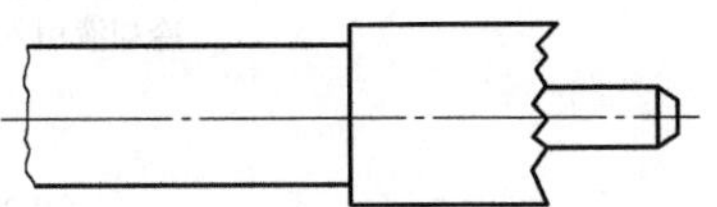
Fig 8.20 Spot facing drill 局部锪面钻

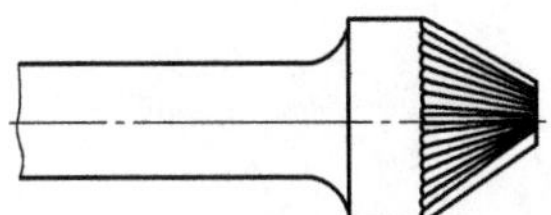
Fig 8.21 Counter sunk drill 倒角锪钻

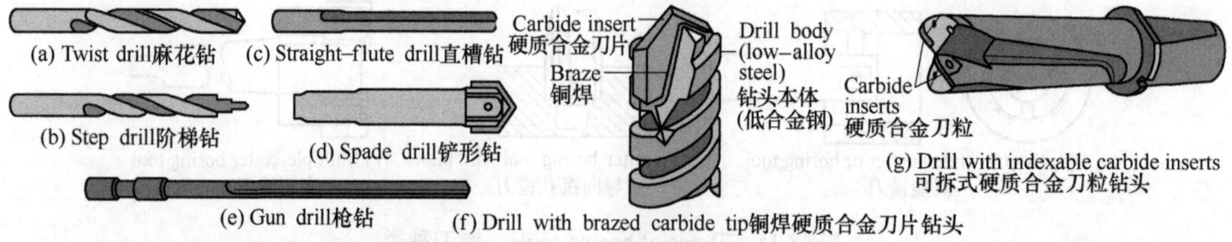

Fig 8.22 Various types of drills 各种钻头

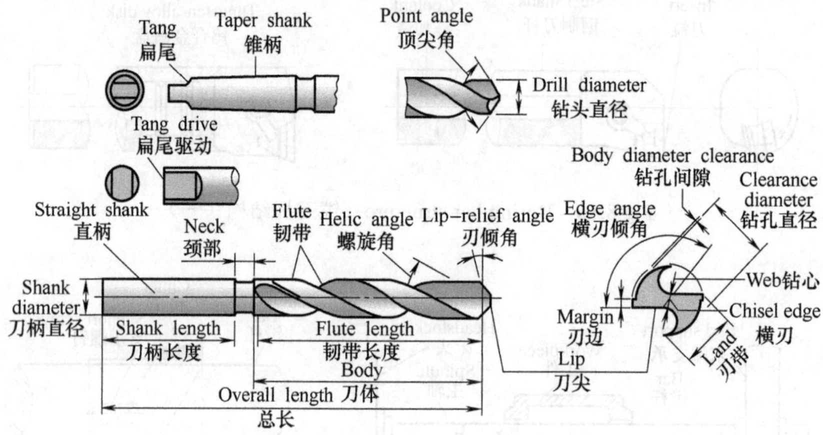

Fig 8.23 Geometry of twist drill 麻花钻结构

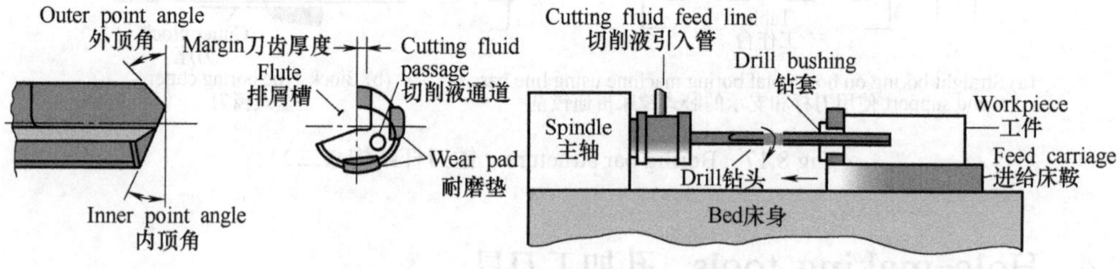

Fig 8.24 A gun drill features and gun-drilling operation 枪钻特征及其应用

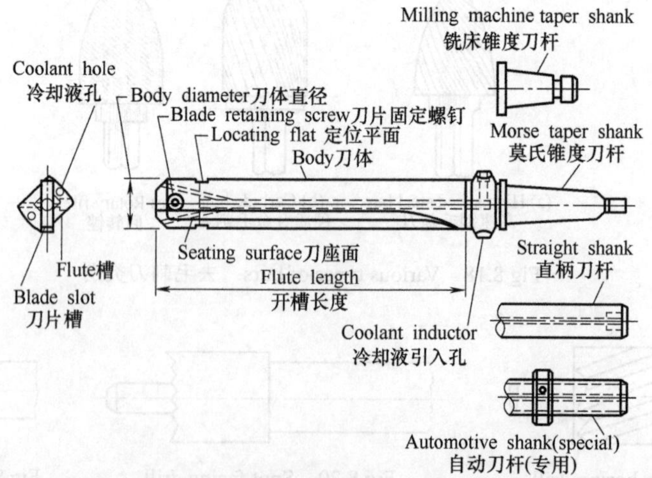

Fig 8.25 Spade drill blade holder 铲形钻刀片固定器

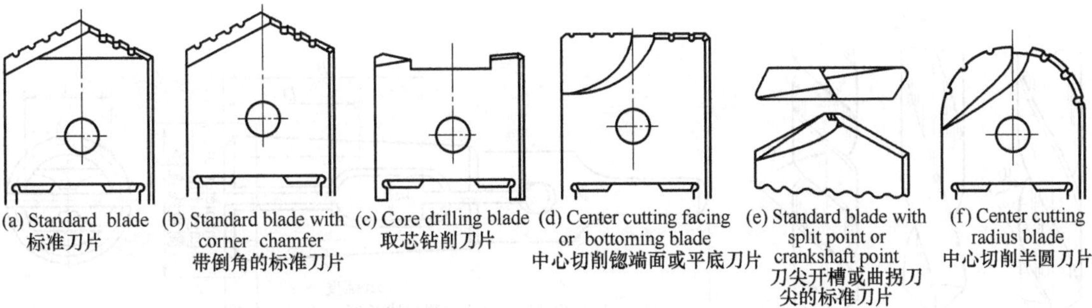

Fig 8.26 Typical spade drill blades 典型铲形钻刀片

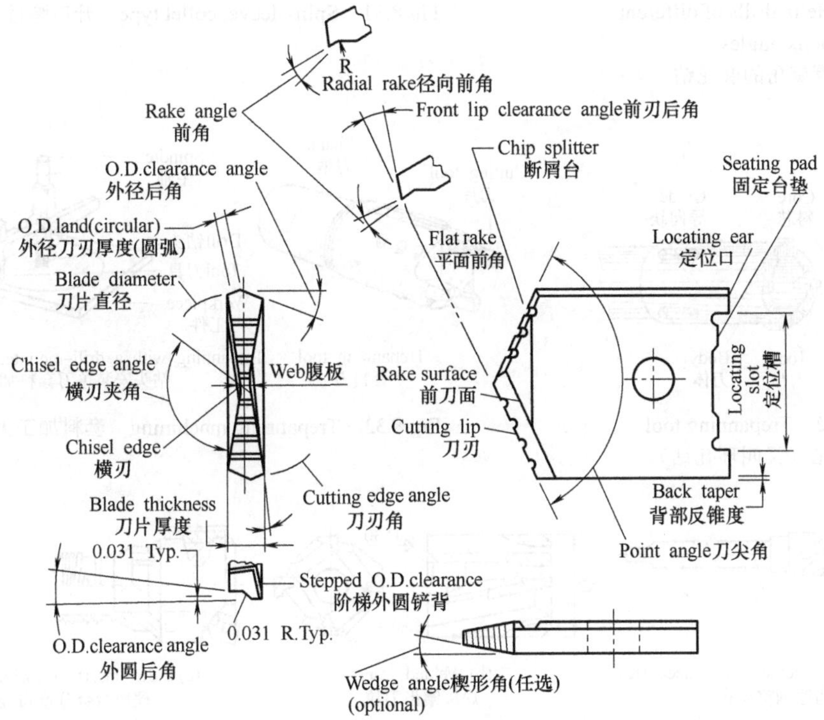

Fig 8.27 Spade drill blade 铲形钻刀片

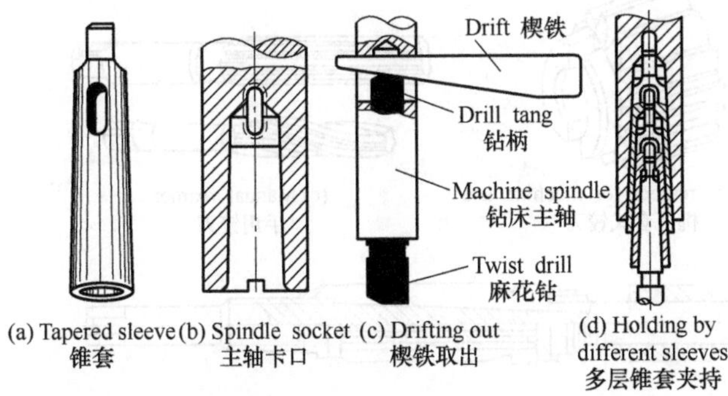

Fig 8.28 Holding drills in spindle socket or sleeves and drifting out from a socket or sleeve
主轴座或套内钻头的夹持及钻头的取出

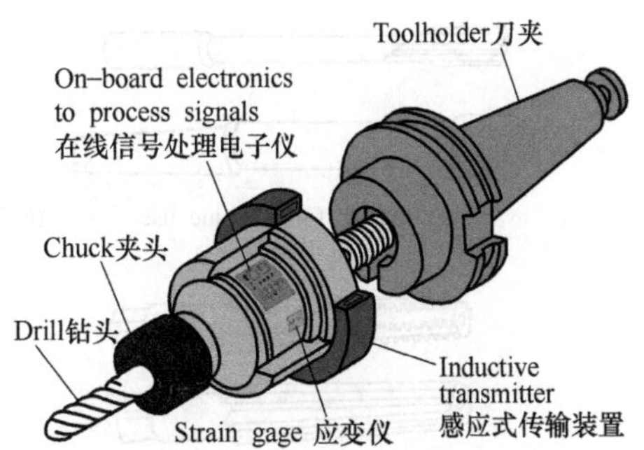

Fig 8.29 A toolholder equipped with thrustforce and torque sensors
装有轴向力传感器和扭力传感器的刀夹

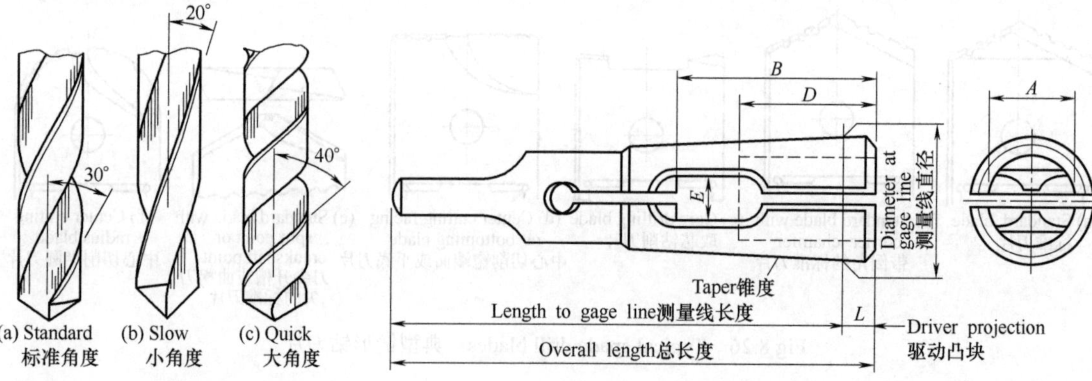

Fig 8.30 Helix drills of different helix angles
不同螺旋角的麻花钻

Fig 8.31 Split-sleeve, collet type 开口弹性套

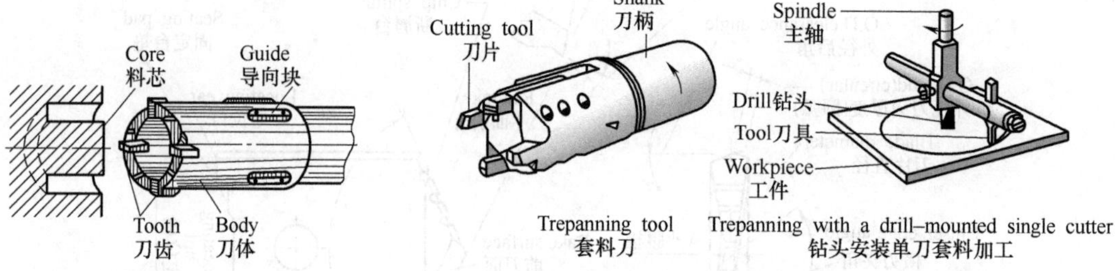

Fig 8.32 Trepanning tool
套料钻（又叫环孔钻）

Fig 8.33 Trepanning machining 套料加工刀具

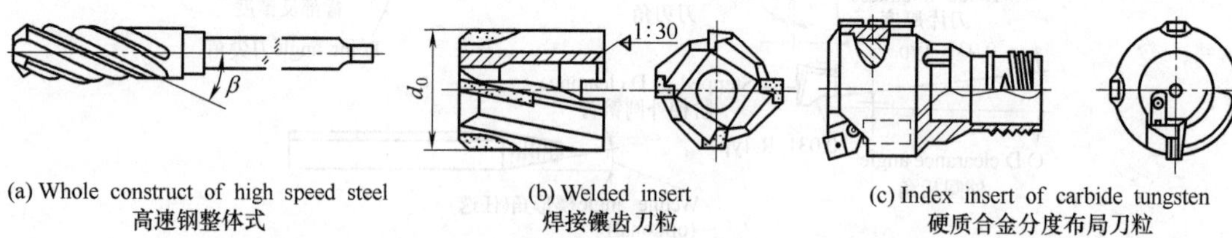

Fig 8.34 Construct of expanding (core) drill (bit) 扩孔钻的结构

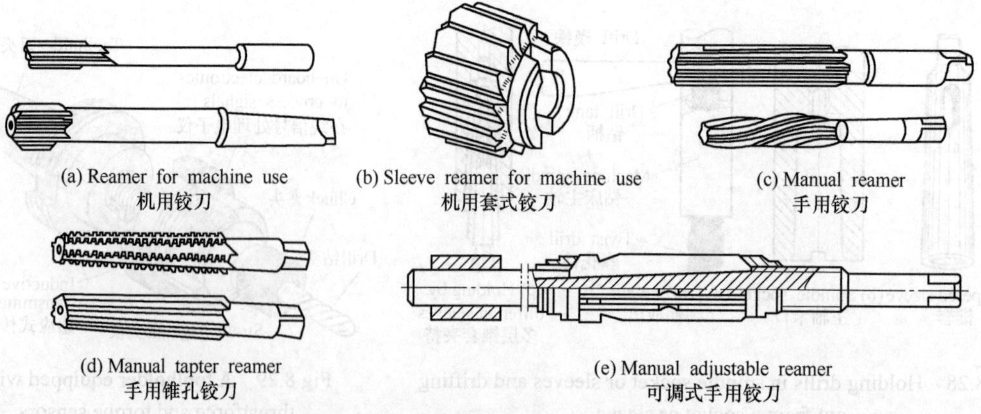

Fig 8.35 Types of reamers 铰刀种类

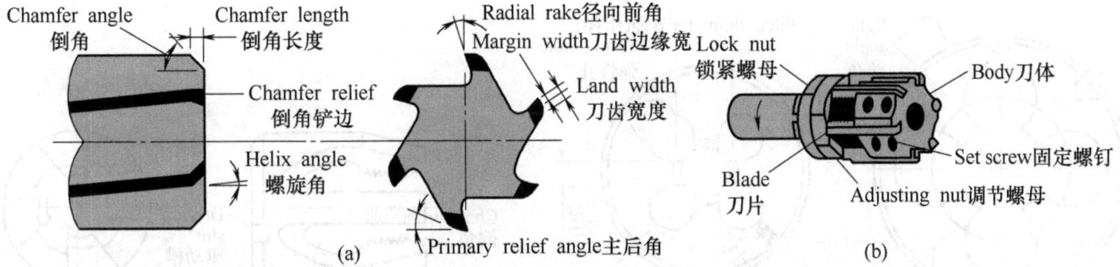

Fig 8.36 Whole construct and assembled reamers 整体铰刀和组装铰刀

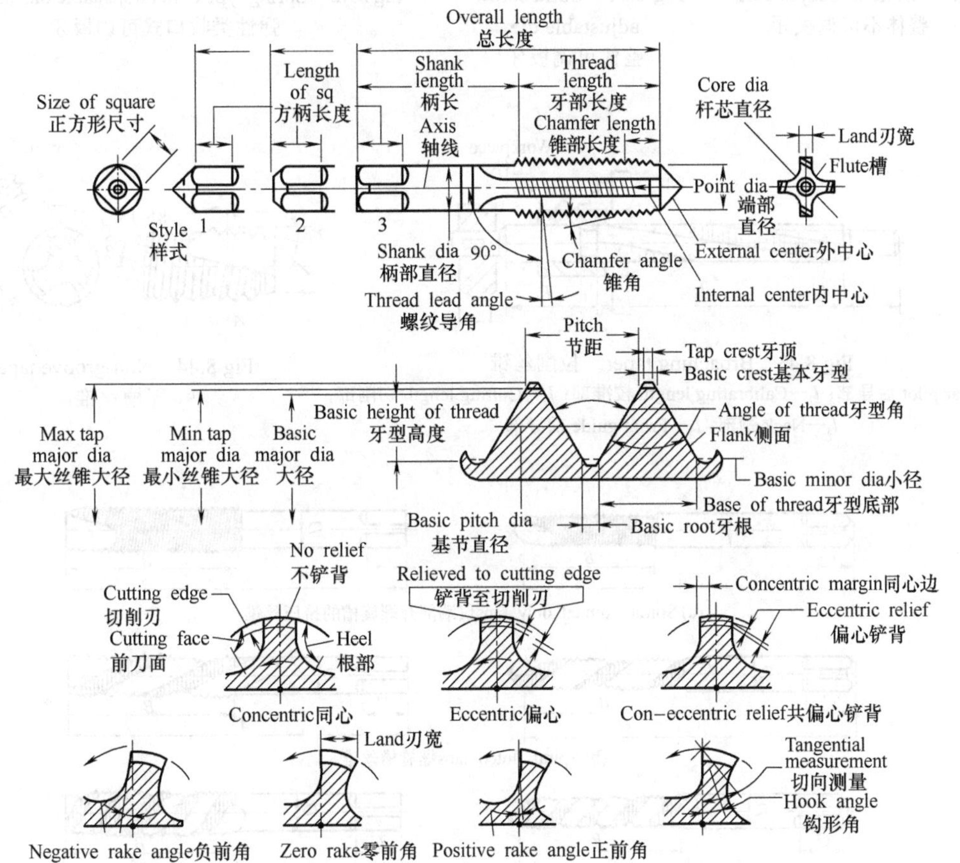

Fig 8.37 Taper terms 丝锥术语

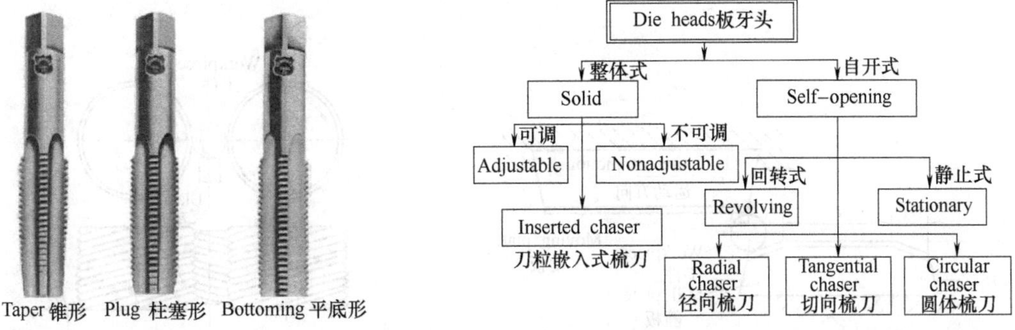

Fig 8.38 Straight flute hand taps 直槽手攻丝锥

Fig 8.39 Classification of threading die heads 板牙头分类

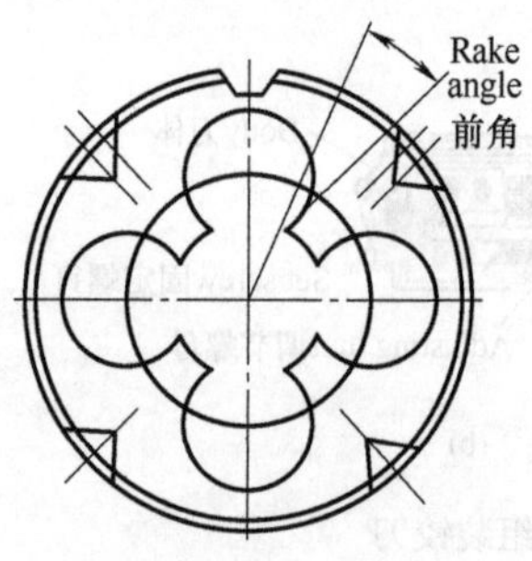

Fig 8.40 Solid nonadjustable die 整体不可调板牙

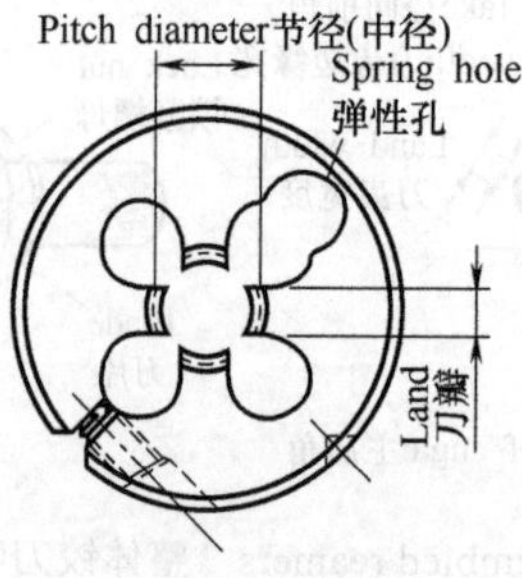

Fig 8.41 Solid screw-adjustable die 整体可调板牙

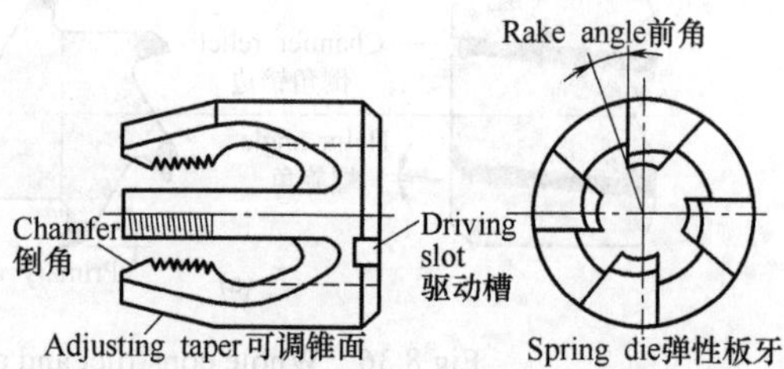

Fig 8.42 Spring type collet-adjustable die and holder 弹性类收口式可调板牙

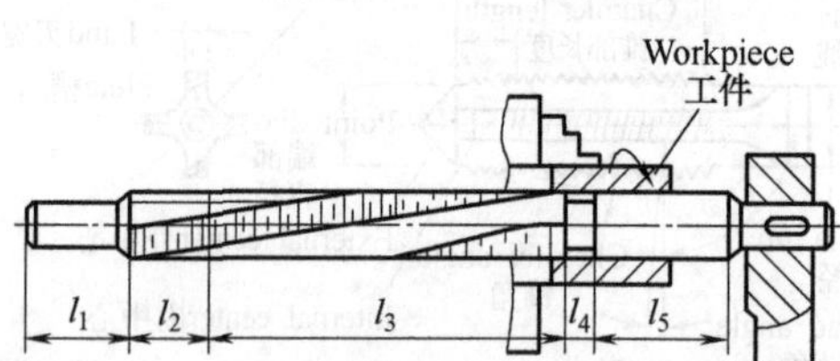

Fig 8.43 Broaching taper 拉削丝锥

l_1—Rear pilot 后导部；l_2—Calibrating length 校准部；l_3—Cutting length 切削部；l_4—Neck 颈部；l_5—Front guide 前导部

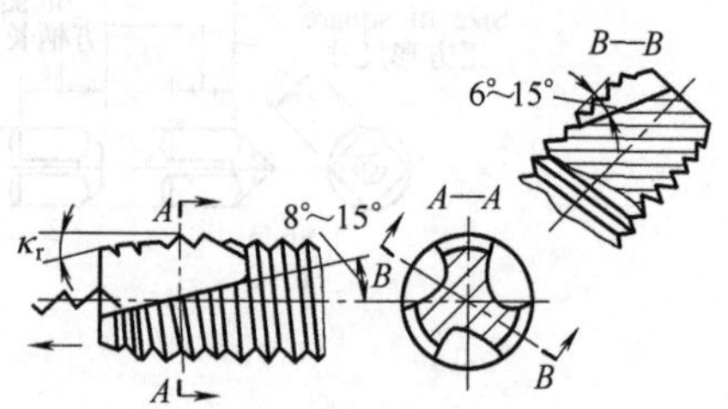

Fig 8.44 Non-groove taper 无槽丝锥

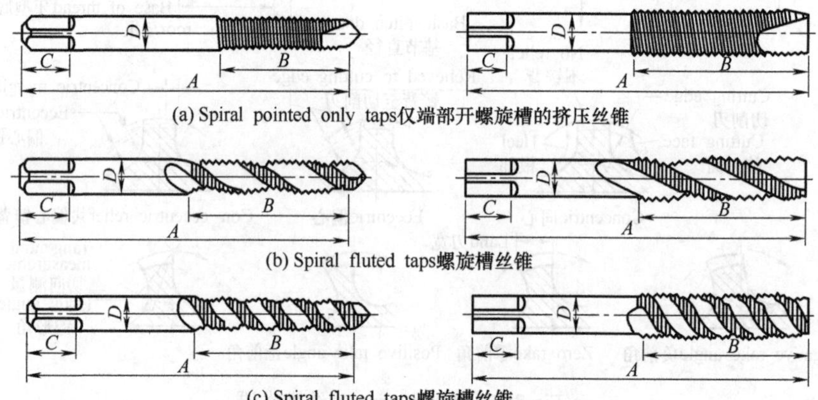

Fig 8.45 Spiral fluted taps 螺旋槽丝锥

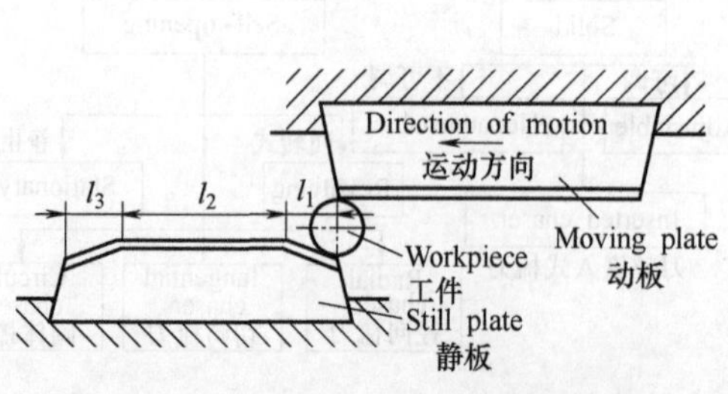

Fig 8.46 Plate die 搓丝板工作情况

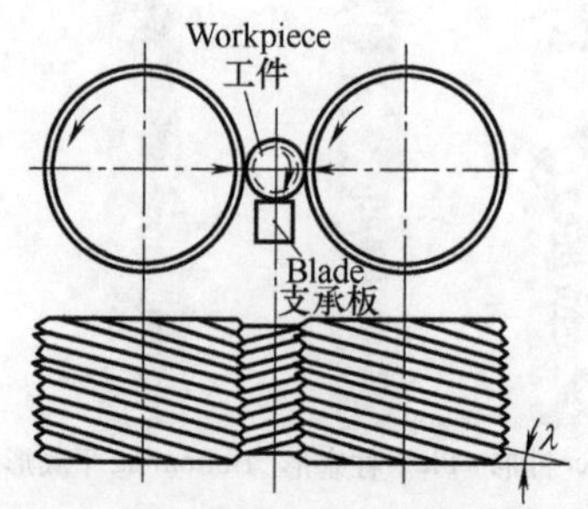

Fig 8.47 Rolled tap 滚丝轮滚压螺纹

8.1.5 Reciprocating process tools 往复运动加工刀具

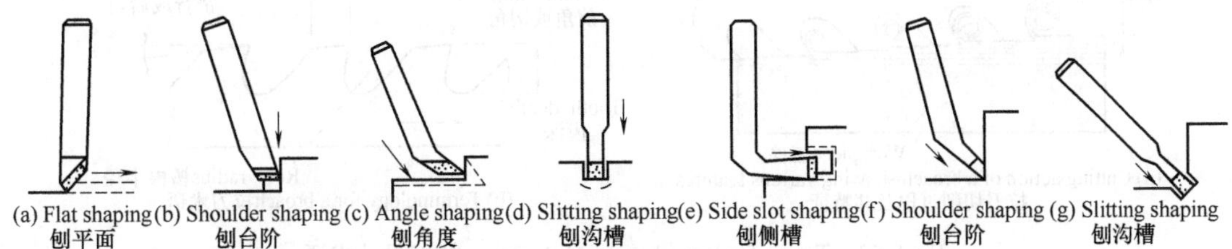

Fig 8.48 Shaping tools and various shaping types 刨刀与刨削加工类型

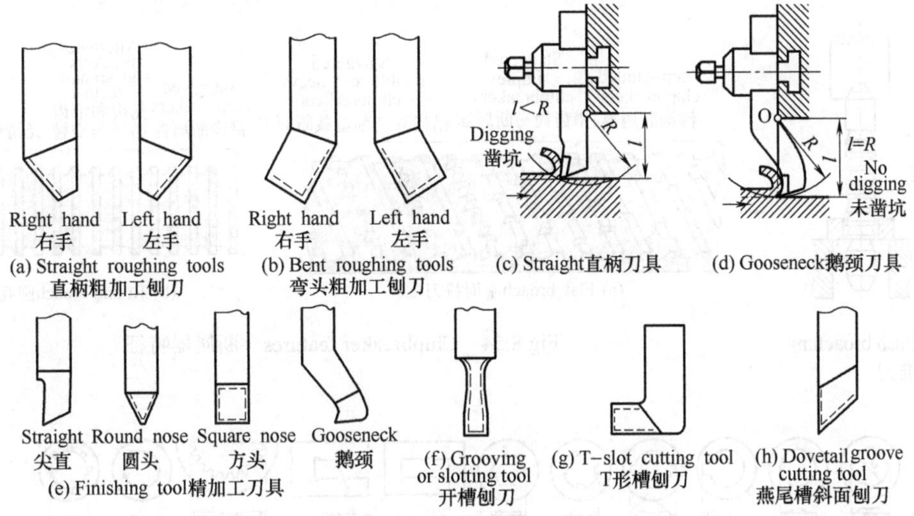

Fig 8.49 Planner tool 龙门刨床用刀具

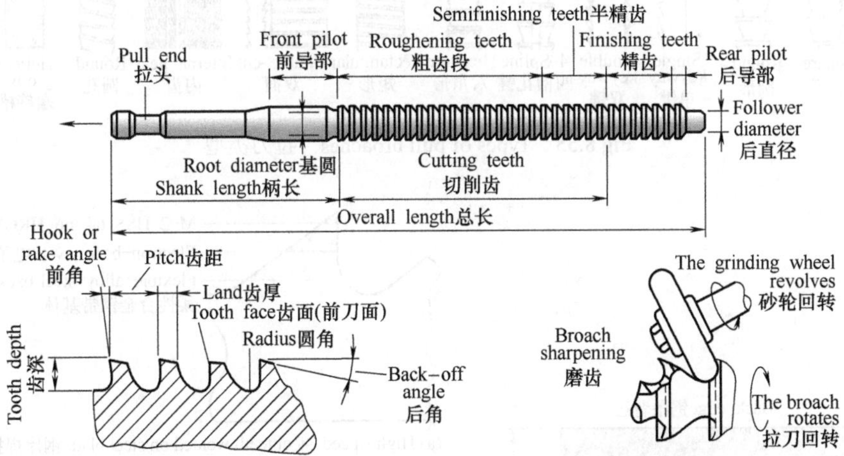

Fig 8.50 Typical construction of a pull broach 拉刀结构

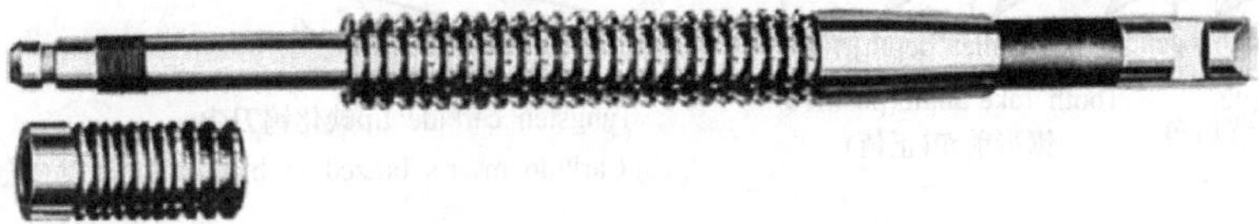

Fig 8.51 Shell broach 套式拉刀

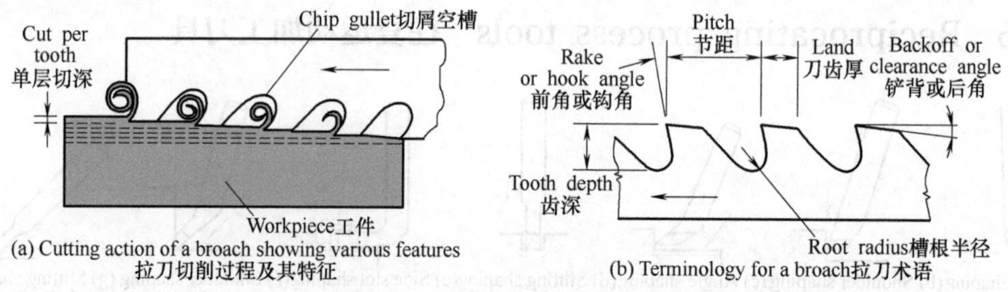

(a) Cutting action of a broach showing various features
拉刀切削过程及其特征

(b) Terminology for a broach拉刀术语

Fig 8.52 Terms for broaching tool design 拉刀设计术语

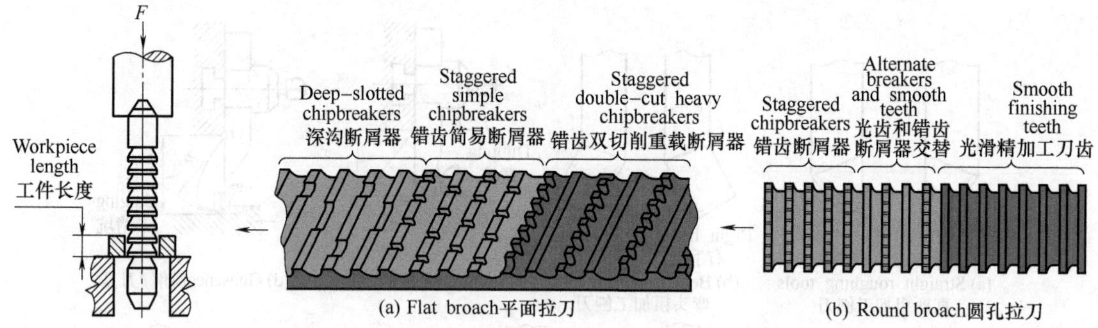

(a) Flat broach平面拉刀

(b) Round broach圆孔拉刀

Fig 8.53 Push broaching
推刀

Fig 8.54 Chipbreaker features 断屑器特征

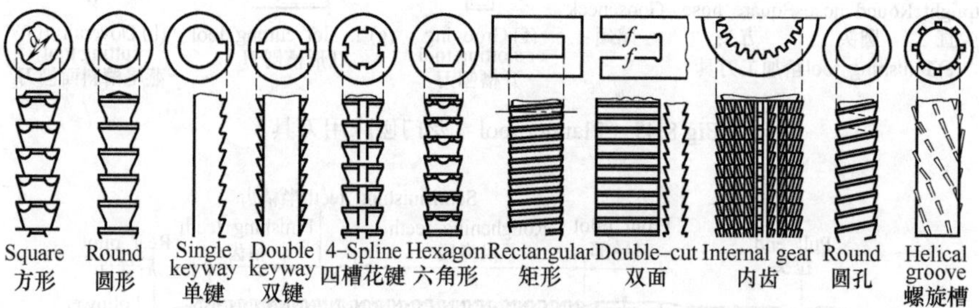

Fig 8.55 Types of pull broaches 拉刀类型

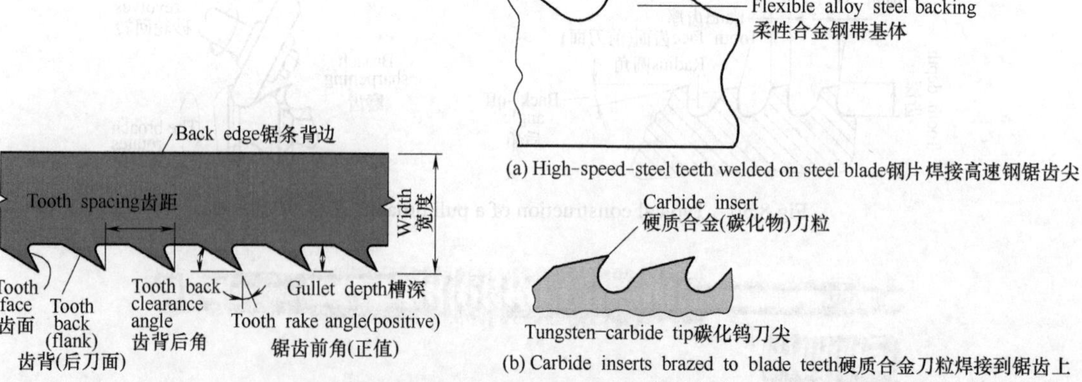

(a) High-speed-steel teeth welded on steel blade钢片焊接高速钢锯齿尖

(b) Carbide inserts brazed to blade teeth硬质合金刀粒焊接到锯齿上

Fig 8.56 Terminology for saw teeth 锯齿术语

Fig 8.57 Welded blade teeth 焊接锯齿

8.1.6 Abrasives 磨具

8.1.6.1 Solid abrasives 固结磨具

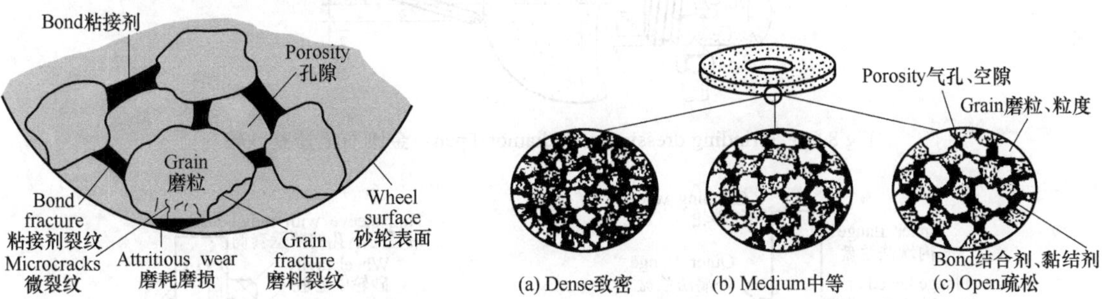

Fig 8.58 Grinding wheel surface 砂轮表面

Fig 8.59 Grinding wheel construct 砂轮组织

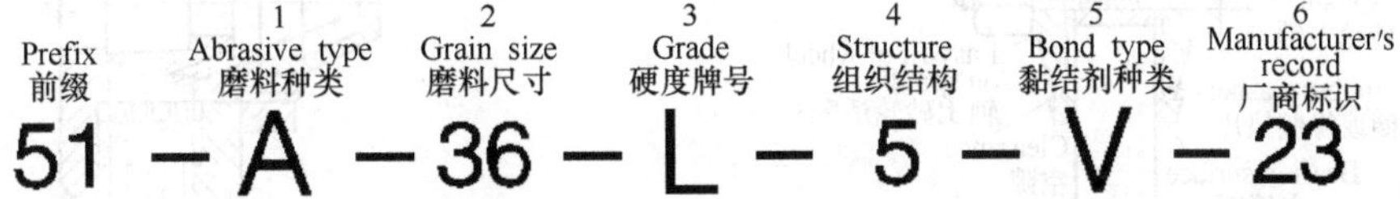

Fig 8.60 Standard grinding wheel marking 标准的砂轮标识

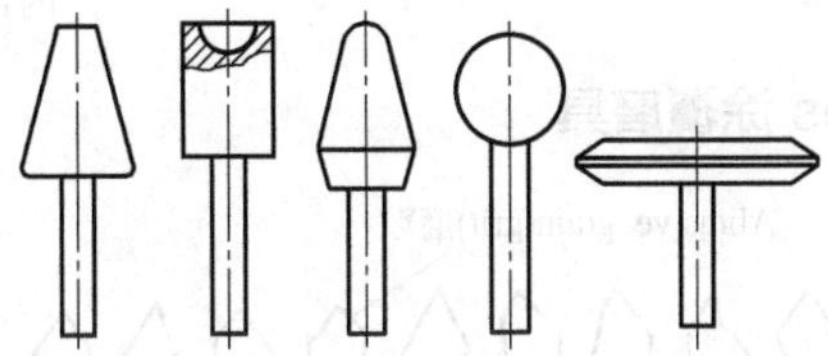

Fig 8.61 Contour grinding wheel 成形磨削砂轮

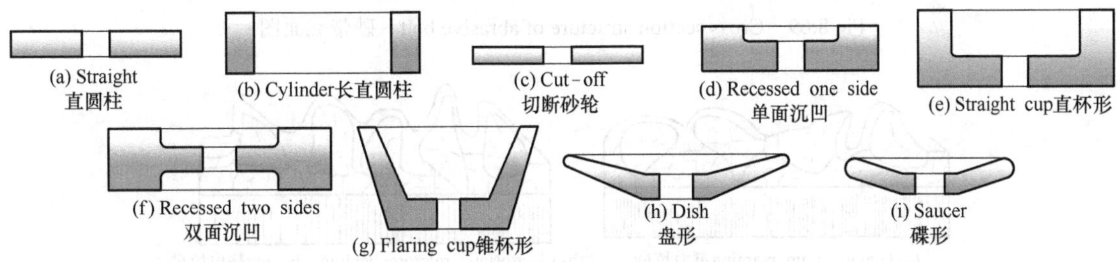

Fig 8.62 Grinding wheel shape 砂轮形状

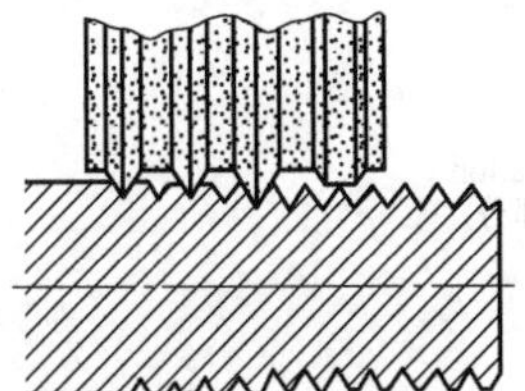

Fig 8.63 Wheel with edges for roughing and finishing 粗精磨削复合砂轮

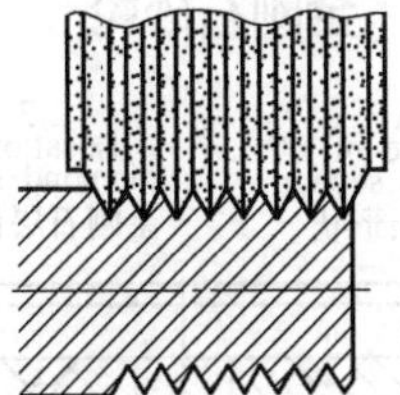

Fig 8.64 Multi-ribbed type of thread-grinding wheel 多牙螺纹磨削砂轮

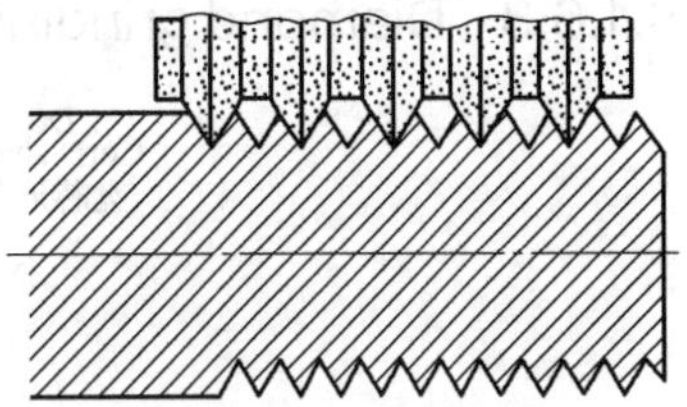

Fig 8.65 Alternate-ribbed wheel for grinding the finer pitches 隔行细牙螺纹磨削砂轮

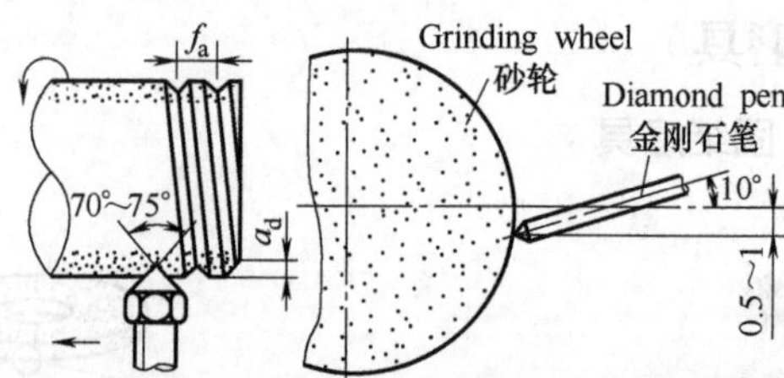

Fig 8.66 Grinding dressing with diamond pen 金刚石笔修整砂轮

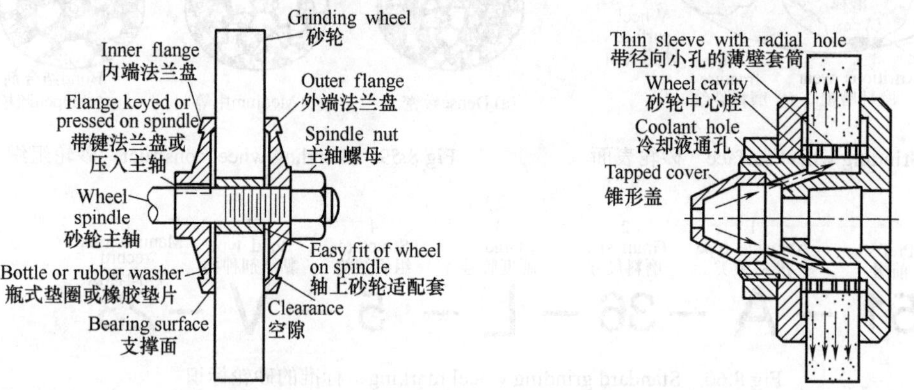

Fig 8.67 Correctly mounted wheel 砂轮的正确安装

Fig 8.68 Internal cooling wheel constructure 内部冷却砂轮结构

8.1.6.2 Coated abrasives 涂覆磨具

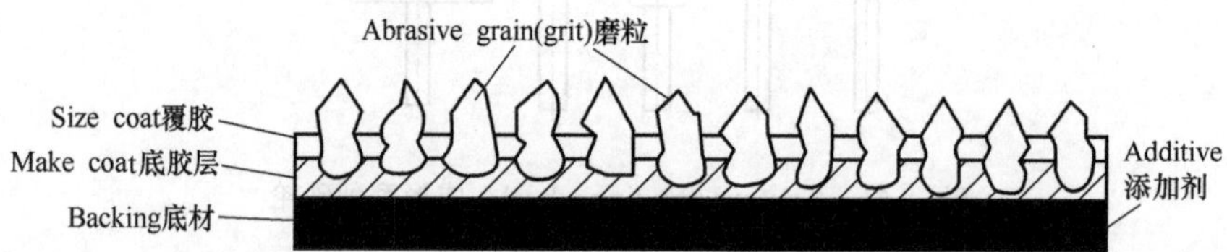

Fig 8.69 Cross section structure of abrasive belt 砂带截面图

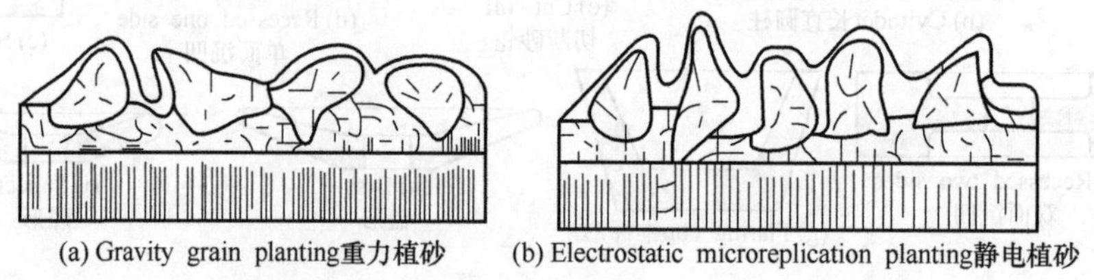

(a) Gravity grain planting重力植砂 (b) Electrostatic microreplication planting静电植砂

Fig 8.70 Comparison of belt structures by different grain planting processes 不同的植砂方式砂带结构比较

8.1.6.3 Diamond grinding wheel 金刚石砂轮

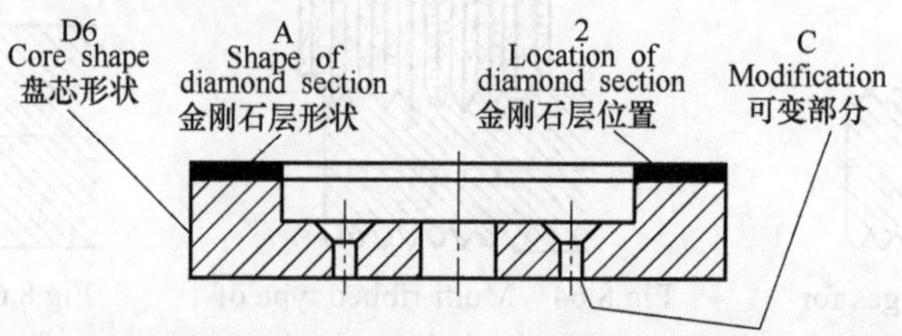

Fig 8.71 A typical diamond wheel shape designation symbol 典型金刚石砂轮形状设计符号

Table 8.1 Designation letters for modifications of diamond wheels 金刚石砂轮改型标识字母

Designation letter 标识字母	Description 描述	Illustration 图解
B—Drilled and counterbored 钻孔并锪沉孔	Holes drilled and counterbored in core 轮毂基体上钻孔并锪沉孔	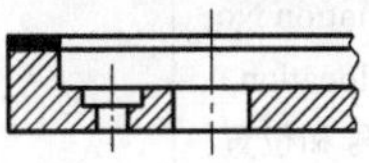
C—Drilled and countersunk 钻孔并锪锥孔	Holes drilled and countersunk in core 轮毂基体上钻孔并锪锥孔	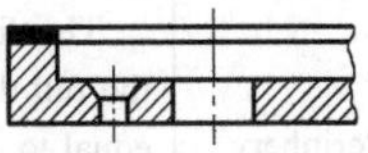
H—Plain hole 钻通孔	Straight hole drilled in core 轮毂基体上钻通孔	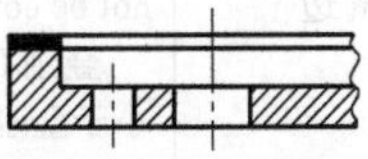
M—Holes plain and threaded 钻通孔和螺纹孔	Mixed holes some plain, some threaded, are in core 轮毂基体上钻孔、螺纹孔	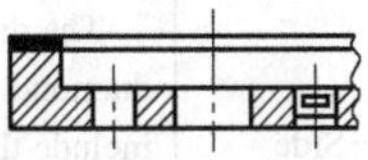
P—Relieved one side 单面内沉	Core relieved on one side of wheel. Thickness of core is less than wheel thickness 砂轮单面内沉，轮中心厚度小于砂轮厚度	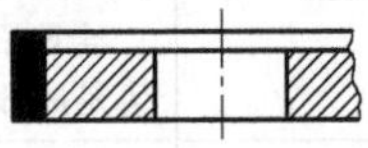
R—Relieved two sides 双面内沉	Core relieved on both sides of wheel. Thickness of core is less than wheel thickness 砂轮双面内沉，轮中心厚度小于砂轮厚度	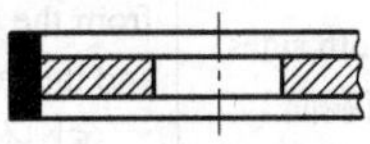
S—Segmented-diamond section 金刚石层开槽分片	Wheel has segmental diamond section mounted on core (Clearance between segments has no bearing on definition) 轮毂上安装开槽分片的金刚石层（片间无支撑）	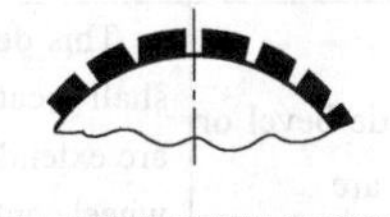
SS—Segmental and slotted 金刚石层分片基体开槽	Wheel has separated segments mounted on a slotted core 开槽轮毂基体上安装金刚石片层	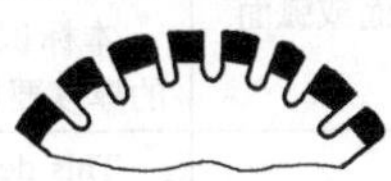
T—Threaded holes 设置螺纹孔	Threaded holes are in core 轮毂基体上设置螺纹孔	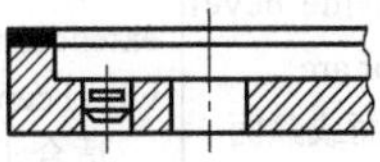
Q—Diamond inserted 金刚石嵌入	Three surfaces of the diamond section are partially or completely enclosed by the core 金刚石剖面的三面全部或部分被轮毂包围	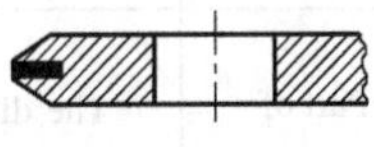
V—Diamond inverted 圆周面金刚石层内凹	Any diamond cross section, which is mounted on the core so that the interior point of any angle, or the con-cave side of any arc, is exposed shall be considered inverted 安装在轮毂上的各种金刚石剖面形状，需要考虑不同角度的内凹点或内凹面能够外露	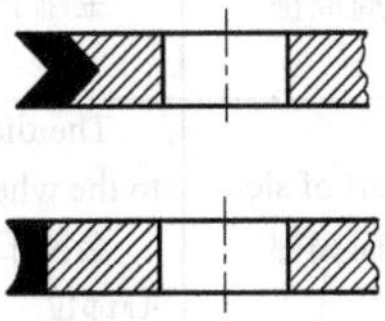

Table 8.2 Designation symbols for composition of diamond and cubic boron nitride wheels 金刚石和立方氮化硼砂轮组分的标识符号

Prefix 前缀	Abrasive type: 磨料种类	Grain size 磨粒尺寸	Grade 硬度	Concentration 集中度	Bond 黏结剂	Bond modification 黏结剂改性	Abrasive depth 磨料层厚度	Manufacturer's ID 制造商识别标记
M	D	120	R	100	B	56	1/8	*

Table 8.3 Designations for location of diamond section on diamond wheel
金刚石砂轮金刚石层剖面及其位置设计

Designation No. and location 设计号和位置	Description 描述	Illustration 图解
1—Periphery 周边	The diamond section shall be placed on the periphery of the core and shall extend the full thickness of the wheel. The axial length of this section may be greater than, equal to, or less than the depth of diamond, measured radially. A hub or hubs shall not be considered as part of the wheel thickness for this definition 金刚石层覆盖圆周及整个宽度，径向测量的剖面轴线长度可大于、等于或小于金刚石砂轮的深度。此定义不包括轮毂的轮厚	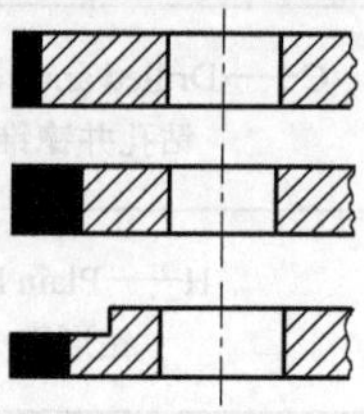
2—Side 端面	The diamond section shall be placed on the side of the wheel and the length of the diamond section shall extend from the periphery toward the center. It may or may not include the entire side and shall be greater than the diamond depth measured axially. It shall be on that side of the wheel which is commonly used for grinding purposes 金刚石层覆盖砂轮的端面并从周边向中心伸展。它可以包括整个端面，也可是局部面。但其轴向长度要等于轮体长度，用于磨削	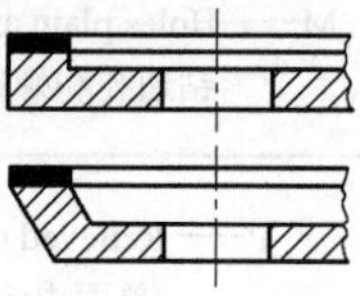
3—Both sides 两端面	The diamond sections shall be placed on both sides of the wheel and shall extend from the peirphery toward the center. They may or may not include the entire sides, and the radial length of the diamond section shall exceed the axial diamond depth 金刚石层从边缘到中心覆盖轮体两个端面，可以是整个表面覆盖，也可以局部表面，其径向长度要大于轴向深度	
4—Inside bevel or arc 内锥面或弧面	This designation shall apply to the general wheel types 2, 6, 11, 12, and 15 and shall locate the diamond section on the side wall, This wall shall have an angle or arc extending from a higher point at the wheel periphery to a lower point toward the wheel center 本标识适合于 2、6、11、12、15 等通用砂轮，金刚石层位于内壁，其周边的厚度要比中心处高	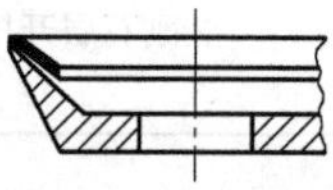
5—Outside bevel or arc 外锥面或弧面	This designation shall apply to the general wheel types 2, 6, 11, and 15 and shall locate the diamond section on the side wall, This wall shall have an angle or arc extending from a lower point at the wheel periphery to a higher point toward the wheel center 适合于 2、6、11、12、15 通用砂轮，金刚石层在位于弧面，中心处厚度高于周边	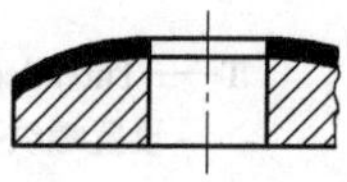
6—Part of periphery 圆周局部	The diamond section shall be placed on the periphery of the core but shall not extend the full thickness of the wheel and shall not reach to either side. 金刚石层位于轮毂周边并不覆盖整个厚度，更不能到达端面	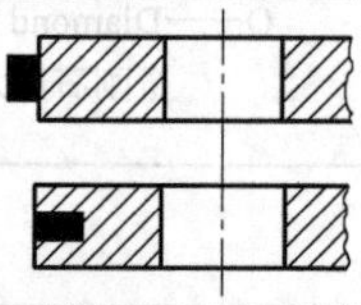
7—Part of side 端面局部	The diamond section shall be placed on the side of the core and shall not extend to the wheel periphery. It may or may not extend to the center 金刚石层位于砂轮局部端面，不能到达周边覆盖整个端面，但可能到达中心部位	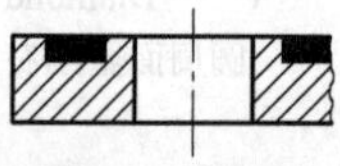
8—Throughout 完全金刚石	Designates wheels of solid diamond abrasive section without cores 整个实体为金刚石磨料，没有轮毂	
9—Corner 角落	Designates a location which would commonly be considered to be on the periphery except that the diamond section shall be on the corner but shall not extend to the other corner 金刚石层位于剖面一个角落，但不宜延伸到另一个角落	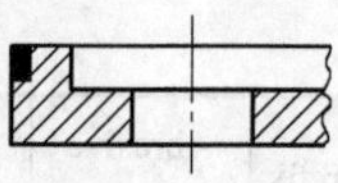
10—Annular 内环面	Designates a location of the diamond abrasive section on the inner annular surface of the wheel 金刚石层位于轮毂内环面上	

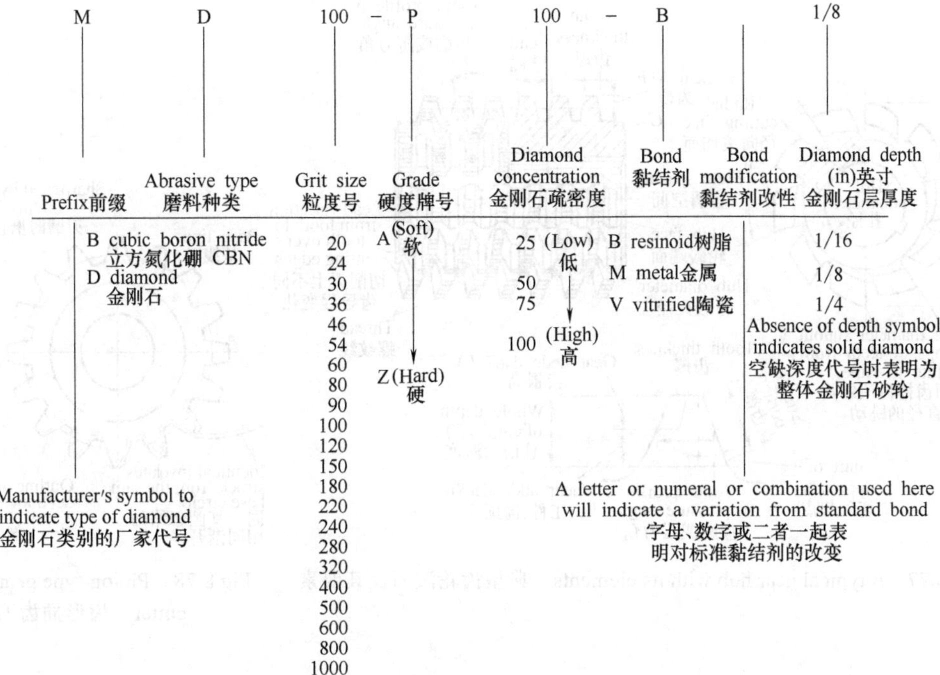

Fig 8.72 Standard marking system for CBN and diamond bonded abrasives
立方氮化硼和金刚石砂轮的标准识别系统

8.1.7 Gear cutting tools 齿轮刀具

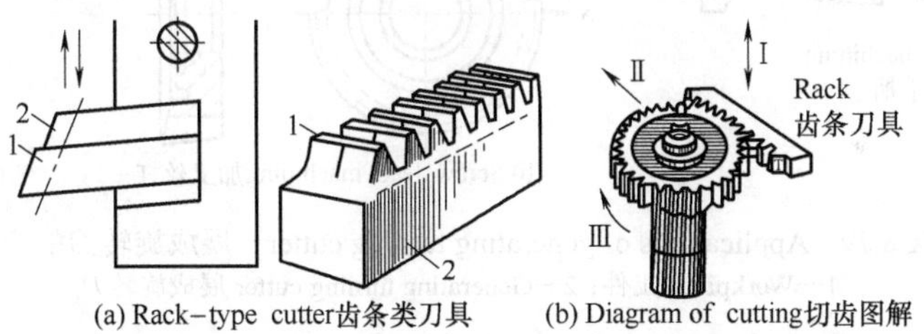

Fig 8.73 Diagram of cutting gears with rack-type cutters 齿条类刀具加工齿轮图解
1—Tool-alloy-steel cutter 合金工具钢；2—Carbon-steel shim 碳钢刀垫

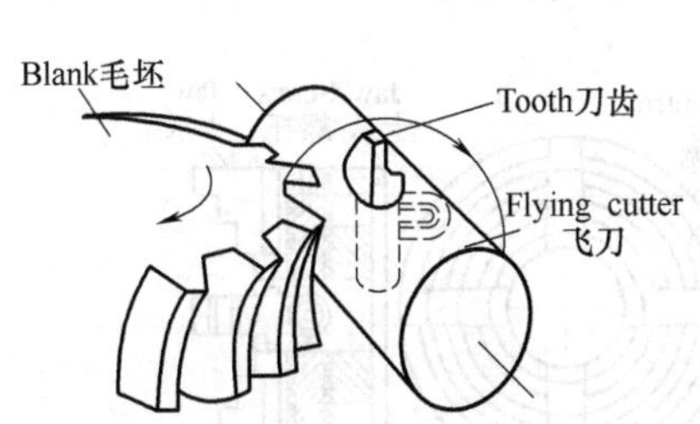

Fig 8.74 Worm wheel cutting with flying cutter 飞刀加工蜗轮

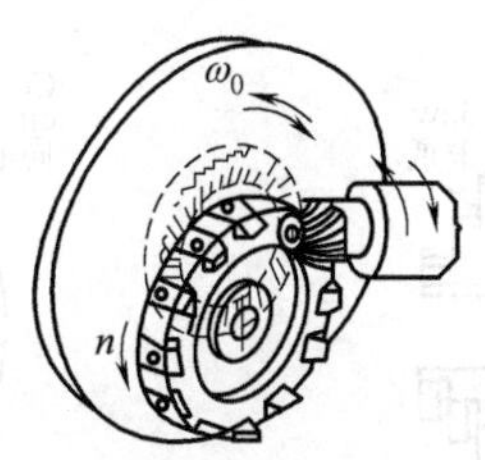

Fig 8.75 Milling cutter bevel-gear of curved tooth
弧齿锥齿铣刀盘

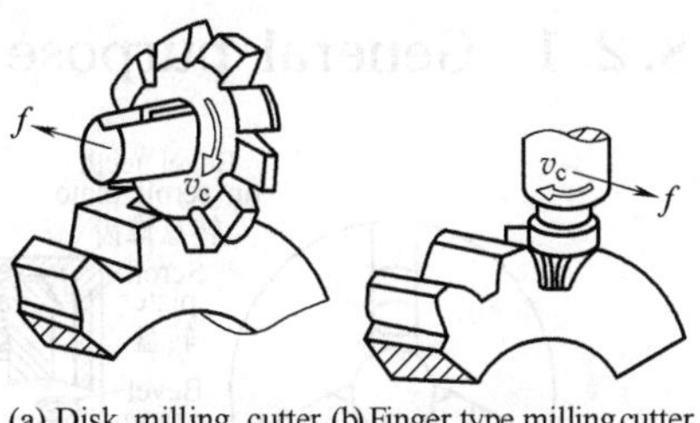

Fig 8.76 Gear form milling
齿轮的成形铣削

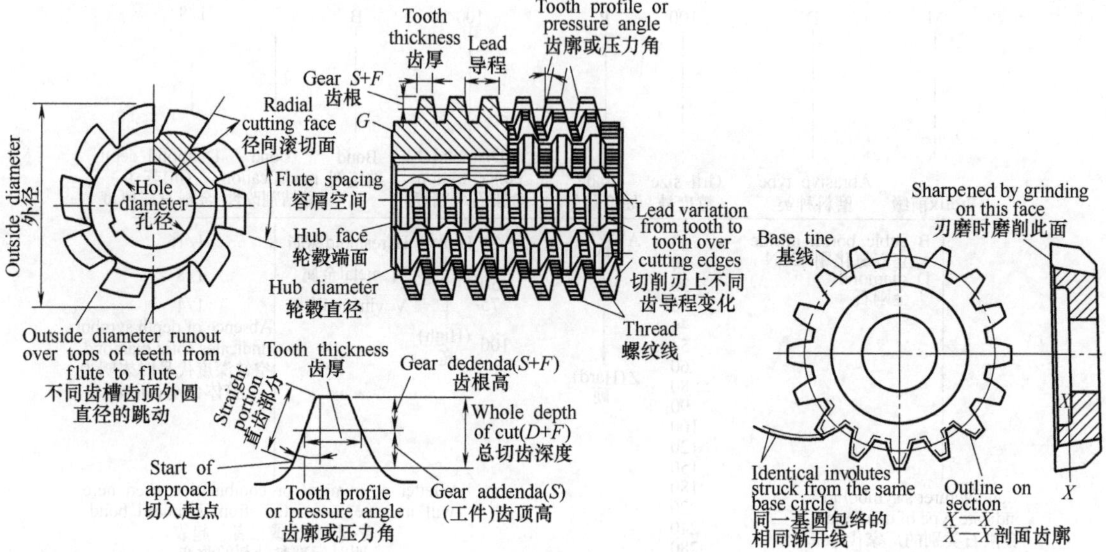

Fig 8.77　A typical gear hob with its elements　典型齿轮滚刀及其要素

Fig 8.78　Pinion-type gear shaper cutter　齿形插齿刀

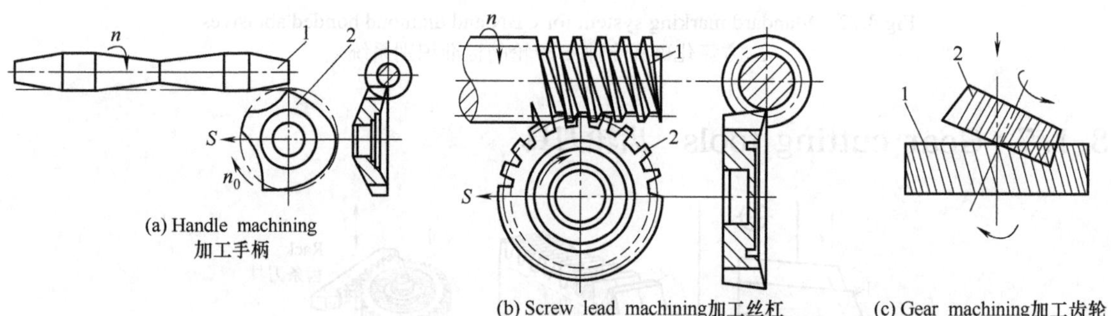

Fig 8.79　Applications of generating turning cutter　展成旋转刀的应用

1—Workpiece 工件；2—Generating turning cutter 展成旋转刀

8.2　Jigs and fixtures, toolings　工装夹具

8.2.1　General purpose jigs and fixtures　通用夹具

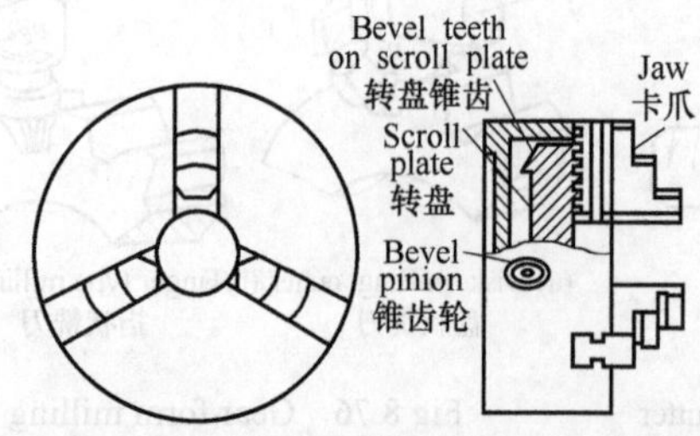

Fig8.80　Schematic layout of a three-jaw chuck

三爪卡盘图解

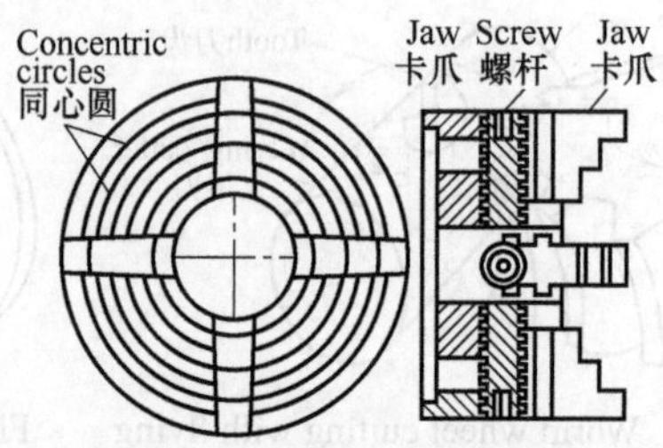

Fig 8.81　Schematic layout of a four-jaw chuck

四爪卡盘图解

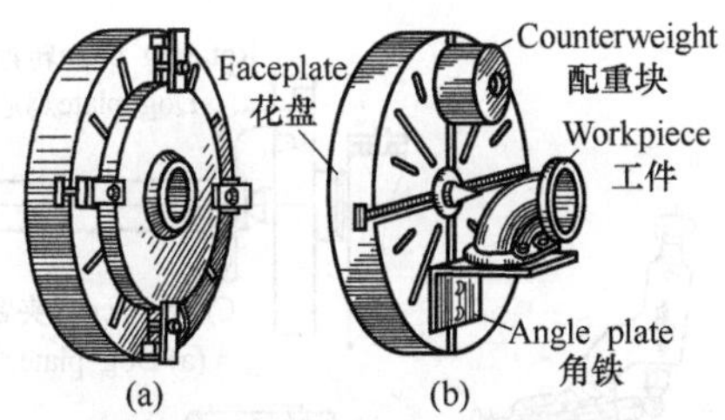

Fig 8.82 Mounting WPs on faceplates 花盘安装工件

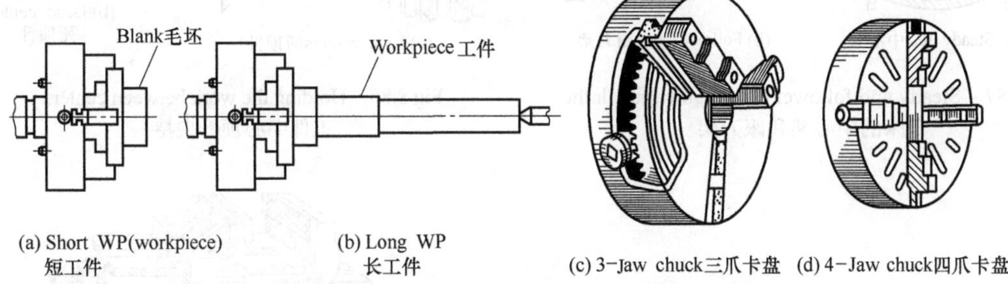

Fig 8.83 Clamping WPs in chucks 卡盘夹持工件

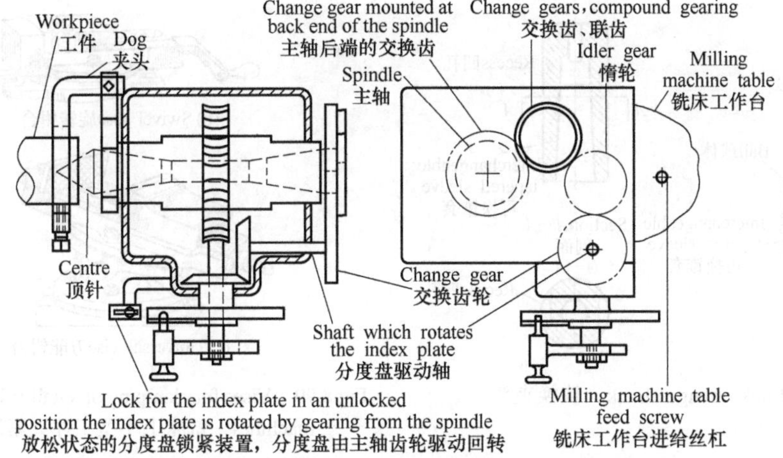

Fig 8.84 Dividing head setup for differential indexing 差分分度头装置

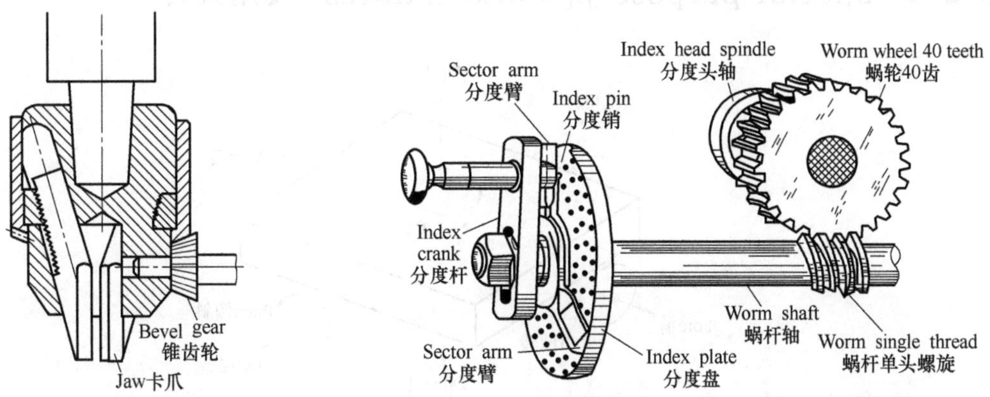

Fig 8.85 Three-jaw drilling chuck 三爪式钻夹头

Fig 8.86 Indexing method of a dividing head 分度头分度方法

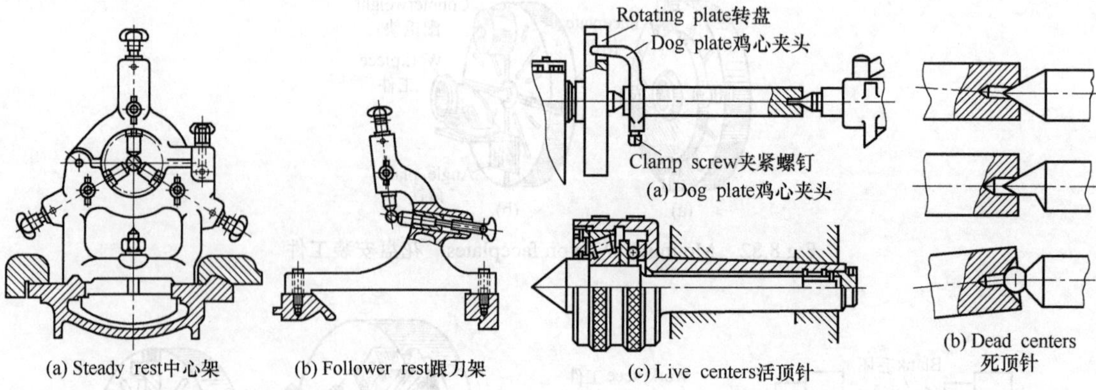

Fig 8.87 Steady and follower rest of an engine lathe
车床的中心架和跟刀架

Fig 8.88 Holding the work between centers
工件中心顶针夹持

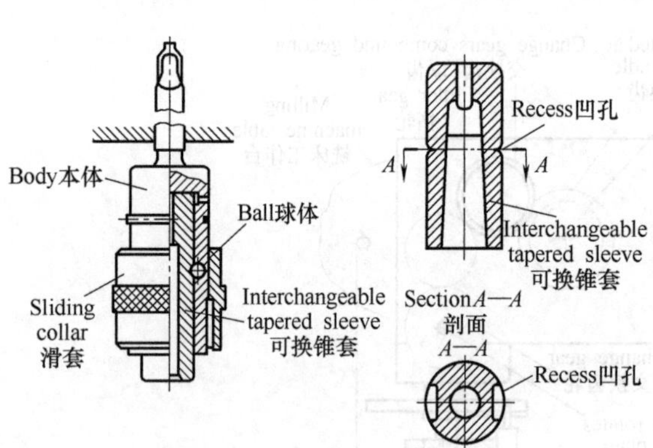

Fig 8.89 Quick-change chuck 快换夹头

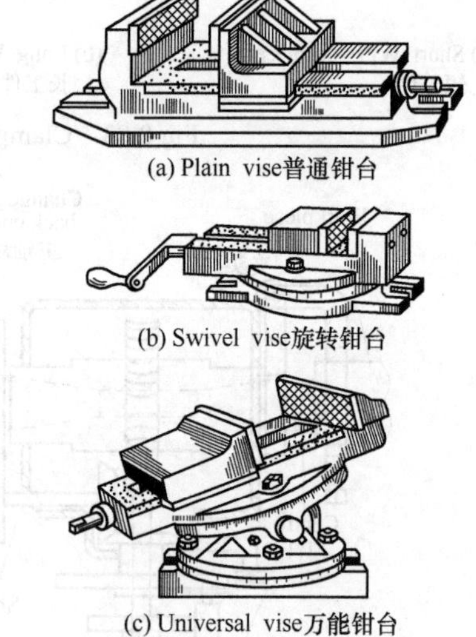

Fig 8.90 Vises for clamping of small WPS on milling machines 铣床上小型工件夹紧用钳台

8.2.2 Special purpose jigs and fixtures 专用夹具

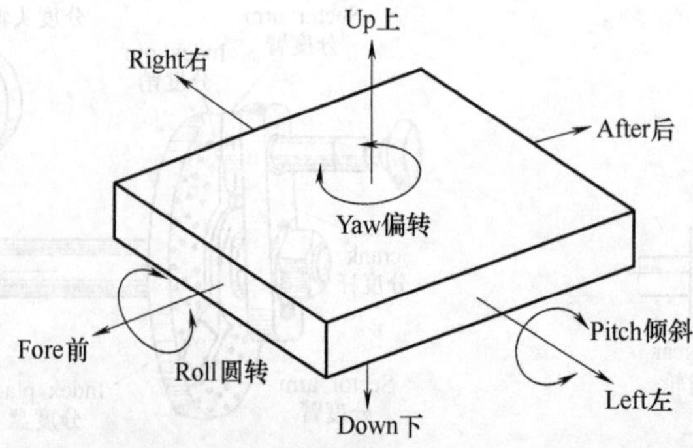

Fig 8.91 Degrees of freedom (movement) that must be controlled 必须控制的自由度（运动）数目

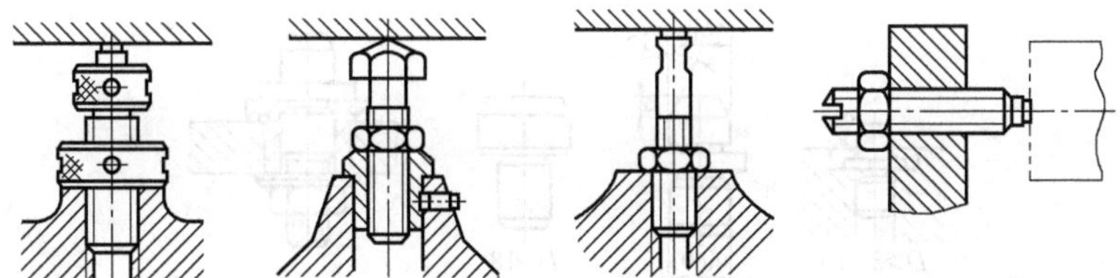

Fig 8.92 Various forms of adjustable supporting pins 可调支承钉的形式

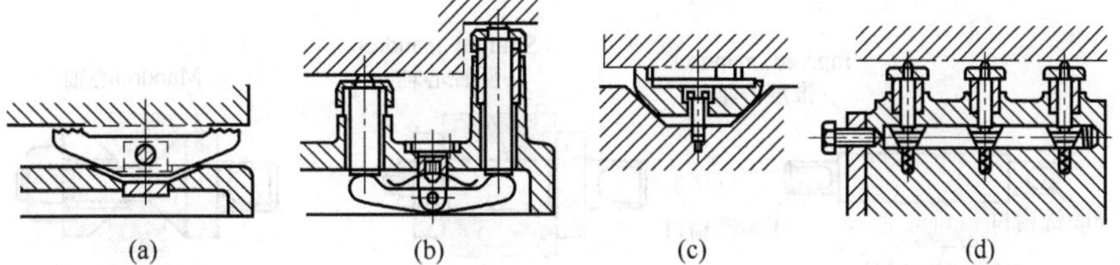

Fig 8.93 Various forms of self-positioning supporting pins 自位支承钉的形式

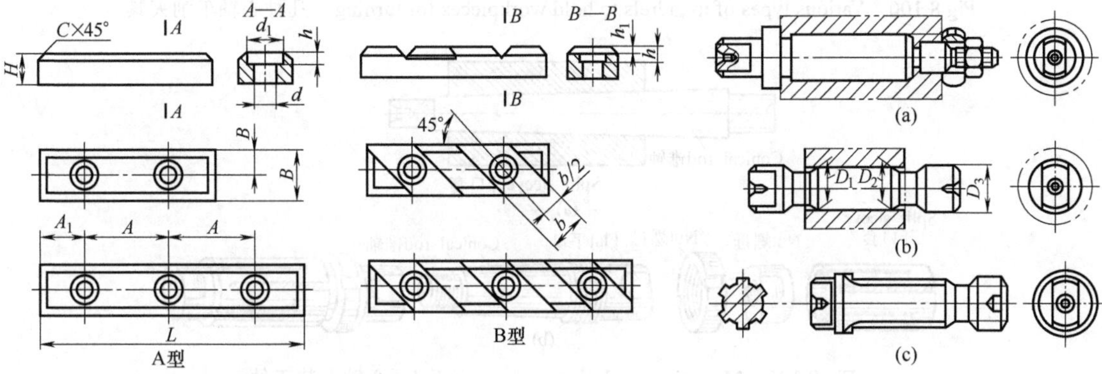

Fig 8.94 Various forms of supporting plates 支承板的形式

Fig 8.95 Cylindrical mandrel 圆柱心轴

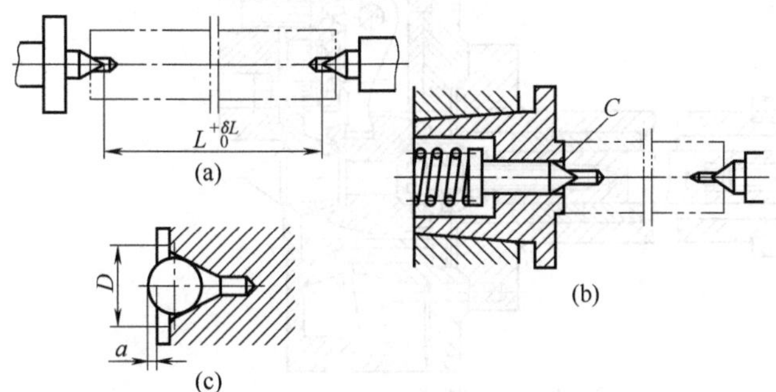

Fig 8.96 Positioning by centre hole 中心孔定位

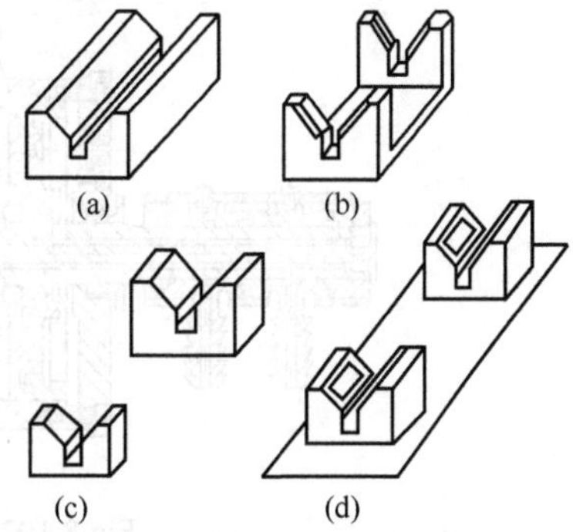

Fig 8.97 V positioning (locating) block V形块

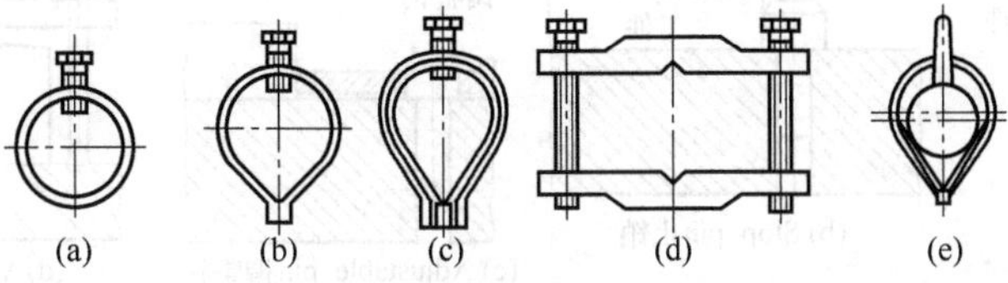

Fig 8.98 Clamping dog 夹头

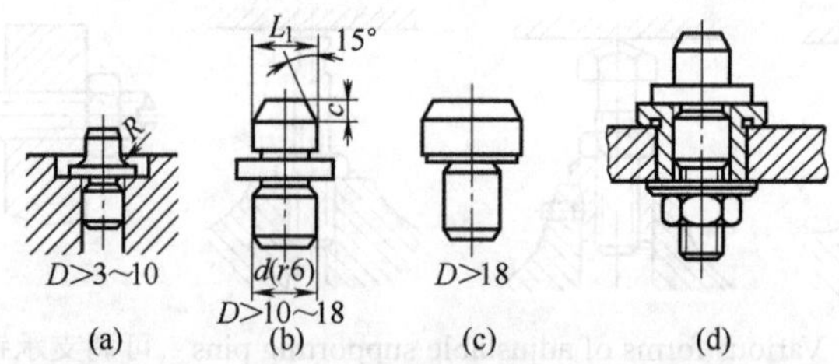

Fig 8.99 Cylindrical positioning pin 圆柱定位销

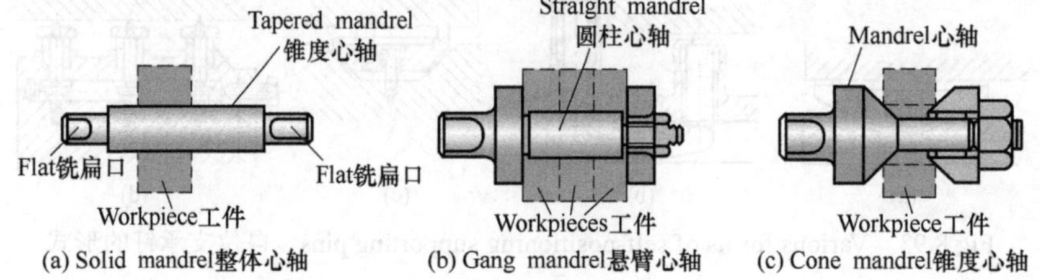

Fig 8.100 Various types of mandrels to hold workpieces for turning 几种心轴车削夹具

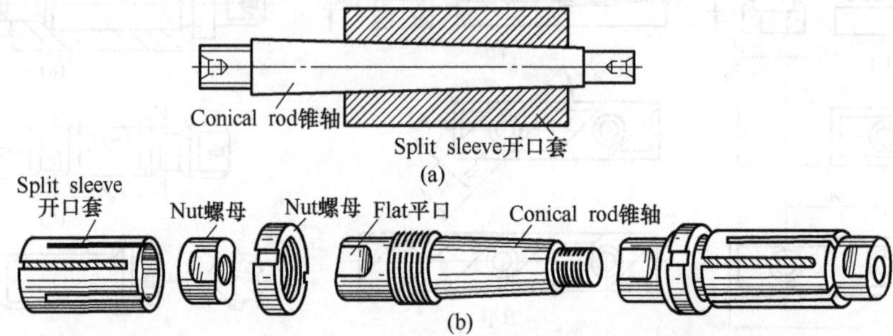

Fig 8.101 Mounting workpieces on a mandrel 心轴安装工件

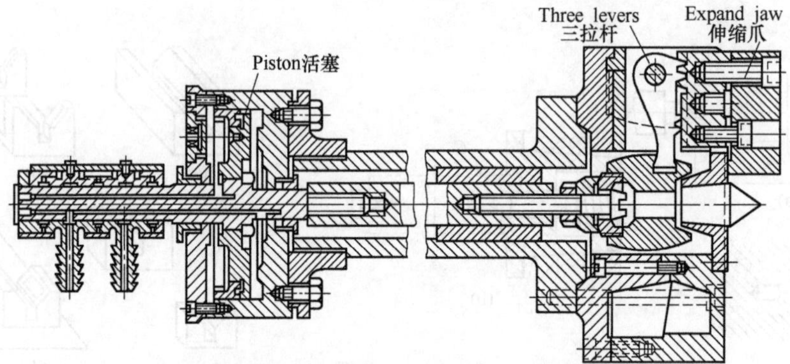

Fig 8.102 Pneumatic chuck 气（风）动夹头

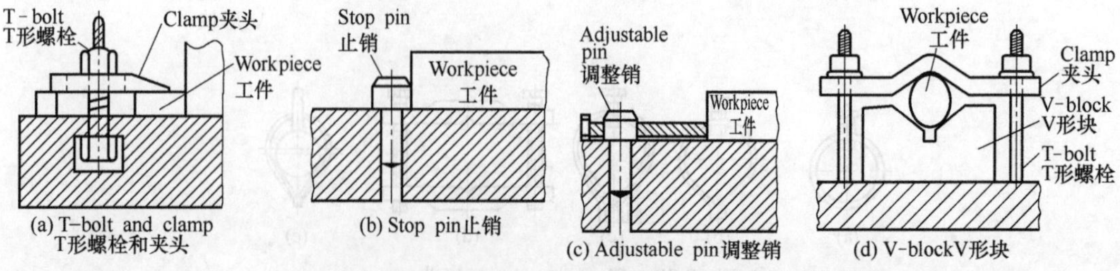

Fig 8.103 Work holding devices for planer 刨床工装夹具

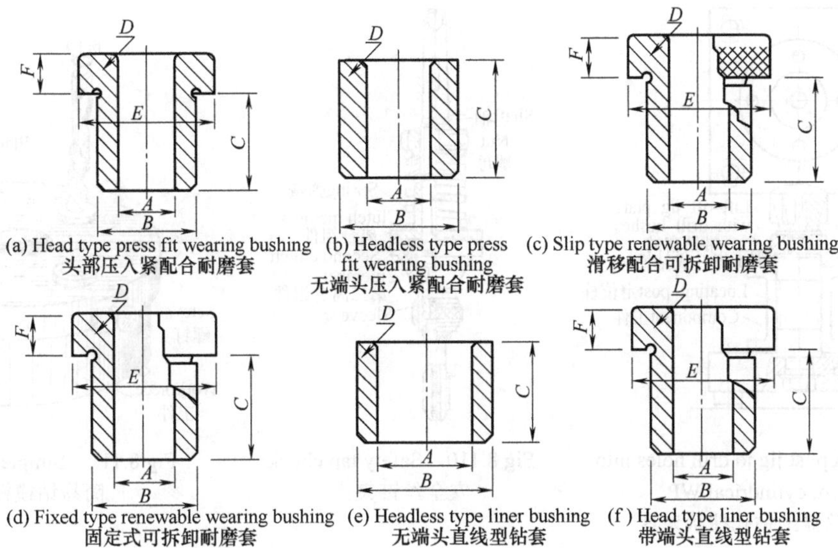

Fig 8.104 Types of jig bushings 夹具衬套（钻套）类别

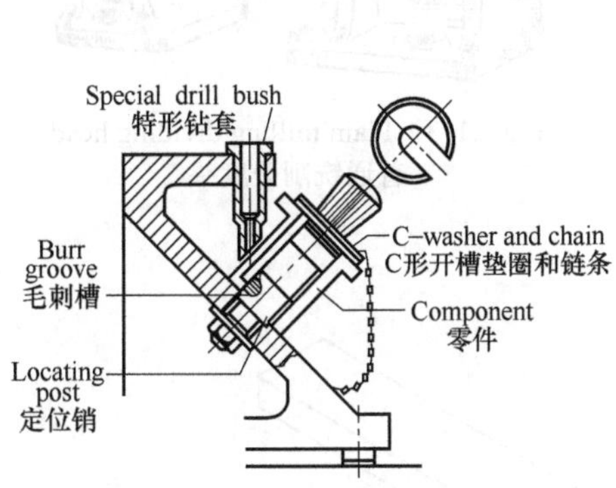

Fig 8.105 Angle drilling jig 角度钻孔夹具

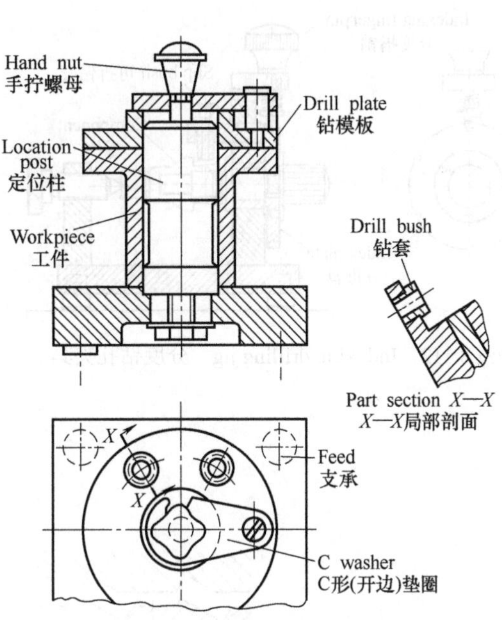

Fig 8.106 Drilling jig 钻削夹具

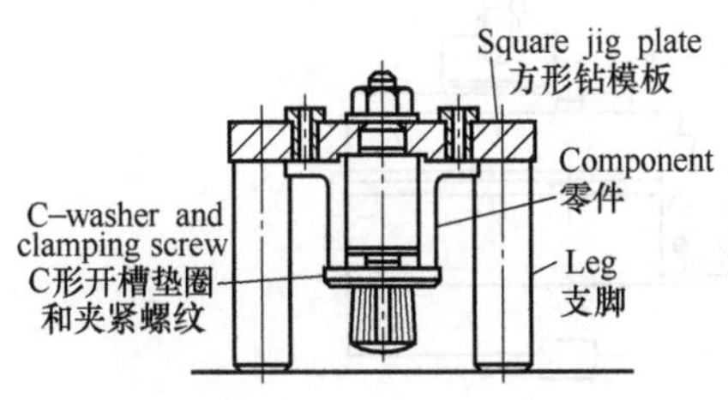

Fig 8.107 Inverted post jig
倒置立柱夹具

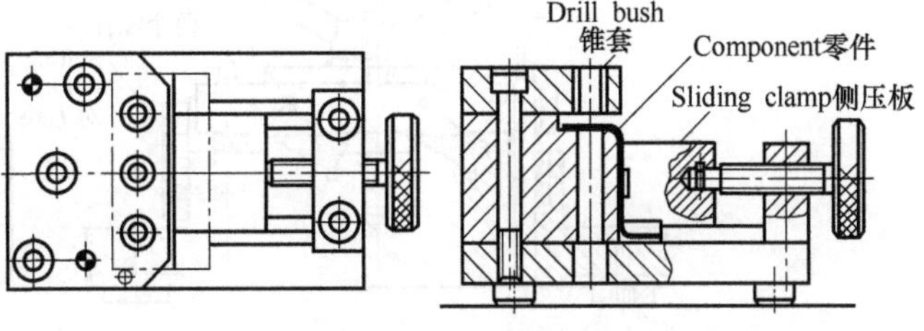

Fig 8.108 Thin plate drilling jig 薄板工件钻夹具

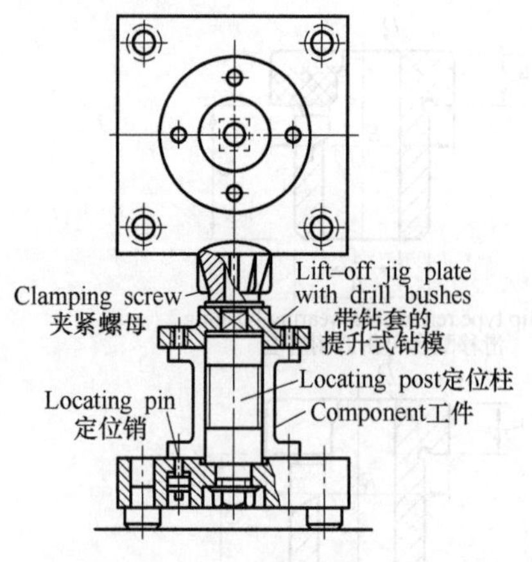

Fig 8.109 A post jig to drill holes into flanged, cylindrical WP
外圆法兰盘上钻孔柱状夹具

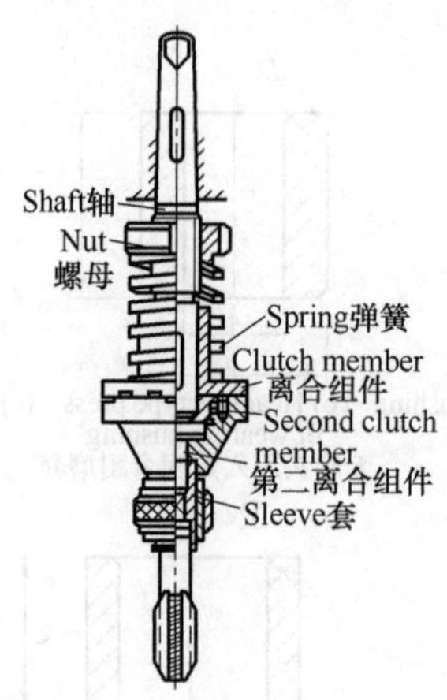

Fig 8.110 Safety tap chuck
安全丝锥夹头

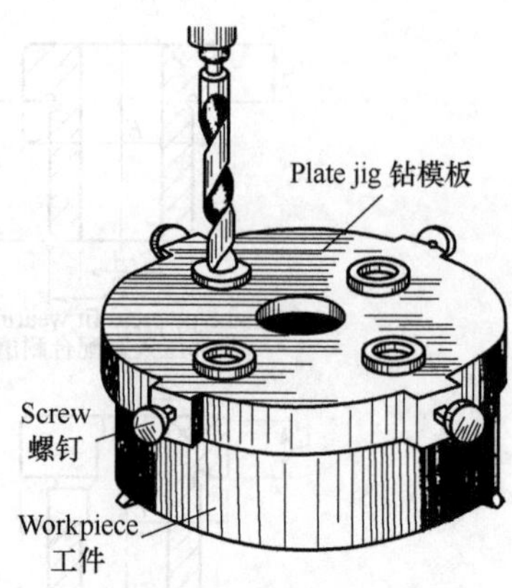

Fig 8.111 Simple plate jig
简易钻模板

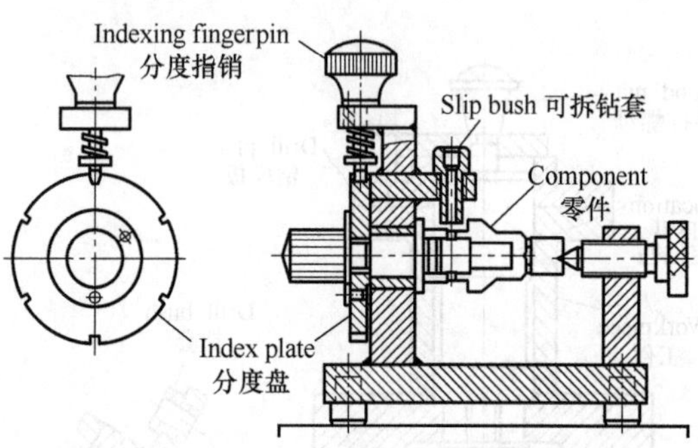

Fig 8.112 Indexing drilling jig 分度钻孔夹具

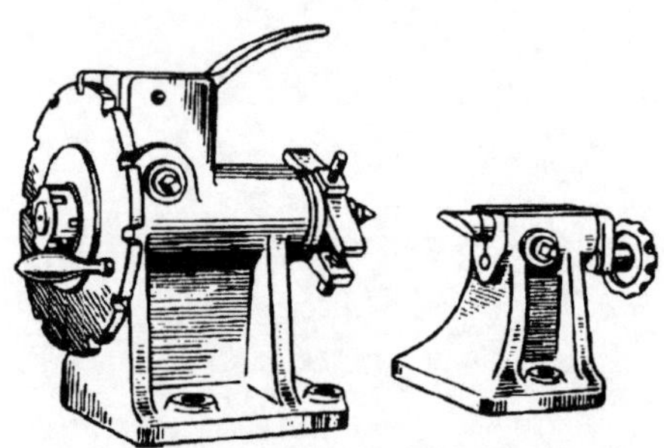

Fig 8.113 Plain milling dividing head
普通铣削分度头

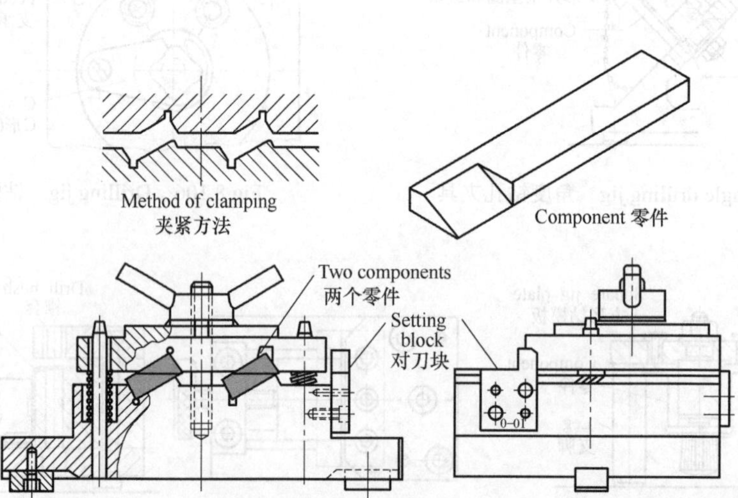

Fig 8.114 A special milling fixture for mounting two rectangular components 两个矩形零件安装专用铣削夹具

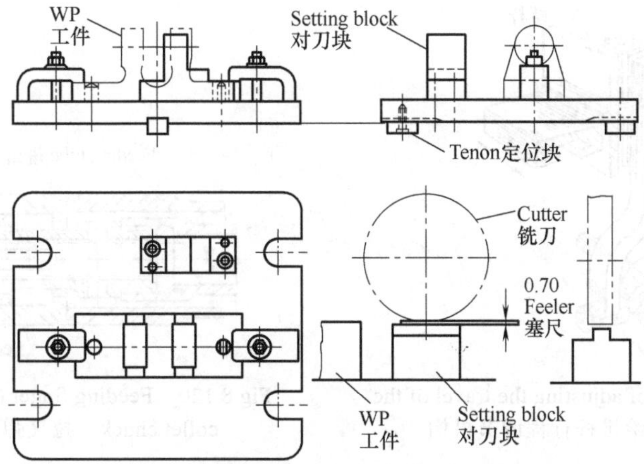

Fig 8.115 Simple milling fixture for a bearing bracket 轴承座简易铣削夹具

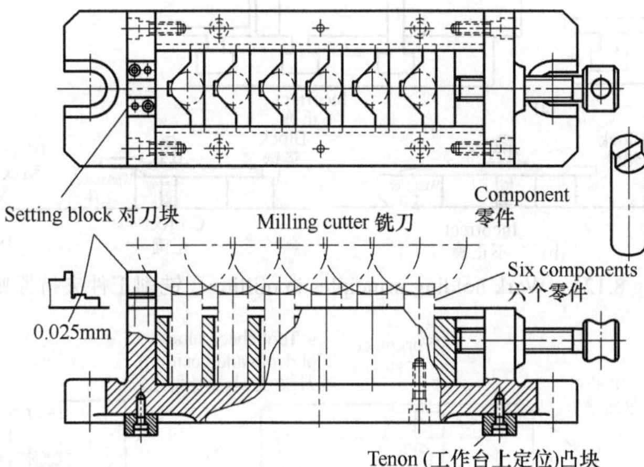

Fig 8.116 Special fixture for milling six cylindrical WPs 六个柱形零件铣削专用夹具

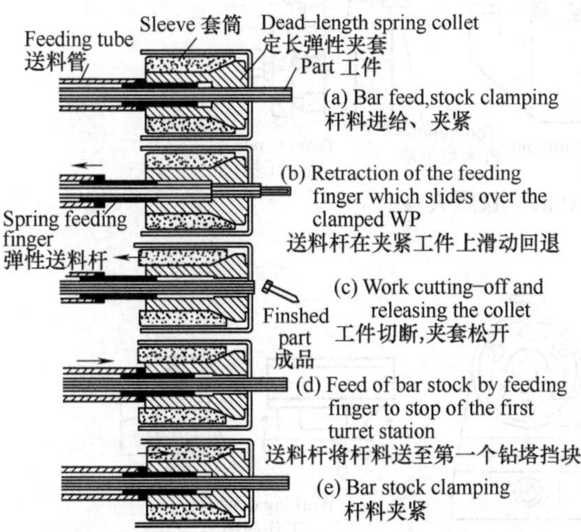

Fig 8.117 Operation of the bar feeding and chucking mechanisms using a dead-length chuck
采用定长夹套实现杆料送进和卡紧机构的动作原理

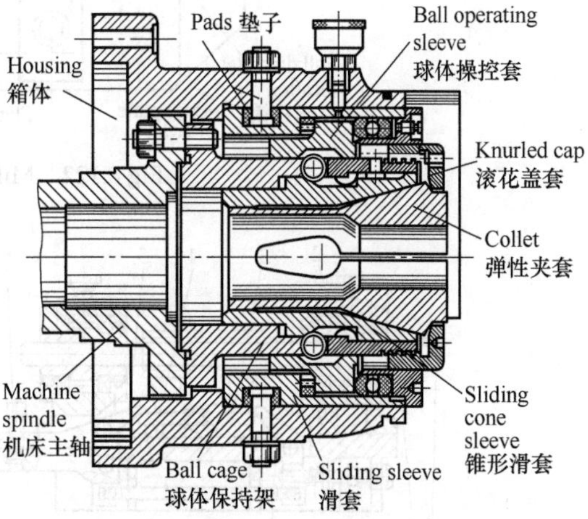

Fig 8.118 Hand-operated collet chuck
手动弹性夹套

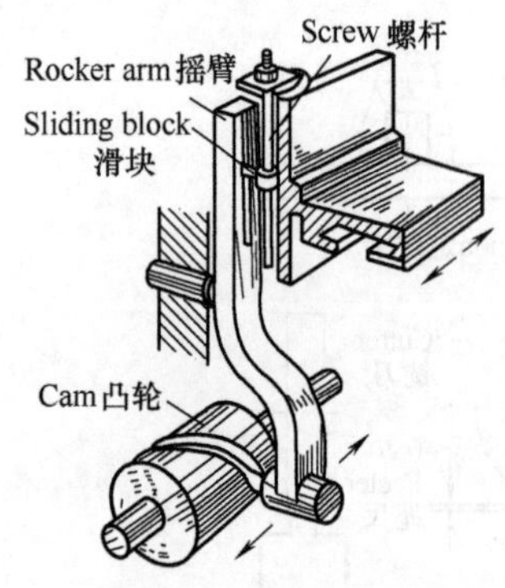

Fig 8.119 Mechanism of adjusting the travel of the feeding finger 进给推杆行程调节机构

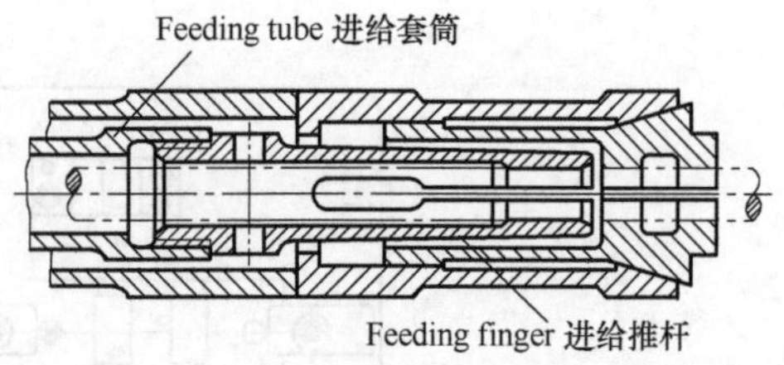

Fig 8.120 Feeding finger (stock pusher) in a pull-in collet chuck 拉夹弹性夹头的进给推杆

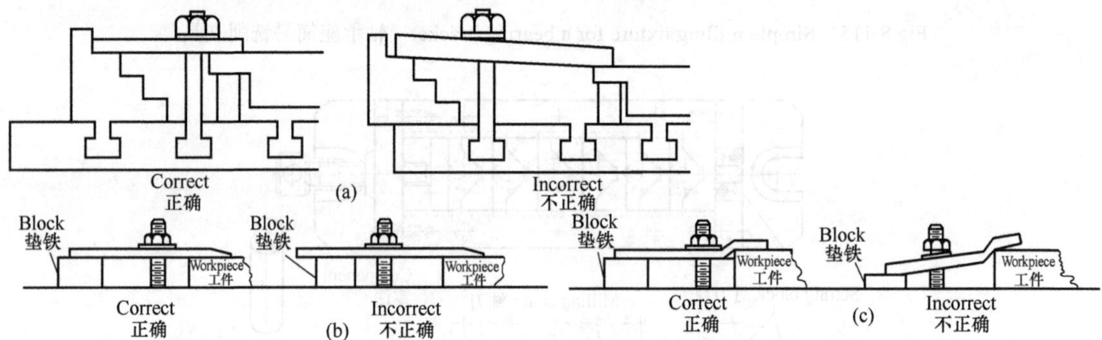

Fig 8.121 Work holding principles in milling 铣削工件夹持原则

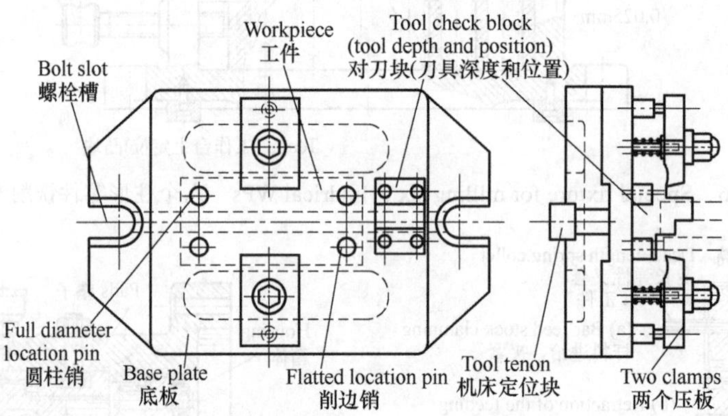

Fig 8.122 Milling fixture 铣床夹具

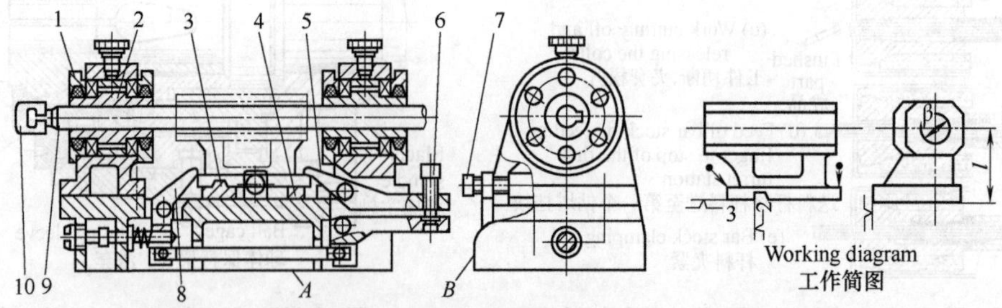

Fig 8.123 Boring jig for tailstock machining 尾座镗削夹具

1—Jig frame 镗模架；2—Boring bushing 镗套；3，4—Positioning plate 定位板；5，8—Clamping bar 压杆；6—Clamping screw 夹紧螺钉；7—Adjustable supporting 可调支承；9—Tool bar 镗刀杆；10—Floating couple 浮动接头

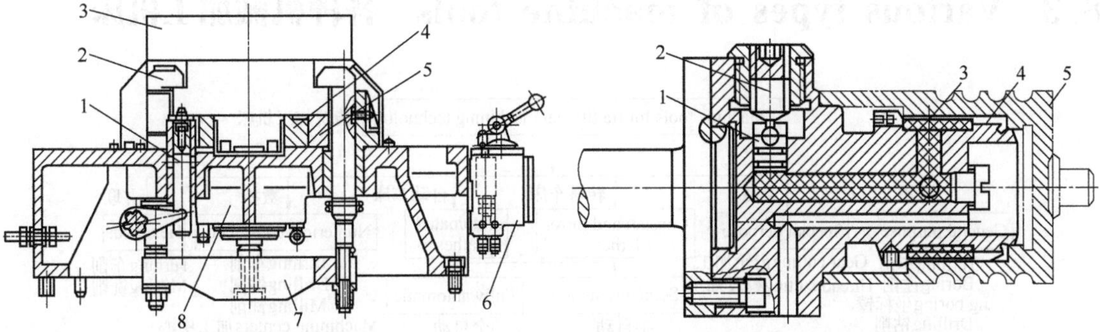

Fig 8.124 Follower jig and fixture on automation line 自动线上随行夹具

1—Slide positioning pin 活动定位销；2—Hook clamping plate 钩形压板；3—Follower jig 随行夹具；4—Convey support 输送支承；5—Positioning support plate 定位支承板；6—Lubricating pump 润滑泵；7—Lever 杠杆；8—Hydraulic cylinder 液压缸

Fig 8.125 Positioning and clamping fixture with liquidized plastics 液性塑料定心夹紧夹具

1—Slide pin 滑柱；2—Clamp bolt 压钉；3—Liquid plastics 液性塑料；4—Thin-wall bushing 薄壁套筒；5—Workpiece 工件

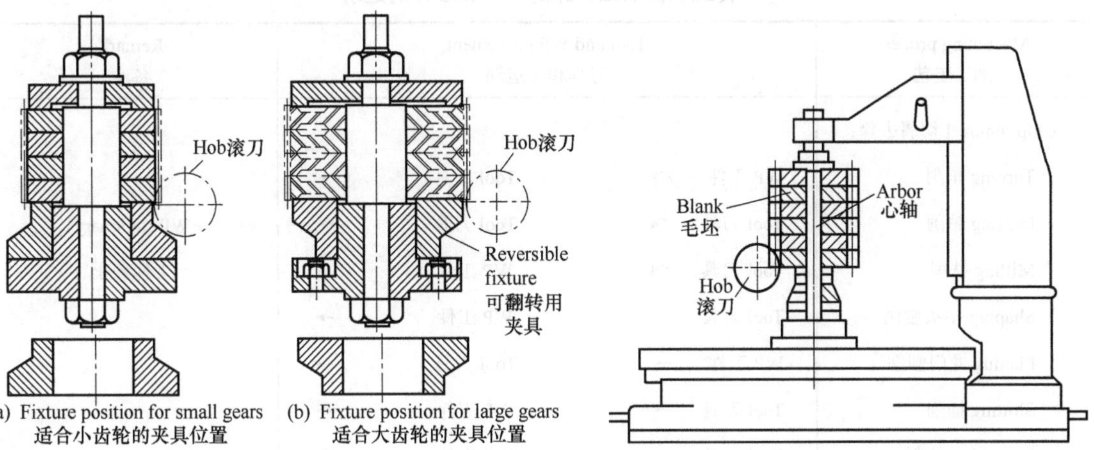

Fig 8.126 Interchangeable hobbing fixture to various size gears 适应不同尺寸滚齿的可换夹具

Fig 8.127 Clamping identical gears in one setup 一次装夹多个相同齿轮

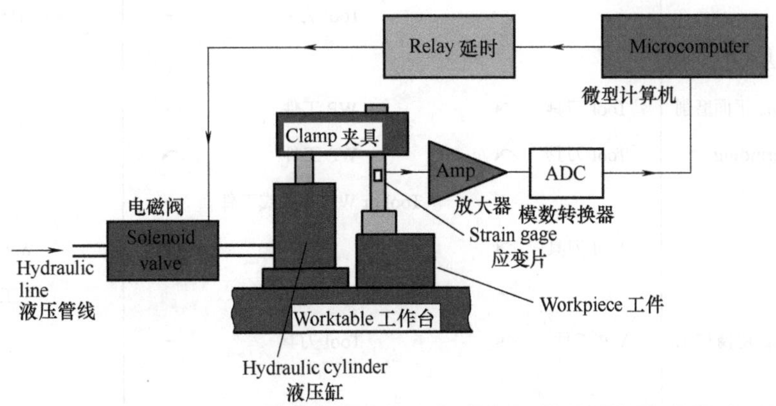

Fig 8.128 A flexible fixturing setup 柔性夹具装置

8.3 Various types of machine tools 各种机械加工机床

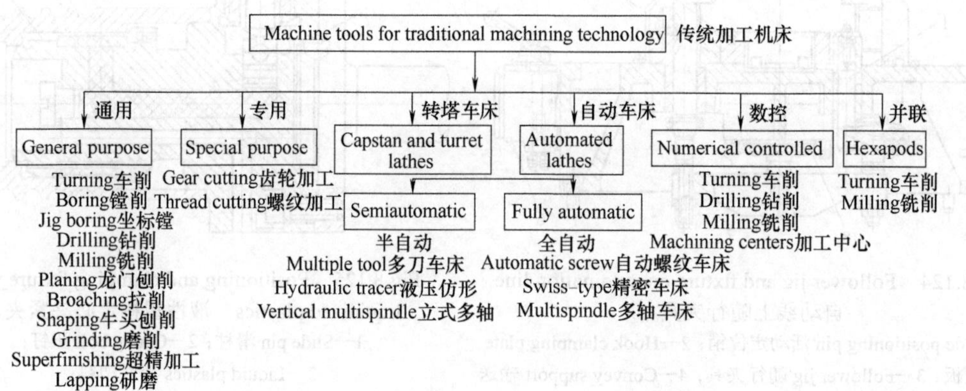

Fig 8.129 Classification of machine tools for traditional machining technology 传统加工机床的分类

Table 8.4 Tool and WP motions for machine tools used for traditional machining
传统机械加工机床的刀具和工件的运动

Machining process 加工工艺	Tool and WP movement 刀具和工运动				Remark 备注
Chip removal 切屑去除：					
Turning 车削	WP 工件	↷	Tool 刀具	→	
Drilling 钻削	Tool 刀具	↷	Tool 刀具	→	WP stationary 工件静止
Milling 铣削	Tool 刀具	↷	WP 工件	→	
Shaping 牛头刨削	Tool 刀具	→	WP 工件	--→	
Planing 龙门刨削	WP 工件	→	Tool 刀具	--→	
Slotting 插削	Tool 刀具	→	WP 工件	--→	
Broaching 拉削	Tool 刀具	→	WP 工件	●	Feed motion is built in the tool 进给运动由刀具结构实现
	WP 工件	→	Tool 刀具	●	
Gear hobbing 滚齿	Tool 刀具	↷	WP 工件	↷	
			Tool 刀具	→	
Abrasion 磨削去除：					
Surface grinding 平面磨削	Tool 刀具	↷	WP 工件	→	
Cylindrical grinding 外圆磨削	Tool 刀具	↷	WP 工件	↷	
			Tool or WP 刀具或工件	→	
Honing 珩磨	Tool 刀具	↷ →		●	WP stationary 工件静止
Superfinishing 超精加工	WP 工件	↷	Tool 刀具	→	

Note 注：↷，Rotation 回转；●，Stationary 静止；→，Linear motion 线性运动；--→，Intermittent 间隙性运动。

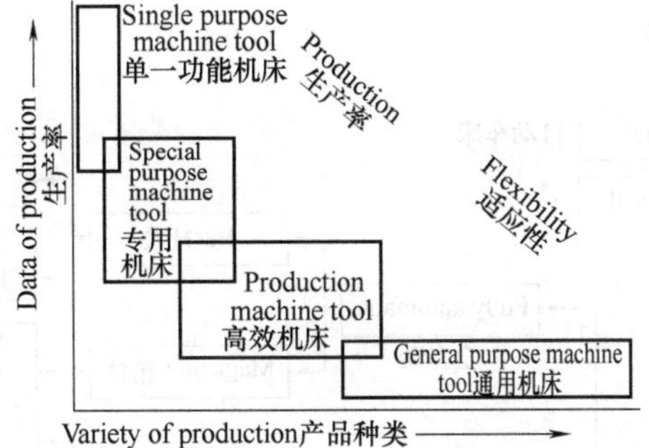

Fig 8.130 Application of machine tools based on their capability 基于生产能力的机床应用

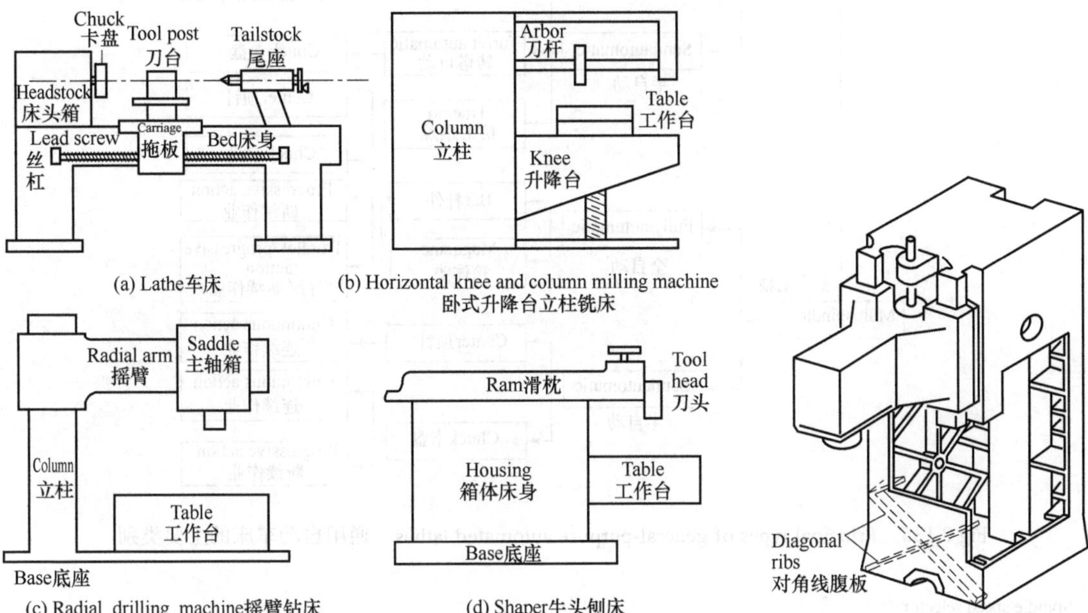

Fig 8.131 Different types of structures found in machine tools 多种常见机床结构

Fig 8.132 An example of a machine tool structure (The boxtype, one-piece design with internal diagonal ribs significantly improves the stiffness of the machine) 机床结构示例（对角线腹板的盒式整体设计明显改善机床刚性）

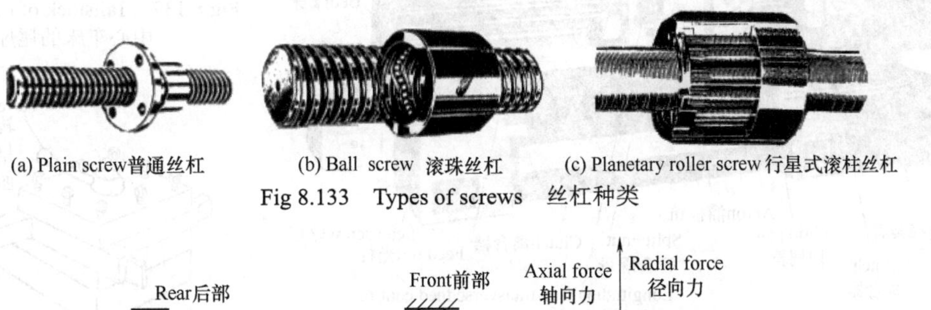

Fig 8.133 Types of screws 丝杠种类

Fig 8.134 Forces acting on machine tool spindle 机床主轴的受力

8.3.1 Lathes 车床

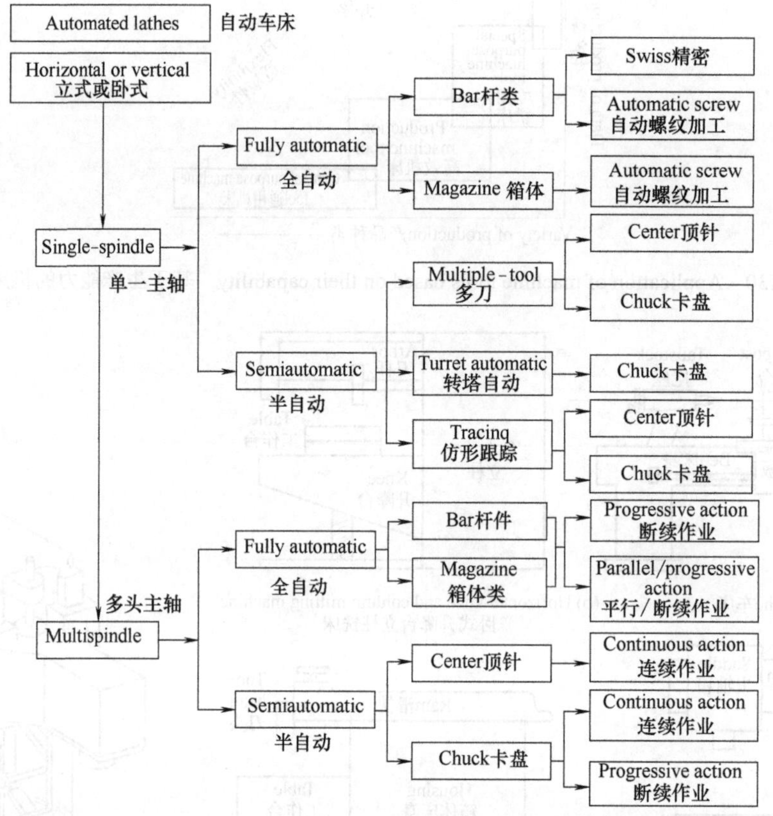

Fig 8.135 Principal types of general-purpose automated lathes 通用自动车床的主要类别

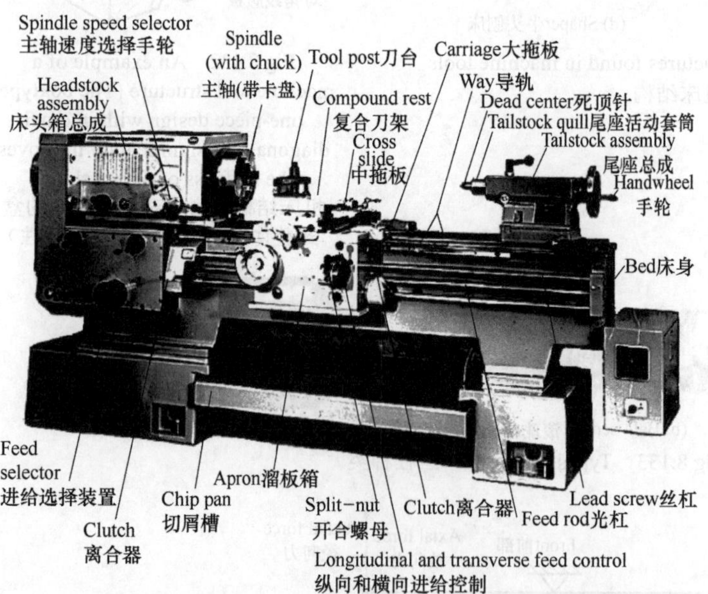

Fig 8.136 Lathe parts 车床部件

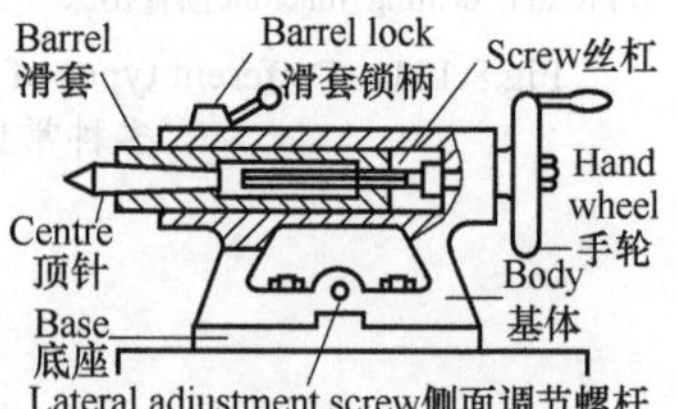

Fig 8.137 Tailstock of centre lathe 中心车床的尾座

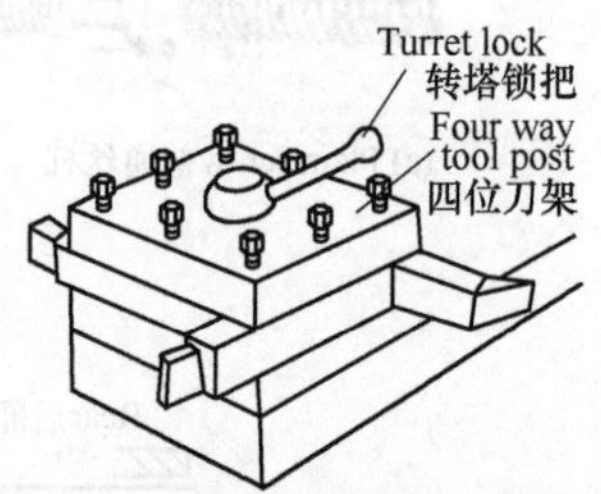

Fig 8.138 Post of centre lathe 中心车床的刀台

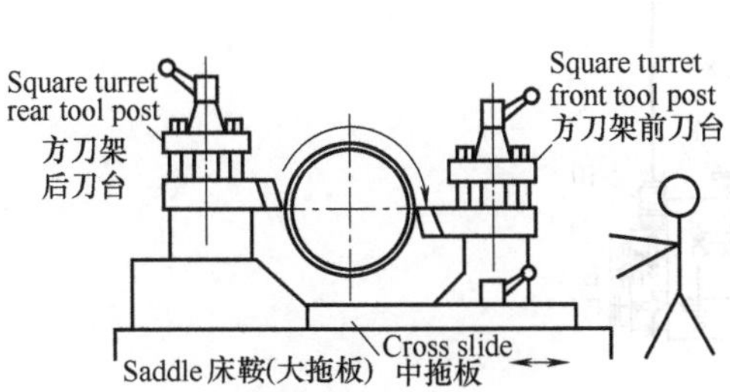

Fig 8.139 Cross slide and square turret tool posts 中拖板与方刀架刀台

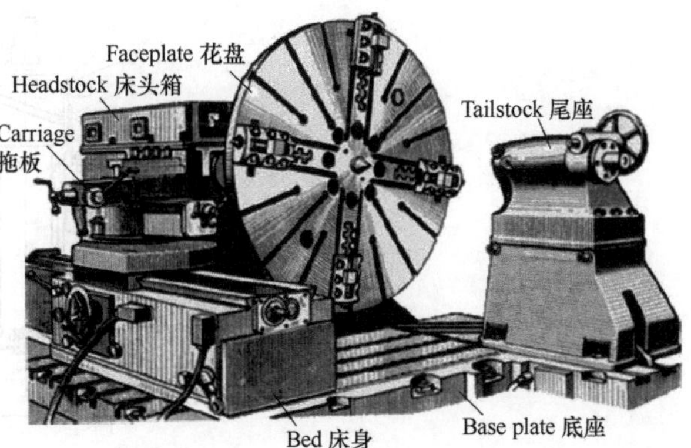

Fig 8.140 Facing lathe 花盘车床

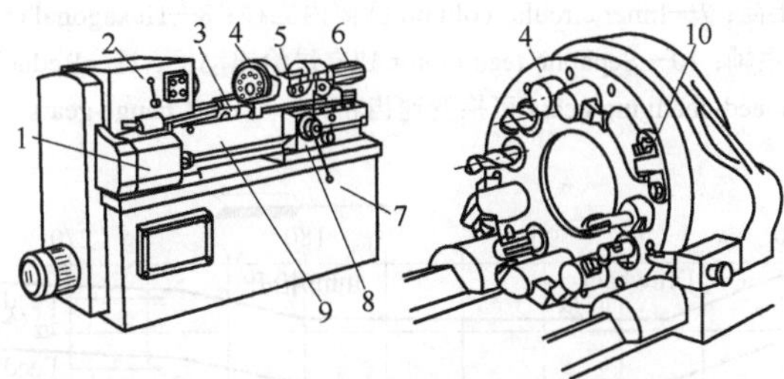

Fig 8.141 Lathe with rotary tool post 回轮车床

1—Feed box 进给箱；2—Headstock 主轴箱；3—Longitudinal stop 纵向挡块；4—Rotary tool post 回转刀架；5—Longitudinal tool carriage 纵向刀具溜板；6—Longitudinal stroke mechanism 纵向定程机构；7—Base 底座；8—Saddle 溜板箱；9—Bed 床身；10—Cross stroke mechanism 横向定程机构

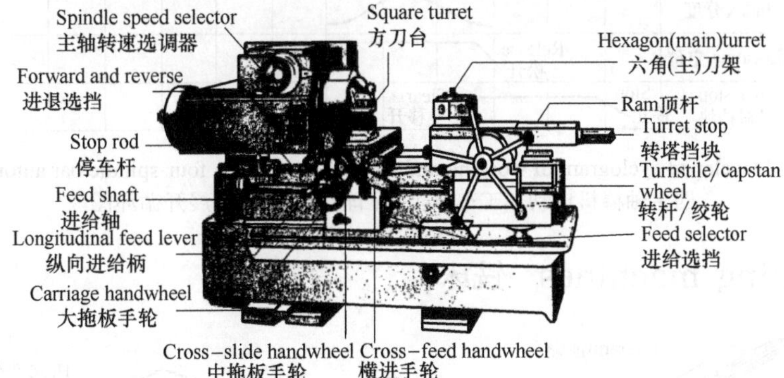

Fig 8.142 Illustration of the components of a turret lathe 转塔车床结构示意图

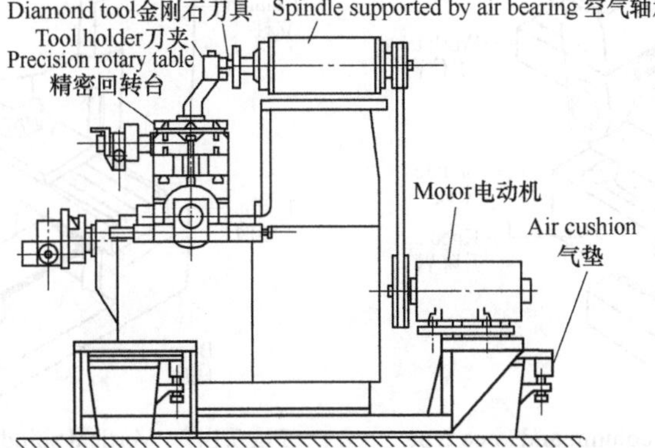

Fig 8.143 Moore precision lathe 摩尔精密车床

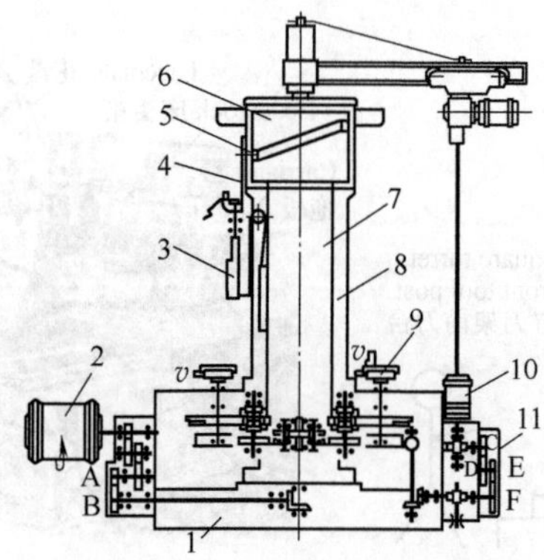

Fig 8.144 Gearing diagram of a vertical multispindle semiautomatic lathe 立式多轴半自动机床传动图

1—Base 基座；2—Spindle motor drive 主轴电动机；3—Tool head 刀头；4—Tie rod 连接杆；5—Roll 滚道；6—Stationary drum 静止筒；7—Inner circular column 圆筒内立柱；8—Hexagonal outer column 六角外柱；9—Work spindle 工件主轴；10—Separate feed motor 独立进给电动机；11—Reducing gear box 减速箱；A，B—Speed-changing gears 可换变速齿轮；C ～ F—Change gears 可换齿轮

0 90 180 270 360

Tool slide 刀具滑块 — Drawback 回退 — Jump抬升 — Feed 进给

Cross slide 横向滑块 — Feed 进给

Bar feed 棒料进给 — Feed 进给 — Return 回退

Chucking 卡紧 — Open 张开 — Close 闭合

Index 分度

Locating 定位 — Release 松开 — Lock 锁紧

Stock stop 材料挡块 — Stop 停止 — Clear 移开

Fig 8.145 Developed cyclogram of working and auxiliary cams of a four-spindle bar automatic lathe
四主轴棒料自动加工车床工作凸轮和副轴凸轮展开循环图

8.3.2 Milling machines 铣床

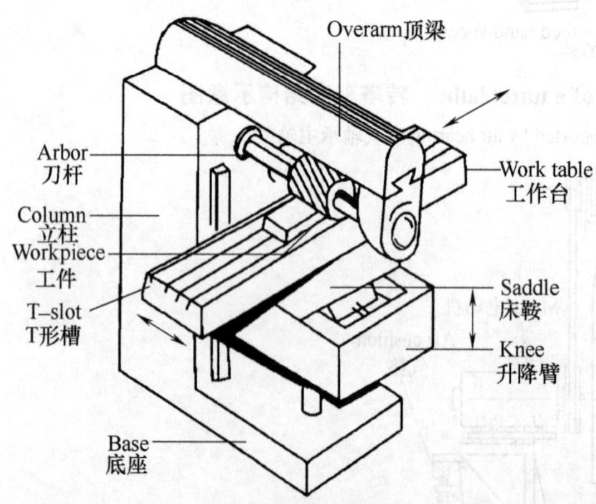

Fig 8.146 Horizontal spindle column and knee type
卧式立柱升降台铣床

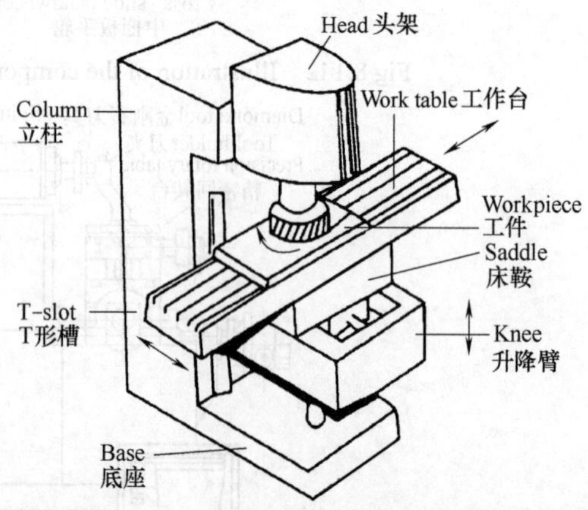

Fig 8.147 Vertical spindle column and knee type
立式立柱升降台铣床

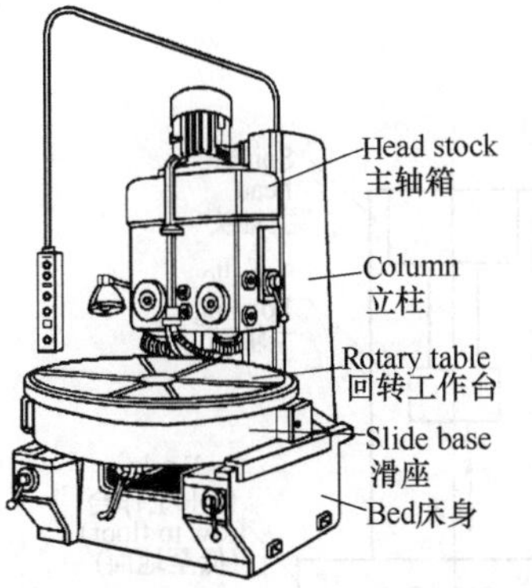

Fig 8.148 Milling machine with rotary table 转台铣床

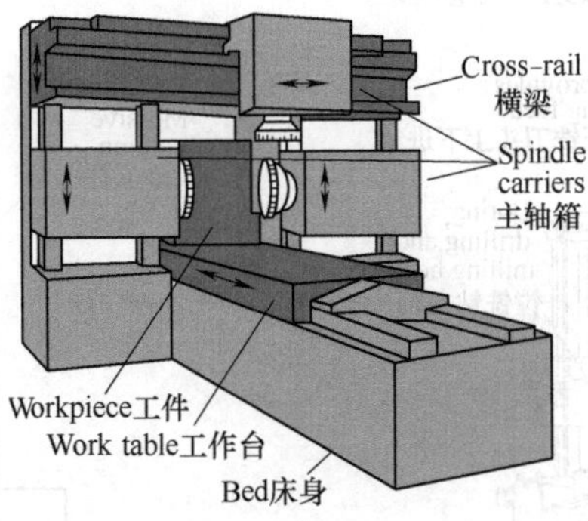

Fig 8.149 A bed-type milling machine 床身式铣床（龙门铣床）

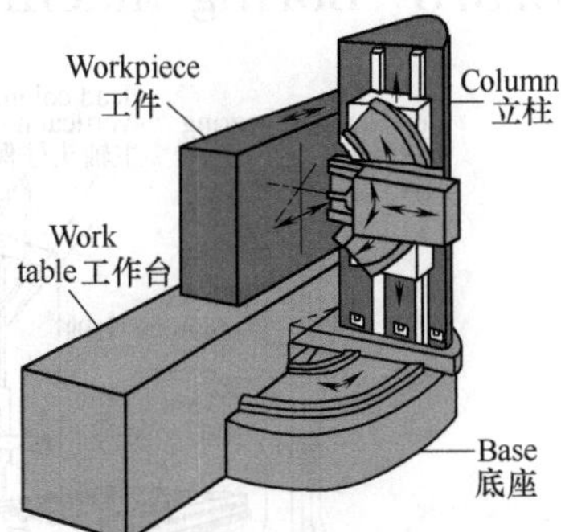

Fig 8.150 A five-axis profile milling machine 五轴轮廓铣床

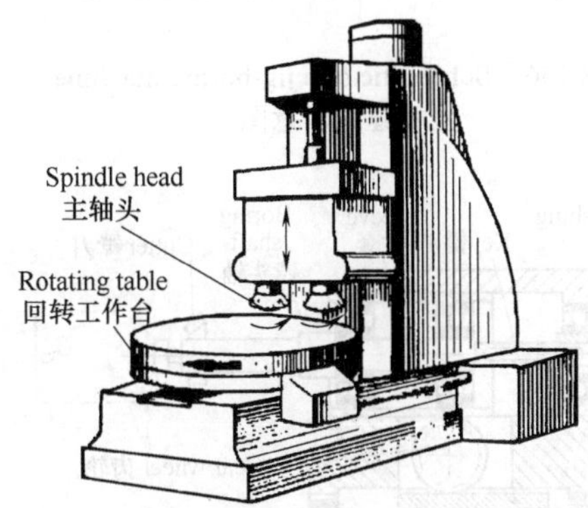

Fig 8.151 Rotary-table milling machine 回转工作台铣床

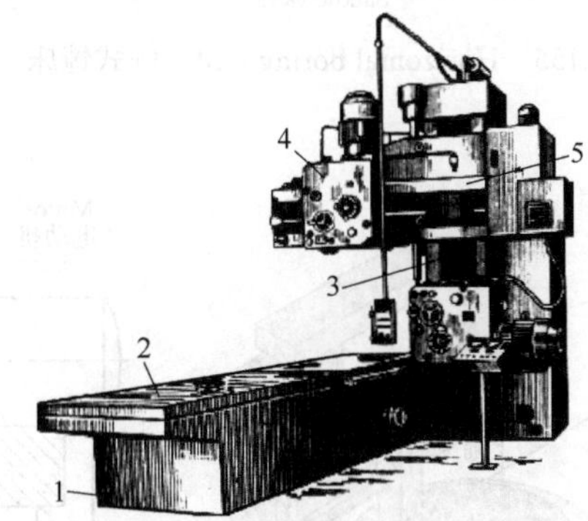

Fig 8.152 Planer-type general-purpose milling machine 龙门式通用铣床

1—Bed 床身；2—Table 工作台；3—Column 立柱；4—Spindle head 主轴头；5—Cross-arm 横梁

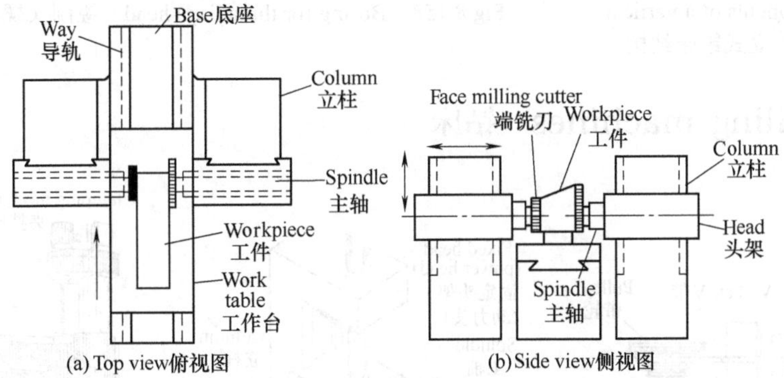

Fig 8.153 Duplex bed type milling machine 双面床身式铣床

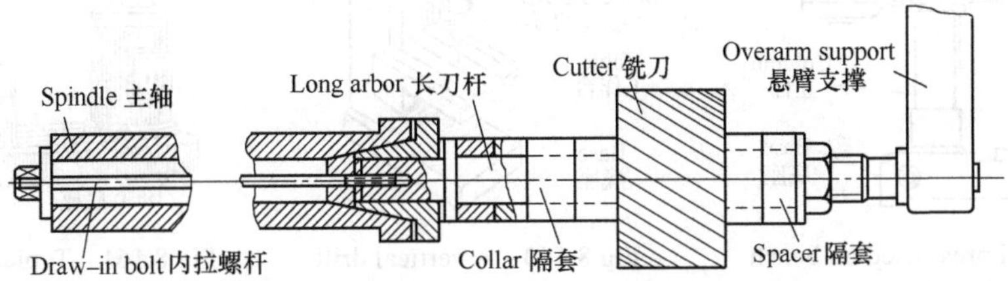

Fig 8.154 Milling machine arbor 铣床刀杆

8.3.3 Boring machines 镗床

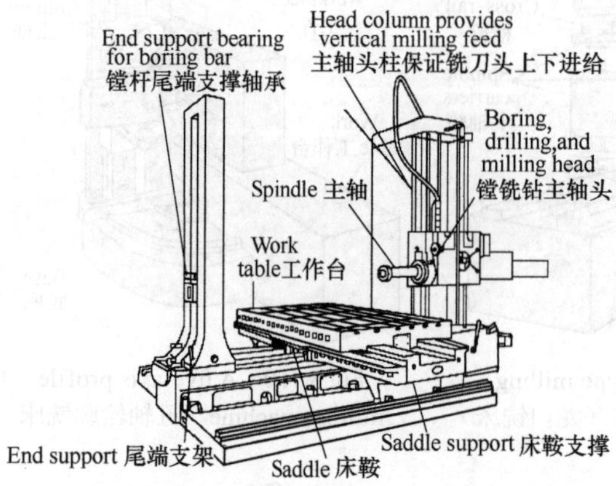

Fig 8.155 Horizontal boring mill 卧式镗床

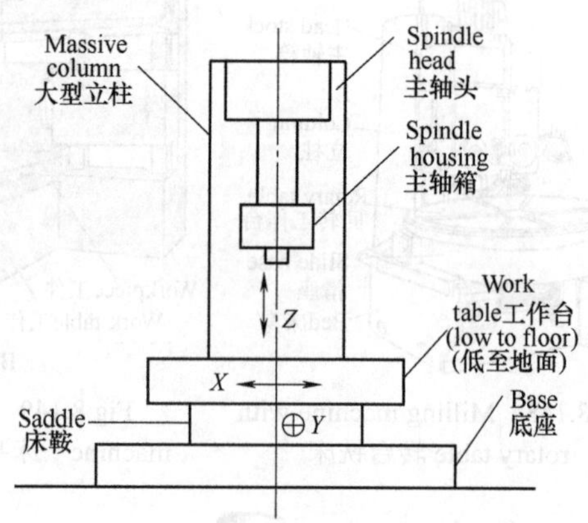

Fig 8.156 Schematic of a jig-boring machine 坐标镗床示意图

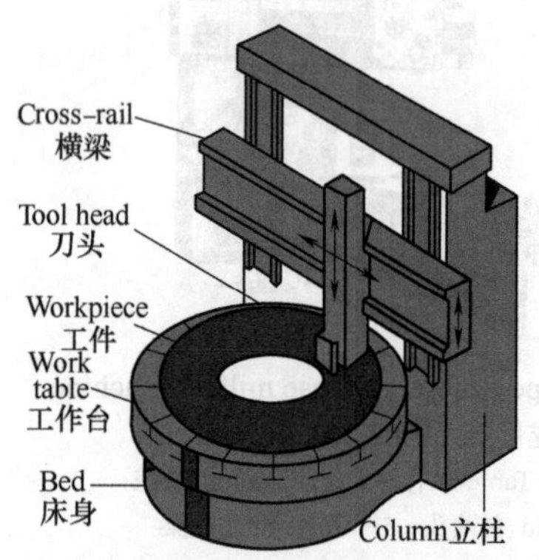

Fig 8.157 Components of a vertical boring machine 立式镗床结构

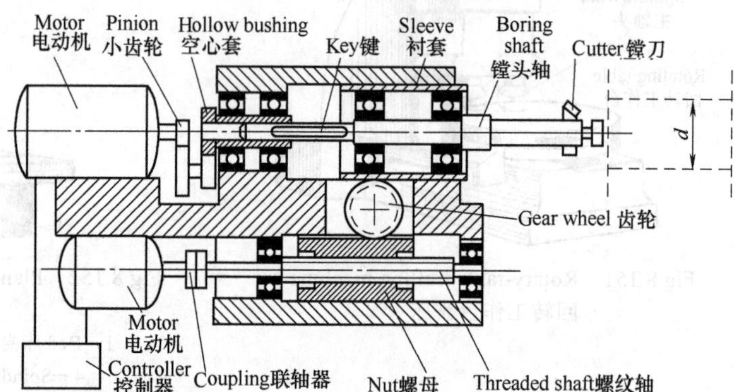

Fig 8.158 Boring (or threading) head 镗削（螺纹加工）头架

8.3.4 Drilling machines 钻床

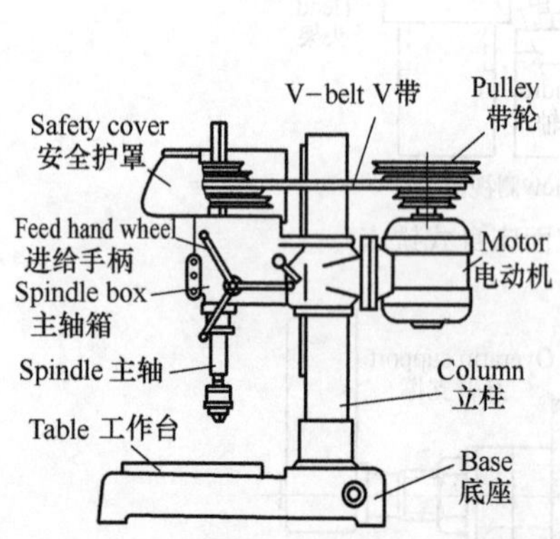

Fig 8.159 Drill press used on bench 台式钻床

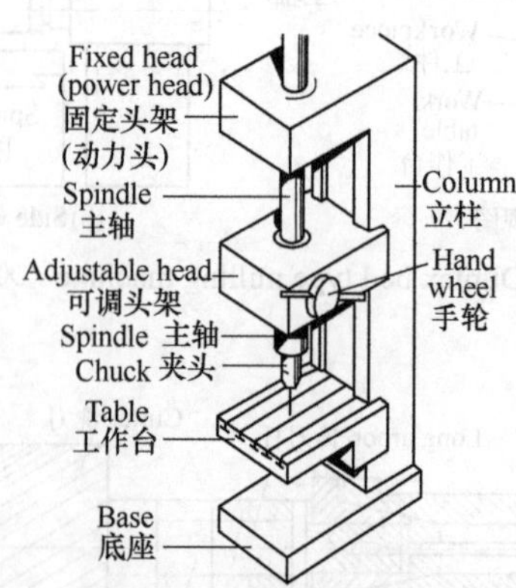

Fig 8.160 A vertical drill press 立式钻床

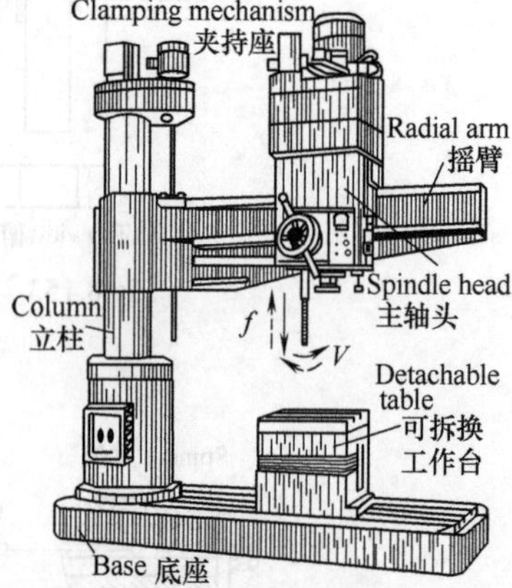

Fig 8.161 Typical radial drilling machine 典型摇臂钻床

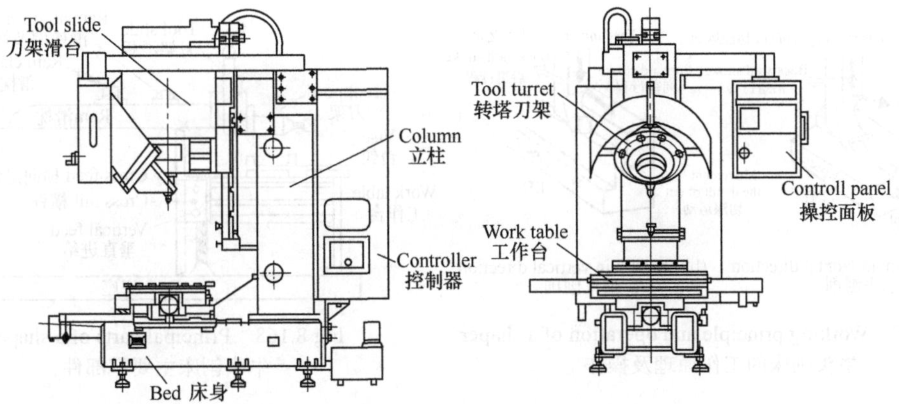

Fig 8.162 NC drilling centre 数控钻削中心

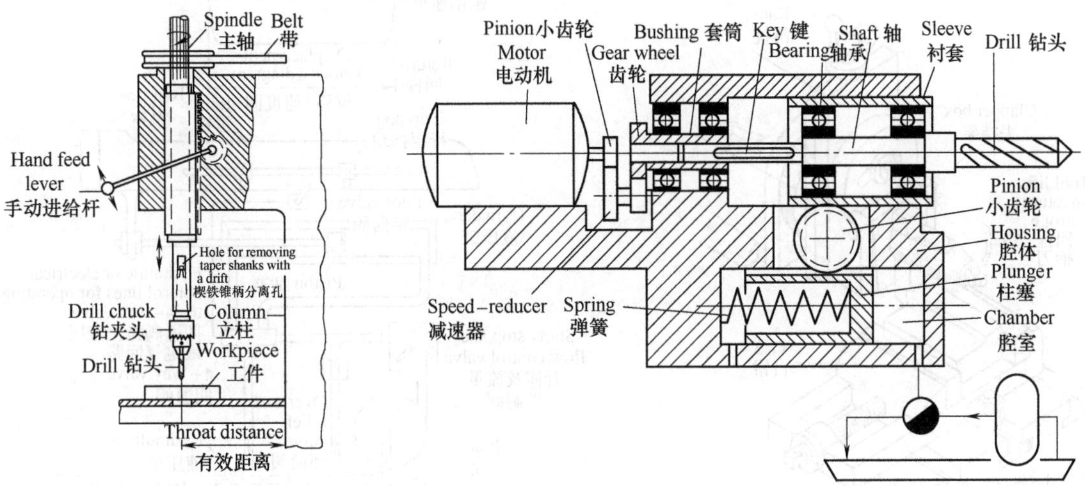

Fig 8.163 The drill press 钻头下压操作机构

Fig 8.164 Drilling head 钻削头架

8.3.5 Reciprocating machines 往复式加工机床

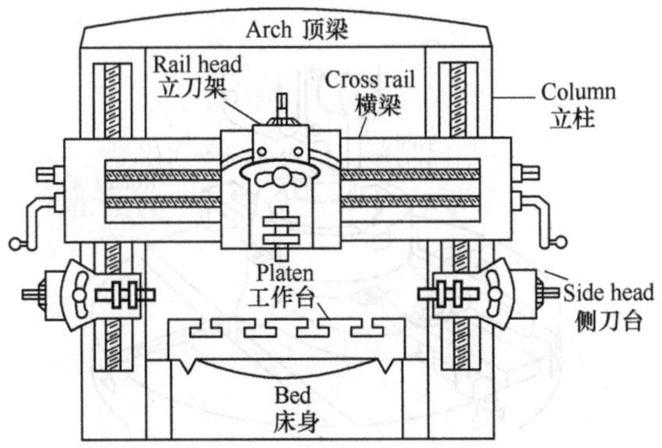

Fig 8.165 Construct of a planning machine with single head 单头龙门刨床结构

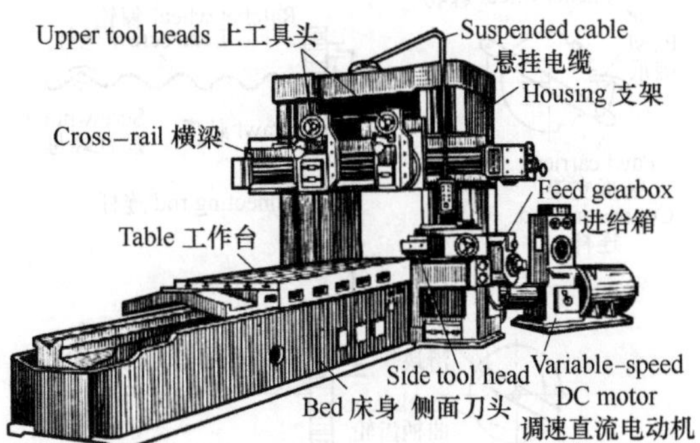

Fig 8.166 Double housing planer 双支架龙门刨床

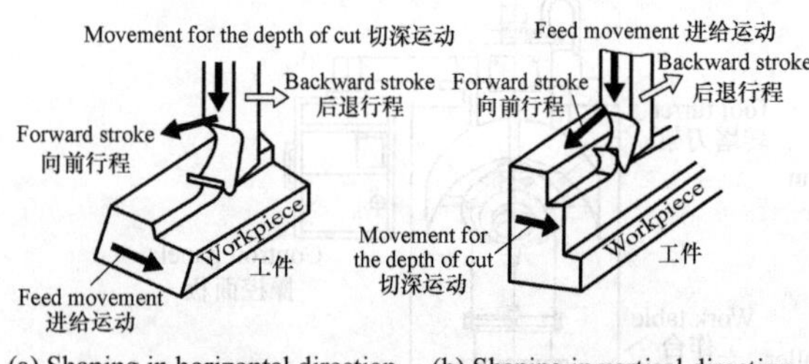

Fig 8.167 Working principle and operation of a shaper
牛头刨床的工作原理及操作

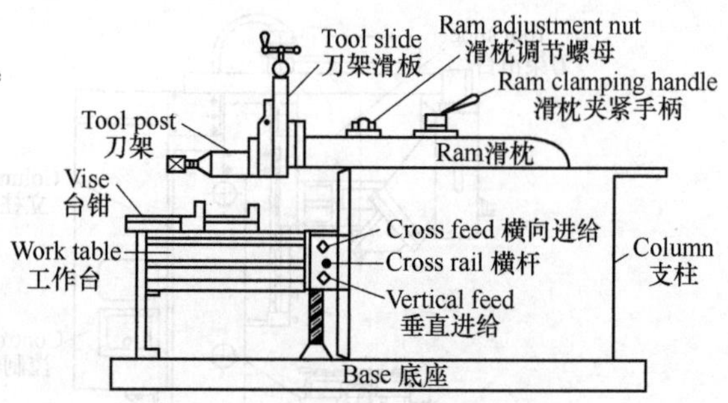

Fig 8.168 Principal parts of a shaper
牛头刨床主要零部件

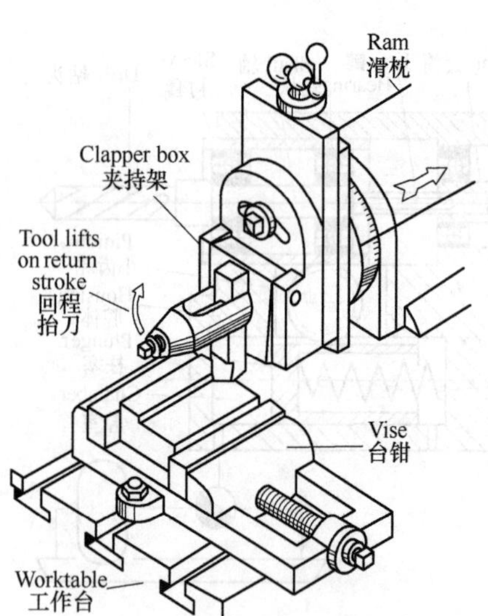

Fig 8.169 Typical arrangement of the workpiece and tool in a shaper
刨床工件和刀具的布置

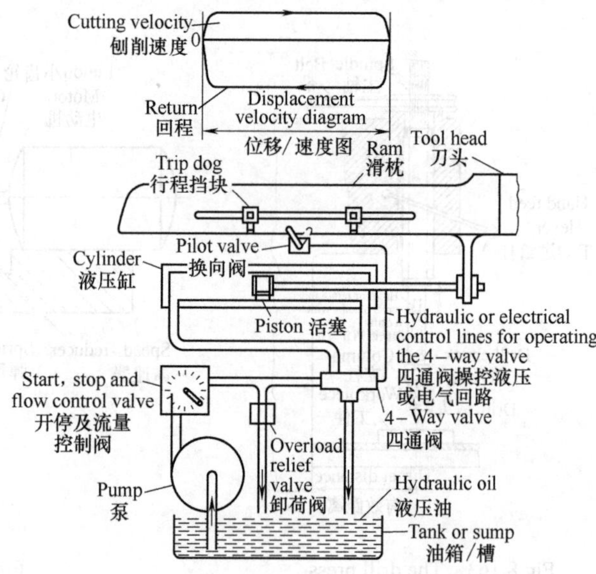

Fig 8.170 Hydraulic shaper 液压牛头刨床

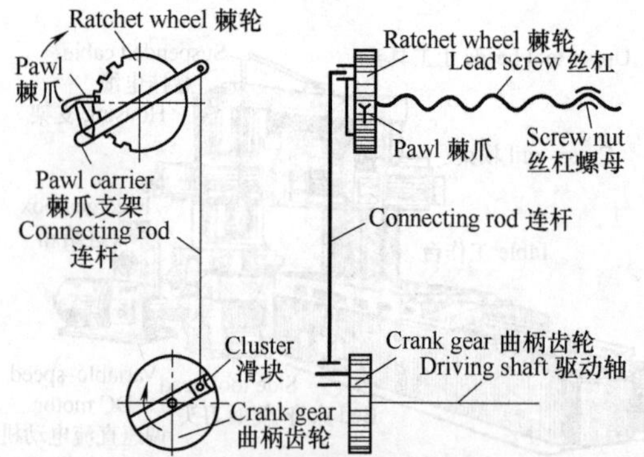

Fig 8.171 Table feed mechanism of a mechanical shaper
牛头刨工作台机械进给机构

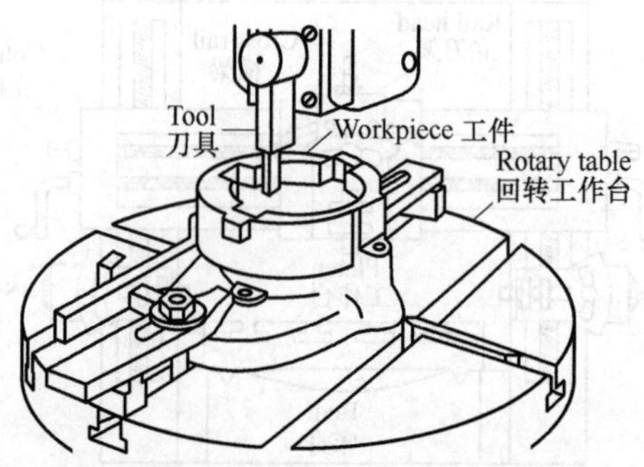

Fig 8.172 A typical component machined in a slotter 插床加工典型工件

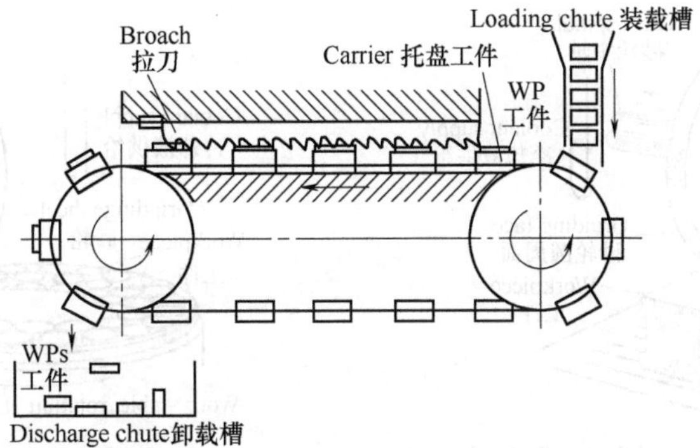

Fig 8.173 Continuous horizontal surface broaching machine 卧式连续平面拉床

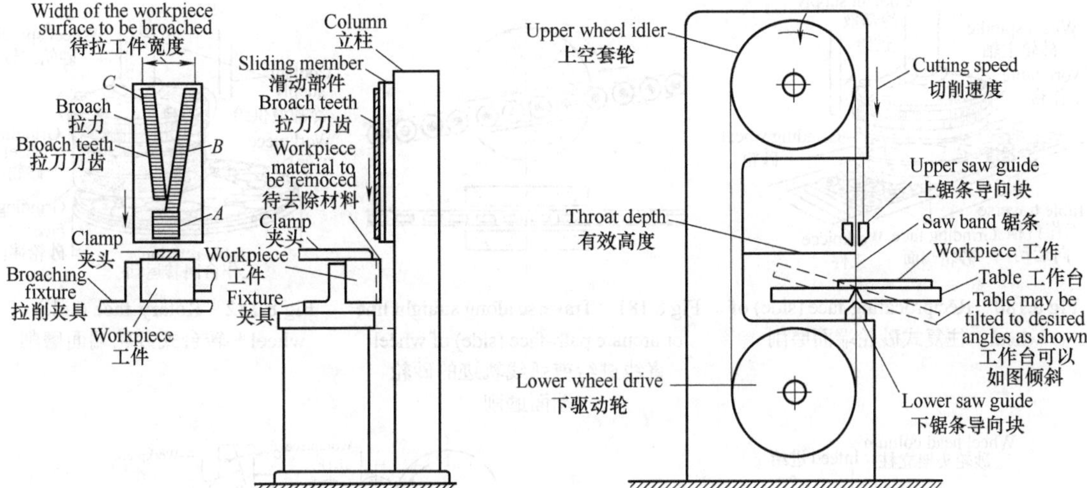

Fig 8.174 Surface broaching machine 平面拉床

Fig 8.175 Vertical band sawing 立式带锯锯床

8.3.6 Grinding machines 磨床

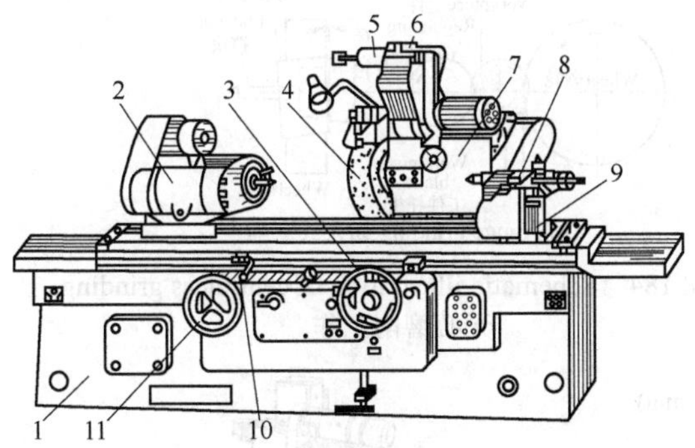

Fig 8.176 Multi-purpose grinding machine for cylinder 万能外圆磨床

1—Bed 床身；2—Headstock 头架；3，11—Hand wheel 手轮；4—Abrasive (grinding)wheel 砂轮；5—Inner surface abrasive 内圆磨具；6—Support frame 支架；7—Wheel supporting frame 砂轮架；8—Tailstock 尾架；9—Work table 工作台；10—Stroke choke 行程挡块

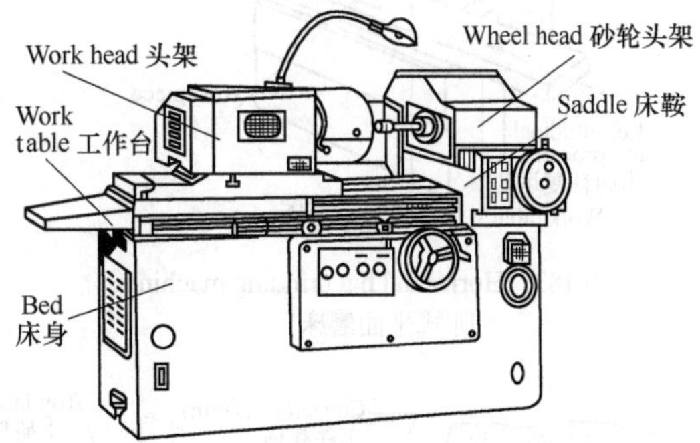

Fig 8.177 Internal surface grinding machine 内圆磨床

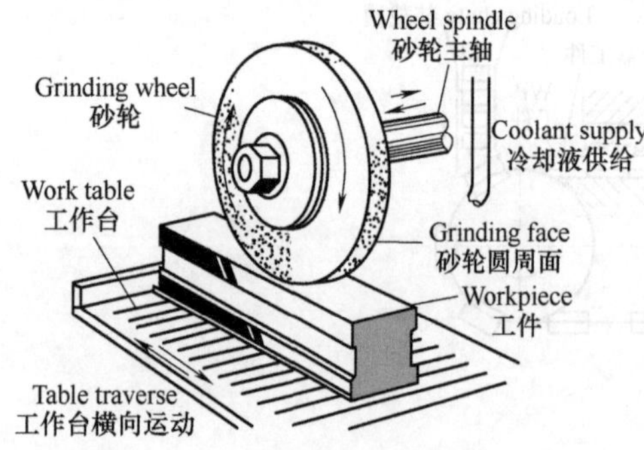

Fig 8.178 Reciprocating-periphery of wheel
往复式砂轮周边磨削

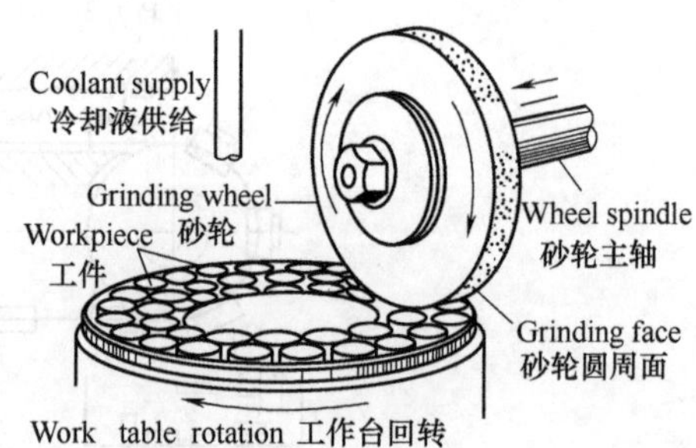

Fig 8.179 Rotary-periphery of wheel
转台式圆周磨削

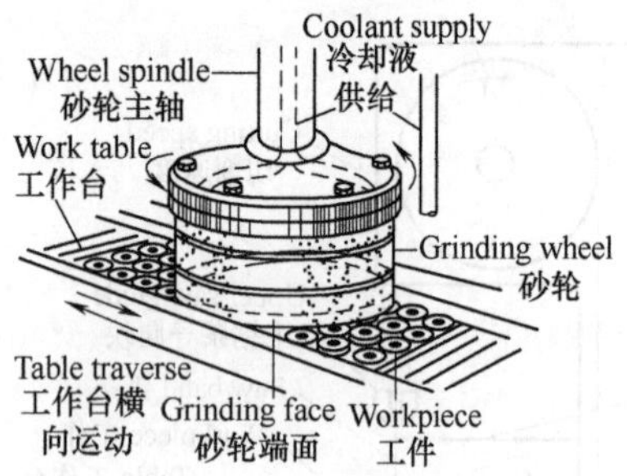

Fig 8.180 Reciprocating-face (side) of wheel 往复式砂轮端面磨削

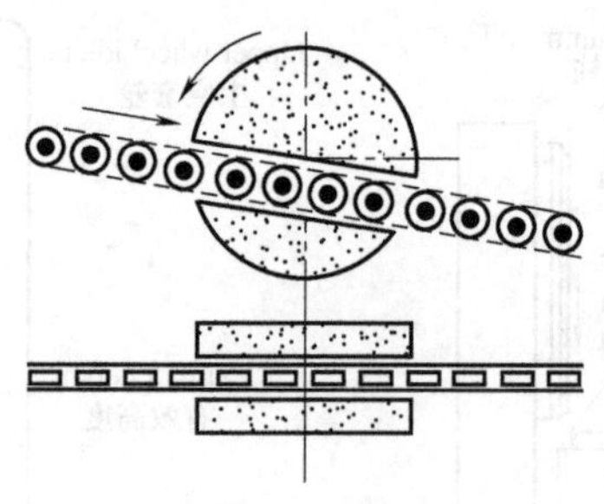

Fig 8.181 Traverse along straight line or arcuate path-face (side) of wheel
直线进给或弧线轨迹的砂轮端面磨削

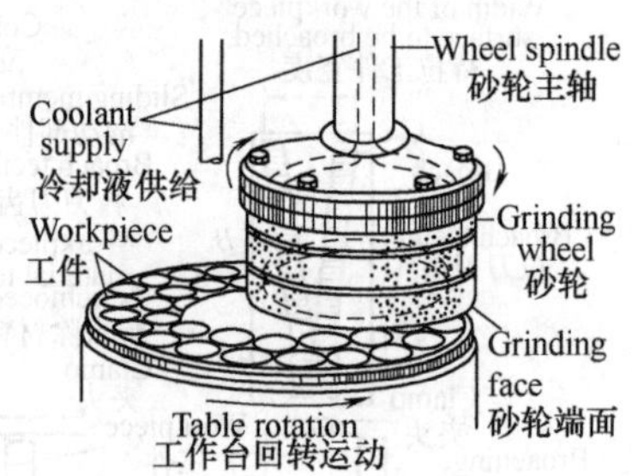

Fig 8.182 Rotary-face (side) of wheel 转台式砂轮端面磨削

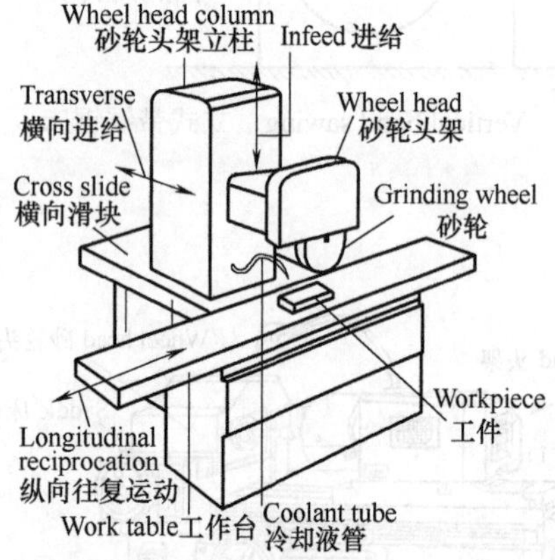

Fig 8.183 Horizental flat grinding machine
卧式平面磨床

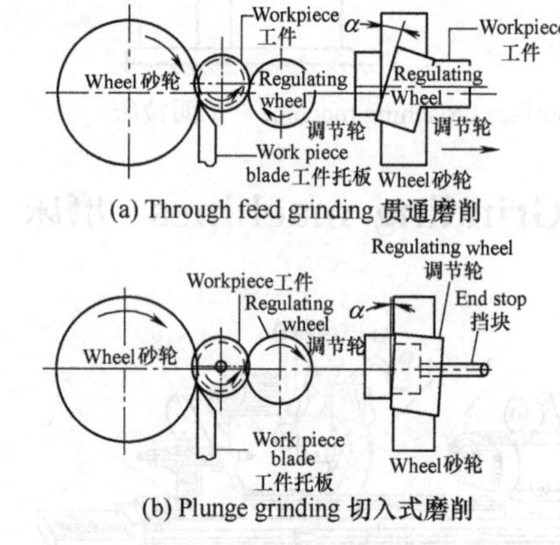

Fig 8.184 Schematic illustration of centreless grinding
无心磨削工作

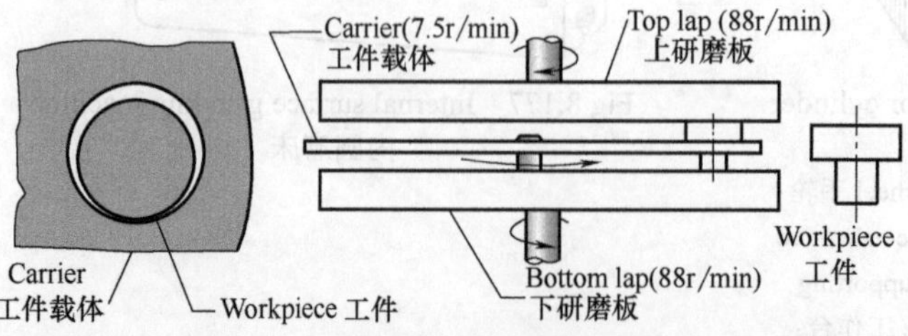

Fig 8.185 Dual-face lapping machine using two bonded abrasive laps
双研磨板双面研磨机

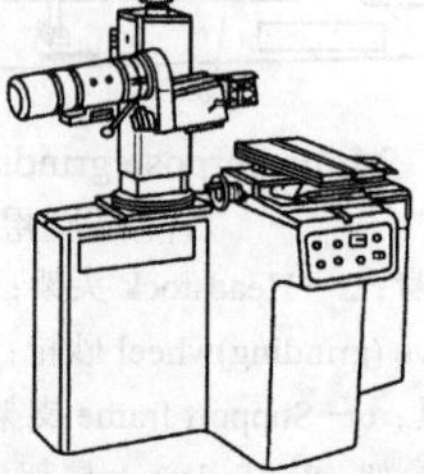

Fig 8.186 Multi-purpose grinding machine for tool 万能工具磨床

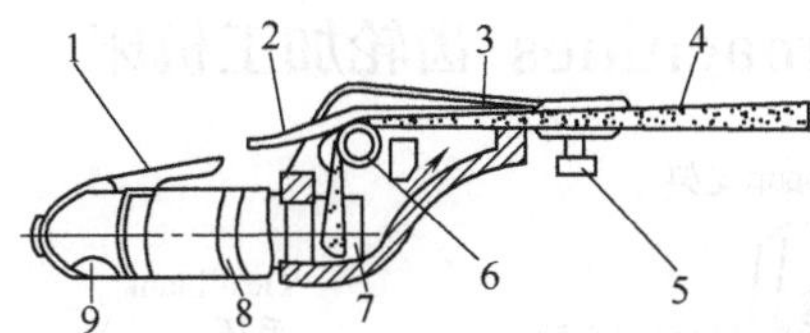

Fig 8.187 Pneumatic pistol-like portable belt grinder 气动手枪式砂带机

1—Switch 开关；2—Tension adjusting bar 调节杆；3—Belt 砂带；4—Contact arm 接触臂；5—Adjusting snob 调节旋钮；6—Idle wheel 惰轮；7—Driving wheel 驱动轮；8—Pneumatic propel wheel 气动叶轮；9—Revolution speed adjust snob 速度调节旋钮

Fig 8.188 Centreless belt grinding machine 无心砂带磨床

Fig 8.189 Portable grinder 手提式砂带机

Fig 8.190 Abrasive disc precision polisher 圆盘精密抛光机

Fig 8.191 Abrasive disc grinder 角磨机

Fig 8.192 Portable welding seam polisher 手提式焊缝打磨机

Fig 8.193 Flap (page) wheel （千）页轮

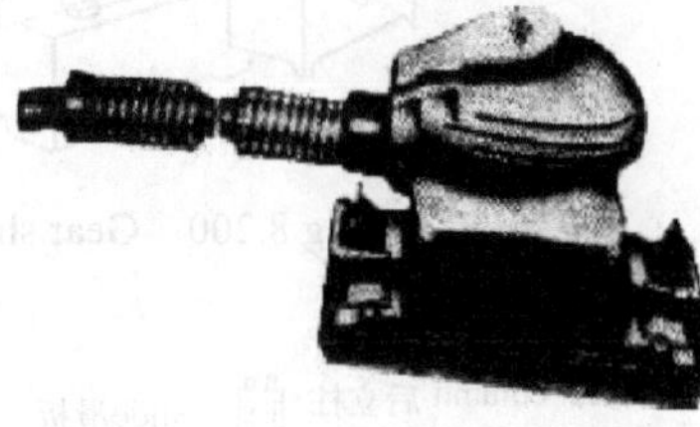

Fig 8.194 Reciprocating vibration flat grinder 往复振动平面磨光机

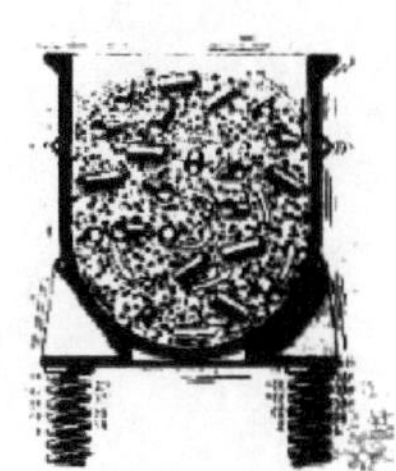

Fig 8.195 Vibration deburring machine 振动去毛刺机

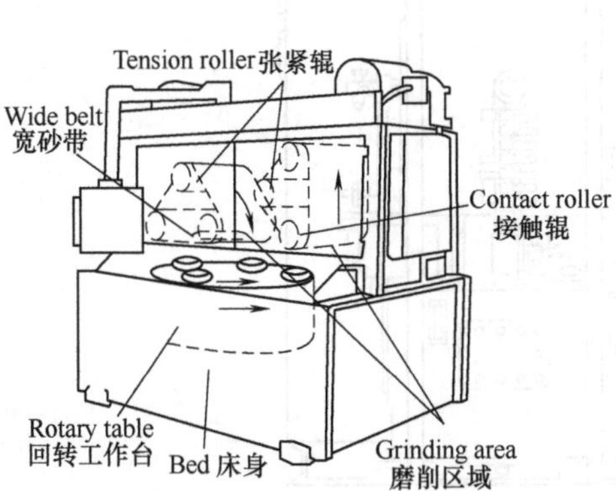

Fig 8.196 Flat surface grinding machine 平面砂带磨床

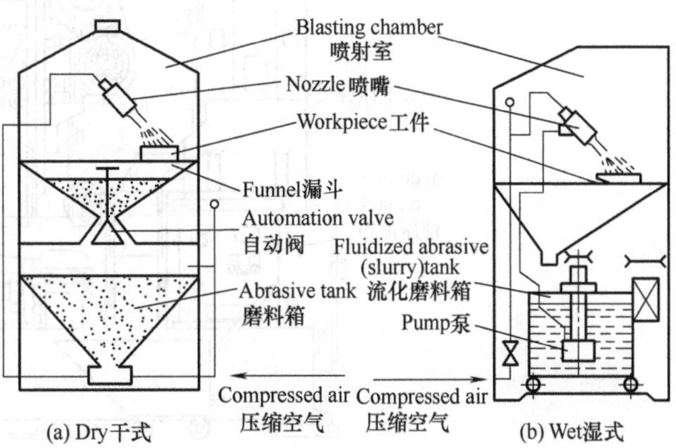

Fig 8.197 Sand blasting (abrasive-jet) setup 喷砂装置

8.3.7 Gear cutting machines 齿轮加工机床

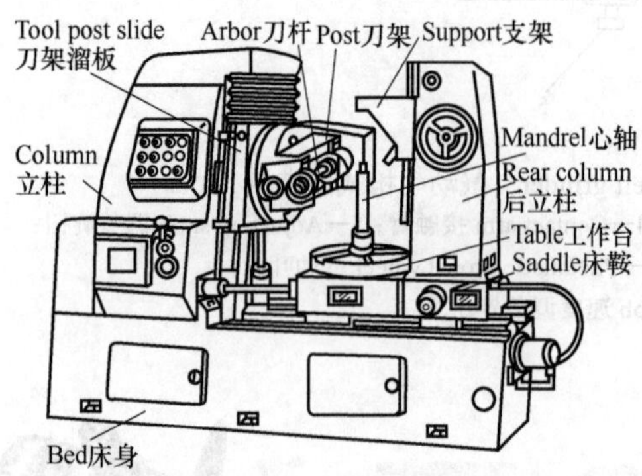

Fig 8.198 Gear hobbing machine 滚齿机

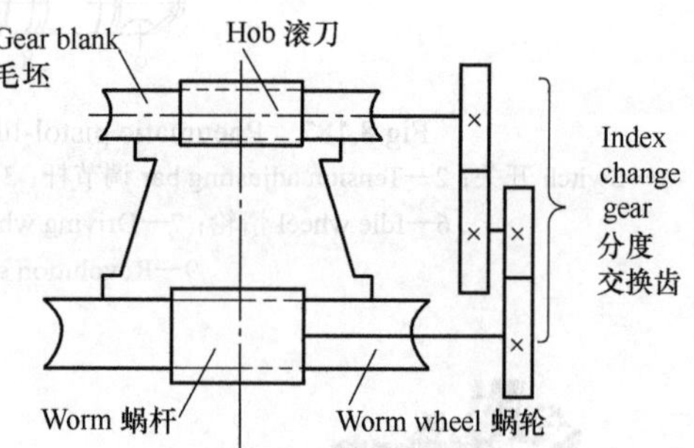

Fig 8.199 Elementary hobbing machine setup
滚齿机基本结构

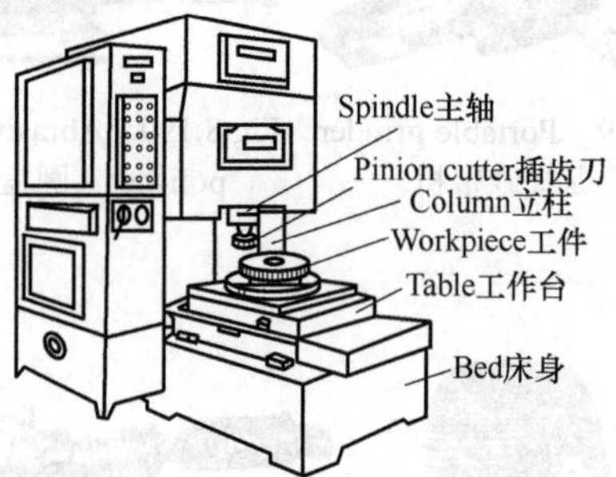

Fig 8.200 Gear shaper 插齿机

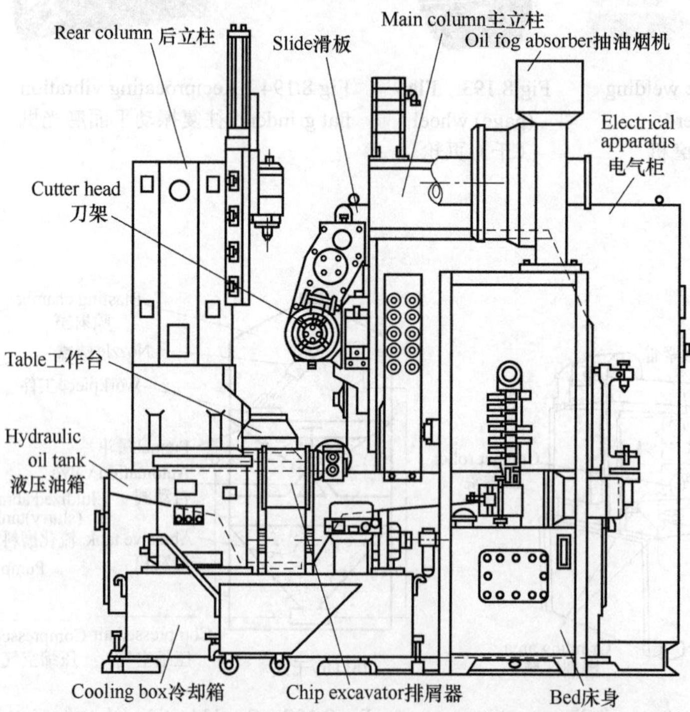

Fig 8.201 NC high speed gear cutting machine 数控高速滚齿机

8.3.8 Automatic screw machines 自动螺纹加工机床

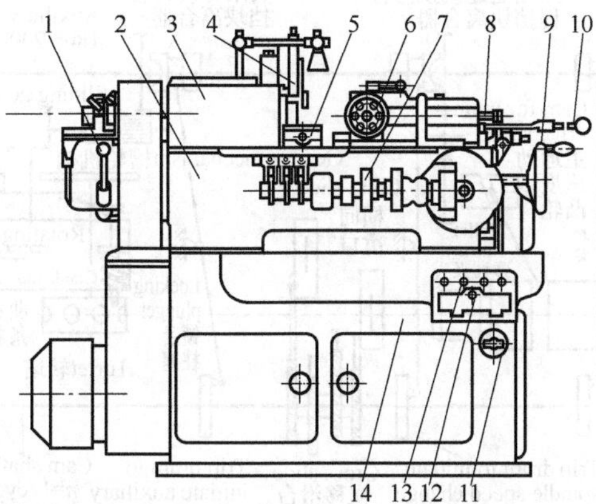

Fig 8.202 General view of the automatic screw machine 自动螺纹加工机外观图

1—Lever to engage auxiliary shaft 辅助轴操控杆；2—Bed 床身；3—Headstock 床头箱；4—Tool slide (vertical) 刀具滑台（立式）；5—Turret-tool slide (horizontal) 转搭刀具滑台（卧式）；6—Turret slide 转塔滑台；7—Main cam shaft 主控凸轮轴；8—Adjustable rod for positioning turret slide with respect to spindle nose 转塔滑台与主轴端调整杆；9—Hand wheel to rotate auxiliary shaft 副轴旋转手轮；10—Lever to traverse turret slide 转塔滑台横移操纵杆；11—Rotary switch 回转开关；12—Console panel for setting up spindle speeds 设置主轴速度操控盘；13—Push button controls of spindle drive 主轴传动控制按钮；14—Base 基座

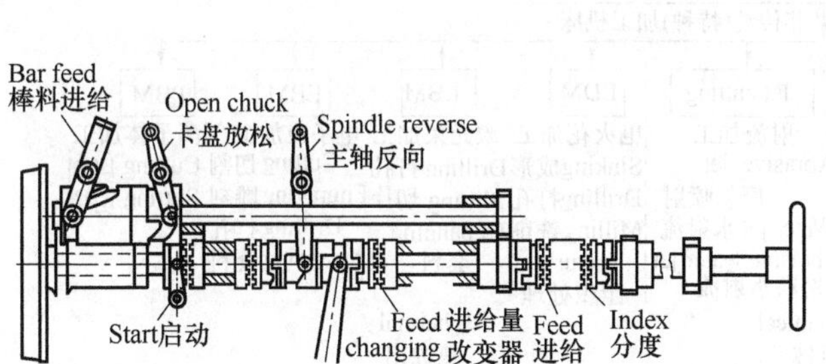

Fig 8.203 The auxiliary shaft assembly 副轴装配图

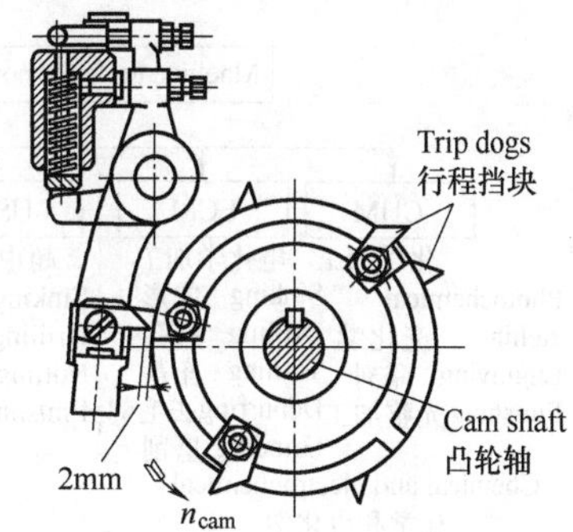

Fig 8.204 Turret indexing trips 转塔分度行程

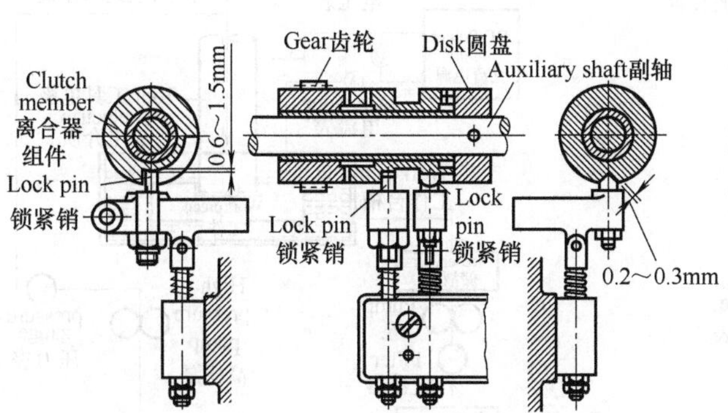

Fig 8.205 Details of dog clutch 挡块离合器详图

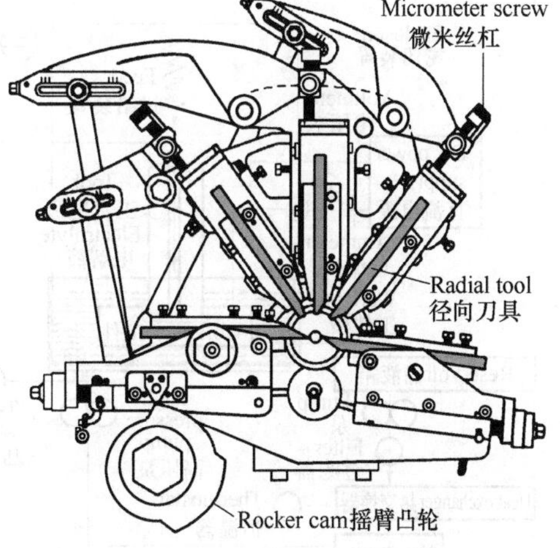

Fig 8.206 Swiss-type automatic screw machine 精密自动螺纹机

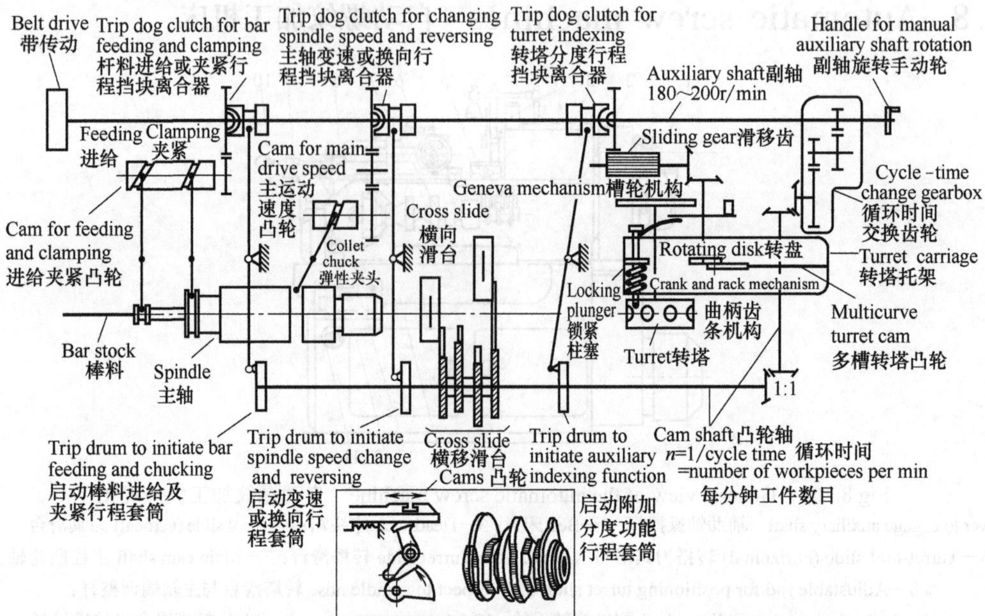

Fig 8.207 Control of automatic screw machine 自动螺纹机床的控制

8.4 Non-traditional machine tools 非传统（特种）加工机床

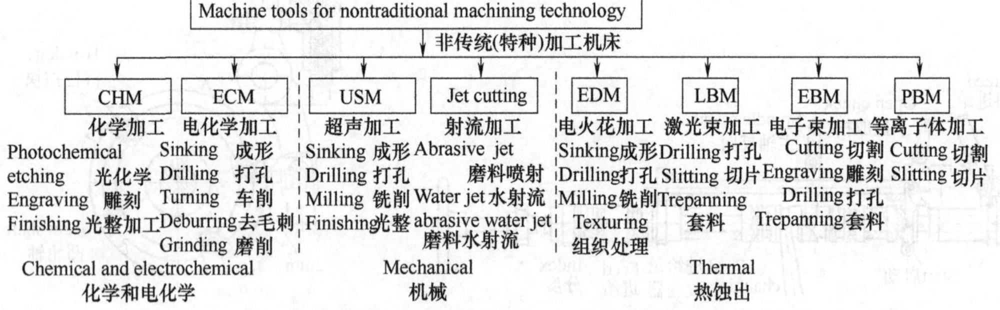

Fig 8.208 Classification of machine tools for non-traditional machining technology
非传统加工（特种加工）技术机床类别

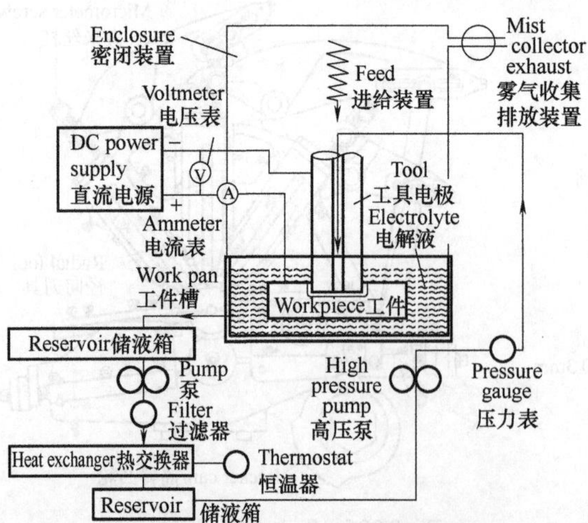

Fig 8.209 Various elements present in a commercial ECM machine 商用电解加工机床构成图解

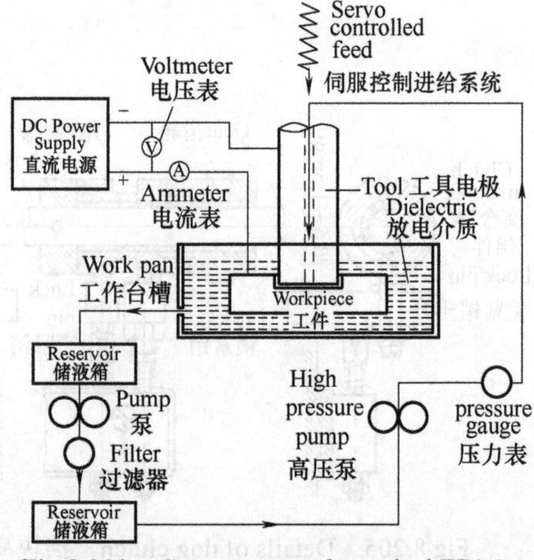

Fig 8.210 Components of a typical EDM machine 典型电火花机床结构

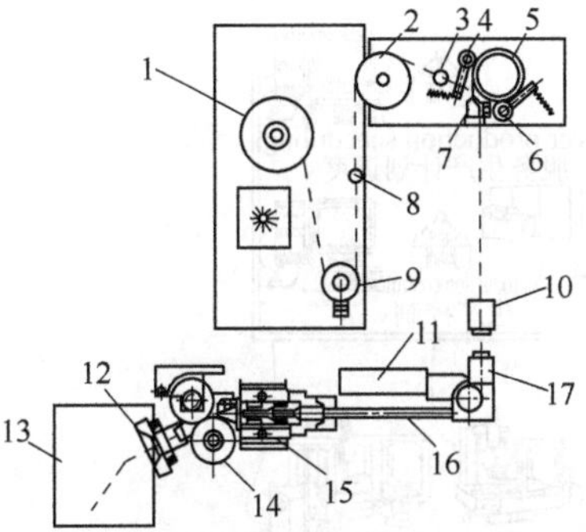

Fig 8.211 Lower speed wire feeding mechanism
低速走丝机构

1—Wire spool 丝筒；2—Pulley 滑轮；3—Felt roller 毡轮；4，6—Press wheel 压轮；5—Tension controller 张紧机构；7—Wire breakage detector 断丝检测器；8—Guide hook 导向钩；9—Spindlebushing 轴套；10—Upper wire guide 上导向器；11—Wire dismantling arm 线丝拆除臂；12—Exit of used wire 线丝排除口；13—Used wire spool 废丝筒；14—Roller 滚轮；15—Pull part 牵引部件；16—Guide tube for wire exit 出丝导管；17—Lower wire guide 下导向器

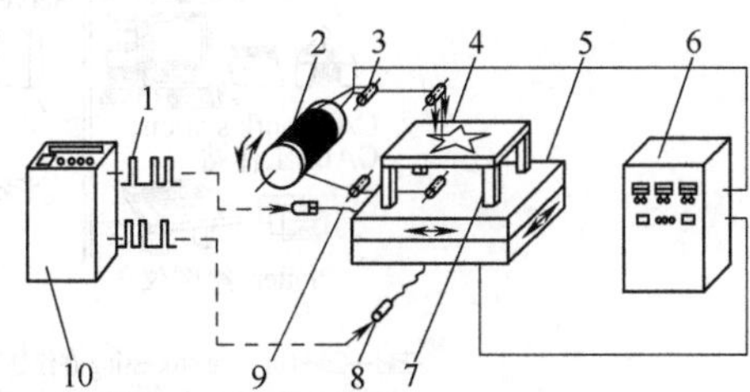

Fig 8.212 Principle of NC wire EDM process
数控线切割工作原理

1—Electric pulse signal 电脉冲信号；2—Wire spool 丝筒；3—Guide wheel 导轮；4—Work piece 工件；5—Work table 工作台；6—Pulse power 脉冲电源；7—Pad 垫铁；8—Stepped motor 步进电动机；9—Lead screw 丝杠；10—Computer controller 计算机控制器

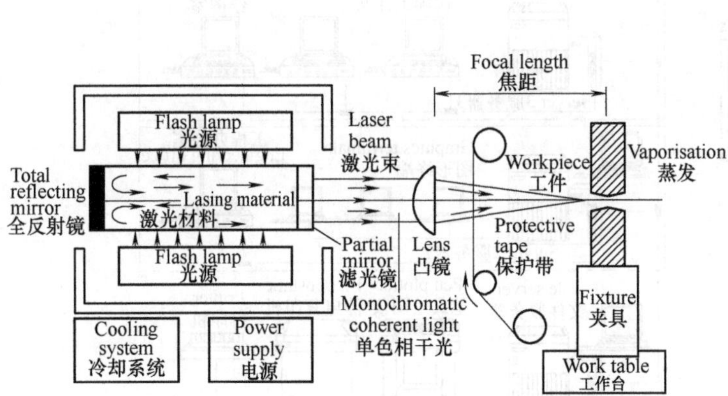

Fig 8.213 Schematic set up of a laser drilling operation
激光钻孔加工装置

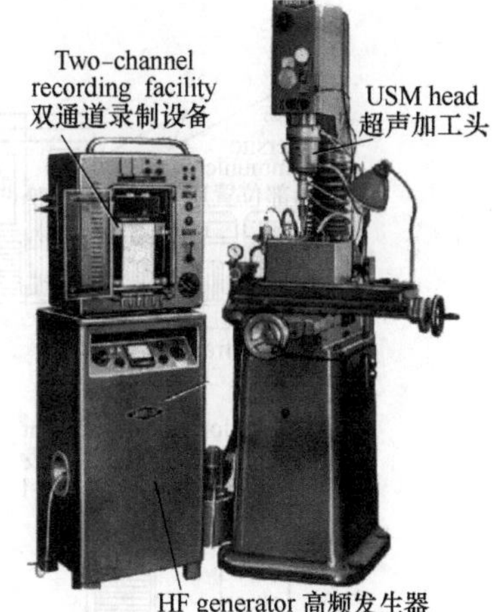

Fig 8.214 USM equipment
超声加工设备

8.5 CAD/CAM/CAPP/FMS/CIMS 与计算机相关的先进系统

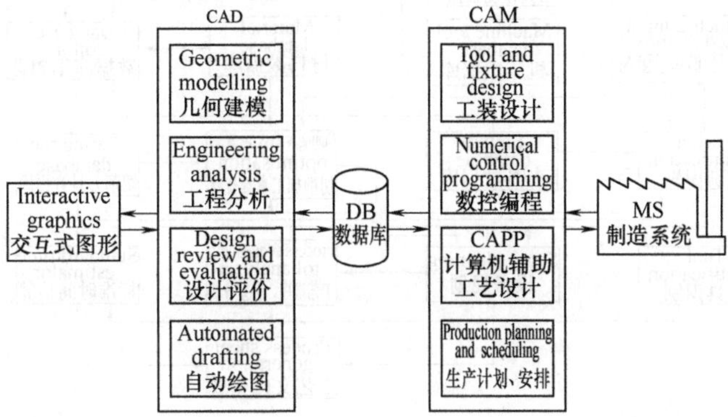

Fig 8.215 Database to CAD/CAM CAD/CAM 数据库

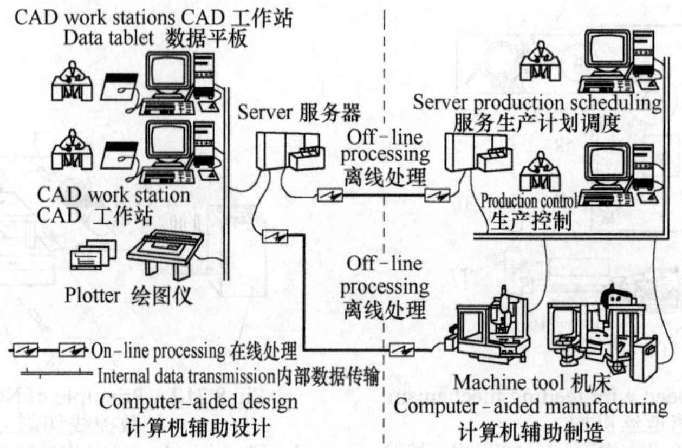

Fig 8.216 Information flow chart in CAD/CAM application CAD/CAM 信息流程图

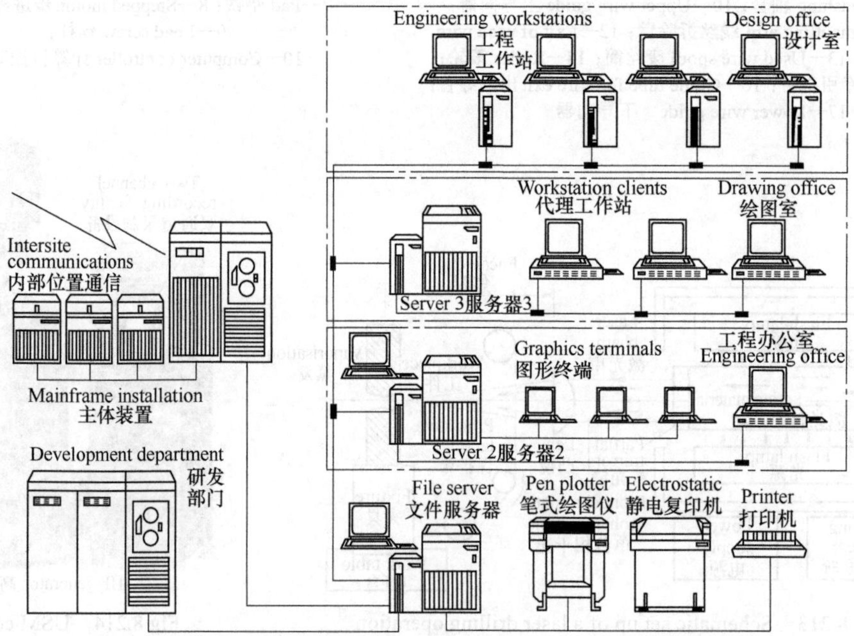

Fig 8.217 An integraled CAD/CAM installation CAD/CAM 集成装置

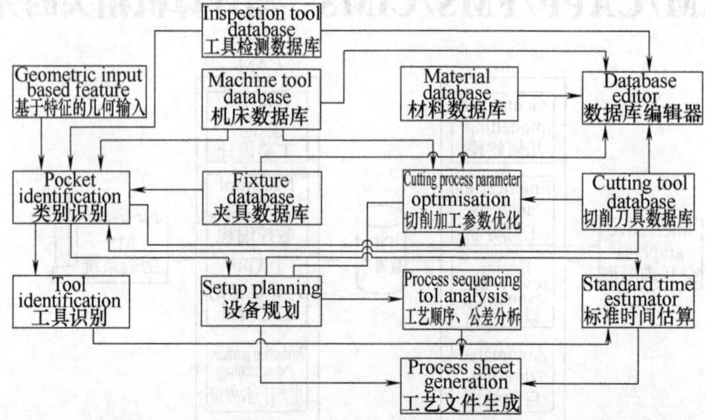

Fig 8.218 Architecture of a CAPP system CAPP 系统构成

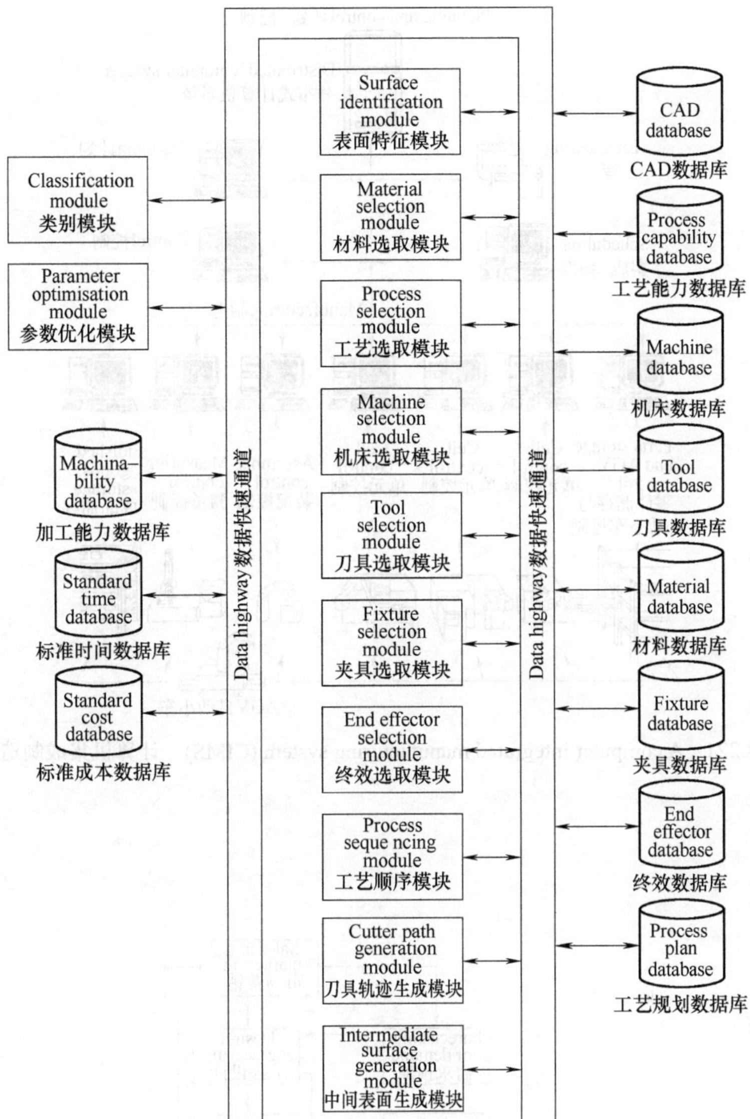

Fig 8.219 Modules and relationship of typical generative CAPP 生成式 CAPP 系统的模块构成及相互关系

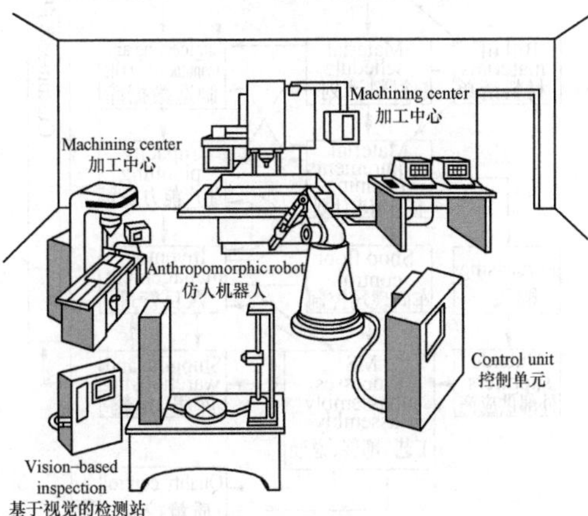

Fig 8.220 Example of FMC composed of two machines, automated part inspection and a serving robot
由两台机床、工件自动检测和伺服机器人组成的柔性制造单元

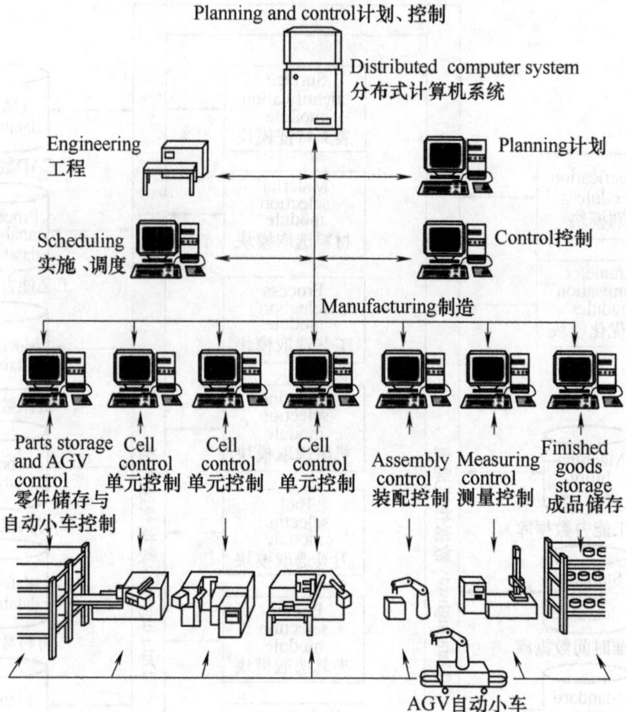

Fig 8.221 A computer integrated manufacturing system (CIMS) 计算机集成制造系统

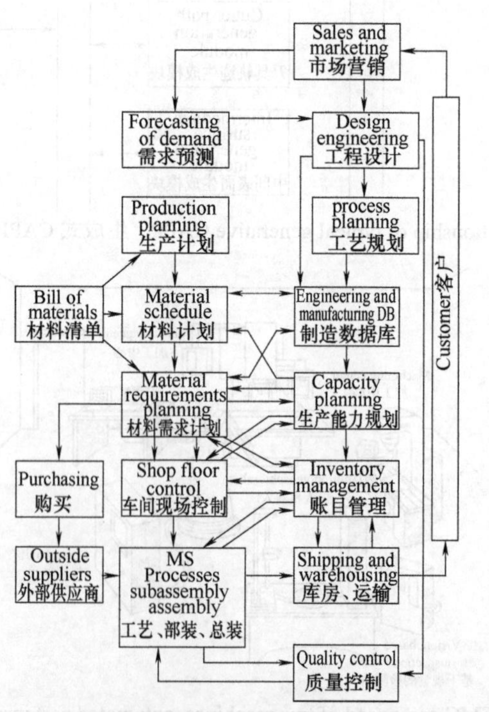

Fig 8.222 Cycles of activities in CIMS 集成制造系统的活动循环

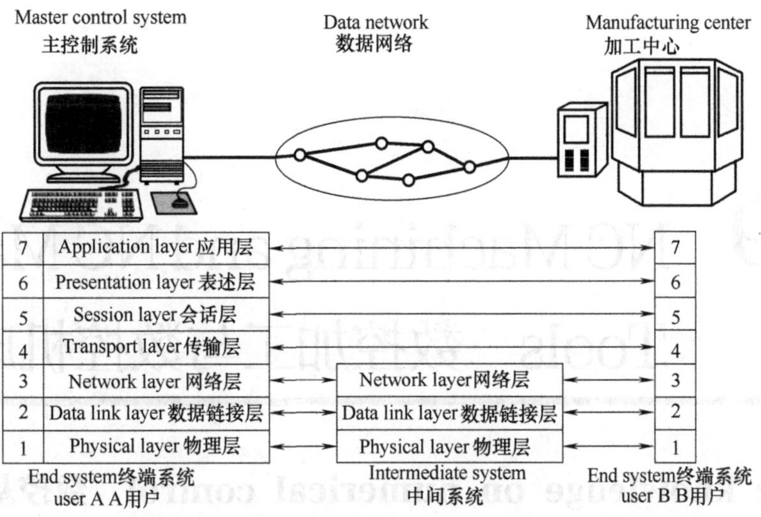

Fig 8.223 The ISO/OSI reference model for open communication 开放式通信 ISO/OSI 参考模型

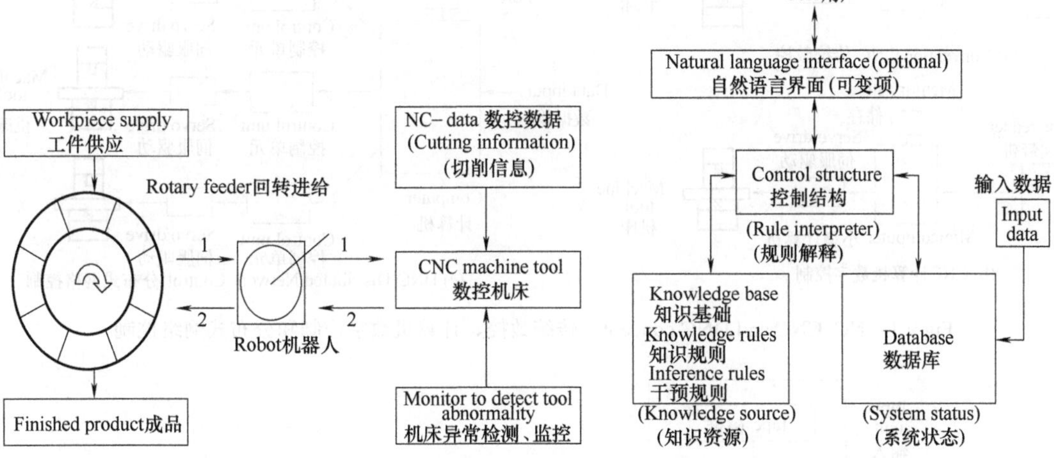

Fig 8.224 Unmanned manufacturing cell 无人制造单元

1—Loading 工件安装；2—Unloading 工件拆卸

Fig 8.225 Basic structure of an expert system 专家系统基本结构

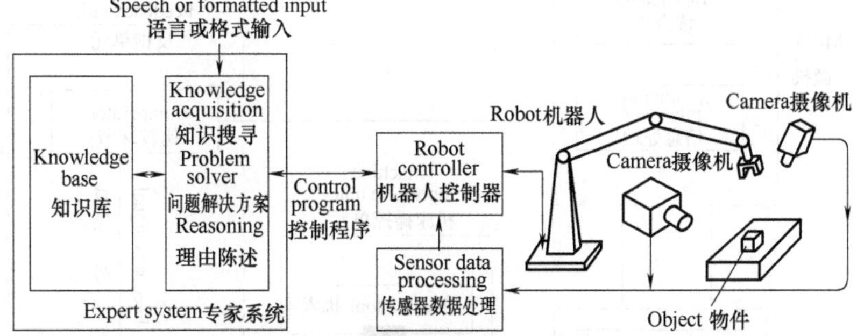

Fig 8.226 Expert system, as applied to an industrial robot guided by machine vision 专家系统用于视觉引导的工业机器人

Unit 9 NC Machining and NC Machine Tools 数控加工与数控机床

9.1 Basic knowledge on numerical control 数控基本知识

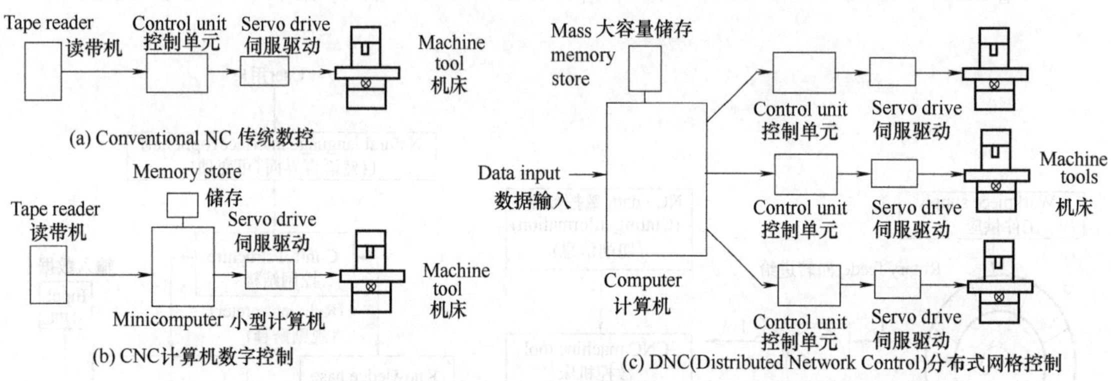

Fig 9.1 NC, CNC and DNC concept 传统数控、计算机数字控制和分布式网络控制

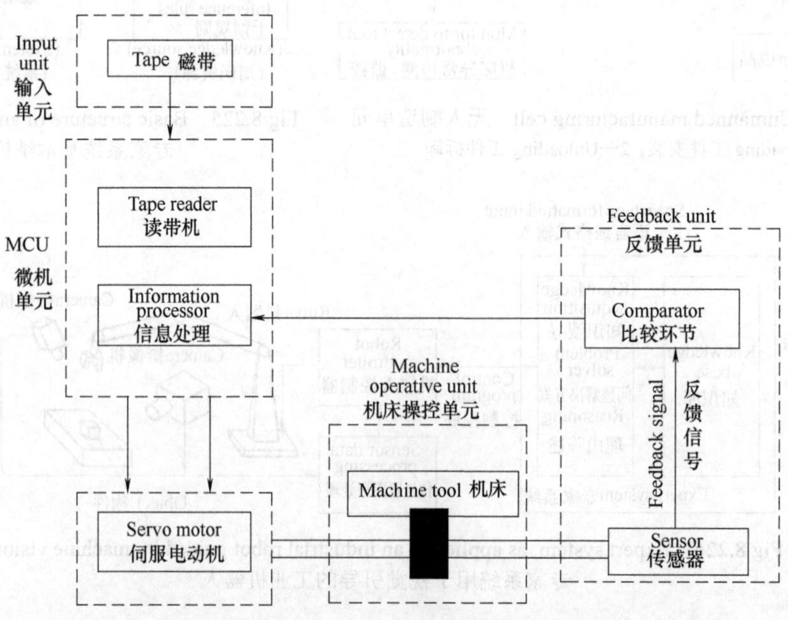

Fig 9.2 Main components of the NC system 数控系统主要组成

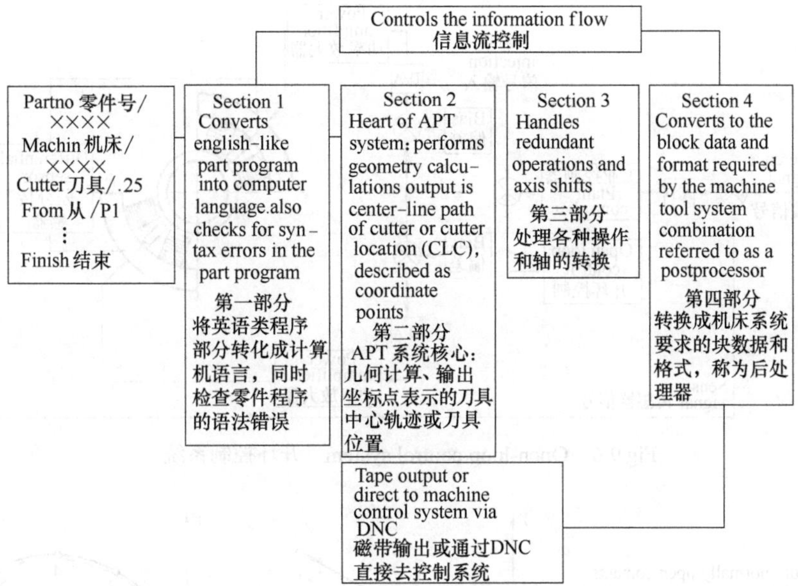

Fig 9.3 Automatically programmed tool (APT) system 自动编程工具（APT）系统

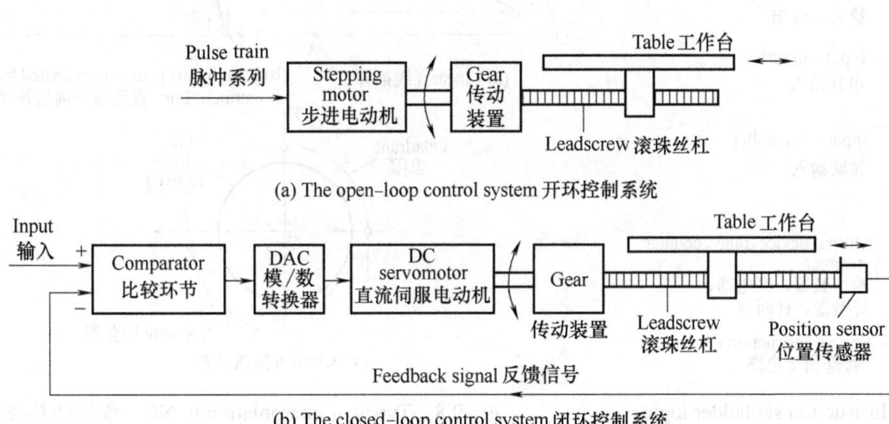

Fig 9.4 The control systems for a NC machine 数控机床的控制系统

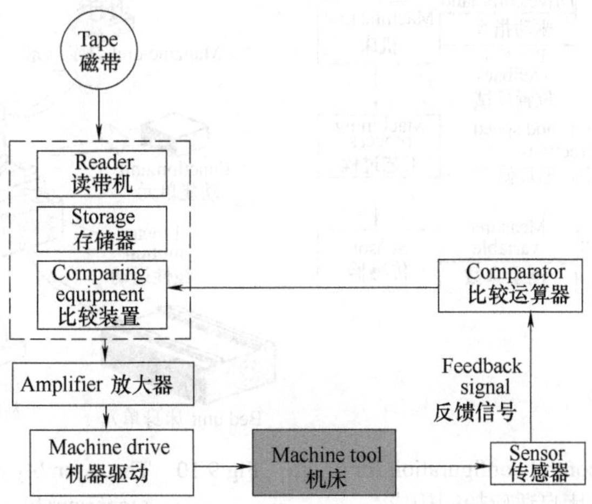

Fig 9.5 Closed-loop control system 闭环控制系统

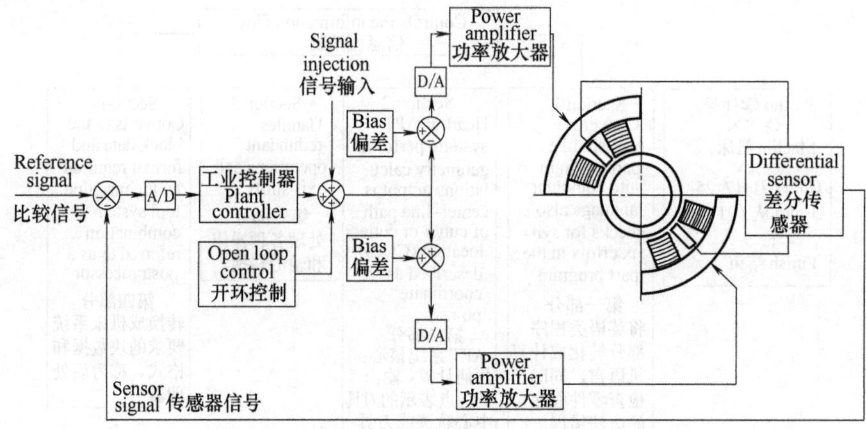

Fig 9.6 Open-loop control system 开环控制系统

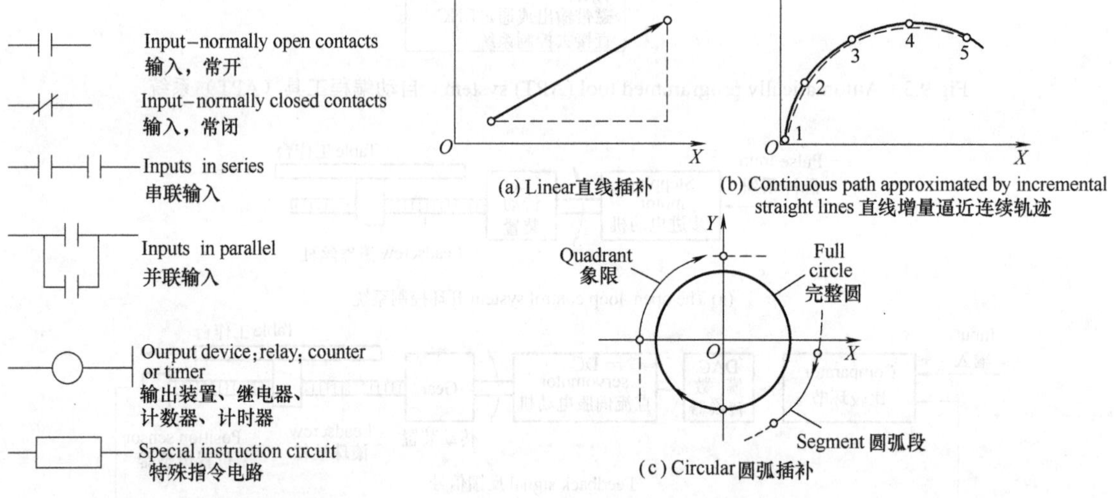

Fig 9.7 Instruction set ladder logic symbols 指令装置梯形逻辑符号

Fig 9.8 Types of interpolation in NC 数控插补类型

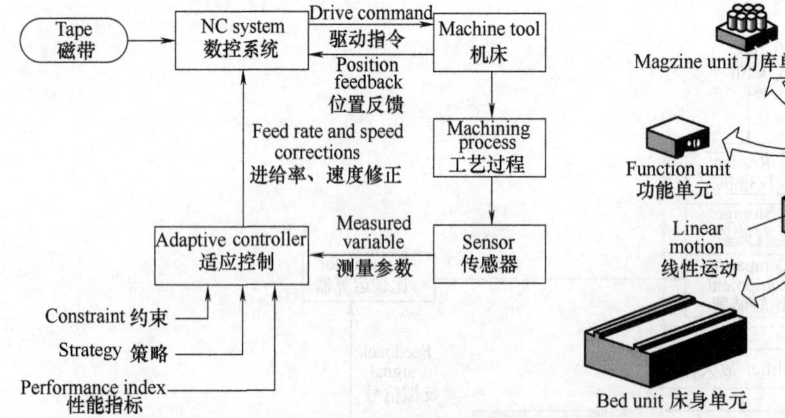

Fig 9.9 Typical adaptive control configuration for a machine tool 典型机床自适应控制构成

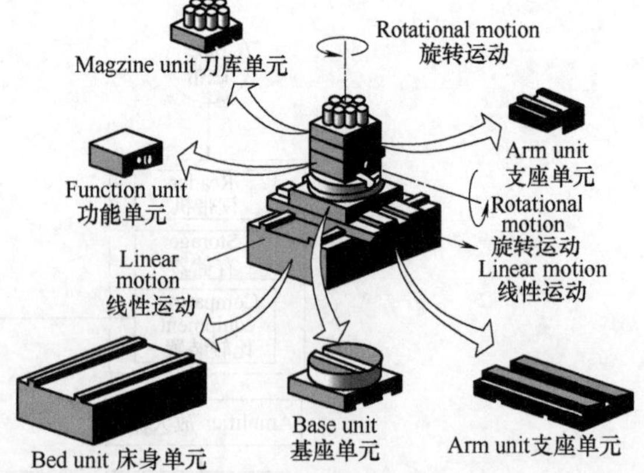

Fig 9.10 The assembly of different components of a reconigurable machining center 可重构的加工中心不同部件的组合

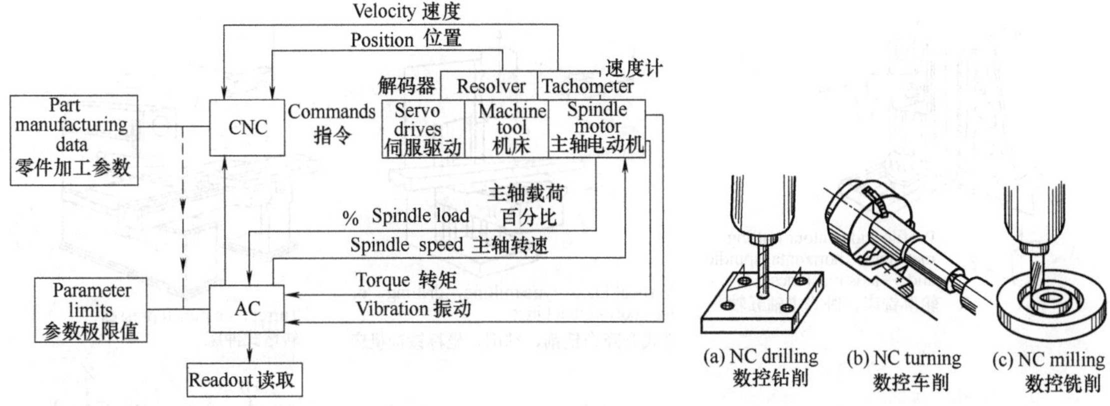

Fig 9.11 Application of adaptive control (AC) for a turning operation 适应控制应用于车削加工

Fig 9.12 Types of NC (numerical control) machining 数控加工的类型（一）

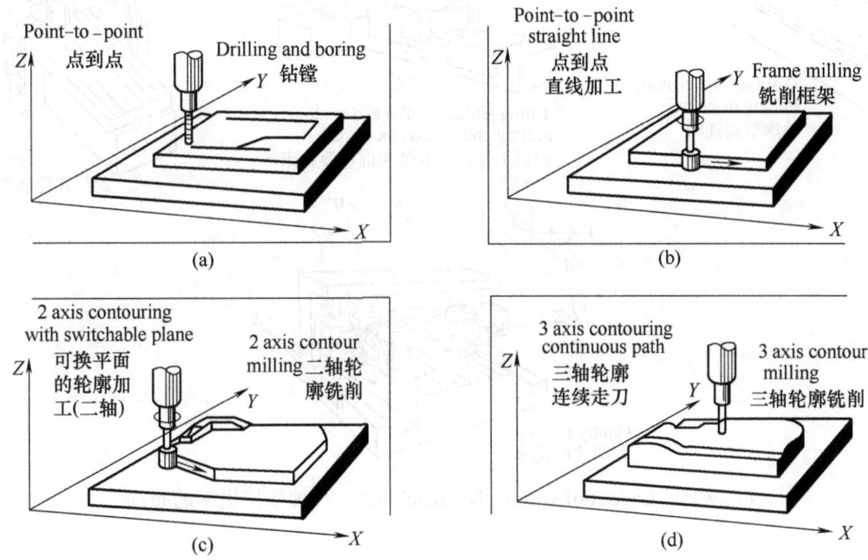

Fig 9.13 Types of NC (numerical control) machining 数控加工的类型（二）

9.2 NC machining tools 数控加工机床

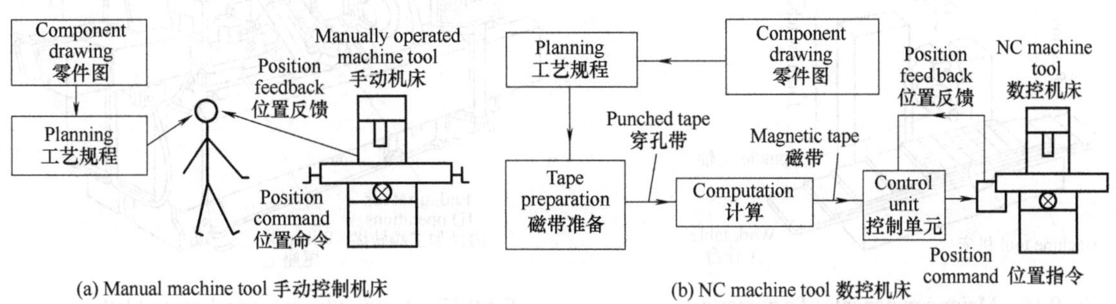

Fig 9.14 Operator-controlled and numerically controlled machine tools 手动机床和数控机床

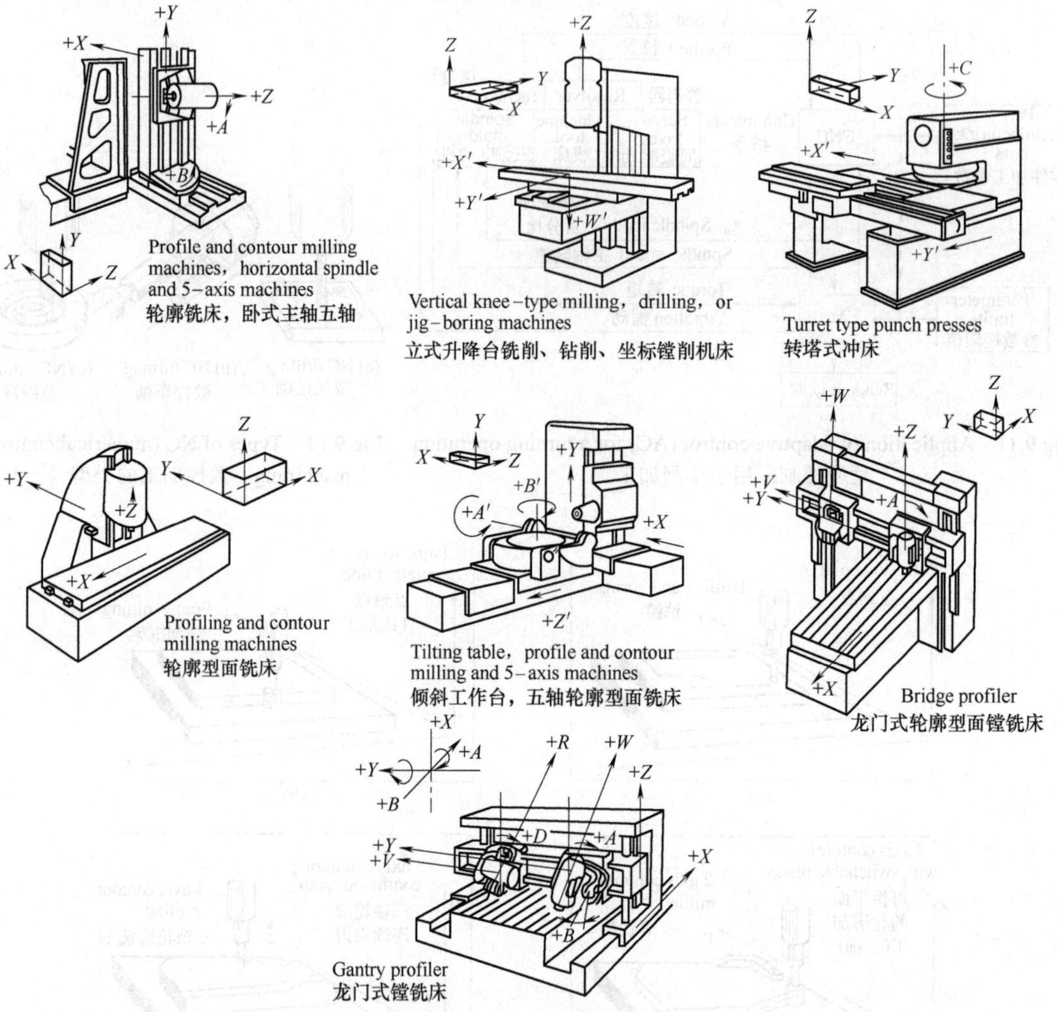

+Y
+X
+Z
+A
+B
Y
X
Z
Profile and contour milling machines, horizontal spindle and 5-axis machines
轮廓铣床，卧式主轴五轴
+Z
Z
Y
X
+X′
+Y′
+W′
Vertical knee-type milling, drilling, or jig-boring machines
立式升降台铣削、钻削、坐标镗削机床
Z
+C
Y
X
+X′
+Y′
Turret type punch presses
转塔式冲床
Z
+Y
Y
X
+Z
+X
Profiling and contour milling machines
轮廓型面铣床
Y
X
Z
+Y
+B′
+A′
+X
+Z′
Tilting table, profile and contour milling and 5-axis machines
倾斜工作台，五轴轮廓型面铣床
+W
Z
Y
X
+Z
+V
+Y
+A
+X
Bridge profiler
龙门式轮廓型面镗铣床
+X
+A
+Y
+B
+R
+W
+Z
+Y
+V
+D
+A
+X
+B
Gantry profiler
龙门式镗铣床

Fig 9.15 Layout of various NC machines 几种数控机床的布局

Computer:
Input commands,
Processing,
Output commands
计算机：输入指令、
工艺规划、输出指令
Position feedback 位置反馈
Drive signals
驱动信号
Limit switches 极限开关
Spindle 主轴
Work table
工作台
Machine tool 机床

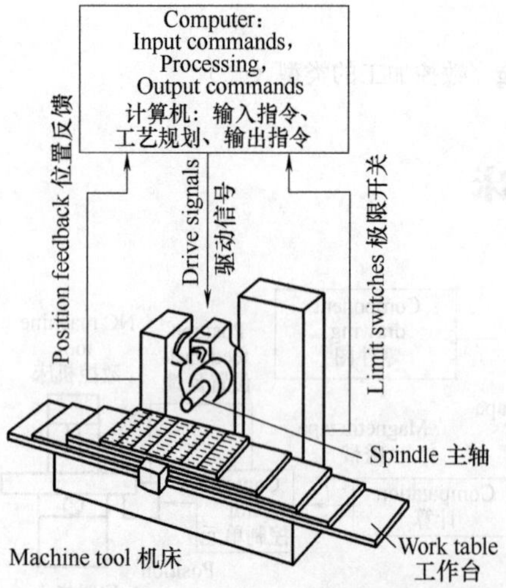

Fig 9.16 Major components of a numerical-control machine tool 数控机床的主要部件

Chuck 卡盘
Round turret for
OD operations
外径加工圆形转塔
CNC unit
数控单元
End turret for
ID operations
内径加工端转塔
Tailstock
尾座

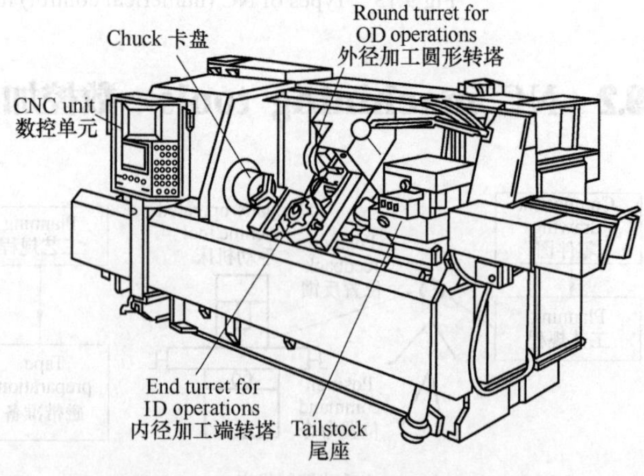

Fig 9.17 A computer numerical control lathe 计算机数控车床

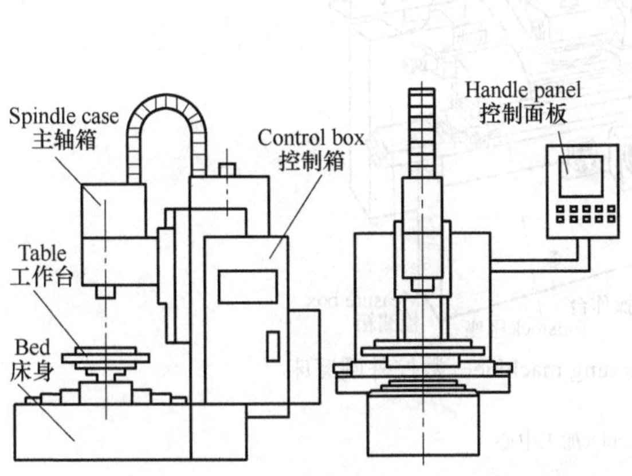

Fig 9.18 Layout of vertical NC milling, boring, drilling complex machine 立式数控铣、镗、钻复合机床

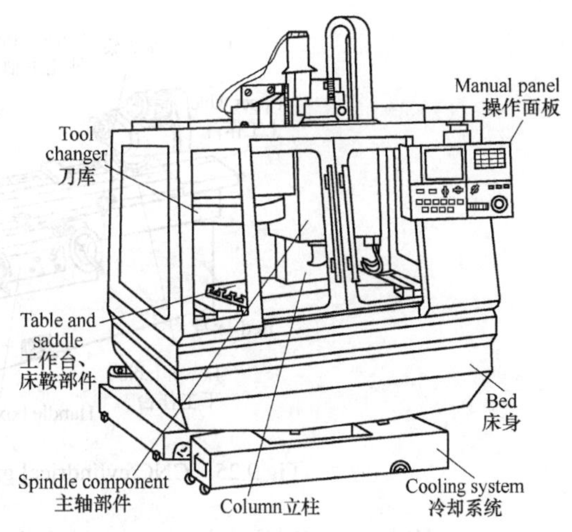

Fig 9.19 Vertical machining centre 立式加工中心

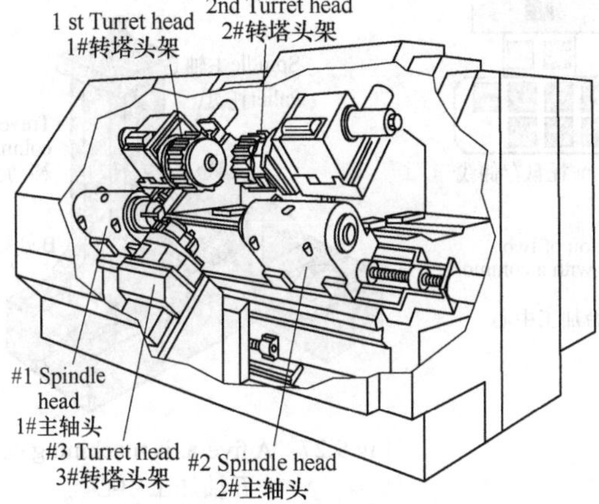

Fig 9.20 A three-turret, two-spindle computer numerical controlled turning center
三个转塔刀架、二轴计算机数控车削中心

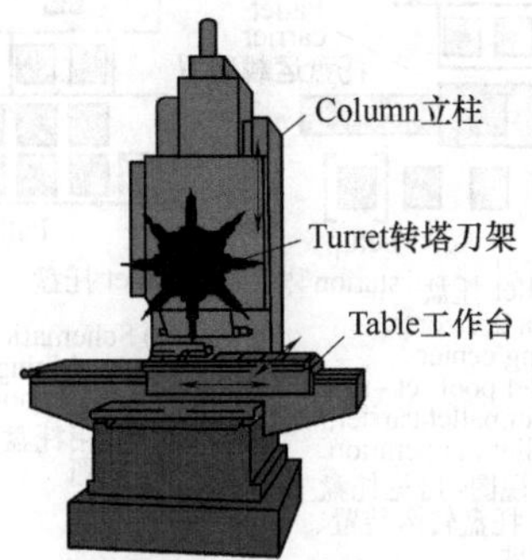

Fig 9.21 A three-axis computer numerical control (CNC) drilling machine 三轴数控钻床

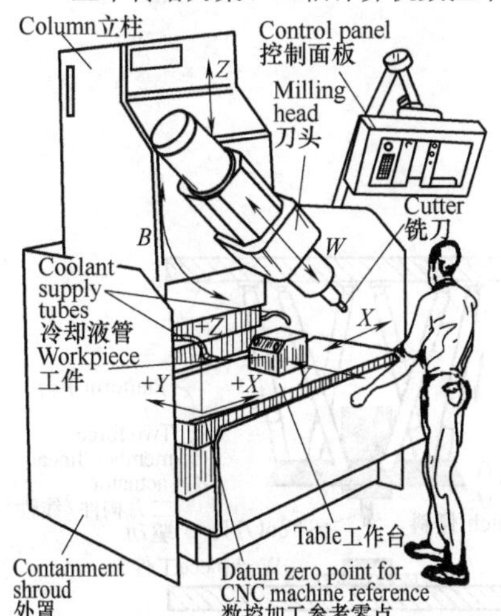

Fig 9.22 CNC milling machine 数控铣床

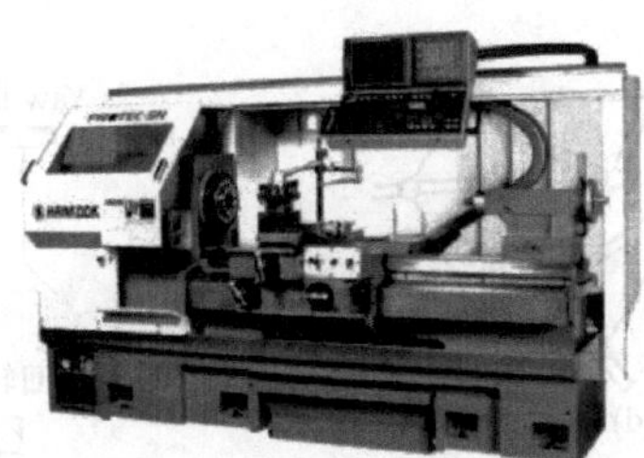

Fig 9.23 CNC horizental lathe 卧式数控车床

Fig 9.24 CNC vertical lathe 立式数控车床

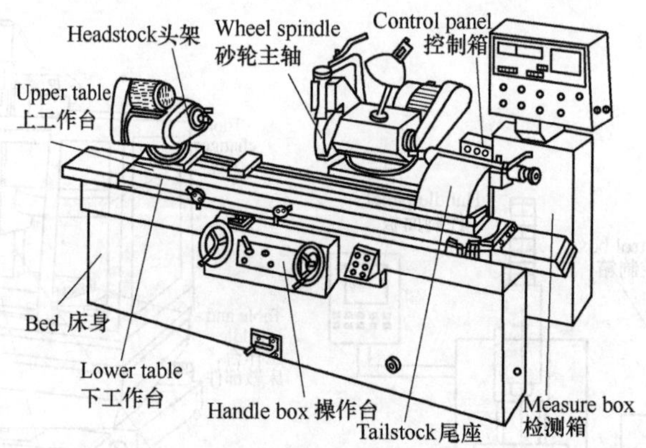

Fig 9.25 CNC cylindrical grinding machine 数控外圆磨床

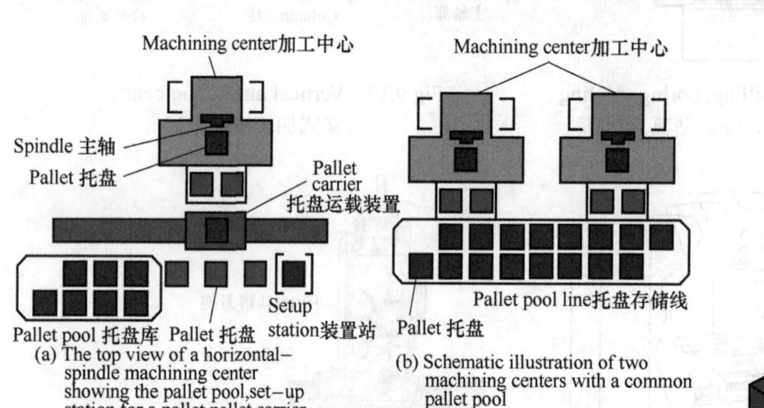

(a) The top view of a horizontal-spindle machining center showing the pallet pool, set-up station for a pallet, pallet carrier, and an active pallet in operation 立式加工中心俯视图：可见托盘库、托盘装置站、托盘转运装置、工作中的活动托盘

(b) Schematic illustration of two machining centers with a common pallet pool 共用托盘库的两台加工中心视图

Fig 9.26 Pallets 托盘装置

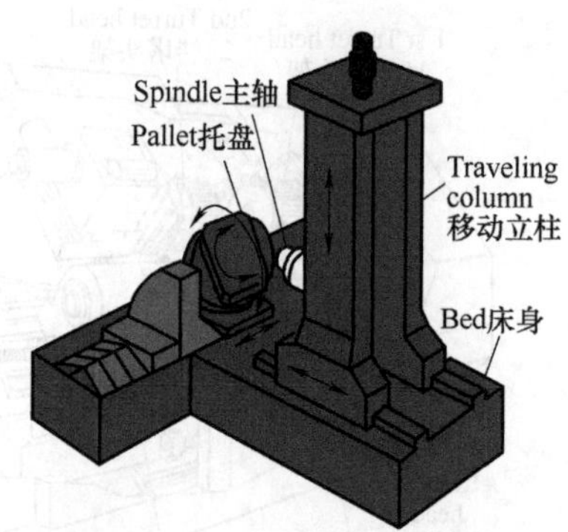

Fig 9.27 A five-axis machining center 五轴加工中心

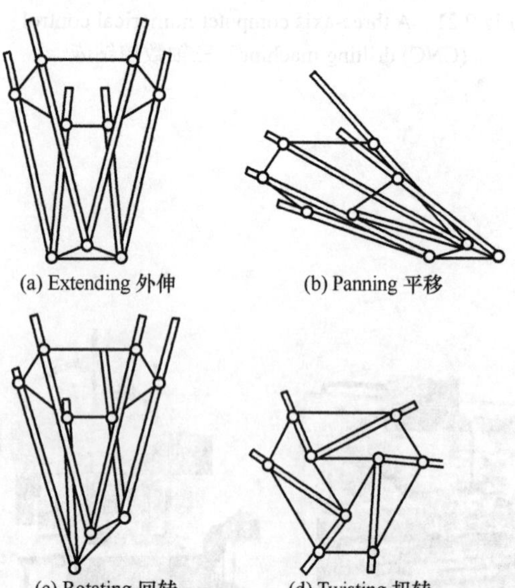

(a) Extending 外伸

(b) Panning 平移

(c) Rotating 回转

(d) Twisting 扭转

Fig 9.28 Models of hexapod movements 六底座运动模型

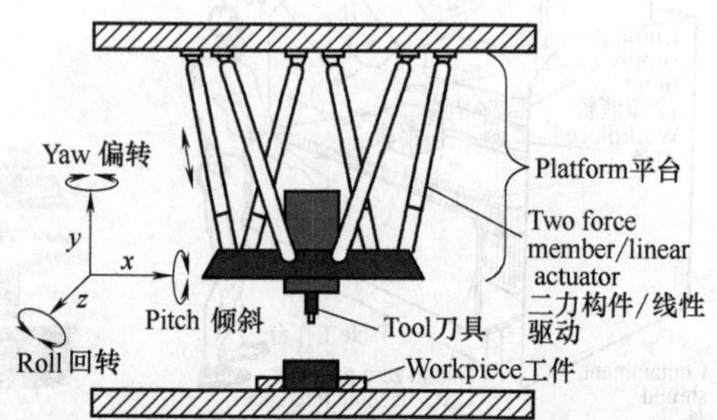

Fig 9.29 Hexapod of telescopic struts, ingersoll system 英格索兰伸缩构件的六底座机构

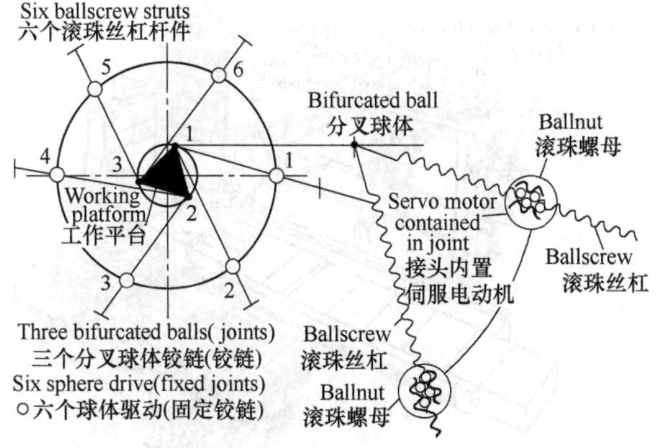

Fig 9.30 Ball screw hexapod 滚珠丝杠六节点底座机床

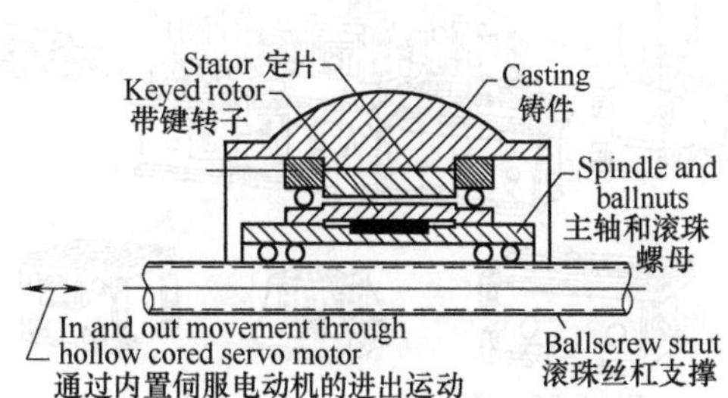

Fig 9.31 Spheredrive 球体传动

(a) Bifurcated ball 二分叉球体

(b) Split bifurcated ball arrangement 剖分二分叉球体

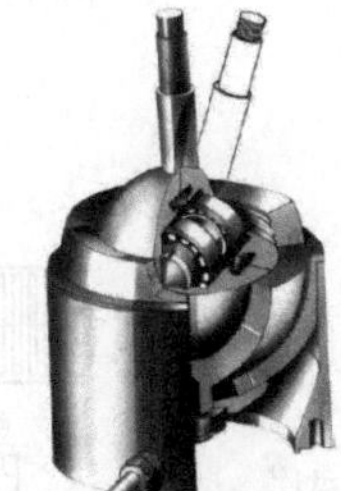

(c) Hydrostatic bifurcated ball joint 静压二分叉球体铰链

(d) Magnetic bifurcated ball joint 磁性二分叉球体铰链

Fig 9.32 Ball joint 球铰链

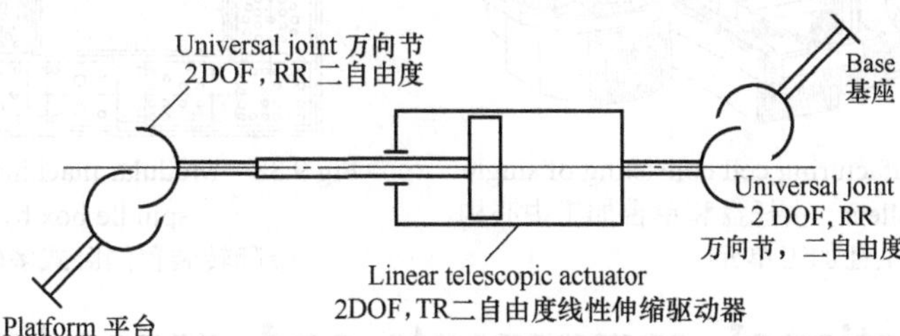

Fig 9.33 Telescopic struts with universal joint 万向节伸缩杆件

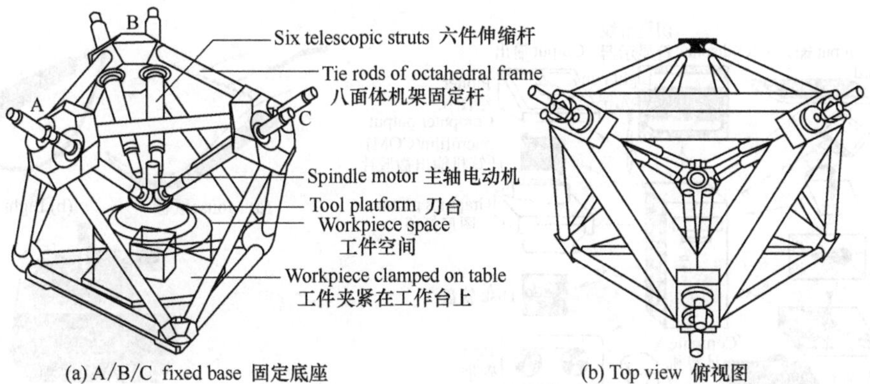

(a) A/B/C fixed base 固定底座

(b) Top view 俯视图

Fig 9.34 Hexpad with rigid frame 刚性结构的六底座

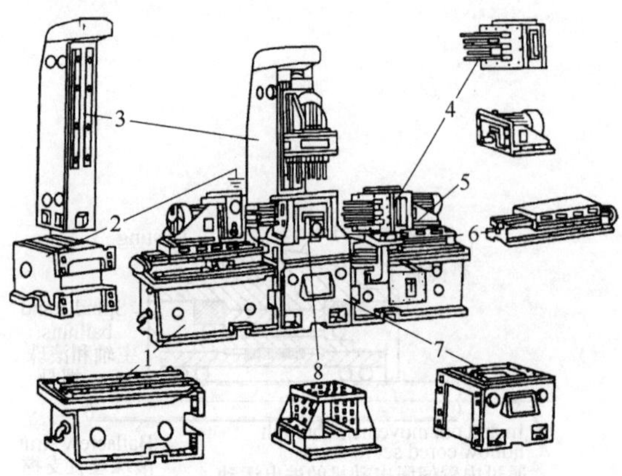

Fig 9.35 Components of modular machine
模块组合机床的组成

1—Left base 左底座；2—Base for column 立柱底座；
3—Right base 右底座；4—Spindle case 主轴箱；
5—Power driving box 动力箱；6—Sliding table 滑台；
7—Central base 中央底座；8—Jig and fixture 夹具

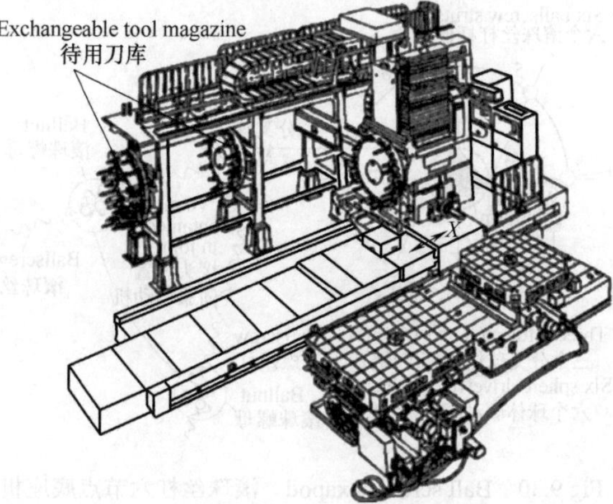

Fig 9.36 Horizontal machining centre with exchangeable tool changer 有可换刀库的卧式加工中心

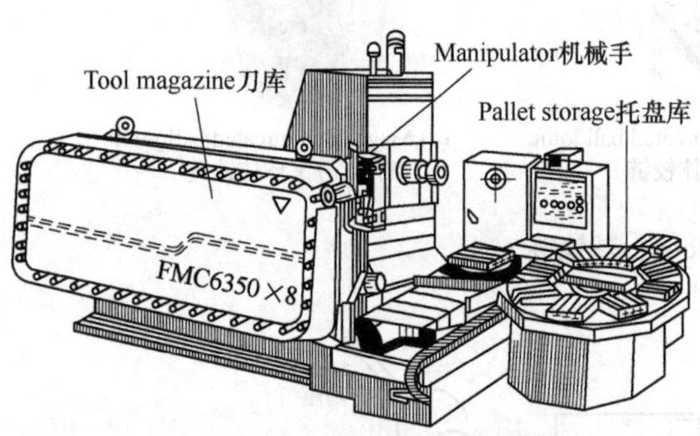

Fig 9.37 Flexible manufacturing cell consisting of single machining center with pallets 由托盘和单台加工中心构成的柔性加工单元

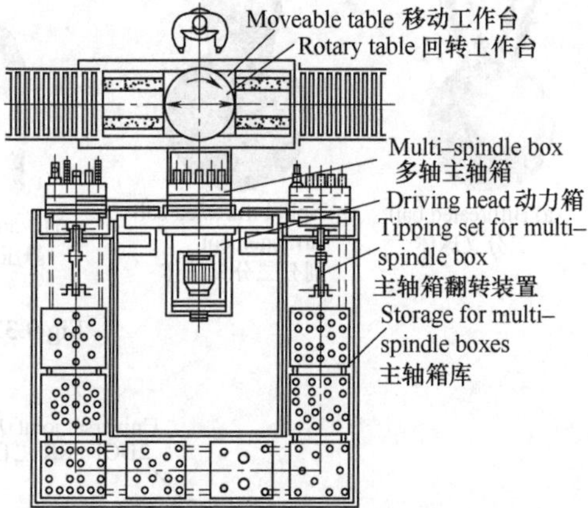

Fig 9.38 Modular machine with changeable multi-spindle box by tipping set
带翻转装置的卧式多轴可换箱组合机床

9.3 NC functional components and appendixes 数控机床功能部件及附件

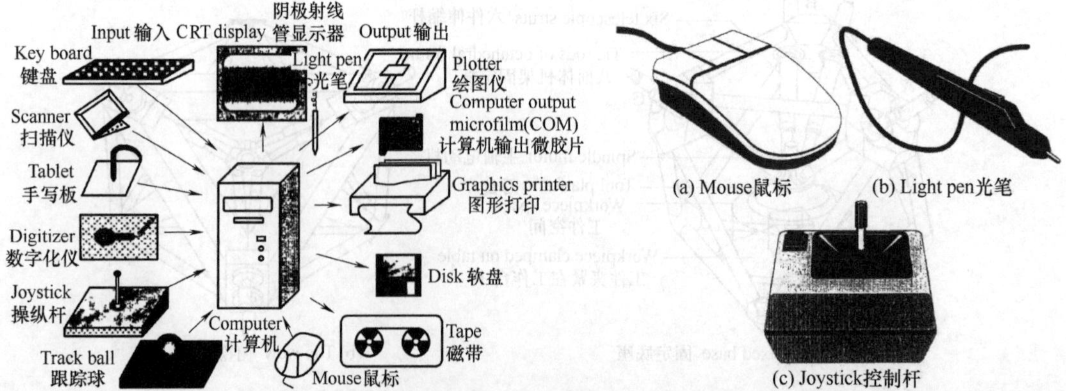

Fig 9.39 I/O devices of a CAD system CAD 系统输入 / 输出装置

Fig 9.40 Interactive devices 互动装置

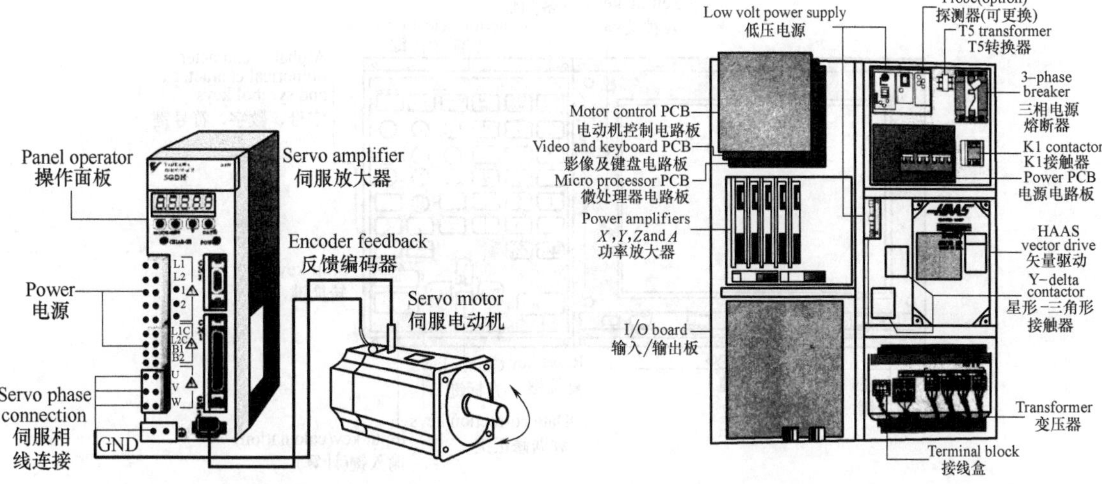

Fig 9.41 Servo driver 伺服驱动器

Fig 9.42 Electrical cabinet 电气柜

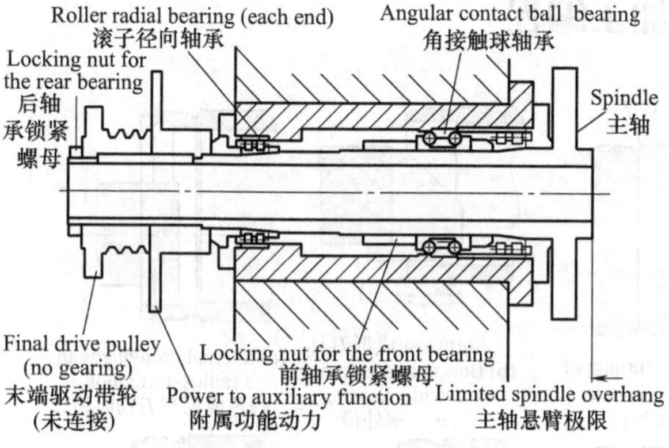

Fig 9.43 Spindle assembly for NC machine tools
数控机床主轴组件

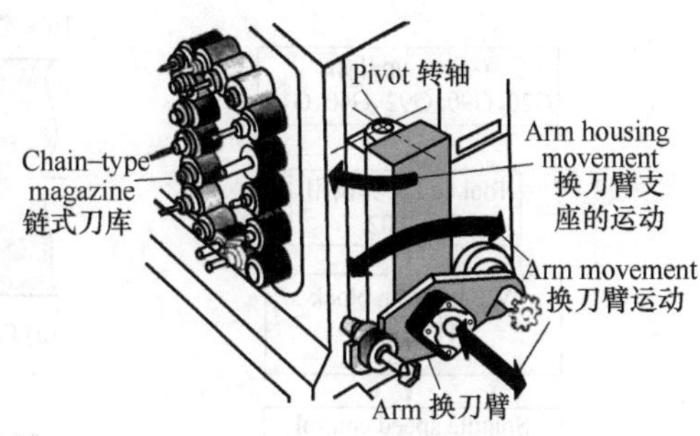

Fig 9.44 Chain-type magazine with automatic tool changer 带自动换刀的链式刀库

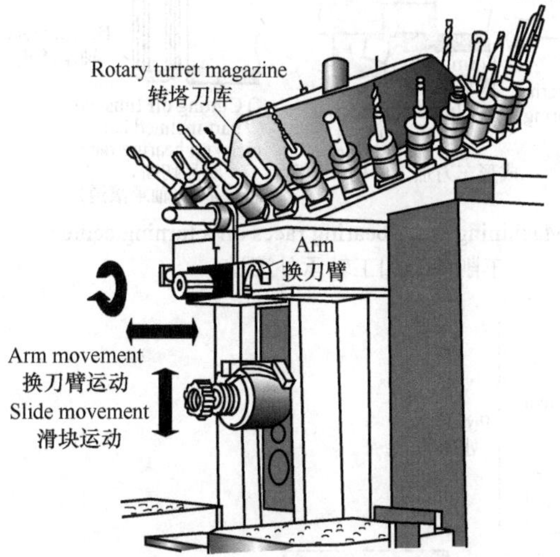

Fig 9.45 Rotary-type magazine with automatic tool changer 带自动换刀装置的回转刀库

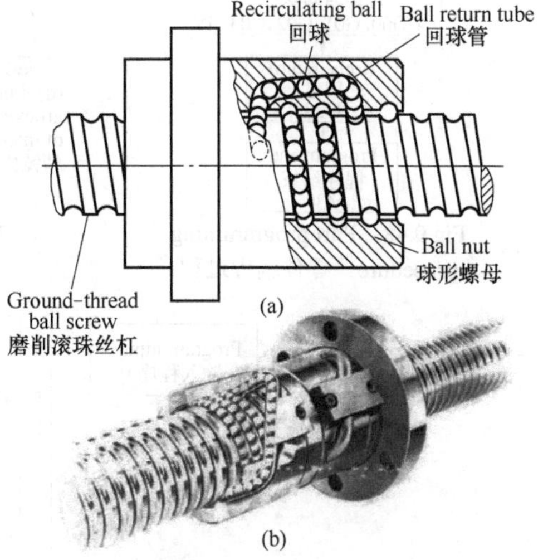

Fig 9.46 External recirculating ball screw
外循环滚珠丝杠

Ready LED
准备闪灯
Setting keys
各种按键
Function selection keys
功能选择键
Alphabet character,
numerical character ,
and symbol keys
字母、数字、符号键
Shift key
转换键
Menu keys
菜单键
Page key
翻页键
Reset key
复位键
Cursor keys
光标键
Data correction keys
数据修正键
Input key(calculation)
输入键(计算)

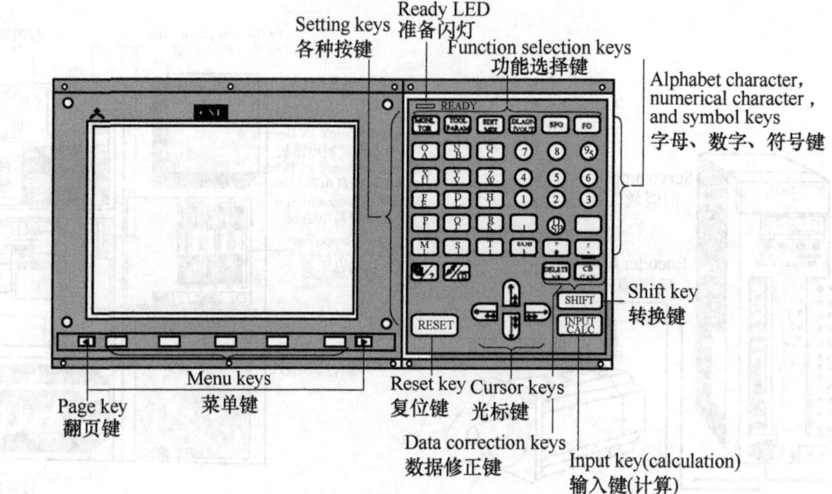

Fig 9.47 CNC control panel 数控控制面板

9.4 NC programming 数控加工编程

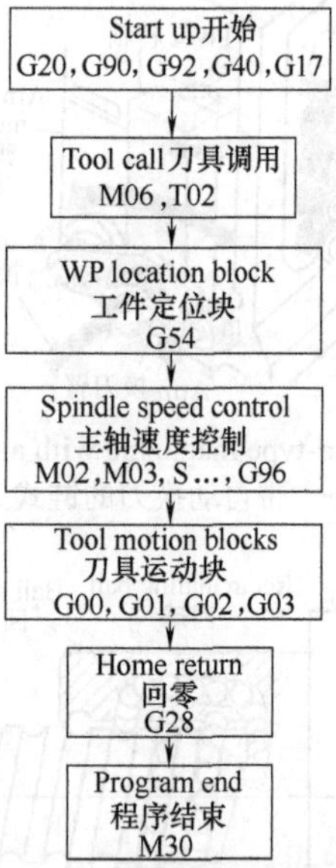

Start up开始
G20, G90, G92, G40, G17
Tool call刀具调用
M06, T02
WP location block
工件定位块
G54
Spindle speed control
主轴速度控制
M02, M03, S…, G96
Tool motion blocks
刀具运动块
G00, G01, G02, G03
Home return
回零
G28
Program end
程序结束
M30

Fig 9.48 Part programming procedure 零件编程过程

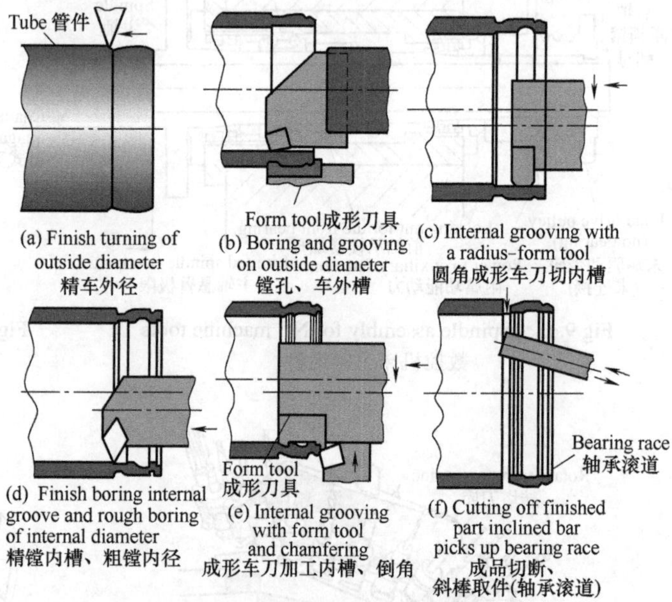

Tube 管件
(a) Finish turning of outside diameter
精车外径
Form tool成形刀具
(b) Boring and grooving on outside diameter
镗孔、车外槽
(c) Internal grooving with a radius-form tool
圆角成形车刀切内槽
(d) Finish boring internal groove and rough boring of internal diameter
精镗内槽、粗镗内径
Form tool
成形刀具
(e) Internal grooving with form tool and chamfering
成形车刀加工内槽、倒角
Bearing race
轴承滚道
(f) Cutting off finished part inclined bar picks up bearing race
成品切断、斜棒取件(轴承滚道)

Fig 9.49 Machining outer bearing races on a turning center 车削中心加工轴承外滚道

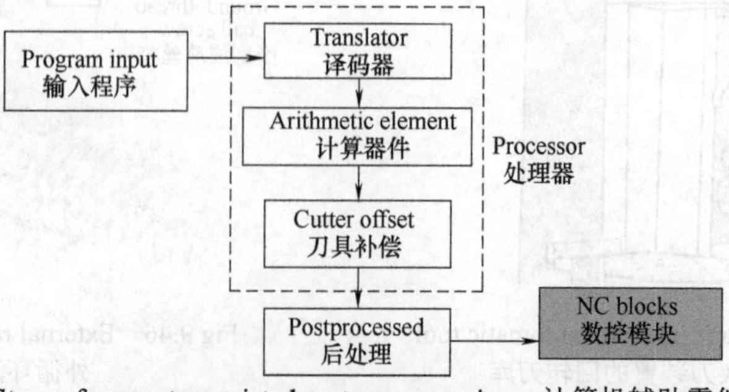

Program input
输入程序
Translator
译码器
Arithmetic element
计算器件
Cutter offset
刀具补偿
Processor
处理器
Postprocessed
后处理
NC blocks
数控模块

Fig 9.50 Steps of computer-assisted part programming 计算机辅助零件编程步骤

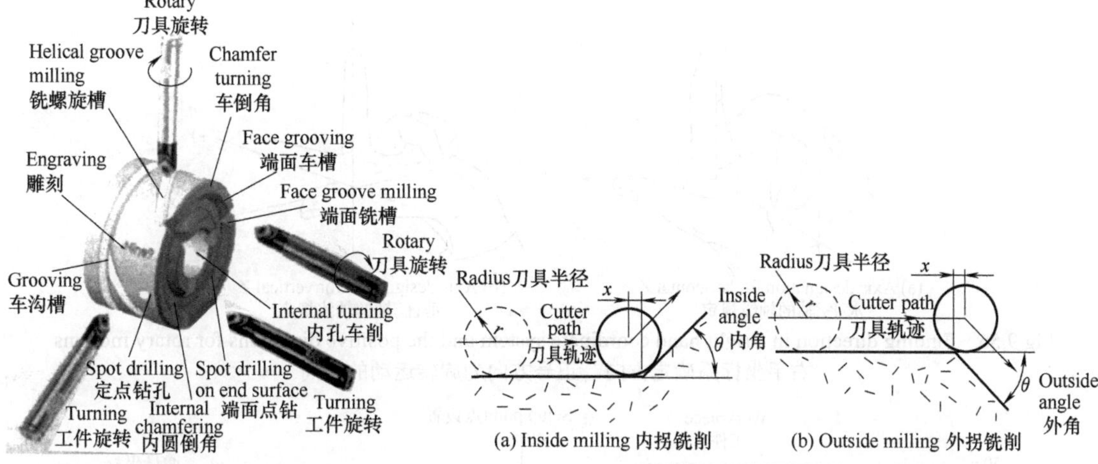

Fig 9.51 Applications on a lathe machining centre 车削加工中心的各种操作

Fig 9.52 Counter Milling 补偿铣削

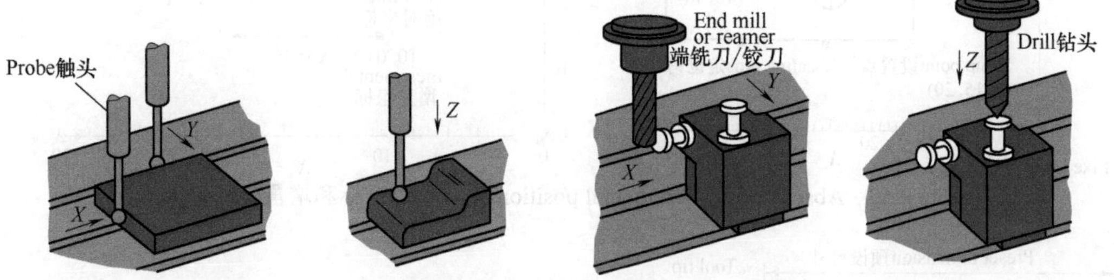

(a) *X*–*Y*direction *X*-*Y*方向

Touch probe determining the *X*–*Y*(horizontal) position of a workpiece 触头确定工件在*X*-*Y*(水平面)位置

(b) *Z* direction *Z*方向

Determining the height of a horizontal surface 确定垂直面的高度

(c) Diameter compensation直径补偿

Determining the planar position of the surface of a cutter (for instance, for cutter-diameter compensation) 确定刀具表面的位置(如,刀具直径补偿)

(d) Tool length offset刀具长度补偿

Determining the length of a tool for tool-length offset 确定刀具补偿长度

Fig 9.53 Touch probes used in machining centers for determining workpiece and tool positions and surfaces relative to the machine table or column

加工中心触头用于确定工件、刀具位置及与机床工作台或立柱的相关表面

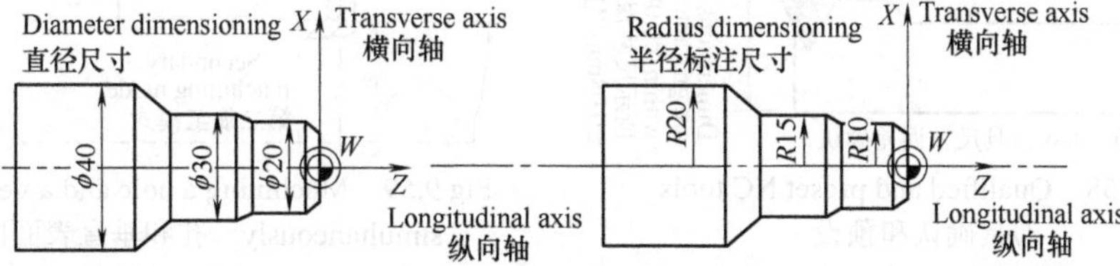

Fig 9.54 Diameter and radius dimensioning for the transverse axis 移动轴的直径与半径尺寸

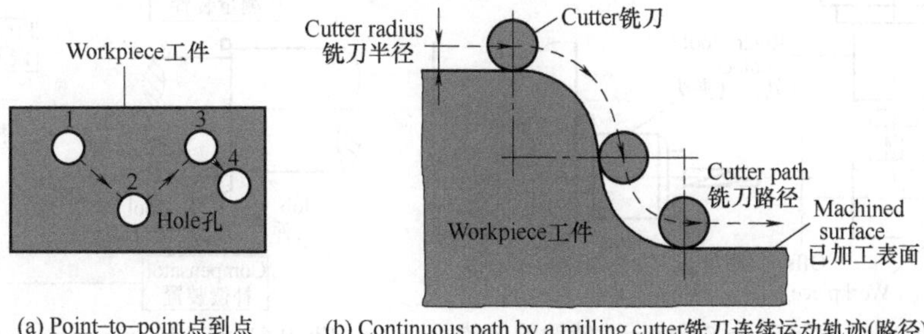

(a) Point-to-point点到点

(b) Continuous path by a milling cutter铣刀连续运动轨迹(路径)

Fig 9.55 Movement of tools in NC machining 数控加工刀具的运动

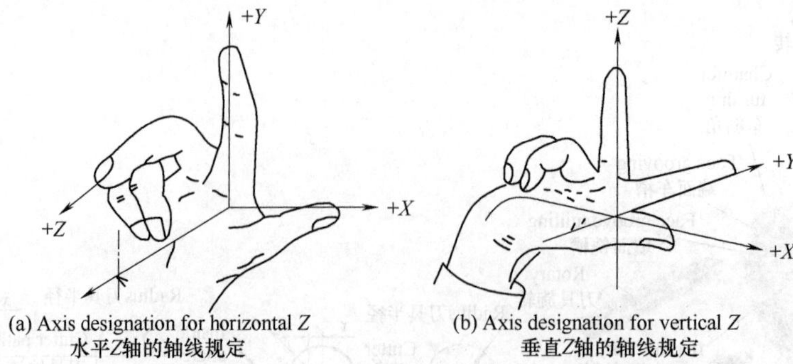

(a) Axis designation for horizontal Z 水平Z轴的轴线规定

(b) Axis designation for vertical Z 垂直Z轴的轴线规定

Fig 9.56 Finding direction in a right hand coordinate system and the positive directions for rotary motions 右手坐标系确定方向，正号方向为旋转运动的方向

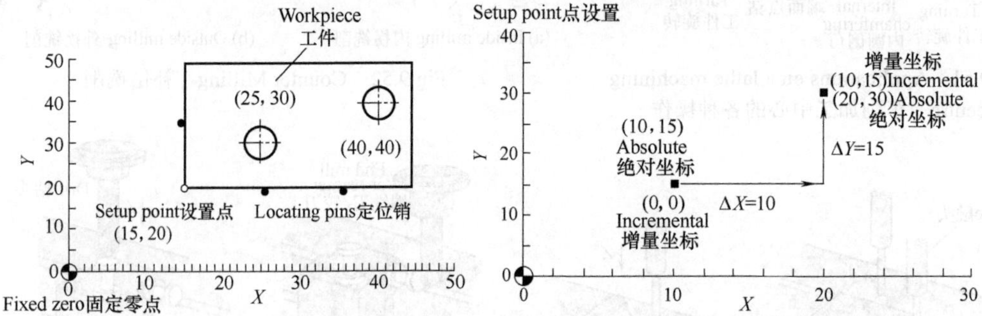

Fig 9.57 Absolute and incremental positioning 绝对坐标和增量坐标

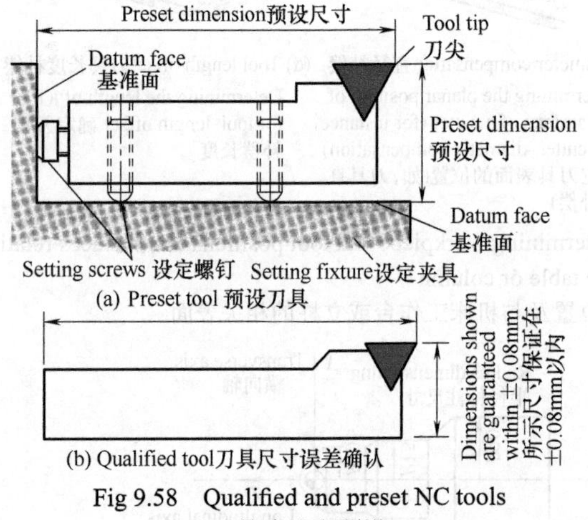

(a) Preset tool 预设刀具

(b) Qualified tool刀具尺寸误差确认

Fig 9.58 Qualified and preset NC tools 刀具确认和预设

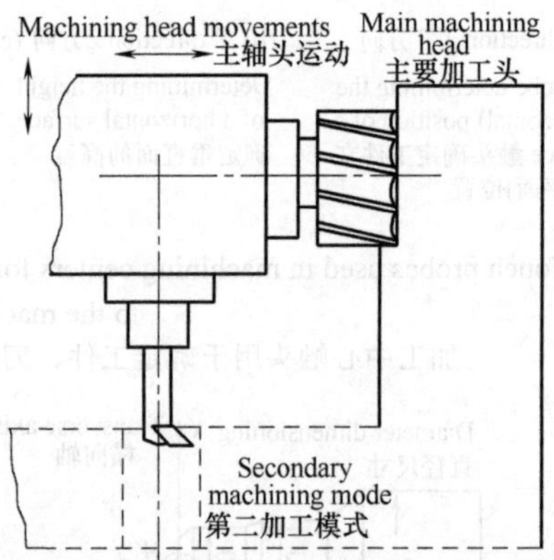

Fig 9.59 Machining a hole and a vertical surface simultaneously 孔和垂直表面同时加工

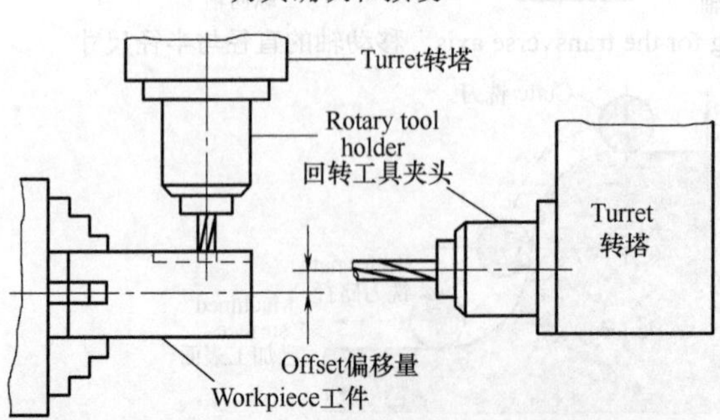

Fig 9.60 Additional tooling facilities 辅助的工装

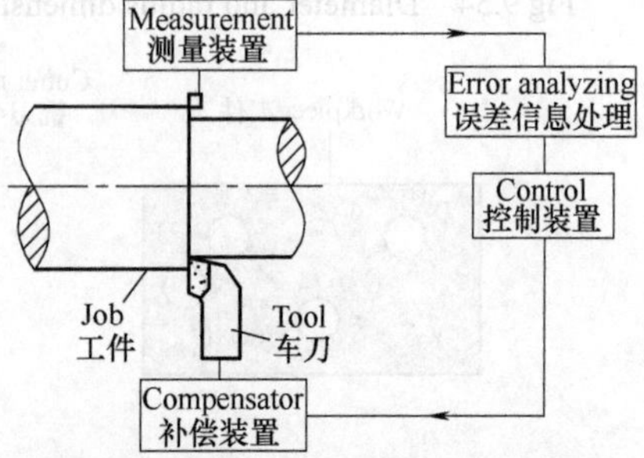

Fig 9.61 Tool compensation system configuration 刀具补偿系统的组成

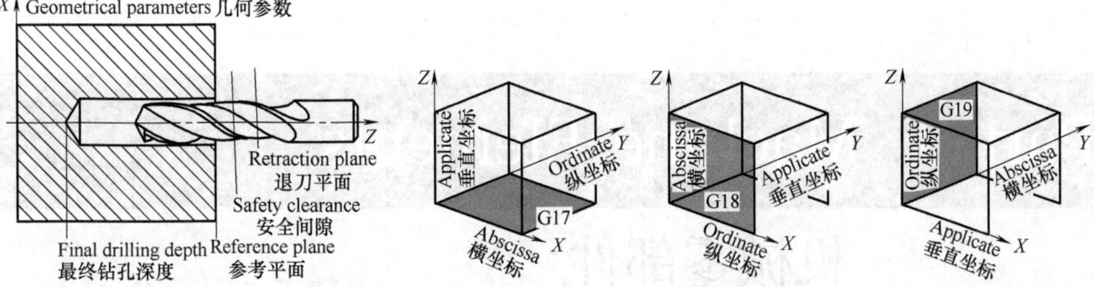

Fig 9.62 Drilling cycles 钻孔循环

Fig 9.63 Plane and axis assignment 平面和轴的分配

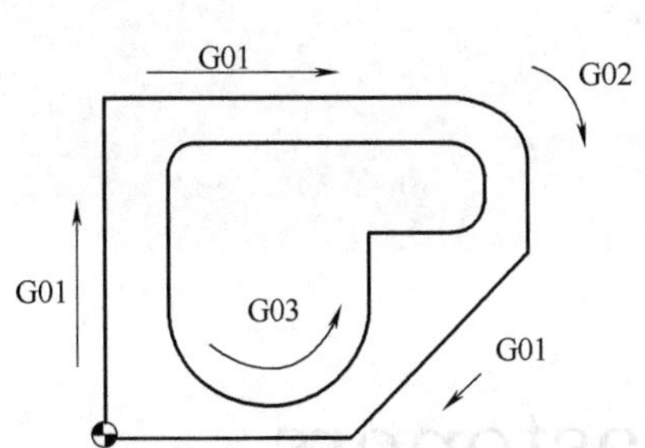

Fig 9.64 Linear interpolation (G01) and circular interpolation (right G02 and left G03) 直线插补（G01）和圆弧插补（右插补 G02，左插补 G03）

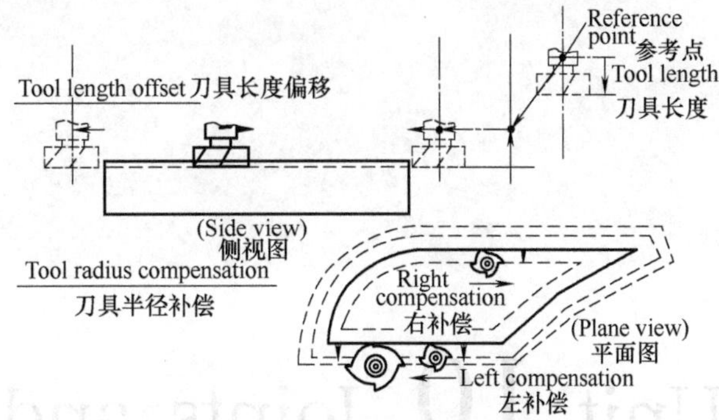

Fig 9.65 Tool offset 刀具偏移

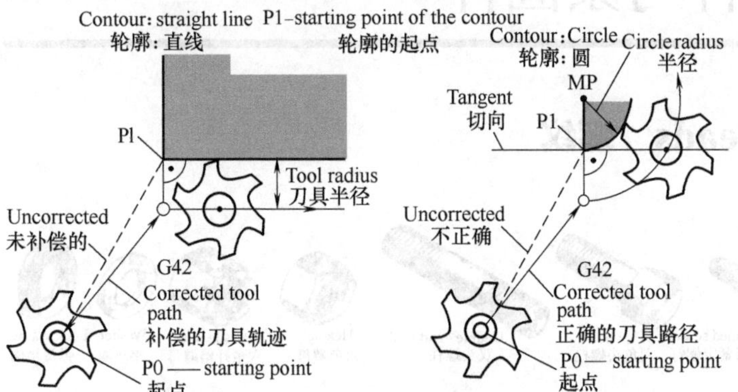

Fig 9.66 Start of the tool radius compensation with G42 以 G42 进行刀具半径补偿

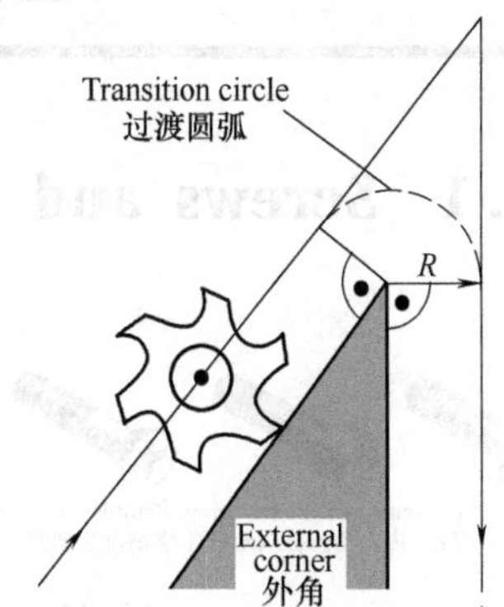

Fig 9.67 Acute contour angle and switching to transition circle 轮廓锐角和过渡圆弧转换

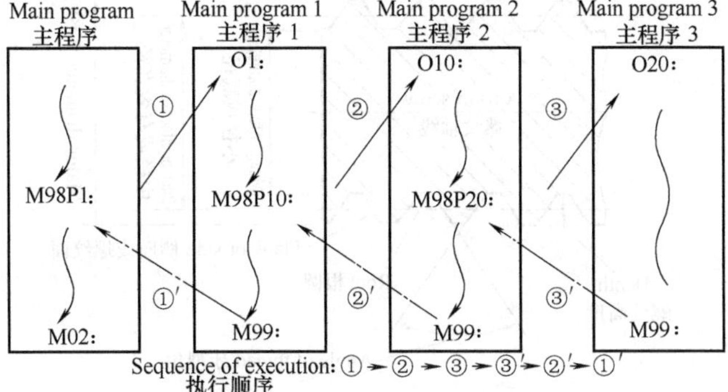

Fig 9.68 Subprogram 子程序

Part 3 Machine Elements

机械零部件

Unit 10 Joints and Fasteners

连接件与紧固件

10.1 Screws and threads 螺纹

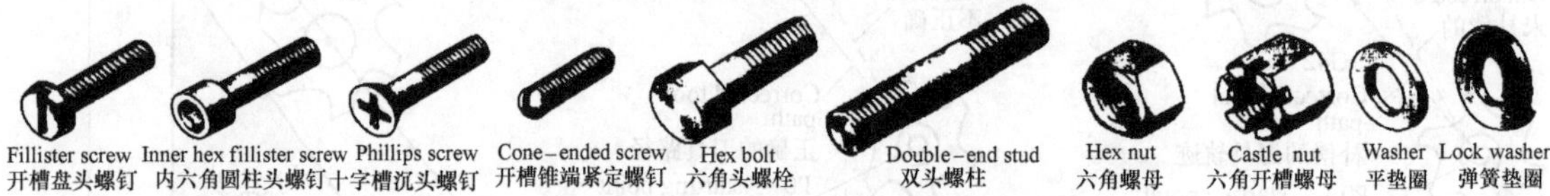

Fig 10.1 Components of bolt coupling 常见螺纹连接构件

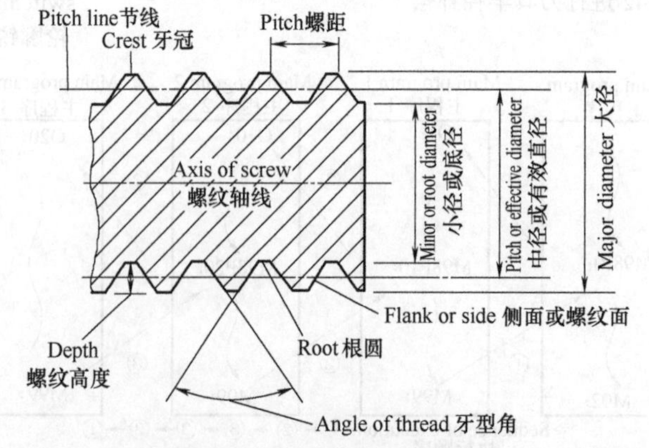

Fig 10.2 Thread terms 螺纹术语

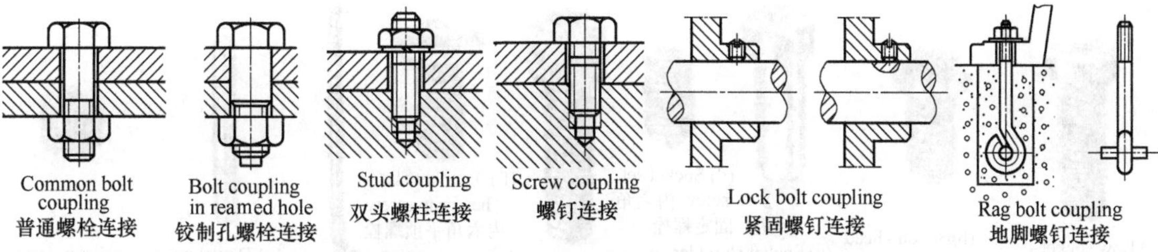

Fig 10.3 Commonly used bolt coupling 常用螺纹连接

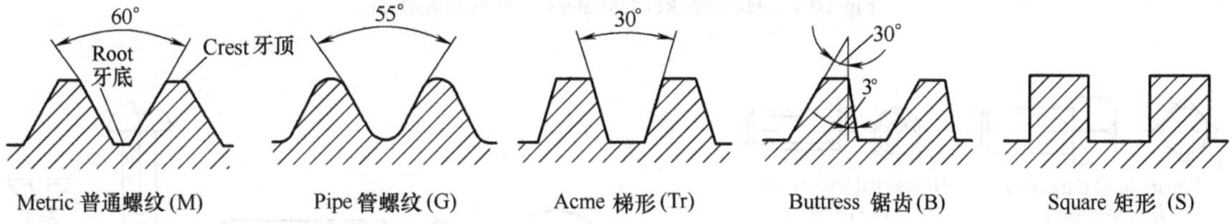

Fig 10.4 Commonly used screw thread 常见的螺纹牙型

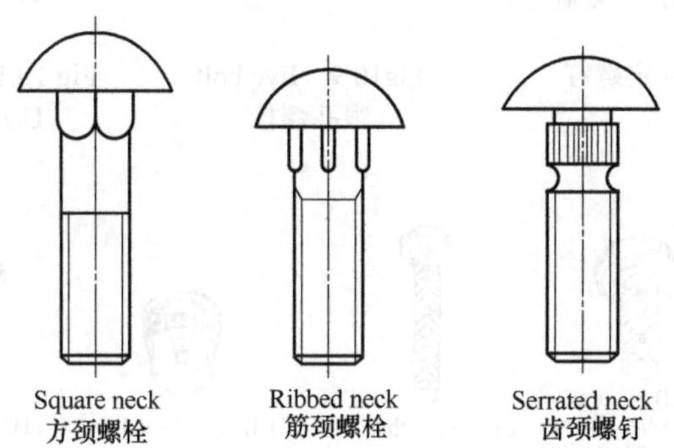

Fig 10.5 Coach bolts (carriage bolts) 方头螺钉（车架螺栓）

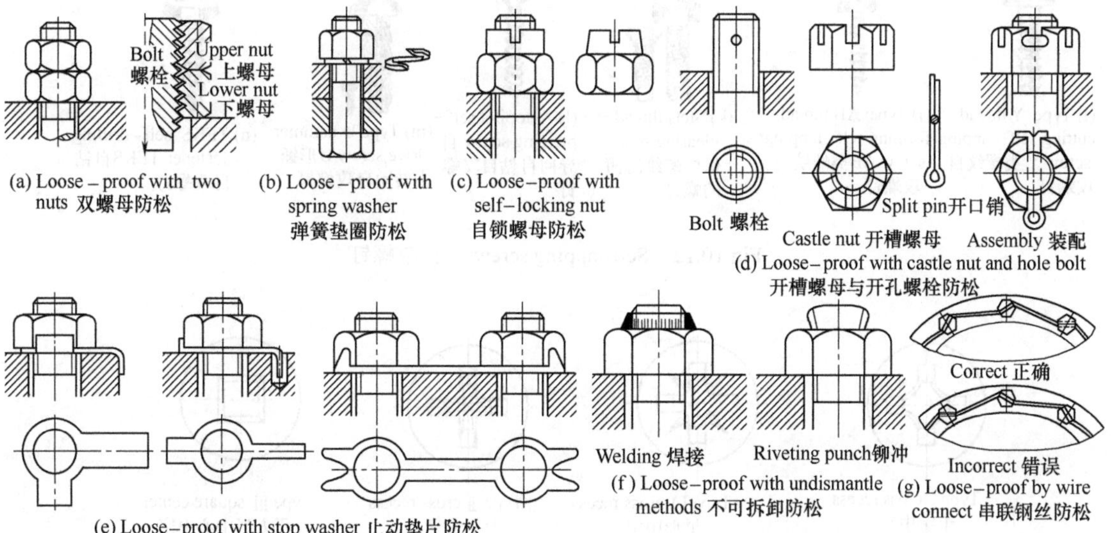

Fig 10.6 Loose-proof of screw coupling 螺纹连接的防松措施

(a) Low socket-head cap screw 低口内六角螺栓
(b)Button-head socket cap screw 纽扣头(圆帽头)内六角螺栓
(c) Socket shoulder screw 内六角轴肩螺栓
(d) Socket set screw 内六角固定螺栓
(e) Socket-head cap screw 内六角头帽螺栓
(f) A socket flat-head cap screw 内六角平顶螺栓
(g) Hexagon key 六角扳手(钥匙)
(h) Dowel pin 销钉
(i) Socket pressure plug 内六角压力螺栓塞

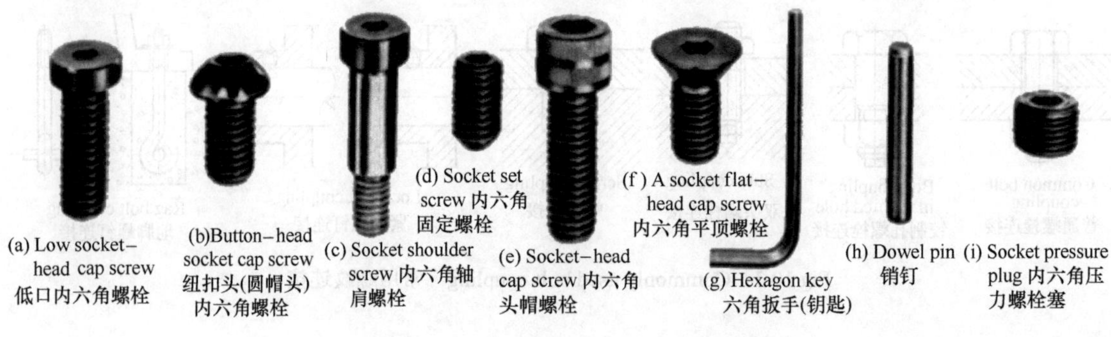

Fig 10.7 Hex-socket fasteners 内六角紧固件

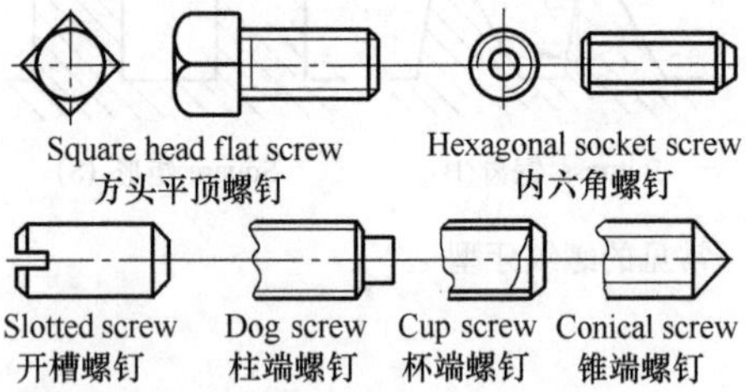

Square head flat screw 方头平顶螺钉
Hexagonal socket screw 内六角螺钉
Slotted screw 开槽螺钉
Dog screw 柱端螺钉
Cup screw 杯端螺钉
Conical screw 锥端螺钉

Fig 10.8 Set screws 锁定螺钉

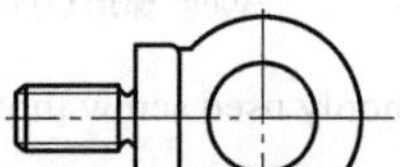

Fig10.9 Eye bolt 眼孔螺栓

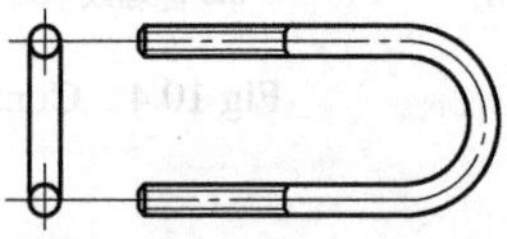

Fig 10.10 U-bolt U 形螺栓

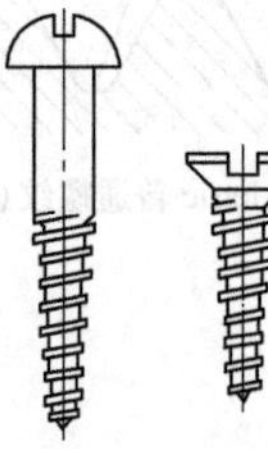

Fig 10.11 Wood screw 木螺钉

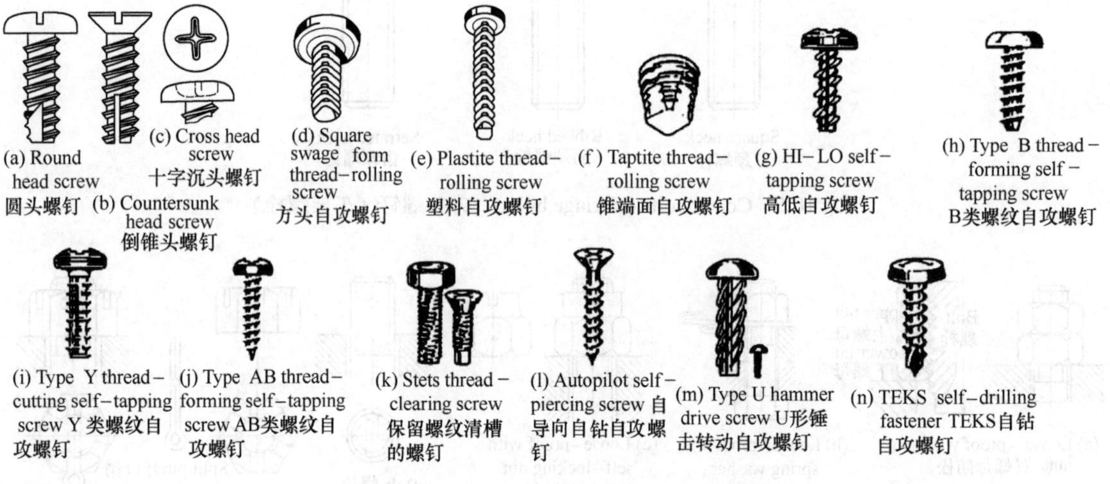

(a) Round head screw 圆头螺钉
(b) Countersunk head screw 倒锥头螺钉
(c) Cross head screw 十字沉头螺钉
(d) Square swage form thread-rolling screw 方头自攻螺钉
(e) Plastite thread-rolling screw 塑料自攻螺钉
(f) Taptite thread-rolling screw 锥端面自攻螺钉
(g) HI-LO self-tapping screw 高低自攻螺钉
(h) Type B thread-forming self-tapping screw B类螺纹自攻螺钉
(i) Type Y thread-cutting self-tapping screw Y 类螺纹自攻螺钉
(j) Type AB thread-forming self-tapping screw AB类螺纹自攻螺钉
(k) Stets thread-clearing screw 保留螺纹清槽的螺钉
(l) Autopilot self-piercing screw 自导向自钻自攻螺钉
(m) Type U hammer drive screw U形锤击转动自攻螺钉
(n) TEKS self-drilling fastener TEKS自钻自攻螺钉

Fig 10.12 Self-tapping screws 自攻螺钉

Type Ⅰ cross recess 十字中空
Type IA cross recess 星形中空
Type Ⅱ cross recess 十字中空
Type Ⅲ square center 正方形中心中空

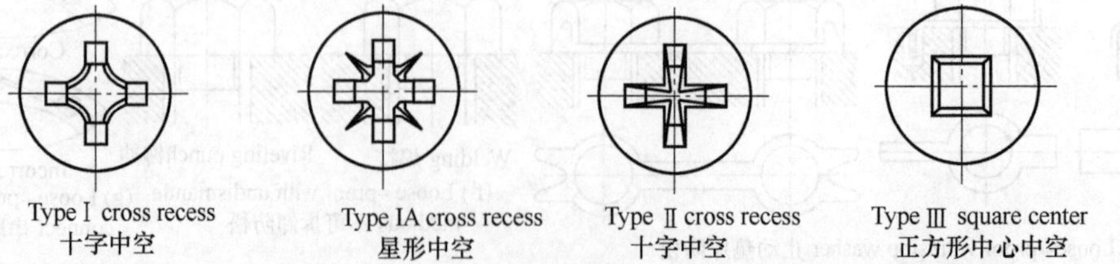

Fig 10.13 Machine screw cross recesses 机制螺钉头交叉凹陷形式

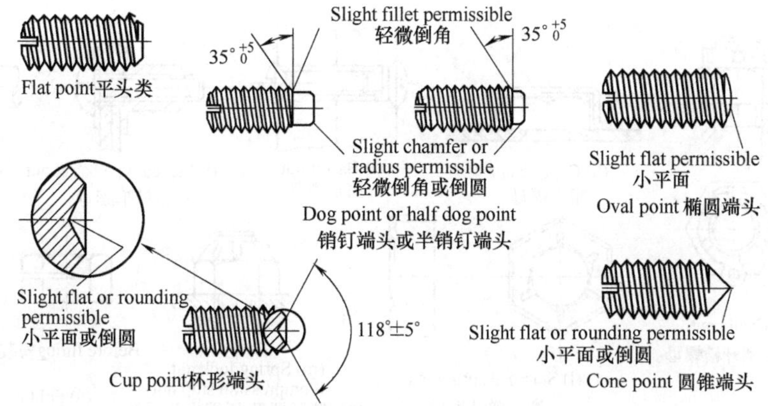

Fig 10.14 Standard slotted headless set screws 标准无头开槽固定螺钉

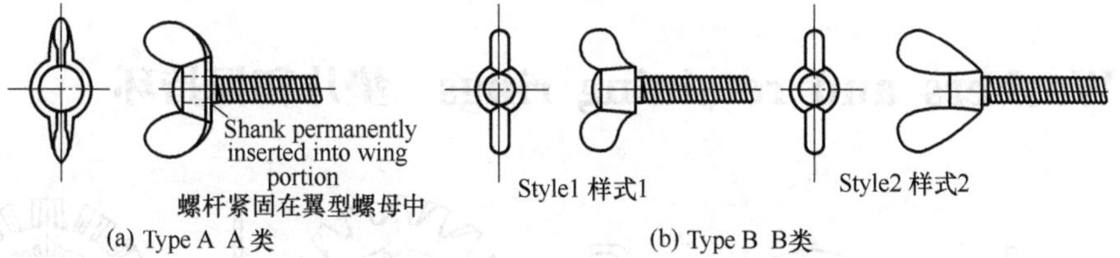

Fig 10.15 Standard wing screw 标准翼型螺钉

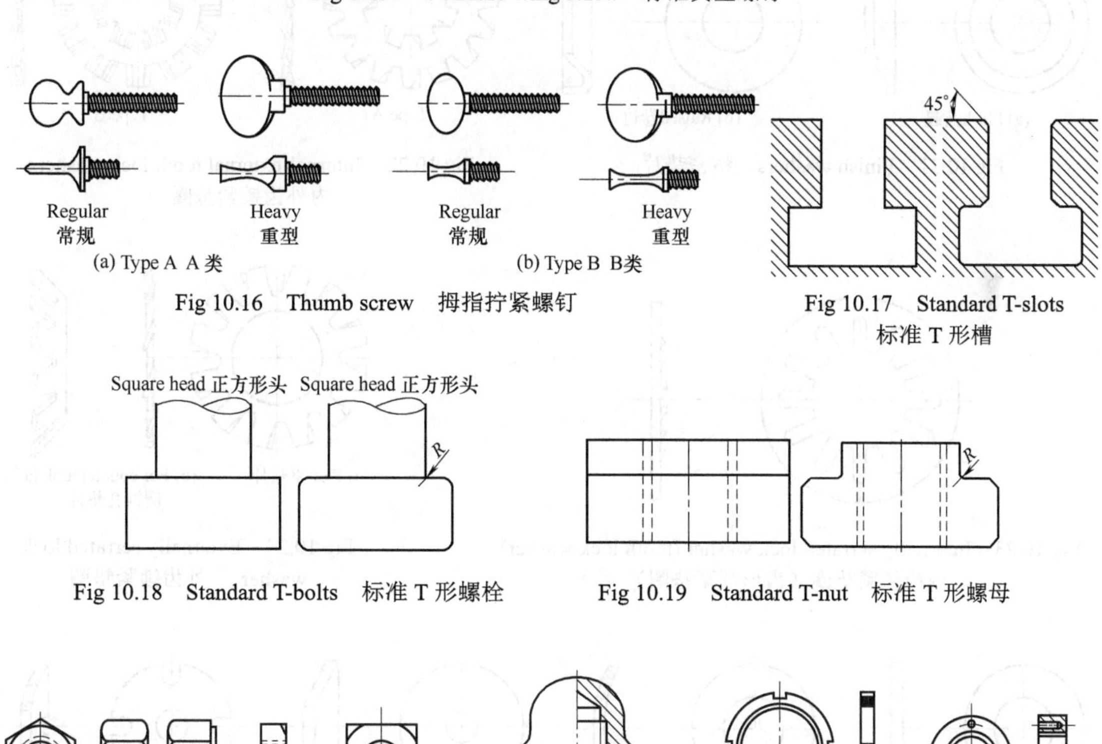

Fig 10.16 Thumb screw 拇指拧紧螺钉

Fig 10.17 Standard T-slots 标准 T 形槽

Fig 10.18 Standard T-bolts 标准 T 形螺栓

Fig 10.19 Standard T-nut 标准 T 形螺母

(a) Metric hex nuts 公制六角螺母

(b) Square nut 方形螺母

(c) Cap nut(crown nut, dome nut) 盖形(冠形、圆顶)螺母

(d) Slotted ring nut 开槽环形螺母

(e) Drilled ring nut 钻孔环形螺母

Fig 10.20

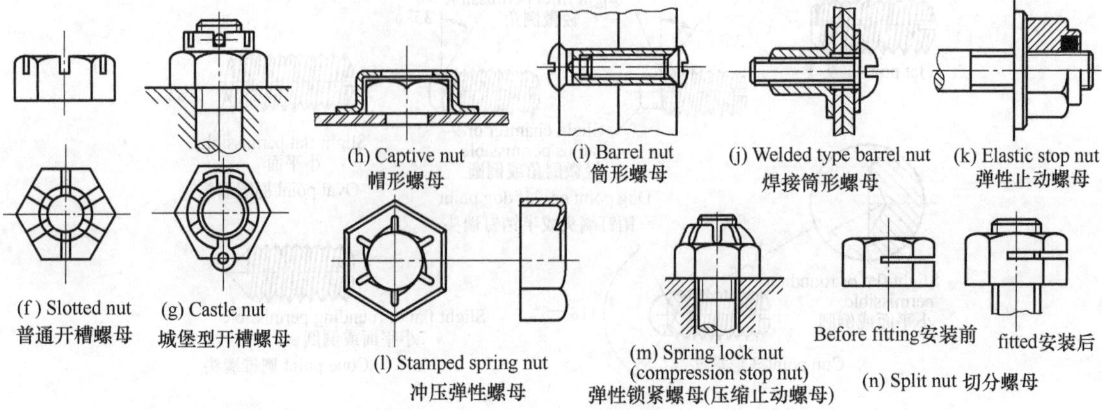

Fig 10.20 Nut 螺母

10.2 Washers and retaining rings 垫片和保持环

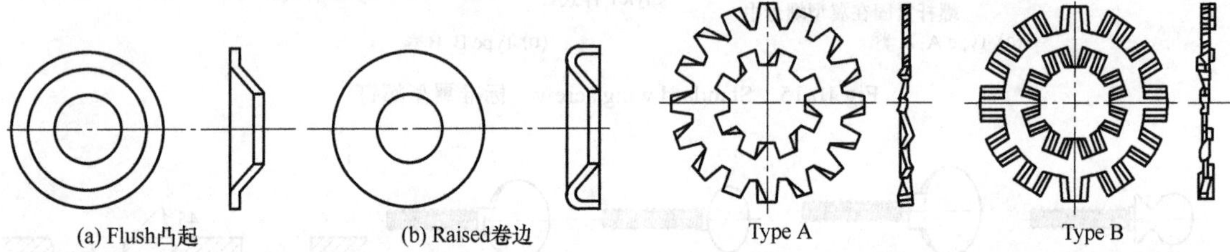

Fig 10.21 Finish washers 精密垫片

Fig 10.22 Internal-external tooth lock washers 内外齿锁紧垫圈

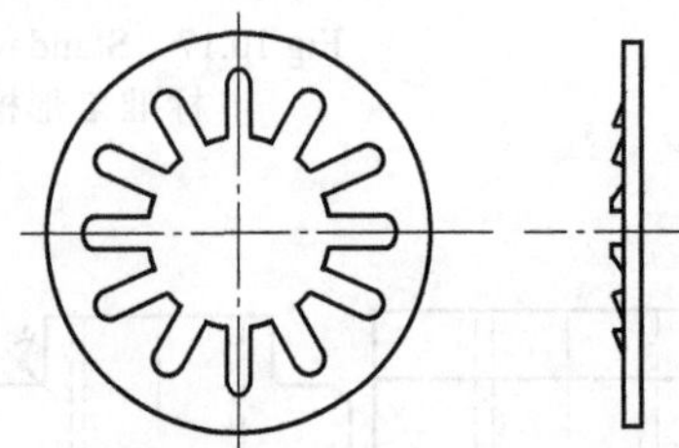

Fig 10.23 Internally serrated lock washer (tooth lock washer) 内齿锁紧垫圈（齿形锁紧垫圈）

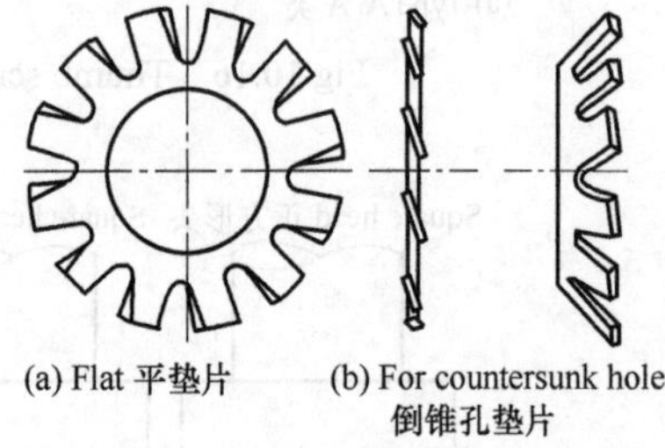

Fig 10.24 Externally serrated lock washer 外齿锁紧垫圈

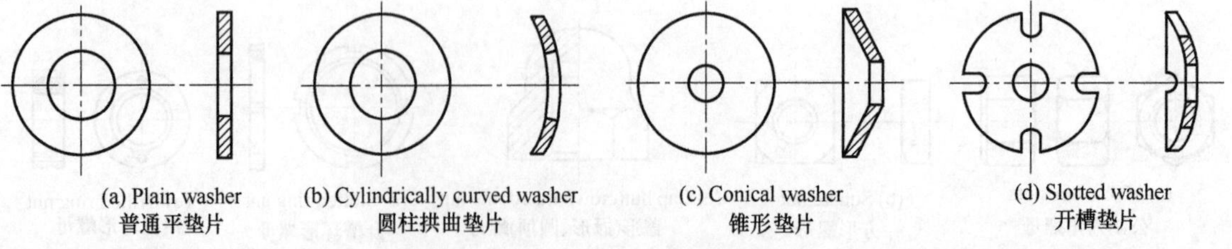

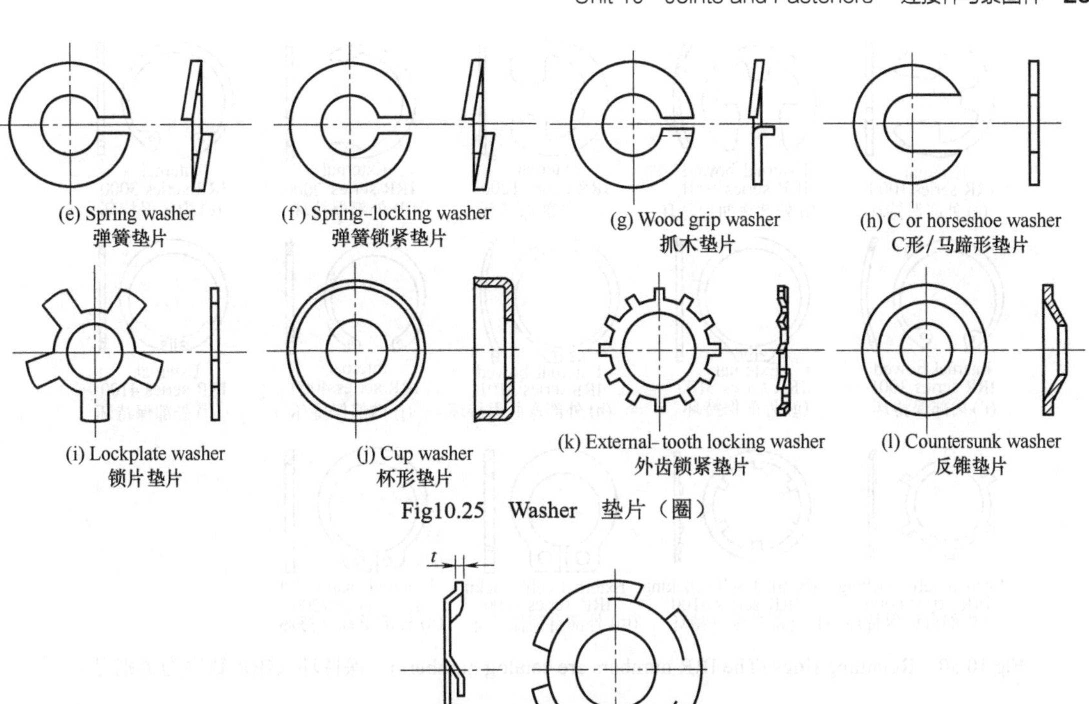

Fig10.25 Washer 垫片（圈）

Fig 10.26 Finger washer 指状垫圈

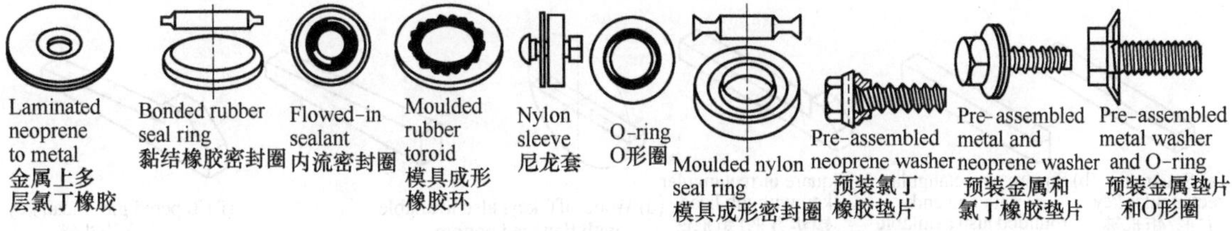

Fig 10.27 Self-sealing fasteners and washers 自封紧闭件和垫圈

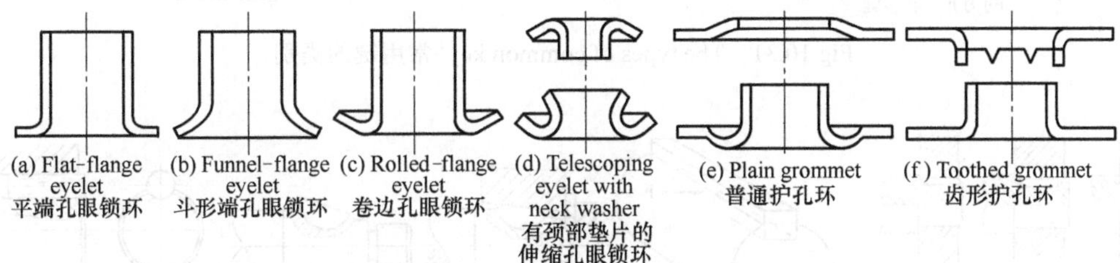

Fig 10.28 Eyelet and grommet 眼锁环和护孔环

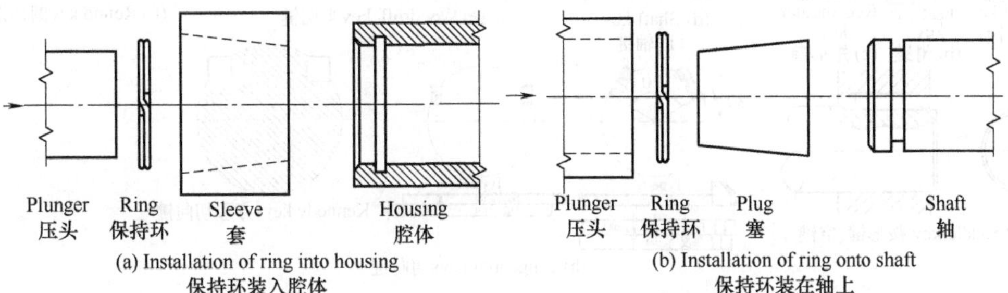

Fig 10.29 Spiral retaining rings 螺旋保持环

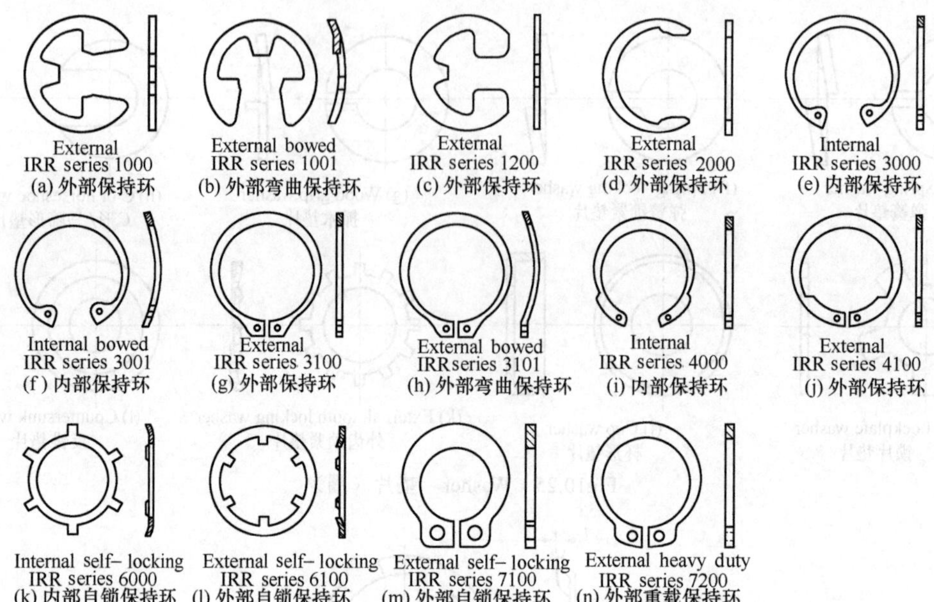

Fig 10.30 Retaining rings (The IRR numbers are catalog numbers) 保持环（IRR 数字为类别号）

10.3 Keys 键

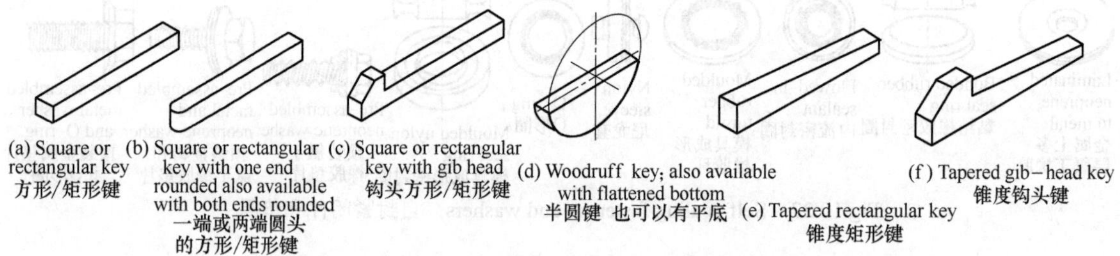

Fig 10.31 The types of common key 常用键的类别

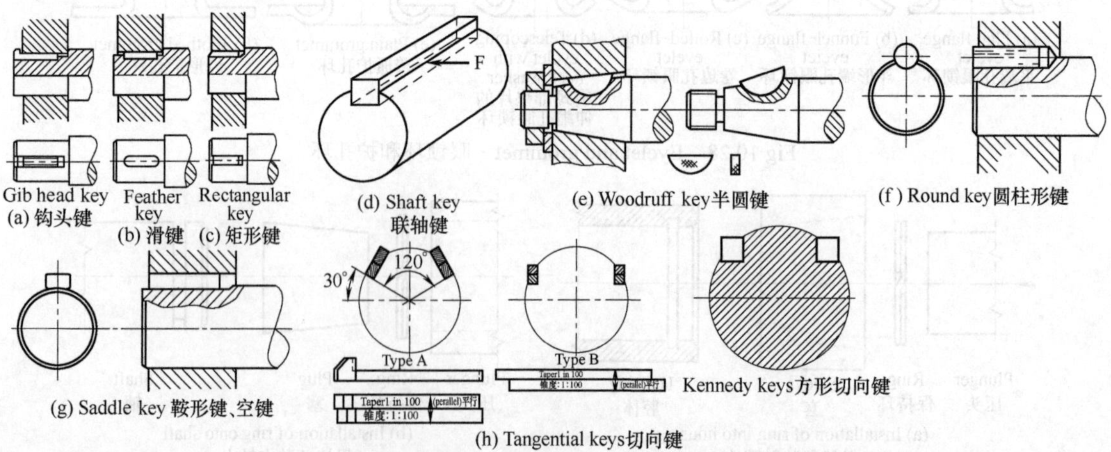

Fig 10.32 Key application 键的应用

10.4 Pins 销

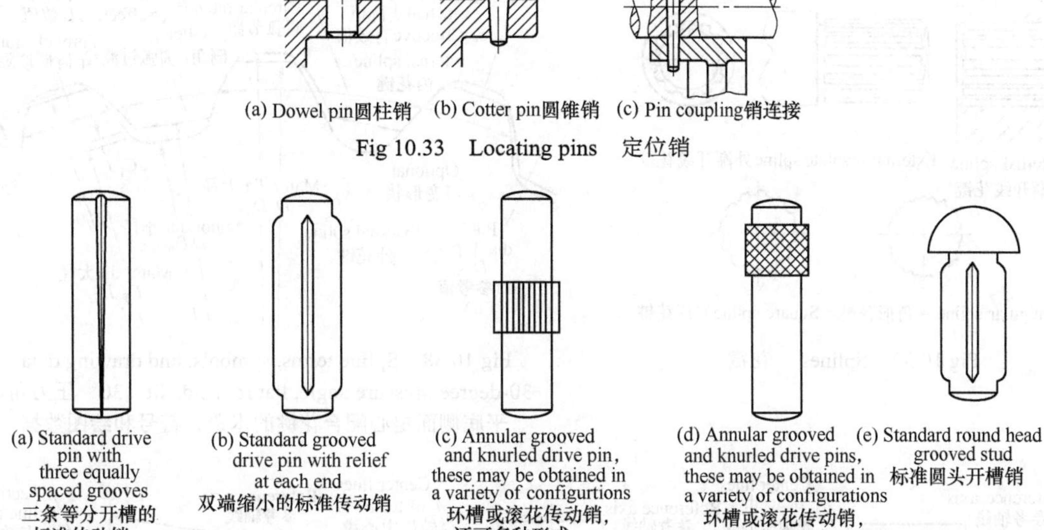

Fig 10.33 Locating pins 定位销

Fig 10.34 An assortment of drive pins 传动销的种类

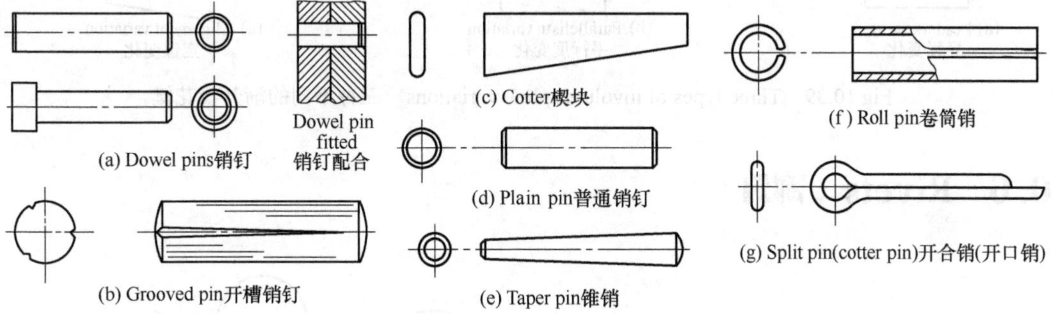

Fig 10.35 Types of pins 销钉的种类

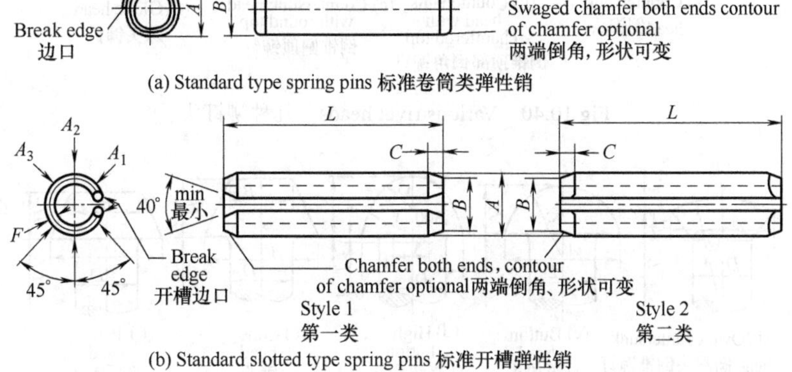

Fig 10.36 Spring pins 弹性销

10.5 Splines 花键

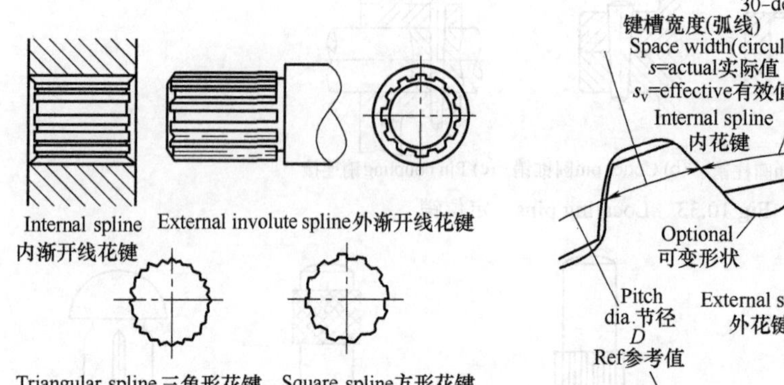

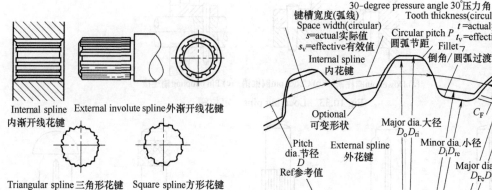

Fig 10.37 Splines 花键

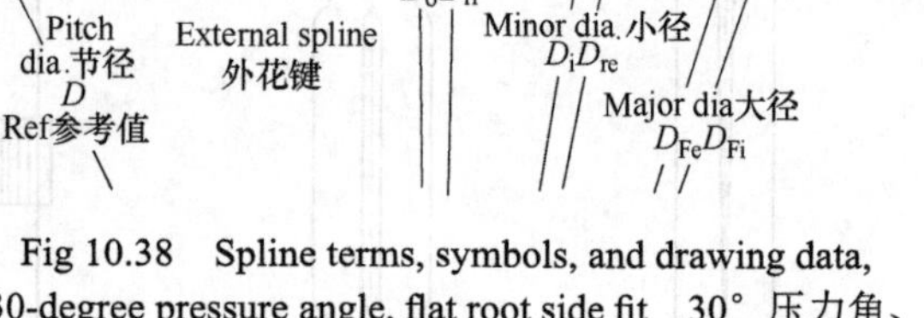

Fig 10.38 Spline terms, symbols, and drawing data, 30-degree pressure angle, flat root side fit 30° 压力角、平底侧面定心配合花键的术语、符号和绘图数据

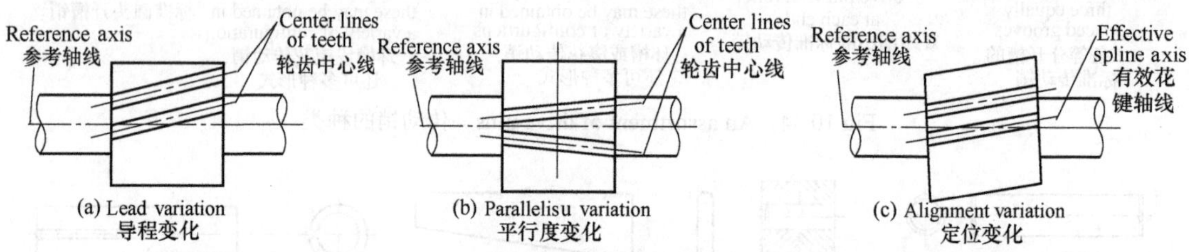

Fig 10.39 Three types of involute spline variations 三种不同的渐开线花键

10.6 Rivets 铆钉

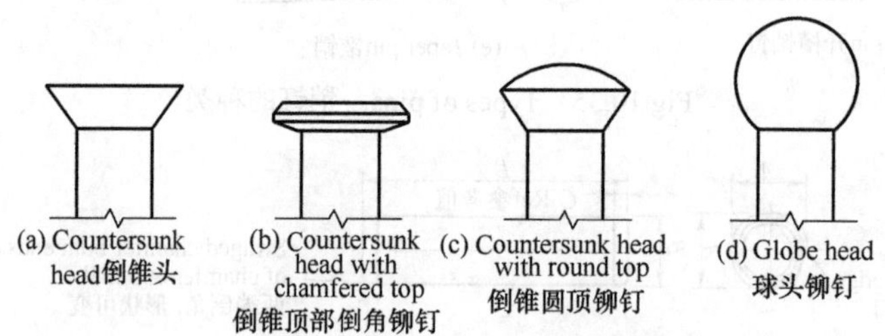

Fig 10.40 Various rivet heads 几种铆钉头

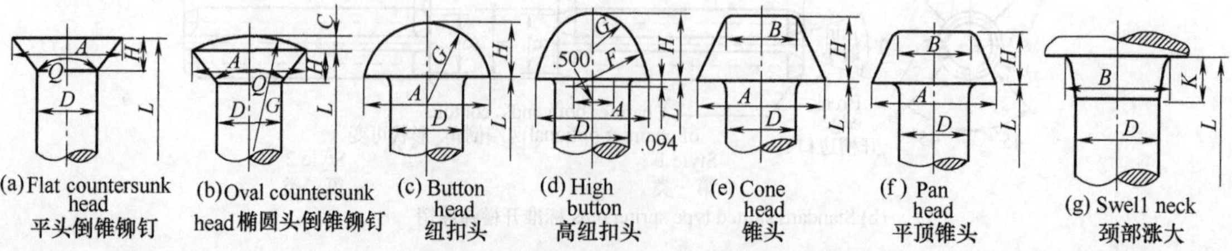

Fig 10.41 Standard large rivets 标准大型铆钉

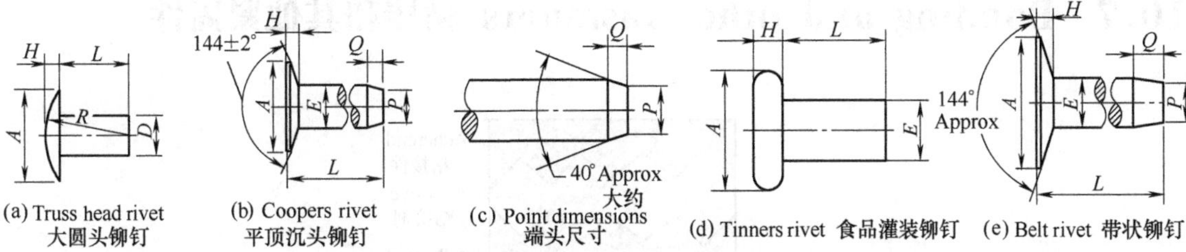

(a) Truss head rivet 大圆头铆钉 (b) Coopers rivet 平顶沉头铆钉 (c) Point dimensions 端头尺寸 (d) Tinners rivet 食品灌装铆钉 (e) Belt rivet 带状铆钉

Fig 10.42 Standard small solid rivets 标准小型实心铆钉

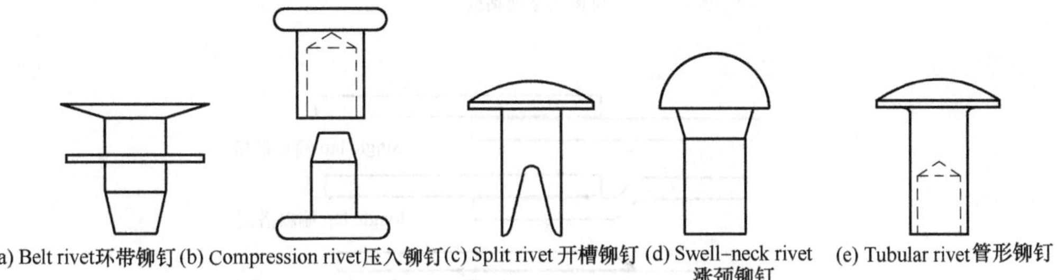
(a) Belt rivet环带铆钉 (b) Compression rivet压入铆钉 (c) Split rivet 开槽铆钉 (d) Swell-neck rivet 涨颈铆钉 (e) Tubular rivet管形铆钉

Fig 10.43 Standard rivet heads with flat bearing surfaces 平底支撑面的标准铆钉头

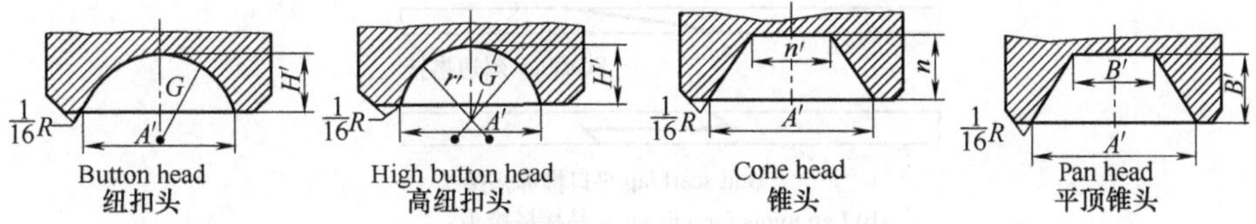

Button head 纽扣头 High button head 高纽扣头 Cone head 锥头 Pan head 平顶锥头

Fig 10.44 Hold-on (dolly bar) and rivet set impression 铆顶棍和铆钉座形状

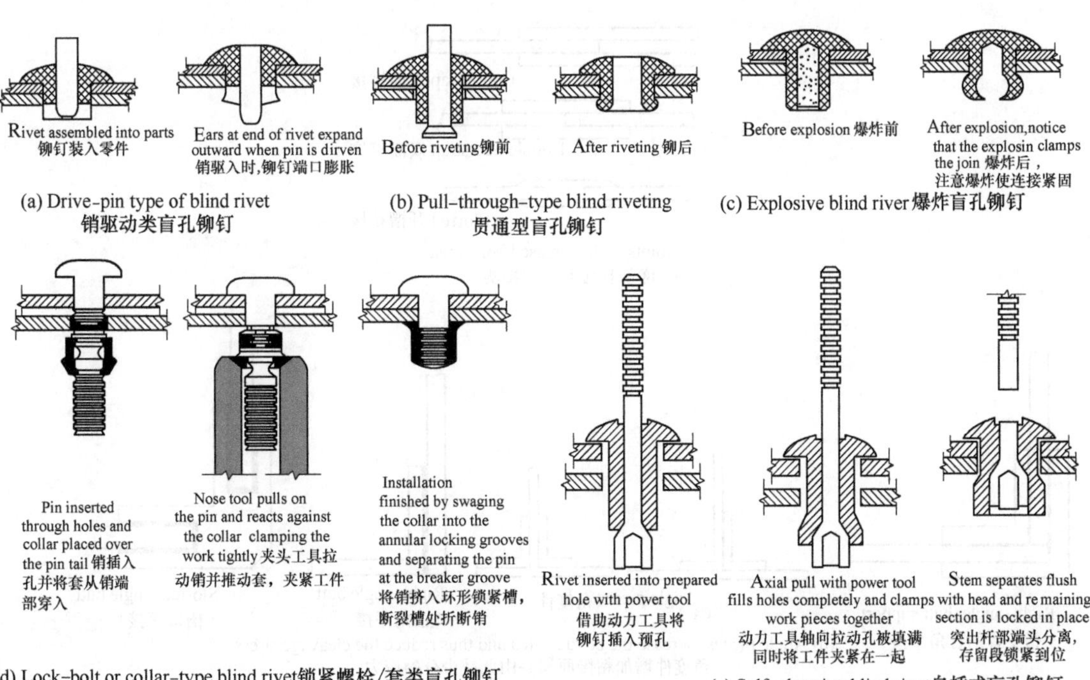

(a) Drive-pin type of blind rivet 销驱动类盲孔铆钉 (b) Pull-through-type blind riveting 贯通型盲孔铆钉 (c) Explosive blind river 爆炸盲孔铆钉

(d) Lock-bolt or collar-type blind rivet锁紧螺栓/套类盲孔铆钉 (e) Self-plugging blind rivet自插式盲孔铆钉

Fig 10.45 Blind rivet 盲孔铆钉

10.7 Bonding and other fasteners 粘接和其他紧固件

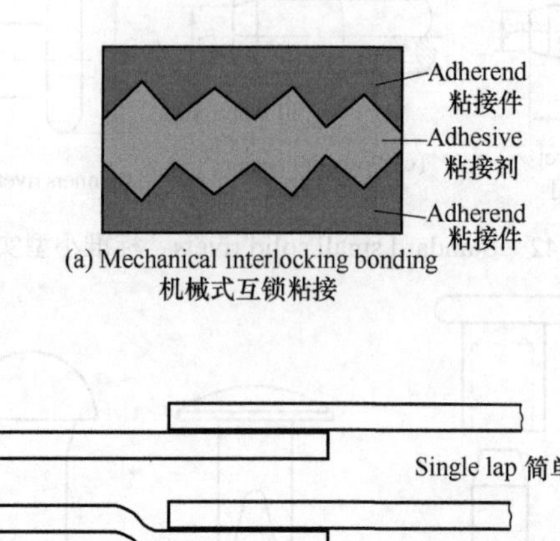

(a) Mechanical interlocking bonding
机械式互锁粘接

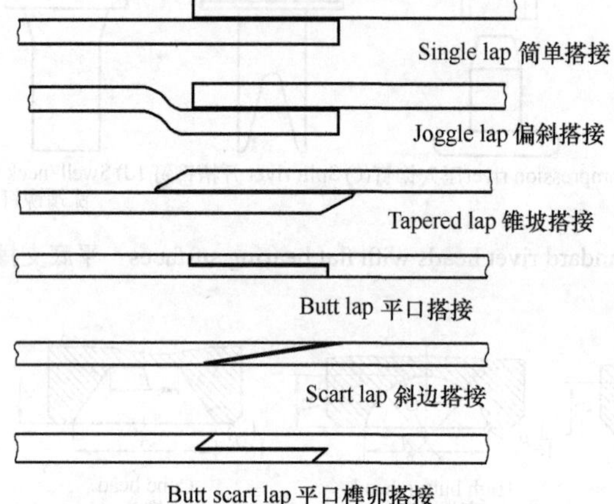

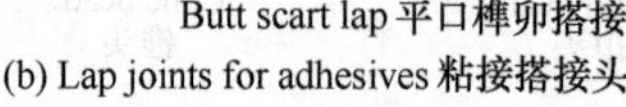

(b) Lap joints for adhesives 粘接搭接头

Double lap 双面搭接

Double butt strap 双面平接

Slotted 开槽粘接

Joints with increased bond area
(c) 增加粘接面积的接头

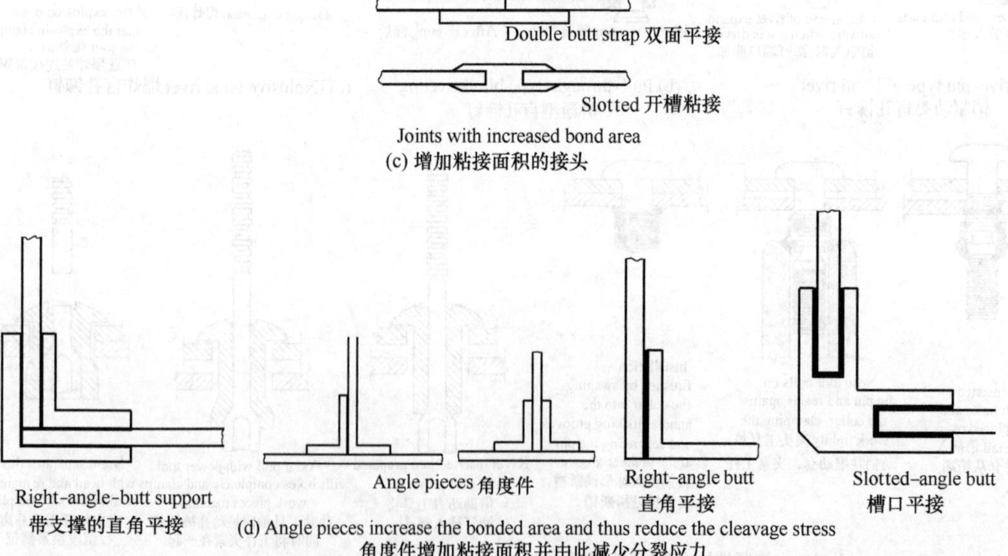

(d) Angle pieces increase the bonded area and thus reduce the cleavage stress
角度件增加粘接面积并由此减少分裂应力

Fig 10.46 Bonding 粘接

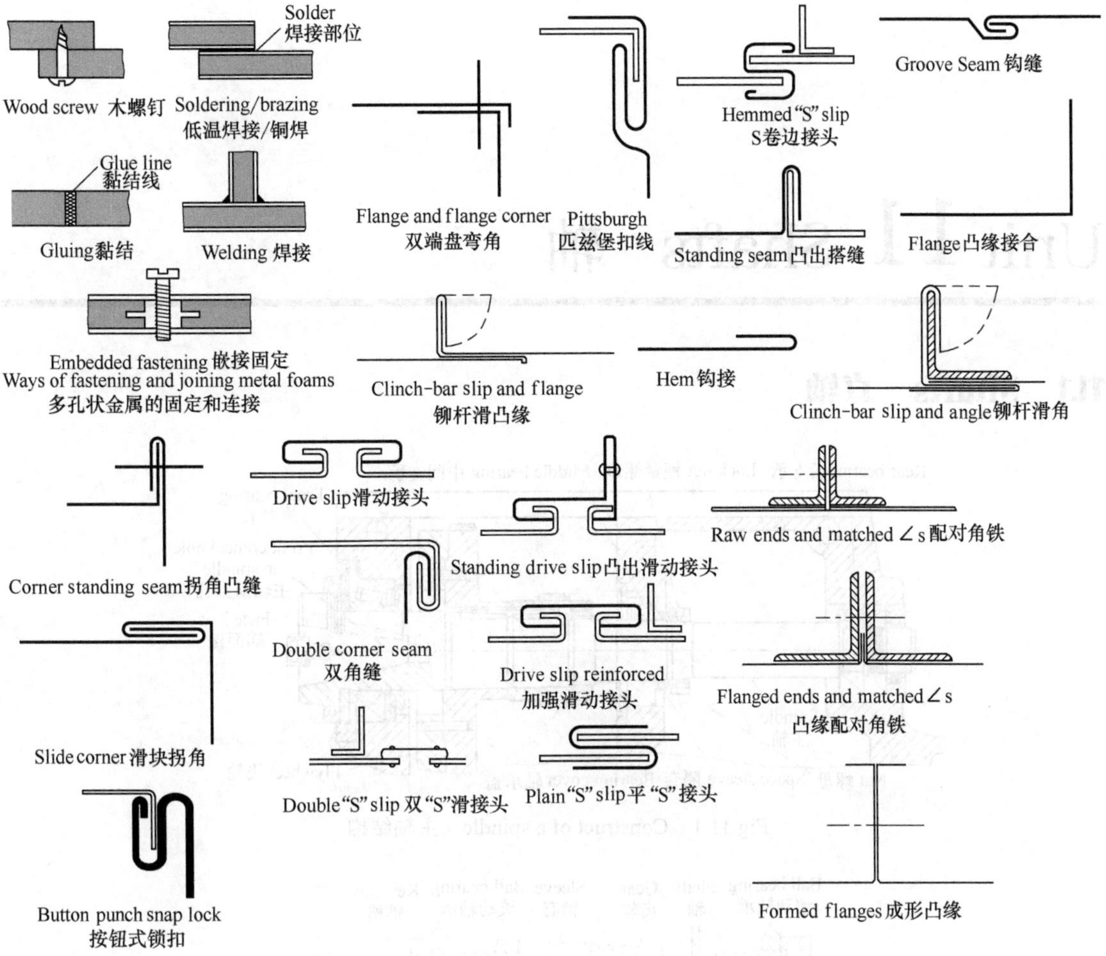

Wood screw 木螺钉
Solder
焊接部位
Soldering/brazing
低温焊接/铜焊
Glue line
黏结线
Gluing黏结
Welding 焊接
Embedded fastening 嵌接固定
Ways of fastening and joining metal foams
多孔状金属的固定和连接
Flange and flange corner
双端盘弯角
Pittsburgh
匹兹堡扣线
Hemmed "S" slip
S卷边接头
Groove Seam 钩缝
Standing seam凸出搭缝
Flange凸缘接合
Clinch-bar slip and flange
铆杆滑凸缘
Hem 钩接
Clinch-bar slip and angle铆杆滑角
Corner standing seam拐角凸缝
Drive slip滑动接头
Standing drive slip凸出滑动接头
Raw ends and matched ∠s 配对角铁
Double corner seam
双角缝
Drive slip reinforced
加强滑动接头
Flanged ends and matched ∠s
凸缘配对角铁
Slide corner 滑块拐角
Double "S" slip 双 "S"滑接头
Plain "S" slip 平 "S" 接头
Button punch snap lock
按钮式锁扣
Formed flanges 成形凸缘

Fig 10.47 Other fasteners 其他紧固件

(a) Dart-type clip
飞镖型锁夹
(b) Nylon expansion nut
尼龙膨胀螺母
(c) Plastic rivet
塑料铆接头
(d) Ratchet rivet
棘齿铆接头
(e) Tri-piece rivet
三组件锁夹铆接头

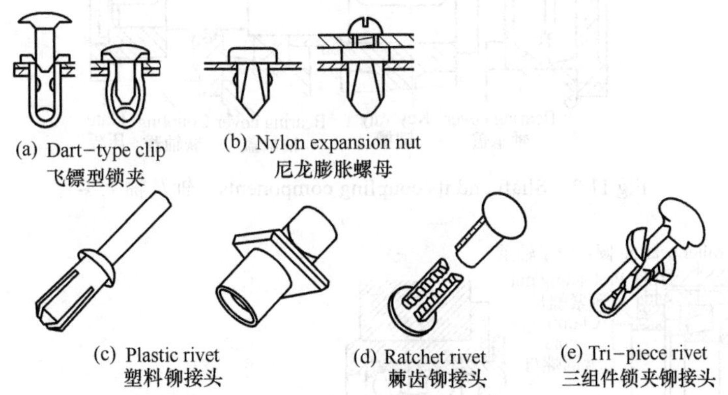

Fig 10.48 Various types of plastics fasteners 各种塑料紧固件

Unit 11 Shafts 轴

11.1 Shafts 直轴

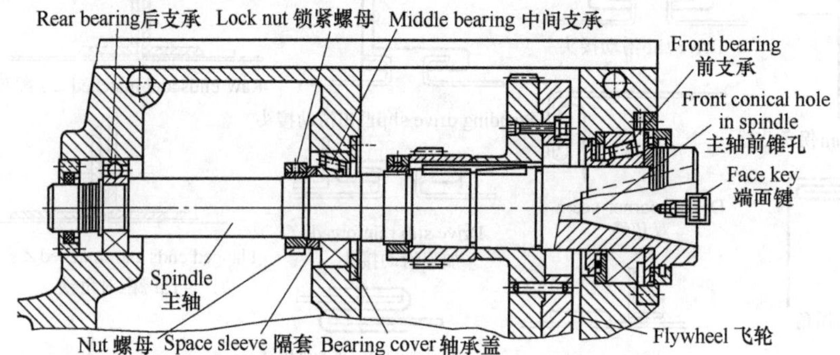

Fig 11.1 Construct of a spindle 主轴结构

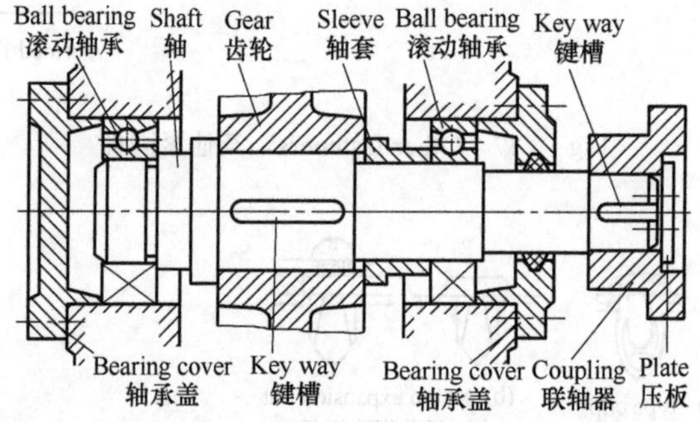

Fig 11.2 Shaft and its coupling components 轴及轴上零件

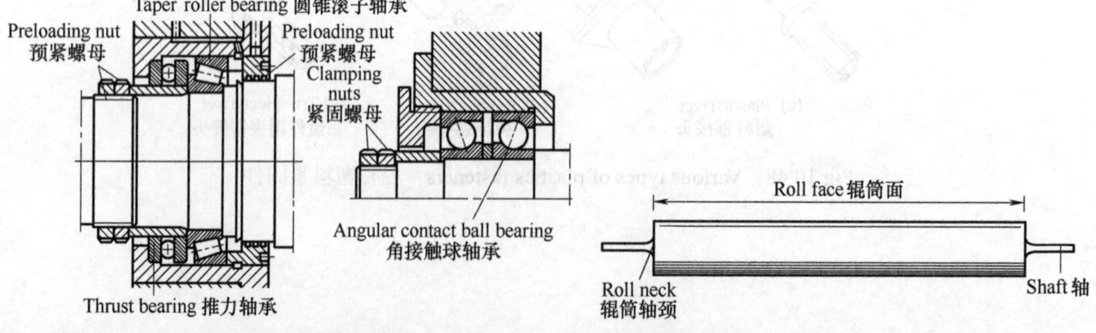

Fig 11.3 Typical spindle-bearing arrangements 典型主轴轴承布局

Fig 11.4 Typical process roll 典型工艺辊

Table 11.1 Good and bad design practices 良好设计和不合理设计示例

Poor fatigue strength 疲劳强度差	Improved fatigue strength 改善疲劳强度	Poor fatigue strength 疲劳强度差	Improved fatigue strength 改善疲劳强度
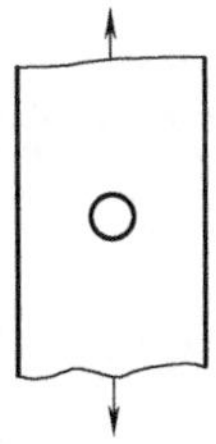	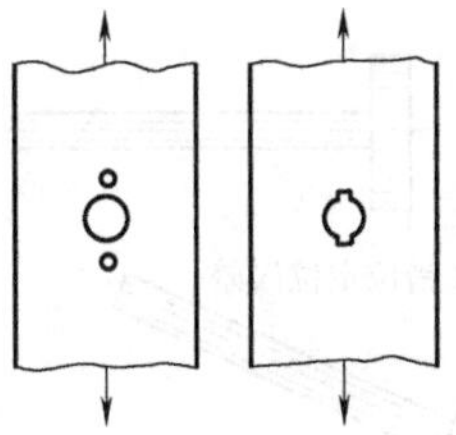	Grooves 沟槽 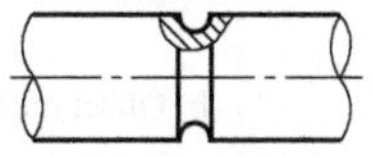	Stress-relieving grooves 应力释放槽 
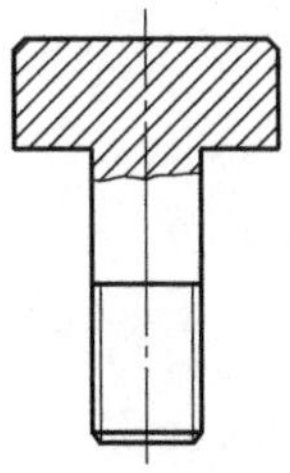	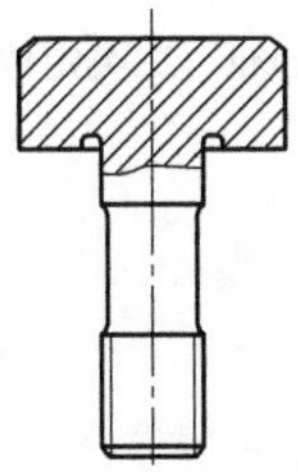	Splines花键 Sharp corner 尖角 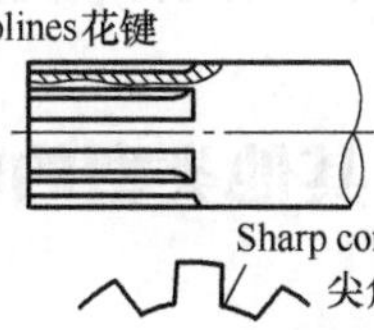	Increased shaft size 增大轴的尺寸 Radiused fillet 圆弧过渡 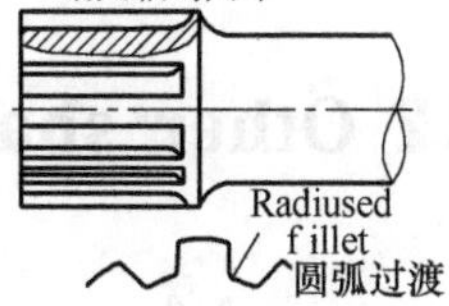
Shoulders 轴肩 Sharp corners 尖角 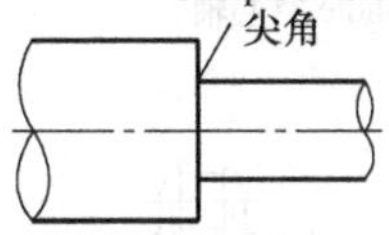	Large fillet radius 大的过渡圆半径 Undercut fillet with fitted collar 开过渡槽和配合套 Undercut radiused fillets 开过渡圆 Stress- relieving grooves 应力释放槽 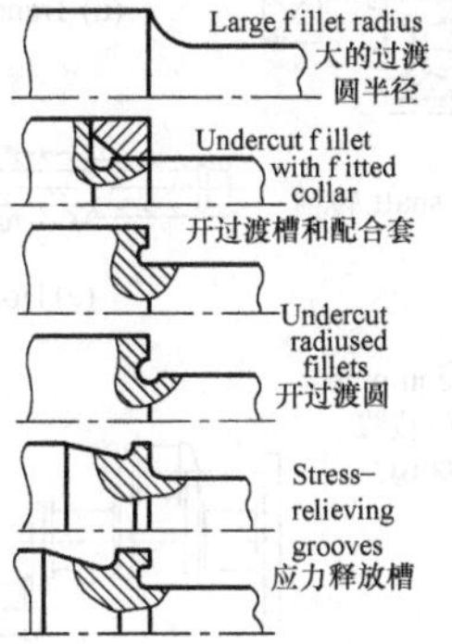	Fitted assemblies 配合装配 Shaft 轴 Wheel、gear、etc 带轮、齿轮等 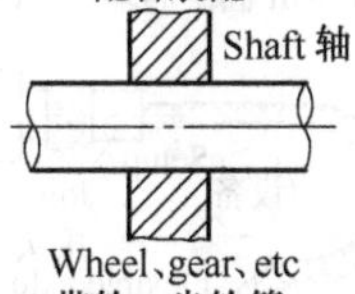	Generous radius 大半径 Increase in journal size 增加轴颈尺寸 Grooves in shaft 轴上开槽 Fillets on hub 轮毂过渡 Grooves in hub 轮毂开槽 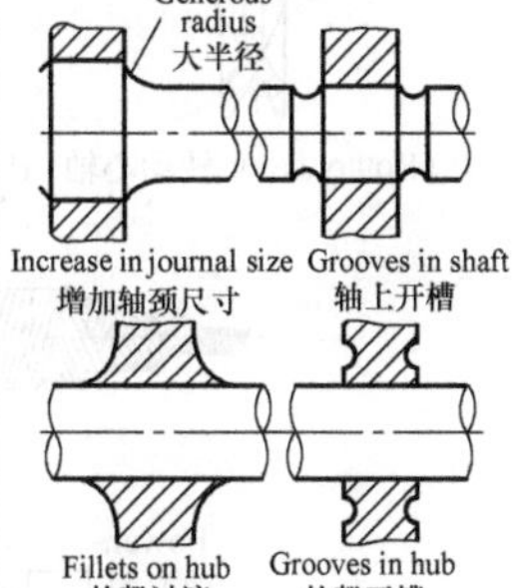
Holes孔眼 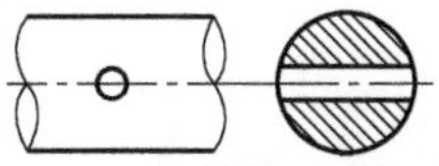	Enlarged section at hole 孔处放大 Stress-relieving grooves 应力释放槽 	Key ways 键槽 Sharp corners 尖角 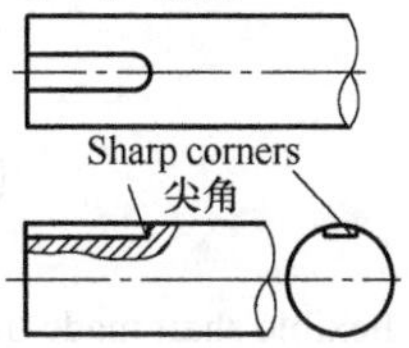	Increased shaft size 增加轴尺寸 Radiused corner 圆弧拐角 Radius半径 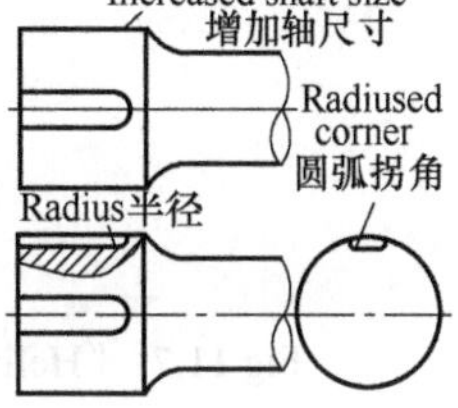

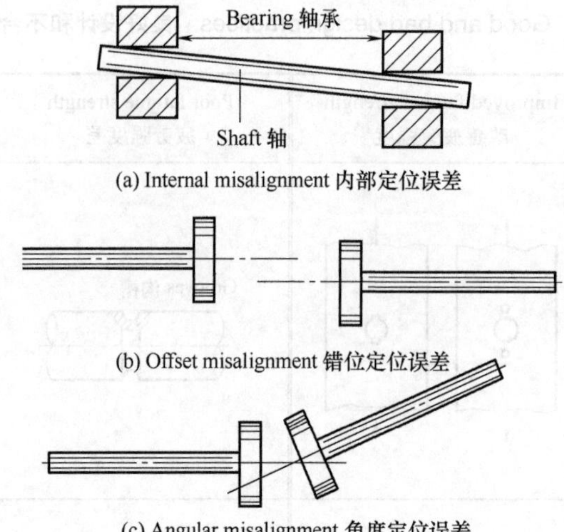

(a) Internal misalignment 内部定位误差

(b) Offset misalignment 错位定位误差

(c) Angular misalignment 角度定位误差

Fig 11.5 Three types of misalignment 三类定位误差

11.2 Other shafts 其他类型的轴

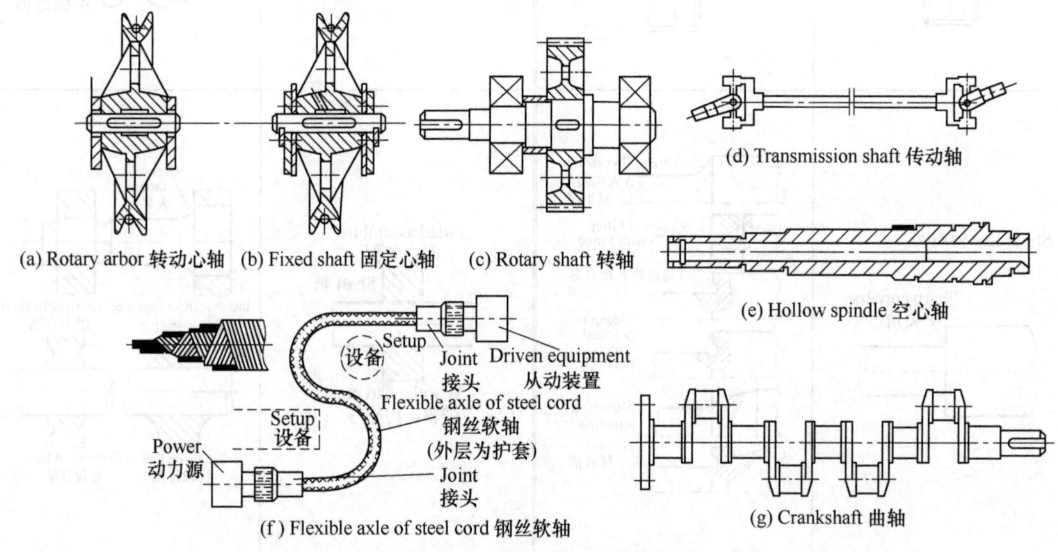

(a) Rotary arbor 转动心轴 (b) Fixed shaft 固定心轴 (c) Rotary shaft 转轴 (d) Transmission shaft 传动轴 (e) Hollow spindle 空心轴 (f) Flexible axle of steel cord 钢丝软轴 (g) Crankshaft 曲轴

Fig 11.6 Types of shafts 轴的种类

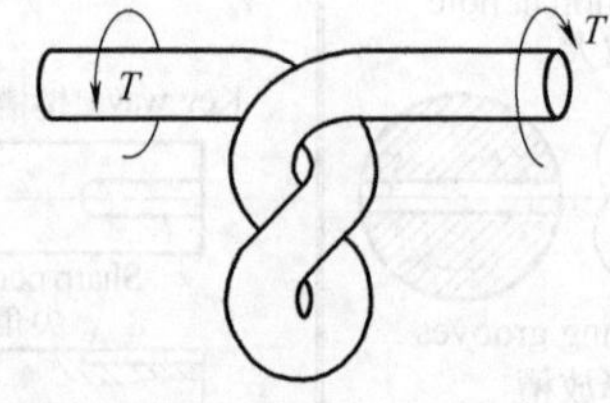

Fig 11.7 "Helixed" flexible shaft made of wire or rope 金属丝或绳制作的“螺旋”柔性轴

Unit 12 Bearings 轴承

12.1 Plain bearings 滑动轴承

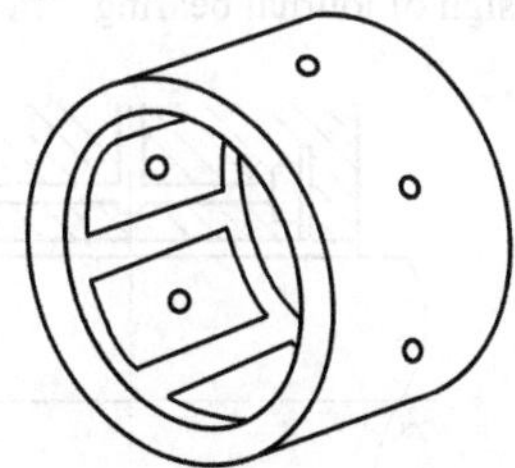

Fig 12.1 Typical hydrostatic journal bearing 典型静压滑动轴承

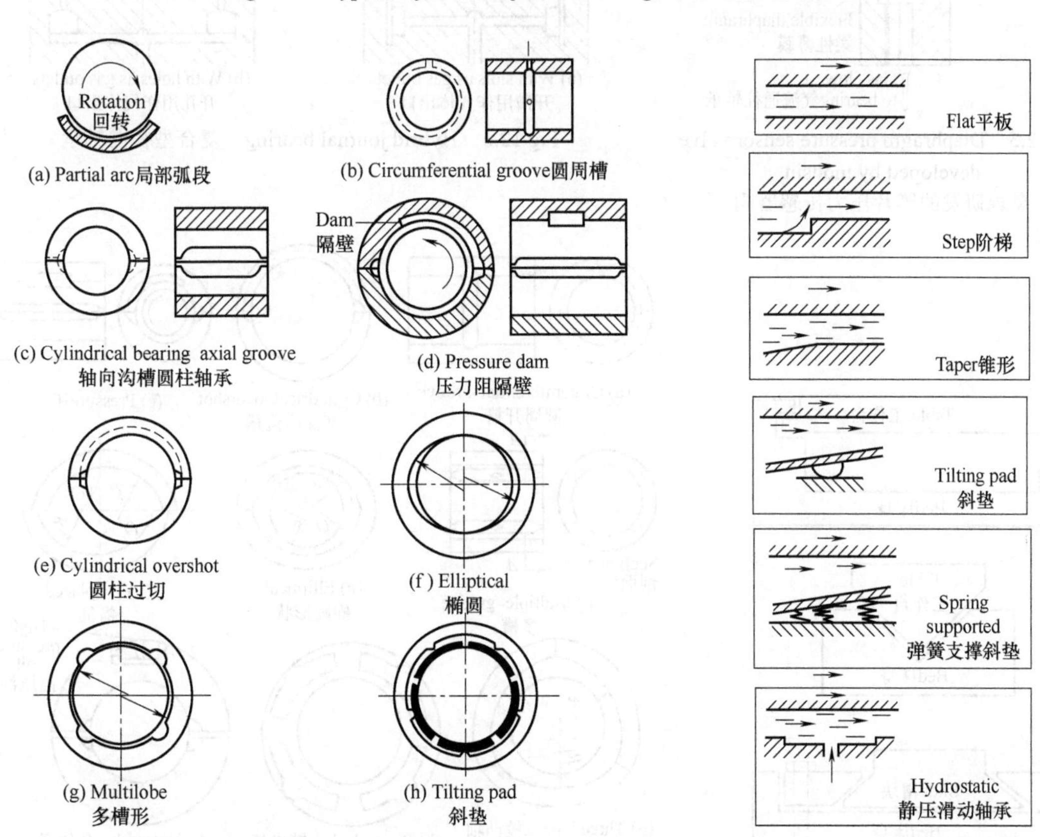

Fig 12.2 Types of pressure-fed journal bearings
动压滑动轴承的类别

Fig 12.3 Common thrust bearings
常用滑动推力轴承

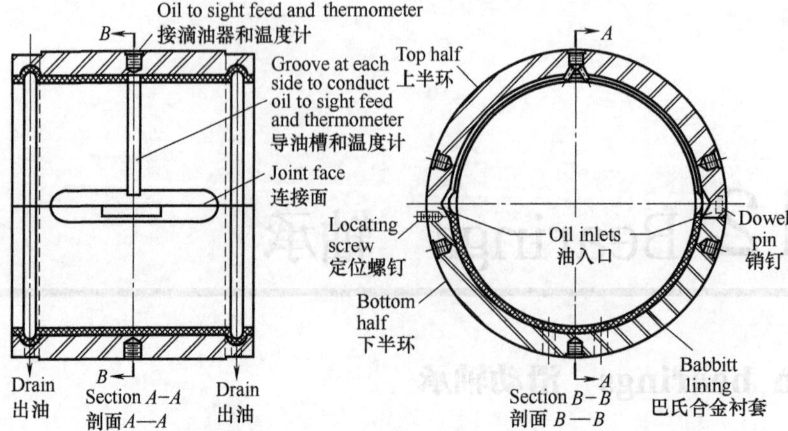

Fig 12.4 Typical design of journal bearing 滑动轴承的典型设计

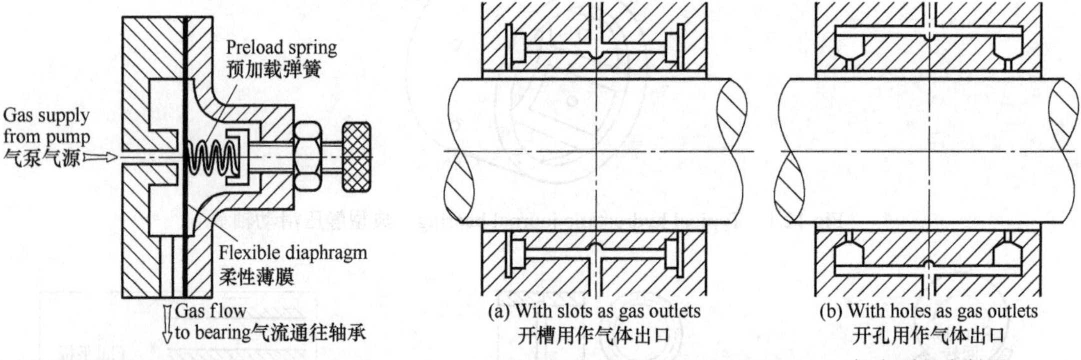

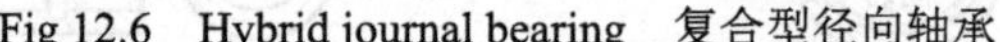

Fig 12.5 Diaphragm pressure sensor valve developed by mohsin
莫森研发的膜片压力传感器阀

Fig 12.6 Hybrid journal bearing 复合型径向轴承

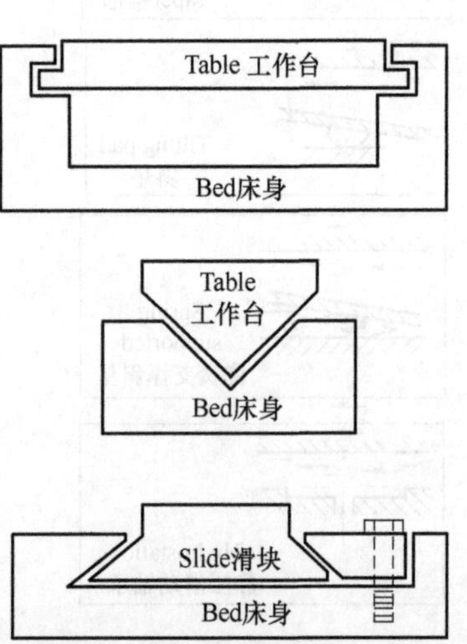

Fig 12.7 Types of guide bearings
导轨轴承的类别

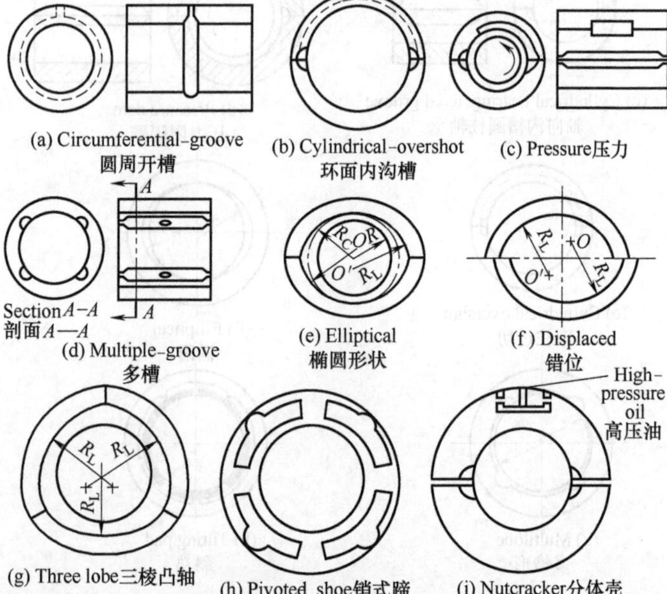

Fig 12.8 Typical shapes of several types of pressure-fed bearings
几种静压轴承的典型结构

Table 12.1 Types of hydrostatic journal bearings 静压轴承类别

Type 类别	Diagram 图例	Type 类别	Diagram 图例	Type 类别	Diagram 图例
Single-pad journal bearing 单轴瓦滑动轴承		Multipad journal bearing 多轴瓦滑动轴承		Multirecess journal bearing 多凹槽滑动轴承	

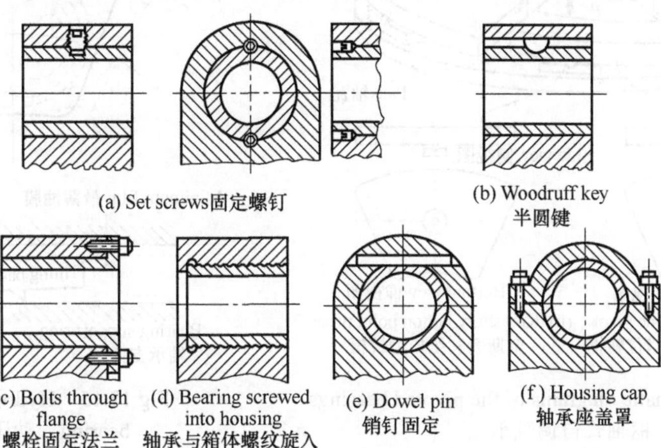

Fig 12.9 Methods of retaining bearings 轴承固定的方法

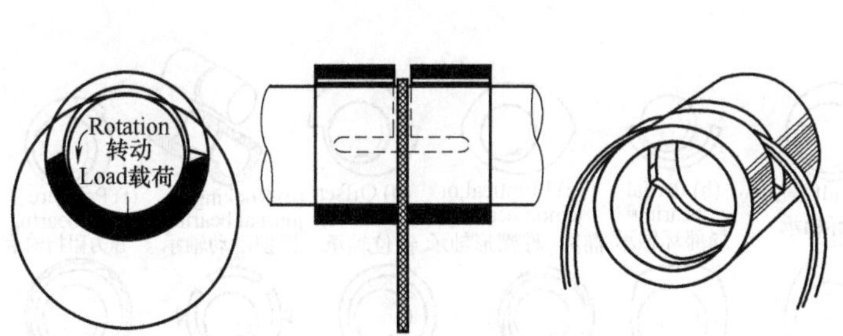

Fig 12.10 Ring oiled bearings 油环轴承

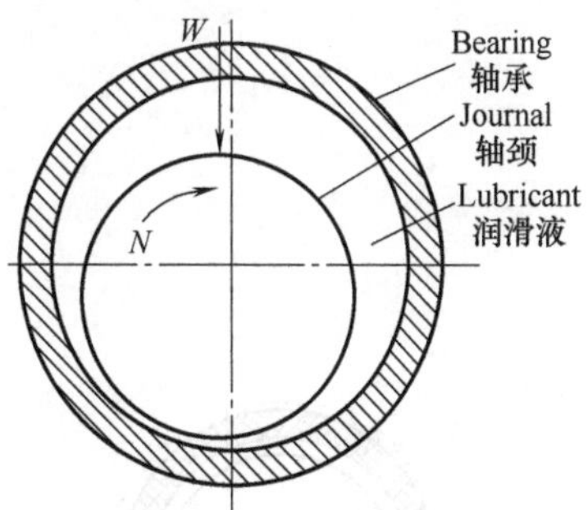

Fig 12.11 Basic components of a journal bearing 滑动轴承基本构成

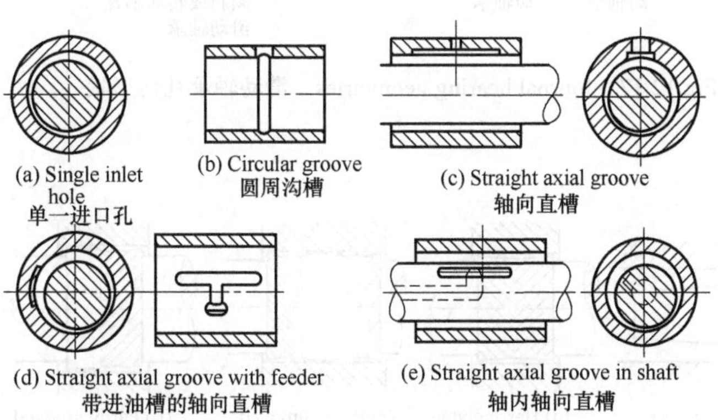

Fig 12.12 Types of journal bearing oil grooving 滑动轴承油槽种类

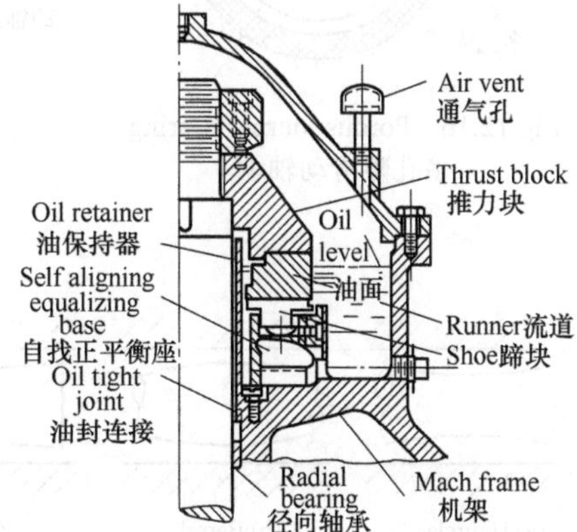

Fig 12.13 Half section of mounting for vertical thrust bearing 立式推力轴承安装的半剖图

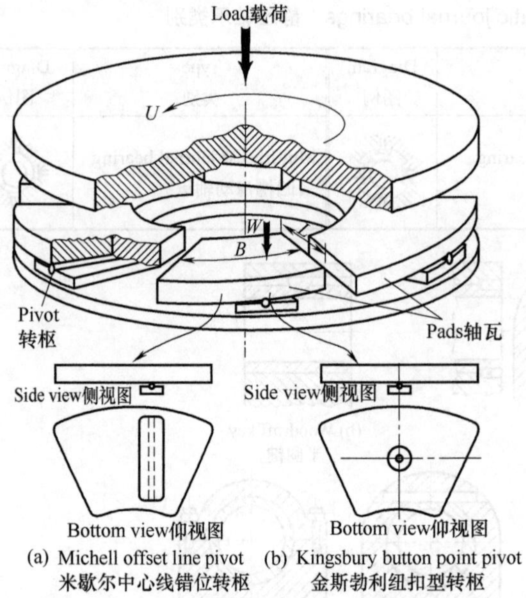

Fig 12.14 Schematic diagrams of the pivoted bearings 枢轴式自调轴承

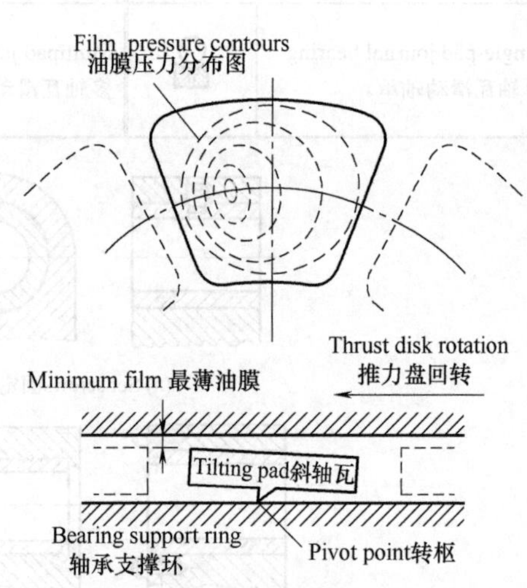

Fig 12.15 Hydrodynamic thrust bearing 动压推力轴承

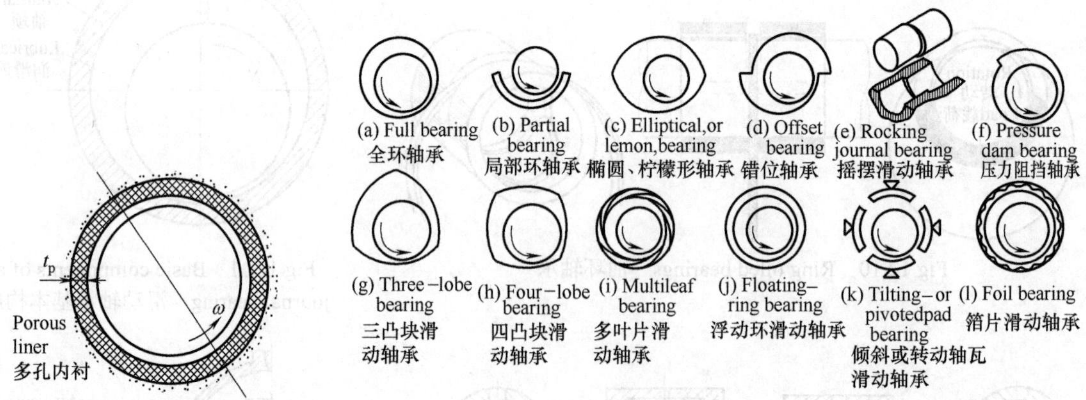

Fig 12.16 Porous journal bearing 多孔隙滑动轴承

Fig 12.17 Journal bearing geometries 滑动轴承几何形状

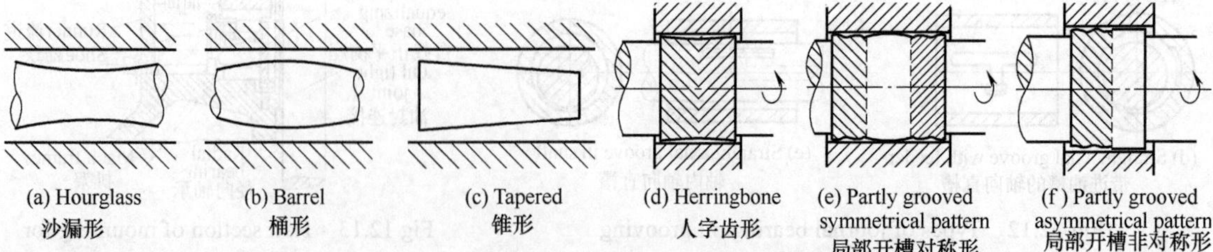

Fig 12.18 Journal shapes 滑动轴承形状

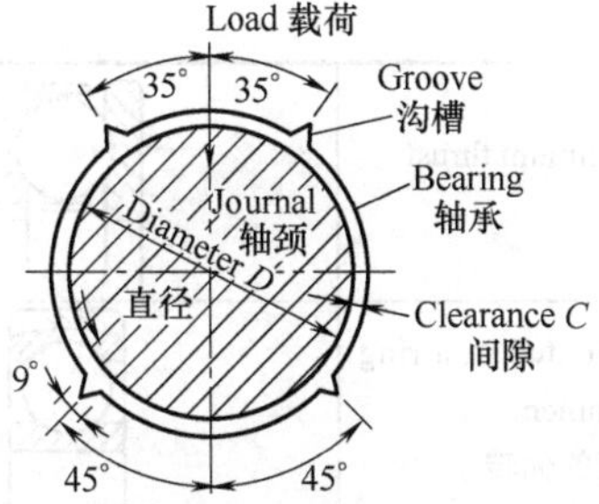

Fig 12.19 Four-axial groove bearing 四轴向槽轴承

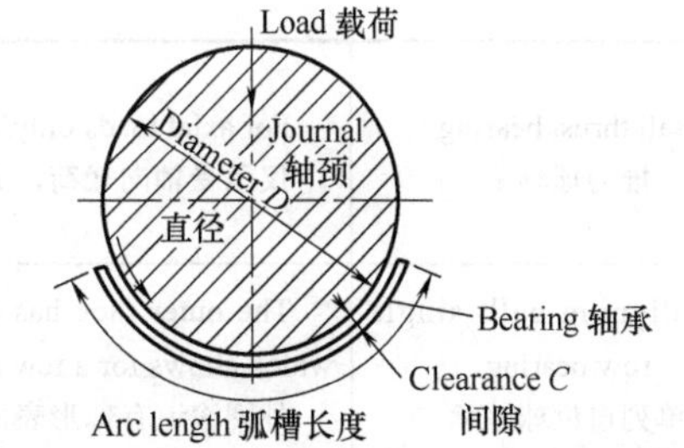

Fig 12.20 Partial arc bearing 局部弧段轴承

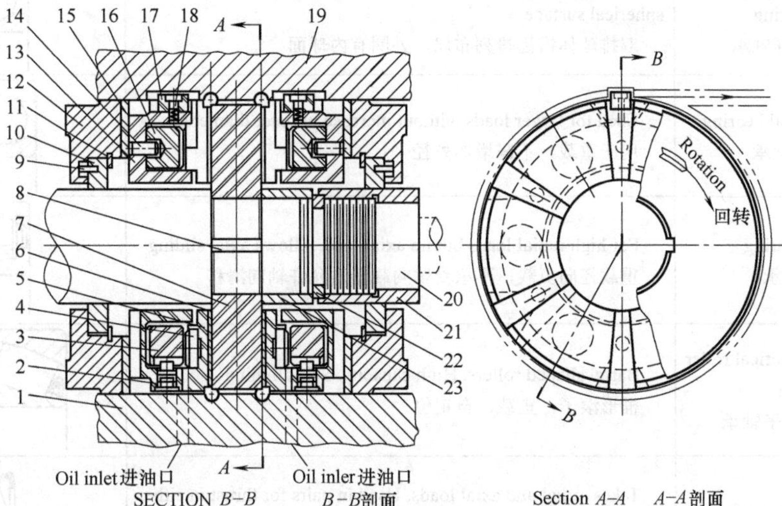

Fig 12.21 Section of tilting-pad thrust bearing 斜板（轴瓦）推力轴承

1—Bearing bracket 轴承架；2—Leveling-plate set-screw 平衡盘固定螺钉；3—Upper leveling plate 上平衡盘；4—Shoe support 蹄块支撑；5—Shoe 蹄块；6—Shoe babbitt 蹄块巴氏合金衬套；7—Collar 套筒；8—Key 键；9—Pin 销；10—Oil guard 油护盖；11—Snap ring 卡环；12—Thrust-bearing ring 推力轴承环；13—Base ring(in halves) 基环（半片）；14—Leveling-plate dowel 平衡板销；15—Shim 底板；16—Lower leveling plate 下平衡盘；17—Base-ring key 基环键；18—Base-ring key screw 基环键螺钉；19—Bearing-bracket cap 轴承支架盖；20—Shaft 轴；21—Outer check nut 外部校正螺母；22—Retaining ring 保持环；23—Inner check nut 内部校正螺母

12.2 Rolling bearings 滚动轴承

Table 12.2 Types of rolling bearings 滚动轴承类别

Ball bearing 球轴承	Used for radial load but will take one third load axially. Deep grooved type now used extensively. Light, medium and heavy duty types available 用于径向支撑，承受三分之一轴向载荷，深沟类应用广泛，有轻型、中型和重型可选	Light 轻型　Medium 中型　Heavy 重型
Angular contact ball bearing 角接触球轴承	Takes a larger axial load in one direction. Must be used in pairs if load in either direction 承受较大的单向轴向力，载荷双向改变的话，必须成对使用	

续表

Ball thrust bearing 推力球轴承	For axial loads only. Must have at least a minimum thrust 仅承受轴向载荷，必须具备最小的推力	
Self-aligning ball, single row bearing 单列自位球轴承	The outer race has a spherical surface mounted in a ring which allows for a few degress of shaft misalignment 外圈滚道有球形表面内环，允许轴有几度的偏转	
Self-aligning ball, double row bearing 双列自位球轴承	Two rows of balls in staggered arrangement. Outer race with spherical surface 双排球体错齿排列布局，外圈有内球面	
Double row ball bearing 双列球轴承	Used for larger loads without increase in outer diameter 用于重载，不需增加外径	
Roller bearing 滚子轴承	For high radial loads but no axial load. Allows axial sliding 重载径向负载，不承受轴向载荷，允许轴向滑移	
Self-aligning spherical roller bearing 自位球面滚子轴承	Barrel shaped rollers. High capacity. Self-aligning 桶形滚子，重载，自定位	
Tapered roller bearing 圆锥滚子轴承	Takes radial and axial loads. Used in pairs for thrust in either direction 承受径向和轴向力，推力改变方向时需成对使用	
Needle roller bearing 滚针轴承	These run directly on the shaft with or without cages. Occupy small space 直接在轴上运转、可以不要保持架，占空间小	
Shields, seals and grooves bearing 防尘、密封沟槽轴承	Shields on one or both sides prevent ingress of dirt. Seals allow packing with grease for life. A groove allows fitting of a circlip for location in bore 一端或两端防止异物进入，密封可以长期保持润滑脂卡槽用于孔里装配定位	Shields 防尘盖 Shields and seals 防尘盖和密封圈 Circlip groove 卡槽

Ball 球体　Cylindrical roller 圆柱滚子　Taper roller 圆锥滚子　Spherical roller 球面滚子　Needle pin 滚针

Fig 12.22　Types of roller　滚子的种类

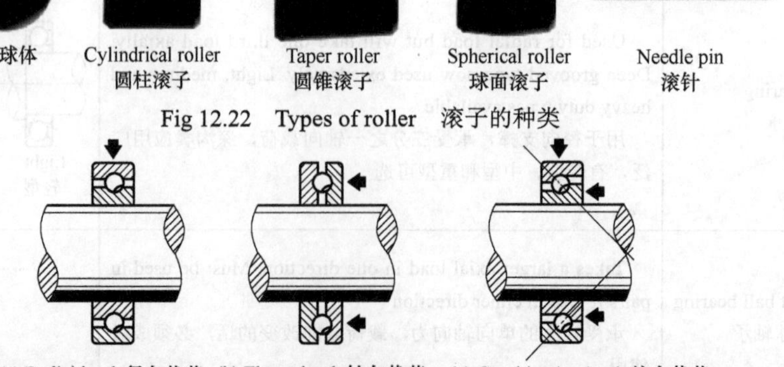

(a) Radial load 径向载荷　(b) Thrust load 轴向载荷　(c) Combination load 综合载荷

Fig 12.23　Three principal types of ball bearing loads　三种主要的轴承载荷

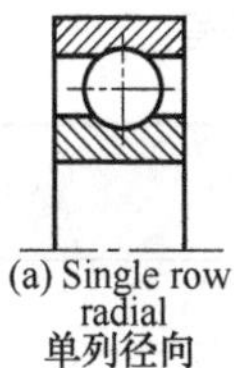
(a) Single row radial
单列径向

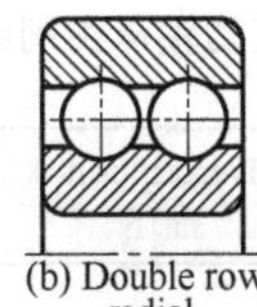
(b) Double row radial
双列径向

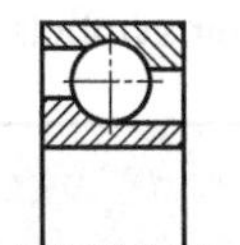

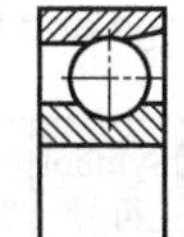
(c) Radial thrust(angular contact)
单列径向推力(角接触)

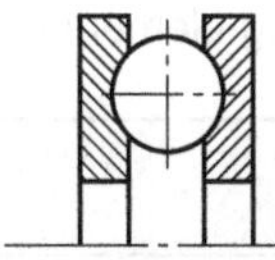
(d) Single row thrust
单列推力

(e) Double thrust
双列推力

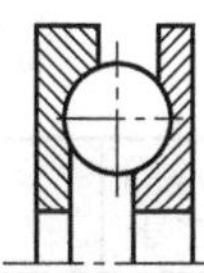
(f) Thrust radial
推力径向

Fig 12.24 Types of ball bearing 球轴承类型

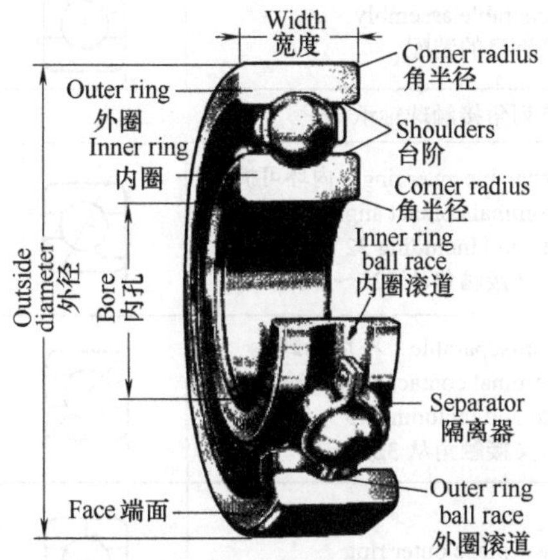

Fig 12.25 Conrad anti-friction ball bearing parts 康拉德防摩擦球轴承部件

ADEC tolerance grade ADEC 公差等级
Bearing number
轴承代号
Burnished dot shows high point of eccentricity
磨光点表示偏心的高点
Apply thrust against this shoulder
对着轴肩承受施加的轴向力
Deep shoulder is stamped side of rearing bearing back
轴承冲压侧面形成深阶梯面(轴承背面)
Deep shoulder 深轴肩
Outer ring
外圈
Outer raceway
外圈滚道
Ball cage
球体保持架
Balls球形滚动体
Inner ring
内圈
Inner raceway
内圈滚道
Counterbored low shoulder bearing face
反镗低轴肩(轴承前面)

Fig 12.26 Photograph of a precision ball bearing of the type generally used in machine-tool applications to illustrate terminology 机床常用精密轴承术语说明

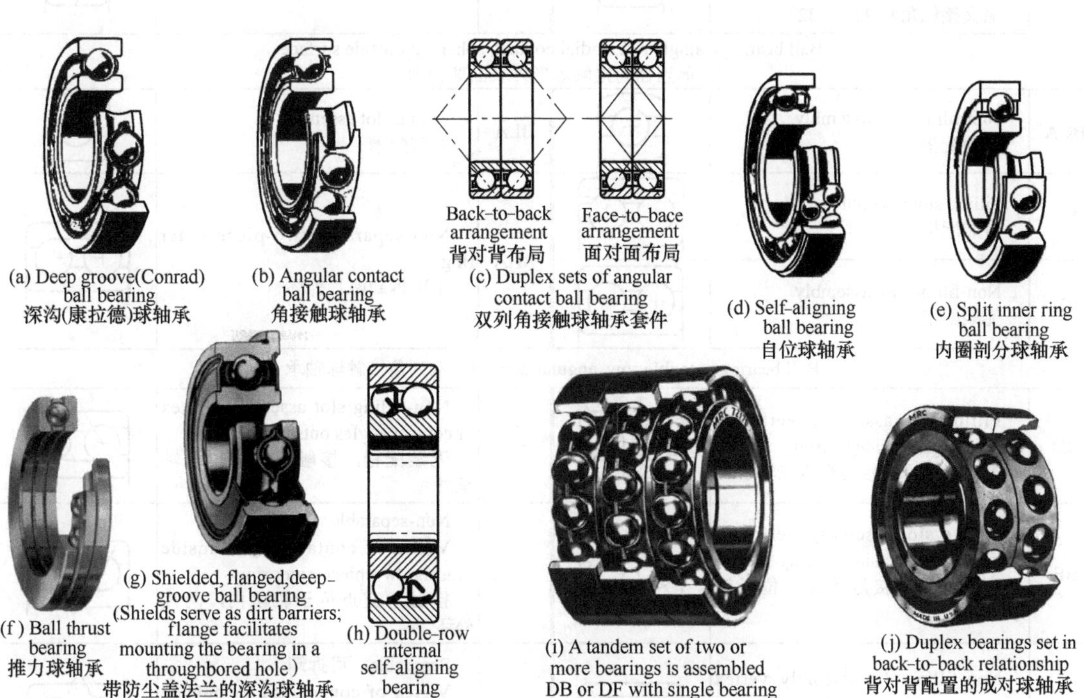

(a) Deep groove(Conrad) ball bearing
深沟(康拉德)球轴承

(b) Angular contact ball bearing
角接触球轴承

(c) Duplex sets of angular contact ball bearing
双列角接触球轴承套件

(d) Self-aligning ball bearing
自位球轴承

(e) Split inner ring ball bearing
内圈剖分球轴承

(f) Ball thrust bearing
推力球轴承

(g) Shielded, flanged, deep-groove ball bearing (Shields serve as dirt barriers; flange facilitates mounting the bearing in a throughbored hole)
带防尘盖法兰的深沟球轴承(防尘盖阻止脏物进入，法兰使得通孔安装轴向定位)

(h) Double-row internal self-aligning bearing
双列内部自位轴承

(i) A tandem set of two or more bearings is assembled DB or DF with single bearing
两个或更多级联轴承组与单个轴承背对背或面对面装配

(j) Duplex bearings set in back-to-back relationship
背对背配置的成对球轴承

Fig 12.27 Ball bearing 球轴承

Table 12.3 Types of rolling element bearings and their symbols 滚动轴承类别和符号

Symbol 符号	Description 描述		Symbol 符号	Description 描述	
	Ball bearings, single row, radial contact 单列径向接触球轴承				
BC	Non-filling slot assembly 无压入槽的装配	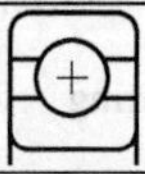	BH	Non-separable counter-bore assembly 不可拆离带锥面的装配	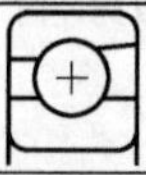
BL	Filling slot assembly 有压入槽的装配	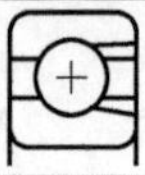	BM	Separable assembly 可拆离的装配	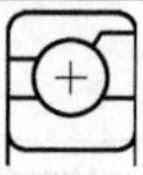
	Ball bearings, single row, angular contact[a] 单列角接触球轴承				
BN	Non-separable 不可拆卸 Nominal contact angle: from above 10° to and including 22° 名义接触角从 10°～22°	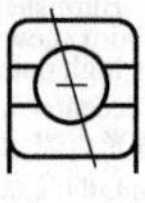	BAS	Separable inner ring 内环可拆卸 Nominal contact angle: from above 22° to and including 32° 名义接触角从 22°～32°	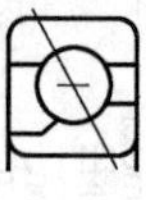
BNS	Separable outer ring 可拆卸 Nominal contact angle: from above 10° to and including 22° 名义接触角从 10°～22°	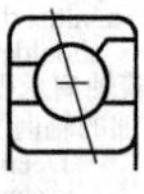	BT	Non-separable 不可拆卸 Nominal contact angle: from above 32° to and including 45° 名义接触角从 32°～45°	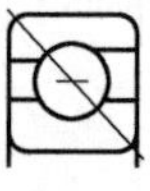
BNT	Separable inner ring 可拆卸 Nominal contact angle: from above 10° to and including 22° 名义接触角从 10°～22°	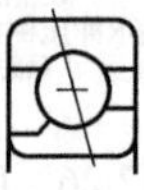	BY	Two-piece outer ring 两片外环	
BA	Non-separable 不可拆卸 Nominal contact angle: from above 22° to and including 32° 名义接触角从 22°～32°	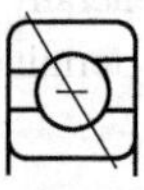	BZ	Two-piece inner ring 两片内环	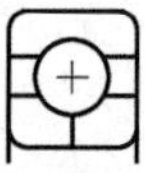
	Ball bearings, single row, radial contact, spherical outside surface 单列径向接触球形外表面球轴承				
BCA	Non-filling slot assembly 无装配槽		BLA	Filling slot assembly 有装配槽	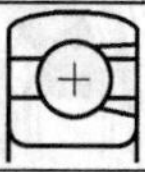
BF	Filling slot assembly 有装配槽	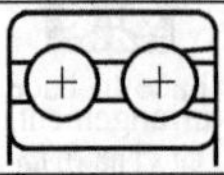	BHA	Non-separable two-piece outer ring 不可拆卸两片外环	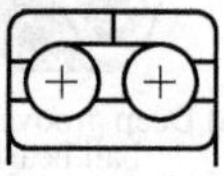
BK	Non-filling slot assembly 无装配槽	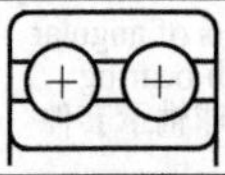			
	Ball bearings, double row, angular contact 双列角接触球轴承				
BD	Filling slot assembly, vertex of contact angles inside bearing 有装配槽，接触角交点在轴承内	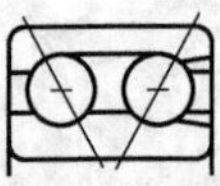	BG	Non-filling slot assembly, vertex of contact angles outside bearing 无装配槽，接触角交点位于轴承外	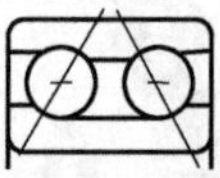
BE	Filling slot assembly, vertex of contact angles outside bearing 有装配槽，接触角交点位于轴承外	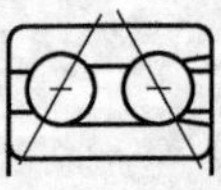	BAA	Non-separable 不可拆卸 Vertex of contact angles inside bearing, two-piece outer ring 接触角交点位于轴承内，两片外环	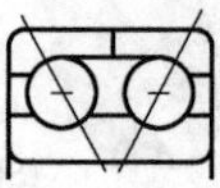
BJ	Non-filling slot assembly, vertex of contact angles inside bearing 无装配槽，接触角交点位于轴承内	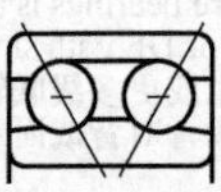	BVV	Separable 可拆卸 Vertex of contact angles outside bearing, two-piece inner ring 接触角交点位于轴承外，两片内环	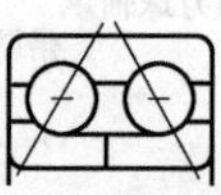

续表

Symbol 符号	Description 描述		Symbol 符号	Description 描述	
Ball bearings, double row, self-aligning 双列自位球轴承					
BS	Raceway of outer ring spherical 外环滚道球形				
Cylindrical roller bearing, single row, non-locating type 单列非定位圆柱滚子轴承					
RU	Inner ring without ribs, double-ribbed outer ring, inner ring separable 内环无肋，外环双肋，内环可拆卸		RNS	Double-ribbed inner ring, outer ring without ribs, outer ring separable Spherical outside surface 内环双肋，外环无肋，外环可拆卸，球形外表面	
RUP	Inner ring without ribs, double-ribbed outer ring with one loose rib, both rings separable 内环无肋，外环双肋，一松肋，内外环可拆卸		RAB	Inner ring without ribs, single-ribbed outer ring, both rings separable 内环无肋，外环单肋，内 / 外环可拆卸	
RUA	Inner ring without ribs, double-ribbed outer ring，inner ring separable Spherical outside surface 内环无肋，外环双肋，内环可拆卸，球形外表面		RM	Inner ring without ribs, rollers located by cage, end-rings or internal snap rings recesses in outer ring, inner ring separable 内环无肋，有保持架，外环内沉，内环可拆卸	
RN	Double-ribbed inner ring, outer ring without ribs, outer ring separable 外环无肋，内环双肋，外环可拆卸		RNU	Inner ring without ribs, outer ring without ribs, both rings separable 内环无肋，外环无肋，内外环都可拆卸	
Cylindrical roller bearings, single row, one-direction-locating type 单列单向定位圆柱滚子轴承					
RR	Single-ribbed inner-ring, outer ring with two internal snap rings, inner ring separable 内环单肋，外环双肋，内卡圈，内环可拆卸		RF	Double-ribbed inner ring, single-ribbed outer ring, outer ring separable 内环双肋，外环单肋，外环可拆卸	
RJ	Single-ribbed inner ring, double-ribbed outer ring, inner ring separable 内环单肋，外环双肋，内环可拆卸		RS	Single-ribbed inner ring, outer ring with one rib and one internal snap ring, inner ring separable 内环单肋，外环单肋和卡圈槽，内环可拆卸	
RJP	Single-ribbed inner ring, double-ribbed outer ring with one loose rib, both rings separable 内环单肋，外环双肋，一松肋，内外环都可拆卸		RAA	Single-ribbed inner ring, single-ribbed outer ring, both rings separable 内环单肋，外环单肋，内外环都可拆卸	
Cylindrical roller bearings, single row, two-direction-locating type 单列双向定位圆柱滚子轴承					
RK	Double-ribbed inner ring, outer ring with two internal snap rings, non-separable 内环双肋，外环双卡槽，不可拆卸		RY	Double-ribbed inner ring, outer ring with one rib and one internal snap ring, non-separable 内环双肋，外环单肋，单卡槽，不可拆卸	

续表

Symbol 符号	Description 描述		Symbol 符号	Description 描述	
Cylindrical roller bearings, single row, two-direction-locating type 单列双向定位圆柱滚子轴承					
RC	Doble-ribbed inner ring, double-ribbed outer ring, non-separable 内环双肋，外环双肋，不可拆卸		RCS	Double-ribbed inner ring, double-ribbed outer ring, non-separable, spherical outside surface 内环双肋，外环双肋，不可拆卸，外表面球形	
RG	Inner ring, with one rib and one snap ring, double-ribbed outer ring, non-separable 内环单肋，单卡槽，外环双肋，不可拆卸				
RP	Double-ribbed inner ring, double-ribbed outer ring with one loose rib, outer ring separable 内环双肋，外环双肋，一松肋，外环可拆卸		RT	Double-ribbed inner ring with one loose rib, double-ribbed outer ring, inner ring separable 内环双肋，一松肋，外环双肋，内环可拆卸	
Cylindrical roller bearings 圆柱滚子轴承					
Double row non-locating type 双列无定位			Double row two-direction-locating type 双列双定位		
RA	Inner ring without ribs, three integral ribs on outer ring, inner ring separable 内环无肋，外环三肋，内环可拆卸		RB	Three integral ribs on inner ring, outer ring without ribs, with two internal snap rings, non-separable 内环三肋，外环无肋，双卡槽，不可拆卸	
RD	Three integral ribs on inner rings, outer ring without ribs, outer ring separable 内环三肋，外环无肋，外环可拆卸		Multi-row non-locating type 多列无定位		
RE	Inner ring without ribs, outer rings without ribs, with two internal snap rings, inner ring separable 内环无肋，外环无肋，两卡槽，内环可拆卸		RV	Inner ring without ribs, double-ribbed outer ring (loose ribs), both rings separable 内环无肋，外环双肋，松肋，内外环都可拆卸	
Needle roller bearings(drawn cup) 滚针轴承（冲压外圈）					
NIB NB	Needle roller bearing, full complement drawn cup, without inner ring 滚针轴承、全接触、冲压外圈、无内环		NIYM NYM	Needle roller bearing, full complement,rollers retained by lubricant, drawn cup, closed end, without inner ring 滚针轴承、全接触、润滑保持滚针、冲压外圈、端头闭合无内环	
NIBM NBM	Needle roller bearing, full complement, drawn cup, closed end without inner ring 滚针轴承、全接触、冲压外圈、闭合端头、无内环		NIH NH	Needle roller bearing, with cage, drawn cup, without inner ring 滚针轴承、有保持架、冲压外圈、无内环	
NIY NY	Needle roller bearing,full complement rollers retained by lubricant, drawn cup, without inner ring 滚针轴承、全接触、滚针靠润滑油保持，冲压外圈、无内环		NIHM NHM	Needle roller bearing, with cage, drawn cup, closed end, without inner ring 滚针轴承、保持架、全接触、冲压外圈、无内环	

续表

Symbol 符号	Description 描述	Symbol 符号	Description 描述
	Needle roller bearings(drawn cup) 滚针轴承（冲压外圈）		
	Needle roller bearings 保持架滚针轴承		Needle roller and cage assemblies 滚针轴承和保持架组件
NIA NA	Needle roller bearing, with cage, machined ring lubrication hole and groove in OD without inner ring 保持架滚针轴承、外圆加工润滑孔和槽、无内环	NIM NM	Needle roller and cage assembly 滚针轴承和保持架组件
	NEEDLE ROLLER BEARING INNER RINGS 滚针轴承内环		Machined ring needle roller bearings Type NIA may be used with inch dimensioned inner rings Type NIR, and Type NA may be used with metric dimensional inner rings Type NR. NIA 类尺寸为英制，NIR 和 NA 可为公制尺寸
NIR NR	Needle roller bearing inner ring, lubrication hole and groove in bore 滚针轴承内环、润滑孔和槽		
	Tapered roller bearings—inch 圆锥滚子轴承（英制）		
TS	Single row 单列	TDI	Two row, double-cone single cups 双列、双锥外环、单内环
TDO	Two row,double-cup single-cone adjustable 双列、双杯内环、单锥可调	TNA	Two row, double-cup single cone nonadjustable 双列、双杯内环、单锥不可调
TQD, TQI	Four row, cup adjusted 四列、外环调整		
	Tapered roller bearings—metric 公制圆锥滚子轴承（公制）		
TS	Single row, straight bore 单列直孔	TSF	Single row ,straight bore, flanged cup 单列直孔、法兰边外环
TDO	Double row, straight bore, two single cones, one double cup with lubrication hole and groove 双列直孔、两个单锥、双面润滑孔和沟槽	2TS	Double row, straight bore, two single cones, two single cups 双列直孔、两个单锥、两个外环
	Thrust tapered roller barings 圆锥滚子推力轴承		
TT	Thrust bearing 推力轴承		
	Self-aligning roller bearings, double row 双列自位滚子轴承		
SD	Three integral ribs on inner ring, race way of outer ring spherical 内环三肋、外环球形滚道	SL	Raceway of outer ring spherical Rollers guided by the cage, two integral ribs on inner ring 外环球形滚道，保持架导引滚动体，内环有两条完整的肋槽
SE	Race way of outer ring spherical, rollers guided by separate center guide ring in outer ring 外环球形滚道、独立中心导环导向		Self-aligning roller bearings single row 单列自位滚子轴承
SW	Race way of inner ring spherical 内环球形滚道	SR	Inner ring with ribs, raceway of outer ring spherical, radial contact 内环带肋、外环球形滚道、径向接触

续表

Symbol 符号	Description 描述	Symbol 符号	Description 描述
	Self-aligning roller bearings, double row 双列自位滚子轴承		
SC	Raceway of outer ring spherical, rollers guided by separate axially floating guide ring on inner ring 外环球形滚道、内环上轴向浮动导环导向	SA	Raceway of outer ring spherical, angular contact 外环球形滚道、角度接触
		SB	Raceway of inner ring spherical, angular contact 内环球形滚道，角接触
	Thrust ball bearings 轴向推力球轴承		
TA TB	Single direction, grooved raceways, flat seats 单向沟槽滚道，平面座	TDA	Double direction, washers with grooved raceways, flat seats 双向垫圈沟槽滚道平面座
TBF	Single direction flat washers, flat seats 单向、平垫圈、平面座		
	Thrust roller bearings 轴向推力滚子轴承		
TS	Single direction, aligning flat seats, spherical rollers 单向导向平面座球面滚子	TPC	Single direction, flat seats, flat races, outside band, cylindrical rollers 单向平面座平滚道外环带圆柱滚子
TP	Single direction, flat seats, cylindrical rollers 单向平面座圆柱滚子	TR	Single direction, flat races, aligning seat with align-ing washer, cylindrical rollers 单向平面滚道垫圈定位圆柱滚子

(a) Single row radial 单列径向　(b) Double row radial 双列径向　(c) Radial thrust 径向推力　(d) Self-aligning 自位　(e) Needle bearing 滚针轴承

Fig 12.28 Types of roller bearing 滚子轴承类型

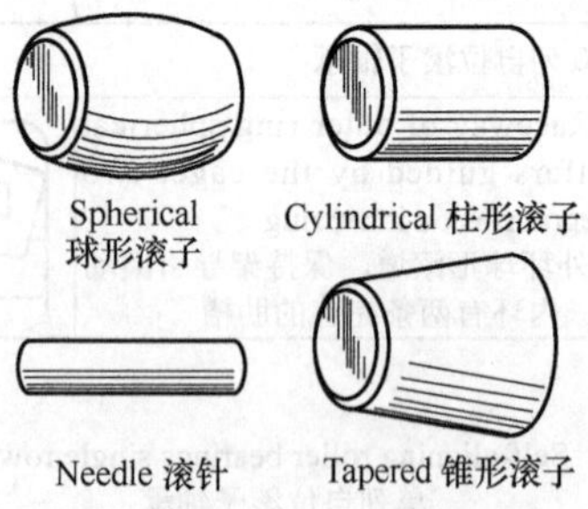

Fig 12.29 Types of roller elements
滚子滚动体元件种类

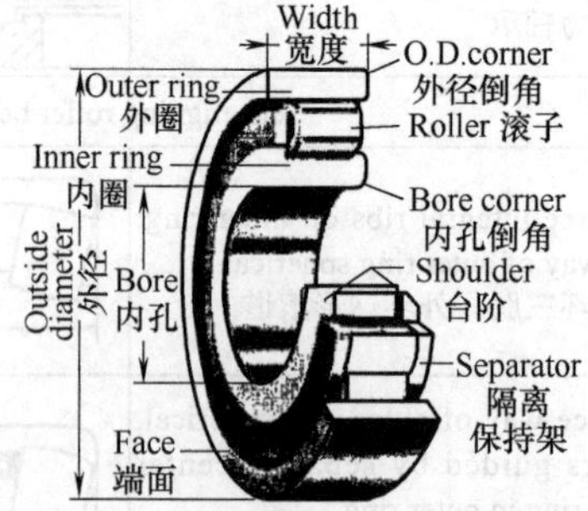

Fig 12.30 Cylindrical roller bearing
柱形滚子轴承

Unit 13 Springs and Flywheels 弹簧和飞轮

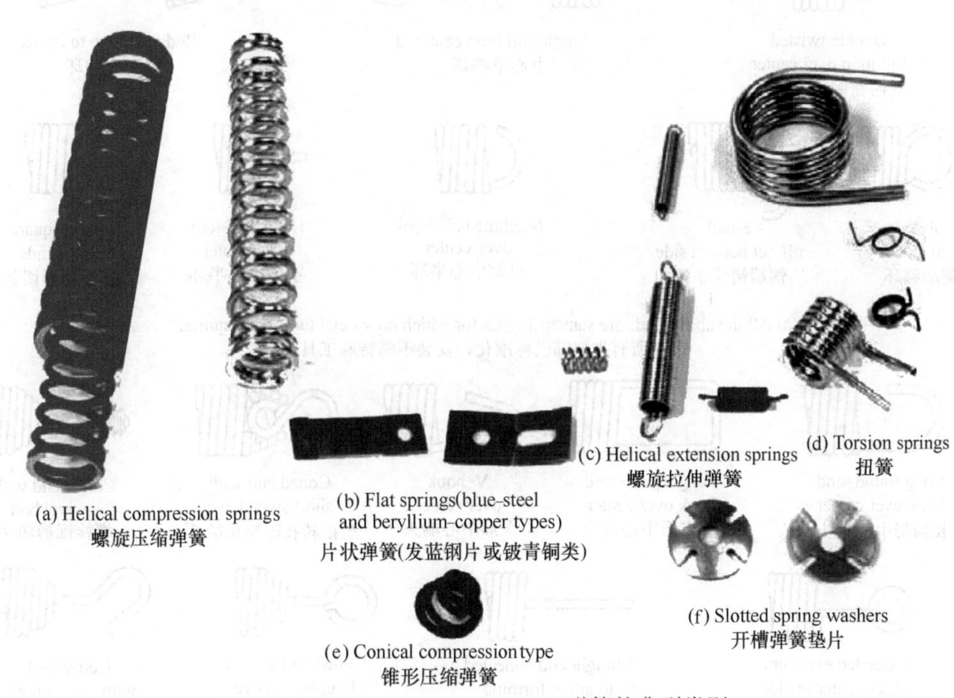

Fig 13.1 Typical types of springs 弹簧的典型类别

13.1 Helical springs 螺旋弹簧

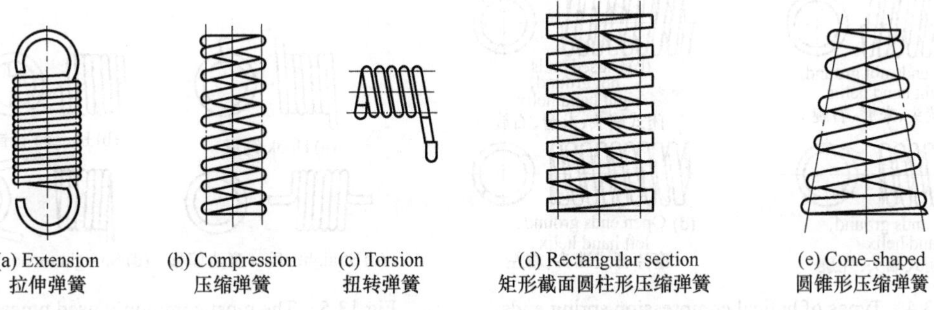

Fig 13.2 Helical spring 螺旋弹簧

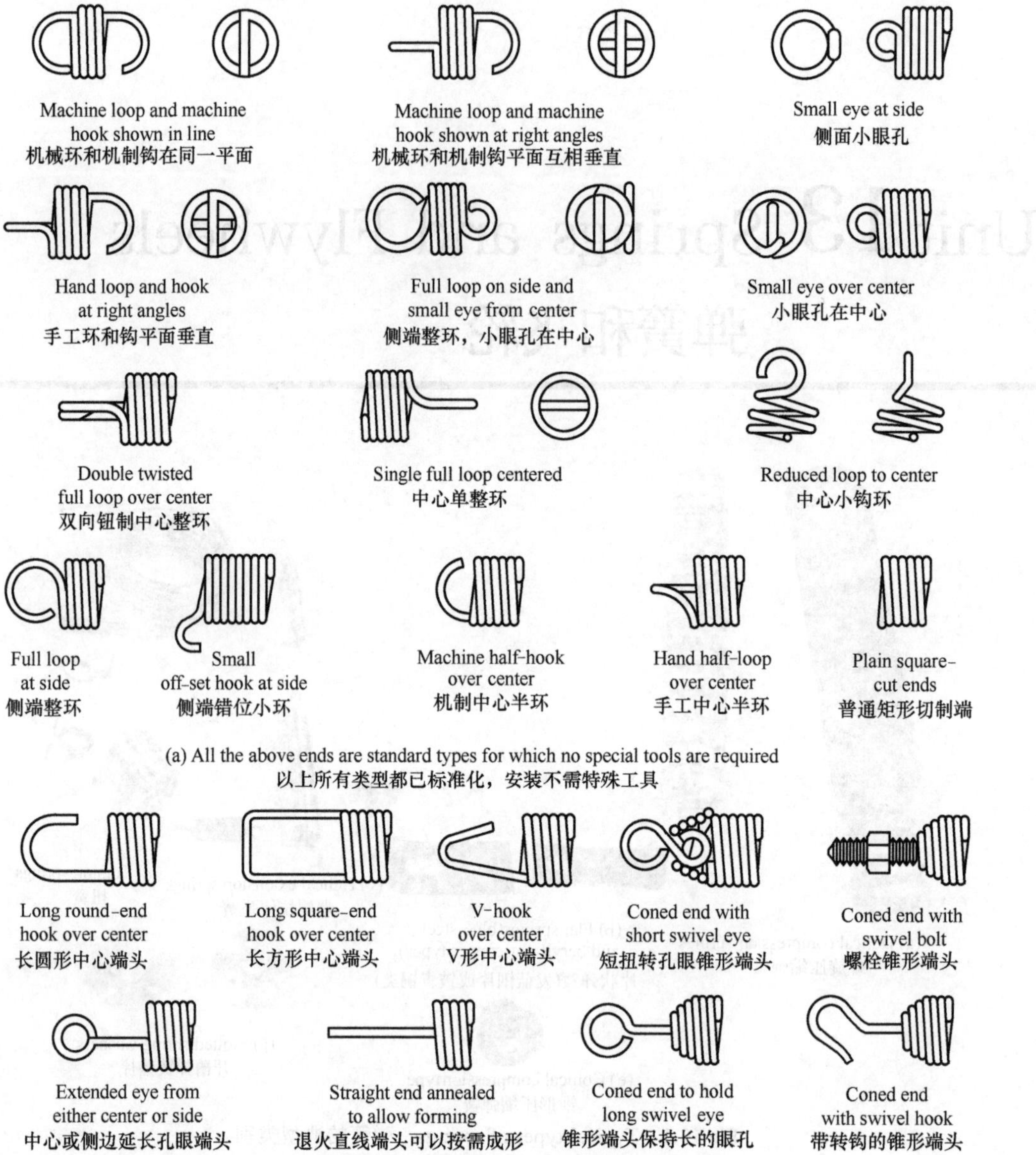

Machine loop and machine hook shown in line
机械环和机制钩在同一平面
Machine loop and machine hook shown at right angles
机械环和机制钩平面互相垂直
Small eye at side
侧面小眼孔
Hand loop and hook at right angles
手工环和钩平面垂直
Full loop on side and small eye from center
侧端整环，小眼孔在中心
Small eye over center
小眼孔在中心
Double twisted full loop over center
双向钮制中心整环
Single full loop centered
中心单整环
Reduced loop to center
中心小钩环
Full loop at side
侧端整环
Small off-set hook at side
侧端错位小环
Machine half-hook over center
机制中心半环
Hand half-loop over center
手工中心半环
Plain square-cut ends
普通矩形切制端
(a) All the above ends are standard types for which no special tools are required
以上所有类型都已标准化，安装不需特殊工具
Long round-end hook over center
长圆形中心端头
Long square-end hook over center
长方形中心端头
V-hook over center
V形中心端头
Coned end with short swivel eye
短扭转孔眼锥形端头
Coned end with swivel bolt
螺栓锥形端头
Extended eye from either center or side
中心或侧边延长孔眼端头
Straight end annealed to allow forming
退火直线端头可以按需成形
Coned end to hold long swivel eye
锥形端头保持长的眼孔
Coned end with swivel hook
带转钩的锥形端头
(b) This group of special ends requires special tools 这组端头需要使用特殊工具安装

Fig 13.3 Types of helical extension spring ends 螺旋拉伸弹簧端头类型

(a) Open ends not ground, right hand helix
开式端头、未磨、右旋
(b) Closed ends not ground, right hand helix
闭式端头、未磨、右旋
(c) Closed ends ground, left hand helix
磨平闭式端头、左旋
(d) Open ends ground, left hand helix
磨片开式端头、右左

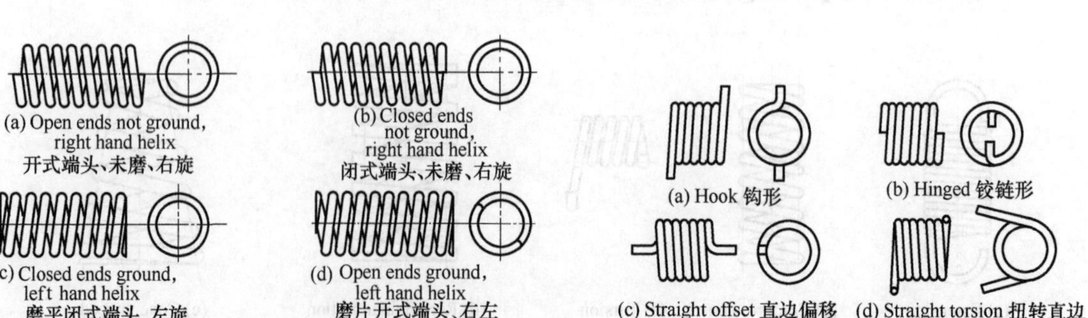

Fig 13.4 Types of helical compression spring ends
螺旋压缩弹簧端头类型

(a) Hook 钩形
(b) Hinged 铰链形
(c) Straight offset 直边偏移
(d) Straight torsion 扭转直边

Fig 13.5 The most commonly used types of ends for torsion springs 扭簧的常用端头类别

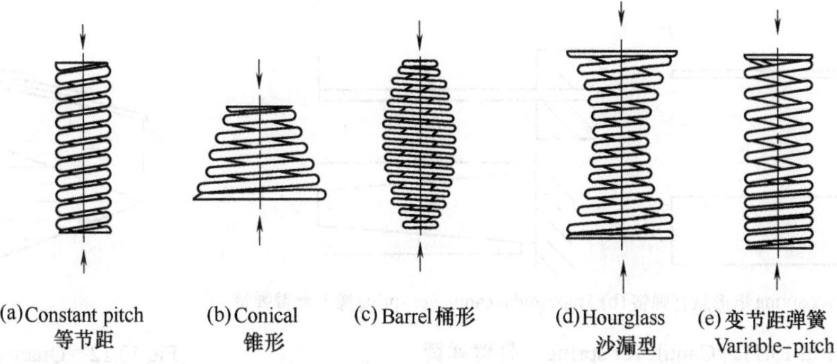

Fig 13.6 Various compression-spring body shapes 几种压缩弹簧外形

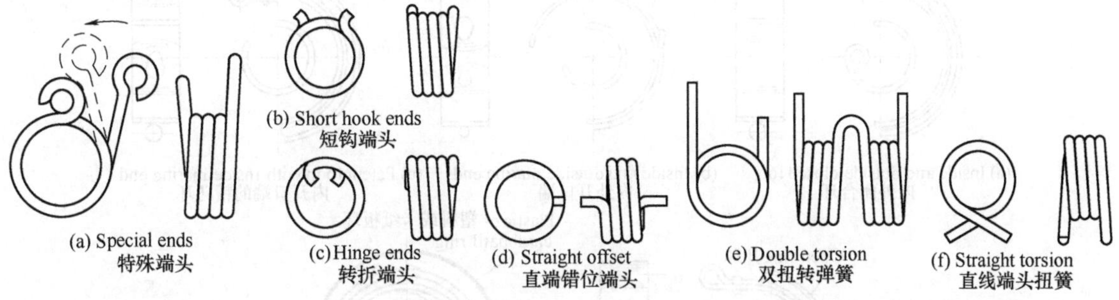

Fig 13.7 Common helical torsion-spring end configurations 常用螺旋扭转弹簧的端部构造

13.2 Other springs 其他弹簧

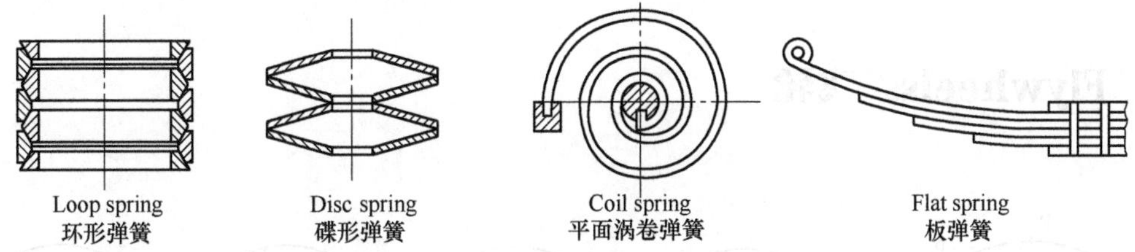

Fig 13.8 Non-helical spring 非螺旋弹簧

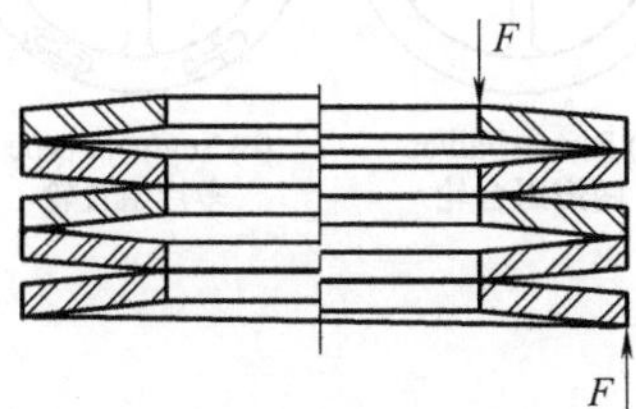

Fig 13.9 Disc springs in series stacking
系列叠片碟簧

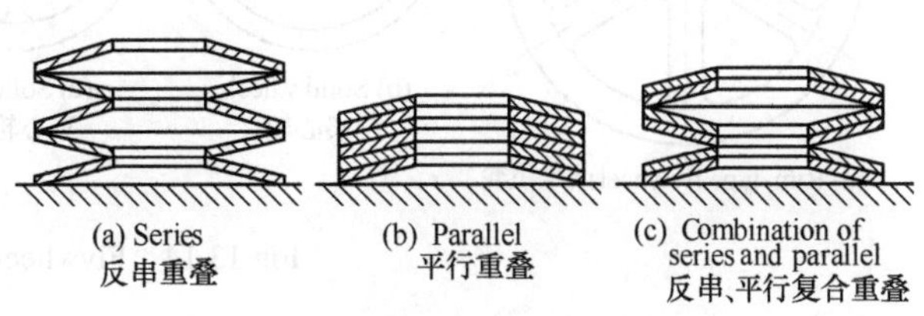

Fig 13.10 Stacks of belleville washers
碟形弹性垫圈的重叠布置

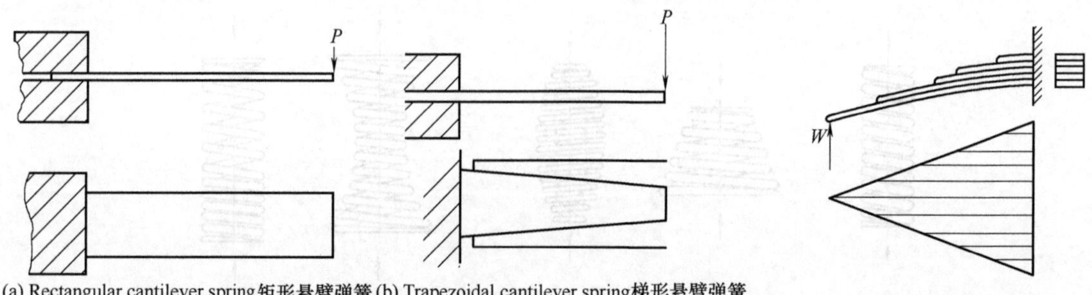

Fig 13.11　Cantilever spring　悬臂弹簧

Fig 13.12　Quarter-elliptic spring　四分之一椭圆叠层弹簧

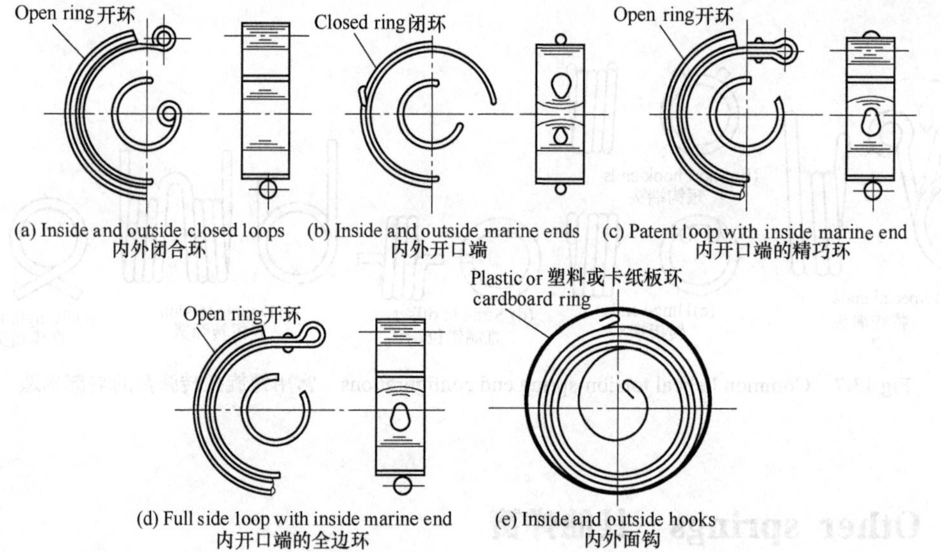

Fig 13.13　Typical power spring retainers and ends　典型动力弹簧的保持器和端头

13.3 Flywheels　飞轮

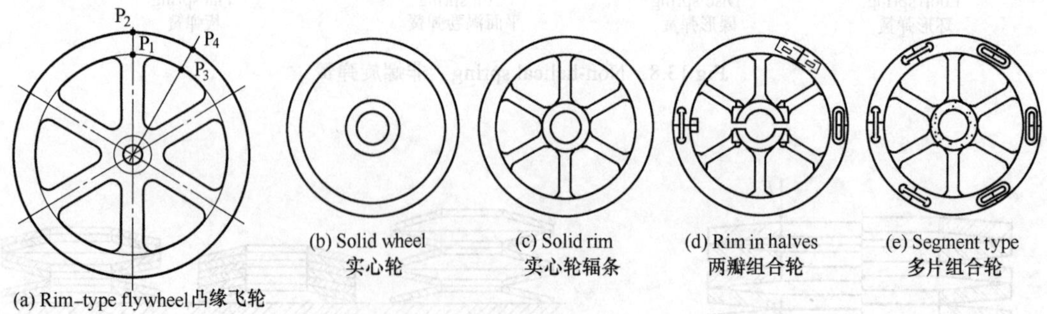

Fig 13.14　Flywheels　飞轮

Unit 14 Lubrication Systems and Sealings 润滑系统与密封件

14.1 Lubrication systems 润滑系统

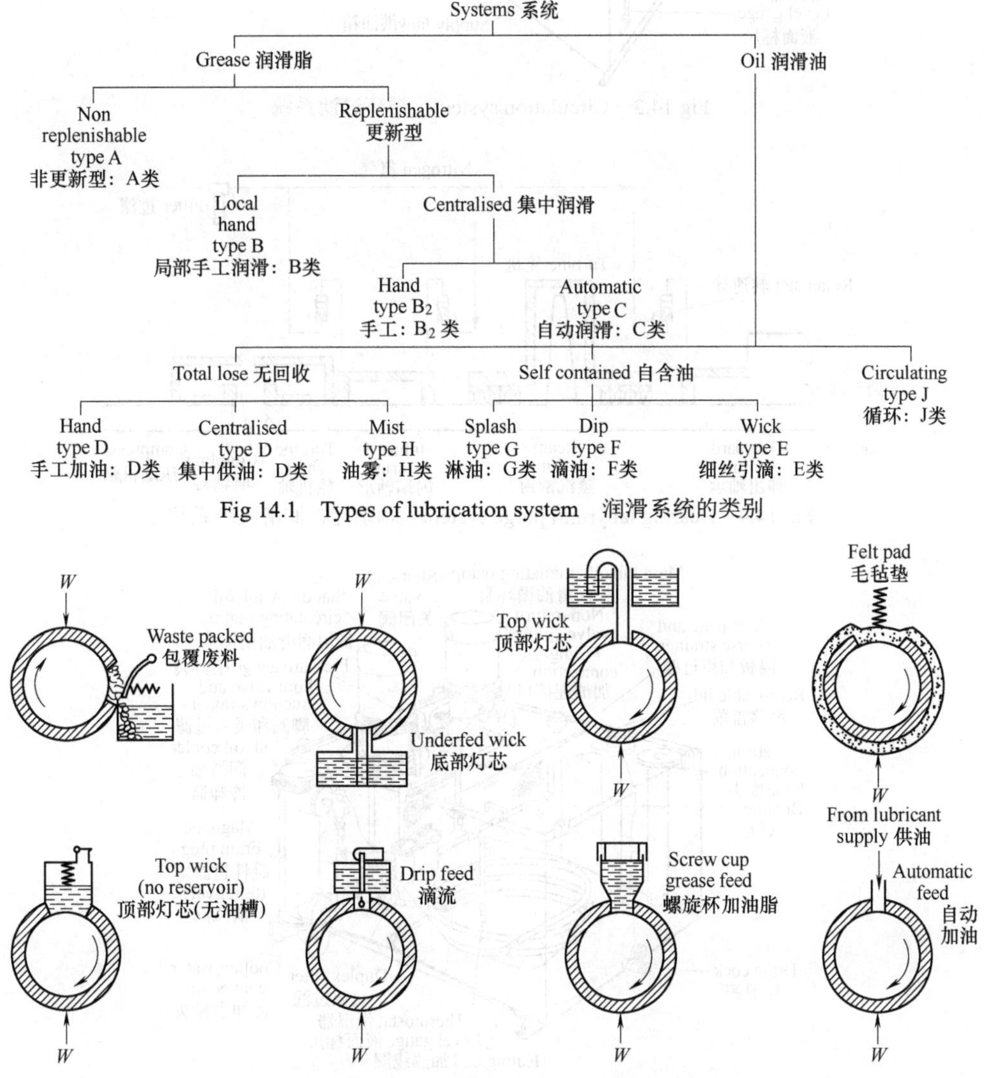

Fig 14.1 Types of lubrication system 润滑系统的类别

Fig 14.2 Typical lubricant feed arrangements 典型的润滑油加注方式设计

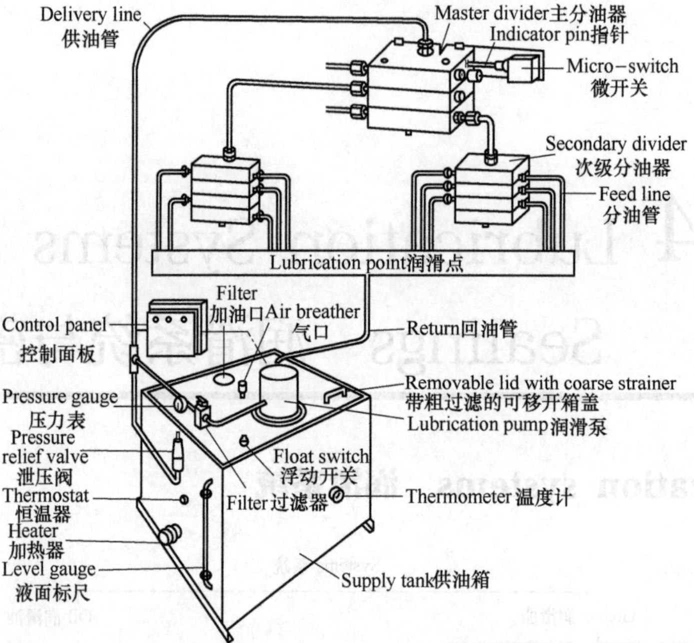

Fig 14.3 Circulation system 循环过滤系统

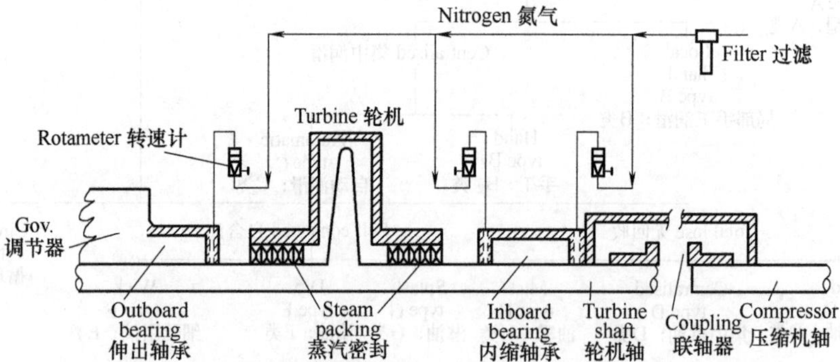

Fig 14.4 Bearing labyrinth purge system 轴承迷宫润滑流动系统

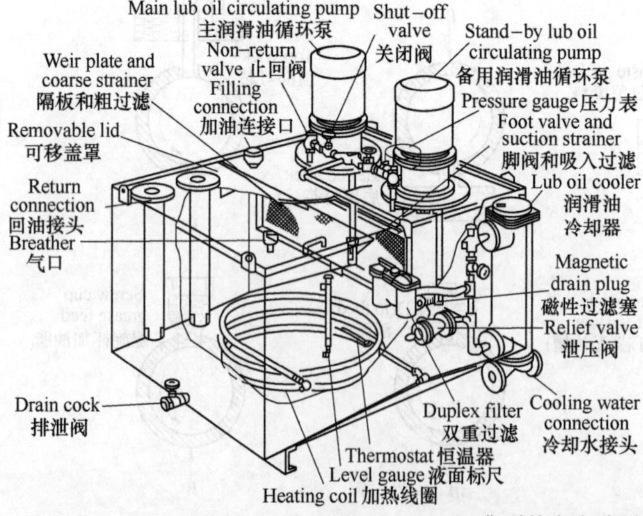

Fig 14.5 A typical self-contained oil circulatory system 典型的自含油循环系统

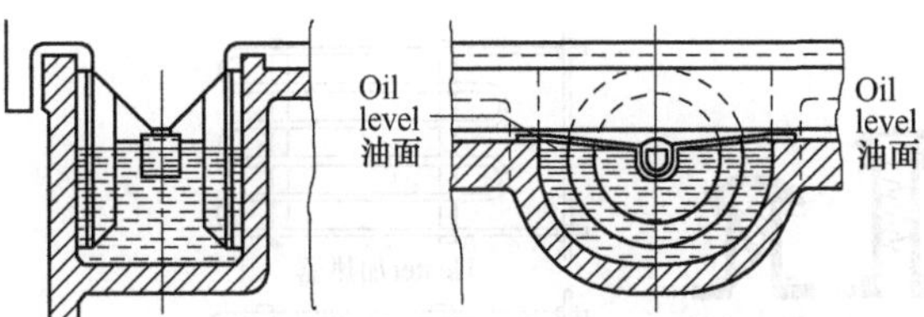

Fig 14.6 Typical roller lubricator arrangement on a machine tool 机床的转动甩油润滑

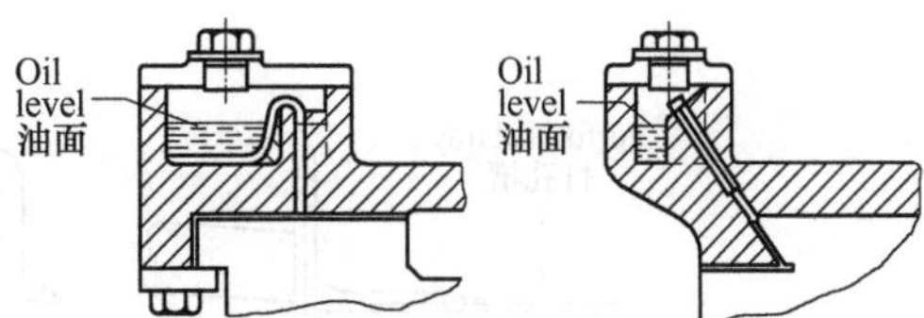

Fig 14.7 Typical wick lubricator arrangement on a machine tool 机床的滴浸润滑

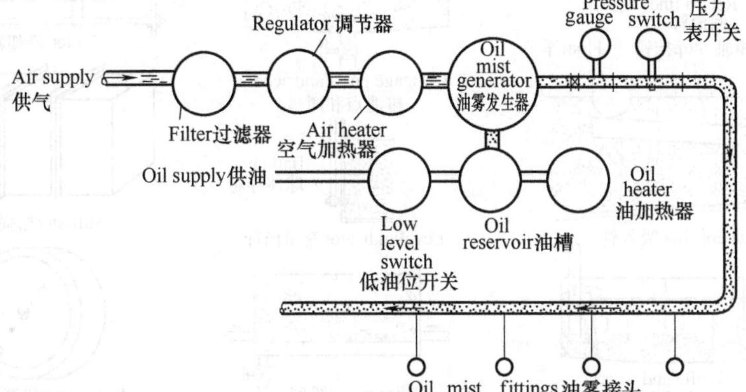

Fig 14.8 Oil mist schematic 油雾流动方案

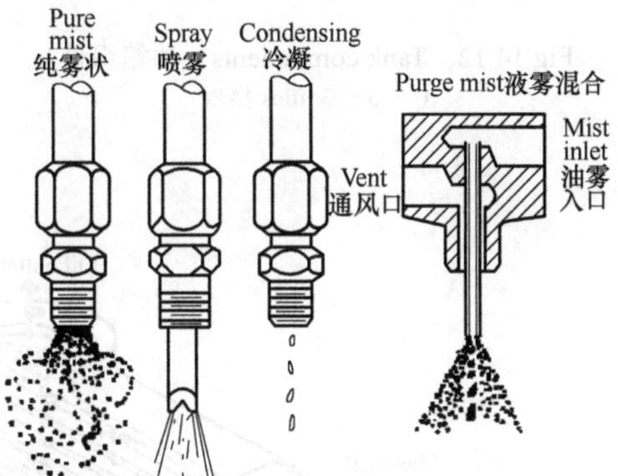

Fig 14.9 Application fittings 实用接头

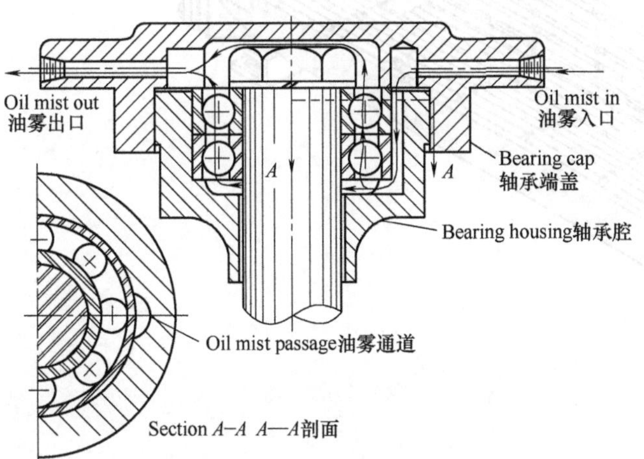

Fig 14.10 Oil mist lubricated thrust bearings on vertical motor 立式电动机推力轴承的油雾润滑

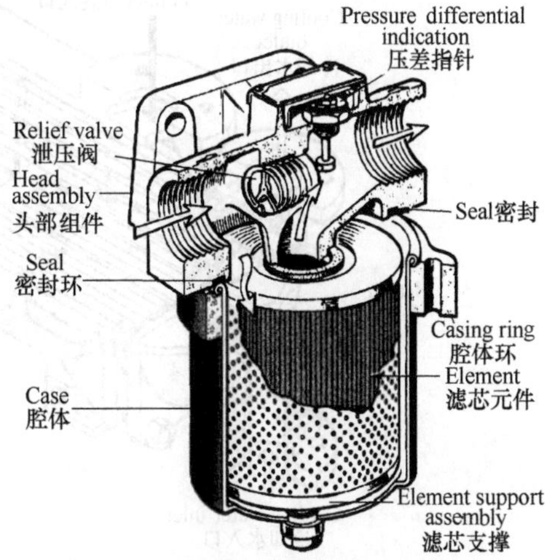

Fig 14.11 Typical full-flow pressure filter with integral bypass and pressure differential indicator 带有整体旁路和压差指示器的典型全流量压力过滤器

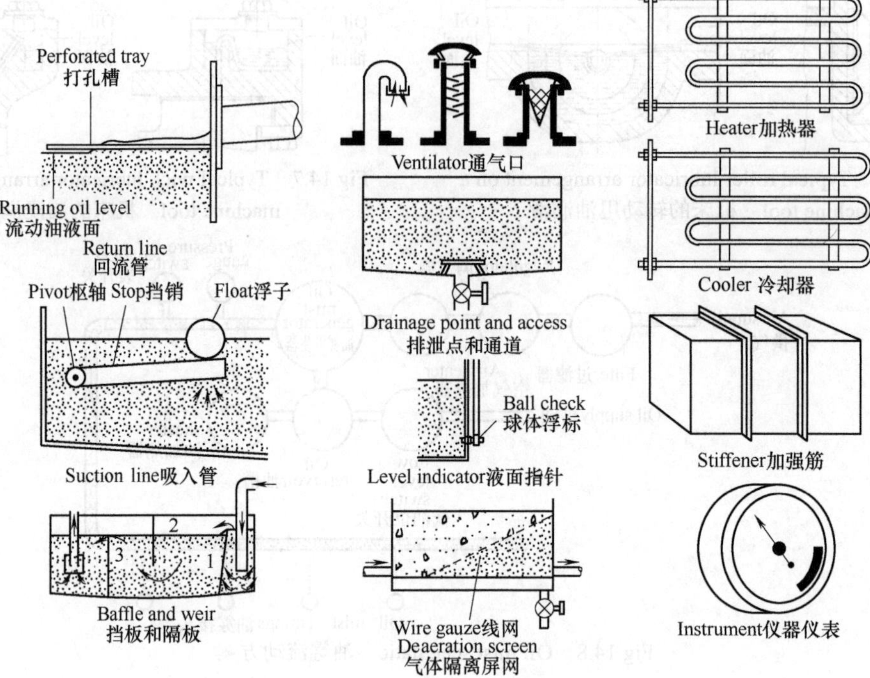

Fig 14.12 Tank components 油箱组件

1 ～ 3—Baffles 挡板

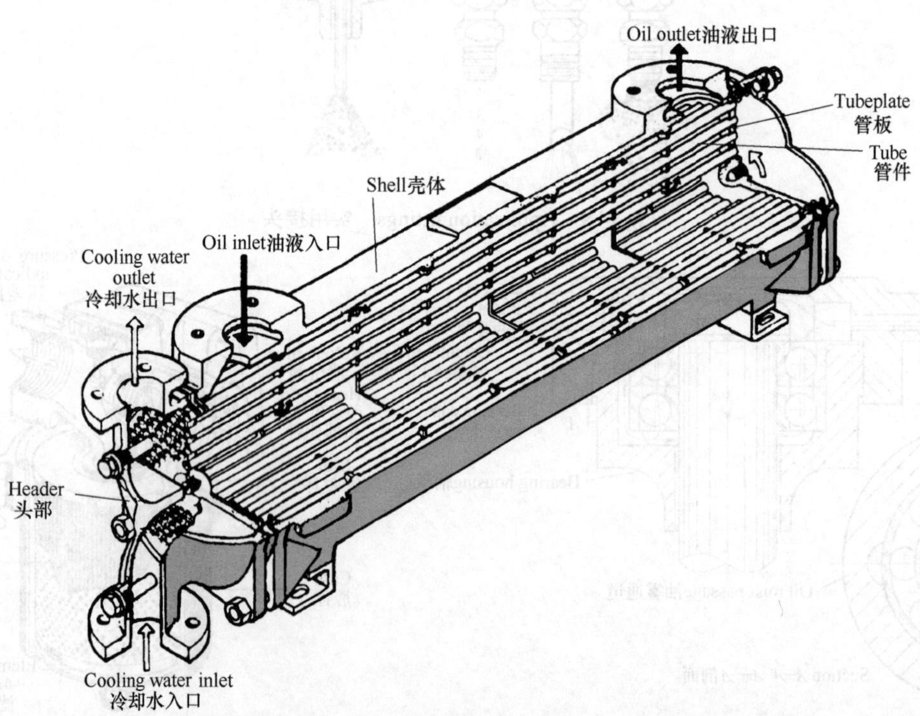

Fig 14.13 Sectional view of a typical oil cooler 典型冷却器剖视图

14.2 Sealings 密封件

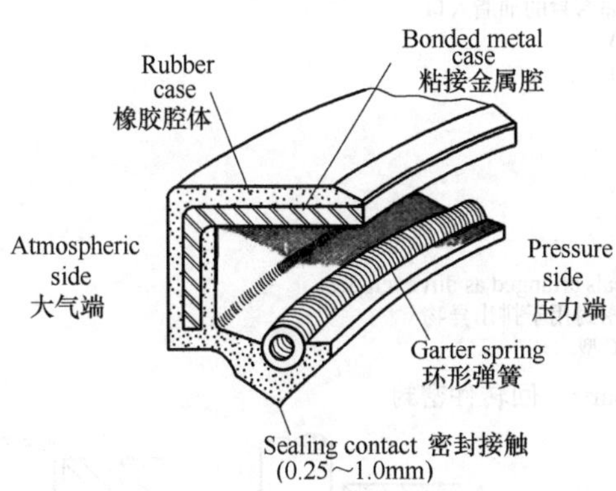

Fig 14.14 Rotary lip seal 回转边缘（唇形）密封

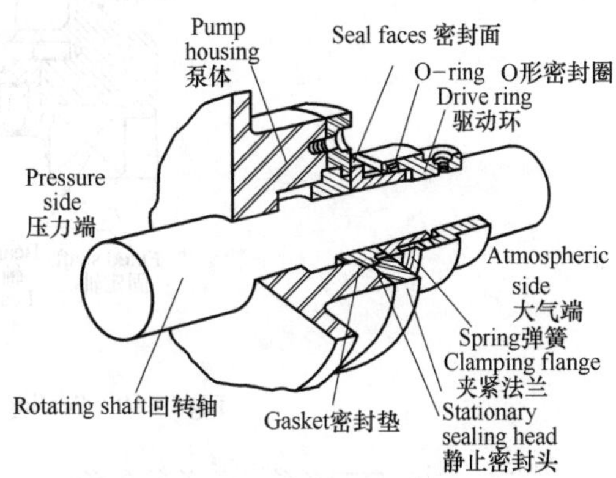

Fig 14.15 Mechanical seal 机械密封

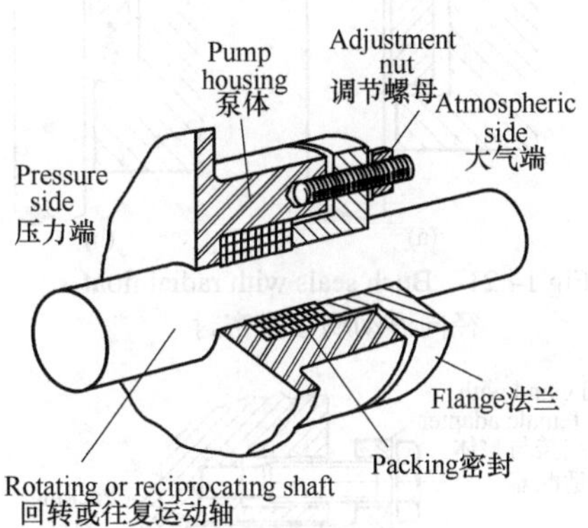

Fig 14.16 Packed gland 填塞密封

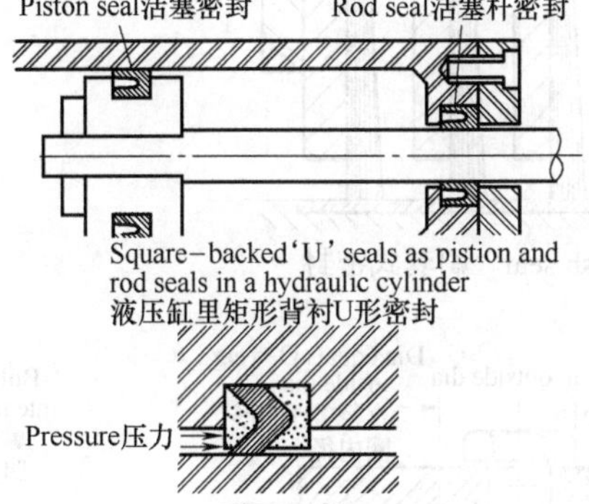

Fig 14.17 Chevron seal with shaped support rings 成形支撑环 V 形密封

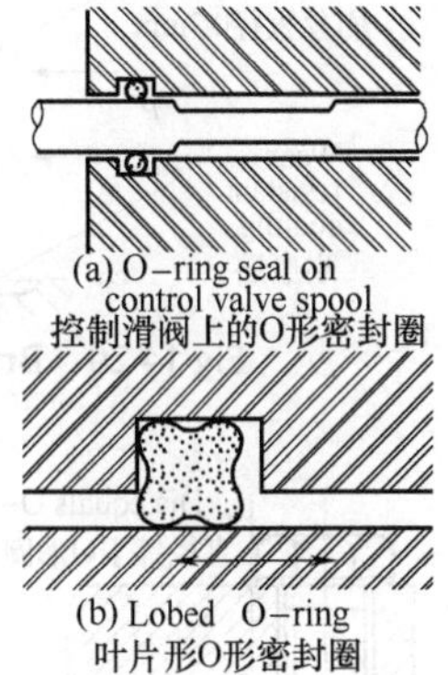

Fig 14.18 O-ring seal O 形密封圈

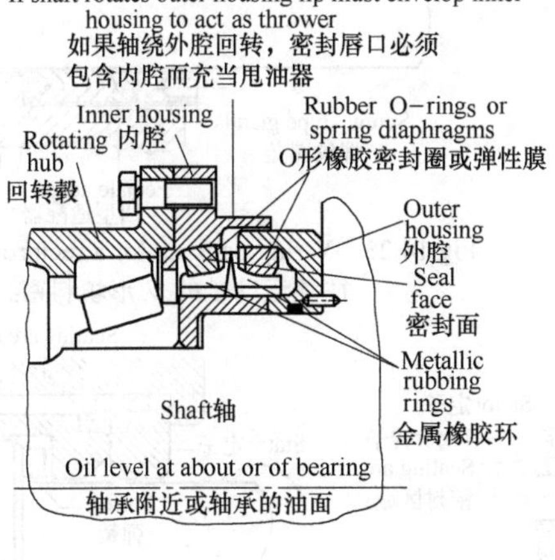

(a) Type A:A型

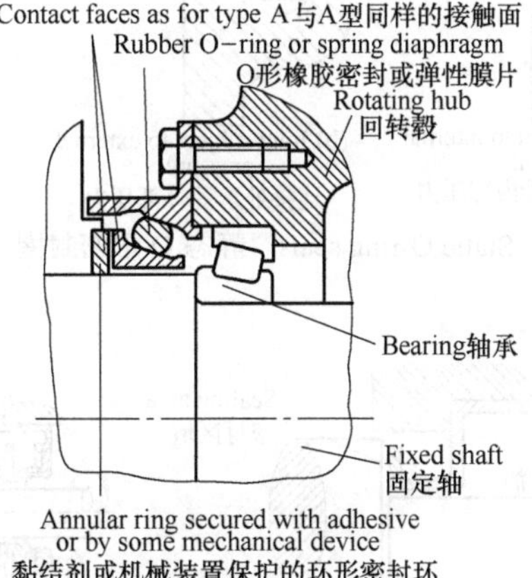

(b) Type B:B型

Fig 14.19

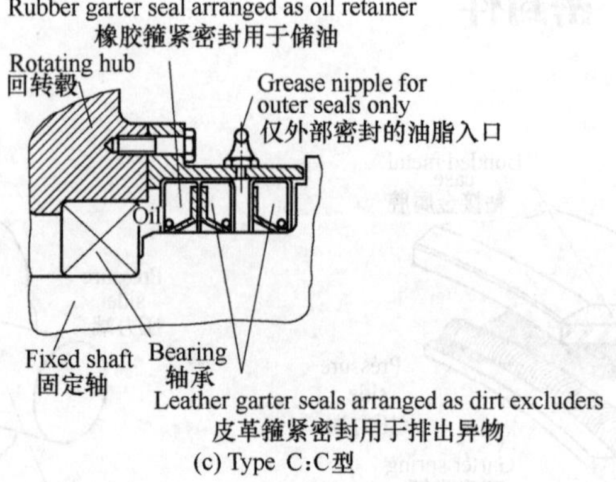

(c) Type C:C型

Fig 14.19 Sealing of rotary parts 回转件密封

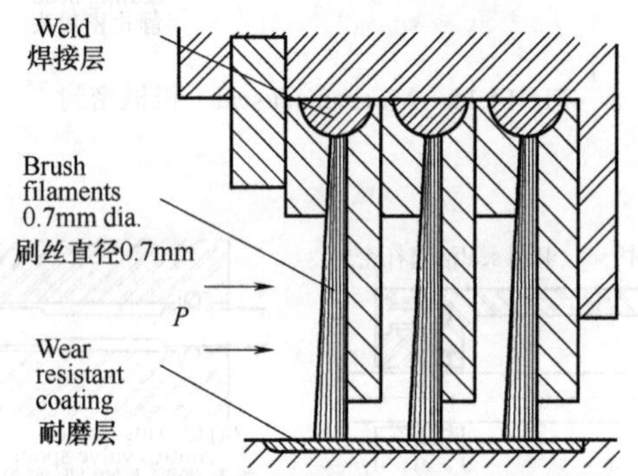

Fig 14.20 Brush seal 刷套式密封

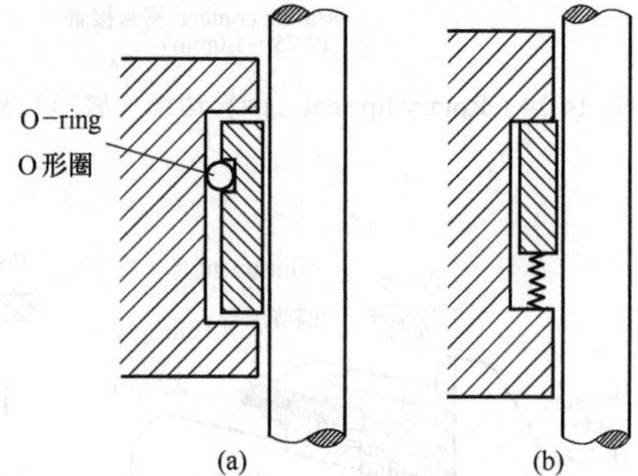

Fig 14.21 Bush seals with radial float 径向浮动刷套式密封

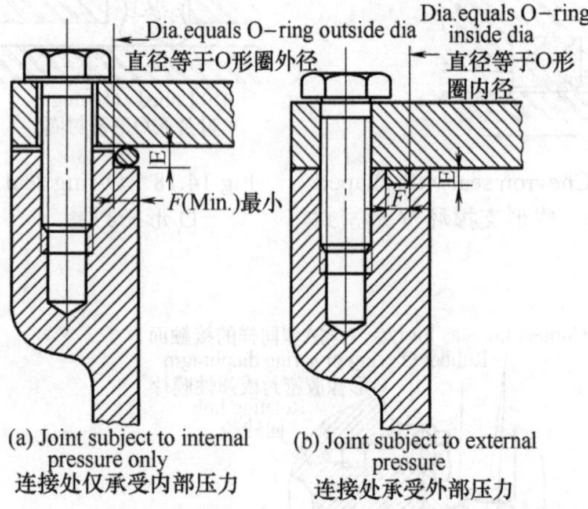

Fig 14.22 Static O-ring seals 静态O形密封圈

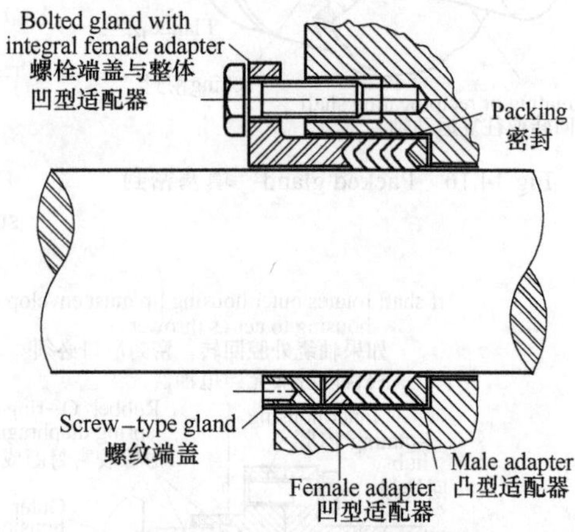

Fig 14.23 V-ring packing for a reciprocating shaft 往复运动轴的V形填料密封

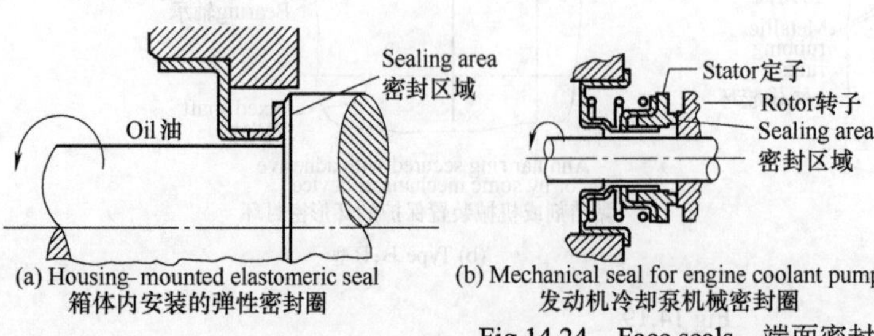

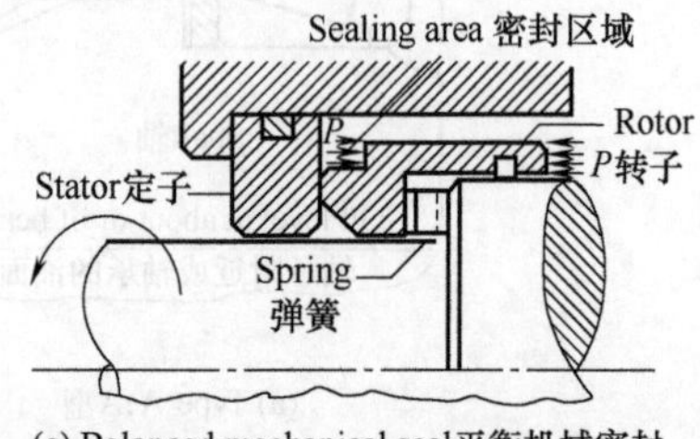

Fig 14.24 Face seals 端面密封

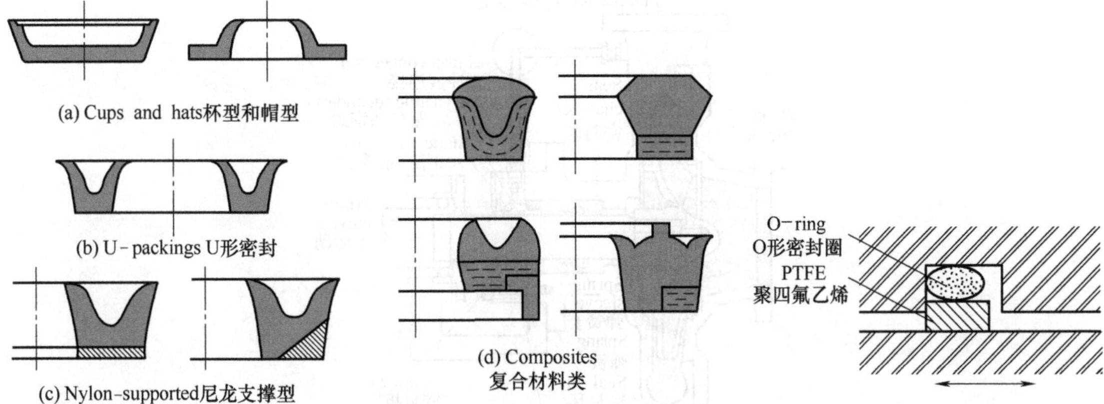

Fig 14.25 Seals for reciprocating shafts 往复运动轴的密封类型

Fig 14.26 Coaxial PTFE seal 共轴聚四氟乙烯密封圈

Table 14.1 Special seals and additional design features 特殊密封及其附加设计

Ceramic/ PTFE seal 陶瓷/聚四氟乙烯密封	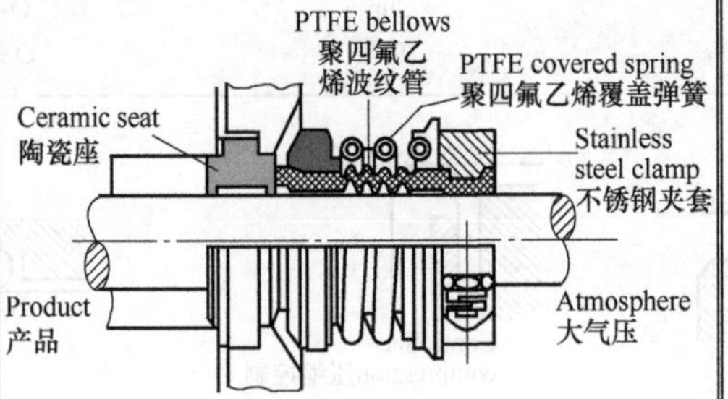 PTFE bellows 聚四氟乙烯波纹管; PTFE covered spring 聚四氟乙烯覆盖弹簧; Ceramic seat 陶瓷座; Stainless steel clamp 不锈钢夹套; Product 产品; Atmosphere 大气压	Flush 冲刷密封	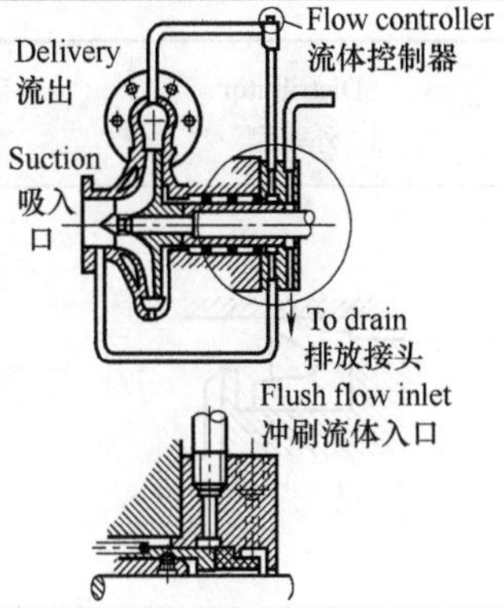 Flow controller 流体控制器; Delivery 流出; Suction 吸入口; To drain 排放接头; Flush flow inlet 冲刷流体入口
Split seal 剖分密封	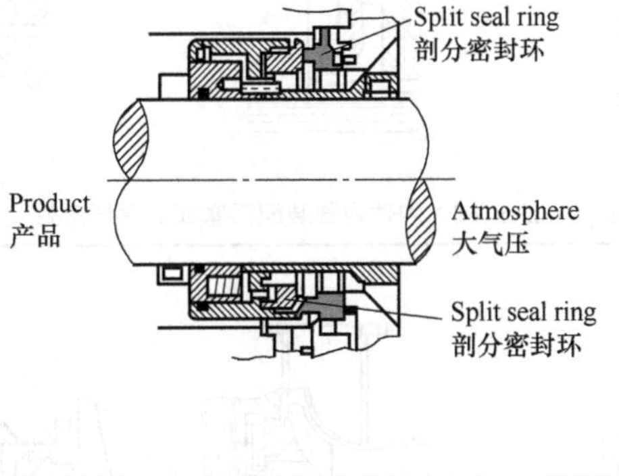 Split seal ring 剖分密封环; Product 产品; Atmosphere 大气压; Split seal ring 剖分密封环	Quench 喷流密封	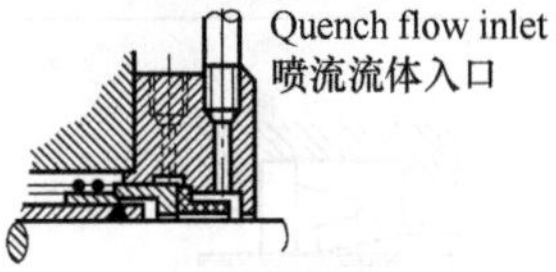 Quench flow inlet 喷流流体入口
		Safety bush 安全套	—
		Auxiliary packing/ back up seal 附属密封/支撑密封	—

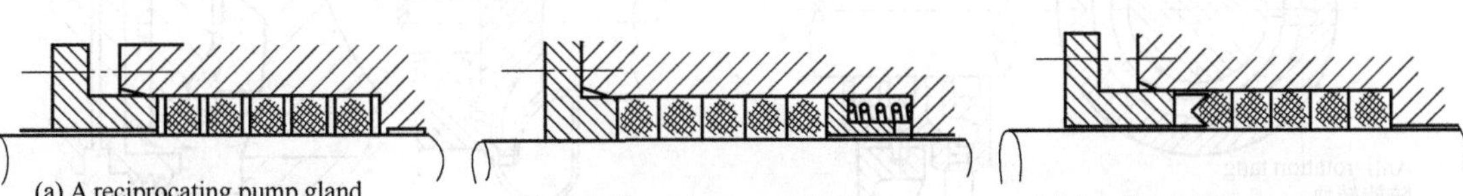

(a) A reciprocating pump gland with PTFE anti extrusion washers between the packing rings
密封环之间带聚四氟乙烯防挤出垫片的往复泵密封

(b) A reciprocating pump gland with internal spring loading to maintain compression of the packing
内部弹簧加载维持密封压缩的往复式泵密封

(c) A reciprocating pump gland with an anti-extrusion moulded hard fabric lip seal
带防挤出的模压成形的硬纤维唇形密封的往复泵密封

Fig 14.27 Reciprocating pump glands 往复泵密封

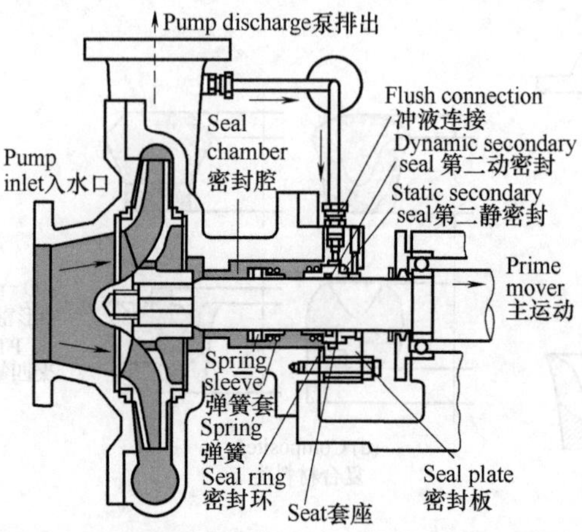

Fig 14.28 A typical mechanical seal installation 典型机械密封装配

Table 14.2 Soft piston seals 软活塞密封

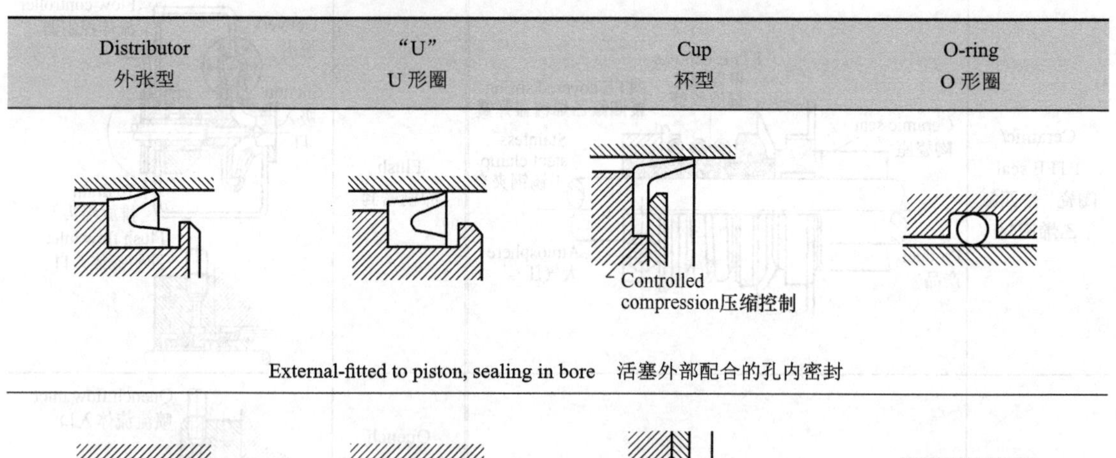

Distributor 外张型	“U” U 形圈	Cup 杯型	O-ring O 形圈
External-fitted to piston, sealing in bore 活塞外部配合的孔内密封			
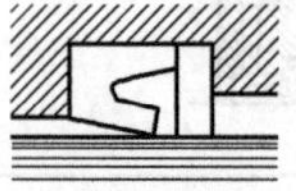	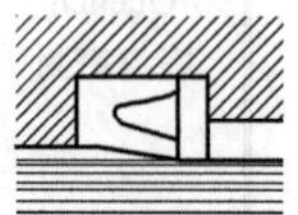	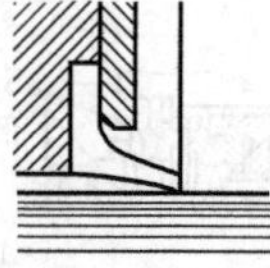	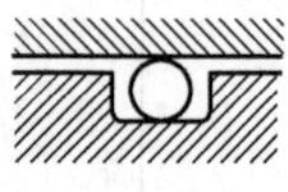
Internal-fitted in housing, sealing on piston or rod 腔体内部装配活塞或活塞杆密封			

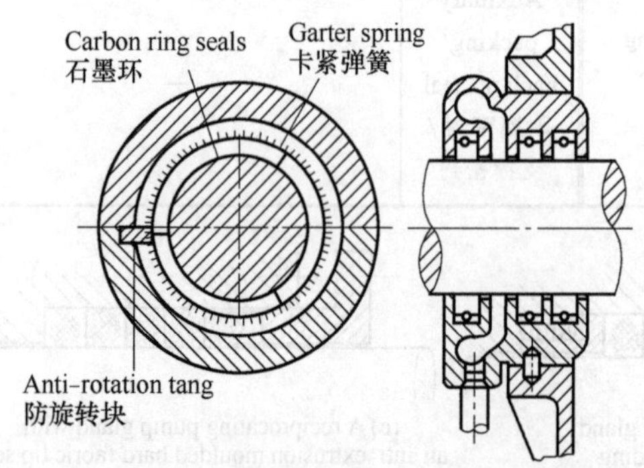

Fig 14.29 Carbon ring gland assembly 石墨环密封的装配

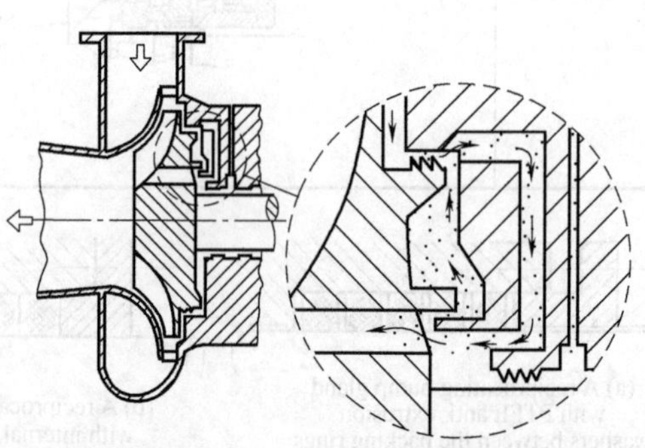

Fig 14.30 Dust-free seal 防尘密封

Table 14.3 Typical gasket designs and descriptions 典型密封垫设计与说明

Type 类型	Cross section 断面	Comments 说明
Flat 平板		Basic form. Available in wide variety of materials. Easily fabricated into different shapes 基本形式，材料来源广泛，易于各种成形制造
Reinforced 强化型		Fabricor metal-reinforced. Improves torque retention and blowout resistance of flat types. Reinforced type can be corrugated 纤维或金属强化，改进转矩保持力，消除平板类阻力，强化型可以做成波纹状
Flat with rubber beads 平板带橡胶珠		Rubber beads located on flat or reinforced material afford high unit sealing pressure and high degree of conformability 平面上的橡胶珠或强化材料承受高的密封压力，密合度好
Flat with metal grommet 平板带金属垫圈		Metal grommet affords protection to base material from medium and provides high unit sealing stress. Soft metal wires can be put under grommet for higher unit sealing stress 金属垫圈保护机体材料免受介质之害，并承受高的密封应力，位于垫片下的软金属线可承受更高应力
Plain metal jacket 普通金属蒙皮		Basic sandwich type. Filler is compressible. Metal affords protection to filler on one edge and across surfaccs 基本夹心层，填料压缩，金属覆盖并保护填料边口
Corrugated or embossed 波纹状或镶嵌		Corrugations provide for increased sealing pressure and higher conformability. Primarily circular. Corrugations can be filled with soft filler 波纹型承受密封压力更高、密合性更好，首圈波纹填入软材料
Profile 型面		Multiple sealing surfaces. Seating stress decreases with increase in pitch. Wide varieties of designs are available 多层密封面，节距增加，安装应力减少，设计形式变化多样
Spiral-wound 螺旋缠绕		Interleaving pattern of metal and filler. Ratio of metal to filler can be varied to meet demands of different applications 金属和填料的交替构形，金属与填料混合比例可随应用场合而变化

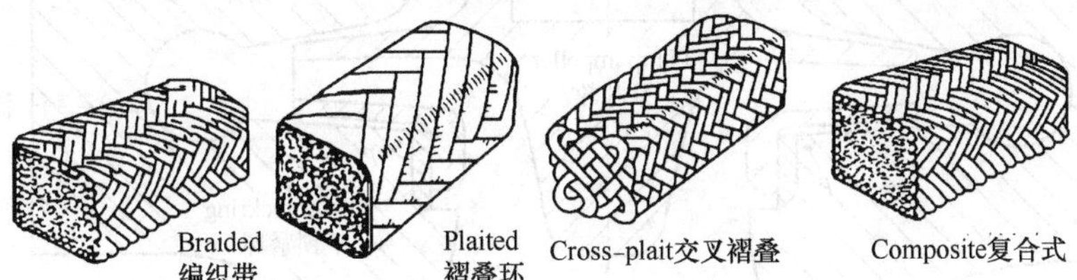

Fig 14.31 Basic packing construction 基本密封结构

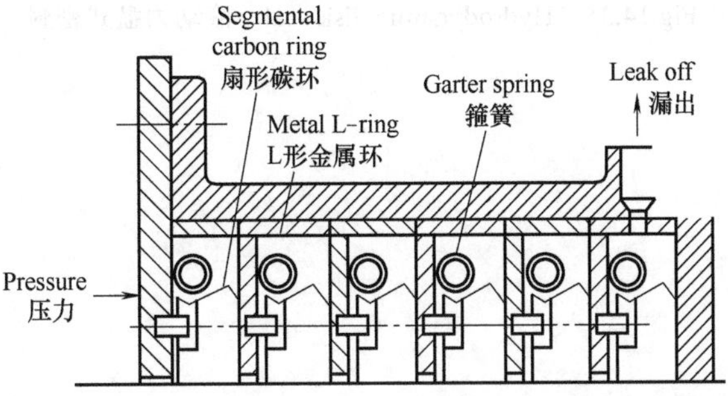

Fig 14.32 Bevel-section carbon gland 锥形碳密封套

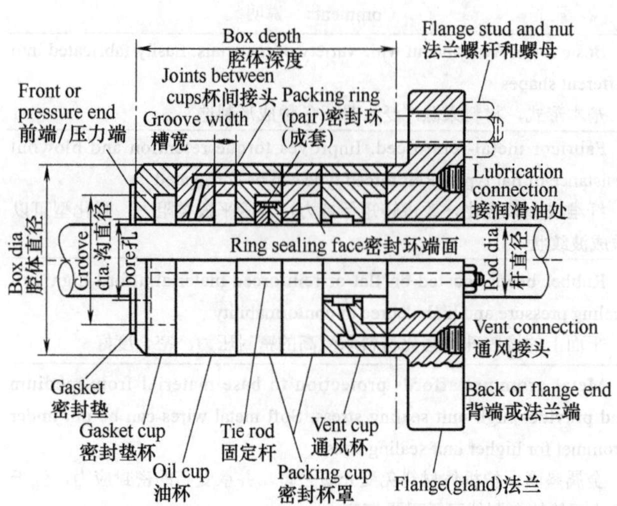

Fig 14.33 General arrangement of a typical mechanical piston rod packing assembly 典型活塞杆机械密封组件的通用设计

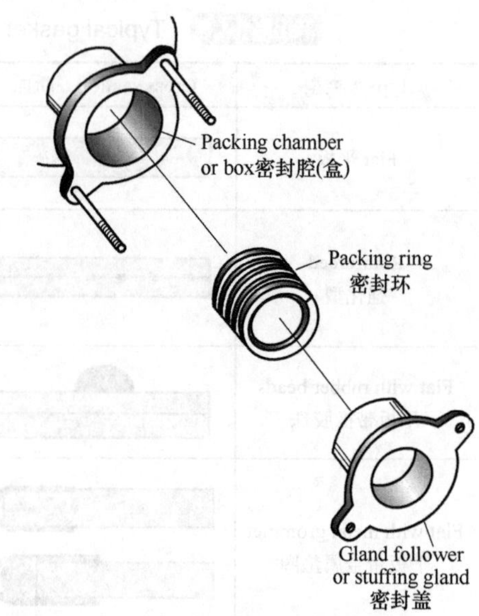

Fig 14.34 Typical packed stuffing box 典型填料密封盒

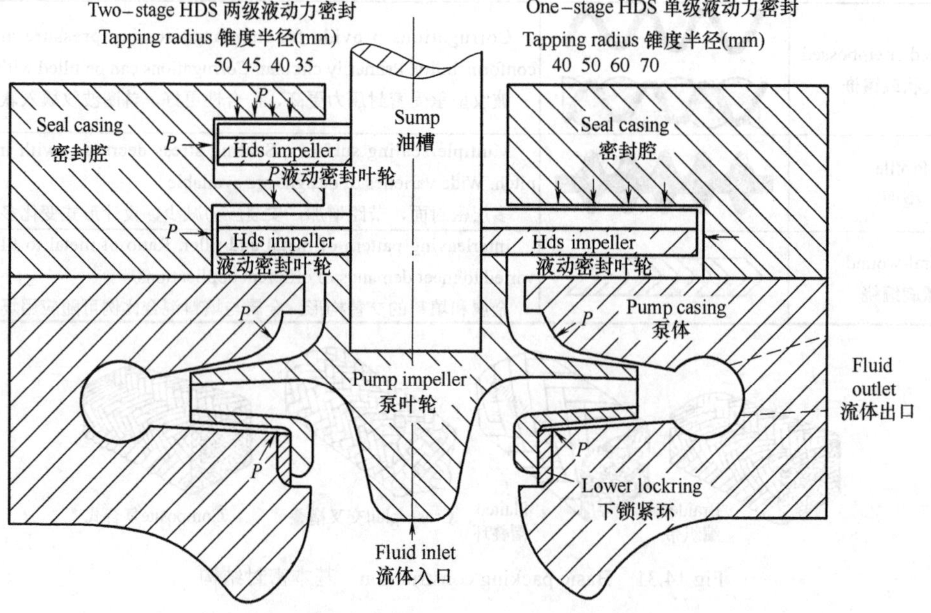

Fig 14.35 Hydrodynamic disk seal 液动力盘式密封

Unit 15 Beds, Frames and Guideways 床身、支架及导轨

15.1 Beds and frames 床身和支架

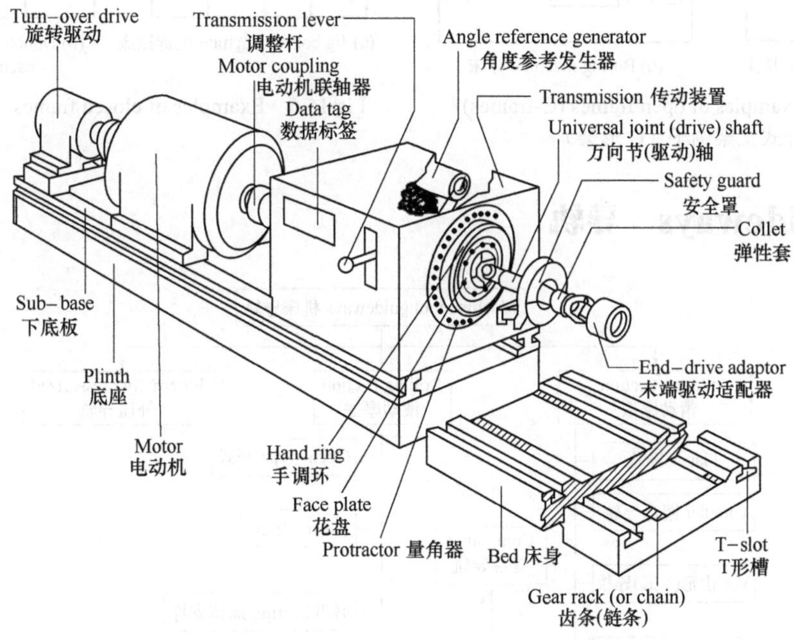

Fig 15.1 V-type headstock with end-drive and bed V 形床头箱、末端驱动和床身

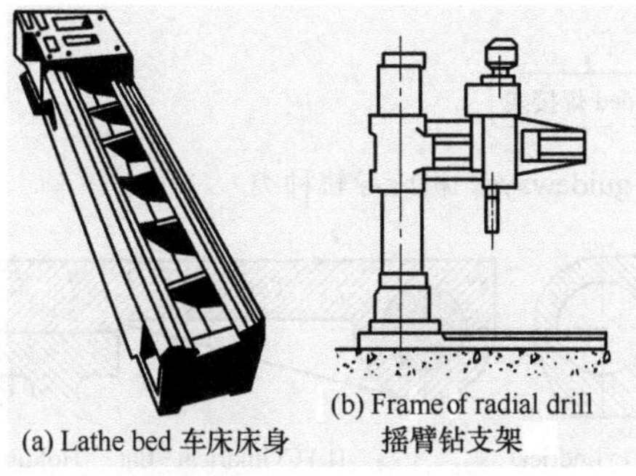

(a) Lathe bed 车床床身

(b) Frame of radial drill 摇臂钻支架

Fig 15.2 Typical bed of center lathe and frame of a drilling machine 典型中心车床床身和钻床支架

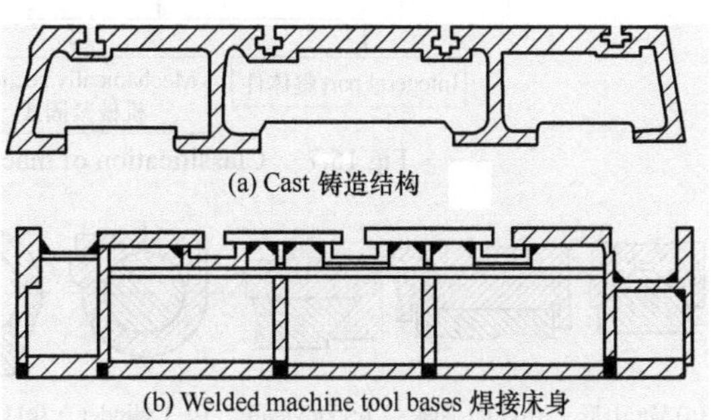

(a) Cast 铸造结构

(b) Welded machine tool bases 焊接床身

Fig 15.3 Cast and fabricated structures 铸造床身结构和组合床身结构

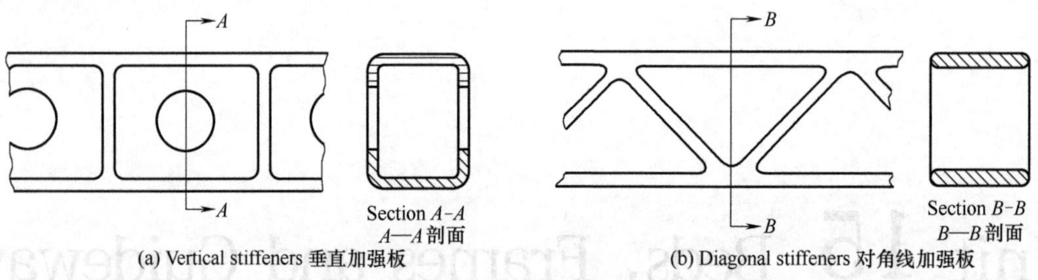

Fig 15.4 Arrangement of stiffeners in machine tool beds 机床床身的加强板布局

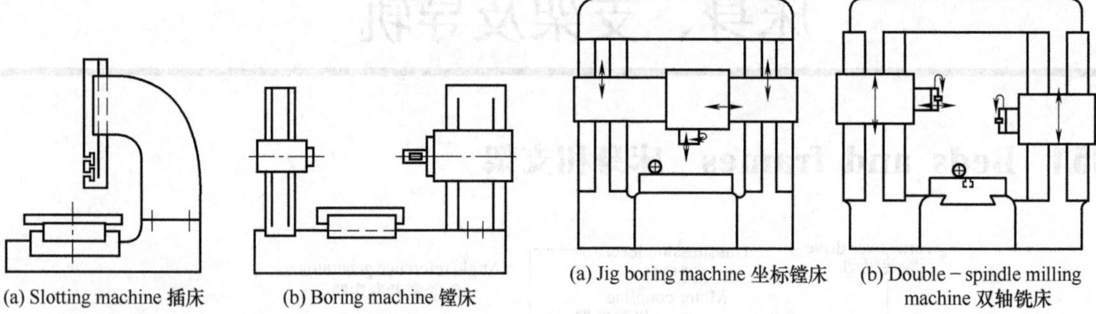

Fig 15.5 Examples of open frames (C-frames) 开式支架例子（C 形架）

Fig 15.6 Examples of closed frames 闭式支架实例

15.2 Guideways 导轨

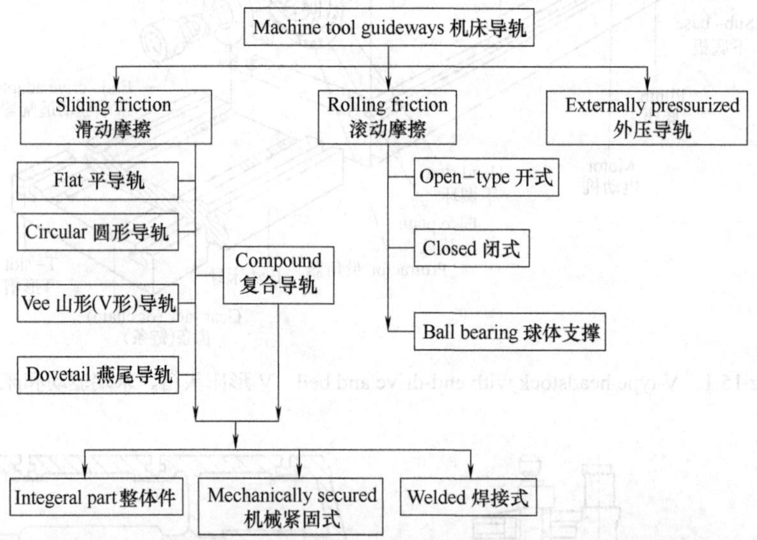

Fig 15.7 Classification of machine tool guideways 机床导轨种类

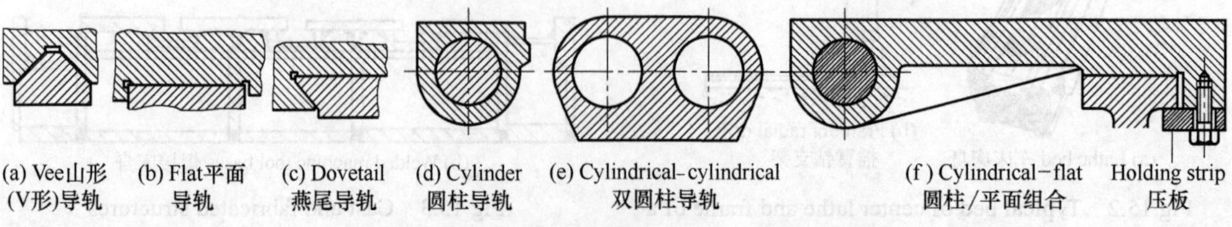

Fig 15.8 Cross sections of guideway 导轨的断面

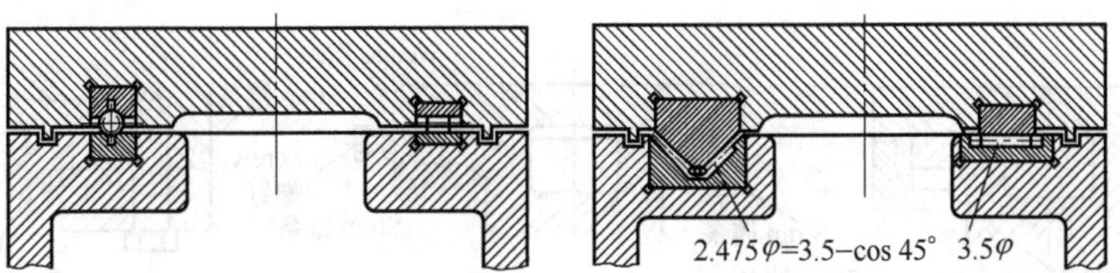

(a) Flat rolling guideway平面滚动导轨 (b) Vee−flat rolling guidewayV形/平面滚动导轨

Fig 15.9 Open-type rolling friction guideways 开式滚动摩擦导轨

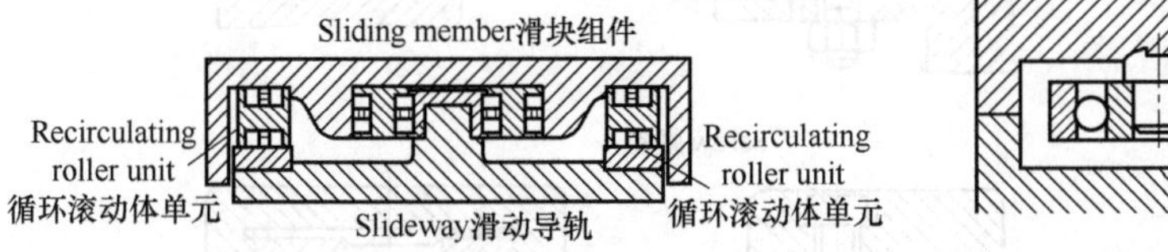

Fig 15.10 Recirculating rolling friction guideway 循环滚动摩擦导轨

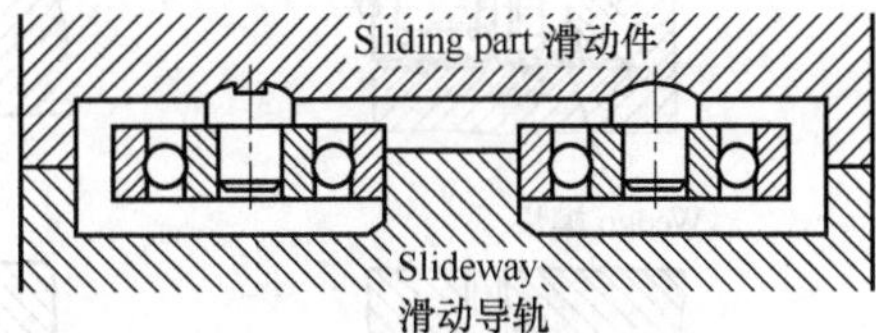

Fig 15.11 Ball bearing guideway 球轴承导轨

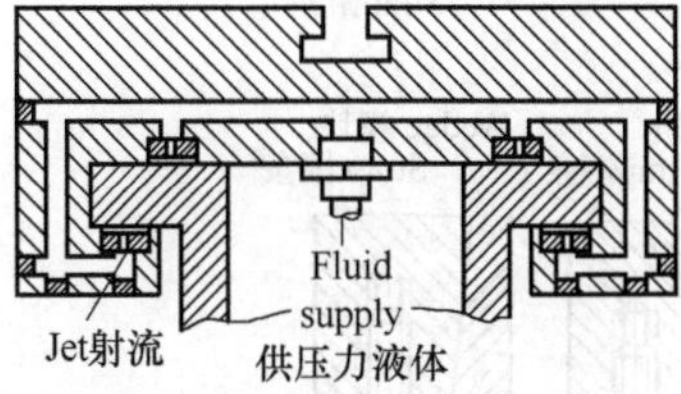

Fig 15.12 Externally pressurized guideway 外部压力导轨

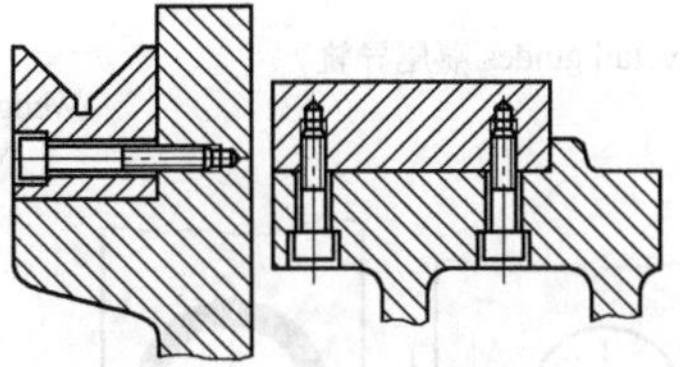

Fig 15.13 Mechanically secured guideway 机械连接导轨

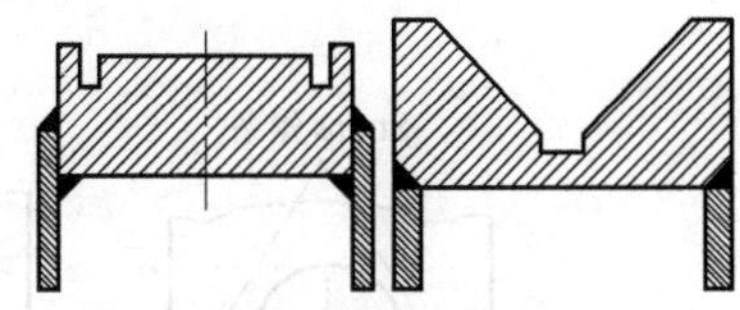

Fig 15.14 Welded guideway 焊接导轨

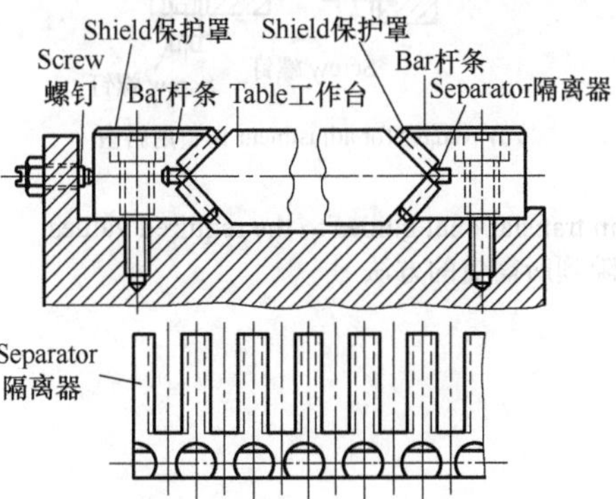

Fig 15.15 Dovetail rolling support 燕尾滚动支撑

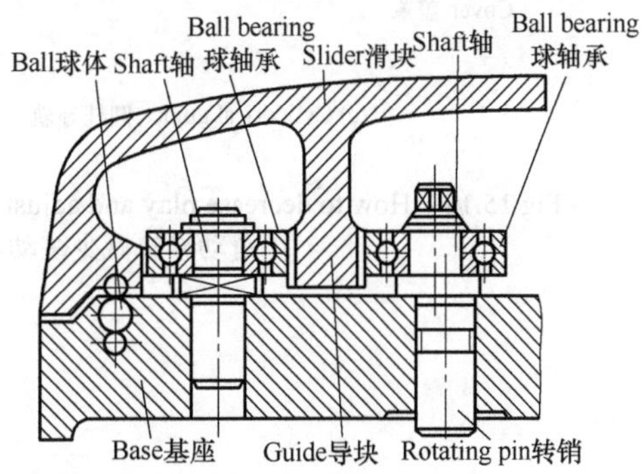

Fig 15.16 Design of a heavy-duty rolling support 重载滚动支撑设计

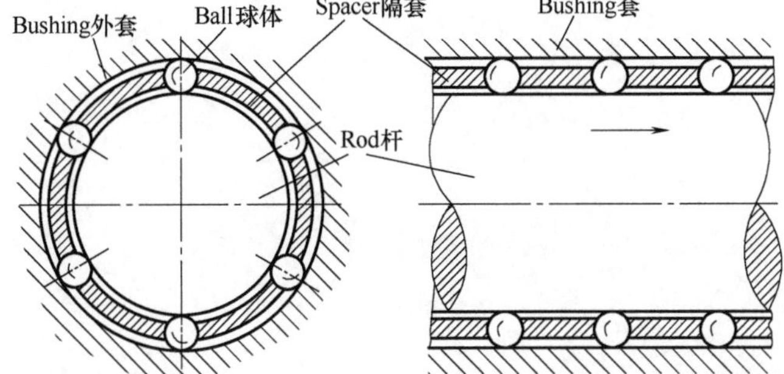

Fig 15.17 Cylindrical rolling support with a separator holding the balls 带隔离保持球体架的圆柱滚子支撑

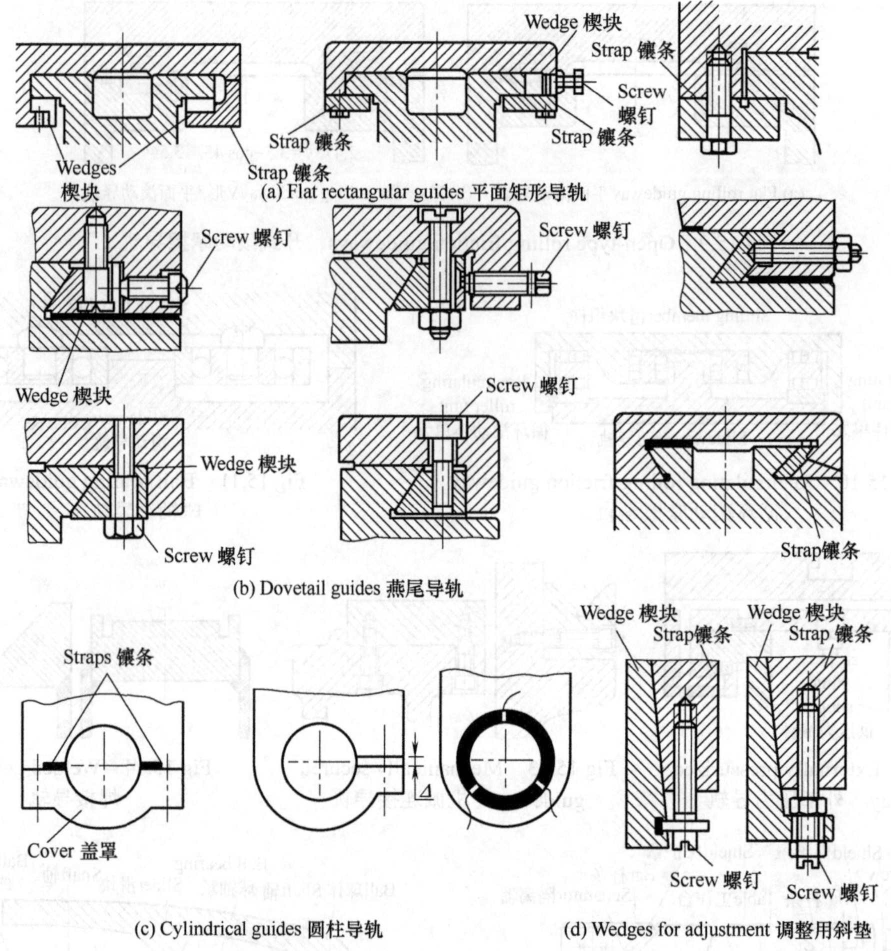

Fig 15.18 How to decrease play and adjust backlash in translational guides to the required values
滑动导轨减少跳动、调整间隙到所需值的方法

Unit 16 Couplings, Clutches and Brakes
联轴器、离合器和制动器

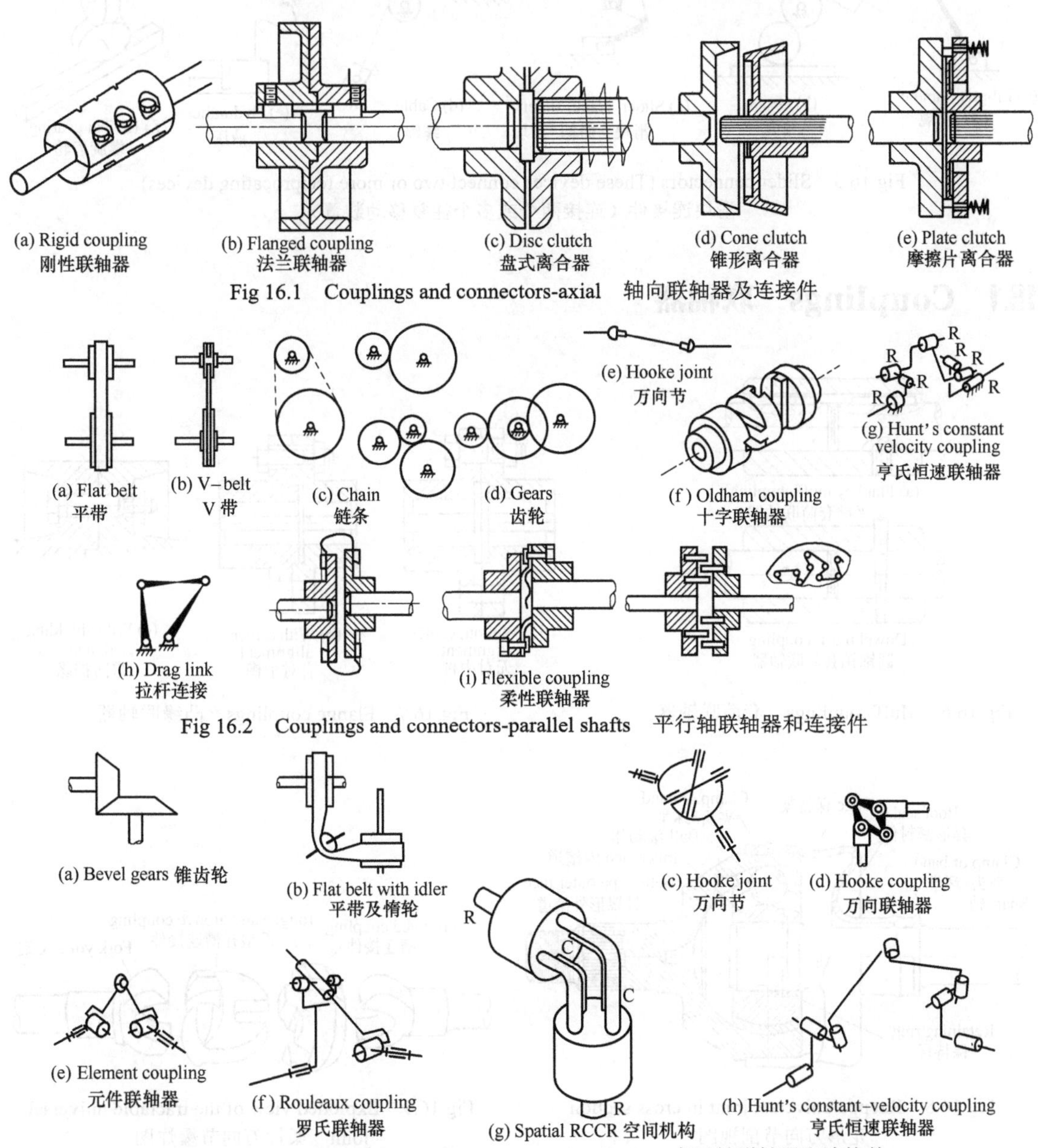

Fig 16.1 Couplings and connectors-axial 轴向联轴器及连接件

Fig 16.2 Couplings and connectors-parallel shafts 平行轴联轴器和连接件

Fig 16.3 Couplings and connectors-intersecting shafts 交叉轴联轴器和连接件

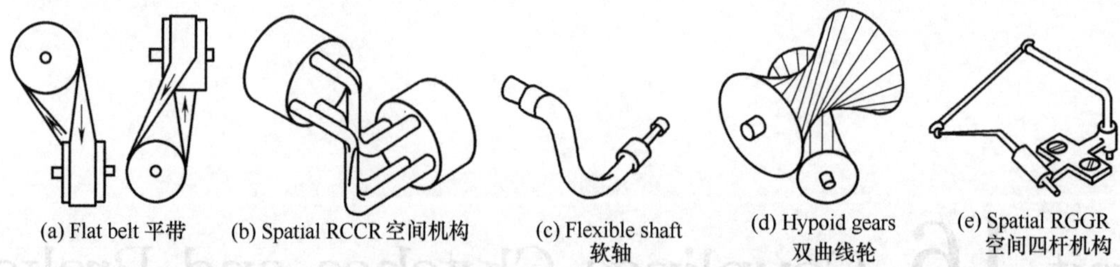

(a) Flat belt 平带 (b) Spatial RCCR 空间机构 (c) Flexible shaft 软轴 (d) Hypoid gears 双曲线轮 (e) Spatial RGGR 空间四杆机构

Fig 16.4 Couplings and connectors—skew shafts 交错轴的联轴装置和连接件

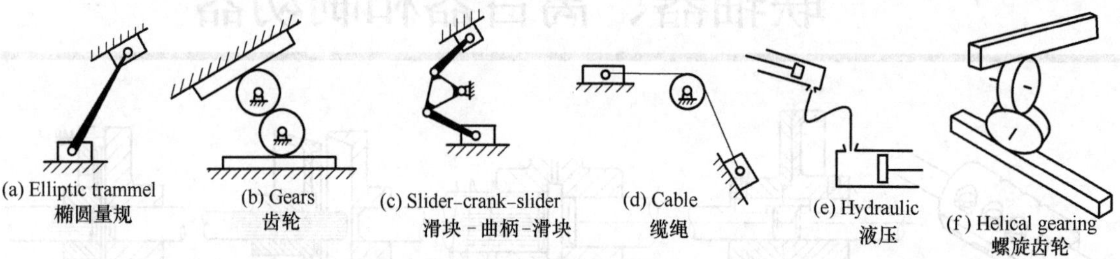

(a) Elliptic trammel 椭圆量规 (b) Gears 齿轮 (c) Slider-crank-slider 滑块-曲柄-滑块 (d) Cable 缆绳 (e) Hydraulic 液压 (f) Helical gearing 螺旋齿轮

Fig 16.5 Slider connectors (These devices connect two or more reciprocating devices) 滑块连接件（连接两个或多个往复移动装置）

16.1 Couplings 联轴器

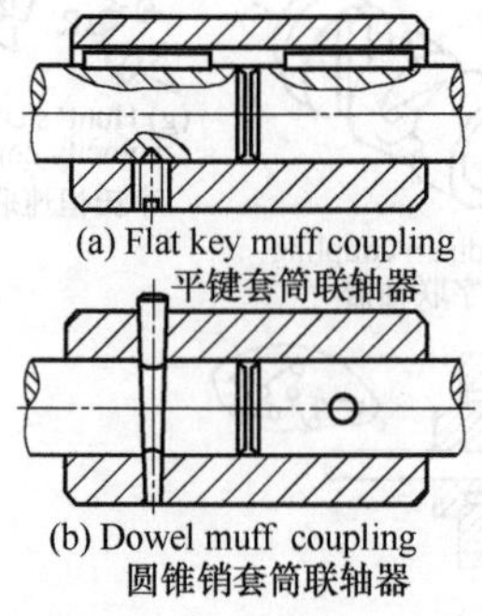

(a) Flat key muff coupling 平键套筒联轴器

(b) Dowel muff coupling 圆锥销套筒联轴器

Fig 16.6 Muff couplings 套筒联轴器

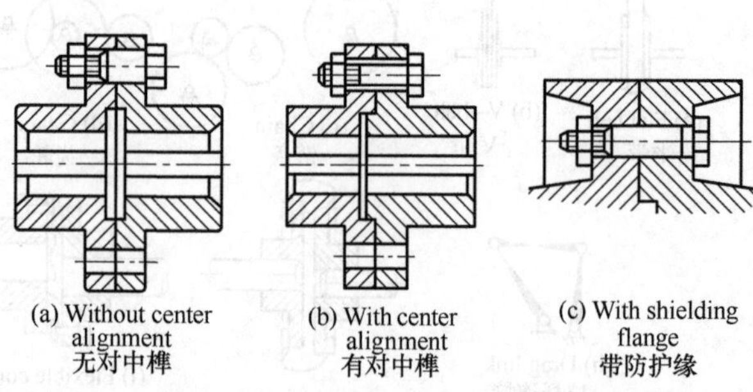

(a) Without center alignment 无对中榫 (b) With center alignment 有对中榫 (c) With shielding flange 带防护缘

Fig 16.7 Flange couplings 凸缘联轴器

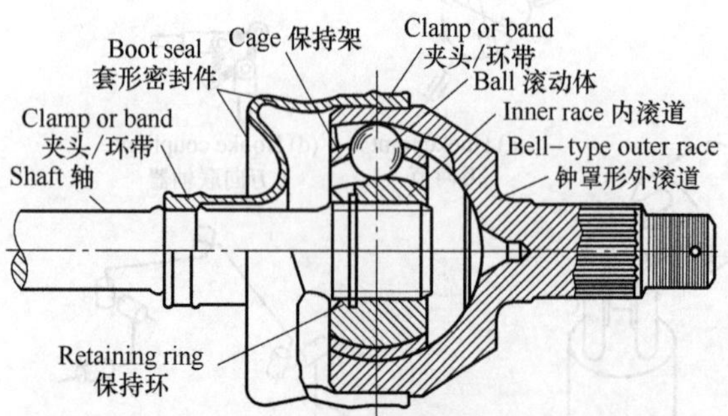

Fig 16.8 Rzeppa universal joint in cross section 笼形万向节剖视图

Fig 16.9 Exploded view of the tractable universal joint 柔性万向节爆炸图

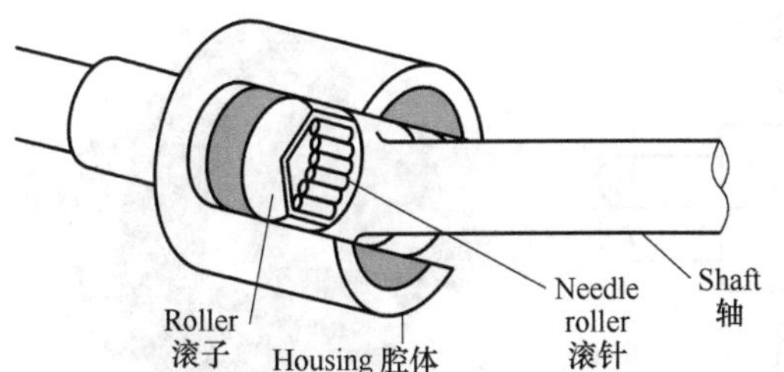

Fig 16.10 Roller trunion universal joint 滚子耳轴万向节

Fig 16.11 Flange-type gear coupling 法兰式齿轮联轴器

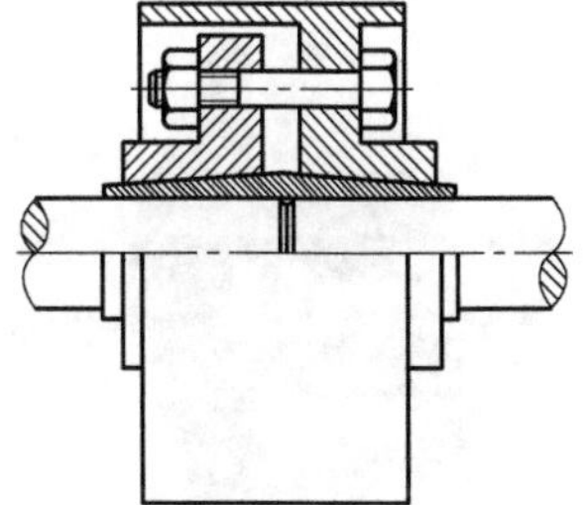

Fig 16.12 Compression coupling 压紧联轴器

Fig 16.13 Silent-chain coupling 无声链联轴器

Fig 16.14 Roller-chain coupling 滚子链联轴器

Fig 16.15 Link coupling 连杆联轴器

Fig 16.16 Metallic grid coupling with cover removed to show grid detail 金属格栅联轴器，盖罩移除便于展示细节

Fig 16.17 Beam coupling 单梁式联轴器

Fig 16.18 Cutaway view of diaphragm coupling assembly showing multiple convoluted diaphragms 膜片式联轴器装配剖切图，可见其多层盘旋膜片

Fig 16.19 Hydraulic coupling (Cutaway shows oil forced between inner and outer tapered sleeves. Note the oil piston chamber at left) 液压联轴器（剖切图可见压力油在内外锥套间流动，油压活塞缸在左边）

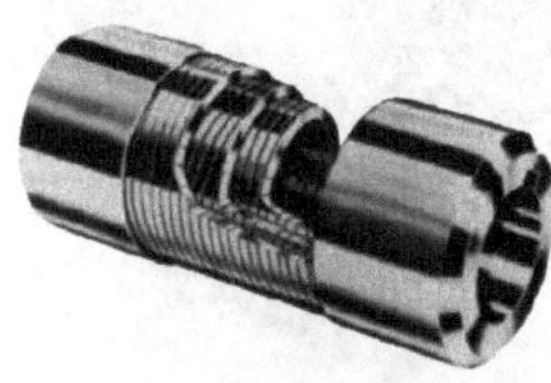

Fig 16.20 Uniflex flexible-spring coupling 万向柔性弹簧联轴器

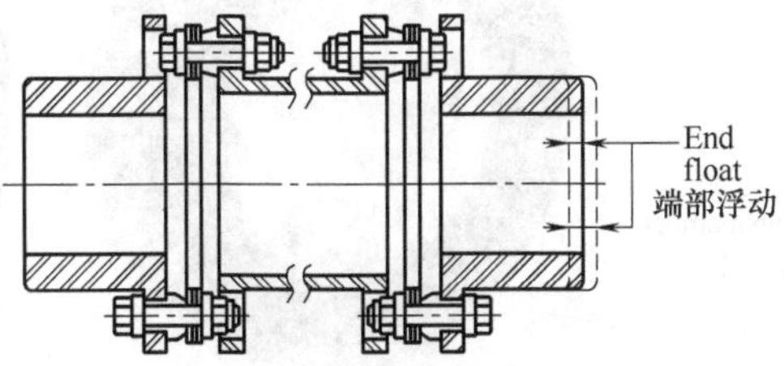

Fig 16.21 Flexible-disk coupling 柔性盘式联轴器

Fig 16.22 Rotating-link coupling 回转连杆联轴器

Fig 16.23 Pressure bushing
压力轴套（联轴器）

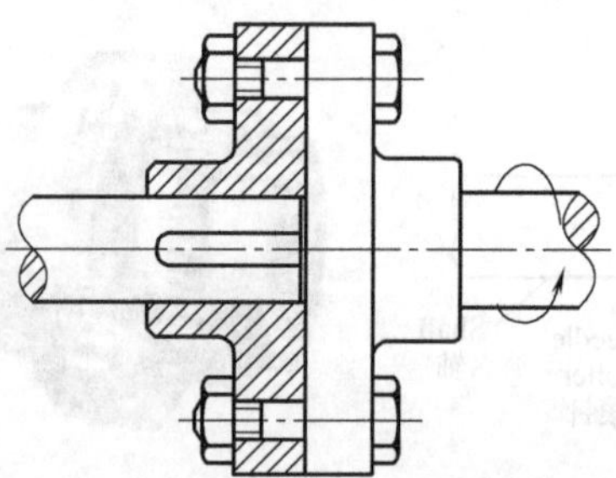
Fig16.24 Solid bolted shaft coupling
刚性螺栓轴联轴器

Fig 16.25 Precompressed axially restrained coupling
预压轴向约束联轴器

Fig 16.26 Bellows coupling
波纹管联轴器

Fig 16.27 Exploded view of jaw-type compression coupling
卡爪式压缩联轴器爆炸图

Fig 16.28 Shear undamped coupling
无阻尼剪切联轴器

Fig 16.29 Tyre coupling
胎式联轴器

Fig 16.30 Torus coupling
环式联轴器

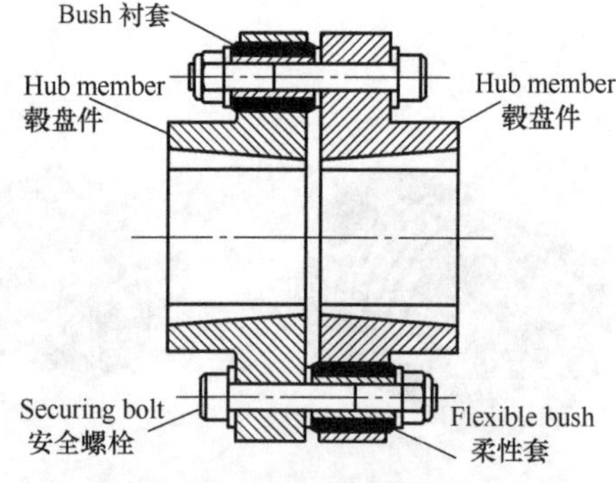

Fig 16.31 Section through pinflex coupling
轴销柔性联轴器剖面图

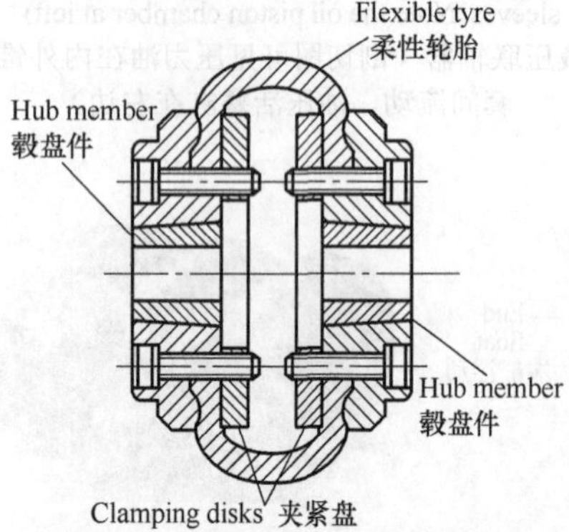

Fig 16.32 Section through a Renold Uratyre coupling 雷诺 U 形轮胎联轴器剖面图

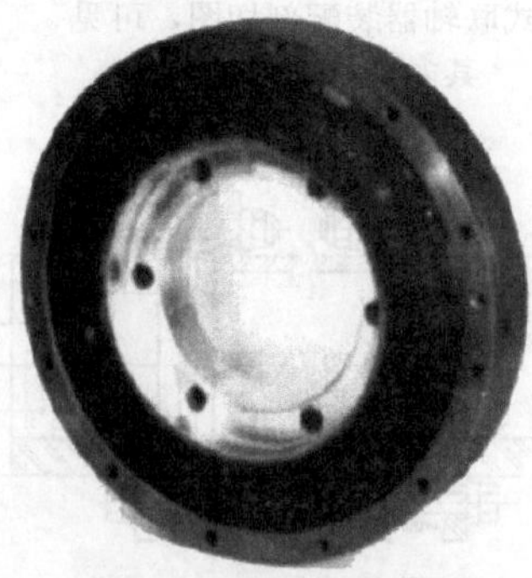
Fig 16.33 Bonded type of a shear undamped coupling
黏结式无阻尼剪切联轴器

Fig 16.34 Tension coupling
张紧式联轴器

Fig 16.35 Precompressed radially restrained coupling
径向预压约束联轴器

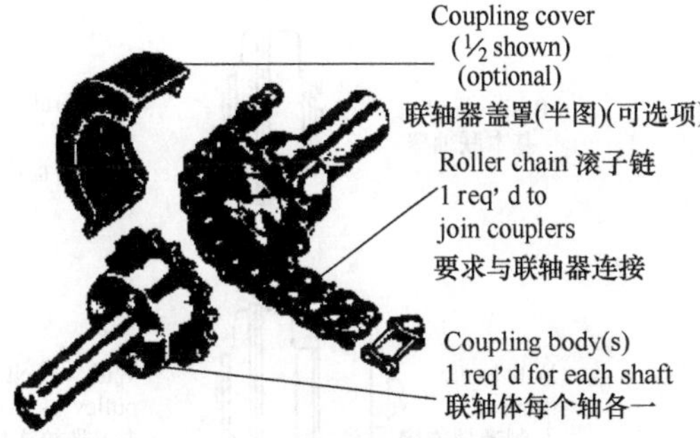

Fig 16.36 Typical chain coupling
典型链条联轴器

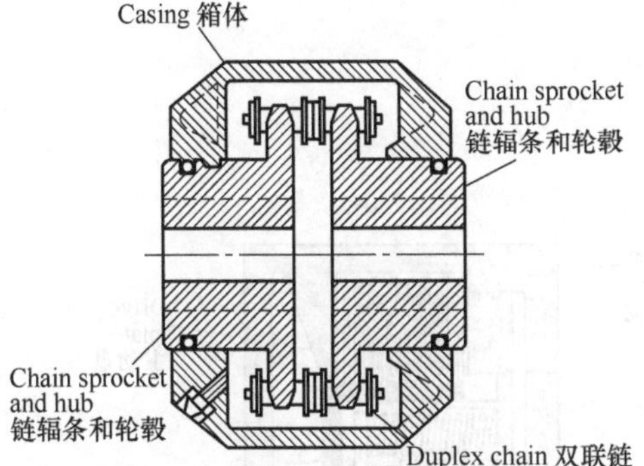

Fig 16.37 Section through a chain type coupling
链式联轴器剖面

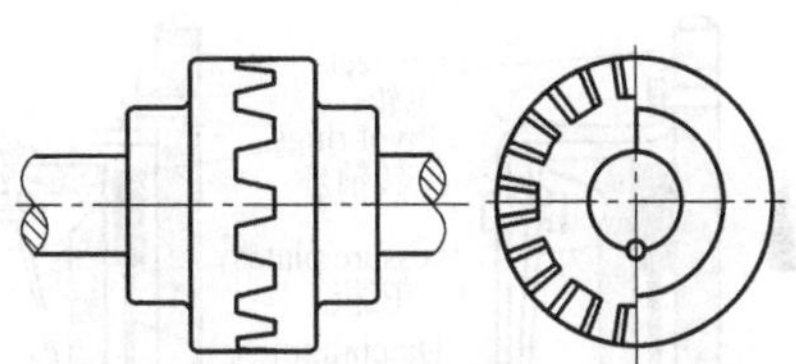

Fig 16.38 Claw coupling 爪型联轴器

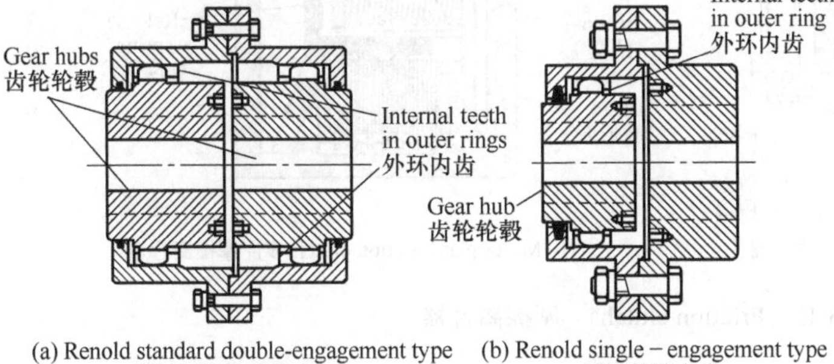

Fig 16.39 Sections through internal gear coupling 内齿联轴器剖面

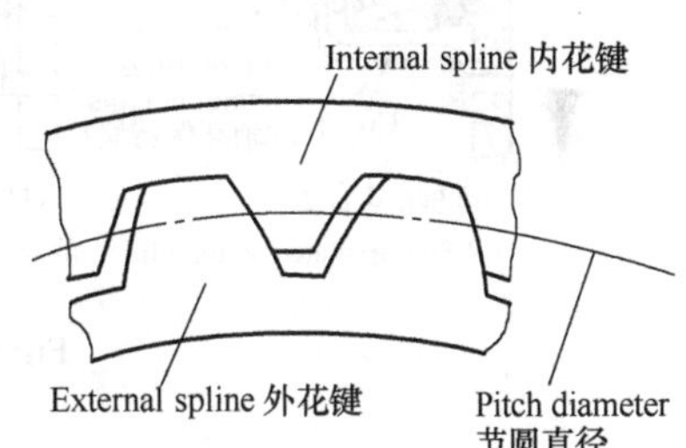

Fig 16.40 Splined coupling
花键联轴器

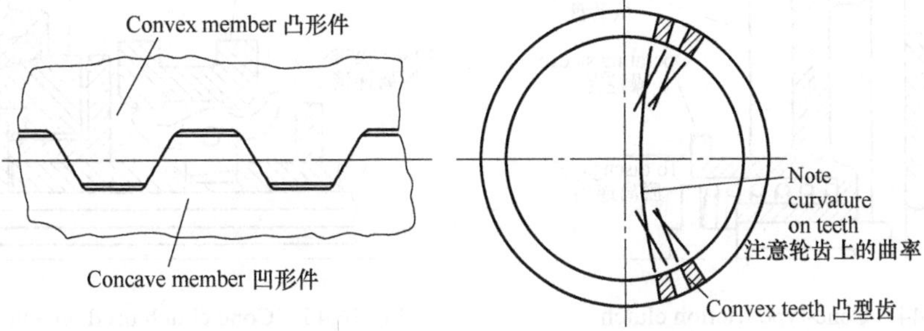

Fig 16.41 Curvic coupling 曲面联轴器

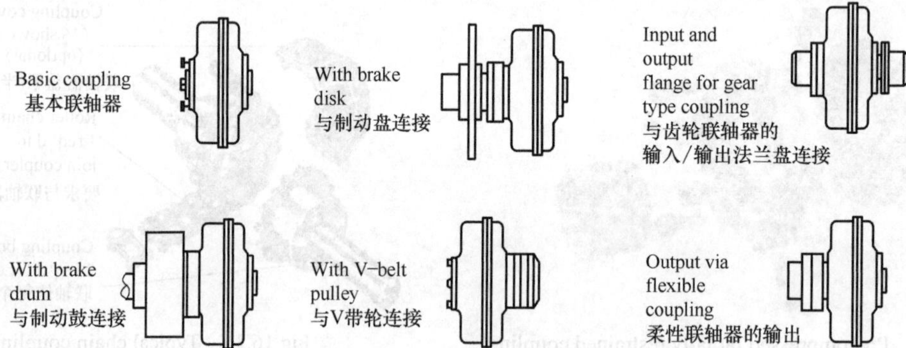

Fig 16.42 Examples of using a fluid coupling in conjunction with other transmission elements
液力联轴器与其他传动元件的连接

16.2 Clutches 离合器

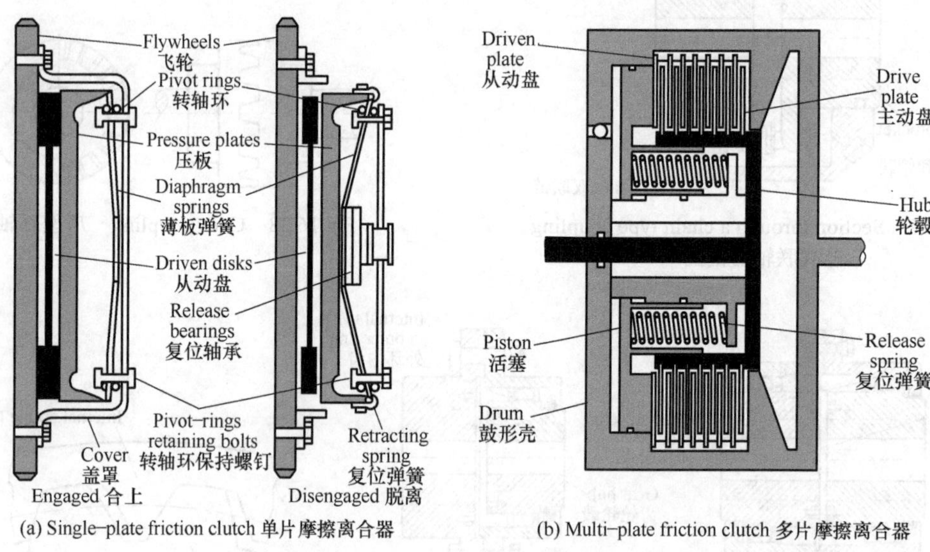

(a) Single-plate friction clutch 单片摩擦离合器 (b) Multi-plate friction clutch 多片摩擦离合器

Fig 16.43 Friction clutchs 摩擦离合器

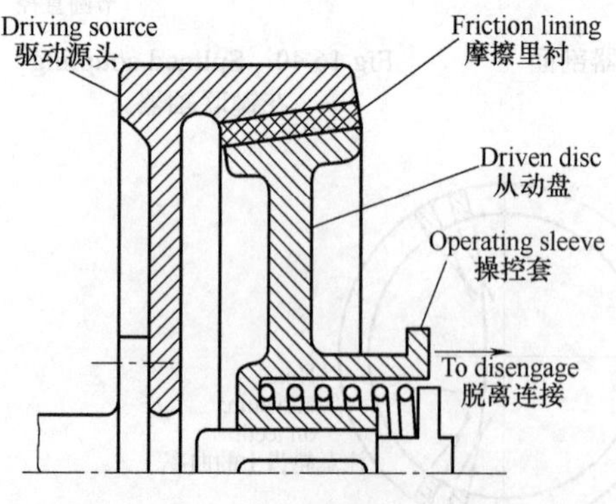

Fig 16.44 Cone-type friction clutch
锥形摩擦离合器

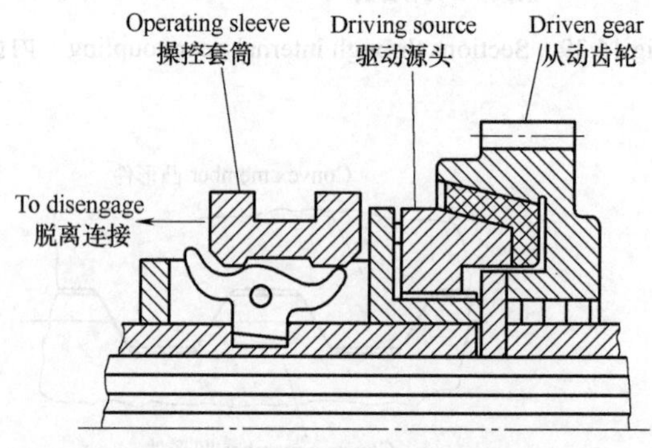

Fig 16.45 Cone clutch used on automatic lathe
自动车床用锥形离合器

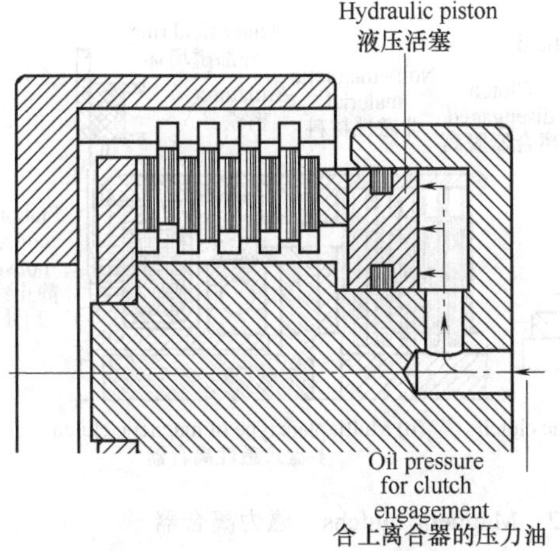

Fig 16.46 Diagram showing operation of hydraulic clutch 液压离合器工作原理图解

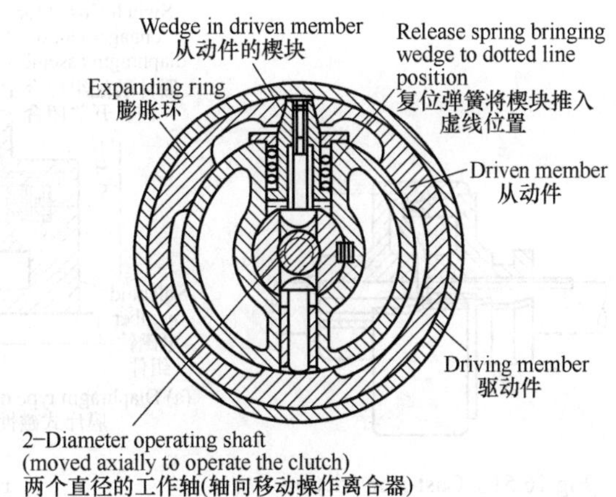

Fig 16.47 Expanding ring type of friction clutch 膨胀环式摩擦离合器

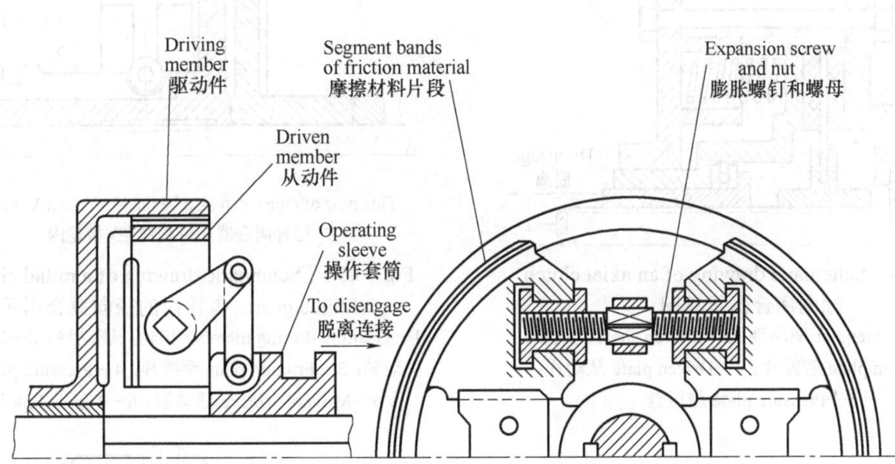

Fig 16.48 Expanding band friction clutch 膨胀环带摩擦离合器

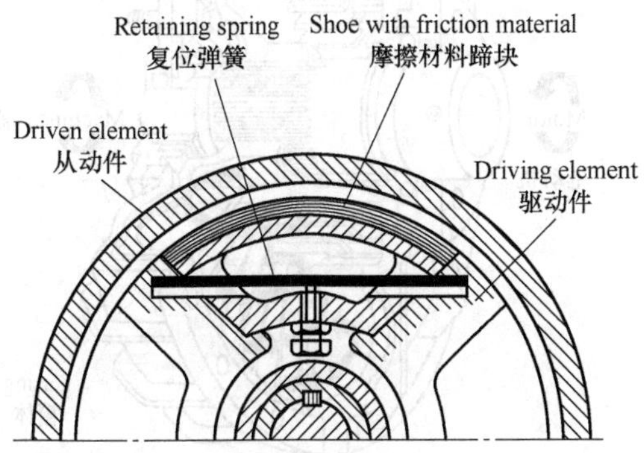

Fig 16.49 Centrifugal clutch with spring control 弹簧控制的离心离合器

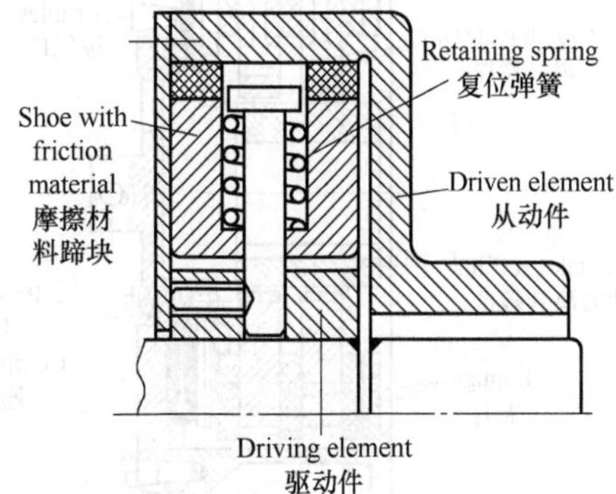

Fig 16.50 Light-duty type of centrifugal clutch 轻载型离心离合器

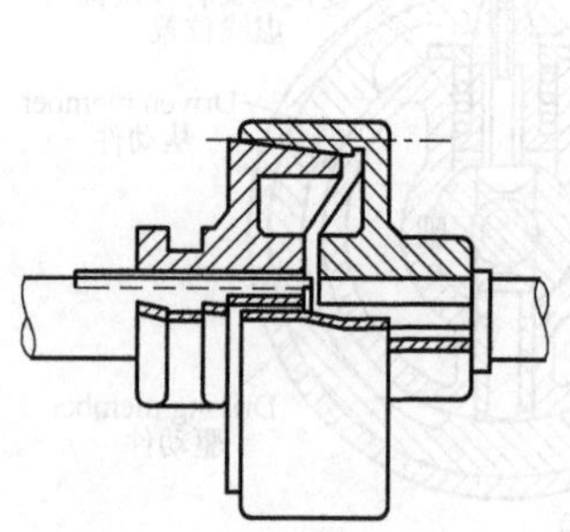

Fig 16.51 Cast-iron friction clutch 铸铁摩擦离合器

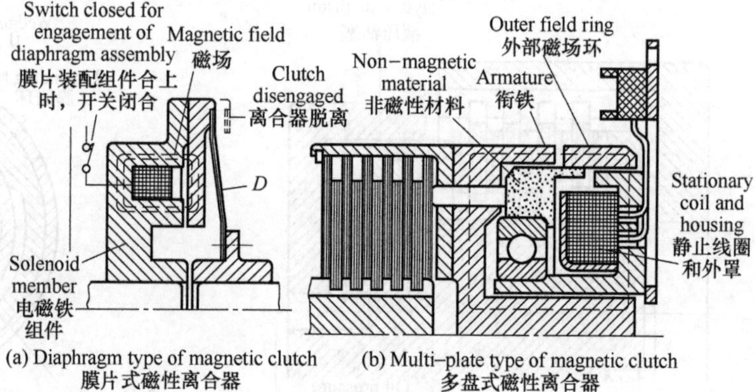

(a) Diaphragm type of magnetic clutch 膜片式磁性离合器

(b) Multi-plate type of magnetic clutch 多盘式磁性离合器

Fig 16.52 Magnetic clutchs 磁力离合器

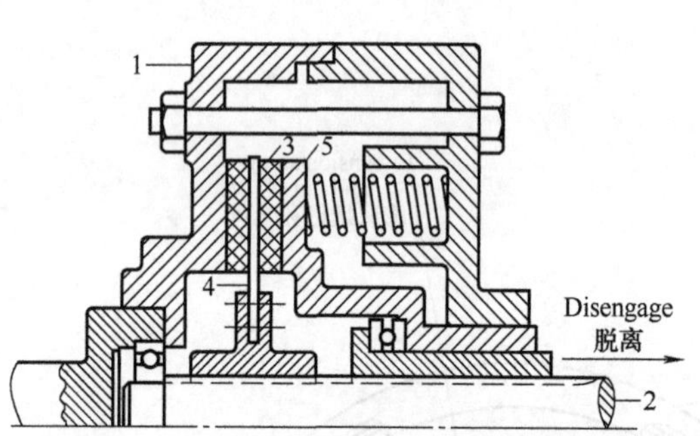

Fig 16.53 Schematic drawing of an axial clutch 轴向离合器示意图

1—Driving member 驱动件；2—Driven shaft 从动轴；3—Friction plate 摩擦片；4—Driven plate 从动片；5—Pressure plate 施压盘

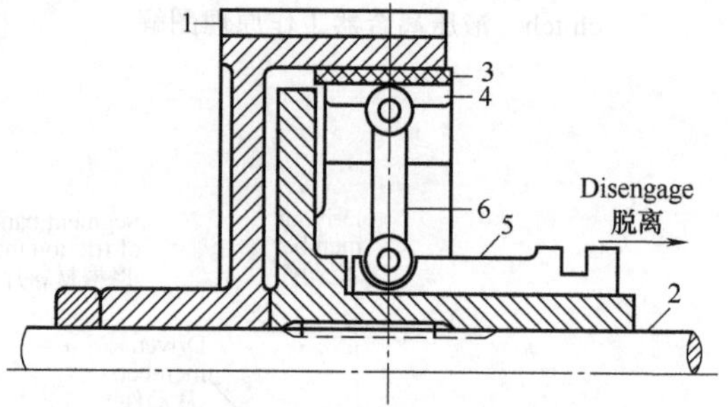

This type of clutch can also be made within a V-belt sheave

这种离合器还可构建在V带轮内

Fig 16.54 Schematic drawing of a radial clutch built within a gear 齿轮内的径向离合器示意图

1—Gear,the driving member 齿轮，驱动件；2—Driven shaft 从动轴；3—Friction plate 摩擦片；4—Pressure plate 施压盘；5—Movable sleeve 活动套；6—Toggle link 肘连接

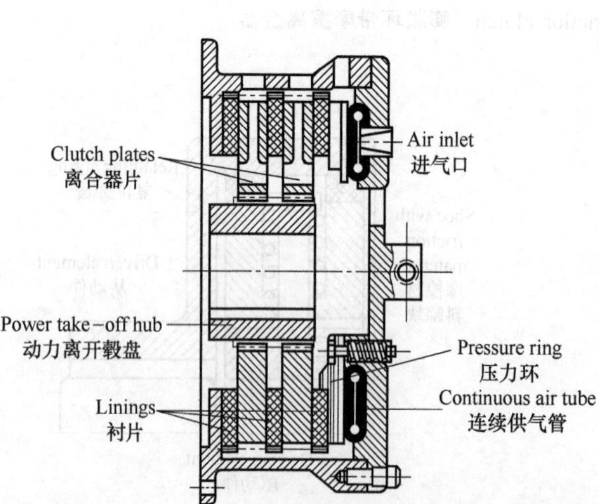

Fig 16.55 Typical air-operated clutch 典型气动离合器

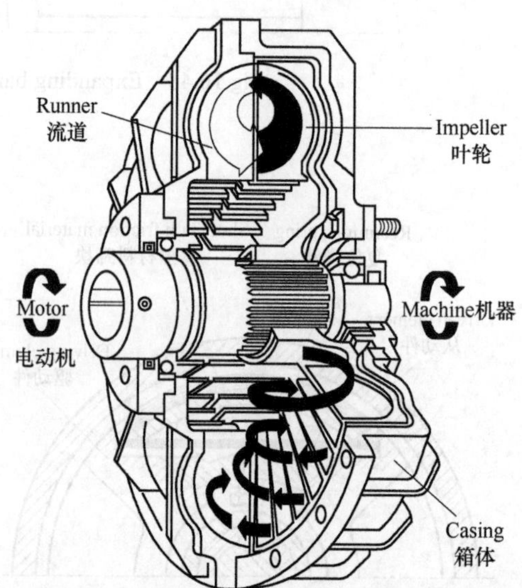

Fig 16.56 The principle of fluid coupling 液力离合器原理

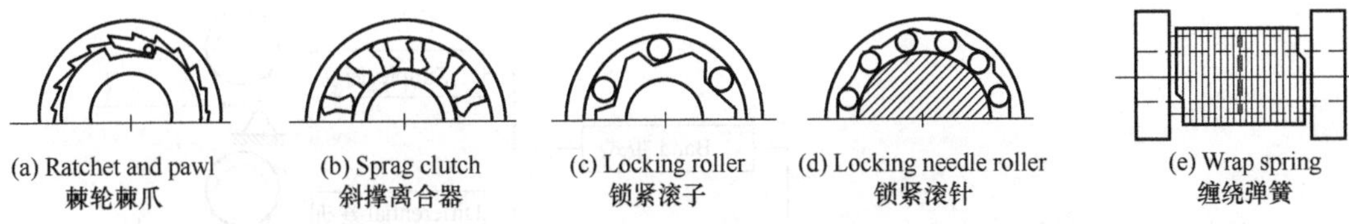

Fig 16.57　Types of one-way clutch (Overrunning clutch)　单向离合器（超越离合器）类型

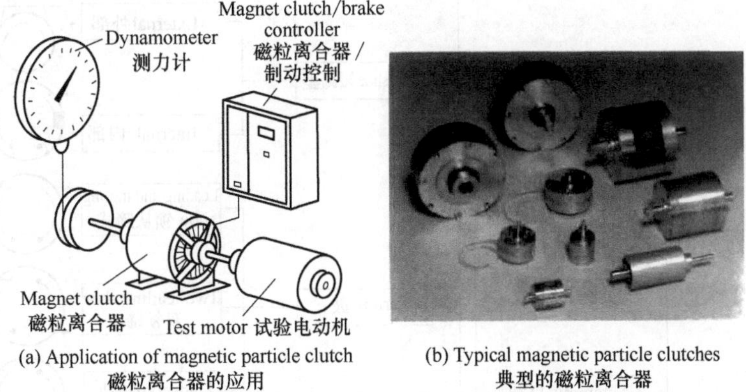

Fig 16.58　Magnetic particle clutch　磁粒离合器

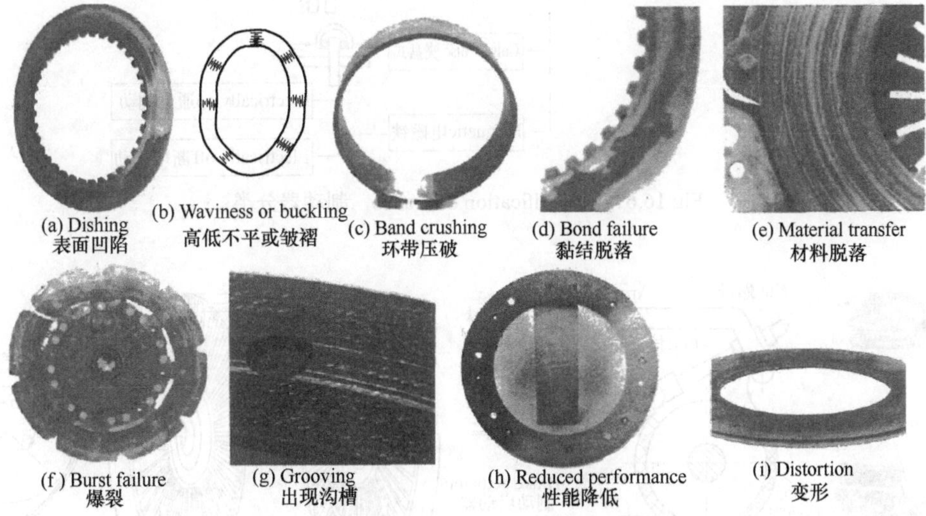

Fig 16.59　Clutch troubles　离合器失效形式

16.3　Brakes　制动器

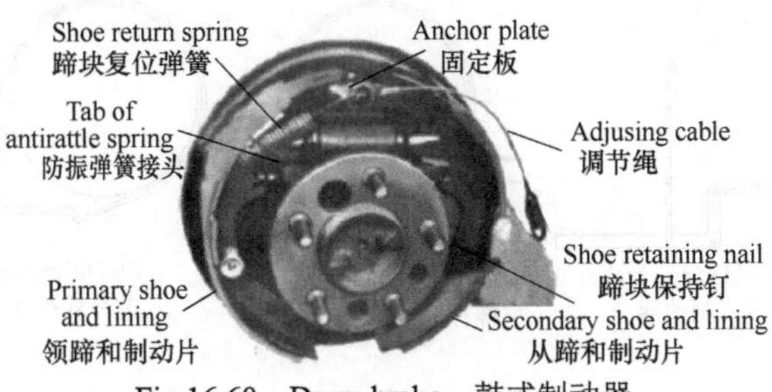

Fig 16.60　Drum brake　鼓式制动器

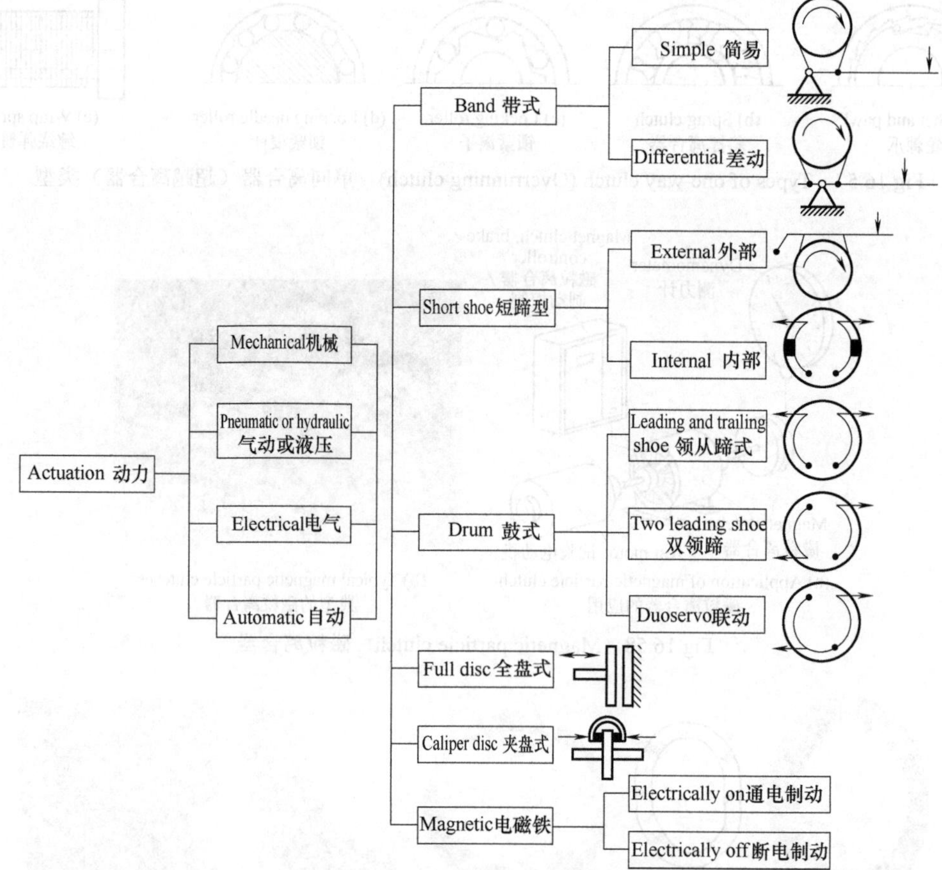

Fig 16.61 Classification of brakes 制动器分类

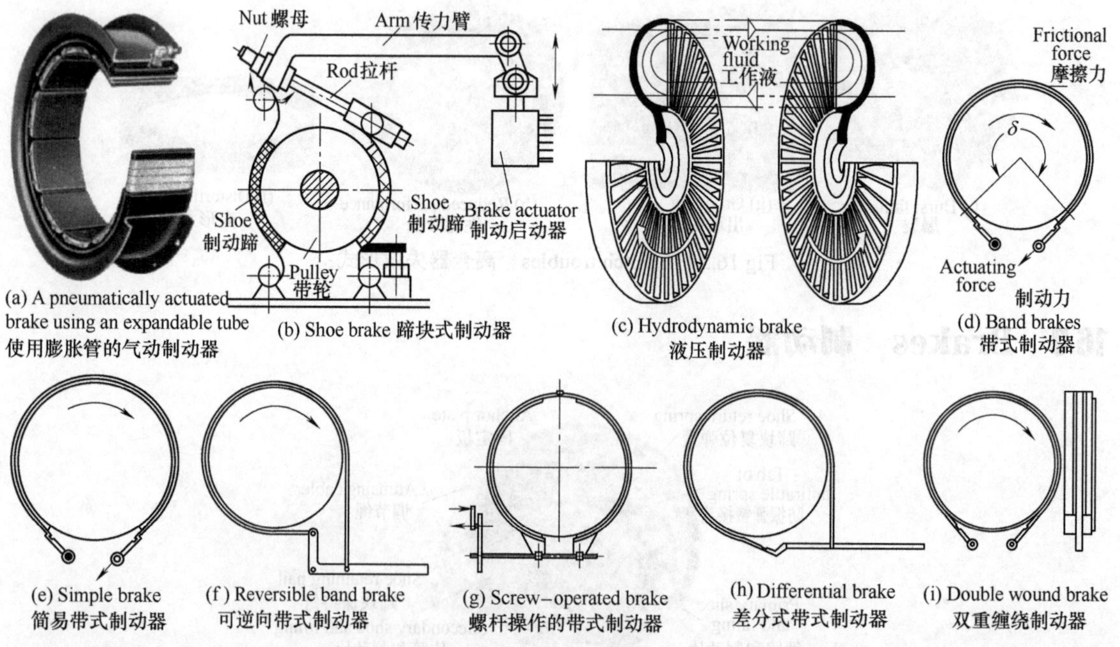

(a) A pneumatically actuated brake using an expandable tube 使用膨胀管的气动制动器
(b) Shoe brake 蹄块式制动器
(c) Hydrodynamic brake 液压制动器
(d) Band brakes 带式制动器
(e) Simple brake 简易带式制动器
(f) Reversible band brake 可逆向带式制动器
(g) Screw-operated brake 螺杆操作的带式制动器
(h) Differential brake 差分式带式制动器
(i) Double wound brake 双重缠绕制动器

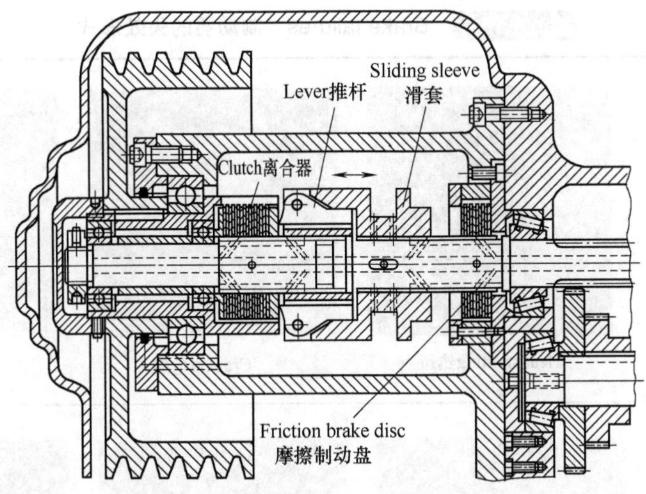

(j) Friction brake 摩擦制动器

Fig 16.62 Types of brake 制动器类型

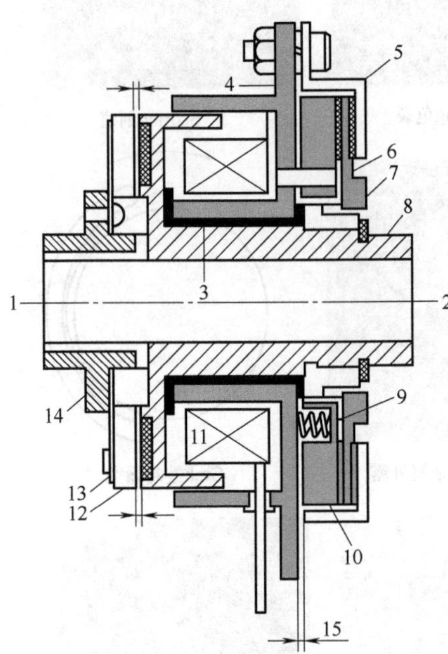

Fig 16.63 Clutch-brake transmission 离合器制动器传动图

1—Input 输入；2—Output 输出；3—Field coil bearing 磁场线圈轴承；4—Field coil assembly 磁场线圈组件；5—Pressure cup 压力杯；6—Brake plate 制动盘；7—Hub spring 轮毂弹簧；8—Rotor assembly 转子组件；9—Brake spring 制动弹簧；10—Brake armature 制动衔铁；11—Field coil 线圈；12—Clutch armature 离合衔铁；13—Clutch spring 离合弹簧；14—Drive plate 驱动盘；15—Air gap 空隙

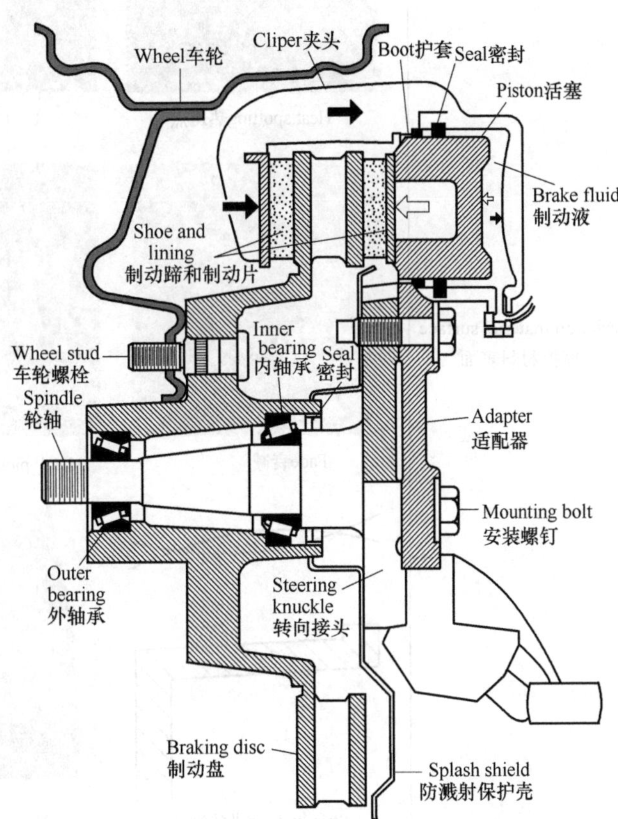

Fig 16.64 Automotive disc brake 汽车盘式制动器

Table 16.1 Brake failures 制动器的失效形式

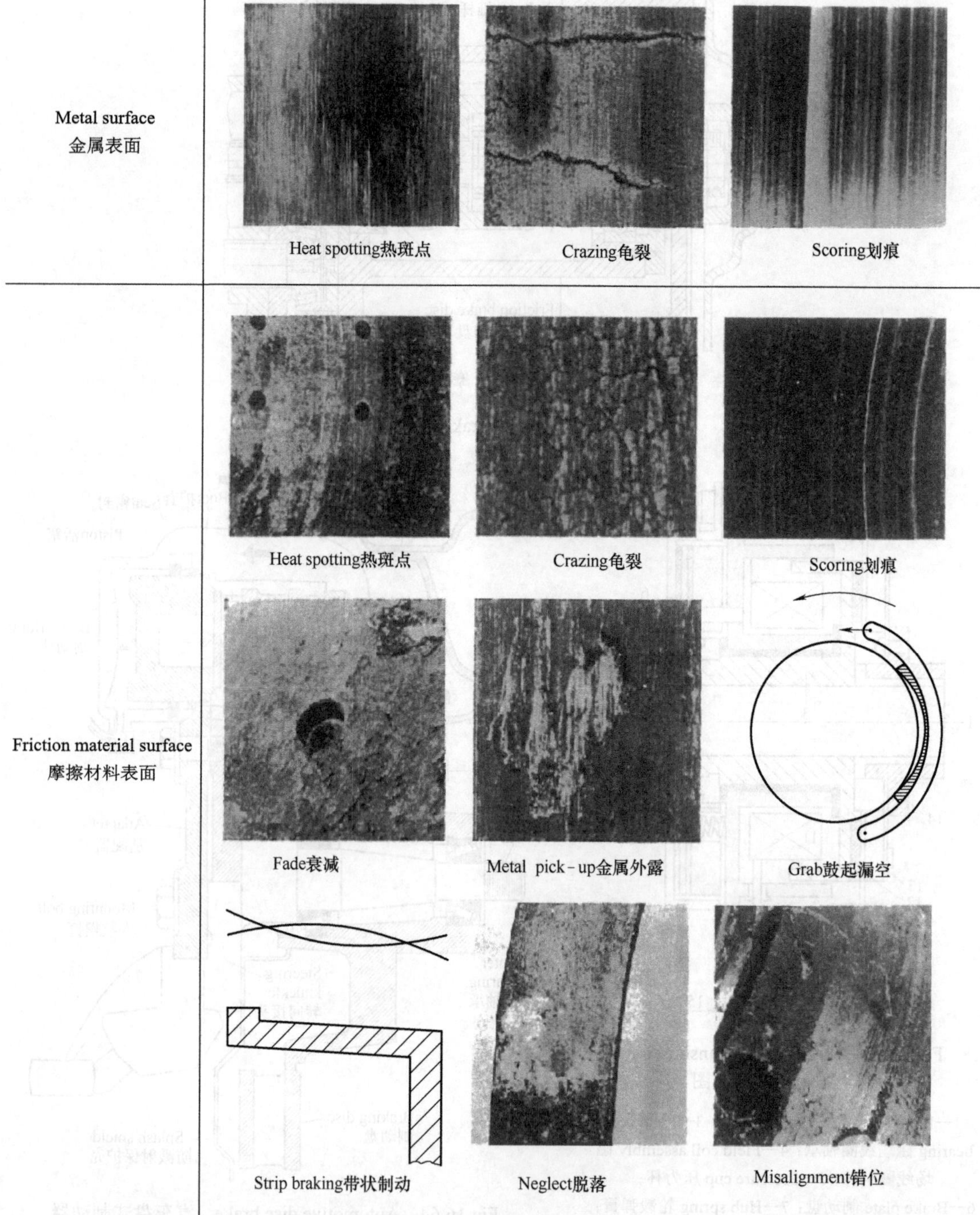

Metal surface 金属表面	Heat spotting热斑点	Crazing龟裂	Scoring划痕
Friction material surface 摩擦材料表面	Heat spotting热斑点	Crazing龟裂	Scoring划痕
	Fade衰减	Metal pick-up金属外露	Grab鼓起漏空
	Strip braking带状制动	Neglect脱落	Misalignment错位

Unit 17 Reductors and Speed Changers 减速器与变速器

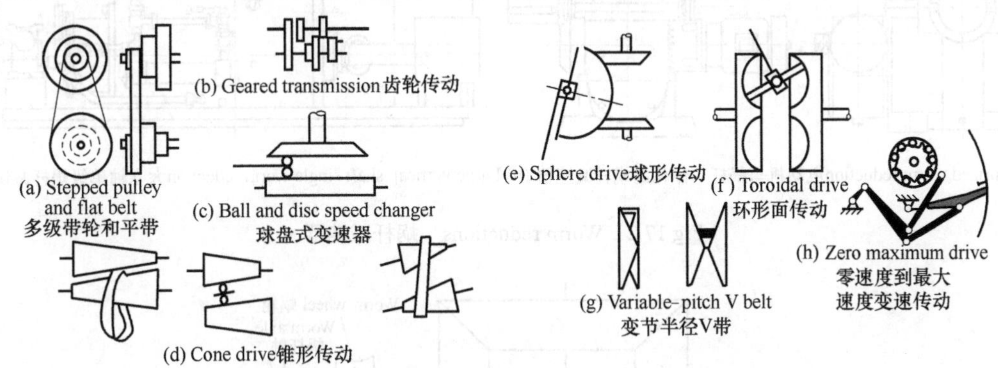

Fig 17.1 Speed-changing mechanisms 变速机构

17.1 Reductors 减速器

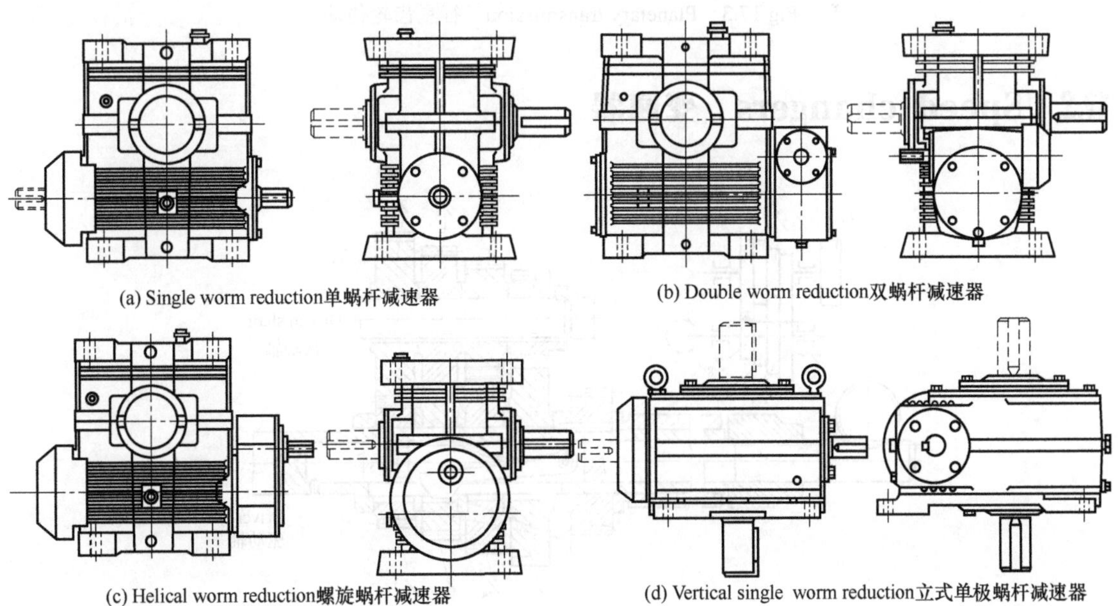

Fig 17.2

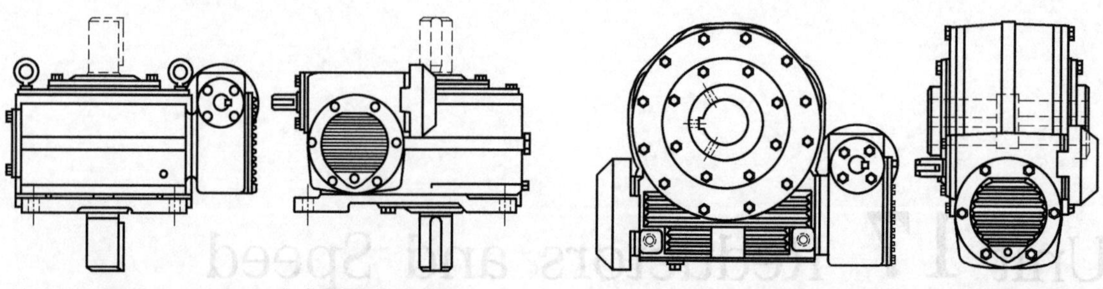

(e) Vertical double worm reduction立式两级蜗杆减速器

(f) Double - worm-reduction shaft-mount unit双蜗杆减速器轴安装单元

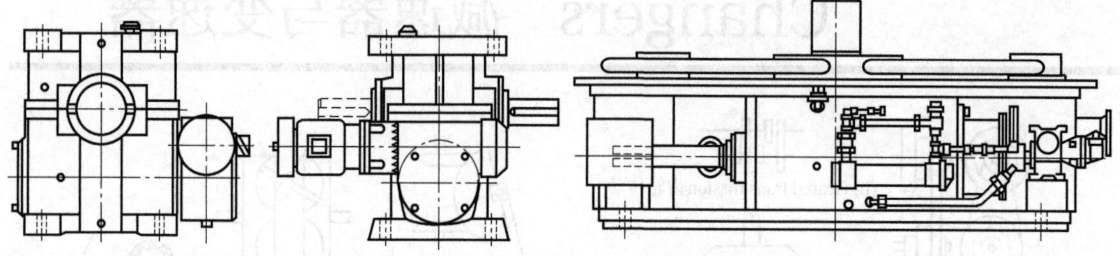

(g) Motorized worm reduction电动机与蜗杆一体化的蜗杆减速器

(h) Large vertical-shaft single worm reduction大立轴单级蜗杆减速器

Fig 17.2 Worm reductions 蜗杆减速器

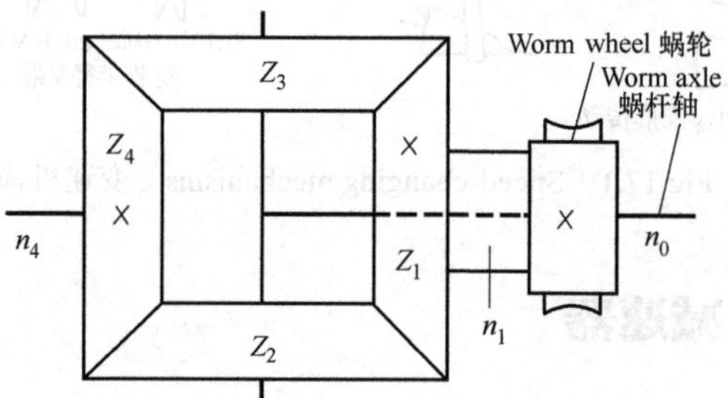

Fig 17.3 Planetary transmission 行星齿轮传动

17.2 Speed changers 变速器

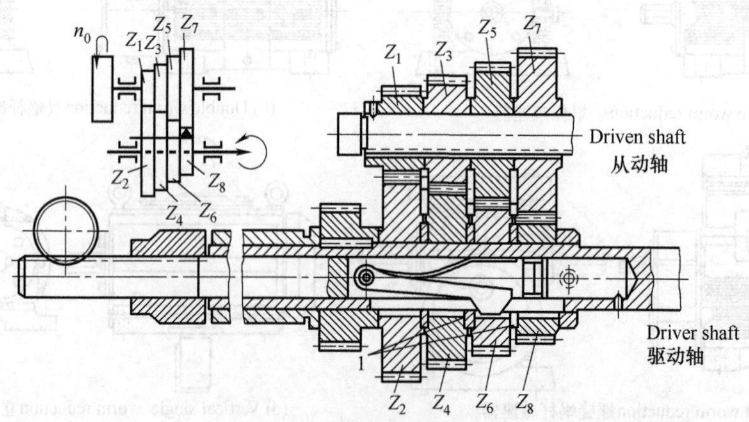

Fig 17.4 Feed gearbox with sliding gears 带滑移齿的变速箱

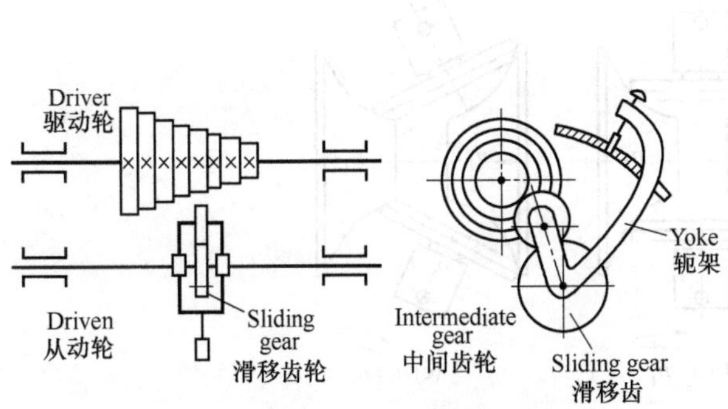

Fig 17.5 Norton gearbox 诺顿齿轮变速箱

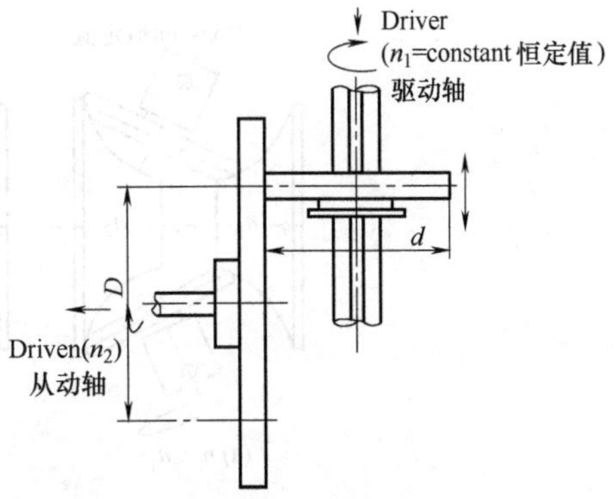

Fig 17.6 Disk-type friction stepless drive 盘式摩擦无级变速传动

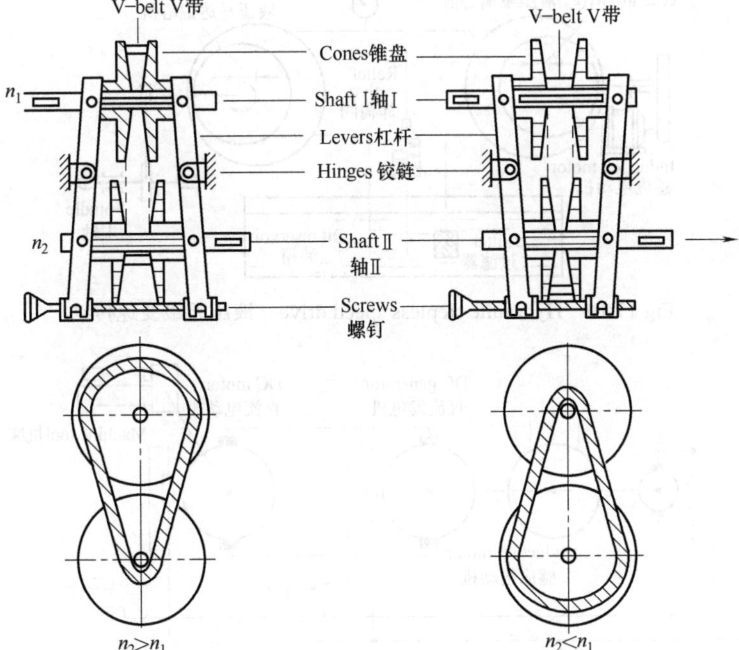

Fig 17.7 Positive infinitely variable drive 强制无级变速传动

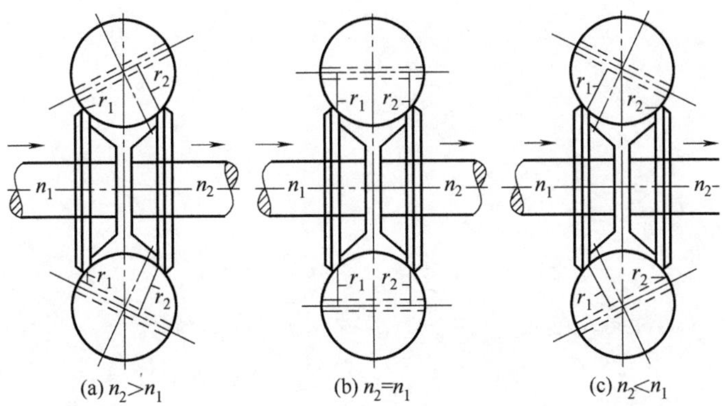

Fig 17.8 Kopp stepless speed mechanisms 科比无级变速机构

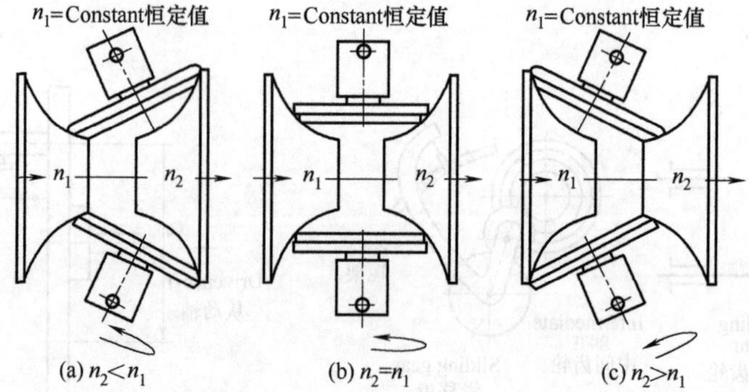

Fig 17.9 Toroidal stepless speed transmissions 锥体无级变速传动

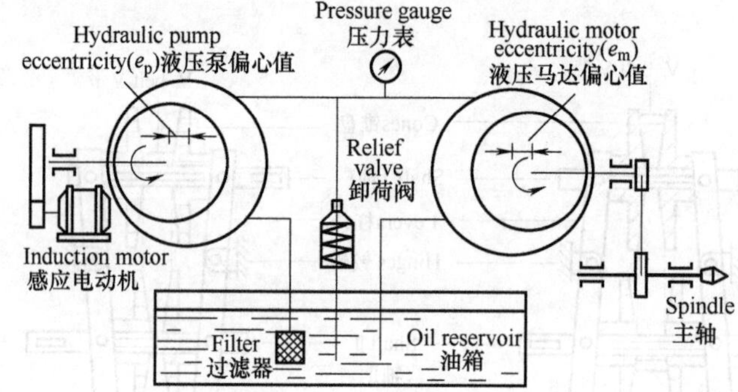

Fig 17.10 Hydraulic stepless speed drive 液压无级变速传动

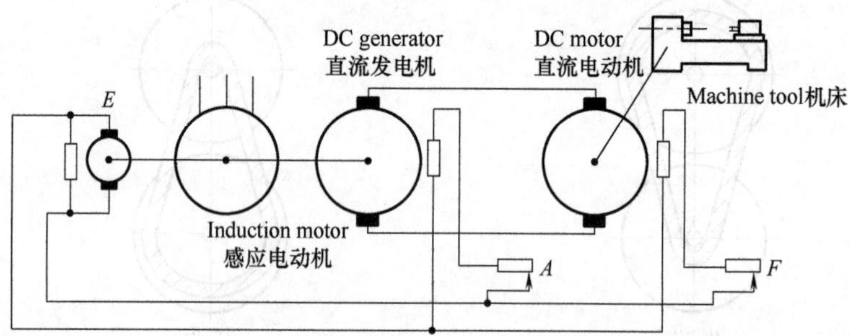

Fig 17.11 Leonard set (electrical stepless speed drive) 发动机 - 电动机机组控制装置（电气无级变速传动）

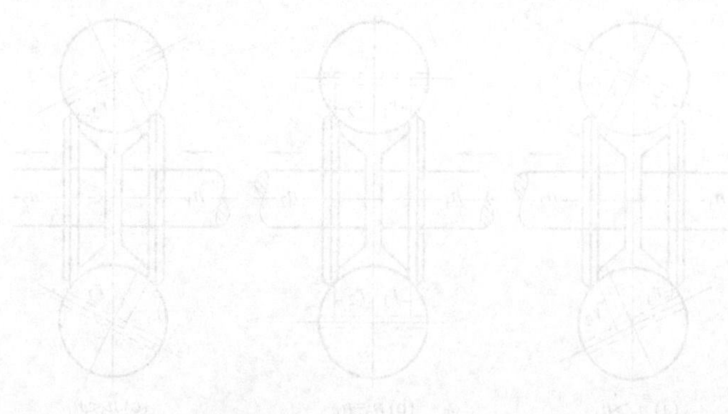

Part 4 Mechanical Equipment for Special Industry

专业机械

Unit 18 Power Generation Equipments

动力设备

18.1 Power generation equipments 多种发电设备

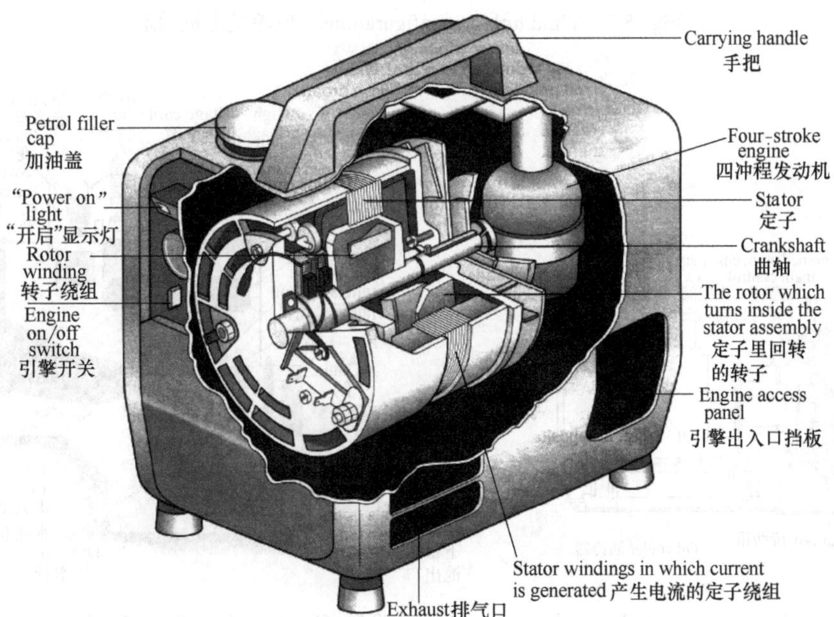

Fig 18.1 Portable generator 手提式发电机

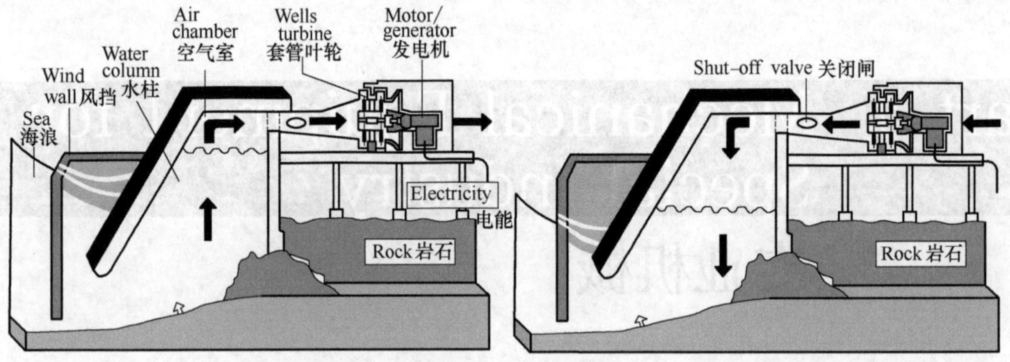

Fig 18.2 Wave power 波浪发电机

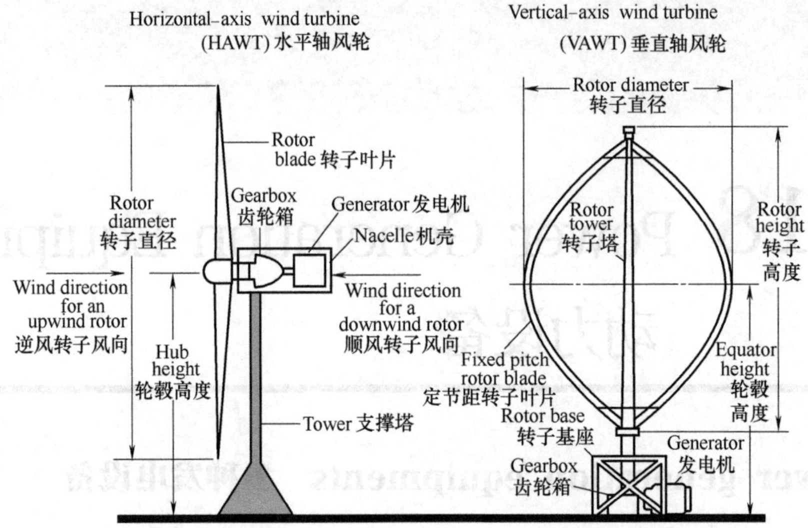

Fig 18.3 Wind turbine configurations 风轮发电机构造

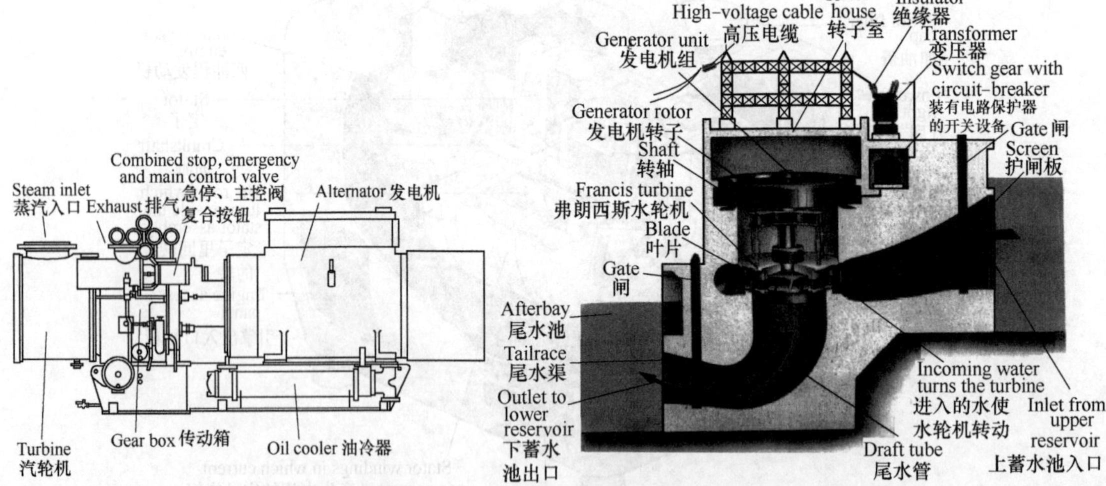

Fig 18.4 A single-stage steam turbine driving an alternator 单级蒸汽汽轮发电机

Fig 18.5 Cross-section of a hydroelectric power station 水力发电站剖面图

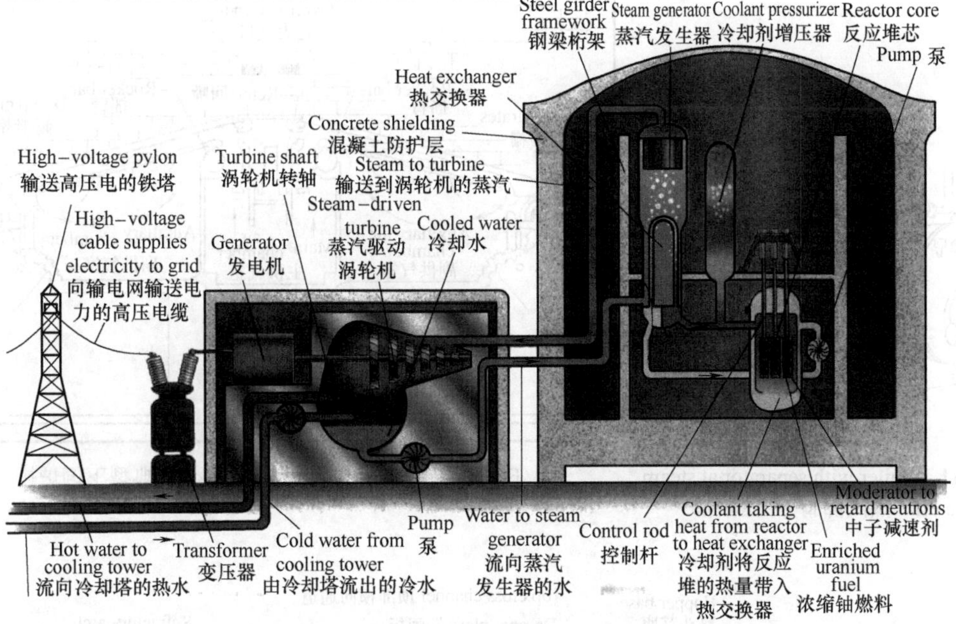

Fig 18.6 Cross-section of a nuclear power 核电站剖面图

18.2 Boilers 锅炉

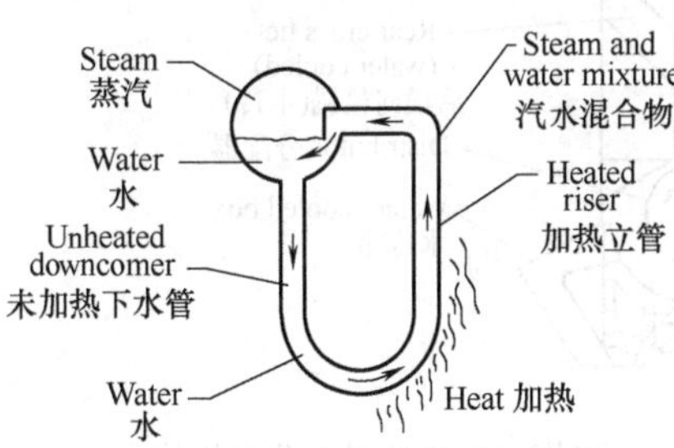

Fig 18.7 Steam drum with tubes 汽包及管路

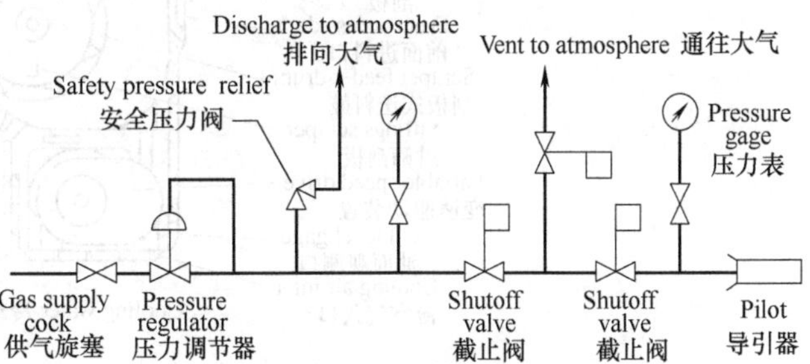

Fig 18.8 Typical pilot and gas piping arrangement 典型导引通气系统

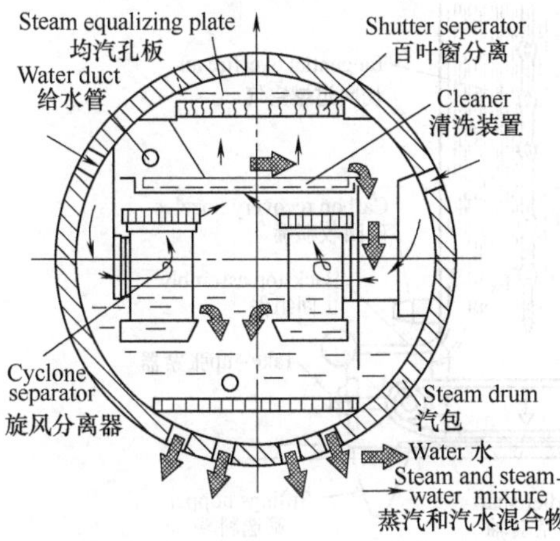

Fig 18.9 Internal setups for steam drum 汽包内部装置

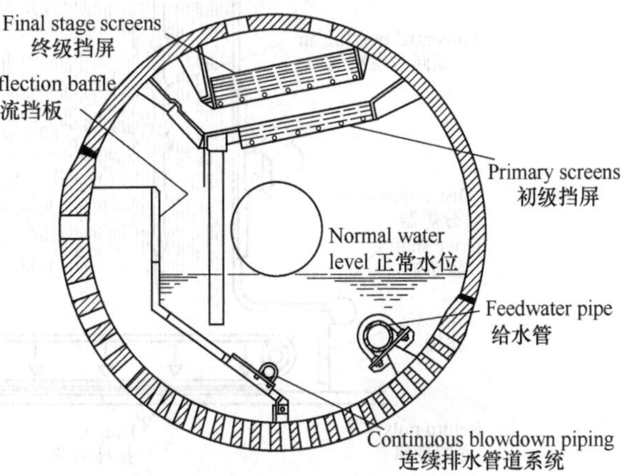

Fig 18.10 Steam drum internals-baffle type 挡板式锅筒内件

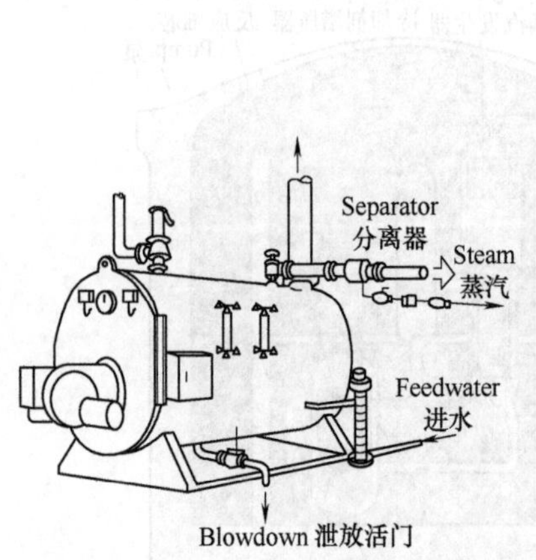

Fig 18.11 Boiler with separator at steam take off 蒸汽沸腾分离器锅炉

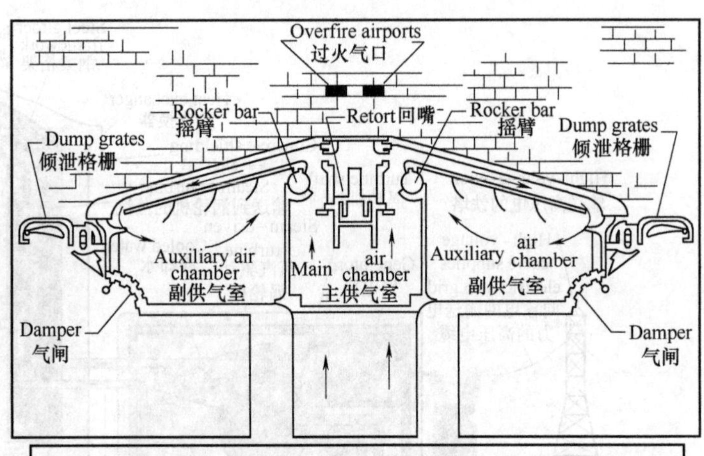

Fig 18.12 Typical underfeed stoker 典型下饲炉排

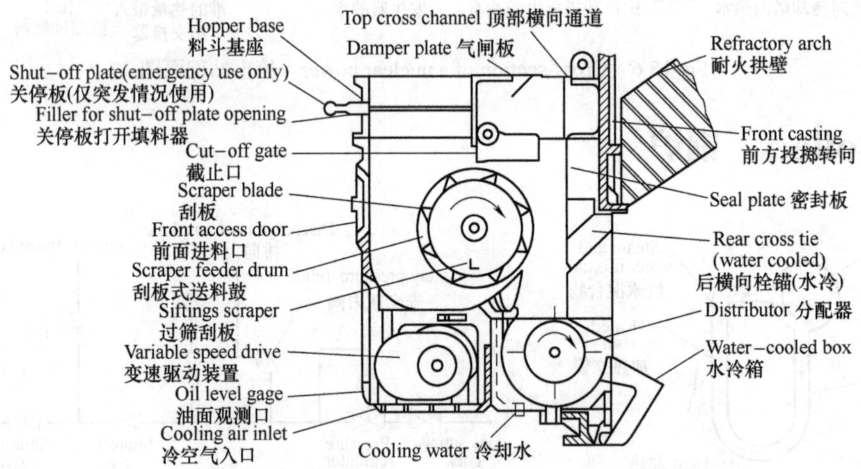

Fig 18.13 Feeder-distributor for firing of coal on a speader stoker 抛煤机炉排的煤火分布机构

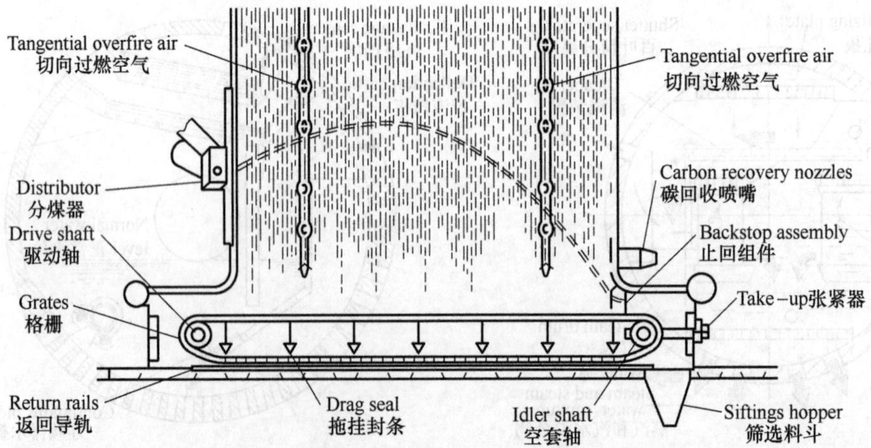

Fig 18.14 Spreader stoker, continuous ash discharge grate 抛煤机，连续抖灰格栅

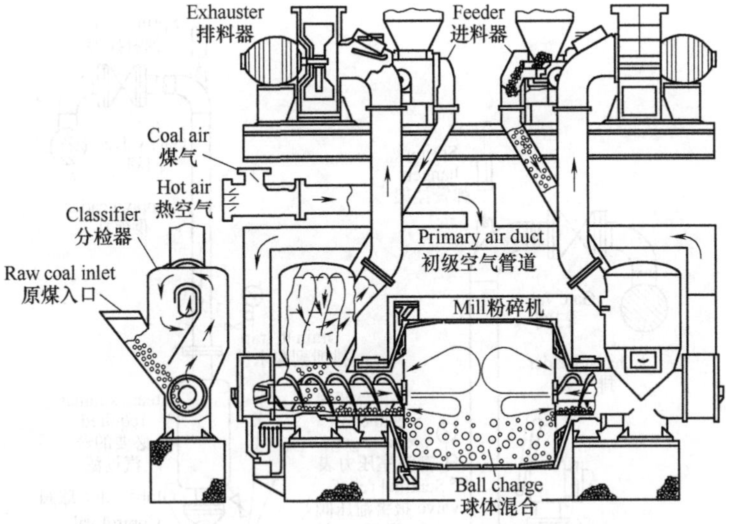

Fig 18.15 Arrangement ball-tube mill 筒式球体磨煤机

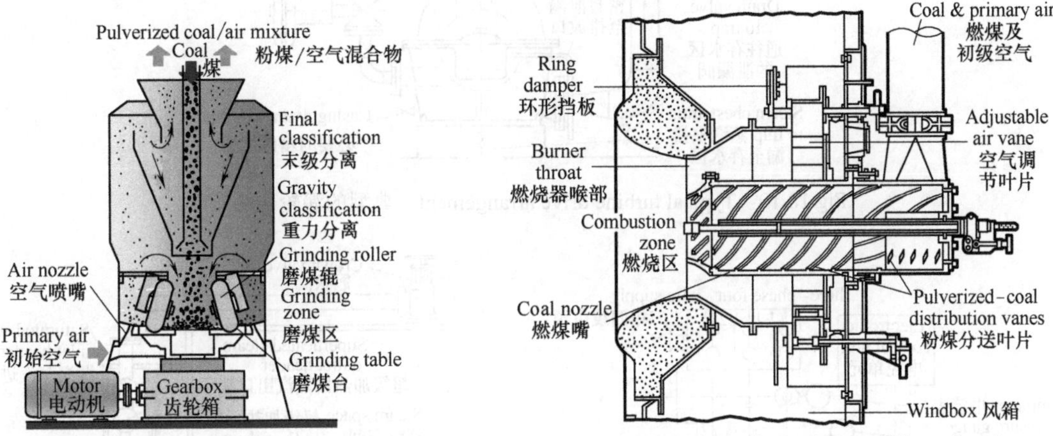

Fig 18.16 Roller medium-speed coal mill
辊辊式中速磨煤机

Fig 18.17 Burner for horizontal firing of coal
燃煤水平点火燃烧器

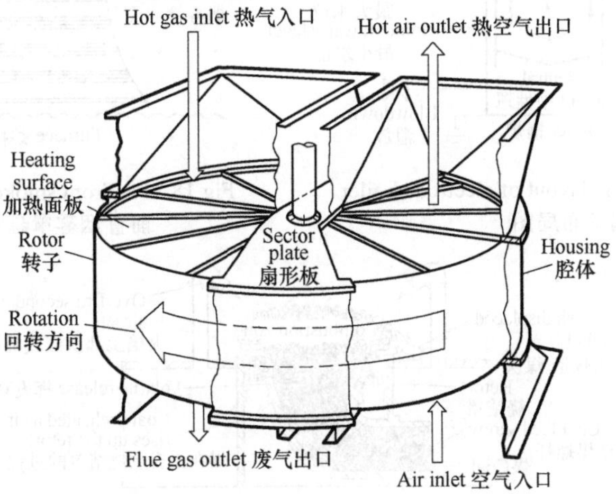

Fig 18.18 Details of regenerative air preheater, bisector type 双扇形再生式空气预热器

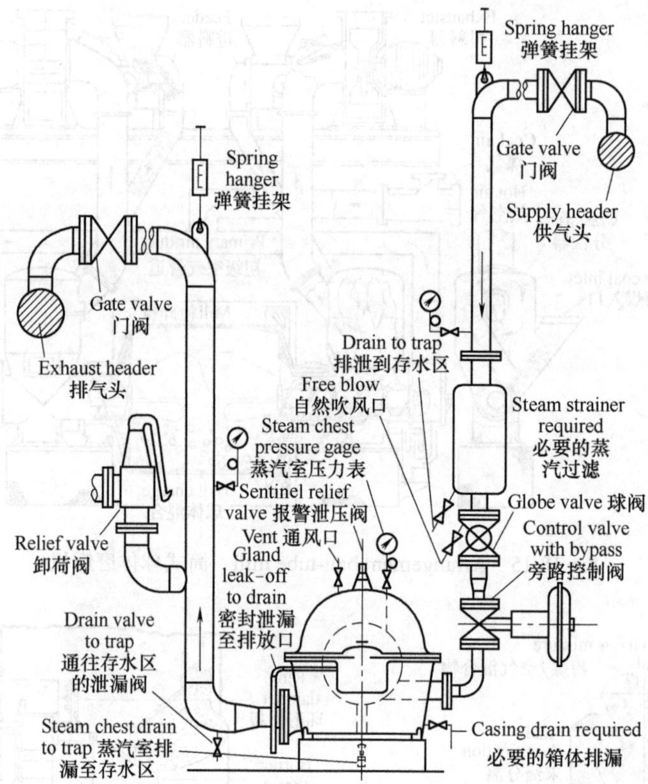

Fig 18.19 Typical turbine drive arrangement 典型的涡轮传动布局

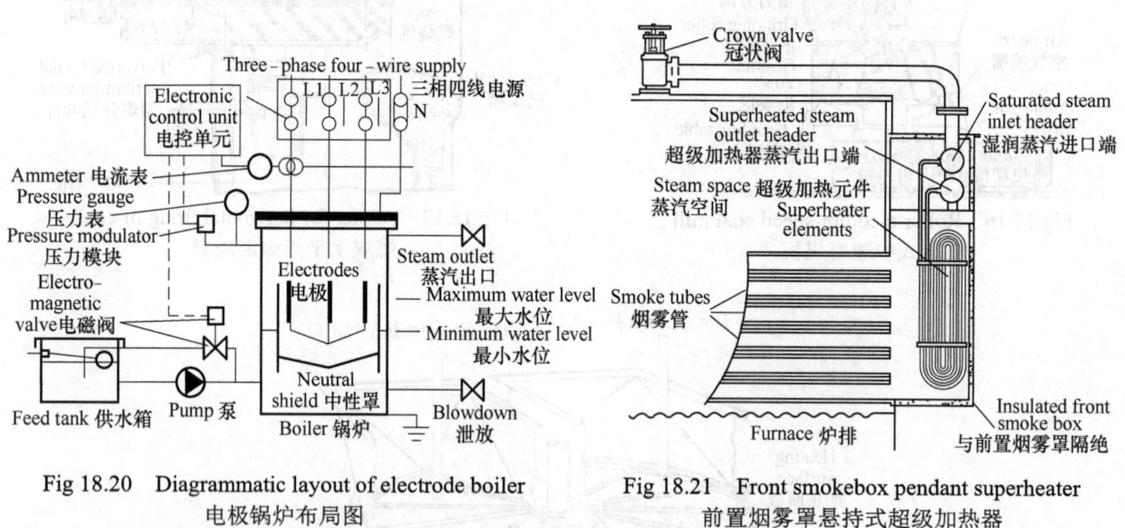

Fig 18.20 Diagrammatic layout of electrode boiler
电极锅炉布局图

Fig 18.21 Front smokebox pendant superheater
前置烟雾罩悬持式超级加热器

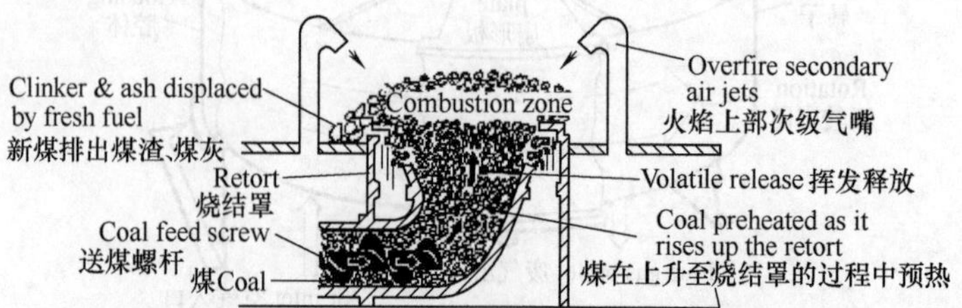

Fig 18.22 Underfeed stoker 底部送进加煤机

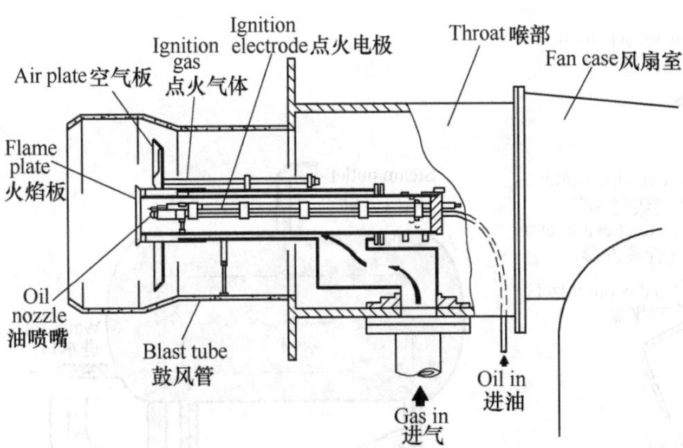

Fig 18.23 Dual-fuel burner based on pressure-jet configuration 基于压力喷射结构的双燃料燃烧器

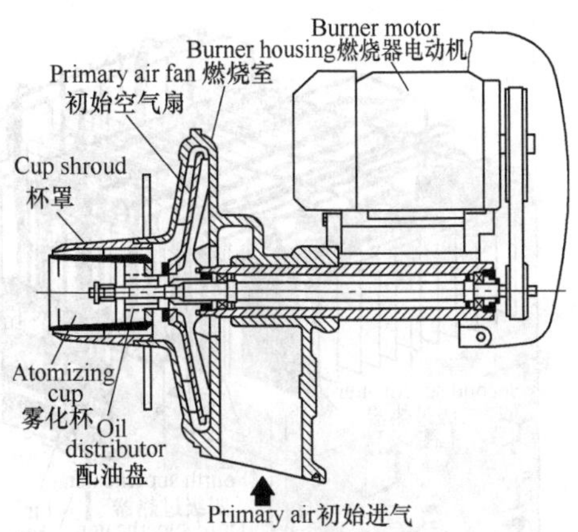

Fig 18.24 Section through rotary cup atomizer 回转杯型雾化器剖视图

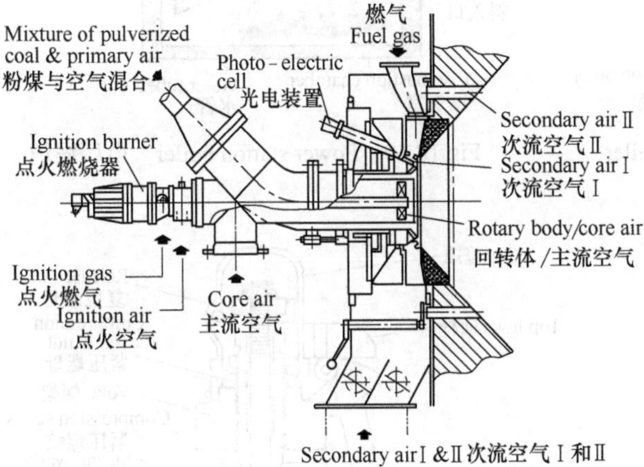

Fig 18.25 Pulverized fuel burner 粉末燃料燃烧器

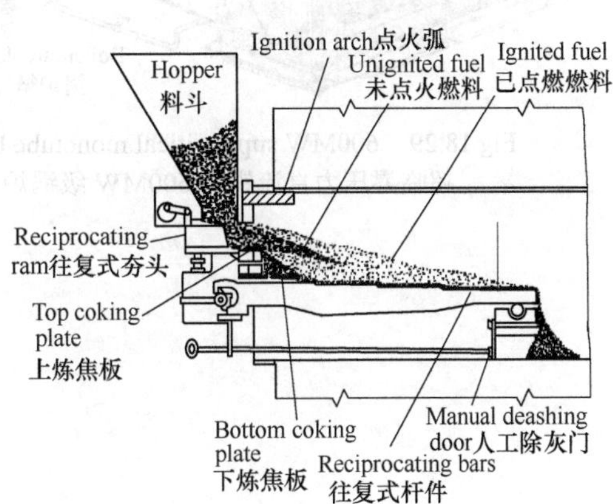

Fig 18.26 Coking stoker 炼焦碎煤机

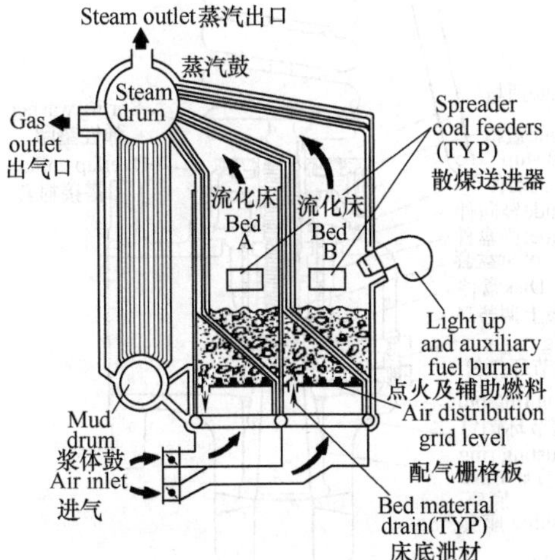

Fig 18.27 Fluid-bed combustion system 流化床燃烧系统

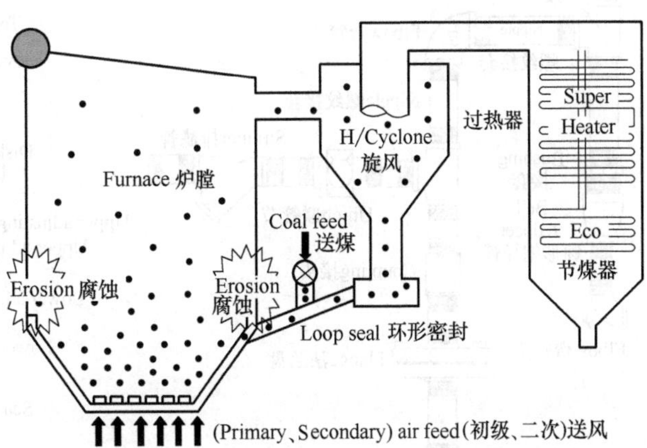

Fig 18.28 Circulating fluidized reboiler 循环流化床锅炉

Steam separator
蒸汽分离器
First superheater
初级过热器
Gas duct evaporator
气管蒸发器
Coal economizer
节煤器
Second superheater
二级过热器
Fourth superheater
四级过热器
First reheater
初级再热器
Third superheater
三级过热器
Furnace炉膛
Second
reheater
二级再热器
Boiler circulation pump
锅炉循环泵

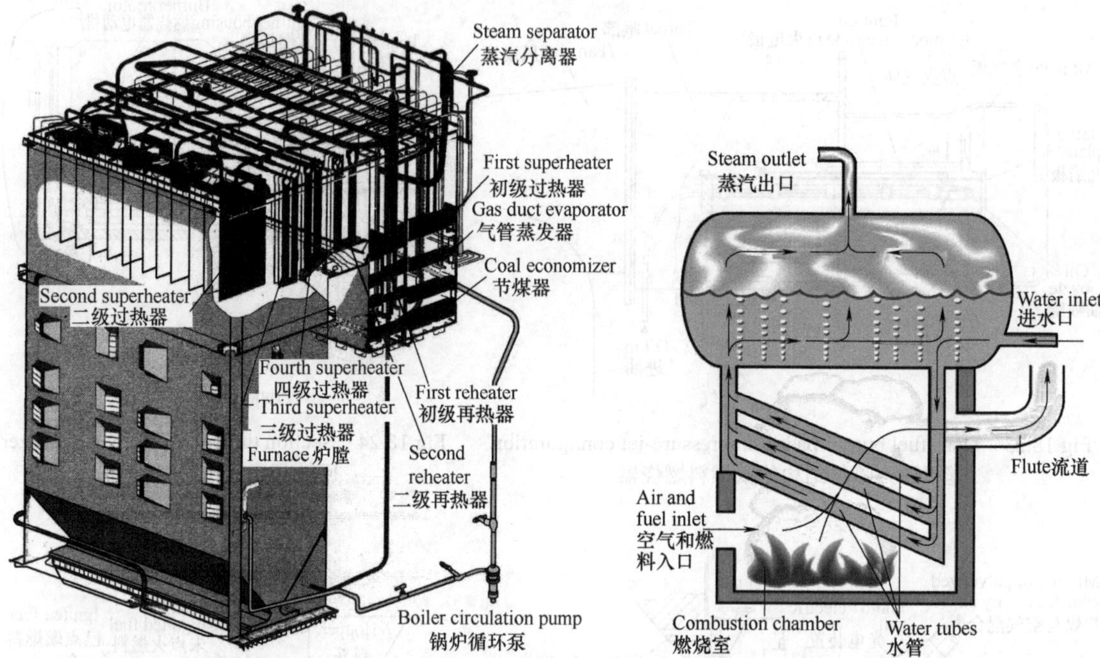

Fig 18.29 600MW supercritical monotube boiler 超临界压力直流锅炉 600MW 级锅炉

Steam outlet
蒸汽出口
Water inlet
进水口
Flute流道
Air and
fuel inlet
空气和燃
料入口
Combustion chamber
燃烧室
Water tubes
水管

Fig 18.30 Power station boiler 电厂锅炉

Cap盖罩
Release nut
复位螺母
Top lever顶杆
Compression
screw nut
紧压螺母
Yoke 轭架
Compression screw
紧压螺纹
Spring washer
弹性垫圈
Drop lever下落杆
Spring 弹簧
Spindle
主轴杆件
Top plate顶板
Spring washer
弹性垫圈
Cover plate盖板
Lift stop
上升止动件
Overlap collar
搭接衬套
Guide导向件
Disk holder固盘件
Disk collar盘套
Disk盘件
Upper adjusting ring上调节环
Upper adjusting ring pin
上调节环销钉
Lower adjusting ring pin
下调节环销钉
Lower adjusting ring
下调节环 Base
底座
Seat bushing 座套
Inlet neck
入口颈部

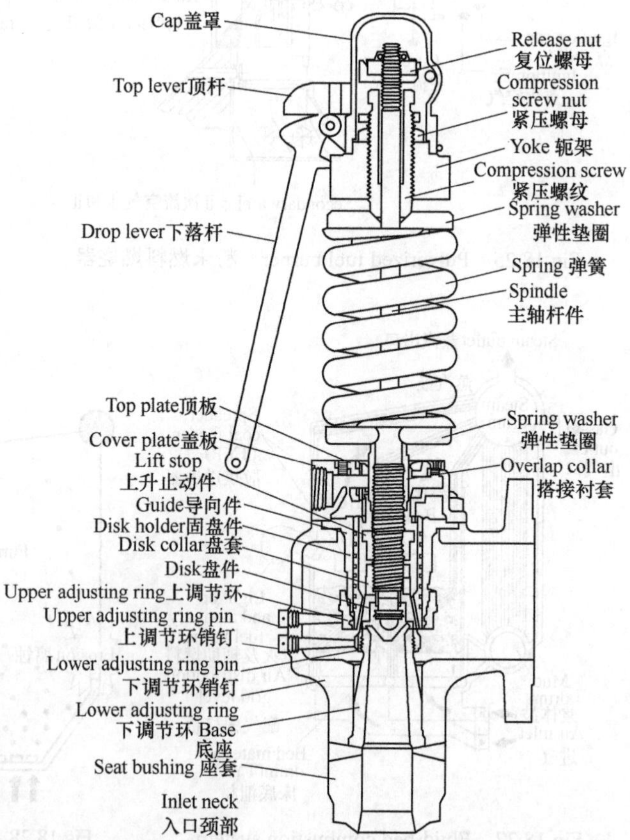

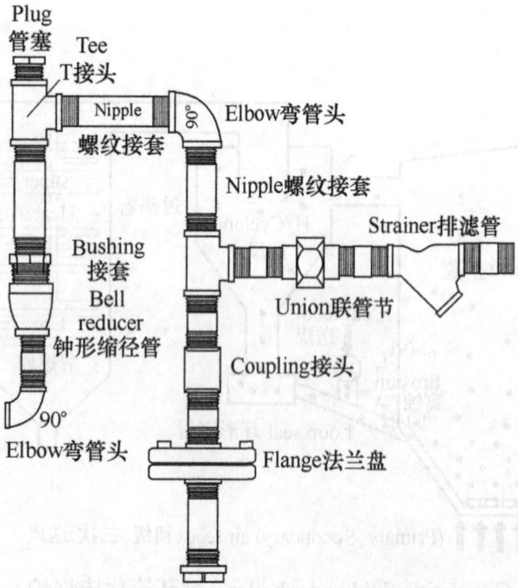

Plug
管塞
Tee
T接头
Nipple
90°
Elbow弯管头
螺纹接套
Nipple螺纹接套
Strainer排滤管
Bushing
接套
Bell
reducer
钟形缩径管
Union联管节
Coupling接头
90°
Elbow弯管头
Flange法兰盘

Fig 18.31 Pipe fittings 管件接头

Fig 18.32 Typical safety valve 典型的安全阀

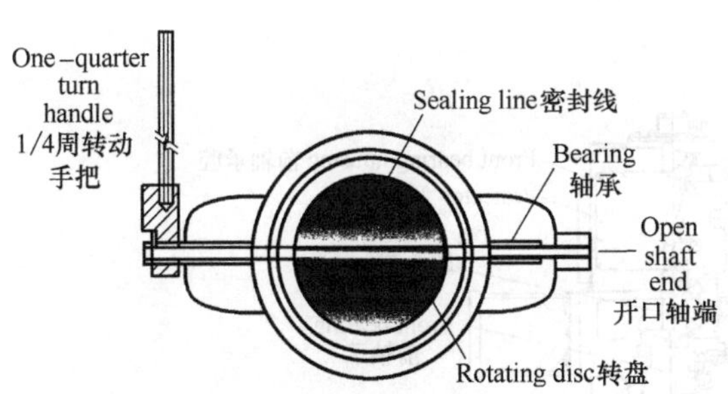

Fig 18.33 Butterfly valve 蝶阀

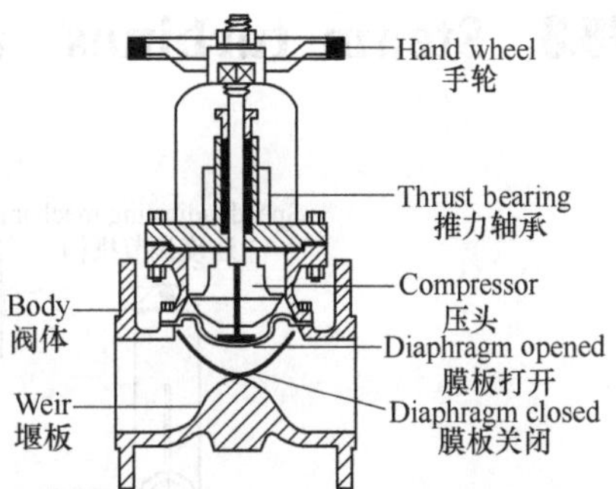

Fig 18.34 Diaphragm valve 膜板阀

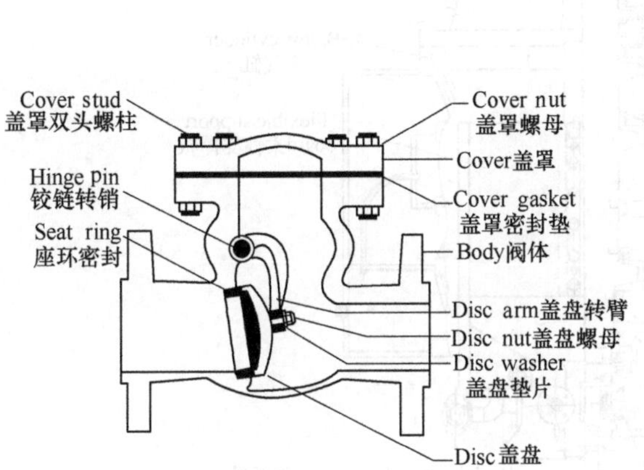

Fig 18.35 Check valve 单向阀

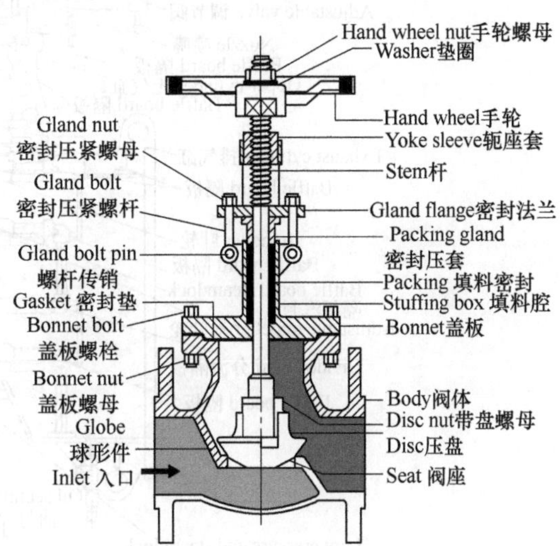

Fig 18.36 Globe valves 球阀

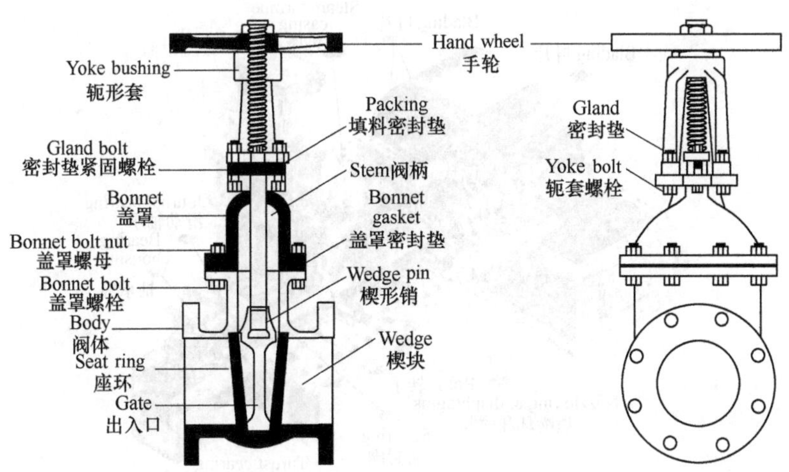

Fig 18.37 Gate valves 闸阀

18.3 Steam turbines 蒸汽轮机

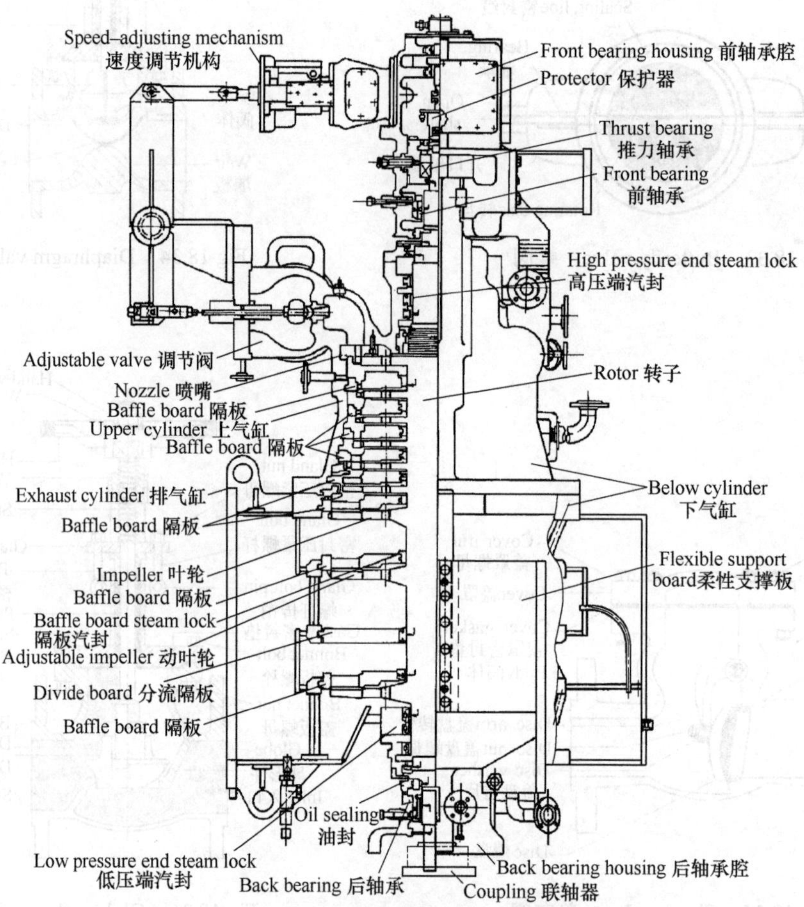

Fig 18.38 Steam turbine 蒸汽轮机

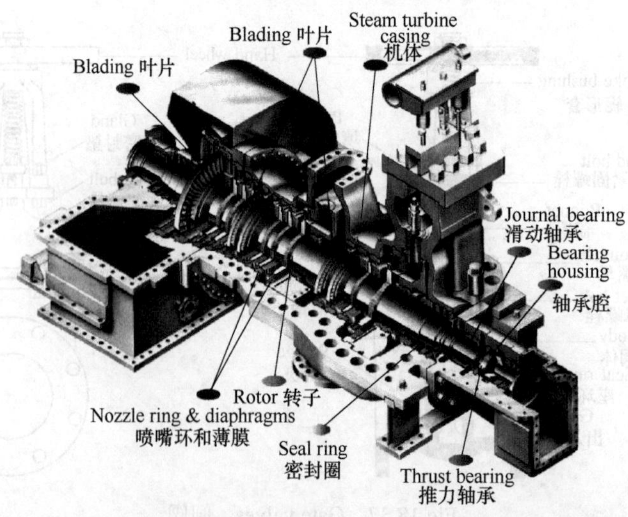

Fig 18.39 Section through a steam turbine 蒸汽轮机剖切图

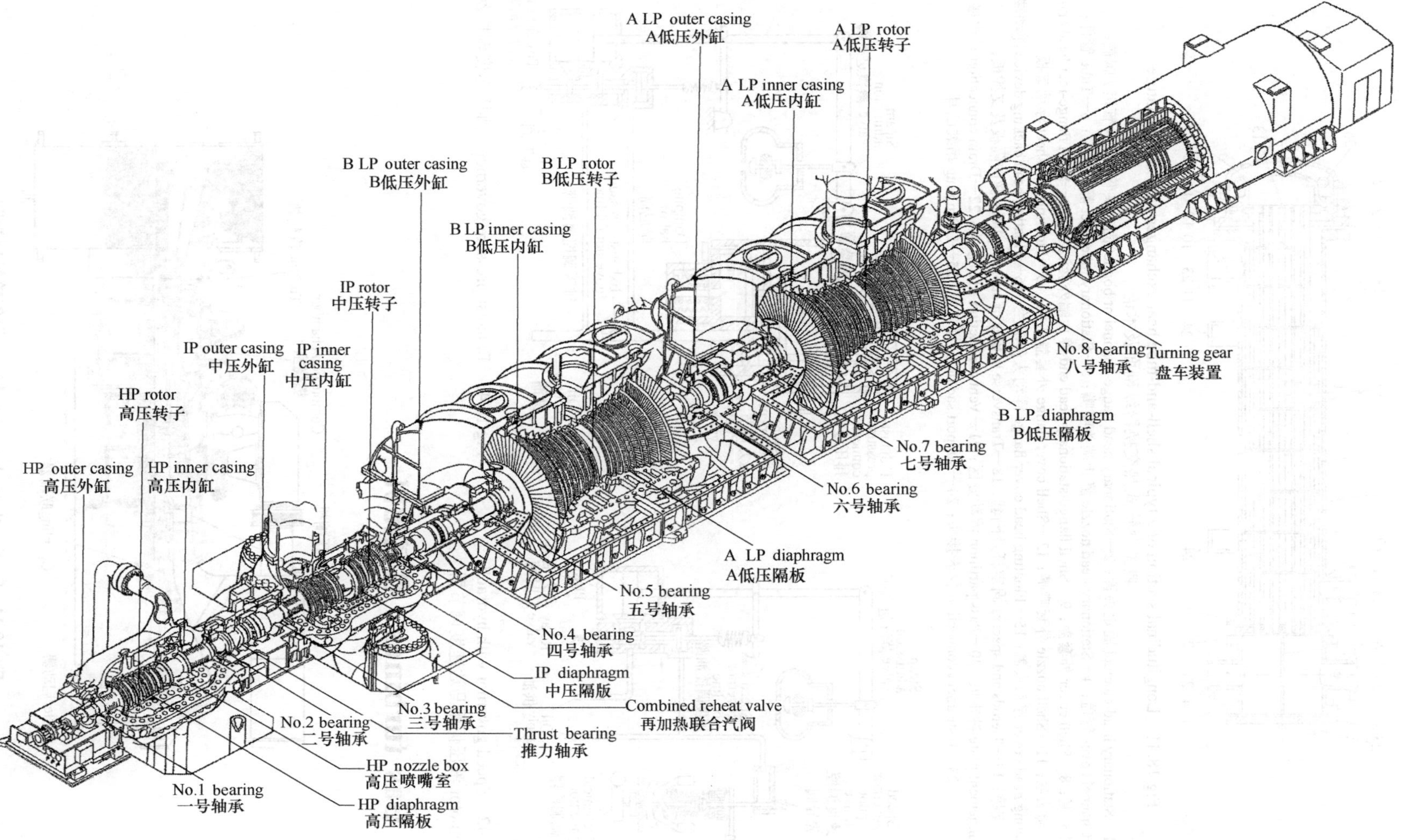

Fig 18.40 Supercritical turbine supassed 1000MW 超临界 1000MW 汽轮机

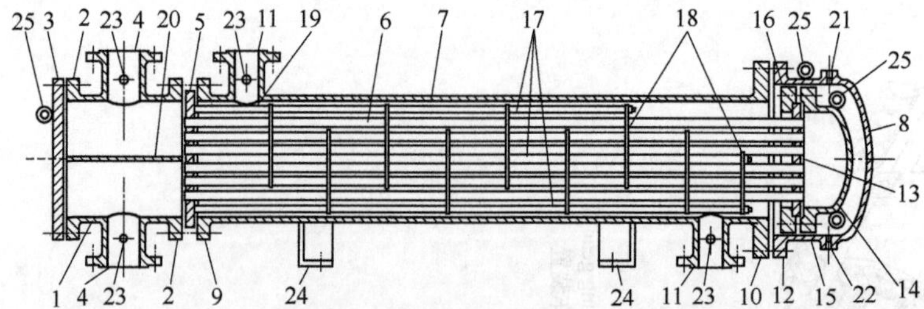

Fig 18.41 Longitudinal section of a typical shell-and-tube heat exchanger with nomenclature
典型套管式热交换器纵剖面及术语

1—Stationary head-channel 固定头管；2—Stationary head flange-channel or bonnet 固定头管法兰 - 管道 / 阀盖；3—Channel cover 管盖；4—Stationary head nozzle 静止端头喷嘴；5—Stationary tubesheet 静止管板；6—Tube 管件；7—Shell 外套；8—Shell cover 外套盖；9—Shell flange-stationary head end 静止端外套法兰；10—Shell flange-rear head end 尾端法兰；11—Shell nozzle 外套喷嘴；12—Shell cover flange 外套盖法兰；13—Floating tubesheet 浮动管板；14—Floating head cover 浮动头盖；15—Floating head cover flange 浮动头盖法兰；16—Floating head backing device 浮动端头支撑；17—Tierods and spacers 固定杆及隔套；18—Transverse baffles or support plates 横向隔板及支撑板；19—Impingement plate 冲击板；20—Pass partition 流动分区；21—Vent connection 出口接头；22—Drain connection 排出接头；23—Instrument connection 仪表接头；24—Support saddle 支撑鞍座；25—Lifting lug 吊装凸耳

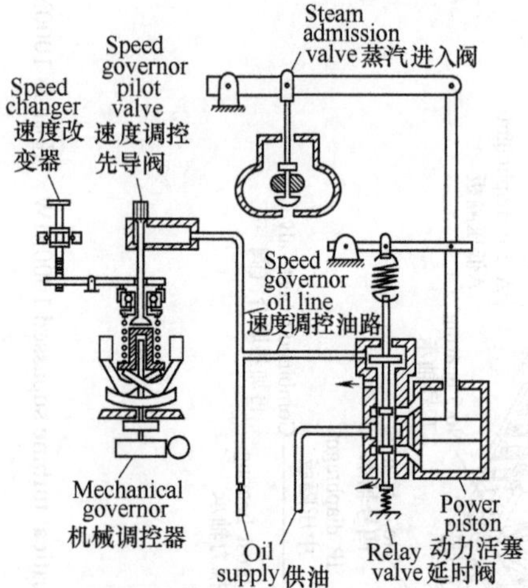

Fig 18.42 Speed governor with hydraulic servomotor 液压伺服马达速度调节器

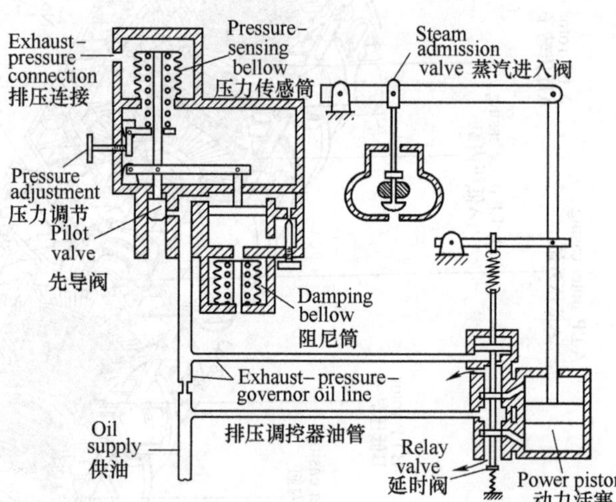

Fig 18.43 Exhaust-pressure governor 排压速度调控器

18.4 Gas turbines 燃气轮机

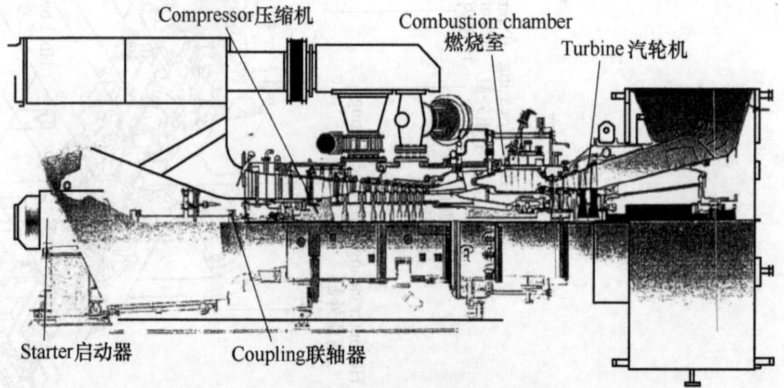

Fig 18.44 Section through a gas turbine 燃气轮机剖视图

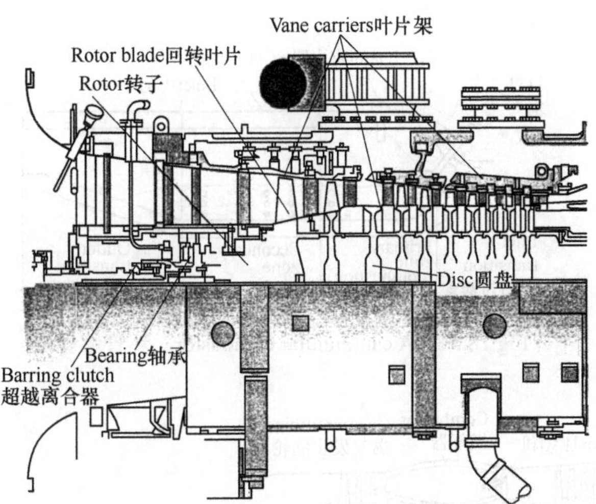

Fig 18.45 Gas turbine compressor 燃气轮机压缩机

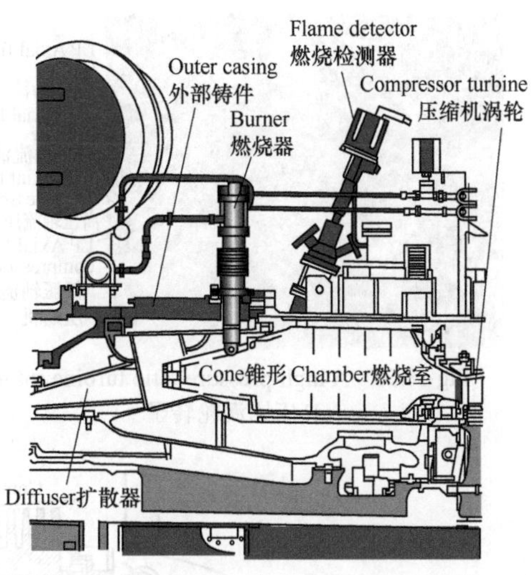

Fig 18.46 Combustion section, gas turbine 燃气轮机燃烧室剖视图

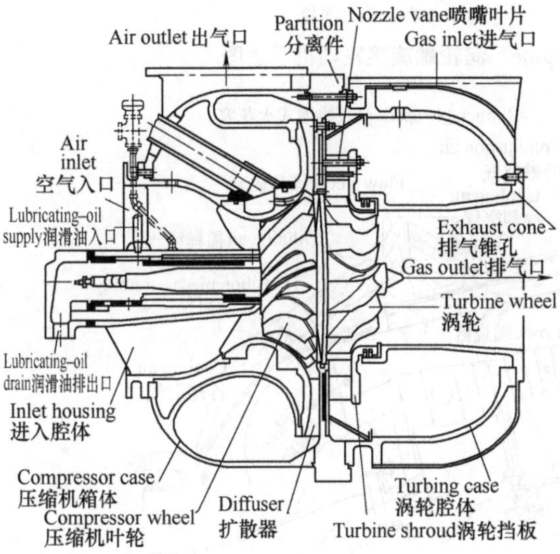

Fig 18.47 Turbocharger 涡轮增压器

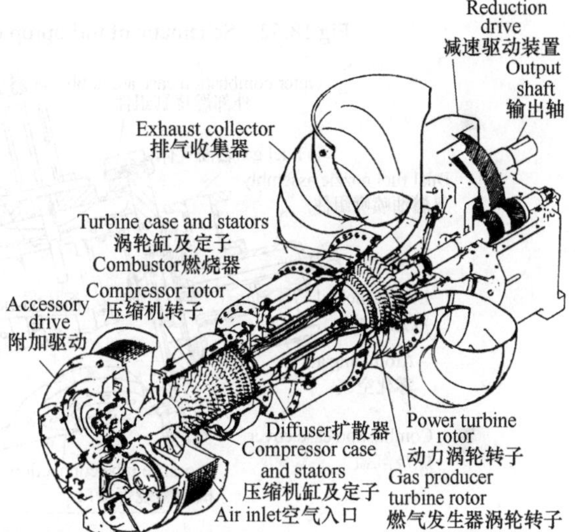

Fig 18.48 Medium size industrial gas turbine 中型尺寸工业燃气轮机

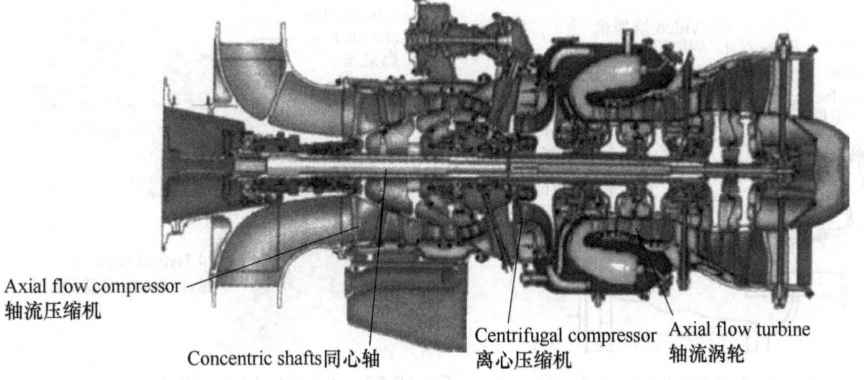

Fig 18.49 Small aeroderivative gas turbine 小型气引导燃气轮机

Fig 18.50 A high-pressure ratio turbine rotor 高压比涡轮转子

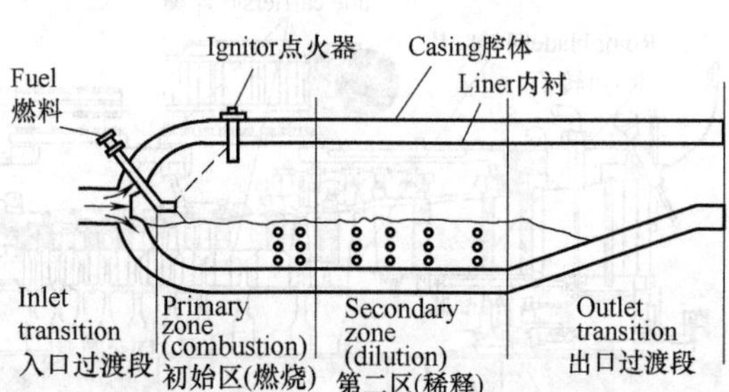

Fig 18.51 A can-annular combustor 环管燃烧器

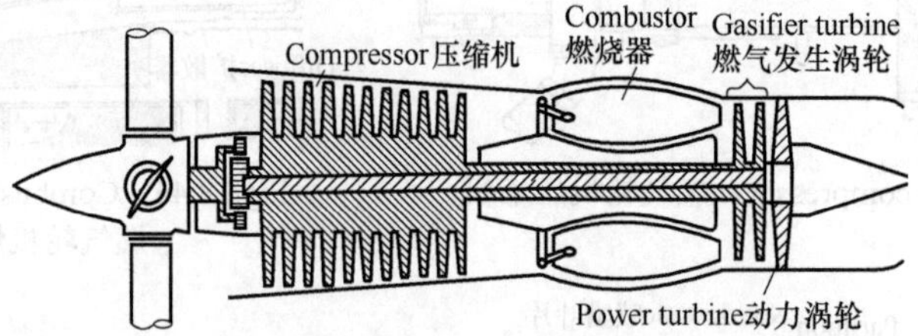

Fig 18.52 Schematic of turboprop engine 涡轮螺旋桨发动机示意图

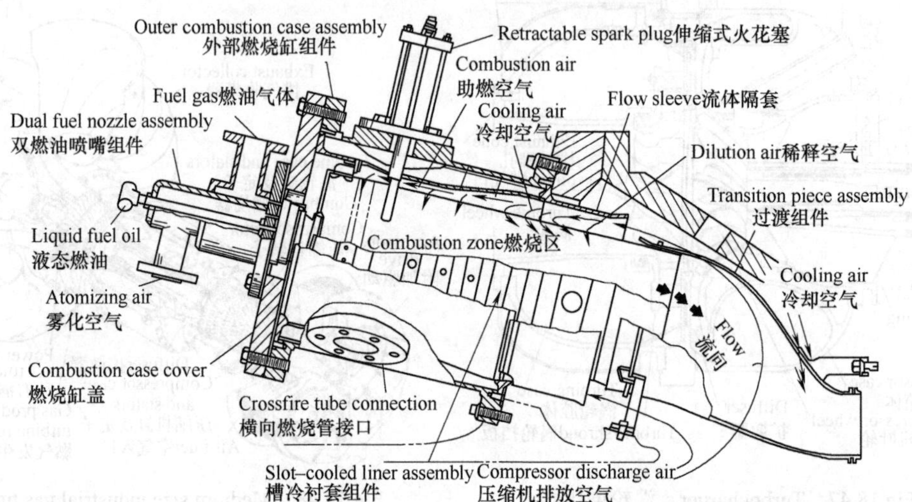

Fig 18.53 A typical reverse flow can-annular combustor 典型的逆流环管燃烧室

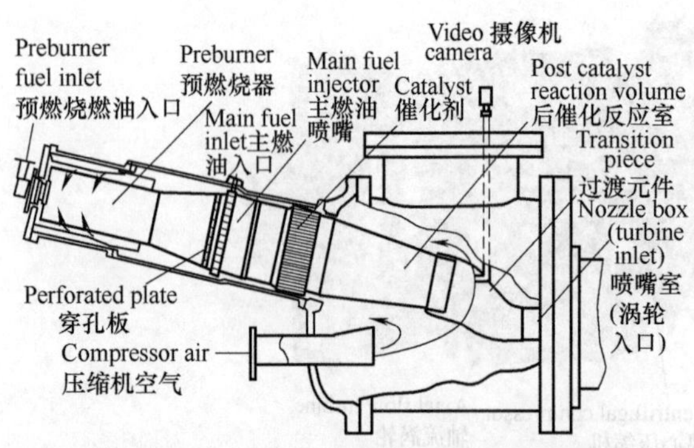

Fig 18.54 Schematic of a full catalytic combustor 完全催化燃烧器

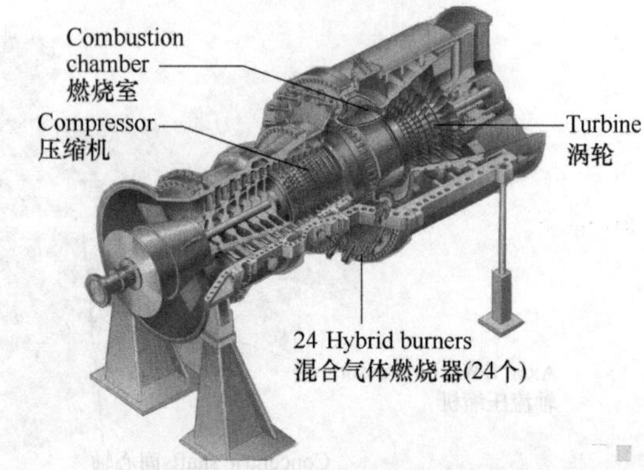

Fig 18.55 A gas turbine with hybid burner ring (HBR) combustion 有混合燃气环布燃烧器的燃气轮机

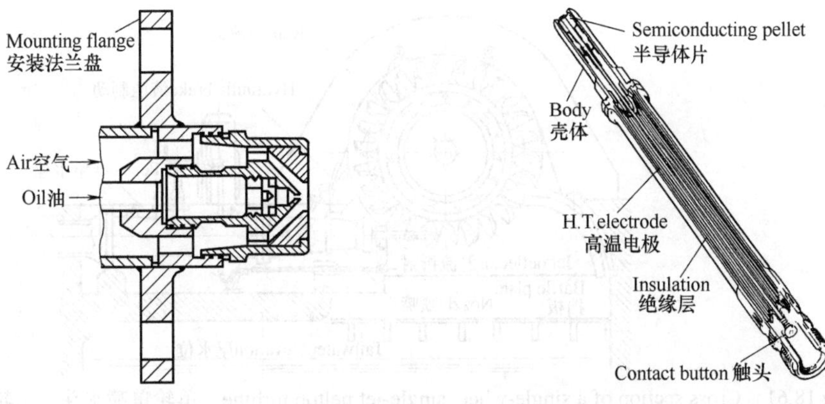

Fig 18.56 Low-pressure air atomizer 低压空气雾化器

Fig 18.57 Ignition plug 点火塞

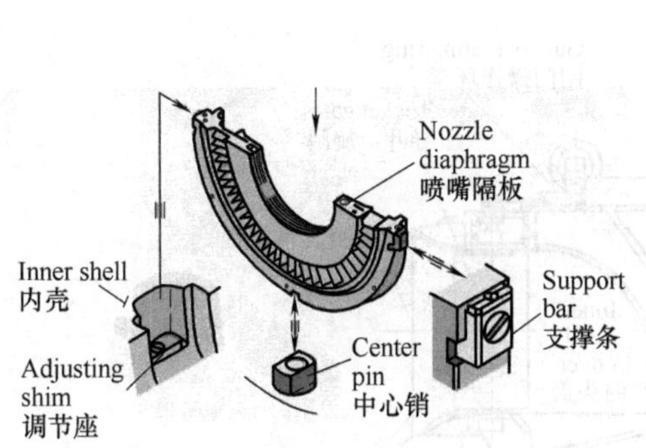

Fig 18.58 Centerline supported diaphragm
中心线支撑的隔板

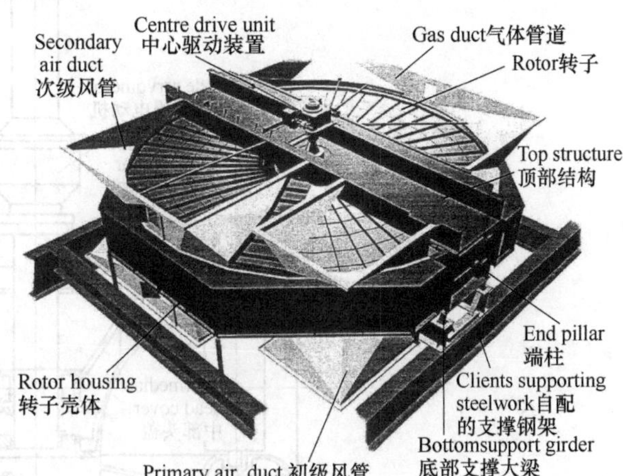

Fig 18.59 Rotary air preheater 回转式空气预热器

18.5 Hydraulic turbines 水轮机

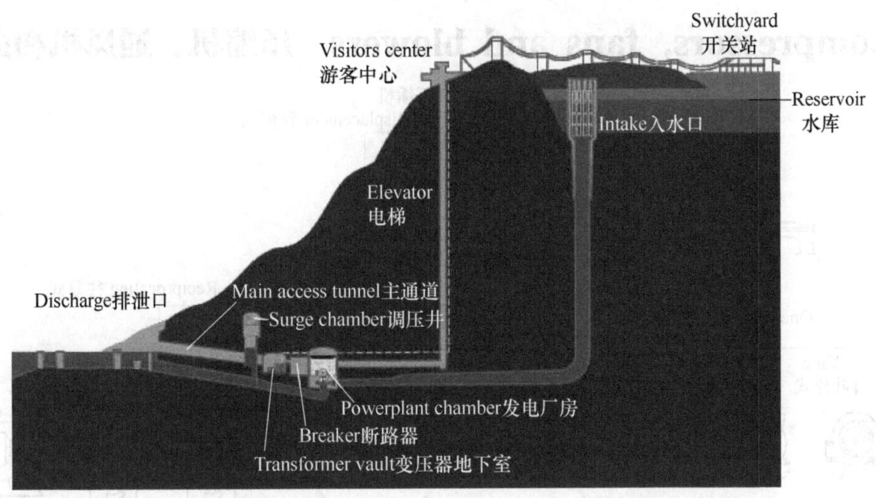

Fig 18.60 Pumped storage power station 抽水蓄能电站

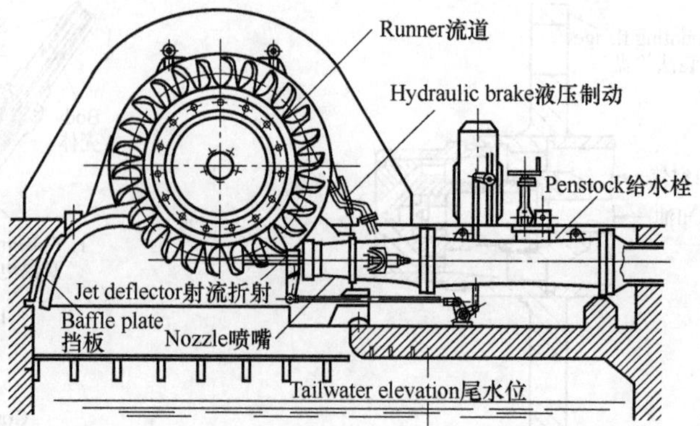

Fig 18.61 Cross section of a single-wheel,single-jet pelton turbine 单轮单喷水斗式水轮机截面图

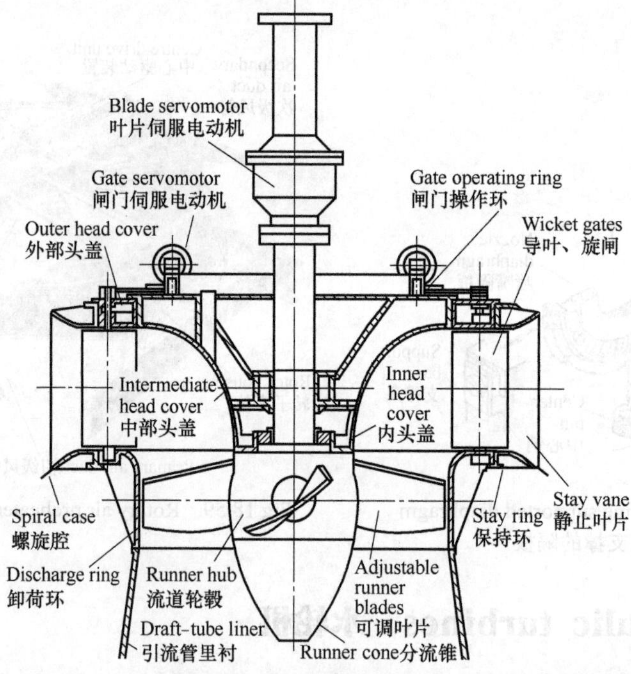

Fig 18.62 Axial-flow turbine with adjustable-pitch runner blades 可调叶片节距的轴流式涡轮机

18.6 Compressors, fans and blowers 压缩机、通风机和鼓风机

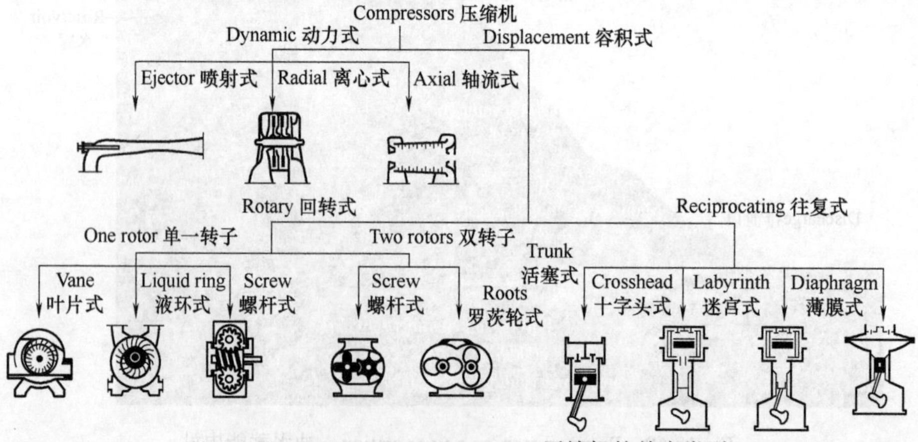

Fig 18.63 Basic compressor types 压缩机的基本类型

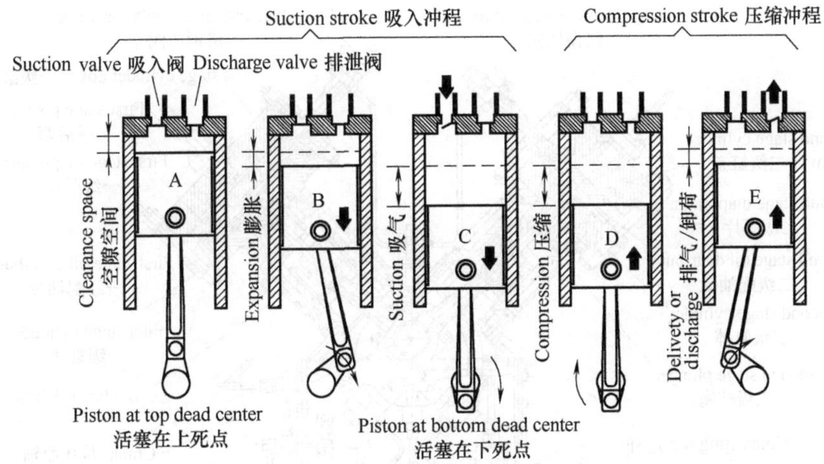

Fig 18.64 Two-cycle,or single-action, air compressor cylinders 双循环单作用空气压缩机缸体

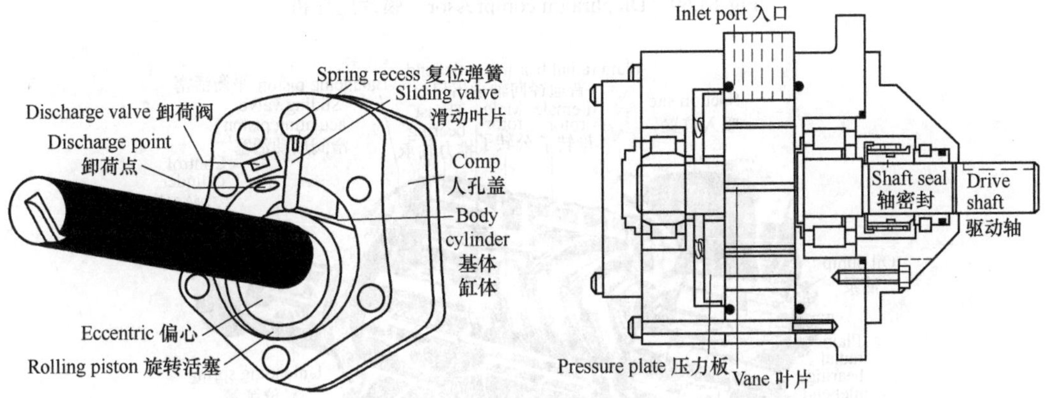

Fig 18.65 Rolling piston compressor
旋转活塞压缩机

Fig 18.66 Rotary vane compressor
回转叶片压缩机

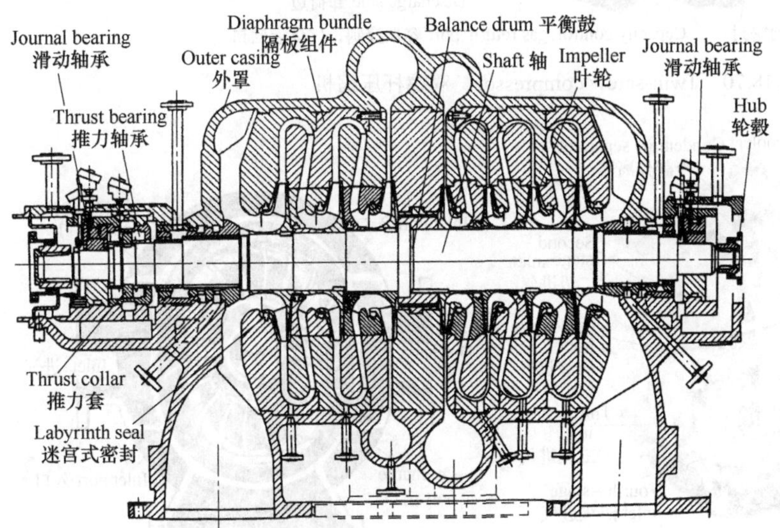

Fig 18.67 Centrifugal compressor 离心压缩机

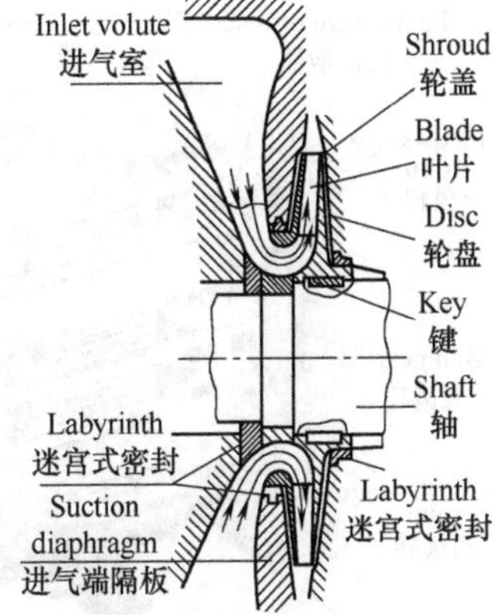

Fig 18.68 Compressor first impeller and inlet volute
压缩机首级叶轮和进气室

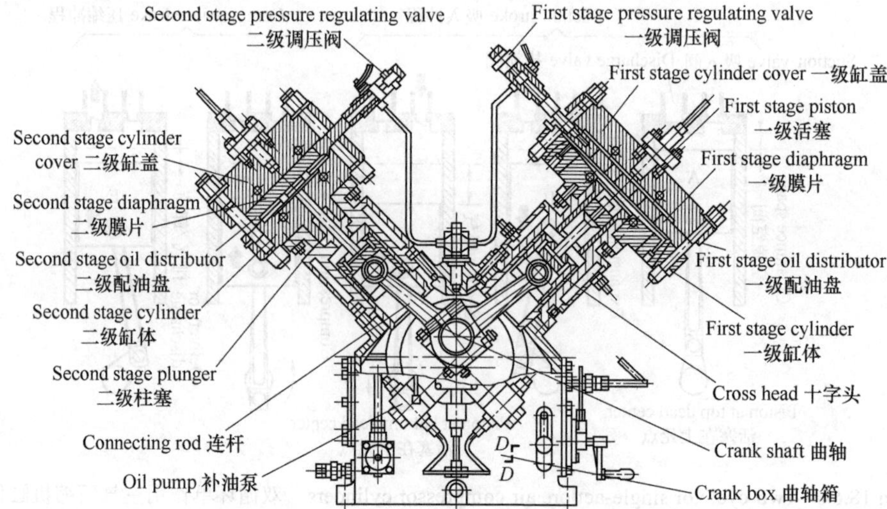

Fig 18.69 Diaphragm compressor 膜片压缩机

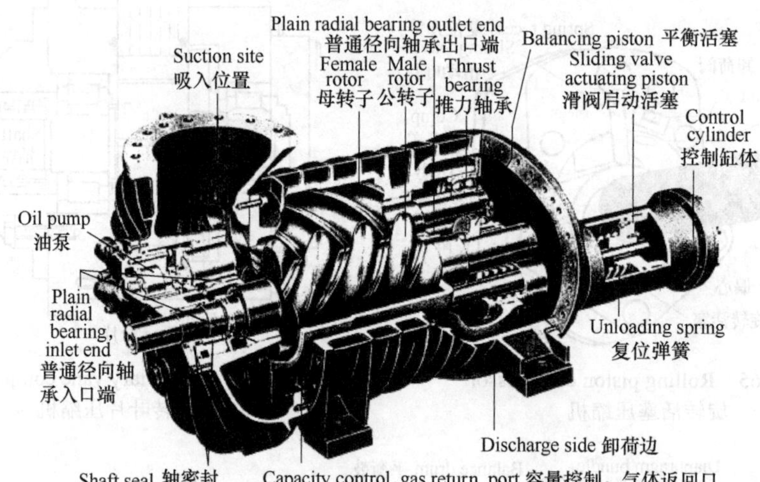

Fig 18.70 Twin-screw compressor 双螺杆压缩机

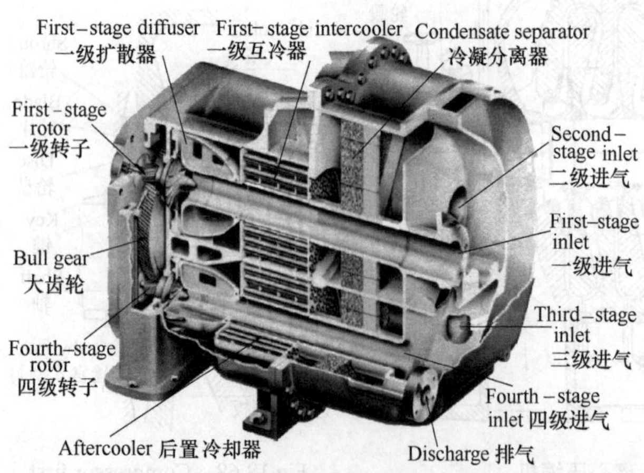

Fig 18.71 Bullgear centrifugal compressor 大齿轮离心压缩机

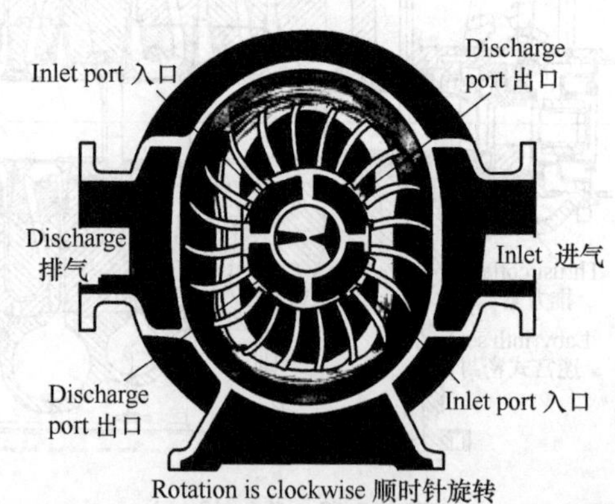

Fig 18.72 Liquid-seal ring rotary air compressor 液体密封环回转压缩机

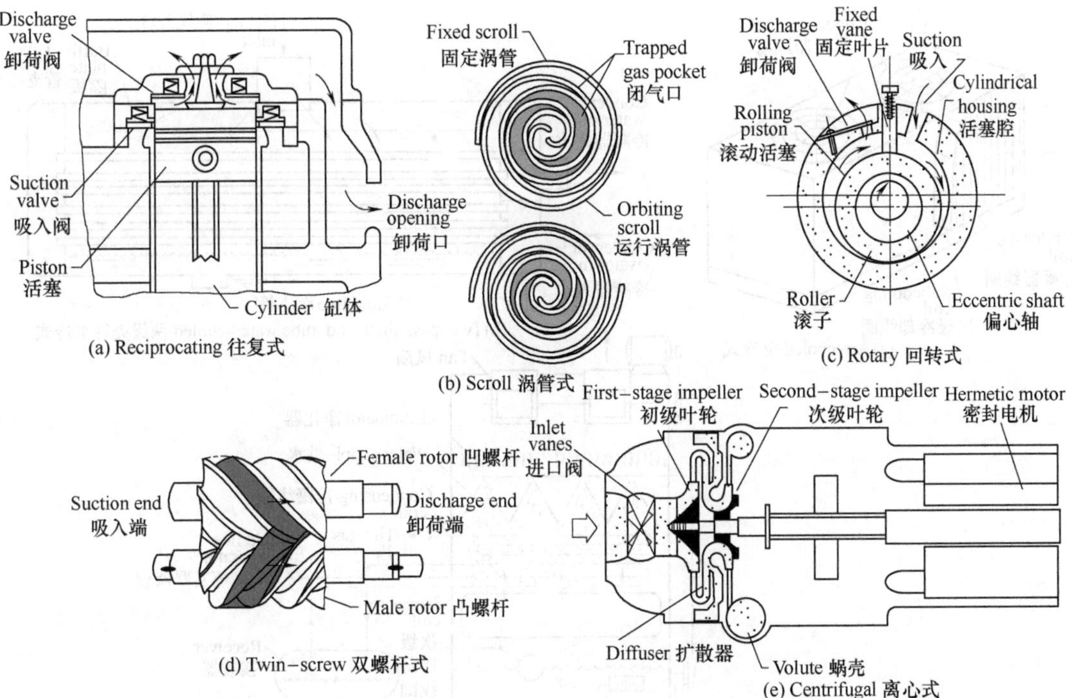

Fig 18.73 Various types of refrigeration compressors 制冷压缩机的类型

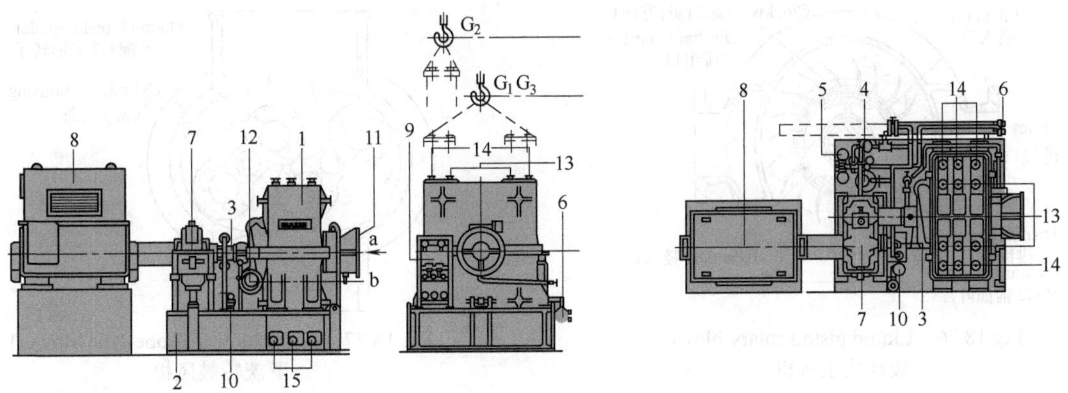

Fig 18.74 Main components and pipe connection for compressor 压缩机的主要部件和管路连接

Main components 主要部件：

1—Compressor 压缩机；2—Base frame with integrated lube oil system 集成润滑箱体机架；
3—Main oil pump 主油泵；4—Auxiliary oil pump 副油泵；
5—Twin oil filter 两级滤油器；6—Oil cooler 油冷却器；7—Gear 齿轮；
8—Motor 电动机；9—Instrument rack 仪表架；10—Oil mist fan 油雾扇；

Pipe connections 管件连接：

11—Compressor suction 压缩机吸入口；12—Compressor discharge 压缩机卸荷口；
13—Cooling-water inlet 冷却水入口；14—Cooling-water outlet 冷却水出口；
15—Compressor condensate drain 压缩机冷凝排出口

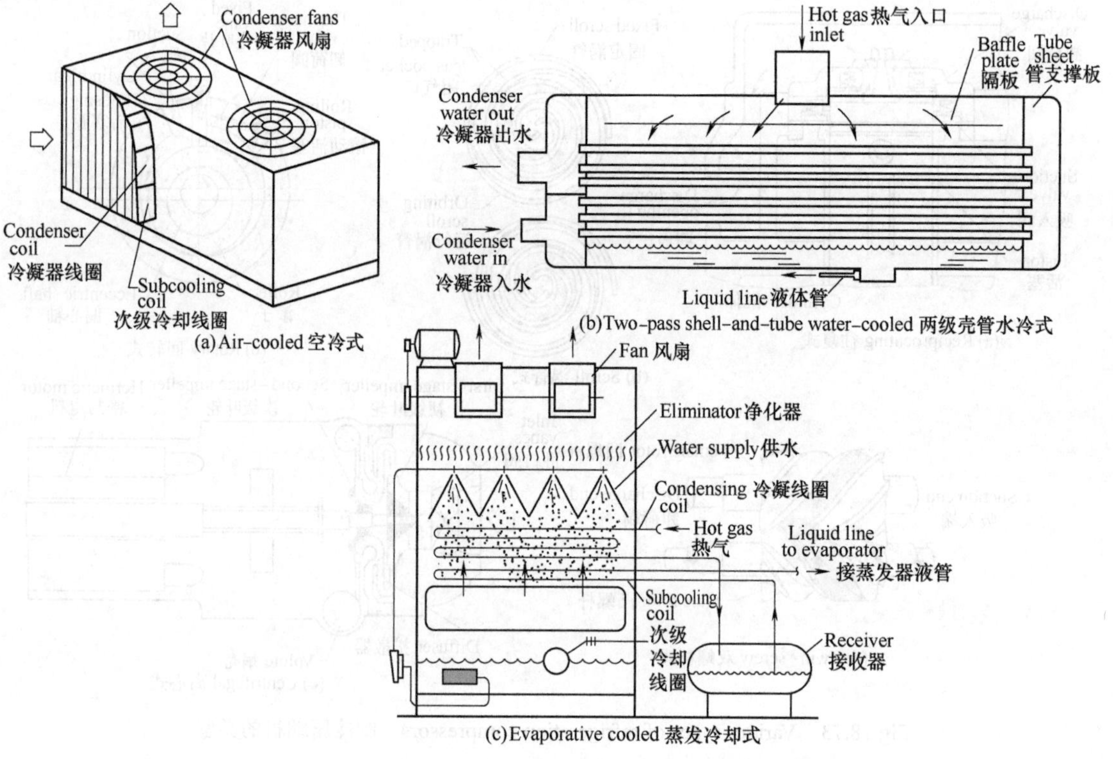

Fig 18.75 Various types of refrigeration condensers 制冷冷凝器类型

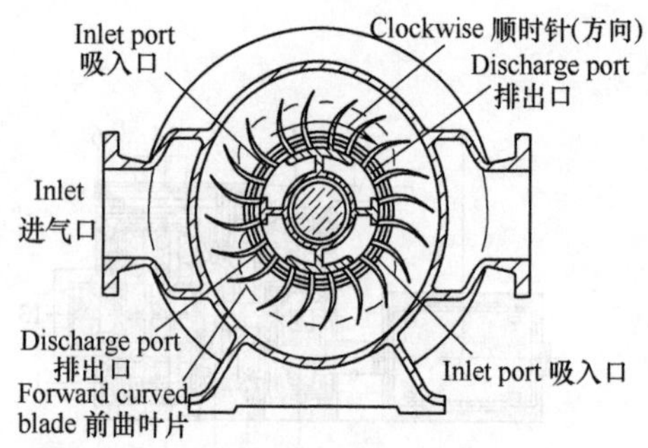

Fig 18.76 Liquid piston rotary blower
液环式压气机

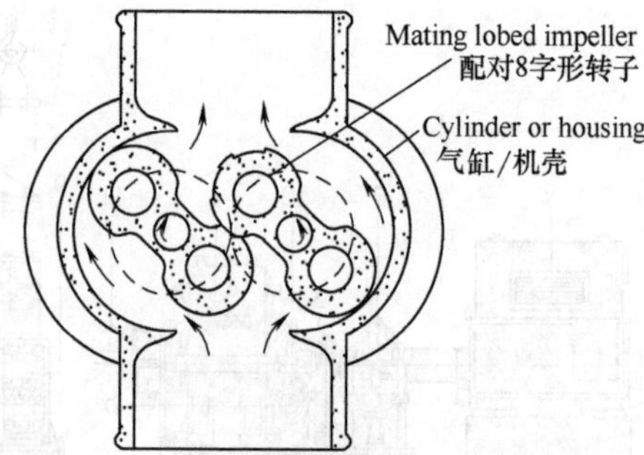

Fig 18.77 Roots blower (Lobe-type blower)
罗茨轮鼓风机

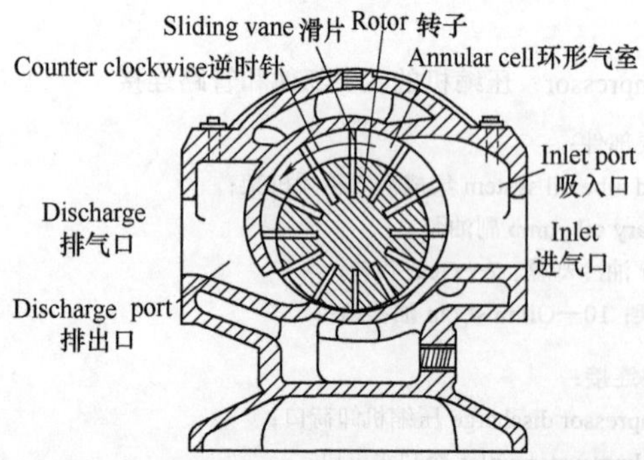

Fig 18.78 Sliding-vane rotary blower
滑片式压气机

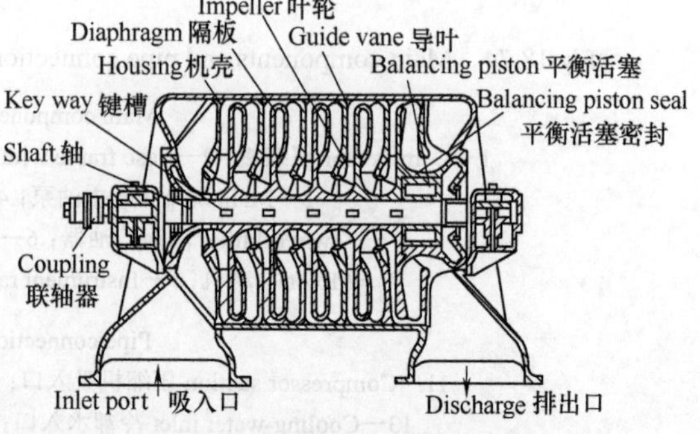

Fig 18.79 Five-stage turbo blower
五级涡轮式压气机

Unit 19 Materials Handling Equipments 运输机械

19.1 Elevators 电梯

19.1.1 The types of elevator 电梯分类

(a) Elevator 垂直电梯 (b) Escalator 扶手电梯 (c) Passenger conveyor 自动人行道

(d) The machineroom elevator 有机房电梯 (e) The machine room- less elevator 无机房电梯 (f) Construction hoist 施工升降机

(g) Passenger elevator 乘客电梯 (h) Sightseeing elevator 观光电梯 (i) Hospital elevator 病床电梯 (j) Freight elevator 载货电梯

Fig 19.1

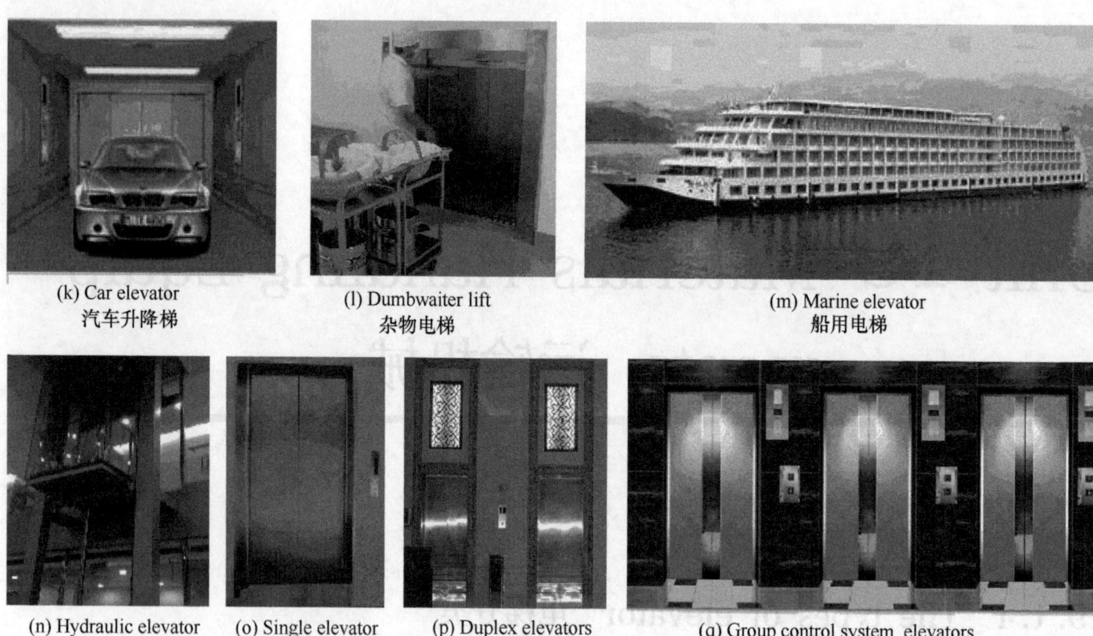

(k) Car elevator
汽车升降梯

(l) Dumbwaiter lift
杂物电梯

(m) Marine elevator
船用电梯

(n) Hydraulic elevator
液压电梯

(o) Single elevator
单控电梯

(p) Duplex elevators
并联电梯

(q) Group control system elevators
群控电梯 (≥3台)

Fig 19.1 The types of elevator 电梯分类

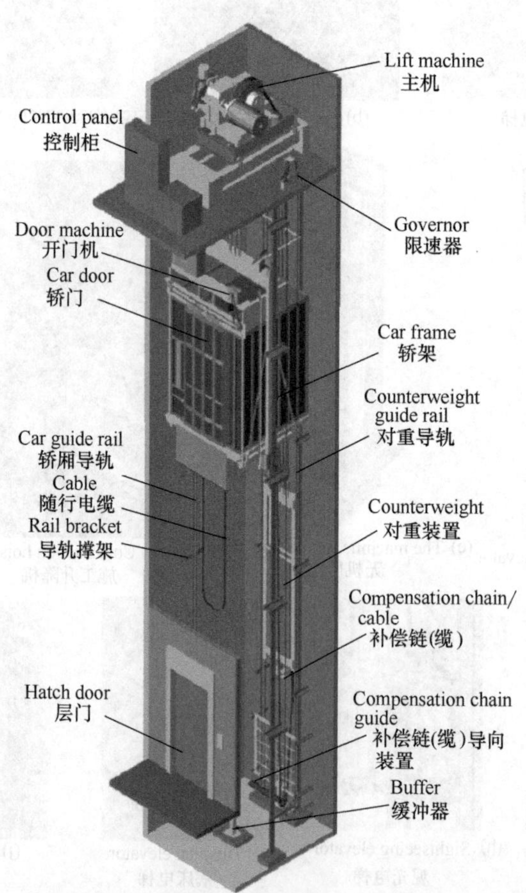

Fig 19.2 Structure of elevator 电梯的结构

19.1.2 The lift machine 电梯主机

Fig 19.3 The lift machine with gear and single bracket 机械变速单臂主机

Fig 19.4 The lift machine with gear and bracket 机械变速双臂主机

Fig 19.5 Permanent magnet synchronous machine 永磁同步主机

Fig 19.6 Permanent-magnet synchronic machine 永磁同步主机

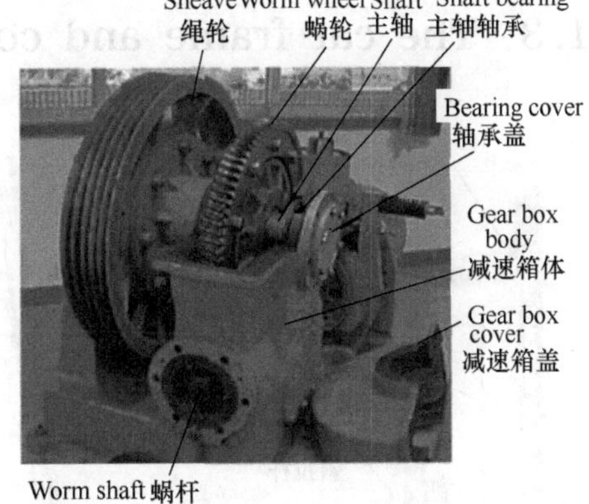

Fig 19.7 The inner structure of the gear box 减速箱内部结构

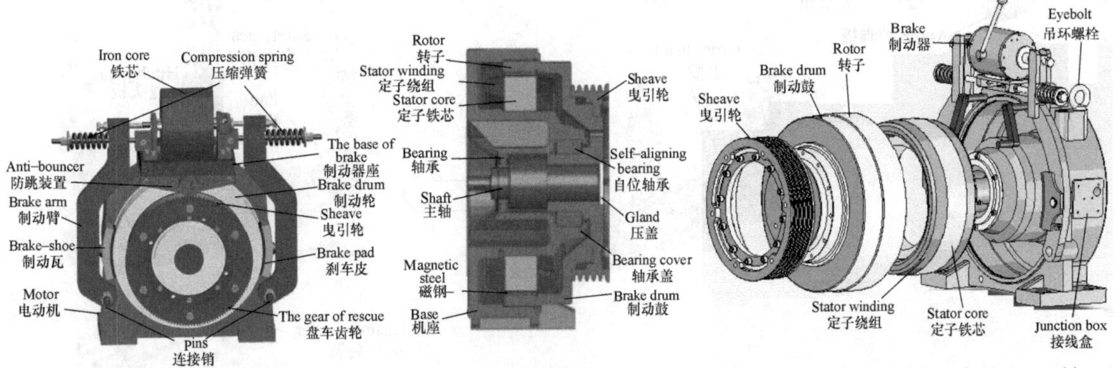

(a) The composition of a synchronous machine 同步主机结构

(b) The profile of a synchronous machine 同步主机剖面图

(c) The inner structure of synchronous machine 同步主机内部结构

Fig 19.8 Synchronous machines 同步主机

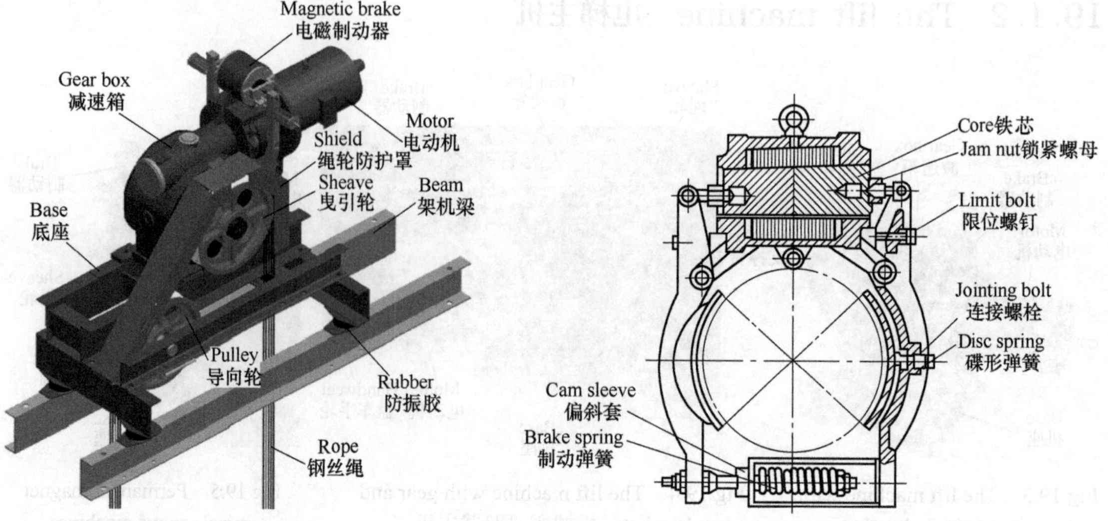

Magnetic brake
电磁制动器
Gear box
减速箱
Motor
电动机
Shield
绳轮防护罩
Sheave
曳引轮
Beam
架机梁
Base
底座
Pulley
导向轮
Rubber
防振胶
Rope
钢丝绳
Fig 19.9　The lift machine in installation　主机及其安装

Core铁芯
Jam nut锁紧螺母
Limit bolt
限位螺钉
Jointing bolt
连接螺栓
Disc spring
碟形弹簧
Cam sleeve
偏斜套
Brake spring
制动弹簧
Fig 19.10　Shoe brake　闸瓦式制动器

19.1.3 The car frame and counterweight　轿架和对重

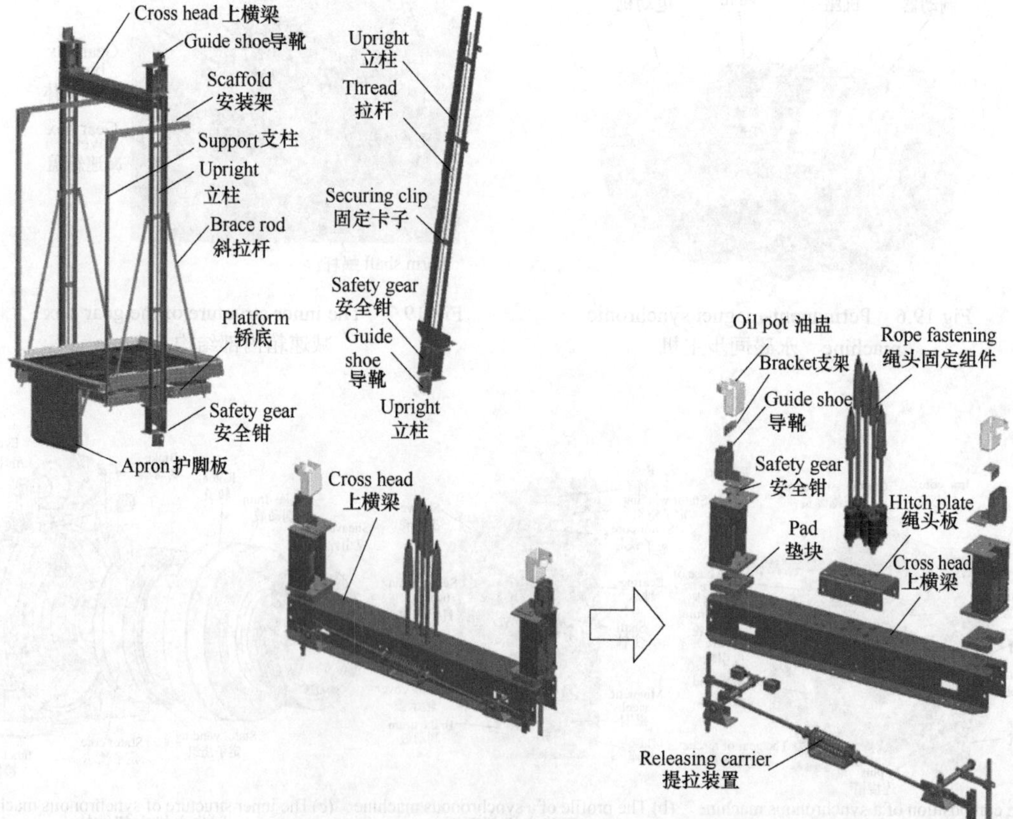

Cross head 上横梁
Guide shoe导靴
Scaffold
安装架
Support支柱
Upright
立柱
Brace rod
斜拉杆
Platform
轿底
Safety gear
安全钳
Apron护脚板
Upright
立柱
Thread
拉杆
Securing clip
固定卡子
Safety gear
安全钳
Guide
shoe
导靴
Upright
立柱
Cross head
上横梁
Oil pot 油盅
Rope fastening
绳头固定组件
Bracket支架
Guide shoe
导靴
Safety gear
安全钳
Hitch plate
绳头板
Pad
垫块
Cross head
上横梁
Releasing carrier
提拉装置
Fig 19.11　The structure of car frame　轿架结构

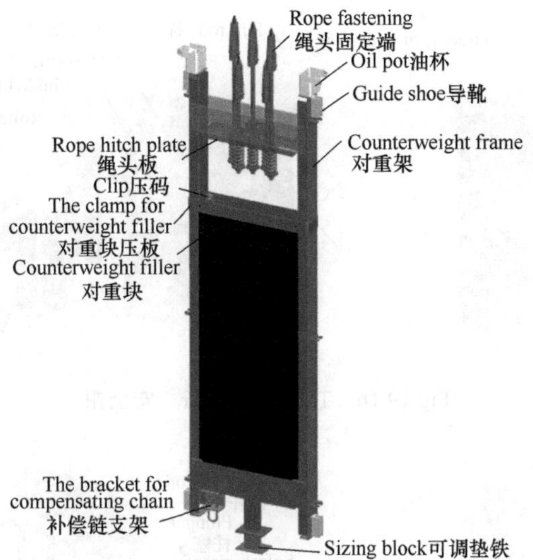

Fig 19.12 The counterweight 对重架

19.1.4 The safety device 安全装置

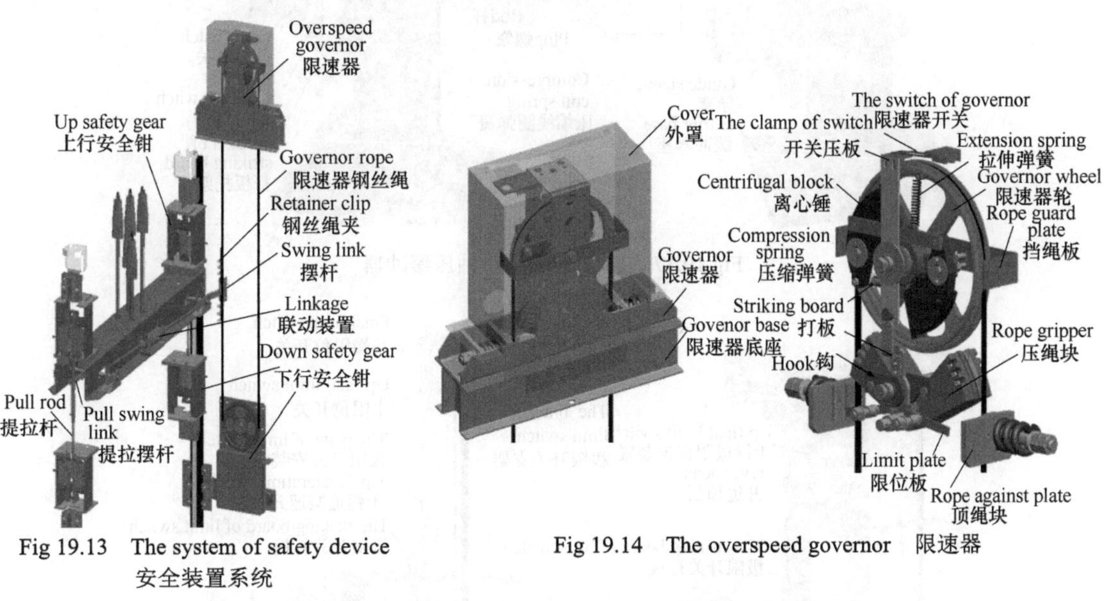

Fig 19.13 The system of safety device 安全装置系统

Fig 19.14 The overspeed governor 限速器

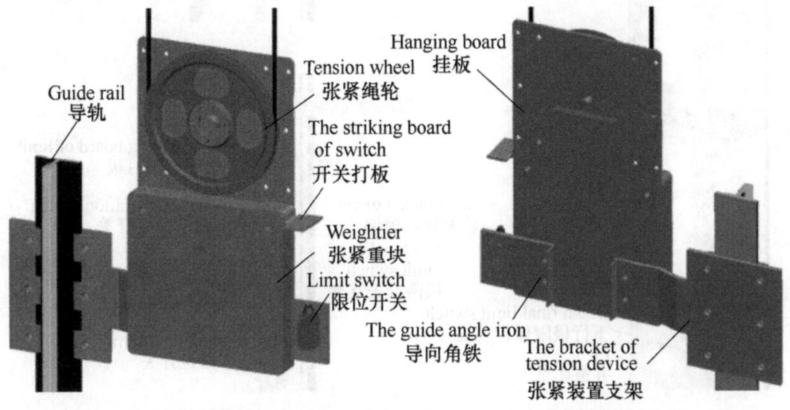

Fig 19.15 The tension device of governor 限速器张紧装置

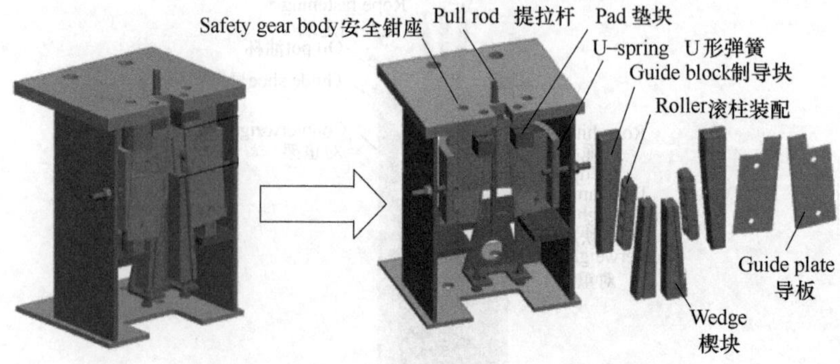

Fig 19.16 The safety gear 安全钳

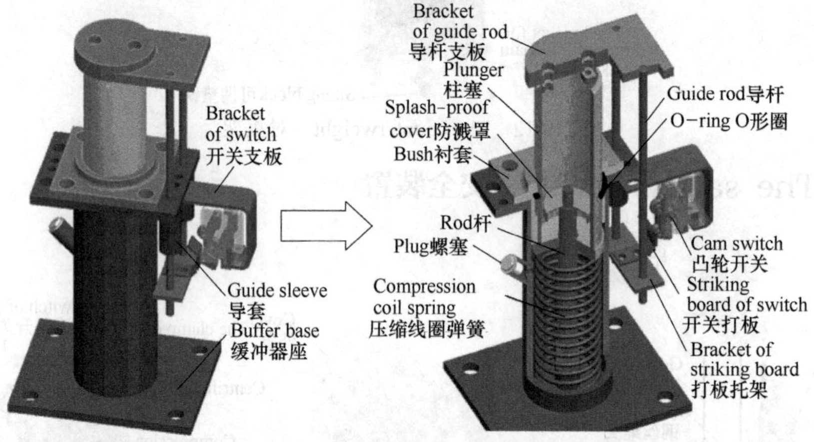

Fig 19.17 The oil buffer 油压缓冲器

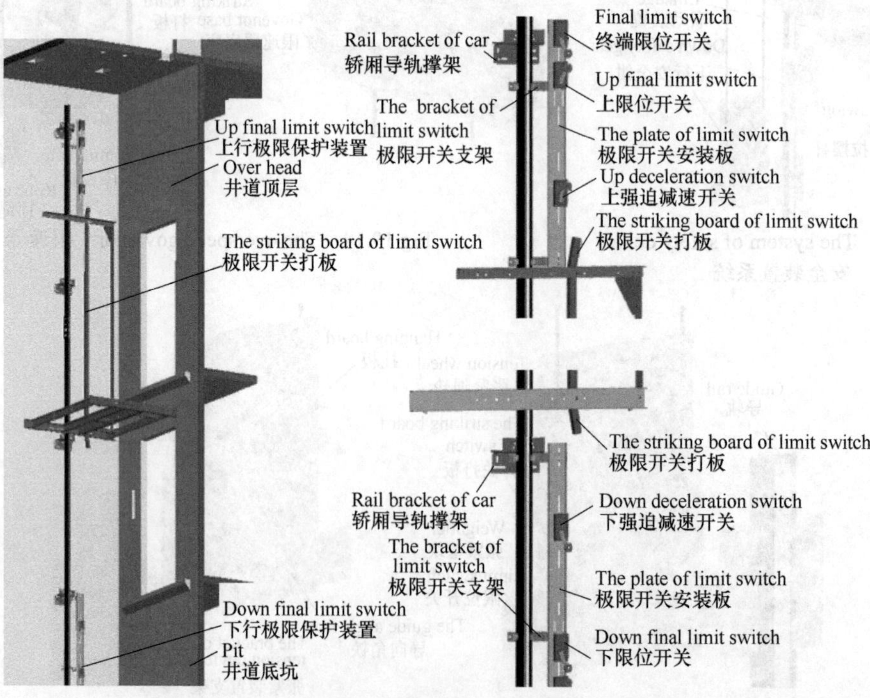

Fig 19.18 The limit switch device 极限开关组件

19.1.5 The hoistway 井道件

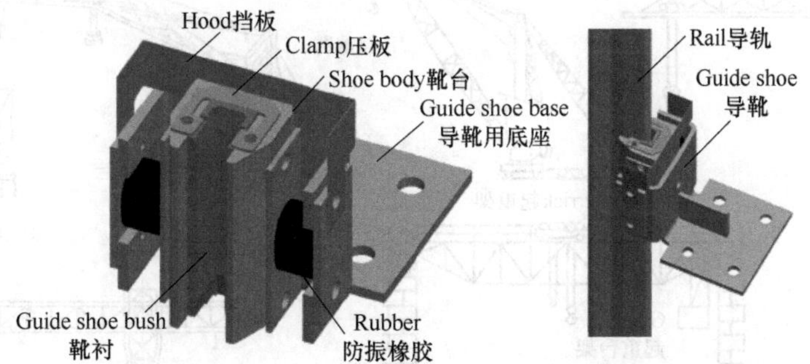

Fig 19.19 The sliding guide shoe 滑动导靴

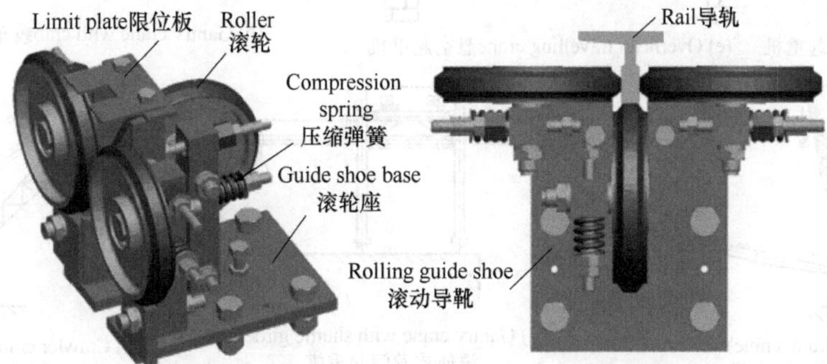

Fig 19.20 The rolling guide shoe 滚动导靴

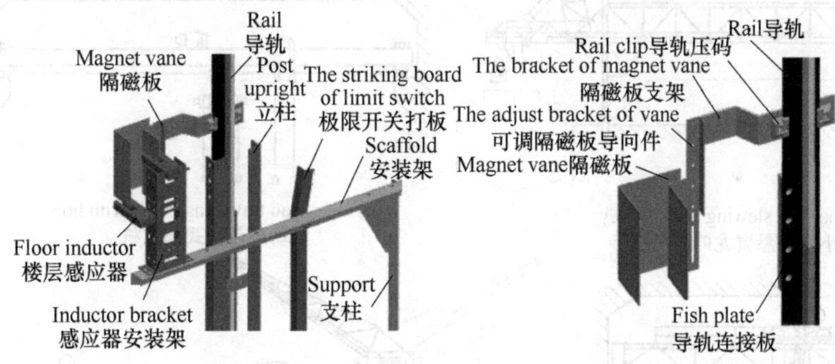

Fig 19.21 The floor inductor and magnet vane 感应器和隔磁板

19.2 Hoisting machinery 起重机械

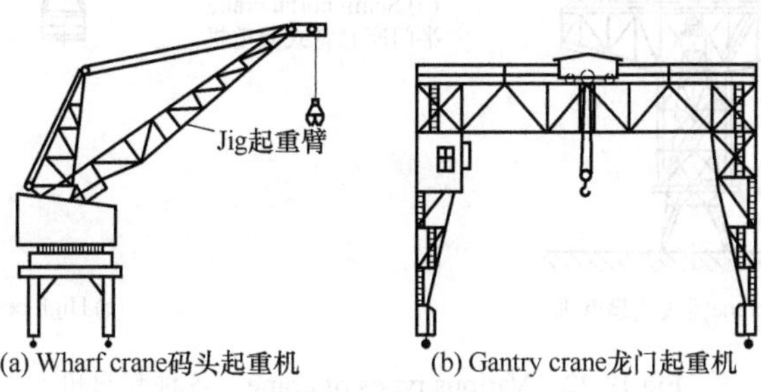

(a) Wharf crane码头起重机 (b) Gantry crane龙门起重机

Fig 19.22

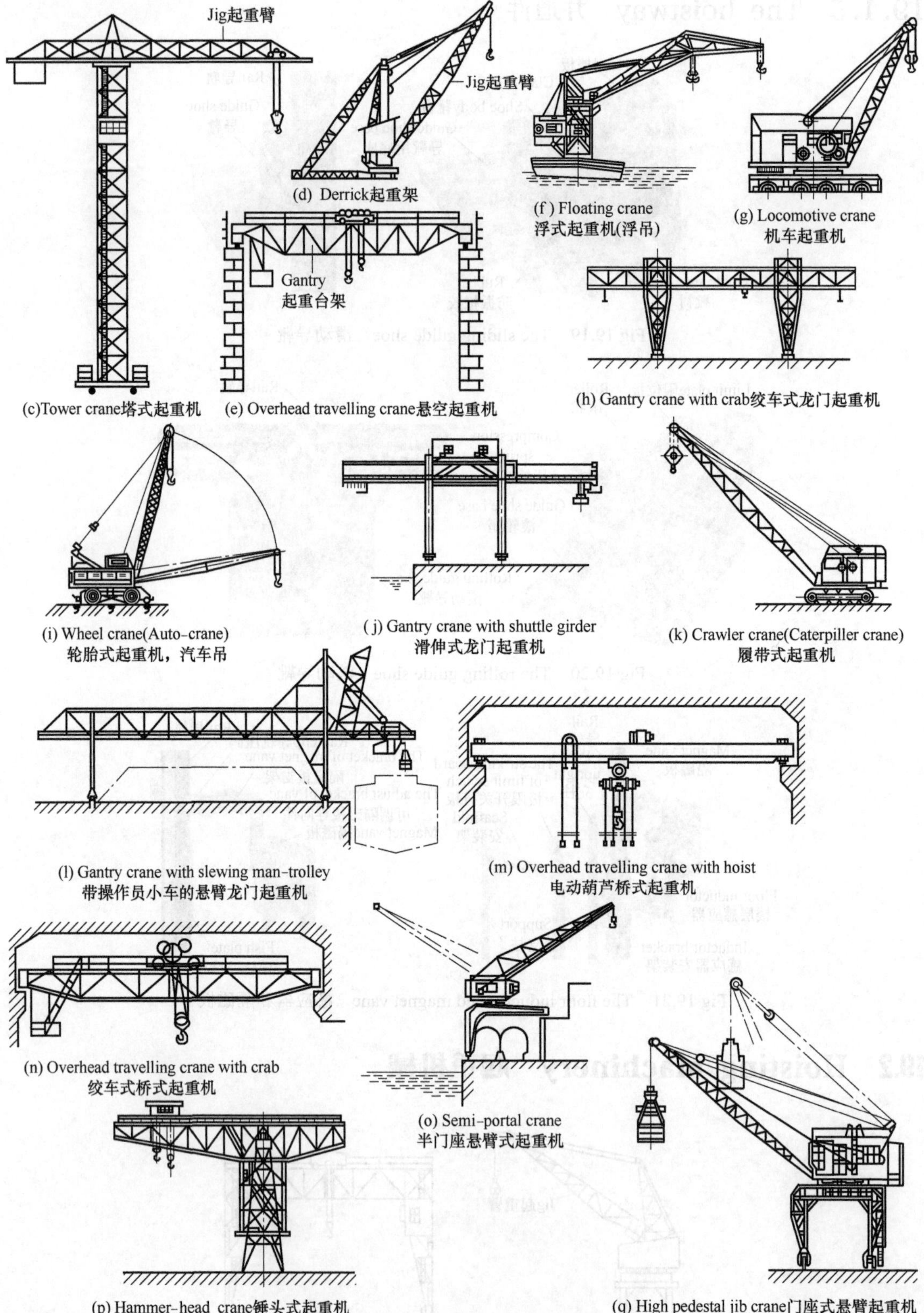

Fig 19.22 Various types of crane 各种起重机

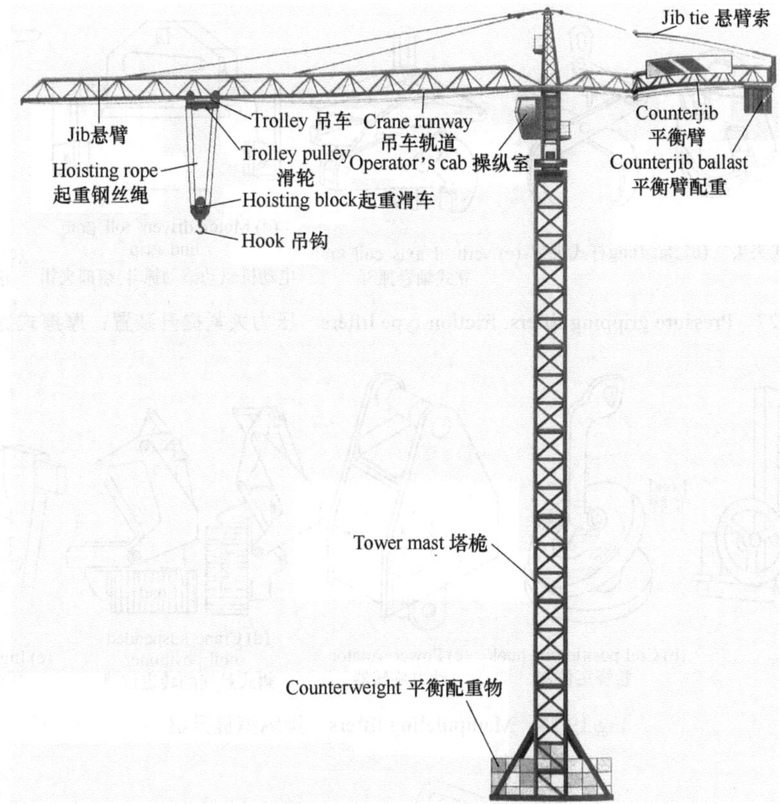

Fig 19.23 Tower crane 塔式起重机

Fig 19.24 Mobile crane 卡车起重机

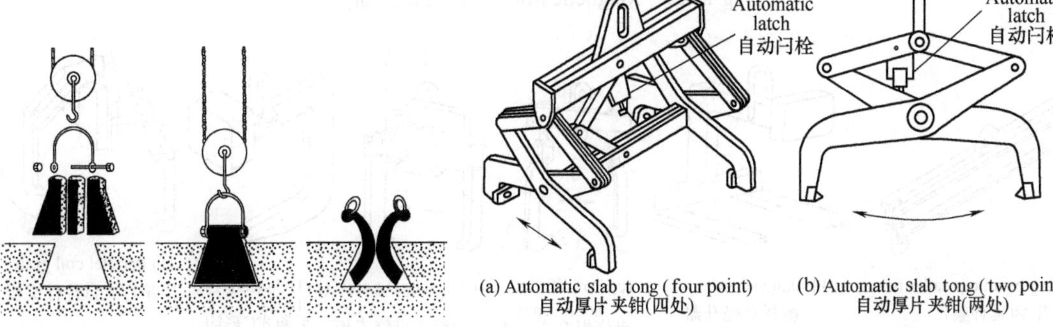

Fig 19.25 Lewis 吊楔（起重钩）

Fig 19.26 Pressure gripping lifters: indentation-type lifters
压力夹紧提升器：压入式提升器

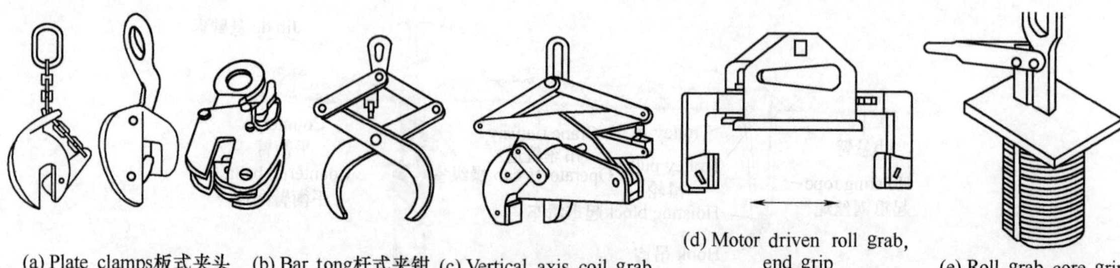

(a) Plate clamps板式夹头 (b) Bar tong杆式夹钳 (c) Vertical axis coil grab 立式轴卷抓斗 (d) Motor driven roll grab, end grip 电动机驱动滚动抓斗,端部夹钳 (e) Roll grab, core grip 滚动抓斗,内部夹钳

Fig 19.27 Pressure gripping lifters: friction-type lifters 压力夹紧提升装置：摩擦式提升器

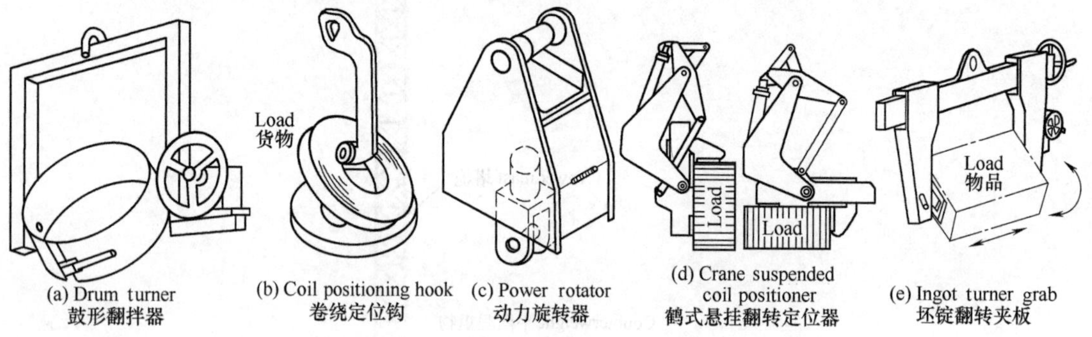

(a) Drum turner 鼓形翻拌器 (b) Coil positioning hook 卷绕定位钩 (c) Power rotator 动力旋转器 (d) Crane suspended coil positioner 鹤式悬挂翻转定位器 (e) Ingot turner grab 坯锭翻转夹板

Fig 19.28 Manipulating lifters 操纵型提升器

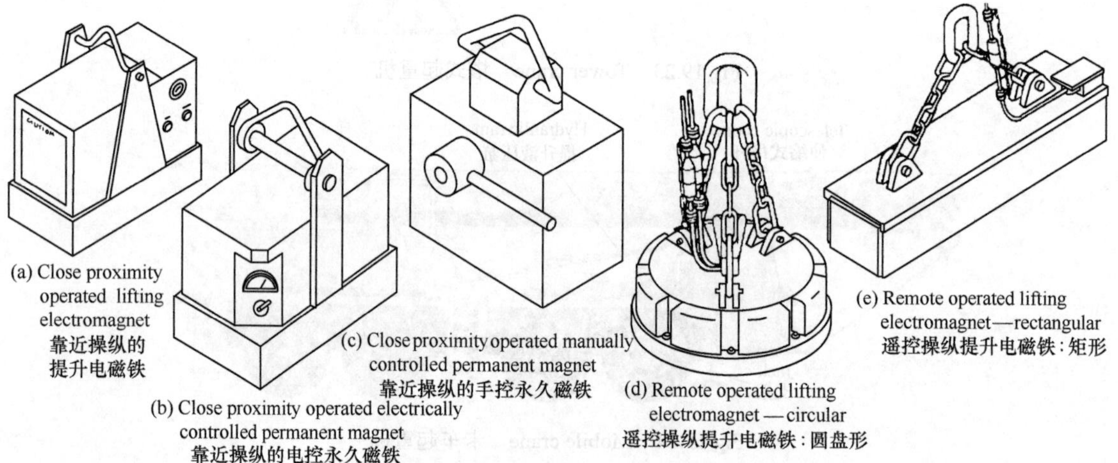

(a) Close proximity operated lifting electromagnet 靠近操纵的提升电磁铁 (b) Close proximity operated electrically controlled permanent magnet 靠近操纵的电控永久磁铁 (c) Close proximity operated manually controlled permanent magnet 靠近操纵的手控永久磁铁 (d) Remote operated lifting electromagnet — circular 遥控操纵提升电磁铁：圆盘形 (e) Remote operated lifting electromagnet—rectangular 遥控操纵提升电磁铁：矩形

Fig 19.29 Magnetic lifters 磁力提升器

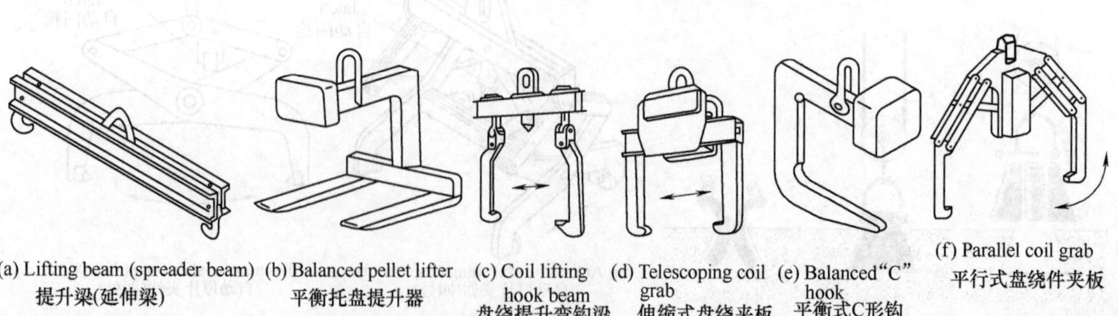

(a) Lifting beam (spreader beam) 提升梁(延伸梁) (b) Balanced pellet lifter 平衡托盘提升器 (c) Coil lifting hook beam 盘绕提升弯钩梁 (d) Telescoping coil grab 伸缩式盘绕夹板 (e) Balanced "C" hook 平衡式C形钩 (f) Parallel coil grab 平行式盘绕件夹板

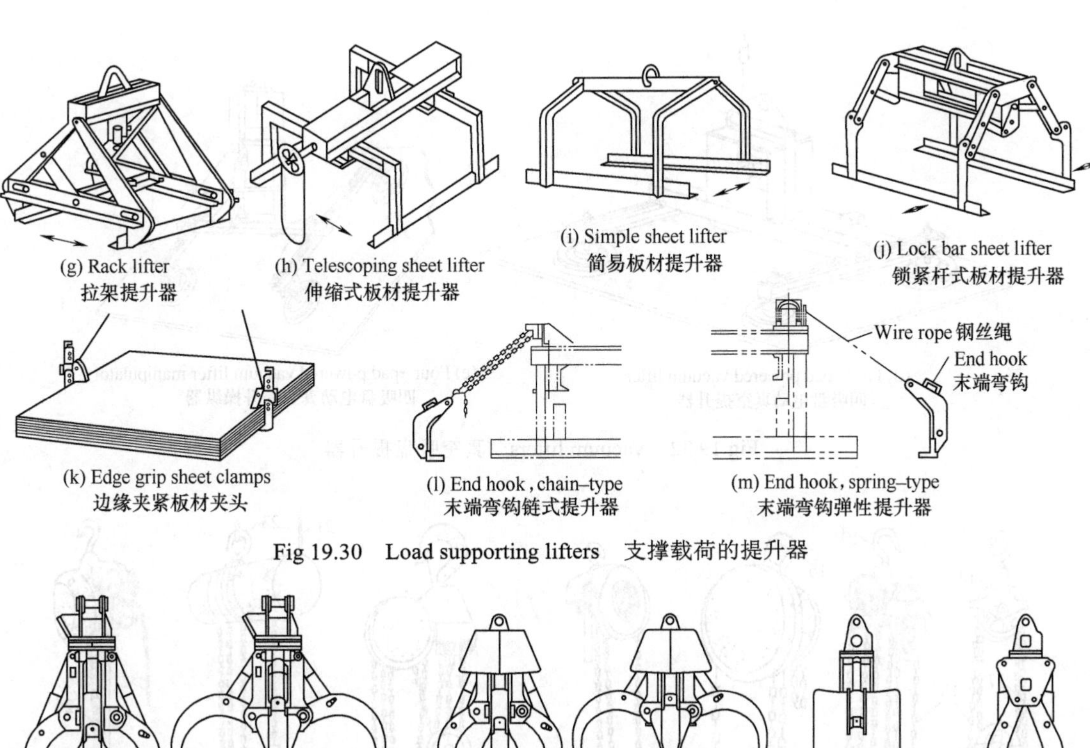

(g) Rack lifter
拉架提升器

(h) Telescoping sheet lifter
伸缩式板材提升器

(i) Simple sheet lifter
简易板材提升器

(j) Lock bar sheet lifter
锁紧杆式板材提升器

(k) Edge grip sheet clamps
边缘夹紧板材夹头

(l) End hook, chain–type
末端弯钩链式提升器

(m) End hook, spring–type
末端弯钩弹性提升器

Fig 19.30 Load supporting lifters 支撑载荷的提升器

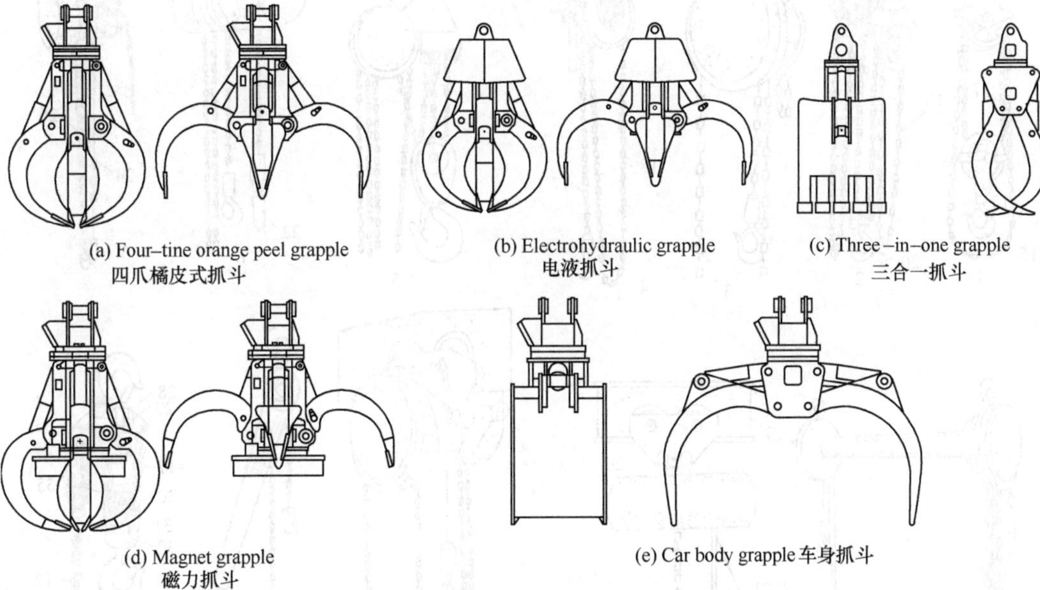

(a) Four–tine orange peel grapple
四爪橘皮式抓斗

(b) Electrohydraulic grapple
电液抓斗

(c) Three–in–one grapple
三合一抓斗

(d) Magnet grapple
磁力抓斗

(e) Car body grapple 车身抓斗

Fig 19.31 Scrap and material-handling grapples 刮动及材料搬运抓斗

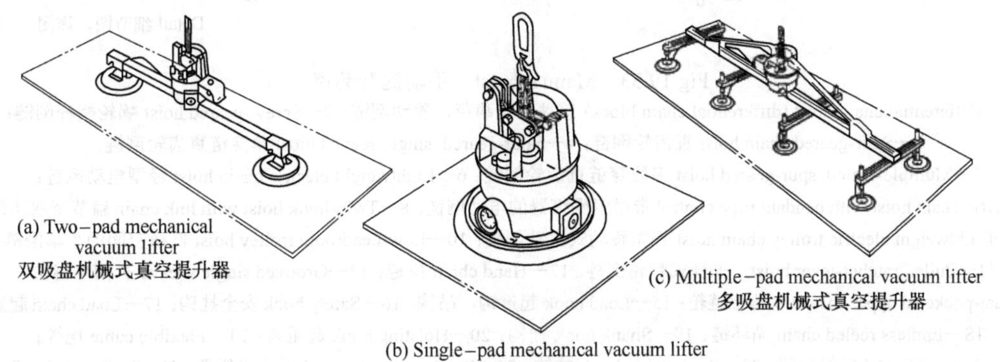

(a) Two–pad mechanical vacuum lifter
双吸盘机械式真空提升器

(b) Single–pad mechanical vacuum lifter
单吸盘机械式真空提升器

(c) Multiple–pad mechanical vacuum lifter
多吸盘机械式真空提升器

Fig 19.32

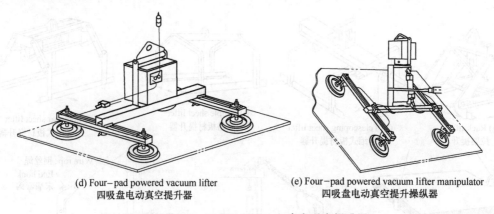

(d) Four-pad powered vacuum lifter
四吸盘电动真空提升器

(e) Four-pad powered vacuum lifter manipulator
四吸盘电动真空提升操纵器

Fig 19.32 Vacuum lifters 真空吸盘提升器

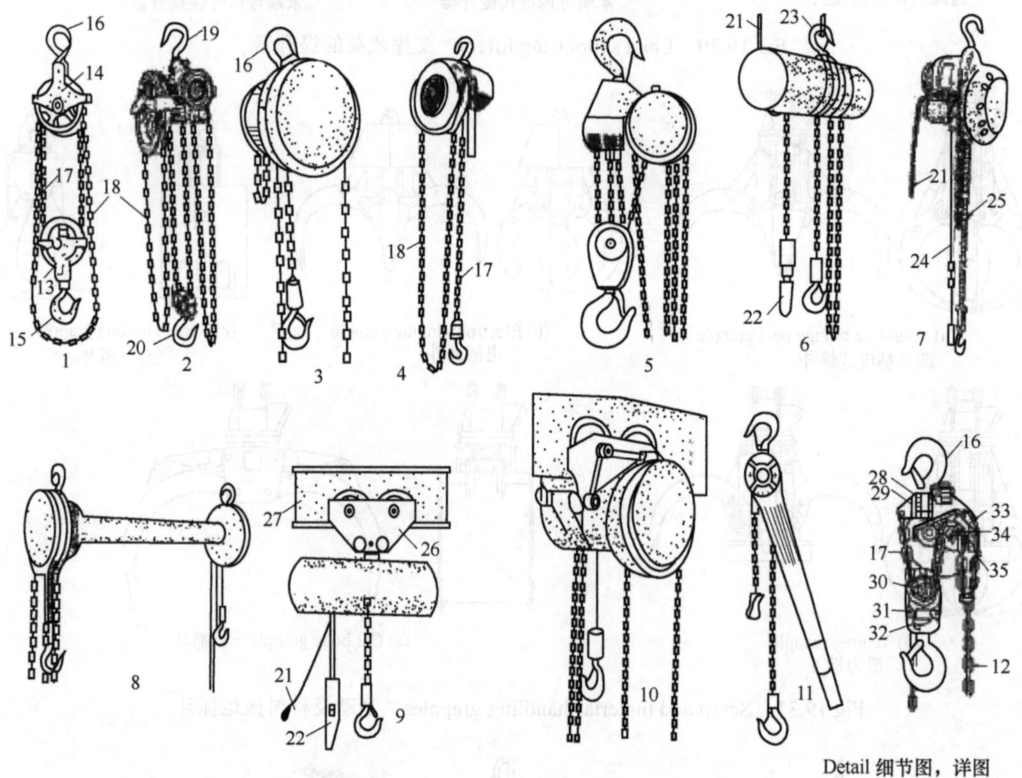

Detail 细节图，详图

Fig 19.33 Manual hoist 手动提升葫芦

1—Differential chain hoist (differential chain block) 差动手拉葫芦，差动倒链；2—Screw-geared hoist 蜗轮蜗杆倒链；3—Spur-geared chain hoist 直齿轮倒链；4—Spur-geared, single-reeved hiost 单穿链直齿轮倒链；5—Multiple-reeled, spur-geared hoist 多股穿链直齿轮倒链；6—Lightweight electric chain hoist 轻型电动倒链；7—Electric chain hoist with pendant rope control 带吊索操纵器的电动倒链；8—Twin-hook hoist with link chain 扁节倒双钩倒链；9—Lightweight electric trolley chain hoist 轻型移动式电动倒链；10—Low-headroom trolley hoist 低净空间的小车倒链；11—Puller, ratchet-lever hoist 推杆式棘轮倒链；12—Hand chain 拉链；13—Grooved single sheave 有槽单链轮；14—Dual-pocketed upper sheave 双槽上链轮；15—Load hook 起重钩，吊钩；16—Safety hook 安全挂钩；17—Load chain 起重链；18—Endless reeled chain 循环链；19—Shank hook 挂钩；20—Hoisting hook 起重钩；21—Flexible cable 电缆；22—Pushbutton control 操纵按钮；23—Suspension hook 挂钩；24—Pendant rope control 吊索操纵器；25—Roller chain 滚子链；26—Trolley 小车；27—Trolley beam 小车梁；28—Hanger 吊架；29—Upper hook collar 上部钩环；30—Hook block sheave 吊钩链轮；31—Hook block 吊钩头；32—Thrust bearing 推力轴承；33—Chain guide 链条导轨；34—Lift wheel 提升轮；35—Loose end 解链端

19.3 Materials handling systems 物料搬运系统

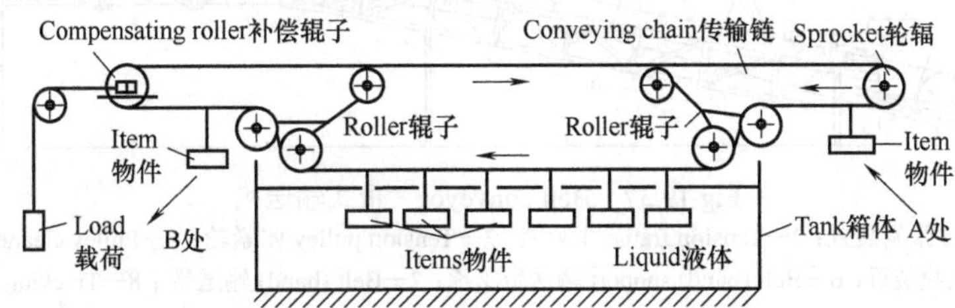

Fig 19.34 Design of a continuous transportation device 连续传输装置设计

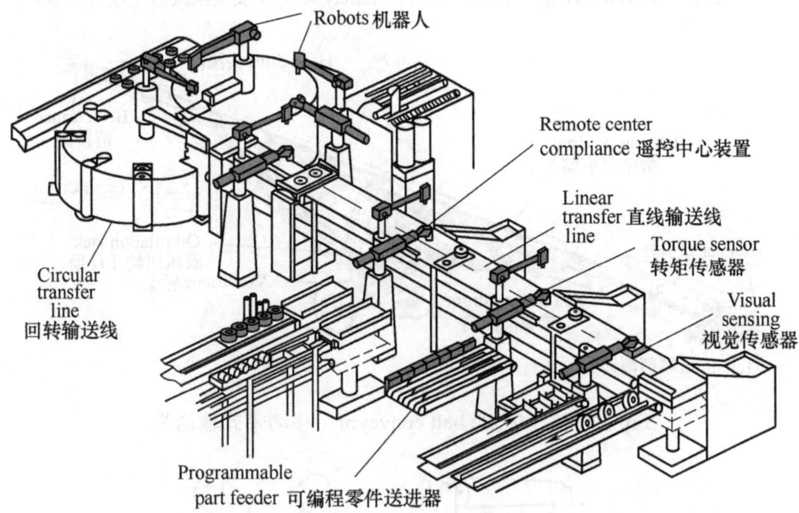

Fig 19.35 Automated assembly operations using industrial robots and circular and linear transfer lines
使用工业机器人、回转输送线和直线输送线实现自动装配作业

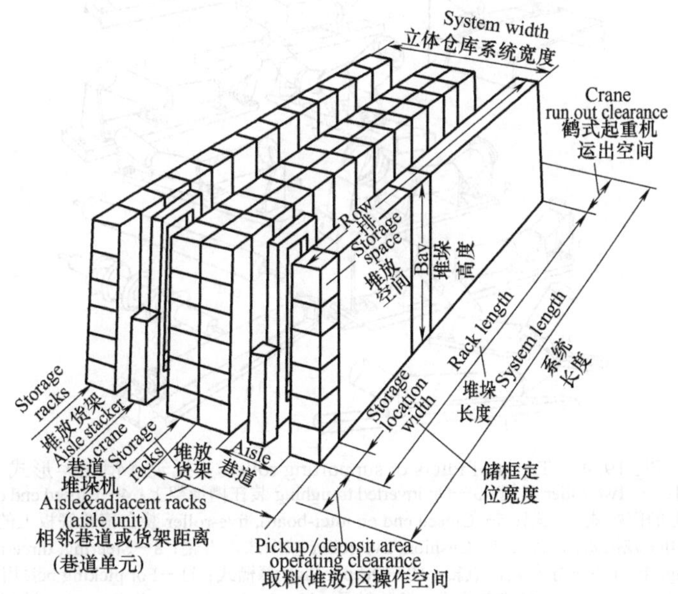

Fig 19.36 Automatic storage houseware and retrieval hardware 自动立体仓库和取存硬件设备

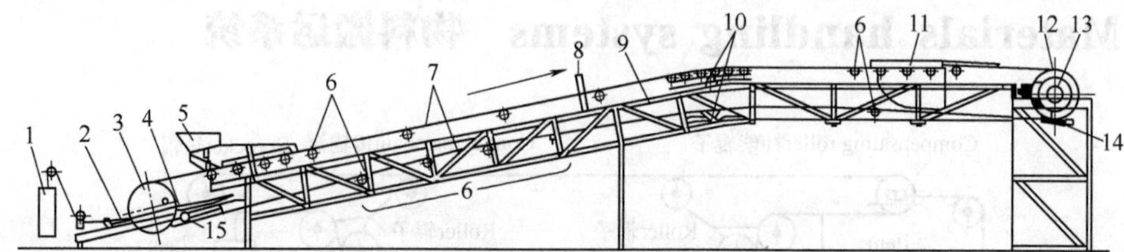

Fig 19.37 Belt conveyor 带式输送机

1—Tension device 张紧装置；2—Tension frame 张紧架；3—Tension pulley 张紧轮；4—Pulley cleaner 带轮清扫器；5—Loading device 装载装置；6—Belt (band) support 输送带支承；7—Belt (band) 输送带；8—Tracking devices 导向装置；9—Conveyor structure 输送机构架；10—Break points and break idler system 转向点及转向惰轮装置；11—Discharge plough 卸料刮板；12—Drive pulley 驱动轮；13—Anti-run back gear 齿轮逆止器；14—Belt (band) cleaner 输送带清扫器；15—Safety scraper 安全刮板（空段清扫器）

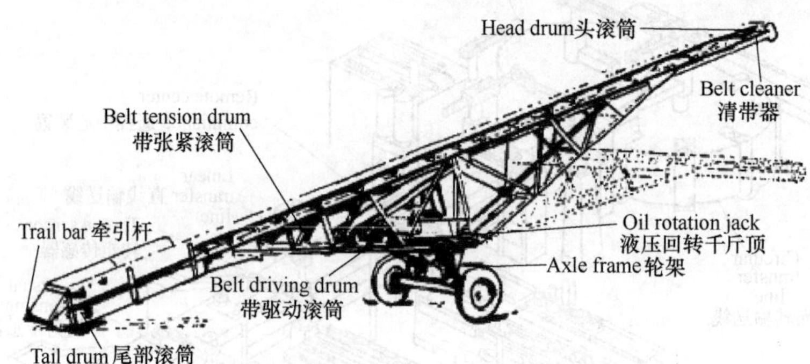

Fig 19.38 Mobile belt conveyor 移动带式输送机

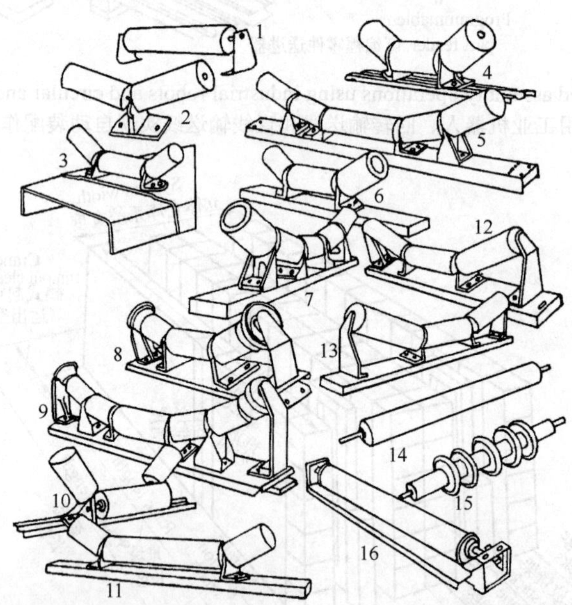

Fig 19.39 Types of idlers or supporting rollers 托辊（惰轮）形式

1—For a flat belt 平带用；2—Two roller 双辊；3—On inverted troughing 装在槽形架上；4—Closed end on idler-board, three-roller 装在惰轮底板上的闭端式，三辊；5—Closed end on idler-board, five-roller 装在惰轮底板上的闭端式、五辊；6—Cushion, three-roller 减振式，三辊；7—Cushion, five-roller 减振式，五辊；8—Steering, three-roller 导向式，三辊；9—Steering, five-roller 导向式，五辊；10—Deep trough 深槽式；11—For picking belt 用于筛分带；12—End supported, roller bearing 端部支承式，滚珠轴承；13—End supported, ball bearing 端部支承式，球轴承；14—Return idler 回程惰轮；15—Disc return 盘式回程（惰轮）；16—Steering return 导向回程（惰轮）

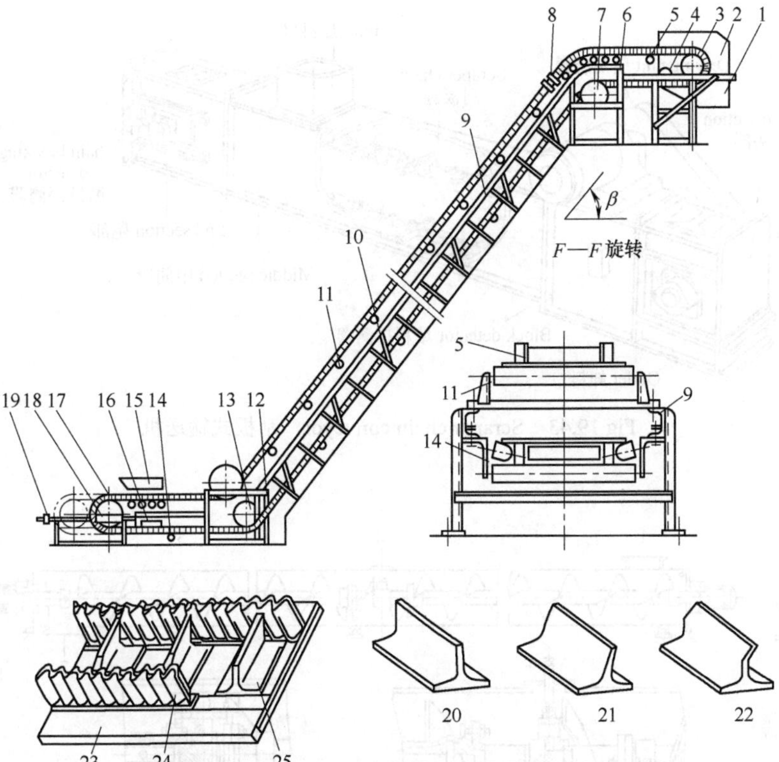

Fig 19.40 Skirt side plate inclined belt convey 裙边挡板斜坡输送机

1—Discharge chute 卸料漏斗；2—Head hood 头部护罩；3—Driving pulley 传动滚筒；4—Beating cleaner 拍打清扫器；5—Skirt side plate 挡边带；6—Conveyor structure for arch section 凸弧段机架；7—Belt pressing wheel 压带轮；8—Steering roller 挡辊；9—Main structure 主机架；10—Leg of main structure 主架支腿；11—Conveying idler 输送托辊；12—Conveyor structure for upturning section 翻转段机架；13—Band pulley 带轮；14—Returning idler 回程托辊；15—Loading channel 上料槽；16—Safety cleaner 空段清扫器；17—Tension pulley 张紧筒；18—Tension device 张紧装置；19—End structure 尾架；20—Type T T 型；21—Type C C 型；22—Type TC TC 型；23—Base belt 基带；24—Skirt side belt 挡边；25—Seperater 隔板

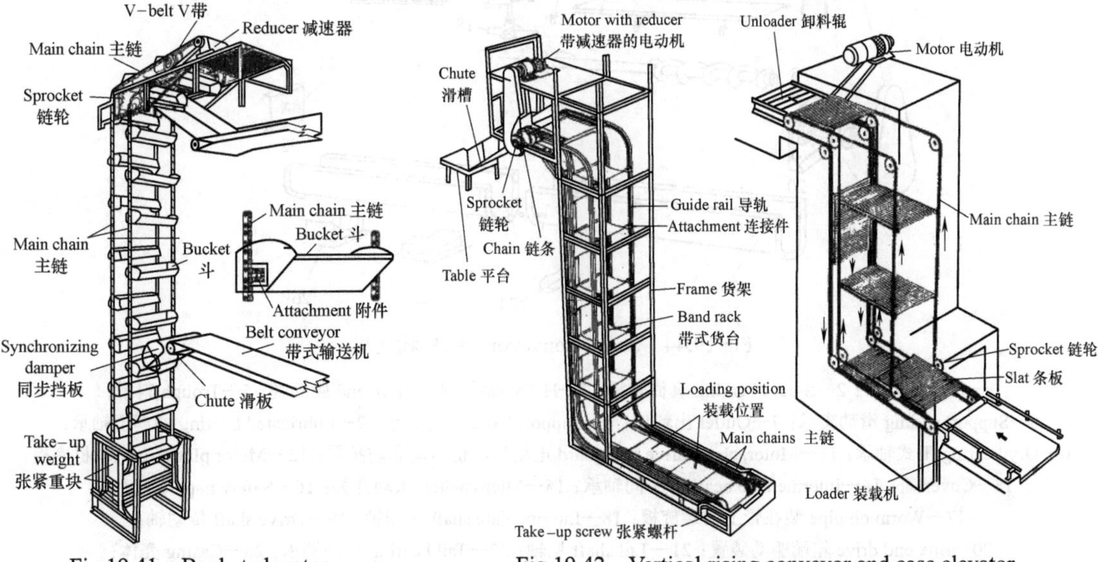

Fig 19.41 Bucket elevator 斗式提升机

Fig 19.42 Vertical rising conveyor and case elevator 垂直提升机和箱体提升机

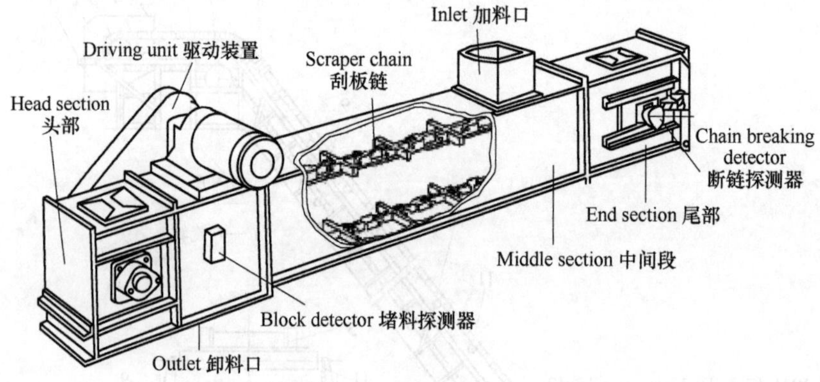

Fig 19.43 Scraper chain conveyor 刮板式输送机

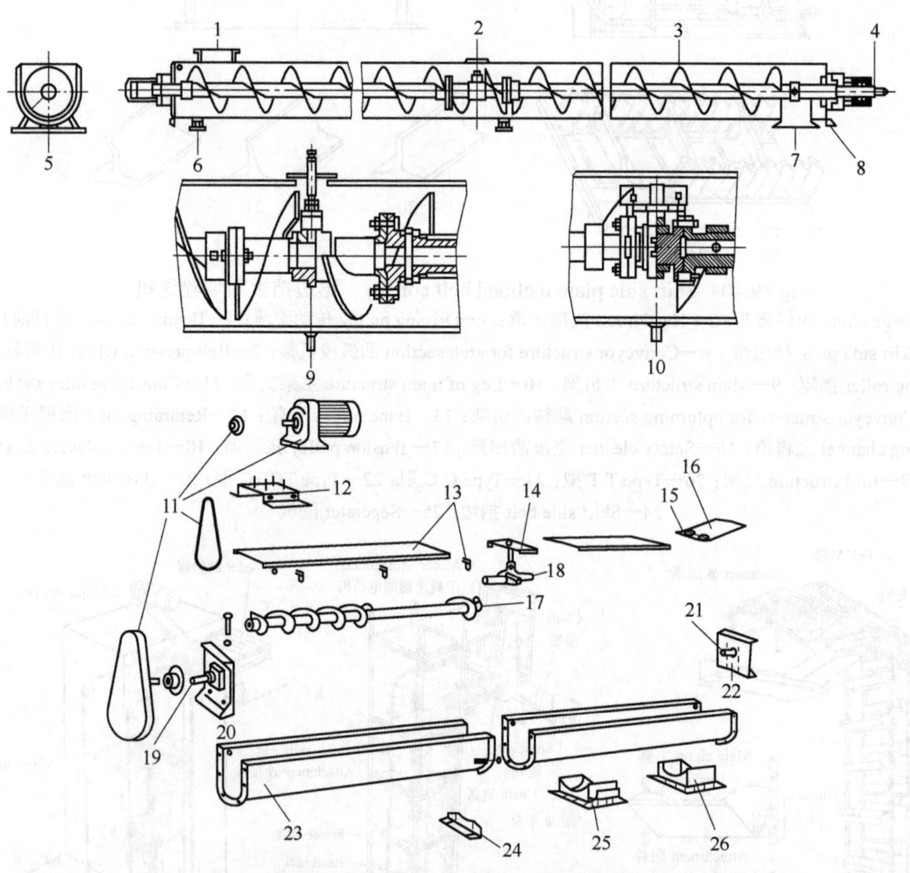

Fig 19.44 Screw conveyor 螺杆输送机

1—Inlet 加料口；2—Sleeve bearing 套筒轴承；3—Helix 螺线；4—Drive end 驱动端；5—Trough 料槽；6—Support sliding 滑动支承；7—Outlet 出料口；8—Support fixed 固定支承；9—Lubricated bearing 油润滑轴承；10—Dry bearing 干式轴承；11—Motor, chain drive and guard 电动机、链传动和防护罩；12—Motor platform 电动机底板；13—Cover 盖；14—Intermediate bearing 中间轴承；15—Microswitch 微动开关；16—Safety flap 安全活板；17—Worm on pipe 装在管上的螺旋板；18—Intermediate shaft 中间轴；19—Drive shaft 传动轴；20—Box end drive 箱端驱动装置；21—Tail shaft 尾轴；22—Tail bearing 尾部轴承；23—Casing 壳体；24—Foot 支座；25—Outlet with slide 带滑板的出料口；26—Outlet 出料口

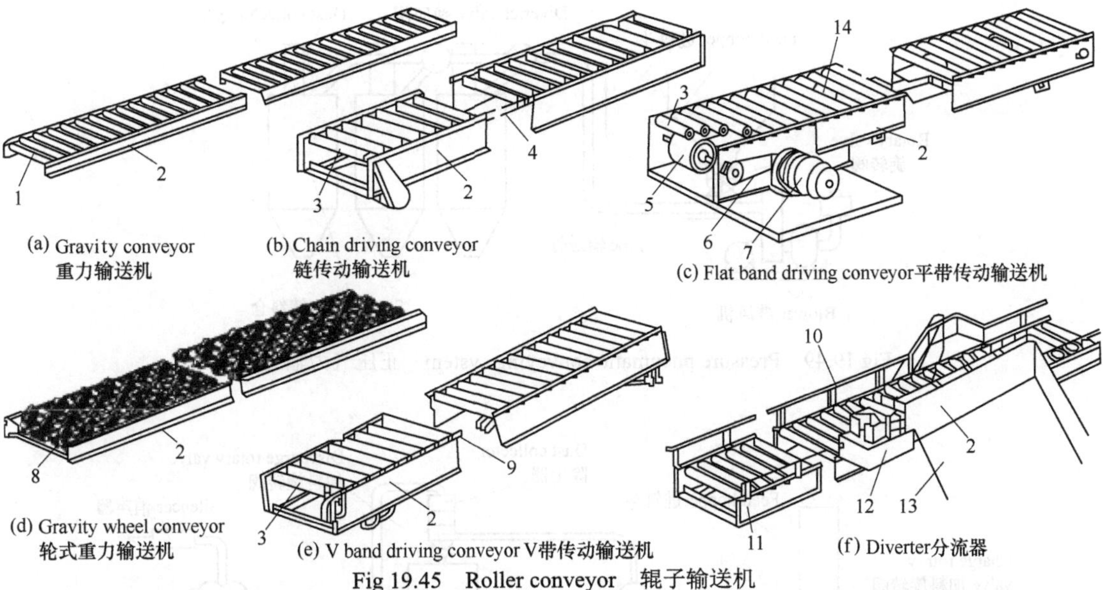

Fig 19.45 Roller conveyor 辊子输送机

1—Idler roller 惰辊；2—Frame 机架；3—Roller 辊子；4—Driving chain 传动链；5—Driving belt 传动带；6—Transmission chain 传动链；7—Drive unit 传动装置；8—Wheel 轮；9—Driving V belt 驱动 V 带；10—Guide rail 导轨；11—Roller conveyor 辊子输送机；12—Driving unit of diverter 分流器的传动装置；13—Chute 溜槽；14—Automatic pressure regulator 自动压力调整器

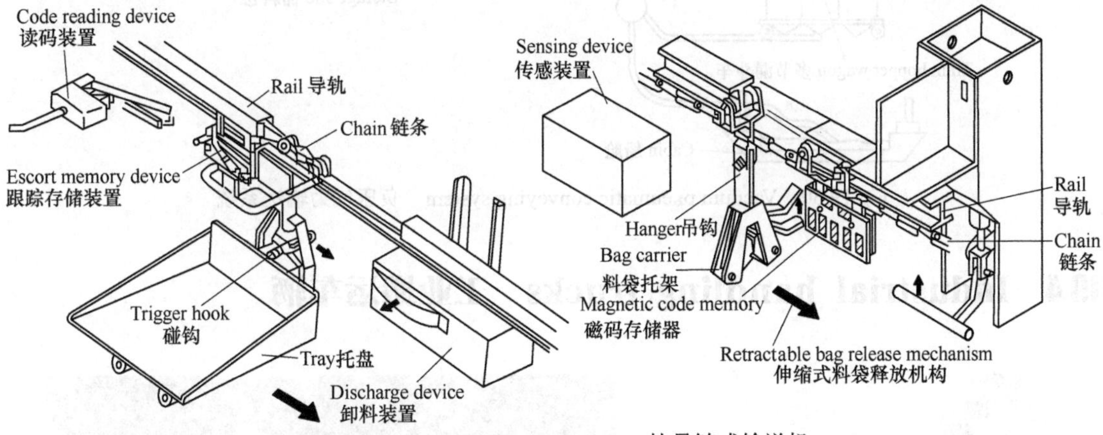

Fig 19.46 Chain trolley conveyor 挂吊链式输送机

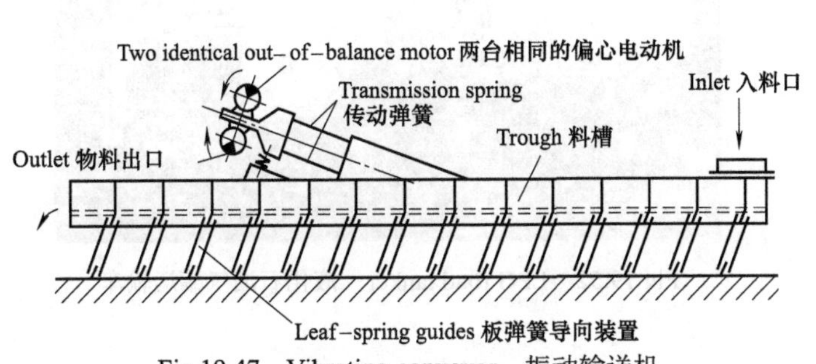

Fig 19.47 Vibrating conveyor 振动输送机

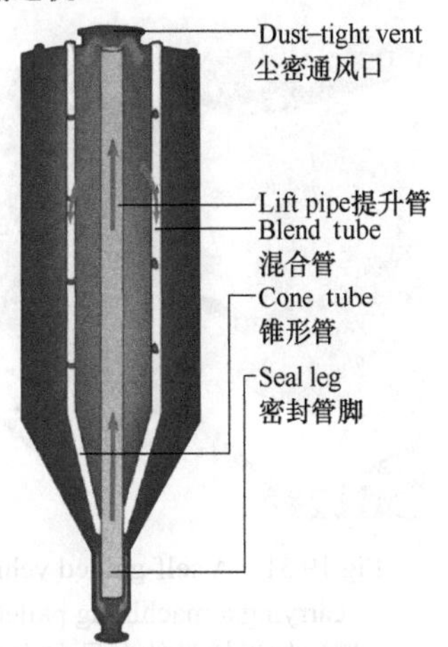

Fig 19.48 Pneumatic conveying fill system 气动传输加料系统

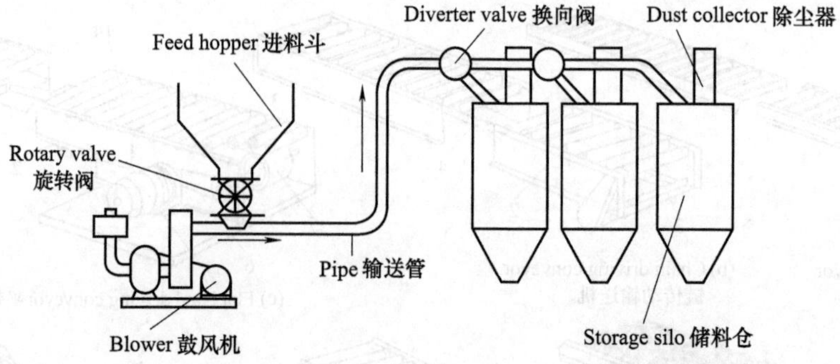

Fig 19.49 Pressure pneumatic conveying system 正压气力输送系统

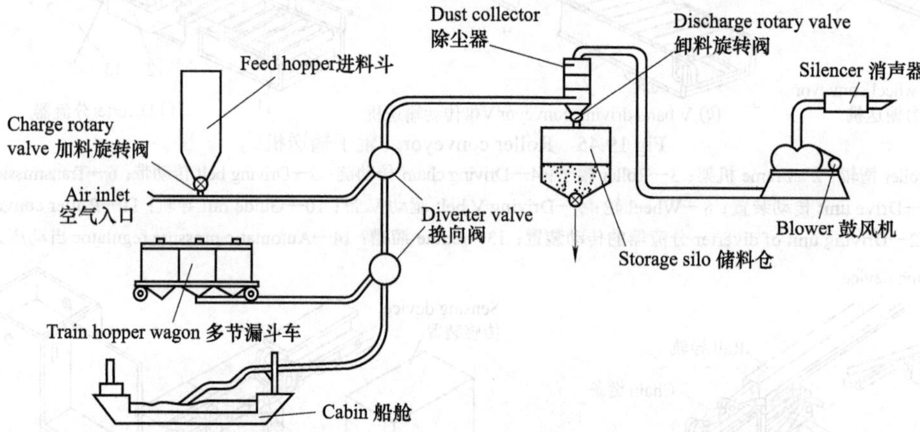

Fig 19.50 Vacuum pneumatic conveying system 负压气力输送系统

19.4 Industrial handling trucks 工业搬运车辆

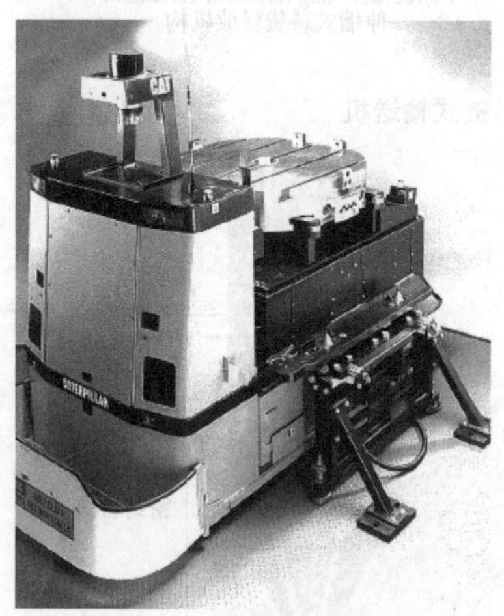

Fig 19.51 A self-guided vehicle carrying a machining pallet 搬运加工托盘的自导向小车

Fig 19.52 LPG road tanker 液化石油气罐运输车

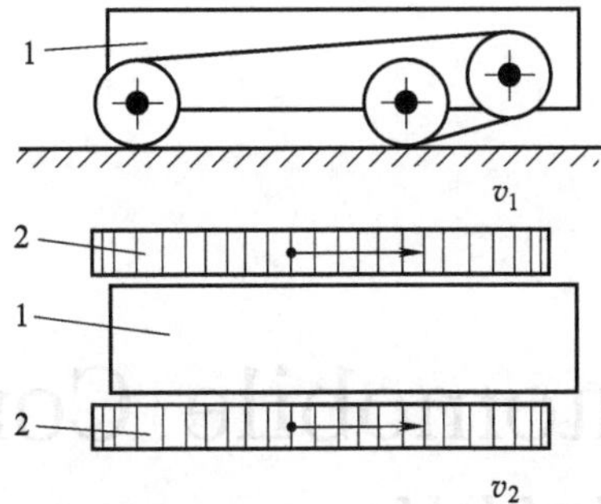

Fig 19.53 Caterpillar-driven vehicle 履带驱动车辆

1—Body 车体；2—Track 履带

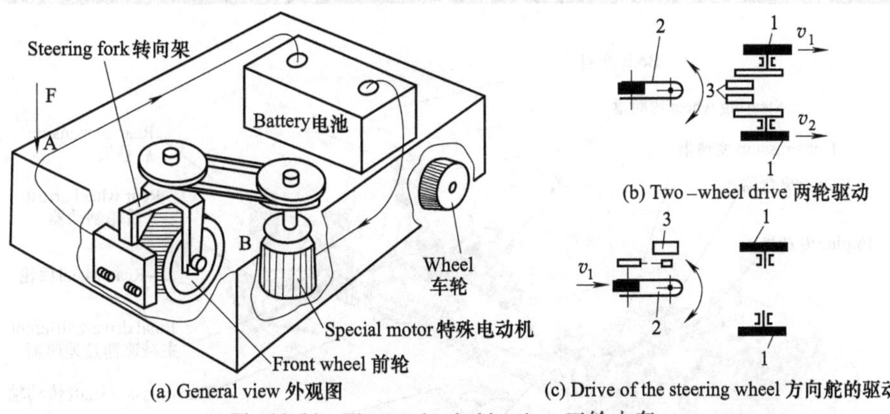

Fig 19.54 Three-wheeled bogie 三轮小车

1—Wheel 车轮； 2—Steering fork 转向架；3—Motor 电动机

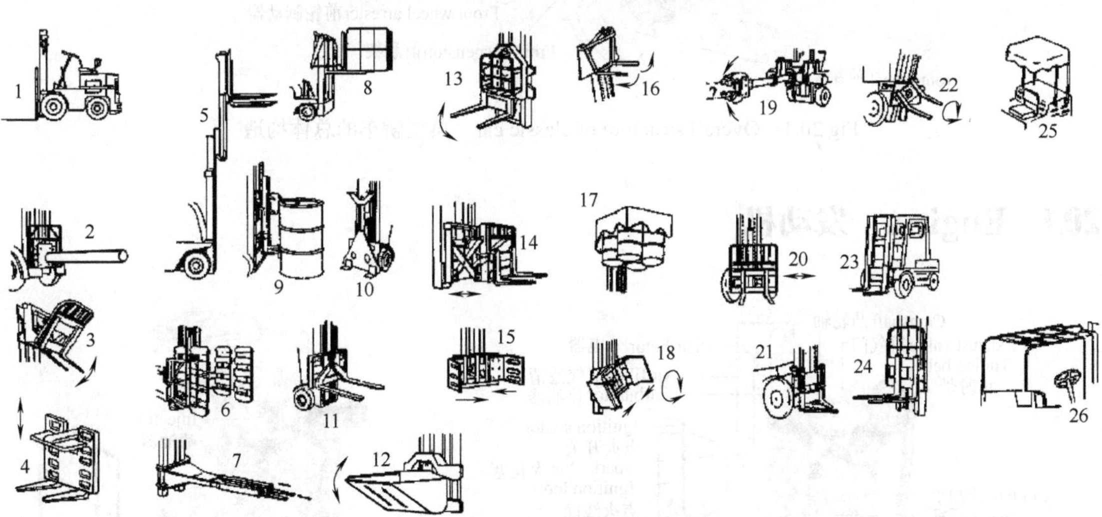

Fig 19.55 Forklift and attachment 叉车及其附件

1—Fork lift truck 叉车；2—Ram 串杆；3—Dumping fork 卸货叉；4—Load stabilizer 货物稳定叉；5—Triple telescoping uprights 三节伸缩式升举器；6—Carton clamp 盒式夹头；7—Charger 装货机构；8—Pusher device 推货设备；9—Drum grab 桶抓；10—Vertical crate grab 立式板条箱抓紧器；11—Gripping forks 夹紧叉；12—Bucket 货斗；13—Hinged fork 转向叉架；14—Reach fork 取货叉；15—Load clamp 货夹；16—Cradle fork 摇动叉；17—Drum carrier 桶架；18—Rotating clamp 回转夹头；19—Manipulator 机械手；20—Side shift 侧向移动；21—Swing fork 旋转叉；22—Rotating fork 转动叉；23—Cab 操纵员室；24—Back rest (load safety rack) 货物安全架；25—Weather shade (awning) 遮阳篷；26—Overhead guard 护顶

Unit 20 Automobile Constructs 汽车结构

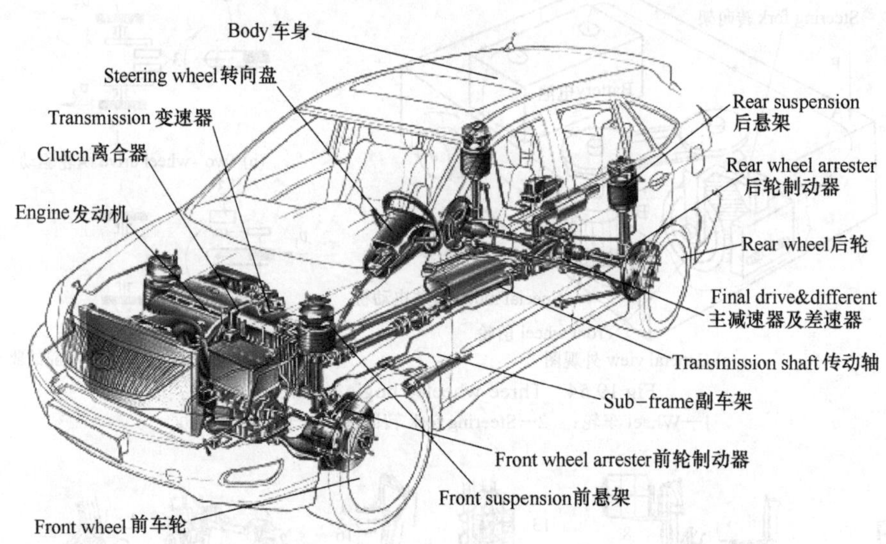

Fig 20.1 Overall structure of classic car 典型轿车的总体构造

20.1 Engine 发动机

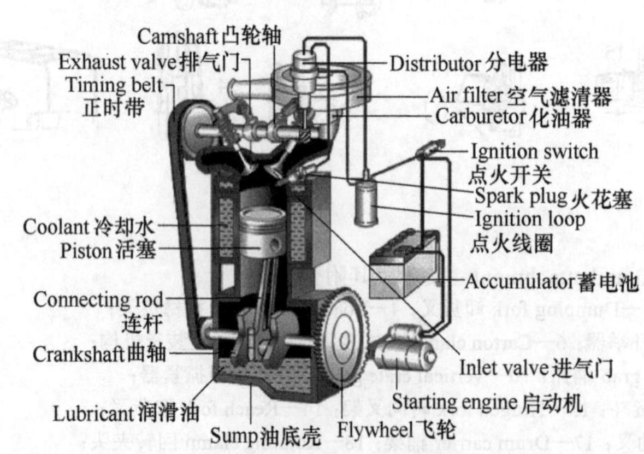

Fig 20.2 Gasline engine appearance structure 汽油发动机外形图

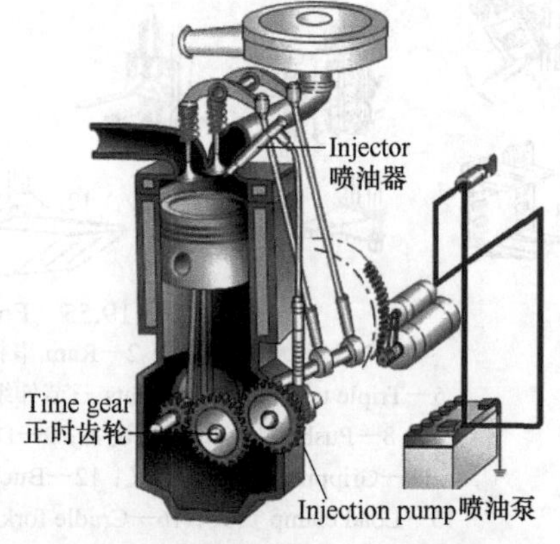

Fig 20.3 Diesel engine appearance structure 柴油发动机外形结构

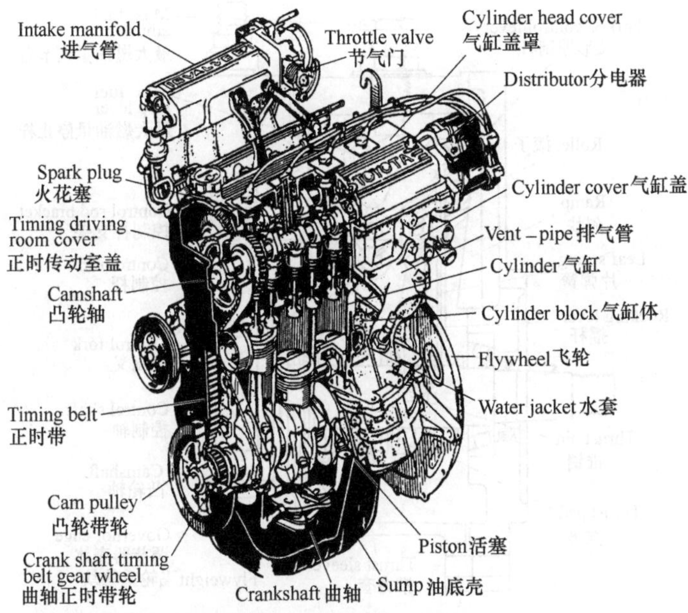

Fig 20.4 Engine construction 发动机构造

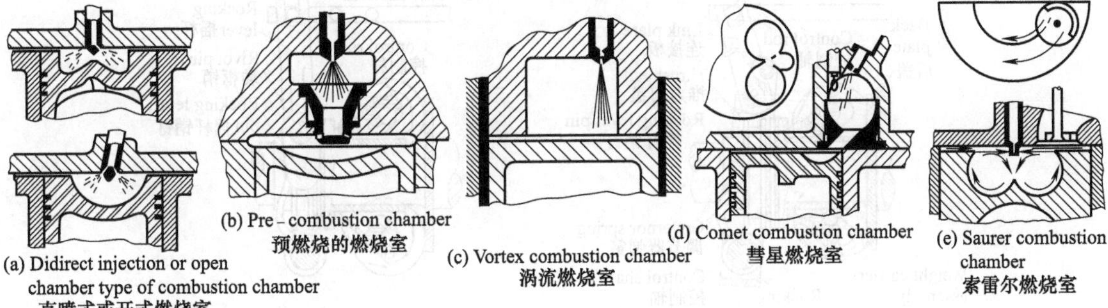

Fig 20.5 Types of chamber 燃烧室类型

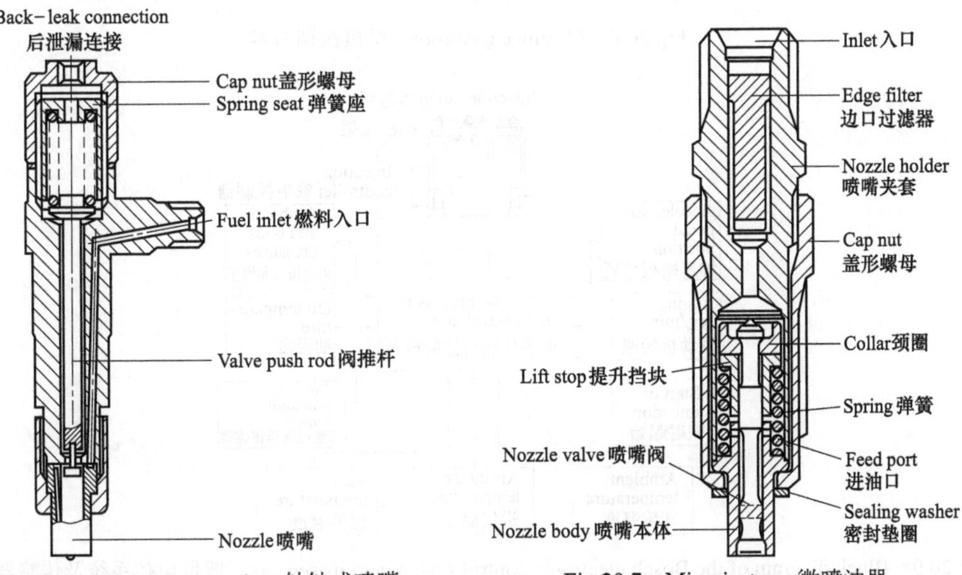

Fig 20.6 Pintle type nozzle 轴针式喷嘴

Fig 20.7 Microjector 微喷油器

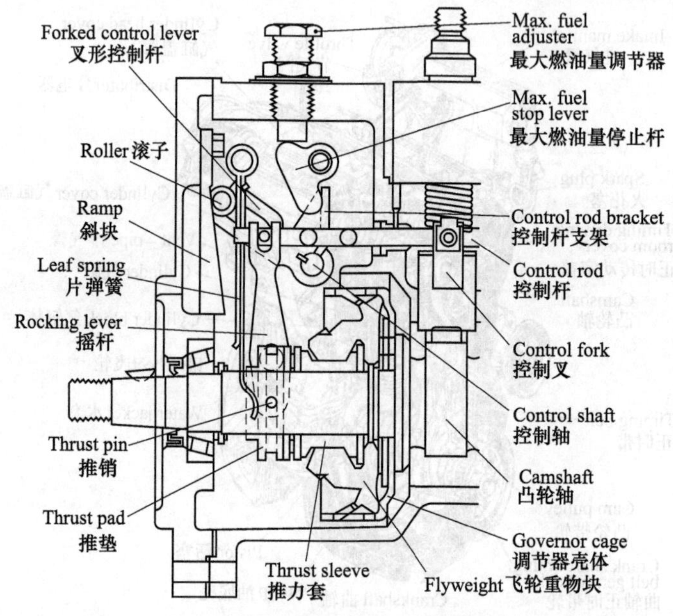

(a) Has a leaf spring 具有片弹簧

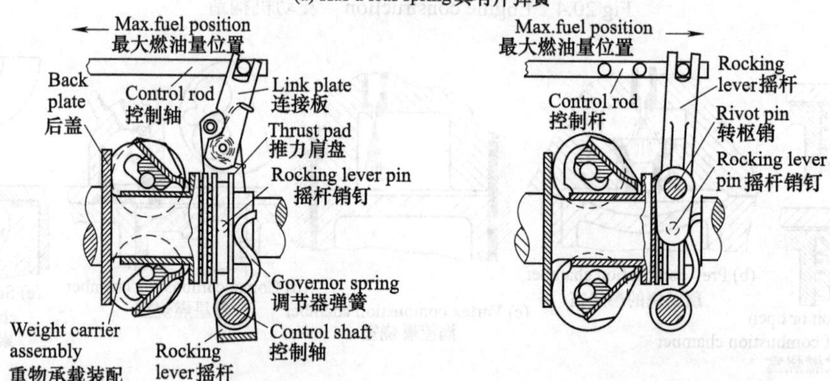

(b) A Minimec governor with a coil spring loaded in tension 线圈弹簧张紧时的微机械调节器

Fig 20.8 Minimec governor 微机械调节器

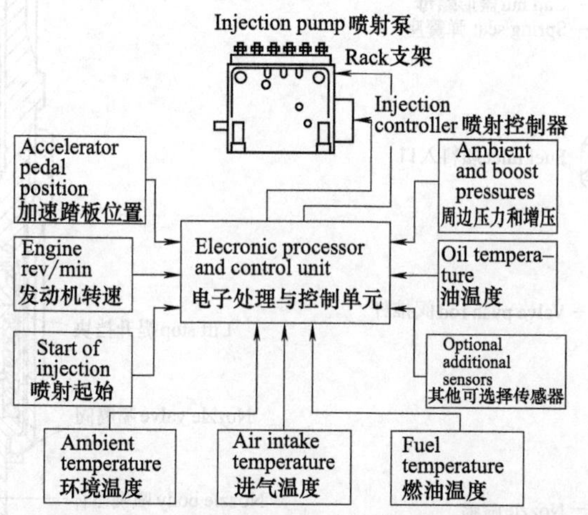

Fig 20.9 Block diagram of the Bosch electronic control system and its sensors 博世电控系统及传感器框图

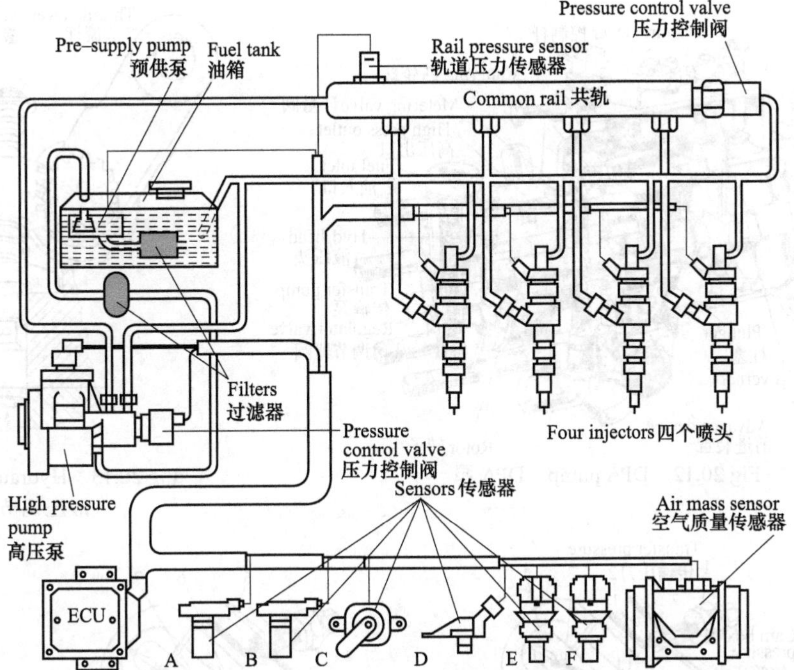

Fig 20.10 Principal components of the Bosch common rail injection system 博世共轨喷射系统的主要部件

A—Crankshaft position 曲轴位置；B—Camshaft position 凸轮轴位置；C—Accelerator pedal 加速踏板；D—Boost pressure 涡轮增压；E—Air temperature 空气温度；F—Coolant temperature 冷却液温度

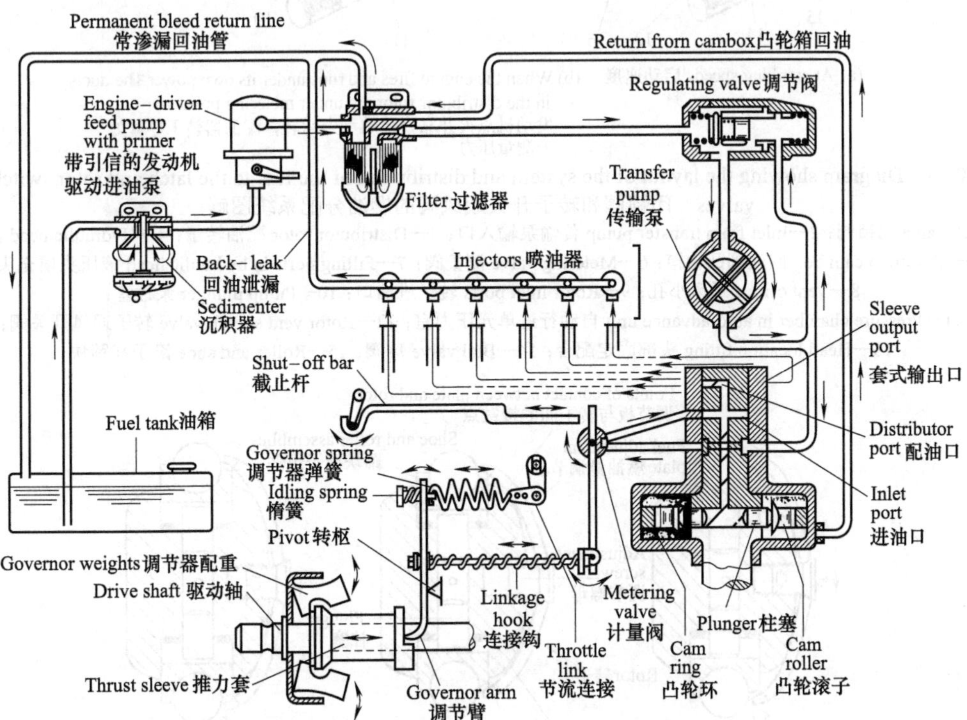

Fig 20.11 Fuel system with DPA pump and mechanical governor

带分配式配油泵和机械调节器的燃油系统

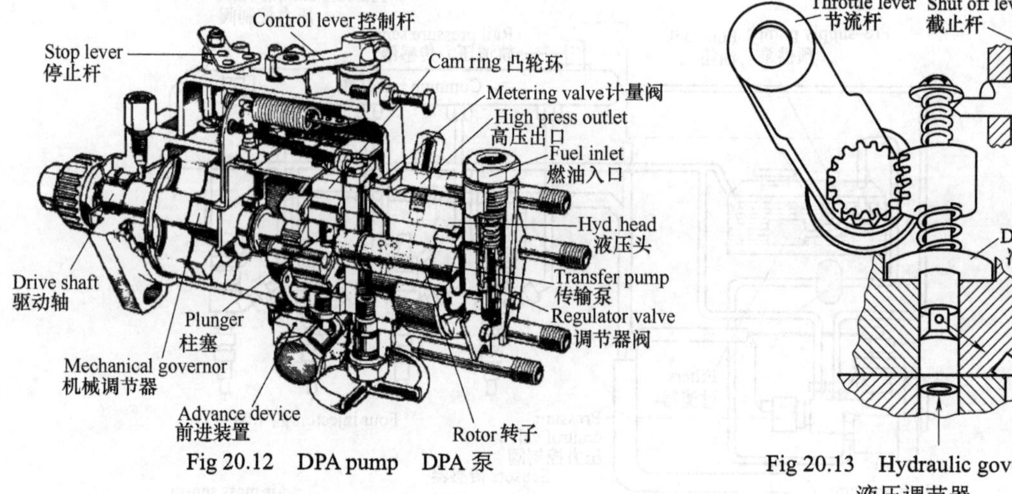

Fig 20.12 DPA pump DPA 泵

Fig 20.13 Hydraulic governor 液压调节器

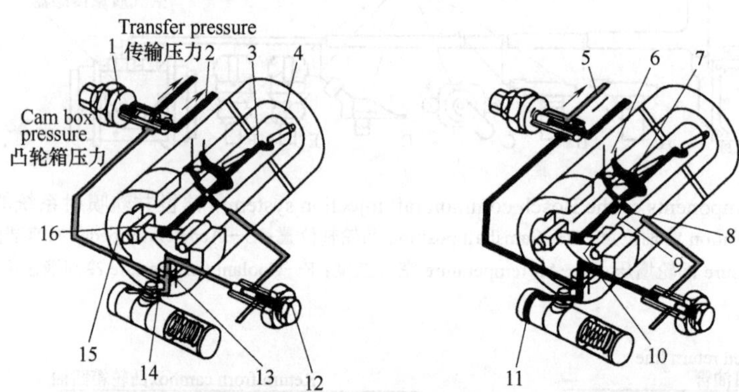

(a) At cranking speeds 启动速度 (b) When the engine fires and runs under its own power.The ducts in the distributor rotor are under metering pressure 发动机点火并靠自己的动力运转，配油器转子导管处于测量压力

Fig 20.14 Diagram showing the layout of the system and distribution of the fuel to the latch and rotor switch vent valves 闭锁阀和转子开关通风阀的燃油分配系统图解

1—Latch valve 闭锁阀；2—Inlet from transfer pump 传输泵输入口；3—Distributor rotor 配油转子；4—Hydraulic head 液压头；5—Return to cam box 回油到凸轮箱；6—Metering valve 测量阀；7—Filling ports in hydraulic head 液压头填充头；8—Vent orifice 通风小孔；9—Rotor inlet ports 转子进油口；10—Pump plunger 泵柱塞；11—Pressure chamber in auto-advance unit 自动行进单元压力室；12—Rotor vent switch valve 转子通风开关阀；13—Head location fitting 头部固定配合；14—Ball valve 球阀；15—Roller and shoe 滚子和蹄块

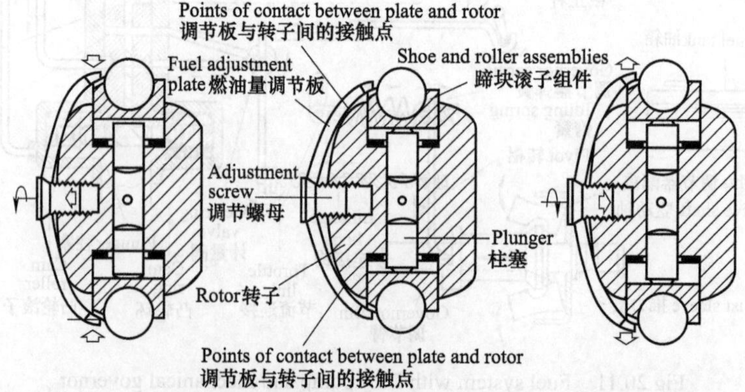

Fig 20.15 Maximum fuel adjustment device 最大燃油量调节装置

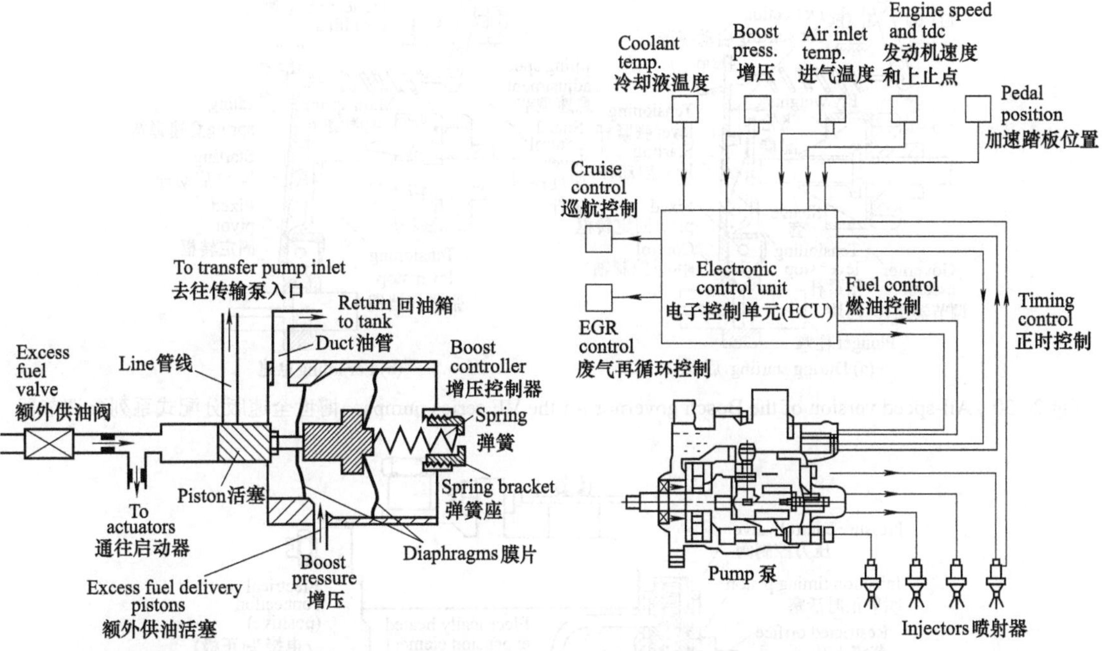

Fig 20.16 Boost controller at rest or idling
处于停止或怠速的增压控制器

Fig 20.17 Electronic control system for indirect injection
间接喷射的电子控制系统

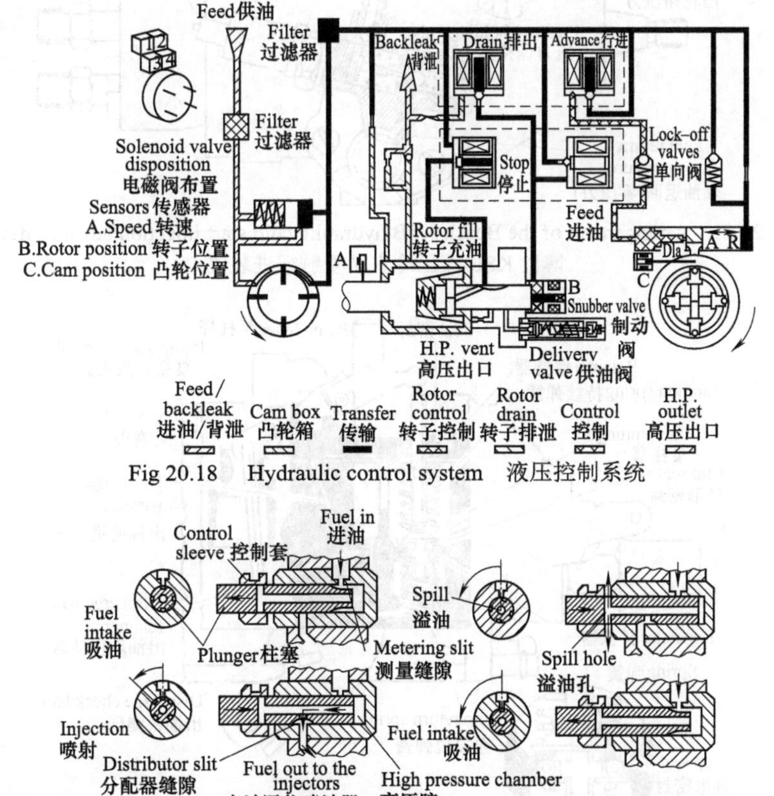

Fig 20.18 Hydraulic control system 液压控制系统

Fig 20.19 Elevation and plan of the plunger and control sleeve of the Bosch VE type pump (left, in the delivery and, right, in the spill condition control, by the governor, over the fuelling is effected by axial movement of the control sleeve to vary the spill point) 调节器对博世分配型泵的柱塞和控制套的提升设计
（左图：供油状态，右图：溢油条件控制。控制套轴向移动改变溢油点从而影响供油量）

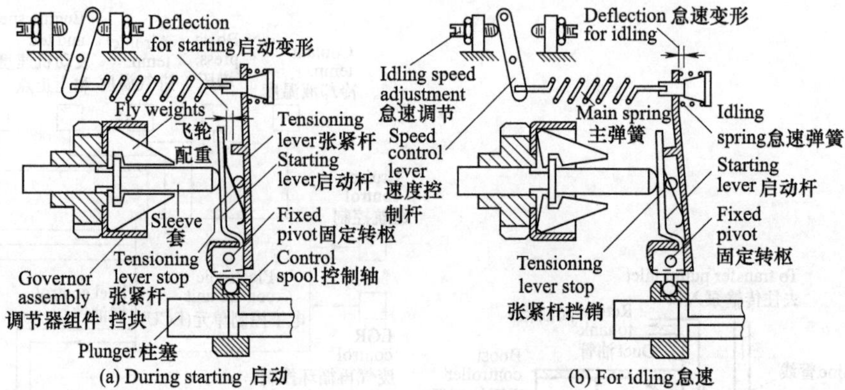

Fig 20.20 All-speed version of the Bosch governor for the VE series pumps 博世全速版分配式系列泵调节器

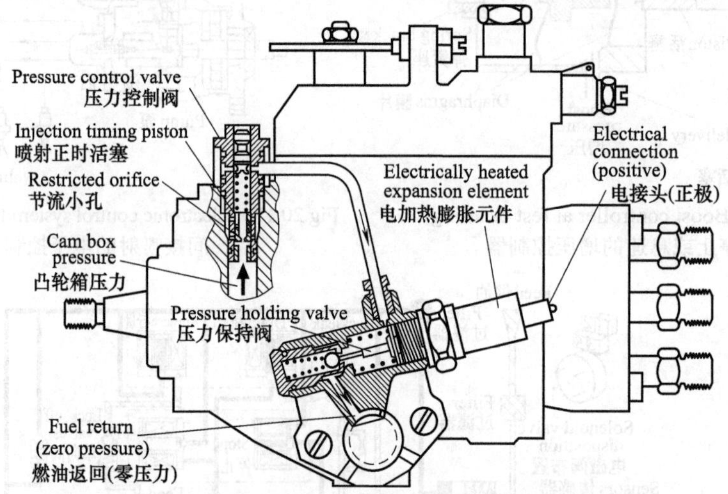

Fig 20.21 Arrangement of the Bosch KSB hydraulic cold start injection advance device
博世 KSB 液压冷启动喷射行进装置

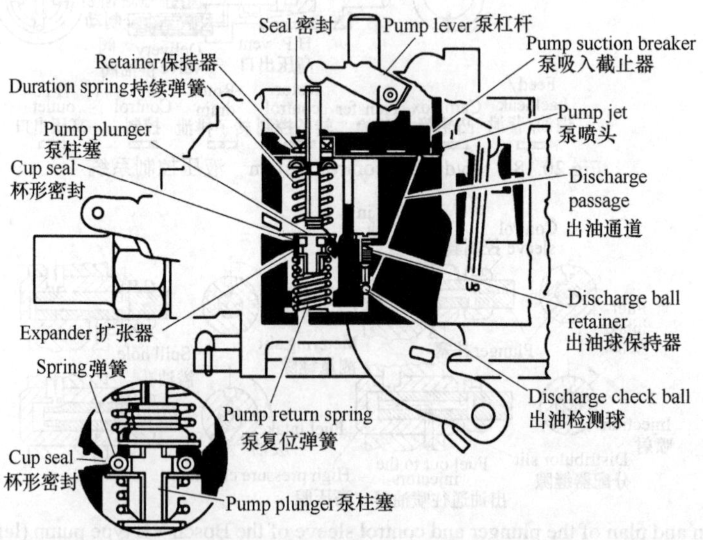

Fig 20.22 Acceleration pump system with, in the scrap view, details of the head of the plunger and its seal
加速泵系统及其拆解图和柱塞、密封放大图

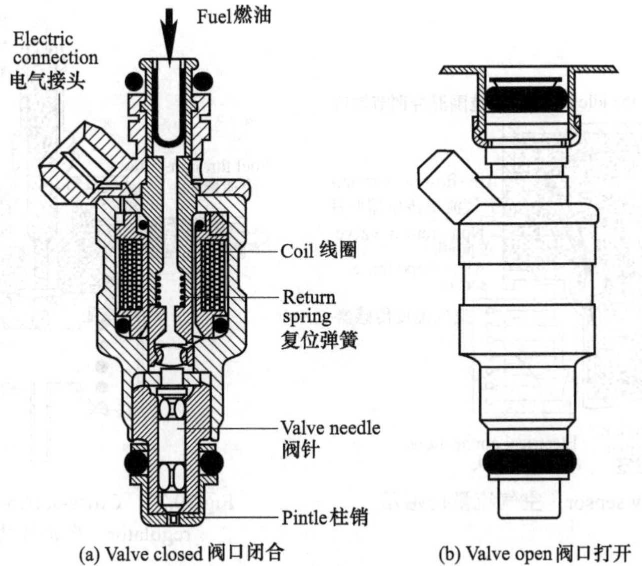

Fig 20.23 Injector for the Bosch K-jetronic system 博世 K 系燃油电喷系统

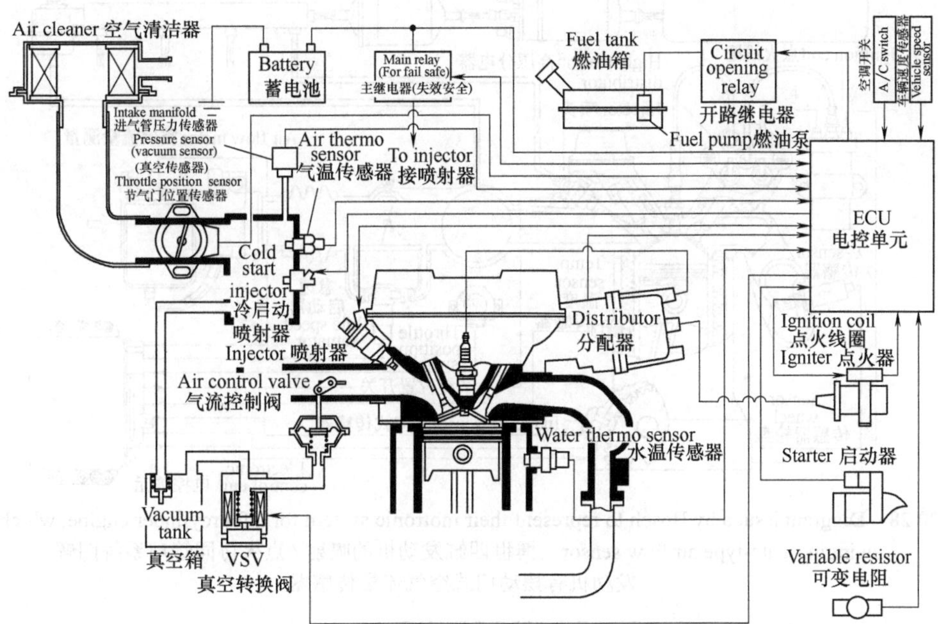

Fig 20.24 Electronic fuel injection system 电子燃油喷射系统

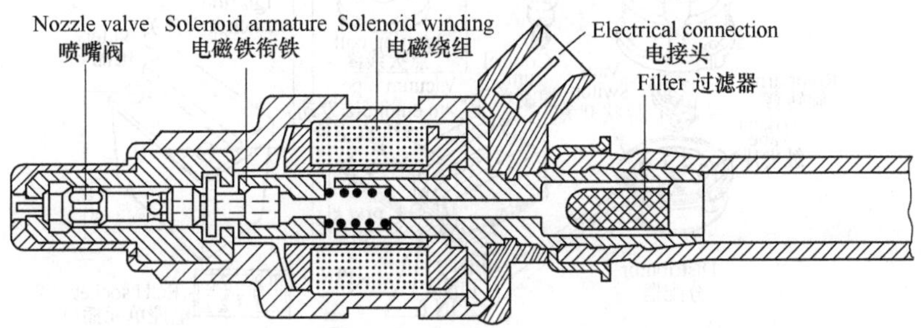

Fig 20.25 Cross-section of the injection valve 喷射阀剖视图

Mixture adjustment screw for the idle range 怠速范围混合调节螺钉
Air-flow sensor flap
空气流量传感器叶片
Non-return valve
单向阀
Air-temperature sensor
空气温度传感器
Compensation flap
补偿叶片
Damping chamber
阻尼室
Electrical connections
电接头

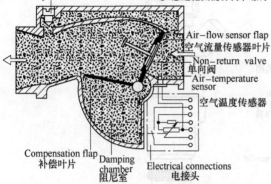

Fig 20.26 Air-flow sensor 空气流量传感器

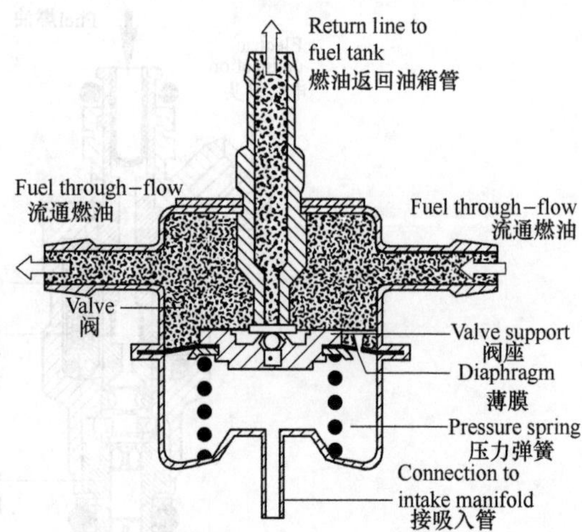

Fig 20.27 Cross-section of the fuel-pressure regulator 燃油压力调节器剖视图

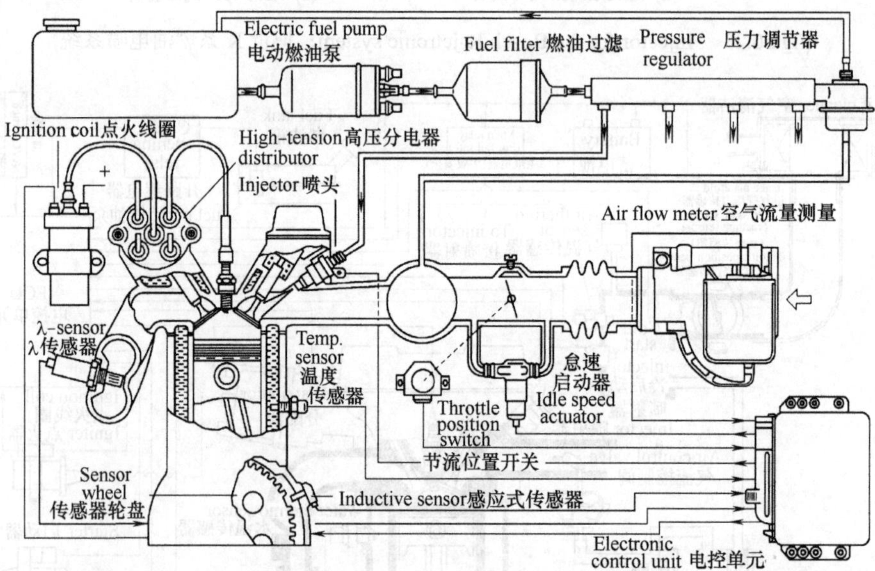

Fig 20.28 Diagram issued by Bosch to represent their motronic system for a fourcylinder engine, which has a swinging-gate-type air flow sensor 博世四缸发动机的喷射 / 点火协同管理系统图解，发动机有摆动门式空气流量传感器

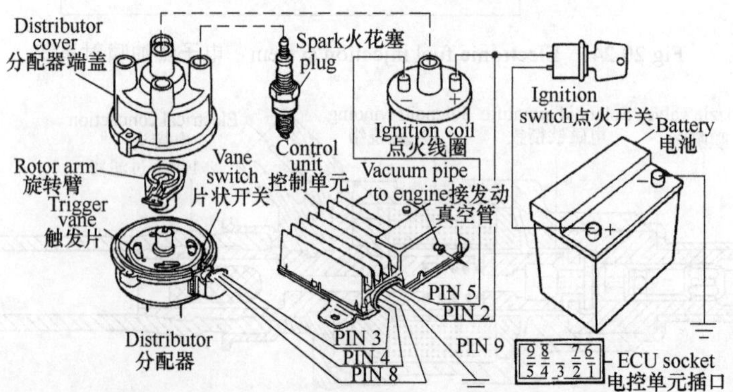

Fig 20.29 Digital ignition system 数字点火系统

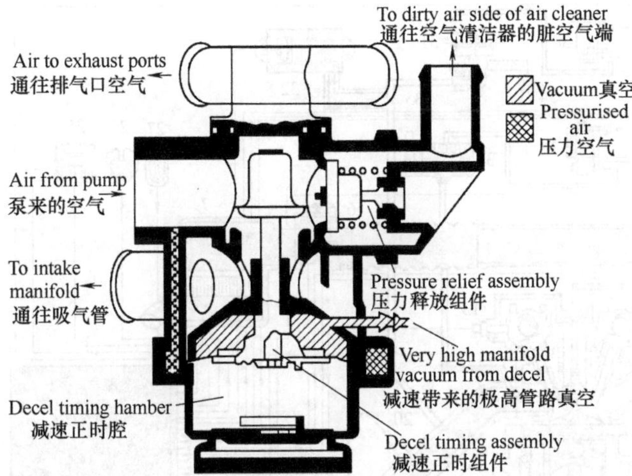

Fig 20.30 Air intake control valve 进气控制阀

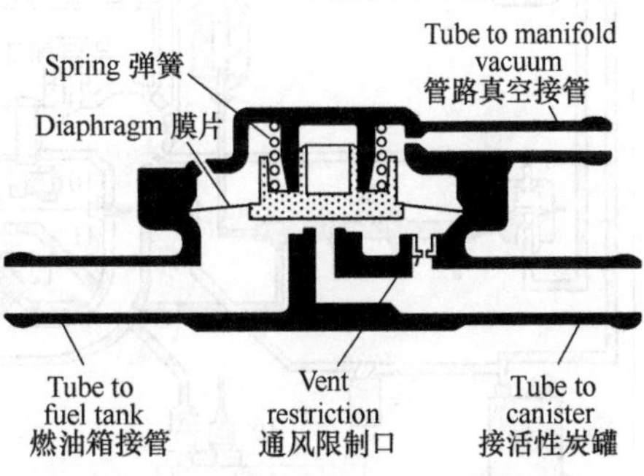

Fig 20.31 Fuel tank pressure control valve 燃油箱压力控制阀

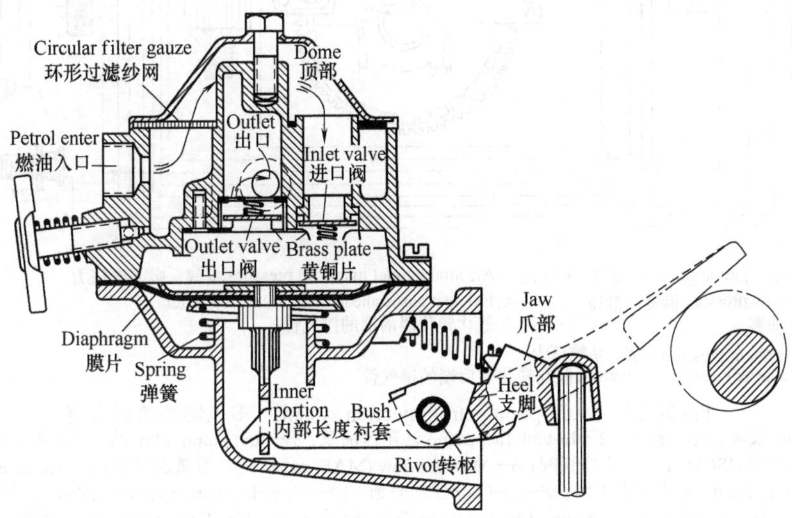

Fig 20.32 AC fuel pump 交流燃油泵

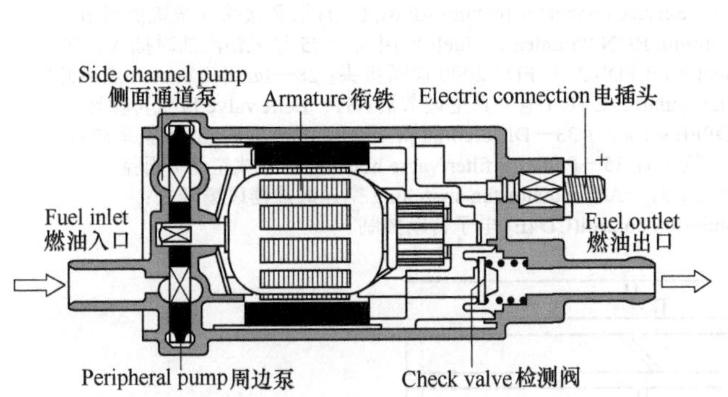

Fig 20.33 Bosch fuel pump for delivery at pressures below 1 bar 博世低压（低于 1bar）燃油供油泵

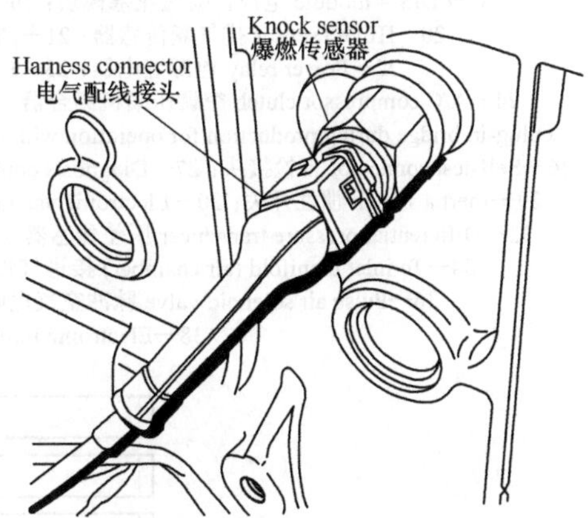

Fig 20.34 The knock sensor on the engine 发动机爆燃传感器

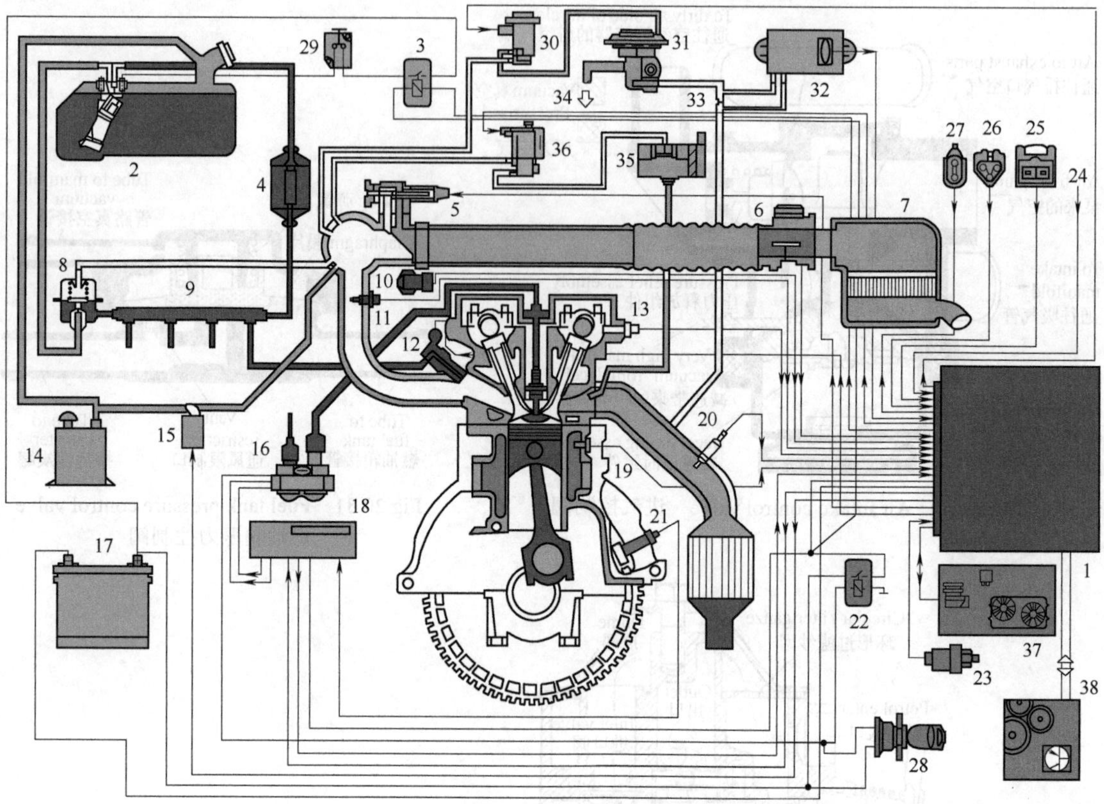

Air intake–atmospheric pressure 吸气 –大气压 Air intake–inlet manifold pressure 吸气 – 吸入管压力
Fuel supply–low pressure 供油 – 低压 Exhaust gases ahead of catalytic converter 催化转化器前端的排气管
Fuel vapour 油雾
Fuel supply–system pressure 供油 –系统压力
Exhaust gases after catalytic converter 催化转化器后端的排气管

Fig 20.35 Multi-point petrol injection system 多点燃油喷射系统

1—EEC IV module 发动机电控模块；2—In-tank fuel pump 油箱内油泵；3—Fuel pump relay 油泵继电器；4—Fuel filter 油滤；5—Idle speed control (ISC) valve 怠速控制阀；6—Mass air flow (MAF) meter 空气质量流量计；7—Air cleaner 空气清洁器；8—Fuel pressure regulator 燃油调压器；9—Fuel rail 燃油轨道；10—Throttle position sensor (TPS) 节气门位置传感器；11—Air charge temperature (ACT) sensor 进气温度传感器；12—Fuel injector 燃油喷嘴；13—Camshaft identification (CID)sensor 凸轮轴识别传感器；14—Carbon canister (EVAP) 活性炭罐；15—Purge solenoid valve (EVAP) 清污电磁阀；16—DIS coil 直喷火花塞线圈；17—Battery 蓄电池；18—EDIS-4 module 电控直喷火花塞模块；19—Engine coolant temperature (ECT) sensor 引擎冷剂温度传感器；20—HEGO sensor 排气氧传感器；21—Crankshaft position/speed (CPS)sensor 曲轴位置 / 速度传感器；22—Power relay 电源继电器；23—Power steering pressure switch (PSPS) 动力转向压力开关；24—A/C compressor clutch 空调压缩机离合器；25—Service connector (octane adjust, OAI) 服务接头（辛烷值调节）（plug-in bridge during production for operation with premium RON 95 unleaded fuell 使用高于 95 号无铅汽油时插入连接）；26—Self-test connector 自检接头；27—Diagnosis connector for FDS 2000 FDS 2000 诊断接头；28—Ignition switch 点火开关；29—Inertia switch 惯性开关；30—Electronic vacuum regulator (EVR) 电子真空调节器；31—EGR valve 废气再循环阀；32—Differential pressure transducer 压差传感器（DPFE sensor）；33—Differential pressure sampling point 压差采样点；34—To inlet manifold (air chamber) 接进气管（气室）；35—Pulse air filter/valve housing 脉冲式空滤 / 阀腔；36—Pulse air solenoid valve 脉冲空气电磁阀；37—A/C radiator fan switching 空调散热器风扇开关；38—Electronic transmission control(CD4E) 电子传动控制

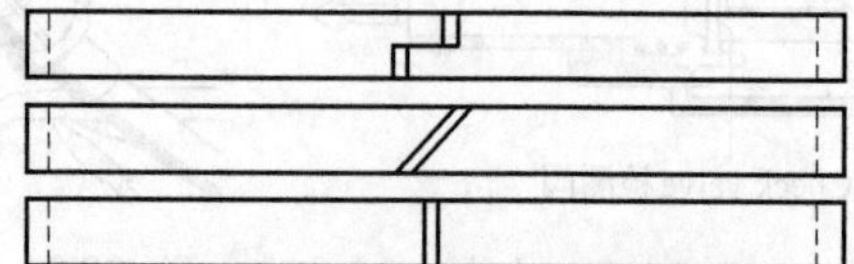

Fig 20.36 Piston rings 活塞环

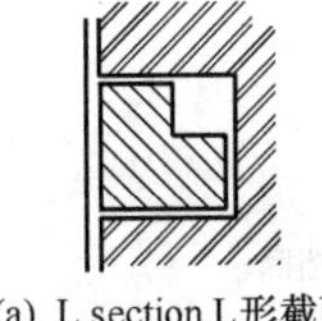
(a) L section L形截面

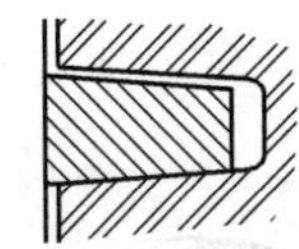
(b) Wedge section 楔形截面

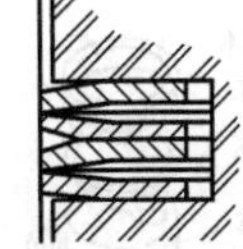
(c) Cords ring 碟形柔性环

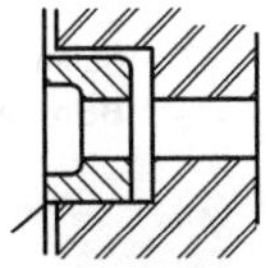
(d) Grooved and slotted rings 槽环

Fig 20.37 Simple rectangular-section rings 矩形截面活塞环简图

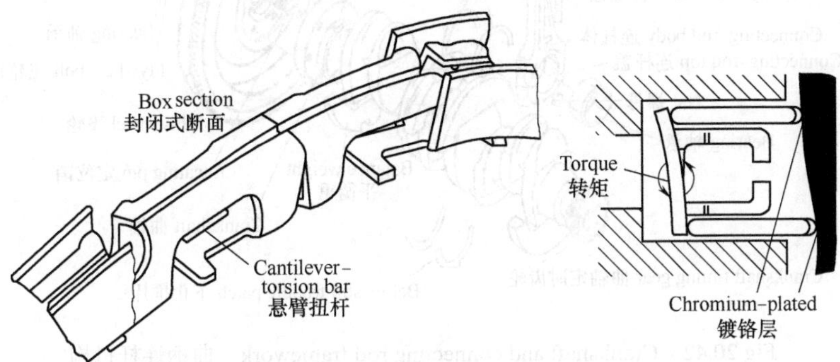

Fig 20.38 Three-piece oil control rings 三组件活塞控油环

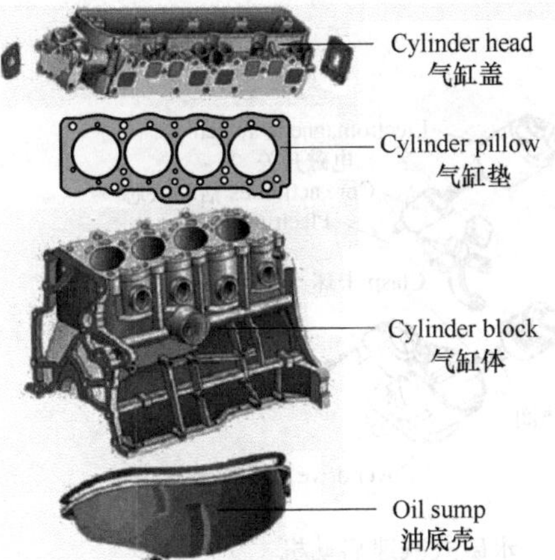

Fig 20.39 Block group 机体组

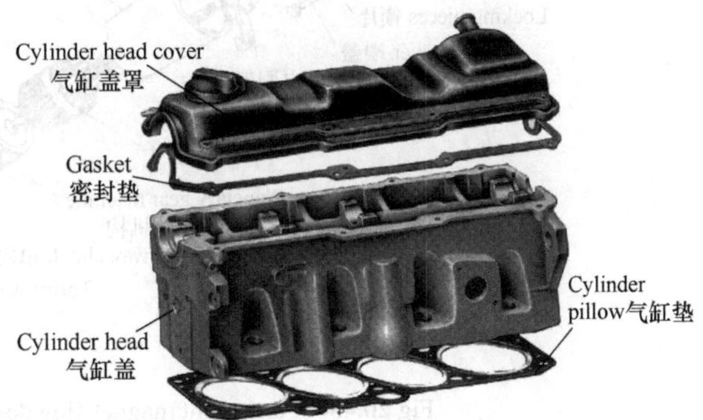

Fig 20.40 Cylinder head and its cover 缸盖及盖罩

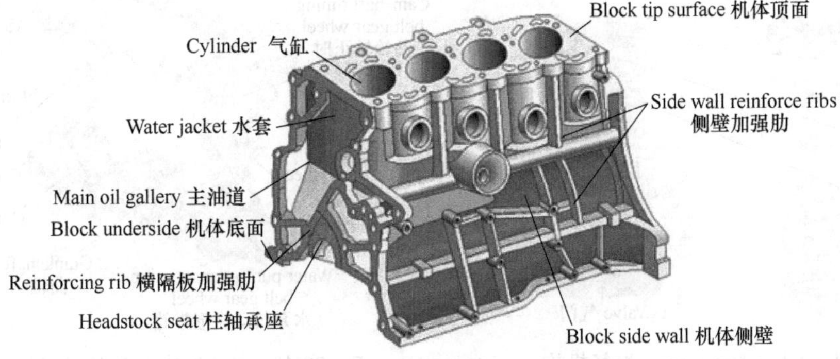

Fig 20.41 Cylinder block 缸体

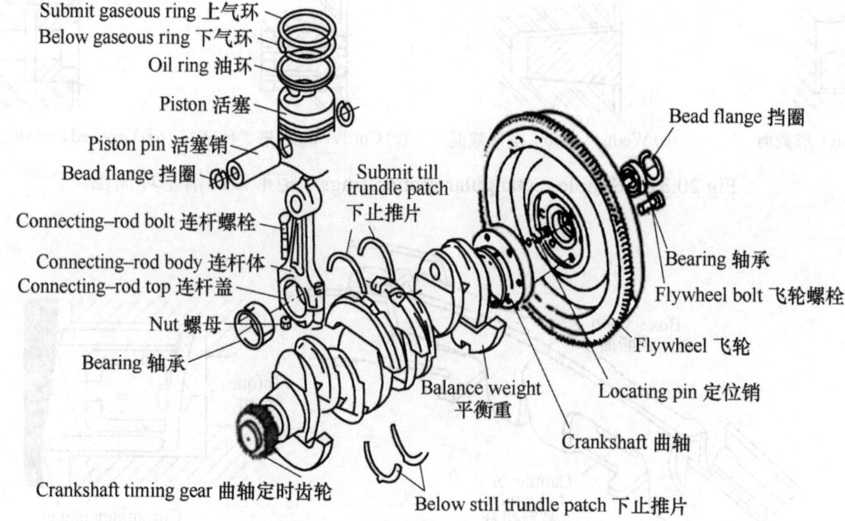

Fig 20.42 Crankshaft and connecting rod framework 曲柄连杆机构

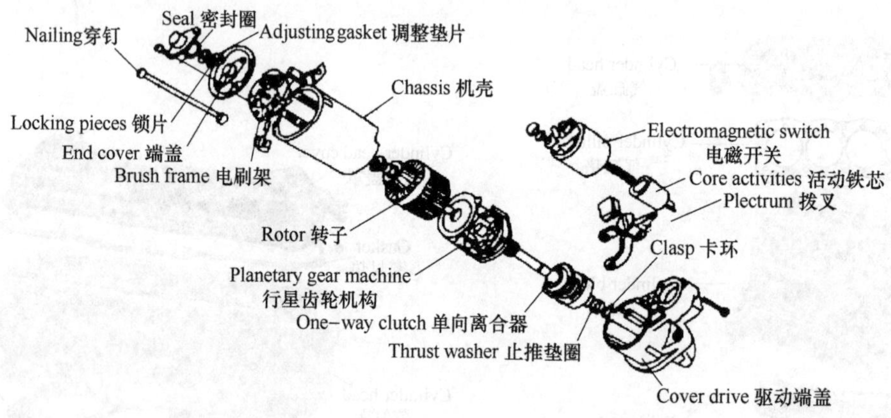

Fig 20.43 Permanent magnet slowdown starter 永磁式减速启动机

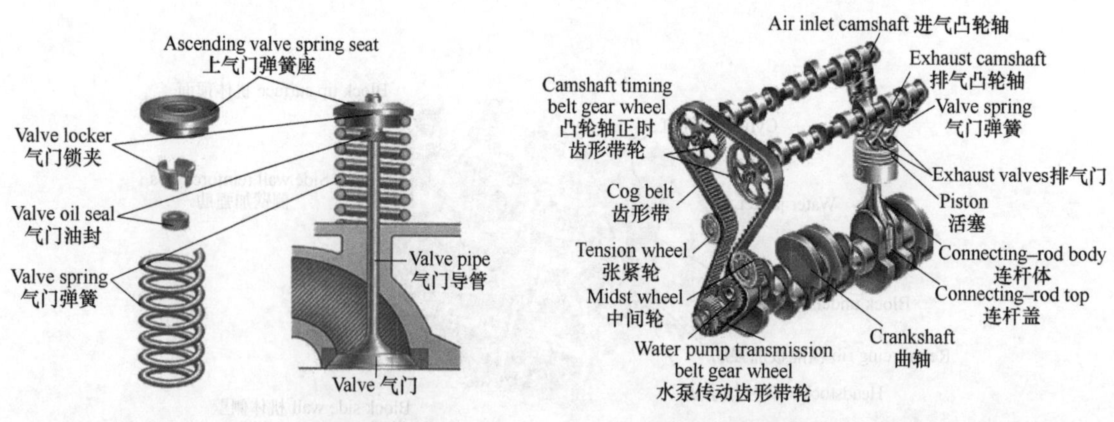

Fig 20.44 Admission gear 配气机构

Fig 20.45 Crankshaft timing belt gear wheel 曲轴正时齿形带轮

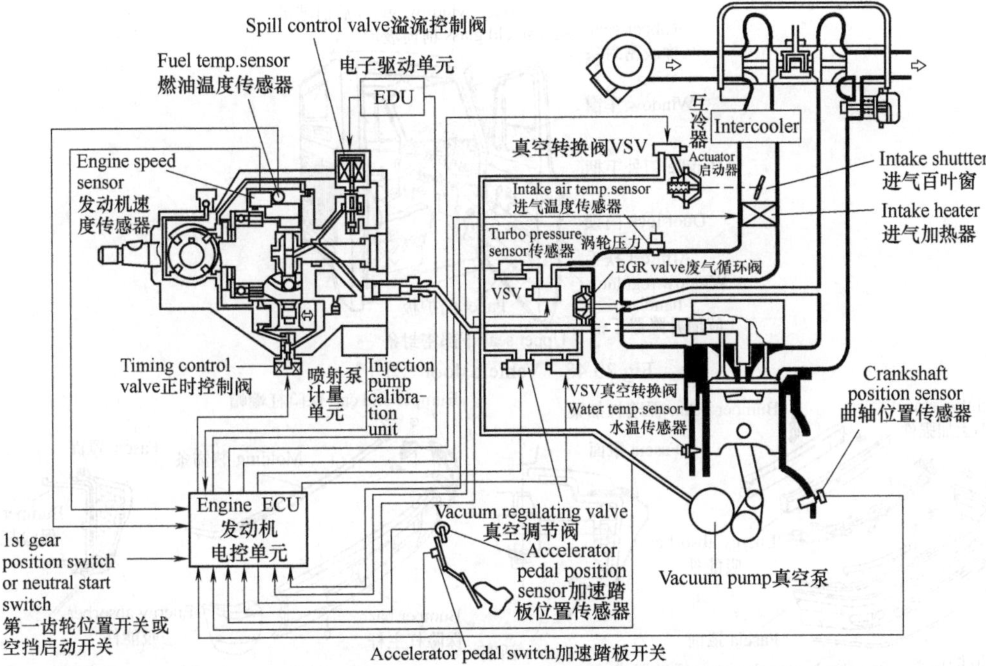

Fig 20.46 Computer controlled diesel engine system 计算机控制的柴油发动机系统

20.2 Body 车身

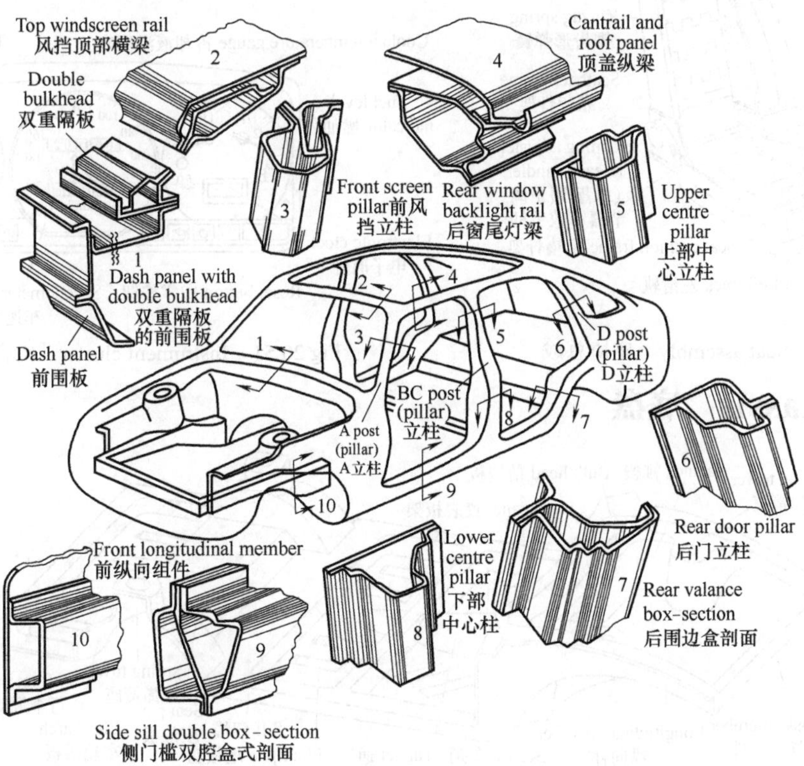

Fig 20.47 Load bearing body box-section members 承载车身腔盒式组件剖面

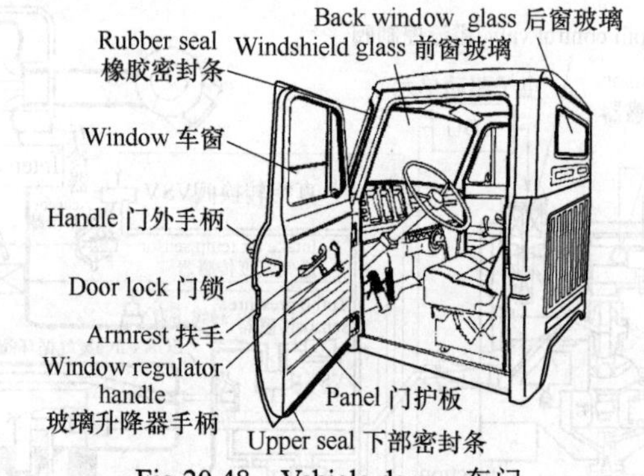

Fig 20.48 Vehicle door 车门

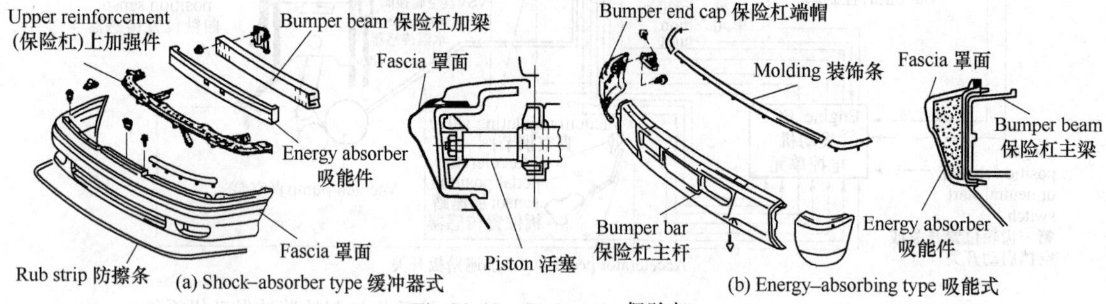

Fig 20.49 Bumper 保险杠

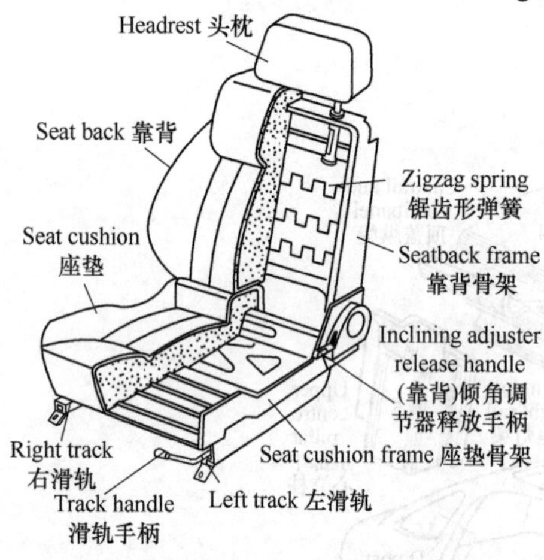

Fig 20.50 Seat assembly 座椅总成

Fig 20.51 Instrument cluster 组合仪表

20.3 Chassis 底盘

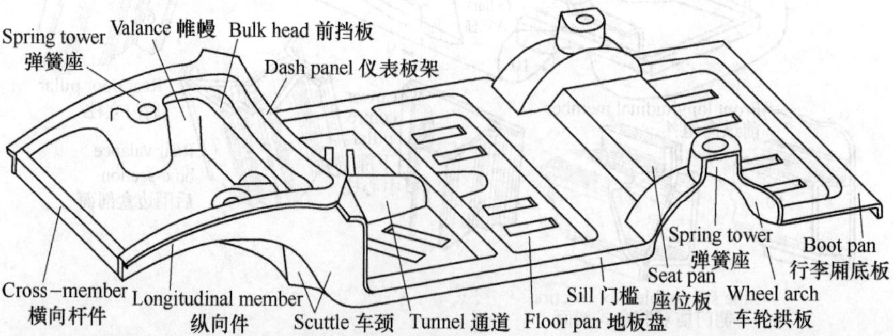

Fig 20.52 Platform chassis (small central tunnel, sills, front valance, rear wheel arches and all round spring towers) 平台式底盘（小中心通道、门槛、前帷幔、后轮拱板和全部圆形弹簧座）

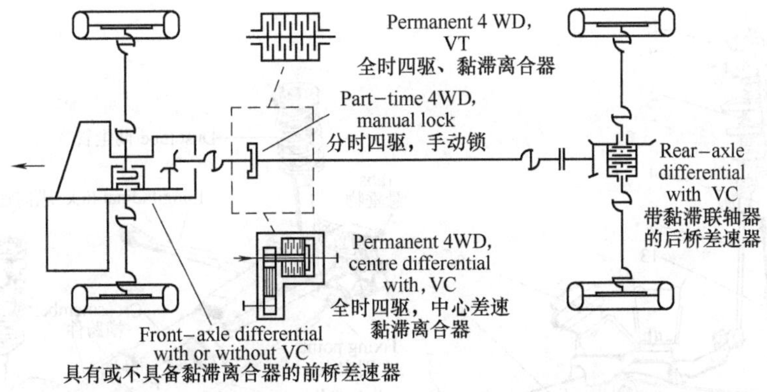

Fig 20.53 Four-wheel-drive systems 四轮驱动系统

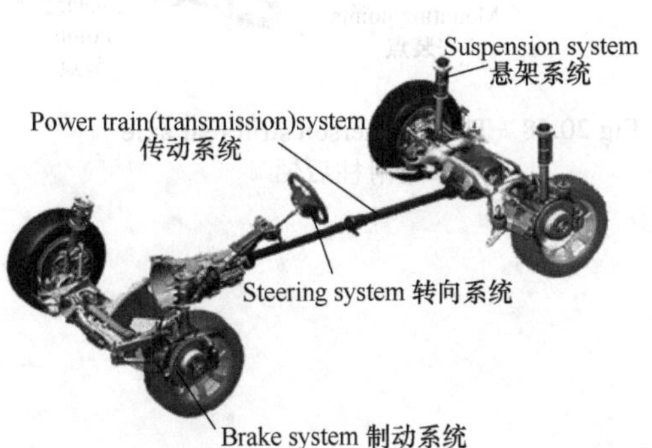

Fig 20.54 Chassis assembly map 底盘总成图

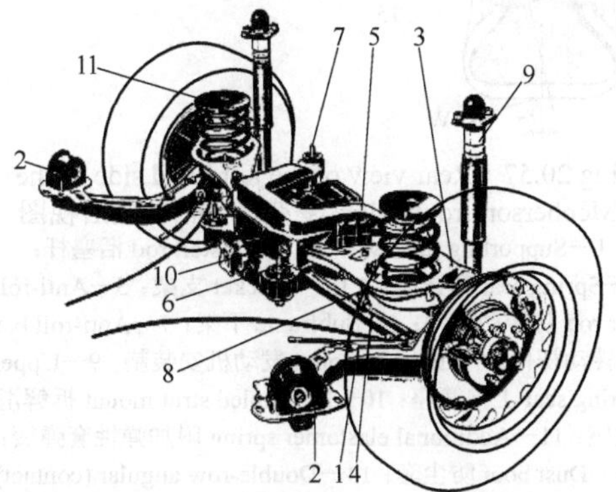

Fig 20.55 Multi-link rear axle 多连杆后轴

1—Trailing arm 领臂；2—Points 节点；3—Upper arm 上臂；
4—Transverse control arm 横向控制臂；5—Subframe 下支架；
6，7—Rubber bushes 橡胶套；8—Anti-roll bar 防转杆；
9—Additional polyurethane spring 聚氨基甲酸乙酯弹簧；
10—Differential 差速器；11—Spring 弹簧

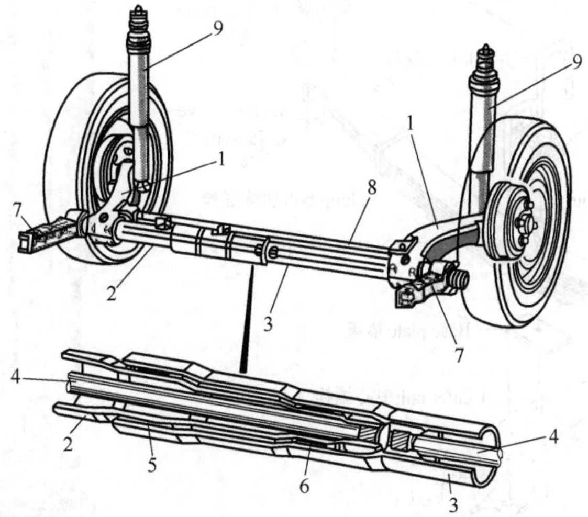

Fig 20.56 Compact trailing arm rear axle 紧凑型领臂后轴

1—Cast trailing arm 铸铁领臂；2，3—Guide tube 导管；4—Torsion bar 扭转杆；5，6—Torsionally elastic bearing 扭转弹性轴承；7—Bracket 支架；8—Anti-roll bar 防转杆；9—Shock absorber 减振器

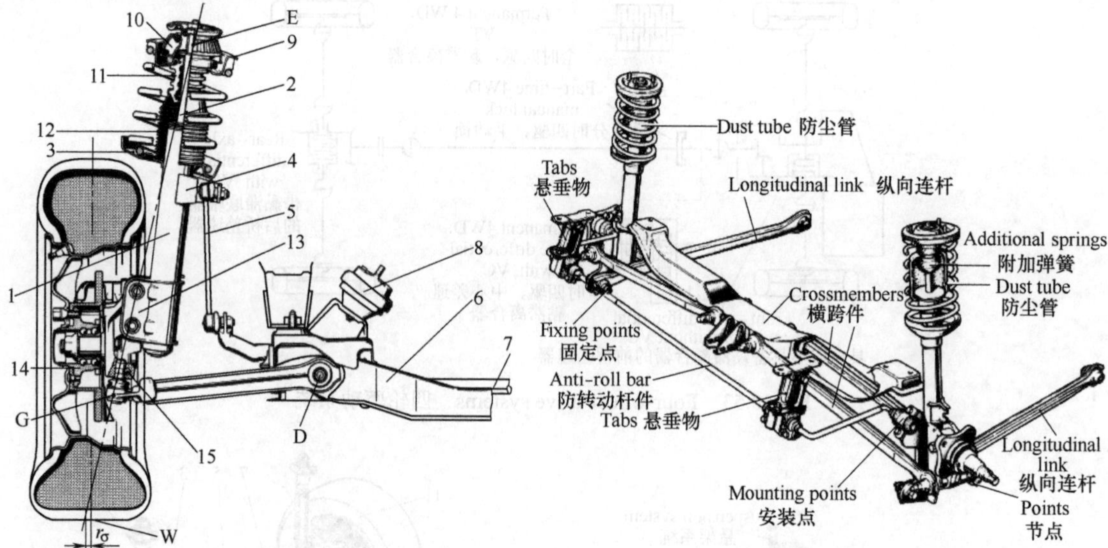

Fig 20.57 Rear view of the left-hand side of the Mcpherson front axle 麦弗逊前轴左侧后视图

1—Supporting tube 支撑管；2—Piston rod 活塞杆；3—Spring seat 弹簧座；4,13—Bracket 支架；5—Anti-roll bar rod 防转动杆件；6—Subframe 下架；7—Anti-roll bar 防转动杆件；8—Engine mount 发动机安装垫；9—Upper spring seat 上弹簧座；10—Decoupled strut mount 拆解滑柱座；11—Additional elastomer spring 附加弹性套弹簧；12—Dust boot 防尘套；14—Double-row angular (contact) ball bearing 双列角接触球轴承；15—Transverse screw 横向螺杆

Fig 20.58 The Mcpherson strut rear axle 麦弗逊滑柱后轴

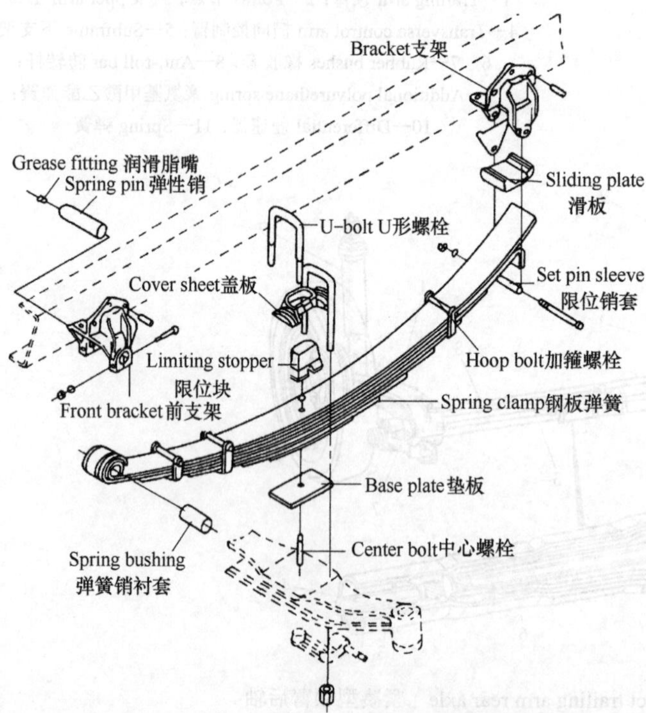

Fig 20.59 Rigid axle suspension 刚性轴悬架

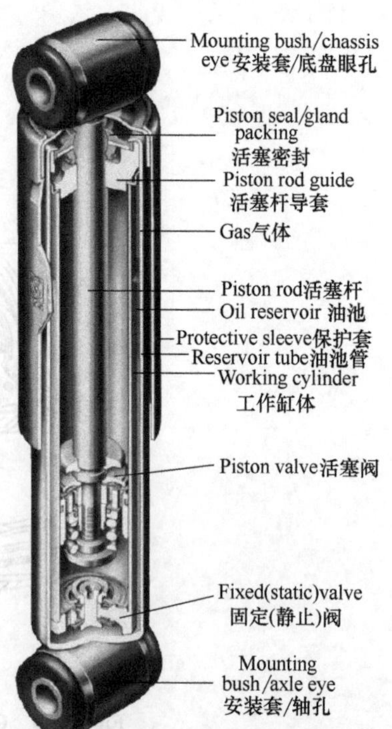

Fig 20.60 Low-pressure twin-tube shock absorber 低压双管减振器

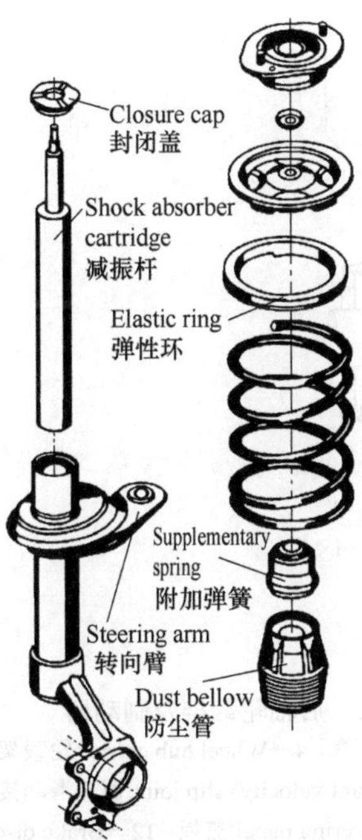

Fig 20.61 Exposive view of damping suspension strut 阻尼悬架柱爆炸图

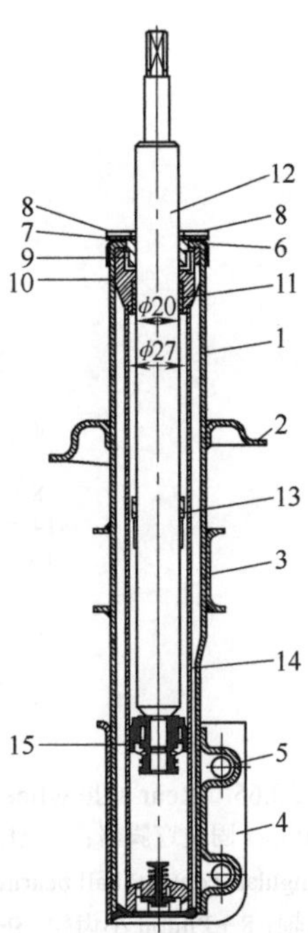

Fig 20.62 Mcpherson strut of the Fiat Panda 菲亚特熊猫的麦弗逊滑柱

1—Outer tube 外管；2—Spring seat 弹簧座；3—Tab 耳片；
4，5—Bracket parts 支架；6—Rolled edge 卷边；7—Stop disc 挡盘；
8—Transverse groove 横槽；9—Seal 密封；
10—Sintering iron rod guide 硬化铸铁活塞杆导套；
11—Bush 衬套；12—Piston rod 活塞杆；13—Bounce stop 振动挡块；
14—Cylinder tube 缸体筒；15—Low friction ring 低摩擦环

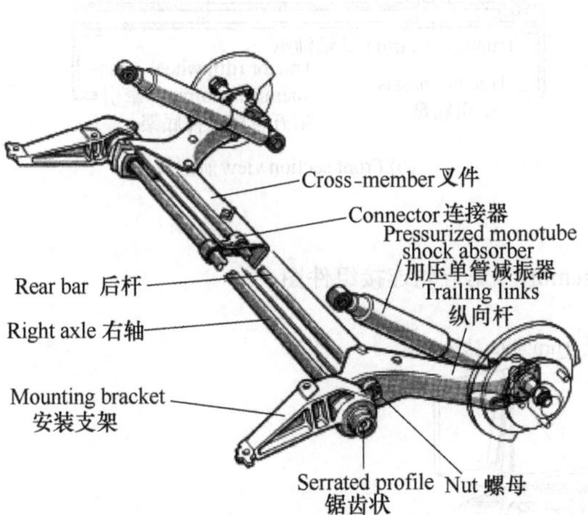

Fig 20.63 An extremely compact four-bar twist beam axle 极其紧凑的四杆扭转梁轴

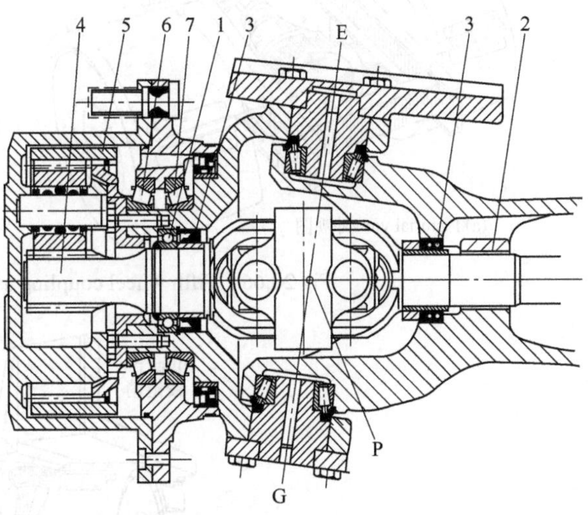

Fig 20.64 Driven, rigid steering axle with dual joint 双节点从动刚性转向轴

1，2—Bearing 轴承；3—Radial sealing ring 径向密封圈；
4—Drive shaft 驱动轴；5—Internal-geared wheel 内齿轮；
6，7—Tapered-roller bearing 圆锥滚子轴承

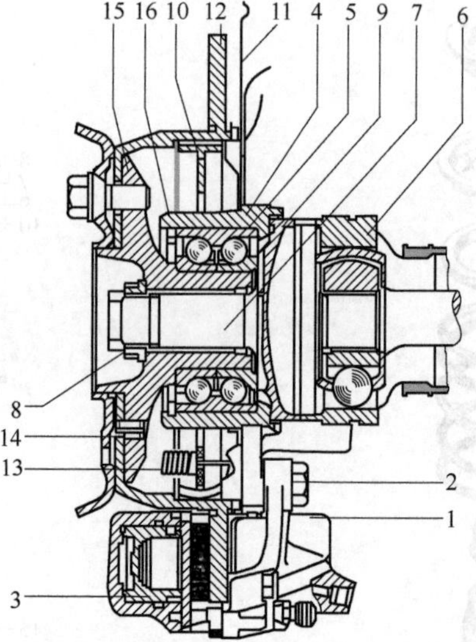

Fig 20.65 Rear axle wheel hub carrier with wheel and brake 后轴轮毂架及制动器

1—Fixed calliper 固定弯脚器；2—Hexagonal bolt 六角螺栓；3—Piston 活塞；4—Wheel hub carrier 轮毂架；5—Two row angular (contact) ball bearing 双列角接触球轴承；6—CV (constant velocity) slip joint 恒速滑动接头；7—Drive shaft 驱动轴；8—Pinion 小齿轮；9—Collar 套；10—Jaw 卡爪；11—Covering panel 盖板；12—Brake disc 制动盘；13—Spring 弹簧；14—Dowel 销；15—Flange 法兰；16—Snap ring 卡圈

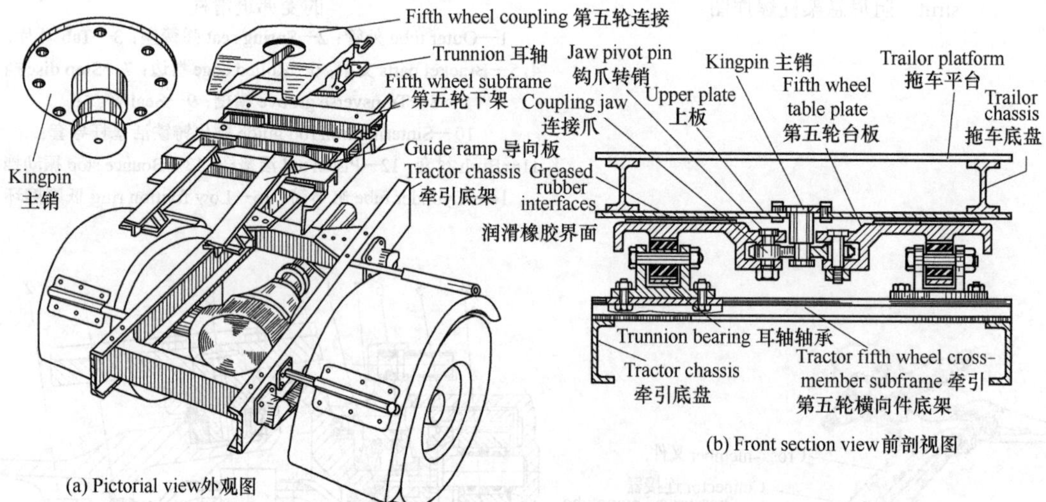

Fig 20.66 Fifth wheel coupling assembly 第五轮连接组件图

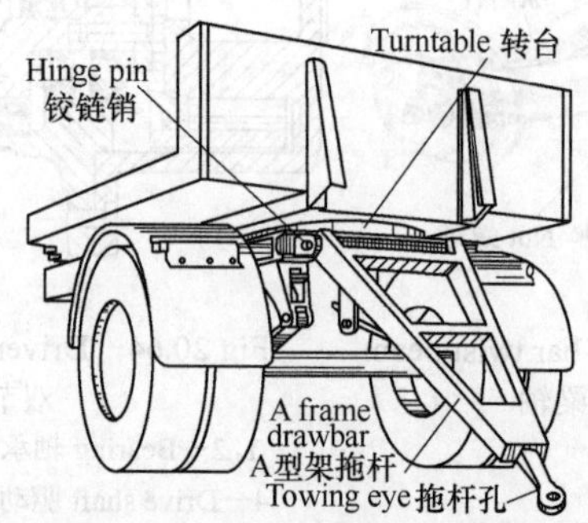

Fig 20.67 Drawbar trailer 拖杆拖车

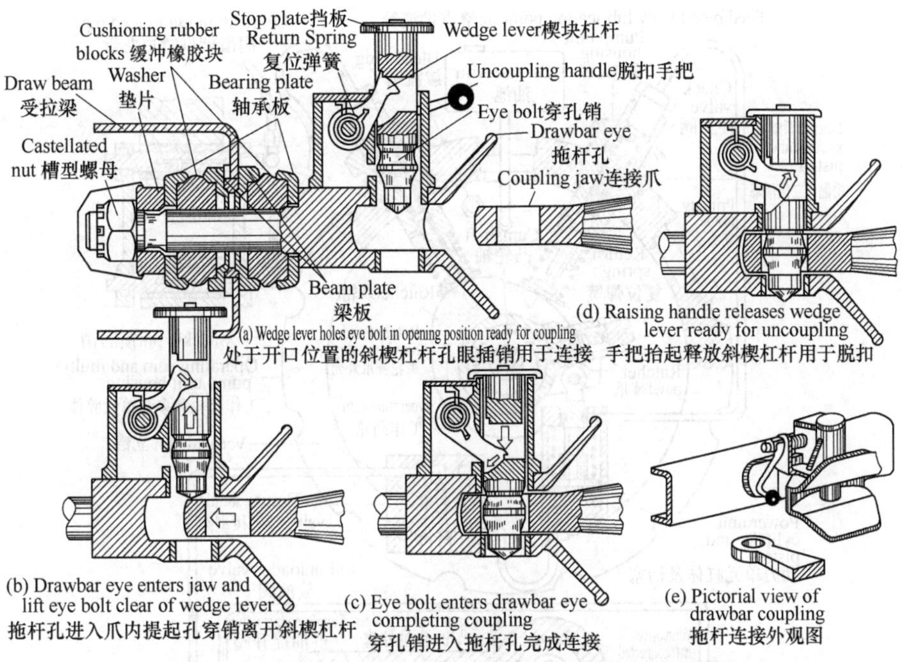

Fig 20.68 Automatic drawbar coupling 自动拖杆的连接

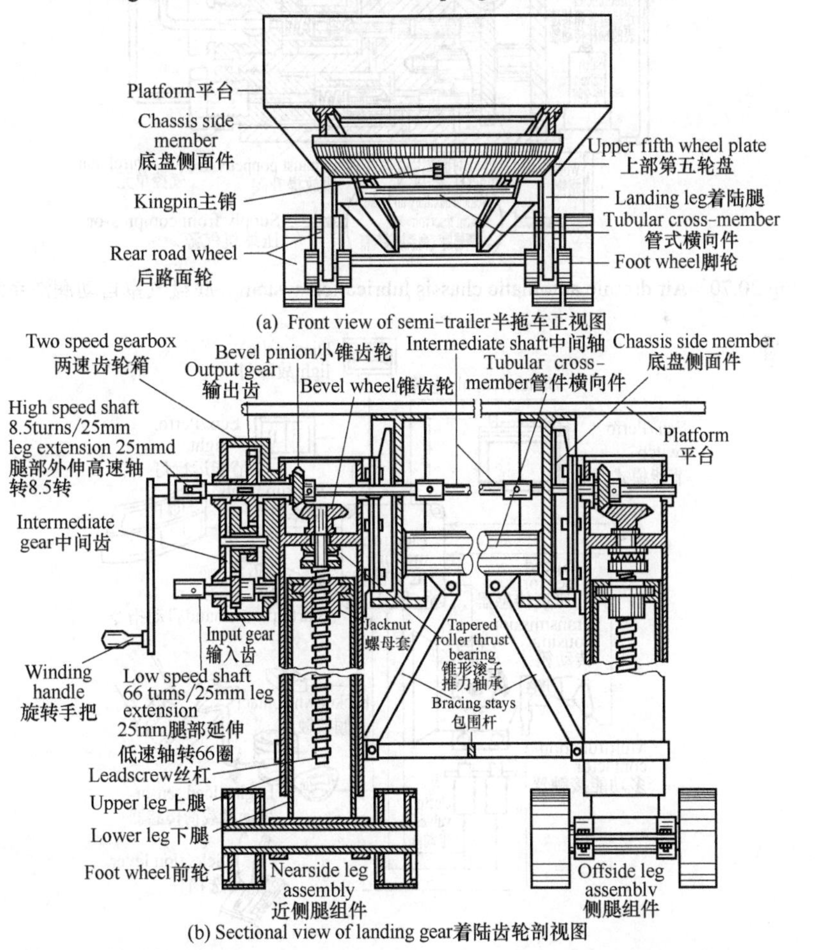

Fig 20.69 Semi-trailer landing gear 半拖车着陆齿轮

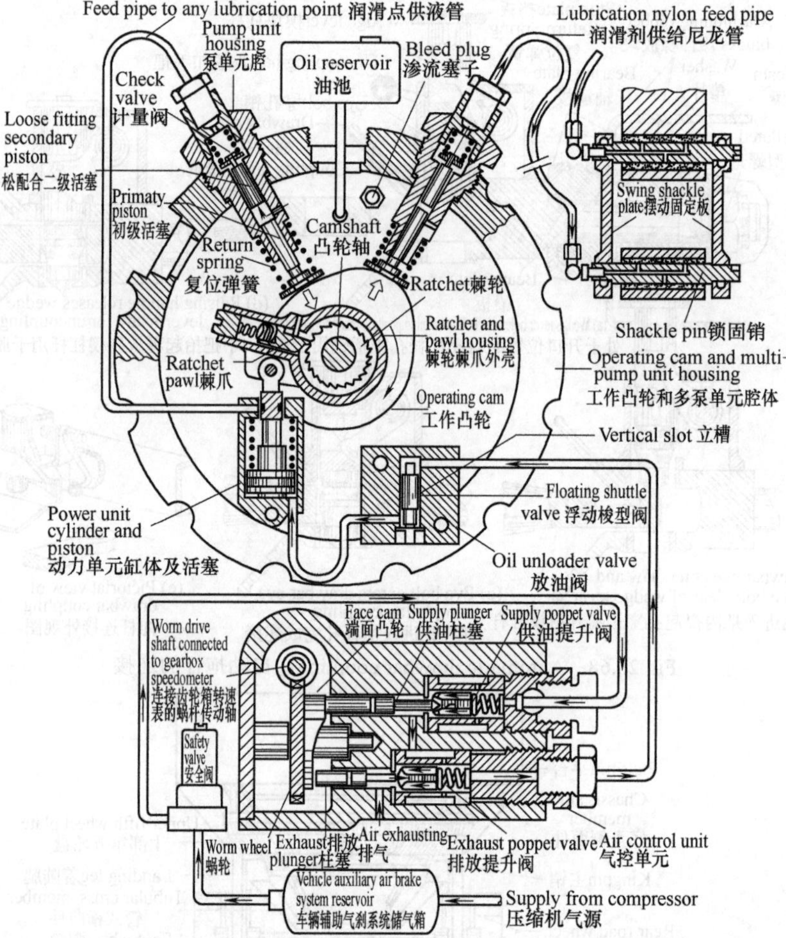

Fig 20.70 Air dromic automatic chassis lubrication system 底盘气室自动润滑系统

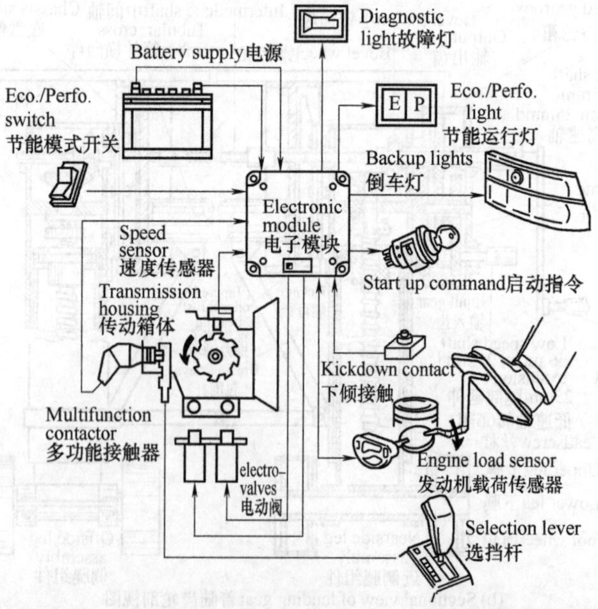

Fig 20.71 A computer controlled transmission system 计算机控制的传动系统

20.4 Functional components 功能部件

Nozzle 喷嘴
Piston 活塞
Rocker-arm shaft 摇臂轴
Injection pump 喷油泵
Connecting rod 连杆
Cam shaft 凸轮轴
Crankshaft 曲轴
Air compressor 空气压缩机
Super charger 增压器
Oil pump 机油泵
Timing gear 正时齿轮
Engine oil cooler 机油冷却器
Engine oil cleaner 机油滤清器
Concentrate colander 集滤器
Pressing limiting valve 限压阀

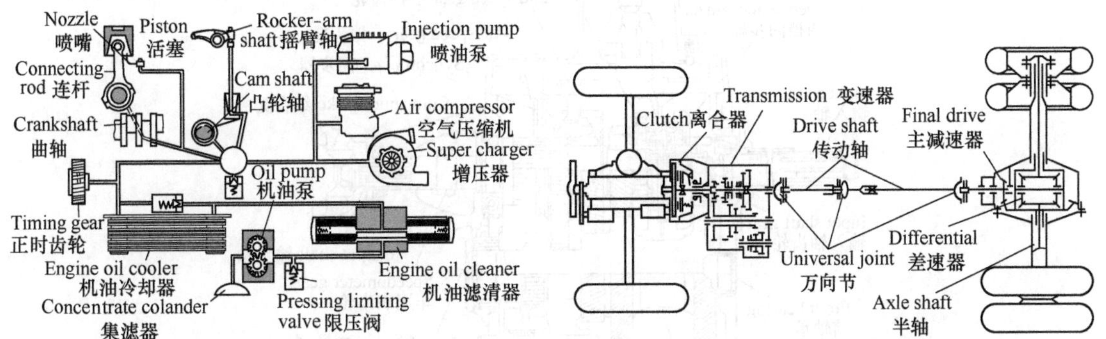

Fig 20.72 Lubrication system oil route 润滑系统油路

Transmission 变速器
Clutch 离合器
Drive shaft 传动轴
Final drive 主减速器
Differential 差速器
Universal joint 万向节
Axle shaft 半轴

Fig 20.73 Power train system assembly 传动系统的组成

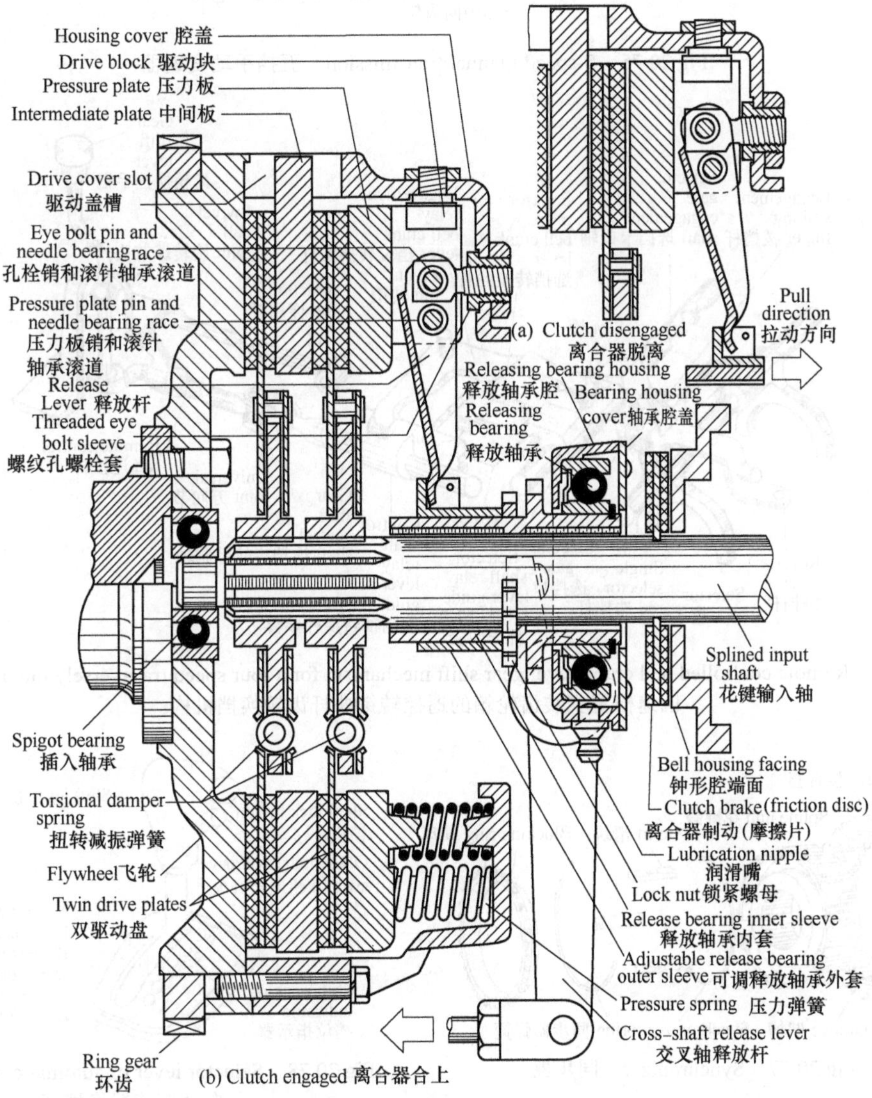

Housing cover 腔盖
Drive block 驱动块
Pressure plate 压力板
Intermediate plate 中间板
Drive cover slot 驱动盖槽
Eye bolt pin and needle bearing race 孔栓销和滚针轴承滚道
Pressure plate pin and needle bearing race 压力板销和滚针轴承滚道
Release Lever 释放杆
Threaded eye bolt sleeve 螺纹孔螺栓套
(a) Clutch disengaged 离合器脱离
Pull direction 拉动方向
Releasing bearing housing 释放轴承腔
Bearing housing cover 轴承腔盖
Releasing bearing 释放轴承
Splined input shaft 花键输入轴
Spigot bearing 插入轴承
Bell housing facing 钟形腔端面
Clutch brake (friction disc) 离合器制动(摩擦片)
Torsional damper spring 扭转减振弹簧
Lubrication nipple 润滑嘴
Flywheel 飞轮
Lock nut 锁紧螺母
Twin drive plates 双驱动盘
Release bearing inner sleeve 释放轴承内套
Adjustable release bearing outer sleeve 可调释放轴承外套
Pressure spring 压力弹簧
Cross-shaft release lever 交叉轴释放杆
Ring gear 环齿
(b) Clutch engaged 离合器合上

Fig 20.74 Twin drive plates pull type clutch 双驱动盘拉引型离合器

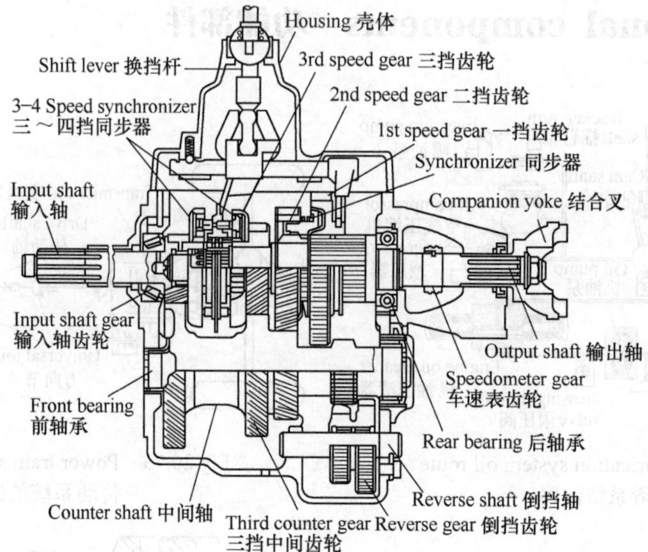

Fig 20.75　5-speed manual transmission　五挡手动变速器

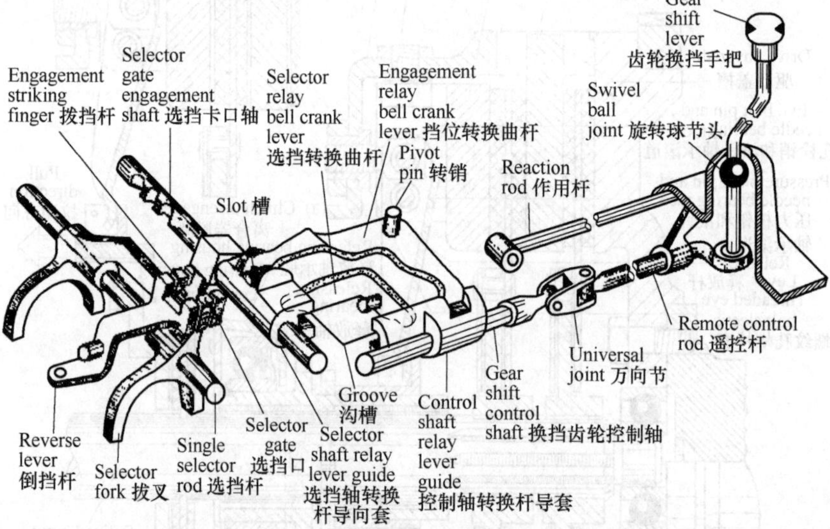

Fig 20.76　Remote controlled bell crank level gear shift mechanism for a four speed transversely mounted gearbox
四速横置安装齿轮箱的遥控转销曲杆齿轮换挡机构

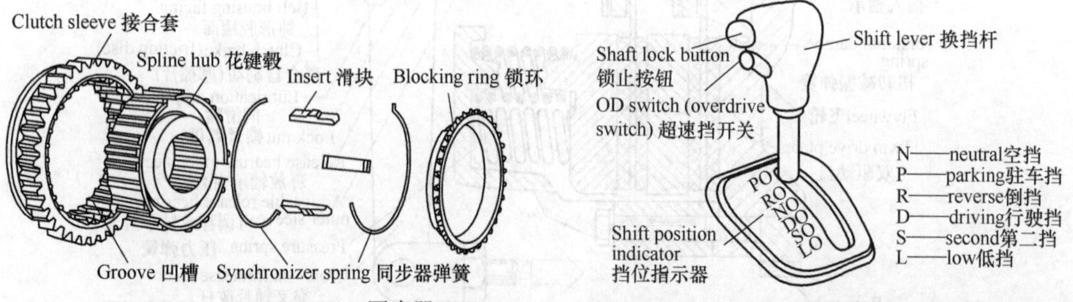

Fig 20.77　Synchronizer　同步器

Fig 20.78　Selector lever of automatic transmission
自动变速器换挡杆

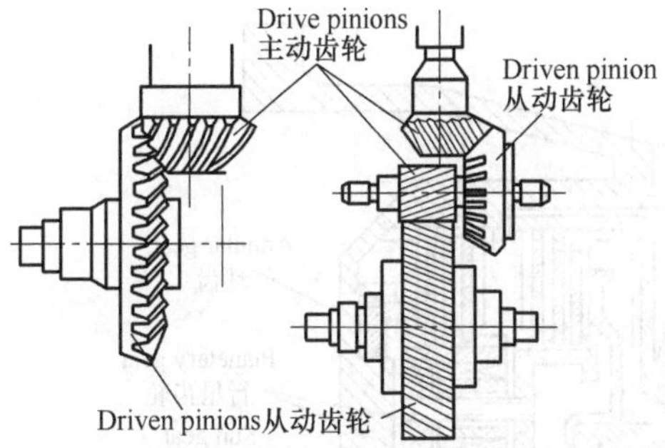

Fig 20.79 Final Drive 主减速器

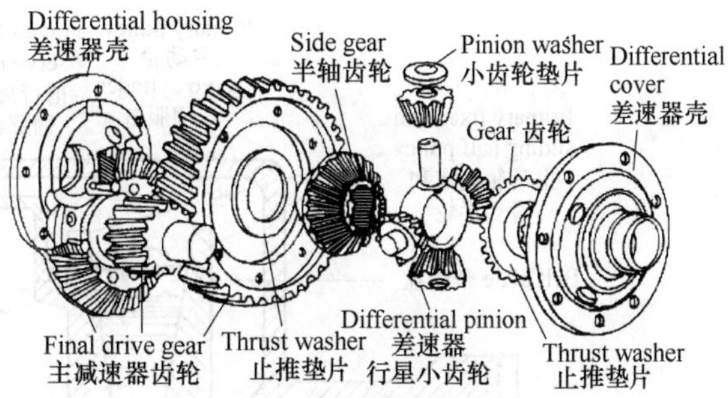

Fig 20.80 Differential 差速器

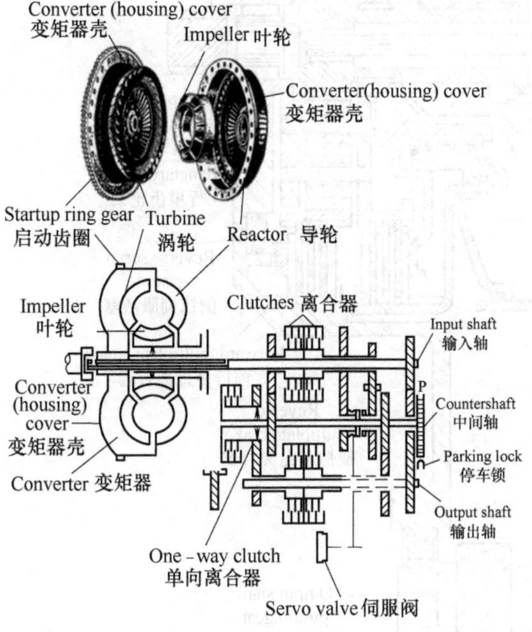

Fig 20.81 Automatic transmission (AT) 自动变速器

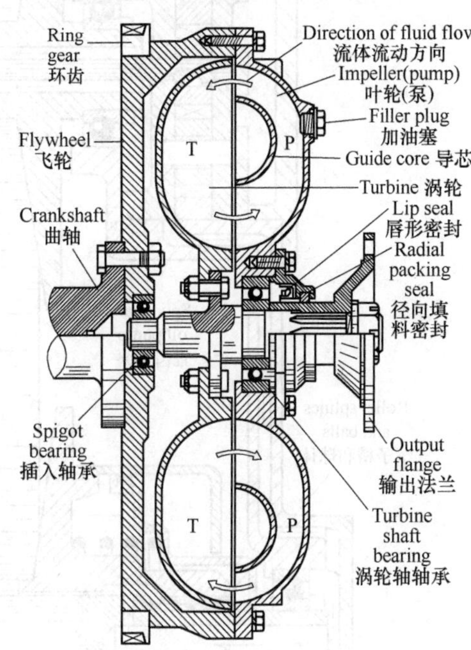

Fig 20.82 Fluid coupling 液力偶合器

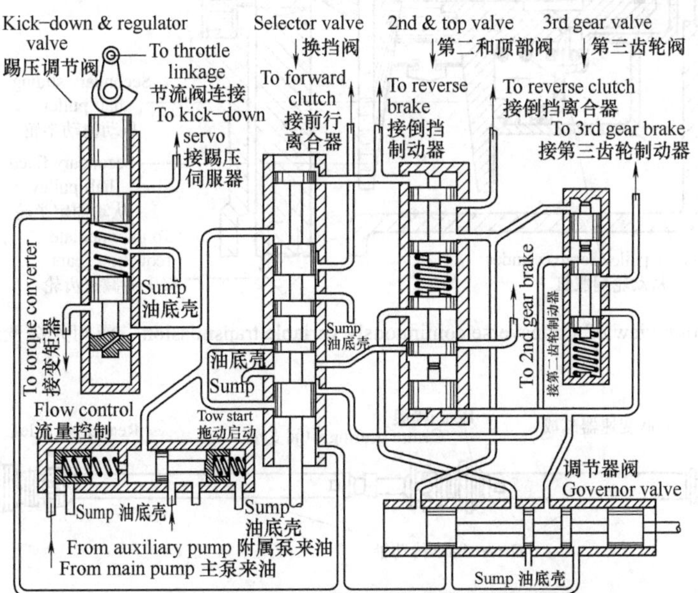

Fig 20.83 AP automatic gearbox control system AP 自动变速箱的控制系统

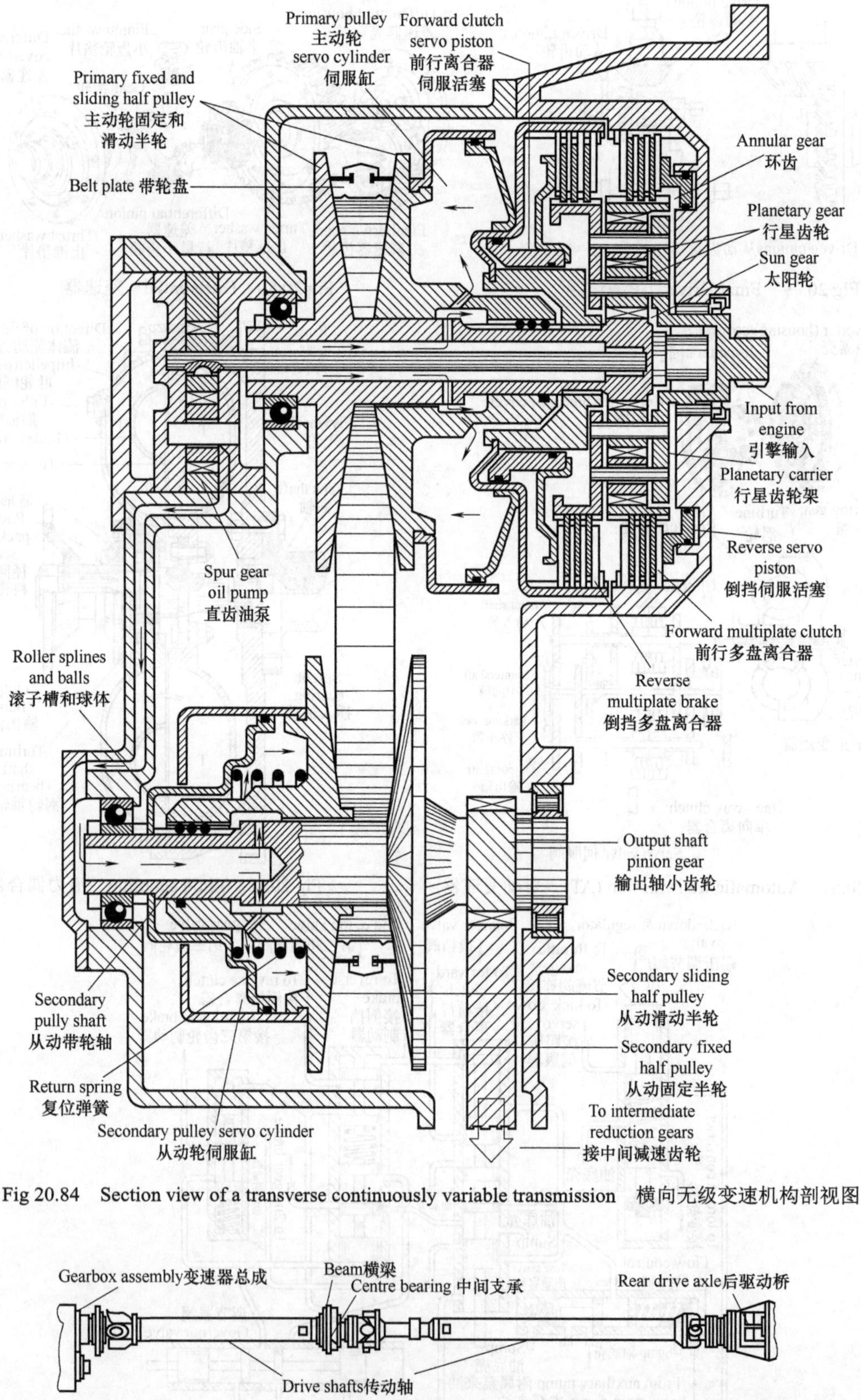

Primary pulley
主动轮
servo cylinder
伺服缸
Forward clutch
servo piston
前行离合器
伺服活塞
Primary fixed and
sliding half pulley
主动轮固定和
滑动半轮
Belt plate 带轮盘
Annular gear
环齿
Planetary gear
行星齿轮
Sun gear
太阳轮
Input from
engine
引擎输入
Planetary carrier
行星齿轮架
Reverse servo
piston
倒挡伺服活塞
Forward multiplate clutch
前行多盘离合器
Reverse
multiplate brake
倒挡多盘离合器
Spur gear
oil pump
直齿油泵
Roller splines
and balls
滚子槽和球体
Output shaft
pinion gear
输出轴小齿轮
Secondary sliding
half pulley
从动滑动半轮
Secondary fixed
half pulley
从动固定半轮
To intermediate
reduction gears
接中间减速齿轮
Secondary
pully shaft
从动带轮轴
Return spring
复位弹簧
Secondary pulley servo cylinder
从动轮伺服缸

Fig 20.84 Section view of a transverse continuously variable transmission 横向无级变速机构剖视图

Gearbox assembly变速器总成
Beam横梁
Centre bearing 中间支承
Rear drive axle后驱动桥
Drive shafts传动轴

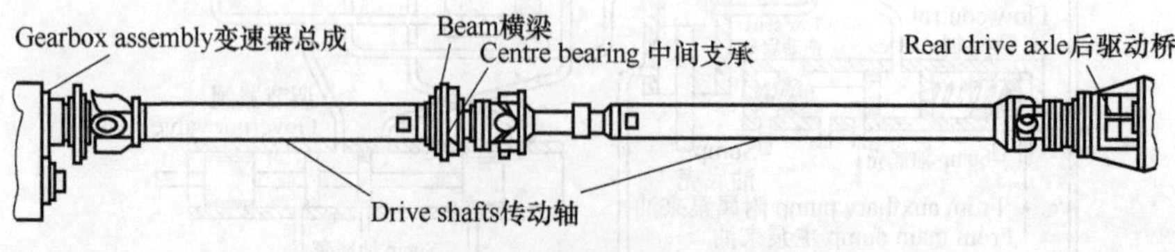

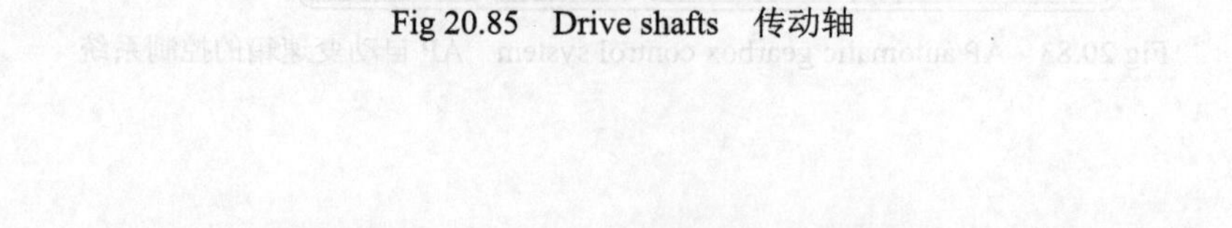

Fig 20.85 Drive shafts 传动轴

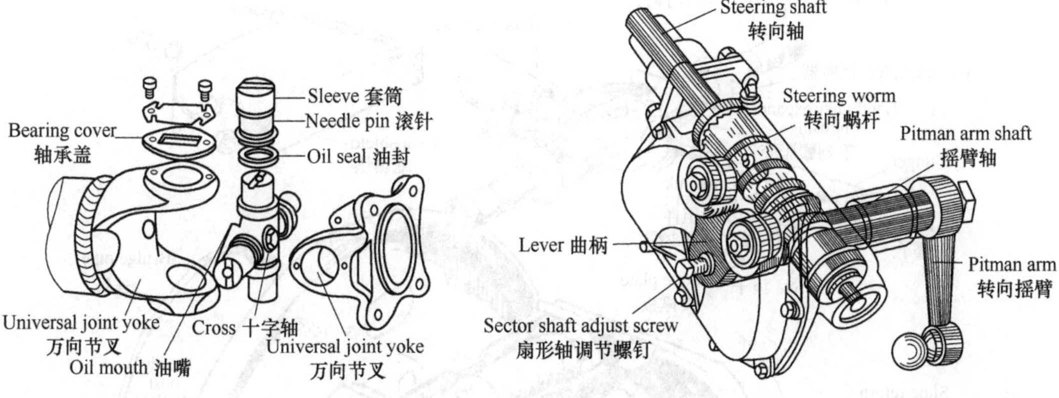

Fig 20.86 Cardan universal joint 十字轴式万向节　Fig 20.87 Cam and lever steering gear 凸轮杠杆转向装置

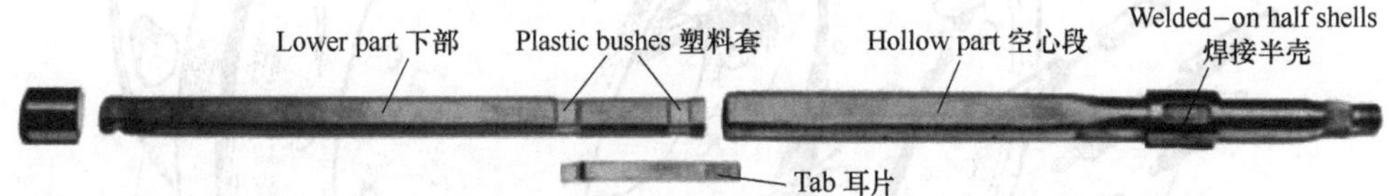

Fig 20.88 Telescopic collapsible steering tubes 伸缩式可拆卸转向管

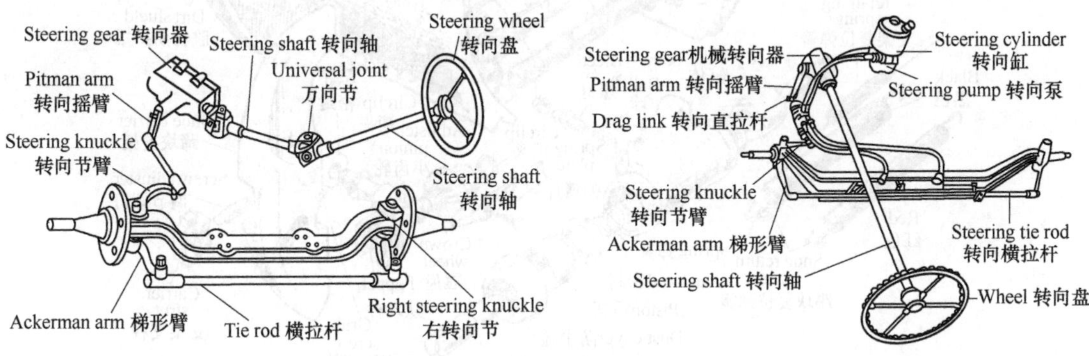

Fig 20.89 Steering system 转向系统　Fig 20.90 Power steering system 动力转向系统

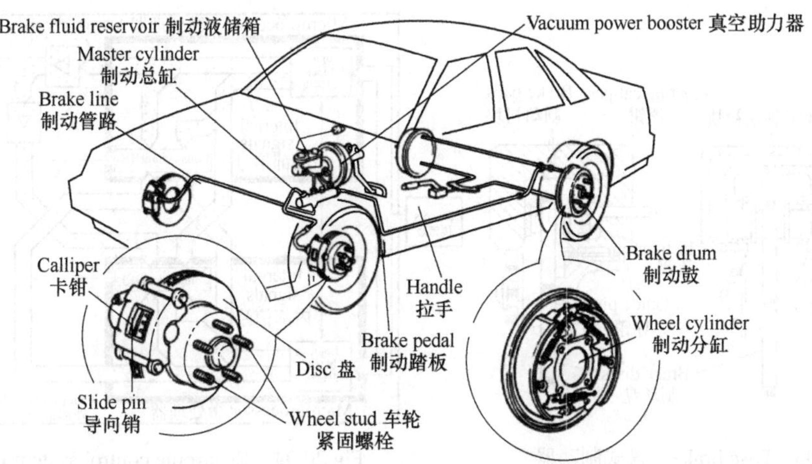

Fig 20.91 Brake system components 制动系统零件

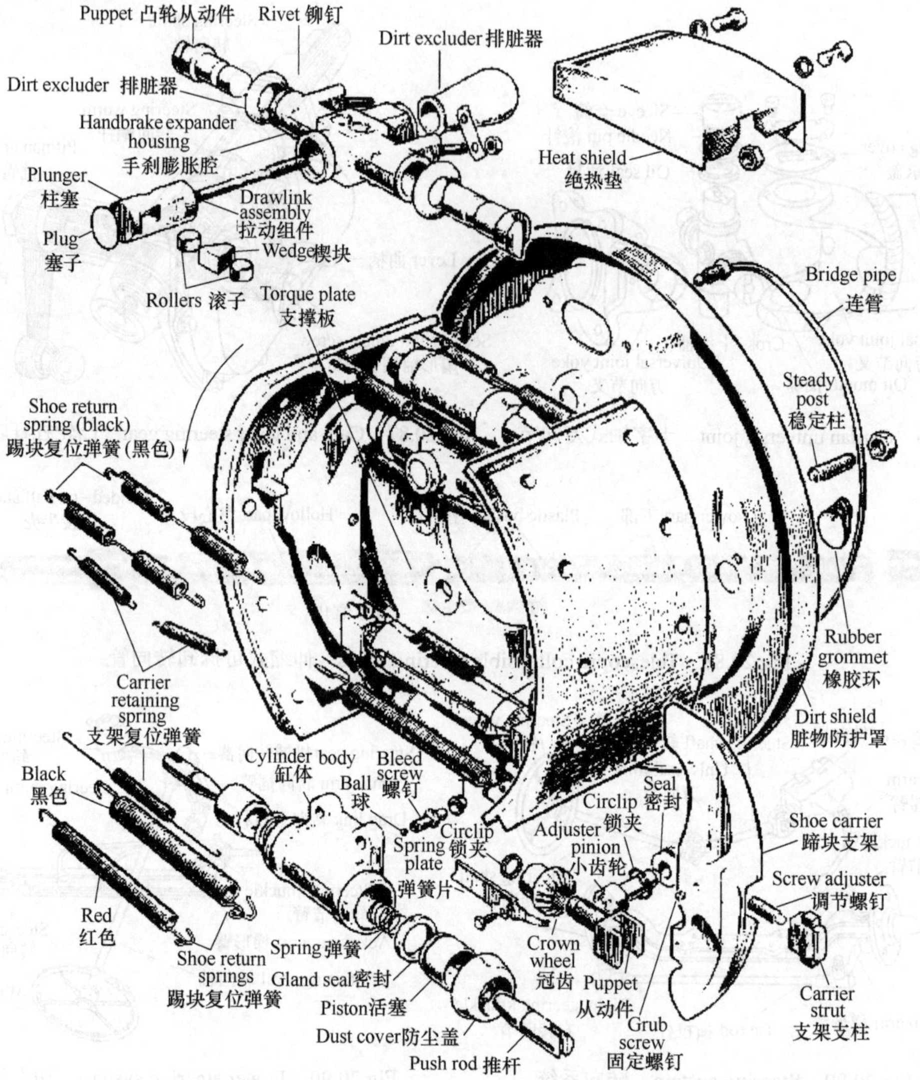

Puppet 凸轮从动件
Rivet 铆钉
Dirt excluder 排脏器
Dirt excluder 排脏器
Handbrake expander housing
手刹膨胀腔
Plunger
柱塞
Drawlink assembly
拉动组件
Plug
塞子
Wedge楔块
Rollers 滚子
Torque plate
支撑板
Heat shield
绝热垫
Bridge pipe
连管
Steady post
稳定柱
Shoe return spring (black)
踢块复位弹簧(黑色)
Rubber grommet
橡胶环
Dirt shield
脏物防护罩
Carrier retaining spring
支架复位弹簧
Black
黑色
Cylinder body
缸体
Bleed screw
螺钉
Ball
球
Seal
密封
Circlip
锁夹
Adjuster pinion
小齿轮
Circlip
锁夹
Shoe carrier
蹄块支架
Spring plate
弹簧片
Screw adjuster
调节螺钉
Red
红色
Shoe return springs
踢块复位弹簧
Spring 弹簧
Gland seal密封
Piston活塞
Dust cover防尘盖
Push rod 推杆
Crown wheel
冠齿
Puppet
从动件
Grub screw
固定螺钉
Carrier strut
支架支柱

Fig 20.92 Heavy-duty drum brake 重载鼓式制动器

Brake pad 制动衬块
Floating caliper
浮钳
Brake pads
制动衬块
Hydraulic pressure
液压
Pistons
活塞
Fixed caliper
固定钳
Guide pin
导销
Piston 活塞
Brake discs
制动盘

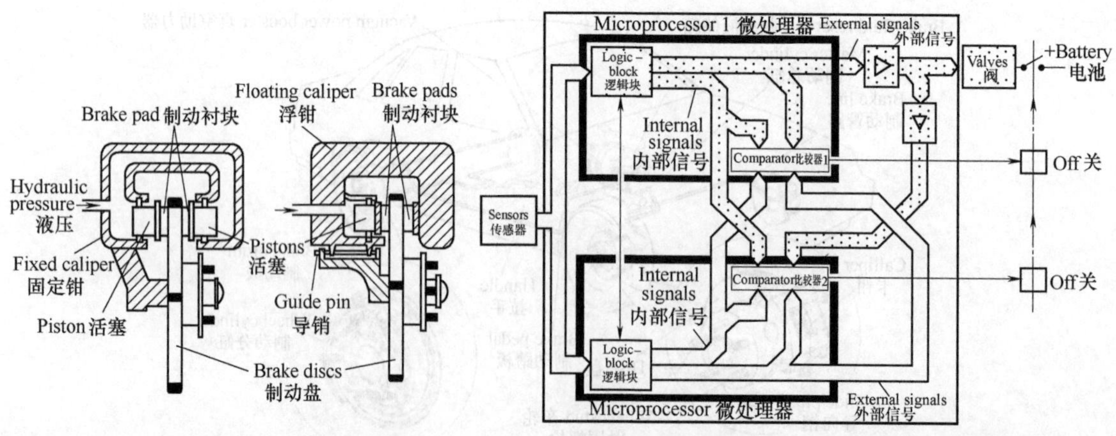

Fig 20.93 Disc brake 盘式制动器

Microprocessor 1 微处理器
External signals
外部信号
Valves
阀
+Battery
电池
Logic-block
逻辑块
Internal signals
内部信号
Comparator比较器1
Off关
Sensors
传感器
Internal signals
内部信号
Comparator比较器2
Off关
Logic-block
逻辑块
External signals
外部信号
Microprocessor 微处理器

Fig 20.94 Electronic control system for ABS
防抱死系统的电控系统

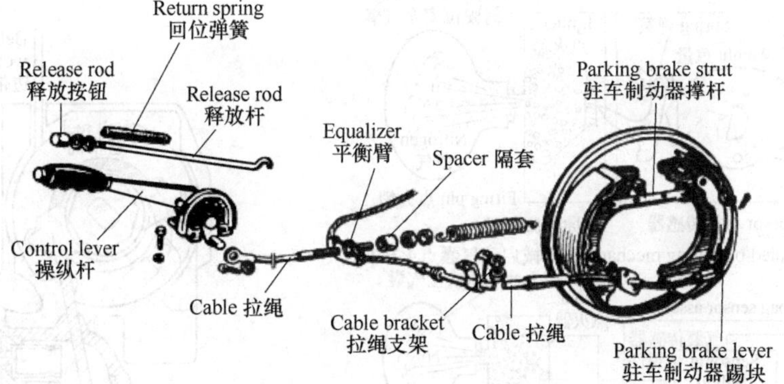

(a) Hand-operated parking brake 手动驻车制动器

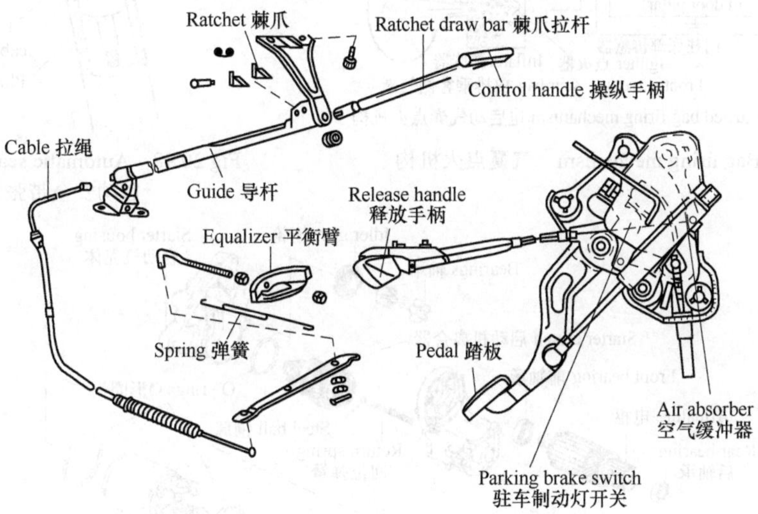

(b) Foot-operated parking brake 脚动驻车制动器

Fig 20.95 **Parking brakes** 驻车制动器

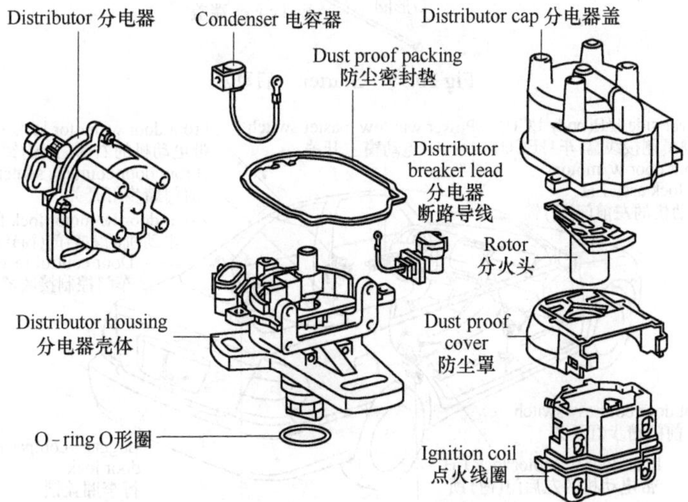

Fig 20.96 **Distributor** 分电器

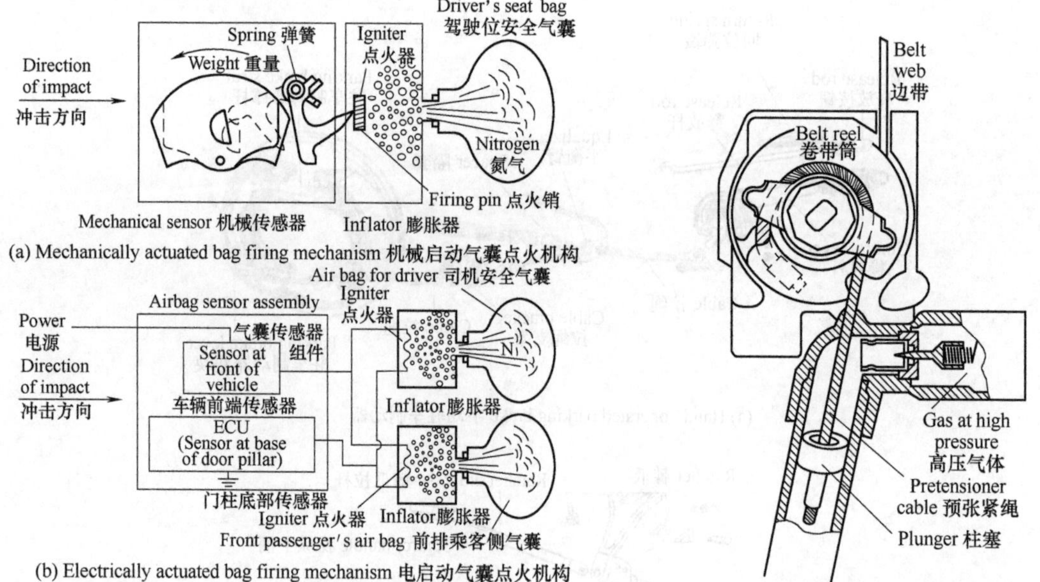

Fig 20.97 Bag firing mechanism 气囊点火机构

Fig 20.98 Automatic seat belt tensioner 自动安全带张紧器

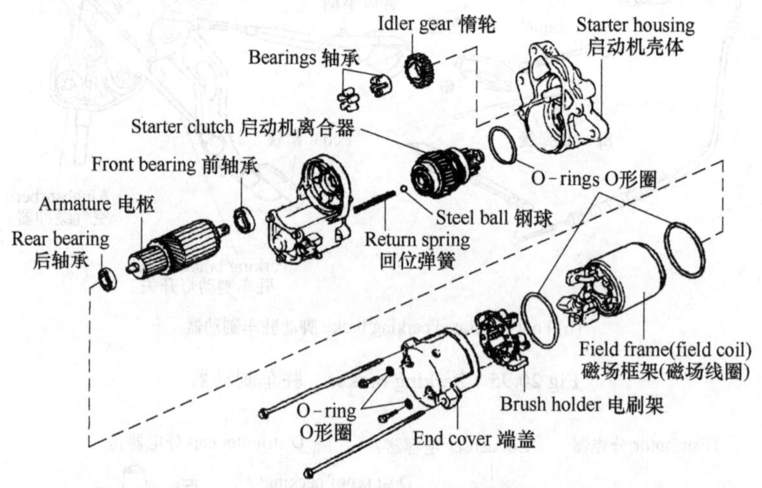

Fig 20.99 Starter 启动机

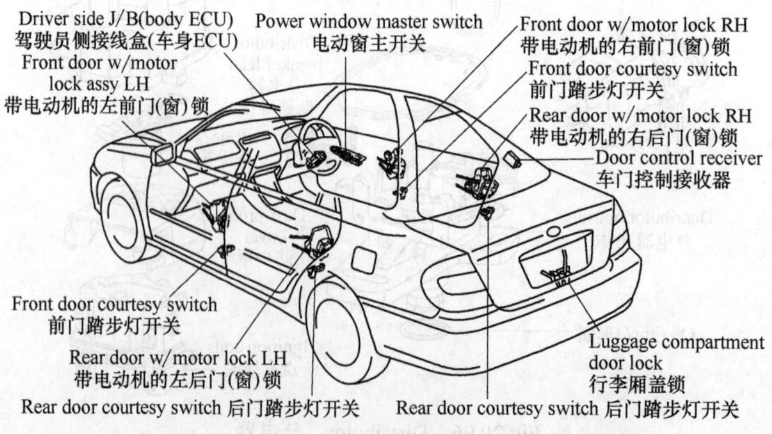

Fig 20.100 Door lock control system 门锁控制系统

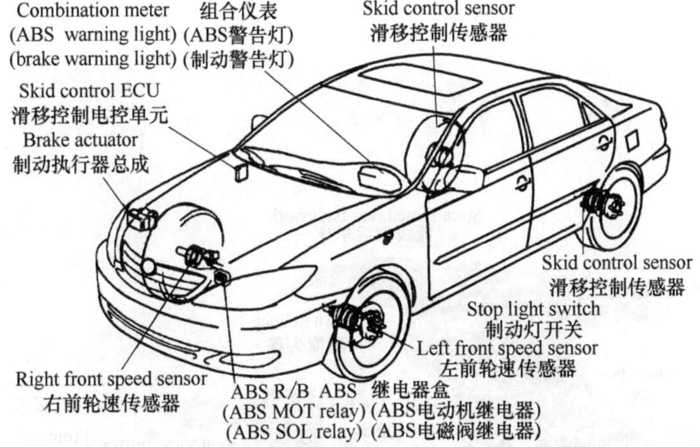

Fig 20.101 ABS & EBD System 防抱死制动系统和电子制动力分配系统

20.5 Special vehicles 特种车辆

Fig 20.102 Amphibian 水陆两用车

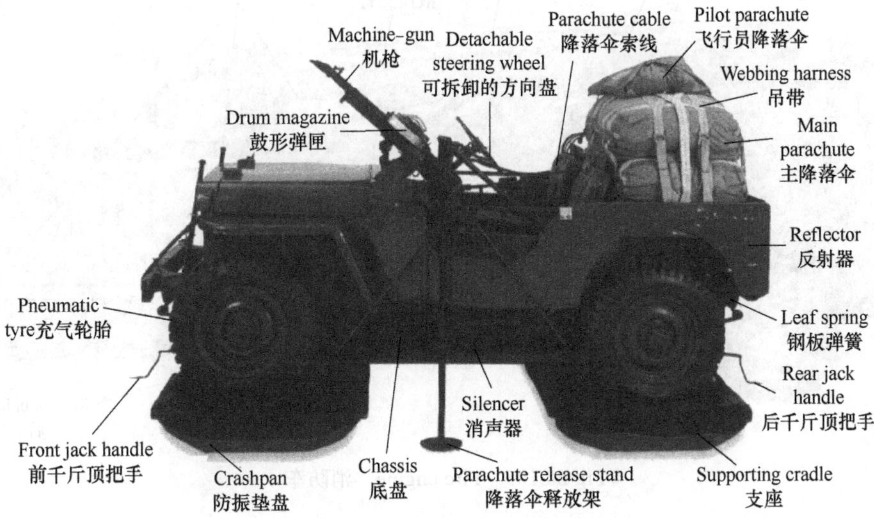

Fig 20.103 Jeep 吉普车

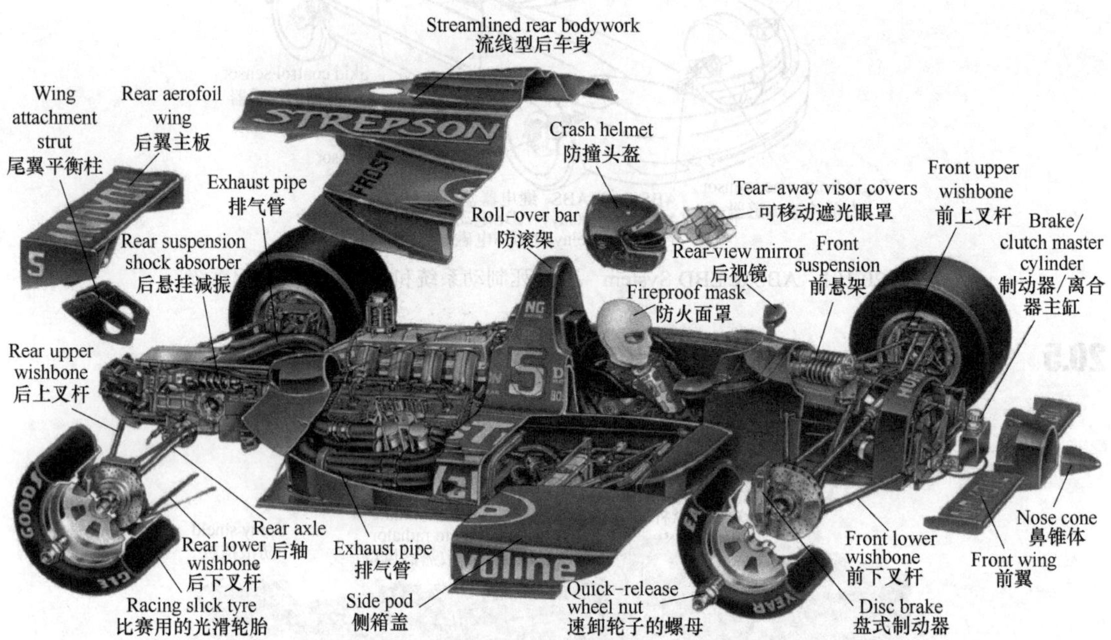

Fig 20.104 Exploded view of a turbocharged, V-8 engine Indycar 涡轮增压 V-8 发动机英迪赛车分解图

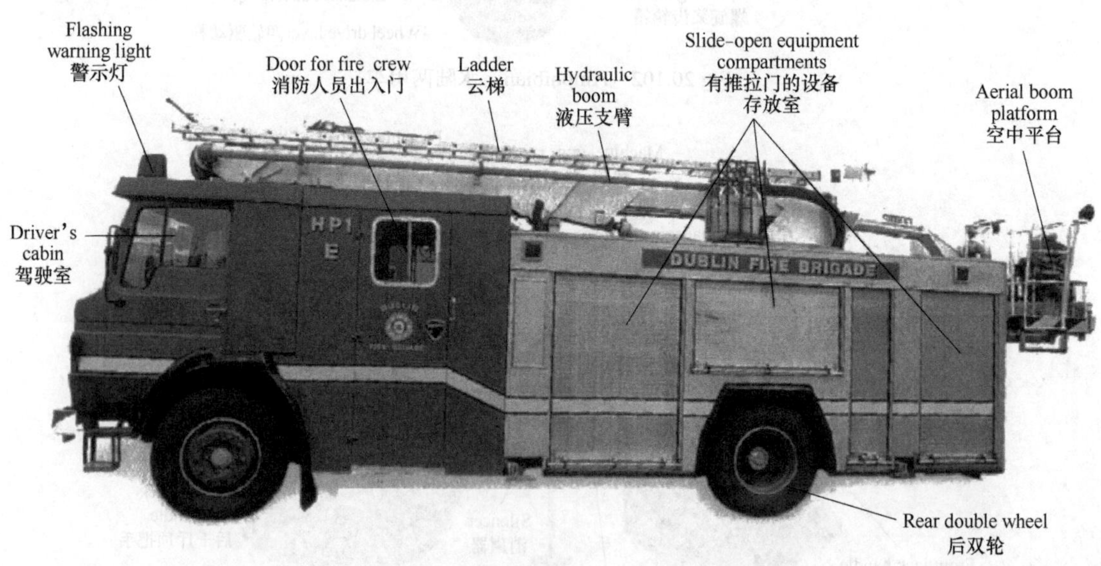

Fig 20.105 Fire engine 消防车

20.6 Electric vehicles 电动车

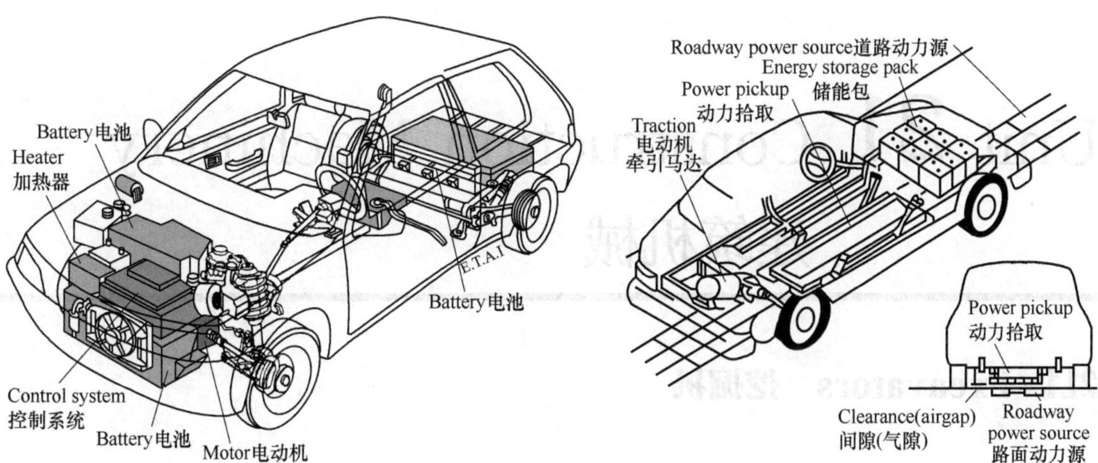

Fig 20.106 Electric car 电动汽车

Fig 20.107 Road-induced electricity 路面感应电动车

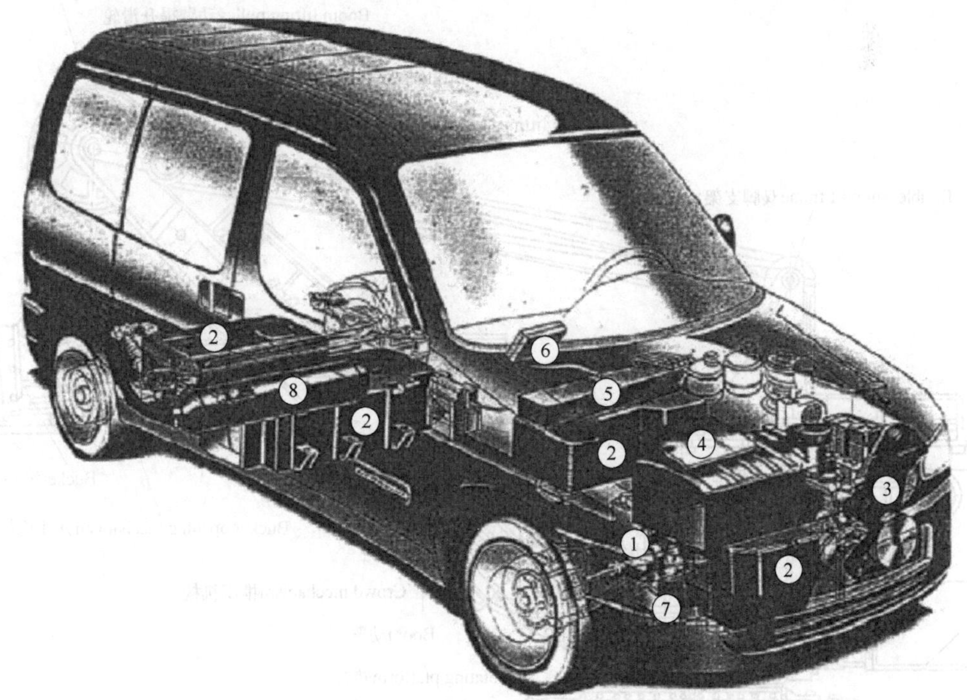

Fig 20.108 Citroen berlingo dynavolt 雪铁龙 Berlingo 电动车

1—Electric motor and drive 电动机和驱动装置；2—Traction battery pack 牵引电池组；3—Generator set 发电机装置；4—Motor controller 电动机控制器；5—Generator controller 发电机控制器；6—Drive programme selector 驱动程序选择面板；7—LPG regulator LPG 调节器；8—LPG storage tank LPG 储存箱

Unit 21 Construction Machinery 建筑机械

21.1 Excavators 挖掘机

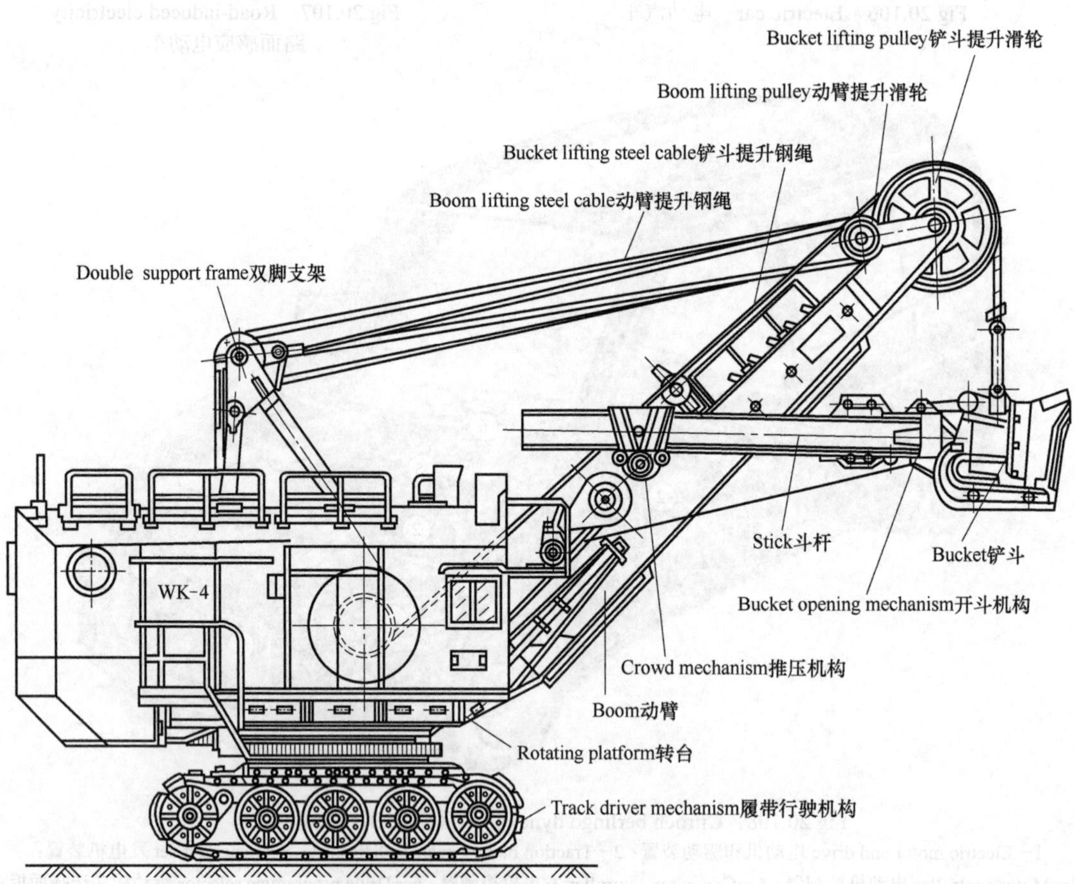

Fig 21.1 Face excavator 正铲挖掘机

Fig 21.2 Tunnel boring machine 隧道盾构机

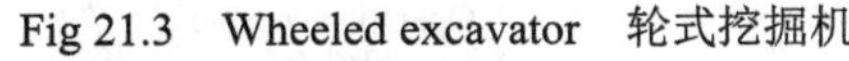
Fig 21.3 Wheeled excavator 轮式挖掘机

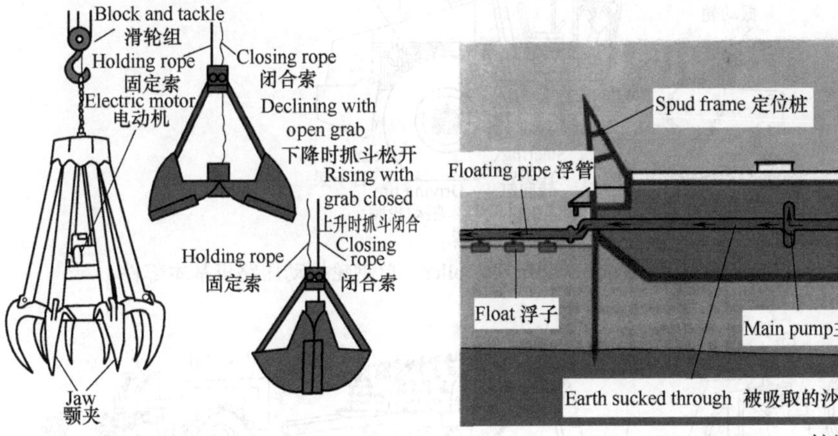

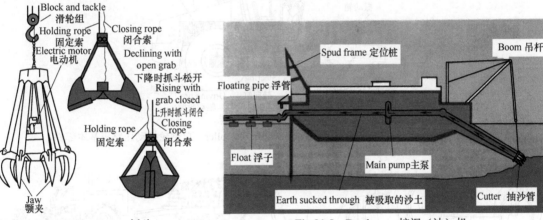

Fig 21.4 Grab 抓斗

Fig 21.5 Dredger 挖泥（沙）机

21.2 Earthmoving machinery 铲土运输机

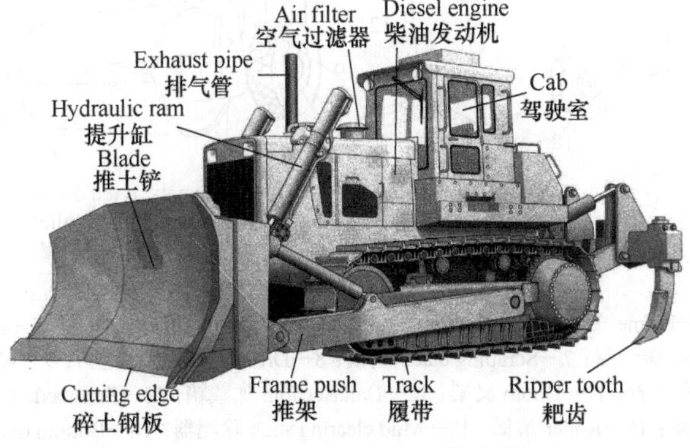

Fig 21.6 Bulldozer 推土机

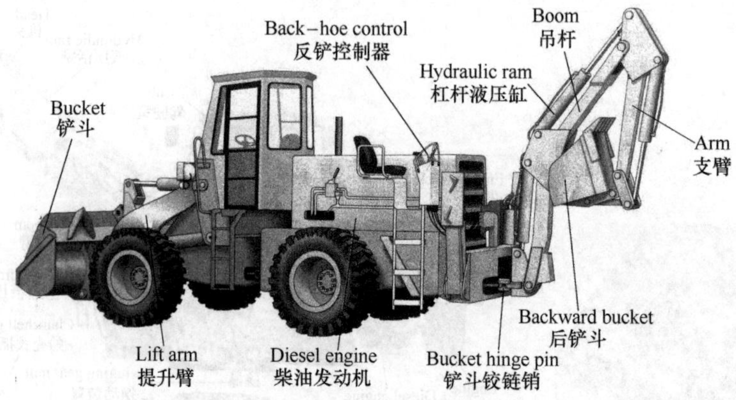

Fig 21.7 Loader 装载机

21.3 Compacting machinery 压实机械

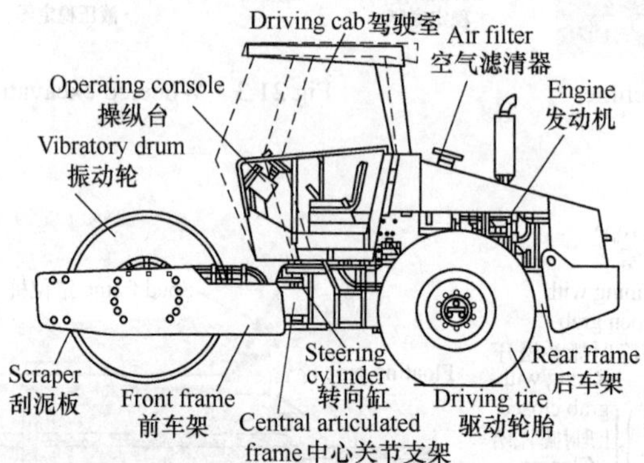

Fig 21.8 Basic components of single drum vibratory roller 单钢轮振动压路机基本部件

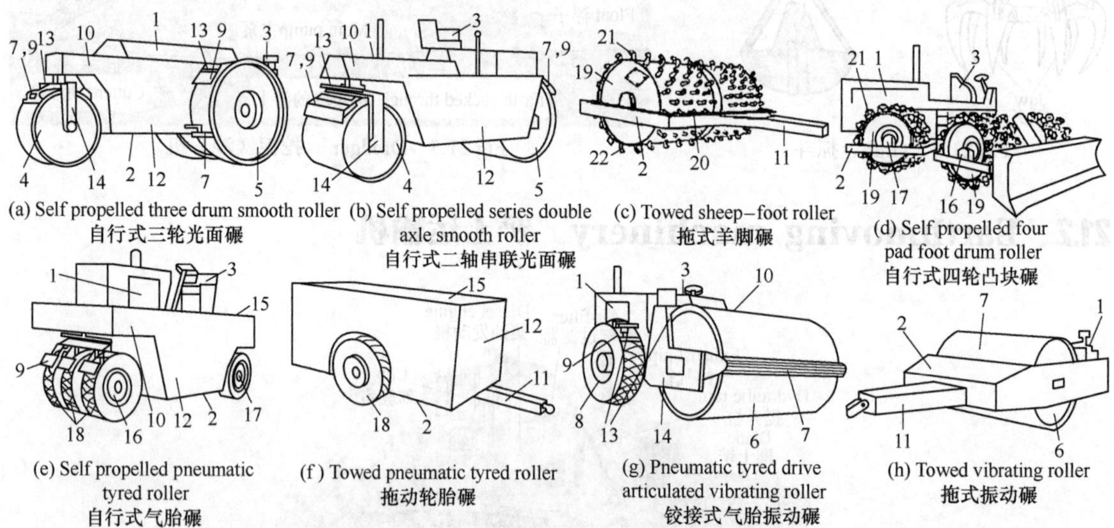

Fig 21.9 Figures of rollers 滚压机外形图

1—Engine 发动机；2—Frame 车架；3—Driver seat 驾驶台；4—Steering roller 转向碾；5—Driving roller 驱动碾；6—Vibrating roller 振动碾；7—Scraping blade 刮板；8—Driving tyre 驱动轮胎；9—Water tank 水箱；10—Sprinkling equipment 喷水装置；11—Boom 叉架；12—Ballast cabin 加载箱；13—Front axle 前轴；14—Behind axle 后轴；15—Pneumatic tyre 气胎；16—Roller 滚筒；17—Mud clearing stick 除泥器；18—Bottom of sheep-foot 羊脚根部；19—Top of sheep-foot 羊脚尖；20—Blade 推土板；21—Edge of blade 刀刃

21.4 Pavement machinery 铺路机械

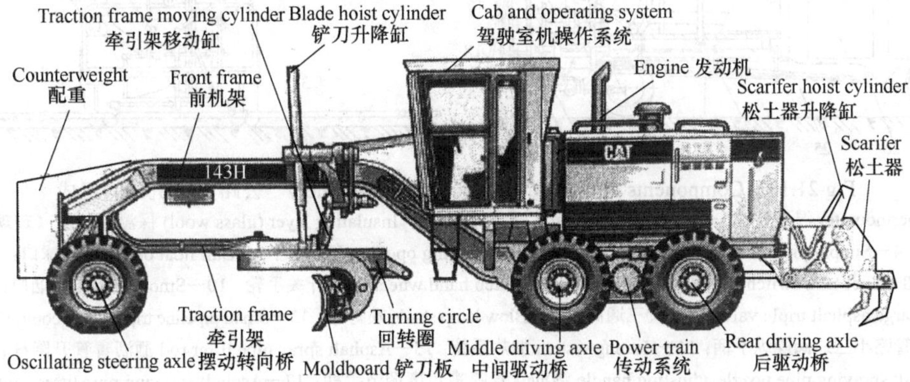

Fig 21.10 Overall structure of grader 平地机的总体构造

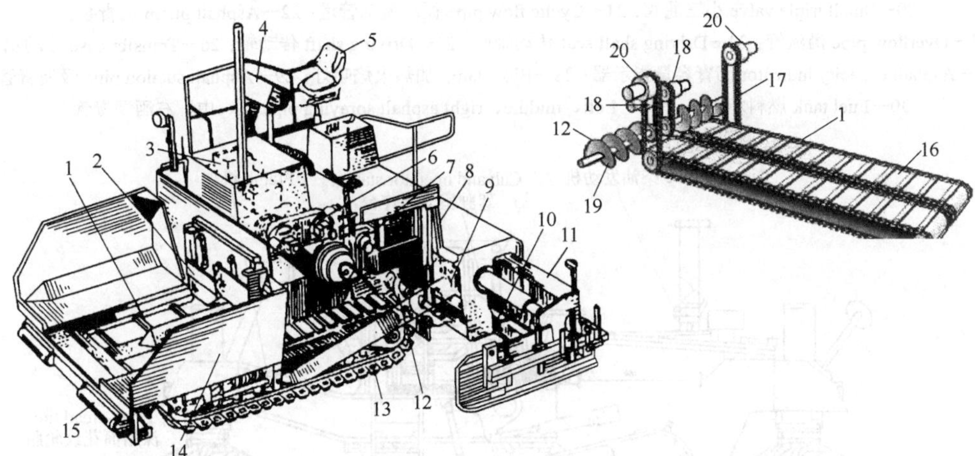

Fig 21.11 Components of asphalt paver 沥青摊铺机的组成

1—Drag conveyer 刮板输送器；2—Gate 闸门；3—Engine 发动机；4—Operator's station 操纵台；5—Operator's seat 驾驶座椅；6—Transmission system 传动系统；7—Frame 机架；8—Lifting cylinder 提升缸；9—Towed arm 牵引臂；10—Tamper 振捣器；11—Screed units 熨平装置；12—Auger conveyer 螺旋分料器；13—Running gear 行走机构；14—Hopper 料斗；15—Push roller 推辊；16—Drag plate 刮板；17—Driving chain 传动链条；18—Drag drive motor 刮板驱动电动机；19—Auger blade 螺旋叶片；20—Auger drive motor 螺旋驱动电动机

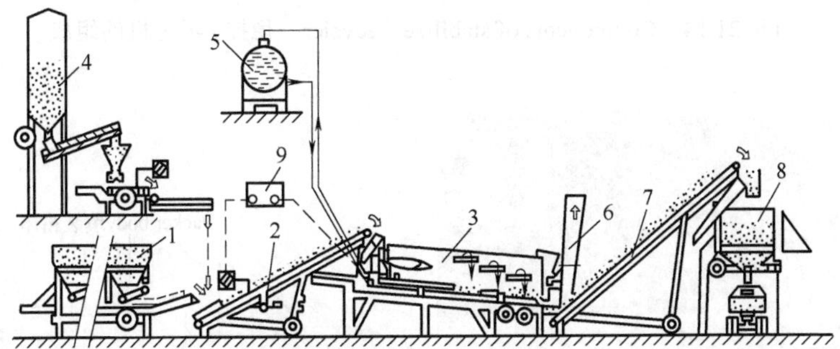

Fig 21.12 Basic components of continuous drum mixing plant 连续滚筒式搅拌设备基本部件

1—Cold aggregate bins and feeder 冷骨料储存和给料器；2—Cold aggregate belt conveyer and weighing system 冷骨料带式输送机及称量系统；3—Dryer-mixer drum 干燥搅拌筒；4—Mineral filler storage silo and supply system 粉料储仓及供给装置；5—Asphalt supply system 沥青供给系统；6—Dust collector 除尘装置；7—Mixed material conveyer 成品料输送机；8—Mixed material storage silo 成品料储仓；9—Control system 控制系统

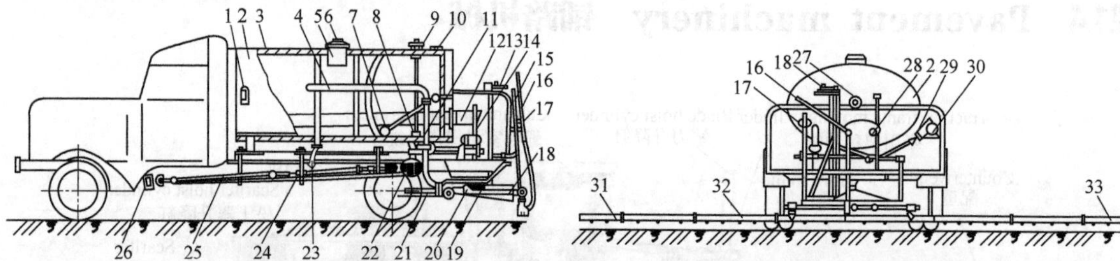

Fig 21.13 Components of highway maintenance machinery 公路养护机械的组成

1—Thermometer 温度计；2—Asphalt storage tank 沥青储箱；3—Insulating layer (glass wool) 保温隔热层（玻璃棉）；4—Supply pipe 进料管；5—Strainer 滤网；6—Charging opening 装料口；7—Ball float opening 浮球口；8—Fuel tank switch 油箱开关；9—Fuel tank switch hand wheel 油箱开关手轮；10—Smoke outlet 排烟口；11—Large asphalt triple valve 沥青大三通阀；12—Blow lamp shade 喷灯罩；13—Small pipeline triple valve control handle 管路小三通阀操纵手柄；14—Heating system 加热系统；15—Asphalt spray pipe lifter rod 洒沥青管升降杆；16—Asphalt spraying pipe nozzle adjusting handle 洒沥青管喷嘴角度调节手柄；17—Asphalt spraying pipe lifter control handle 洒沥青管升降操纵手柄；18—Spherical connecting pipe 球状连接管；19—Asphalt discharging pipe 沥青排放管；20—Small triple valve 小三通阀；21—Cyclic flow pipe 循环流动管道；22—Asphalt pump 沥青泵；23—Overflow pipe 溢流管；24—Driving shaft seat 传动轴座；25—Driving shaft 传动轴；26—Transfer case 分动箱；27—Asphalt capacity indicator 沥青容量指示器；28—Blow lamp 加热火焰喷灯；29—Asphalt suction pipe 吸沥青管；30—Fuel tank 燃料箱；31 ～ 33—Left、middle、right asphalt spraying pipe 左、中、右洒沥青管

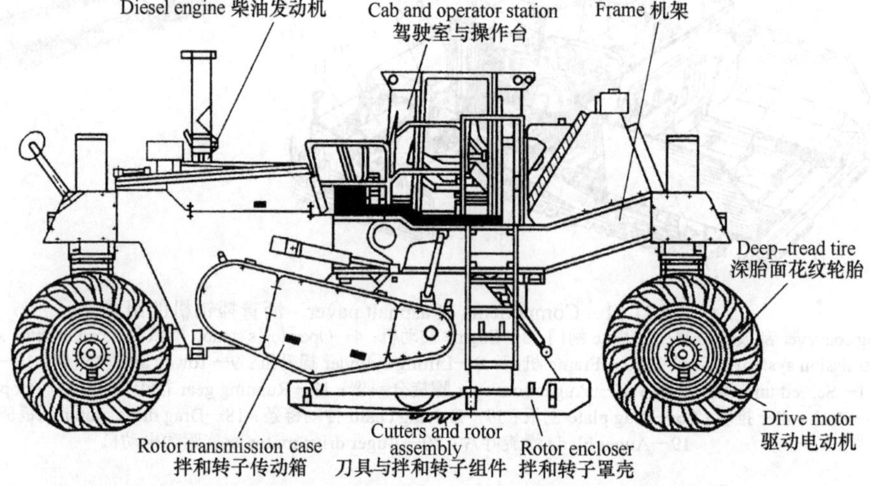

Fig 21.14 Components of stabilizer/recycler 稳拌 / 再生机的组成

Fig 21.15 Road sweeper 扫路机

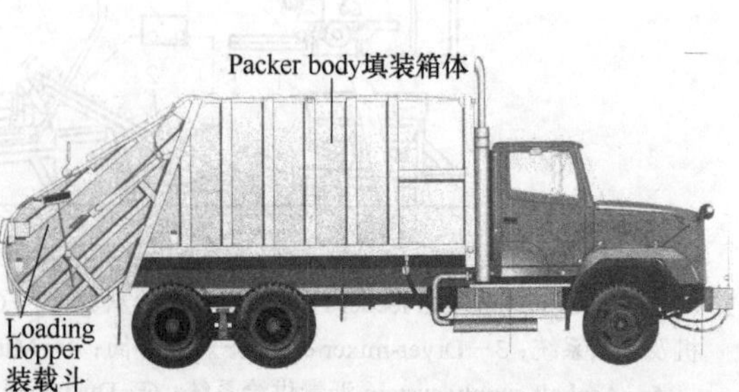

Fig 21.16 Refuse lorry/dustcart 垃圾车

Fig 21.17 Snowblower 吹雪机

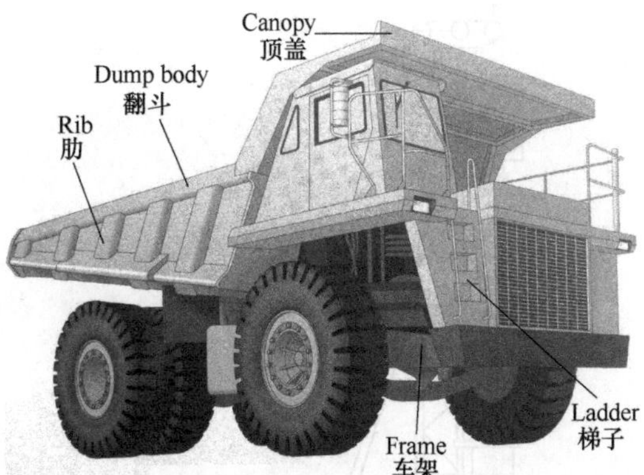

Fig 21.18 Dump truck 自动倾卸车

Fig 21.19 Breakdown lorry/recovery lorry 拖车

21.5 Pile-driving machinery 桩工机械

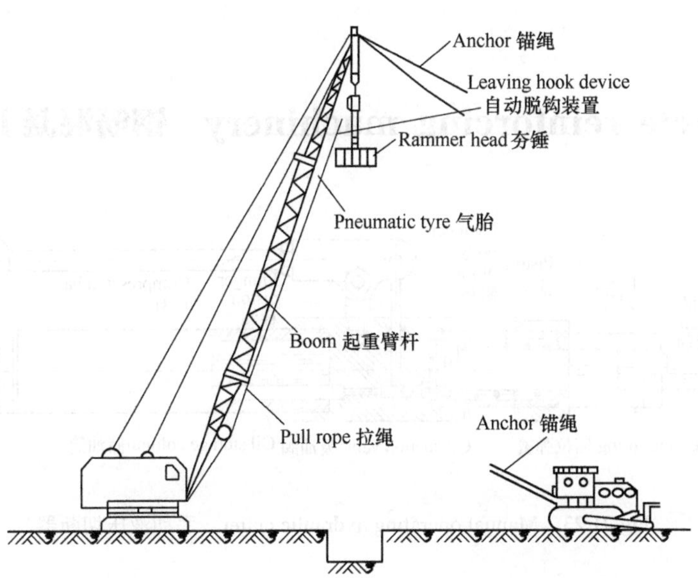

Fig 21.20 Track crane rammer 轨道起重式强夯机

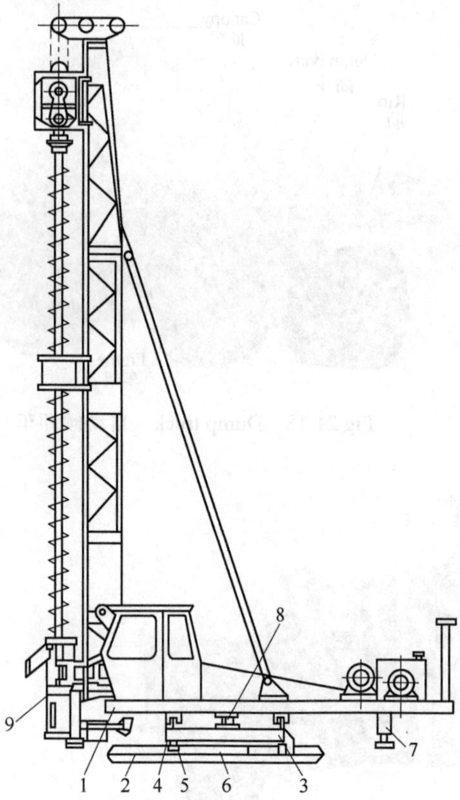

Fig 21.21 Walking drilling machine 步履式钻孔机

1—Upside platform 上盘；2—Lower platform 下盘；
3—Middle platform 中盘；4—Turning truckle 转向滚轮；
5—Walking truckle 行走滚轮；6—Steel cable pulley 钢绳滑轮；
7—Outrigger 支腿；8—Center axle 回转中心轴；
9—Soil discharge hole 出土筒

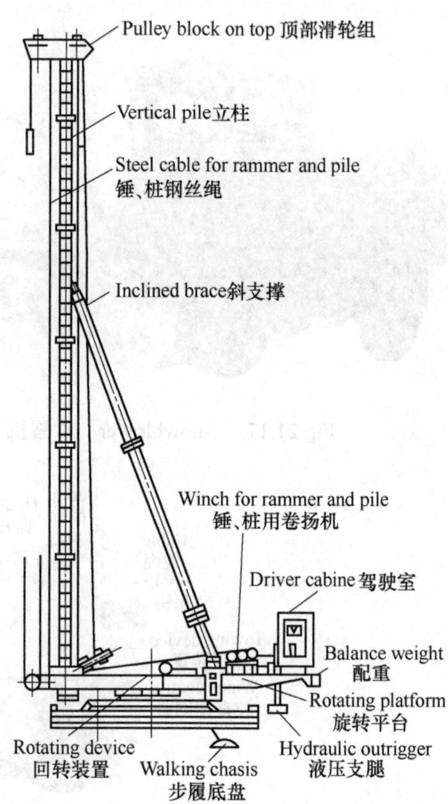

Fig 21.22 Walking pile frame 步履式打桩架

21.6 Concrete reinforcing machinery 钢筋混凝土机械

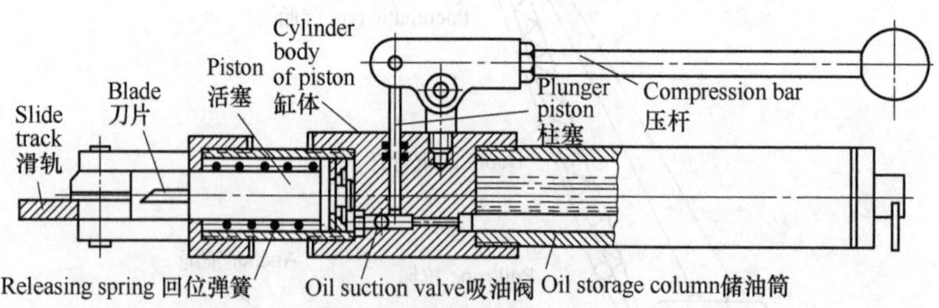

Fig 21.23 Manual operating hydraulic cutter 手动液压切断器

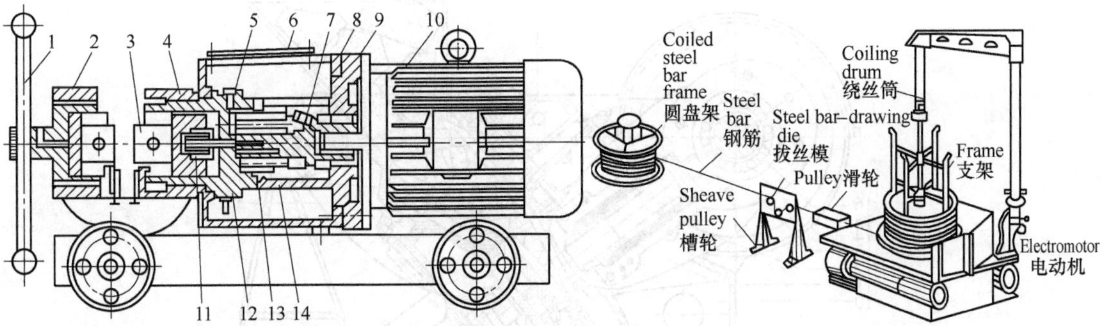

Fig 21.24 Electric hydraulic steel bar cutter 电动液压钢筋切断机

1—Stick 手柄；2—Frame 支架；3—Blade 刀片；4—Piston 活塞；5—Oil drain valve 放油阀；6—Glass window 观察玻璃；7—Eccentric axle 偏心轴；8—Oil tank 油箱；9—Link frame 连接架；10—Electromotor 电动机；11—Rubber cup 皮碗；12—Cylinder body 油缸体；13—Oil cylinder 油缸；14—Plunger piston 柱塞

Fig 21.25 Vertical single drum steel bar stretcher 立式单筒拔丝机

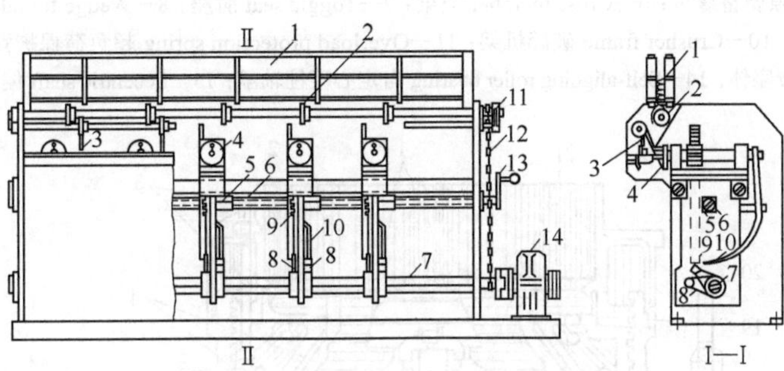

Fig 21.26 Steel bar bender 钢筋弯曲机

1—Steel bar storage vessel 钢筋储槽；2—Steel bar feeding device 拨料轮；3—Push pole for feed 进料推杆；4—Bender device 工作盘；5—Nut for opening and close 锁帽；6—Lead screw 丝杠；7—Driving shaft 传动轴；8，11—Cam 凸轮；9—Gear rack 齿条；10—Connect bar 拔心轴用的连杆；12—Chain 链条；13—Manual operation rotating shaft 手摇轮；14—Reductor 减速器

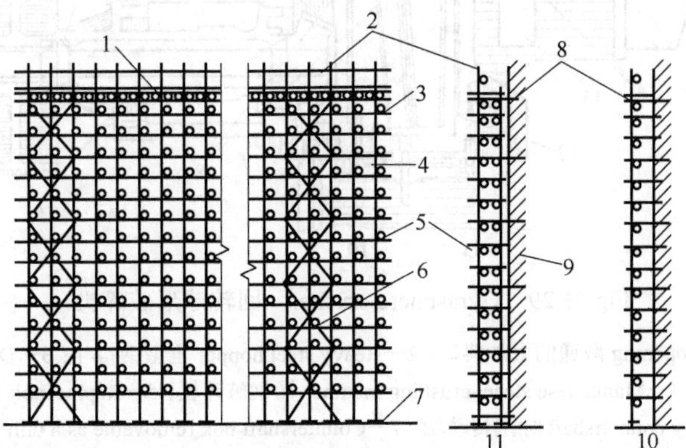

Fig 21.27 Coupler connected steel tube exterior scaffolding 扣件式钢管外脚手架

1—Ledger board 脚手板；2—Baluster 栏杆；3—Runner 大横杆；4—Bearer 小横杆；5—Ricker 立杆；6—Crossing-bracing 剪刀撑；7—Pedestal 底座；8—Tie rod anchored in wall 连墙杆；9—Wall 墙；10—Single row scaffolding 单排脚手架；11—Double row scaffolding 双排脚手架

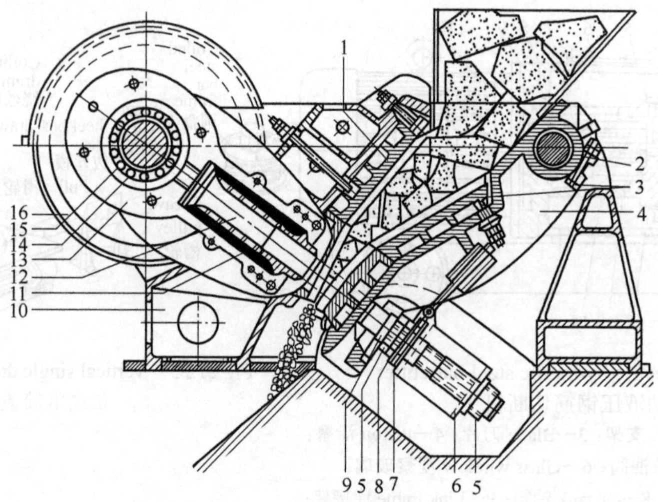

Fig 21.28 Impact jaw crusher 颚式冲击破碎机

1—Stationary jaw 固定颚板；2—Swing jaw axle 摆动颚轴；3—Swing jaw plate 摆动颚板；4—Swing jaw 摆动颚；5—Gap adjusting nut 间隙调整螺母；6—Cross member 横梁；7—Toggle seat 肘座；8—Wedge for adjusting nut 调整螺母的楔；9—Toggle plate 肘板；10—Crusher frame 破碎机架；11—Overload protection spring 超负荷保护弹簧；12—Pitman 连杆；13—Spring housing 弹簧座体；14—Self-aligning roller bearing 自定心滚柱轴承；15—Eccentric shaft 偏心轴；16—Flywheel 飞轮

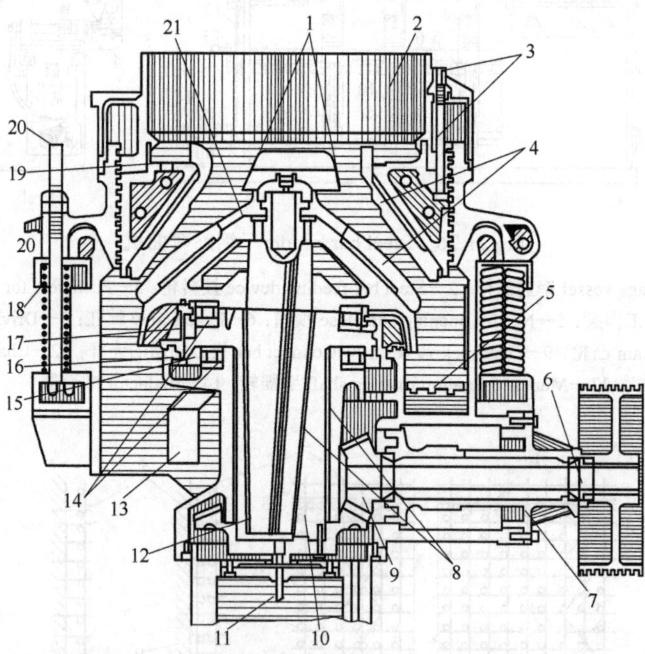

Fig 21.29 Gyrasphere crusher 回转球形破碎机

1—Large unobstructed feed opening 敞通的大加料口；2—Heavy steel hopper 重型钢料斗；3—Discharge opening adjusting mechanism 出料口调节机构；4—Manganese steel crushing member 锰钢破碎板；5—Replaceable countershaft box shield 可更换的主轴箱护盖；6—Heavy countershaft 加重传动轴；7—Countershaft box removable as a unit 可整体拆卸的传动轴箱；8—Large diameter long sleeve eccentric bearing 大直径长套筒偏心轴承；9—Cut steel drive gear 铣齿钢驱动齿轮；10—Large heavy eccentric 重型大偏心轴；11—Main oil supply 主供油口；12—Large diameter heavy duty main shaft 大直径重型主轴；13—Frame arm shield 机架臂护板；14—Heavy duty roller bearing 重型滚柱轴承；15—Cam and lever 凸轮和杠杆；16—Piston ring and labyrinth seal 活塞环和迷宫式密封；17—Rotary seal 回转密封件；18—Spring relief 弹簧保险；19—Gun lock 扣紧枪机；20—Supports for adjusting mechanism 调节机构立柱；21—Spherical shaped crushing head 球形破碎头

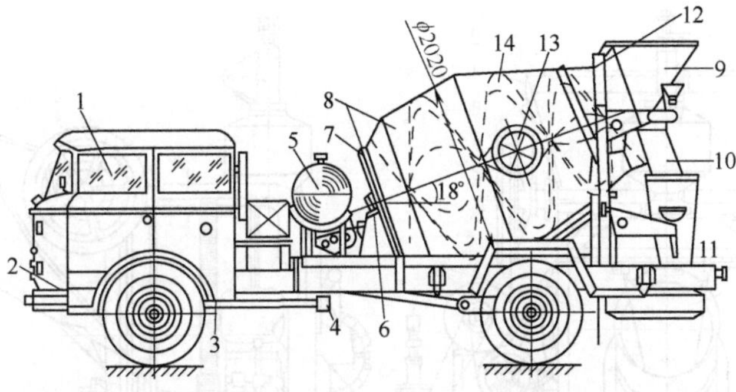

Fig 21.30 Concrete truck mixer 汽车混凝土搅拌运输车

1—Driver's cabin 驾驶室；2—Force dividing mechanism 分力箱；3—Drive shaft 传动轴；4—Gear box 齿轮箱；5—Water tank 水箱；6—Axial direction bearing fixer on mixer 拌筒中心轴承座；7—Chain wheel 链轮；8—Mixer 搅拌筒；9—Feed inlet 进料斗；10—Discharge outlet 出料口；11—Conveying trough 输料槽；12—Frame 机架；13—Man hole 入孔；14—Mixer blade 搅拌叶片

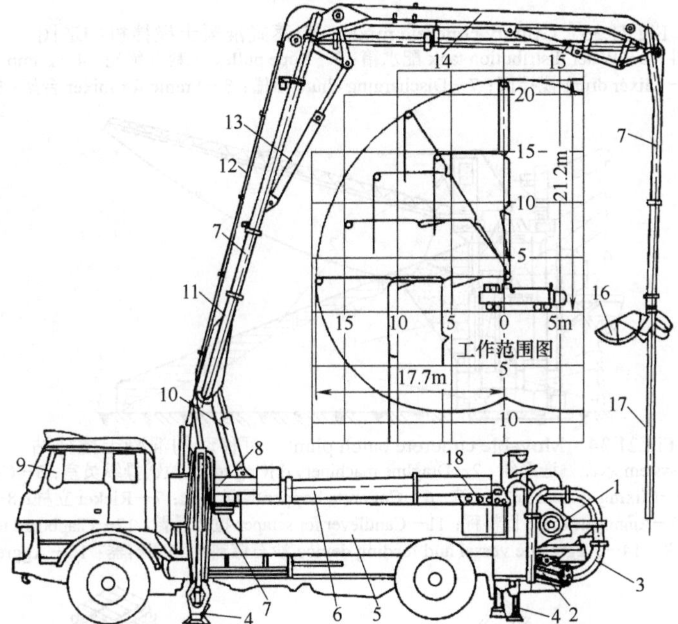

Fig 21.31 Truck with hydraulic concrete pump and placing boom 装有布料臂杆的液压混凝土泵车

1—Hopper and mixer 料斗及搅拌器；2—Concrete pump 混凝土泵；3—Discharge outlet 出料口；4—Hydraulic pressure outrigger 液压外伸支腿；5—Water tank 水箱；6—Standby pipe for concrete 备用混凝土输送管；7—Pipe fixed to rotating platform and along the boom 接入旋转台的和沿臂杆安装的导管；8—Rotating platform 旋转台；9—Driver's cabin 驾驶室；10，13，15—Oil cylinder for folding jib 折叠臂杆用油缸；11，14—Boom 臂杆；12—Oil pipe 油管；16—Support frame of rubber hose 橡胶软管弯曲支管支撑架；17—Rubber hose 橡胶软管；18—Manipulate panel 操纵柜

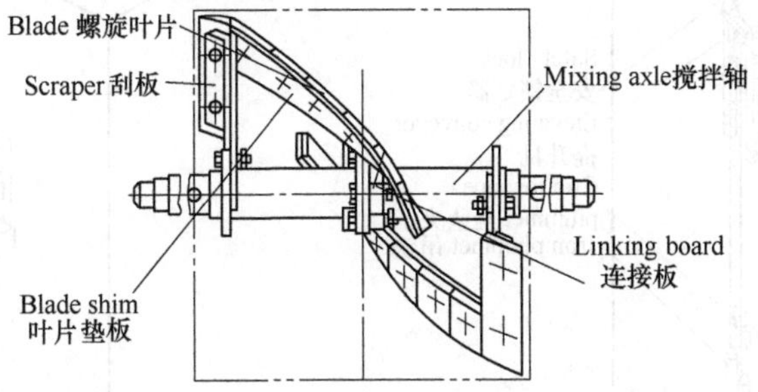

Fig 21.32 Mixing axle of concrete mixer 混凝土搅拌器搅拌轴

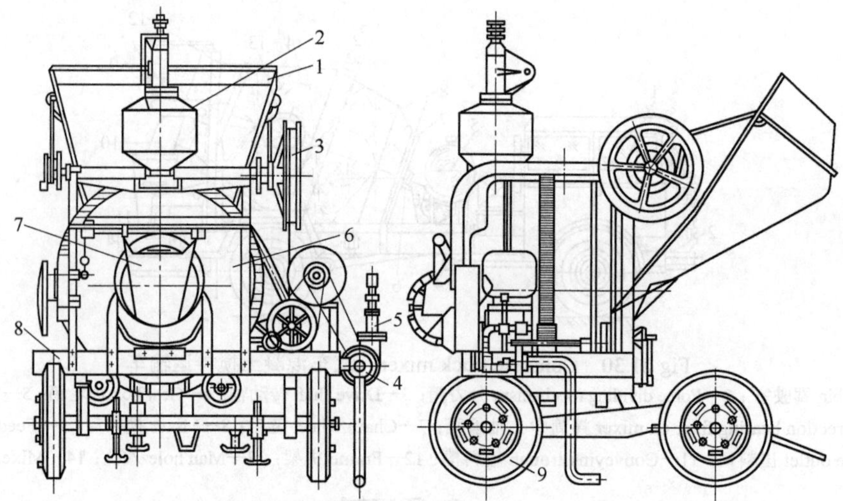

Fig 21.33 Gravity concrete mixer 自落式混凝土搅拌机构造图

1—Feeding hopper 上料斗；2—Water distribution tank 配水箱；3—Rope pulley 上料斗绳轮；4—Pump 水泵；5—Water pump pipe 水泵管路；6—Mixer drum 搅拌筒；7—Discharging chute 溜槽；8—Frame for mixer 台架；9—Wheel 车轮

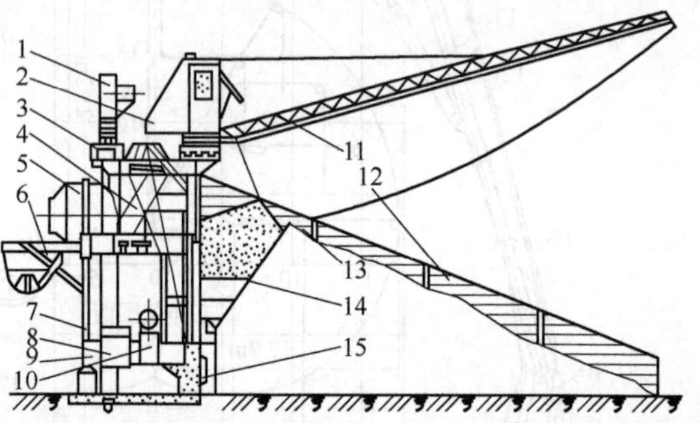

Fig 21.34 Movable concrete batch plant 可移动的混凝土搅拌站

1—Cement transporting system 水泥输送系统；2—Dragline machinery driving room 拉铲操纵员室；3—Cement scale 水泥秤；4—Upper frame 上机架；5—Mixing system 搅拌系统；6—Concrete hopper 混凝土仓；7—Ricker 立柱；8—Electrical bin 电气柜；9—Lower frame 下机架；10—Aggregate scale 骨料秤；11—Cantilever for scraper 拉铲悬臂；12—Clapboard for material 料场隔仓板；13—Dragline bucket 拉铲斗；14—Aggregate vessel and feeding device 沙石储料仓和喂料器；15—Aggregate lifter 沙石提升斗

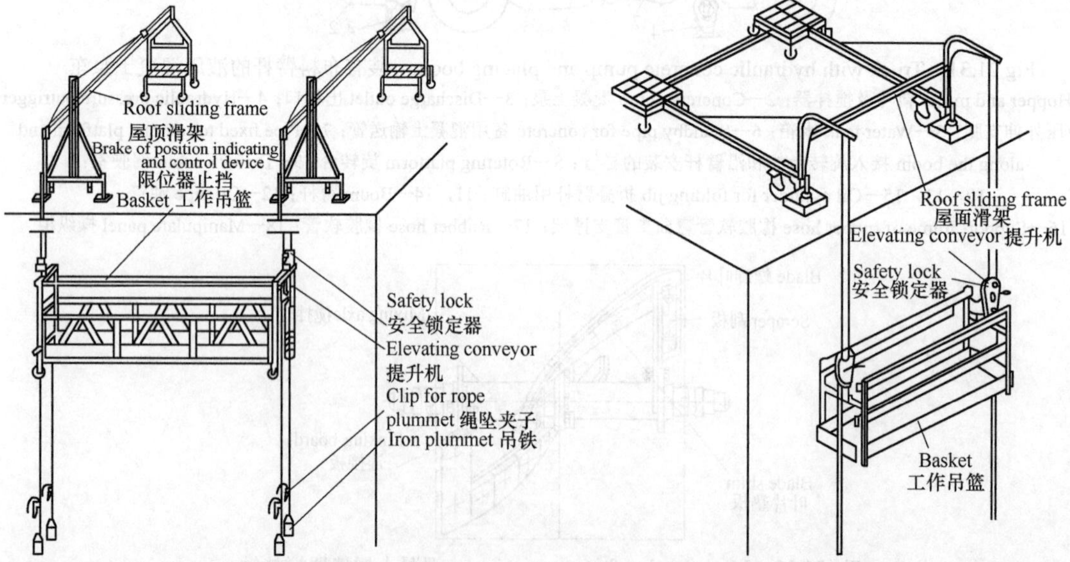

Fig 21.35 Electromotion basket 电动吊篮

Fig 21.36 Manual operation basket 手动吊篮

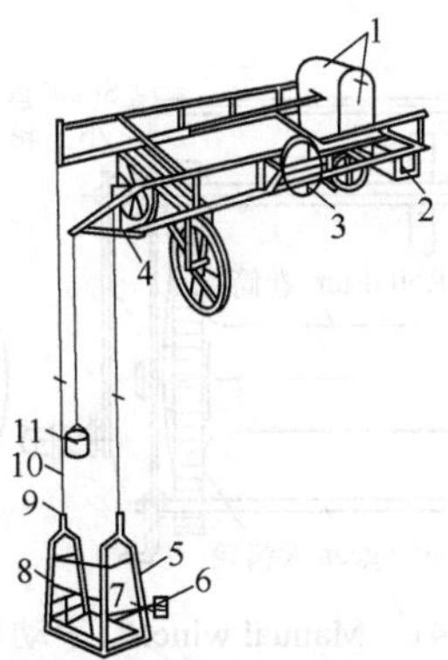

Fig 21.37 Manual operating roof suspension hoist machine 手动屋面移动吊机

1—Gear box 变速箱；2—Balance weight case 配重箱；3—Frame of steel thread drum for hopper 料斗升降钢丝绳卷轮架；4—Movable frame for hopper 摆动送料架；5—Basket (one person) 吊篮（供一人用）；6—Rubber wheel 橡胶轮；7—Handrail of basket 吊篮栏杆；8—Safety hook 安全钩；9—Steel cable 钢丝绳；10—Hopper 料罐；11—Pulley for basket 吊篮升降滑轮

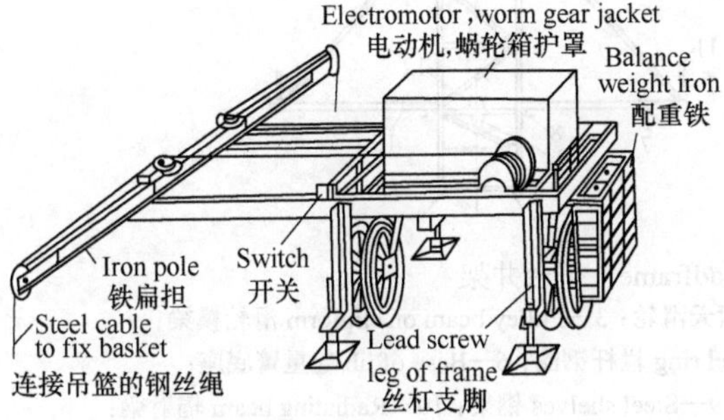

Fig 21.38 Electrocution roof suspension hoist machine 电动屋面移动吊机

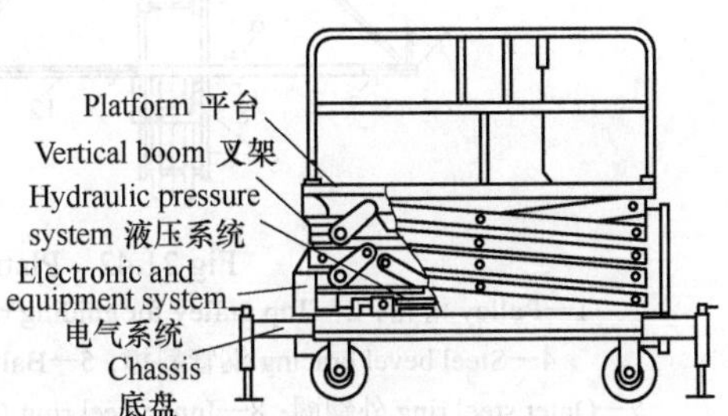

Fig 21.39 Formation and configuration of ZTY hydraulic pressure lift platform ZTY 液压升降平台外形及构造示意图

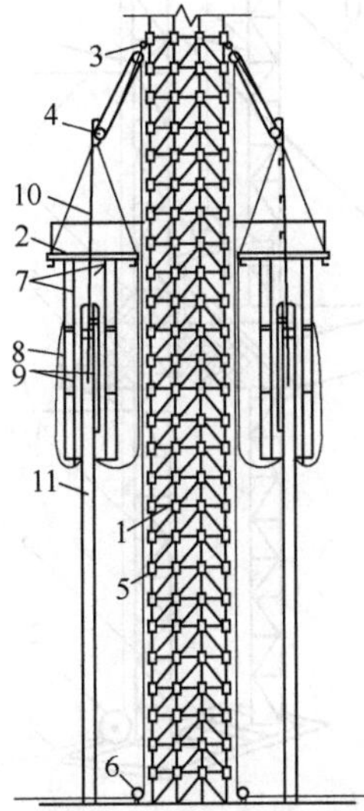

Fig 21.40 Sketch of winch elevating process 卷扬机提升工作示意图

1—Frame of shaft 竖井架；2—Platform 工作台；3—Clamp for steel cable 钢丝绳扣；4—Pulley 滑车；5—Steel cable 钢丝绳；6—Steering pulley 导向滑轮；7—Hoisting inside and outside 内外吊装；8—Safety net 安全网；9—Formwork inside and outside 内外模板；10—Chimney steel cable 烟囱钢缆；11—Chimney 烟囱筒壁

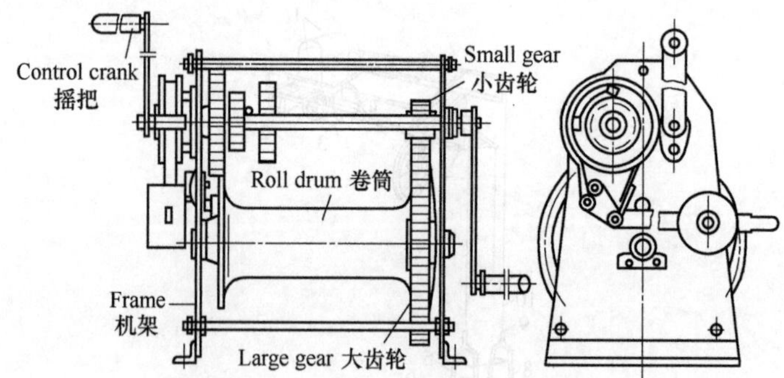

Fig 21.41 Manual winch 手动卷扬机

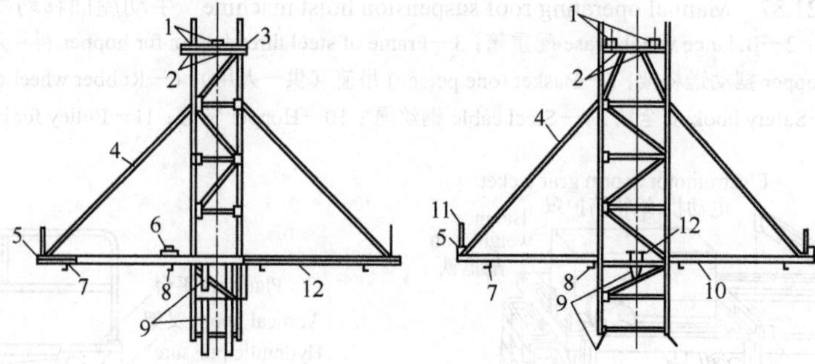

Fig 21.42 Platform headframe 平台井架

1—Pulley 滑轮；2—Top pulley for guiding rope 导索天滑轮；3—Pulley beam on slipform 滑轮模梁；
4—Steel bevel bracing 钢管斜撑；5—Baluster steel ring 栏杆钢圈；6—Base of jib 起重臂底座；
7—Outer steel ring 外钢圈；8—Inner steel ring 内钢圈；9—Steel shelves 钢架；10—Radiating beam 辐射梁；
11—Baluster 栏杆；12—Crossbeam 大梁

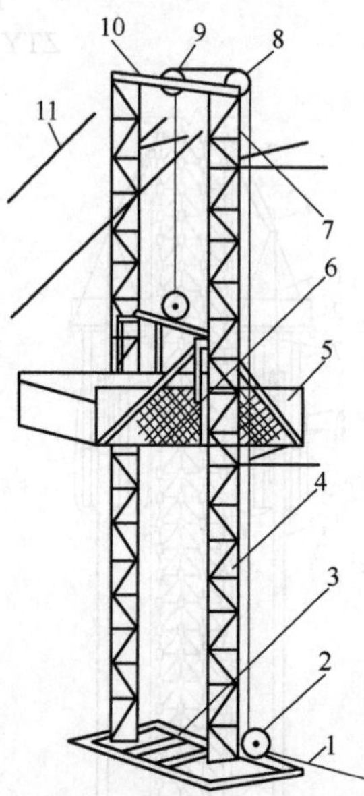

Fig 21.43 MSS-100 frame crane MSS-100 龙门架

1—Elevating steel cable 提升钢丝绳；2—Bottom pulley 地轮；3—Chassis 底盘；4—Vertical shaft column 立柱；
5—Suspension cage 吊笼；6—Load bearing frame 承重架；7—Safety device 安全装置；8—Top pulley 天轮；
9—Steering pulley 导向轮；10—Gantry crane girder 横梁；11—Guy 缆风绳

Unit 22 Petroleum Drilling Machinery and Refining Equipments 石油钻采与炼制机械

22.1 Petroleum drilling rigs 石油钻机

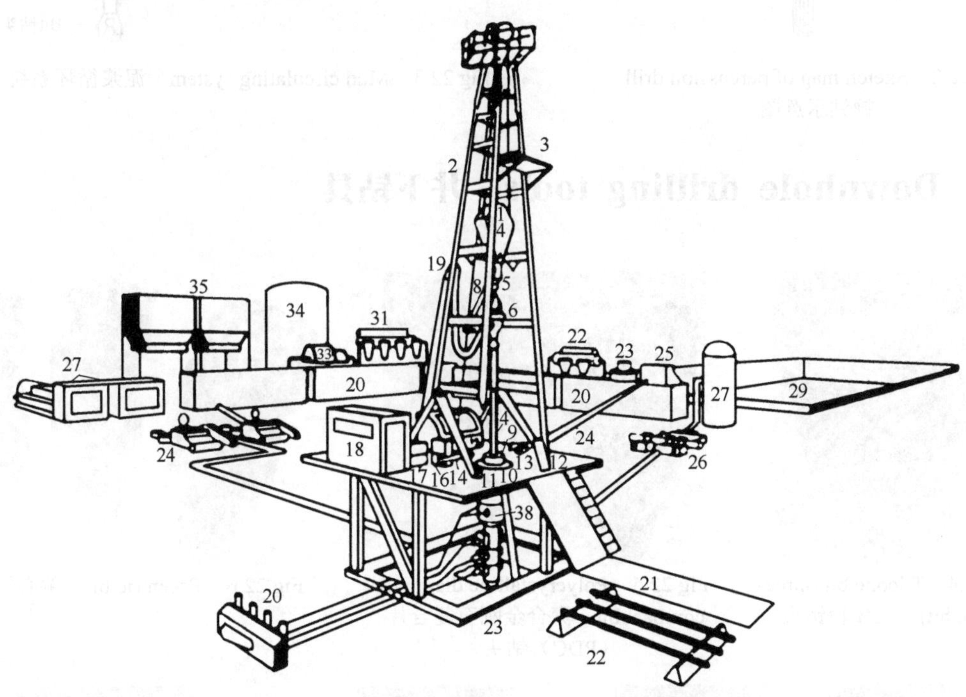

Fig 22.1 Schematic diragram of drilling wells and equipment 钻井井场及设备示意图

1—Crown block 天车；2—Drill derrick 井架；3—Monkey board 二层台；4—Travelling block 游动滑车；5—Hook 大钩；6—Swivel 水龙头；7—Hanger 吊卡；8—Kelly 方钻杆；9—Kelly drive bushing 方钻杆补芯；10—Master bushing 方补芯；11—Mouse hole 小鼠洞（接单根用）；12—Rat hole 大鼠洞（放方钻杆用）；13—Backup tong 固定大钳；14—Make up tongs 接管用大钳；15—Draw-work hoist 绞车；16—Weight indicator 指重表；17—Driller's console 司钻控制台；18—Dog house 井场值班室；19—Rotary hose 水龙带；20—Energy storage 蓄能装置；21—Pipe ramp 管子坡道；22—Pipe rack 管架；23—Derrick base 井架底座；24—Mud return line 泥浆返回管线；25—Mud shaker 泥浆振动筛；26—Choke manifold 节流管汇；27—Mud gas separator 泥浆 - 天然气分离器；28—Degasser 脱气装置；29—Reserve pit 泥浆储备池；30—Mud pit 泥浆池；31—(Deduster) Desilter desander（脱泥机）泥浆过滤设备；32—Desander 脱砂机；33—Centrifuge 离心机；34—Mud pump 泥浆泵；35—Dry cement reserve tank 干水泥储备罐；36—Water tank 储水罐；37—Electric generator 发电机；38—Blowout preventer stack 防喷器组

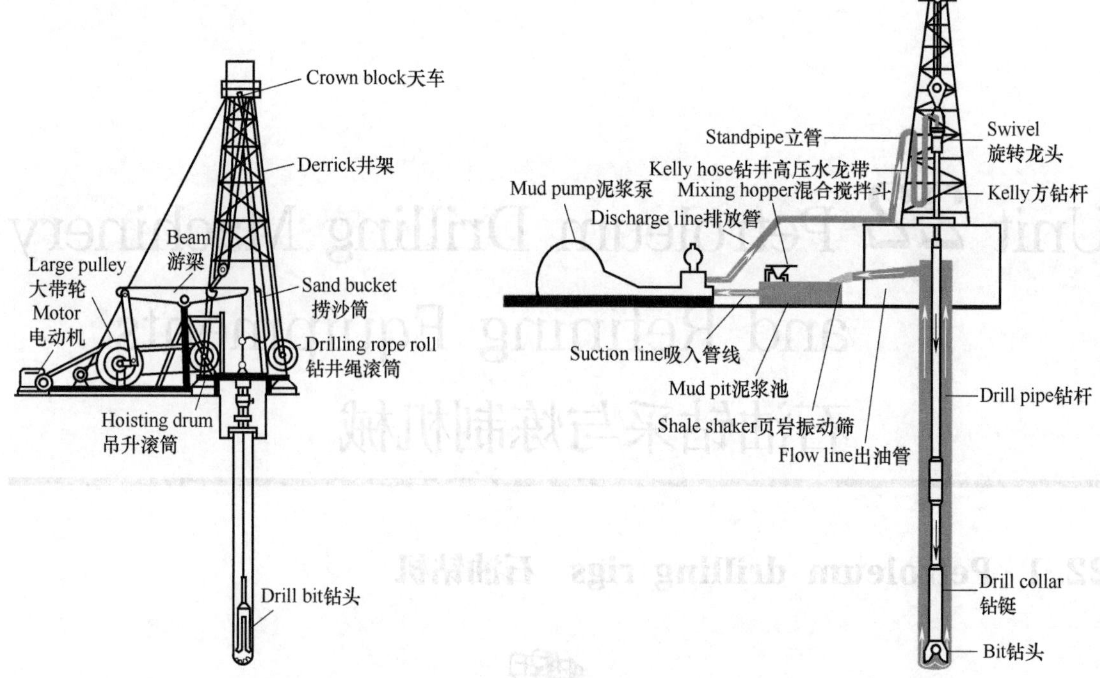

Fig 22.2 Sketch map of percussion drill 顿钻示意图

Fig 22.3 Mud circulating system 泥浆循环系统

22.2 Downhole drilling tools 井下钻具

Fig 22.4 Tricone bit (three-cone bit) 三牙轮钻头

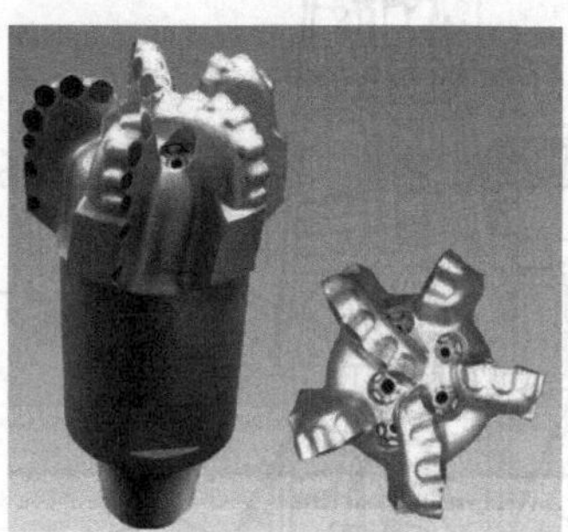

Fig 22.5 Polycrystalline diamond compact bit 聚合金刚石复合片（PDC）钻头

Fig 22.6 Eccentric bit 偏心钻头

Fig 22.7 Drill collars 钻铤

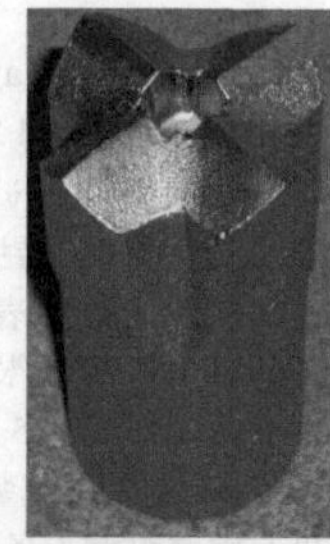

Fig 22.8 Four-wing rotary bit(Cross drill bit) 十字钻头

Fig 22.9 Drag bit (Blade bit) 刮刀钻头

Fig 22.10 Bent housing PDM 弯外壳螺杆钻具

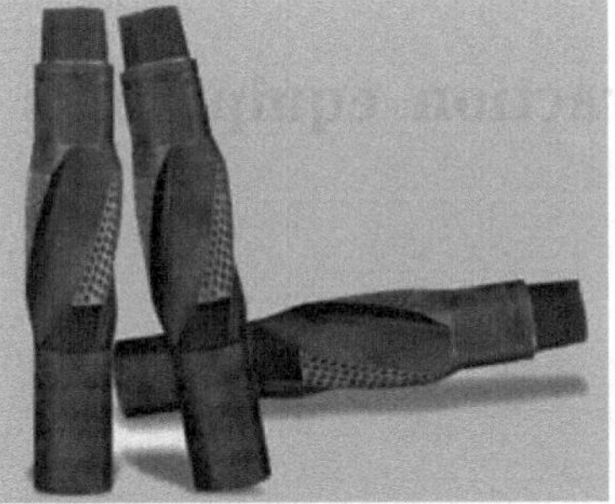

Fig 22.11 Whipstock 导斜器

Fig 22.12 Kelly spinner 方钻杆旋转器

Fig 22.13 Slips （转盘）卡瓦

Fig 22.14 Spinning chain 旋链

Fig 22.15 Core barrel 取芯筒

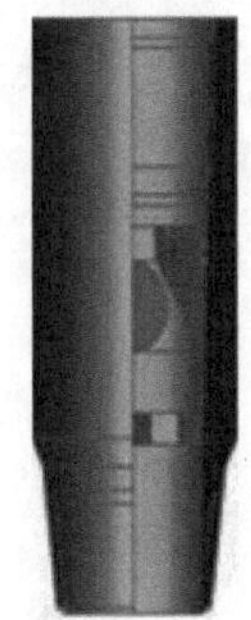

Fig 22.16 Float shoe 浮鞋

Fig 22.17 Float collar 浮箍

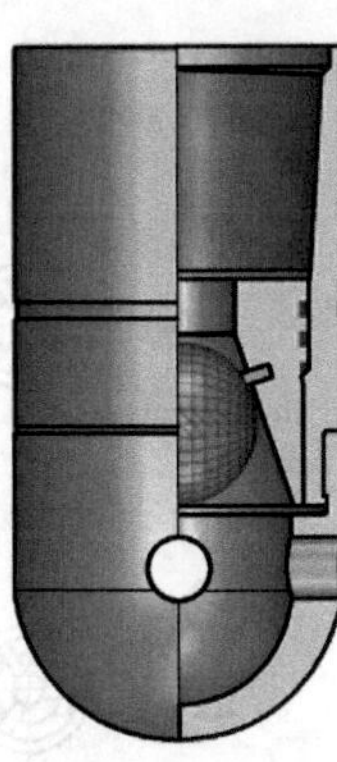

Fig 22.18 Casing shoe 套管鞋

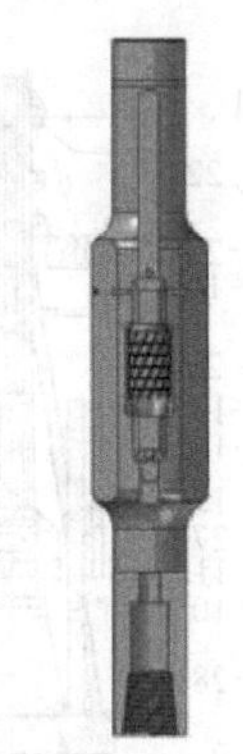

Fig 22.19 Reaming device 划眼器

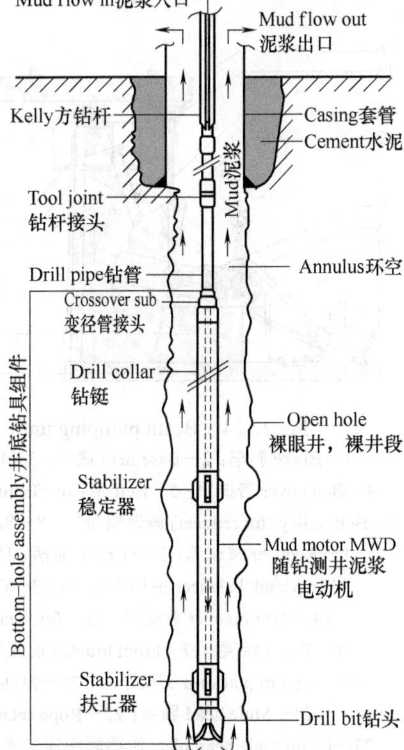

Fig 22.22 Drillstring components 钻柱组件

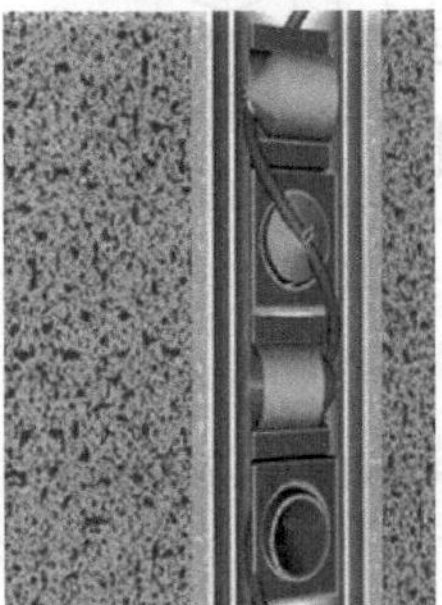

Fig 22.20 Perforating tool 射孔枪

Fig 22.21 Shaped charge 射孔弹

22.3 Petroleum extraction equipments 采油设备

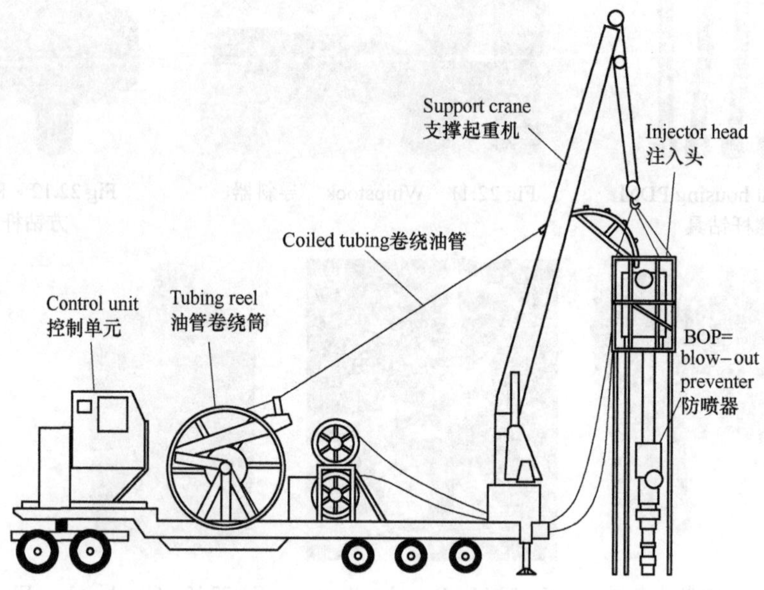

Fig 22.23 Coiled-tubing unit (trailer mounted) 连续油管装置（安装在拖车上）

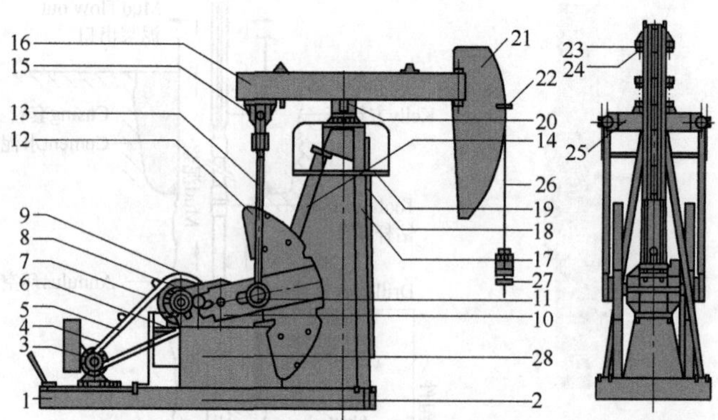

Fig 22.24 Beam pumping unit 游梁式抽油机

1—Brake 刹车；2—Basement 底座；3—Electric motor 电动机；
4—Belt cover 带护罩；5—Belt 带；6—Tubing platform 筒体小平台；
7—Belt pulley (for reducer) 减速器带轮；8—Safety brakes 安全刹车装置；
9—Reducer 减速器；10—Crank 曲柄；11—Crank pin 曲柄销；
12—Crank balancer 曲柄平衡块；13—Connecting rod 连杆；
14—Rear bracket 后支架；15—Tail bearing seat 尾轴承座；
16—Beam 游梁；17—Front bracket 前支架；18—Ladder 梯子；
19—Support platform 支架平台；20—Mid-bearing seat 中轴承座；
21—Mule head 骡头；22—Rope retainer 钢丝绳护圈；
23—Beam-mule head plate 游梁骡头连接板；24—Bearing pin 销轴；
25—Cross beam 横梁；26—Hoist cable 吊绳；
27—Beam hanger 悬绳器；28—Cylinder 筒体

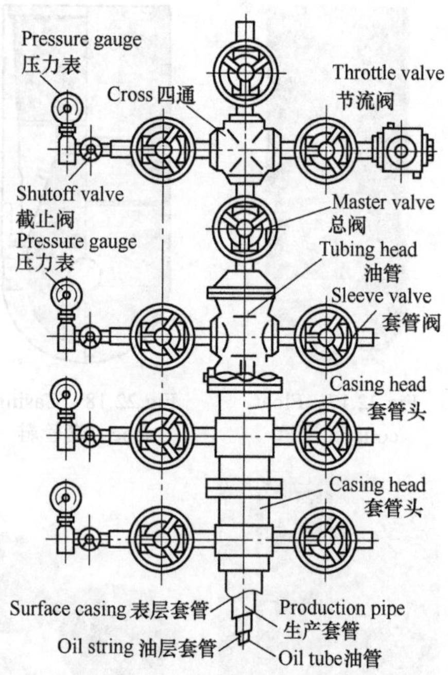

Fig 22.25 Threaded wellhead equipment
螺纹式井口装置

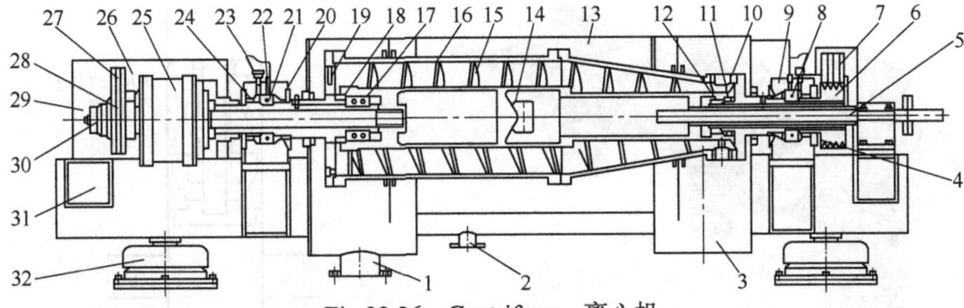

Fig 22.26 Centrifuge 离心机

1—Liquor export 母液排口；2—Residue export 残液排口；3—Filter residue outlet 滤渣排口；4—Main motor 主电动机；5—Feed pipe 进料管；6—Driven wheel 主电动机从动轮；7—Action wheel 主电动机主动轮；8—Front bearing 前主轴承；9—Front bearing base 前轴承座；10—Slag wear set 排渣耐磨套；11—Screw front bearing seal 螺旋前轴密封；12—Screw front bearing 螺旋前轴承；13—Shell 外壳；14—Distributor 布料器；15—Screw push feeder 螺旋推料器；16—Drum components 转鼓组件；17—Screw rear bearing 螺旋后轴承；18—Screw rear bearing seal 螺旋后轴承密封；19—Overflow plate 溢流板；20—Emperor bearing base 主轴承座；21—Emperor bearing 主轴承；22—Temperature sensor 温度传感器；23—Pours the lubricate cap 注油杯；24—Rear bearing seal 后主轴承密封；25—Differential 差速器；26—Differential motor 差速电动机；27—Main differential pulley 主差速带轮；28—Subordinate differential pulley 从差速带轮；29—Belt guard 带保护罩；30—Torque protection system 转矩保护装置；31—Shock absorber 减振器；32—Machine base 机座

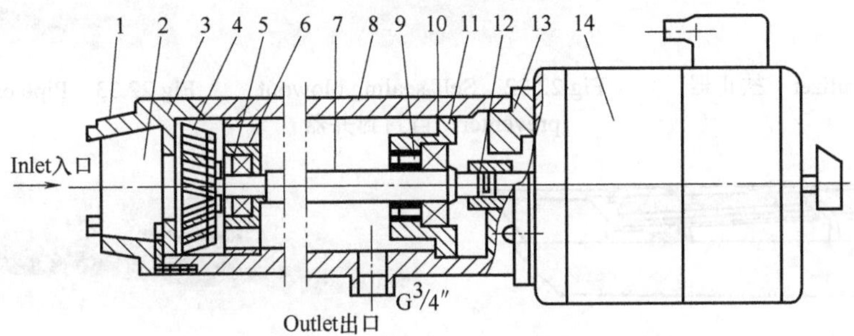

Fig 22.27 (Oil) pump 抽油泵

1—(Pump) cover 泵罩；2—Filter 过滤网；3—Pump nozzle 泵嘴；4—Impeller 叶轮；5—Heteroblade bracket 异叶片支架；6，10—Bearing 轴承；7—Pump shaft 泵轴；8—Pump body 泵体；9—Reinforced seal 骨架密封；11—Bearing seat 轴承座；12—Coupling 联轴器；13—Pump deck 泵座；14—Electric motor 电动机

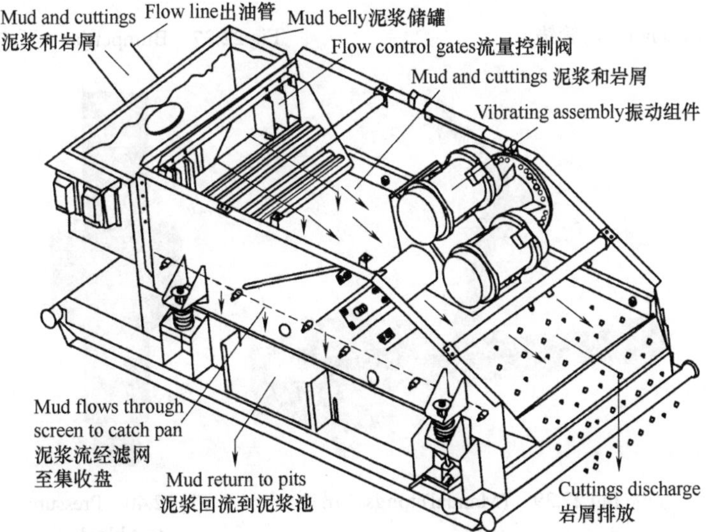

Fig 22.28 Adjustable linear shaker 可调直线型振动筛

Fig 22.29 Venturi nozzle 文丘里喷嘴

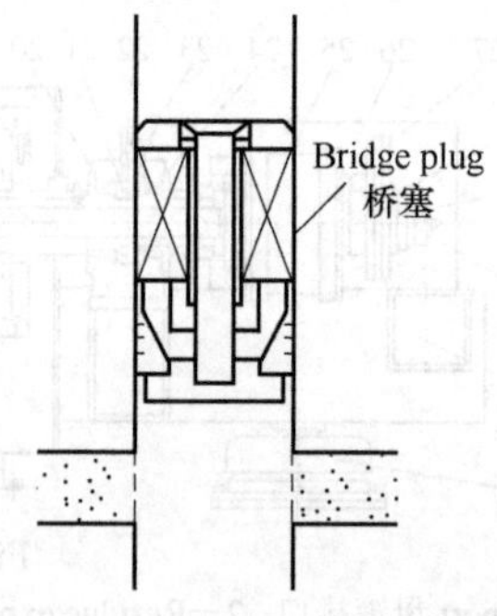

Fig 22.30 Bridge plug 桥塞

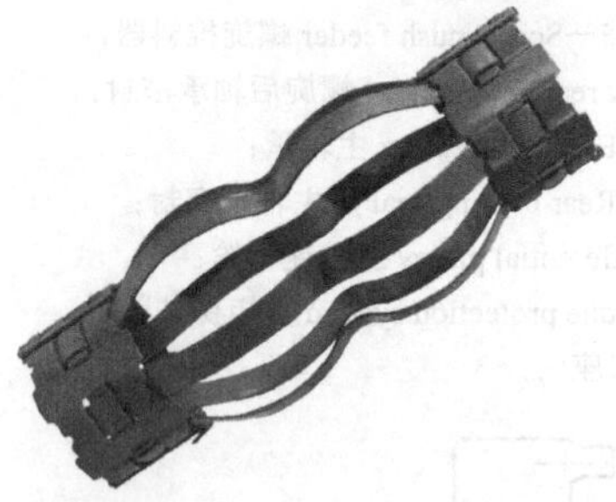
Fig 22.31 Centralizer 扶正器

Fig 22.32 Self-sealing blowout preventer 自封封井器

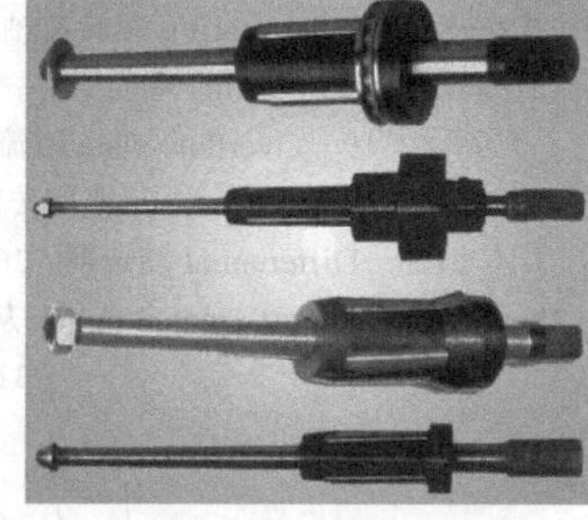
Fig 22.33 Pipe expander 胀管器

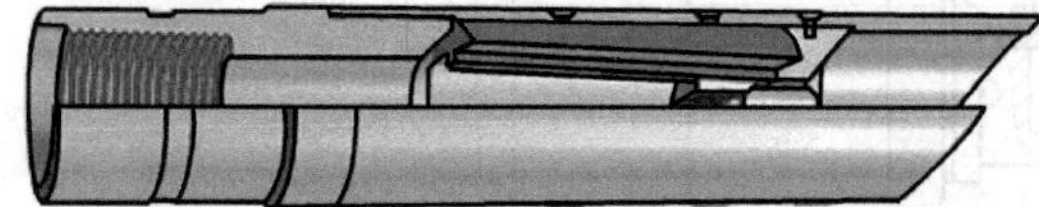
Fig 22.34 Fishing rod tube 抽油杆打捞筒

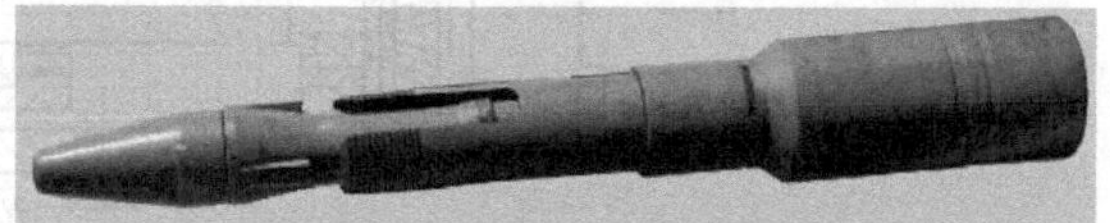
Fig 22.35 Fishing tool 打捞工具

Fig 22.36 Tubing anchor 油管锚

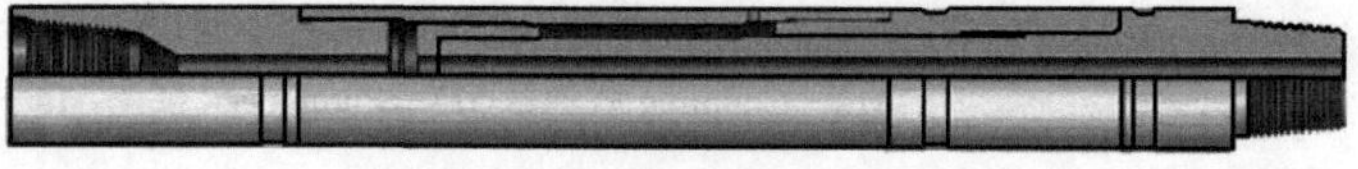
Fig 22.37 Bumper jar 下击器

Fig 22.38 Tube scraper 套管刮削器

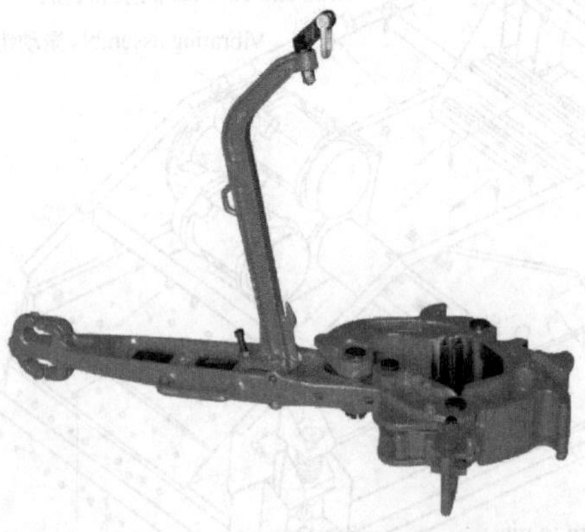
Fig 22.39 (Manual) tongs 吊钳

Fig 22.40 Pressure test block 试压堵塞

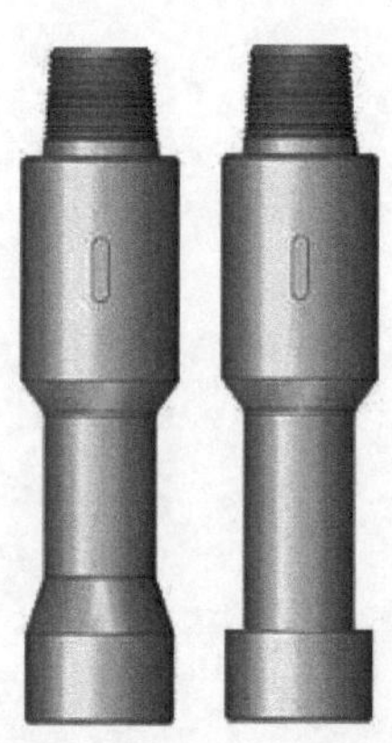
Fig 22.41 Lift sub 提升短节

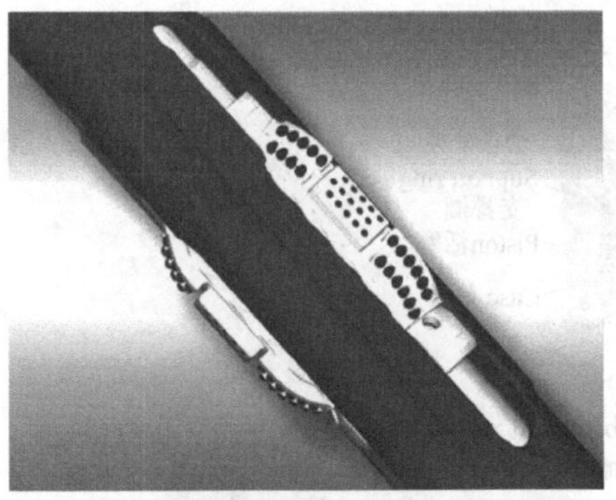

Fig 22.42 Underreamer 扩眼器

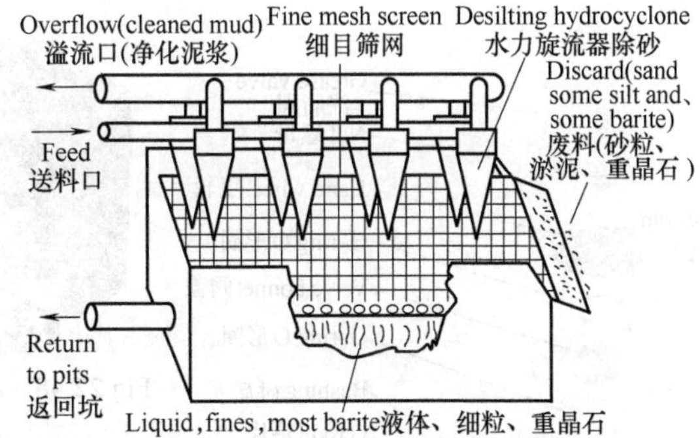

Fig 22.43 Schematic of a mud cleaner 泥浆清洁器示意图

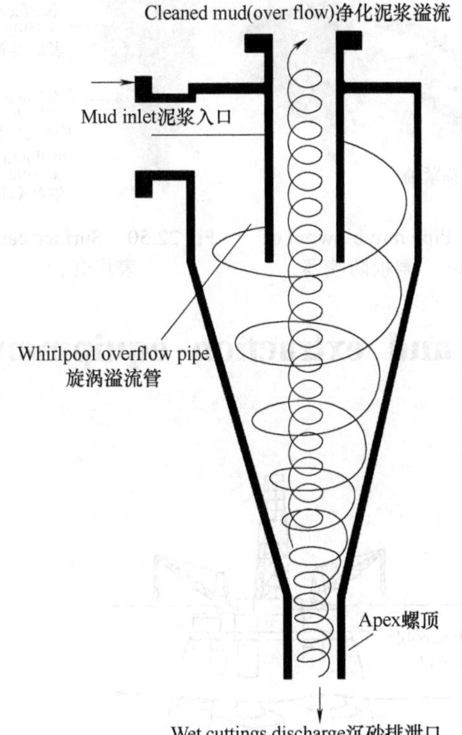

Fig 22.44 Hydroclone schematic 水力旋流器简图

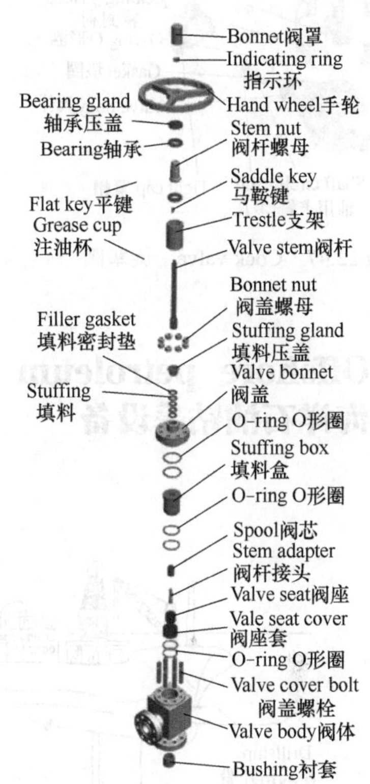

Fig 22.45 Hand throttle valve 手动节流阀

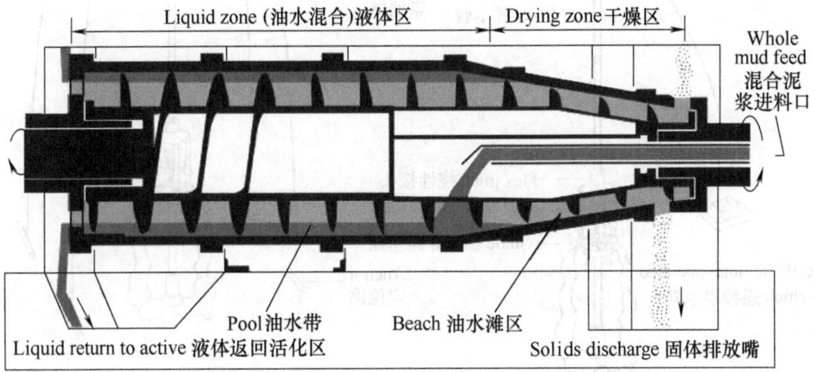

Fig 22.46 Cross section of a decanting centrifuge 沉降式离心分离机截面图

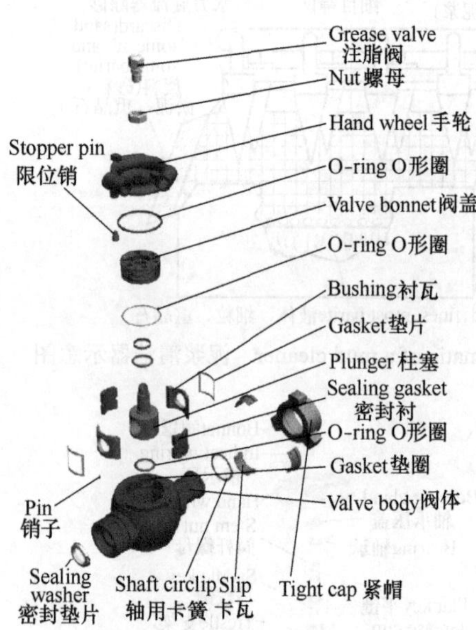

Fig 22.47 Cock valve 旋塞阀

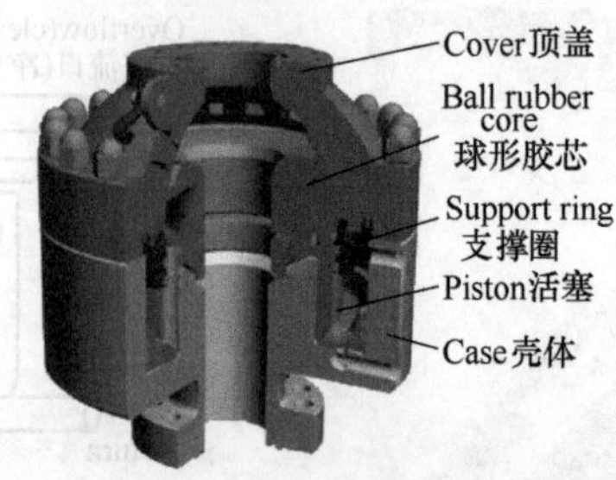

Fig 22.48 Annular blowout preventer 环形防喷器

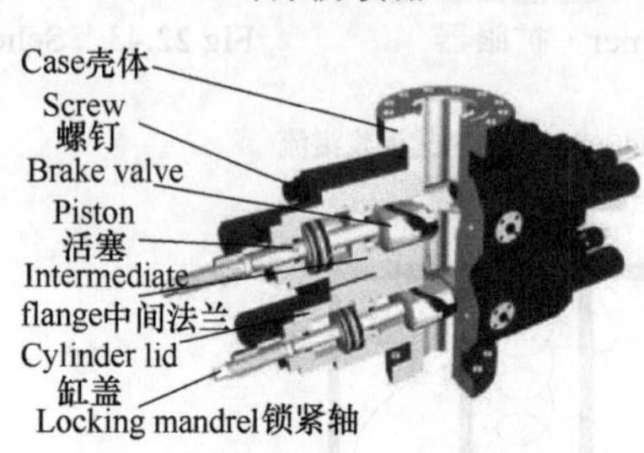

Fig 22.49 Pipe ram blowout of preventer 闸板防喷器

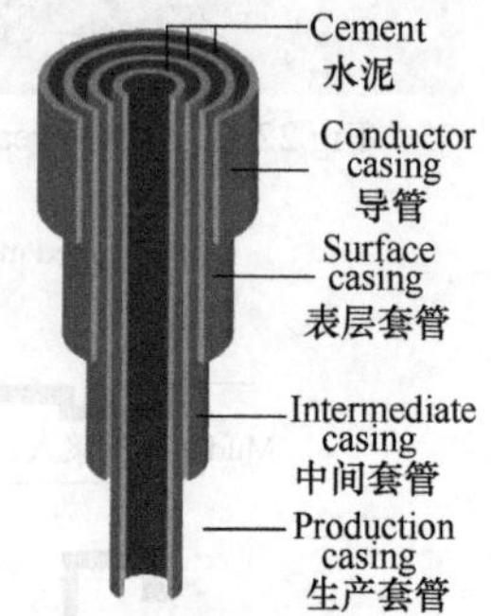

Fig 22.50 Surface casing 表层套管

22.4 Offshore petroleum drilling and extraction equipments 海洋石油钻采设备

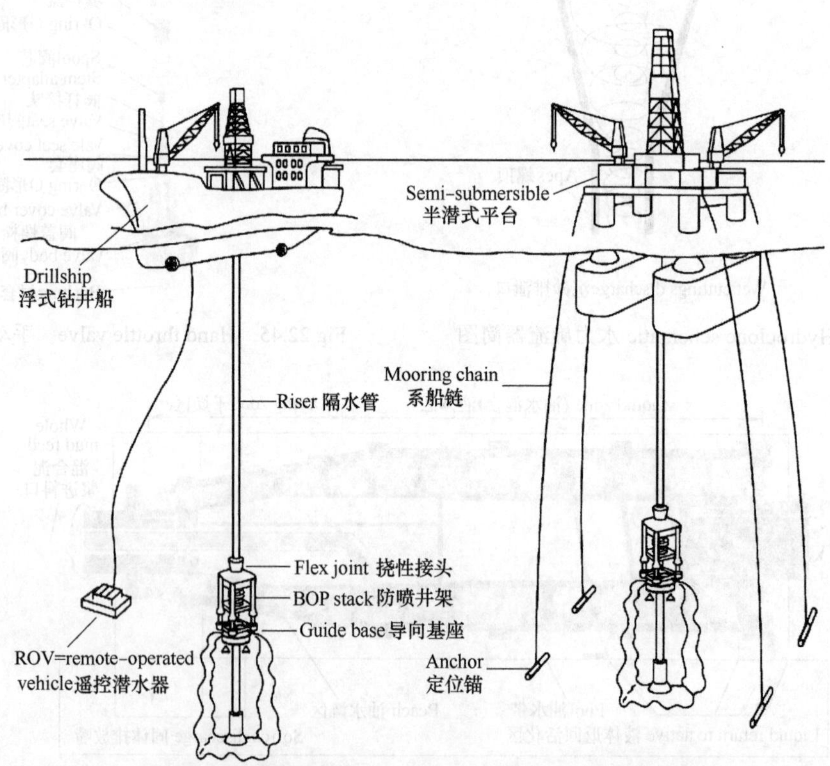

Fig 22.51 Illustration of deep water rigs 深水钻井架示意图

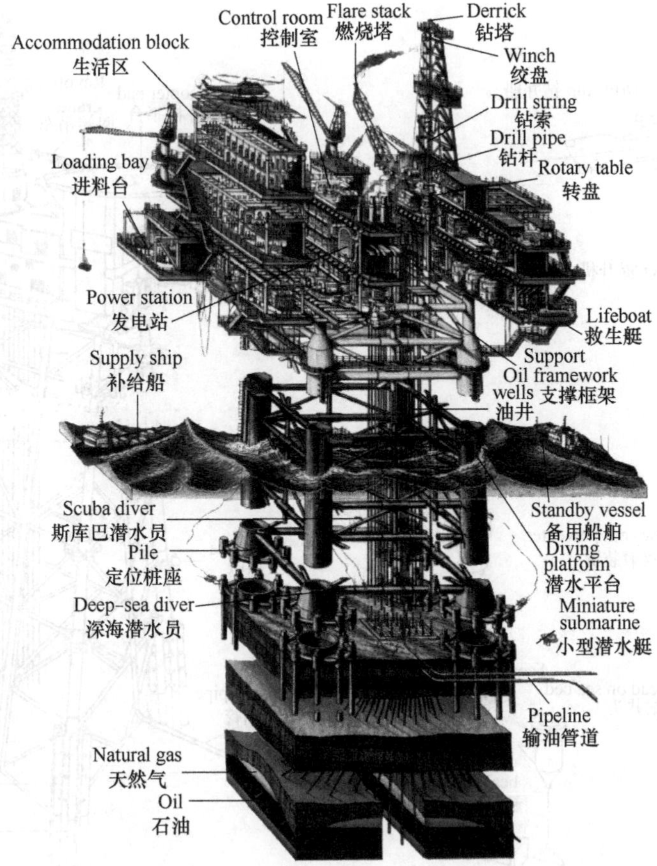

Fig 22.52 Oil platform 海上石油钻进平台

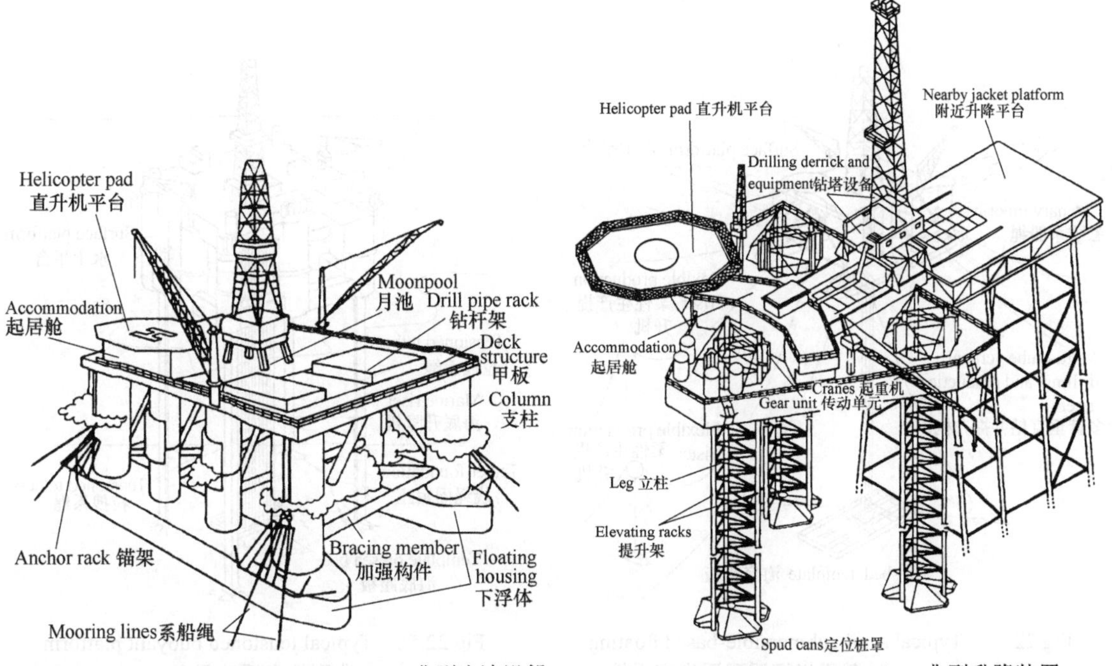

Fig 22.53 Typical semi-submersible vessel 典型半淹没船

Fig 22.54 Typical jack-up rig 典型升降装置

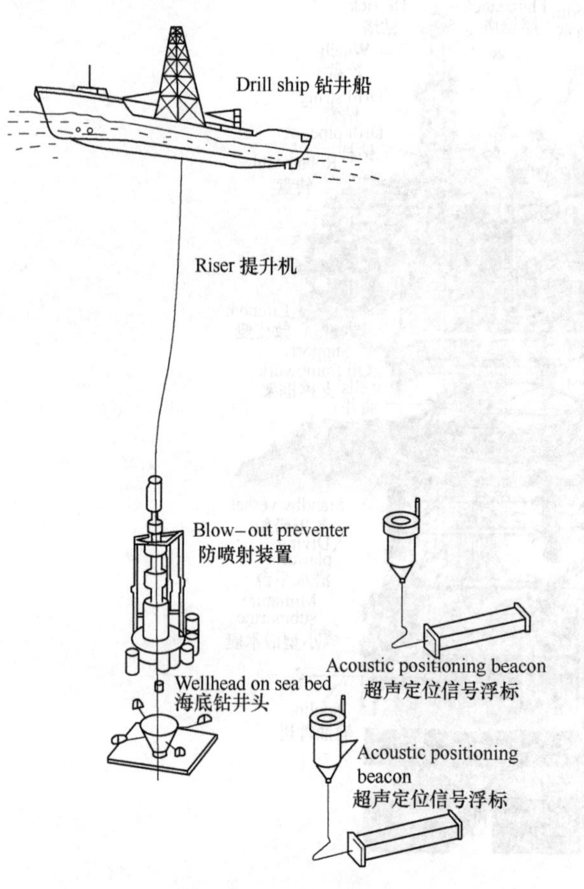

Fig 22.55 Dynamically positioned drill ship
动态停靠钻井船

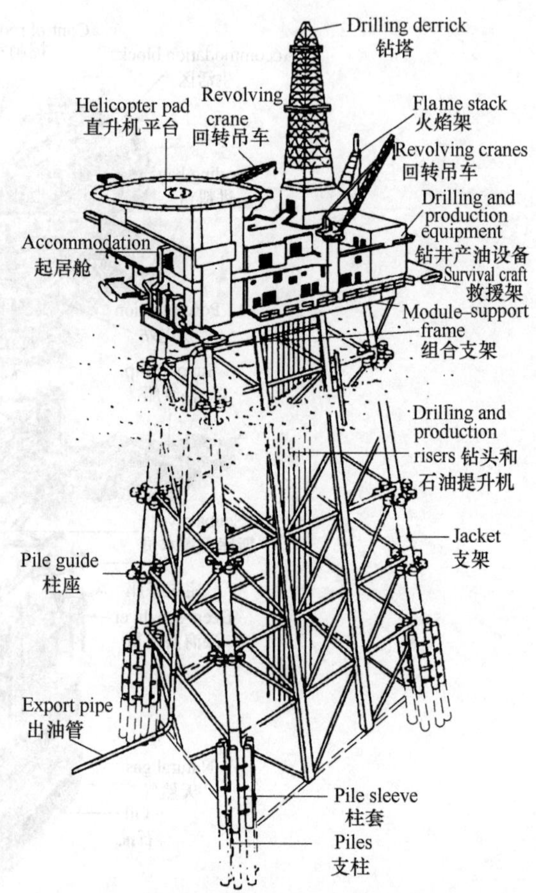

Fig 22.56 Typical offshore drilling and production platform 典型离岸钻井和生产平台

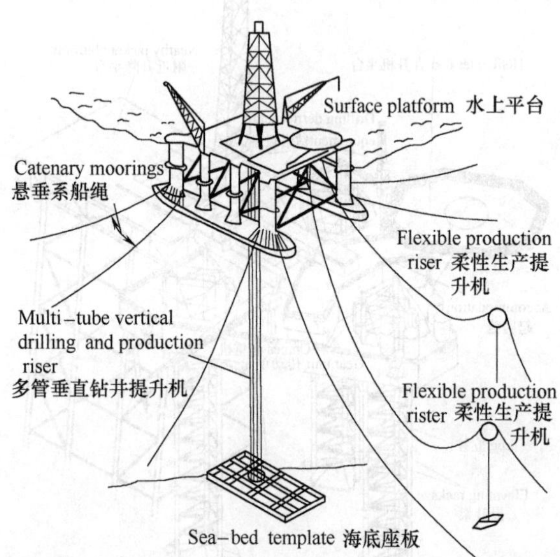

Fig 22.57 Typical semi-submersible-based floating production system 基于半沉浮飘浮生产系统

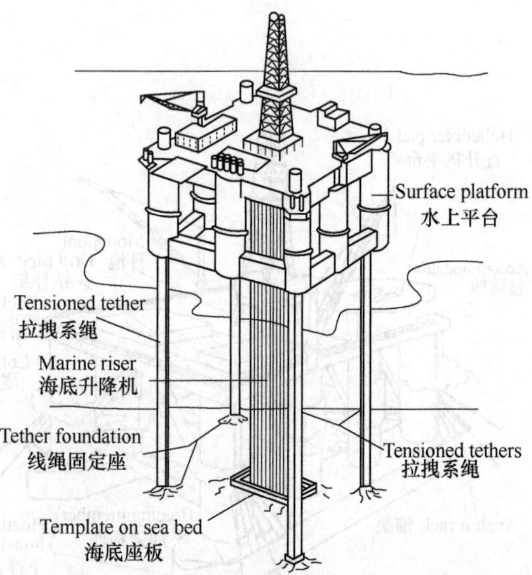

Fig 22.58 Typical tensioned buoyant platform
典型拉曳浮动平台

22.5 Petroleum refining equipments 石油炼制设备

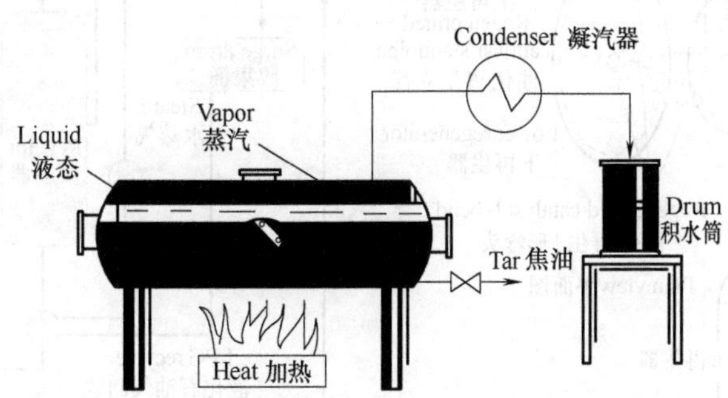

Fig 22.59 Simple batch distillation 简易分批蒸馏装置

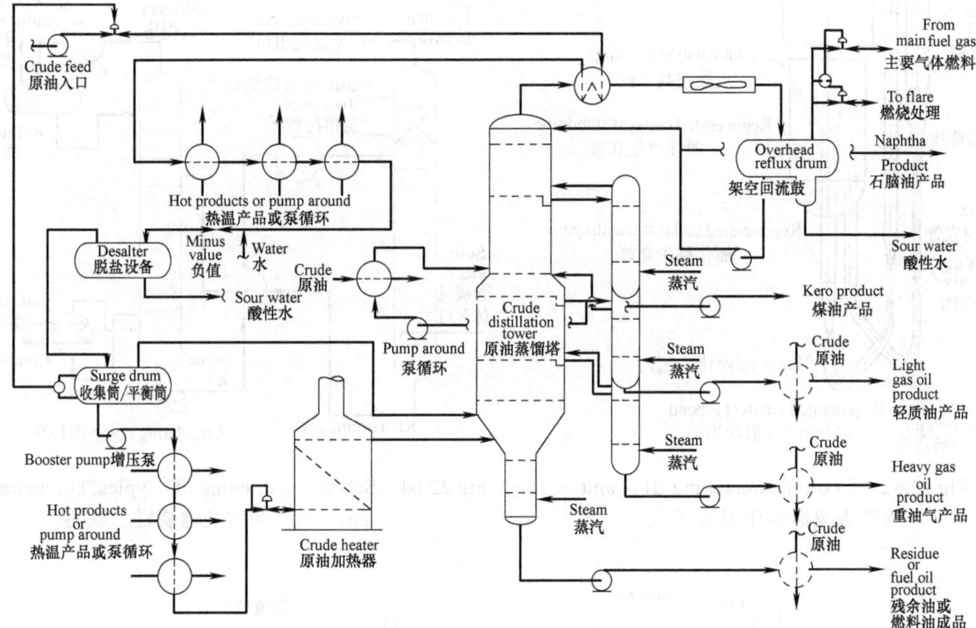

Fig 22.60 A typical atmospheric crude distillation unit 典型的常压原油蒸馏装置

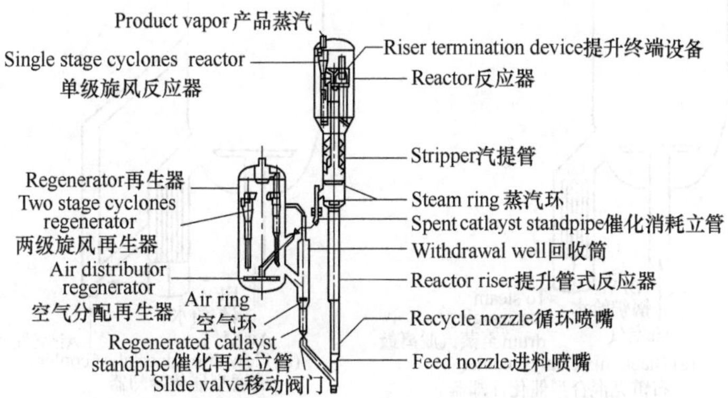

Fig 22.61 Reactor-regenerator of modern FCCU 现代催化裂化装置反应再生器

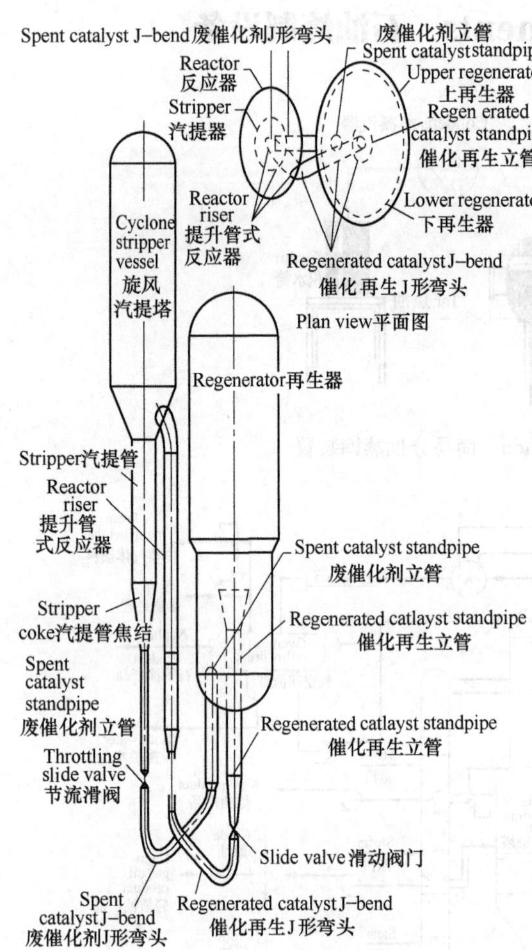

Fig 22.62 Exxon flexicracking Ⅲ R unit
埃克森灵活裂化Ⅲ R 单元

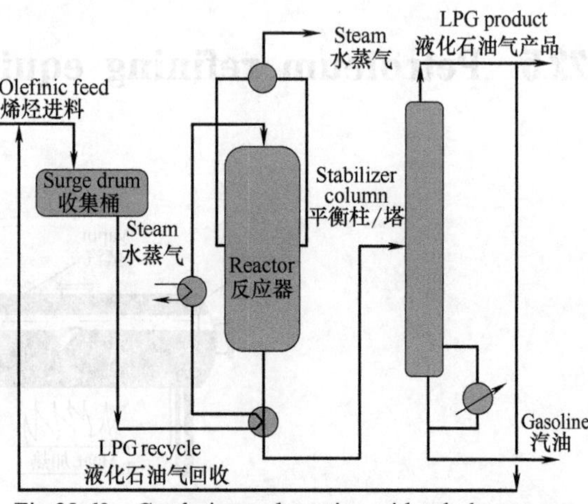

Fig 22.63 Catalytic condensation with tubular reactor
催化缩合与管式反应器

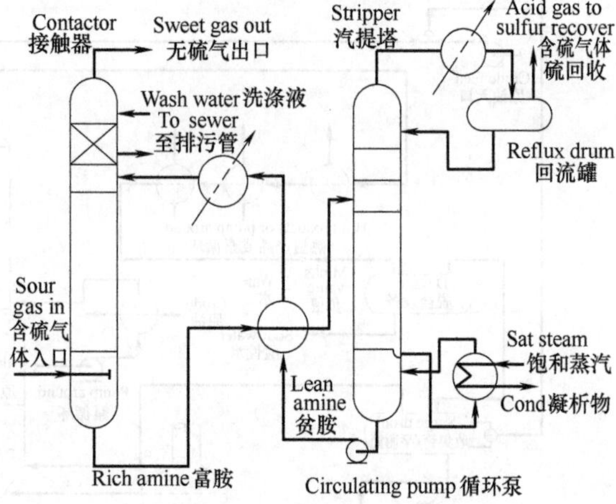

Fig 22.64 Schematic drawing of a typical amine treating unit 典型的胺处理装置示意图

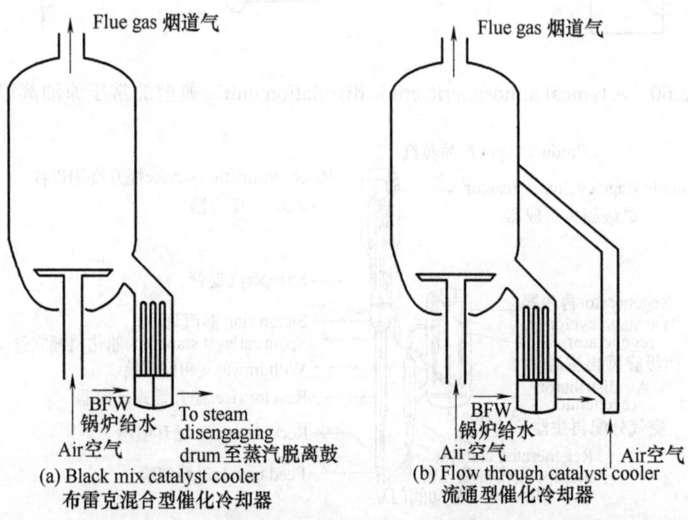

(a) Black mix catalyst cooler
布雷克混合型催化冷却器

(b) Flow through catalyst cooler
流通型催化冷却器

Fig 22.65 Typical catalyst coolers 典型的催化冷却器

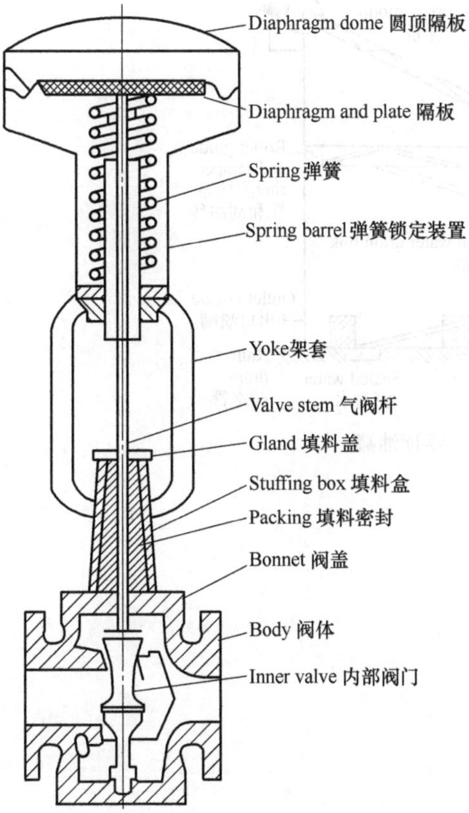

Diaphragm dome 圆顶隔板
Diaphragm and plate 隔板
Spring 弹簧
Spring barrel 弹簧锁定装置
Yoke架套
Valve stem 气阀杆
Gland 填料盖
Stuffing box 填料盒
Packing 填料密封
Bonnet 阀盖
Body 阀体
Inner valve 内部阀门

Fig 22.66 A conventional control valve 传统控制阀

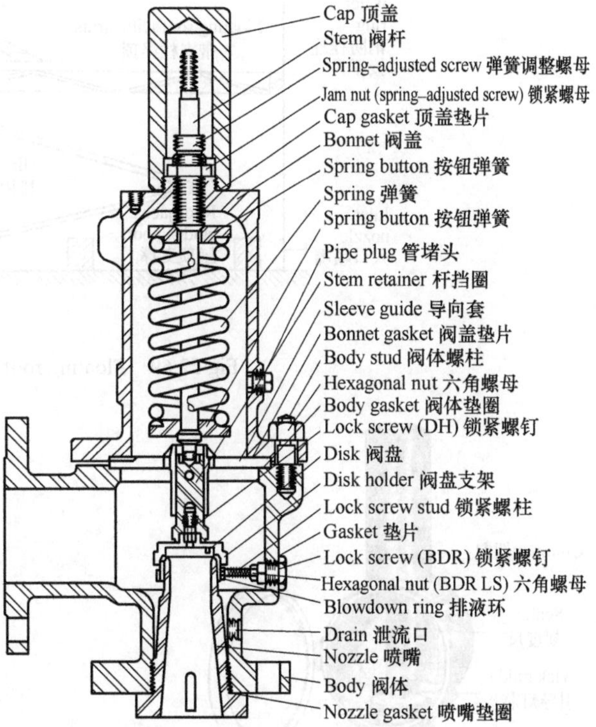

Cap 顶盖
Stem 阀杆
Spring–adjusted screw 弹簧调整螺母
Jam nut (spring–adjusted screw) 锁紧螺母
Cap gasket 顶盖垫片
Bonnet 阀盖
Spring button 按钮弹簧
Spring 弹簧
Spring button 按钮弹簧
Pipe plug 管堵头
Stem retainer 杆挡圈
Sleeve guide 导向套
Bonnet gasket 阀盖垫片
Body stud 阀体螺柱
Hexagonal nut 六角螺母
Body gasket 阀体垫圈
Lock screw (DH) 锁紧螺钉
Disk 阀盘
Disk holder 阀盘支架
Lock screw stud 锁紧螺柱
Gasket 垫片
Lock screw (BDR) 锁紧螺钉
Hexagonal nut (BDR LS) 六角螺母
Blowdown ring 排液环
Drain 泄流口
Nozzle 喷嘴
Body 阀体
Nozzle gasket 喷嘴垫圈

Fig 22.67 A diagram of a conventional safety relief valve 常规安全阀示意图

Vent通风口
Sample hatch样品舱口盖
Platform with fire water ring attached 消防栓连接平台
Inlet nozzle 入口喷嘴
Tank bund 油罐围界
Inspection and maintenance hatch 检查维护舱门
Sump 贮油槽
Outlet to pumps出口泵

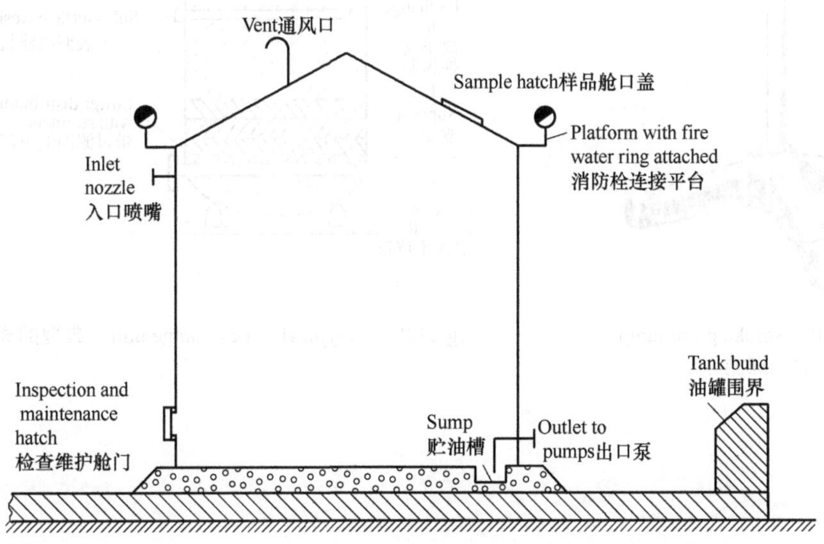

Fig 22.68 Cone roof tank 锥顶油罐

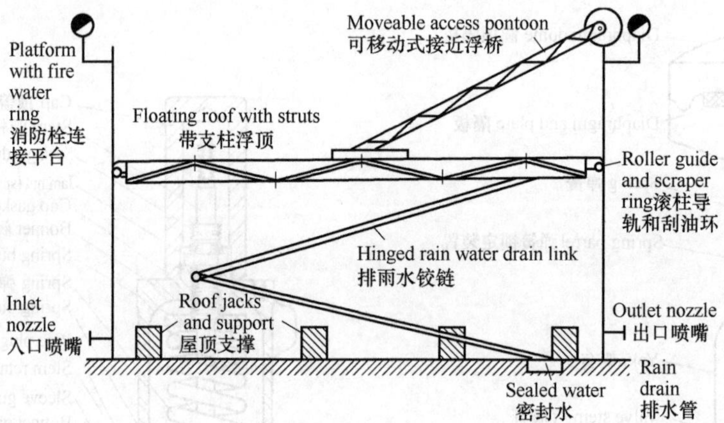

Fig 22.69 Floating roof tank 浮顶油罐

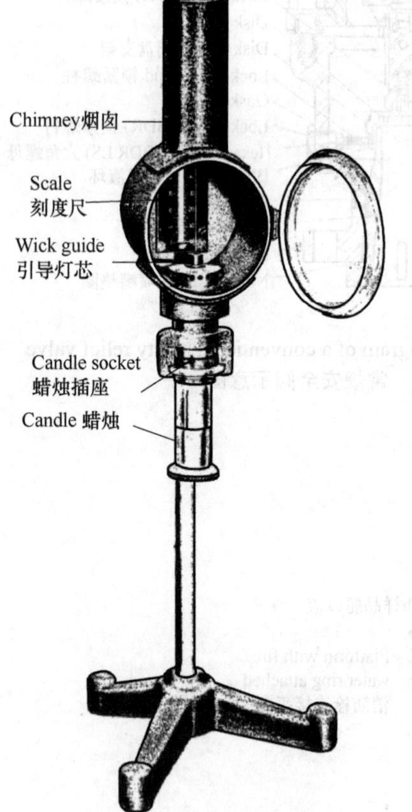

Fig 22.70 Smoke point lamp
烟点探测灯

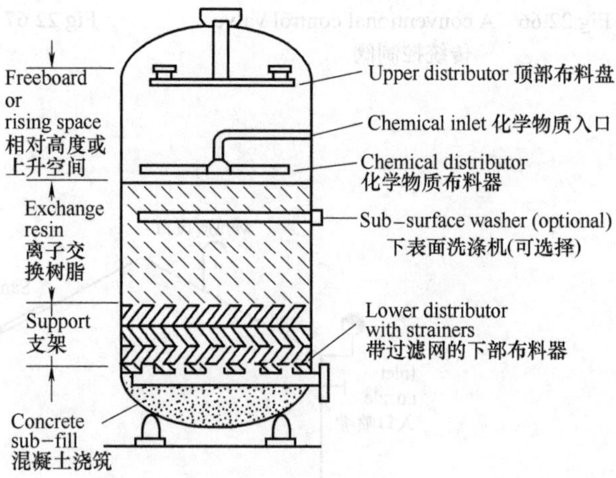

Fig 22.71 A typical ion exchange unit 典型的离子交换装置

Unit 23 Food and Packaging Machinery 食品与包装机械

23.1 Grading and sorting machinery 分级分选机械

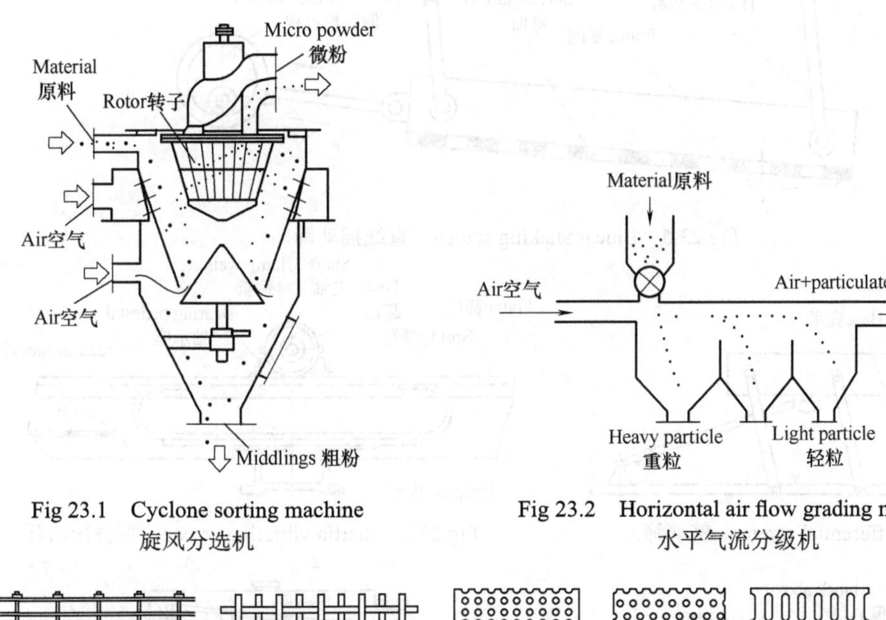

Fig 23.1 Cyclone sorting machine 旋风分选机

Fig 23.2 Horizontal air flow grading machine 水平气流分级机

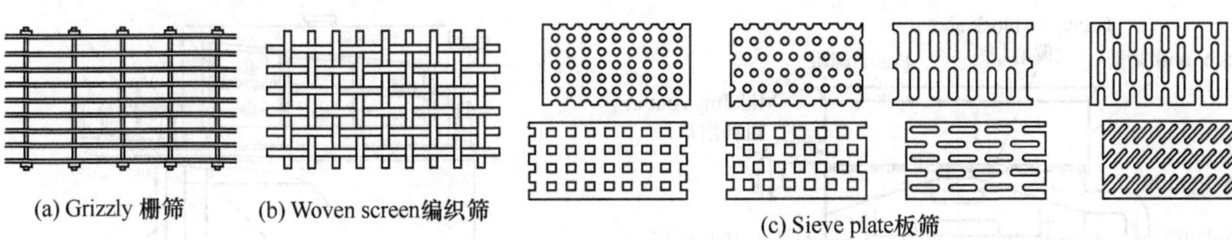

Fig 23.3 Type of screen surface 筛面种类

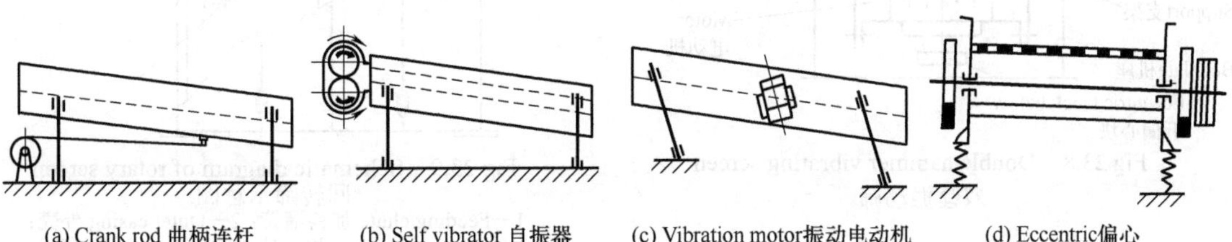

Fig 23.4

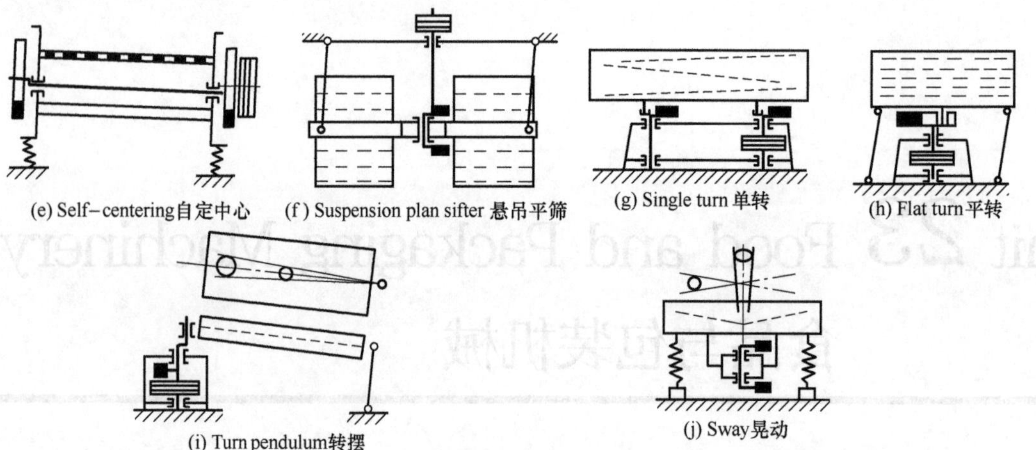

Fig 23.4 Transmission forms of screen 筛子传动方式

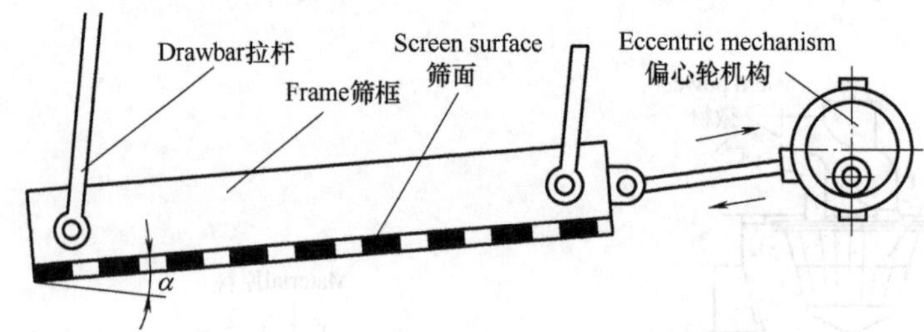

Fig 23.5 Linear shaking screen 直线摇动筛

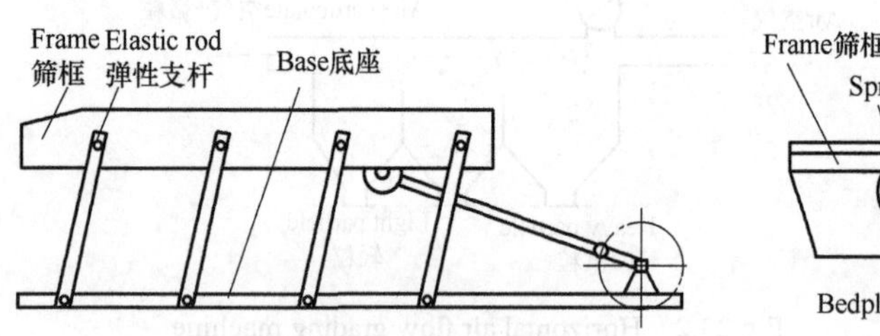

Fig 23.6 Differential screen 差动筛

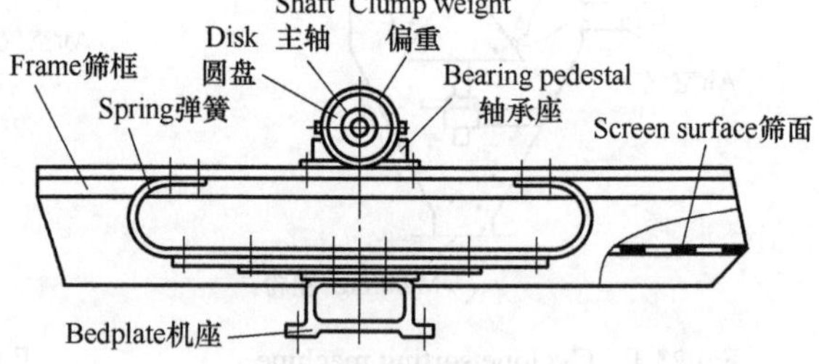

Fig 23.7 Inertia vibrating screen 惯性振动筛

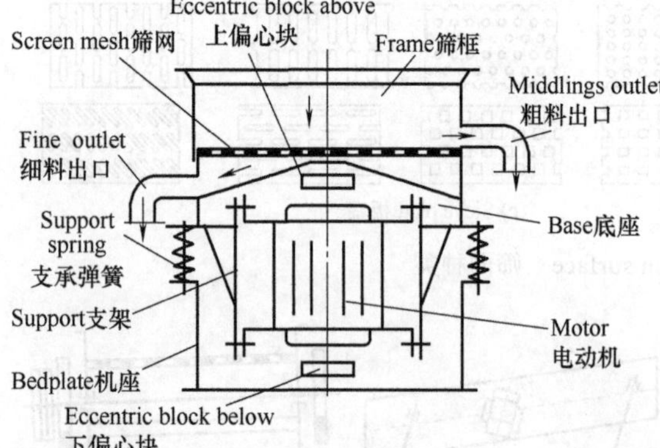

Fig 23.8 Double hammer vibrating screen
双锤振动筛

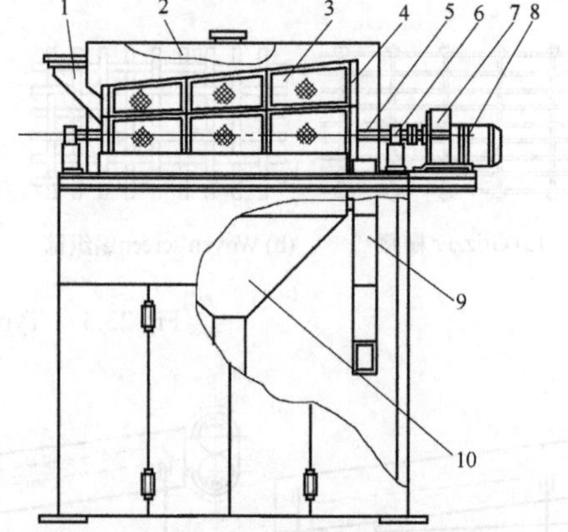

Fig 23.9 Schematic diagram of rotary screen
回转筛示意图

1—Feeding chute 加料溜管；2—Outer casing 外壳；
3—Screen mesh 筛网；4—Rotary drum 回转筒；
5—Spindle 主轴；6—Bearing 轴承；7—Reducer 减速器；
8—Motor 电动机；9—Unloading chute 卸料溜槽；
10—Hopper 料斗

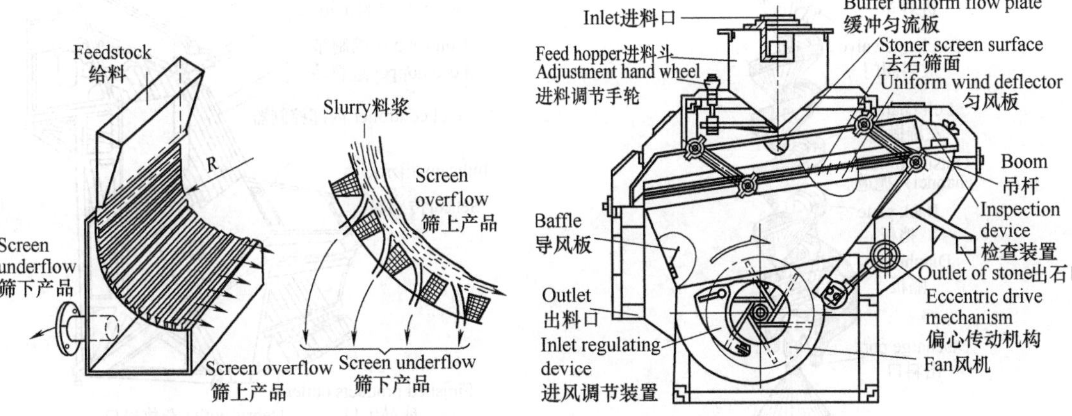

Fig 23.10 Bending screen 曲筛

Fig 23.11 Density stoner 密度去石机

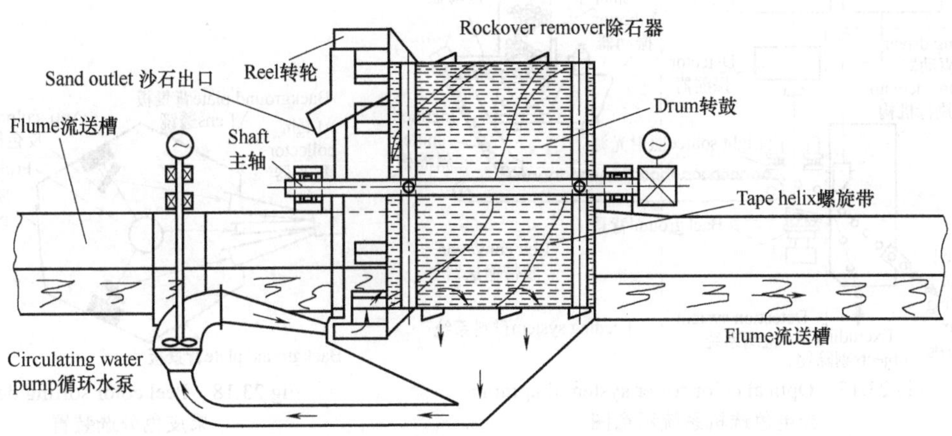

Fig 23.12 Tumble stoner 转筒式除石机

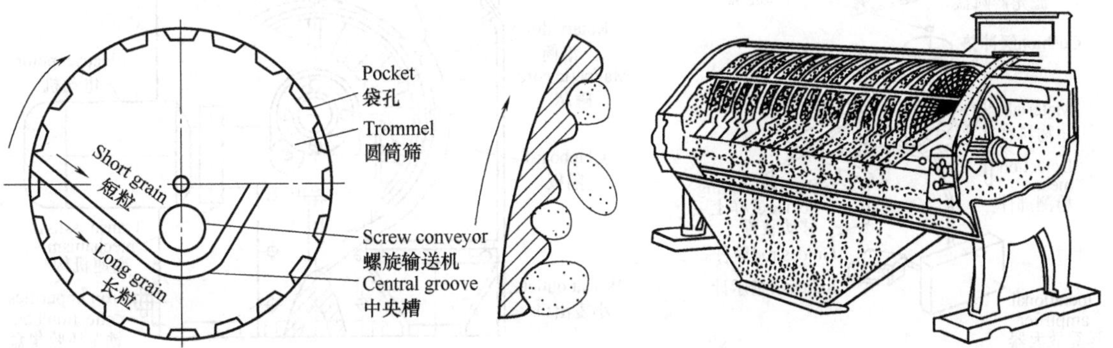

Fig 23.13 Working diagram of drum selection machine 滚筒式精选机工作示意图

Fig 23.14 Disc selection machine 碟片精选机

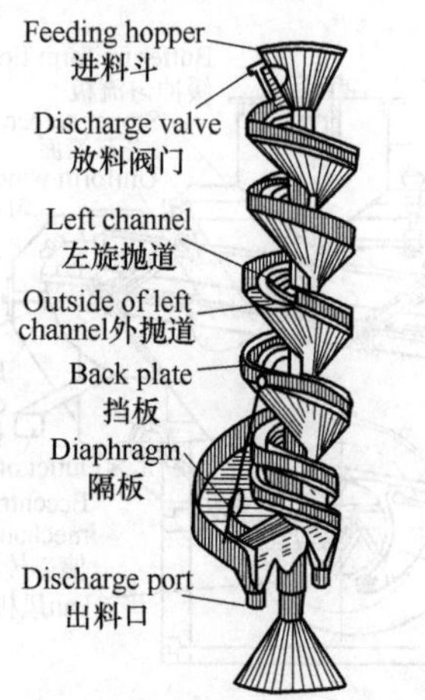

Fig 23.15 Structure of the spiral selection
螺旋精选器的结构

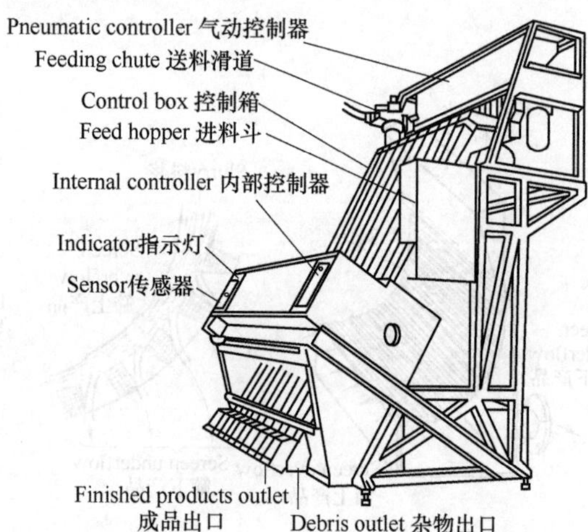

Fig 23.16 Automatic color sorter 全自动色选机

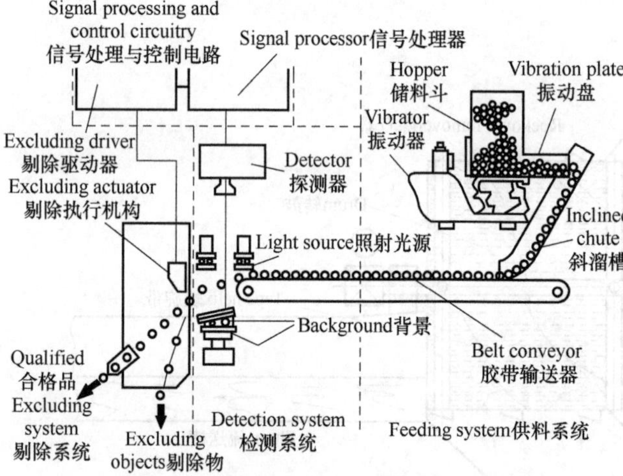

Fig 23.17 Optical color sorter system diagram
光电色选机系统示意图

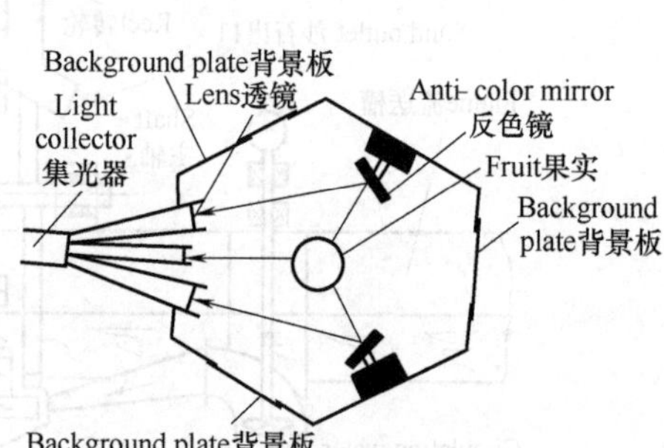

Fig 23.18 Peel color sorting device
果皮色分选装置

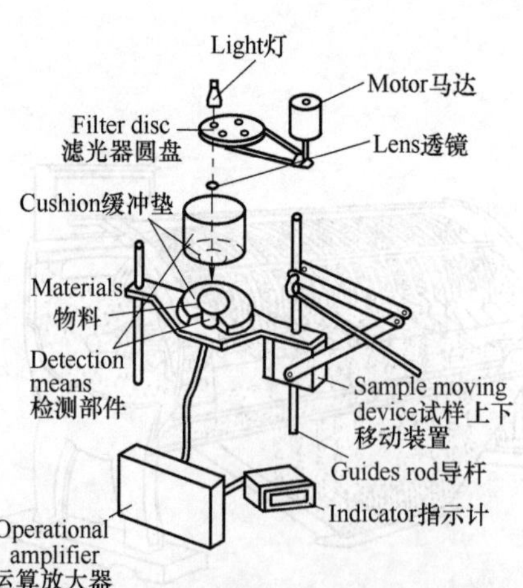

Fig 23.19 Detection device of internal quality of agricultural products
农产品内部品质检测装置

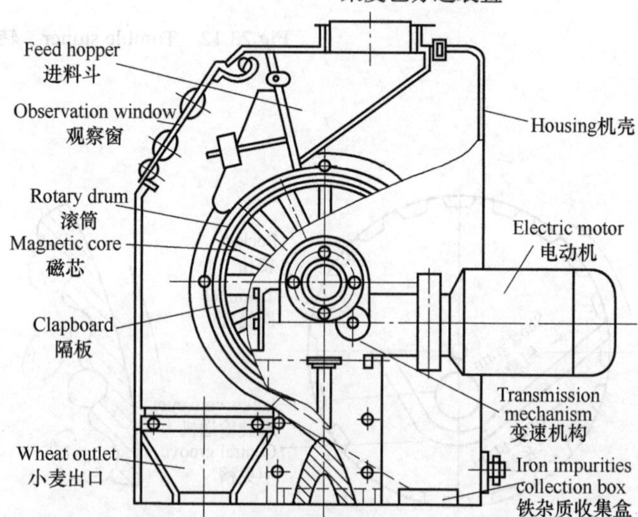

Fig 23.20 Magnetic separation machinery
磁选机械

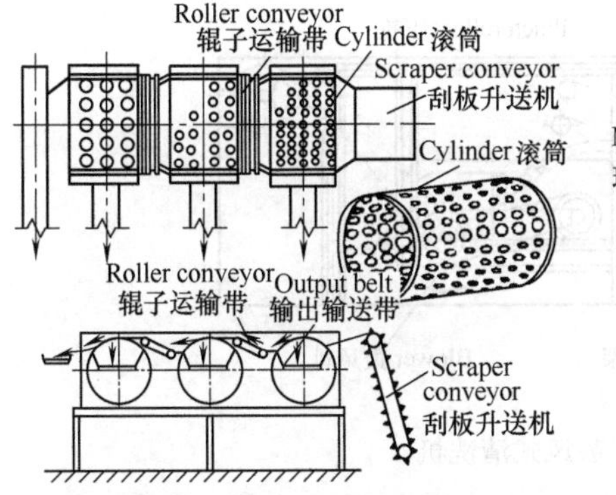

Fig 23.21 Citrus fruit grader with tumble 柑橘用转筒式分级机

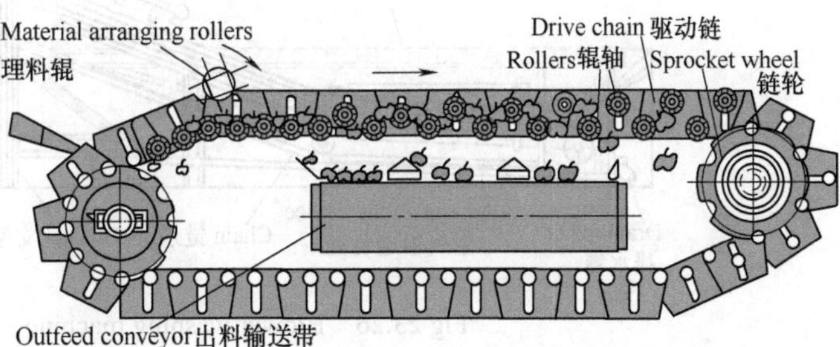

Fig 23.22 Shape grader with roller shaft 辊轴式形状分级机

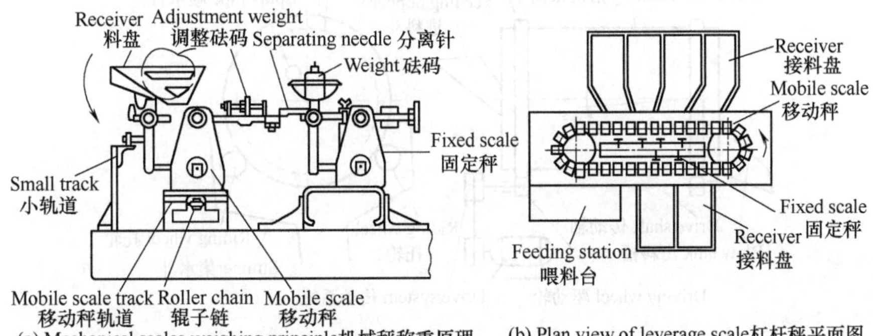

(a) Mechanical scales weighing principle机械秤称重原理 (b) Plan view of leverage scale杠杆秤平面图

Fig 23.23 Weighing-type weight grading machines 称重式重量分级机

23.2 Cleaning machines and devices 清洗机械与设备

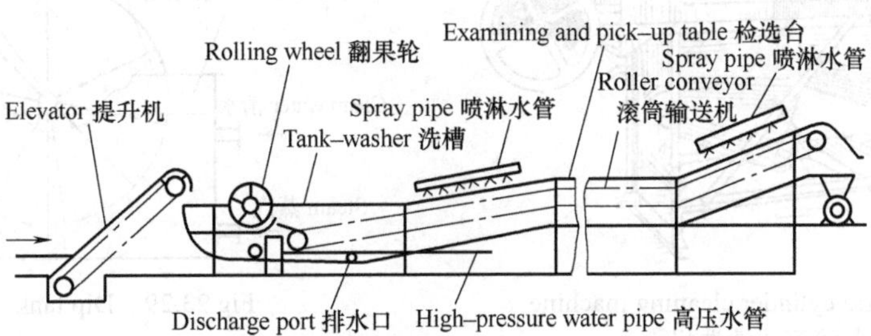

Fig 23.24 Floating washing machine 浮洗机

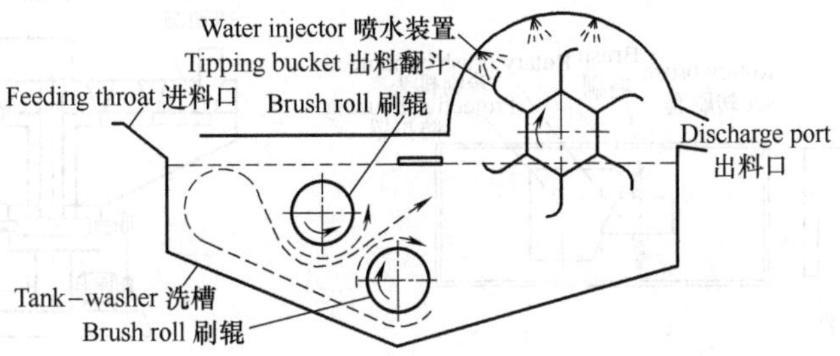

Fig 23.25 Fruit washing machine 洗果机

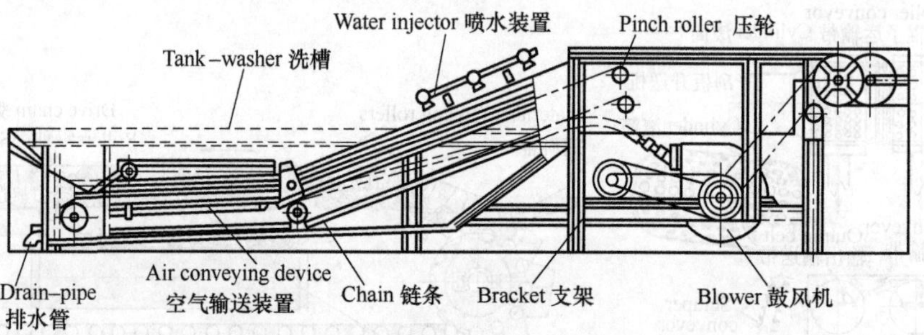

Fig 23.26 Blower washing machine 鼓风式清洗机

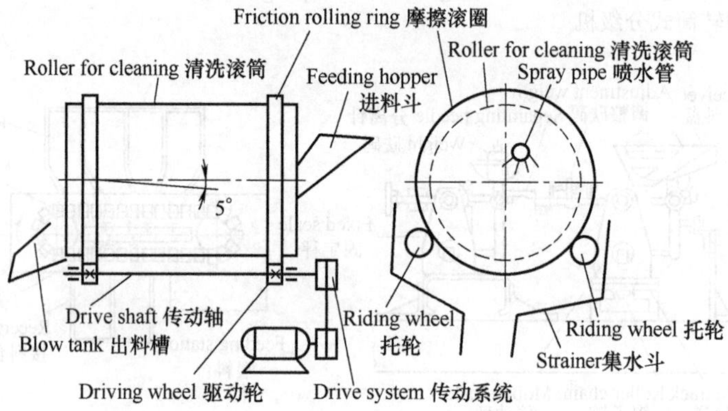

Fig 23.27 Roller washing machine 滚筒式清洗机

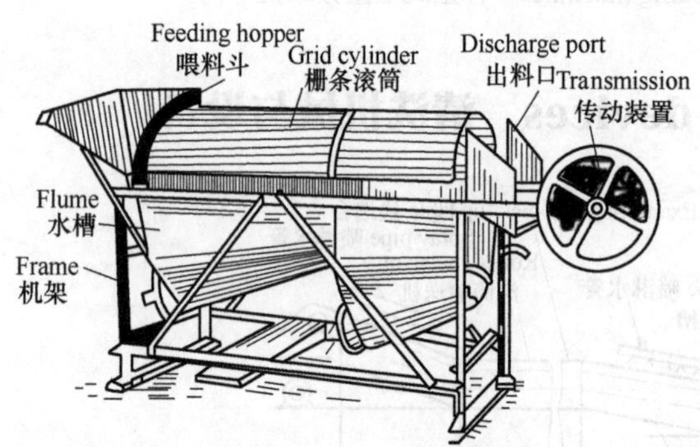

Fig 23.28 Grid cylinder cleaning machine 栅条滚筒式清洗机

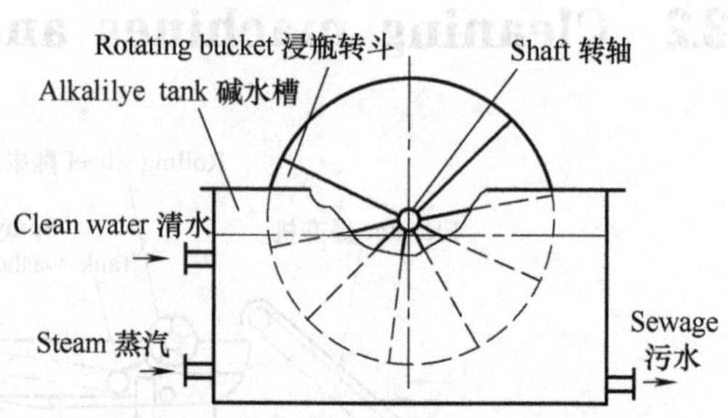

Fig 23.29 Dip tank 浸泡槽

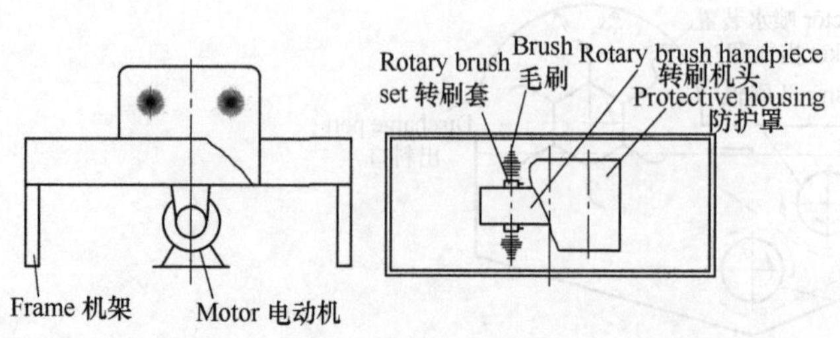

Fig 23.30 Bottle-brushing machine 刷瓶机

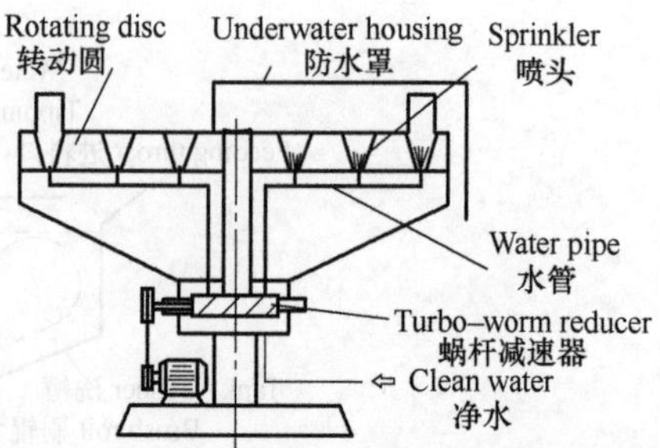

Fig 23.31 Bottle washer 冲瓶机

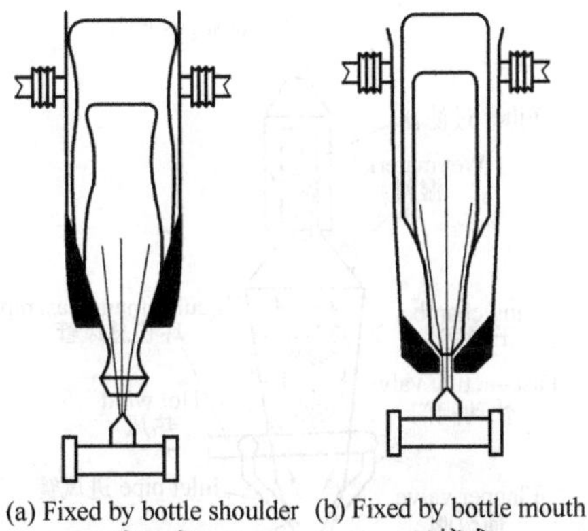

Fig 23.32 Layouts of bottle caps 瓶罩形式

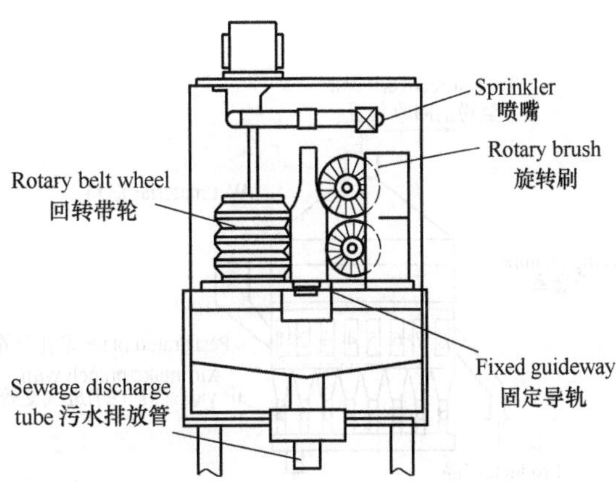

Fig 23.33 External washing machine 外洗机

23.3 Drying equipments 干燥设备

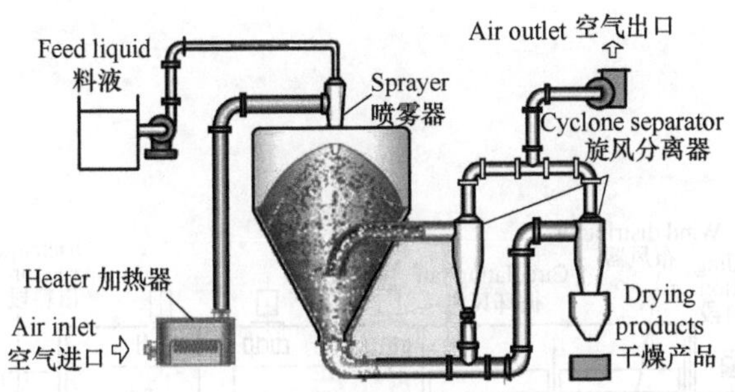

Fig 23.34 Spray drying system 喷雾干燥系统

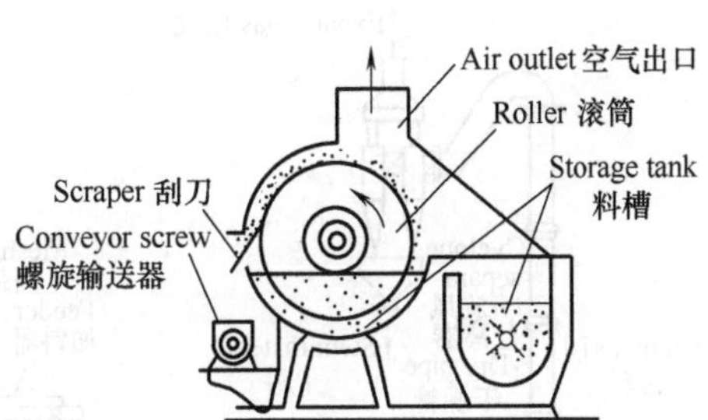

Fig 23.35 Single roller dryer 单滚筒式干燥器

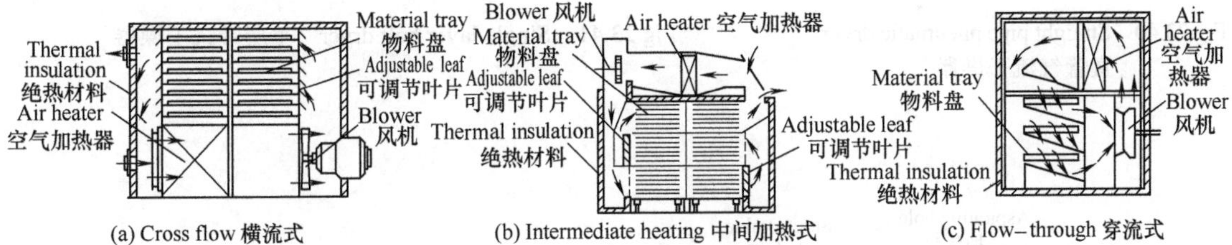

Fig 23.36 Tray dryer 箱式干燥器

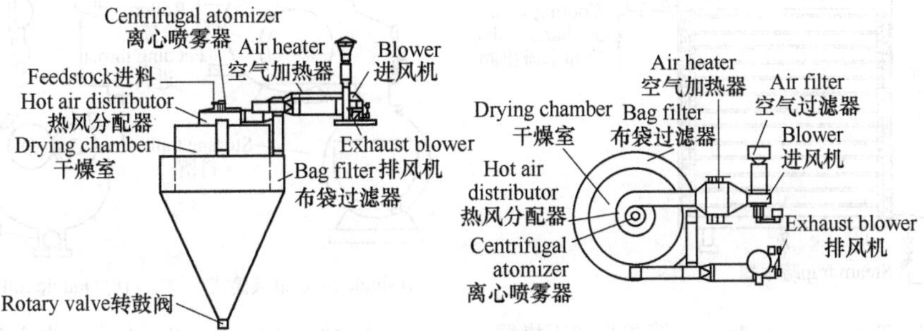

Fig 23.37 Centrifugal spray drying device 离心喷雾干燥装置

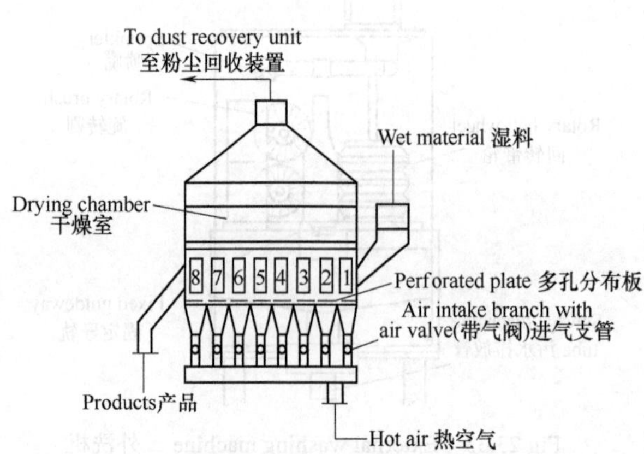

Fig 23.38 Horizontal multi-chamber fluidized bed dryer
卧式多室流化床干燥器

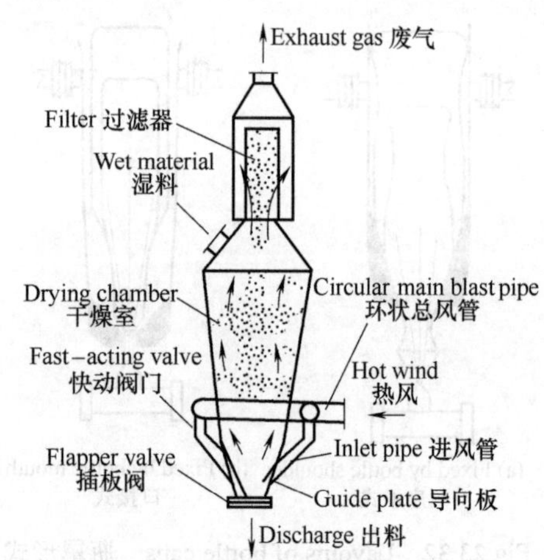

Fig 23.39 Pulsed fluidized bed dryer
脉冲流化床干燥器

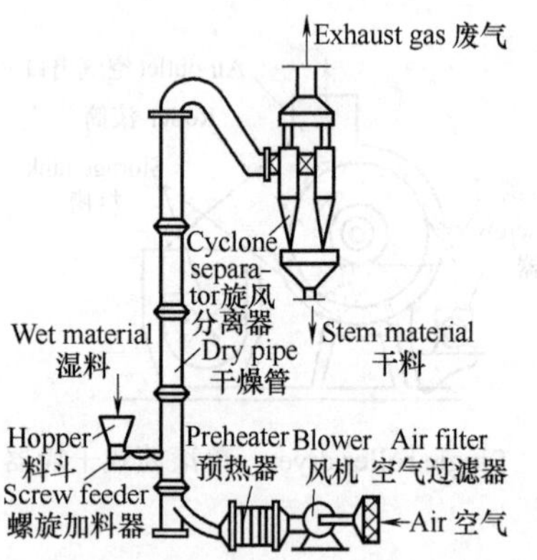

Fig 23.40 Straight pipe pneumatic dryer
直管气流干燥器

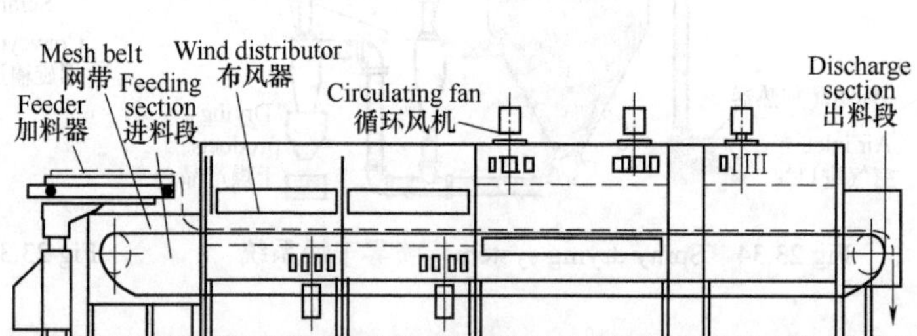

Fig 23.41 Single layer belt dryer 单层带式干燥器

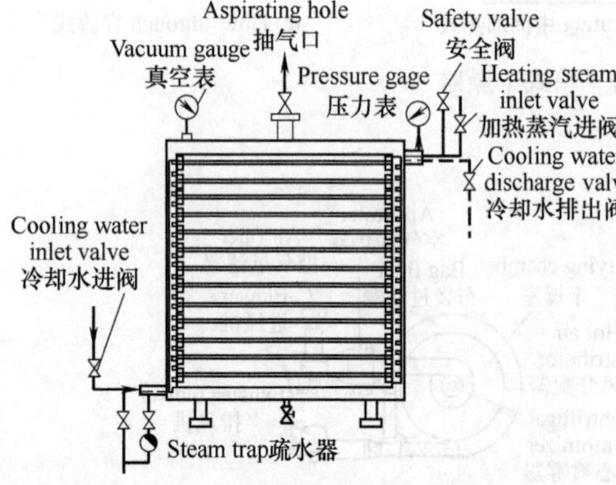

Fig 23.42 Tray vacuum dryer 箱式真空干燥器

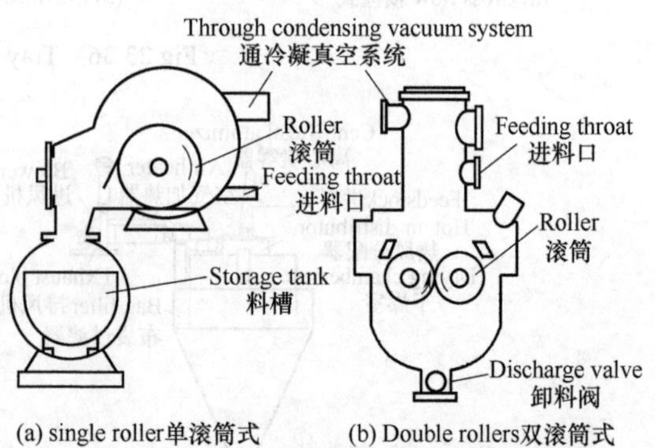

Fig 23.43 Vacuum roller dryer 真空滚筒干燥器

23.4 Crushing and cutting machinery 粉碎切割机械

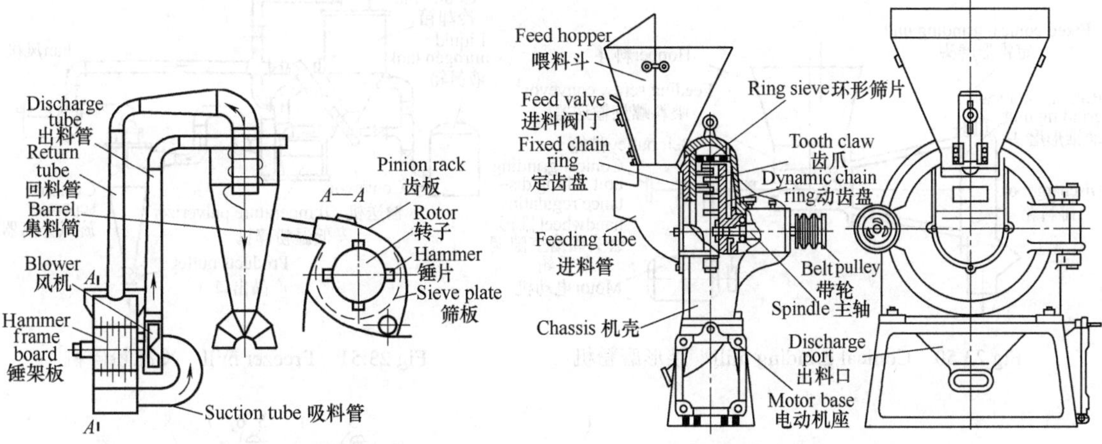

Fig 23.44 Hammer crushing system 锤式粉碎系统

Fig 23.45 Disk mill 齿爪式粉碎机

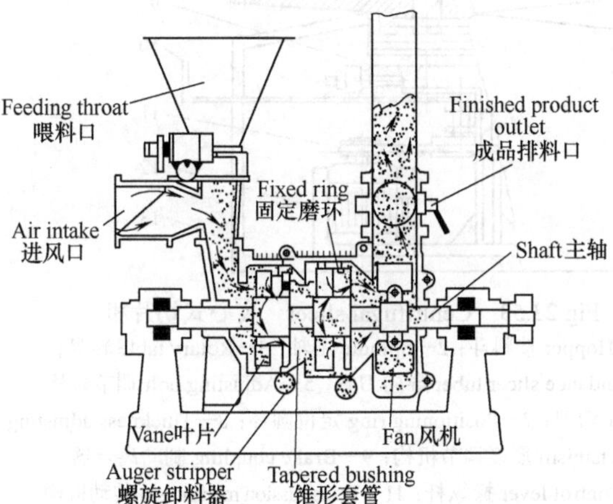

Fig 23.46 Horizontal pulveriser 卧式超微粉碎机

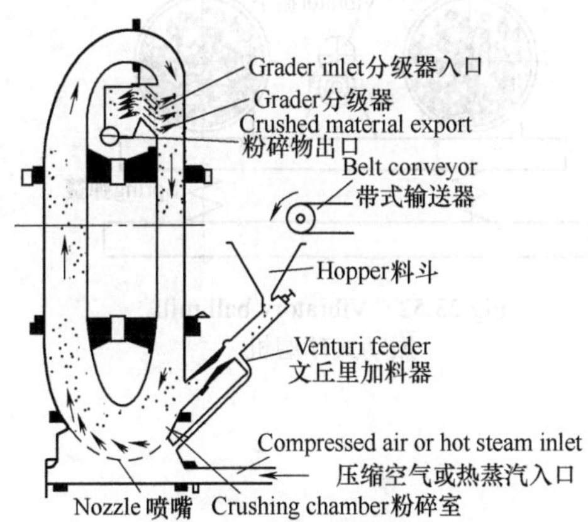

Fig 23.47 Vertical annular jet mill 立式环形喷射式粉碎机

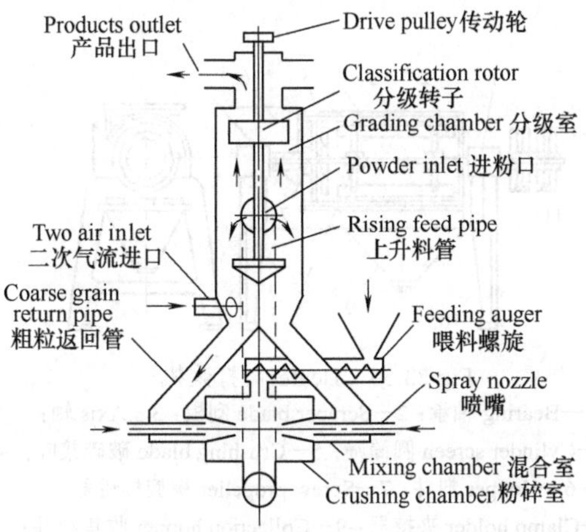

Fig 23.48 Counter-impact jet mill 对撞式气流粉碎机

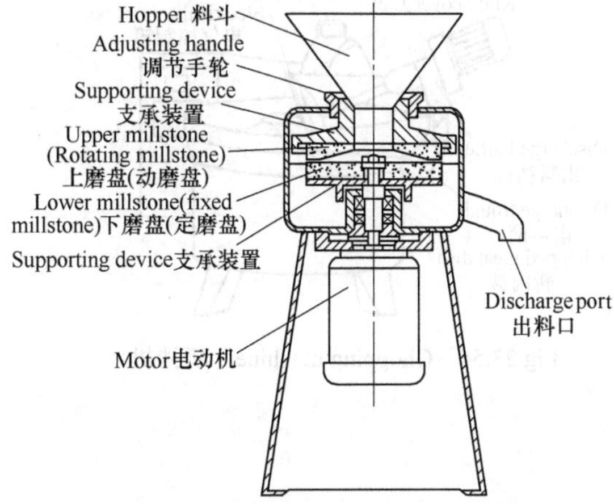

Fig 23.49 Vertical grinding machine 立式磨浆机

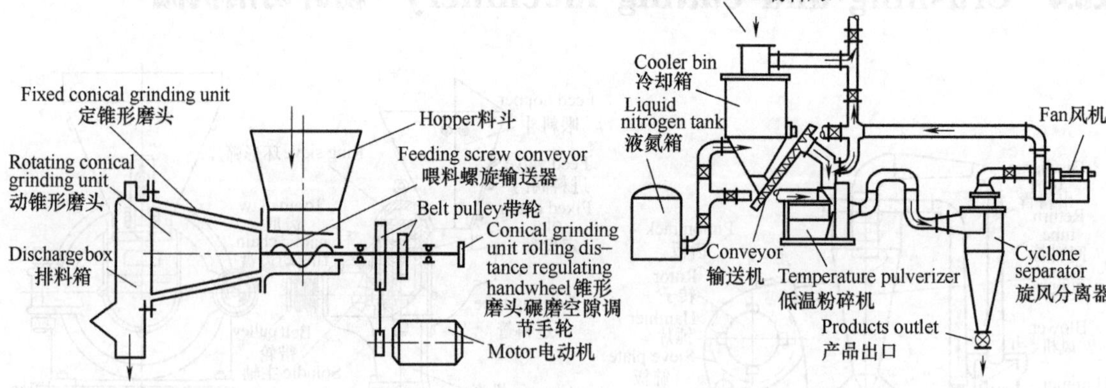

Fig 23.50 Conical grinding mill 锥形磨粉机

Fig 23.51 Freezer mill 冷冻粉碎机

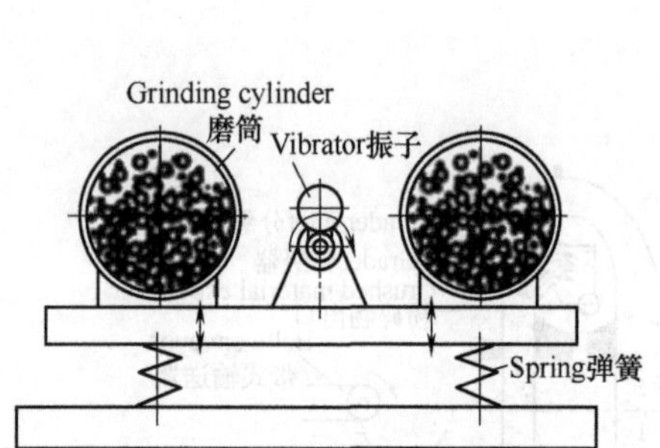

Fig 23.52 Vibratory ball mill 振动式球磨机

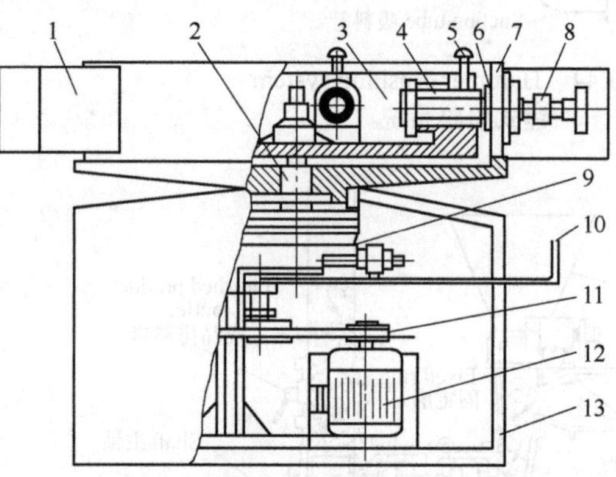

Fig 23.53 Centrifugal slicer 离心式切片机

1—Hopper 接料斗；2—Spindle 主轴；3—Rotary table 转盘；4—Guidance slicer tuber 导向刀管；5—Adjusting bolt 调节螺栓；6—Slicer 切刀；7—Positioning ring 定位圆环；8—Thickness adjusting mechanism 厚度调节机构；9—Brake coupling 制动离合器；10—Control lever 操纵杆；11—Transmission mechanism 传动机构；12—Motor 电动机；13—Chassis 机架

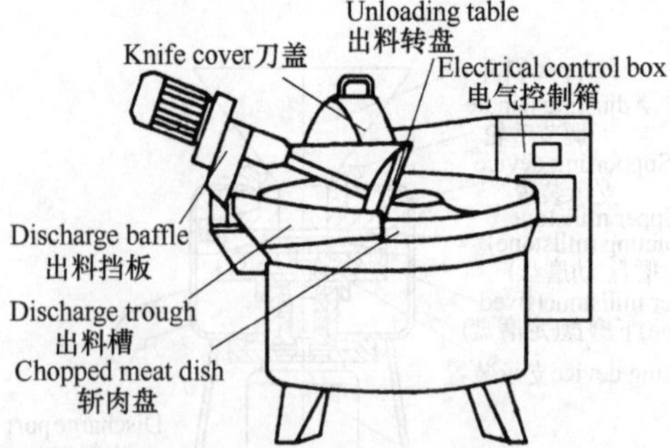

Fig 23.54 Chopping machine 斩拌机

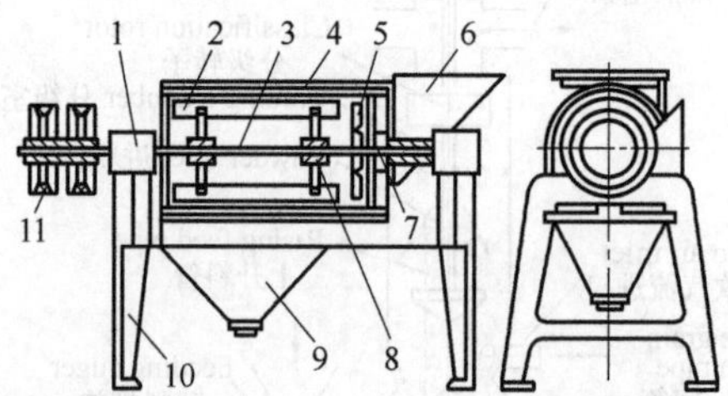

Fig 23.55 Beater 打浆机

1—Bearing 轴承；2—Scraper blade 刮板；3—Axis 轴；4—Cylinder screen 圆筒筛；5—Crushing blade 破碎桨叶；6—Hopper 料斗；7—Screw propeller 螺旋推进器；8—Clamp holder 夹持器；9—Collection hopper 收集料斗；10—Chassis 机架；11—Transmission system 传动系统

23.5 Hot and cold exchange processing machinery 冷热交换处理机械

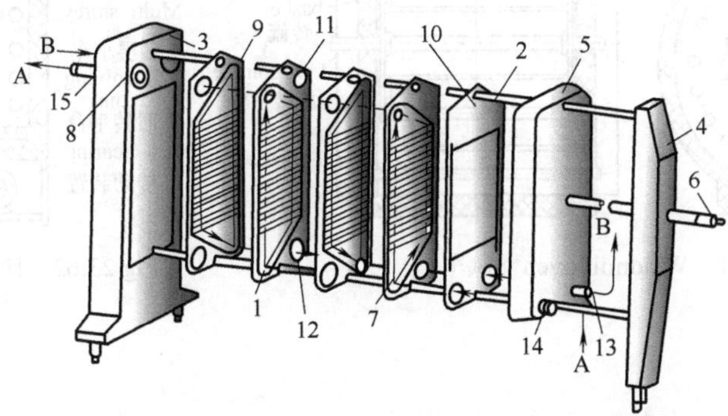

Fig 23.56 Procedural map of plate heat exchanger 片式热交换器流程图

1—Heat transfer plate 传热板；2—Guide bar 导杆；3—Front bracket (dead plate) 前支架（固定板）；4—Rear bracket 后支架；5—Press plate 压紧板后支架；6—Compression screw 压紧螺杆；7—Plate frame rubber washer 板框橡胶垫圈；8—Joint pipe 连接管；9—Upper corner pore 上角孔；10—Separating plate 分界板；11—Ring rubber washer 圆环橡胶垫圈；12—Lower corner pore 下角孔；13 ～ 15—Joint pipe 连接管

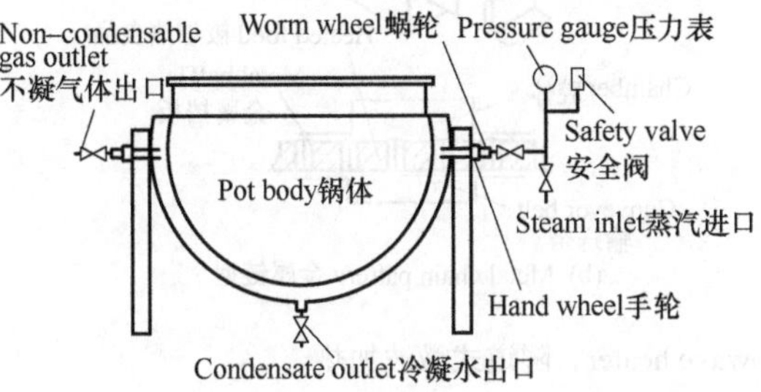

Fig 23.57 Tiltable jacketed pot 可倾式夹层锅

Fig 23.58 Dead jacketed pot 固定式夹层锅

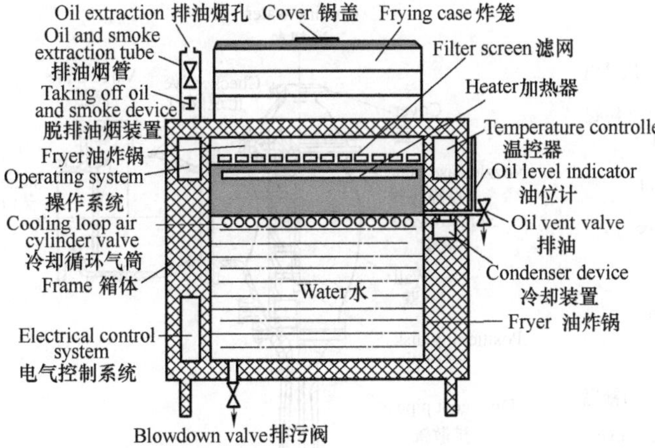

Fig 23.59 Intermittent smoke-free multi-functional frying machine mixing water with oil

间歇式无烟型多功能水油混合式油炸设备

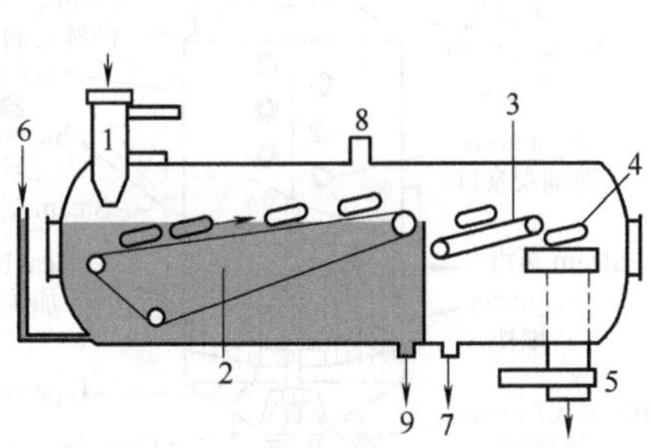

Fig 23.60 Successional vacuum low temperature frying machine 连续式真空低温油炸设备

1—Air-lock valve 闭风器；2—Conveyor 输送器；3，4—Non-oil zone conveyor 无油区输送带；5—Discharge air-lock valve 出料闭风器；6—Oil filler pipe 加油管；7—Residual oil outlet 残油出口；8—Vacuum pump interface 真空泵接口；9—Waste oil outlet 废油出口

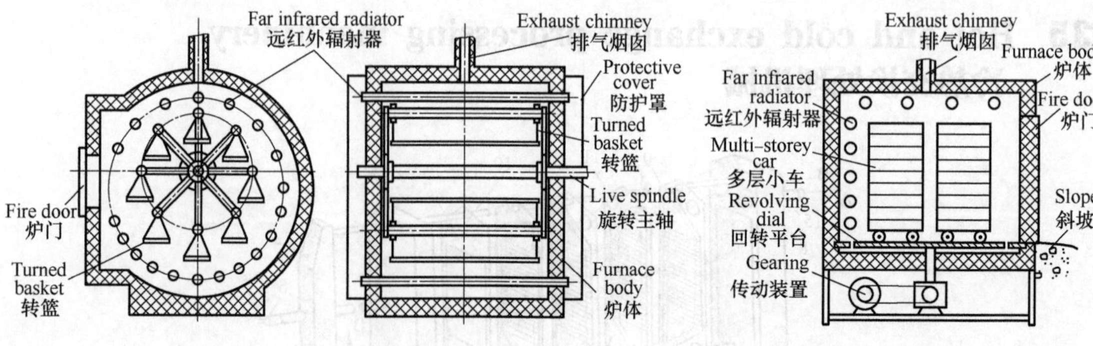

Fig 23.61 Windmill oven 风车炉

Fig 23.62 Horizontal rotary furnace 水平旋转炉

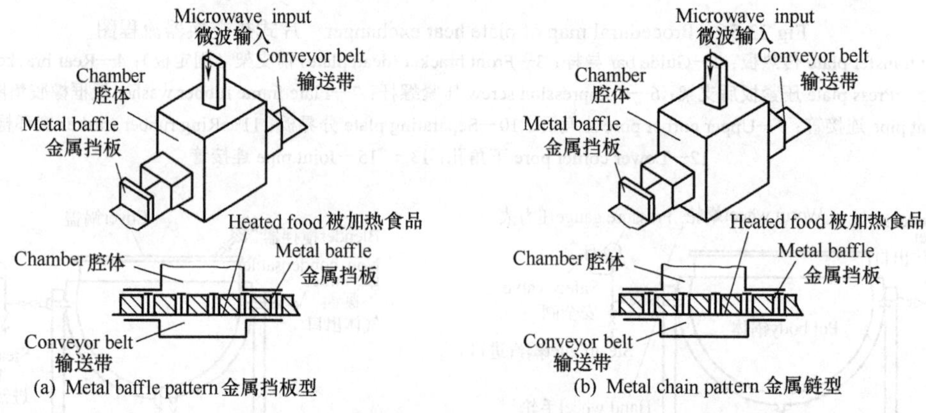

Fig 23.63 Continuous tunnel microwave heater 隧道式微波加热

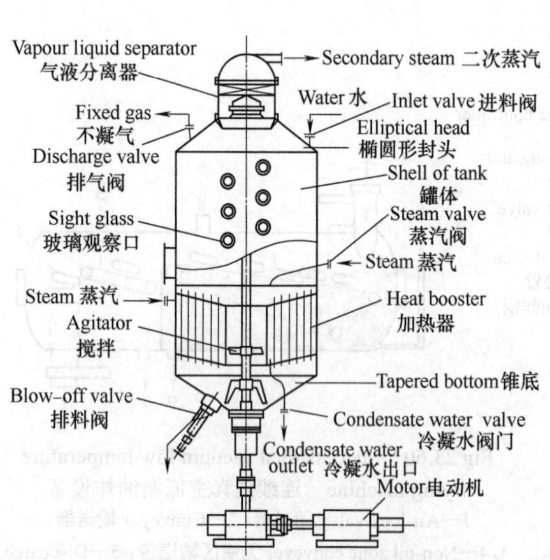

Fig 23.64 Centre cycling evaporator
中央循环式蒸发器

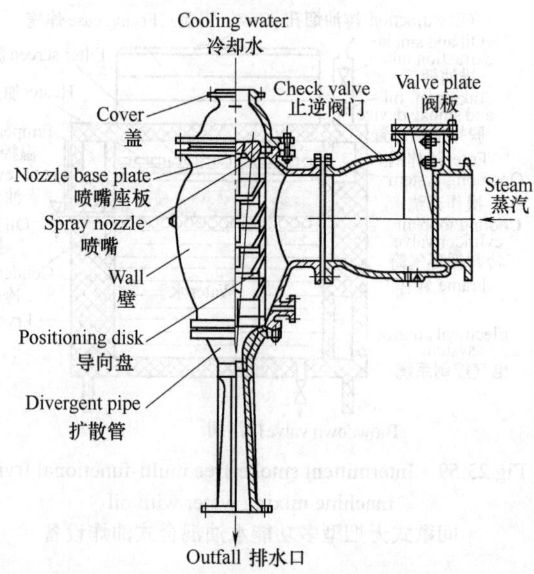

Fig 23.65 Jet condenser 喷射式冷凝器

23.6 Sterilization machinery 杀菌机械

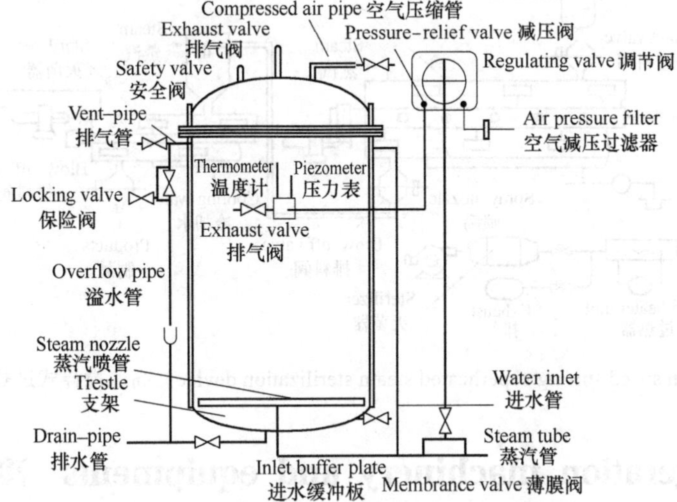

Fig 23.66 Vertical sterilization pot 立式杀菌锅

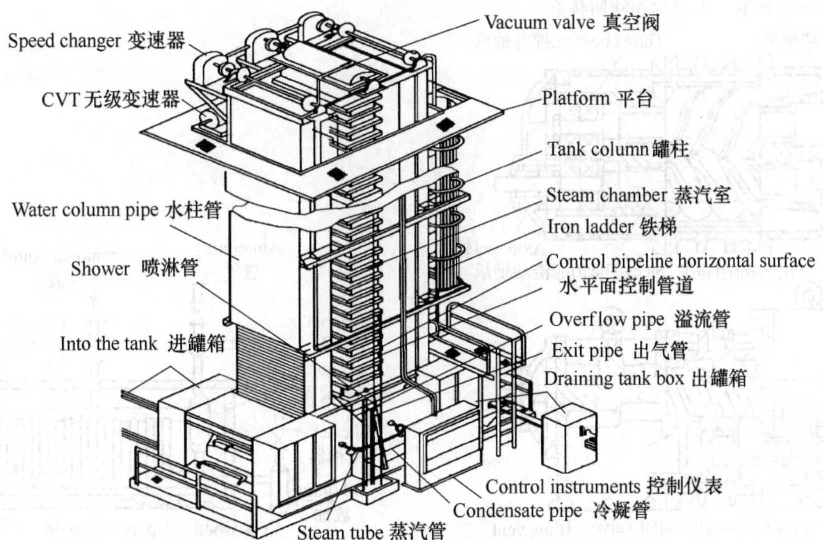

Fig 23.67 Hydrostatic continuous sterilization equipment 静水压连续杀菌设备装置

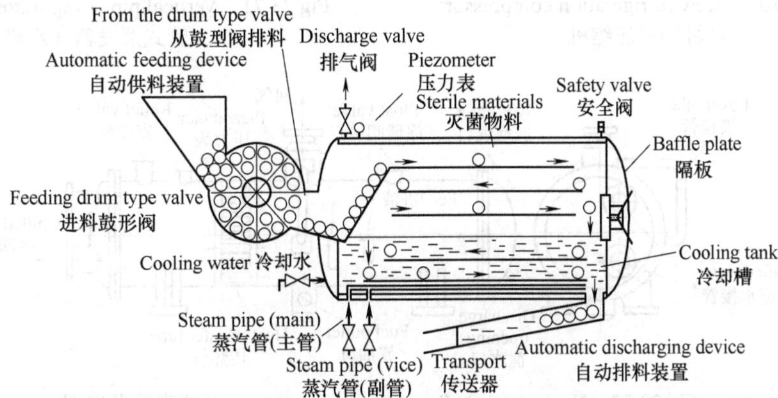

Fig 23.68 Water seal type continuous high pressure sterilization equipment 水封式连续杀菌设备

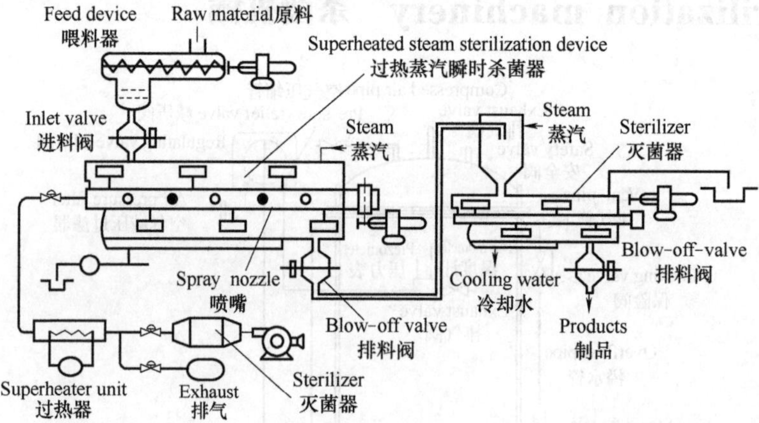

Fig 23.69 High speed stirring superheated steam sterilization device 高速搅拌式过热蒸汽杀菌装置

23.7 Refrigeration machinery and equipments 冷冻机械与设备

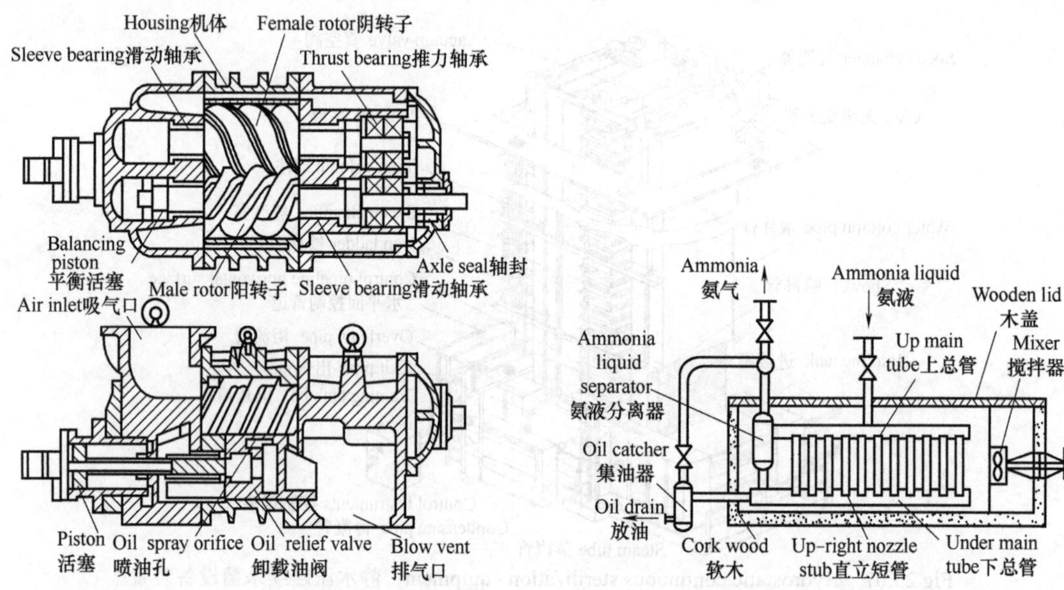

Fig 23.70 Screw refrigeration compressor
螺杆制冷压缩机

Fig 23.71 Vertical pipe evaporator (Water tank)
立管式蒸发器（水箱）

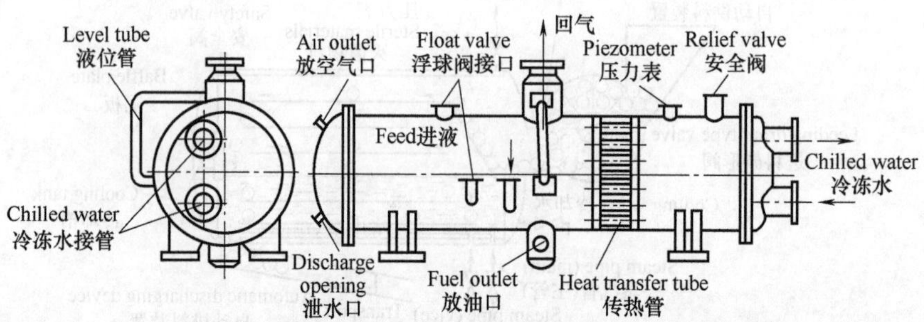

Fig 23.72 Horizontal shell and tube evaporator 卧式壳管蒸发器

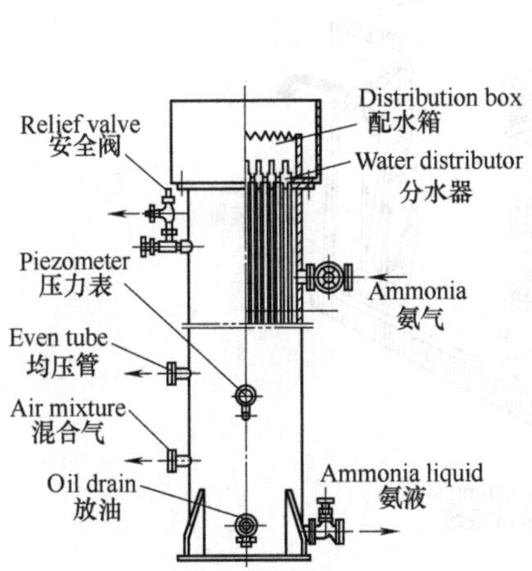

Fig 23.73 Vertical shell and tube condenser 立式壳管冷凝器

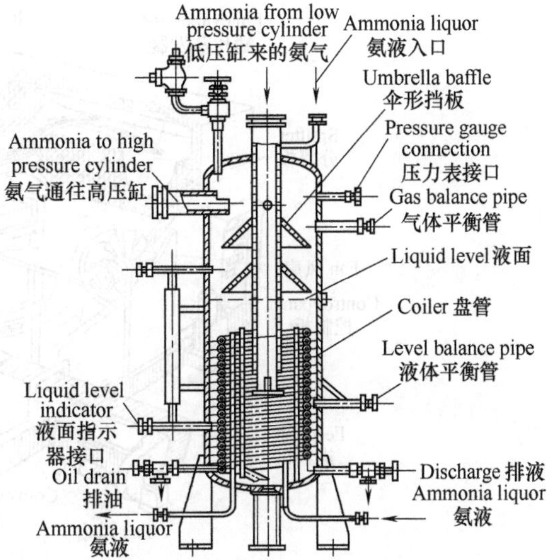

Fig 23.74 Intercooler 中间冷却器

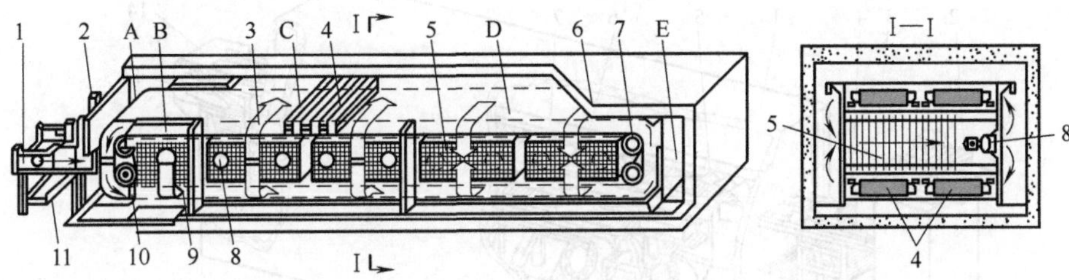

Fig 23.75 Belt freezing tunnel 带式冻结隧道

1—Loading and unloading device 装卸设备；2—Defrost means 除霜装置；3—Air flow direction 空气流动方向；4—Freezing tray 冻结盘；5—Plate-fin type evaporator 板片式蒸发器；6—Insulation housing 隔热外壳；7—Steering device 转向装置；8—Axial fans 轴流风机；9—Light tube evaporator 光管蒸发器；10—Hydraulic drive mechanism 液压传动机构；11—Frozen block conveyor 冻结块输送带；A—Drive chamber 驱动室；B—Water separation chamber 水分分离室；C，D—Freezing chamber 冻结室；E—Bypass 旁路

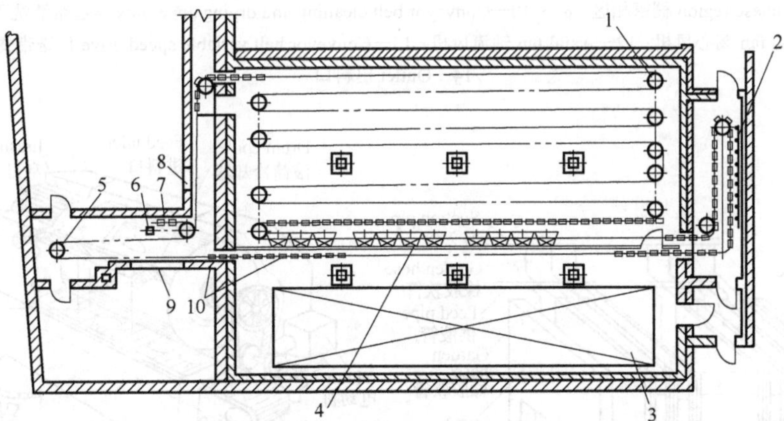

Fig 23.76 Hanging basket type continuous freezer 吊篮式连续冻结装置

1—Transverse wheel 横向轮；2—Ethanol sprinkler system 乙醇喷淋系统；3—Evaporator 蒸发器；4—Axial fan 轴流风机；5—Tension wheel 张紧轮；6—Drive motor 驱动电动机；7—Deceleration device 减速装置；8—Discharge port 卸料口；9—Feed port 进料口；10—Chain plate 链盘

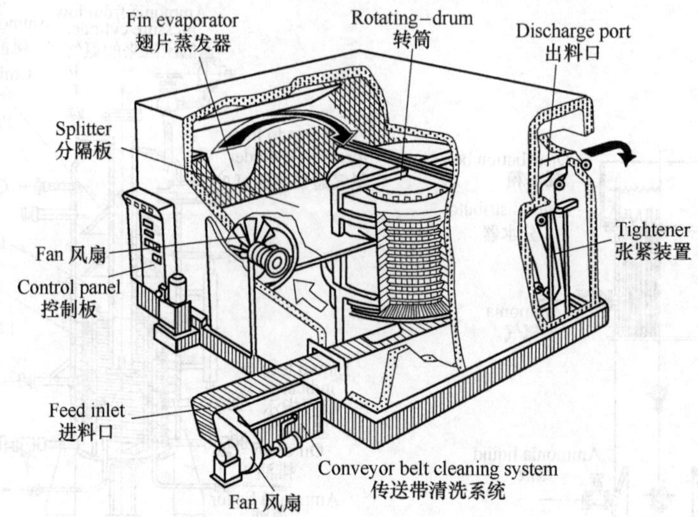

Fig 23.77 Spiral freezer 螺旋式冻结装置

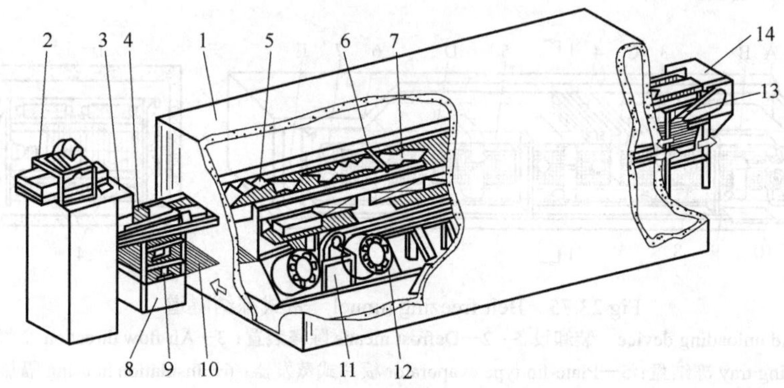

Fig 23.78 One stage belt type fluidization freezer 单级带式流态冻结装置示意图

1—Insulation 隔热层；2—Dehydration oscillator 脱水振荡器；3—Metering funnel 计量漏斗；
4—Variable speed feed belt 变速进料带；5—Loose-phase zone 松散相区；6—Refiner rod 匀料棒；
7—Dense phase region 稠密相区；8 ～ 10—Conveyor belt cleaning and drying apparatus 传送带清洗干燥装置；
11—Centrifugal fan 离心风机；12—Axial fan 轴流风机；13—Conveyor belt variable speed drive 传送带变速驱动装置；
14—Outlet 出料口

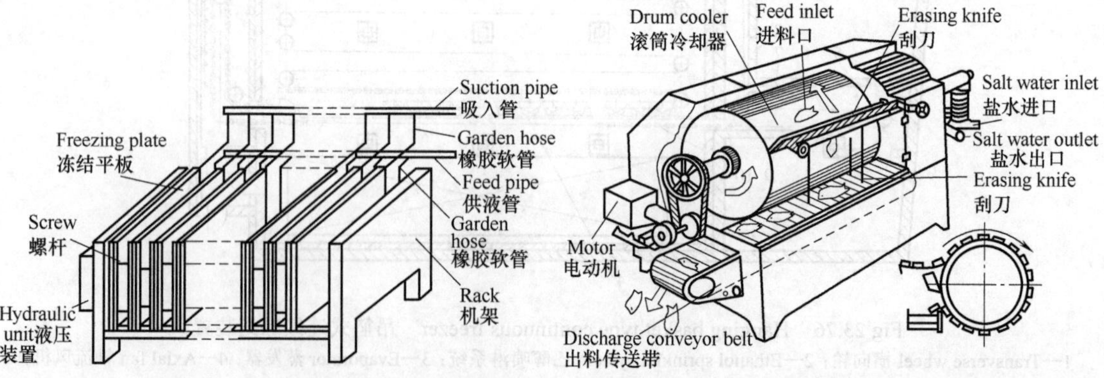

Fig 23.79 Vertical plate freezer 立式平板冻结装置

Fig 23.80 Rotary freezer 回转式冻结装置

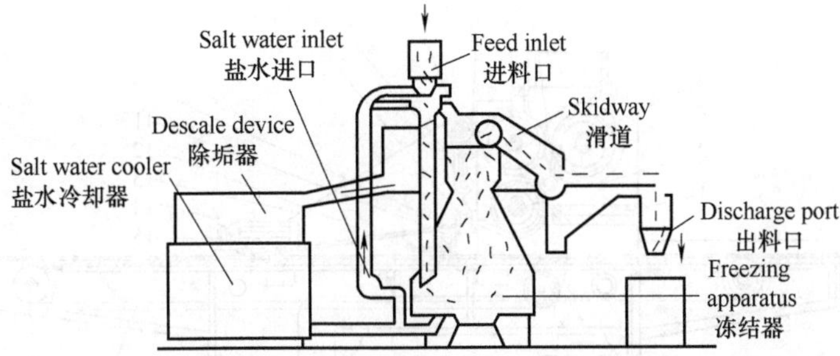

Fig 23.81 Brine continuous immersion freezing device 盐水连续浸渍冻结装置

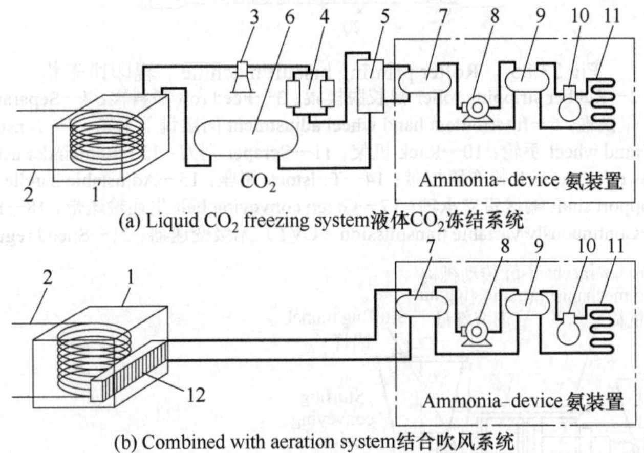

Fig 23.82 CO_2 freezing device CO_2 冻结装置

1—Freezer envelope structure 冻结器围护结构；2—Screw band 螺旋带；3—CO_2 compressor CO_2 压缩机；4—Pairs of CO_2 dryer 成对的 CO_2 干燥机；5—CO_2 condenser CO_2 冷凝器；6—Liquid CO_2 reservoir 液态 CO_2 储液器；7,9—Low or medium pressure reservoir 低压或中压储存器；8,10—Low and high-pressure stage ammonia compressor 低压级与高压级氨压缩机；11—Ammonia condenser 氨冷凝器；12—Air cooler 空气冷却器

23.8 Forming machines 成形机械

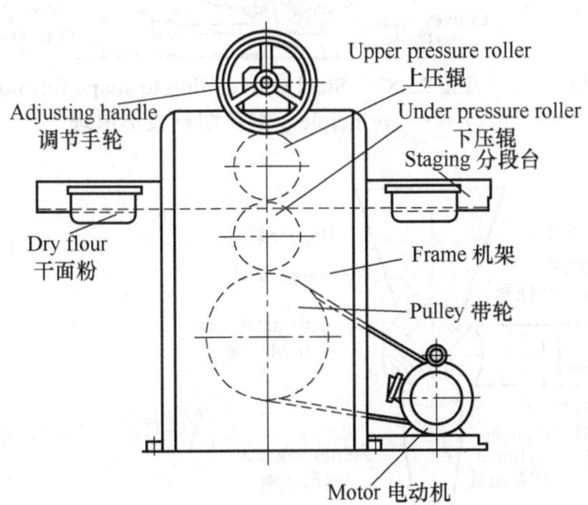

Fig 23.83 Intermittent horizontal calender 间歇卧式压延机

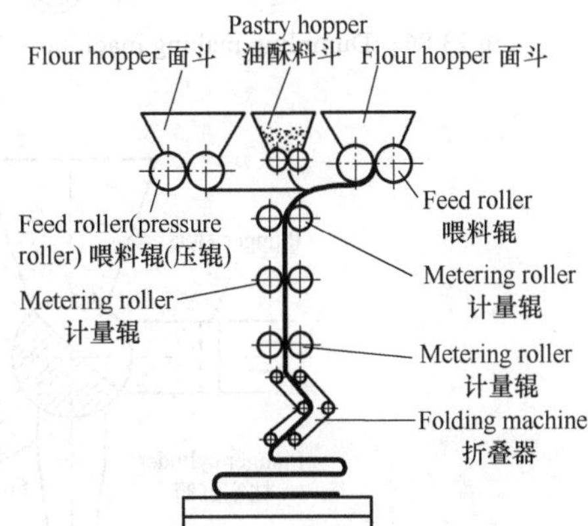

Fig 23.84 Vertical roller press 立式辊压机

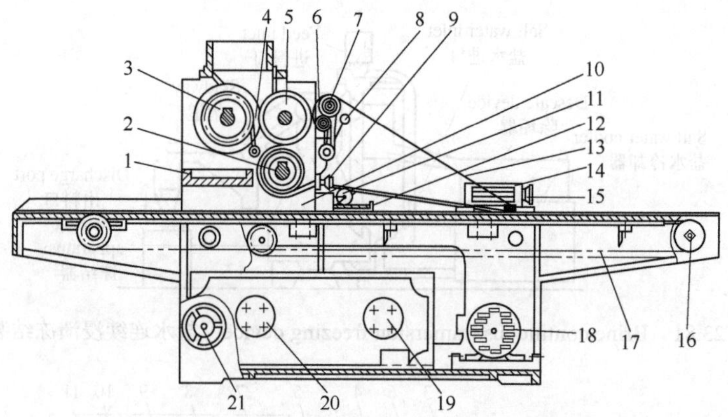

Fig 23.85 Roller printing biscuit machine 辊印饼干机

1—Feed tray 接料盘；2—Rubber stripping roller 橡胶脱模辊；3—Feed roll 喂料辊；4—Separating scraper 分离刮刀；5—Impression roller 印模辊；6—Intermittent hand wheel adjustment 间歇调节手轮；7—Tensioning wheel 张紧轮；8—Handle 手柄；9—Hand wheel 手轮；10—Rack 机架；11—Scraper 刮刀；12—Remainder accepting tray 余料接盘；13—Canvas releasing belt 帆布脱模带；14—Tailstock 尾座；15—Adjustable handle 调节手柄；16—Conveying belt support shaft 输送带支承轴；17—Green conveying belt 生坯输送带；18—Electric motor 电动机；19—Reducer 减速器；20—Continuously variable transmission（CVT）无级变速器；21—Speed regulation handwheel 调速手轮

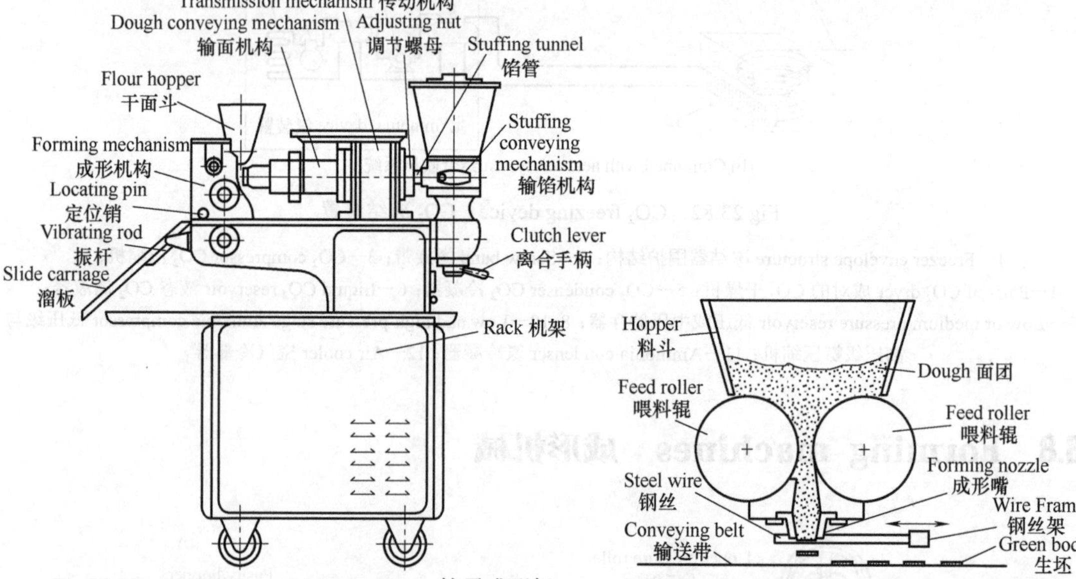

Fig 23.86 Dumpling making machine 饺子成形机

Fig 23.87 Steel wire cutting to shape forming principle 钢丝切割成形原理

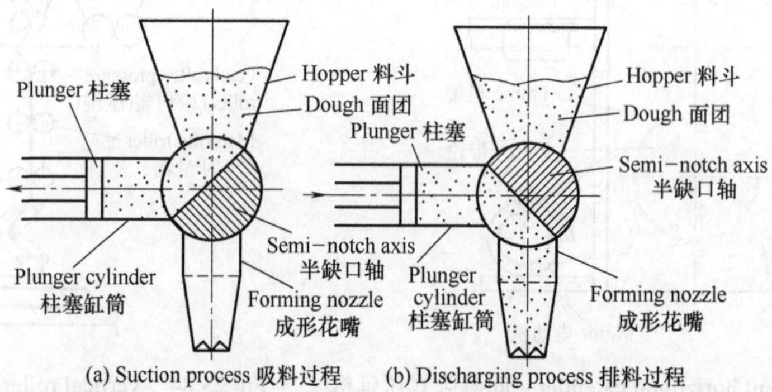

(a) Suction process 吸料过程 (b) Discharging process 排料过程

Fig 23.88 Plunger dosing device 柱塞式定量供料装置

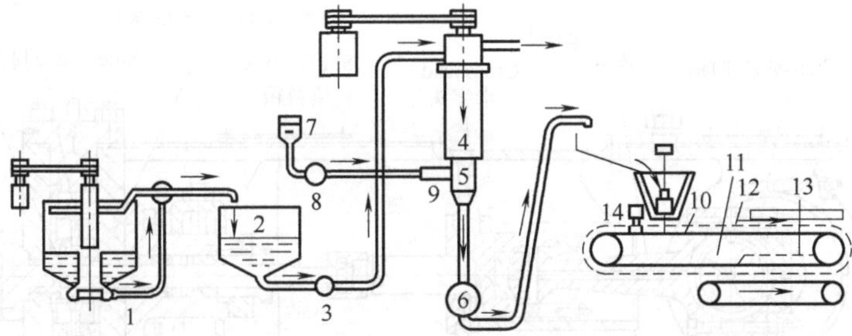

Fig 23.89 Continuous candy pouring molding machine 连续式糖果浇模成形机

1—Pot of melting sugar 化糖锅；2—Syrup storage pot 糖浆储锅；3—Syrup pump 糖浆泵；4—Vacuum chamber of boiling sugar 真空熬糖室；5—Spices mixing chamber 香料混合室；6—Unloading pump 卸料泵；7—Acid, spices, coloring liquid container 酸、香料、色素液容器；8—Metering pumps 计量泵；9—Suction head 吸入头；10—Mold plate 模盘；11—Stripping point 脱模点；12—Airflow above mold plate 模盘上方气流；13—Airflow under mold plate 模盘下方气流；14—Lubricant sprayer 润滑剂喷雾器

23.9 Material conveying machinery 物料输送机械

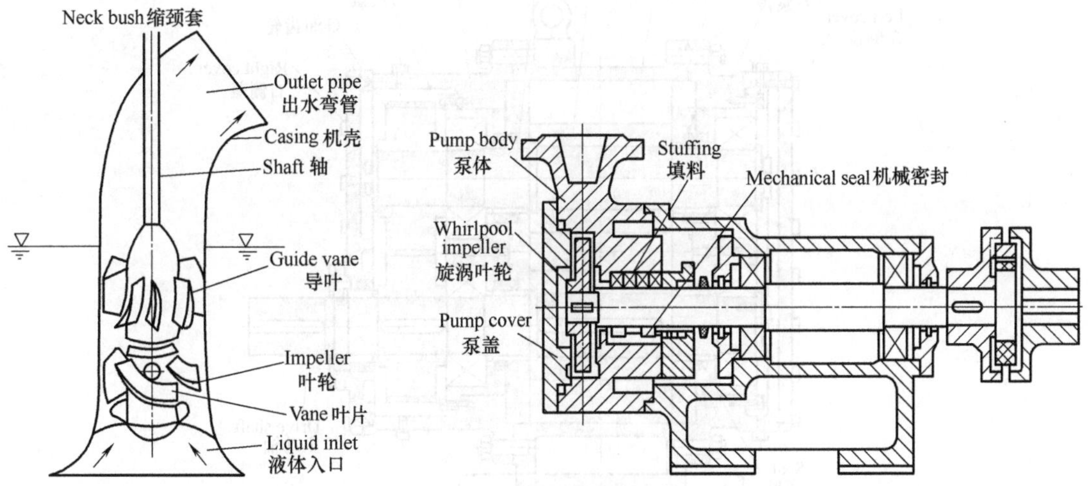

Fig 23.90 Axial flow pump 轴流泵 Fig 23.91 Single-stage cantilevered vortex pump 单级悬臂式旋涡泵

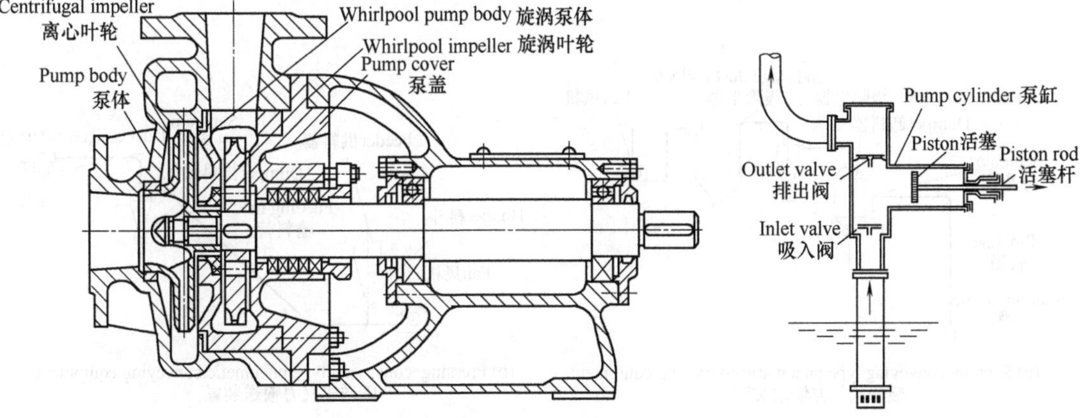

Fig 23.92 Centrifugal vortex pump 离心旋涡泵 Fig 23.93 Reciprocating pump 往复泵

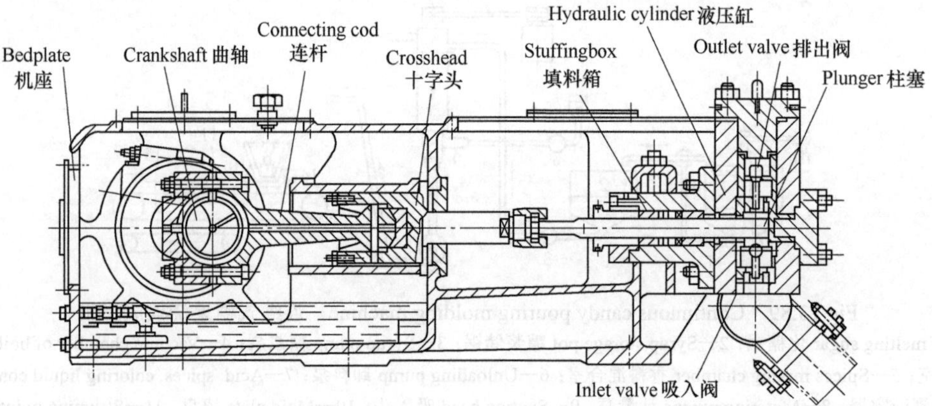

Fig 23.94　Horizontal three plunger high-pressure pump　卧式三柱塞高压泵

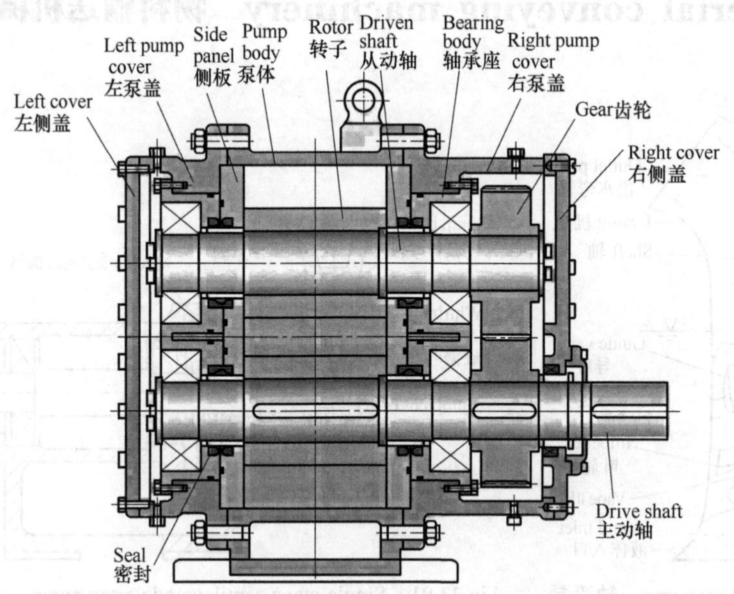

Fig 23.95　Energy saving rotor pump　节能转子泵

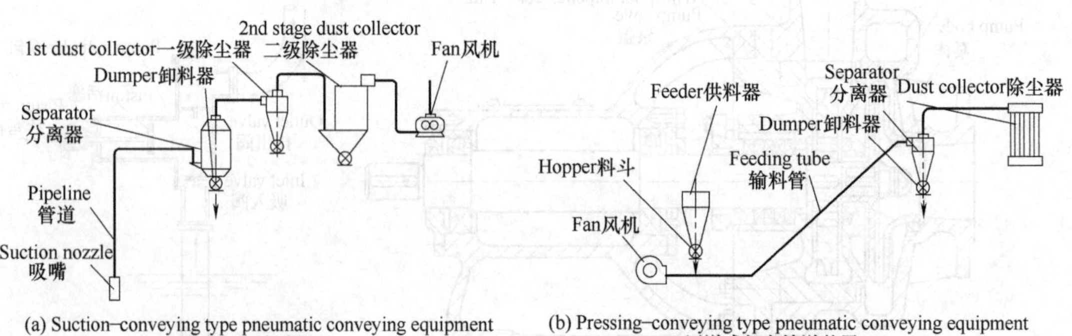

(a) Suction–conveying type pneumatic conveying equipment
吸送式气力输送装置

(b) Pressing–conveying type pneumatic conveying equipment
压送式气力输送装置

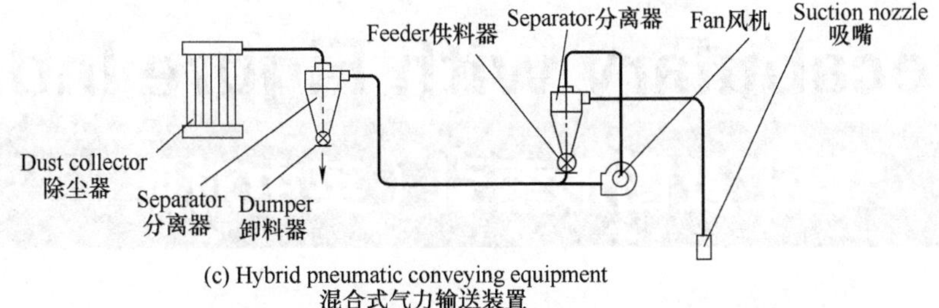

Fig 23.96 Conveying equipment 气力输送装置

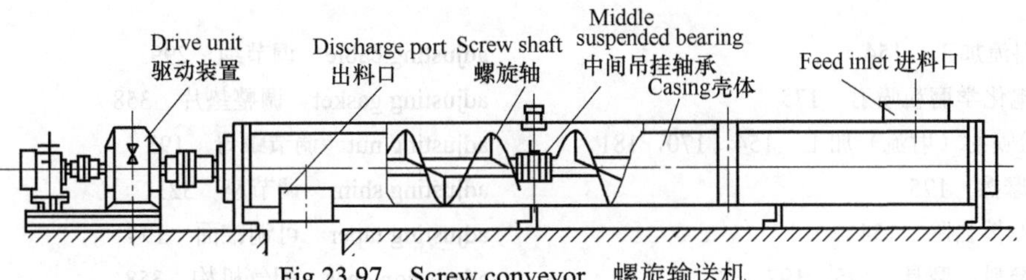

Fig 23.97 Screw conveyor 螺旋输送机

Cover 头罩
Drive roller
传动滚筒
Traction piece
牵引件
Case外壳
Discharge outlet
卸料口
Orifice孔口
Hopper料斗
Traction piece
牵引件
Orifice孔口
Higher feeding inlet
高位装料口
60°
45°
Tension device
张紧装置
Orifice孔口
Lower feeding inlet
低位装料口

Fig 23.98 Hopper elevator 斗式提升机

Traction component
牵引构件
Scraper
刮板
Driving wheel驱动轮
A—A
A
A
Tension pulley
张紧轮
Compression
roller压辊
Supporting
roller托辊
Casing机壳

Fig 23.99 Scraper conveyor 刮板输送机

Vocabulary with Figure Index

词汇及图形索引（英中对照）

A

AFM　磨料流加工　154

AJECM　电化学磨粒喷射　175

AJM　磨粒喷射（射流）加工　154，170，181

abrading　摩擦　175

abrasion　摩擦去除　154

abrasive　磨料，磨具　175，197

abrasive belt　砂带　167

abrasive disc grinder　角磨机　223

abrasive disc precision polisher　圆盘精密抛光机　223

abrasive jet　磨料喷射　226

abrasive paper　砂纸　126

abrasive slurry　研磨膏（浆）　176

abrasive water jet　磨料水射流　226

abrasive water jet machining (AWJM)
磨料水射流加工　170,182

accelerator pedal　加速踏板　349

accommodation　起居舱　399

accumulator　蓄能器储料器，蓄电池　92，175，346

ackerman arm　梯形臂　371

acoustic positioning beacon　超声定位信号浮标　400

actuator　执行元件　93

addendum　齿顶高　66

adherend　粘接件　256

adhesive　黏结剂　19，49，256

adjustable air vane　空气调节叶片　311

adjustable head　可调头架　218

adjustable impeller　动叶轮　316

adjustable linear shaker　可调直线型振动筛　395

adjustable pin　调整销　206

adjustable spanner　活动板手　126

adjustable supporting　可调支承　210

adjustable supporting pin　可调支承钉　205

adjustable valve　调节阀　316

adjusting cable　调节绳　299

adjusting gasket　调整垫片　358

adjusting nut　调节螺母　193

adjusting shim　调节座　321

adjusting taper　可调锥面　194

admission gear　配气机构　358

advance device　前进装置　350

aftercooler　后置冷却器　92，324

air cap　气盖　137

air compressor　空压机　93

air filter　气体过滤器，空气滤清器　92，93

air gap　气隙　12

air inlet camshaft　进气凸轮轴　358

air intake control valve　进气控制阀　355

air jet　气射流　182

air mass sensor　空气质量传感器　349

air valve　气阀　151

aligner　定位器　178

allen key　内六角扳手　126

alternator　发电机　308

ammeter　电流表　312

amphibian　水陆两用车　375

amplifier　放大器　233

anchor　锚绳　383

anchor plate　固定板　299

anchor rack　锚架　399

angle drilling jig　角度钻孔夹具　207

angle milling cutter　角度铣刀　187

angle of thread　牙型角　193

angle pin　斜导柱　111

angle plate　角铁　203

angular contact ball bearing　角接触球轴承　258

angularity　倾斜度　22

annular gear 环齿 370
annulus ring 环套 71
anti-roll bar 防转杆 361
anvil 砧座 123
aperture 小孔 178
apron 护脚板 329
arbor 刀杆 213，216
arc 弧焊，电弧 122，148
arc welding 电弧焊 144
arc welding eqiupment 弧焊设备 147
arch 顶梁 219
armature 衔铁 298，355
arrow head 箭头 21
asphalt paver 沥青摊铺机 381
auger stripper 螺旋卸料器 413
automatic latch 自动闩栓 335
automatic screw machine 自动螺纹加工机床 225
auxiliary oil pump 副油泵 325
auxiliary piston 附加活塞 94
auxiliary shaft 副轴 225
axial flow pump 轴流泵 423
axial flow turbine 轴流涡轮 319
axial piston pump 轴向柱塞泵 82
axial relief angle 轴向铲背面 188
axis 轴线 4，193
axle frame 轮架 340

Babbitt lining 巴氏合金衬套 262
BUE 积屑瘤 155
back bearing housing 后轴承箱 316
backlash 侧隙 66
backup light 倒车灯 366
baffle board 隔板 316
balance drum 平衡鼓 323
balance weight 平衡重，配重 358，384
balanced core 平衡型芯 118
balanced pellet lifter 平衡托盘提升器 336
balancing piston 平衡活塞 326
ball bearing 滚动轴承，球轴承 258，267
ball cage 球体保持架 267
ball screw 滚珠丝杠 213，241
ball thrust bearing 推力球轴承 266
ball transmission 球体传动 54
ball valve 球阀 350
ball-lock ratchet 球体锁紧棘轮 58
baluster 栏杆 390
baluster steel ring 栏杆钢圈 390
bar forming 锻棒料 138
bar tong 杆式夹钳 336
barrel 集料筒 413
barrel nut 筒形螺母 250
barrel finishing 外圆超精加工 31
base 底座 215
base circle 基圆 66
base diameter 基圆直径 66
base for column 立柱底座 240
base metal 基体金属 151
basic major dia 大径 193
basic minor dia 小径 193
basic root 牙根 193
bead flange 挡圈 358
beam 横梁 370
beam compass 长臂划规 127
beam coupling 单梁式联轴器 293
beam deflector 离子束偏转器 178
beam monitor 离子束监控器 178
bearing 轴承 54，219
bearing assembly 轴承组件 82
bearing bracket 轴承架 265
bearing cover 轴承盖 258
bearing housing 轴承腔 279
bearing pedestal 轴承座 406
bearing pin 承载销 77
bearing support ring 轴承支撑环 264
bed 床身 213
bellow 波纹管 54
below cylinder 下气缸 316
below gaseous ring 下气环 358
bellows coupling 波纹管联轴器 294
belt 带 219
belt-drive bracket 带驱动架 74

belt cleaner 清带器 340
belt conveyor 带式输送机 340
belt driving drum 带驱动滚筒 340
belt freezing tunnel 带式冻结隧道 419
belt pulley 带轮 413
belt tension drum 带张紧滚筒 340
bending 折弯 122
bevel 斜角尺 25
bevel gear 锥齿 53～55，203，291
bevel gear transmission 圆锥齿轮传动 65
bevel teeth 锥齿 202
bevel pinion 锥齿轮 202
bilateral tolerance 双向公差 27
bladder-type accumulator 气囊式蓄能器 92
blade 刀片 193
blade slot 刀片槽 190
blading 叶片 316
blank 边料，毛坯 99，102
blank guide 导料板 101
blanking 落料 122
blanking punch 落料凸模，落料冲头 101，105
blast furnace 冲天炉 11
blast tube 鼓风管 313
blasting gun 喷枪 182
blend tube 混合管 343
blind hole 盲孔 163
blind hole with flat bottom 平底盲孔 163
blind riser 封闭冒口 118
blind rivet 盲孔铆钉 255
blister 气孔 34
block 块料 19
block side wall 机体侧壁 357
block tip surface 机体顶面 357
blocking ring 锁环 368
block-type boring cutter 盒式镗刀 189
blow 砂眼 34
body 泵体 26，192
body clearance 副偏角 186
boiler 锅炉 309
bolt 螺栓 26
bonnet bolt 盖板螺栓 315
bonnet nut 盖板螺母 315
boom 起重臂杆 383
boost pressure 涡轮增压，增压 349，351
boring 镗，镗孔，镗削 28，31，154
boring bushing 镗套 210
boring tool 镗刀 189
boss 凸台 35
bottom pulley 地轮 390
bottom view 仰视图 26
box spanner 套筒扳手 126
brace 弓形钻 127
bracing member 加强构件 399
bracket 叉架 39
brake actuator 制动启动器 300
brake armature 制动衔铁 301
brake drum 制动鼓 371
brake system 制动系统 361
branch runner 分流道 110
braze 铜焊 190
breast drill 胸压手摇钻 127
bridge plug 桥塞 396
broaching 拉削 28，31，154，212
broaching taper 拉削丝锥 194
brush frame 电刷架 358
bucket 货斗 345
buffer plunger 缓冲柱塞 92
buffer throttle 缓冲节流阀 92
buffing 擦光 154
bulk head 前挡板 360
bumper 保险杠 360
bumper beam 保险杠主梁 360
bumper jar 下击器 396
burner housing 燃烧室 313
burr 毛刺 36，162
bush 轴套，套筒，衬套 26，77，363
bushing 套筒，衬套 219，397
butt 平接 145
butt lap 平口搭接 256
butterfly valve 蝶阀 315

C

C clamp 钩形夹头 56
C washer C 形垫圈，C 形垫片 207，251
CHM 化学铣削，化学加工 154，170，226
cable 缆绳 383
calliper 卡钳 127
cam clamp 凸轮压头 56
cam operated clamp 凸轮操控压板 56
cam ring 凸轮环 349
cam roller 凸轮滚子 349
cam-lock ratchet 凸轮锁紧棘轮 58
camshaft 凸轮轴 54，346 ～ 348
cantilever-torsion bar 悬臂扭杆 357
cap nut 盖形螺母 249，347
capstan lathe 六角车床 127
captive nut 帽形螺母 250
car elevator 汽车升降梯 328
car frame 轿架 328
car guide rail 轿厢导轨 328
carbide insert 硬质合金刀粒 190
carburetor 化油器 346
carriage 拖板 213
case 箱体 82
casing ring 腔体环 279
cast ingot 铸锭 138
casting 铸造 122
cast-iron friction clutch 铸铁摩擦离合器 298
castle nut 六角开槽螺母，城堡型开槽螺母 246，250
catch basin 接水槽 12
cathode 阴极 165
cavity 孔隙，模腔 7，110
center 顶针 165
center drilling 中心钻 163
center line 中心线 21
center pin 中心销 321
central base 中央底座 240
central brush 中央刷 382
centralizer 扶正器 396
centre cycling evaporator 中央循环式蒸发器 416
centreless belt grinding machine 无心砂带磨床 223
centreless grinding 无心磨削 165
centrifugal block 离心锤 331
centrifugal casting 离心铸造 122
centrifugal compressor 离心压缩机 319
centrifugal impeller 离心叶轮 423
centrifugal pump 离心泵 80
centrifugal slicer 离心式切片机 414
centrifugal spray drying device 离心喷雾干燥装置 411
centrifugal vortex pump 离心旋涡泵 423
centrifuge 离心机 395
chain driving conveyor 链传动输送机 343
chain line 粗点画线 21
chain saw 链锯 126
chain trolley conveyor 挂吊链式输送机 343
chamber 腔室 219
chamfer 倒角 22
chamfer angle 倒角 193
chamfering form tool 倒角成形车刀 186
change gear 交换齿轮 203
chasis 底盘 360
chassis 机壳，底盘 358，389，413
check valve 单向阀，检测阀 315，355
chemical erosion 化学腐蚀 170
chemical milling 化学铣 31
chill 激冷件 35
chip excavator 排屑器 224
chip removal 切屑去除 154
chip splitter 断屑台 191
chipping hammer 除渣锤 147
chisel 凿子 126
choke 节流口 118
chopped fiber sprayer 喷涂法 17
chopping machine 斩拌机 414
chuck 夹头，卡盘 104，127，191，213
circuit breaker 断路器 59
circuit opening relay 开路继电器 353
circular 圆弧插补 234
circular pitch 圆节距 66
circular runout 圆跳动 23
circularity 圆度 22
clamp 压板 187，333

clamp bolt 压钉 211
clamp handle 锁紧手柄 123
clamp screw 压紧螺钉 187
clamping bar 夹压杆 114
clamping disk 夹紧盘 294
clamping mechanism 夹紧机构，夹持座 56，218
clamping nut 紧固螺母 258
clamping surface 夹持面 41
clapboard 隔板 408
clapper box 夹持架 220
clasp 卡环 358
classifier 分检器 311
claw coupling 爪型联轴器 295
claw hammer 羊角锤 124
clearance 顶隙 66
clearance angle 后角 155
clearance locational fit 间隙定位配合 29
clearance surface 容屑表面 188
clearance fits 间隙配合 28
closed die 封闭模 106
closed-loop control system 闭环控制系统 233
closing rope 闭合索 379
clump weight 偏重 406
cluster 滑块 220
clutch 离合器 369
clutch armature 离合衔铁 301
clutch sleeve 接合套 368
coal nozzle 燃煤嘴 311
coarse pitch 大节距 44
coated abrasive 涂覆磨具 198
coated abrasive belt grinding 砂带磨削加工 167
coaxiality 同轴度 23
cock valve 旋塞阀 398
cog belt 齿形带 358
coil positioning hook 卷绕定位钩 336
coil spring 螺旋弹簧，平面涡卷弹簧 127，275
coiler 卷料器 115
coining 精压 122
cold compaction 冷压 143
cold rolling 冷轧 31
cold shut 冷褶痕 34
cold slug well 冷料穴 110
column 立柱 213
combination die 复合模具 99，115
combination plier 组合钳 126
combination spanner 组合扳手 126
combustible gas cylinder 燃气罐 150
combustion chamber 燃烧室 137，318
combustion zone 燃烧区 311
comp 人孔盖 323
compacting machinery 压实机械 380
comparator 比较器，比较运算器 233，372
comparing equipment 比较装置 233
compensating roller 补偿辊子 339
compensation chain/cable 补偿链 328
compensation flap 补偿叶片 354
compression 抗压 3
compression coupling 压紧联轴器 293
compression molding 模压成形 17
compression strength 抗压强度 2
compressive stress 压应力 4
compressor 压缩机，压头 92，315，318，322
compressor turbine 压缩机涡轮 319
compressor wheel 压缩机叶轮 319
concentricity 同心度 23
condensate separator 冷凝分离器 324
condenser 电容器 373
condenser lens 聚焦透镜 178
condensing 冷凝 279
cone 锥盘 305
cone clutch 锥形离合器 291
cone mandrel 锥度心轴 206
cone tube 锥形管 343
cone-ended screw 开槽锥端紧定螺钉 246
cone-type friction clutch 锥形摩擦离合器 296
conical grinding mill 锥形磨粉机 414
conical washer 锥形垫片 250
conical screw 锥端螺钉 248
conical taper 锥度 23
connecting rod 连杆 50，140，220，324
connector 连接器 363
construction hoist 施工升降机 327
contact roller 接触辊 223
contact wheel 接触轮 167
continuous candy pouring molding machine 连续式糖果浇模成形机 423

continuous casting 连铸 10
continuous pultrusion 连续拉拔成形 17
control 控制器 181
control fork 控制叉 348
control lever 控制杆 350
control panel 控制柜 328
control rod 控制杆 348
control rod bracket 控制杆支架 348
control shaft 控制轴 348
control valve with bypass 旁路控制阀 312
controlled radius 受控半径 22
converter 变矩器 369
convey support 输送支承 211
conveying chain 传输链 339
conveying equipment 气力输送装置 425
conveying idler 输送托辊 341
conveyor belt 传输带 176
coolant 冷却液 165，189
coolant temperature gauge 冷却液温度表 360
cooler 冷却器 115，280
cooling box 冷却箱 224
cope 上箱型砂 118
copper electrode 铜电极 173
core 型芯 35，132，329
core activity 活动铁芯 358
core barrel 取芯筒 393
core dia 杆芯直径 193
core drilling 扩孔 163
core rod 芯棒 116
core sand 砂芯 118
corner 角接 145
corrosion resistance 耐蚀性 2
corrugated block 波纹块料 19
corrugated panel 波纹隔板 19
corrugated sheet 波纹板 19
corrugating roll 波纹辊 19
cotter pin 圆锥销 253
counter boring drill 沉孔锪钻 189
counter sunk drill 倒角锪钻 189
counter taper 侧隙锥 36
counterweight 对重装置，配重块 283，328
counterweight frame 对重架 331
counterweight guide rail 对重导轨 328
counterbore 沉孔 22，163
counter-boring tool with pilot 导向沉孔镗刀 189
counterboring 沉孔加工 163
counter-impact jet mill 对撞式气流粉碎机 413
countersink 沉坑、锪锥形沉孔 22，23，163
countersink bit 锪钻、沉头钻 127
countersinking 锪锥面 163
countersunk washer 反锥垫片 251
coupling 联轴器 90，218，258，326
coupling nut 连接螺母 181
cover 盖罩 129，315
cover die 凹模，凹压（定）模 115，132
cover drive 驱动端盖 358
cover gasket 盖罩密封垫 315
cover nut 盖罩螺母 315
cover stud 盖罩双头螺柱 315
cradle fork 摇动叉 345
crane 起重机，鹤式起重机构，鹤式起重机 50，339
crane runway 吊车轨道 335
crank 曲柄 50
crank gear 曲柄齿轮 220
crank pin 曲柄销 168
crank shaft timing belt gear wheel 曲轴正时带轮 347
crankshaft 曲轴 260，307，347，358
crashpan 防振垫盘 375
crawler crane 履带式起重机 334
crazing 龟裂 302
creep 蠕变 3
creep feed grinding 缓进给磨削 167
creep resistance 蠕变性质 2
crisscross shaft gear transmission 交错轴齿轮传动 65
critical stress 临界应力 3
cross 十字轴 371
cross feed 横向进给 166
cross flow 横流 173
cross head 上横梁 329
cross hole 交叉孔 40
cross rail 横梁 219
cross shaft gear transmission 相交轴齿轮传动 65
cross stroke mechanism 横向定程机构 215
cross-arm 横梁 217

crossbeam 大梁 390
cross-feed handwheel 横进手轮 215
crosshead guide 十字滑块 80
cross-member 叉件 363
cross-slide handwheel 中拖板手轮 215
crowd mechanism 推压机构 378
crown block 天车 391
crown valve 冠状阀 312
crucible 坩埚 137，144
crushing chamber 粉碎室 413
cup screw 杯端螺钉 248
cup seal 杯形密封 352
cup washer 杯形垫片 251
curve bevel gear transmission 曲齿圆锥齿轮传动 65
curvic coupling 曲面联轴器 295
cut-away view 剖视图 26
cutting 切割 226
cutting edge 主切削刃 186
cutting plane line 截面线 21
cutting plier 切断钳 126
cutting teeth 切削齿 195
cutting tool 刀片 192
cyclone separator 旋风分离器 309
cylinder 缸体，气缸 51，347
cylinder barrel 气缸筒 82
cylinder block 气缸体 347，357
cylinder cover 气缸盖 347
cylinder head 气缸盖 357
cylinder head cover 气缸盖罩 347
cylinder pillow 气缸垫 357
cylindrical cam 圆柱形凸轮，柱形凸轮 52
cylindrical gear transmission 圆柱齿轮传动 65
cylindrical grinding 圆柱面磨削，外圆磨削 165，212
cylindrical helical gear transmission 圆柱螺旋齿轮传动 65
cylindrical mandrel 圆柱心轴 205
cylindrical positioning pin 圆柱定位销 206
cylindrically curved washer 圆柱拱曲垫片 250
cylindricity 圆柱度 22

DC generator 直流发电机 306
DC motor 直流电动机 306
damper 气闸 310
damping bellow 阻尼筒 318
damping chamber 阻尼室 354
dashpot 冲击口 350
dead jacketed kettle 固定式夹层锅 415
deburring 去毛刺 226
deck structure 甲板 399
dedendum 齿根高 66
demister 除雾器 92
depth gauge 深度尺 128
depth of cut 切深，吃刀深度 161，165
derrick 起重架，井架 334，392
detent 凹槽锁销机构 58
diagnostic light 故障灯 366
diameter 直径 22
diamond boring 金刚石镗削 28
diamond grinding wheel 金刚石砂轮 198
diamond turning 金刚石车削 28
diaphragm 隔板 326
diaphragm compressor 膜片压缩机 324
diaphragm spring 薄板弹簧 296
diaphragm valve 膜板阀 315
die 模具，板牙 107，127
die casting 压铸，压力铸造 28，31
die cavity 模腔 115
die holder 模座 139
die insert 镶块 111
die land 模口 111
die set 下模座 102
dielectric fluid 电火花介质（工作液） 171
diesel engine 柴油发动机 346
die-stock 板牙扳手 127
differential 差速器 361，369
differential brake 差分式带式制动器 300
differential chain hoist 差动手拉葫芦 338
differential hoist 差动提升机 58
differential sensor 差分传感器 234
diffuser 扩散器 319
digital ignition system 数字点火系统 354
dimension figure 尺寸数字 21

dimension line 尺寸线 21
direction of rotation 回转方向 161
dirt excluder 排脏器 372
disc 盘 371
disc arm 盖盘转管 315
disc brake 盘式制动器 301
disc clutch 盘式离合器 291
disc cutter 盘式铣刀 188
disc nut 盖盘螺母 315
disc spring 碟形弹簧 329
disc washer 盖盘垫片 315
discharge chute 卸载槽，卸料漏斗 221，341
discharge point 卸荷点 323
discharge ring 卸荷环 322
discharge tube 出料管 413
discharge valve 排泄阀，卸荷阀 323，325
disk mill 齿爪式粉碎机 413
distribution box 配水箱 419
distributor 分电器，分配器 346，353，373
distributor port 配油口 349
distributor pump assembly (DPA) 分配式配油泵 350
distributor rotor 配油转子 350
diverter 分流器 343
divide board 分流隔板 316
divider 分规 25
dividing head 分度头 57
dog screw 柱端螺钉 248
double cam actuated clamp 双凸轮驱动压头 56
double cam clamp 双凸轮压头 56
double hammer vibrating screen 双锤振动筛 406
double open ended spanner 呆扳手 126
double pawl ratchet 双动式棘轮机构 64
double row ball bcaring 双列球轴承 266
double wedge 双楔块压头 56
double wound brake 双重缠绕制动器 300
double-crank mechanism 双曲柄机构 50
double-end stud 双头螺柱 246
double-ended cutter or boring tool 双端镗刀 189
double-stroke engine 二冲程引擎 60
double-volute pump 双蜗壳泵 80
dovetail guide 燕尾导轨 290
dowel pin 销钉，圆柱销 248，253
down milling 顺铣 161
drag 下箱型砂 118
drag link 转向直拉杆 371
drain 接水槽 181
drain hole 排泄孔 83
drain plug 排泄塞 83
drawbar trailer 拖杆拖车 364
drawing 拉制 31
drawing board 绘图板 25
drift 楔铁 191
drill 钻头 190，219
drill bit 钻头 392
drill bush 钻套 207
drill bushing 钻套 190
drill collar 钻铤 392
drill derrick 井架 391
drill diameter 钻头直径 190
drill pipe rack 钻杆架 399
drill plate 钻模板 207
drill ship 钻井船 400
drill tang 钻柄 191
drilled ring nut 钻孔环形螺母 249
drilling 钻削，钻孔、打孔 28，31，226
drilling jig 钻削夹具 207
drilling machine 钻床 128
drip lubrication 滴漏润滑 78
drive block 驱动块 367
drive cover slot 驱动盖槽 367
drive pin 传动销 253
drive plate 主动盘 296
drive ring 驱动环 281
drive shaft 传动轴 370
driven disk 从动盘 296
driven gear 从动齿轮 296
driven member 从动件 297
driven part 从动件 64
driven ratchet 棘轮 64
driving member 驱动件 297
driving plate 主动轮 64
driving shaft 驱动轴 220
drop core 侧置型芯 119
drop forging 落锻 138
dross 熔渣 179
drum 转鼓 407

drum brake 鼓式制动器 299
drum turner 鼓形翻拌器 336
dry bearing 干式轴承 342
ductility 延展性 2
dumbwaiter lift 杂物电梯 328
dumping fork 卸货叉 345
dumpling making machine 饺子成形机 422
duplex elevator 并联电梯 328
durability 可靠性 3
duration spring 持续弹簧 352
dust bellow 防尘管 363
dust-free seal 防尘密封 284
dynamic chain ring 动齿盘 413
dynamically positioned drill ship 动态停靠钻井船 400
dynamometer 测力计 299

EBM 电子束加工 226
EC deep hole drilling 深孔加工 174
EC hogging 拱起加工 174
EC surfacing 平面加工 174
ECA 电化学摩擦 175
ECAM 电化学电火花 175
ECDM 电火花电化学 175
ECG 电化学磨削 175
ECH 电化学珩磨 175
ECM 电化学加工 154，173，226
ECM of turbine blade 叶片化学加工 174
ECU 电化学超声 175
EDM 电火花加工 154，226
EDMed hole 电火花加工孔 46
EDSCAN machining 电火花电极平动加工 171
ESC 电化学磨料修正 175
earthing clamp 接地夹头 147
eccentric 偏心 323
eccentric bit 偏心钻头 392
eccentric block above 上偏心块 406
eccentric shaft 偏心轴 140，325
edge 边缘平齐 145
edge filter 边口过滤器 347
edge grip sheet clamp 边缘夹紧板材夹头 337
ejector 推杆，顶出杆 99，107
ejector bush 顶件器 101
ejector die 凸模，凸压（动）模 115，132
ejector pin 顶杆，推杆，顶出杆 101，111，115
ejector plate 卸料板，推板 99，115
ejector return pin 复位杆 111
elastic ring 弹性环 71
elastic stop nut 弹性止动螺母 250
electric connection 电插头 355
electric drill 电钻 127
electric furnace 电炉 12
electric screwdriver 电动螺丝刀 127
electrical apparatus 电气柜 224
electrical connection 电刷，电接头 174，352
electrical(connection)brush 电刷 165
electrically heated expansion element 电加热膨胀元件 352
electro magnetic valve 电磁阀 312
electro-chemical 电化学 31
electrode 电极 144，178，312
electrode carrier 电极支架 173
electrode guide tube 电极导管 151
electrode holder 电极夹头 147
electrolysis 电解液 165
electrolyte tank 电解液槽 173
electrolytic grinding 电解磨削 31
electromagnetic switch 电磁开关 358
electron beam 电子束 31，122
electron beam welding 电子束焊接 144
electronic fuel injection system 电子燃油喷射系统 353
electroplating 电镀 174
electro-polish 电抛光 31
electro-slag welding 电渣焊 144
element coupling 元件联轴器 291
elevating steel cable 提升钢丝绳 390
elevator 电梯，垂直电梯 327，328
embedded plate 卸料镶板 99
embossing 模压 122
enclosure 密闭装置 226
encoder feedback 反馈编码器 241
end cover 端盖 358

end cutting edge angle 主切刃角 186
end milling cutter 指状立铣刀 42
end pillar 端柱 321
end relief angle 主后角 186
end seal 端部密封 111
end stop 挡块 222
end surface grinding 端面磨削 166
energy absorber 吸能件 360
energy saving rotor pump 节能转子泵 424
engine 发动机 346
engraving 雕刻 226
enlarging tool 扩孔刀 163
epicyclic gear 行星齿轮 56
erasing knife 刮刀 420
erosion 蚀除 154
escalator 扶手电梯 327
escapements mechanism 擒纵机构 59
excavator 挖土机 379
exhaust camshaft 排气凸轮轴 358
exhaust outlet 排气口 375
exhaust valve 排气门 346
exhauster 排料器 311
exhaust-pressure governor 排压速度调控器 318
expanded panel 膨胀隔板 19
external geneva mechanism 外槽轮机构 63
extension line 尺寸界线 21
external involuter spline 外渐开线花键 254
external-tooth locking washer 外齿锁紧垫片 251
extruding 挤出 31
extrusion 挤出 122
eye bolt 眼孔螺栓 248

F

filament winding 线材缠绕成形 17
face 前刀面，端面 186，267
face cam 端面凸轮 52
face cutter 端面铣刀 187
face excavator 正铲挖掘机 378
face gear 端齿盘 67
face key 端面键 258
face shield 面罩 147
faceplate 花盘 203
fade 衰减 302
fascia 罩面 360
fasten nut 紧固螺母 26
fatigue 疲劳性 2
fatigue strength 疲劳强度 3
feed 进给方向 161
feed box 进给箱 215
feed carriage 进给床鞍 190
feed hand wheel 进给手柄 218
feed nozzle 进料喷嘴 401
feed roller 喂料辊 421
feed valve 进料阀门 413
feeder 进料器 311
feeding finger 进给推杆 210
feeding tube 进给套筒 210
feedwater pipe 给水管 309
feeler gauge 塞尺 128
feether 滑键 252
felt roller 毡轮 227
female rotor 凹螺杆 325
field coil 线圈 301
filler 填料 14
filler plug 加油塞 369
fillister screw 开槽盘头螺钉 246
filter 过滤器 171，226，353
filtering core 滤芯 92
filtering cup 滤杯 92
final drive 主减速器 369
fine adjustment mechanism 微调机构 58
fine pitch 小节距 44
finger washer 指状垫圈 251
finishing 精铣刀，光整 188，226
finishing teeth 精齿 195
fire engine 消防车 376
fixed chain ring 定齿盘 413
fixed core 定模型芯 111
fixed head 固定头架 218
fixed plate 固定板 98
fixed platen 定板 115

fixed ring 固定磨环 413
fixed scroll 固定涡管 325
fixed shaft 固定心轴 260
fixture 夹具 41
flame 火焰 150
flame cutting 火焰切割 31
flange 法兰 87，281
flange coupling 凸缘联轴器 292
flange-type gear coupling 法兰式齿轮联轴器 293
flanged coupling 法兰联轴器 291
flap wheel 飞翼轮，页轮 126，223
flash welding 闪光焊接 144
flask 砂箱 118
flat band driving conveyor 平带传动输送机 343
flat broach 平面拉力 196
flat spring 板弹簧 275
flatness 平面度 22
flatter 平锤 124
flexible arm 柔性臂 181
flexible axle of steel cord 钢丝软轴 260
flexible bush 柔性套 294
flexible coupling 柔性联轴器 291
flexible diaphragm 柔性薄膜 262
flexible shaft 软轴 292
flexible support board 柔性支撑板 316
flexible tyre 柔性轮胎 294
flexible-disk coupling 柔性盘式联轴器 293
flexural 抗弯 3
flexural modulus 抗弯模量 3
float collar 浮箍 393
float shoe 浮鞋 393
floating couple 浮动接头 210
floating crane 浮式起重机(浮吊) 334
floating ring bearing 浮动环滑动轴承 264
floating roof tank 浮顶油罐 404
floor inductor 楼层感应器 333
flour hopper 面斗 421
flow meter 流量计 173
flow plate 导流板 92
flow sleeve 流体隔套 320
flume 流送槽 407
flute 刀齿槽，排屑槽，流道 188，190，314
flux hopper 焊药料斗 149
flyweight 飞轮重物块 348
flywheel 飞轮 140，276，296，347
flywheel bolt 飞轮螺栓 358
foil bearing 箔片滑动轴承 264
folding machine 折叠器 421
follower jig 随行夹具 211
force fits 传力配合 29
forged boring tool 锻制镗刀 188
forging 锻压 31，122，138
fork lift truck 叉车 345
fork yoke 叉架 292
form relieved circular cutter 铲背圆弧成形铣刀 187
formability 可成形性 3
forming 成形 122，143
forming machine 成形机械 421
forward multiplate clutch 前行多盘离合器 370
four lobe bearing 四凸块滑动轴承 264
four-stroke engine 四冲程发动机 307
fracture 断裂口 4
fracture toughness 脆裂强度 2
frame 机架，支架，车架 140，165，383
freeing lever 释放杆 64
freezer mill 冷冻粉碎机 414
freight elevator 载货电梯 327
friction 摩擦 122
friction brake 摩擦制动器 301
friction brake disc 摩擦制动盘 301
friction roller 摩擦滚筒 74
friction clutch 摩擦离合器 296
friction lining 摩擦里衬 296
friction plate 摩擦片 298
friction ratchet mechanism 摩擦棘轮机构 57
friction welding 摩擦焊接 144
front bearing 前支承，前轴承 258，316
front clearance 主后角 186
front lower wishbone 前下叉杆 376
front suspension 前悬架 346
front upper wishbone 前上叉杆 376
front view 主视图 26
front wheel arrester 前轮制动器 346
fuel level indicator 燃油表 360
fuel temp.sensor 燃油温度传感器 359
full bearing 全环轴承 264

function generators 函数发生器 62
funnel 漏斗 223
furnace 炉体，熔炉 108，132
fusion welding 熔焊 144

G clamp 弓形钩，夹钳 127
gang mandrel 悬臂心轴 206
gantry 起重台架 334
gantry crane 龙门起重机 333
gantry crane girder 横梁 390
gantry crane with crab 绞车式龙门起重机 334
gantry crane with shuttle girder 滑伸式龙门起重机 334
garter spring 卡紧弹簧 284
gas control valve 气控阀 150
gas nozzle 气嘴 137
gas regulator 气体调节器 150
gas turbine 燃气轮机 318
gas welding 气焊 144
gasket 垫片，垫圈，密封垫 26，281，315
gasket cup 密封垫杯 286
gasline engine 汽油发动机 346
gate 水口 34
gate valve 门阀，闸阀 312，315
gear 齿轮 26，258
gear blank 齿坯，毛坯 170，224
gear box 传动箱，减速箱 308，329
gear cutting 齿轮加工 154
gear geometry 齿轮啮合 66
gear hobbing 滚齿 168，212
gear pump 齿轮泵 26，80
gear shaping 插齿 168
gear shaving 剃齿 168
gear transmission 齿轮传动 65
gear wheel 齿轮 219
gearbox 齿轮箱 311
geared engine 齿轮传动引擎 60
geared inverted slider crank 倒置齿轮滑块曲柄机构 57
geared slider crank 齿轮滑块曲柄机构 57
geneva intermittent mechanism 槽轮停歇机构 57
gerotor pump 摆线轮液压泵 80
gibhead 钩头键 252
gland 密封垫 315
gland bolt 密封压紧螺杆 315
gland bolt pin 螺杆传销 315
gland flange 密封法兰 315
gland nut 密封压紧螺母 315
globe valve 球阀 312
goggle 护目镜 147
gooseneck 鹅颈管 132
gouge 弧口凿、半圆凿 127
governor 限速器 328
governor arm 调节臂 349
governor cage 调节器壳体 348
governor spring 调节器弹簧 349
governor weight 调节器配重 349
grab 鼓起漏空，抓斗 302，379
grader 平地机 381
grate 格栅 310
gravity conveyor 重力输送机 343
gravity wheel conveyor 轮式重力输送机 343
grinding 磨削 31，154，226
grinding machine 砂轮机、磨床 127
grinding roller 磨煤辊 311
grinding wheel 砂轮 126，156，167
grip 夹具 5
grip segment 夹紧块 108
groove 凹槽 368
grooved wheel 开槽轮 108
guide 导向块 52，192
guide bush 钻套，导套 41，111
guide core 导芯 369
guide hook 导向钩 227
guide pillar 导柱 111
guide pin 导柱 102
guide plate 导板 101
guide rail 导轨 331
guide shoe 导靴 329
guide vane 导叶 326，423
guide wheel 导轮 227
guideway 导轨 288
gun drilling 枪钻 163
gun drill 枪钻 190
guy 缆风绳 390

H

HB 布氏硬度 3
HF generator 高频发生器 227
HV 韦氏硬度 3
hammer-head crane 锤头式起重机 334
hand lay-up 手工涂覆 17
hand throttle valve 手动节流阀 397
hand wheel 手轮 315
handle 拉手 371
hanging core 悬置型芯 119
hardened zone 硬化区 153
hardness 硬度 2
harness connector 电气配线接头 355
head drum 头滚筒 340
head hood 头部护罩 341
headrest 头枕 360
headstock 床头箱，主轴箱 213，215
headstock seat 柱轴承座 357
heat distortion 热变形 2
heat exchanger 换热器，热交换器 173，226，309
heat shield 绝热垫 372
heat spotting 热斑点 302
heat-affected zone(hza) 热影响区 179
heater 加热器 110，280
heavy boring tool 重载镗刀 188
heel 刀棱，根部 188
helical 斜齿 44
helical bevel gear transmission 斜齿圆锥齿轮传动 65
helical gear 斜齿轮，螺旋齿轮 67
helical gear transmission 斜齿圆柱齿轮传动 65
helical peripheral cutter 螺旋圆周铣刀 187
helical spring 螺旋弹簧 273
helicopter pad 直升机平台 399
helix 螺线 342
herring-bone gear 人字齿轮 67
herring-bone gear transmissiom 人字齿轮传动 65
hex bolt 六角头螺栓 246
hex nut 六角螺母 246
hexagon key 六角扳手 248
hexagon(main)turret 六角（主）刀架 215
hex-socket fastener 内六角紧固件 248
hidden line 虚线 21
high pedestal jib crane 门座式悬臂起重机 334
high pressure intensifier 高压增压泵 182
high-torque ratchet 大转矩棘轮 58
hinge 铰链 305
hinge pin 铰链转销 315
hob 滚刀 211，224
hoisting block 起重滑车 335
hoisting drum 吊升滚筒 392
hoisting machinery 起重机械 333
holding bush 固定护套 99
holding pawl 制动棘爪 64
holding plate 夹持板 99
holding rope 固定索 379
hole tolerance 孔公差 28
hollow spindle 空心轴 260
honing 珩磨 31，154，175，212
hood 挡板 333
hook 吊钩 383
hook clamping plate 钩形压板 211
hook spanner 钩形扳手 127
hooke coupling 万向联轴器 291
hooke joint 万向节 291
hopper 料斗 143
hopper elevator 斗式提升机 425
horizontal core 卧式型芯 118
horizontal pulveriser 卧式超微粉碎机 413
horizontal rotary furnace 水平旋转炉 416
horizontal shell and tube evaporator 卧式壳管蒸发器 418
horseshoe washer 马蹄形垫片 251
hose 软管 150
hospital elevator 病床电梯 327
hot compaction 热压 143
hot rolling 热轧制 31
hot roll 热辊 115
hot tear 热裂缝 35
housing 腔体，机壳 219，326
housing cover 腔盖 367
hub member 毂盘件 294
hydraulic amplifier 液压放大器 90

hydraulic boom 液压支臂 376
hydraulic cylinder 液压缸 211
hydraulic elevator 液压电梯 328
hydraulic governor 液压调节器 350
hydraulic head 液压头 350
hydraulic jack 液压千斤顶 79，127
hydraulic outrigger 液压支腿 384
hydraulic shot cylinder 液压注射缸 132
hydraulic transmission 液压传动 54
hydraulic turbine 水轮机 321
hydraulic vane motor 叶片式液压马达 83
hydroclone 水力旋流器 397
hydrodynamic brake 液压制动器 300
hydrostatic extrusion 静压挤出 138

I

idler 惰轮 73
idler gear 惰轮 203
idler pulley 惰轮 74
idler rotor 惰转子 81
idling spring 惰簧 349
igniter 点火器 353
ignition coil 点火线圈 353
ignition electrode 点火电极 313
ignition loop 点火线圈 346
ignition switch 点火开关 346
ignition plug 点火塞 321
impact strength 冲击强度 3
impeller 叶轮 82，316，326，423
impeller pump 叶轮泵 369
index crank 分度杆 203
index head spindle 分度头轴 203
index pin 分度销 203
index plate 分度盘 203
indexing drilling jig 分度钻孔夹具 208
indexing mechanism 分度机构 57
indexing table 分度台 55
indicating line 指示线 21
induction coil 感应线圈 37，137，153
induction motor 感应电动机 306
inertia vibrating screen 惯性振动筛 406
ingot casting 铸锭 10
ingot tumer grab 坯锭翻转夹板 336
injection controller 喷射控制器 348
injection pump 喷油泵 346
injection pump calibration unit 喷射泵计量单元 359
injection timing piston 喷射正时活塞 352
injector 喷油器 346
inlet port 进油口 349
inlet valve 进气门 346
inner flange 内端法兰盘 198
inner gerotor 内摆线轮 80
inner hex fillister screw 内六角圆柱头螺钉 246
inner ring 内圈 267
inner ring ball race 内圈滚道 267
inner-hexagon screw 内六角螺柱 26
insert 刀粒，滑块 162，368
instrumentrack 仪表架 325
insulation 绝缘层 173
insulator 绝缘器 308
intake manifold 进气管 347
intensifier 增压器 181
intensity factor 应力集中系数 3
interactive device 互动装置 240
interchangeable tapered sleeve 可换锥套 204
intercooler 中间冷却器，内冷装置，互冷器 92，359，419
interference 过盈配合 28
interference locational fits 定位过盈配合 29
intermediate gear 中间齿轮 305
intermediate plate 中间板 367
intermittent horizontal calender 间歇卧式压延机 421
internal centerless grinding 内圆无心磨削 167
internal geneva mechanism 内槽轮机构 63
internal spline 内渐开线花键 254
inverted post jig 倒置立柱夹具 207
investment 失模 122
investment casting 熔模铸造 31
involute spur gear 渐开线直齿齿轮 65
ion source 离子源 178

jack 千斤顶 127
jam nut 锁紧螺母 329
jaw 卡爪 202，364
jeep 吉普车 375
jet 射流 289
jet condenser 喷射式冷凝器 416
jet cutting 射流加工 226
jet tube 射流管 171
jig 起重臂 334
jig frame 镗模架 210
joggle lap 偏斜搭接 256
jointing bolt 连接螺栓 329
journal bearing 滑动轴承 316，323
joystick 控制杆 240

kelly 方钻杆 392
kelly spinner 方钻杆旋转器 393
kennedy key 方形切向键 252
kerf 槽 179
key 键 26，219，252
key way 键槽 258
kingpin 主销 365
knee 升降台，升降臂 213，216
knock sensor 爆燃传感器 355
knockout rod 卸料杆，打料杆 105，115

LBM 激光束加工 154，226
LECM 激光电化学 175
labyrinth seal 迷宫式密封 323
ladder 云梯，梯子 376，383
lance 喷管 11
land 刃宽 193
lap 搭结，研具 145，168
lapping 研磨 31，154
lapping sleeve 研磨套 168
large helix angle 大螺旋角 44
laser 激光 31，122
laser welding 激光焊接 144
latch 插销 111
latch valve 闭锁阀 350
latch mechanism 插销机构 58
lateral brush 边刷 382
lead screw 丝杠 213，220
leader 指引线 21
leadscrew 滚珠丝杠 233
leaf spring 钢板弹簧 375
leaf-spring guide 板弹簧导向装置 343
left cap 左端盖 26
left track 左滑轨 360
lever 杠杆，曲柄 211，305，371
lift pipe 提升管 343
lift stop 提升挡块 347
lift sub 提升短节 396
lifting beam（spreader beam） 提升梁（延伸梁） 336
lifting head 直动头 85
light boring tool with bend shank 弯头轻载镗刀 188
limit dimension 极限尺寸 27
limit bolt 限位螺钉 330
linear actuator 线性执行元件 59
linear motion 线性运动 234
linear shaking screen 直线摇动筛 406
line-of-action 啮合线 66
lining 衬片 298
link coupling 连杆联轴器 293
link plate 连接板 77
linkage hook 连接钩 349
lip 刀尖 188
lip angle 刀尖角 188
lip seal 唇形密封 369
liquid alloy 液态合金 7
loading arm 加载臂 5
loading channel 上料槽 341

loading chute 装载槽 221
locating ear 定位口 191
locating mechanism 定位锁紧机构 56
locating pin 定位销 102，358
locating slot 定位槽 191
location post 定位柱 207
lock bar sheet lifter 锁紧杆式板材提升器 337
lock bolt coupling 紧固螺钉连接 247
lock nut 锁紧螺母 129，193
lock washer 锁紧垫圈，弹簧垫圈 87，246
locking bolt 锁紧螺栓 58
locking piece 锁片 358
locking plier 锁紧钳 126
lockpin 锁销 187
lockplate washer 锁片垫片 251
locomotive crane 机车起重机 334
logic element 逻辑元件 93
logic-block 逻辑块 372
longitudinal 纵向 15
longitudinal stop 纵向挡块 215
longitudinal stroke mechanism 纵向定程机构 215
longitudinal tool carriage 纵向刀具溜板 215
loop spring 环形弹簧 275
low level switch 低油位开关 279
lower die 下模 139
lower die shoe 下模座 106
lower horn 下臂 151
lower punch 下冲头 116
lubricated bearing 油润滑轴承 342
lubricating pump 润滑泵 211
lubrication nipple 润滑嘴 367

MAM 磁性磨料加工 154，183
Mcpherson strut rear axle 麦弗逊滑柱后轴 362
machining 加工 122
machining allowance 加工余量 46
magnet clutch 磁粒离合器 299
magnet grapple 磁力抓斗 337
magnet vane 隔磁板 333
magnetic brake 电磁制动器 329
magnetic core 磁芯 408
magnetic lifter 磁力提升器 336
magnetic separation machinery 磁选机械 408
main oil pump 主油泵 325
main parachute 主降落伞 375
major diameter 大径 246
major flank 主后刀面 186
male rotor 凸螺杆 325
mallet 木锤 124
mandrel 心轴 107，112，206
manipulator 机械手 240
manual hoist 手动提升葫芦 338
manual metal-arc welding 手工金属弧焊 144
manual operation lubrication 手工润滑 78
manufacturing tolerance 制造公差 2
marine elevator 船用电梯 328
mask 掩膜 46
material conveying machinery 物料输送机械 423
maximum interference 最大过盈 28
mechanical abrasion 机械磨蚀 170
mechanical governor 机械调控器 318
mechanical seal 机械密封 423
metal active-gas welding MAG 熔焊 144
metal arc welding 金属弧焊 144
metal inert-gas welding MIG 熔焊 144
metal padding 金属垫 35
metering valve 计量阀 349
metric 普通螺纹 247
metric hex nut 公制六角螺母 249
micro EDM 细微电火花加工 171
microjector 微喷油器 347
micrometer screw 微米丝杠 225
microprocessor 微处理器 372
midst wheel 中间轮 358
mill 粉碎机 311
milling 铣削 31，212，226
milling cutter 铣刀 162，188
minimec governor 微机械调节器 348
minimum interference 最小过盈 28
minor cutting edge 副切削刃 186

minor flank 副后刀面 186
minor diameter 小径 246
misalignmen 错位 302
misrun 漏充 34
mitre square 45°尺 25
modulus of elasticity 弹性模量 3
mold 模具 17
mold cavity 模腔 118
mold half 半模 115
molten alloy 熔融合金 7
molten metal 熔融金属 12
molten slag 熔渣 148，151
moonpool 月池 399
motor 电动机 54，180
motor pulley 电动驱动轮 74
mounting flange 安装法兰盘 321
movable clamp 活动钳口 123
movable plate 活动板 98
movable sleeve 活动套 298
movable upper crosshead 活动横梁 4
moving core 活动型芯，动模芯 111，115
moving platen 动板 115
mud drum 浆体鼓 313
muff coupling 套筒联轴器 292
muliple-ply belt 多层带 75
multi flute end mill 多槽端铣刀 187
multi-cavity die 多腔模 115
multi-edges belt 多楔带 75
multileaf bearing 多叶片滑动轴承 264
multi-plate friction clutch 多片摩擦离合器 296
multiple-cutter boring tool 多刃镗刀 189
multiple-thread cutter 多头铣刀 188

NC drilling 数控钻削 235
NC milling 数控铣削 235
NC turning 数控车削 235
nail puller 起钉器 127
nailing 穿钉 358
needle nose plier 尖嘴钳 126
needle pin 滚针 266，371
needle roller bearing 滚针轴承 266
neglect 脱落 302
neutral shield 中性罩 312
noise absorber 消声器 93
non-groove taper 无槽丝锥 194
nose 凸块 52
nose cone 鼻锥体 376
nose radius 刀尖圆角半径 186
notch sensitivity 缺口敏感性 3
note 标注 21
nozzle 浇嘴，喷嘴 47，115，181
nozzle valve 喷嘴阀 347，353
nozzle vane 喷嘴叶片 319
nut 螺母 26，86，250

O-ring O 形密封圈，O 形圈 87，397
objective lens 物镜 178
offset 偏差，偏移量 5，244
offset bearing 错位轴承 264
offset link 错位链节 77
offset link pin 错位链节销 77
offset plate 错位链板 77
oil buffer 油压缓冲器 333
oil cooler 油冷器 308
oil cup 油杯 171，286
oil inlet cavity 回油腔 83
oil mist fan 油雾扇 325
oil mist generator 油雾发生器 279
oil mouth 油嘴 371
oil nozzle 油喷嘴 313
oil pump 补油泵 324
oil reservoir 油槽 279
oil retainer 油保持器 263
oil ring 油杯 358
oil seal 油封 371
oil spray 油雾器 93
oil stone 油石 126

oil stream lubrication 喷油润滑 78
oldham coupling 十字联轴器 291
one-way clutch 单向离合器 358
open die 自由锻 138
open riser 开式冒口 118
open-loop control system 开环控制系统 233
operating sleeve 操控套 296
orbiting scroll 运行涡管 325
orifice 孔口 425
oscillating mechanisms 摆动机构 57
oscillation 往复振动 167
oscillator 振动器 150
outer flange 外端法兰盘 198
outer gerotor 外摆线轮 80
outer ring ball race 外圈滚道 267
output flange 输出法兰 369
overarm 顶梁 216
overflow valve 溢流阀 171
overhanging core 悬臂型芯 36
overhead travelling crane 悬空起重机 334
overhead travelling crane with crab 绞车式桥式起重机 334
overhead travelling crane with hoist 电动葫芦桥式起重机 334
overrunning clutch 超越离合器 58
overspeed governor 限速器 331
oxygen cylinder 氧气罐 150
oxygen lance 氧喷枪 12

P

PBM 等离子加工 154
packing 密封 281
packing chamberor box 密封腔（盒） 286
packing cup 密封杯罩 286
packing gland 密封压套 315
packing ring 密封环 286
pallet 托盘 238
parallel 平行仪 25
parallel shaft gear transmission 平行轴齿轮传动 65
parallelism 平行度 23
partial bearing 局部环轴承 264
parting line 分型线 118
parting-off tool 切断刀 186
parting plane 分型面 132
passenger conveyor 自动人行道 327
passenger elevator 乘客电梯 327
path generator 轨迹生成器 62
pavement machinery 铺路机械 381
pawl 棘爪 58，220
pendulum 摆锤 4
perforating tool 射孔枪 393
peripheral cutting edge 周边铣削刀刃 188
peripheral grinding 周边磨削 166
peripheral pump 周边泵 355
permanent mold casting 永久模型 31
permanent magnet 永磁铁 183
permanently rotating part 常转零件 64
perpendicularity 垂直度 23
phantom line 双点画线 21
phillips screw 十字槽沉头螺钉 246
philips screwdriver 十字螺丝刀 127
photochemical etching 光化学 226
piercing 冲孔 122
piercing punch 冲孔凸模 101
pilot operated relief valve 先导型溢流阀 83
pilot strip 导尺 101
pilot valve 先导阀 318
pin 圆柱销，销钉 26，49
pin coupling 销连接 253
pinch roll 夹压辊筒 12
pinch roller 压轮 410
pinflex coupling 轴销柔性联轴器 294
pinion 小齿轮 219
pintle 柱销 353
pintle type nozzle 轴针式喷嘴 347
pipe 管件，管螺纹 87，247
pipe expander 胀管器 396
pipe vise 管子台虎钳 127
piston 柱塞 82，125，347
piston pin 活塞销 358
piston rod 活塞杆 363
piston shoe 柱塞掌 82
piston-type accumulator 活塞式蓄能器 92
pitch circle 节圆 66

pitch diameter　节圆直径　66
pitch line　节线　66
pitch or effective diameter　中径或有效直径　246
pitch point　节点　66
pitman 连杆　386
pitman arm　转向摇臂　371
pitman arm shaft　摇臂轴　371
pivot　转轴转枢　241，264，349
pivot ring　转轴环　296
pivoted shoe　销式蹄　262
plain bearing　滑动轴承　261
plain screw　普通丝杠　213
plain washer　平垫片　250
planet carrier　行星轮架　71
planet gear　行星齿轮　72
planet wheel　行星齿轮　71
planetary carrier　行星齿轮架　370
planetary gear　行星齿轮　370
planetary gear machine　行星齿轮机构　358
planetary roller screw　行星式滚柱丝杠　213
planetary transmission　行星齿轮传动　304
planing　刨削　31
plasma　等离子　122，177
plasma arc welding　等离子弧焊　144
plate　压板　258
plate clamp　板式夹头　336
plate clutch　摩擦片离合器　291
plate die　搓丝板　194
plate cam　盘形凸轮　52
platform　支撑平台，轿底　180，329
platform headframe　平台井架　390
plectrum　拨叉　358
plumbago crucible　炭精坩埚　132
plunge grinding　切入式磨削　167，222
plunger　活塞，柱塞　132，219
plunger bush　衬套　111
plunger dosing device　柱塞式定量供料装置　422
plunger pump　柱塞泵　80
plunger-detent mechanism　插件凹槽机构　63
pneumatic chuck　气（风）动夹头　206
pneumatic drill　风钻　127
pneumatic hammer　气动锤　124
pneumatic tyre　充气轮胎，气胎　375，383
point angle　顶尖角　190
point punch　样冲　126
polishing　抛光　31，154
polishing pad　抛光垫　176
polishing table　抛光工作台　176
porosity　孔隙度，孔隙　2，35
porous liner　多孔内衬　264
port plate　孔口板　82
portable grinder　手提式砂带机　223
portable welding seam polisher　手提式焊缝打磨机　223
position　位置度　23
positioning support plate　定位支承板　211
pot　熔锅　132
pouring basin (cup)　浇口（杯型）　118
pouring cup　浇口杯　35
powder metal　金属粉末　28
powder metal sintered　金属粉末烧结　28
power amplifier　功率放大器　234
power rotator　动力旋转器　336
power train system　传动系统　361
power turbine　动力涡轮　320
preland　前端模口　111
preload spring　预加载弹簧　262
preloading nut　预紧螺母　258
press wheel　压轮　227
pressure angle　压力角　66
pressure bar　压杆　104
pressure bushing　压力轴套　294
pressure compensated　压力补偿　79
pressure dam bearing　压力阻挡轴承　264
pressure guage　压力表　173，312
pressure holding valve　压力保持阀　352
pressure meter　压力表　171
pressure modulator　压力模块　312
pressure plate　压料板，压力板，压板　99，296，367
pressure ring　压力环　298
pressure sensing bellow　压力传感筒　318
pressure sensor　压力传感器　176
pressure test block　试压堵塞　396
pressure vessel　压力腔　108
pressure-reducing valve　减压阀　84
pressurized oil cavity　压油腔　83
primary circuit　主回路　85

primary shoe 领蹄 299
process control computer 工艺控制计算机 180
profile of a line 线轮廓度 22
profile of a surface 面轮廓度 22
projected tolerance zone 凸带公差带 23
projection welding 凸出焊接 144
propeller 螺旋桨 375
propeller gear box 螺旋桨齿轮箱 375
proportional divider 比例规 25
protractor 量角器，半圆规 25，287
pull rope 拉绳 383
pulley 带轮 218
pulley groove 带轮槽 75
pulse 脉冲 233
pulse power 脉冲电源 227
pulsed fluidized bed dryer 脉冲流化床干燥器 412
pump 泵，气泵，液压泵 173，223
pump cylinder 泵缸 423
pump housing 泵体 281
pump impeller 泵叶轮 286
pump jet 泵喷头 352
pump lever 泵杠杆 352
pump plunger 泵柱塞 350
pump return spring 泵复位弹簧 352
punch 冲头 6，55，107
punch set 上模板 102
punching 冲压 28
puppet 凸轮从动件 372
pusher 推杆 49

quick removable cartridge 快换夹头 182

Roots blower 罗茨轮鼓风机 326
Rouleaux coupling 罗氏联轴器 291
Rzeppa universal joint 笼形万向节 292
rack 机架 50
rack lifter 拉架提升器 337
radial 径向 15
radial arm 摇臂 213
radial engine 径向引擎 60
radiating beam 辐射梁 390
radius 半径 22
radius turning form tool 圆弧成形车刀 186
rag bolt coupling 地脚螺钉连接 247
rail 导轨 333
rail bracket 导轨撑架 328
rail head 立刀架 219
rake face 前刀面 155
ram 压头，滑枕，电极座板 110，173，213
rammer head 夯锤 383
ratchet 棘轮，棘爪 58，373
ratchet brake 棘轮制动器 64
ratchet mechanism 棘轮机构 57
ratchet screwdriver 棘轮式改锥 128
ratchet wheel 棘轮 220
reach fork 取货叉 345
reactor 导轮 369
reamer 铰刀 41，126
reaming 铰孔，铰削 28，31，163
reaming device 划眼器 393
rear aerofoil wing 后翼主板 376
rear double wheel 后双轮 376
rear view 后视图 26
rear-view mirror 后视镜 376
rear bar 后杆 363
receiver 接收器 92
recess 凹孔 204
reciprocating machine 往复式加工机床 219
reciprocating mechanism 往复运动机构 60
reciprocating pump 往复泵 423
reciprocating vibration flat grinder 往复振动平面磨光机 223
rectangular 矩形键 252
rectifier 整流器 171

recycle nozzle 循环喷嘴 401
reductor 减速器 303
reel 卷轮 172
reflector 反射器 375
regulating valve 调节阀 349
regulating wheel 调节轮 85，167，222
reinforcement 强化层 36
reinforcing rib 横隔板加强肋 357
release bearing 复位轴承 296
releasing spring 回位弹簧 384
relief port 溢流口 85
relief valve 卸荷阀 312
reservoir 储液箱 171，226
resin 树脂 17
resin transfer molding 树脂转移模压法 17
resistance 电阻 122
resistance butt welding 电阻对焊 144
resistance welding 电阻焊 144
restraint frame 应变框 5
restricted orifice 节流小孔 352
retainer 压边圈 98
retainer pin 固定销 111
retaining ring 保持环 252
retaining spring 复位弹簧 297
retractable spark plug 伸缩式火花塞 320
returning idler 回程托辊 341
reverse flow 倒流 173
reverse multiplate brake 倒挡多盘离合器 370
reverse servo piston 倒挡伺服活塞 370
reversible band brake 可逆向带式制动器 300
reversible ratchet 可变向棘轮（机构） 58，64
reversing belt 换向带 56
reversing gear 换向齿轮 56
reversing mechanism 反向机构 55
revolving crane 回转吊车 400
ribbed neck 筋颈螺栓 247
rider 护垫 32
right cap 右端盖 26
right steering knuckle 右转向节 371
right track 右滑轨 360
right-angled screwdriver 直角改锥 127
rigid axle suspension 刚性轴悬架 362
rigid coupling 刚性联轴器 291
rigid ring 刚性环 71
rigidity 刚性 3
ring damper 环形挡板 311
ring gear 冠齿轮，环齿 72，367
ring sieve 环形筛片 413
ring spanner 梅花扳手 126
riser 冒口 36
rivet 铆钉 254，372
riveting punch 铆冲 247
rock crusher 摇杆粉碎机构 56
rocker 摇杆 50
rocker arm 摇臂 210
rocker cam 摇臂凸轮 225
rocking journal bearing 摇摆滑动轴承 264
rocking lever 摇杆 348
roll 辊，滚轮，轧辊 19，143，167
roll pin 卷筒销 253
roller 滚筒，辊子 77，339
roller burnishing 精辊 31
roller conveyor 辊子输送机 343
roller follower 滚子从动件 52
roller printing biscuit machine 辊印饼干机 422
roller radial bearing 滚子径向轴承 241
roller trunion universal joint 滚子耳轴万向节 293
roller-chain coupling 滚子链联轴器 293
roller-detent mechanism 滚子凹槽机构 63
rolling 滚压 122
rolling bearing 滚动轴承 265
rolling guide shoe 滚动导靴 333
rolling piston 旋转活塞 323
root diameter 底径 246
rope 钢丝绳 329
rotary air preheater 回转式空气预热器 321
rotary arbor 转动心轴 260
rotary file 回转锉 189
rotary shaft 转轴 260
rotary table 回转工作台 220
rotary tool post 回转刀架 215
rotary-swaging 旋转锻压 140
rotating cylindrical guide 回转导套 49
rotating fork 转动叉 345
rotating hub 回转毂 281
rotating lap 回转研磨板 168

rotating shaft 回转轴 281
rotating table 回转台 55
rotating-link coupling 回转连杆联轴器 293
rotor 转子 282，316
rotor housing 回转腔，转子壳体 81，321
rotor winding 转子绕组 307
roughening teeth 粗齿段 195
roughing 粗铣刀 188
roughness 粗糙度 30
round broach 圆孔拉力 196
round key 圆柱形键 252
rub strip 防擦条 360
rubber cup 皮碗 385
rubber mallet 皮锤 124
rudder 舵 375
runner 流道 35

S

saddle 主轴箱，溜板箱，床鞍 213，215，216
saddle key 鞍形键，空键 252
safety cleaner 空段清扫器 341
safety gear 安全钳 329，333
safety valve 安全阀 314
sand 型砂 118
sand casting 砂型铸造 31
sand mold 型砂 34
sander 打磨机，磨光机 127
saw 锯 126
sawing 锯切 31
scab 夹砂凸起 34
scaffold 安装架 329
scar 斑疤 34
scarifer 松土器 381
scart lap 斜边搭接 256
scoring 划痕 302
scotch yoke 止转棒轭 60
scraper blade 刮板 310
scraper conveyor 刮板输送机 425
screw 丝杠，螺杆 55，202
screw adjustment 调节螺栓 80
screw clamp 螺钉压板 56
screw conveyor 螺旋输送机 425
screw coupling 螺钉连接 247
screw nut 丝杠螺母 220
screw refrigeration compressor 螺杆制冷压缩机 418
screw-operated brake 螺杆操作的带式制动器 300
scroll plate 转盘 202
seal 密封圈，密封 86，352
seal chamber 密封腔 82
seal face 密封面 281
seal leg 密封管脚 343
sealing ring 密封圈 26
seam welding 缝焊 144
seat 刀座 187
seat back 靠背 360
seat cushion 座垫 360
seat ring 密封座环 315
seating pad 固定台垫 191
secondary circuit 次级回路 85
secondary shoe 从蹄 299
section line 剖面阴影线 21
section 剖面 21
sectional view 断面图 26
sector arm 分度臂 203
sector plate 扇形板 311
securing clip 固定卡子 329
sedimenter 沉积器 349
self-positioning supporting pin 自位支承钉 205
self-tapping screw 自攻螺钉 248
semiconducting pellet 半导体片 321
semi-portal crane 半门座悬臂式起重机 334
sensor 传感器 232
separator 隔离器 267
sequence valve 顺序阀 84
serrated neck bolt 齿颈螺钉 247
servo control 伺服控制 171
servo controlled feed system 伺服控制进给系统 226
servo drive 伺服驱动 232
set screw 紧固螺钉，固定螺钉，锁定螺钉 86，193，248
set square 三角板 25
setting block 对刀块 208

shaft 轴 26，219，258，326
shaft key 联轴键 252
shaft tolerance 轴公差 28
shank 刀杆 187
shaped charge 射孔弹 393
shaping 刨削 154
sharp corner 尖角 41
shear 剪切 3
shear undamped coupling 无阻尼剪切联轴器 294
sheave 曳引轮 329
sheet 板材 19
shell 壳体 280
shell broach 套式拉刀 195
shell end mill 中空端面铣刀 187
shielding gas 保护气体 148
shock absorber 减振器 361
shoe plate 掌板 82
short break line 短折线 21
shoulder 台阶 267
shrink fits 压缩过盈配合 29
shrinkage cavity 收缩孔隙 35
shut-off bar 截止杆 349
side clearance 侧后角 186
side cutting edge angle 副切刃角 186
side gear 半轴齿轮 369
side milling cutter 圆盘周铣刀 42
side pod 侧箱盖 376
side rake angle 副前角 186
side relief angle 副后角 186
side view 右视图 26
side wall reinforce rib 侧壁加强肋 357
sieve plate 筛板 413
sightseeing elevator 观光电梯 327
silent chain plate 无声链板 77
silent ratchet mechanism 无声棘轮机构 64
silent-chain coupling 无声链联轴器 293
simple sheet lifter 简易板材提升器 337
single cavity die 单腔模 115
single elevator 单控电梯 328
single layer belt dryer 单层带式干燥器 412
single roller dryer 单滚筒式干燥器 411
single-plate friction clutch 单片摩擦离合器 296
single-stage cantilevered vortex pump 单级悬臂式旋涡泵 423
sinking 成形 226
sintering 烧结 143
sleeve 衬套，套筒，轴套 219，258，371
sleeve bearing 套筒轴承 342
sleeve output port 套式输出口 349
slice 条料 19
sliced 分切 42
slide 滑块 140
slide pin 滑柱，导向销 211，371
slide positioning pin 活动定位销 211
slider crank 滑块曲柄 60
slider-drive crank 滑块驱动曲柄机构 57
sliding bush 活动护套，卸料滑套 99，102
sliding collar 滑套 204
sliding table 滑台 240
sliding valve 滑动叶片 323
slip （转盘）卡瓦 393
slitting 切片 226
slitting saw cutter 锯切铣刀 187
sliding gear 滑移齿 305
slope 斜度 23
slot 切槽 40
slotted 开槽螺钉 248
slotted nut 普通开槽螺母 250
slotted ring nut 开槽环形螺母 249
slotted screwdriver 一字螺丝刀 127
slotted washer 开槽垫片 250
slotting 插削 212
small helix angle 小螺旋角 44
snagging 清铲 31
snap ring 卡圈 364
snap-action mechanism 快动机构 59
snap-action switch 快动开关 59
snap-action toggle switch 快动肘节开关 59
snip 马口铁剪刀 126
socket screwdriver 套筒螺丝刀 127
solder 填（焊）料 151
solder wave 焊波 149
soldering iron 烙铁 151
solenoid armature 电磁铁衔铁 353
solenoid valve 电磁阀 211
solenoid winding 电磁绕组 353

solid abrasive 固结磨具 197
solid alloy 固态合金 7
solid bolted shaft coupling 刚性螺栓轴联轴器 294
solid mandrel 整体心轴 206
solidified metal 固化金属 12
space sleeve 隔套 83，258
spacer 隔套 83，162
spade drill 铲形钻 190
spark plug 火花塞 346
special lock 特形锁扣 54
specimen 试件 5
speed changer 变速器 304
speed-adjusting mechanism 速度调节机构 316
speed-reducer 减速器 219
spherical diameter 球直径 22
spherical radius 球半径 22
spider 十字叉 141
spigot bearing 插入轴承 367
spill control valve 溢流控制阀 359
spindle 主轴 218
spindle box 主轴箱 218
spindle head 主轴头 217
spindle nut 主轴螺母 198
spindle speed selector 主轴转速选调器 215
spinning chain 旋链 393
spiral bevel 螺旋锥齿 44
spiral case 螺旋腔 322
spiral fluted tap 螺旋槽丝锥 194
spiral freezer 螺旋式冻结装置 420
spiral retaining ring 螺旋保持环 251
spiral-bevel set 螺旋锥齿轮副 70
spline 花键 254
spline hub 花键毂 368
splined coupling 花键联轴器 295
splined input shaft 花键输入轴 367
split cotter 开口销 77
split nut 切分螺母 250
split pin 开合销 253
spool 阀芯 397
spot facing 锪孔 163
spot facing drill 局部锪面钻 189
spot welding 点焊 144
spotface 点锪 23
spray 喷雾 279
spray drying system 喷雾干燥系统 411
spring 弹簧 55
spring bracket 弹簧座 351
spring clip 弹簧锁夹 59
spring hanger 弹簧挂架 312
spring lock nut 弹性锁紧螺母 250
spring pin 弹性销 253
spring recess 复位弹簧 323
spring seat 弹簧座 363
spring washer 弹簧垫片 251
spring-locking washer 弹簧锁紧垫片 251
sprocket 轮辐 339
sprue 浇道，浇铸道 35，118
sprue hole 浇孔 115
sprue pin 分流锥 115
spud frame 定位桩 379
spur bevel gear transmission 直齿圆锥齿轮传动 65
spur gear 直齿 55，116
spur gear oil pump 直齿油泵 370
spur gear transmission 直齿圆柱齿轮传动 65
square jig plate 方形铅模板 207
square neck bolt 方颈螺栓 247
square nut 方形螺母 249
square spline 方形花键 254
square turret 方刀台 215
squeeze casting 挤压铸造 16
sraggered tooth cutter 交错齿铣刀 187
stacked and milled 叠层铣削 42
stamped spring nut 冲压弹性螺母 250
star washer 星形垫片 129
star wheel 恒星齿轮 71
starter 启动器，启动机 353
statical tolerance 静公差 23
stator 定子 282，307
stay ring 保持环 322
stay vane 静止叶片 322
steam drum 蒸汽鼓 313
steam turbine 蒸汽轮机 316
steel ball 钢球 86
steel bevel bracing 钢管斜撑 390
steel pipe 钢管 182
steel-ball grinding 钢球磨削 167

steering cylinder 转向缸 371
steering fork 转向架 345
steering gear 转向器 371
steering knuckle 转向节臂 371
steering pulley 导向轮 390
steering pump 转向泵 371
steering shaft 转向轴 371
steering system 转向系统 361
steering wheel 转向盘 371
steering worm 转向蜗杆 371
step drill 阶梯孔 163
step drilling 阶梯钻 163
stepped motor 步进电动机 227
stiffener 加强筋 280
still clamp 固定钳口 123
stop block 限位块 102
stop lever 停止杆 350
stop pin 止销 206
stopcock 旋塞阀 128
straight bevel 直齿锥齿 44
straight bevel gear 直齿锥齿轮 69
straight flow 直流 173
straight pipe pneumatic dryer 直管气流干燥器 412
straight spur 直齿 44
straight-flute drill 直槽钻 190
straightness 直线度 22
strain 应变 5
strain gage 应变仪 191，211
strap 镶条 290
strength 强度 3
stress rupture 断裂 3
stretch gripper 延展夹头 105
strip braking 带状制动 302
stripper 打料杆，汽提管 99，401
stroke choke 行程挡块 221
stud 螺柱 129
stud coupling 双头螺栓连接 247
stuffing 填料 423
stuffing gland 填料压盖 397
stuffing box 填料函 397
stylus 触针 32
submerged-arc welding 埋弧焊 144
submit gaseous ring 上气环 358
substrate 基体 177
suction valve 吸入阀 323
sump 油底壳 347
sun gear 太阳轮 370
sun wheel 太阳轮 71
superfinishing 超精加工 31，154，212
support blade 托板 156
surface broaching machine 平面拉床 221
surface grinding 平面磨削 165，212
suspension system 悬架系统 361
swage 型模 123
swaging 型锻 138
swashplate 旋转斜盘 82
swing fork 旋转叉 345
swing link 摆杆 331
swivel 转环，旋转龙头 77，392
swivel vise 旋转钳台 204
symmetry 对称度 23
synchronizer 同步器 368
synchronous belt 同步带 75
synchrotizing damper 同步挡板 341

T

T-bolt T 形螺栓 206
T-slot T 形槽 162
T-square 丁字尺 25
tab 耳片 363
table 工作台 213
tachometer 转速表 360
tail drum 尾部滚筒 340
tailstock 尾座 213
take-up weight 张紧重块 341
tangential 切向 15
tangential key 切向键 252
tank 油箱 85
tap chuck 丝锥夹头 208
tap crest 牙顶 193
tap wrench 丝锥扳手 127
tape helix 螺旋带 407

tape reader 读带机 232
taper 锥度 22，192
taper mandrel 锥度芯棒 168
taper pin 锥销 253
taper shank 锥柄 190
tapered bushing 锥形套管 413
tapered lap 锥坡搭接 256
tapered mandrel 锥度心轴 206
tapered roller bearing 圆锥滚子轴承 258，266
tapered sleeve 锥套 191
tapered roller thrust bearing 锥形滚子推力轴承 365
tee T 形连接 145
telescoping cover 伸缩盖 173
telescoping sheet lifter 伸缩式板材提升器 337
template 样板，模板 25，46
tensile 抗拉 3
tensile property 拉伸性 2
tensile stress 拉应力 4
tension coupling 张紧式联轴器 294
tension pulley 张紧筒 341
tension roller 张紧辊 223
tension wheel 张紧轮 167，358
tensioned buoyant platform 典型拉曳浮动平台 400
tensioning pulley 张紧轮 74
texturing 组织处理 226
thermostat 恒温器 226
thickness 厚度 22
thin plate drilling jig 薄板工件钻夹具 207
thin-wall bushing 薄壁套筒 211
thread 螺纹，拉杆 86，329
thread cutting tool 螺纹车刀 186
thread lead angle 螺纹导角 193
thread milling 螺纹铣削 188
three-jaw drilling chuck 三爪式钻夹头 203
three-lobe bearing 三凸块滑动轴承 264
three-lobe pump 三凸轮泵 80
throttle lever 节流杆 350
throttle valve 节流阀，节气门 171，347
through feed grinding 贯通磨削 167，222
through hole 通孔 163
thrust bearing 推力轴承 71，86，316
thrust collar 推力套 323
thrust pad 推垫 348
thrust pin 推销 348
thrust sleeve 推力套 348
thrust washer 止推垫圈 358
tie rod 固定杆，横拉杆 286，371
tiltable jacketed kettle 可倾式夹层锅 415
tilting pad 斜轴瓦 264
time gear 正时齿轮 346
timed-screw pump 同步螺杆泵 80
timing belt 正时带 346
timing control 正时控制 351
toggle clamp 肘节夹头 56
toggle link 肘连接 298
toggle mechanism 肘节机构 58
toggle press mechanism 肘节压紧机构 56
toggle ratchet 肘节棘轮 58
tolerance 公差 21
tommy bar 套筒扳手旋转手把 127
tongs 夹钳，吊钳 125
tool cathode 工具电极（阴极） 171
tool face 刀面 186
tool head 刀头 213
tool point 刀尖 186
tool post 刀台 213
tool turret 转塔刀架 219
toolholder 刀夹 191
tooth 刀齿 188
tooth claw 齿爪 413
tooth face 齿面（前刀面） 188
top pulley 天轮 390
torch 火炬 151
torque sensor 转矩传感器 339
torque spanner 扭矩扳手 126
torsion bar 扭转杆 361
torsional damper spring 扭转减振弹簧 367
torsionally elastic bearing 扭转弹性轴承 361
torus coupling 环式联轴器 294
total runout 全跳动 23
toughness 韧性 2
tower crane 塔式起重机 334
tower mast 塔桅 335
track 履带 345
track crane rammer 轨道起重式强夯机 383
track driver mechanism 履带行驶机构 378

track handle 滑轨手柄 360
trailing arm 领臂 361
transfer pump 传输泵 349
transformer secondary 变压器次级线圈 151
transition fits 过渡配合 29
translating cam 移动凸轮 52
transmission gear 传动齿轮 26
transmission shaft 传动轴 260
tray dryer 箱式干燥器 411
tray vacuum dryer 箱式真空干燥器 412
trepanning 套料 226
trestle 支架 397
triangular spline 三角形花键 254
trigger hook 碰钩 343
trigger vane 触发片 354
trolley 吊车 335
trolley pulley 滑轮 335
trommel 圆筒筛 407
trough 料槽 342
trunnion 耳轴，枢轴 135，364
try square 直角尺、矩尺 25，127
tube 管件 280
tube electrode 管状电极 171
tube scraper 套管刮削器 396
tubeplate 管板 280
tubing anchor 油管锚 396
tumble stoner 转筒式除石机 407
tumbler yoke gear 转向扼架齿轮 55
tundish 浇口盘 12
tungsten electrode 钨电极 148
tungsten inert-gas welding TIG 焊接 144
tunnel boring machine 隧道盾构机 379
turbine 汽轮机，涡轮 308，369
turbine shaft 涡轮机转轴 309
turbine shaft bearing 涡轮轴轴承 369
turbine wheel 涡轮 319
turbing case 涡轮腔体 319
turbocharger 涡轮增压器 319
turbo-worm reducer 蜗杆减速器 410
turning 车削 28，212
turning tool 车刀 186
turnstile capstan wheel 转杆 / 绞轮 215
turntable 回转台 105
turret 转塔刀架 237
turret lathe 转塔车床 127
turret stop 转塔挡块 215
twin drive plates 双驱动盘 367
twist drill 麻花钻（头） 163，190
tyre coupling 胎式联轴器 294

U-bolt U 形螺栓 248
U-packing U 形密封 283
USM 超声加工 154，226
ultrasonic machining 电化学超声 175
ultrasonic vibration 超声振动 175
underreamer 扩眼器 397
uniflex flexible-spring coupling 万向柔性弹簧联轴器 293
unilateral tolerance 单向公差 27
unit die 单元模 115
universal joint 万向节 367，371
universal joint yoke 万向节叉 371
universal vise 万能钳台 204
unstacked 拆除叠放 42
untimed-screw pump 非同步螺杆泵 80
up milling 逆铣 161
upper arm 上臂 361
upper cylinder 上气缸 316
upper die 上模 139
upper die shoe 上模座 106
upper horn 上臂 151
upper punch 上冲头 116
upright 立柱 329
upsetting 镦粗 138

V engine V 形引擎 60
Venturi nozzle 文丘里喷嘴 396
VSV 真空转换阀 353
vacuum encapsulate 真空封装 16
vacuum pump 真空泵 359
vacuum regulating valve 真空调节阀 359
vacuum roller dryer 真空滚筒干燥器 412
valve 阀，阀件 125，181
valve base 阀座 83
valve body 阀体 397
valve plate 阀板 82
valve seat 阀座 397
valve spring 气门弹簧 358
valve stem 阀杆 397
vane 叶片 80，423
vane switch 片状开关 354
variable resistor 可变电阻 129，353
vent 通气口，通风口 118，312
vent cup 通风杯 286
vent-pipe 排气管 347
vertical annular jet mill 立式环形喷射式粉碎机 413
vertical axis coil grab 立式轴卷抓斗 336
vertical band sawing 立式带锯锯床 221
vertical boom 叉架 389
vertical core 立式型芯 118
vertical grinding machine 立式磨浆机 413
vertical pile 立柱 384
vertical pipe evaporator 立管式蒸发器 418
vertical roller press 立式辊压机 421
vertical shell and tube condenser 立式壳管冷凝器 419
vertical sterilization pot 立式杀菌锅 417
vertical view 俯视图 26
vibrating conveyor 振动输送机 343
vibratory ball mill 振动式球磨机 414
view 视图 26
vise 虎钳，台钳 56，220
vise grip 夹钳夹头 56
visible line 粗实线 21
visual sensing 视觉传感器 339

WJM 水射流加工 154，170，181
wafer 晶片 176
washer 垫圈，垫片 83，251
water distributor 分水器 419
water exhaust valve 排水阀 92
water jacket 水套 347
water jet 水射流 226
water-cooled copper shoe 水冷铜座 151
waterway 冷却水道 111
waviness 波纹度 30
wear pad 耐磨垫 190
wear resistance 耐磨性 3
wedge 楔块 290
wedge pin 楔形销 315
welded type barrel nut 焊接筒形螺母 250
welding 焊接 122
welding gun 焊枪 150
welding tip 焊接头 150
welding torch 焊炬手把 150
welding wire 焊丝 150
welding with pressure 压力焊接 144
wharf crane 码头起重机 333
wheel crane 轮胎式起重机 334
wheel cylinder 制动分缸 371
wheel spindle 砂轮主轴 198
wheel stud 车轮紧固螺栓 371
whipstock 导斜器 393
whirlpool overflow pipe 旋涡溢流管 397
winch 纹盘 383
windbox 风箱 311
windmill oven 风车炉 416
wing attachment strut 尾翼平衡柱 376
wing screw 翼型螺钉 249
wire 线丝 172
wire brush 钢丝刷 147
wire guide 线丝导轮 172

wire spool 丝筒 227
wood grip washer 抓木垫片 251
wood screw 木螺钉 248
woodruffkey 半圆键 252
workpiece anode 工件电极（阴极） 171
worm 蜗杆 224
worm reduction 蜗杆减速器 304
worm shaft 蜗杆轴 203
worm wheel 蜗轮 224，304

X-ray receiver X 射线接收器 12
X-ray transmitter X 射线传导器 12

yield 屈服 3
yoke 轭架 305
yoke bolt 轭套螺栓 315
yoke bushing 轭形套 315

zero line 零线 63
zigzag spring 锯齿形弹簧 360

Vocabulary with Figure Index

词汇及图形索引（中英对照）

安全阀　safety valve　92，314
安全钳　safety gear　329，333
安装法兰盘　mounting flange　321
安装架　scaffold　329
鞍形键　saddle key　252
凹槽　groove　368
凹槽锁销机构　detent　58
凹孔　recess　204
凹螺杆　female rotor　325
凹模　cover die　115
凹压（定）模　cover die　132

巴氏合金衬套　Babbitt lining　262
摆锤　pendulum　4
摆动机构　oscillating mechanism　57
摆杆　swing link　331
摆线轮液压泵　gerotor pump　80
板式夹头　plate clamp　336
斑疤　scar　34
板材　sheet　19
板弹簧　flat spring　275
板弹簧导向装置　leaf-spring guide　343
板牙　die　127
板牙扳手　die-stock　127
半导体片　semiconducting pellet　321
半径　radius　22
半门座悬臂式起重机　semi-portal crane　334
半模　mold half　115
半圆键　woodruffkey　252
半轴齿轮　side gear　369
薄板弹簧　diaphragm spring　296
薄板工件钻夹具　thin plate drilling jig　207
薄壁套筒　thin-wall bushing　211
箔片滑动轴承　foil bearing　264
保持环　retaining ring，stay ring　252，322
保护气体　shielding gas　148
保险杠　bumper　360
保险杠主梁　bumper beam　360
爆燃传感器　knock sensor　355
杯端螺钉　cup screw　248
杯形垫片　cup washer　251
杯形密封　cup seal　352
泵　pump　173，223
泵复位弹簧　pump return spring　352
泵缸　pump cylinder　423
泵杠杆　pump lever　352
泵喷头　pump jet　352
泵体　body，pump housing　26，281
泵叶轮　pump impeller　286
泵柱塞　pump plunger　350
鼻锥体　nose cone　376
比较器　comparator　372
比较运算器　comparator　233
比较装置　comparing equipment　233
比例规　proportional divider　25
闭合索　closing rope　379
闭环控制系统　closed-loop control system　233
闭锁阀　latch valve　350
边口过滤器　edge filter　347
边料　blank　99

边刷 lateral brush 382
边缘夹紧板材夹头 edge grip sheet clamp 337
边缘平齐 edge 145
变矩器 converter 369
变速器 speed changer 304
变压器次级线圈 transformer secondary 151
标注 note 21
并联电梯 duplex elevator 328
病床电梯 hospital elevator 327
拨叉 plectrum 358
波纹板 corrugated sheet 19
波纹度 waviness 30
波纹隔板 corrugated panel 19
波纹管 bellow 54
波纹管联轴器 bellows coupling 294
波纹辊 corrugating roll 19
波纹块料 corrugated block 19
补偿辊子 compensating roller 339
补偿链 compensation chain/cable 328
补偿叶片 compensation flap 354
补油泵 oil pump 324
布氏硬度 HB 3
步进电动机 stepped motor 227

C 形垫片 C washer 251
擦光 buffing 154
操控套 operating sleeve 296
槽 kerf 179
槽轮停歇机构 geneva intermittent mechanism 57
侧壁加强肋 side wall reinforce rib 357
侧后角 side clearance 186
侧隙 backlash 66
侧隙锥 counter taper 36
侧箱盖 side pod 376
侧置型芯 drop core 119
测力计 dynamometer 299
叉车 fork lift truck 345
叉架 bracket，fork yoke 39，292，389
叉件 cross-member 363
插齿 gear shaping 168
插件凹槽机构 plunger-detent mechanism 63
插入轴承 spigot bearing 367
插销 latch 111
插销机构 latch mechanism 58
插削 slotting 212
差动手拉葫芦 differential chain hoist 338
差动提升机 differential hoist 58
差分传感器 differential sensor 234
差分式带式制动器 differential brake 300
差速器 differential 361
拆除叠放 unstacked 42
柴油发动机 diesel engine 346
铲背圆弧成形铣刀 form relieved circular cutter 187
铲形钻 spade drill 190
常转零件 permanently rotating part 64
超精加工 superfinishing 31，154，212
超声定位信号浮标 acoustic positioning beacon 400
超声加工 USM 154
超声加工电火花 EDM 175
超声振动 ultrasonic vibration 175
超越离合器 overrunning clutch 58
车刀 turning tool 186
车架 frame 383
车轮紧固螺栓 wheel stud 371
车削 turning 28，154，212，226
沉积器 sedimenter 349
沉坑 countersink 22，163
沉孔 counterbore 22，163
沉孔锪钻 counter boring drill 189
沉孔加工 counterboring 163
衬片 lining 298
衬套 bush，bushing，phunger bush sleeve 111，219，363
成形 forming，sinking 122，226
成形锻压 forging 138
成形机械 forming machine 421
承载销 bearing pin 77
城堡型开槽螺母 castle nut 250
乘客电梯 passenger elevator 327
吃刀深度 depth of cut 165
持续弹簧 duration spring 352

尺寸界线 extension line 21
尺寸数字 dimension figure 21
尺寸线 dimension line 21
齿顶高 addendum 66
齿根高 dedendum 66
齿颈螺钉 serrated neck bolt 247
齿轮 gear，gear wheel 26，219
齿轮泵 gear pump 26，80
齿轮传动 gear transmission 65
齿轮传动引擎 geared engine 60
齿轮滑块曲柄机构 geared slider crank 57
齿轮加工 gear cutting 154
齿轮啮合 gear geometry 66
齿轮箱 gearbox 311
齿面 tooth face 188
齿坯 gear blank 170
齿形带 cog belt 358
齿爪 tooth claw 413
齿爪式粉碎机 disk mill 413
冲击口 dashpot 350
冲击强度 impact strength 3
冲孔 piercing 122
冲孔凸模 piercing punch 101
冲天炉 blast furnace 11
冲头 punch 6，55，107
冲压 punching 28
冲压弹性螺母 stamped spring nut 250
充气轮胎 pneumatic tyre 375
出料管 discharge tube 413
除雾器 demister 92
除渣锤 chipping hammer 147
储料器 accumulator 175
储液箱 reservoir 171，226
触发片 trigger vane 354
触针 stylus 32
穿钉 nailing 358
传动齿轮 transmission gear 26
传动系统 power train system 361
传动箱 gear box 308
传动销 drive pin 253
传动轴 transmission shaft，drive shaft 260，370
传感器 sensor 232
传力配合 force fits 29
传输泵 transfer pump 349
传输带 conveyor belt 176
传输链 conveying chain 339
船用电梯 marine elevator 328
床鞍 saddle 216
床身 bed 213
床头箱 headstock 213
垂直电梯 elevator 327
垂直度 perpendicularity 23
锤头 ram 140
锤头式起重机 hammer-head crane 334
唇形密封 lip seal 369
磁力提升器 magnetic lifter 336
磁力抓斗 magnet grapple 337
磁粒离合器 magnet clutch 299
磁芯 magnetic core 408
磁性磨料加工 MAM 154，183
磁选机械 magnetic separation machinery 408
次级回路 secondary circuit 85
从动齿轮 driven gear 296
从动件 driven part，driven member 64，297
从动盘 driven disk 296
从蹄 secondary shoe 299
粗糙度 roughness 30
粗齿段 roughening teeth 195
粗点画线 chain line 21
粗实线 visible line 21
粗铣刀 roughing 188
脆裂强度 fracture toughness 2
搓丝板 plate die 194
错位 misalignmen 302
错位链板 offset plate 77
错位链节 offset link 77
错位链节销 offset link pin 77
错位轴承 offset bearing 264

D

搭结 lap 145
打孔 drilling 226

打料杆　stripper，knockout rod　99，115
打磨机　sander　127
大节距　coarse pitch　44
大径　basic major dia，major diameter　193，246
大梁　crossbeam　390
大螺旋角　large helix angle　44
大转矩棘轮　high-torque ratchet　58
呆扳手　double open ended spanner　126
带　belt　219
带轮　pulley，belt pulley　218，413
带轮槽　pulley groove　75
带驱动滚筒　belt driving drum　340
带驱动架　belt-drive bracket　74
带式冻结隧道　belt freezing tunnel　419
带式输送机　belt conveyor　340
带张紧滚筒　belt tension drum　340
带状制动　strip braking　302
单层带式干燥器　single layer belt dryer　412
单滚筒式干燥器　single roller dryer　411
单级悬臂式旋涡泵　single-stage cantilevered vortex pump　423
单控电梯　single elevator　328
单梁式联轴器　beam coupling　293
单片摩擦离合器　single-plate friction clutch　296
单腔模　single cavity die　115
单向阀　check valve　315
单向公差　unilateral tolerance　27
单向离合器　one-way clutch　358
单元模　unit die　115
挡板　hood　333
挡块　end stop　222
挡圈　bead flange　358
刀齿　tooth　188
刀齿槽　flute　188
刀杆　shank，arbor　187，213
刀夹　toolholder　191
刀尖　tool point，lip　186
刀尖角　lip angle　188
刀棱　heel　188
刀粒　insert　162
刀面　tool face　186
刀片　cutting tool，blade　192，193
刀片槽　blade slot　190
刀台　tool post　213
刀体　body　192
刀头　tool head　213
刀座　seat　187
导板　guide plate　101
导尺　pilot strip　101
导轨　rail，guideway，guide rail　288，331，333
导轨撑架　rail bracket　328
导料板　blank guide　101
导流板　flow plate　92
导轮　guide wheel，reactor　227
导套　guide bush　111
导向沉孔镗刀　counter-boring tool with pilot　189
导向钩　guide hook　227
导向块　guide　52
导向轮　steering pulley　390
导向销　slide pin　371
导斜器　whipstock　393
导芯　guide core　369
导靴　guide shoe　329
导叶　guide vane　326
导柱　guide pin，guide pillar　102，111
倒车灯　backup light　366
倒挡多盘离合器　reverse multiplate brake　370
倒挡伺服活塞　reverse servo piston　370
倒角　chamfer angle　193
倒角成形车刀　chamfering form tool　186
倒角锪钻　counter sunk drill　189
倒流　reverse flow　173
倒置齿轮滑块曲柄机构　geared inverted slider crank　57
倒置立柱夹具　inverted post jig　207
等离子　plasma　122，177
等离子加工　PBM　154，170，226
等离子弧焊　plasma arc welding　144
低油位开关　low level switch　279
滴漏润滑　drip lubrication　78
底盘　chassis　389
底座　base　215
地脚螺钉连接　rag bolt coupling　247
地轮　bottom pulley　390
点焊　spot welding　144
点火电极　ignition electrode　313
点火开关　ignition switch　346

点火器 igniter 353
点火塞 ignition plug 321
点火线圈 ignition loop，ignition coil 346
电插头 electric connection 355
电磁阀 solenoid valve，electro magnetic valve 211，312
电磁开关 electromagnetic switch 358
电磁绕组 solenoid winding 353
电磁铁衔铁 solenoid armature 353
电磁制动器 magnetic brake 329
电动葫芦桥式起重机 overhead travelling crane with hoist 334
电动机 motor 54
电动螺丝刀 electric screwdriver 127
电动驱动轮 motor pulley 74
电镀 electroplating 174
电弧 arc 148
电弧焊 arc welding 144
电化学 electro-chemical，ECM 31，175
电化学超声 ultrasonic machining，ECU 175
电化学电火花 ECAM 175
电化学珩磨 ECH 175
电化学加工 ECM 154
电化学摩擦 ECA 175
电化学磨粒喷射 AJECM 175
电化学磨料修正 ESC 175
电化学磨削 ECG 175
电火花电化学 ECDM 175
电火花电极平动加工 EDScan machining 171
电火花加工 EDM 154
电火花加工孔 EDMed hole 46
电火花介质（工作液） dielectric fluid 171
电极 electrode 144
电极导管 electrode guide tube 151
电极夹头 electrode holder 147
电极支架 electrode carrier 173
电极座极 ram 173
电加热膨胀元件 electrically heated expansion element 352
电接头 electrical connection 352
电解磨削 electrolytic grinding 31
电解液 electrolysis 165
电解液槽 electrolyte tank 173
电流表 ammeter 312
电炉 electric furnace 12
电抛光 electro-polish 31
电气柜 electrical apparatus 224
电气配线接头 harness connector 355
电容器 condenser 373
电刷 electrical(connection)brush 165，174
电刷架 brush frame 358
电梯 elevator 328
电渣焊 electro-slag welding 144
电子燃油喷射系统 electronic fuel injection system 353
电子束 electron beam 31，122
电子束焊接 electron beam welding 144
电子束加工 EBM 226
电阻 resistance 122
电阻对焊 resistance butt welding 144
电阻焊 resistance welding 144
电钻 electric drill 127
垫片 gasket 26
垫圈 gasket，washer 26，83
雕刻 engraving 226
吊车 trolley 335
吊车轨道 crane runway 335
吊钩 hook 383
吊钳 tongs 396
吊升滚筒 hoisting drum 392
叠层铣削 stacked and milled 42
碟形弹簧 disc spring 329
蝶阀 butterfly valve 315
丁字尺 T square 25
顶（出）杆 ejector pin 115
顶尖角 point angle 190
顶件器 ejector bush 101
顶梁 overarm，arch 216，219
顶隙 clearance 66
顶针 center 165
定板 fixed platen 115
定齿盘 fixed chain ring 413
定模型芯 fixed core 111，115
定位槽 locating slot 191
定位过盈配合 interference locational fits 29
定位口 locating ear 191
定位器 aligner 178
定位锁紧机构 locating mechanism 56

定位销　locating pin　102，253，358
定位支承板　positioning support plate　211
定位柱　location post　207
定位桩　spud frame　379
定子　stator　282
动板　moving platen　115
动齿盘　dynamic chain ring　413
动力涡轮　power turbine　320
动力旋转器　power rotator　336
动态停靠钻井船　dynamically positioned drill ship　400
动叶轮　adjustable impeller　316
斗式提升机　hopper elevator　425
读带机　tape reader　232
端部密封　end seal　111
端齿盘　face gear　67
端盖　end cover　358
端面　face　267
端面键　face key　258
端面磨削　end surface grinding　166
端面凸轮　face cam　52
端面铣刀　face cutter　187
端柱　end pillar　321
短折线　short break line　21
断裂　stress rupture　3
断裂口　fracture　4
断路器　circuit breaker　59
断面图　sectional view　26
断屑台　chip splitter　191
锻棒料　bar forming　138
锻压　forging　31，122，138
锻制镗刀　forged boring tool　188
对称度　symmetry　23
对刀块　setting block　208
对重导轨　counterweight guide rail　328
对重架　counterweight frame　331
对重装置　counterweight　328
对撞式气流粉碎机　counter-impact jet mill　413
镦粗　upsetting　138
多槽端铣刀　multi flute end mill　187
多层带　muliple-ply belt　75
多孔内衬　porous liner　264
多片摩擦离合器　multi-plate friction clutch　296
多腔模　multi-cavity die　115
多刃镗刀　multiple-cutter boring tool　189
多头铣刀　multiple-thread cutter　188
多楔带　multi-edges belt　75
多叶片滑动轴承　multileaf bearing　264
舵　rudder　375
惰簧　idling spring　349
惰轮　idler，idler pulley，idler gear　74，203
惰转子　idler rotor　81

鹅颈管　gooseneck　132
轭架　yoke　305
轭套螺栓　yoke bolt　315
轭形套　yoke bushing　315
耳片　tab　363
耳轴　trunnion　364
二冲程引擎　double-stroke engine　60

F

发电机　alternator　308
发动机　engine　346
阀　valve　125
阀板　valve plate　82
阀杆　valve stem　397
阀体　valve body　397
阀芯　spool　397
阀座　valve base，valve seat　83，397
法兰　flange　87，281
法兰联轴器　flanged coupling　291
法兰式齿轮联轴器　flange-type gear coupling　293
反馈编码器　encoder feedback　241
反射器　reflector　375
反向机构　reversing mechanism　55

反锥垫片 countersunk washer 251
方刀台 square turret 215
方颈螺栓 square neck bolt 247
方形花键 square spline 254
方形螺母 square nut 249
方形切向键 kennedy key 252
方形钻模板 square jig plate 207
方钻杆 kelly 392
方钻杆旋转器 kelly spinner 393
防擦条 rub strip 360
防尘管 dust bellow 363
防尘密封 dust-free seal 284
防振垫盘 crashpan 375
防转杆 anti-roll bar 361
放大器 amplifier 233
飞轮 flywheel 140，276，296，347
飞轮螺栓 flywheel bolt 358
飞轮重物块 flyweight 348
飞翼轮 flap wheel 126
非同步螺杆泵 untimed-screw pump 80
分电器 distributor 346
分度臂 sector arm 203
分度杆 index crank 203
分度机构 indexing mechanism 57
分度盘 index plate 203
分度台 indexing table 55
分度头 dividing head 57
分度头轴 index head spindle 203
分度销 index pin 203
分度钻孔夹具 indexing drilling jig 208
分规 divider 25
分检器 classifier 311
分流道 branch runner 110
分流隔板 divide board 316
分流器 diverter 343
分流锥 sprue pin 115
分配器 distributor 353
分配式配油泵 distributor pump assembly (DPA) 350
分切 sliced 42
分水器 water distributor 419
分型面 parting plane 132
分型线 parting line 118
粉碎机 mill 311
粉碎室 crushing chamber 413
风车炉 windmill oven 416
风箱 windbox 311
风钻 pneumatic drill 127
封闭冒口 blind riser 118
封闭模 closed die 106
缝焊 seam welding 144
扶手电梯 escalator 327
扶正器 centralizer 396
浮顶油罐 floating roof tank 404
浮动环滑动轴承 floating ring bearing 264
浮动接头 floating couple 210
浮箍 float collar 393
浮式起重机（浮吊） floating crane 334
浮鞋 float shoe 393
辐射梁 radiating beam 390
俯视图 vertical view 26
附加活塞 auxiliary piston 94
复合模（具） combination die 99，115
复位弹簧 retaining spring，spring recess 297，323
复位杆 ejector return pin 111
复位轴承 release bearing 296
副后刀面 minor flank 186
副后角 side relief angle 186
副偏角 body clearance 186
副前角 side rake angle 186
副切刃角 side cutting edge angle 186
副切削刃 minor cutting edge 186
副油泵 auxiliary oil pump 325
副轴 auxiliary shaft 225

G

盖板螺母 bonnet nut 315
盖板螺栓 bonnet bolt 315
盖盘垫片 disc washer 315
盖盘螺母 disc nut 315
盖盘转管 disc arm 315
盖形螺母 cap nut 249，347

盖罩 cover 129
盖罩螺母 cover nut 315
盖罩密封垫 cover gasket 315
盖罩双头螺柱 cover stud 315
坩埚 crucible 137
杆式夹钳 bar tong 336
杆芯直径 core dia 193
感应电动机 induction motor 306
感应线圈 induction coil 137
干式轴承 dry bearing 342
刚性 rigidity 3
刚性环 rigid ring 71
刚性联轴器 rigid coupling 291
刚性螺栓轴联轴器 solid bolted shaft coupling 294
刚性轴悬架 rigid axle suspension 362
钢板弹簧 leaf spring 375
钢管 steel pipe 182
钢管斜撑 steel bevel bracing 390
钢球 steel ball 86
钢球磨削 steel-ball grinding 167
钢丝软轴 flexible axle of steel cord 260
钢丝绳 rope 329
钢丝刷 wire brush 147
缸体 cylinder 51
杠杆 lever 211
高频发生器 HF generator 227
高压增压泵 high pressure intensifier 182
格栅 grate 310
隔板 baffle board，clapboard，diaphiagm 316，326，408
隔磁板 magnet vane 333
隔离器 separator 267
隔套 space sleeve 83，258
隔套 spacer 83，162
给水管 feedwater pipe 309
根部 heel 193
工件电极 workpiece anode 171
工具电极 tool cathode 171
工艺控制计算机 process control computer 180
工作台 table 213
弓形钩 Gclamp 127
弓形钻 brace 127
公差 tolerance 21
公制六角螺母 metric hex nut 249
功率放大器 power amplifier 234
拱起加工 hogging 174
钩头键 gibhead 252
钩形扳手 hook spanner 127
钩形夹头 clamp 56
钩形压板 hook clamping plate 211
毂盘件 hub member 294
鼓风管 blast tube 313
鼓起漏空 grab 302
鼓式制动器 drum brake 299
鼓形翻拌器 drum turner 336
固定板 fixed plate，anchor plate 98
固定杆 tie rod 286
固定护套 holding bush 99
固定卡子 securing clip 329
固定螺钉 set screw 193
固定磨环 fixed ring 413
固定钳口 still clamp 123
固定式夹层锅 dead jacketed kettle 415
固定索 holding rope 379
固定台垫 seating pad 191
固定头架 fixed head 218
固定涡管 fixed scroll 325
固定销 retainer pin 111
固定心轴 fixed shaft 260
固化金属 solidified metal 12
固结磨具 solid abrasive 197
固态合金 solid alloy 7
故障灯 diagnostic light 366
刮板 scraper blade 310
刮板输送机 scraper conveyor 425
刮刀 erasing knife 420
挂吊链式输送机 chain trolley conveyor 343
观光电梯 sightseeing elevator 327
管板 tubeplate 280
管件 pipe，tube 87，280
管状电极 tube electrode 171
管子台虎钳 pipe vise 127
贯通磨削 through feed grinding 167，222
冠齿轮 ring gear 72
冠状阀 crown valve 312
惯性振动筛 inertia vibrating screen 406
光化学 photochemical etching 226

光整（加工） finishing 226
龟裂 crazing 302
轨道起重式强夯机 track crane rammer 383
轨迹生成器 path generator 62
辊 roll 19
辊印饼干机 roller printing biscuit machine 422
辊子 roller 339
辊子输送机 roller conveyor 343
滚齿 gear hobbing 168
滚刀 hob 211
滚动导靴 rolling guide shoe 333
轴承 ball bearing，rolling bearing 258，265
滚轮 roll 167
滚筒 roller 77
滚压 rolling 122
滚针 needle pin 266，371
滚针轴承 needle roller bearing 266
滚珠丝杠 ball screw，leadscrew 241，233
滚子凹槽机构 roller-detent mechanism 63
滚子从动件 roller follower 52
滚子耳轴万向节 roller trunion universal joint 293
滚子径向轴承 roller radial bearing 241
滚子链联轴器 roller-chain coupling 293
锅炉 boiler 309
过渡配合 transition fits 29
过滤器 filter 171
过盈配合 interference 28

函数发生器 function generator 62
焊波 solder wave 149
焊接 welding 122
焊接筒形螺母 welded type barrel nut 250
焊接头 welding tip 150
焊炬手把 welding torch 150
焊枪 welding gun 150
焊丝 welding wire 150
焊药料斗 flux hopper 149
夯锤 rarnmer head 383
行程挡块 stroke choke 221
行星齿轮 planet wheel，planetary gear，epicyclic gear，planet gear 71，72，370
行星齿轮传动 planetary transmission 304
行星齿轮机构 planetary gear machine 358
行星齿轮架 planetary carrier 370
行星轮架 planet carrier 71
行星式滚柱丝杠 planetary roller screw 213
盒式镗刀 block-type boring cutter 189
鹤式起重机 crane 339
恒温器 thermostat 226
恒星齿轮 star wheel 71
珩磨 honing 31，154，175，212
横隔板加强肋 reinforcing rib 357
横进手轮 cross-feed handwheel 215
横拉杆 tie rod 371
横梁 beam，crossrail，cross-arm，gantry crane girder 217，219，370，390
横流 cross flow 173
横向定程机构 cross stroke mechanism 215
横向进给 cross feed 166
后杆 reat bar 363
后角 clearance angle 155
后冷却器 aftercooler 324
后视镜 rear-view mirror 376
后视图 rear view 26
后双轮 rear double wheel 376
后翼主板 rear aerofoil wing 376
后置冷却器 aftercooler 92
后轴承箱 back bearing housing 316
厚度 thickness 22
弧焊 arc 122
弧焊设备 arc welding eqiupment 147
弧口凿、半圆凿 gouge 127
虎钳 vise 56
互动装置 interactive devices 240
互冷器 intercooler 359
护垫 rider 32
护脚板 apron 329
护目镜 goggles 147
花键 spline 254
花键毂 spline hub 368

花键联轴器 splined coupling 295
花键输入轴 splined input shaft 367
花盘 faceplate 203
划痕 scoring 302
划眼器 reaming device 393
滑动叶片 sliding valve 323
滑动轴承 plain bearing，journal bearing 261，323
滑轨手柄 track handle 360
滑键 feether 252
滑块 cluster，insert 220，368
滑块驱动曲柄机构 slider-drive crank 57
滑块曲柄 slider crank 60
滑轮 trolley pulley 335
滑伸式龙门起重机 gantry crane with shuttle girder 334
滑台 sliding table 240
滑套 sliding collar 204
滑移齿 sliding gear 305
滑枕 ram 213
滑柱 slide pin 211
化学腐蚀 chemical erosion 170
化学加工 CHM 154，226
化学铣削 CHM，chemical milling 170
化油器 carburetor 346
环齿 ring gear，annular gear 367
环式联轴器 torus coupling 294
环套 annulus ring 71
环形弹簧 loop spring 275
环形挡板 ring damper 311
环形筛片 ring sieve 413
缓冲节流阀 buffer throttle 92
缓冲柱塞 buffer plunger 92
缓进给磨削 creep feed grinding 167
换热器 heat exchanger 173
换向齿轮 reversing gear 56
换向带 reversing belt 56
回程托辊 returning idler 341
回位弹簧 releasing spring 384
回油腔 oil inlet cavity 83
回转锉 rotary file 189
回转刀架 rotary tool post 215
回转导套 rotating cylindrical guide 49
回转吊车 revolving crane 400
回转方向 direction of rotation 161
回转工作台 rotary table 220
回转毂 rotating hub 281
回转连杆联轴器 rotating-link coupling 293
回转腔 rotor housing 81
回转式空气预热器 rotary air preheater 321
回转台 rotating table，turntable 55，105
回转研磨板 rotating lap 168
回转轴 rotating shaft 281
绘图板 drawing board 25
混合管 blend tube 343
锪孔 spot facing 163
锪锥面 countersinking 163
锪锥形沉孔 countersink 23
锪钻、沉头钻 countersink bit 127
活动板 movable plate 98
活动板手 adjustable spanner 126
活动定位销 slide positioning pin 211
活动横梁 moveable upper crosshead 4
活动横梁 moving core 115
活动护套 sliding bush 99
活动钳口 movable clamp 123
活动套 movable sleeve 298
活动铁芯 core activity 358
活动型芯 moving core 111
活塞 piston，plunger 82，132
活塞杆 piston rod 363
活塞式蓄能器 piston-type accumulator 92
活塞销 piston pin 358
火花塞 spark plug 346
火炬 torch 151
火焰 flame 150
火焰切割 flame cutting 31
货斗 bucket 345

机车起重机 locomotive crane 334
机架 rack，frame 50，140
机壳 chassis 358
机壳 housing，chassis 326，413

机体侧壁 block side wall 357
机体顶面 block tip surface 357
机械磨蚀 mechanical abrasion 170
机械密封 mechanical seal 423
机械手 manipulator 240
机械调控器 mechanical governor 318
机械转向器 steering gear 371
积屑瘤 BUE 155
基体 substrate 177
基体金属 base metal 151
基圆 base circle 66
基圆直径 base diameter 66
激光 laser 31
激光电化学 LECM 175
激光焊接 laser welding 144
激光束加工 LBM 154
激冷件 chill 35
吉普车 jeep 375
极限尺寸 limit dimension 27
棘轮 ratchet，driven ratchet，ratchet wheel 58
棘轮机构 ratchet mechanism 57
棘轮式改锥 ratchet screwdriver 128
棘轮制动器 ratchet brake 64
棘爪 pawl，ratchet 58，220，373
集料筒 barrel 413
挤出 extrusion 122
挤压铸造 squeeze casting 16
计量阀 metering valve 349
加工 machining 122
加工余量 machining allowance 46
加强构件 bracing member 399
加强筋 stiffener 280
加热器 heater 110
加速踏板 accelerator pedal 349
加油塞 filler plug 369
加载臂 loading arm 5
夹持板 holding plate 99
夹持架 clapper box 220
夹持面 clamping surface 41
夹紧机构，夹持座 clamping mechanism 56，218
夹紧块 grip segment 108
夹紧盘 clamping disk 294
夹具 grip，fixture 5，41
夹钳 tongs 125
夹钳夹头 vise grip 56
夹砂凸起 scab 34
夹头 chuck 104
夹压杆 clamping bar 114
夹压辊筒 pinch roll 12
甲板 deck structure 399
尖角 sharp corner 41
尖嘴钳 needle nose plier 126
间隙定位配合 clearance locational fits 29
间歇卧式压延机 intermittent horizontal calender 421
检测阀 check valve 355
减速器 speed-reducer，reductor 219
减速箱 gear box 329
减压阀 pressure-reducing valve 84
减振器 shock absorber 361
剪切 shear 3
简易板材提升器 simple sheet lifter 337
渐开线直齿齿轮 involute spur gear 65
键 key 26
键槽 key way 258
箭头 arrow head 21
浆体鼓 mud drum 313
交叉孔 cross hole 40
交错齿铣刀 sraggered tooth cutter 187
交错轴齿轮传动 crisscross shaft gear transmission 65
交换齿轮 change gear 203
浇道，浇铸道 sprue 35，118
浇孔 sprue hole 115
浇口 pouring basin 118
浇口杯 pouring cup 35
浇口盘 tundish 12
浇嘴 nozzle 115
角度铣刀 angle milling cutter 187
角度钻孔夹具 angele drilling jig 207
角接 corner 145
角接触球轴承 angular contact ball bearing 241，258，265
角磨机 abrasive disc grinder 223
角铁 angle plate 203
饺子成形机 dumpling making machine 422
绞车式龙门起重机 gantry crane with crab 334
绞车式桥式起重机 overhead travelling crane with

crab 334
铰刀 reamer 41，126
铰孔，铰削 reaming 28
铰链 hinge 305
铰链转销 hinge pin 315
轿底 platform 329
轿架 car frame 328
轿厢导轨 car guide rail 328
阶梯孔 step drill 163
阶梯钻 step drilling 163
接触辊 contact roller 223
接触轮 contact wheel 167
接地夹头 earthing clamp 147
接合套 clutch sleeve 368
接收器 receiver 92
接水槽 catch basin，drain 12，181
节点 pitch point 66
节流阀 throttle valve 171
节流杆 throttle lever 350
节流口 choke 118
节流小孔 restricted orifice 352
节能转子泵 energy saving rotor pump 424
节气门 throttle valve 347
节线 pitch line 66
节圆 pitch circle 66
节圆直径 pitch diameter 66
截面线 cutting plane line 21
截止杆 shut-off bar 349
金刚石车削 diamond turning 28
金刚石砂轮 diamond grinding wheel 198
金刚石镗削 diamond boring 28
金属垫 metal padding 35
金属粉末尺寸 powder metal size 28
金属粉末烧结 powder metal sintered 28
金属弧焊 metal arc welding 144
金属外露 metal pick-up 302
筋颈螺栓 ribbed neck bolt 247
紧固螺钉 set screw 86
紧固螺钉连接 lock bolt coupling bolt 247
紧固螺母 fasten nut，clamping nut 26，258
进给床鞍 feed carriage 190
进给手柄 feed hand wheel 218
进给套筒 feeding tube 210
进给推杆 feeding finger 210
进给箱 feed box 215
进料阀门 feed valve 413
进料喷嘴 feed nozzle 401
进料器 feeder 311
进气管 intake manifold 347
进气控制阀 air intake control valve 355
进气门 inlet valve 346
进气凸轮轴 air inlet camshaft 358
进油口 inlet port 349
晶片 wafer 176
精齿 finishing teeth 195
精辊 roller burnishing 31
精铣刀 finishing 188
精压 coining 122
井架 drill derrick，derrick 391
径向 radial 15
径向引擎 radial engine 60
静公差 statical tolerance 23
静压挤出 hydrostatic extrusion 138
静止叶片 stay vane 322
局部环轴承 partial bearing 264
局部锪面钻 spot facing drill 189
矩形键 rectangular 252
锯 saw 126
锯齿形弹簧 Zigzag spring 360
锯切 sawing 31
锯切铣刀 slitting saw cutter 187
聚焦透镜 condenser lens 178
卷料器 coiler 115
卷轮 reel 172
卷绕定位钩 coil positioning hook 336
卷筒销 roll pin 253
绝热垫 heat shield 372
绝缘层 insulation 173
绝缘器 insulator 308

卡环 clasp 358
卡紧弹簧 garter spring 284
卡盘 chuck 127，213
卡钳 caliper 127，371
卡圈 snap ring 364
卡爪 jaw 202
开槽垫片 slotted washer 250
开槽环形螺母 slotted ring nut 249
开槽轮 grooved wheel 108
开槽螺钉 slotted 248
开槽盘头螺钉 fillister screw 246
开槽锥端紧定螺钉 cone-ended screw 246
开合销 split pin 253
开环控制系统 open-loop control system 233
开口销 split cotter 77
开路继电器 circuit opening relay 353
开式冒口 open riser 118
抗拉 tensile 3
抗弯 flexural 3
抗弯模量 flexural modulus 3
抗压 compression 3
抗压强度 compression strength 2
靠背 seat back 360
壳体 shell 280
可变电阻 variable resistor 129，353
可成形性 formability 3
可变向棘轮（机构） reversible ratchet 58，64
可换锥套 interchangeable tapered sleeve 204
可靠性 durability 3
可逆向带式制动器 reversible band brake 300
可倾式夹层锅 tiltable jacketed kettle 415
可调头架 adjustable head 218
可调支承 adjustable supporting 210
可调支承钉 adjustable supporting pin 205
可调直线型振动筛 adjustable linear shaker 395
可调锥面 odjusting taper 194
空段清扫器 safety cleaner 341
空气滤清器 air filter 93，346
空气调节叶片 adjustable air vane 311
空气质量传感器 air mass sensor 349
空心轴 hollow spindle 260
空压机 air compressor 93
孔公差 hole tolerance 28
孔口 orifice 425
孔口板 port plate 82
孔隙 cavity，porosity 7，35
孔隙度 porosity 2
控制叉 control fork 348
控制杆 joystick，control rod，control lever 240，348，350
控制杆支架 control rod bracket 348
控制柜 control panel 328
控制器 control 181
控制轴 control shaft 348
块料 block 19
快动机构 snap-action mechanism 59
快动开关 snap-action switch 59
快动肘节开关 snap-action toggle switch 59
快换夹头 quick removable cartridge 182
扩孔 core drilling 163
扩孔刀 enlarging tool 163
扩散器 diffuser 319
扩眼器 underreamer 397

拉杆 thread 329
拉架提升器 rack lifter 337
拉伸性 tensile property 2
拉绳 pull rope 383
拉手 handle 371
拉削 broaching 31，154，212
拉削丝锥 broaching taper 194
拉曳浮动平台 tensioned buoyant platform 400
拉应力 tensile stress 4
拉制 drawing 31
栏杆 baluster 390
栏杆钢圈 baluster steel ring 390

缆风绳 guy 390
缆绳 cable 383
烙铁 soldering iron 151
冷冻粉碎机 freezer mill 414
冷料穴 cold slug well 110
冷凝 condensing 279
冷凝分离器 condensate separator 324
冷却器 cooler 115，280
冷却水道 waterway 111
冷却箱 cooling box 224
冷却液 coolant 165
冷却液温度表 coolant temperature gauge 360
冷压 cold compaction 143
冷轧 cold rolling 31
冷褶痕 cold shut 34
离合器 clutch 369
离合衔铁 clutch armature 301
离心泵 centrifugal pump 80
离心锤 centrifugal block 331
离心机 centrifuge 395
离心喷雾干燥装置 centrifugal spray drying device 411
离心式切片机 centrifugal slicer 414
离心旋涡泵 centrifugal vortex pump 423
离心压缩机 centrifugal compressor 319
离心叶轮 centrifugal impeller 423
离心铸造 centrifugal casting 122
离子束监控器 beam monitor 178
离子束偏转器 beam deflector 178
离子源 ion source 178
立刀架 rail head 219
立管式蒸发器 vertical pipe evaporator 418
立式带锯锯床 vertical band sawing 221
立式辊压机 vertical roller press 421
立式环形喷射式粉碎机 vertical annular jet mill 413
立式壳管冷凝器 vertical shell and tube condenser 419
立式磨浆机 vertical grinding machine 413
立式杀菌锅 vertical sterilization pot 417
立式型芯 vertical core 118
立式轴卷抓斗 vertical axis coil grab 336
立柱 column，upright，vertical pile 213，329，384
立柱底座 base for column 240
沥青摊铺机 asphalt paver 381
连杆 connecting rod，pitman 50，386
连杆联轴器 link coupling 293
连接板 link plate 77
连接钩 linkage hook 349
连接螺母 coupling nut 181
连接螺栓 jointing bolt 329
连接器 connector 363
连续拉拔成形 continuous pultrusion 17
连续式糖果浇模成形机 continuous candy pouring molding machine 423
连铸 continuous casting 10
联轴键 shaft key 252
联轴器 coupling 90
链传动输送机 chain driving conveyor 343
链锯 chain saw 126
量角器 protractor 25，287
料槽 trough 342
料斗 hopper 143
临界应力 critical stress 3
零线 zero line 63
领臂 trailing arm 361
领蹄 primary shoe 299
溜板箱 saddle 215
流道 runner，flute 35
流量计 flow meter 173
流送槽 flume 407
流体隔套 flow sleeve 320
六角（主）刀架 hexagon(main)turret 215
六角车床 capstan lathe 127
六角扳手 hexagon key 248
六角开槽螺母 castle nut 246
六角螺母 hex nut 246
六角头螺栓 hex bolt 246
龙门起重机 gantry crane 333
笼形万向节 rzeppa universal joint 292
楼层感应器 floor inductor 333
漏充 misrun 34
漏斗 funnel 223
炉体 furnace 108
轮辐 sprocket 339
轮架 axle frame 340
轮式重力输送机 gravity wheel conveyor 343
轮胎式起重机 wheel crane 334
罗茨轮鼓风机 Roots blower 326

罗氏联轴器 Rouleaux coupling 291
逻辑块 logic-block 372
逻辑元件 logic element 93
螺钉连接 screw coupling 247
螺钉压板 screw clamp 56
螺杆 screw 202
螺杆操作的带式制动器 screw-operated brake 300
螺杆传销 gland bolt pin 315
螺杆制冷压缩机 screw refrigeration compressor 418
螺母 nut 26，86，250
螺栓 bolt 26
螺纹 thread 86
螺纹车刀 thread cutting tool 186
螺纹导角 thread lead angle 193
螺纹铣削 thread milling 188
螺线 helix 342
螺旋保持环 spiral retaining ring 251
螺旋槽丝锥 spiral fluted tap 194
螺旋齿轮 helical gear 67
螺旋带 tape helix 407
螺旋弹簧 coil spring，helical spring 127，273
螺旋桨 propeller 375
螺旋桨齿轮箱 propeller gear box 375
螺旋腔 spiral case 322
螺旋式冻结装置 spiral freezer 420
螺旋输送机 screw conveyor 425
螺旋卸料器 auger stripper 413
螺旋圆周铣刀 helical peripheral cutter 187
螺旋锥齿 spiral bevel 44
螺旋锥齿轮副 spiral-bevel set 70
螺柱 stud 129
落锻 drop forging 138
落料 blanking 122
落料冲头 blanking punch 105
落料凸模 blanking punch 101
履带 track 345
履带行驶机构 track driver mechanism 378
履带式起重机 crawler crane 334
滤杯 filtering cup 92
滤芯 filtering core 92

MAG 熔焊 metal active-gas welding 144
MIG 熔焊 metal inert-gas welding 144
麻花钻 twist drill 163，190
马口铁剪刀 snip 126
码头起重机 wharf crane 333
马蹄形垫片 horseshoe washer 251
埋弧焊 submerged-arc welding 144
麦弗逊滑柱后轴 Mcpherson strut rear axle 362
脉冲 pulse 233
脉冲电源 pulse power 227
脉冲流化床干燥器 pulsed fluidized bed dryer 412
盲孔 blind hole 163
盲孔铆钉 blind rivet 255
毛刺 burr 36，162
毛坯 blank，gear blank 102，224
锚架 anchor rack 399
锚绳 anchor 383
铆冲 riveting punch 247
铆钉 rivet 254，372
冒口 riser 36
帽形螺母 captive nut 250
梅花扳手 ring spanner 126
门阀 gate valve 312
门座式悬臂起重机 high pedestal jib crane 334
迷宫式密封 labyrinth seal 323
密闭装置 enclosure 226
密封 packing，seal 281，32
密封杯罩 packing cup 286
密封垫 gasket gland 281，315
密封垫杯 gasket cup 286
密封法兰 gland flange 315
密封管脚 seal leg 343
密封环 packing ring 286
密封面 seal face 281
密封腔 seal chamber，packing chamberor box 82，286
密封圈 sealing ring 26
密封压紧螺杆 gland bolt 315
密封压紧螺母 gland nut 315
密封压套 packing gland 315
密封座环 seat ring 315

面斗　flour hopper　421
面轮廓度　profile of a surface　22
面罩　face shield　147
模板　template　46
模具　mold，die　17，107
模口　die land　111
模腔　cavity，die cavity，mold cavity　110，115，118
模压　embossing　122
模压成形　compression molding　17
模座　die holder　139
膜板阀　diaphragm valve　315
膜片压缩机　diaphragm compressor　324
摩擦　friction，abrading　122，175
摩擦滚筒　friction roller　74
摩擦焊接　friction welding　144
摩擦棘轮机构　friction ratchet mechanism　57
摩擦离合器　friction clutch　296
摩擦里衬　friction lining　296
摩擦片　friction plate　298
摩擦片离合器　plate clutch　291
摩擦去除　abrasion　154
摩擦制动盘　friction brake disc　301
摩擦制动器　friction brake　301
磨具　abrasive　197
磨粒喷射加工　AJM　154
磨料　abrasive　175
磨料流加工　AFM　154
磨料喷射　abrasive jet　226
磨料射流加工　AJM　181
磨料水射流　abrasive water jet　226
磨料水射流加工　abrasive water jet machining，AWJM　170
磨煤辊　grinding roller　311
磨削　grinding　31，154，226
木锤　mallet　124
木螺钉　wood screw　248

耐磨垫　wear pad　190
耐磨性　wear resistance　3
耐蚀性　corrosion resistance　2
内摆线轮　inner gerotor　80
内槽轮机构　internal geneva mechanism　63
内端法兰盘　inner flange　198
内渐开线花键　internal spline　254
内冷装置　intercooler　92
内六角扳手　allen key　126
内六角紧固件　hex-socket fastener　248
内六角螺柱　inner-hexagon screw　26
内六角圆柱头螺钉　inner hex fillister screw　246
内圈　inner ring　267
内圈滚道　inner ring ball race　267
内圆无心磨削　internal centerless grinding　167
逆铣　Up milling　161
黏结剂　adhesive　19
啮合线　line-of-action　66
扭矩扳手　torque spanner　126
扭转弹性轴承　torsionally elastic bearing　361
扭转杆　torsion bar　361
扭转减振弹簧　torsional damper spring　367

O

O 形密封圈，O 形圈　O-ring　281，397

排料器　exhauster　311
排气管　vent-pipe　347
排气口　exhaust outlet　375
排气门　exhaust valve　346
排气凸轮轴　exhaust camshaft　358
排水阀　water exhaust valve　92
排泄阀　discharge valve　323
排泄孔　drain hole　83

排泄塞 drain plug 83
排屑槽 flute 190
排屑器 chip excavator 224
排压速度调控器 exhaust-pressure governor 318
排脏器 dirt excluder 372
盘 disc 371
盘式离合器 disc clutch 291
盘式铣刀 disc cutter 188
盘式制动器 disc brake 301
盘形凸轮 plate cam 52
旁路控制阀 control valve with bypass 312
抛光 polishing 31，154
抛光垫 polishing pad 176
抛光工作台 polishing table 176
刨削 planing 28，31
配气机构 admission gear 358
配水箱 distribution box 419
配油口 distributor port 349
配油转子 distributor rotor 350
配重 balance weight 384
配重块 counterweight 203
喷管 lance 11
喷枪 blasting gun 182
喷射泵 injection pump 348
喷射泵计量单元 injection pump calibration unit 359
喷射控制器 injection controller 348
喷射式冷凝器 jet condenser 416
喷射正时活塞 injection timing piston 352
喷涂法 chopped fiber sprayer 17
喷雾 spray 279
喷雾干燥系统 spray drying system 411
喷油泵 injection pump 346
喷油器 injector 346
喷油润滑 oil stream lubrication 78
喷嘴 nozzle 47，181
喷嘴阀 nozzle valve 347
喷嘴叶片 nozzle vane 319
膨胀隔板 expanded panel 19
碰钩 trigger hook 343
坯锭翻转夹板 ingot tumer grab 336
皮锤 rubber mallet 124
皮碗 rubber cup 385
疲劳强度 fatigue strength 3
疲劳性 fatigue 2
偏差 offset 5
偏斜搭接 joggle lap 256
偏心 eccentric 323
偏心轴 eccentric shaft 140，325
偏心钻头 eccentric bit 392
偏移量 offset 244
偏重 clump weight 406
片状开关 vane switch 354
平锤 flatter 124
平带传动输送机 flat band driving conveyor 343
平底盲孔 blind hole with flat bottom 163
平地机 grader 381
平垫片 plain washer 250
平垫圈 washer 246
平行度 parallelism 23
平行仪 parallel 25
平行轴齿轮传动 parallel shaft gear transmission 65
平衡鼓 balance drum 323
平衡活塞 balancing piston 326
平衡托盘提升器 balanced pellet lifter 336
平衡型芯 balanced core 118
平衡重 balance weight 358
平接 butt 145
平口搭接 butt lap 256
平面度 flatness 22
平面拉床 surface broaching machine 221
平面拉力 flat broach 196
平面磨削 surface grinding 165，212
平面涡卷弹簧 coil spring 275
平台井架 platform headframe 390
剖面 section 21
剖面阴影线 section line 21
剖视图 cut-away view 26
铺路机械 pavement machinery 381
普通开槽螺母 slotted nut 250
普通螺纹 metric 247
普通丝杠 plain screw 213

起钉器　nail puller　127
启动机，启动器　starter　353，374
起居舱　accommodation　399
起重臂　jig　334
起重臂杆　boom　383
起重滑车　hoisting block　335
起重机　crane　334
起重机械　hoisting machinery　333
起重架　derrick　334
起重台架　gantry　334
气泵　pump　176
汽车升降梯　car elevator　328
气动锤　pneumatic hammer　124
气动夹头　pneumatic chuck　206
气阀　air valve　151
气盖　air cap　137
气缸　cylinder　347
气缸垫　cylinder pillow　357
气缸盖　cylinder cover，cylinder head　347，357
气缸盖罩　cylinder head cover　347
气缸体　cylinder block　347，357
气缸筒　cylinder barrel　82
气焊　gas welding　144
气孔　blister　34
气控阀　gas control valve　150
气力输送装置　conveying equipment　425
汽轮机　turbine　308
气门弹簧　valve spring　358
气囊式蓄能器　bladder-type accumulator　92
气射流　air jet　182
气胎　pneumatic tyre　383
汽提管　stripper　401
气体过滤器　air filter　92
气体调节器　gas regulator　150
气隙　air gap　12
汽油发动机　gasline engine　346
气闸　damper　310
气嘴　gas nozzle　137
千斤顶　jack　127
前挡板　bulk head　360
前刀面　rake face，face　155
前端模口　preland　111
前进装置　advance device　350
前轮制动器　front wheel arrester　346
前上叉杆　front upper wishbone　376
前下叉杆　front lower wishbone　376
前行多盘离合器　forward multiplate clutch　370
前悬架　front suspension　346
前支承　front bearing　258
前轴承　front bearing　316
钳工台虎钳　vise　127
枪钻　gun drilling，gun drill　163，190
腔盖　housing cover　367
腔室　chamber　219
腔体　housing　219
腔体环　casing ring　279
强度　strength　3
强化层　reinforcement　36
桥塞　bridge plug　396
切槽　slot　40
切断刀　parting-off tool　186
切断钳　cutting plier　126
切分螺母　split nut　250
切割　cutting　226
切片　slitting　226
切入式磨削　plunge grinding　167
切深　depth of cut　161
切向　tangential　15
切向键　tangential key　252
切屑去除　chip removal　154
切削齿　cutting teeth　195
擒纵机构　escapements mechanism　59
倾斜度　angularity　22
清铲　snagging　31
清带器　belt cleaner　340
球半径　spherical radius　22
球阀　globe valve，ball valve　312，350
球体保持架　ball cage　267
球体传动　ball transmission　54
球体锁紧棘轮　ball-lock ratchet　58

球直径 spherical diameter 22
球轴承 ball bearing 267
驱动端盖 cover drive 358
驱动盖槽 drive cover slot 367
驱动环 drive ring 281
驱动件 driving member 297
驱动块 drive block 367
驱动轴 driving shaft 220
屈服 yield 3
曲柄 crank 50
曲柄 lever 371
曲柄齿轮 crank gear 220
曲柄销 crank pin 168
曲齿圆锥齿轮传动 curve bevel gear transmission 65
曲面联轴器 curvic coupling 295
曲轴 crankshaft 260，307，358
曲轴正时带轮 crank shaft timing belt gear wheel 347
取货叉 reach fork 345
取芯筒 core barrel 393
去毛刺 deburring 226
全环轴承 full bearing 264
全跳动 total runout 23
缺口敏感性 notch sensitivity 3

燃煤嘴 coal nozzle 311
燃气罐 combustible gas cylinder 150
燃气轮机 gas turbine 318
燃烧区 combustion zone 311
燃烧室 combustion chamber，burner housing 137，313，318
燃油表 fuel level indicator 360
燃油温度传感器 fuel temp.sensor 359
热斑点 heat spotting 302
热变形 heat distortion 2
热辊 hot roll 115
热交换器 heat exchanger 226，309
热裂缝 hot tear 35
热压 hot compaction 143
热影响区 heat-affected zone，HZA 179
热轧制 hot rolling 31
人孔盖 comp 323
人字齿轮 herring-bone gear 67
人字齿轮传动 herring-bone gear transmissiom 65
刃宽 land 193
韧性 toughness 2
容屑表面 clearance surface 188
熔锅 pot 132
熔焊 fusion welding 144
熔炉 furnace 132
熔模铸造 investment casting 31
熔融合金 molten alloy 7
熔融金属 molten metal 12
熔渣 molten slag，dross 148，151，179
柔性薄膜 flexible diaphragm 262
柔性臂 flexible arm 181
柔性联轴器 flexible coupling 291
柔性轮胎 flexible tyre 294
柔性盘式联轴器 flexible-disk coupling 293
柔性套 flexible bush 294
柔性支撑板 flexible support board 316
蠕变 creep 3
蠕变性质 creep resistance 2
软管 hose 150
软轴 flexible shaft 292
润滑泵 lubricating pump 211
润滑嘴 lubrication nipple 367

塞尺 feeler gauge 128
三角板 set square 25
三角形花键 triangular spline 254
三凸块滑动轴承 three-lobe bearing 264
三凸轮泵 three-lobe pump 80
三爪式钻夹头 three-jaw drilling chuck 203
砂带 abrasive belt 167
砂带磨削加工 coated abrasive belt grinding 167
砂轮 grinding wheel 126，156，167
砂轮机 grinding machine 127

砂轮主轴 wheel spindle 198
砂箱 flask 118
砂芯 core sand 118
砂型铸造 sand casting 31，122
砂眼 blow 34
砂纸 abrasive paper 126
筛板 sieve plate 413
闪光焊接 flash welding 144
扇形板 sector plate 311
上臂 upper horn，upper arm 151，361
上冲头 upper punch 116
上横梁 cross head 329
导料槽 loading channel 341
上模 upper die 139
上模板 punch set 102
上模座 upper die shoe 106
上偏心块 eccentric block above 406
上气环 submit gaseous ring 358
上气缸 upper cylinder 316
上箱型砂 cope 118
烧结 sintering 143
射孔弹 shaped charge 393
射孔枪 perforating tool 393
射流 jet 289
射流管 jet tube 171
射流加工 jet cutting 226
伸缩盖 telescoping cover 173
伸缩式板材提升器 telescoping sheet lifter 337
伸缩式火花塞 retractable spark plug 320
深度尺 depth gauge 128
深孔加工 EC deep hole drilling 174
升降臂（台） knee 213，216
失模 investment 122
施工升降机 construction hoist 327
十字槽沉头螺钉 phillips screw 246
十字叉 spider 141
十字滑块 crosshead guide 80
十字联轴器 oldham coupling 291
十字螺丝刀 philips screwdriver 127
十字轴 cross 371
蚀除 erosion 154
试件 specimen 5
试压堵塞 pressure test block 396
视觉传感器 visual sensing 339
视图 view 26
释放杆 freeing lever 64
收缩孔隙 shrinkage cavity 35
手动节流阀 hand throttle valve 397
手动提升葫芦 manual hoist 338
手工金属弧焊 manual metal-arc welding 144
手工润滑 manual operation lubrication 78
手工涂覆 hand lay-up 17
手轮 hand wheel 315
手提式焊缝打磨机 portable welding seam polisher 223
手提式砂带机 portable grinder 223
受控半径 controlled radius 22
枢轴 trunnion 135
输出法兰 output flange 369
输送托辊 conveying idler 341
输送支承 convey support 211
树脂 resin 17
树脂转移模压法 resin transfer molding 17
数控钻削 NC drilling，NC milling，NC turning 235
数字点火系统 digital ignition system 354
衰减 fade 302
双锤振动筛 double hammer vibrating screen 406
双点画线 phantom line 21
双动式棘轮机构 double pawl ratchet 64
双端镗刀 double-ended cutter or boring tool 189
双列球轴承 double row ball bearing 266
双驱动盘 twin drive plates 367
双曲柄机构 double-crank mechanism 50
双头螺栓连接 stud coupling 247
双头螺柱 double-end stud 246
双凸轮驱动压头 double cam actuated clamp 56
双凸轮压头 double cam clamp 56
双蜗壳泵 double-volute pump 80
双向公差 bilateral tolerance 27
双楔块压头 double wedge 56
双重缠绕制动器 double wound brake 300
水口 gate 34
水冷铜座 water-cooled copper shoe 151
水力旋流器 hydroclone 397
水陆两用车 amphibian 375
水轮机 hydraulic turbine 321
水平旋转炉 horizontal rotary furnace 416

水射流 water jet 226
水射流加工 WJM 154，170，181
水套 water jacket 347
顺铣 down milling 161
顺序阀 sequence valve 84
丝杠 lead screw 123，213，227，365
丝杠螺母 screw nut 59，220
丝筒 wire spool 227
丝锥扳手 tap wrench 127
丝锥夹头 tap chuck 208
四冲程发动机 four-stroke engine 307
四凸块滑动轴承 four-lobe bearing 264
伺服控制 servo control 171
伺服控制进给系统 servo controlled feed system 226
伺服驱动 servo drive 232
松土器 scarifer 381
速度调节机构 speed-adjusting mechanism 316
随行夹具 follower jig 211
隧道盾构机 tunnel boring machine 379
锁定螺钉 set screw 248
锁环 blocking ring 368
锁紧垫圈 lock washer 87
锁紧杆式板材提升器 lock bar sheet lifter 337
锁紧螺母 lock nut，jam nut 129，193，329
锁紧螺栓 locking bolt 58
锁紧钳 locking plier 126
锁紧手柄 clamp handle 123
锁片 locking piece 358
锁片垫片 lockplate washer 251
锁销 lockpin 187

T 形螺栓 T-bolt 206
T 形槽 T-slot 162
弹簧 spring 83，219
弹簧垫片 spring washer 251
弹簧垫圈 lock washer 246
弹簧挂架 spring hanger 312
弹簧锁夹 spring clip 59
弹簧锁紧垫片 spring-locking washer 251
弹簧座 spring seat，spring bracket 351，363
弹性环 elastic ring 71
弹性模量 modulus of elasticity 3
弹性锁紧螺母 spring lock nut 250
弹性销 spring pin 253
弹性止动螺母 elastic stop nut 250
塔式起重机 tower crane 334
塔桅 tower mast 335
胎式联轴器 tyre coupling 294
台阶 shoulder 267
台钳 vise 220
太阳轮 sun wheel，sun gear 71，370
炭精坩埚 plumbago crucible 132
镗刀 boring tool 189
镗模架 jig frame 210
镗套 boring bushing 210
镗削 boring 154
套管刮削器 tube scraper 396
套料 trepanning 226
套式拉刀 shell broach 195
套式输出口 sleeve output port 349
套筒 bush，bushing，sleeve 77，219，371
套筒扳手 box spanner 126
套筒扳手旋转手把 tommy bar 127
套筒联轴器 muff coupling 292
套筒螺丝刀 socket screwdriver 127
套筒轴承 sleeve bearing 342
特形锁扣 special lock 54
梯形臂 ackerman arm 371
梯子 ladder 383
提升挡块 lift stop 347
提升短节 lift sub 396
提升钢丝绳 elevating steel cable 390
提升管 lift pipe 343
提升梁 lifting beam 336
剃齿 gear shaving 168
天车 crown block 391
天轮 top pulley 390
填料 filler，solder，stuffing 14，151，423
填料函 stuffing box 397
填料压盖 stuffing gland 397
条料 slice 19

调节臂 governor arm 349
调节阀 adjustable valve，regulating valve 316，349
调节轮 regulating wheel 85，167，222
调节螺母 adjusting nut 193
调节螺栓 screw adjustment 80
调节器弹簧 governor spring 349
调节器壳体 governor cage 348
调节器配重 governor weight 349
调节绳 adjusting cable 299
调节座 adjusting shim 321
调整垫片 adjusting gasket 358
调整销 adjustable pin 206
铁芯 core 329
停止杆 stop lever 350
通风杯 vent cup 286
通风（气）口 vent 118，312
通孔 through hole 163
同步带 synchronous belt 75
同步挡板 synchrotizing damper 341
同步螺杆泵 timed-screw pump 80
同步器 synchronizer 368
同心度 concentricity 23
同轴度 coaxiality 23
铜电极 copper electrode 173
铜焊 braze 190
筒形螺母 barrel nut 250
头部护罩 head hood 341
头滚筒 head drum 340
头枕 headrest 360
凸出焊接 projection welding 144
凸带公差带 projected tolerance zone 23
凸块 nose 52
凸轮操控压板 cam operated clamp 56
凸轮从动件 puppet 372
凸轮滚子 cam roller 349
凸轮环 cam ring 349
凸轮锁紧棘轮 cam-lock ratchet 58
凸轮压头 cam clamp 56
凸轮轴 camshaft 54，346
凸螺杆 male rotor 325
凸模 ejector die 115
凸台 boss 35
凸压模 ejector die 132
凸缘联轴器 flange coupling 292
涂覆磨具 coated abrasive 198
推板 ejector plate 115
推垫 thrust pad 348
推杆 pusher，ejector，ejector pin 49，99，111
推力球轴承 ball thrust bearing 266
推力套 thrust collar，thrust sleeve 323，348
推力轴承 thrust bearing 71，86，258，316，323
推销 thrust pin 348
推压机构 crowd mechanism 378
托板 support blade 156
拖板 carriage 213
托盘 pallet 238
拖杆拖车 drawbar trailer 364
脱落 neglect 302
挖土机 excavator 379

U 形密封 U-packing 283
U 形螺栓 U-bolt 248

V engine V 形引擎 60

外摆线轮 outer gerotor 80
外槽轮机构 external geneva mechanism 63
外齿锁紧垫片 external-tooth locking washer 251
外端法兰盘 outer flange 198
外渐开线花键 external involuter spline 254
外圈滚道 outer ring ball race 267

外圆超精加工 barrel finishing 31
外圆磨削 cylindrical griding 212
弯头轻载镗刀 light boring tool with bend shank 188
万尖圆角半径 nose radius 186
万能钳台 universal vise 204
万向节 hooke joint，universal joint 291，367，371
万向节叉 universal joint yoke 371
万向联轴器 hooke coupling 291
万向柔性弹簧联轴器 uniflex flexible-spring coupling 293
往复泵 reciprocating pump 423
往复式加工机床 reciprocating machine 219
往复运动机构 reciprocating mechanism 60
往复振动 oscillation 167
往复振动平面磨光机 reciprocating vibration flat grinder 223
微处理器 microprocessor 372
微机械调节器 minimec governor 348
微米丝杠 micrometer screw 225
微喷油器 microjector 347
微调机构 fine adjustment mechanism 58
韦氏硬度 HV 3
尾部滚筒 tail drum 340
尾翼平衡柱 wing attachment strut 376
尾座 tailstock 213
位置度 position 23
喂料辊 feed roller 421
文丘里喷嘴 Venturi nozzle 396
纹盘 winch 383
涡轮 turbine wheel，turbine 319，369
涡轮机转轴 turbine shaft 309
涡轮腔体 turbing case 319
涡轮增压 boost pressure 349
涡轮增压器 turbocharger 319
涡轮轴轴承 turbine shaft bearing 369
蜗杆 worm 224
蜗杆减速器 worm reduction，turbo-worm reducer 304
蜗杆轴 worm shaft 203
蜗轮 worm wheel 224
卧式超微粉碎机 horizontal pulveriser 413
卧式壳管蒸发器 horizontal shell and tube evaporator 418
卧式型芯 horizontal core 118
钨电极 tungsten electrode 148
无槽丝锥 non-groove taper 194
无声棘轮机构 silent ratchet mechanism 64
无声链板 silent chain plate 77
无声链联轴器 silent-chain coupling 293
无心磨削 centreless grinding 165
无心砂带磨床 centreless belt grinding machine 223
无阻尼剪切联轴器 shear undamped coupling 294
物镜 objective lens 178
物料输送机械 material conveying machinery 423

X 射线传导器 X-ray transmitter 12
X 射线接收器 X-ray receiver 12
吸能件 energy absorber 360
吸入阀 suction valve 323
铣刀 milling cutter 162
铣削 milling 28，154，212
细微电火花加工 micro EDM 171
下臂 lower horn 151
下冲头 lower punch 116
下击器 bumper jar 396
下模 lower die 139
下模座 die set，lower die shoe 102
下气环 below gaseous ring 358
下汽缸 below cylinder 316
下箱型砂 drag 118
先导阀 pilot valve 318
先导型溢流阀 pilot operated relief valve 83
衔铁 armature 298
限速器 governor，overspeed governor 328，331
限位块 stop block 102
限位螺钉 limit bolt 329
线材缠绕成形 filament winding 17
线轮廓度 profile of a line 22
线圈 field coil 301
线丝 wire 172
线丝导轮 wire guide 172

线性运动　linear motion　234
线性执行元件　linear actuator　59
相交轴齿轮传动　cross shaft gear transmission　65
箱式干燥器　tray dryer　411
箱式真空干燥器　tray vacuum dryer　412
箱体　case　82
镶块　die insert　111
镶条　strap　290
消防车　fire engine　376
消声器　noise absorber　93
销钉　pin，dowel pin　49，248
销连接　pin coupling　253
销式蹄　pivoted shoe　262
小齿轮　pinion　219
小节距　fine pitch　44
小径　basic minor dia　193
小径　minor diameter　246
小孔　aperture　178
小螺旋角　small helix angle　44
楔块　wedge　290
楔铁　drift　191
楔形销　wedge pin　315
斜边搭接　scarf lap　256
斜齿　helical　44
斜齿轮　helical gear　67
斜齿圆柱齿轮传动　helical gear transmission　65
斜齿圆锥齿轮传动　helical bevel gear transmission　65
斜导柱　angle pin　111
斜度　slope　23
斜角尺　bevel　25
斜轴瓦　tilting pad　264
卸荷点　discharge point　323
卸荷阀　relief valve，discharge valve　312，323
卸荷环　discharge ring　322
卸货叉　dumping fork　345
卸料板　ejector plate　99
卸料杆　knockout rod　105
卸料滑套　sliding bush　102
卸料漏斗　discharge chute　341
卸料镶板　embedded plate　99
卸载槽　discharge chute　221
心轴　mandrel　107，206
芯棒　core rod　116
星形垫片　star washer　129
型锻　swaging　138
型模　swage　123
型砂　sand mold，sand　34，118
型芯　core　35，132
胸压手摇钻　breast drill　127
虚线　hidden line　21
蓄电池　accumulator　346
蓄能器　accumulator　92，181
悬臂扭杆　cantilever-torsion bar　357
悬臂心轴　gang mandrel　206
悬臂型芯　overhanging core　36
悬架系统　suspension system　361
悬空起重机　overhead travelling crane　334
悬置型芯　hanging core　119
旋风分离器　cyclone separator　309
旋链　spinning chain　393
旋塞阀　stopcock，cock valve　128，398
旋涡溢流管　whirlpool overflow pipe　397
旋转叉　swing fork　345
旋转锻压　rotary-swaging　140
旋转活塞　rolling piston　323
旋转龙头　swivel　392
旋转钳台　swivel vise　204
旋转斜盘　swashplate　82
循环喷嘴　recycle nozzle　401

压板　clamp，plate，pressure plate　187，258，296，333
压边圈　retainer　98
压钉　clamp bolt　211
压杆　pressure bar　104
压紧联轴器　compression coupling　293
压紧螺钉　clamp screw　187
压力板　pressure plate　367
压力保持阀　pressure holding valve　352
压力表　pressure meter，pressure guage　171
压力补偿　pressure compensated　79

压力传感器 pressure sensor 176
压力传感筒 pressure sensing bellow 318
压力焊接 welding with pressure 144
压力环 pressure ring 298
压力角 pressure angle 66
压力模块 pressure modulator 312
压力腔 pressure vessel 108
压力轴套 pressure bushing 294
压力铸造 die casting 31
压力阻挡轴承 pressure dam bearing 264
压料板 pressure plate 99
压轮 press wheel，pinch roller 227，410
压实机械 compacting machinery 380
压缩过盈配合 shrink fits 29
压缩机 compressor 92，318，322
压缩机涡轮 compressor turbine 319
压缩机叶轮 compressor wheel 319
压头 ram 110
压应力 compressive stress 4
压油腔 pressurized oil cavity 83
压铸 die casting 28
牙顶 tap crest 193
牙根 basic root 193
牙型角 angle of thread 193
延展夹头 stretch gripper 105
延展性 ductility 2
研具 lap 168
研磨 lapping 31
研磨膏（浆） abrasive slurry 176
研磨套 lapping sleeve 168
掩膜 mask 46
眼孔螺栓 eye bolt 248
燕尾导轨 dovetail guide 290
羊角锤 claw hammer 124
仰视图 bottom view 26
氧喷枪 oxygen lance 12
氧气罐 oxygen cylinder 150
样板 template 25
样冲 point punch 126
摇摆滑动轴承 rocking journal bearing 264
摇臂 rocker arm，radial arm 210，213，218
摇臂凸轮 rocker cam 225
摇臂轴 pitman arm shaft 371
摇动叉 cradle fork 345
摇杆 rocker，rocking lever 50，348
摇杆粉碎机构 rock crusher 56
叶轮 impeller 81，298，316，423
叶片 vane，blading 80，316
叶片化学加工 ECM of turbine blade 174
叶片式液压马达 hydraulic vane motor 83
页轮 flap wheel 223
曳引轮 sheave 329
液态合金 liquid alloy 7
液压泵 pump 171
液压传动 hydraulic transmission 54
液压电梯 hydraulic elevator 328
液压放大器 hydraulic amplifier 90
液压缸 hydraulic cylinder 211
液压千斤顶 hydraulic jack 79，127
液压调节器 hydraulic governor 350
液压头 hydraulic head 350
液压支臂 hydraulic boom 376
液压支腿 hydraulic outrigger 384
液压制动器 hydrodynamic brake 300
液压注射缸 hydraulic shot cylinder 132
一字螺丝刀 slotted screwdriver 127
仪表架 instrumentrack 325
移动凸轮 translating cam 52
溢流阀 overflow valve 171
溢流控制阀 spill control valve 359
溢流口 relief port 85
翼型螺钉 wing screw 249
阴极 cathode 165
应变 strain 5
应变框 restraint frame 5
应变片（仪） strain gage 191，211
硬度 hardness 2
硬化区 hardened zone 153
应力集中系数 intensity factor 3
硬质合金刀粒 carbide insert 190
永磁铁 permanent magnet 183
永久模型 perm. mold casting 31
油保持器 oil retainer 263
油杯 oil cup 171，286
油环 oil ring 358
油槽 oil reservoir 279

油底壳　sump　347
油封　oil seal　371
油管锚　tubing anchor　396
油冷器　oil cooler　308
油喷嘴　oil nozzle　313
油润滑轴承　lubricated bearing　342
油石　oil stone　126
油雾发生器　oil mist generator　279
油雾器　oil spray　93
油雾扇　oil mist fan　325
油箱　tank　85
油压缓冲器　oil buffer　333
油嘴　oil mouth　371
右端盖　right cap　26
右滑轨　right track　360
右视图　side view　26
右转向节　right steering knuckle　371
预加载弹簧　preload spring　262
预紧螺母　preloading nut　258
元件联轴器　element coupling　291
圆度　circularity　22
圆弧插补　circular　234
圆弧成形车刀　radius turning form tool　186
圆节距　circular pitch　66
圆孔拉力　round broach　196
圆盘精密抛光机　abrasive disc precision polisher　223
圆盘周铣刀　side milling cutter　42
圆跳动　circular runout　23
圆筒筛　trommel　407
圆柱齿轮传动　cylindrical gear transmission　65
圆柱定位销　cylindrical positioning pin　206
圆柱度　cylindricity　22
圆柱拱曲垫片　cylindrically curved washer　250
圆柱螺旋齿轮传动　cylindrical helical gear transmission　65
圆柱面磨削　cylindrical grinding　165
圆柱销　pin，dowel pin　26，253
圆柱心轴　cylindrical mandrel　205
圆柱形键　round key　252
圆柱形凸轮　cylindrical cam　52
圆锥齿轮传动　bevel gear transmission　65
圆锥滚子轴承　tapered roller bearing　258，266
圆锥销　cotter pin　253
月池　moonpool　399
云梯　ladder　376
运行涡管　orbiting scroll　325

杂物电梯　dumbwaiter lift　328
载货电梯　freight elevator　327
凿子　chisel　126
增压　boost pressure　351
增压器　intensifier　181
轧辊　roll　143
闸阀　gate valve　315
毡轮　felt roller　227
粘接剂　adhesive　49
粘接件　adherend　256
斩拌机　chopping machine　414
张紧辊　tension roller　223
张紧轮　tensioning pulley，tension wheel　74，167，358
张紧式联轴器　tension coupling　294
张紧筒　tension pulley　341
张紧重块　take-up weight　341
长臂划规　beam compass　127
掌板　shoe plate　82
胀管器　pipe expander　396
爪型联轴器　claw coupling　295
罩面　fascia　360
折叠器　folding machine　421
折弯　bending　122
真空泵　vacuum pump　359
真空封装　vacuum encapsulate　16
真空滚筒干燥器　vacuum roller dryer　412
真空调节阀　vacuum regulating valve　359
真空转换阀　VSV　353
砧座　anvil　123
振动器　oscillator　150
振动式球磨机　vibratory ball mill　414
振动输送机　vibrating conveyor　343
蒸汽鼓　steam drum　313
蒸汽轮机　steam turbine　316

整流器 rectifier 171
整体心轴 solid mandrel 206
正铲挖掘机 face excavator 378
正时齿轮 time gear 346
正时带 timing belt 346
正时控制 timing control 351
支撑平台 platform 180
支架 frame，trestle 165，397
执行元件 actuator 93
直槽钻 straight-flute drill 190
直齿 straight spur，spur gear 44，55
直齿油泵 spur gear oil pump 370
直齿圆柱齿轮传动 spur gear transmission 65
直齿圆锥齿轮传动 spur bevel gear transmission 65
直齿锥齿 straight bevel 44
直齿锥齿轮 straight bevel gear 69
直动头 lifting head 85
直管气流干燥器 straight pipe pneumatic dryer 412
直角尺、矩尺 try square 127
直角改锥 right-angled screwdriver 127
直径 diameter 22
直流 straight flow 173
直流电动机 DC motor 306
直流发电机 DC generator 306
直升机平台 helicopter pad 399
直线度 straightness 22
直线摇动筛 linear shaking screen 406
止推垫圈 thrust washer 358
止销 stop pin 206
止转棒轭 scotch yoke 60
指示线 indicating line 21
指引线 leader 21
指状垫圈 finger washer 251
指状立铣刀 end milling cutter 42
制动分缸 wheel cylinder 371
制动鼓 brake drum 371
制动棘爪 holding pawl 64
制动启动器 brake actuator 300
制动系统 brake system 361
制动衔铁 brake armature 301
制造公差 manufacturing tolerance 2
中间板 intermediate plate 367
中间齿轮 intermediate gear 305
中间冷却器 intercooler 419
中间轮 midst wheel 358
中径或有效直径 pitch or effective diameter 246
中空端面铣刀 shell end mill 187
中拖板手轮 cross-slide handwheel 215
中心线 center line 21
中心销 center pin 321
中心钻 center drilling 163
中性罩 neutral shield 312
中央底座 central base 240
中央刷 central brush 382
中央循环式蒸发器 centre cycling evaporator 416
重力输送机 gravity conveyor 343
重载镗刀 heavy boring tool 188
周边泵 peripheral pump 355
周边环套 annulus ring 71
周边磨削 peripheral grinding 166
周边铣削刀刃 peripheral cutting edge 188
轴 shaft 26
轴承 bearing 54
轴承盖 bearing cover 258
轴承架 bearing bracket 265
轴承腔 bearing housing 279
轴承支撑环 bearing support ring 264
轴承组件 bearing assembly 82
轴承座 bearing pedestal 406
轴公差 shaft tolerance 28
轴流泵 axial flow pump 423
轴流涡轮 axial flow turbine 319
轴套 bush，sleeve 26，258
轴线 axis 4，193
轴向铲背面 axial relief angle 188
轴向柱塞泵 axial piston pump 82
轴销柔性联轴器 pinflex coupling 294
轴针式喷嘴 pintle type nozzle 347
肘节机构 toggle mechanism 58
肘节棘轮 toggle ratchet 58
肘节夹头 toggle clamp 56
肘节压紧机构 toggle press mechanism 56
肘连接 toggle link 298
主动盘 drive plate 296
主动轮 driving plate 64
主后刀面 major flank 186

主后角　front clearance，end relief angle　186
主回路　primary circuit　85
主减速器　final drive　369
主降落伞　main parachute　375
主切刃角　end cutting edge angle　186
主切削刃　cutting edge　186
主视图　front view　26
主销　kingpin　365
主油泵　main oil pump　325
主轴　spindle　218，241，258
主轴螺母　spindle nut　198
主轴头　spindle head　217
主轴箱　saddle，headstock，spindle box　213，215，218
主轴转速选调器　spindle speed selector　215
柱端螺钉　dog screw　248
柱塞　plunger　219
柱塞泵　plunger pump　80
柱塞式定量供料装置　plunger dosing device　422
柱塞掌　piston shoe　82
柱销　pintle　353
柱形凸轮　cylindrical cam　57
柱轴承座　headstock seat　357
铸锭　ingot casting，cast ingot　10，138
铸铁摩擦离合器　cast-iron friction clutch　298
铸造　casting　122
抓斗　grab　379
抓木垫片　wood grip washer　251
转动叉　rotating fork　345
转动心轴　rotary arbor　260
转杆 / 绞轮　turnstile capstan wheel　215
转鼓　drum　407
转环　swivel　77
转矩传感器　torque sensor　339
转盘　scroll plate　202
转盘卡瓦　slip　393
转枢　pivot　264
转速表　tachometer　360
转塔车床　turret lathe　127
转塔挡块　turret stop　215
转塔刀架　tool turret，turret　219，237
转筒式除石机　tumble stoner　407
转向泵　steering pump　371
转向传动轴　steering shaft　371
转向扼架齿轮　tumbler yoke gear　55
转向缸　steering cylinder　371
转向架　steering fork　345
转向节　right steering knuckle　371
转向节臂　steering knuckle　371
转向盘　steering wheel　371
转向器　steering gear　371
转向蜗杆　steering worm　371
转向系统　steering system　361
转向摇臂　pitman arm　371
转向直拉杆　drag link　371
转向轴　steering shaft　371
转销　pivot pin　101
转轴　pivot，rotary shaft　241，260
转轴环　pivot ring　296
转子　rotor　282，316，358
转子壳体　rotor housing　321
转子绕组　rotor winding　307
装载槽　loading chute　221
锥柄　taper shank　190
锥齿　bevel gear，bevel teeth　55，202
锥齿轮　bevel gear，bevel pinion　55，202，203，291
锥度　taper，conical taper　22，192
锥度心轴　cone mandrel，tapered mandrel　206
锥度芯棒　taper mandrel　168
锥端螺钉　conical screw　248
锥盘　cone　305
锥坡搭接　tapered lap　256
锥套　tapered sleeve　191
锥销　taper pin　253
锥形管　cone tube　343
锥形滚子推力轴承　tapered roller thrust bearing　365
锥形垫片　conical washer　250
锥形离合器　cone clutch　291
锥形摩擦离合器　cone-type friction clutch　296
锥形磨粉机　conical grinding mill　414
锥形套管　tapered bushing　413
自动螺纹加工机床　automatic screw machine　225
自动人行道　passenger conveyor　327
自动闩栓　automatic latch　335
自攻螺钉　self-tapping screw　248
自位支承钉　self-positioning supporting pin　205
自由锻　open die　138

纵向 longitudinal 15
纵向挡块 longitudinal stop 215
纵向刀具溜板 longitudinal tool carriage 215
纵向定程机构 longitudinal stroke mechanism 215
阻尼室 damping chamber 354
阻尼筒 damping bellow 318
组合扳手 combination spanner 126
组合钳 combination plier 126
组织处理 texturing 226
钻柄 drill tang 191
钻床 drilling machine 128
钻铤 drill collar 392
钻杆架 drill pipe rack 399
钻井船 drill ship 400
钻孔 drilling 31
钻孔环形螺母 drilled ring nut 249
钻模板 drill plate 207
钻套 guide bush，drill bushing，drill bush 41，190，207
钻头 drill，drill bit 190，392
钻头直径 drill diameter 190
钻削 drilling 28
钻削夹具 drilling jig 207
最大过盈 maximum interference 28
最小过盈 minimum interference 28
左端盖 left cap 26
左滑轨 left track 360
座垫 seat cushion 360

附录

附录 1　One-hand alphabet　单手势字母表

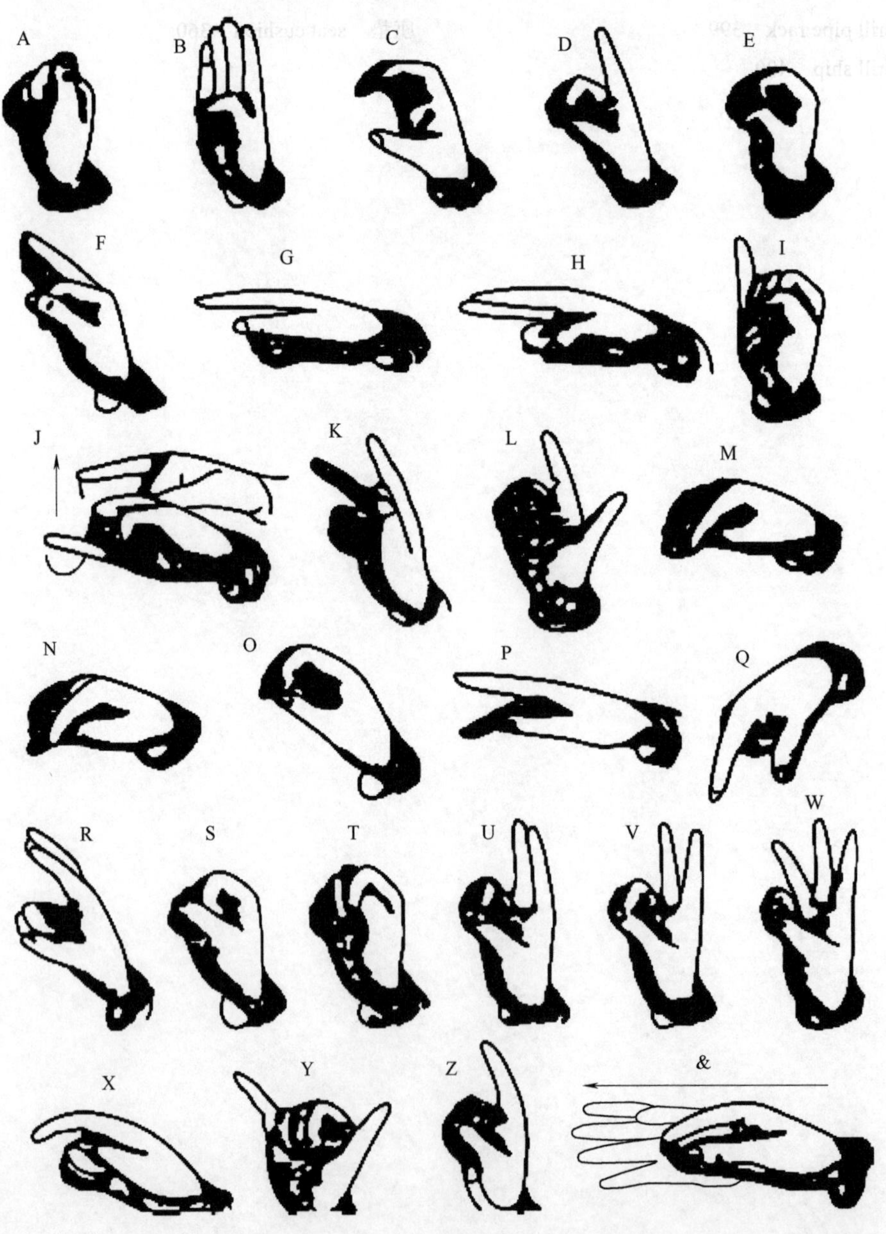

附录 2 Deaf-and-dump alphabet 手语字母表

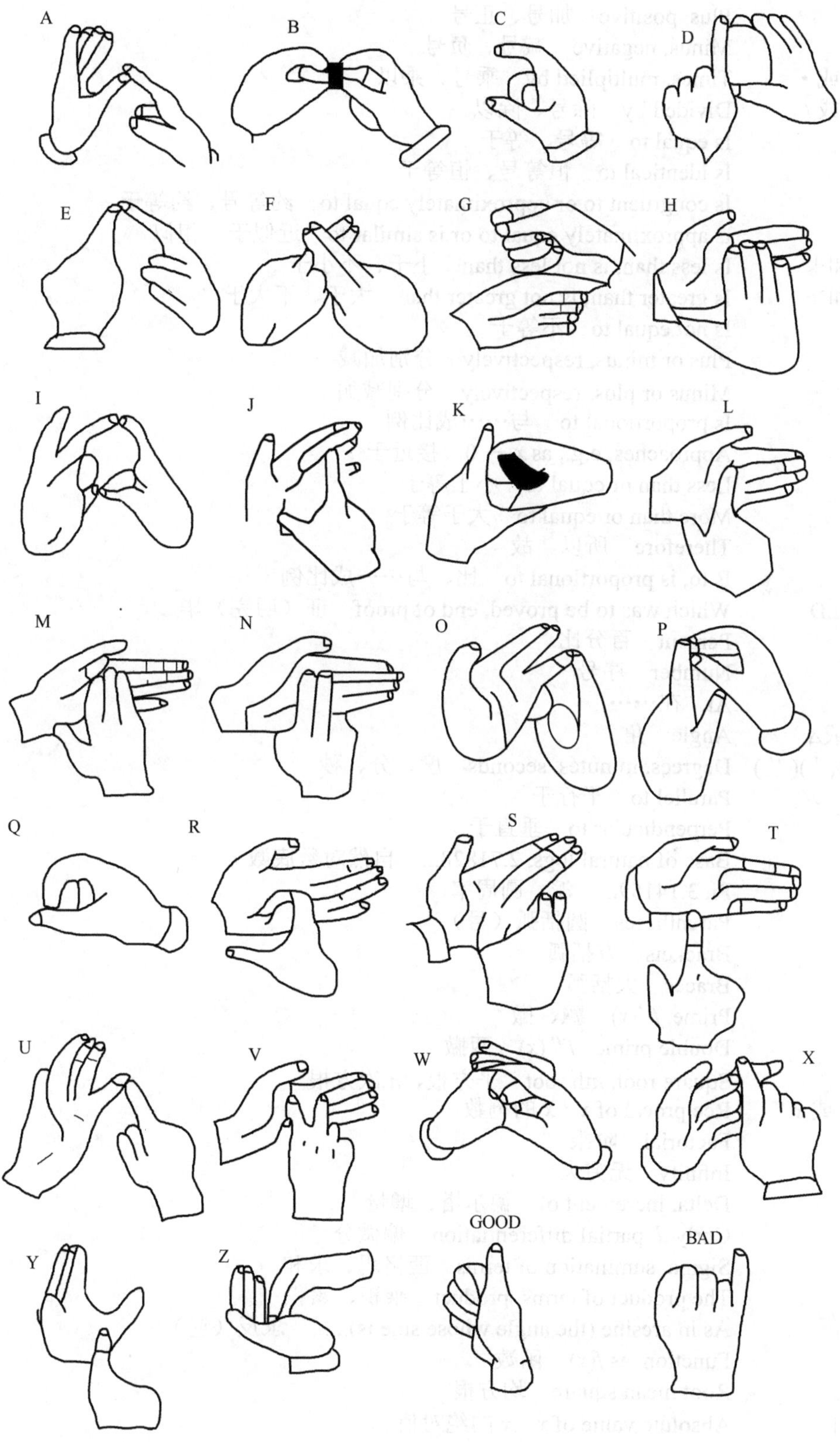

附录 3 Mathematical signs and symbols 数学标识和符号

+	Plus, positive 加号、正号
−	Minus, negative 减号、负号
× 或 ·	Times, multiplied by 乘号、乘以
÷ 或 /	Divided by 除号、除以
=	Is equal to 等号、等于
≡	Is identical to 恒等号、恒等于
≌	Is congruent to or approximately equal to 约等号、约等于
∼	Is approximately equal to or is similar to 近似于、相似
< 和 ≮	Is less than, is not less than 小于、不小于
> 和 ≯	Is greater than, is not greater than 大于、不大于
≠	Is not equal to 不等于
±	Plus or minus, respectively 分别加减
∓	Minus or plus, respectively 分别减加
∝	Is proportional to 与……成比例
→	Approaches, e.g., as $x \to 0$ 接近于
⩽	Less than or equal to 小于等于
⩾	More than or equal to 大于等于
∴	Therefore 所以、故
:	Is to, is proportional to 比、与……成比例
Q.E.D.	Which was to be proved, end of proof 证（明完）毕
%	Percent 百分比
#	Number 序号
@	At 在……
∠或∡	Angle 角
(°)(′)(″)	Degrees, minutes, seconds 度、分、秒
‖ , //	Parallel to 平行于
⊥	Perpendicular to 垂直于
e	Base of natural logs, 2.71828... 自然对数底数
π	Pi, 3.14159... 派、圆周率
( )	Parentheses 圆括弧（号）
[ ]	Brackets 方括弧
{ }	Braces 大括弧
′	Prime, $f'(x)$ 飘、撇
″	Double prime，$f''(x)$ 两撇
$\sqrt{}$, $\sqrt[n]{}$	Square root, nth root 平方根、n 次方根
$1/x$ 或 x^{-1}	Reciprocal of x x 的倒数
!	Factorial 阶乘
∞	Infinity 无穷大
Δ	Delta, increment of 德尔塔、增量
∂	Curly d, partial differentiation 偏微分
∑	Sigma, summation of terms 西格玛、求和
∏	The product of terms, product 乘积、累积
arc	As in arcsine (the angle whose sine is)...... 求反（逆）
f	Function, as $f(x)$ 函数
rms	Root mean square 均方根
$\lvert x\rvert$	Absolute value of x x 的绝对值

参 考 文 献

[1] 朱派龙主编．机械专业英语图解教程．北京：北京大学出版社，2008.

[2] 朱派龙主编．机械工程专业英语图解教程．第2版．北京：北京大学出版社，2013.

[3] 朱派龙主编．图解机械制造专业英语．北京：化学工业出版社，2009.

[4] 朱派龙主编．图解机械制造专业英语（增强版）．北京：化学工业出版社，2014.

[5] 朱派龙，孙永红主编．机械制造工艺装备．西安：西安电子科技大学出版社，2006.

[6] 黄云，朱派龙编著．砂带磨削原理及其应用．重庆：重庆大学出版社，1993.

[7] 曾励，朱派龙等，机电一体化系统设计．北京：高等教育出版社，2004.

[8] 吴永平，陈波，赵利军编著．图解工程机械英语词汇．北京：化学工业出版社，2009.

[9] 加拿大QA国际图书出版公司．学生英汉百科图解词典．北京：外语教学与研究出版社，2005.

[10] 中国寰球工程公司编．英汉化学工程图解词汇．北京：化学工业出版社，2003.

[11] 英国DK公司编．牛津英汉双解大词典（插图版）．北京：外语教学与研究出版社，2005.

[12] 牛津当代百科大辞典．北京：中国人民大学出版社，1995.

[13] Serope Kapakjian, Stever R.Schmid, *Manufacturing Engineering and Technology* (*Fifth Edition*)．制造工程与技术．北京：清华大学出版社，2006.

[14] P N Rao *Manufacturing Technology-Metal Cutting & machine tool*．制造技术——金属切削与机床．北京：机械工业出版社，2003.

[15] Carvill, James *Mechanical Engineer' s Data Handbook*, Butterworth-Heinemann, 2003.

[16] Ronald A. Washer *Handbook of Machining and Metalworking and calculations*, McGraw-Hill, 2001.

[17] Heinz P. Bloch and Fred K. Geitner. *Major process equipment maintenance and repair*, Gulf Publishing.

[18] Heinz P. Bloch and Fred K. *Machinery Component Maintenance and Repair*, Gulf Publishing Company, 1999.

[19] Ben-Zion Sandier, *ROBOTICS Designing the Mechanisms for Automated Machinery* (Second Edition), ACADEMIC PRESS, 1999.

[20] John S. Oakland, *Statistical Process Control* (Fifth Edition), Butterworth-Heinemann, 2003.

[21] R. Keith Mobley, *ROOT CAUSE FAILURE ANALYSIS*, Butterworth-Heinemann, 1999.

[22] Ioan D. Marinescu et al. *Tribology of abrasive machining processes*, William Andrew, Inc. 2004.

[23] Thomas Childs, *Metal Machining Theory and Applications*, Arnold, 2000.

[24] Dan B. Marghitu, *Mechanical Engineer' s Handbook*, ACADEMIC PRESS, 2001.

[25] Edward H. Smith, *Mechanical Engineer' s Reference Book*, Butterworth-Heinemann, 2000.

[26] U.K. Singh, *MANUFACTURING PROCESSES* (Second Edition), New Age International (P) Ltd, 2009.

[27] Joseph E. Shigley, *STANDARD HANDBOOK OF MACHINE DESIGN*, McGraw-Hill, 1996.

[28] ERIK OBERG , *27th Edition Machinery' s Handbook*, 2004 by Industrial Press Inc.

[29] Youssef, Helmi A. *Machining technology: machine tools and operations*, CRC Press Taylor & Francis Group, 2008.

[30] Prof. J.S. Colton, *Manufacturing Processes and Engineering*, GIT 2009.

[31] Bob Mercer, *Industrial Control Wiring Guide* (Second edition), Newnes,2001.

[32] Eric H. Glendinning, *Oxford.English for.Electrical and.Mechanical Engineering*, Oxford University Press, 1995.

[33] John Ridley and John Channing, *Safety at Work* (Sixth edition) , Butterworth-Heinemann, 2003.

[34] Andreas W. Momber, *Hydrablasting and Coating of Steel Structures*, 2003 Elsevier Science Ltd.

[35] John Campbell, *Castings*, Butterworth-Heinemann, 2003.

[36] Ei-ichi YASUDA Michio INAGAKI, *CARBON ALLOYS Novel Concepts to Develop Carbon Science and Technology*, 2003 Elsevier Science Ltd.

[37] J. Edward Pope, *RULES OF THUMB FOR MECHANICAL ENGINEERS*, 1997 by Gulf Publishing Company.

[38] R. Keith Mobley, *AN INTRODUCTION TO PREDICTIVE MAINTENANCE* (Second Edition), Butterworth-Heinemann ,2002.

[39] M. J. NEALE, *THE TRIBOLOGY HANDBOOK* (Second edition), Butterworth-Heinemann , 2001.

[40] Soares, Claire. *Process engineering equipment handbook*, McGraw-Hill, 2002.

[41] Frank Kreith, *The Mechanical Engineering Handbook Series*, CRC Press, 2005.

[42] Mobley, R. Keith , *Plant engineering-Handbooks, manuals*, Butterworth-Heinemann, 2001.

[43] Gwidon W. Stachowiak, *ENGINEERING TRIBOLOGY*, Butterworth-Heinemann, 2000.

[44] Eric H. Glendinning, Oxford English for Electrical and Mechanical Engineering, Oxford University Press, 1995.

[45] ASME, RECOMMENDED GUIDELINES FOR THE CARE OF POWER BOILERS, ASME, 2013.

[46] ASME, BELOW-THE-HOOK LIFTING DEVICES, ASME, 2003.

[47] David S.J.,Handbook of petroleum processing, Springer, 2006.

[48] Ron Hodkinson and John Fenton, Lightweight Electric/Hybrid Vehicle Design, Butterworth-Heinemann, 2001.

[49] Marios Sideris, METHODS FOR MONITORING AND DIAGNOSING THE EFFICIENCY OF CATALYTIC CONVERTERS, ELSEVIER, 1998.

[50] Prof. Dipl.-Ing. Jörnsen Reimpell, The Automotive Chassis: Engineering Principles, SECOND EDITION, Butterworth-Heinemann, 2001.

[51] T.K. GARRETT,The Motor Vehicle,Thirteenth Edition, Butterworth-Heinemann, 2001.

[52] Heinz, Heisler, Advanced vehicle technology, SECOND EDITION, Butterworth-Heinemann, 2002.

[53] Allan W. M. Bonnick, Automotive Computer Controlled Systems :Diagnostic tools and techniques, Butterworth-Heinemann, 2001.